S0-CAS-407

Supplement D

The chemistry of **halides, pseudo-halides and azides** Part 1

THE CHEMISTRY OF FUNCTIONAL GROUPS

A series of advanced treatises under the general editorship of Professor Saul Patai

The chemistry of alkenes (2 volumes)
The chemistry of the carbonyl group (2 volumes)
The chemistry of the ether linkage
The chemistry of the amino group
The chemistry of the nitro and nitroso groups (2 parts)
The chemistry of carboxylic acids and esters
The chemistry of the carbon–nitrogen double bond
The chemistry of amides
The chemistry of the cyano group
The chemistry of the hydroxyl group (2 parts)
The chemistry of the azido group
The chemistry of acyl halides
The chemistry of the carbon–halogen bond (2 parts)
The chemistry of the quinonoid compounds (2 parts)
The chemistry of the thiol group (2 parts)
The chemistry of the hydrazo, azo and azoxy groups (2 parts)
The chemistry of amidines and imidates
The chemistry of cyanates and their thio derivatives (2 parts)
The chemistry of diazonium and diazo groups (2 parts)
The chemistry of the carbon–carbon triple bond (2 parts)
The chemistry of ketenes, allenes and related compounds (2 parts)
The chemistry of the sulphonium group (2 parts)
Supplement A: The chemistry of double-bonded functional groups (2 parts)
Supplement B: The chemistry of acid derivatives (2 parts)
Supplement C: The chemistry of triple-bonded functional groups (2 parts)
Supplement D: The chemistry of halides, pseudo-halides and azides
Supplement E: The chemistry of ethers, crown ethers, hydroxyl groups and their sulphur analogues (2 parts)
Supplement F: The chemistry of amino, nitroso and nitro compounds and their derivatives (2 parts)

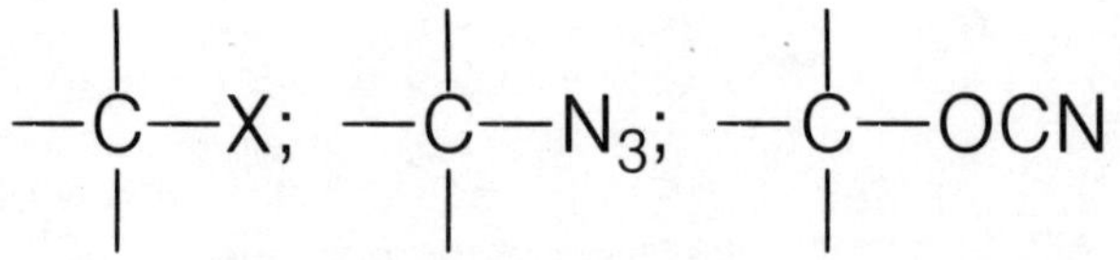

$$-\overset{|}{\underset{|}{C}}-X;\quad -\overset{|}{\underset{|}{C}}-N_3;\quad -\overset{|}{\underset{|}{C}}-OCN$$

Supplement D

The chemistry of **halides, pseudo-halides and azides** Part 1

Edited by

SAUL PATAI

and

ZVI RAPPOPORT

The Hebrew University, Jerusalem

1983

JOHN WILEY & SONS

CHICHESTER – NEW YORK – BRISBANE – TORONTO – SINGAPORE

An Interscience ® Publication

CHEMISTRY

6827-4324

Copyright © 1983 by John Wiley & Sons Ltd.

All rights reserved.

No part of this book may be reproduced by any means, nor transmitted, nor translated into a machine language without the written permission of the publisher.

Library of Congress Cataloging in Publication Data:

The Chemistry of halides, pseudo-halides, and azides.
(The Chemistry of functional groups. Supplement: D)
'An Interscience publication.'
Includes bibliographical references and indexes.
1. Halides. 2. Azides. I. Patai, Saul. II. Rappoport, Zvi. III. Series.
QD165.C48 1983 546′.73 82-23908
ISBN 0 471 10089 7 (set)
ISBN 0 471 10087 0 (v. 1)
ISBN 0 471 10088 9 (v. 2)

British Library Cataloguing in Publication Data:

Supplement D: the chemistry of halides, pseudohalides and azides.—(The Chemistry of functional groups)
1. Halogens 2. Chemistry, Organic
I. Patai, Saul II. Rappoport, Zvi III. Series
547′.02 QD165

ISBN 0 471 10089 7 (set)
ISBN 0 471 10087 0 (v. 1)
ISBN 0 471 10088 9 (v. 2)

Typeset by Preface Ltd., Salisbury, Wiltshire, and printed in Great Britain.

QD165
C48
1983
v.1
CHEM

Contributing authors

E. Baciocchi	Department of Chemistry, University of Perugia, Perugia, Italy
J. Y. Becker	Department of Chemistry, Ben-Gurion University of the Negev, Beer Sheva, Israel
K. Berei	Central Research Institute for Physics, PO Box 49, H-1525 Budapest, Hungary
H. Bock	Institute of Inorganic Chemistry, Johann Wolfgang Goethe University, Niederurseler Hang, D-6000 Frankfurt (M) 50, West Germany
J. M. Brittain	Department of Chemistry, University of Auckland, Private Bag, Auckland, New Zealand
N. De Kimpe	Laboratory of Organic Chemistry, Faculty of Agricultural Sciences, State University of Gent, Coupure 533, B-9000 Gent, Belgium
P. B. D. de la Mare	Department of Chemistry, University of Auckland, Private Bag, Auckland, New Zealand
J.-M. Dumas	Laboratoire de Physico-Chimie des Diélectriques, Faculté des Sciences Fondamentales et Appliquées, Université de Poitiers, 86022 Poitiers Cedex, France
L. K. Dyall	Department of Chemistry, University of Newcastle, New South Wales 2308, Australia
A. Foucaud	Department of Crystal Physics and Structural Chemistry, University of Rennes I, Avenue du Général Leclerc, 35042 Rennes Cedex, France
D. M. Goldish	Department of Chemistry, California State University, Long Beach, California 90840, USA
M. Gomel	Laboratoire de Physico-Chimie des Diélectriques, Faculté de Sciences Fondamentales et Appliquées, Université de Poitiers, 86022 Poitiers Cedex, France
M. Guerin	Laboratoire de Physico-Chimie des Diélectriques, Faculté de Sciences Fondamentales et Appliquées, Université de Poitiers, 86022 Poitiers Cedex, France
R. R. Gupta	Department of Chemistry, University of Rajasthan, Jaipur 302004, India
E. Hadjoudis	Solid State Chemistry Laboratory, Department of Chemistry, Greek Atomic Energy Commission, Nuclear Research Centre 'Demokritos', Aghia Paraskevi, Attiki, Athens, Greece

A. Horowitz	Soreq Nuclear Research Centre, Yavne, Israel
M. Hudlicky	Department of Chemistry, Virginia Polytechnic Institute and State University, Blacksburg, Virginia 24061, USA
T. Hudlicky	Department of Chemistry, Virginia Polytechnic Institute and State University, Blacksburg, Virginia 24061, USA
T. R. B. Jones	Department of Chemistry, Brock University, St Catherines, Ontario, Canada L2S 3A1
M. Kaftory	Department of Chemistry, Technion-Israel Institute of Technology, Haifa, Israel
D. N. Kevill	Department of Chemistry, Northern Illinois University, DeKalb, Illinois 60115, USA
G. F. Koser	Department of Chemistry, University of Akron, Akron, Ohio 44325, USA
G. Lodder	Gorlaeus Laboratories, University of Leiden, Leiden, The Netherlands
A. E. C. Lucken	Physical Chemistry Department, 30 Quai E. Ansermet, 1211 Geneva, Switzerland
G. Marchese	Istituto di Chimia Organica, Università di Bari, Bari, Italy
A. Y. Meyer	Department of Organic Chemistry, Hebrew University, Jerusalem, Israel
J. M. Miller	Department of Chemistry, Brock University, St Catherines, Ontario, Canada L2S 3A1
H. W. Moore	Department of Chemistry, University of California, Irvine, California 92717, USA
F. Naso	Istituto di Chimia Organica, Università di Bari, Bari, Italy
R. K. Norris	Department of Organic Chemistry, University of Sydney, Sydney, New South Wales 2006, Australia
B. E. Smart	Central Research and Development Department, Experimental Station, E. I. du Pont de Nemours & Co., Wilmington, Delaware 19898, USA
L. Vasáros	Central Research Institute for Physics, PO Box 49, H-1525 Budapest, Hungary
R. Verhé	Laboratory of Organic Chemistry, Faculty of Agricultural Sciences, State University of Gent, Coupure 533, B-9000 Gent, Belgium
P. Weyerstahl	Technische Universität Berlin, Institut für Organische Chemie, D-1000 Berlin 12, German Federal Republic
K. Wittel	Institute of Inorganic Chemistry, Johann Wolfgang Goethe University, Niederurseler Hang, D-6000 Frankfurt (M) 50, West Germany
M. Zupan	Department of Chemistry and 'Jožef Stefan' Institute, 'E. Kardelj' University of Ljubljana, Murnikova 6 – PO Box 537, 61001 Ljubljana, Yugoslavia

Foreword

This Supplement D contains material on halides, pseudo-halides and azides. The same functional groups have been treated previously in the following main volumes of the Chemistry of the Functional Groups series:

The Chemistry of the Azido Group (1971)
The Chemistry of the Carbon–Halogen Bond (2 parts, 1973)
The Chemistry of Cyanates and their Thio Derivatives (2 parts, 1977)

Chapters which were also intended to appear in this volume, but did not materialize were the following: "Advances in the preparation and uses of azides"; "Recent advances in biological reactions involving halides and pseudo-halides"; and "Syntheses and uses of isotopically labelled halides and azides". We hope that these chapters will be included in a future supplementary volume, which should be published in several years' time when the amount of new and unreviewed material justifies this.

The present volume concludes the first set of supplementary volumes (Supplements A, B, C, D, E and F) which cover among themselves all the subjects treated in the Functional Groups series.

We will be very grateful to readers who would call our attention to omissions or mistakes in this and other volumes in the series.

SAUL PATAI
ZVI RAPPOPORT

Jerusalem, October 1982

The Chemistry of Functional Groups
Preface to the series

The series 'The Chemistry of Functional Groups' is planned to cover in each volume all aspects of the chemistry of one of the important functional groups in organic chemistry. The emphasis is laid on the functional group treated and on the effects which it exerts on the chemical and physical properties, primarily in the immediate vicinity of the group in question, and secondarily on the behaviour of the whole molecule. For instance, the volume *The Chemistry of the Ether Linkage* deals with reactions in which the C—O—C group is involved, as well as with the effects of the C—O—C group on the reactions of alkyl or aryl groups connected to the ether oxygen. It is the purpose of the volume to give a complete coverage of all properties and reactions of ethers in as far as these depend on the presence of the ether group but the primary subject matter is not the whole molecule, but the C—O—C functional group.

A further restriction in the treatment of the various functional groups in these volumes is that material included in easily and generally available secondary or tertiary sources, such as Chemical Reviews, Quarterly Reviews, Organic Reactions, various 'Advances' and 'Progress' series as well as textbooks (i.e. in books which are usually found in the chemical libraries of universities and research institutes) should not, as a rule, be repeated in detail, unless it is necessary for the balanced treatment of the subject. Therefore each of the authors is asked *not* to give an encyclopaedic coverage of his subject, but to concentrate on the most important recent developments and mainly on material that has not been adequately covered by reviews or other secondary sources by the time of writing of the chapter, and to address himself to a reader who is assumed to be at a fairly advanced post-graduate level.

With these restrictions, it is realized that no plan can be devised for a volume that would give a *complete* coverage of the subject with *no* overlap between chapters, while at the same time preserving the readability of the text. The Editor set himself the goal of attaining *reasonable* coverage with *moderate* overlap, with a minimum of cross-references between the chapters of each volume. In this manner, sufficient freedom is given to each author to produce readable quasi-monographic chapters.

The general plan of each volume includes the following main sections:

(a) An introductory chapter dealing with the general and theoretical aspects of the group.

(b) One or more chapters dealing with the formation of the functional group in question, either from groups present in the molecule, or by introducing the new group directly or indirectly.

(c) Chapters describing the characterization and characteristics of the functional groups, i.e. a chapter dealing with qualitative and quantitative methods of determination including chemical and physical methods, ultraviolet, infrared, nuclear magnetic resonance and mass spectra: a chapter dealing with activating and directive effects exerted by the group and/or a chapter on the basicity, acidity or complex-forming ability of the group (if applicable).

(d) Chapters on the reactions, transformations and rearrangements which the functional group can undergo, either alone or in conjunction with other reagents.

(e) Special topics which do not fit any of the above sections, such as photochemistry, radiation chemistry, biochemical formations and reactions. Depending on the nature of each functional group treated, these special topics may include short monographs on related functional groups on which no separate volume is planned (e.g. a chapter on 'Thioketones' is included in the volume *The Chemistry of the Carbonyl Group*, and a chapter on 'Ketenes' is included in the volume *The Chemistry of Alkenes*). In other cases certain compounds, though containing only the functional group of the title, may have special features so as to be best treated in a separate chapter, as e.g. 'Polyethers' in *The Chemistry of the Ether Linkage*, or 'Tetraaminoethylenes' in *The Chemistry of the Amino Group*.

This plan entails that the breadth, depth and thought-provoking nature of each chapter will differ with the views and inclinations of the author and the presentation will necessarily be somewhat uneven. Moreover, a serious problem is caused by authors who deliver their manuscript late or not at all. In order to overcome this problem at least to some extent, it was decided to publish certain volumes in several parts, without giving consideration to the originally planned logical order of the chapters. If after the appearance of the originally planned parts of a volume it is found that either owing to non-delivery of chapters, or to new developments in the subject, sufficient material has accumulated for publication of a supplementary volume, containing material on related functional groups, this will be done as soon as possible.

The overall plan of the volumes in the series 'The Chemistry of Functional Groups' includes the titles listed below:

The Chemistry of Alkenes (two volumes)
The Chemistry of the Carbonyl Group (two volumes)
The Chemistry of the Ether Linkage
The Chemistry of the Amino Group
The Chemistry of the Nitro and Nitroso Groups (two parts)
The Chemistry of Carboxylic Acids and Esters
The Chemistry of the Carbon–Nitrogen Double Bond
The Chemistry of the Cyano Group
The Chemistry of Amides
The Chemistry of the Hydroxyl Group (two parts)
The Chemistry of the Azido Group
The Chemistry of Acyl Halides
The Chemistry of the Carbon–Halogen Bond (two parts)
The Chemistry of Quinonoid Compounds (two parts)
The Chemistry of the Thiol Group (two parts)
The Chemistry of Amidines and Imidates
The Chemistry of the Hydrazo, Azo and Azoxy Groups (two parts)
The Chemistry of Cyanates and their Thio Derivatives (two parts)
The Chemistry of Diazonium and Diazo Groups (two parts)
The Chemistry of the Carbon–Carbon Triple Bond (two parts)
Supplement A: The Chemistry of Double-bonded Functional Groups (two parts)

The Chemistry of Ketenes, Allenes and Related Compounds (two parts)
Supplement B: The Chemistry of Acid Derivatives (two parts)
Supplement C: The Chemistry of Triple-Bonded Groups (two parts)
Supplement D: The Chemistry of Halides, Pseudo-halides and Azides (two parts)
Supplement E: The Chemistry of Ethers, Crown Ethers, Hydroxyl Groups and their Sulphur Analogues (two parts)
The Chemistry of the Sulphonium Group (two parts)
Supplement F: The Chemistry of Amino, Nitroso and Nitro Groups and their Derivatives (two parts)

Titles in press:

The Chemistry of Peroxides
The Chemistry of Organometallic Compounds
The Chemistry of Organic Se and Te Compounds

Advice or criticism regarding the plan and execution of this series will be welcomed by the Editor.

The publication of this series would never have started, let alone continued, without the support of many persons. First and foremost among these is Dr Arnold Weissberger, whose reassurance and trust encouraged me to tackle this task, and who continues to help and advise me. The efficient and patient cooperation of several staff-members of the Publisher also rendered me invaluable aid (but unfortunately their code of ethics does not allow me to thank them by name). Many of my friends and colleagues in Israel and overseas helped me in the solution of various major and minor matters, and my thanks are due to all of them, especially to Professor Z. Rappoport. Carrying out such a long-range project would be quite impossible without the non-professional but none the less essential participation and partnership of my wife.

The Hebrew University
Jerusalem, ISRAEL

SAUL PATAI

Contents

The Chemistry of Functional Groups, Supplement D
Edited by S. Patai and Z. Rappoport
© 1983 John Wiley & Sons Ltd

CHAPTER **1**

Molecular mechanics and conformation

A. Y. MEYER
Department of Organic Chemistry, Hebrew University, Jerusalem, Israel

I. INTRODUCTION

In writing this chapter I tried not to duplicate material in existing reviews and to avoid altogether topics that had been discussed elsewhere. Even these limitations left grounds too wide to cover in the allotted space. I chose to concentrate on aspects of molecular mechanics that relate to functionalized hydrocarbons, halides in particular

(Sections I–IV), and on topics in the conformation of halides that have attracted or are attracting now the molecular mechanist's attention (Sections V–IX). Material that did not link up with this choice and, admittedly, some material that did, was left out. Section III is the exception: since its subject matter is controversial, and the controversy neither starts nor stops with halides, I thought it advisable to make the coverage wider than called for.

Pertinent reviews are cited at the beginnings of sections and subsections. Further systematized information concerning organic halogen compounds may be found in the following sources: barriers to internal rotation[1]; conformational analysis of halocyclohexanes[2]; energies associated with conformational change[3]; molecular geometry[4]; thermochemistry[5]; geometric and dynamic structures of fluorocarbons and related compounds[6]; structural data[7]; enthalpies of formation[8]; structure of molecules with large-amplitude motions[9].

In preference to the 'PST notation' [10] rotamers are here codified by specifying groups *anti* (*a*), *gauche* (*g*) or *syn* to each other, e.g. Cl/Cl-*a* (**1**), Cl/Cl-*g* (**2** or **3**), Cl/Cl-*syn* (**4**) and Cl/H-*syn* (**5**) in 1,2-dichloroethane. In cyclohexane derivatives, the abbreviations *ax*, *eq*, *aa*, *ae*, and *ee* stand, respectively, for axial, equatorial, diaxial, axial-equatorial and diequatorial orientations. In citing the results of calculations, E_t is the total computed energy. The components of E_t are E_i, and contributions to a given E_i are denoted e_i, such that

$$\sum_j e_{i,j} = E_i \quad \text{and} \quad \sum E_i = E_t.$$

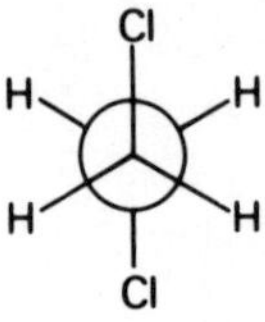

(**1**) 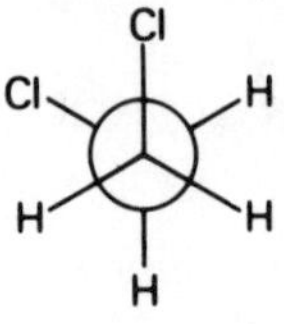 (**2**) 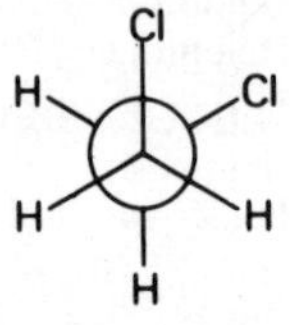 (**3**)

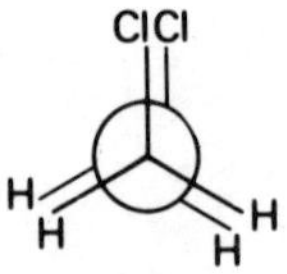

(**4**) 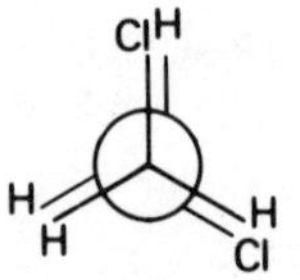 (**5**)

Expressions such as 'conformational energy', 'conformational preference', 'relative stability', refer here to differences in energy (or enthalpy), *not* in free energy. Unless stated otherwise, numbers cited refer to the gas phase. Note that these usages are not accepted by all. In particular, the term 'conformational energy' refers frequently in the literature to free-energy differences in solution[11].

In 1946 Hill[12] wrote:

> 'The forces involved in steric effects are well known: (1) groups or atoms may repel each other if close together; (2) in order to decrease this interaction, the groups or atoms will tend to move apart, but this will generally require the stretching or bending of bonds with a related increase in energy. The final configuration will thus be the result of a compromise between the two types of force, and will be the configuration of minimum energy'.

Twenty years later, a new branch of science was in existence:

'... In the second approach, a molecule is viewed as a system of particles assumed to be held together by classical forces. The energy differences between molecular systems may then be estimated by classical mechanical means, thus avoiding the complexity of quantum-mechanical treatments. The artificiality of this model restricts its use to the evaluation of relative quantities; without further adjustment and modification the method is incapable of yielding total or absolute energies'[13].

In this chapter we concentrate on the 'second approach', called *molecular mechanics* or force field calculation, and on its application to organic halogen compounds. As it happens, organic halides are serving as testing grounds for force field calculations of functionalized hydrocarbons, in particular as regards the electrostatic component in the total energy. The axial symmetry and high group moment of the C–Hal bond are two obvious reasons, as well as the traditional interest in the dipole moments of halides[14].

Quantum chemistry considers molecules in terms of nuclei and electrons. Molecular mechanics, excepting developments of as yet unknown scope[15], concentrates on atoms and bonds. In quantum-chemical work it is expected that improvement in technique will improve the results and thereby permit more profound interpretations of observed phenomena; in molecular mechanics the reproduction of experimental data is in general not problematic, but this does not guarantee interpretation in terms of fundamental physical concepts. In both approaches one strives towards an 'energy-component analysis', that is, breakdown of energy differences into separate contributions. A quantum-mechanical breakdown could include the core energy, electronic energy, core-electron interaction and the kinetic energy[16]. In distinction, the molecular-mechanical components are more akin to those in the classical equations of motion[17]: here one would consider terms due to stretching of bonds, deformation of angles, attraction and repulsion of non-bonded atoms, electrostatic interactions, and more. Most of the force fields described heretofore, but not all[15,18], require also an internal mechanism for correction, the *intrinsic torsion potential* or *torsional term*. This is because inaccuracies unavoidably pile up through the multitude of terms[19], non-bonded and other explicit strains account only partially for barrier heights[20,21], and certain quantum-chemical effects (resonance!) have no immediate counterpart in molecular mechanics. Truly, the framework hardly fits into what has come to be called 'first principles'[22]. Rather, it is a technique for analysing the properties of some molecules by using data on others.

An example – extreme, no doubt – will illustrate these points. In 1,2-difluoroethene (FCH=CHF) the *cis*-isomer is more stable than the *trans*[23] with $\Delta H^\circ = 0.93$ kcal mol^{-1}. This 'counter-intuition' relationship has prompted quite a few *ab initio*[24–26] and semi-empirical[27–29] quantum-chemical studies. Among the *ab initio* calculations, only one[24] (with polarization orbitals) predicts $E_{trans} > E_{cis}$. Of the semi-empirical methods, only INDO performs acceptably[28]. Nonetheless, the computed numbers suggest certain interpretations. To cite a few, *cis* may be endowed of a greater correlation energy[24]; or correspond to a higher two-electron stabilization and lower four-electron destabilization, between the F···F moiety and the double bond[30a]; or enjoy a more stabilizing interaction between the core of each fluorine and the electronic shell of the other[27]. The choice between these alternatives, and others perhaps, is still pending, but the examination has at least been cast in clear-cut physical terms.

A molecular-mechanical analysis of fluorinated olefins has been reported[27]. Here FCH=CHF was taken as the starting point and the corrective term *adjusted* to reproduce the measured difference in enthalpy. The component analysis came out as follows:

	$E_{trans} - E_{cis}$, kcal mol^{-1}
Skeletal strain	−0.012
Non-bonded strain (ΔE_{nb})	−0.083
Electrostatic strain (ΔE_{es})	−0.373
Corrective term (ΔE_{tor})	+1.400
Total (ΔE_t)	+0.93

This build-up, inelegant as it is ($\Delta E_{tor} > \Delta E_t$!), is not unhelpful. One sees that skeletal and non-bonded effects are insignificant and that electrostatic interactions actually favour the *trans* isomer – which is not obvious[25]. Also, explication is not the sole end of computation: energy differences, geometrical detail, dipole moments, thermodynamic quantities, are required *per se*. And the same set of parameters, of which those required for FCH=CHF forms a subset, fits olefins in general[31] and applies to a variety of fluorinated olefins[27].

II. THE FORCE FIELD CALCULATION

Westheimer is considered as the originator of molecular-mechanical (as distinct from spectroscopic[17]) force field calculations[32]. Hendrickson[33] and Wiberg[34] then took over. Subsequent developments have been critically examined[13,19,35,36] and reviewed[37,38]. Westheimer's work initiates also the application to organic halogen compounds. Systematic studies of halides have not been numerous[27,39–48], but papers dealing with a small number of molecules occur in the literature frequently.

In the force field method, any geometry of a given molecule defines a potential energy E_t ('t' for total) which is made up as sum of components. The formulation of E_t, and hence its value for a given structure, may vary from one field to another, but differences in E_t between structural variants (i.e. stereoisomers and such position-isomers that have identical 'structural units' (Ref. 35, p. 39) are meant to represent differences in internal energy which are independent of computational detail.

In current force fields, some or all of the following components, and sometimes others[15], are included:

$$E_t = E_s + E_b + E_{nb} + E_{es} + E_{tor} + \text{cross terms.}$$

Here E_s (stretch) is the energy due to stretching or compression of bonds; E_b (bend) refers to the opening or closing of bond angles; E_{nb} (non-bonded) represents attraction and repulsion between non-bonded non-geminal atoms; E_{es} (electrostatic) stands for intramolecular electrostatic interaction; E_{tor} (torsion) is the corrective term (Section I). Each of these components is itself a sum, wherein each contribution depends on an internal coordinate and gets its zero at some value – called *reference value* – of that coordinate. For example,

$$E_s(\text{molecule}) = \sum_{\substack{j \\ \text{(all bonds)}}} e_{s,j}$$

where each $e_{s,j}$ alludes to a bond length and vanishes at the reference value ascribed to that type of bond (say[49] 1.523 Å for C—C). Computation consists of a first guess at the molecular geometry, followed by systematic variation[19,36] such that E_t achieves a minimum. This provides the equilibrium geometry which can itself be used to derive further quantities[37]. Geometries obtained by this procedure are considered as more trustworthy than the computed energy differences[50,51].

Stretching (e_s) is usually expressed as a Hooke's law harmonic function in the bond length l, but sometimes a cubic term is added,

$$e_s = \tfrac{1}{2}k_s(l - l_0)^2 + k'_s(l - l_0)^3.$$

The force constants k_s and k'_s, and the reference bond length l_0, are best viewed as adjustable parameters and not identified with their spectroscopic or other counterparts. Bending (e_b) is defined analogously in the valence angle θ. Interaction between non-bonded atoms (e_{nb}) is taken[13] as a pairwise sum of atom–atom interactions, excepting geminal pairs (A···B in >C<(A)(B)). Lennard-Jones and Buckingham potentials have been used, as well as the Hill potential (equation 1)[20],

$$e_{nb} = \epsilon\left[A\,e^{-Br/r^*} - \frac{C}{(r/r^*)^6}\right], \qquad (1)$$

variants thereof[43], and other formulations[52]. In equation (1), r is the interatomic distance or an effective distance[53]. For a pair $i \cdots j$, the parameters of energy (ϵ_{ij}) and of distance (r^*_{ij}) are usually[54], but not always[49], made to depend on atomic parameters. With Hill's original constants ($A = 8.28 \times 10^5$, $B = 1/0.0736$, $C = 2.25$), the minimum occurs at $r/r^* = 1.0070$, whereat $e_{nb} = -1.2115$; also, e_{nb} vanishes at $r/r^* = 0.8935$. The latter point defines the transition from the zone of *non-bonded repulsion* to that of *non-bonded attraction* (Figure 1).

E_{es} is taken as a pairwise sum of interactions between charges[55] ('monopole approximation') or between dipolar bonds[56] ('point-dipole approximation'). The charge formula is given in equation (2),

$$e_{es,ij} = \frac{q_i q_j}{d r_{ij}}, \qquad (2)$$

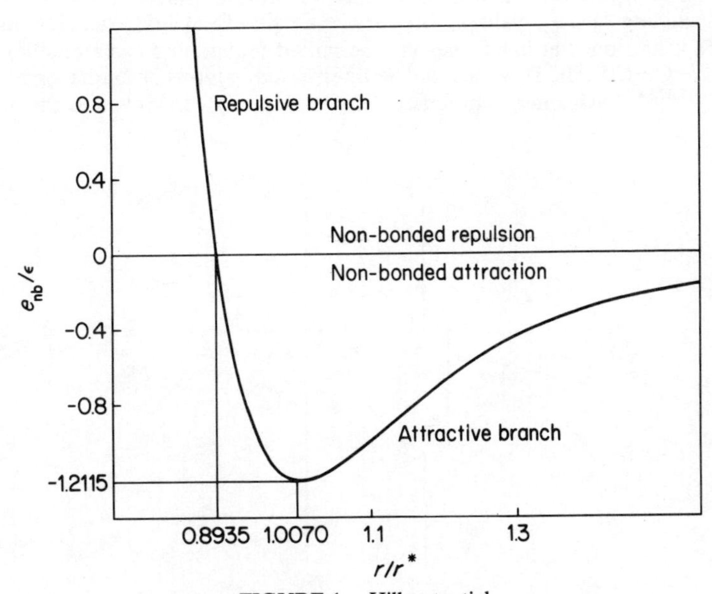

FIGURE 1. Hill potential.

and the dipole formula[57], in scalar form and convenient notation[58], is given in equation (3):

$$e_{es,ij} = \frac{\mu_i \mu_j}{dr^3}(\cos \chi_{ij} - 3 \cos \alpha_i \cos \alpha_j). \quad (3)$$

In these equations d is the dielectric constant, q_i represents a point charge and μ_i the moment of a bond. In practice these are put, respectively, at the atom or at the midpoint of the bond. r_{ij} is the distance between q_i and q_j or μ_i and μ_j, and the angles α_i, α_j and χ_{ij} (dipole angle) are, respectively, between μ_i and r_{ij}, μ_j and r_{ij}, μ_i and μ_j (Figure 2).

The torsional term does not occur in spectroscopic force fields and was not considered by Westheimer. Hendrickson put it in[33] and subsequent fields[34] used a one-term cosine function $e_{tor} = \frac{1}{2}V_n(1 + s \cos n\omega)$, where ω is the dihedral angle between planes ABC and BCD of a four-point sequence A—B—C—D, n is the periodicity of internal rotation[59], and s is plus or minus one. In ethane and ethene n equals 3 and 2, respectively, and it became customary to take $\frac{1}{2}V_3(1 + \cos 3\omega)$ and $\frac{1}{2}V_2(1 - \cos 2\omega)$ for rotation about single and double bonds[31]. Yet, reproduction of the change in total energy (E_t), along the path of internal rotation, requires a more-than-one-term periodic function[60–65]. It was found helpful[66] also to expand e_{tor} in a series, e.g. equation (4):[49]

$$e_{tor} = \tfrac{1}{2}V_1(1 + \cos \omega) + \tfrac{1}{2}V_2(1 - \cos 2\omega) + \tfrac{1}{2}V_3(1 + \cos 3\omega). \quad (4)$$

The constants V_i should not be confused with constants (say, U_k) that occur in the periodic development of E_t itself (equation 5):

$$E_t = \tfrac{1}{2} \sum_k U_k(1 - \cos k\omega). \quad (5)$$

For example, an analysis of internal rotation in $ClCH_2CH_2Cl$ could start with the estimation of E_t for several rotamers. Each of these E_t values is the sum of many terms, including nine e_{tor} values (Cl—C—C—Cl, Cl—C—C—H, etc. cf. equation 4). These E_t values once at hand, they can be united (equation 5) as a function in, say, ω(Cl—C—C—Cl). The U_k values can be interpreted in terms of fundamental physical effects[62,63,67,68]. Such interpretations do not necessarily carry over to the V_i values,

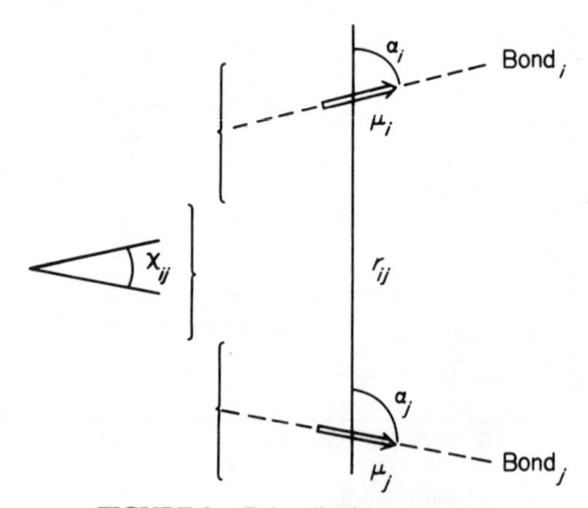

FIGURE 2. Point-dipole parameters.

inasmuch as most current force fields have been parameterized by trial-and-error searches and the uniqueness of the ensuing parameters not ascertained[66].

Cross-terms evolve automatically in the mathematical derivation[17,36] but are only partially retained in molecular-mechanical force fields. The stretch–bend interaction term[49,69] (e_{sb}) is introduced to reproduce the elongation of C—C bonds that accompanies the closure of θ(CCC) in cycloalkanes[53]. Quantum-chemically, the interdependence is ascribable to variation in the hybridization at the central carbon[70] or to repulsion between its substituents[71].

What about C—Hal bond lengths? In CH_3CH_2Br, l(C—Br) ~ 1.950 Å, determined by microwaves (MW)[72]; it is ~1.975 Å in $(CH_3)_3CBr$ (MW)[73]. In an attempt to identify the cause for elongation within the molecular-mechanical model, calculations were conducted for $(CH_3)_3CBr$ with various interactions 'turned off'[48]. First, the non-bonded interactions of bromine were omitted by setting ϵ(Br) = 0 (equation 1). Optimization under this constraint led to l(C—Br) = 1.947 (too short) and θ(CCBr) = 107.2° (probably too small). The sum of non-bonded interactions at this geometry (from a separate calculation) would amount to 2.80 kcal mol^{-1}, 2.24 of which due to H···Br. Next, ϵ(Br) was revived but e_{sb} eliminated. This led to l(C—Br) = 1.972 (almost correct) and θ(CCBr) = 108.4°. With all constraints removed, l(C—Br) = 1.975 (correct) and θ(CCBr) = 108.4°. The sum of non-bonded H···Br interactions at this geometry is 0.735, while the corresponding number in CH_3CH_2Br is only 0.137 kcal mol^{-1}. One concludes that e_{sb} has very little effect[74], and that the primary cause for elongation in this case is H···Br repulsions.

In the example cited, a slight expenditure of stretching energy relieves several times its value of non-bonded strain: recall how steep is the repulsive branch of e_{nb} (Figure 1). Bond angles are more pliable than lengths, and their deformation can become even more beneficial to the molecular energy[32]. Still, force fields have been described that constrain C—Hal lengths and even bond angles to fixed values[15,42,75,76].

It has been estimated[77] that the C—H bond length in ethane changes by 0.019 Å on going from the staggered to the eclipsed rotamer. In halides, rotamerization is expected likewise to be attended by changes in bond lengths and angles. The energetic implications may be illustrated by the following results[42] for barriers to internal rotation in CCl_3CH_2Cl, CCl_3CHCl_2 and CCl_3CCl_3. The barriers computed for rigid rotation (that is, without allowing relaxation in bond lengths and angles) were grossly exaggerated: 17.7, 29.4 and 49.5 kcal mol^{-1}, versus the experimental 10.0, 14.2 and 17.5[60]. Yet, when θ(CCCl) was increased by 1° in eclipsed CCl_3CCl_3, the computed barrier diminished to 25.1 kcal mol^{-1} – a saving of about 24 kcal in E_{nb} at an expense estimated as no more than 1.5 kcal in E_b!

Hydrogen bonding leaves its mark on the stereochemistry of some halides. Thus[78], the major conformations of 2-fluoroethylamine in the gas phase (**6**, **7**) have a F/NH_2

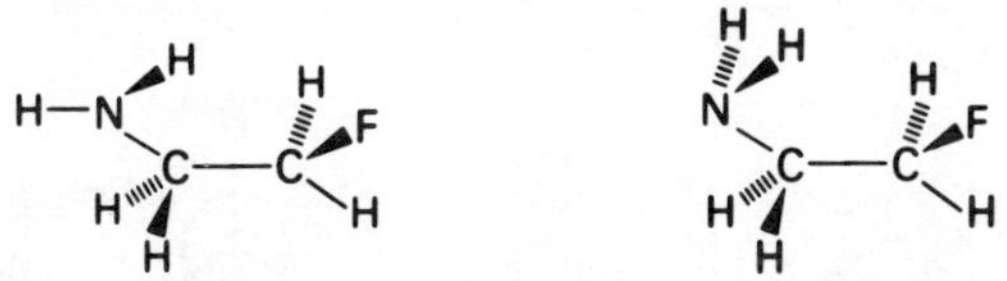

(**6**) (**7**)

skew-relationship. In 2-fluoroethanol the *gauche* conformer is exclusive in the vapour and predominant in the liquid[79] (Section IX). Ways to incorporate hydrogen bonding in the force field procedure have been developed[80,81] but not applied as yet to simple halides.

The opinion is sometimes expressed[82] that 'with so many components in E_t and so many parameters in these components, anything on Earth can be shown'. The following two cases, where both quantum chemistry[30] and molecular mechanics are still at sea, are of some interest in this regard.

A. The 1-Halopropane Problem

There is evidence[61,83–85] that the *gauche* form in 1-halopropanes (**8**) is somewhat more stable (not just more abundant) than the *anti* form (**9**). Dipole–dipole interaction was invoked as an interpretation[42,86] as well as X···CH_3 attraction[27]. In one study[42], the effect was reproduced in $CH_3CH_2CH_2F$ but not in the chlorine and bromine derivatives. The differences, $\Delta E_t = E_{t,9} - E_{t,8}$, came out in that study as follows (all values in kcal mol^{-1}):

	Fluoride	Chloride	Bromide
ΔE_{nb}	−0.06	−0.48	−0.18
ΔE_{es}	0.49	0.37	0.14
ΔE_{tor}	−0.01	−0.18	−0.12
ΔE_t	+0.44	−0.29	−0.16

This suggests that ΔE_{nb} and ΔE_{es} act in opposition, the former favouring X/CH_3-*a* (**9**), the latter favouring X/CH_3-*g* (**8**), such that ΔE_t reflects a very delicate balance. What still displeases the stereochemist is not that $E_{t,9}$ exceeds $E_{t,8}$. Rather[87], as stressed in Figure 3, it is that *gauche*-$CH_3CH_2CH_2X$ is the structural analogue of axial-$C_6H_{11}X$, *anti*-$CH_3CH_2CH_2X$ the analogue of equatorial-$C_6H_{11}X$, and in monohalogenated cyclohexanes the equatorial conformer is the more stable[11]. Consequently, a force field that fits 1-halopropanes would hardly fit halocyclohexanes, and vice versa. In recent force fields parameters were chosen so as to fit to halocyclohexanes[47,48].

Conversely, one wonders what would happen if a field was parametrized to fit

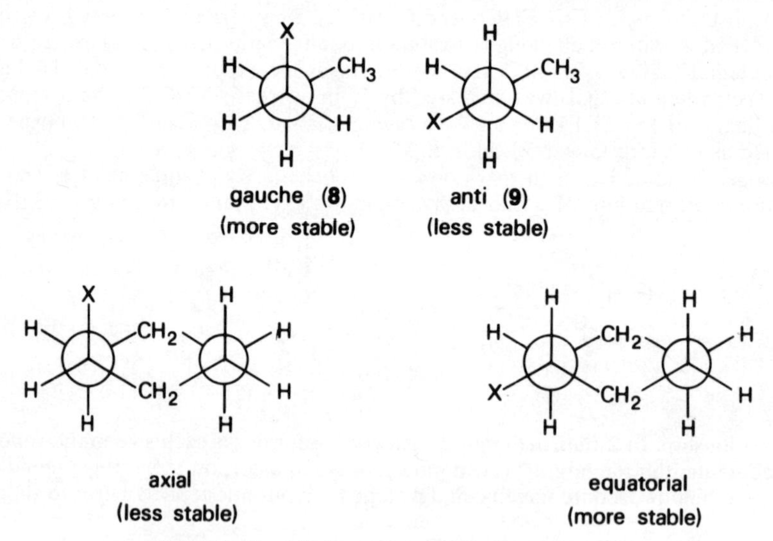

FIGURE 3. 1-Halopropanes and halocyclohexanes.

1-halopropanes and then applied, say, to 1-fluoro-2-propanol. Because of hydrogen-bonding, F/OH-*a* (**10**) can be excluded, and analogy with $CH_3CH_2CH_2F$ suggests that F/H-*a* (**11**) would be preferred to F/CH_3-*a* (**12**). A microwave study, however, indicates that the predominant rotamer is **12**, not **11**[88]. The same goes for epifluorohydrin[89], where the more stable conformer is F/CH_2-*a* (**13**), not F/O-*a* (**14**) nor F/H-*a* (**15**).

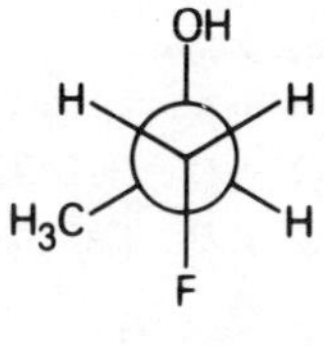

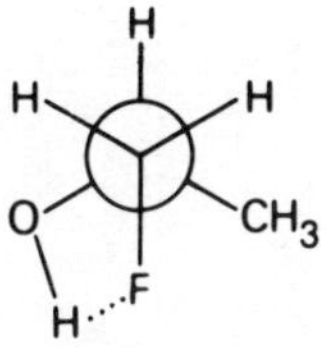

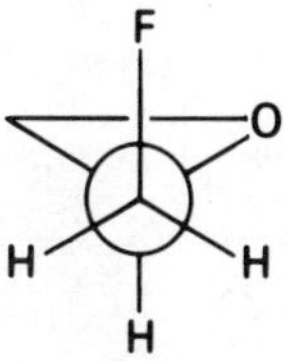

B. The 1-Halopropene Problem

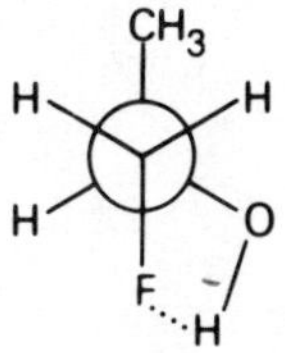

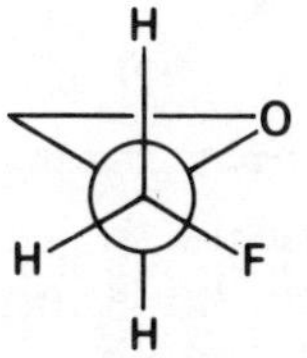

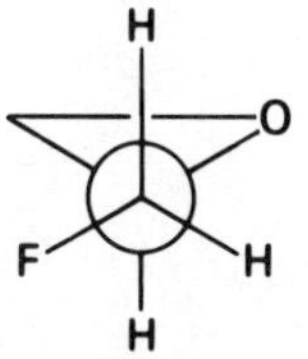

Another problem, and an obvious test case for both quantum chemistry[90] and molecular mechanics, has to do with *cis–trans* isomerism and the barrier to methyl rotation in 1-halopropenes, $XCH{=}CHCH_3$. The pertinent numbers are assembled in Figure 4. In brief, the *cis*-form (**16**) is more stable than the *trans* (**18**), and the barrier is higher in *trans* (**19**) than in *cis* (**17**).

The problem came up in a molecular-mechanical study of internal rotation in substituted propenes[94], where the energy equation did produce somewhat lower barriers in the *cis* isomer (for X = Cl, Br). However, the order of isomer stabilities was reversed, with *cis* being calculated as less stable than *trans*. More recently, calculations of fluoro-olefins were reported[27] wherein an acceptable value was obtained for ΔE(*trans–cis*) in $FCH{=}CHCH_3$. This time, however, the barriers to internal rotation in *cis* and in *trans* came out too close.

III. INTRAMOLECULAR ELECTROSTATICS

The estimation of intramolecular electrostatic interactions (E_{es}) is inseparable from the computation of molecular dipole moments. In reviews of the force field method the topic is touched at only briefly[98], probably because of the tendency to ignore E_{es} and dipole moments in dealing with saturated hydrocarbons[35].

We denote by μ(CX) the moment of bond C—X, and by μ the overall moment of a molecule. The organic chemist's 'Jeans formula'[99] (for Böttcher's 'ideal dipole'[100a]) is herein called 'the point-dipole equation'. The *local* (or 'microscopic') dielectric constant, a parameter attributed to intramolecular space, is denoted d; ϵ refers to the *bulk* dielectric constant of a liquid or a vapour.

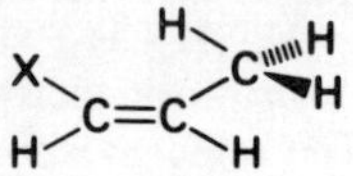

(16)
cis, minimum
$E = 0$

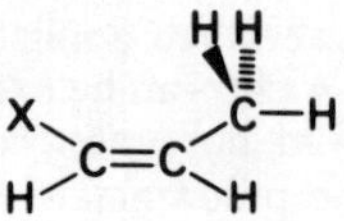

(17)
cis, maximum
E (above *16*):
1.06 (X = F, Ref. 91)
0.62 (X = Cl, Ref. 92)
0.42 (X = Br, Ref. 93)

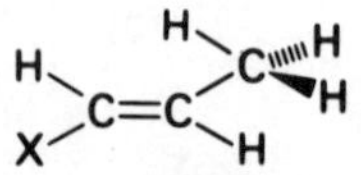

(18)
trans, minimum
E (above *16*, Ref. 95):
0.75 (X = F)
0.76 (X = Cl)
0.73 (X = Br)

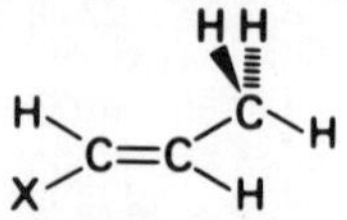

(19)
trans, maximum
E (above *18*)
2.20 (X = F, Ref. 96)
2.17 (X = Cl, Ref. 97)

FIGURE 4. Energy relationships (in kcal mol^{-1}) in 1-halopropenes.

A. Bond Moments

The justification for ignoring polarization in C—C and C—H bonds is that long-range electrostatic interactions in saturated hydrocarbons are of little consequence, short-range interactions still small and accountable through e_{tor} and e_b, and dipole moments frequently of no concern. The dipole moments of unsubstituted alkenes and alkynes are not insignificant, as Table 1 shows[101]. Still, they can be compounded to a

TABLE 1. Dipole moment (in debyes) of some hydrocarbons

Compound	Dipole moment, D	Reference
$CH_3CH_2CH_3$	0.08	101
$(CH_3)_3CH$	0.13	101
$CH_2{=}CHCH_3$	0.36	102
$CH_2{=}C(CH_3)_2$	0.50	102
$CH_2{=}C(CH_3)CH{=}CH_2$	0.25	102
$(CH_3)_2CHCH{=}CH_2$	0.3–0.4	65
$CH_3CH_2C{\equiv}CH$	0.75	103
$CH_3(CH_2)_2C{\equiv}CH$	0.81	104
$CH_3(CH_2)_3C{\equiv}CH$	0.83	104
$CH_3(CH_2)_4C{\equiv}CH$	0.85	104
$(CH_3)_2CHC{\equiv}CH$	0.72	103
$(CH_3)_3CC{\equiv}CH$	0.66	103
Vinylcyclopropane	0.50	102
Ethynylcyclopropane	0.89	103

fair approximation from C ↛ C= and C ↛ C≡ bond moments, with μ(CH) and paraffinic μ(CC) kept at zero [31,105]. Monosubstituted alkanes can be treated analogously, by ascribing a moment only to the bond from carbon to the substituent. E_{es} is then zero[47]. E_{es} would come out non-zero if the molecule contains two or more bonds that are not C—H or paraffinic C—C (e.g. $CH_3C{\equiv}CCl$, $ClCH_2CH_2Cl$, but not $CH_3CH_2CH_2Cl$). In most applications the interaction between geminal dipolar bonds (e.g. in CH_3CHCl_2) is also ignored.

However, experimental dipole moments are not constant even in simple series[106]: for example[101], CH_3Cl, 1.89; CH_3CH_2Cl, 2.04; $CH_3CHClCH_3$, 2.17 D. So, if theoretical dipole moments are to be compounded from bond moments, either μ(CX) is to be admitted wild or else the polarization in C—C and C—H bonds is to be included. Furthermore, C—H bonds in alkanes, alkenes and alkynes are electrically dissimilar[107] and their moments cannot be neglected simultaneously. A striking illustration[108] of what happens when μ(CC) and μ(CH) are ignored is reproduced in Table 2.

A reasonable way to proceed for a given type of skeleton (say, alkane) and substituent (say, chlorine) could be as follows[45]. Select a value for μ(CH), either from calculation[107] or experiment[109,110], by arbitrary choice or some roundabout means[111]. Use the experimental geometry and dipole moment of CH_3X to estimate μ(CX) by vectorial decomposition. Proceed to CH_3CH_2X and derive $\mu(C_\beta C_\alpha)$. Experience shows that the polarization in C—C bonds further away can be ignored. Then turn to disubstituted molecules of unambiguous conformation, such as *cis*-1,2 (**20**) or *cis*-1,4 (**21**)

(20) **(21)**

disubstituted cyclohexane, and use their dipole moments to estimate the *apparent* bond moment μ^*(CX), that is, the bond moment of C—X when close to another strong dipole[112]. Do this for several molecules and formulate μ^*(CX) as a function of the distance and relative orientation of C—X with respect to other strong dipoles in the molecule. This is the approach of Miyagawa (1954), put to the test (1964) by Quivoron and Néel[111], and recently revived[27,45,46,48]. We return to it in Section III.D.

A sample of what can be done is given in Table 3. The last entry refers to the optimized geometry of 1a,2e,3e,4e,5a,6a-hexachlorocyclohexane (gammexane), a

TABLE 2. Apparent bond monents (in debyes)[108]

Compound	Dipole moment, D	Apparent C—Cl moment[a], D
CH_3Cl	1.86	1.86
CH_2Cl_2	1.60	1.39
$CHCl_3$	0.95	0.95
CH_3CH_2Cl	2.03	2.03
CH_3CHCl_2	2.05	1.78
CH_3CCl_3	1.77	1.77
CH_2ClCCl_3	1.39	1.24
$CHCl_2CCl_3$	0.92	0.92

[a]Assuming $\mu(CH) = \mu(CC) = 0$.

TABLE 3. Dipole moments (in debyes) of saturated chlorides[45,101]

Compound	Dipole moments, D	
	Calculated	Experimental
CH_3CH_2Cl	2.04	2.04
$CH_3CH_2CH_2Cl$	2.05	2.05
$CH_3CHClCH_3$	2.14	2.17
$(CH_3)_3CCl$	2.22	2.15
Chlorocyclohexane	2.15	2.1–2.2
cis-1,2-Dichlorocyclohexane	3.13	3.12
cis-1,4-Dichlorocyclohexane	2.87	2.89
trans-1,2-Dichlorocyclohexane		
ax,ax	0.95	1.07[a]
eq,eq	3.24	3.32[a]
2β,3α-Dichloro-5α-androstane	1.23[b,c]	1.27[116]
2α,3β-Dichloro-5α-androstane	3.22[b,d]	3.44[116]
1a,2e,3e,4e,5a,6a-Hexachloro-cyclohexane (gammexane)	2.62[b]	2.82[111]

[a]Estimated independently of bond-moment summation[113].
[b]Calculated[114] as prescribed in Ref. 45.
[c]$\omega(Cl_{ax}CCCl_{ax})$, 153° (exptl[115], 156°).
[d]$\omega(Cl_{eq}CCCl_{eq})$, 61° (exptl[115], 61°).

molecule with six interacting strong dipoles. This geometry is far from standard but still not as far as the geometry reported for the crystal[117]. *Inter alia*, the $C^1C^6C^5$ valency angle is computed (113.5°) larger than other CCC angles but not as wide as reported for the crystal (123°). θ(CCCl), 110.3–112.0° (versus 103–116° in the crystal); ω(ClCCCl), 159.2° (*ax*, *ax*), 51.6° (*ax*, *eq*), 54.6° (*eq*, *eq*). When the calculation was repeated with chlorine interactions 'turned off', angles relaxed to a more standard geometry, with θ(CCCl) in the range 108.6–109.7° and ω(ClCCCl), 166.9° (*ax*, *ax*), 50.6° (*ax*, *eq*), 62.4° (*eq*, *eq*). The computed moment (2.79 D) became closer to the experimental. An earlier calculation[111] assumed standard geometry (tetrahedral valency angles, dihedral angles of 60° or 180°) and produced 2.84 D. Details on the two calculations are given in Table 4.

The data in Table 4 show that the quality of the geometry that one assumes cannot be tested by the quality of the dipole moment that one computes, and that the set of

TABLE 4. Two estimations of bond moments and the dipole moment (in debyes) of gammexane

Geometry	Early calculation[111] Standard	New calculation[114] Optimized
μ°(CCl)	2.05	1.60
μ(CH)	0.35 (H negative)	0.33 (H positive)
$\mu(C^1Cl)$	1.91	1.03
$\mu(C^2Cl)$	1.70	1.00
$\mu(C^3Cl)$	1.70	1.02
$\mu(C^4Cl)$	1.70	1.00
$\mu(C^5Cl)$	1.91	1.03
$\mu(C^6Cl)$	2.58	1.17
Dipole moment	2.84	2.62

bond moments that reproduces a given dipole moment is not unique. Since dipole–dipole interactions are taken proportional to bond moments and made to depend on the geometry (equation 3), it follows that the electrostatic component in E_t could vary appreciably from one force field to another.

A further difficulty concerns the moment ascribed to the C—H bond. If it is to be picked up at will, one may well ask whether $\mu(CH) = 0$ is a good choice. Test cases are rare and not obvious, since C—H moments are small anyhow, often cancel (e.g. *ortho* and *meta* C—H moments in chlorobenzene) or can be incorporated in the moments of other bonds (*para* C—H in chlorobenzene). In a recent study of organic amines[118], moments were alloted only to C ↛ N (0.04 D), H ↛ N (0.76), N ↛ lone pair (0.60), but not to C—H bonds. As a consequence, the computed dipole moment depended only on the pattern of substitution, not on the nature of the alkyl groups: one obtained 1.33 D for monoalkylamines, 1.10 for dialkylamines, 0.64 for trialkylamines. This is quite representative as open-chain amines go, but the number 0.64 D was by necessity obtained also for quinuclidine (**22**), where the experimental value is much higher: 1.17 in C_6H_{12} or 1.22 D in C_6H_6[119]. The discrepancy (~0.5 D) is just of the order[120] of $\mu(CH)$, and indicates that the moment of bond C^4—H ought not to be omitted.

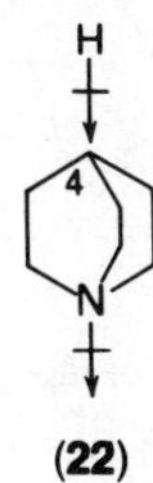

(**22**)

Finally, a famous paper by Coulson[121] led to confusion as to the direction of the C—H bond moment[122]. It seems to be accepted nowadays[120,123a] that the proper sense should be $^-CH^+$.

B. Point-dipole Approach

It is helpful to recall the derivation[57] of the point-dipole equation (Section II, equation 3) by using Figure 5.

Let a dipole of moment $\mu_i = ql$ be defined by charges $-q$ and $+q$ at a distance l. The potential created by this dipole at a point $P(r, \theta)$ is

$$V\,(\text{at P}) = \frac{q}{r_i} - \frac{q}{r_j} = q\,\frac{r_j - r_i}{r_i r_j}.$$

If the distance l is sufficiently small with respect to r, $r_j - r_i \sim l \cos\theta$ and $r_i r_j \sim r^2$, so that $V = \mu_i \cos\theta / r^2$ and the components of the field $\mathscr{E}$ at P become $\mathscr{E}_r = 2\mu_i \cos\theta / r^3$ and $\mathscr{E}_\theta = \mu_i \sin\theta / r^3$. On introducing unit vectors and further development one obtains an expression for $\mathscr{E}$(at P), and then an expression for the potential energy of a dipole μ_2 placed at P, $e_{es} = -\boldsymbol{\mu}_j \cdot \mathscr{E}$. Division by the dielectric constant d provides the relation sought,

$$e_{es} = \mu_i \mu_j (\cos\chi - 3\cos\alpha_i \cos\alpha_j)/dr^3.$$

Strictly speaking, then, equation (3) applies intramolecularly only if the lengths of the interacting dipoles are 'small enough' [124] with respect to their distance. Removal of this limitation is possible[100a] but has not been attempted. Other weak points are that bond moments are not unequivocal (Section III.A) and that it is hard to pinpoint the

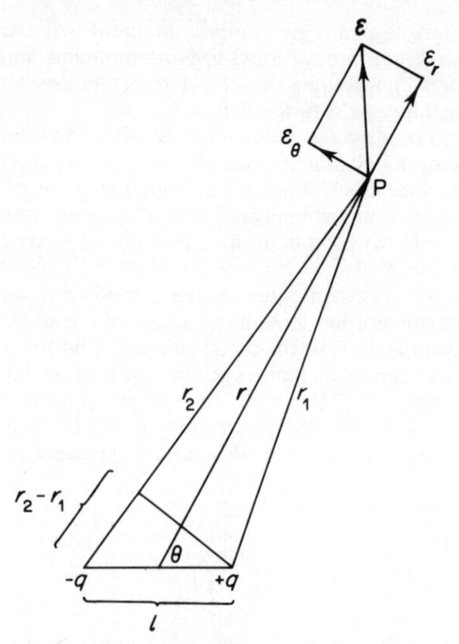

FIGURE 5. Field of dipole ql at point P.

centre of the dipole[58]. In practice one usually places μ(XY) at the midpoint of bond X—Y. Because of the uncertainties, conclusions as to the reliability of the equation have to be based upon its performance. In particular, it is relevant to check whether results by equation (3) compare with the sum of Coulomb interactions, as given by

$$e_{\mathrm{es}} = \sum_{s} \sum_{t} \frac{q_s q_t}{d r_{st}}. \tag{6}$$

This equation is a variant of equation (2), where q_s are the two charges that define μ_i and q_t the two charges that define μ_j. A literature scan produced meagre findings: for most molecules that have been calculated by both formulations, the moments used with equation (3) were inconsistent with the charges used with equation (6).

One case in which equation (3) was checked against equation (6) is the *E–Z* equilibrium in 2-halobenzoic esters[125] (**23**, **24**, with R = CH_3 or CH_3CH_2 and X = F,

E **(23)** Z **(24)**

Cl, Br, I). By dipole moments (in benzene solution) and the carbonyl stretching frequency (in decalin solution) at room temperature, conformation *E* prevails for X = F (~64%, R = CH_3), but its abundance diminishes as X grows (~45% for X = I, R = CH_3). In the calculations, moments were first assigned to bonds, and the charges on individual atoms were then obtained by dividing the bond moment by its length and summing at each atom. Predictions by the charge formula and point-dipole equation different at most by 6% in the population, being 3% on the average. The charge calculation yielded 68% *E* for X = F, and 62% *E* for X = I (R = CH_3).

A statement to the contrary was enunciated by Stolow in the late 1960s[126]. The conformational equilibria (chlorine equatorial ⇌ axial) in 4-chlorocyclohexanone (**25**) and its ethylene ketal, 8-chloro-1,4-dioxaspiro[4.5]decane (**26**) were under scrutiny.

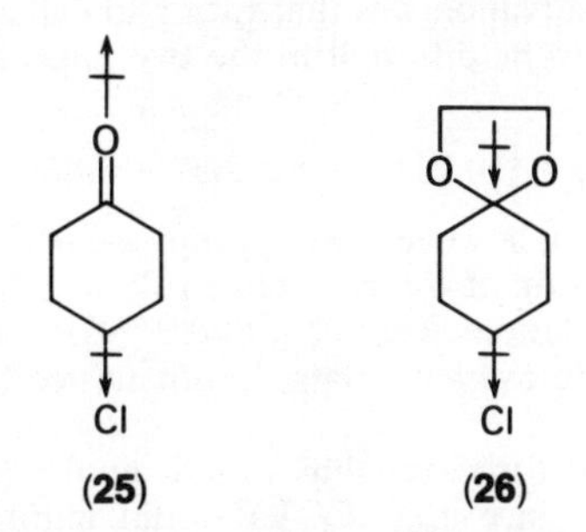

By NMR there is an axial excess in both compounds: ketone, 67% axial in CCl_4 solution, 31°C; ketal, 58% axial in $CFCl_3$, −74°C. Stolow argues that whichever conformation (axial or equatorial) happens to be computed as the more stable in **25** would be computed as the less stable in **26**, since in **25** the moment of the carbonyl points *away* from the six-membered ring while in **26** the dioxolane moment points *towards* it. Put differently, cos χ in equation (3) changes sign on passing from **25** to **26**, and so would the computed e_{es}.

In Stolow's reasoning, the entire *eq*-*ax* energy difference is ascribed to electrostatic effects and variations in steric strain are ignored. Also, the interaction between two polar groups only (C=O and C—Cl in **25**, dioxolane and C—Cl in **26**) is taken to represent the entire electrostatic term. In addition, effects of solvent are ignored. Even so, one is curious to know how badly such an approximate calculation for the ketal could turn out since cos χ in the ketone has the right sign to make the axial conformation more stable. To check this point[127] the geometry of the conformers was optimized by standard procedures[128], standard bond moments were then ascribed to C—Cl (1.7 D[129]) and to the dioxolane (1.47 D[130], pointing towards the six-membered ring) and the interaction was computed by equation (3). With d = 1 this gave $e_{es,eq} - e_{es,ax} = -0.19$ kcal mol^{-1}, whereas experimentally $-\Delta G° = +0.13$ kcal mol^{-1} in solution. The sign is wrong, of course, but agreement is still to within 0.3 kcal mol^{-1} (or better, if $d > 1$).

Optimization of **26** revealed, incidentally, that it is not a 'reversed counterpart' of **25**. The dipole angle χ(C=O/C—Cl) differs markedly in the two conformations of the ketone, being 108° in the axial and 133° in the equatorial form[47]. In the ketal, by contrast, χ(dioxolane/C—Cl) comes out as 52° in the axial and 54° in the equatorial, thus virtually independent of conformation.

One has even contested the applicability of equation (3) to the ketone **25**[131]. In a confrontation of equations (3) and (6), using charges consistent with bond moments, the C=O and C—Cl dipole lengths were taken as 1.22 and 1.5 Å (equation 3) and the charges distanced at 1.22 and 1.56 Å (equation 6). It is claimed that the computed

Δe_{es} was acceptable only with the charge formula while the point-dipole prediction deviated by 0.47 kcal mol^{-1}. The geometrical parameters that computation requires (r, χ, α_1, α_2, see Figure 2) were measured on a Dreiding model. They are not cited in the publication and cannot be checked against the optimized numbers[47].

The two approaches were applied to *trans*-1,4-dichlorocyclohexane[132], but with inconsistent charges and bond moments. The same was done for vicinal dibromides[47]. In $BrCH_2CH_2Br$, for example, *gauche* was computed less stable than *anti* by 1.68 (dipole) or 1.67 kcal mol^{-1} (charge), versus the experimental[133,134] 1.7 kcal mol^{-1} or more. In *trans*-1,2-dibromocyclohexane *ee* was obtained above *aa* by 1.65 or 1.49 kcal mol^{-1} (experimental[113,135] > 1.5 kcal mol^{-1}); in *meso*-$CH_3CHBrCHBrCH_3$, Br/Br-*g* is above Br/Br-*a* by 1.73 or 1.54 kcal mol^{-1} (experimental[136,137] ~1.5 kcal mol^{-1}). An interesting observation was that fitting to experimental energies required e_{tor}(BrCCBr) to be taken to be different in the two approaches: in conjunction with equation (3),

$$e_{tor}(\mathrm{BrCCBr}) = \tfrac{1}{2}[0.71(1 - \cos 2\omega) - 0.83(1 + \cos 3\omega)];$$

when point-charges are used, however, the appropriate position was $e_{tor}(\mathrm{BrCCBr}) = 0$. This is another demonstration of the corrective role of E_{tor} (Section I).

Some point-dipole calculations were reported[46,48] for molecules with dipoles at a 1,3-relation (XCCCY). The available data do not suffice for stringent checks of the results.

Whether equation (3) performs reliably for 1,2-dipoles (XCCY) is a question that can hardly be answered, since e_{tor}(XCCY) would automatically smooth out any insufficiencies. As to 1,1-dipoles (XCY), their energy of interaction need not be computed at all: most of it cancels when taking differences (e.g. when comparing eclipsed with staggered CH_3CHXY), and the remainder is implicitly incorporated in e_b(XCY) during the process of parameter-fitting.

A final remark concerns the local dielectric constant. Since d occurs in the denominator of interaction terms, choosing a high value for this constant is tantamount to de-emphasizing intramolecular electrostatic effects. In the early days of molecular mechanics one looked up to theory to supply appropriate values for d. Kirkwood and Westheimer's work[138] was taken as model[139]. The model has since been further developed[140,141] but nowadays one tends to regard d as one of the adjustable constants in the force field[142] or as a parameter to be selected on pragmatic grounds[18]. Table 5 gives an idea, not too gratifying, of the present status of the problem.

C. Point-charge Approach

In this alternative, charges are distributed at certain points in the molecule – actually at the atoms – and the electrostatic interaction is given in equation (7);

$$E_{es} = \sum_{\substack{i,j \\ (j \neq i)}} \frac{q_i q_j}{d r_{ij}}. \tag{7}$$

Here q_i is the excess charge (positive or negative) associated with atom i, r_{ij} is the distance between point charges i and j. The components of the dipole moment are given by equation (8),

$$\mu_u = \sum_i q_i u_i, \tag{8}$$

where $u = x$, y or z, and the total moment is

$$\mu(\mu_x^2 + \mu_y^2 + \mu_z^2)^{1/2}.$$

TABLE 5. Local dielectric constant

Year	d	Context	Reference	Justification given
1970	1	Halides	42	$d \neq 1$ is tantamount to including d in quantum-chemical Hamiltonian
1971	1	Substituted cyclohexane	143	$d \neq 1$ is nonsense
1975	1	EPEN force field	15	Better than guess
1968	1 or 2	Anomeric effect	144	Conformation-dependent
1978	1.5	Diketone in gas phase	145	Distinction from solution
1980	1.5	Dihalides	47	$1 < d < \epsilon$
1980	1.96	Chrysanthemic acid	142	Optimization
1958	2	Haloketones	139	Close to ϵ in hydrocarbon solvents
1978	2	Dihalides, haloketones	146	Molecule (not vacuum) between dipoles
1977	2.27	2-Halobenzoic acids in benzene	125	$d = \epsilon$
1979	3	Alcohols and ethers	18	With small d, optimization all but impossible
1980	3	Carbohydrates	147	Geometries better than with $d < 3$
1965	3.5	Polypeptides	148	As measured for solids
1971	3.5	Sugars	149	As advocated[148]
1974	3.5	Ethanediol and ethers	55	Close to ϵ of solvents
1966	4	Polypeptides	150	Close to value advocated[148]
1978	4	Halides	46	With smaller d, E_{es} dominates E_t
1980	4	[2.2]paracyclophanes	151	By fitting
1969	5	Bicyclo[2.2.2]octanoic acids	152	Theoretical considerations
1980	5	Ethers	153	By fitting and analogies
1962	5.5	Bicyclo[2.2.1]heptadiene	154	As in diamond
1978	8	Diketone dissolved in benzene	145	Effective ϵ of solvent
1976	Very large	Bicyclo[2.2.2]octanedioic acid	141	Theoretical considerations
1979	Very large	Histamine in solution	155	To eliminate E_{es}

Various schemes are being used to partition the charge. Charges were taken to comply with bond moments[55,106,156], or treated as atomic constants[80] (sometimes[18] readjusted further to preserve $\sum q_i = 0$), or the product $q_i q_j$ considered a pair-constant[42]. More frequently, however, charges are derived by a preliminary quantum-chemical calculation[128,142,157,158]. This brings in a dilemma of consistency[157]: unlike their employment in molecular mechanics, the quantum-chemical 'charges' do not relate to the dipole moment through equation (8).

In brief, let $\varphi_k = \Sigma(\mu) c_{k\mu} \chi_\mu$ be a molecular orbital, where χ_μ represents an atomic orbital tied at atom M by normalization,

$$\begin{aligned} \langle \varphi_k | \varphi_k \rangle &= \sum_\mu \sum_\nu C_{k\mu} C_{k\nu} S_{\mu\nu} \\ &= \sum_\mu C_{k\mu}^2 + 2 \sum_{\substack{\mu,\nu \\ (\nu \neq \mu)}} C_{k\mu} C_{k\nu} S_{\mu\nu} = 1. \end{aligned} \tag{9}$$

If overlap integrals are ignored or not explicited,

$$\sum_\mu C_{k\mu}^2 = 1, \tag{10}$$

and one may identify $c_{k\mu}^2$ with that portion of the charge in χ_μ that is associated with electrons in φ_k. Let n_k denote the number of electrons in φ_k (usually $n_k = 2$). The gross charge on atom M is

$$q_{\mathrm{M}}^{\mathrm{G}} = \sum_{\mu \in \mathrm{M}} \sum_{k\ \mathrm{occ.}} n_k c_{k\mu}^2. \tag{11}$$

By subtracting the number of electrons that atom M donates, one obtains the excess charge (or simply, the charge) q_{M} on M. These q_{M} values fit equation (7) and are consistent[159] with equation (8). Hückel π-charges fall in this category. They have been used widely in the past to compute dipole moments of unsaturated hydrocarbons[160] and, in conjunction with σ-charges, of unsaturated heterocycles[161].

When overlap integrals are not suppressed, equation (9) does not reduce to equation (10). Following Mulliken[162], one lets μ and ν share equally the (μ, ν)-component in the right-hand term of equation (9), which leads to a gross atomic population:

$$\begin{aligned} q_{\mathrm{M}}^{*\mathrm{G}} &= \sum_{\mu \in \mathrm{M}} \sum_{k\ \mathrm{occ.}} n_k \left(C_{k\mu}^2 + \sum_{\nu \neq \mu} C_{k\mu} C_{k\nu} S_{\mu\nu} \right) \\ &= q_{\mathrm{M}}^{\mathrm{G}} + \sum_{\mu \in \mathrm{M}} \sum_{k\ \mathrm{occ.}} n_k \sum_{\nu \neq \mu} C_{k\mu} C_{k\nu} S_{\mu\nu}. \end{aligned}$$

Now the gross charge $q_{\mathrm{M}}^{\mathrm{G}}$ is but one of two addends. A complex expression ensues for the dipole moment[123,163], equation (8) does not apply, and it is not clear[164] what to pose in equation (7). Furthermore, Mulliken's partition of charge is itself arbitrary, and not necessarily the most appropriate choice[165].

A very simple way out is to deal with saturated molecules, or with the σ-skeleton of unsaturated molecules, in a manner analogous to the Hückel method. The method of Del Re[166], which has been adapted to halogen compounds[161], is now making its entry into molecular mechanics[155]. Only bond orbitals are here explicited, overlap is suppressed, and extensive parametrization compensates for the missing factors. Consider a bond based on $\chi_\mu \in \mathrm{M}$ and $\chi_\nu \in \mathrm{N}$. The doubly-occupied molecular orbital is

$$\varphi_{\mu\nu} = C_\mu \chi_\mu + C_\nu \chi_\nu$$

with $c_\mu^2 + c_\nu^2 = 1$. The net charge q_μ in χ_μ and the net charge Q_{M} of atom M are, respectively,

$$q_\mu = 1 - 2c_\mu^2 = c_\nu^2 - c_\mu^2 \equiv Q_{\mu\nu},$$
$$q_M = \sum_{\substack{\nu \text{ bonded} \\ \text{to M}}} Q_{\mu\nu},$$

where q_M is the quantity that equations (7) and (8) require. Approximately,

$$Q_{\mu\nu} = \frac{\delta_\nu - \delta_\mu}{2\epsilon_{\mu\nu}},$$

where $\epsilon_{\mu\nu}$ is now a constant of the bond and the constants δ_μ depend through further parameters $\gamma_{\mu\nu}$ on all atoms to which atom M is bonded (equation 12):

$$\delta_\mu = \delta_\mu^0 + \sum_{\substack{\nu \text{ bonded} \\ \text{to M}}} \gamma_{\mu\nu}\delta_\nu. \qquad (12)$$

Thus, the molecular connectivity dictates a system of linear equations in δ_μ and their solution provides $Q_{\mu\nu}$ and hence Q_M. An obvious shortcoming is that the computed charges depend only on the ordering of bonds in the molecule and are not sensitive to rotamerization.

Smith and Eyring's partition of charge[167] suffers in its original formulation from the same deficiency. Here one expresses the moment of bond M—N as a function of its polarizability, length, and of the effective charges and covalent radii of atoms M and N. This provides the charge on M due to bond M—N and, by summation, q_M. Some of the required parameters were adapted from other work, others were assigned by fitting, and the scheme was simplified[168] by putting $\mu(CH) = 0$. The method has been applied to alkyl halides[169], α,ω-dihaloalkanes[170], formyl and carbonyl halides[171], alkenyl halides[172], to fragments derived from alkyl halides[173] and, very recently[174], to 1-chloro-4,4-difluorocyclohexane and to 1,1-*bis*(trifluoromethyl)-4-chlorocyclohexane.

A novel development is the 'modified Smith–Eyring' method (MSE) of Allinger and Wuesthoff[175], where the charge on M is made to depend on all M···N pairs, not just M—N bonds. MSE is sensitive to geometry and has been used in the theoretical analysis of mono- and dihaloalkanes and haloketones[47,146,175].

D. Variability in Bond Moments

In the point-charge methodology, the Smith–Eyring and Del Re constructions operate by 'sigma induction' [176]: a polar bond affects other bonds in the molecule by successive polarization of the intervening bonds. MSE allows also for through-space interaction. In the point-dipole approach, the effect of one polar bond on the moment of another is introduced through 'field induction' [100b,112,177]. Applications, other than in molecular mechanics, have been reviewed[178a,179a,180]. Here we dwell on the practical aspects of computation.

A dipole adjacent to a polarizable group induces a moment in the same direction but of reversed sense from its own. Hence two dipoles in a molecule affect each other in a way to modify – usually reduce – their respective moments. The energy of induction[100b] is amall enough to be ignored in present-day applications, but changes in bond moments and the ensuing changes in e_{es} and E_{es} are significant. For example[45], the dipole moments of monochloroalkanes can be fitted by the set $\mu(CCl) = 1.60$, $\mu(CH) = 0.33$ and $\mu(C_\beta C_\alpha) = 0.28$ D. With these bond moments one would calculate $\mu = 3.49$ D for *cis*-1,2-dichlorocyclohexane, whereas the experimental value is only

3.12 D – corresponding to μ(CCl) = 1.38 D. The C—Cl bond moment is thus reduced by 0.22 D, and e_{es}(C—Cl/C—Cl) by a factor of $(1.38/1.60)^2 = 0.74$. If μ(CH) and $\mu(C_\beta C_\alpha)$ are taken to be zero, μ(CCl) has to be lowered in this particular case by *c*. 0.33 D[181]. This shows not only that the reduction is appreciable but also that its magnitude depends on the computational framework.

The inductive interaction between two dipoles depends on their polarizability, distance, and relative orientation in space[100b]. These factors were taken into account empirically in early studies of 2-haloketones[99]. Thus[182], in the axial conformer of 2-fluoro-3-cholestanone (**27**) the dipole angle χ(C=O/C—F) is large and induction expected to be small; in the equatorial conformer (**28**) the angle is small and the effect large. In calculating dipole–dipole interaction, therefore, μ(C=O) was taken high (3.01 D) in **27** and low (2.84 D) in **28**.

(27) **(28)**

In a study of polychlorinated alkanes, Miyagawa[182a] proposed to reduce the C—Cl bond moment by 0.46 D for each geminal and by 0.24 D for each vicinal C—Cl bond (see Ref. 111). Smyth gives the following table of moment reduction for geminal dihaloalkanes: C—Cl by C—Cl, 0.36; C—Cl by C—F, 0.56; C—F by C—Cl, 0.11; C—F by C—F, 0.21 D[179b].

Let μ_0 be the reference bond moment (before induction is taken into account), μ^* the apparent moment (with induction), and $\Delta\mu = \mu^\circ - \mu^*$. By equation (3) (Section II), equation (13) is obtained:

$$e_{es} = \mu_1 \mu_2 F, \tag{13}$$

where $F = (\cos\chi - 3\cos\alpha_1\cos\alpha_2)/dr^3$. Denoting by λ_2 the longitudinal polarizability of bond 2, the longitudinal change in μ_2 ($\Delta\mu_2$) due to induction by μ_1 is given by equation (14):

$$\Delta\mu_2 = \mu_2^\circ - \mu_2^* = \lambda_2 \mu_1 F. \tag{14}$$

In a pioneering investigation, Quivoron and Néel derived μ^* for C—Cl bonds in a number of 1,2,3,4,5,6-hexachlorocyclohexanes, assuming idealized geometries[111]. For example, 1,3-diaxial C—Cl bonds were taken parallel, so that $\cos\chi = 1$, $\cos\alpha_1 = \cos\alpha_2 = 0$ (Figure 2). With $d = 1$,

$$\mu^* = \mu^\circ - \frac{\lambda\mu^*}{r^3}$$

or

$$\mu^* = \frac{\mu^\circ}{1 + \lambda/r^3}$$

In this manner, overall dipole moments were expressed as function of μ°(CCl), μ(CH) and λ, and these parameters were estimated by comparison of the resulting expressions with experimental dipole moments.

The approach has been made compatible with the force field procedure and applied to fluoro-[46], chloro-[45] and bromoalkanes[48], and to fluorinated olefins[27]. Only the induction in C—Hal bonds is considered: bonds of other types (C—C, C=C, C—H) are either not polarizable enough and/or not polarizing enough, and their moments are small anyhow. For C—Hal bonds, one approximates equation (14) further to get

$$\mu_2^* = \mu_2^\circ - \lambda_2\mu_1 F \sim \mu_2^\circ - \lambda_2\mu_1^\circ F$$
$$= \mu_2^\circ - (\lambda_2/\mu_2^\circ)\mu_1^\circ\mu_2^\circ F = \mu_2^\circ - k_2 e_{es}^\circ$$

or

$$\Delta\mu_2 = k_2 e_{es}^\circ$$

Here, e_{es}° is the interaction (equation 13) at the reference values of bond moments, and k_2 is taken as a constant of the bond. If the molecule contains several C—X bonds, equation (15) is obtained:

$$\Delta\mu(CX^i) = \sum_{n \neq i} k_{i(n)} e_{es}^\circ(C—X^i/C—X^n). \tag{15}$$

In the force field calculation, a first optimization is performed with the reference bond moments μ°. This provides e_{es}° for all pairs, at a geometry that is virtually final. Equation (15) then furnishes the apparent moments μ^* and the calculation is repeated to provide the final dipole moment and the final $E_{es} = \sum e_{es}$. As an example, the apparent C—Cl bond moments in gammexane were given in Table 4.

IV. SOME HALIDE FORCE FIELDS

It is advantageous to base the force field for a group of functionalized hydrocarbons on a well tested force field for unsubstituted hydrocarbons. This helps when effects of substitution are to be analysed, or when parameters for dissimilar substituents have to be used conjointly (e.g. halogen and carbonyl in haloketones). Many of the results cited in this chapter were obtained by an extension[27,45,46,48] of Allinger's 1973 force field[183]. In this scheme, E_t is the sum of E_s, E_b, E_{sb}, E_{nb}, E_{tor}, E_{tb} (torsion–bend cross-term), and E_{es}. For E_{es} the point-dipole approach is used, with $\mu(^-CH^+$, paraffinic) = 0.33, $\mu(^-CH^+$, olefinic) = 0.52 D, and $d = 4$.

In 1977 Allinger came up with the MM2 program[49] which was later extended to a variety of functional groups, including halides[47]. In this set-up, $d = 1.5$ and E_{es} is evaluated at choice, either by the point-dipole formula with $\mu(CH) = 0$ or through MSE and the point-charge formula (Section III.C). The computer program also furnishes the heat of formation of monohaloalkanes[44]. The point-charge option has been rendered compatible with the computation of solvation energies.

In conjunction with electron diffraction studies, Stølevik and coworkers calculate conformational energies, geometries, barriers to internal rotation and torsional force constants[184]. Their energy is given by equation (16):

$$E_t = E_s + E_b + E_{tor} + E_r, \tag{16}$$

where $e_s = \frac{1}{2}k_s(l - l_0)^2$, $e_b = \frac{1}{2}k_b(\theta - \theta_0)^2$, $e_{tor} = \frac{1}{2}V_3[1 + \cos 3(\omega - \omega_0)]$, and e_r includes van der Waals' repulsion and attraction and the charge interaction. Applications centre about polyhalo-derivatives of open-chain alkanes, e.g. $ClCF_2CF_2CF_2Cl$[185] and $Cl_2CHCH_2CHCl_2$[186].

The last two terms in equation (16) define Abraham and Parry's pioneering force field for halides[42], wherein flexibility in bond lengths and valency angles was not permitted and dipole moments were not sought. Results by this field for a large number of mono- and polyhalogenated alkanes[42,187], and for halogenated aldehydes and acyl halides[188], have been reported.

V. FLUORINE COMPOUNDS

Fluorine seems sometimes to have a logic of its own. Of the two conformations available to compounds of type XCH_2CH_2X, *gauche* (**29**) and *anti* (**30**), fluorine prefers the *gauche*[189–191] (by 1–2 kcal mol^{-1}) while other halogens prefer the *anti*[192a] (by 1–2 kcal mol^{-1}). *Gauche* corresponds also to the global minimum[193] in FOOF and FSSF, as it does in the parent hydrides HOOH and HSSH. The tendency of fluorines to approach each other is further illustrated by FCH_2COF: the conformers here are[194] F/F-*a* (**31**) and F/F-*syn* (**32**), while these are Cl/Cl-*a* (**31**) and Cl/Cl-*g* (**33**) in $ClCH_2COCl$[195]. F_2CHCOF is a mixture of F/H-*g* (**34**) and F/H-*a* (**35**), of which **35**, the conformer with clustered fluorines, is lower in energy by 0.26 kcal mol^{-1}[196]. Fluorine crowding in this molecule is assisted by the tendency of vicinal C—H and C=O bonds to eclipse[197].

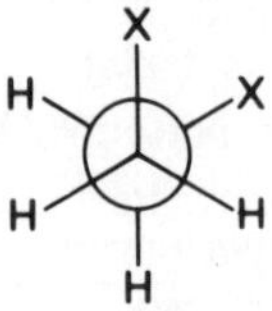

(**29**) 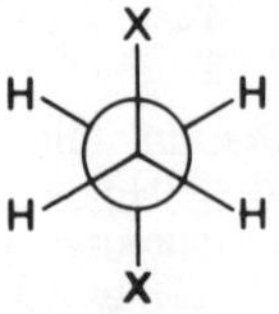 (**30**)

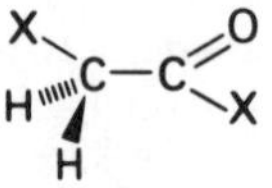

(**31**) 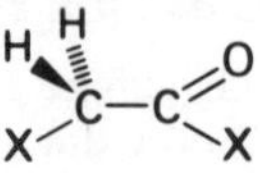 (**32**) 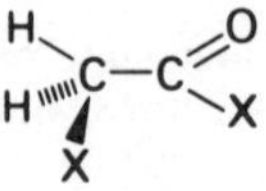 (**33**)

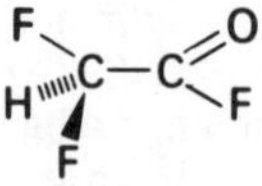

(**34**) 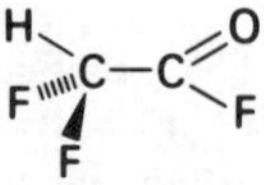 (**35**)

Cases are known, on the other hand, where the conformation of lower energy has the fluorines farther apart than in the alternative. These are not devoid of their own peculiarities. In 1,1,2,2-tetrafluoroethane[198] conformation **36** is stabler than **37**, as it is in the chlorine and bromine analogues; the reported energy difference (~1.2 kcal mol^{-1}), however, exceeds the values reported[199,200] for $Cl_2CHCHCl_2$ (0.5–0.8), $Br_2CHCHBr_2$ (0.6–1.1), Cl_2CHCHF_2 and Br_2CHCHF_2. In 1,1,2-trifluoroethane conformation **38** is more stable than **39**, but now the energy difference[42,201] (0.9–1.4 kcal mol^{-1}) is lower than in the chlorine (1.8–2.0) and bromine analogues[199]. Strangely enough, quantum-chemical reasoning[30a] prefers **37** to **36** and **39** to **38**, and actual

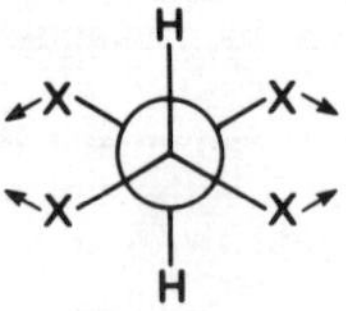

(**36**) 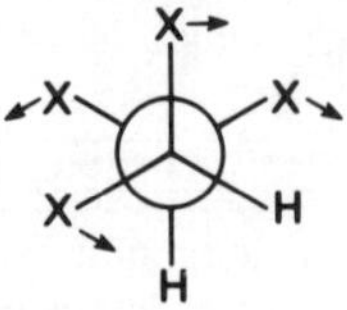 (**37**)

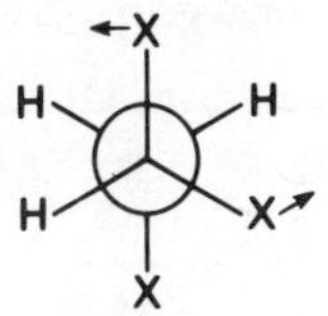

(38) 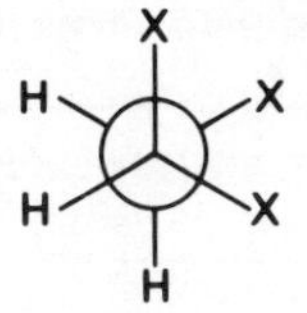 **(39)**

computation[202] makes **37** either too stable or too unstable. Other examples are N_2F_4 and P_2F_4 which are *anti*[203], the presence of fluorines thus overriding the '*gauch* effect' that the parent hydrides manifest[193,204] – unlike in FOOF and FSSF. *Trans* is also the more stable isomer of 1,2-difluorocyclopropane[205].

Flourine tends more than other halogens to approach localities of unsaturation. Benzyl halides ($PhCH_2X$) provide an example. When X = Cl, Br or I, the C—X bond lies in a plane perpendicular to the aromatic ring; fluorine, like hydroxy[206], prefers a periplanar orientation[207]. In $CH_2{=}CHCH_2F$, F/C=C-*syn* (**40**, X = F) is preferred to H/C=C-*syn* (**41**, X = F) by 0.17 kcal mol^{-1} [208]; solution data suggest a reversed order of stability in the chlorine and bromine compounds[209]. Another type of oddity is discerned when one passes from $CH_2{=}CHCH_2F$ to $CH_2{=}CFCH_2F$: the gap between **42** (low) and **43** (high) widens from 0.17 to 0.42 kcal mol^{-1} [210]. With 1,2-difluoroethane in mind, one would expect the gap to close rather than open. It is still lower than the gap in $CH_2{=}CClCH_2Cl$ (about 0.9 kcal mol^{-1} [211]).

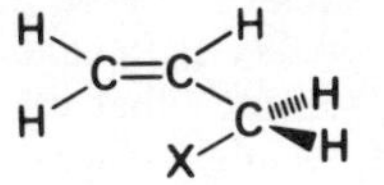

(40) 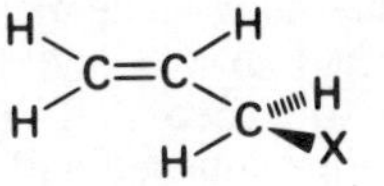 **(41)**

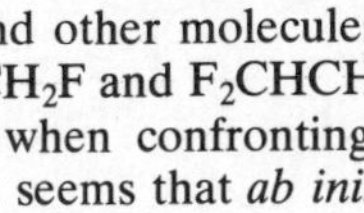

(42) 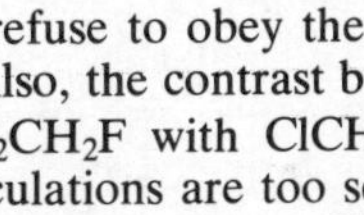 **(43)**

Such peculiarities have prompted extensive efforts along quantum-chemical lines[24–26,30,202,212,213], but results for some molecules are numerically unsatisfying (e.g.[62] FCH_2CH_2F) and other molecules just refuse to obey the quantum chemist's insight (e.g.[16a] F_2CHCH_2F and F_2CHCHF_2). Also, the contrast between fluorine and other halogens – e.g. when confronting FCH_2CH_2F with $ClCH_2CH_2Cl$ – does not come out distinctly. It seems that *ab initio* calculations are too sensitive to the basis set[24,214] and input geometry[215], and that the correlation energy differs too significantly in the structural variants[24]. An extreme illustration is provided by oxalyl fluoride, FCOCOF. Using the same basis set (4-31G) but somewhat different geometries, the secondary minimum was obtained in one study[214] as F/F-*syn*, with energy 0.54 kcal mol^{-1} above *anti*, and in another[216] as *gauche* ($\omega \sim 15°$) with energy 3.93 kcal mol^{-1} above *anti*!

Very little, by contrast, has been attempted along molecular-mechanical lines. An interesting observation concerning X_2CHCHX_2 and related molecules was made by Miyagawa and coworkers[217] in 1957. In geminal dichlorides, θ(XCX) would exceed the tetrahedral value (it is about 112° in CH_2Cl_2[218]) which shifts the Cl atoms of $Cl_2CHCHCl_2$ as indicated by arrows in formulae **36** and **37**. This aggravates the two

TABLE 6. Energy-component analyses (in kcal mol^{-1}) for X_2CHCHX_2 and X_2CHCH_2X (X = F, Cl)[42]

	E(**37**) − E(**36**) X = F	E(**37**) − E(**36**) X = Cl	E(**39**) − E(**38**) X = F	E(**39**) − E(**38**) X = Cl
Non-bonded	−0.06	−0.52	−0.06	0.99
Torsion	0.01	0.06	−0.01	−0.11
Electrostatic	1.36	0.75	1.53	1.06
Total	1.31	0.29	1.46	1.94
Skew ω(XCCX)	62°	66°	62°	68°

skew Cl···Cl interactions in **36** but decreases one of the three Cl···Cl skews in **37**. In geminal difluorides[6], on the other hand, θ(FCF) is smaller than tetrahedral. It is about 108.3° in CH_2F_2[198], 107.4° in CH_3CHF_2[201], 106.8° in F_2CHCH_2F[219] and 107.3° in F_2CHCHF_2[220]. For X = F, then, the arrows in **36** and **37** should be reversed, which implies a rise in conformational energy on passing from $Cl_2CHCHCl_2$ to F_2CHCHF_2[192c]. As for Cl_2CHCH_2Cl, internal rotation (shown in the formula by arrows) can relieve some strain in conformation **38** but not in **39**. In F_2CHCH_2F, FCF-closing could push conformation **38** even lower.

1,1,2-Trihalo- and 1,1,2,2-tetrahaloethanes count among the first halides calculated by the force field method[42]. The field used stresses non-bonded and electrostatic interactions (d = 1) but ignores relexation in bond lengths and valency angles. The component analyses are cited in Table 6. One notes that changes in electrostatic interaction are more pronounced with fluorine, while changes in non-bonded interaction are more pronounced in the chlorine analogues. The preference in F_2CHCH_2F and F_2CHCHF_2 is thus interpreted as determined by ΔE_{es}, while it is the balance of ΔE_{es} and ΔE_{nb} that counts in Cl_2CHCH_2Cl and $Cl_2CHCHCl_2$.

An alternative force field for fluorides[46] de-emphasizes electrostatic interaction (d = 4) and relegates to ΔE_{tor} a higher share in ΔE_t; it does include stretching and bending terms. Results for F_2CHCH_2F and F_2CHCHF_2 are similar to the above, except obviously that ΔE_{tor} is substantial in the sum $\Delta E_{tor} + \Delta E_{es}$. Consider, however, results by this field[114] for the more crowded molecule $CH_3CF_2CF_2CH_3$ (**44**, **45**):

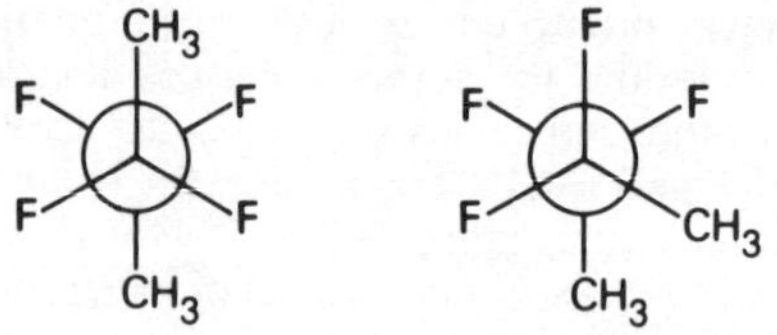

(**44**) (**45**)

Energy component	E(**45**) − E(**44**), kcal mol^{-1}
Steric (mainly bending)	0.476
Non-bonded	−0.226
Torsion	1.563
Electrostatic	0.405
ΔE_t	2.2

The computed ΔE_t exceeds in this instance the quantum-chemical estimate[202] (<1.8, probably around 1.4 kcal mol^{-1}) and might be overestimated. But the analysis recalls attention[32] to bending strain: *inter alia*, CCC angles are forced to open and FCF angles to close in Me/Me-*g* (**45**). Furthermore, it illustrates that E_{nb} can be smaller in the conformer of higher energy; one should not confuse ΔE_{nb} with $|\Delta E_t$[221]. By computation, all anti-interactions in $CH_3CF_2CF_2CH_3$ and also the Me/Me skew-interaction fall in the *attractive branch* of the Hill potential (Figure 1). The average values in **44** and **45** are obtained as: Me/F-anti, −0.128; Me/F-skew, +0.065; Me/Me-anti, −0.273; Me/Me-skew, −0.224; F/F-anti, −0.060; F/F-skew, +0.018 kcal mol^{-1}. The Me/Me interaction energy is similar in the two conformers, as is the sum of F/F interaction energies. In comparison with **44**, **45** enjoys then the more favourable Me/F anti-interactions. This, however, does not suffice to make it more stable than **44**.

With this perspective, let us refer back to Table 6 and note than E_{nb} is not a simple measure of the number of skew X···X interactions[35a,222]. One anti-interaction is replaced by skew both on going from **36** to **37** and from **38** to **39** but, for X = Cl, ΔE_{nb} comes out negative in the former transition and positive in the latter. Whether non-bonded interactions are additive at all is questionable[13], but in molecular mechanics they are taken to be so. The example of $CH_2CF_2CF_2CH_3$ serves to show that all pairs – anti and skew – should be taken in consideration, not just the count of skew-interactions or of X···X skew-interactions (see Ref. 45, p. 135).

To conclude this section let us recall that in 1,2-difluoroethene the *cis* isomer is more stable than the *trans*[23] (Section I). A preference for *cis* isomers is also manifested by FCH=CHCl, FCH=CHBr, FCH=CHI, FCH=CFCl, ClCH=CHCl, and BrCH=CHBr[23,223,224]. Unlike the case in diimide[225] (HN=NH), *cis* is the more stable isomer of difluorodiazine[223] (FN=NF) and, in distinction from $CH_3N{=}NCH_3$[226] and $CF_3N{=}NCH_3$, the more stable isomer of $CF_3N{=}NCF_3$ is almost certainly *cis*[227].

VI. 'NON-BONDED ATTRACTION'

It happens in some molecules that a conformation with large substituents close to each other is stabilized with respect to other conformations, or is relatively more stable than a corresponding conformation in a less crowded analogue[228]. For example, $\Delta E(= E_g - E_a)$ is *c*. 0.97 kcal mol^{-1} in $CH_3CH_2CH_2CH_3$, 0.81 in $(CH_3)_2CHCH_2CH_3$, but only 0.05 in $(CH_3)_2CHCH(CH_3)_2$[222,229]; it is positive in $(CH_3)_3CCH_2SiH_2CH_3$ but negative[230] in $(CH_3)_2CHSiH(CH_3)_2$. In solutions of *meso*-$(CH_3)_3CCHClCHClC(CH_3)_3$ there is a higher percentage of *gauche* forms[231] than in solutions of $ClCH_2CH_2Cl$[232]. The *anti* conformation is not the global minimum in $BrCH_2C{\equiv}CCH_2Br$[233] and is not the predicted global minimum in $FCH_2C{\equiv}CCH_2F$[63]. Hexafluorobutadiene[234] and hexachlorobutadiene[235] prefer a skewed conformation ($F_2C{=}CH{-}CH{=}CF_2$ is *s-trans*[236]).

The phenomenon is sometimes referred to as *non-bonded attraction*, even if this is a biased molecular-mechanical term. The term implies that E_{nb} is (or is assumed to be) less positive or more negative in the crowded arrangement, and that ΔE_{nb} dominates ΔE_t. In some instances computation supports this interpretation. A now classical example is 1,3,5-tri(neopentyl) benzene, where the predominant rotamer has all three *tert*-butyl groups on the same side of the aromatic ring[237]. Another example[238] is provided by *endo*-isomers of 1,2,3,4,5,6-hexamethylbicyclo[2.2.0]hexane (e.g. **46**) that are more stable than the corresponding *exo*-isomers (e.g. **47**). A third is 1,1,2,2-tetracyclohexylethane which prefers the *gauche* conformation in solution and also in the crystal[239]. Quantum-chemistry has its own methodology for dealing with crowdedness[30,240] – as does inorganic chemistry, e.g.[241] in interpreting the eclipsed conformation of $(Cl_4Re{-}ReCl_4)^{-2}$.

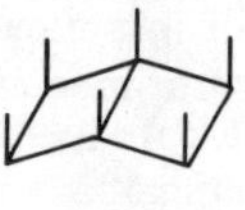

(**46**) 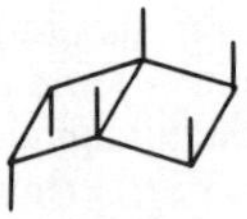 (**47**)

Among halogen compounds[242], best known is the case of 2,2′-dihalobiphenyls[9a]. The conformation with close halogens (**48**, X, Y = Hal) prevails, in the gas at high

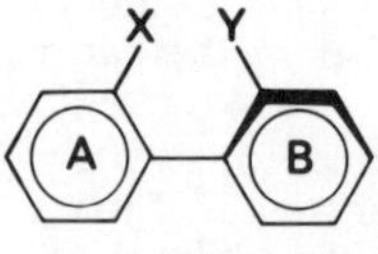

(**48**)

temperatures, to the extent that no other rotamer is observable. In the crystal, the halogens approach each other closer than to the sum of Pauling's contact radii. This is not unexpected, since halogen Y would be attracted not only to X but also to other atoms in ring A, as would X to atoms in ring B[243].

Another instance has to do with 1,5-hexadiyne and its 1,6-dibromo analogue[242]. Electron-diffraction studies detected 75% *anti* in the former (**49**, X = H) but only 47% *anti* in the latter (**49**, X = Br), and the trend (not the numbers) is reproduced by computation[48]. An electron-diffraction study[244] of $BrCH_2CH_2CH_2Br$, at a nozzle temperature of 65°C, detected 67% of *gg* (**50**) with only 30% of *ag* (**51**) and 3% of *aa* (**52**). Molecular-mechanical results[48] correspond to 56, 24 and 19%, respectively.

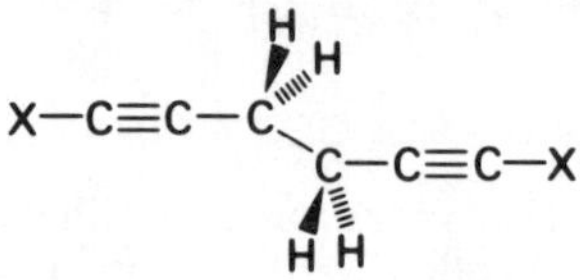

(**49**)

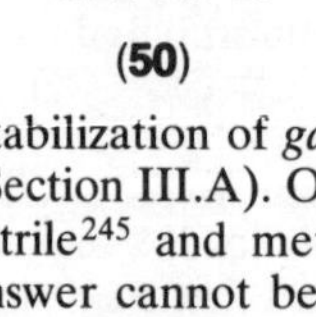

(**50**) 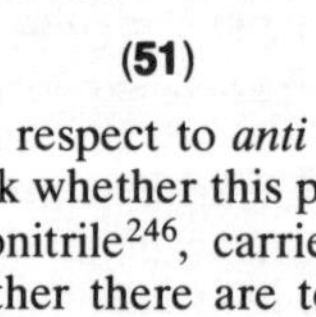 (**51**) 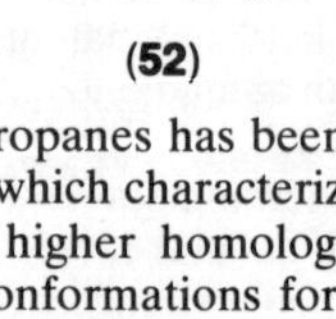 (**52**)

The stabilization of *gauche* with respect to *anti* in 1-halopropanes has been noted above (Section III.A). One may ask whether this preference, which characterizes also butyronitrile[245] and methoxyacetonitrile[246], carries over to higher homologues. A sharp answer cannot be given: either there are too many conformations for a safe analysis or, when a few, there are ponderant factors on top of halide preferences. The conformational energies in 1-bromobutane have been calculated by a field that *does not* reproduce the *gauche*-preference in 1-bromopropane[48]. The computed relative E_t values were: *aa*, 0; *ag*, 0.37; *ga*, 0.60; *gg*, 0.86 kcal mol^{-1}, where the left symbol refers to the conformation of the sequence C—C—C—C and the right symbol to that of Br—C—C—C. These numbers *do* seem representative, judging by the agreement of the derived populations (vapour, 25°C) with the experimental[247]: *aa*, 30% (experimental 36%); *ag*, 33 (24); *ga*, 22 (24); *gg*, 15 (16).

In $(CH_3)_2CHCH_2Br$, the conformation with methyls flanking the bromine has been reported as somewhat less stable than the alternative[248]. It is, however, reported[249] that in $(CH_3)_2CBrCH_2CH_3$ the conformation with one Me/Me-skew and one Br/Br-skew is appreciably preferred to the conformation with two Me/Me-skews. In dissolved $(CH_3)_2CBrCH(CH_3)_2$[250], the conformation with a flanked bromine (X/H-*a*, **53**) is preferred to the alternative (X/Me-*a*, **54**): $\Delta G^\circ \sim 0.39$, $\Delta H^\circ \sim 0.11$ kcal mol^{-1}.

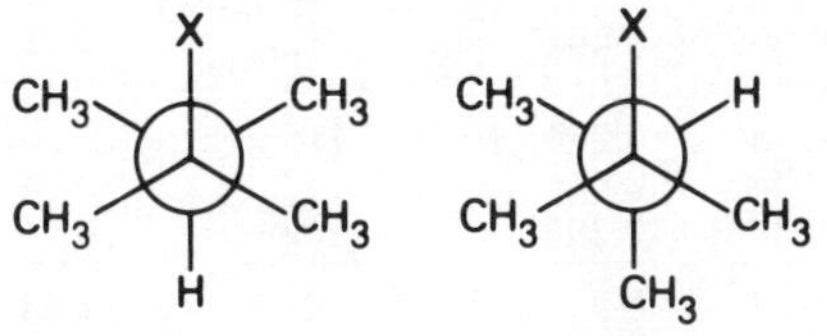

(**53**) (**54**)

The theoretical counterpart[48] of the latter number (vapour) has the following components:

Component	E(**54**) − E(**53**) kcal mol^{-1}
ΔE_s	0.006
$\Delta E_b + \Delta E_{sb}$	−0.342
ΔE_{nb}(1–4)	0.600
ΔE_{nb}(long range)	−0.758
$\Delta E_{tor} + \Delta E_{tb}$	0.669
ΔE_{es}	−0.020
ΔE_t	0.16

The preference is thus interpreted as reflecting a balance of opposing factors, not domination of ΔE_t by ΔE_{nb}. One may add that the chloride and iodide also prefer **53**, but **54** is preferred in the fluoride and the hydrocarbon itself shows no bias[250,251].

Evidence for non-bonded attraction should also be sought in the molecular geometry. A case in point is *trans*-1-chloro-1,3-butadiene (**55**) where the C^2—H bond is stretched to 1.129 Å and the angle C^1=C^2—H is compressed to 114.6°, which enable Cl and H to approach each other[252]. In the *cis* isomer (**56**) one has discerned an analogous interaction, between chlorine and the hydrogen atom four bonds away[253]. Cognate phenomena have been encountered in other instances[254].

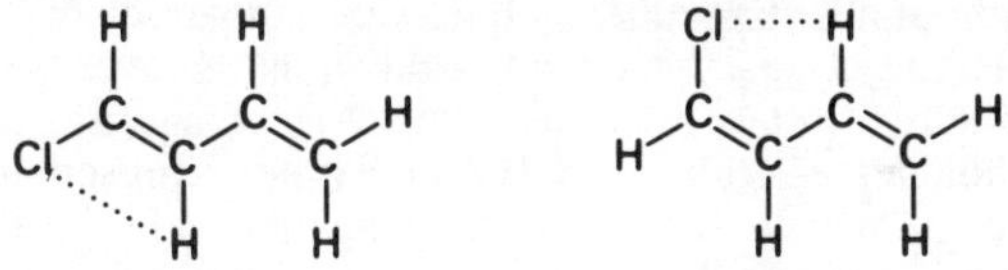

(**55**) (**56**)

VII. 1,2-DIHALOETHANES

1,2-Dihaloethanes, obvious prototypes for vicinal disubstitution, have been repeatedly studied since the early days of conformational analysis[192a,255]. Recent reports are available on FCH_2CH_2Br[256], FCH_2CH_2Cl, FCH_2CH_2I, $ClCH_2CH_2Br$[257], FCH_2CH_2F[190,191], $ClCH_2CH_2Cl$[258], $BrCH_2CH_2Br$[134], ICH_2CH_2I[259], as well as on the cyano analogues $ClCH_2CH_2CN$[260], $BrCH_2CH_2CN$[261], $NCCH_2CH_2CN$[262]. As already

TABLE 7. Conformational energies ($E_g - E_a$, kcal mol^{-1}) and dipole moments (in debyes) of 1,2-dihaloethanes

	FCH_2CH_2F	FCH_2CH_2Cl	$ClCH_2CH_2Cl$	$ClCH_2CH_2Br$	$BrCH_2CH_2Br$
Component analysis					
ΔE_s	0.009	0.001	0.012	0.012	0.015
$\Delta E_b + \Delta E_{sb}$	0.041	0.015	0.086	0.124	0.199
ΔE_{nb}	0.086	0.044	0.115	0.167	0.258
ΔE_{tor}	−0.742	0.016	0.733	0.912	1.198
ΔE_{es}	0.363	0.292	0.252	0.211	0.178
ΔE_t(cald.)	−0.24	0.37	1.20	1.43	1.85
ΔE (exptl)	~−2.0	0.3–0.5	1.2	1.4	~1.8
$\omega(XCCY)_g$	67°	65°	65°	69°	73°
$\Delta\mu(C—X)_g$	0.14	0.13, 0.22	0.21	0.18, 0.17	0.15
$\Delta\mu(C—X)_a$	0.09	0.09, 0.15	0.15	0.14, 0.12	0.12
μ(calc.)	2.66[a]	1.95[b]	1.24	1.15[d]	0.85[e]
μ(exptl)	2.67[a]	1.85[b]	1.12[c]	1.09[d]	0.81[e]

[a] *Gauche* form, Ref. 263.
[b] At 33°C, Ref. 264.
[c] Ref. 255.
[d] At 66°C, Ref. 265.
[e] At 66°C, Ref. 266.

mentioned (Section V) the *gauche* conformation is much more stable in FCH_2CH_2F than the *anti* form. In the other molecules the *anti* form is favoured. Solvation and liquefaction stabilize *gauche* with respect to *anti* (Section IX.B), which reverses in some cases the order of stability[232,257].

A molecular-mechanical view, by the field that de-emphasizes electrostatic interactions[45,46,48] ($d = 4$), is offered in Table 7. Results for the symmetrical molecules (XCH_2CH_2X) were obtained by adjusting constants in $e_{tor}(XCCX)$ such that ΔE_t ($E_g - E_a$) fit the experimental. In FCH_2CH_2F, the calculated ΔE_t is not negative enough: when the field was developed, the extent of *gauche*-stabilization in this molecule had not been appreciated, and later attempts to improve the computed ΔE_t ($= E_g - E_a$) without affecting other output were not successful[114]. In the unsymmetrical cases (XCH_2CH_2Y), $e_{tor}(XCCY)$ was taken as the average $\frac{1}{2}[e_{tor}(XCCX) + e_{tor}(YCCY)]$ and not readjusted. One sees that ΔE_t comes out appropriately for FCH_2CH_2Cl even though the result for FCH_2CH_2F is not representative. This implies that some effect of consequence, specific to the combination F···F, is not contained in the molecular-mechanical set-up. We return to the problem presently, but would like to note here also that *ab initio* calculations cannot cope with 1,2-difluoroethane[62,68]. CNDO/2, INDO and PCILO can[267], and Extended Hückel Theory produced in 1970 the best ever prediction[268] – at the time it must have been considered a failure of the method!

Table 7 contains the following information on the intramolecular electrostatics: the difference in E_{es} between the two conformers (ΔE_{es}); the reduction in the C—X bond moments ($\Delta\mu$); the dipole moment (μ) calculated for the conformational mixture, as estimated from ΔE_t and the moments computed for the separate conformers. The table shows that dipole–dipole interaction favours the *anti* conformation in all cases, its significance diminishing with decreasing electronegativity. Non-bonded interaction also favours *anti*, and its importance increases as the halogen grows. In this particular force field most of the strain resides in ΔE_{tor} and, in the terminology of Section I, is by necessity viewed as 'unexplained'. The computed dipole moments illustrate the smooth performance of the algorithm for μ^*(C—X) (Section III.D). The *gauche*-angle $\omega(XCCY)_g$ is quite open. There are indications[259] that in ICH_2CH_2I it opens up to 85°.

An alternative view[42], by the field that emphasizes electrostatic interaction ($d = 1$) but avoids computation of geometries and dipole moments, is given in Table 8. The results – even if numerically less satisfying than before – corroborate the trends: ΔE_{es} decreases and ΔE_{nb} increases as the halogen becomes bigger.

The period of internal rotation about bond A—B is $N_a N_b/J$, where N_a is the number of symmetry planes in A that intersect the bond A—B, N_b the corresponding number for group B, and J is the maximum number of such planes in A and in B that can be brought simultaneously into coincidence[59]. The potential of internal rotation is given by equation (5) (Section II): $E_t(\omega) = \frac{1}{2} \Sigma U_k(1 - \cos k\omega)$. Experimental or theoretical information on internal rotation in a given molecule can serve to solve for the constants U_k. The magnitude and relative magnitude of these constants may then shed light on the physical effects that underlie the dependence of E_t on ω. For example, U_3 is frequently the leading term in expansions for derivatives of ethane, even in cases where the period of internal rotation differs from 3 (e.g.[269] in $CH_3CH_2CH_2C{\equiv}CH$ where the period is 1). Hence the prominent disturbance in such molecules is of threefold periodicity, as it is in ethane itself. This suggests that the main disturbance is similar in origin to that in ethane itself, and has little to do with the nature of the substituents.

The magnitude and sign of the constants U_3, U_2 and U_1 have been interpreted as reflecting three intramolecular effects[270]. The first effect (U_3, threefold) is some form of bond–bond repulsion. The second (U_2, twofold) is the stabilizing influence of back-donation from lone-pair orbitals at one end of the molecule into antibonding σ-orbitals at the other, or of π-conjugation[271]. The third effect (U_1, onefold) is the interaction of dipoles at the two ends of the bond and/or non-bonded interaction[215].

Let us consider the three-term expansions for $CH_3CH_2CH_2CH_3$, $ClCH_2CH_2Cl$ and FCH_2CH_2F. Data on internal rotation in butane are available[272]. For a rough appreciation of 1,2-dichloroethane one can use the old estimates[273] of 2.8 and 4.5 kcal mol^{-1} (above Cl/Cl-*a*) for the two barriers, together with data on the conformation Cl/Cl-*g* (Table 7). The expansion for 1,2-difluoroethane has been derived from electron-diffraction data (model CP in Table VI of Ref. 191). One has (in kcal mol^{-1}):

TABLE 8. Results of 'unrelaxed optimization' for 1,2-dihaloethanes ($E_g - E_a$, kcal mol^{-1})[42]

	FCH_2CH_2F	FCH_2CH_2Cl	$ClCH_2CH_2Cl$	$ClCH_2CH_2Br$	$BrCH_2CH_2Br$	ICH_2CH_2I
Component analysis						
ΔE_{nb}	−0.04	−0.09	0.49	0.46	0.68	1.41
ΔE_{tor}	−0.07	0.01	0.18	0.18	0.18	0.34
ΔE_{es}	1.43	1.17	0.83	0.77	0.70	0.46
ΔE_t	1.32[a]	1.09	1.50	1.41	1.56	2.21
ω(XCCY)	64°	62°	70°	70°	70°	74°

[a]Diminishes to 0.6 on geometric relexation[187].

	$CH_3CH_2CH_2CH_3$	$ClCH_2CH_2Cl$	FCH_2CH_2F
U_1	2.38	2.0	0.67
U_2	−0.95	−0.3	−2.53
U_3	3.62	2.5	2.21

The three molecules have then in common positive U_1 and U_3, and a negative U_2. Traces of $1 - \cos \omega$, $-(1 - \cos 2\omega)$, and $1 - \cos 3\omega$ are shown in Figure 6. It is seen that a positive U_1 reflects destabilization (dipole–dipole and non-bonded) of X/X-*syn* with respect to X/X-*a*; the negative U_2 reflects destabilization (electronic) of the coplanar with respect to the orthogonal arrangement of X—C—C—X; the positive U_3 reflects destabilization (bond–bond) of eclipsed with respect to staggered rotamers.

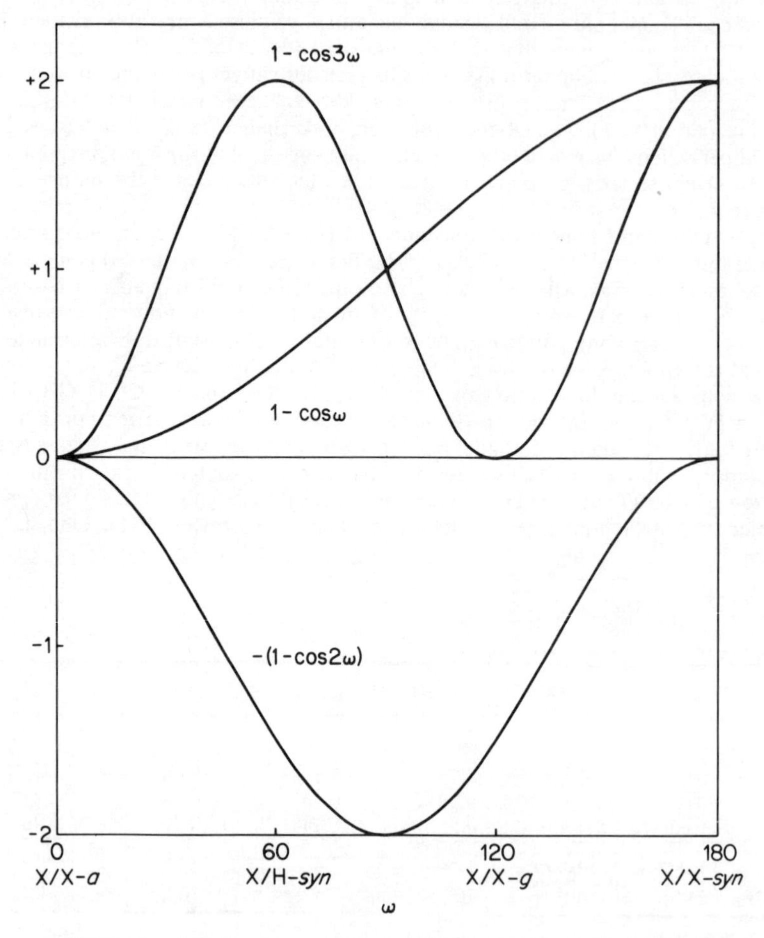

FIGURE 6. Angular change in the components of the torsional potential.

In FCH_2CH_2F the term in U_2 is leading, which reflects a drive of the sequence F—C—C—F to adopt the orthogonal arrangement (**57**). It has been pointed out[62,66] that such an arrangement enables the two fluorines to partake simultaneously of hyperconjugative interaction with the vicinal methylene grouping (schematized in **57**

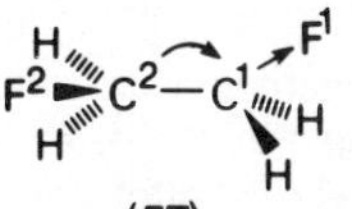

(**57**)

by the electron-shift arrows $C^2 \rightarrow C^1$ and $C^1 \rightarrow F^1$; the corresponding arrows that involve F^2, namely $C^1 \rightarrow C^2$ and $C^2 \rightarrow F^2$, are not marked). This could then be the quantum-chemical effect that molecular mechanics misses in the combination F···F. It is of little consequence in F···Cl and in other X···Y combinations. For $ClCH_2CH_2Cl$, U_2 is small though still negative. As in $CH_3CH_2CH_2CH_3$, the global minimum at Cl/Cl-*a* is interpreted as due to the drive to decrease dipole–dipole, non-bonded, and bond–bond interactions.

VIII. CYCLIC HALIDES

Conformational analysis, in a way, started with the six-membered ring[274], and conformations of halogenated cyclohexanes played an important role in its development. The accumulating material was reviewed in 1962[275], in 1965[99], in 1971[276], and in 1974[232]. Halogenated cyclobutanes[277,278] and cyclopentanes[278,279] were reviewed recently.

Substituted cyclohexanes are mostly being studied in solution and there still seems to be some disagreement as to the entropic contribution to the free-energy difference between conformers[280]. Extrapolation of solution data to the vapour requires corrective factors[232] (Section IX) and here again there is room to mix up free energy with enthalpic quantities[281]. The theoretician finds himself in the peculiar position of not knowing which numbers he is expected to reproduce or to predict. Consequently, component analyses can become less meaningful for cyclic than for acyclic halides. For example, $-\Delta G^\circ(ee\text{–}aa)$ for *trans*-1,2-dibromo-1-methylcyclohexane, as measured in CCl_4, is 1.45 kcal mol^{-1} [113]; a calculated value for ΔE_t is 1.73 kcal mol^{-1} [48]. Is this a 'good' calculation? Could one infer from its details that a diaxial conformer predominates because the compound is a cyclic analogue of $BrCH_2CHBrCH_3$ (wherein Br/Br-*a* is computed more stable than Br/H-*a* or Br/Me-*a* by 1.7 kcal mol^{-1})? It is correct then to conclude that ring-formation is but of little consequence? Such questions still rest unanswered.

As in other known monosubstituted cyclohexanes[11] (excluding mercury compounds[282]) the equatorial form of a cyclohexyl halide is lower in free energy and enthalpy than the axial. Reported energy differences are 0.17 kcal mol^{-1} for F (gas)[283]; 0.45 for Cl (ΔH°, $CFCl_3$); 0.52 (gas)[284], 0.45 (ΔH°, CS_2)[285] and 0.47 (ΔH°, $CFCl_3$)[280] for Br; the iodide would be close in value to the bromide[286]. ΔH° is lowered by 4,4-dimethylation[287] (Cl, Br), and calculation consistently predicts[47,48] that a 4-*eq*-*tert*-butyl group also stabilizes an axial halogen. Though small (~0.08 kcal in the bromide[48]), the effect seems real, and is attributable to a non-bonded attractive interaction between the axial halogen and the bulky alkyl group; an equatorial halogen is too far to be affected.

In halocyclobutanes the halogen is predominantly equatorial[277]. In cyclopentanes, there is some preference for the axial position[288]. Calculations for envelope–cyclopentane, with chlorine at the prow-carbon, suggest that bending deformations are responsible: the angles C^1C^2Cl and $C^1C^2C^3$ are computed as 113.2° and 103.0° in the equatorial conformer, versus values that are closer to normal, 110.9° and 104.6°, in the axial[47]. In these calculations, however, C—H bond moments were ignored, and the extent of eventual attraction[288] – between axial C—Cl and bonds C^3—H and C^4—H – could not be assessed.

The foregoing citations demonstrate that the equatorial preference in halides is relatively weak. It can be enhanced by the drive to avoid 1,3-diaxial interaction, as in *cis*-1,3-dibromocyclohexane[289]. The equatorial drive may yield, however, to features of conflicting requirements, as in *cis*-1-bromo-4-methylcyclohexane[290], where *ax*-Br-*eq*-CH_3 (**58**) is favoured over *eq*-Br-*ax*-CH_3 (**59**). Other examples are the preference of axial (**60**) to equatorial (**61**) in 1-chloro-4,4-difluorocyclohexane ($\Delta H^\circ = -0.68$ kcal mol^{-1} in $CFCl_3$ solution), and in 1-chloro-4,4-*bis*(trifluoromethyl) cyclohexane ($\Delta H^\circ = -0.59$ in $CFCl_3$)[174].

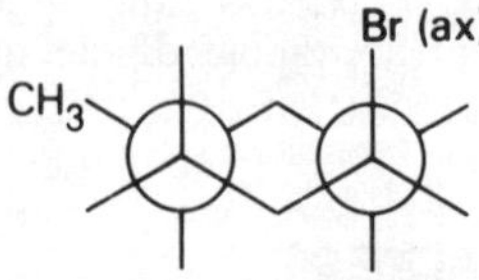

Br (ax) CH3 CH3 Br (eq)

(**58**) 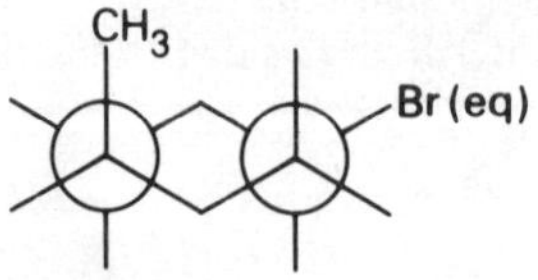 (**59**)

Cl H F F F F Cl H

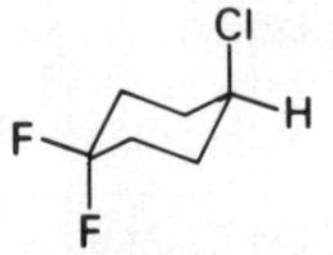

(**60**) 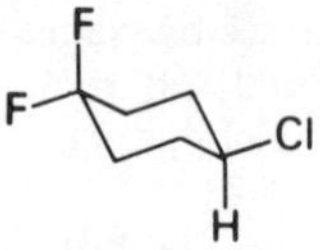 (**61**)

One may recall here the anomeric effect[291,292]. In particular, the halogen in 2-chloro- and 2-bromotetrahydropyran is axial (as in **62**, not as in **63**). The energy difference has been estimated[293] as 2.7 (Cl) and >3.2 kcal mol^{-1} (Br). The quantum-chemical analysis[292] is in terms of interaction between the polar C—X bond and the non-bonding oxygen electrons (marked in **62**), or, alternatively, between the halogen non-bonding electrons and bonds C^2—O and C^6—O. Such effects have no explicit counterpart in the molecular-mechanical model, where one would start by examining the C^6—O/C^2—X electrostatic repulsion[294].

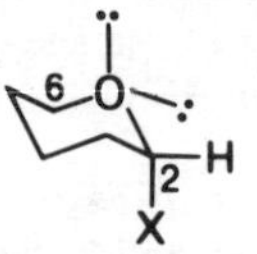

6 O H 2 X H 6 2 O X

(**62**) 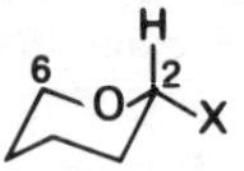 (**63**)

A pioneering estimation of ΔE_{es} by the point-charge approach, with $d = 1$ in **62** and $d = 2$ in **63**, was reported[144] in 1968. It favoured the axial form by 2.8 kcal mol^{-1} in the chloride and by 2.6 in the bromide. True[126,291], readjustment of d is inelegant and the number for the bromide is too low. But the point is made that ΔE_{es} is substantial in ΔE_t. Nowadays the calculation could be improved without affecting this conclusion by supplementing torsional potentials[128] (X—C—O—C, X—C—C—C, etc.). Torsion would serve as the molecular-mechanical guise of the quantum-chemical factor that favours the axial conformation: resonance involving C—X and the axial lone pair on oxygen[295].

The situation comes to a head in *trans*-1,2-dihalocyclohexanes: the preference of each separate halogen for the equatorial position, which implies a diequatorial conformation, **64**, is counteracted by the drive of the entire vicinal combination to assume an *anti* relationship, i.e. the diaxial conformation, **65**. The dibromide has recently been calculated[48]. By calculation, compromise in this compound consists of a conformational mixture wherein a 'quasi-diaxial' form is more stable; that is, a structure derived from **65** by rotating the C—X bonds so as to close the X—C—C—X dihedral angle. Concurrent changes must take place, as indicated in formula **65** by arrows. The calculated dihedral angles in the dibromide are marked in formula **66**; the

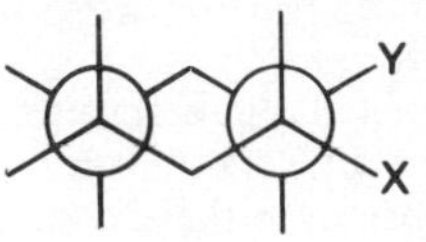

(**64**) 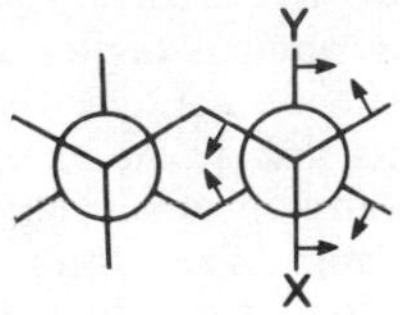 (**65**)

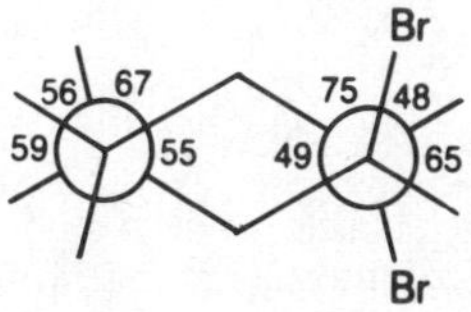

(**66**)

Br—C—C—Br dihedral angle is 161°. Calculation apart, this interpretation is supported by the crystal structure of *trans*-2,3-dichloro-1,4-dioxane[296], the high dipole moments[101] of *trans*-2,3-dibromodecalin (1.15 D), 4-*tert*-butyl-*trans*-1,2-dibromo- and -dichlorocyclohexane (1.2 D), and of *trans*-diaxial-1,2-dichlorocyclohexane[113] (1.07 D). It is also corroborated by the crystal structure of $2\beta,3\alpha$-dichloro-5α-cholestane[115] and the dipole moments of several $2\beta,3\alpha$- and $5\alpha,6\beta$-dihalo-5α-cholestanes[116] (halogen = Br, Cl).

Experimentally, a diaxial preference has been noted to varying extents in dissolved dibromo-, dichloro-, diiodo-, bromochloro- and chloroiodo- cyclohexanes, but not in the fluoroiodo derivative[297]. In *trans*-1,2-dichloro- and -dibromo-1-methylcyclohexane, the diaxial combination (where methyl is equatorial, i.e. **67**, not **68**; R = H) is further stabilized; in *trans*-1,2-dichloro-1,4,4-trimethylcyclohexane it is destabilized, even if the methyl at position 1 is now forced into an axial orientation (**68**, not **67**; R = CH_3)[113].

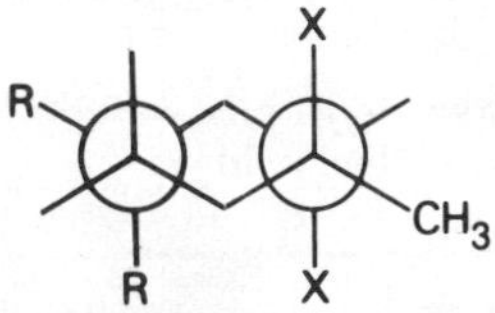

(**67**) 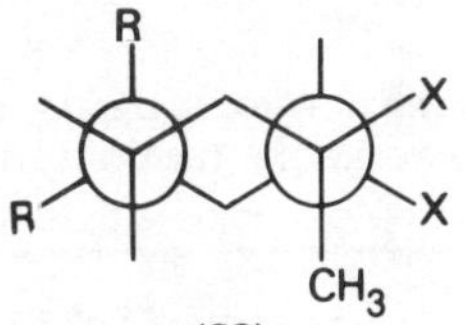 (**68**)

One has wondered at the 'effect of the disappearance of effects' [298] in cyclic dihalides and in cognate molecules[297] and a 'repulsive *gauche* effect' was invoked to cover the phenomena[299]. It seems that the situation is puzzling only as long as one views

$C_6H_{10}XY$ as a superposition of $C_6H_{11}X$ and $C_6H_{11}Y$. It ceases to be so if the molecule is viewed as a cyclic derivative of XCH_2CH_2Y, and its conformational composition as a compromise between the spatial demands of XCH_2CH_2Y (more often strong) and those of $C_6H_{11}X$ and $C_6H_{11}Y$ (always weak). In $BrCH_2CH_2Br$ *anti* is much more stable than *gauche* (~1.8 kcal mol^{-1}) and the preference, diminished as expected (down to about 0.8 kcal mol^{-1}) is conserved in $C_6H_{10}Br_2$. In FCH_2CH_2I, by contrast, *anti* would be only somewhat more stable than *gauche*[257]; correspondingly, the diaxial form of $C_6H_{10}FI$ is not favoured[300].

Whether this sort of argument is extensible to other groups of molecules remains to be checked. It should become unsafe when the opposing tendencies are not markedly unbalanced. An obvious test case which has not been considered as yet is the confrontation of haloacetaldehydes with 2-halocyclanones[301].

Solvent effects obscure the interplay of intramolecular interactions. For 2-halocyclohexanones, an attempt was reported to extricate information on the free molecules from solution data[47]. The conformational equilibrium in these compounds, *eq*(**69**) ⇄ *ax*(**70**) shifts to the left as the medium becomes more polar[302]. Calculation consisted

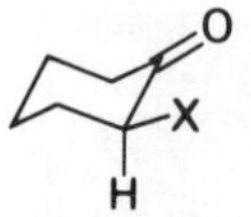

(**69**) 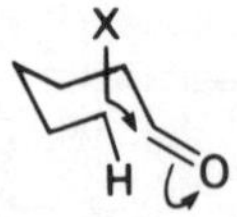 (**70**)

of searching such values for $E_t(eq)$ and $E_t(ax)$ that would, when corrected for the effect of various solvents, yield $E_t(eq) - E_t(ax)$ close to the experimental $\Delta G°$; then, of searching torsional constants for the sequence O=C—C—X such that $E_{tor}(eq)$ and $E_{tor}(ax)$ would produce, when taken together with other strains in each conformation, energies close to the required $E_t(eq)$ and $E_t(ax)$. Matters are complicated by the sensitivity of computed strains to geometrical details which, in turn, are sensitive to the torsional constants. As an illustration of what can be done, the energy/free energy differences (*eq* − *ax*, in kcal mol^{-1}) in 2-chlorocyclohexanone are as follows:

Phase	Measured[302]	Computed[47]
Free molecule	—	0.70 (ΔE_t)
C_6H_{12}	0.74	0.54
CCl_4	0.58	0.49
$CHCl_3$	0.12	0.12
C_6H_6	0.00	−0.03
CH_3CN	−0.30	−0.38

The computed ΔE_t values ($E_t(eq) - E_t(ax)$, in kcal mol^{-1}) for 2-fluoro, 2-chloro-, and 2-bromocyclohexanone turn out to be compounded as follows:

	Fluoride	Chloride	Bromide
ΔE_{st}	−3.18	−1.99	−0.12
ΔE_{es}	+3.04	+2.69	+1.58
ΔE_t	−0.14	+0.70	+1.46

Thus, steric effects favour the equatorial disposition, electrostatic effects favour the axial, and effects of both types decrease from F to Cl to Br. Also ΔE_{st} becomes progressively less important with respect to ΔE_{es}. Hence, the equatorial excess in 2-fluorocyclohexanone is not interpreted as due to a low electrostatic repulsion (C=O/C—F) in the equatorial conformer; the repulsion is therein high, but so turns out to be the torsional strain (O=C—C—F) in the axial conformer. Here, too, one suspects that the adjusted torsional term covers up for some quantum-chemical effect (cf. discussion of FCH_2CH_2F, Section VII, and of 2-halotetrahydropyrans above). This could be 'no-bond resonance' of the type depicted by arrows in formula **70**[303]. Such an electron shift is more likely with chlorine and bromine than with fluorine, which explains why the steric strain comes out relatively high in the axial form of the latter. In other words[304], the stabilizing $n(X) \rightarrow \pi^*(CO)$ charge donation is easier with the larger halogens than with F. Using a still different parlance[305], the axial tendency of a halogen at position alpha to the carbonyl augments with its polarizability.

2-Halocyclohexanones constitute but one class of 2-substituted cyclohexylidene derivatives[306] (**71**). Studies of nitronate salts, certain oximes, iminoxyl radicals and

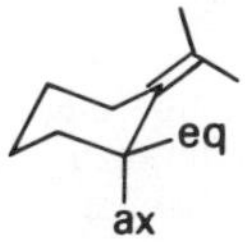

eq

ax

(**71**)

o-methoxyphenylhydrazones of cyclohexanones suggest that the axial preference of a substituent at position alpha is the rule[307].

IX. SOLUTION EQUILIBRIA

Table 9 cites free-energy differences in various solvents between conformers of three molecules. It is evident that these differences are solvent-dependent; that is, the conformers of a given molecule are stabilized to different extents by solvation. This is the solvent effect on conformational equilibria[308]. A change in medium can affect conformational equilibria considerably, because free-energy differences between rotamers and solvation enthalpies of dipolar solutes frequently compare in magnitude. In the last two examples of Table 9 even the order of stability depends on the solvent.

The first two cases in Table 9 illustrate the dictum that 'solvation stabilizes better the conformer of higher dipole moment'. In the third molecule (*trans*-1,4-dichlorocyclohexane) both conformers are devoid of dipole moment but solvation has still a sizeable outcome. All three cases illustrate the dictum that 'the effect of solvent increases with its polarity'. In consequence, the dipole moment of a solute would generally increase on going from a less polar to a more polar solvent – if several conformations that differ in dipole moment are allowable. The same applies to liquefaction. Solvation can also modify geometrical detail[309,310].

Clearly, data on a dissolved compound have to be processed prior to confrontation with theoretical numbers. The obtention of theoretical numbers requires itself a force field constructed by fitting to processed data. It cannot be by mere chance, for example, that molecular-mechanical studies of organic iodides have been so fragmentary[42,75,311]: vapour-phase data are yet to be reported[191] even for the key molecule ICH_2CH_2I[259].

Organic halides have also played a role in the development of theories of solvation. Early work was discussed by Mizushima[255]. Subsequent results were reviewed by Laszlo[312] and by Abraham and Bretschneider[232], and examined within a wider context

TABLE 9. Solvent effect on conformational energies

Solvent	ϵ^b	Conformational energy, kcal mol^{-1} [a]		
		$ClCH_2CH_2Cl$ [c]	$BrCH_2CHO$ [d]	$ClCH(CH_2CH_2)_2CHCl$ [e] *trans*
C_6H_{12}	2.0	0.91	+0.04	
CCl_4	2.2		0	
$CFCl_3$	2.5			0.14
CS_2	2.6	0.87		0.08
$CHCl_3$	4.7		−0.10	−0.03
CH_3COCH_3	20.7	0.26	−0.43	−0.43
CH_3CN	36	0.12	−0.50	
$(CH_3)_2NCHO$	50		−0.57	−0.60

[a] From NMR measurements.
[b] Dielectric constant of solvent.
[c] Reported as ΔE (Ref. 232). *Gauche* (**72**) above *anti* (**73**).
[d] $\Delta G°$ (Ref. 301). Br/H-*a* (**74**) less abundant than H/H-*a* (**75**) in C_6H_{12}.
[e] ΔG at −65°C (Ref. 132). Diequatorial (**76**) less abundant than diaxial (**77**) in $CFCl_3$.

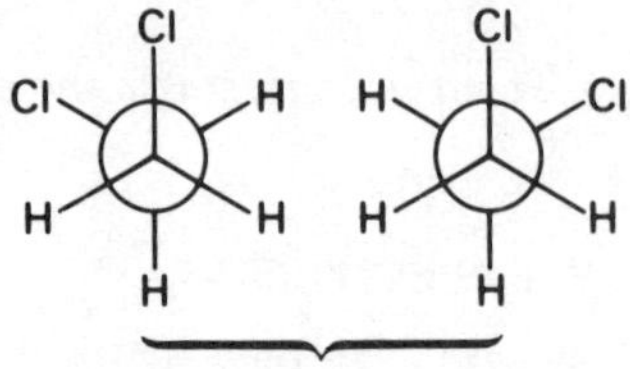

(**72**) 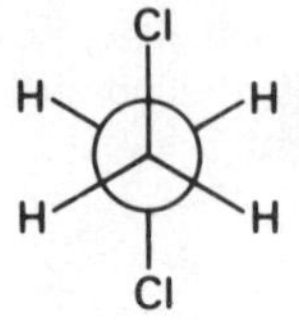 (**73**)

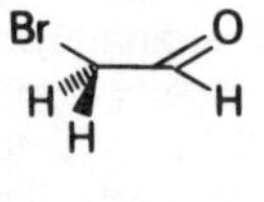

(**74**) 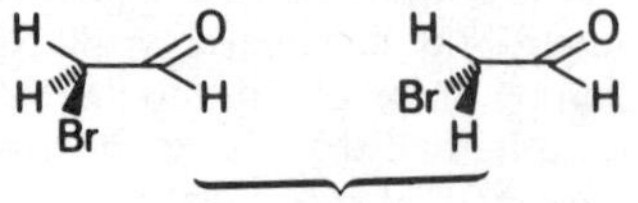 (**75**)

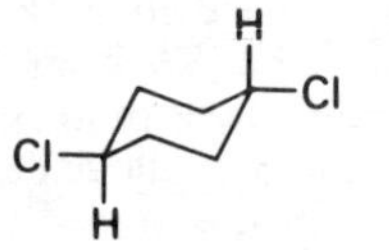

(**76**) 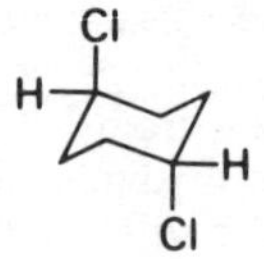 (**77**)

by Reichardt[313] and by other authors[314]. For discussion of halides see, in particular Refs 2 and 232.

A. Formulation

In applications to organic halides, it is still customary to retain only the electrostatic component in the expression for the solvation energy[315].

By Onsager's model[316], if a rigid dipole of permanent moment μ is introduced into a cavity of radius a in an unpolarized, isotropic and continuous medium of dielectric constant ϵ, dipoles are induced in the medium. These dipoles exert in turn a reaction field $R\mu$ on the immersed dipole. Mizushima[317] made the electrostatic stabilization of the dipole, taken as a point in the centre of the cavity, amount to

$$-\int_0^\mu R\mu \, \mathrm{d}\mu.$$

With Onsager's expression for R one obtains equation (17):

$$E_D = \frac{\epsilon - 1}{2\epsilon + 1} \cdot \frac{\mu^2}{a^3}, \tag{17}$$

where E_D is the *lowering* of energy due to interaction of the dipole with the solvent. To obtain E_D in kcal mol^{-1}, with μ in debye's and a in ångstroms multiply by 14.3942. E_D is but one term in the electrostatic contribution to the solvation energy (equation 18):

$$E_S = E_D + E_Q + \ldots, \tag{18}$$

where S stands for solvation, D for dipole, and Q for quadrupole. In Abraham's formulation[232]

$$E_D = \frac{\mu^2}{a^3} \cdot \frac{x}{1 - lx}, \tag{19}$$

$$E_Q = \frac{3}{2a^5} \cdot \frac{3x}{5 - x} q^2 \tag{20}$$

with $x = (\epsilon - 1)(2\epsilon + 1)$, $l = 2(n_D^2 - 1)/(n_D^2 + 2)$, and q is the quadrupole moment of the solute molecule, placed at its centre. For very polar solvents an additional term is required in E_S (equation 18)[318]. For alternative formulations see Ref. 312.

In evaluating ΔE_S, the dipole moment of a given conformation may be estimated by calculation, by analogies, or from vapour-phase data[255]. It is customary, even though not evident[319], to identify the volume of the cavity with the dimension of the solute molecule. The effective radius (a) is equated to the radius of a sphere of volume equal to the molecular volume, that is, $a^3 = 3M/4\pi\rho N_0$ (for a in ångstroms and ρ in g cm^{-3}, $a^3 = 0.3964M/\rho$). M is the molecular weight, N_0 – Avogadro's number, ρ – the density. If the density is not known from measurement, the molar volume, $V_M = M/\rho$, may be estimated by assuming additivity[320] in closely related substances. Additivity can also be assumed in the molar refraction[321], which leads to the refractive index, $[R]_M = (n_D^2 - 1)V_M/(n_D^2 + 2)$. In calculating E_Q (equation 20) a centre has to be chosen in the solute molecule. There is still disagreement as to the preferable choice[232b,315]. The centre of mass and the centre of volume have both served in recent applications.

Now, if ΔE^v represents the energy difference between rotamers in the vapour, the corresponding difference in solution is

$$\Delta E^s = \Delta E^v - \Delta E_S. \tag{21}$$

For a neat liquid one has

$$\Delta E^l = \Delta E^v - \Delta E_L, \tag{22}$$

where L (liquid) replaces S. Once ΔE_S (or ΔE_L) has been computed, equation (21) (or equation 22) can be used to relate theory with experiment. The computed ΔE^s

(or ΔE^{l}) lead to the predicted populations of rotamers in solution (or in the liquid) and, hence, to a predicted value of the overall dipole moment. The 'experimental' dipole moment of a neat liquid can be derived from its dielectric constant by a formula due to Onsager[100c]. With numerical values for the constants and appropriate conversion factors, this formula is given by equation (23):

$$\mu^2 = 1.4757 \times 10^{-4} \frac{(\epsilon - n_D^2)(2\epsilon + n_D^2)}{\epsilon(n_D^2 + 2)^2} \cdot \frac{MT}{\rho} \tag{23}$$

where μ is in debyes, ρ is in g cm^{-3}, and T is in kelvins.

One sometimes encounters calculations that bypass equation (18)[145,322]. In these the solvent is considered to penetrate the solute molecule and affect its energy by modifying the *intra*molecular electrostatic strain, $E_{es} = \sum e_{es}$. The contributions e_{es} are computed by equation (2) or equation (3) as they are for the vapour (Section II), except that ϵ replaces d. If $\epsilon > d$ (i.e. if the dielectric constant of the solution exceeds the value assigned to the parameter d in vapour-phase calculations), and if $E_{es} > 0$ (which is determined by the molecular structure), one gets indeed a reduction in the computed E_{es}, and hence in E_t. To meet the first requirement ($\epsilon > d$) it is customary in this approach to assign a low value to d, lower even than the dielectric constants of hydrocarbon solvents. The second requirement ($E_{es} > 0$) cannot hold in all cases: E_{es} is quite frequently negative. However, it probably holds if one ignores C—H bond moments, eliminating thereby from the summation attractions of the type C^+X^-/C^-H^+. The interactions conserved, C^+X^-/C^+Y^-, are usually repulsive and add up to a positive E_{es}.

B. Computational Pathways

Our first example is a Mizushima-type passage from ΔE^{v} to ΔE^{l} in 1,2-dichloroethane. The two coexisting conformers are Cl/Cl-*g* (**72**) and Cl/Cl-*a* (**73**). From the temperature-dependence of the dipole moment in the gas phase[255a,323], $\Delta E^{v} = E^{v}(g) - E^{v}(a) \sim 1.22$ kcal mol^{-1} and $\mu_g \sim 2.62$ D; μ_a is virtually zero. One assumes[324] that the bulk density ($\rho = 1.2462$ g cm^{-3})[325] and the bulk dielectric constant ($\epsilon = 10.66$) characterizes each of the two conformers. The density leads then to $a^3 = 31.475$ Å^3; equations (17) and (22) then furnish $E_{D,g} = 1.36$, $E_{D,a} = 0$, $\Delta E_L = E_{L,g} - E_{L,a} = 1.36$, and $\Delta E^{l} = E^{l}(g) - E^{l}(a) \sim -0.1$ kcal mol^{-1}. The two conformers of $ClCH_2CH_2Cl$ are indeed of roughly equal energy in the liquid[133].

As a second example[114] consider an Abraham-type calculation for *meso*-$CH_3CHBrCHBrCH_3$, where Br/Br-*a* and Br/Br-*g* coexist. Force-field calculations[48] provide $\Delta E^{v} = 1.57$ kcal mol^{-1}, $\mu_g = 2.96$, $\mu_a = 0$ D, and the bulk liquid constants are known from measurement[326]: $\rho = 1.7747$ g cm^{-3}, $n_D = 1.5091$, $\epsilon = 6.245$ at 25°C. When these numbers and the theoretical geometry (required to evaluate q in E_Q, equation 20) are put in equations (19) and (20), one obtains: $E_{D,g} = 1.322$, $E_{Q,g} = 0.374$, $E_{D,a} = 0$, $E_{Q,a} = 0.986$, $\Delta E_D = E_{D,g} - E_{D,a} = 1.322$, $\Delta E_Q = E_{Q,g} - E_{Q,a} = -0.612$ and $\Delta E_S = 0.71$ kcal mol^{-1}. One notes that ΔE_Q is sizeable in this case and opposite in sign to ΔE_D. By equation (22), $\Delta E^{l} = 0.86$, which corresponds to ~68% Br/Br-*a* in the liquid and an overall dipole moment of 1.68 D. The experimental counterparts are *c*. 70% Br/Br-*a* (NMR[327]) and 1.65 D (by equation 23).

Abraham himself used the formulation in order to derive ΔE^{v} values from data on organic halides in solution[132,200], and Ivanov applied it to 1,2-dihalo-1,2-diphenylethanes, predicting the trends of changing composition as the polarity of the medium grows[328]. A modification due to Allinger was applied to halogenated cyclohexanes and cyclohexanones (Section VIII), this time for converting calculated ΔE^{v} into ΔE^{s} values[47,146]. Thus, a multi-purpose tool is at hand. Under current analysis are the

TABLE 10. Energy differences[a] (in kcal mol^{-1}) in 2-haloethanols[79]

Compound	Vapour	Liquid
FCH_2CH_2OH	>2.00	1.44
$ClCH_2CH_2OH$	1.20	0.28
$BrCH_2CH_2OH$	1.45	0.09
ICH_2CH_2OH	—	0.00

[a] *Anti* above *gauche*.

distinction between enthalpic and free energy quantities[281], the limits of the two-term expansion in equation (18) and of neglecting non-electrostatic contributions to E_S[315], and the assumption that all conformers have the same molar volume[329].

Hydrogen bonding evades equation (18). 2-Haloethanols provide a case in point. As Table 10 shows, the *gauche* conformation **(78)** is definitely more stable in the gas but liquefaction stabilizes the *anti* form **(79)**. Since **78** and **79** would have very similar

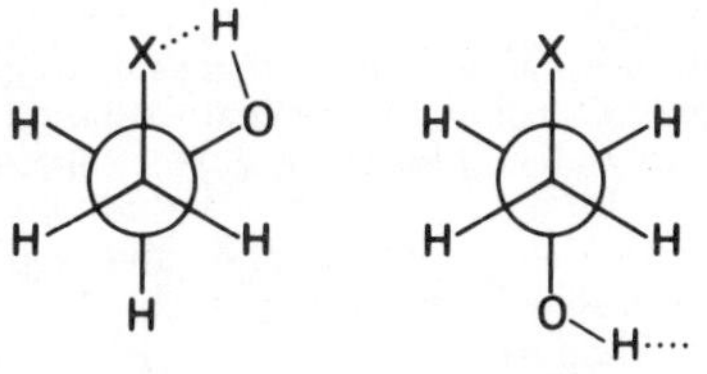

dipole moments[330], it is not ΔE_D that makes ΔE^l differ so much from ΔE^v. The change is interpreted in terms of hydrogen bonding. In the free molecule *intra*molecular bonding accounts – at least in part[331] – for the stabilization of *gauche*[79]; on liquefaction, *inter*molecular bonding can come into effect and stabilize *anti*[330].

X. CONCLUSION

When I started collecting material for this chapter, I presumed that a review of knowledge would ensue. Now that it is written, it looks more like a questionnaire. What are the secrets that the torsional term underlies? Why do the 1,2-dihaloethenes, the 1-halopropanes and the 1-halopropenes make the wrong stereochemical decisions? Are the terms in the energy equation too many[332] or perhaps too few? What makes some 4,4-disubstituted cyclohexyl halides prefer the axial conformation? Is the point-dipole equation applicable in the intramolecular context, and should the charge distribution reproduce the dipole moment? Is there a way to account simultaneously for the conformational bias in FCH_2CH_2F, FCH_2CHF_2 and F_2CHCHF_2? What does the intramolecular dielectric constant stand for, and might not its nature dictate its value? And what is the matter with cyclopentyl halides?

More questions are listed in the preceding pages and there are still more that have not found their way into the chapter. Some regard specific compounds. How come that both θ(HCH) and θ(ClCCl) in CH_2Cl_2 exceed the tetrahedral value[333]? Why is the C—X bond comparatively short in 1-haloadamantanes[73]? What about the twisted conformation[202] of perfluorinated polyethylene? Others regard the model itself. How far can one go assuming additivity in non-bonded interactions? How far can one go ignoring interaction between geminal substituents[71]? Can one go further letting the torsional term take care of itself[66]? When would molecular mechanics and quantum chemistry meet and, more intriguing, where[334]?

XI. ACKNOWLEDGEMENT

I wish to express my indebtedness to Professor N. L. Allinger who has introduced me to, and guided my way about, molecular mechanics.

XII. REFERENCES

1. J. P. Lowe, *Prog. Phys. Org. Chem.*, **6**, 1 (1968).
2. H. Booth, *Prog. NMR Spect.*, **5**, 149 (1969); p. 318.
3. E. Wyn-Jones and R. A. Pethrick, *Topics Stereochem.*, **5**, 205 (1970).
4. J. Trotter in *The Chemistry of the Carbon–Halogen Bond* (Ed. S. Patai), Part 1, Wiley, London, 1973, p. 49.
5. R. Shaw in *The Chemistry of the Carbon–Halogen Bond* (Ed. S. Patai), Part 2, Wiley, London, 1973, p. 1049.
6. A. Yokozeki and S. H. Bauer, *Topics Curr. Chem.*, **53**, 72 (1975).
7. K. H. Hellwege (Ed.), *Landolt-Börnstein Zahlenwerte und Funktionen*, New Series, Vol. II/7, Springer, Berlin, 1976.
8. J. B. Pedley and J. Rylance, *Sussex NPL Computer-analysed Thermochemical Data: Organic and Organometallic Compounds*, University of Sussex, Brighton, 1977.
9. O. Bastiansen, K. Kveseth and H. Møllendal, *Topics Curr. Chem.*, **81**, 99 (1979); (a) pp. 136–141.
10. F. R. Dollish, W. G. Fateley and F. F. Bentley, *Characteristic Raman Frequencies of Organic Compounds*, Wiley, New York, 1974, p. 14.
11. J. A. Hirsch, *Topics Stereochem.*, **1**, 199 (1967).
12. T. L. Hill, *J. Chem. Phys.*, **14**, 465 (1946).
13. J. E. Williams, P. J. Stang and P. von R. Schleyer, *Ann. Rev. Phys. Chem.*, **19**, 531 (1968).
14. P. Debye (Ed.), *The Dipole Moment and Chemical Structure*, English translation, Blackie & Son, London and Glasgow, 1931; P. Debye, *Polar Molecules*, Dover Publications, New York, 1929 Chap. 3.
15. L. L. Shipman, A. W. Burgess and H. A. Scheraga, *Proc. Nat. Acad. Sci.*, **72**, 543 (1975).
16. L. G. Csizmadia, G. Theodorakopoulos, H. B. Schlegel, M. H. Whangbo and S. Wolfe, *Canad. J. Chem.*, **55**, 986 (1977).
17. E. B. Wilson, J. C. Decius and P. C. Cross, *Molecular Vibrations*, McGraw-Hill, New York, 1955.
18. S. Melberg and K. Rasmussen, *J. Mol. Struct.*, **57**, 215 (1979).
19. C. Altona and D. H. Faber, *Topics Curr. Chem.*, **45**, 1 (1974).
20. T. L. Hill, *J. Chem. Phys.*, **16**, 399 (1948).
21. T. W. Bates, *Trans. Faraday Soc.*, **63**, 1825 (1967).
22. R. G. Woolley, *J. Am. Chem. Soc.*, **100**, 1073 (1978), footnote 16.
23. N. C. Craig and E. A. Entemann, *J. Am. Chem. Soc.*, **83**, 3047 (1961).
24. J. S. Binkley and J. A. Pople, *Chem. Phys. Lett.*, **45**, 197 (1977).
25. N. D. Epiotis, R. L. Yates, J. R. Larson, C. R. Kirmaier and F. Bernardi, *J. Am. Chem. Soc.*, **99**, 8379 (1977).
26. D. J. Mitchell and S. Wolfe, *J. Am. Chem. Soc.*, **100**, 3698 (1978); F. Bernardi, A. Bottoni, N. D. Epiotis and M. Guerra, *J. Am. Chem. Soc.*, **100**, 6018 (1978); A. Skancke and J. E. Boggs, *J. Am. Chem. Soc.*, **101**, 4063 (1979).
27. A. Y. Meyer, *J. Comput. Chem.*, **1**, 111 (1980).
28. N. D. Epiotis and W. Cherry, *JCS Chem. Commun.*, 278 (1973).
29. R. C. Bingham, M. J. S. Dewar and D. H. Lo, *J. Am. Chem. Soc.*, **97**, 1307 (1975).
30. N. D. Epiotis, W. R. Cherry, S. Shaik, R. L. Yates and F. Bernardi, *Topics Curr. Chem.*, **70**, 1 (1977); (a) p. 192.
31. N. L. Allinger and J. T. Sprague, *J. Am. Chem. Soc.*, **94**, 5734 (1972).
32. F. H. Westheimer in *Steric Effects in Organic Chemistry* (Ed. M. S. Newman), Wiley, New York, 1956.

33. J. B. Hendrickson, *J. Am. Chem. Soc.*, **83**, 4537 (1961); **84**, 3355 (1962); **86**, 4854 (1964).
34. K. B. Wiberg, *J. Am. Chem. Soc.*, **87**, 1070 (1965).
35. N. L. Allinger, *Adv. Phys. Org. Chem.*, **13**, 1 (1976); (a) p. 32.
36. O. Ermer, *Struct. Bonding*, **27**, 161 (1976).
37. M. B. Hursthouse, G. P. Moss and K. D. Sales, *Ann. Rep. Prog. Chem.*, **75B**, 23 (1978).
38. D. N. J. White in *The Chemical Society, Specialist Periodical Reports, Molecular Structure by Diffraction Methods*, **6**, 38 (1978).
39. N. L. Allinger, J. A. Hirsch, M. A. Miller and I. J. Tyminski, *J. Am. Chem. Soc.*, **91**, 337 (1969).
40. G. Heublein, R. Kühmstedt, P. Kadura and H. Dawczynski, *Tetrahedron*, **26**, 81 (1970).
41. G. Heublein, R. Kühmstedt, H. Dawczynski and P. Kadura, *Tetrahedron*, **26**, 91 (1970).
42. R. J. Abraham and K. Parry, *J. Chem. Soc. B*, 539 (1970).
43. A. Goursot-Leray and H. Bodot, *Tetrahedron*, **27**, 2133 (1971).
44. A. Y. Meyer and N. L. Allinger, *Tetrahedron*, **31**, 1971 (1975).
45. A. Y. Meyer, *J. Mol. Struct.*, **40**, 127 (1977).
46. A. Y. Meyer, *J. Mol. Struct.*, **49**, 383 (1978).
47. A. Y. Meyer, N. L. Allinger and Y. Yuh, *Israel J. Chem.*, **20**, 57 (1980).
48. A. Y. Meyer and N. Ohmichi, *J. Mol. Struct.*, **73**, 145 (1981).
49. N. L. Allinger, *J. Am. Chem. Soc.*, **99**, 8127 (1977).
50. D. A. Dougherty and K. Mislow, *J. Am. Chem. Soc.*, **101**, 1401 (1979).
51. S. Abdi, I. Yavari and M. Askari, *J. Mol. Struct.*, **76**, 37 (1981).
52. M. Cignitti and T. L. Allen, *J. Chem. Phys.*, **43**, 4472 (1965).
53. N. L. Allinger, M. T. Tribble, M. A. Miller and D. H. Wertz, *J. Am. Chem. Soc.*, **93**, 1637 (1971).
54. K. V. Mirskaya, *Tetrahedron*, **39**, 679 (1973).
55. F. Podo, G. Nemethy, P. L. Indovina, L. Radics and V. Viti, *Mol. Phys.*, **27**, 521 (1974).
56. C. P. Smyth, R. W. Dornte and E. B. Wilson, *J. Am. Chem. Soc.*, **53**, 4242 (1931); E. J. Corey, *J. Am. Chem. Soc.*, **75**, 2301 (1953).
57. Organic chemists give as reference J. H. Jeans, *Mathematical Theory of Electricity and Magnetism*, 5th edn, Cambridge University Press, Cambridge, 1933, p. 377; for a modern textbook derivation see, for example, M. Alonso and E. J. Finn, *University Physics*, Vol. II, Addison-Wesley, Reading, Mass., 1967, p. 477.
58. J. M. Lehn and G. Ourisson, *Bull. Soc. Chim. Fr.*, 1113 (1963).
59. J. R. Letelier, Y. N. Chiu and G. Zon, *J. Mol. Struct.*, **63**, 273 (1980).
60. G. Allen, P. N. Brier and G. Lane, *Trans. Faraday Soc.*, **63**, 824 (1967).
61. E. Hirota, *J. Chem. Phys.*, **37**, 283 (1962).
62. L. Radom, W. A. Lathan, W. J. Hehre and J. A. Pople, *J. Am. Chem. Soc.*, **95**, 693 (1973).
63. L. Radom, P. J. Stiles and M. A. Vincent, *J. Mol. Struct.*, **48**, 259 (1978).
64. J. R. Durig and A. W. Cox, *J. Mol. Struct.*, **38**, 77 (1977).
65. R. A. Creswell, M. Pagistas, P. Shoja-Chagheravand and R. H. Schwendeman, *J. Phys. Chem.*, **83**, 1427 (1979).
66. L. S. Bartwell, *J. Am. Chem. Soc.*, **99**, 3279 (1977); N. L. Allinger, D. Hindman and H. Hönig, *J. Am. Chem. Soc.*, **99**, 3282 (1977).
67. L. Radom, W. J. Hehre and J. A. Pople, *J. Am. Chem. Soc.*, **94**, 2371 (1972).
68. K. Kveseth, *Acta Chem. Scand.*, **A32**, 51 (1978).
69. J. D. Andose and K. Mislow, *J. Am. Chem. Soc.*, **96**, 2168 (1974).
70. I. A. Mills, *Spectrochim. Acta*, **19**, 1585 (1963).
71. C. Glidewell, *Educ. Chem.*, **16**, 146 (1979).
72. C. Flanagan and L. Pierce, *J. Chem. Phys.*, **38**, 2963 (1963).
73. A. C. Legon, D. J. Millen and A. Samson-Baktiari, *J. Mol. Struct.*, **52**, 71 (1979).
74. G. Giunchi and L. Barino, *Gazz. Chim. Ital.*, **110**, 395 (1980).
75. S. Sýkora, *Coll. Czech. Chem. Commun.*, **33**, 3514 (1968).
76. D. W. Aksnes and J. Støgård, *Acta Chem. Scand.*, **27**, 3277 (1973).
77. A. Veillard, *Chem. Phys. Lett.*, **3**, 128 (1969).
78. K. M. Marstokk and H. Møllendal, *Acta Chem. Scand.*, **A34**, 15 (1980).
79. D. Davenport and M. Schwartz, *J. Mol. Struct.*, **50**, 259 (1978).
80. D. Poland and H. A. Scheraga, *Biochemistry*, **6**, 3791 (1967).
81. E. Giglio, *Nature*, **222**, 339 (1969); A. I. Kitaigorodsky, *Chem. Soc. Revs*, **7**, 133 (1978).

82. Referee's comment on the first version of Ref. 46.
83. C. Komaki, I. Ichishima, K. Kuratani, T. Miyazawa, T. Shimanouchi and S. Mizushima, *Bull. Chem. Soc. Japan*, **28**, 330 (1955).
84. Y. Ogawa, S. Imazeki, H. Yamaguchi, H. Matsuura, I. Harada and T. Shimanouchi, *Bull. Chem. Soc. Japan*, **51**, 748 (1978).
85. R. Müller, J. Fruwert and G. Geisler, *J. Mol. Struct.*, **70**, 145 (1981).
86. E. B. Wilson, *Chem. Soc. Revs*, **1**, 293 (1972), p. 315.
87. E. L. Eliel and R. J. L. Martin, *J. Am. Chem. Soc.*, **90**, 697 (1968).
88. K. M. Marstokk and H. Møllendal, *J. Mol. Struct.*, **40**, 1 (1977).
89. F. G. Fujiwara, J. L. Painter and H. Kim, *J. Mol. Struct.*, **41**, 169 (1977).
90. M. H. Whangbo, D. J. Mitchell and S. Wolfe, *J. Am. Chem. Soc.*, **100**, 3698 (1978).
91. R. A. Beaudet and E. B. Wilson, *J. Chem. Phys.*, **37**, 1133 (1962).
92. R. A. Beaudet, *J. Chem. Phys.*, **40**, 2705 (1964).
93. R. A. Beaudet, 1972, cited in Ref. 94.
94. H. Dodziuk, *J. Mol. Struct.*, **21**, 29 (1974).
95. P. I. Abell and P. K. Adolf, *J. Chem. Therm.*, **1**, 333 (1969).
96. S. Siegel, *J. Chem. Phys.*, **27**, 989 (1957).
97. R. A. Beaudet, *J. Chem. Phys.*, **37**, 2398 (1962).
98. J. F. Stoddart, *MTP Intern. Rev. Sci., Org. Chem., Series 1*, **1**, 1 (1973), p. 22, and **7**, 1 (1973), p. 7.
99. E. L. Eliel, N. L. Allinger, S. J. Angyal and G. A. Morrison, *Conformational Analysis*, Interscience, New York, 1965.
100. C. J. F. Böttcher, *Theory of Electric Polarization*, Elsevier, Amsterdam, 1952; (a) p. 14; (b) p. 147; (c) p. 323.
101. Unless stated otherwise, dipole moments are from A. L. McClellan, *Tables of Experimental Dipole Moments*, Vol. 1, Freeman, San Fransisco, 1963, and Vol. 2, Rahara Enterprises, El Cerrito, Calif., 1974.
102. E. G. Codding and R. H. Schwendeman, *J. Mol. Spectr.*, **49**, 226 (1974).
103. A. R. Mochel, A. Bjørseth, C. O. Britt and J. E. Boggs, *J. Mol. Spect.*, **48**, 107 (1973).
104. H. J. G. Hayman and S. Weiss, *J. Chem. Phys.*, **42**, 3701 (1965).
105. N. L. Allinger and A. Y. Meyer, *Tetrahedron*, **31**, 1807 (1975).
106. L. G. Groves and S. Sugden, *J. Chem. Soc.*, 158 (1937).
107. K. B. Wiberg, *Comput. Chem.*, **1**, 221 (1977).
108. J. R. Thomas and W. D. Gwinn, *J. Am. Chem. Soc.*, **71**, 2785 (1949).
109. A. R. H. Cole and A. J. Michell, *Spectrochim. Acta*, **20**, 739 (1964).
110. D. E. Williams, *Acta Cryst.*, **30A**, 71 (1974).
111. C. Quivoron and J. Néel, *J. Chim. Phys.*, **61**, 554 (1964).
112. J. Cantacuzene, *J. Chim. Phys.*, **59**, 186 (1962).
113. C. Altona, H. R. Buys, H. J. Hageman and E. Havinga, *Tetrahedron*, **23**, 2265 (1967).
114. A. Y. Meyer, unpublished results.
115. H. J. Geise, C. Romers and W. E. M. Rutten, *Acta Cryst.*, **20**, 249 (1966); H. J. Geise and C. Romers, *Acta Cryst.*, **20**, 257 (1966).
116. H. J. Geise, A. Tieleman and E. Havinga, *Tetrahedron*, **22**, 183 (1966).
117. G. W. van Vloten, C. A. Kruissink, B. Strijk and J. M. Bijvoet, *Acta Cryst.*, **3**, 139 (1950).
118. S. Profeta, *Conformational Analysis of Amines by Force Field and Quantum Mechanical Methods*, Dissertation, University of Georgia, Athens, Ga., 1978.
119. A. C. Vandenbroucke, R. W. King and J. G. Verkade, *Rev. Sci. Instrum.*, **39**, 558 (1968).
120. G. Riley, S. Suzuki, W. J. Orville-Thomas and B. Galabov, *JCS Faraday Trans.*, **2**, 1947 (1978).
121. C. A. Coulson, *Trans. Faraday Soc.*, **38**, 433 (1942).
122. R. E. Hiller and J. W. Straley, *J. Mol. Spectr.*, **5**, 24 (1966).
123. W. Kutzelnigg, G. Del Re and G. Berthier, *Topics Curr. Chem.*, **22**, 1 (1971); (a) pp. 104–105.
124. K. B. Wiberg, *Physical Organic Chemistry*, Wiley, New York, 1964, p. 134.
125. O. Exner and Z. Friedl, *Coll. Czech. Chem. Commun.*, **42**, 3030 (1977).
126. R. D. Stolow in *Conformational Analysis, Scope and Present Limitations* (Ed. G. Chiurdoglu), Academic Press, New York, 1971.
127. U. Burkert and A. Y. Meyer, unpublished results.
128. U. Burkert, *Tetrahedron*, **33**, 2237 (1977); **35**, 209, 1945 (1979).

129. O. Exner, *Dipole Moments in Organic Chemistry*, Thieme, Stuttgart, 1975, p. 33.
130. R. A. Y. Jones, A. R. Katritzky, P. G. Lehman, K. A. F. Record and B. B. Shapiro, *J. Chem. Soc. B*, 1302 (1971).
131. L. J. Collins and D. N. Kirk, *Tetrahedron Lett.*, 1547 (1970).
132. R. J. Abraham and Z. L. Rossetti, *JCS Perkin 2*, 582 (1973).
133. K. Tanabe, *Spectrochim. Acta*, **28A**, 407 (1972).
134. L. Fernholt and K. Kveseth, *Acta Chem. Scand.*, **A32**, 63 (1978).
135. F. T. Chau and C. A. McDowell, *J. Mol. Struct.*, **34**, 93 (1976).
136. S. Kondo, E. Tagami, K. Iimura and M. Takeda, *Bull. Chem. Soc. Japan*, **41**, 790 (1968).
137. P. J. D. Park and E. Wyn-Jones, *J. Chem. Soc. A*, 422 (1969).
138. J. G. Kirkwood and F. H. Westheimer, *J. Chem. Phys.*, **6**, 506, 513 (1938).
139. J. Allinger and N. L. Allinger, *Tetrahedron*, **2**, 64 (1958).
140. J. T. Edward, P. G. Farrell and J. L. Job, *J. Chem. Phys.*, **57**, 5251 (1972).
141. S. Ehrenson, *J. Am. Chem. Soc.*, **98**, 7510 (1976).
142. G. Castellani, R. Scordamaglia and C. Tosi, *Gazz. Chim. Ital.*, **110**, 457 (1980).
143. M. J. T. Robinson, *Pure Appl. Chem.*, **25**, 635 (1971).
144. C. B. Anderson and D. T. Sepp, *Tetrahedron*, **24**, 1707 (1968).
145. N. L. Allinger, L. Došen-Mićović, J. H. Viskocil and M. T. Tribble, *Tetrahedron*, **34**, 3395 (1978).
146. L. Došen-Mićović and N. L. Allinger, *Tetrahedron*, **34**, 3385 (1978).
147. G. A. Jeffrey and R. Taylor, *J. Comput. Chem.*, **1**, 99 (1980).
148. D. A. Brant and P. J. Flory, *J. Am. Chem. Soc.*, **87**, 2791 (1965).
149. V. S. R. Rao, K. S. Vijayalakshmi and P. R. Sundararajan, *Carbohydrate Res.*, **17**, 341 (1971).
150. R. A. Scott and H. A. Scheraga, *J. Chem. Phys.*, **45**, 2091 (1966).
151. S. Acevedo and K. Bowden, *JCS Chem. Commun.*, 608 (1977).
152. R. B. Hermann, *J. Am. Chem. Soc.*, **91**, 3152 (1969).
153. M. J. Bovill, D. J. Chadwick, I. O. Sutherland and D. Watkin, *JCS Perkin 2*, 1529 (1980).
154. R. B. Hermann, *J. Org. Chem.*, **27**, 441 (1962).
155. E. A. Jauregui, M. R. Estrada, L. S. Mayorga and G. M. Ciuffo, *J. Mol. Struct.*, **54**, 257 (1979).
156. G. L. Henderson, *J. Phys. Chem.*, **83**, 856 (1979).
157. A. T. Hagler and A. Lapiccirella, *Biopolymers*, **15**, 1167 (1967).
158. A. A. Lugovski and V. G. Dashevsky, *J. Struct. Chem.* (English translation), **13**, 105, 112 (1972).
159. R. S. Mulliken, *J. Chem. Phys.*, **46**, 539 (1949).
160. B. Pullman and A. Pullman, *Les théories electroniques de la chimie organique*, Masson, Paris, 1952, Chap. VII.
161. H. Berthod and A. Pullman, *J. Chim. Phys.*, **62**, 942 (1965); H. Berthod, C. Giessner-Prettre and A. Pullman, *Theoret. Chim. Acta*, **8**, 212 (1967).
162. R. S. Mulliken, *J. Chem. Phys.*, **23**, 1833 (1955).
163. For formulation and sample calculations see J. E. Bloor, B. R. Gilson and F. P. Billingsley, *Theoret. Chim. Acta*, **12**, 360 (1968); J. A. Pople and G. A. Segal, *J. Chem. Phys.*, **43S**, 136 (1965).
164. G. G. Hall, *Chem. Phys. Lett.*, **20**, 501 (1973); A. D. Tait and G. G. Hall, *Theoret. Chim. Acta*, **31**, 311 (1973).
165. A. F. Marchington, S. C. R. Moore and W. G. Richards, *J. Am. Chem. Soc.*, **101**, 5529 (1979).
166. G. Del Re, *J. Chem. Soc.*, 4031 (1958); G. Del Re, G. Berthier and J. Serre, *Electronic States of Molecules and Atom Clusters*, Lecture Notes in Chemistry No. 13, Springer, Berlin, 1980, pp. 49–58.
167. R. P. Smith, T. Ree, J. L. Magee and H. Eyring, *J. Am. Chem. Soc.*, **73**, 2263 (1951).
168. R. P. Smith and H. Eyring, *J. Am. Chem. Soc.*, **74**, 229 (1952).
169. R. P. Smith and E. M. Mortensen, *J. Am. Chem. Soc.*, **78**, 3932 (1956).
170. R. P. Smith and J. J. Rasmussen, *J. Am. Chem. Soc.*, **83**, 3785 (1961).
171. L. J. Bellamy and R. L. Williams, *J. Chem. Soc.*, 4294 (1957).
172. H. Bodot and J. Jullien, *Bull. Soc. Chim. Fr.*, 1488 (1962).
173. P. A. Bouffier and P. Federlin, *Bull. Soc. Chim. Fr.*, 4079 (1967).

174. R. D. Stolow, P. W. Samal and T. W. Giants, *J. Am. Chem. Soc.*, **103**, 197 (1981).
175. N. L. Allinger and M. T. Wuesthoff, *Tetrahedron*, **33**, 3 (1977).
176. W. F. Reynolds, *JCS Perkin 2*, 985 (1980).
177. H. M. Smallwood and K. F. Herzfeld, *J. Am. Chem. Soc.*, **52**, 1919 (1930).
178. J. W. Smith, *Electric Dipole Moments*, Butterworths, London, 1955; (a) p. 175; (b) p. 181.
179. C. P. Smyth, *Dielectric Behavior and Structure*, McGraw-Hill, New York, 1955; (a) p. 235; (b) p. 238.
180. N. E. Hill, W. E. Vaughan, A. H. Price and M. Davies, *Dielectric Properties and Molecular Behaviour*, Van Nostrand, London, 1969, p. 248; V. I. Minkin, O. A. Osipov and Y. A. Zhdanov, *Dipole Moments in Organic Chemistry*, Plenum Press, New York, 1970, p. 191.
181. W. D. Kumler and A. C. Huitric, *J. Am. Chem. Soc.*, **78**, 3369 (1956).
182. (a) N. L. Allinger, H. M. Blatter, M. A. DaRooge and L. A. Freiberg, *J. Org. Chem.*, **26**, 2550 (1961).
(b) I. Miyagawa, *J. Chem. Soc. Japan*, **75**, 1061 (1954).
183. D. H. Wertz and N. L. Allinger, *Tetrahedron*, **30**, 1579 (1974).
184. R. Stølevik, *Acta Chem. Scand.*, **A28**, 299 (1974).
185. L. Fernholt, R. Seip and R. Stølevik, *Acta Chem. Scand.*, **A32**, 225 (1978).
186. M. Braathen, D. H. Christensen, P. Klaeboe, R. Seip and R. Stølevik, *Acta Chem. Scand.*, **A33**, 437 (1979).
187. R. J. Abraham and P. Loftus, *Chem. Commun.*, 180 (1974).
188. D. Bhaumik and G. S. Kastha, *Indian J. Pure Appl. Chem.*, **15**, 820 (1977).
189. E. J. M. van Schaick, H. J. G. Geise, F. C. Mijlhoff and G. Renes, *J. Mol. Struct.*, **16**, 23 (1973).
190. L. Fernholt and K. Kveseth, *Acta Chem. Scand.*, **A34**, 163 (1980).
191. D. Friesen and K. Hedberg, *J. Am. Chem. Soc.*, **102**, 3987 (1980).
192. J. Hine, *Structural Effects in Organic Chemistry*, Wiley–Interscience, New York, 1975; (a) p. 125.
193. S. Wolfe, *Accts Chem. Res.*, **5**, 102 (1972).
194. E. Saegebarth and E. B. Wilson, *J. Chem. Phys.*, **46**, 3088 (1967).
195. O. Steinnes, Q. Shen and K. Hagen, *J. Mol. Struct.*, **64**, 217 (1980).
196. B. P. van Eijck, *J. Mol. Struct.*, **37**, 1 (1977).
197. D. D. Danielson and K. Hedberg, *J. Am. Chem. Soc.*, **101**, 3730 (1969).
198. D. E. Brown and B. Beagley, *J. Mol. Struct.*, **38**, 167 (1977).
199. F. Heatley and G. Allen, *Mol. Phys.*, **16**, 77 (1969).
200. R. J. Abraham, M. A. Cooper, T. M. Siverns, P. E. Swinton, H. G. Weder and L. Cavalli, *Org. Magn. Reson.*, **6**, 331 (1974).
201. B. Beagley and D. E. Brown, *J. Mol. Struct.*, **54**, 175 (1979).
202. S. Scheiner, *J. Am. Chem. Soc.*, **102**, 3723 (1980).
203. E. L. Wagner, *Theoret. Chim. Acta*, **23**, 115, 127 (1971).
204. J. O. Williams, J. N. Scarsdale, L. Schäfer and H. J. Geise, *J. Mol. Struct.* (*Theochem.*), **76**, 11 (1981).
205. N. C. Craig, T. N. Hu Chao, E. Cuellar, D. E. Hendriksen and J. W. Koepke, *J. Phys. Chem.*, **79**, 2270 (1975).
206. M. Traetteberg, H. Østensen and R. Seip, *Acta Chem. Scand.*, **A34**, 449 (1980).
207. R. J. Abraham, R. A. Hearmon, M. Traetteberg and P. Bakken, *J. Mol. Struct.*, **57**, 149 (1979).
208. E. Hirota, *J. Chem. Phys.*, **42**, 2071 (1965).
209. D. S. Stephenson and G. Binsch, *Org. Magn. Reson.*, **14**, 226 (1980).
210. A. D. English, L. H. Scharpen, K. M. Ewool, H. L. Strauss and D. O. Harris, *J. Mol. Spectr.*, **60**, 210 (1976).
211. E. B. Whipple, *J. Chem. Phys.*, **35**, 1039 (1961).
212. T. K. Brunck and F. Weinhold, *J. Am. Chem. Soc.*, **101**, 1700 (1979).
213. M. H. Whangbo and S. Wolfe, *Israel J. Chem.*, **20**, 36 (1980).
214. J. Tyrell, *J. Am. Chem. Soc.*, **98**, 5456 (1976).
215. J. Tyrell, *J. Am. Chem. Soc.*, **101**, 3766 (1979).
216. A. J. P. Devaquet, R. E. Townshend and W. J. Hehre, *J. Am. Chem. Soc.*, **98**, 4068 (1976).
217. I. Miyagawa, T. Chiba, S. Ikeda and Y. Morino, *Bull. Chem. Soc. Japan*, **30**, 218 (1957).

218. D. Chadwick and D. J. Millen, *Trans. Faraday Soc.*, **67**, 1539 (1971); *J. Mol. Struct.*, **25**, 216 (1975).
219. E. Hirota, T. Tanaka, K. Sakkibara, Y. Ohashi and Y. Morino, *J. Mol. Spectr.*, **34**, 222 (1970).
220. B. Beagley, M. O. Jones and N. Houldsworth, *J. Mol. Struct.*, **62**, 105 (1980).
221. R. T. Morrison and R. N. Boyd, *Organic Chemistry*, 3rd edn, Allyn and Bacon, London, 1973, pp. 78–79.
222. L. S. Bartell and T. L. Boates, *J. Mol. Struct.*, **32**, 379 (1976).
223. N. C. Craig, L. G. Piper and V. L. Wheeler, *J. Phys. Chem.*, **75**, 1453 (1971).
224. H. G. Viehe, *Chem. Ber.*, **93**, 1697 (1960).
225. J. W. Nibler and V. E. Bondeybey, *J. Chem. Phys.*, **60**, 1307 (1974).
226. E. Flood, P. Pulay and J. E. Boggs, *J. Mol. Struct.*, **50**, 355 (1978).
227. C. H. Chang, R. F. Porter and S. H. Bauer, *J. Am. Chem. Soc.*, **92**, 4313 (1970).
228. Y. Kodama, K. Nishihata, S. Zushi, M. Nishio, J. Uzawa, K. Sakamoto and H. Iwamura, *Bull. Chem. Soc. Japan*, **52**, 2661 (1979).
229. L. S. Bartell, *J. Am. Chem. Soc.*, **99**, 3279 (1977).
230. R. J. Ouellette and S. H. Williams, *J. Am. Chem. Soc.*, **93**, 466 (1971).
231. D. C. Best, G. Underwood and C. A. Kingsbury, *Chem. Commun.*, 627 (1969).
232. R. J. Abraham and E. Bretschneider in *Internal Rotation in Molecules* (Ed. W. J. Orville-Thomas), Wiley, London, 1974; (a) p. 515; (b) p. 510.
233. O. H. Ellestad and K. Kveseth, *J. Mol. Struct.*, **25**, 175 (1975).
234. C. H. Chang, A. L. Andreassen and S. H. Bauer, *J. Org. Chem.*, **36**, 920 (1971).
235. R. Aroca Muñoz and Yu. N. Panchenko, *Spectrochim. Acta*, **28A**, 27 (1972).
236. R. M. Conrad and D. A. Dows, *Spectrochim. Acta*, **21**, 1039 (1965).
237. R. E. Carter, B. Nilsson and K. Olsson, *J. Am. Chem. Soc.*, **97**, 6155 (1975); R. E. Carter and P. Stilbs, *J. Am. Chem. Soc.*, **98**, 7515 (1976); B. Aurivillius and R. E. Carter, *JCS Perkin 2*, 1033 (1979).
238. J. M. A. Baas, B. van de Graaf, F. van Rantwijk and A. van Veen, *Tetrahedron*, **35**, 421 (1979).
239. S. G. Baxter, H. Fritz, G. Hellmann, B. Kitschke, H. J. Lindner, K. Mislow, C. Rüchardt and S. Weiner, *J. Am. Chem. Soc.*, **101**, 4493 (1979).
240. A. Liberles, A. Greenberg and J. E. Eilers, *J. Chem. Educ.*, **50**, 676 (1973).
241. F. A. Cotton and G. Wilkinson, *Advanced Inorganic Chemistry*, 3rd edn, Interscience, New York, 1972, p. 552 and p. 980.
242. M. Traetteberg, P. Bakken, R. Seip, S. J. Cyvin, B. N. Cyvin and H. Hopf, *J. Mol. Struct.*, **55**, 199 (1979).
243. N. L. Allinger, M. A. Miller, F. A. VanCatledge and J. A. Hirsch, *J. Am. Chem. Soc.*, **89**, 4345 (1967).
244. P. E. Farup and R. Stølevik, *Acta Chem. Scand.*, **A28**, 680 (1974). For discussion see J. E. Gustavsen, P. Klaeboe and R. Stølevik, *J. Mol. Struct.*, **50**, 285 (1978).
245. S. W. Charles, F. C. Cullen and N. L. Owen, *J. Mol. Struct.*, **34**, 219 (1976).
246. R. Kewley, *Canad. J. Chem.*, **52**, 509 (1974).
247. F. A. Momany, R. A. Bonham and W. H. McCoy, *J. Am. Chem. Soc.*, **85**, 3077 (1963).
248. E. Wyn-Jones and W. J. Orville-Thomas, *Trans. Faraday Soc.*, **64**, 2907 (1968).
249. P. J. D. Park and E. Wyn-Jones, *J. Chem. Soc. A*, 2944 (1968).
250. W. Freitag and H. J. Schneider, *Israel J. Chem.*, **20**, 153 (1980).
251. J. E. Anderson, C. W. Doecke and H. Pearson, *JCS Perkin 2*, 336 (1979).
252. P. Cederbalk and F. Karlsson, *Acta Chem. Scand.*, **A34**, 541 (1980).
253. F. Karlsson and Z. Smith, *J. Mol. Spectr.*, **81**, 327 (1980).
254. M. Hayashi, H. Kato and M. Oyamada, *J. Mol. Spectr.*, **83**, 408 (1980).
255. S. Mizushima, *Structure of Molecules and Internal Rotation*, Academic Press, New York, 1954.
256. M. F. El Bermani, C. J. Vear, A. J. Woodward and N. Jonathan, *Spectrochim. Acta*, **24A**, 1251 (1968).
257. M. F. El Bermani and N. Jonathan, *J. Chem. Phys.*, **49**, 340 (1968).
258. K. Kveseth, *Acta Chem. Scand.*, **A28**, 482 (1974); **A29**, 307 (1975).
259. K. Tanabe, *Spectrochim. Acta*, **30A**, 1901 (1974).
260. K. Tanabe, *J. Mol. Struct.*, **25**, 259 (1975).

261. K. Tanabe, *J. Mol. Struct.*, **28**, 329 (1975).
262. L. Fernholt and K. Kveseth, *Acta Chem. Scand.*, **A33**, 335 (1979).
263. S. S. Butcher, R. A. Cohen and T. C. Rounds, *J. Chem. Phys.*, **54**, 4123 (1971).
264. A. DiGiacomo and C. P. Smyth, *J. Am. Chem. Soc.*, **77**, 1361 (1955).
265. C. T. Zahn, *Phys. Rev.*, **40**, 291 (1932).
266. G. I. M. Bloom and L. E. Sutton, *J. Chem. Soc.*, 727 (1941).
267. D. Bhaumik and G. S. Kastha, *Indian J. Phys.*, **51B**, 202 (1977).
268. K. G. R. Pachler and J. P. Tollenaere, *J. Mol. Struct.*, **8**, 83 (1971).
269. F. J. Wodarczyk and E. B. Wilson, *J. Chem. Phys.*, **56**, 166 (1972).
270. L. Radom, W. J. Hehre and J. A. Pople, *J. Am. Chem. Soc.*, **94**, 2371 (1972).
271. I. Fischer-Hjalmars, *Tetrahedron*, **19**, 1805 (1963).
272. N. L. Allinger and S. Profeta, *J. Comput. Chem.*, **1**, 181 (1980).
273. H. J. Bernstein, *J. Chem. Phys.*, **17**, 262 (1949).
274. See papers by D. H. R. Barton and by O. Hassel, republished in *Topics Stereochem.*, **6**, 1 (1971).
275. E. L. Eliel, *Stereochemistry of Carbon Compounds*, McGraw-Hill, New York, 1962.
276. F. R. Jensen and C. H. Bushweller, *Adv. Alicycl. Chem.*, **3**, 139 (1971).
277. R. M. Moriarty, *Topics Stereochem.*, **8**, 271 (1974).
278. A. C. Legon, *Chem. Rev.*, **80**, 231 (1980).
279. B. Fuchs, *Topics Stereochem.*, **10**, 1 (1978).
280. D. Höfner, S. A. Lesko and G. Binsch, *Org. Magn. Reson.*, **11**, 179 (1978).
281. J. J. Moura Ramos, L. Dumont, M. L. Stien and J. Reisse, *J. Am. Chem. Soc.*, **102**, 4150 (1980).
282. F. A. L. Anet, J. Krane, W. Kitching, D. Dodderel and D. Praeger, *Tetrahedron Lett.*, 3255 (1974).
283. P. Anderson, *Acta Chem. Scand.*, **16**, 2337 (1962).
284. J. Reisse, J. C. Celotti and G. Chiurdoglu, *Tetrahedron Lett.*, 397 (1965).
285. J. Reisse, M. L. Stien, J. M. Gilles and J. F. M. Oth, *Tetrahedron Lett.*, 1917 (1969).
286. F. R. Jensen, C. H. Bushweller and B. H. Beck, *J. Am. Chem. Soc.*, **91**, 344 (1969).
287. J. Reisse in *Conformational Analysis* (Ed. G. Chiurdoglu), Academic Press, New York, 1971.
288. R. C. Loyd, S. N. Mathur and M. D. Harmony, *J. Mol. Spect.*, **72**, 359 (1978).
289. B. Franzus and B. E. Hudson, *J. Org. Chem.*, **28**, 2238 (1963).
290. E. L. Eliel, *Chem. Ind. (London)*, 568 (1959).
291. J. F. Stoddart, *Stereochemistry of Carbohydrates*, Wiley–Interscience, New York, 1971.
292. W. A. Szarek and D. Horton (Eds), *Anomeric Effect*, American Chemical Society, Washington, D.C., 1979.
293. C. B. Anderson and D. T. Sepp, *J. Org. Chem.*, **32**, 607 (1967).
294. J. T. Edward, *Chem. Ind. (London)*, 1102 (1955).
295. N. L. Allinger and D. Y. Chung, *J. Am. Chem. Soc.*, **98**, 6798 (1976).
296. C. Altona and C. Romers, *Rec. Trav. Chim. Pays Bas*, **82**, 1080 (1963); see also K. Fukushima and S. Takeda, *J. Mol. Struct.*, **49**, 259 (1978).
297. N. S. Zefirov, L. G. Gurvich, A. S. Shashkov, M. Z. Krimer and E. A. Vorob'eva, *Tetrahedron*, **32**, 1211 (1976).
298. N. S. Zefirov, *Tetrahedron*, **33**, 3193 (1977).
299. E. L. Eliel and E. Juaristi in *Anomeric Effect* (Eds W. A. Szarek and D. Horton), American Chemical Society, Washington, D.C., 1979.
300. L. D. Hall and D. L. Jones, *Canad. J. Chem.*, **51**, 2914 (1973).
301. G. J. Karabatsos and D. J. Fenoglio, *J. Am. Chem. Soc.*, **91**, 1124 (1969).
302. Y. H. Pan and J. B. Stothers, *Canad. J. Chem.*, **45**, 2943 (1967).
303. E. J. Corey and H. J. Burke, *J. Am. Chem. Soc.*, **77**, 5418 (1955).
304. M. Laudet, F. Metras, J. Petrissans and G. Pfister-Guillouzo, *J. Mol. Struct.*, **29**, 263 (1975).
305. J. Cantacuzene, R. Jantzen and D. Ricard, *Tetrahedron*, **28**, 717 (1972).
306. F. Johnson and S. K. Malhotra, *J. Am. Chem. Soc.*, **87**, 5492 (1965).
307. W. R. Bowman, B. T. Golding and W. P. Watson, *JCS Perkin 2*, 731 (1980).
308. J. E. Leffler and E. Grunwald, *Rates and Equilibria of Organic Reactions*, Wiley, New York, 1963.

309. K. G. R. Pachler and P. L. Wessels, *J. Mol. Struct.*, **68**, 145 (1980).
310. O. Tapia and B. Silvi, *J. Phys. Chem.*, **84**, 2646 (1980).
311. C. Serboli and B. Minasso, *Spectrochim. Acta*, **24A**, 1813 (1968).
312. P. Laszlo, *Prog. NMR Spectr.*, **3**, 231 (1967).
313. C. Reichardt, *Solvent Effects in Organic Chemistry*, Verlag Chemie, Weinheim, 1979.
314. B. Pullman (Ed.), *Environmental Effects on Molecular Structure and Properties*, Reidel, Dordrecht, 1976.
315. I. Tvaroška and T. Bleha, *Coll. Czech. Chem. Commun.*, **45**, 1883 (1980).
316. L. Onsager, *J. Am. Chem. Soc.*, **58**, 1486 (1936).
317. S. Mizushima and H. Okazaki, *J. Am. Chem. Soc.*, **71**, 3411 (1949).
318. R. J. Abraham, *J. Phys. Chem.*, **73**, 1192 (1969).
319. J. Powling and H. J. Bernstein, *J. Am. Chem. Soc.*, **73**, 1815, 4353 (1951).
320. J. R. Partington, *An Advanced Treatise on Physical Chemistry*, Vol. 2, Longmans, London, 1951, pp. 17–28.
321. A. I. Vogel, *Practical Organic Chemistry*, 3rd edn, Longmans, London, 1957, pp. 1034–1035.
322. M. K. Kaloustian. *J. Chem. Educ.*, **51**, 777 (1974).
323. Y. Morino, I. Miyagawa, T. Haga and S. Mizushima, *Bull. Chem. Soc. Japan*, **28**, 165 (1955).
324. S. D. Christian, J. Grundnes, P. Klaeboe, E. Tørneng and T. Woldbaek, *Acta Chem. Scand.*, **A34**, 391 (1980).
325. E. Wilhelm, J. P. E. Grolier and M. H. Karbalai Ghassemi, *Monats. Chem.*, **109**, 369 (1978).
326. S. Winstein and R. E. Wood, *J. Am. Chem. Soc.*, **62**, 548 (1940).
327. F. A. Bothner-By and C. Naar-Colin, *J. Am. Chem. Soc.*, **84**, 743 (1962).
328. P. M. Ivanov and I. G. Pojarlieff, *J. Mol. Struct.*, **64**, 67 (1980).
329. T. Bhela, J. Gajdos and I. Tvaroška, *J. Mol. Struct.*, **68**, 189 (1980).
330. G. S. Kastha, S. B. Roy and S. K. Nandy, *Indian J. Phys.*, **46**, 293 (1972).
331. Y. Hopilliard and D. Solgadi, *Tetrahedron*, **36**, 377 (1980).
332. T. Sundius and K. Rasmussen, *Commentat. Phys.-Math. Soc. Sci. Fenn.*, **47**, 91 (1977).
333. D. Chadwick and D. J. Millen, *J. Mol. Struct.*, **25**, 216 (1975).
334. K. Kovačević, M. Eckert-Maksić and Z. B. Maksić, *Croat. Chem. Acta*, **46**, 249 (1974).

The Chemistry of Functional Groups, Supplement D
Edited by S. Patai and Z. Rappoport
© 1983 John Wiley & Sons Ltd

CHAPTER 2

Diamagnetic behaviour of compounds containing carbon–halogen bonds

R. R. GUPTA
Department of Chemistry, University of Rajasthan, Jaipur 302004, India

I. INTRODUCTION

Much research work has been reported on the diamagnetic behaviour of organic and organometallic compounds in attempts to find applications of diamagnetic measurements in structural chemistry. Systematic studies on the diamagnetic behaviour of organic compounds containing carbon–halogen bonds were initiated by French and Trew[1] and anomalous diamagnetic behaviour was reported for halomethanes. Gupta and Mital[2] have also reported anomalous diamagnetic behaviour of compounds containing carbon–halogen bonds in comparison to analogous compounds of other series and ascribed this to the strong interactions between the carbon and halogen atoms. To account for such anomalous diamagnetic behaviour several theories were developed and methods were worked out to estimate diamagnetic susceptibilities of such compounds. None of these was satisfactory till 1964, when Heberditzl[3] reported a method for the calculation of diamagnetic susceptibilities of organic compounds. Gupta and Mital[4] have modified the Haberditzl method to calculate the diamagnetism of di- and polysubstituted benzenes and their derivatives, accounting for interactions between substituents. By now, a large number of organic compounds have been studied and it has been established that such investigations are useful in studying the nature of bonding in carbon–halogen bonds. No review has appeared on this topic and an account of diamagnetic behaviour of organic compounds containing carbon–halogen bonds is presented here.

II. MEASUREMENT OF DIAMAGNETIC SUSCEPTIBILITY

A. Gouy Method

The Gouy method[5] is the most widely used technique for the measurement of the diamagnetic susceptibilities of organic compounds. It is discussed here briefly and for details Refs 6–18 should be consulted.

1. Theory

The substance is considered in the form of a rod and to satisfy this requirement a glass tube (known as a Gouy tube) is used. This is filled with the substance and suspended vertically from one arm of a balance maintaining its lower end near the middle point of the field between the poles of the magnet and the upper end above the pole gap, out of the influence of the magnetic field. Bates[18] has shown that the force ($\mathrm{d}F$) acting in an homogeneous field (H gauss) on an element ($\mathrm{d}V$, $\mathrm{d}m$) of the material along the gradient ($\mathrm{d}H/\mathrm{d}X$) is given by equation (1):

$$\mathrm{d}F = Hk\,\mathrm{d}V\frac{\mathrm{d}H}{\mathrm{d}X}, \tag{1}$$

where k is the volume susceptibility of the substance. The ends of the Gouy tube are in fields of H and H_0 gauss and the force acting on the tube can be expressed by equation (2):

$$F = \frac{A}{2}k(H^2 - H_0^2), \tag{2}$$

where A is the area of cross-section of the Gouy tube.

The Gouy tube comprises a form of hollow specimen and a force is exerted on the tube in the field. This force δg (δ is the apparent change in the weight of the empty Gouy tube in the field) should be subtracted from the force acting on the Gouy tube when it is filled with the substance in order to get the force acting on the substance. The force acting on the Gouy tube is negative, since it is made of glass. When the Gouy tube is filled with the substance, the air, which possesses a fair amount of magnetic susceptibility and contributes to the force acting on the Gouy tube in the field, is displaced, and correction is applied for this while calculating the force acting on the substance in the field. The force acting on the substance alone in the field can be expressed by equation (3):

$$F = \frac{A}{2}(k - k_1)(H^2 - H_0^2) + \delta g, \quad (3)$$

where k_1 is the volume susceptibility of air. Since the upper end of the Gouy tube is out of the influence of the field, H_0 may be neglected. Equation (3) is reduced to equation (4):

$$F = \frac{A}{2}H^2(k - k_1) + \delta g, \quad (4)$$

where $F = g\mathrm{d}w$ ($\mathrm{d}w$ is the apparent change in the weight of the substance in the field and g is the gravitational constant) and $k = \chi_s(m/V)$ (χ_s, m and V are the mass or volume susceptibility, the mass and the volume of the substance, respectively). Equation (4) can be transformed into equation (5):

$$\chi_s = \frac{1}{m}\left[Vk_1 + \frac{2Vg}{AH^2}(\mathrm{d}w - \delta)\right], \quad (5)$$

where Vk_1 is constant and can be obtained by multiplying the volume susceptibility of air and the volume of the substance in the tube. Let $Vk_1 = \alpha$. The second term, $2Vg/AH^2$, is also constant (because A and V are constant for the given Gouy tube, g is constant for a particular place and H is kept constant for a set of measurements). Let $2Vg/AH^2 = \beta$. Equation (5) then takes the form of equation (6),

$$\chi_s = \frac{\alpha + \beta(\mathrm{d}w - \delta)}{m}, \quad (6)$$

$\mathrm{d}w$ and δ are in milligrams and m is in grams. β is known as the tube calibration constant. By determining the values of α and β for a given Gouy tube under a set of conditions and substituting these values in equation (6), χ_s is calculated.

2. *Instrumentation*

The Gouy balance consists of three parts.

a. Balance. A semi-micro balance is used to measure the apparent change in the weight of the substance in the magnetic field. The Gouy tube is hung from one of the arms, with its lower end at the centre of the pole gap (in the maximum field) and its upper end out of the magnetic field. To exclude air draughts the Gouy tube is surrounded by a glass jacket.

b. Gouy tube. The Gouy tube is generally made of Pyrex glass. It is 6–10 cm in length and 3–9 mm in diameter, depending on the nature of the substance under investigation.

c. Magnet. Generally a magnet capable of producing a field strength of 5000–12 000 gauss is employed. A pole diameter of approximately 3.5 cm is capable of producing a pole gap of 1.5–2.5 cm, which is sufficient for measurements at room temperature. Both permanent magnets and electromagnets are used. Permanent magnets produce fields up to 8000 gauss, but the field strength for a given pole gap cannot be changed. Electromagnets with a current regulator can produce field strengths up to 15 000 gauss, which can be changed for given pole gaps by varying the current. A plot between current and field strength provides the field strength for a given pole gap.

B. Faraday Method

The force given in equation (1) can be transformed to that in equation (7):

$$\mathrm{d}F = \frac{k}{d} d\,\mathrm{d}V \frac{H\,\mathrm{d}H}{\mathrm{d}X} = \chi_s\,\mathrm{d}m \frac{H\,\mathrm{d}H}{\mathrm{d}X} \tag{7}$$

where dm is the mass corresponding to the volume $\mathrm{d}V$ of the substance. Since the volume of the substance is small, $H\,\mathrm{d}H/\mathrm{d}X$ is fairly constant and $\mathrm{d}F$ is proportional to χ_s and forms the basis for the Faraday method[19]. To measure force ($\mathrm{d}F$), either a torsion head[20–24] or an optical[6,16,25] arrangement is employed.

1. *Torsion head arrangement*

A torsion balance is employed to measure the force. The sample is suspended in the pole gap by a quartz torsion fibre and is free to move horizontally. The sample is displaced from the zero position when the field is switched on and the weights are adjusted to twist the torsion head till it returns to the original position. This twist of the torsion head measures the force required to balance the magnetic force at the zero position. Measurements are carried out for a standard substance of known susceptibility and for the substance under investigation. χ_s of the substance is calculated from equation (7).

2. *Optical arrangement*

To measure the force, an optical system is used. The sample (generally placed in a fused quartz bucket of about 1 mm internal diameter) is suspended by a quartz fibre from a phosphor-bronze ring fitted with an optical arrangement. On switching on the field the sample is displaced and the displacement is magnified several hundred times by this arrangement. The balance, suspension and sample systems are enclosed. Let θ_b, θ_r and θ_s be the respective deflections in the field for the empty bucket, for the bucket containing the standard reference substance and for the bucket containing the test sample. It can be shown that the deflection is proportional to the force experienced by the sample in the field and, from equation (7), equation (8) can be obtained:

$$\mathrm{d}F = \chi_s\,\mathrm{d}m \frac{H\,\mathrm{d}H}{\mathrm{d}x}. \tag{7}$$

Put $H\,\mathrm{d}H/\mathrm{d}x = C_1$ (a constant), then

$$\mathrm{d}F/C_1 = \chi_s\,\mathrm{d}m;$$

again put $\mathrm{d}F = C_2\theta$ (C_2 is a constant) and then

$$C_2\theta/C_1 = \chi_s\,\mathrm{d}m,$$

where C_2/C_1 is again a constant, say C. Hence

$$C\theta = \chi_s \, dm. \tag{8}$$

If χ_{s_r} is the susceptibility of the reference substance, equation (8) is reduced to equation (9):

$$C(\theta_r - \theta_b) = \chi_{s_r}(dm_r - dm_b). \tag{9}$$

Similarly,

$$C(\theta_s - \theta_b) = \chi_s(dm_s - dm_b), \tag{10}$$

where dm_b is the mass of the empty bucket, dm_r is the mass of the bucket containing the reference substance and dm_s is the mass of the bucket containing the test sample.

On dividing equation (10) by equation (9), equation (11) is obtained:

$$\frac{(\theta_s - \theta_b)}{(\theta_r - \theta_b)} = \frac{\chi_s}{\chi_{s_r}} \frac{(dm_s - dm_b)}{(dm_r - dm_b)}, \tag{11}$$

$$\chi_s = \chi_{s_r} \frac{(\theta_s - \theta_b)}{(\theta_r - \theta_b)} \frac{(dm_r - dm_b)}{(dm_s - dm_b)}. \tag{12}$$

By noting the successive deflections for the empty bucket, the bucket containing the standard reference substance, and the bucket containing the test sample and substituting the values of θ_s, θ_b, and θ_r in equation (12), χ_s can be calculated.

C. Nuclear Magnetic Resonance (NMR) Method

A method based on chemical shifts in NMR spectra has been reported by Evans[26]. This can be applied only to paramagnetic substances and it will not be discussed here. Frei and Bernstein[27] reported on an NMR method based on the difference in chemical shifts arising from differences in the shape factors for a sphere and a cylinder. They inserted spherically and cylindrically shaped reference tubes (each containing distilled water as a reference material) in a conventional spinning sample tube of 4 mm (outer diameter) and obtained two sharp signals corresponding to two different reference tubes. The separation of signals has been found to be linearly dependent on the volume susceptibility of the sample in the sample tube. Volume susceptibility has been calculated from equation (13):

$$\delta\text{cyl(ref)} - \delta\text{sph(ref)} = [g_{\text{cyl}} - g_{\text{sph}}][K(\text{ref}) - K(\text{sample})] \tag{13}$$

where δ is the chemical shift and δcyl(ref) $-$ δsph(ref) is obtained from the NMR spectra, g is the geometrical constant and for ideal geometry $(g_{\text{cyl}} - g_{\text{sph}}) = 2\pi/3 \approx 2.095$. Frei and Bernstein[27] have found this factor to be 2.058 from a slope of least squares treatment for 15 compounds. By substituting values of different constants and factors, K(sample) is calculated from equation (13), and is converted into χ_s by dividing K (volume susceptibility) by the density of the sample.

III. THEORETICAL CALCULATION OF DIAMAGNETIC SUSCEPTIBILITY

Methods used to calculate the diamagnetic susceptibilities of organic compounds are discussed here in brief.

A. Pascal Method

This method was developed by Pascal[28–39] and is based on the atomic concept of susceptibility. Molecular diamagnetic susceptibilities (χ_M) are considered to be

TABLE 1. Atomic susceptibility data[6,7]

Atom A	χ_A
C	6.00
H	2.93
O in alcohols and ethers	4.61
O in aldehydes and ketones	−1.72
O in carboxylic acids	3.36
N in monoamides	1.54
N in diamides and imides	2.11
N in ring	4.61
N in open chain	5.55
F	11.50
Cl	20.11
Br	30.64
I	44.64

composed of atomic susceptibilities of atoms constituting the molecules. It is expressed by equation (14):

$$\chi_M = \chi_A + \sum \lambda. \tag{14}$$

χ_A is the atomic susceptibility and λ represents constitutive corrections for structural factors. Pascal has derived susceptibility values for several atoms and the constitutive

TABLE 2. Constants for constitutive corrections[6,7]

Structural unit	Constitutive correction, λ
C=C	−5.47
C≡C	−0.77
C=C—C=C	−10.60
C≡N	−0.77
C=N	−8.15
N=N	1.83
N=O	−1.73
C—Cl	−3.07
Cl—C—C—Cl	−4.30
C—Cl	−1.44
C—Br	−4.10
Br—C—C—Br	−6.25
C—I	−4.10
C in monocyclic systems such as benzene, cyclohexane, etc.	+0.24
C in bicyclic systems such as two of the carbon atoms in naphthalene	+3.10
C in tricyclic systems such as two of the carbon atoms in perylene	+4.03
$C^{\alpha}_{(3)}$	+1.30
$C^{\beta}_{(3)}$	+0.50
$C^{\gamma}_{(3)}$, $C^{\delta}_{(3)}$, $C^{\epsilon}_{(3)}$	+1.30
$C^{\alpha}_{(4)}$	+1.55
$C^{\beta}_{(4)}$	+0.50
$C^{\gamma}_{(4)}$, $C^{\delta}_{(4)}$, $C^{\epsilon}_{(4)}$	+1.55

$C_{(3)}$ is a tertiary carbon atom and α, β, γ, δ, ϵ represent its position in a molecule with respect to a functional group. Similarly, $C_{(4)}$ is a quaternary carbon atom.

corrections for different structural factors. These values have been modified and corrected and the values relevant to the present work are included in Tables 1 and 2. Diamagnetic susceptibilities of several compounds containing carbon–halogen bonds have been calculated by this method and the calculated χ_M values are summarized in Table 3. All susceptibility values are expressed in c.g.s. units $\times 10^{-6}$ throughout.

TABLE 3. χ_M values calculated by the Pascal method

Compound	χ_M
o-Bromoaniline	86.3
m-Bromoaniline	86.3
p-Bromoaniline	86.3
Bromobenzene	78.5
p-Bromo-*N*,*N*-dimethylaniline	108.6
o-Bromophenyl methyl ether	94.7
m-Bromophenyl methyl ether	94.7
p-Bromophenyl methyl ether	94.7
o-Bromothiophenol	93.9
m-Bromothiophenol	93.9
p-Bromothiophenol	93.9
o-Chloroaniline	77.70
m-Chloroaniline	77.70
p-Chloroaniline	77.70
Chlorobenzene	69.2
m-Chloro-*N*,*N*-dimethylaniline	99.7
p-Chloro-*N*,*N*-dimethylaniline	97.7
Chloroethylene	35.4
o-Chloronitrobenzene	75.10
m-Chloronitrobenzene	75.10
p-Chloronitrobenzene	75.10
o-Chlorophenyl methyl ether	85.8
m-Chlorophenyl methyl ether	85.8
o-Chlorophenyl methyl sulphide	97.4
m-Chlorophenyl methyl sulphide	97.4
p-Chlorophenyl methyl sulphide	97.4
o-Chlorothiophenol	85.0
m-Chlorothiophenol	85.0
p-Chlorothiophenol	85.0
1-Chloro-1,2,2-trifluoroethylene	61.1
o-Dichlorobenzene	83.50
m-Dichlorobenzene	83.50
p-Dichlorobenzene	83.50
1,1-Dichloro-2,2-difluoroethylene	69.7
Dichlorodifluoromethane	69.2
Dichlorofluoromethane	60.6
1,1-Dichloroethylene	52.66
cis-1,2-Dichloroethylene	52.6
trans-1,2-Dichloroethylene	52.6
Difluorochloromethane	52.0
1,1-Difluoro-2,2-dichloroethyl butyl ether	133.11
1,1-Difluoro-2,2-dichloroethyl ethyl ether	109.39
1,1-Difluoro-2,2-dichloroethyl methyl ether	97.53
1,1-Difluoro-2,2-dichloroethyl amyl ether	144.97
1,1-Difluoro-2,2-dichloroethyl propyl ether	121.25
m-Chloroaniline	66.3

TABLE 3. *continued*

Compound	χ_M
p-Fluoroaniline	66.3
p-Fluoroanisole	74.7
Fluorobenzene	58.3
m-Fluoro-*N*,*N*-dimethylaniline	88.6
p-Fluoro-*N*,*N*-dimethylaniline	88.6
m-Iodoaniline	99.9
p-Iodoaniline	99.9
p-Iodoanisole	108.3
Iodobenzene	92.5
m-Iodo-*N*,*N*-dimethylaniline	122.5
p-Iodo-*N*,*N*-dimethylaniline	122.5
2-Methyl-4-bromothiophenol	108.2
o-Nitrobromobenzene	85.3
m-Nitrobromobenzene	85.3
p-Nitrobromobenzene	85.3
m-Nitrofluorobenzene	65.3
p-Nitrofluorobenzene	65.3
o-Nitroiodobenzene	98.9
m-Nitroiodobenzene	98.9
p-Nitroiodobenzene	98.9
Tetrachloroethylene	86.9
Tetrachloromethane	86.4
1,2,2-Trichloroethylene	69.7
1,2,2-Trichloro-1-fluoroethylene	78.3
Trichlorofluoromethane	77.8

B. Pascal, Pacault and Hoarau Method

Pascal and coworkers[40] have revised the Pascal atomic susceptibility data in order to reduce the number of constitutive corrections and minimize the error. In the modified method a constitutive correction for methyl groups has also been introduced. The revised atomic susceptibility data and constitutive corrections reported in this method are summarized here in Tables 4 and 5 respectively. Diamagnetic susceptibilities calculated for compounds containing carbon–halogen bonds are summarized in Table 6. Results obtained by this method are better than those obtained by the Pascal method.

TABLE 4. Atomic susceptibility data[40]

Atom A	χ_A
C	7.36
H	2.00
O (alcohols)	5.30
C=O (aldehydes and ketones)	6.40
C=O (acids)	15.13
N	9.00
Cl	18.50
Br	27.80
I	42.20
S	16.90

TABLE 5. Constants for constitutive corrections[40]

Structural unit	Constitutive correction, λ
C=C	−5.5
C≡C	−0.80
C=C—C=C	−10.6
CH_3	+0.85
Benzene ring	+15.10
Furan ring	+10.92
Thiophene ring	+14.22

TABLE 6. χ_M values calculated by Pascal, Pacault and Hoarau method

Compound	χ_M
i-Amyl β-chlorovinyl ketone	98.62
o-Bromophenyl β-chlorovinyl ketone	116.68
p-Bromophenyl β-chlorovinyl ketone	116.68
Butyl bromide	76.05
Butyl chloride	66.79
i-Butyl β-chlorovinyl ketone	87.26
Butyl iodide	91.65
Carbon tetrabromide	128.40
Carbon tetrachloride	81.36
Chloroform	64.86
o-Chlorophenyl β-chlorovinyl ketone	107.38
p-Chlorophenyl β-chlorovinyl ketone	107.38
n-Decyl β-chlorovinyl ketone	154.57
1,4-Dichloro-2-butyne	73.64
1,1-Dichloroethane	59.72
Dichloromethane	78.36
Diiodomethane	101.10
Dodecyl β-chlorovinyl ketone	177.29
Ethyl chloride	32.17
Ethyl β-chlorovinyl ketone	63.69
n-Heptyl β-chlorovinyl ketone	120.49
n-Hexyl β-chlorovinyl ketone	109.13
i-Hexyl β-chlorovinyl ketone	109.98
Methyl chloride	32.71
Methyl β-chlorovinyl ketone	52.33
p-Methoxyphenyl β-chlorovinyl ketone	107.54
n-Nonyl β-chlorovinyl ketone	143.21
n-Octyl β-chlorovinyl ketone	131.85
n-Pentadecyl β-chlorovinyl ketone	211.37
Propyl chloride	55.43
n-Propyl β-chlorovinyl ketone	75.05
i-Propyl β-chlorovinyl ketone	75.90
Phenyl β-chlorovinyl ketone	90.88
o-Tolyl β-chlorovinyl ketone	102.24
m-Tolyl β-chlorovinyl ketone	102.24
p-Tolyl β-chlorovinyl ketone	102.24
Tridecyl β-chlorovinyl ketone	188.64
Trifluoroacetic acid	40.66
Trifluoromethyl chloride	42.76
Undecyl β-chlorovinyl ketone	165.93

C. Baudet and Tillieu Wave-mechanical Method

Baudet and coworkers[41–45] introduced the bond concept of susceptibility and developed a method in which molecular susceptibility is considered to be contributed to by inner shell electrons (ISE), bonding electrons (BE), non-bonding electrons (NBE) and π-electrons. χ_M is expressed by equation (15):

$$\chi_M = \chi_{ISE} + \chi_{BE} + \chi_{NBE} + \chi_{\pi\text{-electrons}}. \quad (15)$$

Bond susceptibility data for a number of bonds have been calculated wave-mechanically by using wave functions given by Slater[46] and their parameters given by Coulson[47] and Craig[48]. Contributions of inner shell electrons and non-bonding electrons for different atoms, and bonding electrons and π-electrons for different bonds, calculated wave-mechanically by Baudet and coworkers, are summarized in Tables 7, 8, 9, and 10 respectively. Diamagnetic susceptibilities calculated from these data and which are relevant to the present review are included in Table 11.

TABLE 7. Contribution of inner shell electrons to molecular susceptibility[41–45]

	χ									
Atom	$1s^2$	$2s^2$	$2p^6$	$3s^2$	$3p^6$	$3d^{10}$	$4s^2$	$4p^6$	$4d^{10}$	Total
C	0.15									0.15
N	0.105									0.105
O	0.08									0.08
S	0.019	0.34	0.53							0.889
F	0.06									0.06
Cl	0.017	0.29	0.46							0.767
Br	0.004	0.05	0.08	0.36	0.54	3.07				4.104
I	0.002	0.01	0.03	0.11	0.17	0.92	6.43	0.77	1.13	9.572

TABLE 8. Contributions of non-bonding electrons to molecular susceptibility[41–45]

Atom	μ	χ_{NBE}	Atom	μ	χ_{NBE}
			F	1732	1.37
				0	1.76
N	1	2.22	Cl	0	5.37
	1732	2.43		1732	4.13
				∞	2.58
O	1414	1.73	Br	0	8.45
	1732	1.78		1732	7.46
				∞	3.85
S	1414	0.91	I	0	14.19
	1732	5.14		1732	11.03
				∞	6.34

μ expresses the state of hybridization of the atom: $\mu = \infty$ represents pure p orbitals; $\mu = 1732$ represents sp^3 hybridization; $\mu = 1414$ represents sp^2 hybridization; $\mu = 1$ represents sp hybridization; $\mu = 0$ represents pure s orbitals. μ_A and μ_B express the states of hybridization of the atoms A and B.

TABLE 9. Contribution of bonding electrons to molecular susceptibility[41–45]

	Hybridization		
Bond	μ_A	μ_B	χ_{BE}
H—H	0	0	3.94
C—H	1	0	3.33
	1414	0	3.74
	1732	0	4.05
C—C	1	1	2.38
	1414	1	2.47
	1732	1	2.91
	1414	1414	2.60
	1732	1414	3.01
	1732	1732	3.10
C—N	1	1	2.37
	1732	1732	3.15
C—O	1414	1414	2.55
	1414	1732	2.51
	1732	1732	2.73
C—S	1732	1732	4.83
C—F	1732	1732	2.51
	1732	∞	2.44
C—Cl	1732	1732	4.28
	1732	∞	4.24
C—Br	1732	1732	7.41
	1732	∞	7.69
C—I	1732	1732	7.88
	1732	∞	7.63
N—H	1732	0	3.63
N—N	1	1	2.18
O—H	1732	0	3.34
S—H	1732	0	5.28
S—S	1732	1732	5.29

See notes to Table 8.

TABLE 10. Contribution of π-electrons to molecular susceptibility[41–45]

Bond	$\chi_{\pi\text{-electrons}}$
C=C	3.42
C=O	3.05
C≡C	4.94
C≡N	3.44
N≡N	2.35

D. Pascal, Gallais and Labarré Method

Pascal and coworkers[49] also realized the importance of the bond concept of susceptibility and worked out bond susceptibility data for a number of bonds. Bond susceptibility data are summarized in Table 12 and diamagnetic susceptibility values of a large number of compounds are calculated using these data (Table 13).

TABLE 11. χ_M values calculated by Baudet and Tillieu wave-mechanical method

Compound	χ_M
i-Amyl chloride	75.14
i-Amyl β-chlorovinyl ketone	103.90
Butyl bromide	74.49
Butyl chloride	63.79
Butyl β-chlorovinyl ketone	92.55
Butyl iodide	90.74
Carbon tetrachloride	69.91
Chloroacetic acid	47.58
Chloroacetone	53.40
Chloral	68.61
Chloroform	56.52
1-Chloro-2,3-dihydroxypropane	63.68
2-Chloropropylene	45.51
n-Decyl β-chlorovinyl ketone	160.65
1,2-Dichloroethane	54.48
1,2-Dichloroethylene	48.60
1,3-Dichloro-2-hydroxypropane	71.45
Dichloromethane	43.13
n-Dodecyl β-chlorovinyl ketone	183.35
Ethyl chloride	41.09
Ethyl chloroacetate	70.62
Ethyl β-chlorovinyl ketone	69.85
n-Heptyl β-chlorovinyl ketone	126.60
Hexachloroethane	108.04
n-Hexyl β-chlorovinyl ketone	115.25
Methyl chloride	29.74
Methyl β-chlorovinyl ketone	58.50
n-Nonyl β-chlorovinyl ketone	149.30
n-Octyl β-chlorovinyl ketone	137.95
Octyl chloride	109.14
Pentadecyl β-chlorovinyl ketone	217.40
Pentachloroethane	94.65
n-Propyl β-chlorovinyl ketone	81.20
i-Propyl β-chlorovinyl ketone	81.20
1,1,2,2-Tetrachloroethane	81.26
n-Decyl β-chlorovinyl ketone	194.70
Trifluoroacetic acid	40.25
Trifluoromethyl chloride	36.25
n-Undecyl β-chlorovinyl ketone	172.00

E. Haberditzl Method

Haberditzl[3,50] has shown that the use of one-electron approximations in calculating magnetic susceptibility is inadequate and that the diamagnetic contribution of an electron pair forming a bond depends on the nature and number of other atoms and groups attached to these atoms. In the Haberditzl method molecular susceptibility is considered to be contributed to by inner shell core electrons, bonding electron increments and π-electrons and expressed by equation (16):

TABLE 12. Bond susceptibility[49]

Bond	χ
C—H	4.25
C—C (alkanes)	2.90
C=C (alkenes)	3.60
C≡C (alkynes)	10.30
C—O (alcohols, ethers and acids)	4.10
C=O (aldehydes)	3.30
C=O (ketones)	3.50
C=O (acids)	7.00
C—N (amines)	3.42
C—S	10.20
O—H (alcohols and acids)	5.60
S—H	12.60
S—S	15.40
C—Cl	20.00
C—Br	29.30
C—I	46.10
Constitutive correction for CH_3 group	$\lambda = 1.2$

$$\chi_M = \chi_{\text{inner shell core electrons}} + \chi_{\text{bonding electron increments}} + \chi_{\pi\text{-electrons}}. \quad (16)$$

In this approach due consideration is given to structural parameters under the heading neighbouring-bond effects and it is the best method reported at the time of

TABLE 13. χ_M values calculated by Pascal, Gallais and Labarré method

Compound	χ_M
Carbon tetrachloride	80.00
Chloroacetic acid	48.10
Chloroacetone	51.75
1-Chloro-2,3-dihydroxypropane	66.45
Chloroform	64.25
Chloropropylene	48.95
1,2-Dichloroethylene	52.10
1,3-Dichloro-2-hydroxypropane	76.75
Dichloromethane	48.50
Ethyl chloride	45.35
Ethyl chloroacetate	71.95
Ethyl trichloroacetate	103.45
Hexachloroethane	102.90
2-Amyl chloride	80.75
Methyl chloride	33.95
Octyl chloride	113.75
1,1,1,2,2-Pentachloroethane	107.15
n-Propyl chloride	56.75
1,1,2,2-Tetrachloroethane	91.40
Trichloroacetaldehyde	70.65

TABLE 14. Magnetic susceptibility contributions of inner shell core electrons[3,50]

Atom A	χ
C	0.15
N	2.40
O	3.60
F	5.00
Cl	12.50
Br	20.00
I	36.00

TABLE 15. Magnetic susceptibility contributions of bonds[3,50]

Bond B	χ_B	Bond B	χ_B
$C_{(1)}$—H	4.20	$C_{(2)}$—Cl	8.10
$C_{(2)}$—H	3.80	C^+—Cl	8.10
$C_{(3)}$—H	3.50	$C_{(2)}(Cl_2)$—Cl	6.90
$C^+_{(1)}$—H	3.60	$C_{(3)}(Cl_3)$—Cl	5.90
$C^+_{(2)}$—H	3.20	$C_{(4)}(Cl_4)$—Cl	4.20
$C_{(0)}$—H	3.60	$C_{(2)}(Cl)$—$C_{(2)}(Cl)$	3.10
$C_{(1)}$—$C_{(1)}$	3.60	$C_{(1)}$—Br	8.10
$C_{(1)}$—$C_{(2)}$	3.60	$C_{(2)}$—Br	8.10
$C_{(1)}$—$C_{(3)}$	3.60	$C_{(1)}$—I	8.10
$C_{(1)}$—$C_{(4)}$	3.40	$C_{(2)}$—I	8.10
$C_{(2)}$—$C_{(2)}$	3.60	$C(Cl_2)$—F	1.50
$C_{(2)}$—$C_{(3)}$	3.40	C(Cl)—F	3.10
$C_{(2)}$—$C_{(4)}$	3.30	$C_{(1)}$—$N_{(1)}$	4.80
$C_{(3)}$—$C_{(3)}$	3.30	$C_{(2)}$—$N_{(1)}$	2.70
$C_{(3)}$—$C_{(4)}$	3.30	$C_{(2)}$—$N_{(2)}$	2.20
$C_{(4)}$—$C_{(4)}$	3.30	$C_{(2)}$—$N_{(3)}$	2.20
C—C^+	2.60	$N_{(0)}$—H	5.30
C^+—C^+	2.40	$N_{(1)}$—H	3.50
C—$O_{(1)}$	1.80	$O_{(1)}$—H	3.30
C—$O_{(2)}$	1.70	O(C=O)—H	3.60
C—$C_{(2)}(O_{(1)})$	4.60		
C—$C_{(3)}(O_{(1)})$	4.60		
$C(O_{(1)})$—$C(O_{(1)})$	5.70		
$C^+_{(1)}$=O^+	4.30		
$C^+_{(2)}$=O^+	0.60		
$C^+_{(3)}$=O^+	0.20		
C^+—$O_{(1)}$	1.50		
$C^+(O)$—O	4.00		
$C^+_{(2)}(O)$=O^+	1.80		
$C^+_{(3)}(O)$=O^+	1.80		

$C_{(1)}$, $C_{(2)}$, $C_{(3)}$, $C_{(4)}$ indicate primary, secondary, tertiary and quaternary carbon atoms respectively; $C_{(0)}$ indicates a methane carbon atom. $N_{(1)}$, $N_{(2)}$, $N_{(3)}$ indicate primary, secondary and tertiary nitrogen atoms respectively; $N_{(0)}$ indicates an ammonia nitrogen atom. $O_{(0)}$ and $O_{(1)}$ indicate oxygen atoms attached to two and one hydrogen atoms respectively.

TABLE 16. Magnetic susceptibility contributions of π-electrons[3,50]

π-electrons	χ
$C_{(1)}\ \pi\ C_{(2)}$	3.40
$C_{(2)}\ \pi\ C_{(2)}$	2.20
$C_{(1)}\ \pi\ C_{(3)}$	2.20

TABLE 17. χ_M values calculated by Haberditzl method

Compound	χ_M
i-Amyl β-chlorovinyl ketone	99.70
Butyl bromide	75.60
n-Butyl chloride	67.40
i-Butyl β-chlorovinyl ketone	88.35
Butyl iodide	91.60
Carbon tetrachloride	66.98
Chloroform	58.55
n-Decyl β-chlorovinyl ketone	155.75
1,4-Dichloro-2-butyne	73.50
1,2-Dichloroethane	59.50
Dichloromethane	46.55
n-Dodecyl β-chlorovinyl ketone	178.45
Ethyl chloride	44.70
Ethyl β-chlorovinyl ketone	64.95
n-Hexyl β-chlorovinyl ketone	110.35
i-Hexyl β-chlorovinyl ketone	111.05
n-Heptyl β-chlorovinyl ketone	121.70
Methyl chloride	32.15
Methyl β-chlorovinyl ketone	53.60
n-Nonyl β-chlorovinyl ketone	144.40
n-Octyl β-chlorovinyl ketone	133.05
n-Pentadecyl β-chlorovinyl ketone	212.50
n-Propyl chloride	56.03
n-Propyl β-chlorovinyl ketone	76.30
i-Propyl β-chlorovinyl ketone	77.20
n-Tridecyl β-chlorovinyl ketone	189.80
Trifluoroacetic acid	43.50
Trifluoromethyl chloride	45.05
n-Undecyl β-chlorovinyl ketone	167.10

writing. Magnetic susceptibility contributions for inner shell core electrons, bonding electron increments and π-electrons calculated by Haberditzl on a quantum-mechanical basis are presented in Tables 14, 15 and 16, respectively. Diamagnetic susceptibilities calculated by this method and relevant to the present work are presented in Table 17.

F. Modified Haberditzl Method

Gupta and coworkers[4] have modified and improved the Haberditzl method to calculate molecular susceptibilities of di- and polysubstituted benzenes by accounting for additional interactions between the substituents which affect diamagnetism. This

TABLE 18. χ_M values calculated by modified Haberditzl method

Compound	χ_M
o-Bromoaniline	87.28
m-Bromoaniline	83.72
p-Bromoaniline	85.12
o-Bromotoluene	79.73
m-Bromotoluene	88.17
p-Bromotoluene	89.57
o-Chlorotoluene	84.23
m-Chlorotoluene	80.67
p-Chlorotoluene	82.07

provides different values of diamagnetic susceptibilities of *o*-, *m*- and *p*-disubstituted benzenes and fits experimental observations. χ_M values calculated by the modified Haberditzl method are included in Table 18.

V. RESULTS AND DISCUSSION

In order to make possible a systematic discussion of the results of diamagnetic susceptibilities of compounds containing carbon–halogen bonds they have been classified into five series, as detailed below.

A. Halogenated Hydrocarbons

Diamagnetic susceptibilities of both aliphatic and aromatic halogenated hydrocarbons have been studied extensively. Diamagnetic susceptibilities of chloro-, bromo-

TABLE 19. Diamagnetic susceptibilities of alkyl and aryl halides[1]

Halide	χ_M	
	Experimental	Theoretical
CH_3Cl	32.0	34.9
CH_2Cl_2	46.6	52.1
$CHCl_3$	58.3	69.2
CCl_4	66.8	86.4
CH_3Br	42.8	45.4
CH_2Br_2	65.9	73.1
$CHBr_3$	82.2	100.7
CBr_4	93.7	128.4
CH_3I	57.2	59.4
CH_2I_2	93.5	101.1
CHI_3	117.3	142.7
CI_4	135.6	184.4
C_2H_5Cl	45.0	46.8
C_2H_5Br	55.5	57.2
C_2H_5I	68.5	71.2
C_6H_5Cl	69.9	70.7
C_6H_5Br	78.9	81.2
C_6H_5I	91.3	95.2

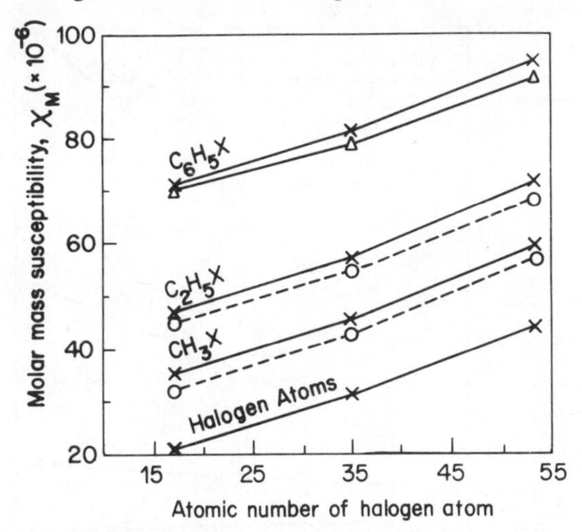

FIGURE 1. Plot of χ_M (×, theoretical; △ and ○, experimental) against atomic number of halogen atoms for the series CH_3X, C_2H_5X and C_6H_5X[1].

and iodoalkanes have been reported[51] to deviate considerably from experimental values. This type of discrepancy is enhanced in dichloro, dibromo and diiodo compounds when both the halogen atoms are attached to the same carbon atom and it is attributed to the interaction between the halogen atoms. Such abnormal diamagnetic behaviour is less marked in fluoro compounds. This type of abnormality in diamagnetic behaviour of C_2H_4, $C_2H_4Cl_2$, $C_2H_4Br_2$, $C_2H_4I_2$ and CCl_4 has been explained on the basis of valency links[52].

French and coworkers[1] have made systematic and exhaustive diamagnetic investigations on alkyl and aryl halides and found that their experimental diamagnetic susceptibilities deviate considerably from the corresponding theoretical values calculated by the Pascal method (see Section III.A). Diamagnetic susceptibilities (experimental and theoretical) are summarized in Table 19 for comparison and discussion. The plots (Figure 1) have been reported for diamagnetic susceptibilities (experimental and theoretical) of CH_3X (X = Cl, Br, I), C_2H_5X and X versus atomic number of halogen atoms. All the curves possess characteristic inflection points. The curves of experimental and theoretical values run almost parallel for the CH_3X and C_2H_5X series, and a divergence is noted for the C_6H_5X series. Such an observation shows that the deviation between experimental and theoretical susceptibilities increases on passing from aryl chloride to iodide while it is almost of the same order for chloro, bromo and iodo compounds of methane and ethane. The deviations in diamagnetic susceptibilities of CH_3X, C_2H_5X and C_6H_5X have been ascribed to the interaction between hydrogen and halogen, which produces an effect on the diamagnetic susceptibility comparable to that caused by the formation of a double bond. It has been further reported that the hydrogen and halogen interaction in aryl halides increases on passing from a chloro to an iodo derivative and an additional divergence is introduced which enhances the lowering in diamagnetism of aryl iodide in comparison to aryl chloride. In order to study additional interactions in dihalo and polyhalo methanes and ethanes which cause lowering in molecular diamagnetism the graphs between the deviations in diamagnetic susceptibilities and atomic number of halogen atoms

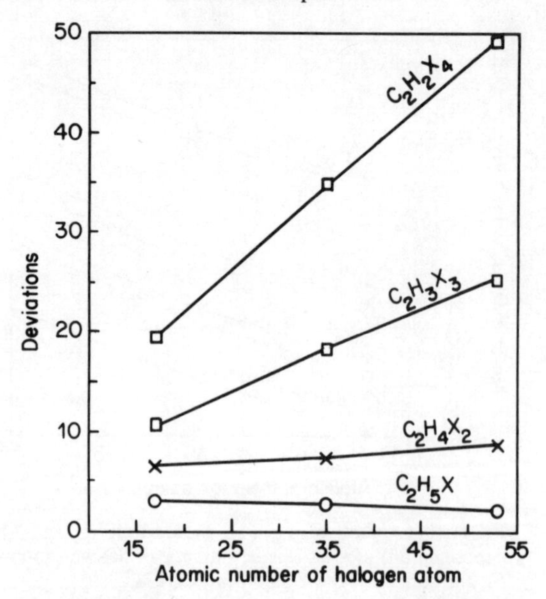

FIGURE 2. Plots of deviation against atomic number of halogen atoms for the series C_2H_5X, $C_2H_4X_2$, $C_2H_3X_3$ and $C_2H_2X_4$[1].

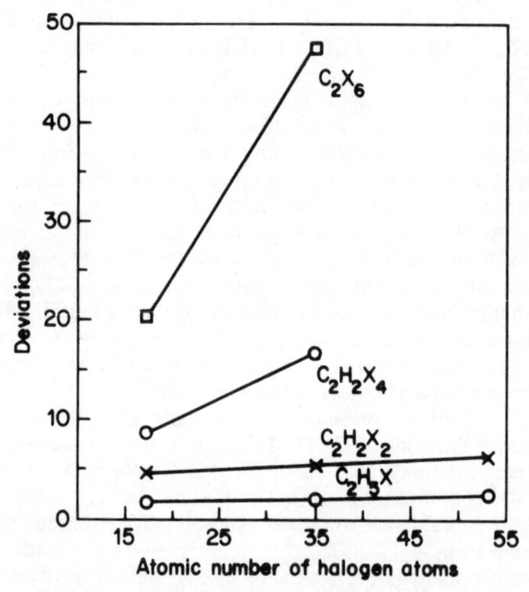

FIGURE 3. Plots of deviations against atomic number of halogen atoms for the series C_2H_5X, $C_2H_4X_2$, $C_2H_2X_4$ and C_2X_6[1].

have been investigated thoroughly. The graphs for mono-, di-, tri-, tetra- and hexahaloethanes are shown in Figures 2 and 3, respectively. The deviation curves reveal that in the dihalide the hydrogen–halogen interaction is approximately double that in the monohalide series and there is a small but systematic rise in deviation on passing from chloro to iodo compounds. It has been further reported that in dihalides, in addition to hydrogen–halogen interaction, there is a deformation of halogen atomic fields due to the polarization effect. The latter causes further lowering in diamagnetism. The curves demonstrate the polarization effect which is in accordance with the order of polarizability of iodine, bromine and chlorine, resulting in more depression in diamagnetism for iodo compounds than for chloro compounds. The increasing slopes of the deviation curves for tri- and tetrahalomethanes and tetra- and hexahaloethanes reveal that the polarization effect increases with an increase in the number of halogen atoms in the molecule. The straight line character of the deviation curves also indicates the additive nature of these deviations which are proportional to the number of halogen atoms producing them. Hence on increasing the number of halogen atoms in a molecule the polarization effect is enhanced and the depression in molecular diamagnetism should be pronounced, as is indeed experimentally found. It has also been reported[1] that the depression in the molecular susceptibility is greater when two or more halogen atoms are attached to the same carbon atom and has been explained using the Pauling view[53], which states that in compounds containing only one halogen atom there is no possibility of resonance to give a double-bonded structure since there are only four orbitals for bond formation but in compounds containing two or more halogen atoms the resonance occurs as shown in **1**, which causes additional lowering in diamagnetism.

$$\begin{array}{c} X^- \\ R\text{—}\underset{\displaystyle \overset{|}{H}}{C}\text{=}X^+ \end{array}$$

(**1**)

Lacher and coworkers[54] have reported diamagnetic studies of chloro and fluoro derivatives of methane and ethylene and the experimental data have been compared with those calculated theoretically by the Pascal method (see Section III.A). It has also been reported that the *trans* isomers have lower values of diamagnetic susceptibility than the corresponding *cis* isomers. Lacher and coworkers[55] have also studied the diamagnetic behaviour of mixed (Cl, Br, F) halomethanes and they observed that tetrahedral interactions contribute to the diamagnetism. Tetrahedral interaction increments have also been reported for different bonds and are summarized in Table 20.

Dorfman[56] has studied polarity of carbon–halogen bonds using the diamagnetic susceptibility data and expressed the diamagnetic susceptibility of a compound by equation (17):

$$\chi_M = \chi_d + \chi_p, \tag{17}$$

where χ_d is the diamagnetic susceptibility contribution to the molecular susceptibility and χ_p is the temperature-independent Van Vleck's paramagnetism. The values of χ_p per carbon–halogen bond for a number of molecules are summarized in Table 21. A critical examination of these values reveals that the value of χ_p per carbon–halogen bond increases with an increase in the number of halogen atoms in a molecule and consequently there is a drop in the polarity of carbon–halogen bonds.

Gupta[57] has reported large bond–bond interactions in chloroalkanes by diamagnetic

TABLE 20. Tetrahedral interaction increments for bonds[54]

Bond	Value
H—H	0.2
F—F	0.8
Cl—Cl	3.4
Br—Br	6.4
I—I	9.4
H—F	1.4
H—Cl	0.1
H—Br	0.2
H—I	0.2
F—Cl	2.4
F—Br	3.6
F—I	5.1
Cl—Br	4.9
Cl—I	6.4
Br—I	7.9

TABLE 21. Values of χ_p per carbon–halogen bond in halogenated hydrocarbons[56]

Compound	χ_p per carbon–halogen bond
Methyl chloride	2.00
Methylene chloride	2.40
Chloroform	2.83
Carbon tetrachloride	4.67
1,2-Dichloroethane	2.00
1,1-Dichloroethane	3.00
1,1,2,2-Tetrachloroethane	2.90
Pentachloroethane	3.20
Hexachloroethane	3.70
Methyl bromide	6.20
Bromoform	10.13
Carbon tetrabromide	10.82
Methyl iodide	11.80
Methylene iodide	13.45
Iodoform	31.00

studies. Studies on organic iodides[58] have been carried out at various temperatures and increase in diamagnetism with increase in temperature has been observed.

B. Disubstituted Benzenes with Halo Substituents

The diamagnetic behaviour of a large number of disubstituted benzenes with one or two halo atoms has been studied extensively. Gupta and coworkers[59] have reported systematic studies on the diamagnetic behaviour of such compounds in order to generalize the effect of substitution in (mono-) halobenzenes. They have reported the values of χ_{Cl} in chloroanilines, chlorotoluenes, chlorobenzaldehydes and chloro-

TABLE 22. Values of χ_{Cl} and χ_{Br} in disubstituted benzenes[59]

Compound	χ_{Cl} or χ_{Br}
o-Chloroaniline	20.48
o-Chlorotoluene	20.82
o-Chlorobenzaldehyde	19.08
o-Chloroacetophenone	18.46
m-Chloroaniline	16.92
m-Chlorotoluene	17.31
m-Chlorobenzaldehyde	21.33
m-Chloroacetophenone	21.38
p-Chloroaniline	18.32
p-Chlorotoluene	18.69
p-Chlorobenzaldehyde	18.23
p-Chloroacetophenone	17.56
o-Bromoaniline	28.82
o-Bromotoluene	28.81
o-Bromoacetophenone	29.31
m-Bromoaniline	25.21
m-Bromotoluene	25.38
m-Bromoacetophenone	32.99
p-Bromoaniline	26.63
p-Bromotoluene	26.70
p-Bromoacetophenone	27.03

acetophenones and χ_{Br} in bromoanilines, bromotoluenes and bromobenzaldehydes. Values of χ_{Cl} and χ_{Br} are included in Table 22. The diamagnetic susceptibility contribution of the substituent depends on its position, the values being *ortho* > *para* > *meta*. Such an observation has also been reported by French[60] in dichlorobenzenes. Balliah and coworkers[61] have correlated the diamagnetic behaviour with the nature of the other substituents. They have reported that the diamagnetism follows the order *ortho* > *meta* > *para* and *meta* > *ortho* > *para* according as to whether the other substituent is electron releasing or electron withdrawing. Their explanation is based on effective conjugative interaction of the other substituent and the halo atom through the benzene nucleus. Extensive diamagnetic studies[62] have also been reported for *meta* and *para* chloro-, bromo- and iodoanilines and -*N*,*N*-dimethylanilines and *p*-chloro, *p*-bromo-, *p*-iodo- and *p*-fluoroanisole. Diamagnetic susceptibilities of such compounds have also been calculated theoretically and compared with measured values.

In haloanilines a small positive divergence has been reported for fluoro compounds. However, negative divergence has been observed in chloro-, bromo-, and iodoanilines following the order iodoanilines > bromoanilines > chloroanilines. These marked deviations have been ascribed to vacant d orbital participation of valence shells in resonance as shown below:

$$X-C_6H_4-NR_2 \longleftrightarrow \bar{X}=C_6H_4=\overset{+}{N}R_2$$

R = H or CH_3

X = F, Cl, Br or I

The behaviour of fluoroaniline is different from that of other haloanilines because fluorine does not have a vacant d orbital in the valence shell and there is no such

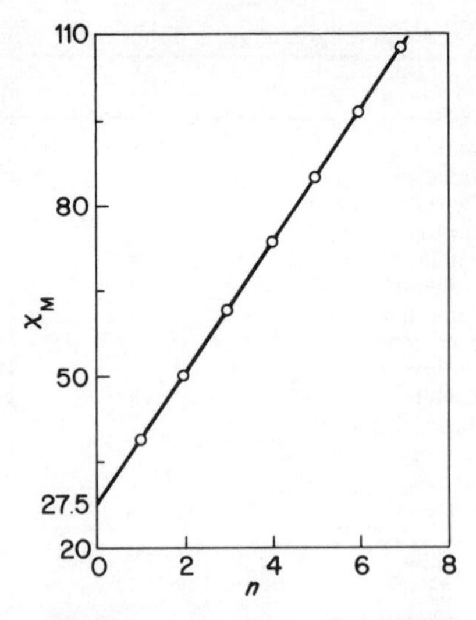

FIGURE 4. Plot of χ_M against n for the series $H(CH_2)_n$ COCl[66].

participation in the resonance. Even greater deviations have been reported for chloro-, bromo- and iodo-*N,N*-dimethylanilines and have been attributed to the greater effectiveness of $-N(CH_3)_2$ than $-NH_2$ in causing d orbital resonance. Diamagnetic behaviour of chloro- and bromothiophenols has also been reported[63]. Appreciable divergences have been observed in halo thiophenols but no explanation has been given. Diamagnetic studies[64] have also been reported for *meta* and *para* halophenyl methyl sulphides, sulphoxides and sulphones.

Diamagnetic measurements[65] have also been used to study chelation and hydrogen bonding in disubstituted benzenes with one chlorine atom.

C. Acyl Chlorides

Diamagnetic measurements[66] have been used to study the behaviour of the carbonyl group in acid chlorides to examine whether it behaves similarly or differently as com-

TABLE 23. Values of χ_{CO} and χ_O for acyl chlorides, aldehydes, ketones, acids and esters[66,67]

	Average value	
Series	χ_{CO}	χ_O
Acyl chlorides	7.0	−0.40
Aldehydes	5.9	−1.50
Ketones	6.7	−0.65
Acids	11.17	+3.87
Acetates	10.69	+3.29
Propionates	10.74	+3.34
Methyl esters	10.71	+3.31

TABLE 24. Values of $\chi_{CO^{2+}}$ for acyl chlorides, aldehydes, ketones and esters[66,69]

Series	$\chi_{CO^{2+}}$
Acid chlorides	2.90
Aldehydes	0.80
Ketones	0.40
Benzaldehydes	2.10
Acetophenones	1.00
Acids	2.83
Acetates	3.56
Propionates	3.57
Methyl esters	3.56

pared to that of aldehydes, ketones, carboxylic acid and esters. The plot of χ_M versus n (Figure 4) for the series $H(CH_2)_nCOCl$ has been used to obtain the value of $H(CH_2)_0COCl$ from the axial intercept. From the latter, diamagnetic contributions of the CO group (χ_{CO} and the oxygen atom (χ_O) have been calculated and compared with those reported[67] for aldehydes, ketones, carboxylic acids and esters (Table 23). The diamagnetic susceptibility contribution of the carbonyl ion ($\chi_{CO^{2+}}$) has been calculated by Yang's method[68] and has been compared with values for other series[69]. Values of $\chi_{CO^{2+}}$ for different series are summarized in Table 24. Tables 23 and 24 reveal that χ_O, χ_{CO} and $\chi_{CO^{2+}}$ are not fixed for all series, but their values depend on the environments in which the CO group is present. The diamagnetic behaviour of acyl chlorides has shown that the CO group in acyl chlorides behaves differently from those of aldehydes, ketones, acids and esters and this shows that different electronic environments prevail in these compounds.

Diamagnetic studies[70] have also been reported for chloro-, dichloro- and trichloroacetyl chlorides. They have been found to possess abnormal diamagnetic behaviour

TABLE 25. Magnetochemical data and resonance energy of alkyl β-chlorovinyl ketones[73]

Compound	χ_M	$\chi_\parallel$	ΔK	Density, d, g cm^{-3}	Resonance energy kcal mol^{-1}
Methyl β-chlorovinyl ketone	54.30	52.33	5.91	1.1300	13.76
Ethyl β-chlorovinyl ketone	65.71	63.69	6.06	1.0702	13.36
n-Propyl β-chlorovinyl ketone	77.02	75.05	5.91	1.0396	12.67
i-Propyl β-chlorovinyl ketone	78.01	75.90	6.33	1.0317	13.45
i-Butyl β-chlorovinyl ketone	89.30	87.26	6.12	1.0117	12.76
i-Amyl β-chlorovinyl ketone	100.60	98.62	5.94	0.9930	12.15
n-Hexyl β-chlorovinyl ketone	111.07	109.13	5.82	0.9871	11.83
i-Hexyl β-chlorovinyl ketone	111.99	109.98	6.03	0.9750	12.10
n-Heptyl β-chlorovinyl ketone	122.42	120.49	5.79	0.9733	11.61
n-Octyl β-chlorovinyl ketone	133.75	131.85	5.70	0.9695	11.39
n-Nonyl β-chlorovinyl ketone	145.15	143.21	5.82	0.9701	11.63
n-Decyl β-chlorovinyl ketone	156.46	154.57	5.67	0.9773	11.54
n-Undecyl β-chlorovinyl ketone	167.84	165.93	5.73	0.9797	11.53
n-Dodecyl β-chlorovinyl ketone	179.15	177.29	5.58	0.9959	11.45
n-Tridecyl β-chlorovinyl ketone	190.56	188.65	5.73	0.9801	11.55
n-Pentadecyl β-chlorovinyl ketone	213.40	211.37	6.09	0.9657	12.12

in the sense that their experimental values of diamagnetic susceptibility deviate considerably from corresponding theoretical values and abnormality increases with increase in the number of chlorine atoms attached to the methyl group.

D. Alkyl and Aryl β-Chlorovinyl Ketones

The diamagnetic behaviour of a number of alkyl and aryl β-chlorovinyl ketones has been studied by Gupta and Mital[71]. An increase has been observed in the molecular susceptibilities of alkyl β-chlorovinyl ketones with chain length. This has been attributed to the mesomeric effect which operates due to conjugation of the carbonyl with the C=C double bond, as shown below:

$$-\underset{\underset{O}{\|}}{C}-CH=CH- \longleftrightarrow -\underset{\underset{\underline{O}^{-}}{|}}{C}=CH-\overset{+}{C}H-$$

It has been further observed that the relative increase in molecular susceptibility decreases on passing from methyl β-chlorovinyl ketone to higher homologues, i.e. on lengthening the alkyl chain, and thus the increase is associated with the COCH=CH—Cl unit of the molecule. *Iso* isomers have been reported to be more diamagnetic than *n*-isomers, and this has been explained by Angus and Hill[72], who postulated that free rotation of alkyl chains lowers diamagnetism.

Increase in diamagnetism[71] has also been observed in aryl β-chlorovinyl ketones. It ranges from 2.92 to 6.82 c.g.s. units and increases with substitution in the benzene ring. This additional increase has been attributed to the interactions introduced by the substitution. *Meta* isomers are more diamagnetic than *ortho* and *para* isomers.

Mital and Gupta[73] have used diamagnetic susceptibility data to calculate the resonance energy of alkyl β-chlorovinyl ketones, expressed by equation (18):

$$E = 2.6\,\Delta K d \times 10^6, \qquad (18)$$

where E is resonance energy, d is the density of the ketone and K is the magnetic anisotropy of the ketone, which is related to the molecular diamagnetism by equation (19),

$$\Delta K = 3(\chi_M - \chi_\parallel), \qquad (19)$$

where $\chi_\parallel$ is the principal susceptibility with the molecular plane parallel to the magnetic field.

$\chi_\parallel$ has been calculated following the Pascal additivity rule. Values of χ_M, $\chi_\parallel$, ΔK, d and E for alkyl β-chlorovinyl ketones are summarized in Table 25. The resonance energy rising from 11.39 to 13.76 kcal mol^{-1} has been considered to be responsible for the increase in molecular diamagnetism of β-chlorovinyl ketones.

E. Miscellaneous Halo Compounds

The diamagnetic behaviour of 1,1-difluoro-2,2-dichloroalkyl ethers has been reported[74]. Deviations have been observed from Pascal's additivity rule and have been attributed to interaction forces between non-bonded atoms directed along the edges of a tetrahedron. The diamagnetic behaviours of 1-(*o*-chlorophenylazo)-2-naphthol and 1-(*p*-chlorophenylazo)-2-naphthol have shown that they exist in the azophenol form[75]. The molecular compound of aniline with 2,4-dinitrochlorobenzene has a lower diamagnetism than the value of χ_M given by the sum of the χ_M value of its components[76].

Mixtures of acetone with chloroform and trichloro-*n*-butyl alcohol have been reported[77] to deviate from Pascal's additivity rule. Diamagnetic behaviours of azeotropic binary liquid mixtures containing organic halo compounds have also been studied[78]. They obey the additivity rule and the magnetic data are not affected by the azeotropic nature of the mixtures.

V. ACKNOWLEDGEMENTS

I gratefully acknowledge the granting of permission by the American Chemical Society, the American Institute of Physics and the Chemical Society (London) to reproduce data from their respective journals. Professor W. Haberditzl is also thanked for permission to reproduce data from his contributions. Professors R. C. Mehrotra and K. C. Joshi are thanked for encouragement. My sincere thanks are due to Dr B. P. Bachlas, Dr K. G. Ojha and Mr M. Kumar for help rendered in numerous ways during the preparation of the manuscript. Mr Jean-Paul Mengolli is sincerely thanked for typing the manuscript. Finally, I acknowledge the never failing aid from my wife, Vimla.

VI. REFERENCES

1. C. M. French and V. C. G. Trew, *Trans. Faraday Soc.*, **41**, 429 (1945).
2. R. R. Gupta and R. L. Mital, *Ann. Soc. Scient. Bruxelles*, **T81** (1), 81 (1967).
3. W. Haberditzl, *Sitzber. Deut. Akad. Wiss. Berlin, Kl. Chem. Geol. Biol. (2)*, 100 (1964).
4. R. L. Mital and R. R. Gupta, *Indian J. Pure Appl. Phys.*, **8**, 177 (1970).
5. L. G. Gouy, *Compt. Rend.*, **109**, 935 (1889).
6. P. W. Selwood, *Magnetochemistry*, 2nd edn, Interscience, New York, 1951.
7. S. S. Bhatnagar and K. N. Mathur, *Physical Principles and Applications of Magnetochemistry*, Macmillan, London, 1935.
8. R. S. Nyholm, *Quart. Rev.*, 377 (1953).
9. A. Earnshaw, *Lab. Pract.*, **10**, 157 (1961).
10. B. N. Figgis and J. Lewis in *Modern Coordination Chemistry* (Ed. J. Lewis and R. G. Wilkins), Interscience, New York (1960).
11. A. Earnshaw, *Lab. Pract.*, **10**, 294 (1961).
12. L. N. Mulay, *Magnetic Susceptibility*, Interscience, New York (1966).
13. A. Earnshaw, *Lab. Pract.*, **10**, 294 (1961).
14. L. N. Mulay and I. L. Mulay, *Anal. Chem.*, **36**, 404 (1964).
15. L. N. Mulay, *Anal. Chem.*, **34**, 343 (1962).
16. A. Earnshaw, *Introduction to Magnetochemistry*, Academic Press, London (1968).
17. E. Mueller, A. Rieker, K. Scheffler and A. Moosmayer, *Angew. Chem. Int. Ed. Engl.*, **5**, 6 (1966).
18. L. F. Bates, *Modern Magnetism*, 4th edn, Cambridge University Press, London (1961).
19. M. Faraday, *Exptl. Res. (London)*, **7**, 27 (1855).
20. L. Sacconi and R. Cini, *J. Sci. Inter.*, **31**, 1142 (1960).
21. E. Wilson, *Proc. Roy. Soc. (London) A*, **96**, 429 (1920).
22. M. C. Day, L. D. Hulett and D. E. Wellis, *Rev. Sci. Instrum.*, **31**, 1142 (1960).
23. A. E. Oxley, *Phil. Trans. Roy. Soc. (London) A*, **215**, 79 (1915).
24. C. Cheneveau, *Phil. Mag.*, **20**, 357 (1910).
25. W. Sucksmith, *Phil. Mag.*, **8**, 158 (1929).
26. D. F. Evans, *J. Chem. Soc.*, 2003 (1950).
27. K. Frei and H. J. Bernstein, *J. Chem. Phys.*, **37**, 1891 (1962).
28. P. Pascal, *Compt. Rend.*, **147**, 56, 242 and 742 (1908).
29. P. Pascal, *Ann. Chim. Phys.*, **19**, 5 (1910).
30. P. Pascal, *Compt. Rend.*, **150**, 1167 (1910).
31. P. Pascal, *Compt. Rend.*, **152**, 862 and 1010 (1911).
32. P. Pascal, *Ann. Chim. Phys.*, **28**, 218 (1913).

33. P. Pascal, *Compt. Rend.*, **156**, 323 (1913).
34. P. Pascal, *Ann. Chim. Phys.*, **25**, 289 (1912).
35. P. Pascal, *Compt. Rend.*, **148**, 413 (1909).
36. P. Pascal, *Compt. Rend.*, **158**, 37 (1914).
37. P. Pascal, *Compt. Rend.*, **176**, 765 and 1887 (1923).
38. P. Pascal, *Compt. Rend.*, **173**, 144 (1921).
39. P. Pascal, *Compt. Rend.*, **180**, 1596 (1925).
40. P. Pascal, A. Pacault and J. Hoarau, *Compt. Rend.*, **233**, 1078 (1951).
41. J. Baudet, J. Tillieu and J. Guy, *Compt. Rend.*, **244**, 2920 (1957).
42. J. Baudet, *J. Chim. Phys.*, **58**, 228 (1961).
43. J. Tillieu, *Ann. Phys.*, **2**, 631 (1957).
44. J. Baudet, J. Tillieu and J. Guy, *Compt. Rend.*, **224**, 1756 (1957).
45. J. Baudet, J. Guy and J. Tillieu, *J. Phys. Rad.*, **21**, 600 (1960).
46. J. C. Slater, *Phys. Rev.*, **36**, 75 (1930).
47. C. A. Coulson, *Trans. Faraday Soc.*, **33**, 338 and 1479 (1937).
48. D. P. Craig, *Proc. Roy. Soc. (London) A*, **210**, 272 (1949–50).
49. P. Pascal, F. Gallais and J. F. Labarré, *Compt. Rend.*, **256**, 335 (1963).
50. W. Haberditzl, *Angew. Chem. Int. Ed. Engl.*, **5**, 288 (1966).
51. P. Pascal, *Bull. Soc. Chim., France*, **4**, 159 (1912).
52. V. I. Vidyanathan and B. Singh, *Indian J. Phys.*, **7**, 19 (1932).
53. L. Pauling, *The Nature of the Chemical Bond*, 3rd edn, Cornell University Press, Ithaca, N.Y., 1960.
54. J. R. Lacher, R. E. Scruby and J. D. Park, *J. Am. Chem. Soc.*, **71**, 1797 (1949).
55. J. R. Lacher, R. E. Scruby and J. D. Park, *J. Am. Chem. Soc.*, **72**, 333 (1950).
56. Y. G. Dorfman, *Diamagnetism and the Chemical Bond*, Edward Arnold, London, 1965, pp. 93–96.
57. R. R. Gupta, *Indian J. Chem.*, **15A**, 353 (1977).
58. M. L. Khanna, *J. Sci. Ind. Res. India*, **6**, 10 (1947).
59. R. D. Goyal, R. R. Gupta and R. L. Mital, *Indian J. Chem.*, **9**, 696 (1971).
60. C. M. French, *Trans. Faraday Soc.*, **41**, 676 (1945).
61. V. Balliah and M. M. Abubucker, *Indian J. Chem.*, **9**, 963 (1971).
62. V. Balliah and C. Srinivasan, *Indian J. Chem.*, **9**, 217 (1971).
63. V. Balliah and C. Srinivasan, *Indian J. Chem.*, **9**, 215 (1971).
64. V. Balliah, C. Srinivasan and M. M. Abubucker, *Indian J. Chem.*, **8**, 981 (1970).
65. P. Rumpf and M. Seguin, *Bull. Soc. Chim., France*, 366 (1949).
66. R. L. Mital and R. R. Gupta, *J. Chem. Phys.*, **54**, 3230 (1971).
67. R. R. Gupta, S. K. Jain and K. G. Ojha, *J. Chem. Phys.*, **66**, 4961 (1977).
68. T. Yang, *J. Chem. Phys.*, **16**, 865 (1948).
69. S. K. Jain, K. G. Ojha and R. R. Gupta, *Ann. Soc. Sci. Bruxelles*, **T91** (3), 179 (1977).
70. A. Fava, G. Geacometti and A. Ilieceto, *Ric. Sci.*, **22**, 1201 (1952).
71. R. R. Gupta and R. L. Mital, *Ann. Soc. Sci. Bruxelles*, **T81** (2), 183 (1967).
72. W. R. Angus and W. K. Hill, *Trans. Faraday Soc.*, **51**, 241 (1955).
73. R. L. Mital and R. R. Gupta, *Bull. Soc. Polon. Sci., Ser. Sci. Chim.*, **17**, 45 (1969).
74. J. D. Park, C. M. Snow and J. R. Lacher, *J. Am. Chem. Soc.*, **73**, 861 (1951).
75. M. Matsunaga, *Bull. Chem. Soc. Japan.*, **30**, 429 (1957).
76. B. Puri, R. C. Sahney, M. Singh and S. Singh, *J. Indian Chem. Soc.*, **24**, 409 (1947).
77. B. Cabrera and A. Madineveitia, *Ann. Soc. Espan. Quim.*, **30**, 528 (1932).
78. K. Venkateswarlu and S. Sriraman, *Bull. Chem. Soc. Japan*, **31**, 211 (1958).

The Chemistry of Functional Groups, Supplement D
Edited by S. Patai and Z. Rappoport
© 1983 John Wiley & Sons Ltd

CHAPTER 3

The mass spectra of azides and halides

JACK M. MILLER and TIMOTHY R. B. JONES
Department of Chemistry, Brock University, St Catharines, Ontario, Canada L2S 3A1

I. INTRODUCTION

Mass spectrometry has long been a useful tool in the study and analysis of halides. Positive ion studies of chlorides and chlorine isotopes were reported by Aston as early as 1919[1,2] with negative ion spectra being investigated a year later[3], the studies being primarily concerned with the isotopic nature of chlorine. The active use of mass spectrometry to solve 'relevant' analytical problems involving halides continues today, with mass spectrometry being the preferred technique for analysis of the ubiquitous tetrachlorodioxins at the part per trillion level[4,5].

The situation concerning azides is not nearly so favourable. Studies of the mass spectra were instituted initially as part of the examination of the gas-phase decomposition of these compounds, electron impact being considered one way of imparting energy to the molecule. Thus we get the early reports of the mass spectra of triarylmethyl azides[6] in 1966 and phenyl azides[7,8] in 1967 and 1968, all three papers comparing electron impact-induced fragmentation to thermal or photochemical breakdown of azides.

In consideration of the lengthy history of halide mass spectrometry, studies in the mass spectra of azides represent a relatively new field, never before reviewed and one for which much of the data in the literature involves incompletely tabulated mass spectra or uninterpreted raw data. It is thus our intention to report in some detail on the current state of the art of mass spectrometry as applied to azides. However, since halides have been extensively reviewed elsewhere, we will report some of the recent developments of interest in the mass spectra of halogenated compounds, selectively, in a second section of this chapter.

II. THE MASS SPECTRA OF AZIDES

A. Introduction

As indicated in the previous section, study of the mass spectra of azides is a relatively recent endeavour. Researchers would do well to contemplate the possible reasons for this delay in application of mass spectrometry to azide chemistry. Surely chemists studying azides are not more resistant than others to the application of new techniques? Was it perhaps the mass spectroscopist's fear of damage to the expensive mass spectrometer from 'explosive' azide compounds? No use was made of the 'oil company' mass spectrometers in the 1940s. Even with the development of the modern 'organic' mass spectrometer and inlet systems in the late 1950s and early 1960s, almost a decade passes before spectra begin to be reported. This fear on the part of mass spectrometer operators appears to be more severe than that which delayed the application of mass spectrometry to organometallic chemistry[9]. This hesitation continues. Thus, we have not yet seen the application of the newer 'soft' ionization techniques such as field desorption, chemical ionization or fast atom bombardment (FAB) to azides. All the work done is with standard electron impact ionization sources. Similarly, only one partial negative ion study is reported, nor has extensive use been made of new techniques for studying metastable ions. If we were to make a prediction, it would be that the use of mass spectrometry, at the least for molecular weight characterization of azides, will increase vastly once field desorption instruments are more common, and the fragility of these instruments is disproven by use. The newly developed 'FAB' sources introduced by two mass spectrometer manufacturers (Kratos and VG) appear to be much easier to use than field desorption, and could easily become the favoured source for azide work if they are widely adopted, no direct heating of the sample being required. Various security agencies are experimenting with atmospheric pressure ionization sources for the monitoring of explosives, and in the course of their work it is suspected that they have looked at the mass spectra of azides in this fashion, but nothing has yet been reported to our knowledge. Since HPLC is far superior to gas chromatography as a separation tool for azides, a third area from which advances in the study of azide mass spectra will come, is the interfacing of liquid chromatographs to mass spectrometers.

It is to be hoped that one outcome of this review will be an increase in the determination and reporting of the mass spectra of azides. The operators' fears are groundless. We have not seen any report in the literature of source explosions from the attempted volatilization of unstable azides into a mass spectrometer. Careful control of temperature and small amounts of the samples on a direct insertion probe will minimize any possibility of damage, but these are the same precautions also taken for organometallic compounds. It is interesting to note that the first people to report on the mass spectra of azides were those already studying their thermolysis[7,8].

In the next sections we discuss the important aspects of the mass spectra of azide derivatives of various compound classifications, and tabulate the compounds for which full (or partial) mass spectra data are available. It is interesting to note that few of these compounds appear in any of the computerized libraries of mass spectral data.

B. Aromatic Azides and Benzoyl Azides

Studies of the mass spectra of aromatic azides were first reported in 1967 by Crow and Wentrup[7], who report on the electron impact mass spectrum of phenyl azide, as part of an investigation into the generation of phenylnitrenes. This was followed a year later by a more complete study by Hedaya and coworkers[8]. The relationship of pyrolysis and photolysis studies to the mass spectra of azides will be discussed in detail in Section D. These studies were followed by more comprehensive mass spectral investigations of phenyl azide beginning in 1970[10,11], a series of aromatic azides[12–14], and partial spectra reported as part of further pyrolysis, photolytic or other spectroscopic studies[15–21]. In Table 1 we summarize aryl and aroyl azides and related compounds for which mass spectral data are available and note the completeness or incompleteness of the data and their interpretation.

TABLE 1. Aryl azides, benzoyl azides and related compounds

No.	Compound	Completeness of data	Reference
1	$C_6H_5N_3$	Interpreted spectrum	7, 13
1	$C_6H_5N_3$	Pyrolysis data	8
1	$C_6H_5N_3$	Isotope labelling, fragmentation rearrangements	10, 11
2	2-$NCC_6H_4N_3$	Spectra and discussion	12
3	3-$NCC_6H_4N_3$	Spectra and discussion	12
4	4-$NCC_6H_4N_3$	Spectra and discussion	12
5	2-$ClC_6H_4N_3$	Spectra and discussion	12
6	3-$ClC_6H_4N_3$	Spectra and discussion	12
7	4-$ClC_6H_4N_3$	Spectra and discussion	12
8	2-$O_2NC_6H_4N_3$	Spectra and discussion	12, 13
9	3-$O_2NC_6H_4N_3$	Spectra and discussion	12, 13
10	4-$O_2NC_6H_4N_3$	Spectra and discussion	12, 13
11	2-$CH_3OC_6H_4N_3$	Spectra and discussion	12
12	3-$CH_3OC_6H_4N_3$	Spectra and discussion	12
13	4-$CH_3OC_6H_4N_3$	Spectra and discussion	12
14	2-$C_6H_5C_6H_4N_3$	Spectra and discussion	12
15	3-$C_6H_5C_6H_4N_3$	Spectra and discussion	12
16	4-$C_6H_5C_6H_4N_3$	Spectra and discussion	12
17	2-$CH_3C_6H_4N_3$	Isotope labelling, discussion	12, 13
18	3-$CH_3C_6H_4N_3$	Isotope labelling, discussion	12, 13
19	4-$CH_3C_6H_4N_3$	Isotope labelling, discussion	12, 13

TABLE 1. *continued*

No.	Compound	Completeness of data	Reference
20	2-O_2N-6-$CH_3C_6H_3N_3$	Spectra and discussion	12
21	2-O_2N-4-$CH_3C_6H_3N_3$	Spectra and discussion	12, 13
22	3-O_2N-4-$CH_3C_6H_3N_3$	Spectra and discussion	13
23	3-O_2N-6-$CH_3C_6H_3N_3$	Spectra and discussion	13
24	2,6-$(CH_3)_2C_6H_3N_3$	Spectra and discussion	13
25	2-$BrC_6H_4N_3$	Spectra and discussion	13
26	3-$BrC_6H_4N_3$	Spectra and discussion	13
27	4-$BrC_6H_4N_3$	Spectra and discussion	13
28	$C_6H_5CON_3$	Spectra and discussion	13, 14
29	4-$O_2NC_6H_4CON_3$	Spectra and discussion	13, 14
30	4-$CH_3OC_6H_4CON_3$	Spectra and discussion	13, 14
31	4-$CH_3C_6H_4CON_3$	Spectra and discussion	13, 14
32	4-$ClC_6H_4CON_3$	Spectra and discussion	14
33	4-$BrC_6H_4CON_3$	Spectra and discussion	14
34	4-$C_6H_5C_6H_4CON_3$	Spectra and discussion	14
35	$C_6H_5CH{=}CHCON_3$	Spectra and discussion	14
36	4-$CH_3OC_6H_4CH{=}CHCON_3$	Spectra and discussion	14
37	2,4,6-$(O_2N)_3C_6H_2N_3$	Spectra and discussion	19
38	3-Azido-2-naphthoyl azide	Partial data	15
39	2,3-Diazidonaphthalene	Partial data	15
40	Methyl 3-azido-2-naphthoate	Partial data	15
41	2,2′-Diazidobiphenyl	Partial data	6
42	2,2′-Diazido-6,6′-dimethylbiphenyl	Partial data	6
43	4-Azidocarbazole	Partial data	6
44	4(2′-Azido-2-biphenylazo)carbazole	Partial data	6
45	2,2′-*bis*(*o*-Azidophenyl)azobenzene	Partial data	6
46	2,3-Diazido-1,4-dimethoxynaphthalene	Partial data	17
47	2-Azido-2′-phenylazobiphenyl	Partial data and interpretation	18
48	2-Azido-2′-(4-tolylazo)biphenyl	Partial data	18
49	2-Azido-2′-(2′-biphenylazo)biphenyl	Partial data	18
50	2-Azido-2′-(4-chlorophenylazo)biphenyl	Partial data	18
51	2,4-$Br_2C_6H_3N_3$	Partial data	21
52	2-N_3-3,5-$Br_2C_6H_2CN$	Partial data	21
53	2-N_3-3,5-$Br_2C_6H_2N_2C_6H_5$	Partial data	21
54	2-N_3-3,5-$Br_2C_6H_2COC_6H_5$	Partial data	21
55	2-N_3-3,5-$Br_2C_6H_2CHO$	Partial data	21
56	2-Iodo-2′-azidobiphenyl	Partial data	22
57	1-Azido-8-iodonaphthalene	Partial data	22

All 57 compounds listed in Table 1 exhibit one important feature in their 70 eV EI mass spectra: a significant molecular ion is present in each case. Parent ion intensities vary from 1–2% to 30% and average about 10% of the base peak. At 17 eV the molecular ion of phenyl azide increases to a conspicuous 70% of the base peak, carrying over 20% of the total positive ion current. Thus, despite concerns over the stability which may have precluded study of azide mass spectra, this is certainly not the situation for these aromatic derivatives. Neither instrument nor inlet system and temperature have significant effect on the observed spectra, though some difference in the metastable ions supporting fragmentation have been observed[12–14] for some azides on changing instruments. This behaviour is not limited to azides alone. Even compounds with molecular weights over 400 gave good spectra and are thus amenable to mass spectrometric identification. Since aryl azides readily thermolyse in the 140–170°C range, by using 3-cyanophenylazide as a model azide, Abramovitch and coworkers[12] studied the effect of ion source temperature over the range 165–250°C. Spectra were identical except for a change in ratio of $M^{+\cdot}/(M\text{-}N_2)^{+\cdot}$ which changed from about 0.22 to 0.17 over the temperature range. This results from the nitrene or azepinium ion being formed both by fragmentation of the parent ion and by ionization of aryl nitrene produced by thermolysis. As ionization electron energy was reduced from 70 eV, little change was observed in the above ratio until the potential had been reduced below 12 eV.

The base peak in the spectrum of phenyl azide corresponds to losses of N_2 from the molecular ion[7,11]. This mode of fragmentation is quite common, the $(M\text{-}N_2)^{+\cdot}$ ion being, if not the base peak, an ion of significant abundance for most of the azides discussed in this section. It has been shown that loss of N_2 is not due to decomposition in the ion source.

If we look first at the spectrum of phenyl azide[7], and both the ^{13}C-labelled[11] and ^{2}H-labelled[10] compounds, the fragmentation can be well understood. The observation of metastable ions for the loss of N_2 to give $C_6H_5N^+$ (formally ionized phenyl nitrene or on rearrangement [azepinium-hydrogen]$^+$) prompted the labelling studies; *m/z* 90 then showed a metastable ion supported loss of both acetylene and hydrogen cyanide[7,10–12] as shown in Scheme 1, while the complete spectrum is shown in Figure 1. (Throughout this paper an asterisk (*) in a fragmentation scheme indicates that the transition is supported by the observation of metastable ions.) This loss is reminiscent of that observed in the spectrum of aniline. However, in contrast to the $C_6H_7N^{+\cdot}$ odd

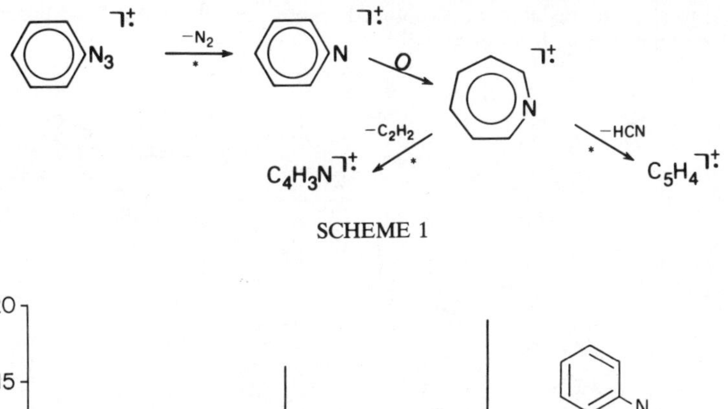

SCHEME 1

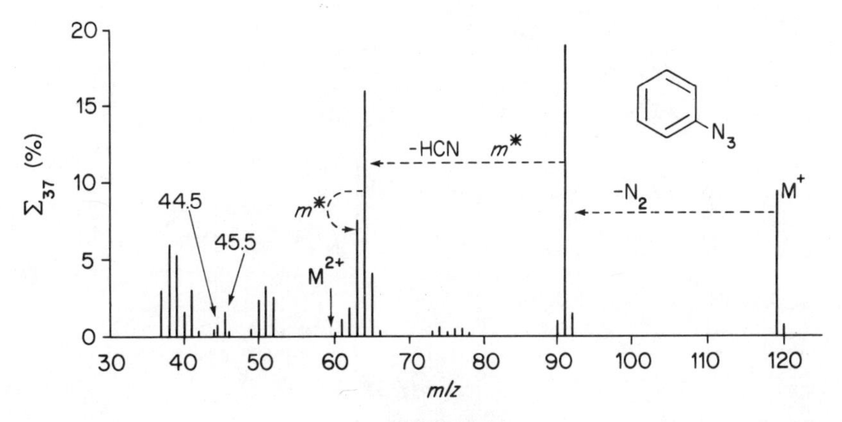

FIGURE 1.

electron molecular ion of aniline, the odd electron $C_6H_5N^{+\cdot}$ species from phenyl azide has been shown to undergo ring expansion with the insertion of nitrogen typical of carbenoid species[23,24] as proven by ^{13}C labelling at the N_3 carbon[11]. This rearrangement precedes expulsion of hydrogen cyanide or acetylene. The data for loss of acetylene suggests that a C—N bond is broken and that insertion of nitrogen was specific, rather than random. The HCN loss is less specific in indicating the insertion site. A study of 2,4,6-d_3-phenyl azide shows complete scrambling of hydrogen and deuterium prior to the loss of HCN[18]. Abramovitch reported a general scheme by which the 'normal' fragmentation of many substituted azides could be described (Scheme 2), which was analogous to that of phenyl azide itself (Scheme 1). This scheme was appropriate to eight of 18 monosubstituted aryl azides studied, i.e. the three chloro and three cyano derivatives plus the 3- and 4-nitro derivatives. These 'normal' fragmentations are characterized by relatively high ion current for $(M\text{-}N_2)^{+\cdot}$, and/or 90 or 89, 63, 62 and 39–37. There is only low ion current at 77, 76 and 52–50 for these azides, suggesting strongly that ring fragmentation is determined by either the substituent X or by the ring expansion and nitrogen insertion; i.e. we do not get fragmentation based on the free phenyl ion. Abramovitch suggested that route *a*, i.e. loss of HCN from $(M\text{-}N_2)^{+\cdot}$ is usually rare, but is observed for chlorophenyl azides[12] in the case of 'normal' fragmentation. However, Fraser and collaborators[13], as well as Abramovitch, report significant peaks and metastable ion-supported loss of HCN for systems which are exceptions to the 'normal' behaviour. These exceptions usually involve fragmentations of the functional group 'X' which introduces added complexity to the spectra. Another system which could be categorized as obeying 'normal' fragmentations is that of the three bromo derivatives[13], but unlike the chlorides these do not exhibit HCN loss, bromine loss being preferred. These bromo derivatives were the first to show metastable ion-supported loss of the substituent from the molecular ion, in competition with N_2 loss.

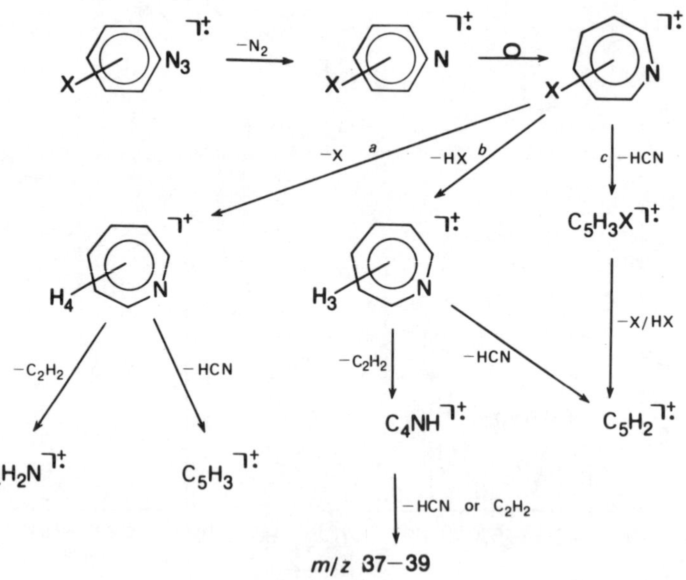

SCHEME 2

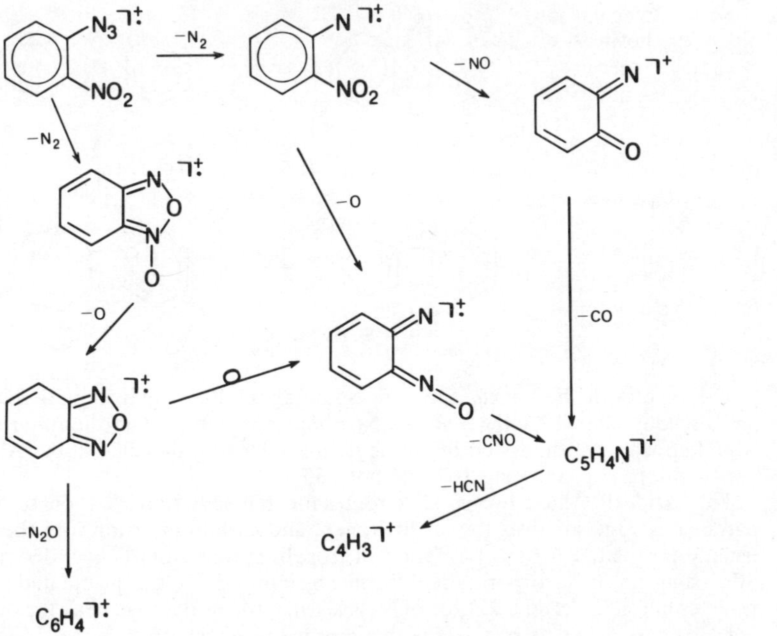

SCHEME 3

Looking at exceptions to normal behaviour, the first example to become apparent, that of 2-nitrophenyl azide, displays peaks due to a pronounced *ortho* effect (m/z 120, 78, 76 and 51) in addition to the expected m/z 90, 63 and 39. These have been rationalized as shown in Scheme 3.

Abramovitch[16] postulated the two routes to m/z 78 to account for the different 76/78 ratio in the spectrum of this compound, compared to that of benzofuroxan (a thermolysis product), while Fraser[13] shows the third route as well.

In the methoxyphenyl azides, one has competing CH_3 and CO losses in addition to normal fragmentation. Again, however, *ortho* effects are observed for 2-methoxyphenyl azide, with a large m/z 120 being observed, rather than 121, the former presumably being a benzo species. The *p*-methoxy derivative, unlike the other two, has large peaks at m/z 80 and 52, i.e.

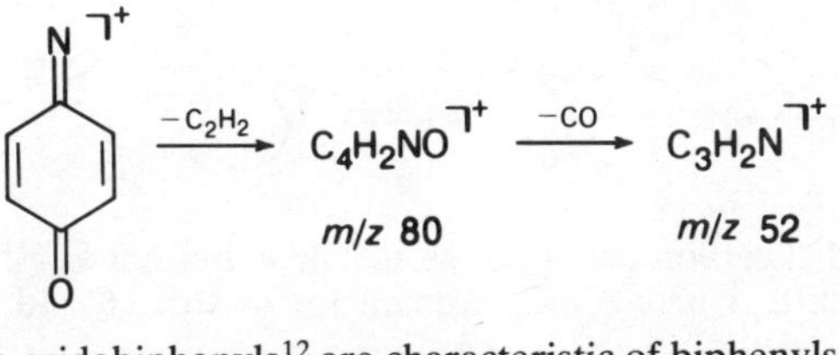

$$\xrightarrow{-C_2H_2} C_4H_2NO^{+} \xrightarrow{-CO} C_3H_2N^{+}$$

m/z 80 $\quad$ m/z 52

The spectra of the azidobiphenyls[12] are characteristic of biphenyls in having most of the ion current carried by high m/z ions. These ions have been postulated to fragment as shown in Scheme 4 below, with the intermediate proposed for the *ortho* derivative being analogous to those proposed for fragmentation of di- and triphenyl derivatives of phosphorus, arsenic and antimony[9,25].

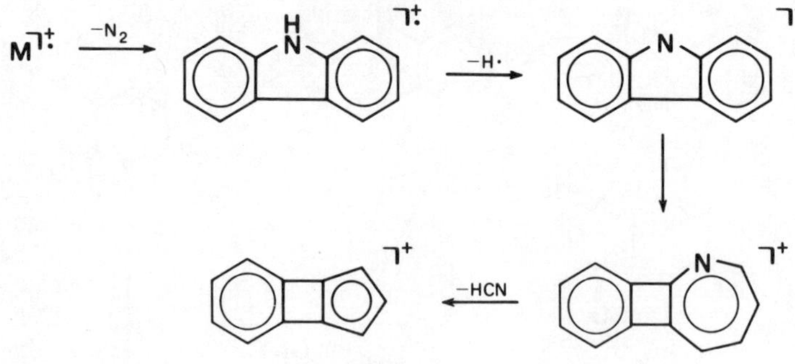

SCHEME 4

The spectra of these *meta* and *para* isomers are almost indistinguishable, with the *ortho*-azidobiphenyl having a somewhat simpler spectrum than the other isomers. It would appear that the spectrum of the *para* isomer is mislabelled in the Abramovitch paper, the base peak being 167 and not 165.

Tolyl azides[12,13] have two possible routes for fragmentation after loss of N_2 from the parent ion. One involves loss of hydrogen, and carbon insertion into the ring, as is usually postulated for the formation of tropylium ions from benzyl derivatives. The other suggests the incorporation of the nitrogen into the ring as postulated for 'normal' fragmentation in Scheme 2. Loss of N_2 gives m/z 105 as the base peak for the *meta* and *para* isomers, whereas m/z 104 is the base peak for the *ortho* isomer, though, in all three cases, loss of hydrogen, m/z 105 → 104, is supported by metastable ions. The *ortho* isomer also has a strange m/z 79 (loss of C_2H_2 from m/z 105) and shows a metastable ion for the transition m/z 78 → 77. High resolution[12] showed that m/z 107 is due both to $(M\text{-}C_2H_2)^{+\cdot}$ and to the more usual primary amine cation[12], formed by loss of N_2 from the molecular ion, followed by hydrogen abstraction in the ion source. On the basis of varying m/z 104/105 ratios, Fraser[13] suggests that ring expansion is unlikely. Abramovitch[12], using $^{13}CH_3$-labelled $2\text{-}^{13}CH_3C_6H_4N_3$ unambiguously established that $C_7H_7N^{+\cdot}$ (m/z 105) fragments by loss of neutral C_2H_2, HCN and H_2CN; giving 79^+, 78^+ and 77^+ and following loss of $H^\cdot$; $C_7H_6N^+$. M/z 104 fragments exclusively by loss of HCN and H_2CN, yielding ions at m/z 77 and 76. Thus, ions at mass 77 arise via two routes. It would be interesting to see some of this work repeated using the full metastable ion studies possible with modern instrumentation. Since the carbon label is lost preferentially, and unexpected loss of H occurs, it has been postulated[6] that the following ion structures are responsible:

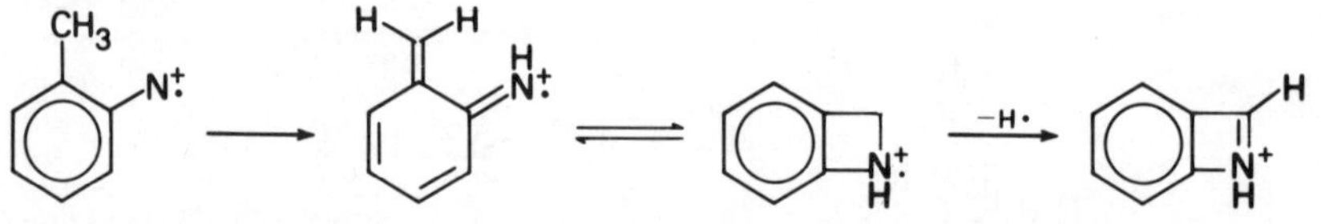

the labelled methyl carbon being lost as acetylene but not as $H^{13}CN$ from the second structure. The third structure may account for loss of $H^\cdot$ and the small amount of $H^{13}CN$ lost.

A weak loss of acetylene from the *meta* and *para* isomers may suggest that the quinoid structure alone might be more important than ring expansion involving either C or N.

Dimethylphenyl azides[13] were shown to behave similarly to tolyl azides. Both show impurity peaks due to scavenging of hydrogen from the walls of the ion source by

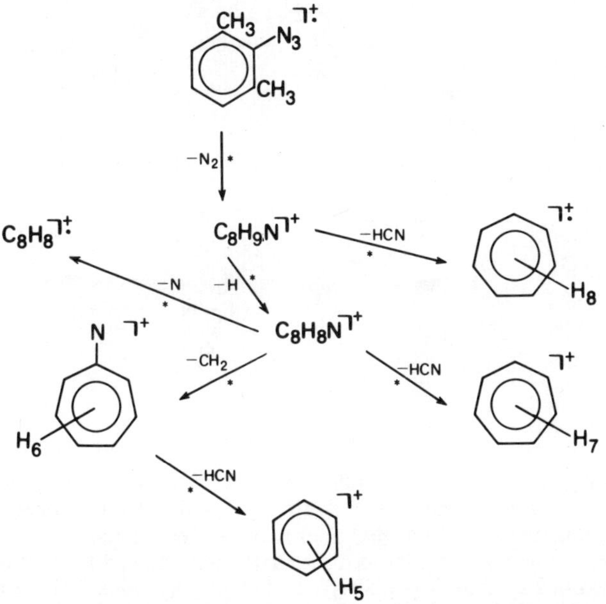

SCHEME 5. Reproduced from Fraser *et al.*[13] by permission of Heyden & Sons Ltd.

the nitrenium ion. As shown in Scheme 5, the ion at m/z 104 is a doublet arising from two processes.

The spectra of azidonitrotoluenes combine the features of both tolyl azides and nitrophenyl azides[12,13]. An *ortho* effect appears to be operative for these compounds, the $M^{+\cdot}/(M\text{-}N_2)^{+\cdot}$ ratio being much larger in the *ortho* case, suggesting perhaps that the *o*-methyl group hinders the concerted cyclization–elimination of nitrogen, giving the proposed benzofuroxan (Scheme 3).

Another well studied group of azides related to the simple aromatic azides are benzoyl azide, and a series of its *para*-substituted derivatives, first studied by Fraser's group[13] and then examined more extensively by Matjeka[14]. Benzoyl azide did not eliminate CO from the molecular ion, as do so many other benzoyl compounds, the usual azide fragmentation involving loss of N_2 from $M^{+\cdot}$ predominating. Fraser[13] suggested that the $(M\text{-}N_2)^{+\cdot}$ ion did not rearrange to an isocyanate ion, since, as shown below, three modes of loss were observed from this ion; i.e. rearrangement is

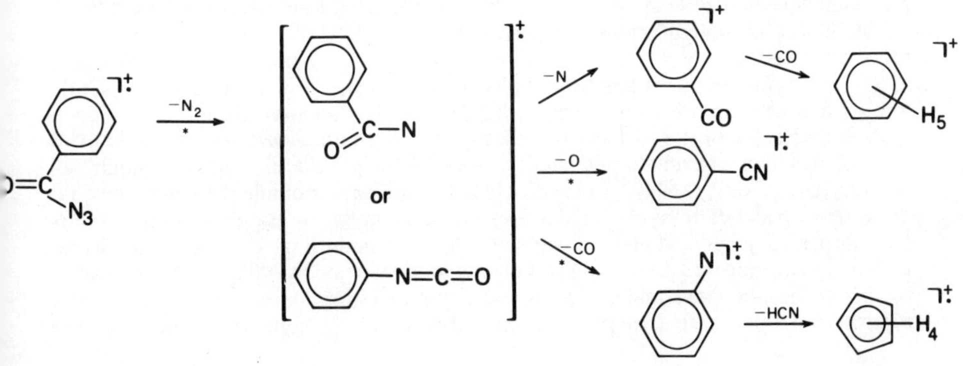

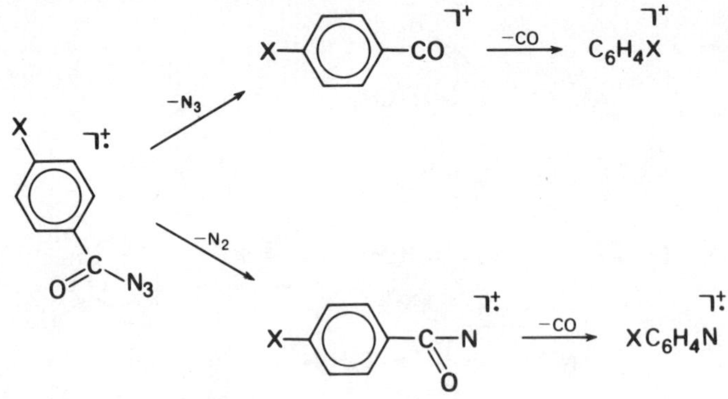

SCHEME 6

obviously necessary for CO elimination. Matjeka[14], unlike Fraser and coworkers[13], observed a metastable ion for loss of N_3 from $M^{+\cdot}$ and did not observe loss of $O^{\cdot}$ from the $(M-N_2)^{+\cdot}$ ion. Thus the isocyanate ion may be a possibility.

Unlike *p*-nitrophenyl azide, the nitrobenzoyl azide showed the metastable ion that confirmed loss of $NO^{\cdot}$ from the $(M-N_2)^{+\cdot}$ and $(M-N_2-N)^+$ ions[13]. However, Matjeka[14] again reported a difference from Fraser's data, namely NO_2 rather than the NO species being lost from the two species above, and the $(M-N_3)^+$ ion was shown to be formed directly from the parent ion. The reason for these differences is not apparent.

The remaining 10 substituted benzoyl azides (CH_3O, CH_3, Cl, Br and Ph substituents) all follow a standard route[14] (Scheme 6) with $(M-N_3)^+$ being the base peak in most cases. The $XC_6H_4N^+$ ion has most likely undergone ring expansion with the N being included in the ring (Scheme 2). The methyl derivative shows a large ion at *m/z* 134 corresponding to a loss of molecular nitrogen and a methyl radical. The resulting ion is postulated as a conjugated isocyanate ion by analogy with the species observed in the spectrum of *p*-methyldiazoacetophenone[14]; i.e.

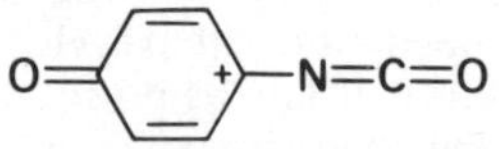

Again, like *p*-phenyldiazoacetophenone, *p*-phenylbenzoyl azide fragments with loss of N_2 and CO from $M^{+\cdot}$ at the expense of $(M-N_3)^+$. The resulting *m/z* 167 ion is presumably similar to that observed for azidobiphenyl[12], where it is the base peak.

Two other carbonyl azides, $XC_6H_4CH{=}CHC(O)N_3$ (X = H or OCH_3) were reported in the same work[14], fragmenting by analogy to Scheme 6, i.e. loss of N_3 and CO, or N_2, rearrangement and loss of CO. HCN was also lost from the $(M-N_2-CO)^{+\cdot}$ ion; its fragmentation is characteristic of X, as was its daughter ion.

Peek and collaborators[15] have reported partial data on 3-azido-2-naphthoyl azide and 2,3-diazidonaphthalene and methyl 3-azido-2-naphthoate. Not surprisingly, all showed parent ions, $(M-N_2)^{+\cdot}$ ions and for the diazido compounds $(M-2N_2)^{+\cdot}$ ions, i.e. both the ring and carbonyl bonded azides break down by the standard route – loss of N_2. Surprisingly, no $(M-N_3)^+$ is reported for the naphthoyl derivative, a distinct difference from that observed for benzoyl derivatives. The naphthoyl derivative readily decomposed thermally to 3-azido-2-naphthoyl isocyanate, a fact which perhaps supports an isocyanate structure for the $(M-N_2)^{+\cdot}$ ion in benzoyl compounds. The naphthoyl derivatives also showed peaks at m/z 169 corresponding to $(M\text{-}N_2\text{-}CH_2O)^{+\cdot}$ and m/z 140, i.e. $(NC_{10}H_6)^+$ results by loss of HC=O from 169.

Yabe and Honda[16] report partial data for several diazido and azido compounds (Table 1, compounds **41–45**). All showed molecular ions and $(M\text{-}N_2)^{+\cdot}$ ions, while diazido derivatives showed $(M\text{-}2N_2)^{+\cdot}$ as well. Yabe[17] found similar results for a diazidodimethoxynaphthalene.

Spagnolo and coworkers[18] have reported partial uninterpreted mass spectral data on four 2-azido-2′-arylazobiphenyls, where the aryl group is one of phenyl, 4-tolyl, 2″-biphenyl and 4-chlorophenyl. It is clear that all four show loss of N_2 and N_3 from the parent ions. Like the simple biphenyls themselves[12], these more complex compounds also show loss of $H^\cdot$ from the $(M\text{-}N_2)^+$. But loss of H is also seen from the $(M\text{-}N_3)^+$ ion in the first two cases. A peak is also observed at m/z 180 which could be assigned to the following species, which can eliminate N_2 as shown, producing the

N=N $\xrightarrow{-N_2}$

m/z 180 $\qquad$ m/z 152

biphenylene ion. The data also suggest that the $(M\text{-}N_3)^+$ ion may eliminate CN_2 with rearrangement.

Muller has recently reported the mass spectrum of 2,4,6-trinitrophenyl azide.[19] The base peak was the NO^+ at m/z 30, but significant peaks were assigned to $(M\text{-}N_2)^{+\cdot}$, and losses of NO_2, NO and NO_2. This is not consistent with the fragmentation expected for nitroazides[12,13] where loss of N_2 and $O^\cdot$ are observed. Major rearrangement would also be required for the loss of four nitrogen–oxygen species. One might, perhaps, have expected loss of $O^\cdot$ and N_2O from $(M\text{-}N_2)^{+\cdot}$ to account for m/z 150, but in the absence of the complete spectrum of metastable ion evidence, definitive assignment cannot be made.

Most recently, Dickson and Dyall[21] have reported partial mass spectral data for a series of *o,p*-dibromophenyl azides (Table 1, compounds **51–55**). There was no pattern for these derivatives, with no common base peak. All show $M^{+\cdot}$ and $(M\text{-}N_2)^{+\cdot}$ ions. The 2,4-dibromoazidobenzene has as its base peak the $(M\text{-}N_2\text{-}Br)^+$ ion, a 'normal' behaviour. For **52** the base peak again was $(M\text{-}N_2\text{-}Br)^+$. Further fragmentation occurs by loss of the second bromine and HCN. Significant HBr^+ was reported. For **53**, as a result of ring expansion with a nitrogen of the azobenzene substituent, the base peak is m/z 91. No species were reported which correspond to single loss of one or two bromines. Compound **54** again shows $(M\text{-}N_2\text{-}Br)^+$, with a single phenyl ion m/z 77 being the base peak. The azido benzaldehyde (**55**) shows $(M\text{-}N_2\text{-}CHO)^+$ ion as well as $(M\text{-}N_2)^{+\cdot}$ and significant $(M\text{-}N_2\text{-}Br)^+$. This latter ion also shows loss of HCN and CO, i.e. ions at m/z 142 and 140. The base peak at m/z 87 results from loss of Br and HCO from the $(M\text{-}N_2\text{-}Br)^+$ ion. Thus the basic fragmentations are what would be expected from the substituted aromatic azides. For neither of these carbonyl compounds are ions observed which correspond to the loss of N_3 from the molecular ion, as was observed in the case of benzoyl azides. It is thus not the presence of a carbonyl but rather the azide attached to the carbonyl that favours N_3 loss.

Benati and coworkers[22] have reported that 2-iodo-2′-azidobiphenyl shows a strong molecular ion and that fragmentation begins with loss of N_2. Similar behaviour is reported for 1-azido-8-iodonaphthalene.

C. Aliphatic Azides

The first report of mass spectra of azides was in 1966. When Moriarty and Kirkien-Konasiewicz[6] studied the electron impact decomposition of triarylmethyl azides (i.e. *p*-tolyl-, *p*-chlorophenyl- and *p*-nitrophenyl diphenylmethyl azides) only weak parent ions were observed. The base peak in all cases was $(M\text{-}N_3)^+$, i.e. the triarylmethyl carbonium ion, reminiscent of that observed[14] for benzoyl azides. $(M\text{-}N_2)^{+\bullet}$ intensities varied with substituent, but no evidence was presented favouring either of the two possible structures, i.e. the nitrene $Ar_3C{-}N^{+\bullet}$, or a rearranged arylimino ion $(Ar_2\overset{+}{C}{=}N{-}Ar)^{+\bullet}$. Lowering the ionizing voltage from 70 eV to 50 eV had little effect, which the authors took as suggesting nitrene formation by pyrolysis in the source prior to ionization. However, without reducing the ionizing voltage to the 15–20 eV range, such a conclusion is invalid. No metastable ions were reported for either loss of N_2 or loss of N_3 from $M^{+\bullet}$. In Scheme 7, the major decompositions from the $(M\text{-}N_2)^{+\bullet}$ and $(M\text{-}N_3)^+$ ions are shown for the *p*-tolyldiphenylmethyl azide. The ion at *m/z* 180 is preferentially assigned to $(C_6H_5{-}C{\equiv}N{-}C_6H_5)^+$ on the basis of a similarity to the spectrum of benzophenone *N*-phenylimine. Since the stability of the triarylmethyl carbonium ion appeared so important in controlling the electron impact fragmentation and since the $((C_6H_5)_2CCOOH)^+$ ion should be much less stable than

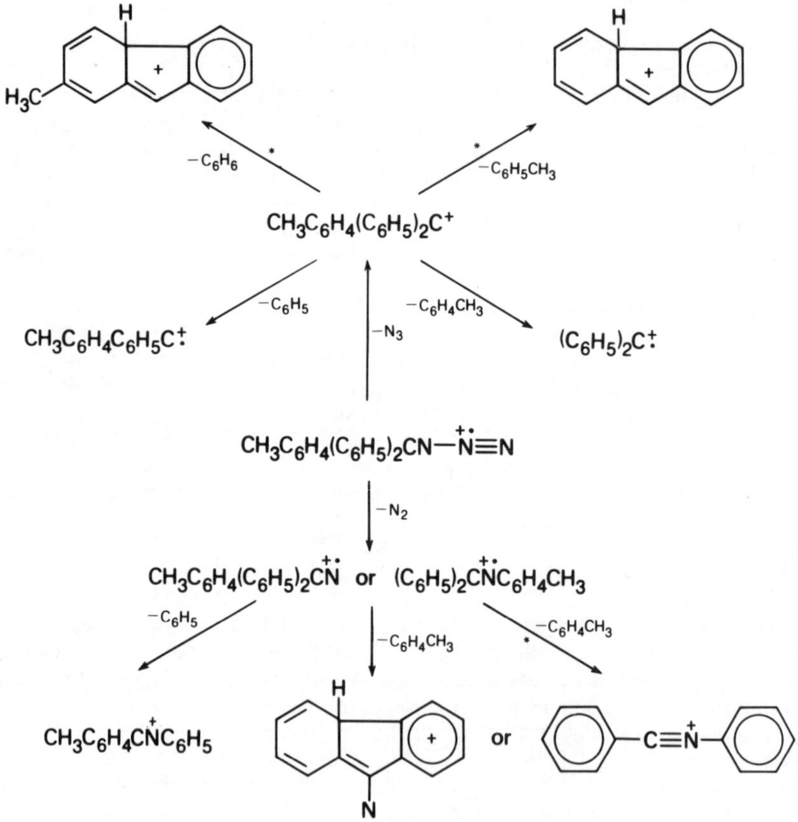

$CH_3C_6H_4(C_6H_5)_2C^+$

$-C_6H_6$ $-C_6H_5CH_3$ $-C_6H_5$ $-C_6H_4CH_3$

$CH_3C_6H_4C_6H_5C^{+\bullet}$ $(C_6H_5)_2C^{+\bullet}$

$-N_3$

$CH_3C_6H_4(C_6H_5)_2CN{-}\overset{+\bullet}{N}{\equiv}N$

$-N_2$

$CH_3C_6H_4(C_6H_5)_2C\overset{+\bullet}{N}$ or $(C_6H_5)_2C\overset{+\bullet}{N}C_6H_4CH_3$

$-C_6H_5$ $-C_6H_4CH_3$ $-C_6H_4CH_3$

$CH_3C_6H_4C\overset{+}{N}C_6H_5$ or $C{\equiv}\overset{+}{N}$

SCHEME 7

$(C_6H_5)_3C^+$, the spectrum of azidodiphenylacetic acid was determined[6]. The base peak for the acid now corresponded to $((C_6H_5)_2CN_3COOH—N_2—CO_2)^{+\cdot}$, the same species as that which resulted from solution photolysis.

Curci and coworkers[26] report the unassigned spectrum of 1,2,3-tri-*t*-butyl-3-azidocyclopropene. As one might expect, the *t*-butyl ion forms the base peak. The primary mode of fragmentation is loss of N_3 from the molecular ion. This contrasts with the corresponding cyano derivative which loses Me_3C rather than CN.

Reed and Lwowski[27] report partial data for 1-azidonorbornane. A parent ion is observed, which then shows loss of either N_2 or C_2H_4, which cannot be distinguished in the absence of high resolution data. Loss of 29, i.e. N_2H or C_2H_5 is also observed and is of greater intensity than $(M\text{-}28)^{+\cdot}$. m/z 108 in fact becomes the base peak in the 13 eV spectrum. A second loss of 28 from m/z 107, or 27 from m/z 108 is observed, presumably either C_2H_4 or CH_2N from m/z 107 and HCN from m/z 108. The base peak is observed at m/z 69, presumably either $C_6H_4^+$ or $C_4NH_7^+$. No loss of N_3 is reported.

Barone and collaborators[28] have reported partial data for a series of α-azidonitriles and α-azidoaldehydes. These compounds **(62–73)** are shown in Table 2. Parent ions

TABLE 2. Aliphatic azides

No.	Compound	Completeness of data	Reference
58	$4\text{-}CH_3C_6H_4(C_6H_5)_2CN_3$	Spectra and discussion	6
59	$4\text{-}ClC_6H_4(C_6H_5)_2CN_3$	Spectra and discussion	6
60	$4\text{-}O_2NC_6H_4(C_6H_5)_2CN_3$	Spectra and discussion	6
61	$(C_6H_5)_2(CO_2H)CN_3$	Comment only	6
62	1,2,3-Tri-*t*-butyl-3-azidocyclopropene	Uninterpreted spectrum	26
63	1-Azidonorbornane	Uninterpreted spectrum	27
64	1-Azidocyclododecane-1-carboxaldehyde	Partial data	28
65	$(CH_2)_5C(CHO)N_3$	Partial data	28
66	$(CH_2)_7C(CHO)N_3$	Partial data	28
67	$CH_3(C_6H_5)C(CHO)N_3$	Partial data	28
68	$(C_6H_5CH_2CH_2)_2C(CHO)N_3$	Partial data	28
69	$(c\text{-}C_5H_9)CH_3C(CHO)N_3$	Partial data	28
70	1-Azidocyclododecane	Partial data	28
71	$(CH_2)_5C(CN)N_3$	Partial data	28
72	$(CH_2)_7C(CN)N_3$	Partial data	28
73	$CH_3(C_6H_5)C(CN)N_3$	Partial data	28
74	$(C_6H_5CH_2CH_2)_2C(CN)N_3$	Partial data	28
75	$(c\text{-}C_5H_9)CH_3C(CN)N_3$	Partial data	28
76	*trans*-2-Azidocyclohex-3-enyl acetate	Partial data	29
77	$(C_6H_5)_3CN_3$	Spectra and partial interpretation	30
78	$2\text{-}C_6H_5C_6H_4(C_6H_5)_2CN_3$	Spectra and partial interpretation	30
79	$C_6H_5CH(CH_3)N_3$	Spectra and partial interpretation	30
80	2-Phenylisopropyl azide	Spectra and partial interpretation	30
81	2-(4′-Biphenyl)isopropyl azide	Spectra and partial interpretation	30
82	2-(2′-Biphenyl)isopropyl azide	Spectra and partial interpretation	30
83	2-Azido-2,3,3-trimethylbutane	Spectra and partial interpretation	30
84	4-Azidoheptane	Spectra and partial interpretation	30
85	(Tri-*n*-pentyl)methyl azide	Spectra and partial interpretation	30
86	2-Azido-2-methyl-4-phenylbutane	Spectra and partial interpretation	30
87	*cis*-1-Azido-2-chlorocyclopentane	Spectra and partial interpretation	31
87a	Tri(*o*-azidothioacetophenone)	Partial data	33b
88	*cis*-1-Azido-2-chlorocycloheptane	Spectra and partial interpretation	31

were not observed for the aldehydes listed except that an azido steroid, to be described in Section E, showed a strong parent ion. Loss of N_3 or N_2 was general in some cases, but there was no constancy in the mass losses suggesting a complex fragmentation scheme which would require high resolution data to resolve. However, many of the nitriles did show reasonable intensity parent ions, or in their absence, $(M-N_3)^+$ ions. As the primary loss is that of N_3, there seems to be no particularly interesting fragmentation in these compounds except for simple phenyl derivatives for which $(M-N_2)^{+\cdot}$ ions are important. This suggests nitrogen incorporation here, as observed for aromatic derivatives.

Becsi and Zbiral[29] observed a weak parent ion for *trans*-2-azidocyclohex-3-enyl acetate, with consequent loss of N_3 or CH_3COOH with significant intensity due to loss of both N_3 and CH_3COOH.

The most varied and comprehensive study of aliphatic azides has been that of Abramovitch and Kyba[30], who studied nine compounds (Table 2, **75–84**). These were divided into three categories. The first, consisted of triphenylmethyl azide and 2-biphenyldiphenyl methyl azide, for which no parent ion were observed – surprising in view of Moriarty's results[6]. C—N bond cleavage is most important, with N—N cleavage being minimal, fragmentation of the biphenyl proceeding as shown in Scheme 8, the base peak being presumably a 9-phenylfluorenyl cation.

A second group consisted of α-methylbenzyl azide, 2-phenylisopropyl azide and 2-(4′-biphenylyl)isopropyl azide, these having only one aryl group attached to the azide-bearing carbon. Each showed a molecular ion, with C—N cleavage predominating. The 2-(2′-biphenylyl)isopropyl azide showed no parent ion and had as its base peak the 9-methylfluorene cation. The last group, **81–84**, with no aryl group on the same carbon as the azide, showed no molecular ion even at low ionizing electron energy, and the $(M-N_2)^+$ and $(M-N_3)^+$ ions are weaker, the latter, however, being 10–100 times the intensity of the former. There is no evidence of HCN loss from any $(M-N_2)^+$ ions, which probably confirms that rearrangements involving nitrogen have not occurred for these derivatives.

Schweng and Zbiral[31] report partial data on two *cis*-1-azido-2-chlorocycloalkanes, these being characterized by weak or absent molecular ions. The main loss is that of N_2, which is then followed by loss of chlorine.

Thus, although we can draw no general scheme for the fragmentation of aliphatic azides, parent ions can be observed in many cases. Certainly with modern soft ionization techniques one should be able to observe a molecular ion or quasi-molecular ion for aliphatic azides. The most general mode of decomposition,

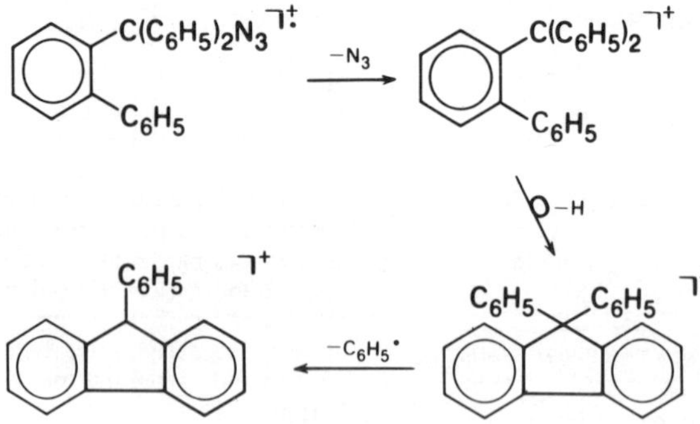

$C(C_6H_5)_2N_3$, C_6H_5 $\xrightarrow{-N_3}$ $C(C_6H_5)_2$, C_6H_5 $\xrightarrow{-H}$ C_6H_5, C_6H_5 $\xrightarrow{-C_6H_5^\cdot}$ C_6H_5

SCHEME 8

like benzoyl derivatives, involves loss of N_3. Therefore rearrangement, with the incorporation of nitrogen into the remaining structure, as is so common for aromatic derivatives, is missing, except where there are nearby aromatic substituents such as in the triphenylmethyl azides[6,30] or the azidoaldehydes and nitriles[28].

D. Pyrolysis, Photolysis and the Mass Spectra of Azides

As indicated previously, most of the early mass spectral studies of azides were related to studies of pyrolysis and photolysis, or the partial spectra are simply reported as part of the characterization. Thus the earliest report of Moriarty and Kirkien-Konasiewicz[6] described the decompositions of triarylmethyl azides on electron impact in order to compare it with the thermal[32] and photochemical[33a] decompositions. Both proceeded via loss of N_2, with subsequent aryl migration from carbon to nitrogen, to yield a Shiff's base, the thermal migration being selective while the photochemical is statistical. The $(M\text{-}N_3)^+$ ion is contrary to anything observed in the thermal or photochemical processes. Azidodiphenylacetic acid, with a less stable carbonium ion, however[6], did give the same mode of fragmentation as solution photolysis. The authors[6] were able to conclude that the similarity of the EI mass spectra of the azides they studied to thermolysis or photolysis was extremely dependent on the specific compound.

Aromatic azides have received more attention, starting with Crow and Wentrup[7], who found little similarity in the thermolysis and EI mass spectrum of phenylazide. Hedaya and coworkers[8] carried out thermolysis of phenyl azide in the mass spectrometer inlet system. Initially, they observed $M^{+\cdot}$ (m/z 119), $C_6H_5NH_2^{+\cdot}$ (m/z 93) formed by reaction of the nitrene produced with materials in the inlet system and $C_6H_5N^{+\cdot}$ (m/z 91). $M^{+\cdot}$ decreased as the sample was heated above 300°C, while m/z 91 decreased until 500°C and then increased to a maximum at 800°C, m/z 93 increasing continuously. The increase in m/z 90 above 500°C and the decrease of dimeric species suggested that the phenylnitrene was isomerizing, perhaps to cyanocyclopentadiene.

Ashby and Suschitzky[33b] produced tri(*o*-azidothioacetophenone) (**87a**) for further pyrolysis study. While its mass spectrum showed no parent ion, the base peak was due to loss of N_2 from the monomer, *o*-azidothioacetophenone; not surprising, since, at its melting point (147°C), it yielded monomer and then evolved N_2.

Abramovitch and Kyba[30] have studied the thermolysis of alkyl azides, and for the five tertiary azides observed equal amounts of C—N and N—N bond fission, with alkyl nitrenes shown to be intermediates by the isolation of intramolecular aromatic substitution products. Other azides did not show this behaviour. The substituted fluoroenes obtained in thermolysis of biphenyl-substituted methyl azides corresponds to species observed in their mass spectra.

Unlike his phenyl azide study[12], Abramovitch[30] found that the mass spectrum of **81**, 2-$C_6H_5C_6H_4C(CH_3)_2N_3$, was very dependent on the inlet system. The heated batch inlet (200°C) gave a spectrum quite different from the direct probe results at ambient temperature, the former being rationalized as a composite of the thermolysis products (azide **81** begins to decompose at 185°C). The thermal decomposition products are characterized by imine formation or loss of N_2, not inconsistent with weak peaks in the azide mass spectra.

Peek and collaborators[15] report the flash photolysis of 2,3-diazidonaphthalene, but since their report is limited to partial mass spectra, it is impossible to compare EI mass spectral rearrangement with the photolysis. Yabe and Honda[16] report loss of N_2 from 2,2′-diazidobiphenyls on photolysis, consistent with their partial mass spectral data, suggesting nitrogen incorporation via ring formation. They reported no loss of $N^{\cdot}$ in the mass spectra, contrary to a major photolysis product. Yabe[17] also studied

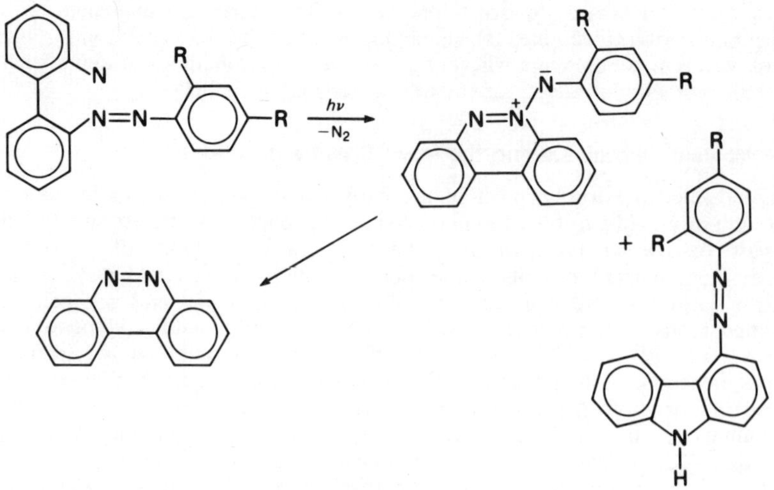

SCHEME 9

photolysis of 2,3-diazidonaphthalenes, where successive loss of N_2 corresponded to the mass spectrum.

Spagnolo and coworkers[18] has reported photolysis and thermolysis of some 2-azido-2′-arylazobiphenyls. Loss of N_2 resulted in formation of the first two product species shown in Scheme 9, and it is easy to see how the third also arises.

The mass spectra are consistent with this behaviour, m/z 180 corresponding to the third species. The absence of m/z 166 in any of the mass spectra suggests that the ion corresponding to the first species probably does not form, since it can be shown that this molecule eliminates the arylazo fragment to give m/z 166 as the base peak. However, although the fragmentations in the mass spectra of the second species, benzo[*c*]cinnoline *N*-arylimides correspond to species in the mass spectrum of the parent azide; they do not account for all ions, perhaps indicating that more than a single thermolysis is occurring.

Dickson and Dyall[21] studied *ortho* effects in the pyrolysis of aryl azides, and found N_2 elimination proceeded via an electrocyclic mechanism with formation of a new heterocyclic ring as shown below:

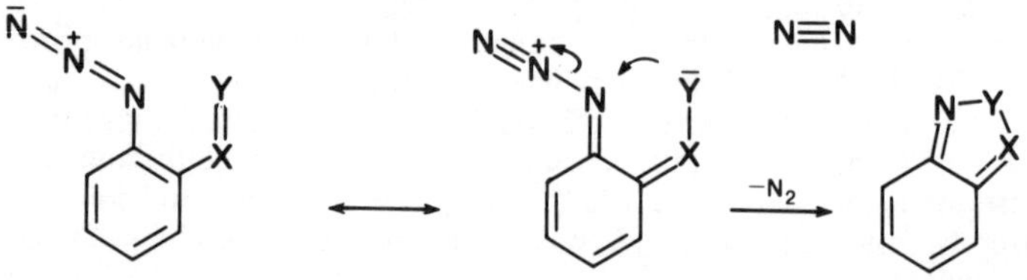

where ring formation is impossible, as for **51** the nitrene abstracted hydrogen to form an amine. The partial mass spectral data do not appear to show similar amine formation. There is insufficient mass spectral data to verify the analogous process in the electron impact mass spectra.

McDonald and Chowdhury[20] have generated a phenylnitrene anion radical, $C_6H_5N^{-\bullet}$ in a flowing afterglow apparatus and studied its ion molecule reactions with phenyl azide. This is the only report we have discovered in which any negative ion data for azide mass spectra are reported.

In conclusion, we can say that, although there are many similarities observed in the mass spectral fragmentation and thermolysis of photolysis of azides, one has to take care in drawing conclusions from any one set of data. The mass spectra must always be checked to eliminate the possibility of generation of the detected species by inlet or source pyrolysis. Possibly the best way to do this is to observe a metastable ion supporting its unimolecular formation from the molecular ion, or to carry out a variable temperature mass spectral study.

E. Steroid and Alkaloid Azides

Katzenellenbogen and coworkers[34] were the first to report mass spectral data on azidosteroids. 2-Azidoestrone and -estradiol, and the model compounds 3-azido- and 3,3′-diazidohexestrol, i.e. *meso*-3,4-*bis*(4′-hydroxyphenyl)hexane (Table 3), all gave parent ions as expected of an azido group attached to an aromatic ring. The base peak in the first three cases was $(M\text{-}26)^{+\bullet}$ while m/z 28, presumably N_2, formed the base peak in the latter three derivatives with the unusual loss of 26 still being observed for the 2-azido-estradiol. The species of mass 26 cannot be identified from their data, but would appear to be products of loss of CN as a result of major rearrangement. However, C_2H_2 cannot be eliminated as a contributing species since $(M\text{-}25)^+$ is also observed. The four steroids also show the expected $(M\text{-}28)^{+\bullet}$ ion. From the partial data presented, no loss of either N_2 or N_3 is seen for the hexestrols, the diazido compound **94** showing an ion corresponding either to symmetrical cleavage of the molecule, or to a very high intensity for the doubly charged ion. The monoazido derivative shows similar cleavage between C_3 and C_4, the non-azido side giving an ion at m/z 135 almost as intense as the base peak. The ion corresponding to the azido side of the molecule was not reported, but there is an ion at m/z 150 which could correspond to loss of 26 mass units from this species, analogous to the observation for the four steroids. No loss of N_3 was observed from any of these compounds.

Astier and coworkers[35,36] report partial mass spectral data on some pregnane, **95**, **96**, **98** and androstane derivatives **97** and candenolides **99–102**. Weak molecular ions are observed in most cases, with $(M\text{-}N_2)^{+\bullet}$ and $(M\text{-}N_3^+)$ in most cases being a common factor; however, in cases where $(M\text{-}42)^+$ is observed and in some where it is not, an ion corresponding to $(M\text{-}43)^+$ is also reported. This could be due to loss of HN_3 or alternately to loss of the common C_3H_7 hydrocarbon fragment, or C_2H_3O oxygen containing fragments, or, since we have an azide, the possibility of 43 being CH_3N_2 exists. high resolution studies would be required to resolve these species. This $(M\text{-}43)^+$ species is the base peak in some cases. For the diacetoxypregnane **98**, loss of OH or (44 + 17) are also observed from the parent ion.

Barone and collaborators[28] report data for an azidopregnane aldehyde (**103**) and an azidopregnane nitrile (**104**), both showing good intensities ($\simeq$10% and 50% respectively) for the molecular ions. Neither show loss of N_2 or N_3, the fragmentation being dominated by the steroid. Loss of 55 and 57 atomic mass units (a.m.u.) occur for the former compound with 14, 31 and 54 a.m.u. losses being observed for the latter. Again, without high resolution or metastable ion data it is impossible to assign these losses definitely.

Schweng and Zbiral[31] report partial spectral data on 2-azido-2-cholestene. They did not observe a molecular ion, but the $(M\text{-}28)^+$ ion was confirmed by exact mass measurement to correspond to loss of N_2. Similarly, loss of 43 from the parent ion was assigned to loss of HN_3.

Cambie's group[37,38] report a series of iodoazido steroids (**106–109**) and related

TABLE 3. Azido steroids, alkaloids and related compounds

No.	Compound	Completeness of data	Reference
89	4-Azidoestrone	Uninterpreted spectrum	34
90	4-Azidoestradiol	Uninterpreted spectrum	34
91	3-Azidohexestrol	Uninterpreted spectrum	34
92	2-Azidoestrone	Uninterpreted spectrum	34
93	2-Azidoestradiol	Uninterpreted spectrum	34
94	3,3′-Diazidohexestrol	Uninterpreted spectrum	34
95	14-Azido(5α,14β,17α)pregnane	Partial data	35
96	20*S*,3β,20-diacetoxy-14-azido(5α,14β,17α)-pregnane	Partial data	35
97	14-Azido(5β,14β)androstane	Partial data	35
98	20*S*,3β,20-diacetoxy-14-azido-5β-pregnane	Partial data	35
99	3β-Acetoxy-14-azido(5β,14β)canden-20(20)olide	Partial data	36
100	3β-Acetoxy-14-azido(5β,14β,17α)canden-20(22)-olide	Partial data	36
101	3β-Acetoxy-14-azido(5α,14β,17α)canden-20(22)-olide	Partial data	36
102	3β,19-Diacetoxy-5-hydroxy-14-azido(5β,14β,17α)-canden-20(22)olide	Partial data	36
103	20-Azido-6β-methoxy-3,5α-cyclopregnane-20-carboxaldehyde	Partial data	28
104	20-Azido-3β-hydroxy-5-pregnane-20-carbonitrile	Partial data	28
105	2-Azido-2-cholestene	Partial data	31
106	15β,17-Diazido-16α-iodo-13β-kuarang	Partial data	37
107	2β-Azido-3α-iodo-3-methyl-5α-androstane	Partial data	38
108	3β-Azido-2α-iodo-3-methyl-5α-androstane	Partial data	38
109	2α-(1-Azido-1-iodoethyl)-5α-A-norandrostane	Partial data	38
110	2-(1-Azidoethylidene)-5α-A-norandrostane	Partial data	38
111	2β-Azido-3α-acetyl-3β-methyl-5α-androstane	Partial data	38
112	*r*-1-Azido-*trans*-2-iodo-*trans*-3-*t*-butylcyclohexane	Partial data	39
113	*r*-1-Azido-*trans*-2-iodo-*cis*-3-*t*-butylcyclohexane	Partial data	39
114	*r*-1-Azido-*trans*-2-iodo-*trans*-3-methoxycyclohexane	Partial data	39
115	*r*-1-Azido-*trans*-2-iodo-*cis*-3-methoxycyclohexane	Partial data	39
116	6-Deoxy-6-azidodihydroisomorphine	Interpreted data	40
117	6-Deoxy-6-azidodihydroisocodeine	Interpreted data	40
118	3-*O*-Ethyl-6-deoxy-6-azidodehydroisomorphine	Interpreted data	40
119	6-Deoxy-6-azido-14-hydroxydehydroisomorphine	Interpreted data	40
120	6-Deoxy-6-azido-14-hydroxydihydroisocodeine	Interpreted data	40
121	6-Deoxy-8-azidopseudomorphine	Interpreted data	40
122	6-Deoxy-8-azidopseudocodeine	Interpreted data	40
123	6-Deoxy-8-azido-14-hydroxypseudocodeine	Interpreted data	40
124	6-Deoxy-6-azido-14-hydroxyisocodeine	Interpreted data	40

cyclohexanes[39] (**112–115**). Weak parent ions are usually observed, but the most interesting species appear to involve the loss of IN_2, IN_3, and IN_4 species. Those losses are shown in Table 4. Without metastable ion evidence, we cannot know if these are one-step, e.g. IN_4, or multi-step, e.g. $IN_2 + N_2$ or $N_2 + N_2 + I$, losses.

The only other observation related to these iodonitrogen compounds is the loss of Cl and N_2 reported by Schweng and Zbiral[31] for 1-azido-2-chlorocyclopentane (**87**) and -cycloheptane (**88**). These halo compounds warrant proper metastable ion study.

TABLE 4. Interesting losses from iodo azides

No.	Losses from molecular ion	Reference
106	IN_2, IN_3, IN_4	37
107	N_2, N_3, IN_2	38
108	N, N_3, (CH_3N_3), IN_2	38
109	HN_2, N_3, IN_2	38
112	N_3, $CH(CH_3)_3$, I, IN_2	39
113	IN_2	39
114	I, HI, (I + HN_3)	39
115	HI, (HI + HN_3)	39

A comprehensive mass spectrometic study of nine azidomorphine derivatives (**116–124**) has been carried out by Tamás and coworkers[40], who reported unusual fragmentations which differ from those of either morphine or azides. All nine derivatives gave molecular ions of good intensity, although weaker than the molecular ion in the corresponding morphine alkaloid. Peaks characteristic of the morphine skeleton such as $(M\text{-}57)^+$ are observed, i.e. loss of C_3H_7N, along with fragmentation of the amine bridge (*m/z* 70, 59, 58) and a peak at *m/z* 115. Most of the abundant fragments are derived from azido group fragmentation. Compounds **116–120** fragment primarily by loss of N_2, and **121–124** by loss of N_3. Thus we have characteristics similar to both aromatic and aliphatic azides. Actually, with respect to the base peak,

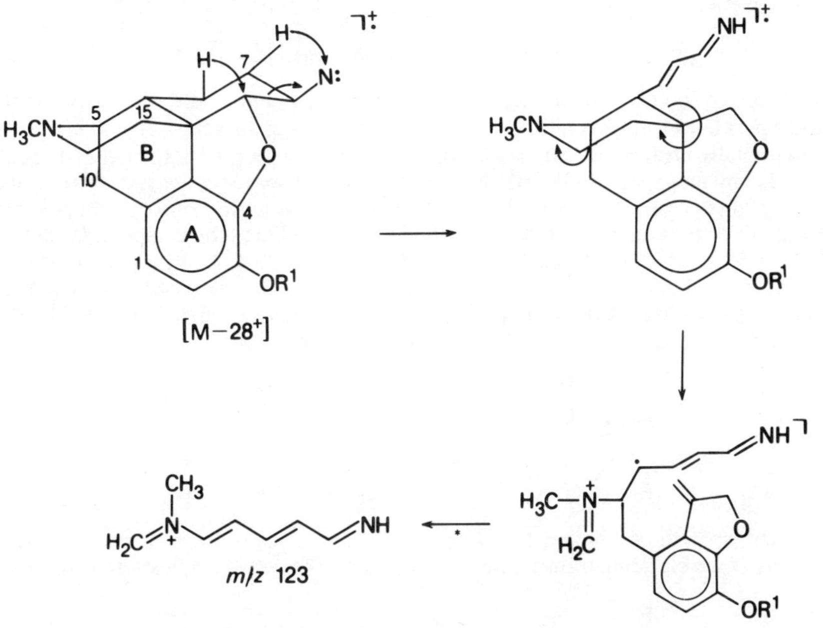

(116) $R^1 = H$ (117) $R^1 = CH_3$ (118) $R^1 = C_2H_5$

SCHEME 10. Reproduced from Tamás *et al.*[40] by permission of Heyden & Sons Ltd.

$(M\text{-}N_2)^{+\cdot}$ ions are of similar intensity for all nine compounds, but $(M\text{-}N_3)^+$ varies between 3% and 8% for the first five, not much different from the $(M\text{-}N_2)^{+\cdot}$ intensity, but it becomes the base peak for the last four compounds. The first four compounds also show $(M\text{-}57)^+$ ion, i.e. $(M\text{-}C_3H_7N)^+$, and from this ion the formation of $(M\text{-}128)^+$ and $(M\text{-}141)^+$ ions, the former requiring a hydrogenation at C-14.

For compounds **116–118** the base peak is unexpectedly m/z 123, (shown by exact mass measure to be $C_7H_{11}N_2^+$) which does not appear in the spectra of the other derivatives. Metastable ion scan techniques show that m/z 123 arises for $(M\text{-}N_2)^{+\cdot}$ as indicated by Tamás[40] and shown in Scheme 10. Cleavage of at least four bonds is required for any reasonable generation of m/z 123 from $(M\text{-}N_2)^{+\cdot}$. The presence of the 14-hydroxy group in **119** and **120** causes a difference, $(M\text{-}112)^+$ ion being formed from $(M\text{-}28)^+$, this involving cleavage of C5—C6 and C13—C14 bonds along with C9—C14, i.e. a bond β to both the OH group and the bridging nitrogen atom.

Compounds **121–124** have a double bond in ring C, and show simpler spectra with fewer peaks, the base peak being due to $(M\text{-}N_3)^+$, as mentioned earlier. Tamás[40] attributes this to an allylic effect of the double bond, which weakens the polar C—N_3 bond and stabilizes the $(M\text{-}N_3)^+$ ion. This is consistent with the thermally induced allylic rearrangement in solution[41].

F. Heterocyclic Azides

Azide–tetrazole tautomerism in tetrazoloazines has been investigated by Wentrup[42] who, from mass spectra, showed that for the labile tetrazoles, azide tautomerism followed by nitrene formation is the first step in the gas-phase pyrolysis. The same basic fragmentation,

$$M^{+\cdot} \xrightarrow{m^*} (M\text{-}N_2)^{+\cdot} \xrightarrow{m^*} (M\text{-}N_2\text{-}HCN)^{+\cdot},$$

is common to both tetrazoloazines and azidoazines. For both thermally stable tetrazolopyridines and -pyrimidines the $(M\text{-}N_2)^{+\cdot}$ ion was usually the base peak, but for those thermally unstable or existing as azides at room temperature, the $(M\text{-}N_2)^{+\cdot}$ peak was weak, and an apparent $(M\text{-}26)^+$ ion was observed. These ions are shown to be due to a thermal process $((M\text{-}N_2) + 2H)^+$ in the mass spectrometer source, only possible for the azide tautomer. This has been verified by ^{15}N labelling. The compounds studied (**125–137**) are listed in Table 5. It is interesting to note that, even for these azido-substituted tetrazolopyrimidines, there is no loss of N_3 observed from the parent ion, i.e. they are characteristic of aromatic azides. Fragmentation is dominated by the tetrazolo part of the molecule tautomerizing, as shown below:

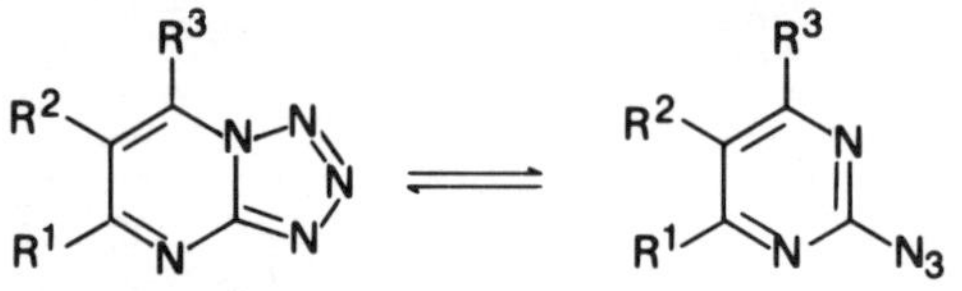

R3 R3
R2 N–N N R2 N
R1 N N R1 N N3

The mass spectra of **130** and **131** were studied under several sets of mass spectral conditions (variable temperature and volatage) and the results indicate that thermal

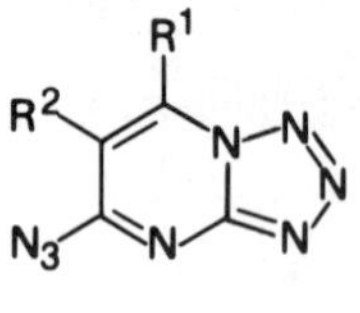

R1
R2 N–N N
N3 N N

(**130**) $R^1 = CH_3$, $R^2 = H$
(**131**) $R^1 = H$, $R^2 = CH_3$

TABLE 5. Heterocyclic azides

No.	Compound	Completeness of data	Reference
125	Tetrazolo[1.5.a]pyrimidine	Data and interpretation	42
126	5,7-Dimethyltetrazolo[1.5.a]pyrimidine	Data and interpretation	42
127	5-Chloro-7-methyltetrazolo[1.5.a]pyrimidine	Data and interpretation	42
128	5-Mcthoxy-7-methyltetrazolo[1.5.a]pyrimidine	Data and interpretation	42
129	5-Methoxy-7-methylmercaptotetrazolo[1.5.a]-pyrimidine	Data and interpretation	42
130	5-Azido-7-methyltetrazolo[1.5.a]pyrimidine	Data and interpretation	42
131	5-Azido-6-methyltetrazolo[1.5.a]pyrimidine	Data and interpretation	42
132	5,7-Dimethyltetrazolo[1.5.c]pyrimidine	Partial data	42
133	5,7,8-Trimethyltetrazolo[1.5.c]pyrimidine	Partial data	42
134	4-Azido-2,6-dimethoxypyrimidine	Partial data	42
135	7-Methyl-5-methylmercaptotetrazolo[1.5.c]-pyrimidine	Partial data	42
136	Tetrazolo[1.5.a]pyrazine	Raw data	42
137	Tetrazolo[1.5.6]pyridazine	Raw data	42
138	4-Azido-2,3,5,6-tetrafluoropyridine	Partial data	43
139	4-Azido-3-chloro-2,5,6-trifluoropyridine	Partial data	43
140	4-Azido-3,5-dichloro-2,6-difluoropyridine	Partial data	43
141	*C*-Methyl-*C*-azido-1,2,4-triazole	Spectra and discussion	44
142	*C*-Methyl-2*N*-methyl-*C*-azido-1,2,4-triazole	Spectra and discussion	44
143	*C*-Dimethylamino-*C*-azido-1,2,4-triazole	Spectra and discussion	44
144	*C*-Methoxy-*C*-azido-1,2,4-triazole	Spectra and discussion	44
145	*C*-Methylthio-*C*-azido-1,2,4-triazole	Spectra and discussion	44
146	*C*-Methylthio-*N*-methyl-*C*-azido-1,2,4-triazole	Spectra and discussion	44
147	8-Azido-5-methyldibenz[*c*,*e*]azocine	Partial data	45

reactions are reponsible for the formation of the ions $(M\text{-}N_2 + 2)^{+\bullet}$, and $(M\text{-}2N_2 + 2)^{+\bullet}$.

Banks and Sparkes[43] have reported on the mass spectra of tetrahalopyridines (Table 5, **138–140**). Intense molecular ions are observed. Loss of N_2 but not N_3 is observed, characteristic of aromatic azides. The base peak for the azidotetrafluoropyridine corresponds to C_4F_3N formed by loss of FCN from the $(M\text{-}N_2)^{+\bullet}$ ion. The base peak of the dichloro derivative derives from CIF loss from the $(M\text{-}N_2)^{+\bullet}$ ion.

C-Azido-1,2,4-triazoles have been studied in detail. Heitke and McCarty[44] used low resolution ^2H isotope labelling and 'normal' metastable ion techniques (Table 5, **141–146**). They observed only weak peaks for loss of N_2 or N_3 from the parent ion even though a metastable ion is observed for $(M\text{-}N_2)^+$ formation. Rather, a strong parent ion is seen leading to the base peak, due to metastable ion-supported loss of an RN_4 species shown in Scheme 11.

Although it is not discussed by Heitke[44], there is evidence for thermal reactions similar to those described by Wentrup[42]. The presence of (M-26) ions, which were proposed to arise from H-transfer, i.e. $(M\text{-}N_2 + 2H)$, suggests that such effects deserve careful consideration when proposals are made for fragmentation pathways or, in the simplest case, when fingerprinting spectra.

The partial mass spectra of 8-azido-5-methyldibenz[*c*,*e*]azocine has also been reported[45]. The large peak is due to loss of nitrogen from the molecular ion with no N_3 loss observed.

In summary, the mass spectral behaviour of heterocyclic aliphatic and aromatic azides under electron impact shows no significant deviation from patterns observed for

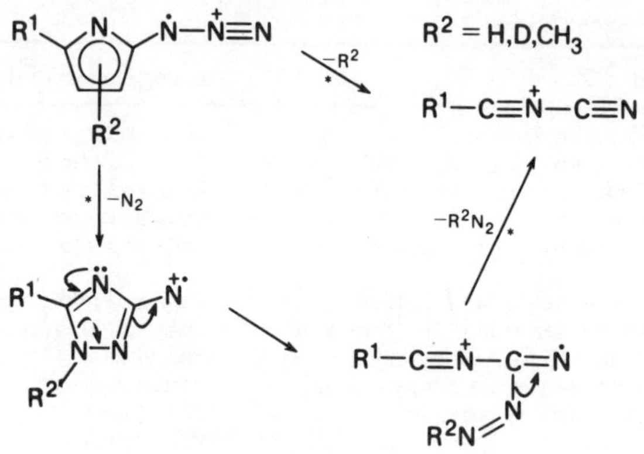

SCHEME 11

the organic compounds. The examples available are comprised entirely of N-heterocycles. Of these, the pyridine[43] and tetrazole[42] azides exhibit the 'normal' pattern of fragmentation which is typified by a reasonably strong molecular ion and a base peak resulting from expulsion of a nitrogen molecule by the parent ion. Only the triazole azides studied by Heitke and McCarty[44] exhibit 'abnormal' behaviour. In the latter case, the apparent driving force is a stable nitrillium ion formed via direct loss of HN_4 by the parent compound in each case studied.

G. Sulphonyl Azides, Phosphine Azides and Inorganic Systems

The mass spectra of the six alkyl and aryl sulphonyl azides, described by Campbell and Dunn[46] (Table 6, **148–153**), provide a direct contrast to the normal carbon–azide patterns.

Departing from the expected pattern of N_2 loss, all of these compounds show a distinct preference for N_3 loss, and in addition, for the isopropyl derivative (**153**), we have a significant *ortho* effect with the base peak corresponding to CH_3N_2 loss. The *ortho* effect has been noted for many substituted aromatics, in particular by Smith and coworkers[47], who have studied a wide variety of isopropyl derivatives which showed a preferred *ortho* effect with the branched chain substituents. High resolution confirms this loss as CH_3N_2, not HN_3 and leads us to view with some uncertainty the nature of supposed HN_3 losses reported for other azide derivatives for which exact mass measurements are not reported.

Abramovitch and Holcomb[48] report very few details of the mass spectra of some ferrocene sulphonyl azides and it is likely that these decompose by simple N_2/N_3 losses. Loss of 26 from the molecular ion was observed as the base peak **155**, which could be due either to hydrogen abstraction in the ion source by the $(M\text{-}N_2)^{+\cdot}$ ion to give amine ions as noted previously (in this case the disulphamyl species) or loss of C_2H_2, which has been noted for some aryl azides[12] and is also common in metal cyclopentadienyl[49]. Compound **156** showed the same behaviour. The azepine **157** had as its base peak *m/e* 92 corresponding to $C_6H_6N^+$, the azatropylium ion.

Reports of mass spectra of compounds involving $P\text{—}N_3$ groups are limited to the works of Schröder and Müller[50,51]. The compounds studied were of the dialkylphosphonyl class and consisted of R_2PXN_3, where R = CH_3, CD_3, C_2H_5 and X = O, S, Se (**158–165**) and three diphenyl derivatives (**166–168**). The compounds are described as stable, monomeric and non-explosive.

TABLE 6. Sulphonyl, phosphorus and inorganic azides

No.	Compound	Completeness of data	Reference
148	$CH_3SO_2N_3$	Partial data	46
149	$C_6H_5SO_2N_3$	Partial data	46
150	4-$CH_3C_6H_4SO_2N_3$	Partial data	46
151	4-$O_2NC_6H_4SO_2N_3$	Partial data	46
152	2,4,6-$(CH_3)_3C_6H_2SO_2N_3$	Partial data	46
153	2,4,6-(Isopropyl)$_3C_6H_2SO_2N_3$	Partial data	46
154	Ferrocene 1,1′-disulphonyl azide	Partial data	48
155	1′-Sulphonylferrocene sulphonyl azide	Partial data	48
156	1′-(*N*-Phenylsulphonyl)ferrocene sulphonyl azide	Partial data	48
157	*N*-(1′-Sulphonylazido)ferrocene sulphonyl azepine	Partial data	49
158	$(CH_3)_2P(O)N_3$	Spectrum and interpretation	50
159	$(C_2H_5)_2P(O)N_3$	Spectrum and interpretation	50
160	$(CH_3)_2P(S)N_3$	Spectrum and interpretation	50
161	$(C_2H_5)_2P(S)N_3$	Spectrum and interpretation	50
162	$(CD_3)_2P(O)N_3$	Spectrum and interpretation	51
163	$(CD_3)_2P(S)N_3$	Spectrum and interpretation	51
164	$(CH_3)_2P(Se)N_3$	Spectrum and interpretation	51
165	$(C_2H_5)_2P(Se)N_3$	Spectrum and interpretation	51
166	$(C_6H_5)_2P(O)N_3$	Spectrum and interpretation	51
167	$(C_6H_5)_2P(S)N_3$	Spectrum and interpretation	51
168	$(C_6H_5)_2P(Se)N_3$	Spectrum and interpretation	51
169	$ClSO_2N_3$	Partial spectrum and interpretation	54
170	$H_2NSO_2N_3$	Partial spectrum and interpretation	54
171	*n*-$BuNHSO_2N_3$	Partial spectrum and interpretation	54
172	$F_2P(:S)N_3$	Spectrum	55
173	WF_5N_3	Interpretation	56

The mass spectra of this series appears to be dominated by two major considerations. The first is the relative strength of the P—X bond, an important factor noted in the mass spectra of R_2PXY compounds[52] where Y is halide (F, Cl, Br). The second consideration is the 'normal' azide tendency for fragmentation to be dominated by initial loss of nitrogen. The dimethyl compounds $(CH_3)_2PXN_3$ (X = O, S, Se) demonstrate these effects dramatically. The oxide spectrum gives no evidence for direct loss of molecular nitrogen, preferentially breaking down via N_3 and HN_3 loss (m/z 77 $(M-N_3)^+$ and m/z 43 (HN_3^+) are base peaks). The mass spectra of the sulphide and selenide both indicate that N_3 loss is still common, yet N_2 loss also becomes significant, while HN_3^+ is completely absent from the spectra of both compounds. This behaviour, which is confirmed in the spectra of the d_3-methyl compounds, supports the premise concerning the importance of the P—X bond strength. Thus, unlike the sulphide and selenide, the oxide gives rise to very few fragment ions.

Similar pattern is observed with the diethyl compounds where the preference for N_2 loss is X = Se > X = S > X = O and the HN_3^+ ion is large for the oxide and very weak in the sulphide spectrum. A rather anomalous situation arises for the selenide where, despite the presence of a reasonable molecular ion, there are no ions corresponding to N_2 or N_3 losses and no other fragments resulting from small mass losses. Of notable fact, however, is that, unlike the dimethyl derivative, HN_3^+ has become the base peak for this compound. It appears that direct HN_3 loss is the major fragmentation mode for this particular compound, with secondary pathways arising out of P—Se cleavage.

In summary, when other factors such as P—O bond strength are taken into account, the fragmentation of azides of this type differs little from the expected modes. The oxidation state of the phosphorus atom in these compounds could be an important influence on the P—N_3 interaction and the procurement of data related to the corresponding phosphinic azides $R_2P(N_3)$ would be a useful exercise.

For the aromatic S and Se derivatives **167** and **168** there is a greater stability of the $(C_6H_5)P^+$ and $(C_6H_4)_2P^+$ ions; the latter of which may well have rearranged to a dibenzphosphole species[9,52]. As usual, such species are not seen for the P=O compounds.

The selenides appear to show loss of SeN_2, though there is little data to indicate whether or not this is a one-step process.

Although a large number of stable volatile main group organometallic azides have been prepared[53], most of this work was done in the mid 1960s when few organometallic chemists had access to mass spectrometers. As a result, there is little mass spectral data available on these azides.

Shozda and Vernon[54] report the spectra of sulphuryl azide chloride and two sulphamoyl azides (**169–171**). These give parent ions, the former showing loss of chloride or azide in almost equal amounts, while for the sulphamoyl derivatives fragmentation occurs between the SO_2 and N_3 moieties. N^+ ions are observed.

Colburn and coworkers[55] report on the spectrum of thiophosphonyldifluoride azide, $F_2P(:S)N_3$. A large parent ion is observed along with loss of both N_2 and N_3. There is also evidence for SF loss with the azide remaining attached to the phosphorus. SN^+ is also formed by rearrangement.

The most recent work is that of Fawcett and collaborators[56] who reported the partial specta of azidopentafluorotungsten(VI). The molecular ion may lose N_3 to give the base peak, which then shows successive fluorine losses or, alternatively, loss of a fluorine atom (67% of base peak), which then shows loss of N_2. The WF_4N^+ ion then fragments either by loss of nitrogen or by successive loss of fluorine down to a WN^+ species. This spectrum had to be obtained at a source temperature of 58°C to prevent decomposition. Thus, even somewhat unstable azides give reasonable parent ions and spectra.

H. Conclusions

Thus, we can conclude that the mass spectra of azides are not particularly mysterious. Of the 173 compounds listed, most give parent ions and there are no reports in the literature of particular difficulty or hazard in handling these azides in the mass spectrometer.

III. THE MASS SPECTRA OF HALIDES

A. Introduction

Recently the emphasis on studies of the mass spectra of various halides has been on the analytical use of mass spectrometry. Researchers have taken advantage of the potentially high specificity and sensitivity of high resolution mass spectrometry, high resolution gas chromotagraphy/mass spectrometry or collisional activation mass spectrometry in order to detect and quantify organic halides, both in the environment and in biological systems. Other areas in which advances have occurred in the mass spectrometry of halides have involved the postulation of halonium ion structures to explain observed intensities in halide mass spectra and in the study of halogen migrations in the mass spectra of organometallic compounds. Here, halide migration occurs from carbon to the central metal, some observed and unobserved transitions being explained in terms of hard/soft acid–base theories.

B. Environmental Applications – Dioxins

A very extensive literature on the mass spectrometric analysis of halogenated compounds has developed over the past few years. Papers are to be routinely found not only in mass spectrometry and analytical chemistry journals, but also in agricultural, food, environmental, toxicological, biological and medical journals, reflecting the concern about man-made contaminants in the environment and their effect on human health. Using the tetrachlorodibenzo-*p*-dioxins (TCDD) and, in particular, the most hazardous isomer, 2,3,7,8-TCDD, as an example, and drawing only from a small part of the chemical literature, we will attempt to illustrate the scope of the problems and the wide range of mass spectrometric solutions proposed.

Firestone's group[57] were the first to detect chlorodioxin in commercial chlorophenols. In early studies, high resolution and time-averaging techniques preceded by tedious separation techniques gave 10^{-12} g detection limits and the ability to measure TCDD at parts per trillion levels in fish[58]. Early attempts to improve the sensitivity of low resolution gas chromatograph/mass spectrometry for TCDD involved use of negative chemical ionization mass spectrometry[59,60]. However, one result of the low resolution studies was the demonstration of interferences in 2,4,5-trichlorophenols (the precursors to TCDD) from other chlorinated species with peaks at the same nominal masses and with similar retention times to TCDDs[61].

During the clean-up and environmental monitoring after the industrial accident at Seveso, Italy, in 1976[62], low resolution gas chromatography together with multiple ion detection mass spectrometry were the principal techniques used to monitor 2,3,7,8-TCDD down to the part per trillion level. Typically, m/z 320, 322 and 324 were monitored, the 77 : 100 : 49 intensity ratio being taken as evidence for the dioxin.

However, as noted above, other interferences are possible. Lamparski and coworkers[63], using a low resolution ($M/\Delta M = 400$) quadrapole mass spectrometer, were able to detect TCDD in fish at levels of 10–100 parts per trillion. They also note that, at higher resolution, ($M/\Delta M = 1000$) additional specificity was obtained with lower limits of detection, since, even at 1000 resolution, much better separation is obtained from hydrocarbon background. In this day of computerized mass spectrometers, with digital plots of nominal masses rather than oscillographic recorder traces, fewer scientists realize that, even at low resolution, polychlorinated or polybrominated derivatives, with their negative mass defect, are often clearly separated from hydrocarbon background peaks which have a positive mass defect. It was also necessary to prepare and identify all 22 TCDD isomers[64] as part of the analytical methods development.

Hummel and Shadoff[65] severely criticized di Domenico and coworkers[62] for giving the impression that low resolution gas chromatography–low resolution mass spectrometry (LRGC–LRMS) can specifically detect and quantify TCDD without a prior specific cleanup. Although in their study LRGC–LRMS may have been specific, it was noted[65] that when this specific cleanup is lacking, high resolution ($M/\Delta M = 9000$) mass spectrometry must be used. Hummel and Shadoff conclude from the EPA study of beef fat samples analysed for TCDD that LRGC–LRMS is a satisfactory screening procedure for TCDD in beef fat above 20 parts per trillion, though many false positive results are obtained. As a result, LRGC–HRMS must be used to eliminate the false positives; but even in this case several false positive results were still obtained.

Phillipson and Puma[66] have reported the interference of chlorinated methoxybiphenyls with the GC/MS determination of TCDD at low resolution. These methoxy derivatives must be removed by the clean-up procedure, or they must be separated by the GC from the TCDD, or the mass spectrometer must be able to resolve the overlapping species. For the tetrachloro compounds a resolution of 8800 is needed. Gas chromatographic separations do not appear to be possible, since there are 198 possible tetrachloromethoxybiphenyl isomers, with some having retention times almost identical to 2,3,7,8-TCDD. They suggest monitoring not only $M^{+\cdot}$ but also $(M\text{-}COCl)^+$ to be certain of dioxin identification at low resolution. Despite the above cautions, however, low resolution data continue to be used for dioxins in human milk[5]. In particular[67], another recent study uses capillary GC combined with atmospheric pressure, negative chemical ionization mass spectrometry for TCDD determination without polychlorinated biphenyl (PCB) interference[68].

The use of high resolution mass spectrometry (HRMS), together with high resolution GC columns, and multiple peak monitoring is by far the most sensitive specific test for TCDDs. The high resolution work also permits the use of an isotopically labelled internal standard without the complication of differential interference for the standard and unknown[69]. High resolution ($M/\Delta M$ = 10 000) single ion monitoring has proven to be exceedingly sensitive[70]. A recent study[71] showed a detection limit of about 5 fg for 2,3,7,8-TCDD. Harless and collaborators[72] use capillary column GC–HRMS to quantify 2,3,7,8-TCDD in human milk and in fish at the parts per trillion level, and down to the 0.03 parts per trillion level for water and sediment samples.

Chess and Gross[4] have used metastable decomposition monitoring of the loss of COCl from TCDD. Selectivity as good as GC–HRMS was noted, but the accuracy was poorer. This approach is basically what is becoming known as mass spectrometry–mass spectrometry (MS/MS). MS/MS is carried out both on sector instruments, as above[4], or with dual or triple tandem low resolution quadrupoles[73], where the second analyser is used to monitor a specific fragment or fragments from a parent or daughter ion selected by the first mass analyser. These granddaughter ions can be uniquely characteristic of a molecule. For example, using the TAGA™ 6000 MS/MS system, the parent ion of TCDD (m/z 322) is tuned on the first analyser, collision occurs with argon in a second quadrupole region, and the third quadrupole mass spectrometer is used for selective ion monitoring of m/z 259 and 196, i.e. $(M\text{-}COCl)^+ + (M\text{-}2COCl)^+$. They were able to detect 2,3,7,8-TCDD in fish at the 0.07 parts per trillion level.

Thus, from the examples selected, we see the rapid progress in improving the selectivity and detection limits for the analysis of 2,3,7,8-TCDD from a wide variety of materials. There is little new work on the interpretation of the mass spectra of the chlorinated dioxins, but rather, given the known spectra, instrumentation and mass spectrometric techniques are being pushed to and beyond their limits in order to meet all the needs of society in monitoring an extremely toxic material.

C. Halonium Ions

The participation of halides in electron impact-induced fragmentation processes is a well documented fact as evidenced by Refs 74–78. This influence is evident for alkyl, alkenyl and aromatic systems and proceeds, more or less, by the classical McLafferty mechanism:

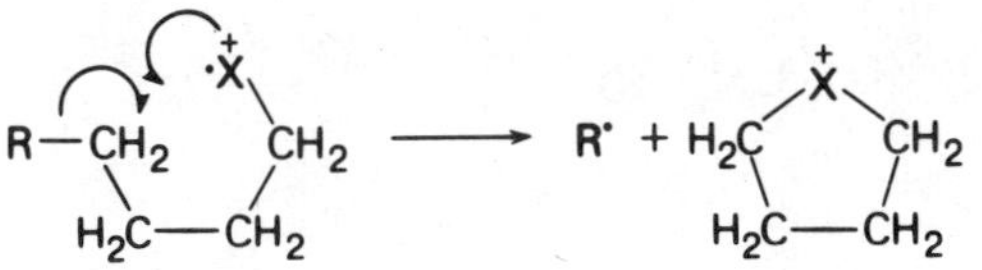

Stahl and Gäumann[74] reported a thorough investigation of the EI mass spectum of 1-heptyl iodide using ^{13}C-labelled compounds to derive fragmentation modes. It is clear from their work that iodine-containing ions are not the major driving force towards rearranged products. They report that $(CH_2)_mI^+$ (m = 1–4) species carry only 2% of the total ion current and play a much more subtle role than was observed in the case of the chloride[79]. Van de Sande and McLafferty[79] have studied the tetramethylenechloronium ion formed from both 1-chloroheptane and 1-chlorohexane. The similarity of the collisional activation spectra of this ion when dideuterated at the 1,1- or 4,4-positions is taken as evidence for the cyclic structure. An open chain would not have the necessary symmetry.

Arseniyadis and coworkers described the mass spectra of some halogenated alkenes[75] and alkynes[76]. In both cases, the compounds exhibited a concerted rearrangement involving the halide and the unsaturated end of the molecule.

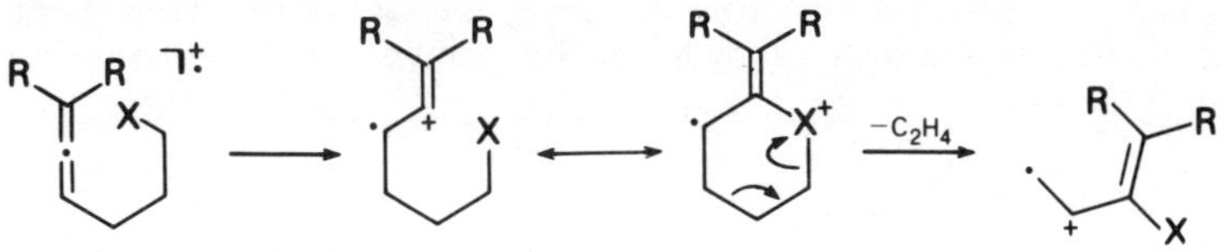

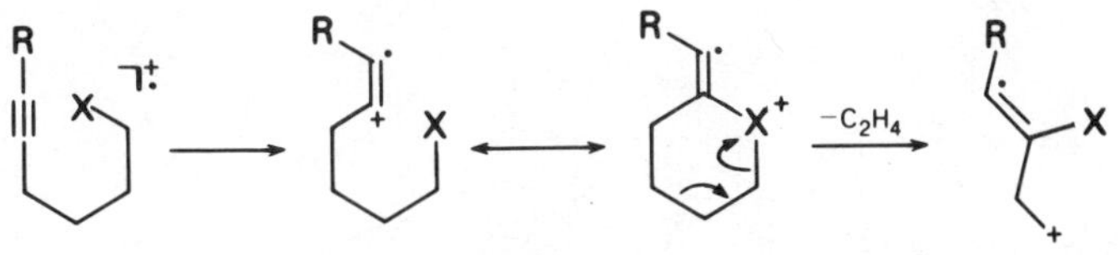

Being more concerned with an acetylene–allene isomerization, the authors do not offer a comparison of the effect of halide substituent on the mode of fragmentation observed. Other than an absence of a molecular ion for the bromine derivatives, the spectra of these compounds are virtually identical, suggesting that little difference exists in the relative abilities of the Cl and Br towards halonium ion formation under these conditions. Though such ambiguity is not consistent with this kind of mechanism, the role of the unsaturated function may be contributing to the stability of the resultant halonium species.

In studies of some phenyl propionyl halides, Hittenhausen-Gelderblom and coworkers[77] found the fragmentation processes of the compounds to be dominated by rearrangements involving the acidic α-carbonyl protons. This acidity is also a function of the electronegativity of the halogen atoms. The major rearrangement involving the halogen atoms results in formation of the m/z 55 ion $C_3H_3O^{+\cdot}$. This proceeds as shown below:

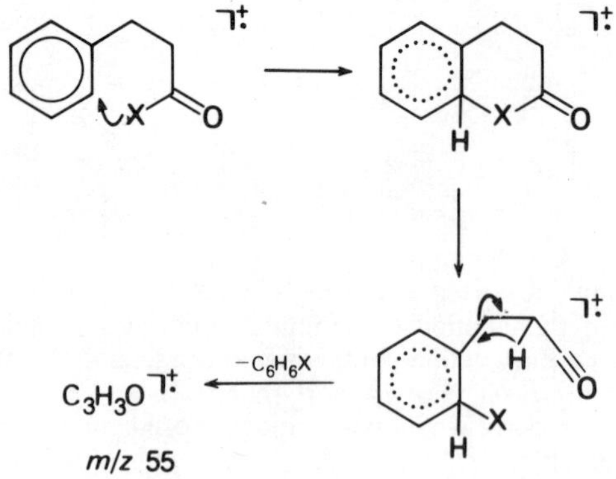

$C_3H_3O^{+\cdot}$ ← $-C_6H_6X$

m/z 55

The intensity of the *m/z* 55 ion reflects the relative ability of the halogen to form a halonium intermediate, i.e. X = Br > Cl > F.

Levy and Oswald[78] have observed the behaviour of a number of PCBs under electron impact. They were able to explain the lack of intense $(M\text{-}Cl)^+$ ions in the spectra of most PCBs by the fact that the pattern of *ortho*-chlorine substitution determined the level of ring stability achievable under conditions of electron impact. They proceeded to show that the most intense $(M\text{-}Cl)^+$ ions were obtained when the PCB contained three *ortho* chloro subitituents. Because of the steric interaction of these *ortho* chlorines, a significantly better configuration could be attained if a single chlorine atom was lost:

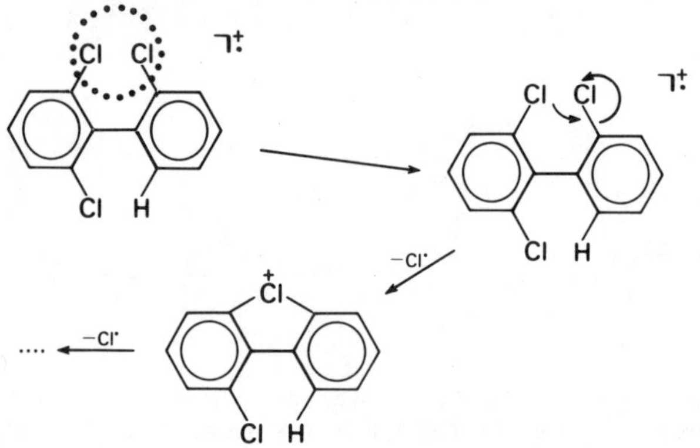

$-Cl^\cdot$

.... ← $-Cl^\cdot$

As shown in the figure, the authors suggested that displacement of the chlorine occurs via a chloronium intermediate. These results are also verified by Shushan and Boyd[80].

Though this appears to be an excellent descriptive mechanism for the variety of compounds described here, it is nevertheless surprising that a far greater number of cases have not been considered along these lines. It may be that the bulk of recent halide data has not made the most of the techniques available to study mechanistic pathways.

D. Halide Migrations and Hard/Soft Acid–Base Theory

The transfer of a halogen atom from an organic substituent to the central atom is a phenomenon which has been thoroughly studied in our laboratory with respect to many main group[81–92] and transition metal elements[81,90,92].

In general, the most interesting mass spectral behaviour along these lines occurs with the perfluoroaromatic systems. The figure below describes two sequences which are fairly typical within this class of compounds when subjected to electron-impact mass spectrometry:

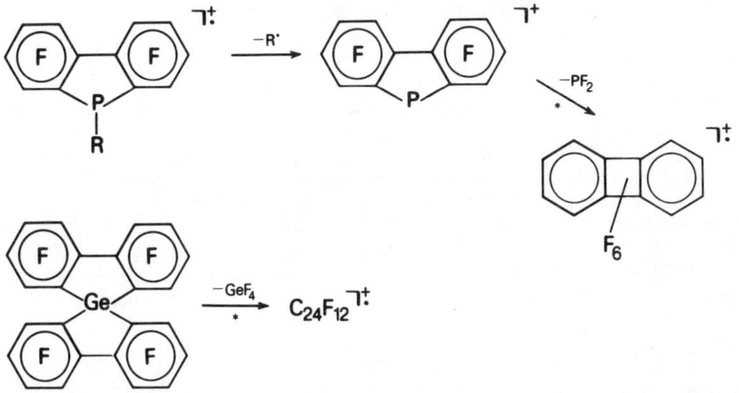

These losses of central atom/halide usually represent significant breakdown pathways in the overall fragmentation process, and have been of predictive value for synthetic purposes. Of the groups of compounds studied to date there seem to be only two critical factors relating to the observation of this phenomenon. The first, and this has mechanistic implications, is that the central atom requires empty orbitals capable of overlap with the filled halogen p orbitals prior to migration. It is noteworthy from this aspect, that no halide transfer is observed with compounds of nitrogen or oxygen which do not have suitable orbitals of the type mentioned above.

Secondly, patterns of halogen migrations observed in the mass spectra of organometallic and coordination compounds have been explained by Miller, Jones and Deacon[93] in terms of hard/soft acid–base theory (HSAB). Although the HSAB concept[94] is a relatively crude one, it is clearly seen that 'hard' fluorine atoms migrate to 'hard' metals and metaloids such as silicon, germanium, tin, lead, iron, etc., but not to 'soft' metals such as mercury, gold or platinum. Softer halides such as bromide do migrate from carbon to mercury. Morris and Koob[95] have extended this approach to the study of fluorinated β-diketonate complexes. Fluoride migration to the metal was observed for Mn(II), Al(III), Cr(III), Fe(III), Fe(II), Co(II) and Ni(II) but not for Zn(II) or Cu(I). However, if, rather than using hexafluoroacetylacetone, a β-diketone is used, with one CF_3 replaced by a more electron-rich substituent, the acidity of the metal is softened, and fluorine rearrangement is less prevalent. As the metal softens with change of substituent, HF loss begins to compete with fluorine migration to the metal. As yet this phenomenon has not been studied for chlorinated or brominated β-diketones.

E. Conclusions

In our brief consideration of the mass spectra of halides we have shown through several examples the extent and depth to which their study has progressed. It is a 'mature' area of study compared to the azides in the previous section, with many applications, as well as more basic studies underway. It is clear that future work will be concentrated in the analytical applications, and the mechanistic implications of the observed spectra.

IV. REFERENCES

1. F. W. Aston, *Nature*, **104**, 393 (1919).
2. F. W. Aston, *Phil. Mag.*, **39**, 620 (1920).
3. F. W. Aston, *Mass Spectra and Isotopes*, Edward Arnold, London (1942), p. 142.
4. E. K. Chess and M. L. Gross, *Anal. Chem.*, **52**, 2057 (1980).
5. M. L. Longhurst and L. A. Shadoff, *Anal. Chem.*, **52**, 2037 (1980).
6. R. M. Moriarty and A. M. Kirkien-Konasiewicz, *Tetrahedron Letters*, 4123 (1966).
7. W. D. Crow and C. Wentrup, *Tetrahedron Letters*, 4379 (1967).
8. E. Hedaya, M. E. Kent, D. W. McNeil, F. P. Lossing and T. McAllister *Tetrahedron Letters*, 3415 (1968).
9. J. M. Miller and G. L. Wilson, *Adv. Inorg. Chem. Radiochem.* **18**, 229 (1966).
10. D. G. I. Kingston and J. D. Henion, *Org. Mass Spect.*, **3**, 413 (1970).
11. P. O. Woodgate and C. Djerassi, *Tetrahedron Letters*, 1875 (1970).
12. R. A. Abramovitch, E. P. Kyba and E. F. V. Scriven, *J. Org. Chem.*, **36**, 3796 (1971).
13. R. T. M. Fraser, N. C. Paul and M. J. Bagley, *Org. Mass Spectr.*, **7**, 83 (1973).
14. E. R. Matjeka, *Part I: Mass Spectrometry of Organic Azides and Diazocompounds*, Ph.D. thesis, Iowa State University, (1974), pp. 3–47; Xerox University Microfilms 75–10, 489.
15. M. E. Peek, C. W. Rees and R. C. Storr, *JCS Perkin I* 1260 (1974).
16. A. Yabe and K. Honda, *Bull. Chem. Soc. Japan*, **49**, 2496 (1976).
17. A Yabe, *Bull. Chem. Soc. Japan*, **52**, 789 (1979).
18. P. Spagnolo, A. Tundo and P. Zanirato, *J. Org. Chem.* **42**, 292 (1977).
19. J. Muller, *Z. Naturforsch.*, **346**, 437 (1979).
20. R. N. McDonald and A. K. Chowdhury, *J. Amer. Chem. Soc.*, **102**, 5118 (1980).
21. N. J. Dickson and L. K. Dyall, *Aust. J. Chem.* **33**, 91 (1980).
22. L. Benati, P. C. Montevecchi and P. Spagnolo, *Tetrahedron Letters*, 815, (1978)
23. C. Wentrup, *Chem. Commun.*, 1386 (1969).
24. W. D. Crowe and C. Wentrup, *Chem. Commun.*, 1387 (1969).
25. A. T. Rake and J. M. Miller, *J. Chem. Soc. A*, 1881 (1970).
26. R. Curci, V. Lucchini, G. Modena, P. J. Korienski and J. Ciabattoni, *J. Org. Chem.*, **38**, 3149 (1973).
27. J. O. Reed and W. Lwowski, *J. Org. Chem.*, **36**, 2864 (1971).
28. A. D. Barone, D. L. Snitman and D. S. Watt, *J. Org. Chem.*, **43**, 2066 (1978).
29. F. Becsi and E. Zbiral, *Monats. Chem*., **110**, 955 (1979).
30. R. A. Abramovitch and E. P. Kyba, *J. Amer. Chem. Soc.*, **96**, 480 (1974).
31. J. Schweng and E. Zbiral, *Liebigs Ann. Chem.*, 1089 (1978).
32. W. H. Saunders Jr. and J. C. Ware, *J. Amer. Chem. Soc*., **80**, 3328 (1958).
33. (a) W. H. Saunders Jr. and E. A. Caress, *J. Amer. Chem. Soc*., **86**, 861 (1964).
    (b) J. Ashby and H. Suschitzky, *Tetrahedron Letters*, 1315 (1971).
34. J. A. Katzenellenbogen, H. N. Myers and H. J. Johnson, Jr, *J. Org. Chem.*, **38**, 3525 (1973).
35. A. Astier, Q. Khong-Huu and A. Pancrazi, *Tetrahedron*, **34**, 1481 (1978).
36. A. Astier, A. Pancrazi and Q. Khong-Huu, *Tetrahedron*, **34**, 1487 (1978).
37. R. C. Cambie, R. C. Hayward, P. S. Rutledge, T. Smith-Palmer and P. D. Woodgate, *JCS Perkin I*, 840 (1976).
38. R. C. Cambie, P. S. Rutledge, T. Smith-Palmer and P. D. Woodgate, *JCS Perkin I*, 2250 (1977).
39. R. C. Cambie, P. S. Rutledge, T. Smith-Palmer and P. D. Woodgate, *JCS Perkin I*, 997, (1978).
40. J. Tamás, M. Mak and S. Makleit, *Org. Mass Spect.* **9**, 845 (1974).
41. S. Makleit, L. Radics, R. Bognár, T. Mile and E. Oláh, *Acta Chim. Acad. Sci. Hung.*, **74**, 99 (1972); *Magy. Kém. Folyoirat*, **78**, 223 (1972).
42. C. Wentrup, *Tetrahedron*, **26**, 4969 (1970).
43. R. E. Banks and G. R. Sparkes, *JCS Perkin I*, 2964 (1972).
44. B. T. Heitke and C. G. McCarty, *Canad. J. Chem.*, **52**, 2861 (1974).
45. A. Padwa, A. Ku, H. Ku and A. Mazzu, *J. Org. Chem.*, **43**, 66 (1978).
46. M. M. Campbell and A. D. Dunn, *Org. Mass Spect.*, **6**, 599 (1972).
47. J. G. Smith, G. L. Wilson and J. M. Miller, *Org. Mass Spect.*, **10**, 5 (1975).

48. R. A. Abramovitch and W. D. Holcomb, *J. Org. Chem.*, **41**, 491 (1976).
49. T. R. B. Jones, J. M. Miller, S. A. Gardner and M. D. Rausch, *Canad. J. Chem.*, **57**, 335 (1979).
50. H. Fr. Schröder and J. Müller, *Z. Anorg. Chem.*, **418**, 247 (1975).
51. H. Fr. Schröder and J. Müller, *Z. Anorg. Chem.*, **451**, 158 (1979).
52. T. R. B. Jones, J. M. Miller, and M. Fild, *Org. Mass Spect.*, **12**, 317 (1977).
53. J. S. Thayer and R. West, *Adv. Organomet. Chem.*, **5**, 169 (1967).
54. R. J. Shozda and J. A. Vernon, *J. Org. Chem.*, **32**, 2876 (1967).
55. C. B. Colburn, W. E. Hill and D. W. A. Sharp, *J. Chem. Soc. A*, 2221 (1970).
56. J. Fawcett, R. D. Peacock, and D. R. Russell, *JCS Dalton*, 2294 (1980).
57. D. Firestone, J. Ross, N. L. Brown, R. P. Barron and J. N. Daminco, *J. Assoc. Off. Anal. Chem.*, **55**, 85 (1972).
58. R. Baughman and M. Meselman, *Adv. Chem. Soc.*, **120**, 92 (1973).
59. D. F. Hunt, T. M. Harvey and J. W. Russell, *JCS Chem. Commun.*, 152 (1975).
60. J. R. Hass, M. D. Friesen, D. J. Horvan and C. E. Parker, *Anal. Chem.*, **50**, 1474 (1978).
61. L. A. Shadoff, W. W. Blaser, C. W. Koches and H. G. Fravel, *Anal. Chem.*, **50**, 1474 (1978).
62. A. di Domenico, F. Ment, L. Bomiforti, I. Comoni, A. DiMuccio, F. Taggi, L. Vergori, G. Colli, E. Elli, A. Gormi, P. Grossi, G. Invernizzi, A. Jemma, L. Luciani, F. Cattaleni, T. DeAnglis, G. Galli, C. Chiabrando and R. Fanelli, *Anal. Chem.*, **51**, 735 (1979).
63. L. C. Lamparski, T. J. Nestrick and R. H. Stehl, *Anal. Chem.*, **51**, 1453 (1979).
64. T. J. Nestrick, L. C. Lamparski and R. H. Stehl, *Anal. Chem.*, **51**, 2273 (1979).
65. R. A. Hummel and G. A. Shadoff, *Anal. Chem.*, **52**, 191 (1980).
66. D. W. Phillipson and B. J. Puma, *Anal. Chem.*, **52**, 2328 (1980).
67. L. L. Lamparski and T. J. Nestrick, *Anal. Chem.*, **52**, 2045 (1980).
68. R. K. Mitchum, G. F. Moler and W. A. Korfmacher, *Anal. Chem.*, **52**, 2278 (1980).
69. R. D. Craig, R. H. Bateman, B. N. Green and D. S. Millington, *Phil. Trans. Roy. Soc.* (London) *A* **293**, 135 (1979).
70. T. A. Gough, K. S. Welb, A. Carick and D. Hazelby, *Anal. Chem.*, **51**, 989 (1979).
71. K. T. Taylor and G. Gooch, *Kratos Data Sheet* 121, Kratos Instruments, Manchester (1980).
72. R. L. Harless, E. O. Oswald, M. K. Wilkinson, A. E. Deprey, Jr, D. D. McDaniel and H. Tai, *Anal. Chem.*, **52**, 1239 (1980).
73. J. A. Buckley, J. B. French, B. H. Dawson and N. M. Reid, paper presented at MS/MS Conference, Asilomar, Calif., Sept. 1980 and private communication from N. M. Reid, Sciex Inc., Toronto, Ontario, to J. M. Miller, Jan. 1981.
74. D. Stahl and T. Gäumann, *Org. Mass Spect.*, **12**, 761 (1977).
75. S. Arseniyadis, J. Gore and M. L. Roumestant, *Org. Mass Spect.*, **12**, 262 (1977).
76. S. Arseniyadis, J. Gore and M. L. Roumestant, *Org. Mass Spect.*, **13**, 54 (1978).
77. R. Hittenhausen-Gelderblom, A. Venema and N. M. M. Nibbering, *Org. Mass Spect.*, **9**, 878 (1979).
78. L. A. Levy and E. O. Oswald, *Biomed. Mass Spect.* **3**, 88 (1976).
79. C. C. Van de Sande and F. W. McLafferty, *J. Amer. Chem. Soc.*, **97**, 2298 (1975).
80. B. I. Shushan and R. K. Boyd, *Biomed. Mass Spect.*, in press.
81. J. M. Miller, *J. Chem. Soc. A*, 828 (1967),
82. J. M. Miller, *Canad. J. Chem.*, **47**, 1612 (1969).
83. A. T. Rake and J. M. Miller, *J. Chem. Soc. A*, 1881 (1970).
84. A. T. Rake and J. M. Miller, *Org. Mass Spect.*, **3**, 237 (1970).
85. T. Chivers, G. F. Lanthier and J. M. Miller, *J. Chem. Soc. A*, 2556 (1971).
86. G. F. Lanthier and J. M. Miller, *Org. Mass Spect.*, **6**, 89 (1972).
87. G. F. Lanthier, J. M. Miller and A. J. Oliver, *Canad. J. Chem.*, **51**, 1945 (1973).
88. S. C. Cohen, A. G. Massey, G. F. Lanthier and J. M. Miller, *Org. Mass Spect.*, **6**, 373 (1972).
89. S. C. Cohen, A. G. Massey, G. F. Lanthier and J. M. Miller, *Org. Mass Spect.*, **8**, 235 (1974).
90. T. R. B. Jones, J. M. Miller, J. L. Peterson and D. W. Meek, *Canad. J. Chem.*, **54**, 1478 (1976).

91. T. R. B. Jones, J. M. Miller and M. Fild, *Org. Mass Spect.*, **12**, 317 (1977).
92. T. R. B. Jones, J. M. Miller, S. A. Gardner and M. D. Rausch, *Canad. J. Chem.*, **57**, 335 (1979).
93. J. M. Miller, T. R. B. Jones and G. B. Deacon, *Inorg. Chem. Acta*, **32**, L75 (1979).
94. R. G. Pearson, *J. Chem. Educ.*, **45**, 581 and 643 (1968).
95. M. L. Morris and R. D. Koob, *Inorg. Chem.*, **20**, 2737 (1981).

The Chemistry of Functional Groups, Supplement D
Edited by S. Patai and Z. Rappoport
© 1983 John Wiley & Sons Ltd

CHAPTER **4**

Nuclear quadrupole resonance of carbon-bonded halogens

E. A. C. LUCKEN

Physical Chemistry Department, 30 Quai E. Ansermet, 1211 Geneva, Switzerland

I. THE THEORETICAL FOUNDATIONS OF NUCLEAR QUADRUPOLE COUPLING

A. Introduction

Although the existence of nuclear quadrupole moments has been known since 1935[1], it is only since 1950, when Dehmelt and Kruger developed the technique of pure nuclear quadrupole resonance (NQR) spectroscopy[2], that nuclear quadrupole coupling constants became readily accessible experimental quantities. Compounds containing halogen constitute the class for which the greatest number of coupling constants have been measured, so it is particularly appropriate that this subject should be reviewed here.

The theory of the interaction of nuclear quadrupole moments with the electrostatic field produced by the surrounding electrons has been treated in detail in a number of monographs[3–11] and this treatment will not be repeated here. Enough of the basic results will be given, however, to make this chapter reasonably self-contained.

B. The Nuclear Quadrupole Moment

The electrical behaviour of a charge distribution is usually best described in terms of its various multipole moments. The most familiar of these is the dipole moment but since nuclei have a centre of symmetry they cannot have electric dipoles and their first non-trivial moment is a quadrupole. This is a second-rank tensor quantity whose elements are defined in terms of the scalar quadrupole moment Q:

$$Q = \frac{1}{e}\int (3z^2 - r^2)\rho(x, y, z)\, \mathrm{d}x\, \mathrm{d}y\, \mathrm{d}z. \tag{1}$$

TABLE 1. Properties of the stable halogen nuclei

Isotope	Relative abundance	Nuclear spin	Magnetic moment[a]	Quadrupole moment[b]
^{35}Cl	75.4	3/2	0.82089	-8.0×10^{-2}
^{37}Cl	24.6	3/2	0.68329	-6.3×10^{-2}
^{79}Br	50.57	3/2	2.0990	0.33
^{81}Br	49.43	3/2	2.2626	0.28
^{127}I	100	5/2	2.7939	−0.75

[a]In nuclear magnetons.
[b]In barns (10^{-28} m^2).

In this formula the z-axis is not in fact the symmetry axis of the stationary nucleus but the effective symmetry axis corresponding to the maximum projection, $M_J = I$, of the nuclear spin angular momentum, I. This has the consequence that nuclear quadrupole moments are zero unless the nuclear spin quantum number is greater than or equal to unity. Thus, of the four halogens, the only stable isotope of fluorine, ^{19}F, has a spin of one-half and has no quadrupole moment whereas the other three halogens all have abundant stable nuclei possessing quadrupole moments. Their properties are summarized in Table 1.

Nuclear quadrupole coupling constants for fluorine have been obtained either by using the 197 keV excited state of ^{19}F, the perturbed angular correlation method[12], or by studying the decay of the polarized unstable isotopes[13], ^{17}F and ^{20}F. The number of compounds which have been so studied is very small; they are mainly inorganic compounds and these fluorine nuclear quadrupole coupling constants will not be discussed further.

C. Nuclear Quadrupole Coupling Constants

The advantage of the multipole moment description is that each multipole moment interacts only with the corresponding derivatives of the electrostatic potential. In this case quadrupole moments interact only with the second derivatives of the electrostatic potential $(\partial^2 V/\partial z^2)$, etc. These derivatives likewise form a symmetric second rank tensor and thus in an appropriate axis system the tensor is diagonal. The energy of interaction between the electric quadrupole and the potential derivatives may then be thought of as depending on the angles between the principal axes of the two tensors. It is this feature which, via the space quantization of the nuclear spin angular momentum, gives rise to the quantification of the nuclear quadrupole interaction.

Unlike the nuclear quadrupole moment, the potential field produced by the electrons does not necessarily have cylindrical symmetry. When it does, however – and as we will see below this is often the case for organohalogen compounds – a simple analytical expression may be given for the interaction energy, E_Q, with, as usual, the z-axis being chosen as the symmetry axis:

$$E_Q = e^2Q\left(\frac{\partial^2 V}{\partial z^2}\right)\frac{(3m^2 - I)(I + 1)}{4I(2I - 1)}. \quad (2)$$

In this expression I and m are the usual nuclear spin quantum numbers. The potential derivatives are often measured in units of the electronic charge, this, together with the convention that Q has the dimensions of an area, is responsible for the factor e^2. A form in which the field gradient is measured in macroscopic units, thus having only the factor 'e', is also found. Finally the term $(\partial^2 V/\partial z^2)$ is often represented by the symbol 'q' or, if there is a need to distinguish the different derivatives, by the symbol q_{zz}.

The quantity e^2Qq/h, now measured in hertz, is called the nuclear quadrupole coupling constant. For a given nucleus it is thus a function of the nuclear environment, reflected by the value of q.

In pure quadrupole resonance we observe direct transitions between the levels, E_Q. The levels are characterized by the absolute value of 'm' and are thus doubly degenerate. For the two cases which interest us here, $I = 3/2$ and $I = 5/2$, we have

$I = 3/2$:

$$\nu_{1/2 \leftrightarrow 3/2} = \frac{e^2Q}{2h} q. \tag{3}$$

$I = 5/2$:

$$\left.\begin{aligned} \nu_{1/2 \leftrightarrow 3/2} &= \frac{3e^2Q}{20h} q, \\ \nu_{3/2 \leftrightarrow 5/2} &= \frac{3e^2Q}{10h} q. \end{aligned}\right\} \tag{4}$$

If the charge distribution around the nucleus does not have cylindrical symmetry these formulae become more complex. The lack of cylindrical symmetry implies $q_{xx} \neq q_{yy}$ but the three derivatives are not independent since, like the quadrupole moment, the corresponding tensor is traceless (the Laplace equation):

$$q_{xx} + q_{yy} + q_{zz} = 0. \tag{5}$$

With the convention $|q_{zz}| \geqslant |q_{yy}| \geqslant |q_{xx}|$ it is convenient to define the asymmetry parameter, η:

$$\eta \equiv \frac{q_{xx} - q_{yy}}{q_{zz}}.$$

The asymmetry parameter thus lies in the range $0 \leqslant \eta \leqslant 1$.

For half-integer nuclear spin the presence of an asymmetry parameter does not remove the degeneracy of the $\pm m$ levels but produces a shift. An expression for the corresponding resonance frequencies may be given exactly for $I = 3/2$ but only in the form of a polynomial in η for $I = 5/2$.

$$\nu_{1/2 \leftrightarrow 3/2} = \frac{e^2Q}{2h} q(1 + \eta^{2/3})^{1/2} \tag{6}$$

$I = 5/2$:

$$\nu_{1/2 \leftrightarrow 3/2} = \frac{3e^2Q}{20h} q(1 + \tfrac{5}{4}\eta^2 + \ldots),$$

$$\nu_{3/2 \leftrightarrow 5/2} = \frac{3e^2Q}{10h} q(1 - \tfrac{11}{54}\eta^2 + \ldots).$$

Tabulated solutions are available for the $I = 5/2$ case at intervals of 0.01 in η [5].

It is apparent from the above that whereas the single resonance frequency for an $I = 3/2$ nucleus yields no information concerning the asymmetry parameter, for $I = 5/2$ both the asymmetry parameter and the coupling constant may be obtained from the two resonance frequencies. It is unfortunate that the most easily studied halogen nucleus, ^{35}Cl, has a spin of 3/2 and thus the vast majority of the pure quadrupole resonance frequencies do not give separate values of e^2Qq and η. However, for $\eta < 0.2$, as is true for almost all organochlorine compounds, the factor $(1 + \eta^2/3)^{1/2}$ produces an effect of less than 1% so that e^2Qq is still essentially given by twice the resonance frequency.

D. The NQR Zeeman Effect: The Measurement of Asymmetry Parameters for Nuclei with $I = 3/2$

The effect of a weak (tens of gauss) magnetic field on the nuclear quadrupole energy levels is to remove the degeneracy of the $\pm m$ levels. The magnitude of the effect, however, depends both on the strength of the field and on its orientation with respect to the field gradient principal axis system. The study of a single crystal thus reveals the directions of these field gradient principal axes and likewise furnishes the value of the asymmetry parameter. These measurements require large (~1 g) single crystals showing reasonably strong resonance lines, and thus only a very small number of compounds have been studied in this way. Several attempts have been made at removing the necessity for the preparation of a single crystal by simulating the line-shape of the resonance for a polycrystalline sample in the presence of a magnetic field[14–16]. Although the first of these[14] could be criticized on the grounds that only approximate methods were used to calculate and analyse the line-shape, the most recent are irreproachable from the theoretical point of view, but the results of such methods are unreliable, possibly because a truly isotropic polycrystalline sample is difficult to prepare.

E. The NQR Stark Effect

The application of intense uniform electric fields to crystalline samples induces a modification of the NQR spectrum. The effect is due to polarization of the electrons in the molecule and hence to a modification of the electronic structure in the vicinity of the nucleus under study. Much of the experimental work has been concerned with halogen resonances and the observed frequency shifts can be correlated with the polarizability of the system. The experimentally observed shifts have been used in attempts to estimate the magnitude of solid state shifts (Section I.F) produced by *internal* electric fields.

The subject has been reviewed by Ainbinder and Svetlov and an English translation is available[18].

F. The Effect of Temperature and Crystalline Environment on NQR Frequencies[19]

Quadrupole coupling constants may in principle be determined for isolated molecules in the gas phase by analysis of the hyperfine structure that they produce in the molecule's rotational spectrum. The difficulties with this technique are first that the precision of the measurements is limited by the line-width of the rotational bands and thus the uncertainty in the value of the coupling constant may be as much as 1%. An improvement in accuracy may be obtained using techniques such as double resonance and simple molecules may also be studied using the various molecular beam methods, but they are not generally applicable. For polyatomic molecules even more serious is the fact that the coupling constants are determined with respect to the inertial axes which will not in general coincide with the field gradient axes. Since the off-diagonal elements of the field gradient tensor in the inertial framework only appear in the second-order terms, they are often not known experimentally and the transformation to the field gradient principal axis system can then only be made by supposing, for example, that the principal field gradient z-axis lies along the carbon–halogen bond.

The field gradient at the nucleus, and hence the nuclear quadrupole coupling constant, is a function of the coordinates of the atoms making up the molecule and hence the observed coupling constant is averaged over the vibrational motion and dependent upon the vibrational state. This dependence has been studied for a large number of diatomic molecules, usually by molecular beam techniques, and is usually expressed in the form

$$e^2Qq_v = e^2Qq_0 + e^2Qq_I(v + \tfrac{1}{2}) + \dots \qquad (7)$$

The ratio e^2Qq_I/e^2Qq_0 is of the order of 10^{-2} and both positive and negative signs are encountered with equal probability. For polyatomic molecules this expression can be generalized:

$$e^2Qq_v = e^2Qq_0 + \sum_i e^2Qq_i(v_i + \tfrac{1}{2}) + \dots \qquad (8)$$

Table 2 shows the vibrational dependence of the coupling constant of some simple organic halides.

The dependence on vibrational averaging also gives rise to an isotope effect. This takes two forms, the first being illustrated by the data in Table 3, where the halogen coupling constant depends on the isotopic composition of the rest of the molecule. The second form arises when we compare the coupling constants of two isotopes of the same atom, for example ^{35}Cl and ^{37}Cl. Although the static field gradient, q_0, is constant for the two isotopes, the vibrational field gradients, q_i, are mass-dependent so that the ratios of the ^{35}Cl and ^{37}Cl coupling constants vary slightly from one molecule to another. This is not, however, very important in the present context.

Only a very small number of organohalogen compounds have been studied in the gas phase and most of the data which will be discussed below are derived from measurements in the solid state. We are usually interested, however, in interpreting the coupling constants in terms of the structure of the isolated molecule, so it is essential to understand what happens when the isolated molecule is transferred to the crystalline state.

In the first place a molecule in a crystal has additional vibrational modes over and above those of the isolated molecule and in the second the neighbouring molecules will produce an additional field gradient which will also be vibration dependent. We thus generalize equation (8) even further:

$$e^2Qq = e^2Qq_0 + \sum_i e^2Qq_i(v_i + \tfrac{1}{2}) + \sum_j e^2Qq_j(v_j + \tfrac{1}{2}) + e^2Qq'_0 + \sum_k e^2Qq'_k(v_k + \tfrac{1}{2}), \qquad (9)$$

where e^2Qq_j, and v_j are the vibrational dependence and vibrational quantum number of the crystalline vibrational modes and e^2Qq' and $e^2Qq'_k$ are the static and vibrationally dependent terms of the intermolecular field gradients.

Separate coupling constants corresponding to different vibrational states are not, however, observable in the solid state. Instead we see a thermal average over the populations of the various vibrational states. If the vibrations are treated as harmonic

TABLE 2. Vibrational dependence of quadrupole coupling constants in linear and symmetric top molecules

Molecule	Nucleus	Vibrational state	e^2Qq, MHz	Reference
HC≡CCl	^{35}Cl	$v = 0$	−79.67	20
		$v_5 = 1, l = 1$	−79.15	
		$v_5 = 2, l = 2$	−78.90	
		$v_5 = 2, l = 0$	−78.00	
		$v_4 = 1, l = 1$	−79.15	
		$v_3 = 1, l = 0$	−79.47	
HC≡CBr	^{79}Br	$v = 0$	648.00	21
		$v_5 = 1, l = 1$	644.90	
		$v_5 = 2, l = 2$	644.9	
		$v_5 = 2, l = 0$	646	
		$v_4 = 1, l = 1$	644.9	
		$v_3 = 1$	646	
CH_3Cl	^{35}Cl	$v = 0$	−74.7533 ± 0.0020	22
		$v_2 = 1$	−74.51 ± 0.1	
		$v_3 = 1$	−74.87 ± 0.1	
		$v_5 = 1$	−75.05 ± 0.3	
		$v_6 = 1$	−74.90 ± 0.1	
CH_3Br	^{79}Br	$v = 0$	577.1 ± 1	23
		$v_2 = 1$	578.2 ± 1	
		$v_3 = 1$	577.0 ± 1	
		$v_3 = 2$	577.4 ± 1	
		$v_5 = 1$	580.0 ± 3	
		$v_6 = 1$	579.8 ± 1	
CH_3I	^{127}I	$v = 0$	−1933.93 ± 0.1	24
		$v_2 = 1$	−1936.83 ± 0.8	
		$v_3 = 1$	−1932.43 ± 0.8	
		$v_3 = 2$	−1933.98 ± 0.1	
		$v_3 = 3$	−1932.7 ± 1	
		$v_5 = 1$	−1941.61 ± 0.02	
		$v_6 = 1$	−1939.3 ± 0.1	
		$v_6 = 2$	−1947.3 ± 0.6	
		$v_3 = 1, v_6 = 1$	−1943.2 ± 2	
		$v_3 = 1, v_3 = 1$	−1940.4 ± 2	
BrCN	^{79}Br	$v = 0$	685.6 ± 0.4	25, 26, 27
		$v_1 = 1$	687.9 ± 0.6	
		$v_2 = 1$	681.6 ± 0.7	
ICN	^{127}I	$v = 0$	−2420.5 ± 0.1	27, 28, 29
		$v_1 = 1$	−2426.9 ± 0.2	
		$v_1 = 2$	−2429 ± 2	
		$v_2 = 1, l = 1$	−2411.5 ± 0.1	
		$v_2 = 2, l = 0$	−2404.8 ± 0.4	
		$v_2 = 2, l = 2$	−2402.6 ± 0.5	
		$v_2 = 3, l = 1$	−2395 ± 2	
		$v_2 = 3, l = 3$	−2395.0 ± 0.4	
		$v_1 = 1, v_2 = 1$ $l = 1$	−2420 ± 1	

TABLE 3. Isotope effects in gas-phase quadrupole coupling constants of symmetric tops and linear molecules

Molecule	Nucleus	Coupling constant, MHz	Reference
CH_3Cl	^{35}Cl	-74.7533 ± 0.0020	22
CD_3Cl		-74.5731 ± 0.001	22
$^{13}CD_3Cl$		-73.2 ± 1.2	30
$HC{\equiv}CCl$	^{35}Cl	-79.67	20
$DC{\equiv}CCl$		-79.66	20
CH_3Br	^{79}Br	577.1 ± 1	23
CD_3Br		575.4 ± 1	23
$^{13}CD_3Br$		573.9 ± 2	23
$HC{\equiv}CBr$	^{79}Br	648.00	21
$DC{\equiv}CBr$		647.96	21
CH_3I	^{127}I	-1933.93 ± 0.1	24
CD_3I		-1929 ± 1	31
$^{13}CH_3I$		-1932.84 ± 2	24

oscillators, we get the standard result:

$$e^2Qq = e^2Q(q_0 + q'_0) + \frac{1}{2}\sum_i e^2Qq_i \coth\left(\frac{h\nu_i}{2kT}\right). \quad (10)$$

In this expression we have lumped together all the different vibrational dependencies q_i, ν_i being the corresponding vibrational frequencies. (This equation, which corresponds to the temperature dependence at constant volume, must be modified to give the temperature dependence at constant pressure. This aspect is not of real concern here.)

One of the crystalline vibrational modes is that of molecular libration. It may easily be demonstrated that any libration about an axis not coincident with the principal field gradient z-axis will decrease the magnitude of the field gradient, i.e. $e^2Qq'_{\mathrm{LIB}}/e^2Qq_0$ is negative. Since the vibrational modes usually have low frequencies associated with them, these terms usually dominate equation (10), and hence most coupling constants and resonance frequencies decrease with increasing temperature. A typical curve is shown in Figure 1. A further solid state effect is that any phase change produces a discontinuity in the temperature dependence (Figure 2) and different crystalline phases have different coupling constants (Figure 3).

Finally the coupling constants in the solid state, even if an attempt is made to extrapolate the temperature dependence to absolute zero, are different from those of the isolated molecule. This is illustrated in Table 4 by the data for a few simple organic molecules. It is remarkable that such large differences, both positive and negative, are observed, and it is possible that they may not be typical. Thus the methyl halides may well have considerable librational movements in the solid state while the highly polar cyanogen halides may exhibit strong intermolecular association. It is clear, however, that the uncertainties introduced by what we may call a solid state effect imply that firm conclusions should not be drawn from a study of just one molecule but where possible a series of closely related systems should be studied.

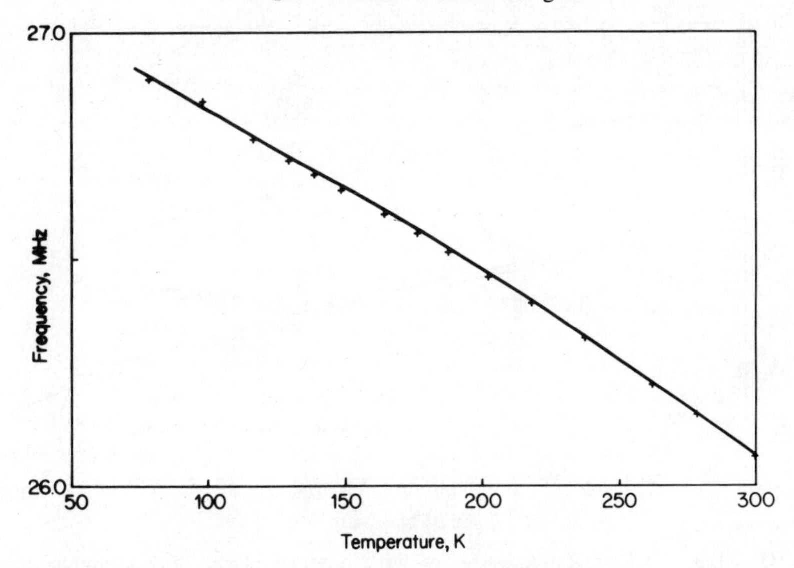

FIGURE 1. The temperature variation of a ^{35}Cl NQR frequency of $CH_3CO_2SbCl_4$ (unpublished observations made in the author's laboratory).

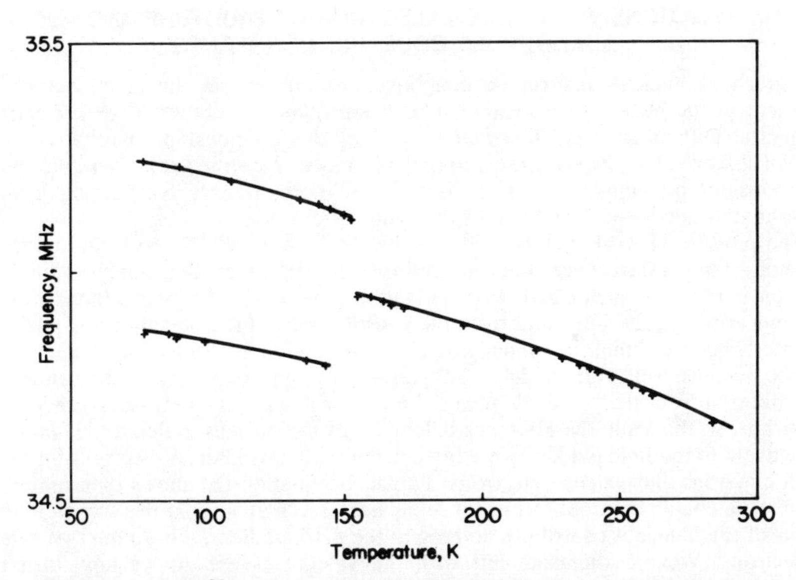

FIGURE 2. The ^{35}Cl NQR resonance frequencies of the *p*-Cl substituent in 1,3-*bis*(trichloromethyl)-5-*p*-chlorophenyl-*s*-triazine (unpublished observations made in the author's laboratory) illustrating the effects of phase changes.

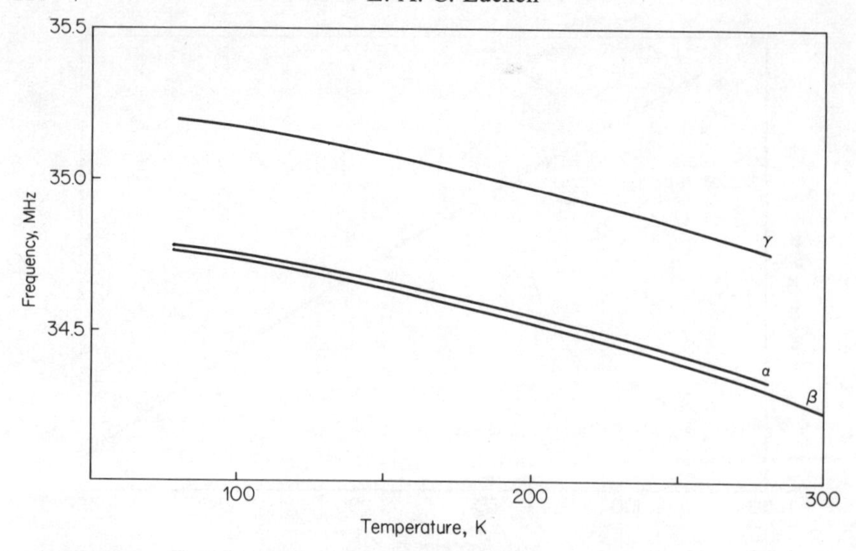

FIGURE 3. ^{35}Cl NQR frequencies of three different phases of *p*-dichlorobenzene (drawn using data published by G. G. Moross and H. S. Story, *J. Chem. Phys.*, **45**, 3370 (1966)).

II. THE RELATIONSHIP BETWEEN ELECTRONIC STRUCTURE AND NUCLEAR QUADRUPOLE COUPLING CONSTANTS

Inasmuch as nuclear quadrupole coupling constants reflect the electrostatic field gradient at the nucleus it is clear that they are related to the molecule's electronic structure. Difficulties arise, however, in making this relationship quantitative in the sense (observed coupling constant) to (derived electronic structure). These difficulties are considerably diminished if, as is the case here, we are considering a singly coordinated nucleus always bound to the same atom.

The pioneering work in this field was due to Townes and Dailey[48] and, although several attempts have been made to improve the details of their analysis, the basic approach remains unchanged. In a polyatomic molecule the field gradient at the nucleus arises on the one hand from the contributions of the neighbouring positively charged nuclei and their accompanying electrons, and on the other from the electrons on the nucleus itself. If the neighbouring atoms are approximately electroneutral, the contribution from the positively charged nuclei will approximately cancel that of the electrons, so that only the electrons belonging to the nucleus under study contribute effectively to the field gradient. We can conveniently divide these electrons into inner shell electrons and valence electrons. Laplace's equation (5) shows that spherically symmetric charge distributions produce no field gradient, so that the complete inner shells of the halogens contribute nothing to the field gradient, nor do the two valence s-electrons. We are therefore left with the valence p-shell so we now turn to a discussion of the field gradient produced by a p-electron. Since p-orbitals have cylindrical symmetry the corresponding field gradient tensor also has cylindrical symmetry. Thus, using the superscripts x, y, z to distinguish the three p-orbitals, we

TABLE 4. Gas-phase and solid state (77 K) quadrupole coupling constants. The asymmetry parameters of the solid state coupling constants are assumed to be zero unless otherwise stated

Molecule	Nucleus	Gas phase		Solid state at 77 K			
		e^2Qq/h, MHz	Ref.	e^2Qq/h, MHz	Ref.	$\Delta\nu$	$100\Delta\nu/e^2Qq$ (gas)
CH_3Cl	^{35}Cl	−74.7533	22	(−)68.06	32	6.69	8.9
CH_3CH_2Cl	^{35}Cl	−68.80	33	(−)65.4	32	3.4	4.9
$H_2C{=}CHCl$	^{35}Cl	−70.16	34	(−)66.88	35	3.28	4.7
$HC{\equiv}CCl$	^{35}Cl	−79.67	20	(−)78.42	36	1.25	1.6
$N{\equiv}CCl$	^{35}Cl	−83.39	37	(−)83.47	36	−0.08	−0.1
C_6H_5Cl	^{35}Cl	−71.09 ($\eta = 0.074$)	38	(−)69.32 ($\eta = 0.08$)	32	1.77	2.5
CH_3Br	^{79}Br	577.1	23	529.0	39	48.1	8.3
CH_3CH_2Br	^{79}Br	538.0	40	497.5	39	40.5	7.5
$N{\equiv}CBr$	^{79}Br	685.6	25	709.27	41	−23.67	−3.4
C_6H_5Br	^{79}Br	558.9 ($\eta = 0.046$)	38	538.0 ($\eta = 0.66$)	42	20.9	3.7
CH_3I	^{127}I	−1933.93	24	(−)1765.85	39	168.1	8.7
CH_3CH_2I	^{127}I	−1771	43	(−)1649.0	39	122.0	6.9
$CH_2{=}CHI$	^{127}I	−1892.1 ($\eta = 0.03$)	44	(−)1822.01 ($\eta = 0.06$)	45	70.1	3.7
C_6H_5I	^{127}I	−1877 ($\eta = 0.056$)	46	(−)1847 ($\eta = 0.042$)	47	30.0	1.6

have the result for the field gradient diagonal elements produced by one electron:

$$\left.\begin{aligned} q_{xx}^{x} &= -2q_{yy}^{x} = -2q_{zz}^{x} = q_0, \\ q_{yy}^{y} &= -2q_{xx}^{y} = -2q_{zz}^{y} = q_0, \\ q_{zz}^{z} &= -2q_{xx}^{z} = -2q_{yy}^{z} = q_0, \end{aligned}\right\} \tag{11}$$

For Hartree–Fock type orbitals given by the product of a radial function and the appropriate p-orbital spherical harmonic, it may be shown that the value q_0 is given by the expression

$$q_0 = -\frac{4}{5}\left\langle\frac{1}{r^3}\right\rangle. \tag{12}$$

The value of q_0 may thus be obtained, in principle, from appropriate atomic wave-functions, but, since nuclear quadrupole moments are not in general known independently of the nuclear quadrupole coupling constant, this equation serves more to determine the value of the nuclear quadrupole moment from the observed coupling constant of a simple system.

The above results for a singly occupied p-orbital may be used to calculate the field gradient tensor components for an arbitrarily occupied p-shell. If the populations of the p_x, p_y and p_z orbitals are a, b and c respectively, we have

$$\left.\begin{aligned} q_{xx} &= \left(a - \frac{b+c}{2}\right)q_0, \\ q_{yy} &= \left(b - \frac{a+c}{2}\right)q_0, \\ q_{zz} &= \left(c - \frac{a+b}{2}\right)q_0. \end{aligned}\right\} \tag{13}$$

The labels for the x, y and z axes must of course be chosen to correspond to the field gradient axes with the usual convention:

$$|q_{zz}| \geqslant |q_{yy}| \geqslant |q_{xx}|. \tag{14}$$

The quadrupole coupling tensor only provides, at most, two experimental quantities, and three are required to specify the p-orbital population. Assumptions as to at least one of these are therefore necessary. In the case of the organic halides this usually means setting one or more of the populations to 2, for example in an aliphatic halide such as methyl chloride. In this case the field gradient z-axis is along the C—Cl bond and we have:

$$\left.\begin{aligned} q_{zz} &= -(2-c)q_0, \\ q_{xx} &= q_{yy} = -\tfrac{1}{2}q_{zz}. \end{aligned}\right\} \tag{15}$$

It will not, however, always be possible to assume a population of 2. For example in chloroacetylene there is undoubtedly some delocalization of the lone pair electrons to the π-electron system and we have

$$q_{zz} = -(b-c)q_0. \tag{16}$$

The coupling constant then only yields the difference in the populations.

When a halogen is attached to a planar conjugated system, where it is reasonable to assume that only one of the two lone pair orbitals participates in the π-bonding, a complete determination is again possible. The orbital participating in π-bonding is the

p_x orbital and replacing its population, b, by the symbol 2π and, assuming that the p_y orbital's population is 2, we have[49]

$$\left.\begin{aligned} q_{zz} &= -\left(2 - c - \frac{\pi}{2}\right) q_0, \\ \eta q_{zz} &= -\tfrac{3}{2} q_0. \end{aligned}\right\} \tag{17}$$

The effect of π-bonding is thus to decrease the coupling constant and its extent may be directly determined if the asymmetry parameter has been measured (see, however, Section III.E). It should be stressed that these equations are only approximate and many of the difficulties and approximations have been glossed over. Even aside from the sweeping assumptions leading to a consideration of only the valence shell, there is the assumption that the parameters of the valence p orbital, and hence the value of q_0, are the same whatever the nature of the molecule containing the atom. Likewise it is assumed that the values of the parameters for each of the three p orbitals are identical. For the present purposes, however, where the immediate environment of the halogen is fairly constant and where we are mainly interested in following the changes in this environment, the use of the above equations in a differential form is much more likely to yield reliable results. Nevertheless, the relationship between partial double-bond character and asymmetry parameter should be treated with some reserve.

It remains to discuss the value of q_0 or, more exactly, e^2Qq_0. For all three halogens the value of the coupling constant for the free atom has been determined by atomic beam measurements. Since here we know that the populations of the p_x, p_y and p_z. orbitals are 2, 2 and 1 respectively, the experimental coupling constant gives the value of e^2Qq_0. This value does not take into account changes in the p orbital brought about by bond formation. A possible solution to this is to use the coupling constants for the molecular halogen molecules – where again the populations are 2, 2 and 1 – but, unfortunately, it is known that in the solid state there is considerable intermolecular bonding, and the gas-phase values are unknown since the molecule does not possess a dipole moment. An estimate of the coupling constant of the free molecular halogen can, however, be obtained from the coupling constant of the gaseous interhalogen molecules[50]. Table 5 shows the values of e^2Qq for the halogen nuclei obtained from both methods and it will be seen that, except for ^{127}I, there is very little difference between the two. In all calculations given below the values derived from the atomic couplings have been employed. The above discussion has concentrated on the field gradients produced by p-electrons. It can be shown, both theoretically and

TABLE 5. Valence shell p-orbital coupling constants (in megahertz) derived from atomic coupling constants or from the interpolated values for the molecular halogens

	e^2Qq_0/h, MHz	
Nucleus	Atomic	Molecular
^{35}Cl	109.75	115
^{37}Cl	86.51	91
^{79}Br	−769.76	−798
^{81}Br	−643.30	−667
^{127}I	2292.7	2510

experimentally, that the field gradients produced by d- or f-electrons are much smaller than those produced by p-electrons of the same principal quantum number. For the halogens, where valence d-orbital contributions are likely to be small, the p-electron terms dominate and the neglect of these terms in this approximate analysis is totally justified.

III. THE CALIBRATION OF THE METHOD: NQR FREQUENCIES IN SIMPLE SYSTEMS

A. Introduction

The vast majority of halogen NQR frequencies are concerned with the ^{35}Cl nucleus. The trends exhibited here are usually likewise manifest in the corresponding ^{79}Br or ^{127}I resonances and this is particularly true for organic halides where the change from one halide to another is not normally accompanied by any fundamental structural difference. For this reason most of what follows will be concerned with ^{35}Cl resonances, though the ^{79}Br and ^{127}I resonances will be discussed or referred to when this is possible and appropriate.

In many of the tables references are not given to measurements of individual compounds but to secondary sources such as review papers or articles devoted to correlating previously published data. In particular the monograph of Semin, Babushkina and Yakobson[5] contains an exhaustive list of resonance frequencies published prior to 1970.

Many of the early systematic studies of halogen quadrupole resonance frequencies of organic halides were concerned with the effects of substituents in simple, well studied systems and indeed this question has continued to attract attention at intervals right to the present day. These studies serve several purposes. In the first place, by showing a relationship with other measures of substituent properties, they show that resonance frequencies do indeed reflect the electronic structure of the carbon–halogen bond. Unlike many substituent parameters such as the Hammett σ or Taft σ^*, the resonance frequencies are determined only by the properties of the molecule in its unperturbed ground state, so that the relationship between the substituent parameter and the effect of the substituent on a resonance frequency throws light on the relative importance of the various effects which determine the value of that substituent parameter and of the chemical reactivity of which it is a measure. Finally, the goodness or otherwise of these correlations indicates the extent of the scatter which is brought about by solid state effects and indicates the limits of the reliability and significance of quadrupole resonance frequencies as an index of electronic structure.

B. NQR Frequencies and the Polarity of the Carbon–Halogen Bond

In much of what follows we will be concerned with the effects of substituents or other structural details on the electronic character of the carbon–halogen bond. In this section we attempt to use the quadrupole coupling data to establish the basic electronic structure of an 'unsubstituted' organic halide for which the only variable is the coordination number of the carbon atom. The results of this analysis will serve as reference points for the discussion which follows and for this reason we confine our attention to coupling constants derived from gas-phase coupling constants.

Table 6 shows the data for the series CH_3X, CH_3CH_2X, $CH_2{=}CHX$ and $HC{\equiv}CX$. They have been analysed according to equations (15) and (17). For the acetylenes, where the extent of π-bonding cannot be obtained directly, it is assumed that it occurs to the same extent as in the vinyl halides for each of the lone pair orbitals.

The figures shown are a very satisfactory confirmation of the expected trends in this

TABLE 6. Quadrupole coupling constants (in megahertz) and bond populations in haloalkanes, haloalkenes and haloalkynes

	e^2Qq/h	η	Sigma density, c	π (%)	Charge	Ref.
CH_3Cl	−74.75	0	1.320	0	−0.320	22
CH_3CH_2Cl	−68.80	0.035	1.367	(0)	−0.367	33
$CH_2{=}CHCl$	−70.16	0.141	1.332	6.0	−0.299	34
$HC{\equiv}CCl$	−79.7	0	1.215	12.0	−0.155	20
CH_3Br	577.1	0	1.250	0	−0.250	23
CH_3CH_2Br	538	0.005	1.298	(0)	−0.298	40
$CH_2{=}CHBr$	543.8	0.066	1.278	3.1	−0.262	51
$HC{\equiv}CBr$	648	0	1.141	7.0	−0.106	21
CH_3I	−1933.9	0	1.159	0	−0.159	24
CH_3CH_2I	−1771	0	1.230	0	−0.230	43
$CH_2{=}CHI$	−1877	0.056	1.169	3.0	−0.154	44

series. Thus for any given molecular type the negative charge on the halogen decreases regularly on going from chlorine to iodine. On going from the methyl halide to the ethyl halide there is an increase in charge owing to the electron-releasing effect of the methyl group, and for the series haloethane, haloethylene and haloacetylene the charge decreases owing to the increase in effective electronegativity of carbon with its increasing participation of the 2s orbital in the σ bond. Finally, in the vinyl halides the extent of π-bonding decreases on going from chlorine to iodine.

It remains to situate the figures for the carbon–halogen bond with respect to those of the halogen bound to the neighbours of carbon in the Periodic Table. For comparison Table 7 shows the coupling constants and derived charges for the silyl halides, the hydrogen halides and chloramine. Once again the expected trends are seen, i.e. the highly covalent character of the N—Cl bond and the highly ionic nature of the silicon–halogen bond. The rather similar ionic characters of the halogen–hydrogen and halogen–carbon bonds are a striking illustration on the influence of solvation on the ionic dissociation of the hydrogen halides.

C. Substituted Methyl Halides

The literature is full of correlations of the resonance frequencies of substituted chloromethanes embracing a gradually increasing range of compounds. Probably the first study of the suitability of resonance frequencies for investigating the electronic structure of the carbon–halogen bond was that of Livingstone[32] who showed, *inter alia*, that the ^{35}Cl resonance frequency increased regularly in the series $CH_3Cl < CH_2Cl_2 < CHCl_3 < CCl_4$ and decreased in the series $CH_3Cl > CH_3CH_2Cl > (CH_3)_2CHCl > (CH_3)_3CCl$. Table 8 shows the ^{35}Cl resonance frequencies of a selection of monosubstituted chloromethanes, RCH_2Cl. It can be seen that the nature of the single substituent produces a range of frequencies from 30 to 39 MHz, a range of 26% about the median value of 34.5 MHz. On the basis of equation (15) these limits represent a charge on the chlorine atom from 0.45 to 0.28, a range of 0.17 electrons.

The most obvious substituent parameter to use for correlation purposes in these circumstances where no conjugation is possible is the Taft σ^* polar substituent constant[58]. The results of extensive investigations of this kind have been summarized

TABLE 7. Quadrupole coupling constants (in megahertz) and halogen atom charge for methyl halides and their neighbours in the periodic table

	^{35}Cl			^{79}Br			^{127}I		
Molecule	e^2Qq/h	Ref.	Charge	e^2Qq/h	Ref.	Charge	e^2Qq/h	Ref.	Charge
H—X	−67.619	52	−0.385	−532.304	53	−0.309	−1828.29	54	−0.205
H_2N—X	−98.1	55	−0.102	–	–	–	–	–	–
H_3C—X	−74.753	22	−0.320	577.1	23	−0.250	−1933.9	24	−0.159
H_3Si—X	−39.74	56	−0.639	335.1	57	−0.565	−1245.1	57	−0.459

TABLE 8. Halogen resonance frequencies (in megahertz at 77 K) of monosubstituted halomethanes, RCH_2X [3]

R	^{35}Cl	^{79}Br	$^{127}I_{(1/2 \leftrightarrow 3/2)}$
—H	34.029	264.508	265.102
$—CH_3$	32.646, 32.759	248.78	247.374
$—C_6H_5$	33.627		
$—COCH_3$	35.075, 35.484		
—COOH	36.131, 36.429	284.760	298.642
		287.014	299.483
—COONa	34.794		
—CN	38.122		
$—NO_2$	37.635		
$—OCH_3$	29.817, 30.206		
—F	33.799		
—Cl	35.991		
—Br	36.14		
—I	36.498		
$—CF_3$	37.387		
$—CCl_3$	36.40		
$—SiCl_3$	36.786		
$—GeCl_3$	37.926		
$—SnCl_3$	37.870		
$—PCl_2$	36.09		
$—P(O)Cl_2$	38.224, 38.286		
$—SO_2Cl$	38.612, 30.039		

by Biryukov and Voronkov[59] and by Voronkov and Feshin[8]. As can be seen from Table 9 there may well be a departure from additivity inasmuch as the parameters of the correlation are not the same for the series RCH_2Cl, $RCCl_3$, $R^1R^2CCl_2$ and $R^1R^2R^3CCl$. It should be noted, however, that for the first two series there is necessarily only one compound corresponding to each substituent whereas the last two series permit numerous combinations of the substituents and thus a more widespread statistical analysis and so the apparent lack of additivity may be an artefact. Following on the work of Weiss and coworkers[60–66] on the chlorobenzenes, Ramanamurti, Venkatacharyulu and Premaswarup[67] have fitted the resonance frequencies of substituted chloromethanes to the equation

$$\nu^{77} = 33.983 + \sum_i m_i \tag{18}$$

TABLE 9. The relationship between the ^{35}Cl resonance frequency at 77 K and the Taft polar substituent, σ^*, for chloromethanes of various types; $\nu = \nu_0 + \rho\sigma^*$ [38]

Molecular type	ν_0	ρ	Standard deviation	Correlation coefficient
R—Cl	33.120	2.849	0.27	0.93
RCH_2Cl	33.107	1.214	0.23	0.99
$R^1R^2R^3CCl$	32.05	1.019	0.27	0.96
$RCCl_3$	38.045	0.933	0.06	0.96
$R^1R^2CCl_2$	35.215	0.959	0.27	0.93

TABLE 10. Substituent parameters (m, σ_m, σ_v^*) derived from ^{35}Cl quadrupole resonance frequencies and their comparison with the Taft substituent parameter (σ^*) and the group electronegativity (χ)

Substituent	m	σ_m	σ_v^*	σ^*	χ
$-CCl_3$	2.86	3.30	2.75	2.65	3.0
$-CHCl_2$	1.27	1.74	1.90	1.94	2.8
$-CH_2Cl$	0.38	0.86	–	1.05	2.75
$-C_6H_5$	−0.13	0.36	0.80	0.60	3.0
$-H$	0.0	−0.49	0.40	0.49	2.28
$-CH_3$	−0.86	−0.36	0.0	0.00	2.3
$-C(CH_3)_3$	0.62	1.10	–	−0.30	–
$-F$	1.14	1.61	–		3.95
$-Cl$	2.19	2.64	2.7		3.03
$-Br$	2.10	2.55	–		2.80
$-I$	2.21	2.66	–		2.47
$-CN$	3.68	4.10	3.7		3.3
$-NO_2$	3.22	3.65	–		3.4
$-CF_3$	2.52	2.96	–		3.35
$-COOH$	1.96	2.41	–		2.84
$-COCl$	2.76	3.19	2.5		–
$-CONH_2$	1.14	1.61	1.4		–
$-COONa$	0.67	1.15	–		2.92
$-CHO$	0.72	1.20	–		–
$-OCH_3$	−3.97	−3.41	–		3.7
$-SCH_3$	−0.88	−0.37	–		–
$-SO_2Cl$	−2.07	−1.54	–		–
$-SCl$	1.34	1.80	1.6		–

and hence obtained values of the substituent coefficients m_i. The values of m_i are shown in Table 10 and compared to the σ^* values where available[58]. The standard deviation for the frequencies predicted by this equation is 0.405 MHz. It should be noted that the reference compound in the above study is methyl chloride so that m for —H is 0.00. For a direct comparison with the σ^* values they should therefore be divided by 1.019 (the slope of the σ^* correlation for the $R^1R^2R^3CCl$ series in Table 10) and 0.49, the σ^* value of H, should be added. The values thus calculated, σ_m, are also shown in the above table. It can be seen that for most of the substituents originally considered by Taft there is a satisfactory agreement between the two and indeed Biryukov and Voronkov[59] used a technique of this sort to obtain values of σ^* for groups for which Taft's original method had not been applied. These values, σ_v^*, are likewise shown in the table. A final point of comparison is provided by the group electronegativities given by Wells[68].

The parallelism between the m values and other measures of the polarity of a substituent is conspicuously absent from the first-row substituents F— and CH_3O—. Indeed these substituents were specifically excluded by Biryukov and Voronkov from their correlations between resonance frequencies and σ^*. Various suggestions have been made as to the reasons for this discrepancy and these are discussed in Section IV.D.

D. Substituted Halobenzenes

One of the earliest attempts at systematically applying NQR spectroscopy to the problem of the electronic structure of organic compounds concerned the correlation between the effect of substituents on the reactivity of a particular position in a substituted benzene derivative, quantified via the well known Hammett equation, and the resonance frequency of a halogen nucleus attached at that same position. The earliest measurements could be said to have been undertaken more to check that quadrupole resonance frequencies did indeed reflect the electronic structure of the C—X bond[69], but as the available data became more extensive it became clear that the NQR frequencies could be used both to obtain Hammett constants for substituents which were difficult to study by more conventional methods and to separate the different types of electronic displacements which substituents could produce.

Table 11 shows the ^{35}Cl resonance frequencies of several monosubstituted chlorobenzenes. Qualitatively the frequencies follow the pattern which one would have expected from the nature of the substituent, although inductive effects seem to dominate and the shifts produced by a particular substituent are in the order *ortho* > *meta* > *para*. However, the scatter produced by intermolecular fields makes it necessary to extend these studies as much as possible and to include multiply-substituted chlorobenzenes.

The first large scale systematic study of this sort was undertaken by Bray and Barnes[70], who correlated the ^{35}Cl resonance frequencies of 52 substituted chlorobenzenes with the Hammett constants of the substituents. For multiply-substituted derivatives the effective substituent constant was taken as the sum

TABLE 11. ^{35}Cl resonance frequencies (in megahertz at 77 K) of monosubstituted chlorobenzenes, RC_6H_4Cl. Data have been taken from Refs 5, 70 and 60 except those for NH_3^+ from Ref. 66

R	*o*	*m*	*p*
—H	34.622	34.622	34.622
—CH_3	34.19	34.36	34.53
—CCl_3			34.564
—CF_3	35.633	35.073	
—COOH	36.305	34.766	34.673
		35.227	34.700
—CN	35.410		35.150
	35.453		
	35.500		
	35.541		
—NH_2	33.953	34.388	34.146
		34.468	
—$\overset{+}{N}H_3Cl^-$	35.916	34.854	34.619
			34.886
—NO_2	37.260	35.457	34.88
—NCO	34.635	34.653	35.259
—OH	35.37	34.766	34.700
		34.825	34.945
—F	36.294	34.97	34.818
		35.05	35.226
			35.286
—Cl	35.580	34.809	34.799
	35.755	34.875	
	35.824	35.030	
—Br		34.85	34.81
		34.99	

of the constants for each substituent. The correlation thus obtained was

$$\nu^{77} = 34.826 + 1.024\sum_i \sigma_i \pm 0.360. \tag{19}$$

Using this equation they obtained the values of 29 previously unknown substituent constants. In particular they showed that the appropriate *ortho* substituent parameter appeared to be that defined by the ratio of the dissociation constants of the substituted and unsubstituted acids in water at 25°C.

These results were sporadically extended by various workers and have been summarized by Voronkov and Feshin[10]. If the *ortho, meta* and *para* positions are considered together, the correlation remains essentially unchanged, but if each position is considered separately somewhat different correlations are obtained[71]. Thus for *meta* substituents, where the substituent effect is essentially inductive, we have

$$\nu_m^{77} = 34.40 + 1.223\,\sigma_m. \tag{20}$$

For *para* substituents there are deviations from this simple relationship by amounts that are either positive or negative according to whether the substituents have conjugative electron-accepting or electron-donating abilities as measured by the Taft resonance parameter, σ_R. Thus we here have the two-parameter equation

$$\nu_p^{77} = 34.40 + 1.223\,\sigma_p - 1.446\,\sigma_R \pm 0.07. \tag{21}$$

Ortho substituents behave as usual in a much less systematic manner and the 'explanations' of this are no more precise than they are for other measures of the properties of *ortho* substituents. Similar studies on more restricted series of compounds have been made for bromobenzenes and iodobenzenes. For ^{79}Br resonance frequencies we have[72,73]

$$\nu^{77} = 271.6 + 9.14\sum_i \sigma_i \tag{22}$$

while for iodobenzenes[74]

$$e^2Qq^{298} = 1837 + 112\sum_i \sigma_i. \tag{23}$$

Note that here we have the quadrupole coupling constants and not the resonance frequencies since the two resonance frequencies of the ^{127}I nucleus allow one to separate the coupling constant and asymmetry parameter.

In order to compare the three series we express these relationships in a rearranged form:

$$\left.\begin{aligned} \nu_{Cl}^{77} &= 34.826\left(1 + 0.030\sum_i \sigma_i\right), \\ \nu_{Br}^{77} &= 271.6\left(1 + 0.034\sum_i \sigma_i\right), \\ e^2Qq_{I}^{298} &= 1837\left(1 + 0.061\sum_i \sigma_i\right). \end{aligned}\right\} \tag{24}$$

From this it would appear that the iodobenzenes are the most sensitive to the effect of substituents. This may, however, be an artefact of the much smaller sample of substituted iodobenzenes.

An alternative rearrangement, which yields the same conclusion, is to express the relationship in terms of the corresponding atomic parameters:

$$\left.\begin{aligned} \nu_{\mathrm{Cl}}^{77} &= 54.875\left(0.6248 + 0.0186 \sum_i \sigma_i\right), \\ \nu_{\mathrm{Br}}^{77} &= 384.88\left(0.7057 + 0.0237 \sum_i \sigma_i\right), \\ e^2Qq_{\mathrm{I}}^{298} &= 2292.7\left(0.8012 + 0.0489 \sum_i \sigma_i\right). \end{aligned}\right\} \tag{25}$$

Weiss and his coworkers have adopted a somewhat different approach[60–66]. Considering that Hammett substituent constants reflect the overall effect of a number of different types of electronic displacement and that the relative weight of these different contributions are not necessarily the same in their effect on the nuclear quadrupole coupling constant, they have proposed correlating the coupling constants of substituted chlorobenzenes according to a one-parameter equation where the substituent parameters are chosen to give the best statistical fit:

$$\nu^{77} = 34.622 + \sum_i K_i. \tag{26}$$

Subsequently of course the resultant parameters can be compared with other more traditional substituent parameters such as the Hammett parameter discussed above. The K-values and the related σ parameter are shown in Table 12. On the basis of these studies Weiss concluded that in general the ^{35}Cl resonance frequencies were mainly determined by inductive effects and that steric effects were not important for *ortho* substituents.

TABLE 12. Substituent parameters (K_i, equation 26) for substituted chlorobenzenes. Values of the Hammett sigma parameter, σ, are shown for comparison

		K_i	σ
$-CH_3$	*o*	-0.392 ± 0.054	
	m	-0.207 ± 0.161	-0.069
	p	-0.004 ± 0.126	-0.170
$-CF_3$	*o*		
	m	0.611 ± 0.147	0.415
	p	0.740 ± 0.207	0.551
$-COOH$	*o*	1.704 ± 0.162	
	m	0.377 ± 0.153	0.355
	p	0.409 ± 0.148	0.265
$-NH_2$	*o*	-0.534 ± 0.119	
	m	-0.103 ± 0.130	-0.16
	p	-0.119 ± 0.133	-0.66
$-NMe_2$	*o*	0.336 ± 0.652	
	m	-0.380 ± 0.066	-0.21
	p	-0.550 ± 0.150	-0.60
$-NH-C(=NH)NH_2$	*o*	~ 0	
	m	~ 0	
	p	~ 0	

TABLE 12. *continued*

		K_i	σ
$-NHCOCH_3$	*o*	0.667 ± 0.101	
	m	0.343 ± 0.106	0.21
	p	0.056 ± 0.162	−0.02
$-\overset{+}{N}H_3$	*o*	1.213 ± 0.047	
	m	0.705 ± 0.065	0.64
	p	0.469 ± 0.062	0.49
$-\overset{+}{N}Me_3$	*o*	1.56	
	m	0.679 ± 0.142	0.90
	p	0.495 ± 0.070	0.86
$-NH-C(NH_2)_2^+$	*o*	0.838 ± 0.075	
	m	0.494 ± 0.109	
	p	0.282 ± 0.162	0.38
$-\overset{+}{N}\equiv N$	*o*	3.204 ± 0.095	
	m	1.910 ± 0.092	1.8
	p	0.824 ± 0.112	1.9
$-NO_2$	*o*	2.096 ± 0.087	
	m	1.069 ± 0.093	0.710
	p	0.607 ± 0.125	0.778/1.270[a]
$-OH$	*o*	0.843	
	m	0.545	+0.10
	p	−0.114 ± 0.143	−0.18
$-OCH_3$	*o*	0.917 ± 0.172	
	m		
	p	0.122 ± 0.145	−0.268
$-OC_2H_5$	*o*	0.793 ± 0.172	
	m		
	p	−0.027 ± 0.157	−0.250
$-OCOCH_3$	*o*	1.015 ± 0.144	
	m		
	p	0.329 ± 0.154	
$-OCOCH_2Cl$	*o*	1.528 ± 0.194	
	m		
	p	0.348 ± 0.154	
$-OCOCHCl_2$	*o*	1.395 ± 0.194	
	m		
	p	0.505 ± 0.194	
$-OCOCCl_3$	*o*	1.429 ± 0.194	
	m		
	p	0.169 ± 0.194	
$-OCOCF_3$	*o*	1.812 ± 0.194	
	m		
	p	0.751 ± 0.194	
$-OCH_2COOH$	*o*		
	m		
	p	−0.266 ± 0.233	

TABLE 12. *continued*

		K_i	σ
$-SO_3Na$	*o*	1.403 ± 0.240	
	m	0.398 ± 0.233	
	p	0.804 ± 0.235	
$-Cl$	*o*	1.206 ± 0.047	
	m	0.499 ± 0.035	0.373
	p	0.329 ± 0.072	0.227
$-Br$	*o*	0.975 ± 0.204	
	m		
	p	0.312 ± 0.173	0.232
$-I$	*o*	0.917 ± 0.196	
	m		
	p	0.199 ± 0.173	0.276

[a] σ^- value.

E. Asymmetry Parameters

As discussed in Section II, equation (17), the asymmetry parameter of a halogen atom bonded to a conjugated system is used as a measure of the partial double bond character of the carbon–halogen bond. Values so obtained have already been discussed for chloroethylene and are mentioned below in Sections III and IV for various chlorinated heterocyclic compounds. Here the limitations of the use of equation (17) are discussed.

As noted in Section II, equations based on the Townes and Dailey method[48] are most likely to give valid results if they are used to compare similar systems since it is only then that the terms of the contributions to the field gradient which are neglected in this theory are likely to remain constant and thus cancel out in a comparison. In the present context the assumption that the field gradient is dominated by the valence p-orbitals means that, in a localized bond description of a molecule, all aliphatic halides should have an asymmetry parameter of zero. Table 13 shows a selection of the asymmetry parameters of such compounds which, although small, overlap with those measured for halobenzenes. Local 'non-bonding' interactions have made non-negligible contributions to the asymmetry parameter so that even if no partial

TABLE 13. Halogen coupling constants and asymmetry parameters in aliphatic halides

Compound	Nucleus	e^2Qq, MHz	η	Ref.
CH_2Cl_2	^{35}Cl	72.04	0.068	75
$CHCl_2COOH$	^{35}Cl	70.30	0.05	76
		71.56	0.11	
p-$ClC_6H_4CCl_3$	^{35}Cl	76.90	0.040	77
		78.57	0.023	
		78.59	0.045	
CH_3I	^{127}I	1765.85	0.0293	78, 79
CH_2I_2	^{127}I	1897.37	0.0272	80
CH_3CH_2I	^{127}I	1647.0	0.016	78
$CH_3CH_2CH_2I$	^{127}I	1666.9	0.052	78
CH_2ICOOH	^{127}I	1978	0.061	78
		1988	0.046	

double bonding occurred the asymmetry parameter of, say, chlorobenzene would not be zero. Thus absolute double-bond characters cannot be calculated from equation (17) nor can two systems be compared whose local environments are not similar – such as 1,3,5-trichlorobenzene and 1,4,6-*s*-triazine. Note, too, the asymmetry parameter of methyl iodide, which arises entirely from crystal field effects.

In the substituted halobenzenes these considerations immediately preclude the study of *ortho* substituents – at least if the estimation of partial double bond character is the object of the study. Furthermore, *meta* substituents are not likely to produce any marked conjugative effects so that the most interesting series is that of the *para*-substituted halobenzenes.

The values for ^{35}Cl, ^{79}Br and ^{127}I for the *para*-substituted halobenzenes are shown in Tables 14, 15 and 16. The substituents are listed in the order of the Hammett σ constant[80]. The values shown in parentheses have been estimated from those for similar groups. Positive σ constants would be expected to favour partial double-bond character and hence increase η while negative σ constants should have the reverse effect. While it is clear that this general trend is respected, there are a number of glaring anomalies. Thus the asymmetry parameter of ^{35}Cl in 4,4′-dichlorodiphenylsulphone seems much too high, but since this is derived from Zeeman studies there is a possibility of error if by chance there were two almost parallel C—Cl vectors in the crystal. This cannot be the case for *p*-fluoroiodobenzene and may indicate that either the σ constant is too low or that it is not a good measure of the conjugative power of a *para* substituent. One must not indeed forget that Hammett

TABLE 14. ^{35}Cl resonance frequencies (in megahertz) and asymmetry parameters for *p*-substituted chlorobenzenes

Substituents	ν, MHz	η	Hammett σ	Ref.
$-SO_2C_6H_4Cl$-*p*	34.33	0.22 ± 0.03	0.73	81
—Cl	3	0.067 ± 0.001	0.23	82
—OH	34.12	0.071 ± 0.001	−0.37	83
	34.35	0.096 ± 0.001		
$-NH_2$	33.77	0.048 ± 0.001	−0.66	84

TABLE 15. ^{79}Br resonance frequencies (in megahertz) and asymmetry parameters for *p*-substituted bromobenzenes

Substituents	ν, MHz	η	Hammett σ	Ref.
$-NH_3^+$	273.6	0.105	(0.82)[a]	85
$-SO_2Cl$	278.9	0.065	(0.73)[a]	85
$-COOCH_3$	272.18	0.094	0.45	86
—Br	267.89	0.045	0.23	87
$-SC_6H_4Br$-*p*	272.18	0.080	0	88
$-OC_6H_4Br$-*p*	273.15	0.064	−0.32	88
—OH	264.78	0.053	−0.37	89
$-NH_2$	265.57	0.029	−0.66	90

[a]The Hammett σ constants shown in parentheses in the fourth column are estimated from the Hammett σ constants of related substituents, namely $-\overset{+}{N}Me_3$ and $-SO_2CH_3$ respectively.

TABLE 16. ^{127}I coupling constants and asymmetry parameters for *p*-substituted iodobenzenes

Substituents	e^2Qq	η	Hammett σ	Ref.
$-\overset{+}{N}H_3Cl^-$	1909.5	0.0771	0.82	67
$-NO_2$	1887.7	0.063	0.78	67
$-COOH$	1814.8	0.0476	0.45	67
$-COOCH_3$	1882.7	0.0726	0.31	67
$-I$	1834.7	0.0477	0.18	67
	1839.4	0.0365		
$-H$	1820.4	0.069	0	5
	1823.7	0.069		
$-NHCOCH_3$	1820.9	0.0449	0	67
$-F$	1847.3	0.112	0.06	5
$-OH$	1823.1	0.029	−0.37	5
	1830.2	0.028		
$-NH_2$	1738.1	0.039	−0.66	67

σ constants and other measures of substituent effects are themselves only approximate quantities and are the result of a number of quite different effects.

F. Conclusion

The investigations discussed above show clearly that chlorine resonance frequencies do indeed reflect the nature of the carbon–chlorine bond and far less extensive studies of bromine and iodine resonances point to the same conclusion. The fluctuations introduced by the intermolecular fields of the solid state are, however, by no means negligible and as a rule of thumb we must conclude that, for ^{35}Cl resonance frequencies, a difference in resonance frequency between two different chlorine nuclei of 0.5 MHz or less is by no means evidence of a difference in electronic environment, and only when this difference gets to 1.0 MHz or more can we conclude that a difference in electronic structure almost certainly exists. The best way of overcoming this difficulty is by studying series of related compounds so as to drown the fluctuations in a mass of statistics.

IV. NQR AS A TOOL

A. The Attenuation of the Effect of Substituents with Chain Length

We have already seen that for the halobenzenes the effect of a substituent is largely determined by its inductive effect, which is rapidly attenuated as the distance between the two centres increases. This question has also been studied in aliphatic systems, typically by measuring the halogen quadrupole resonance frequency in the series $R(CH_2)_nX$.

A selection of such studies is shown in Table 17 for, as usual, ^{35}Cl resonances. In Figure 4 a plot of the data for two of these series where complete coverage is available illustrates strikingly the rapid exponential attenuation of the inductive effect with chain length and the oscillation of the resonance frequency about a value of 33 MHz characteristic of a long-chain aliphatic primary chloride. These oscillations provide, incidentally, a measure of the crystal field scatter, here only of the order of 0.2 MHz.

TABLE 17. Attenuation of the substituent effect with chain length. The ^{35}Cl resonance frequencies (in megahertz at 77 K) of $R(CH_2)_nCl$ compounds

n	$H{-}(CH_2)_n{-}Cl$[91]	$Cl{-}(CH_2)_n{-}Cl$[92]	$Cl_3C{-}(CH_2)_n{-}Cl$[93]	$NC{-}(CH_2)_n{-}Cl$[5]	$HOOC{-}(CH_2)_n{-}Cl$[94]	$ClCO{-}(CH_2)_n{-}Cl$[94]
1	34.029	35.991	36.40	38.122	36.131	37.517
					36.429	
2	32.646	34.361	34.629	34.110	33.948	34.017
					33.978	
3	32.968	32.868				
		32.967				
4	33.255	32.850	33.345	32.844	33.547	
5	33.103	33.222				
6	32.958	33.008			32.980	33.155
					33.228	
7	33.106	32.604				
		32.988				
8	33.009	33.084	33.384	33.198	33.650	33.149
9	33.150	32.550				
		33.000				
10	33.024	33.120				

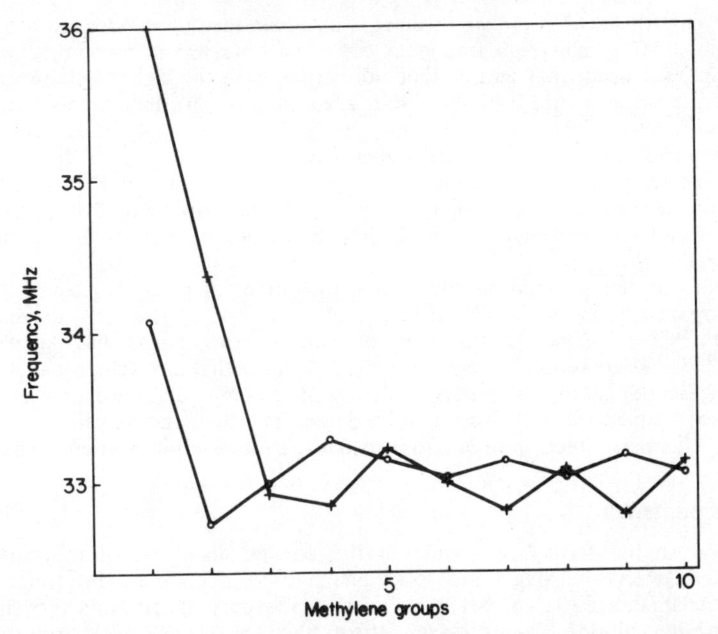

FIGURE 4. Average ^{35}Cl NQR frequencies as a function of the number of methylene groups in $Cl(CH_2)_nCl$ (+) and $H(CH_2)_nCl$ (○).

Very approximately the inductive effect is decreased from one-half to one-third for every additional $-CH_2-$ group. This is the order of magnitude of attenuation which had been estimated from purely chemical studies.

B. Alicyclic Compounds: The Cyclopropanes

Table 18 shows the ^{35}Cl resonance frequencies of a series of monochloro- and *gem*-dichlorocycloalkanes. For both series the frequency is essentially independent of ring size – and equal to that of the open chain analogue – except when we arrive at the

TABLE 18. ^{35}Cl resonance frequencies (in megahertz at 77 K) of alicyclic chlorides. The data have been taken from Ref. 5 except the two marked with an asterisk which are unpublished data, measured by S. Arjomande in the author's laboratory

n	$(CH_2)_n$ CHCl (ring)		$(CH_2)_n$ CCl_2 (ring)
1	33.411		36.258, 36.524, 36.873
2	34.063		36.610
3	31.752*		34.710, 35.358
4	32.125*		34.984, 35.060
5	–		34.904, 34.989
$(CH_3)_2CHCl$	31.939	$(CH_3)_2CCl_2$	34.883

chlorocyclopropanes. At this point there is a sudden increase in frequency of a little less than 2 MHz. This high frequency has been observed in numerous studies of substituted cyclopropanes and undoubtedly arises from the higher s-character of the two extracyclic sp hybrid orbitals of the carbon atoms. This increased s-character is incidentally also reflected in the opening up of the ClCCl angle to $114°38' \pm 15'$[95] compared to the $112° \pm 18'$ in methylene dichloride[96,97].

In Table 18 are also included the 'two-membered rings': the chloroethylenes. Their frequencies are remarkably similar to those of the corresponding cyclopropanes, as indeed would be expected if the higher frequency of the cyclopropanes is a hybridization effect.

In view of the postulation of 'homoconjugation' and of suggestions that the cyclopropane ring has some sort of unsaturated character, the effects of substituents on the ^{35}Cl resonance frequencies of chlorocyclopropanes has been widely studied[98–103]. There is not, however, any clear evidence that the cyclopropane ring is in any way exceptional in this respect. It should be remembered, however, that halogen resonance frequencies seem largely to be determined by inductive effects.

Stereochemical effects in the cyclopropanes are discussed in Section IV.K.

C. Adamantanes

Several studies have been made of the adamantanes and of transmission of substituent effects through the rigid adamantane skeleton. The frequency of 1-chloroadamantane (31.442 MHz) is typical of a tertiary chloride and very similar to that of *t*-butyl chloride. The substituent effects have almost exclusively been studied in the series 3-*R*-1-chloroadamantane but, as we have seen, the attenuation of inductive effects at this distance is already considerable so that no very useful conclusions can be drawn from these studies[104,105].

D. Geminal Interactions in Methyl Halides

1. *Introduction*

As has been mentioned in the previous section, the usually close correlation between the effect of a substituent on a halogen resonance frequency and the various parameters such as electronegativity or Taft σ^*, which provide a semiquantitative measure of the properties of that substituent, break down completely for the first-row substituents F— and RO—, the resonance frequencies being much lower than would be expected. On the other hand, second-row substituents often have markedly greater resonance frequencies than their electronegativities would lead one to expect. These points are illustrated by the general data for monosubstituted chloromethanes in Tables 8 and 10 and in Table 19 further data point them out specifically.

2. *First-row substituents*

For RO— and F— there is clearly present an electron-releasing interaction, which renders the chlorine atom more ionic than the simple electronegativity of the substituent would predict. It is interesting to note in this context that the corresponding R_2NCH_2Cl compounds ionize spontaneously even in the solid state[106] and exist as $(R_2N{=}CH_2)^+Cl^-$.

This spontaneous ionization of the α-aminochloromethanes immediately suggests an explanation of the effect. This is simply that substituents already participate in the

TABLE 19. The effects of fluoro and methoxy substituents on the ^{35}Cl resonance frequencies (in megahertz at 77 K) of substituted chloromethanes

Compounds	ν^{77}	Ref.
CH_3Cl	34.029	32
CH_2Cl_2	35.991	32
CH_2FCl	33.797[a]	32
CCl_4	40.6[b]	32
CCl_3F	39.161, 39.517, 39.703	32
CCl_2F_2	38.450	32
$CClF_3$	38.089	32
CCl_3COOH	39.964, 40.160, 40.236	32
$CFCl_2COOH$	39.594	107
$CF_2ClCOOH$	37.485	107
CH_3OCH_2Cl	30.181	107
CH_3SCH_2Cl	33.104	107
$ClCH_2OCH_2Cl$	32.381, 32.587	32
$ClCH_2SCH_2Cl$	34.526, 34.749	107
$(CH_3)_3CCH_2Cl$	33.015	114
$(CH_3)_3SiCH_2Cl$	34.32	114
$(CH_3)_3GeCH_2Cl$	34.25	114
$(CH_3)_3SnCH_2Cl$	33.936	114
$ClP(CH_2Cl)_4$	38.927, 39.263, 39.445, 35.480	5
$ClP(CH_3)(CH_2Cl)_3$	38.245, 38.486, 38.780	5
$ClP(CH_3)_2(CH_2Cl)_2$	37.436, 37.590, 37.814	5

[a] At 20 K.
[b] Average of 15 resonances.

formation of an incipient double bond and thus drive electrons towards the halogen atom[107]:

$$\overset{\delta+}{X}\cdots\!=\!CH_2\cdots\overset{\delta-}{Cl}$$

A second explanation invokes the internal electric field at the halogen atom produced by the neighbouring dipole in the C—X bond. The direction of this field is such that it increases the ionicity of the carbon–halogen bond and thus decreases the halogen resonance frequency. It is supposed that the effect is particularly marked for the two substituents in question since the dipole moments of the C—X bond will be high and the short C—X distance brings it closer to the halogen nucleus[106,109]. This is essentially an extension of Scrocco's calculations on the shifts produced by intra- and intermolecular fields in polychlorobenzenes[110]. Although the proposed mechanism has been subjected to the test of a calculation of the magnitude of the frequency shift and apparently gives reasonable results, there is a considerable degree of arbitrariness since there is (a) choice in the location of the dipole or dipoles connected with the first-row substituent and (b) an interpretation of the results of the Stark effect in NQR (Section I.E).

It may well be that a detailed analysis of these two mechanisms would show them to amount to the same thing, inasmuch as the structure of atoms and molecules is a result of an interplay of the electrostatic forces between the particles of the system and a change in structure is the result of a change in the internal electrostatic fields, but the first explanation has the advantage that it is expressed in terms of a chemical model and has, moreover, stereochemical consequences which can be easily verified. Thus for the chloro ethers there is only one lone pair π-orbital which can participate in the bonding and the extent to which it affects the chlorine atom will depend on the dihedral angle, θ, between the nodal plane of this orbital and the OCCl plane:

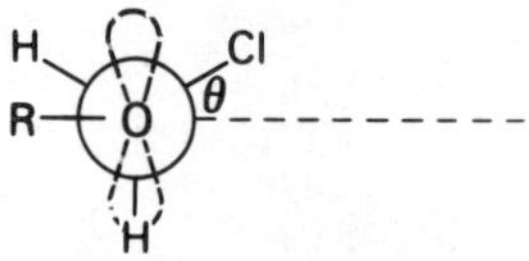

When θ is 90° the interaction will be a maximum and hence the ionicity of the carbon–chlorine bond a maximum with the resonance frequency a minimum, whereas when $\theta = 0°$ there will be no first-order interaction and the resonance frequency will increase. This prediction of the model is completely born out by experience for a large number of cyclic α-chloro ethers of known structure, the average frequency difference between chloro substituents in equatorial positions ($\theta \approx 0°$) and axial positions ($\theta \approx 60°$) being more than 2 MHz; much greater than any 'crystal field' scatter (Section IV.K). Although such a conformational dependence could no doubt be built into the internal field model, for example by adding a point dipole corresponding to the lone pair, this would certainly be done only *after* the experimental discovery of the conformational dependence.

As remarked above, the spontaneous ionization of the α-chloroamines both make it impossible to study the nitrogen analogues of the α-chloro ethers and may be said to support the incipient π-bond model. Unsaturated analogues are readily available, however, for example 2-chloropyridine, and indeed its resonance frequency is much lower than would have been expected. This difference was ascribed to large partial double bonding in the 2-chloropyridine derivative but although the asymmetry parameter of 2-chloropyridine is a little greater than that of 3-chloropyridine the difference in asymmetry parameter is not enough to explain the frequency difference. This question is discussed in more detail in Section IV.F, but it is possible that a similar lone pair interaction may be occurring here and also in the carbonyl chlorides where the frequencies are again very low (Section IV.G).

There is finally another observation which can be explained in terms of the above interaction. The ^{35}Cl resonance frequency of $ClCH_2CH_2CH_2Cl$ (32.952, 33.130 MHz) is almost identical to that of $CH_3CH_2CH_2Cl$ (32.968 MHz) since, as seen in Section IV.A, two methylene groups almost completely attenuate the effect of the chlorine substituent. For the analogous CH_3OCH_2Cl compared with $ClCH_2OCH_2Cl$ the frequency difference is 2.2 MHz[111], greater than that between CH_3Cl and CH_2Cl_2 (1.97 MHz). This would be explained in terms of the lone pair donation hypothesis as competition between the two CH_2Cl groups for the lone pair electrons on the oxygen atom. This cannot, however, be the whole story since a similar difference, 1.54 MHz, occurs between CH_3SCH_2Cl and $ClCH_2SCH_2Cl$ while the frequency of CH_3SCH_2Cl itself is much as would be expected. It is perhaps the methylene group, which is an exceptionally poor transmitter of substituent effects.

3. *Non-first-row substituents*

This phenomenon has been most extensively studied by Voronkov and his coworkers for Group IV substituents[8,10,112–114]. In almost all cases the ^{35}Cl resonance frequency of the compound $R^1R^2R^3MCH_2Cl$ is greater for a given set of substituents

R^1, R^2 and R^3 when M = Si, Ge or Sn than when M = C, despite the fact that the electronegativity of carbon is about 0.7 units higher than that of the other three atoms. Conversely, however, the effect of the substituents R^1, R^2 and R^3 is transmitted more effectively through a carbon atom than through a silicon atom. The dependence of ^{35}Cl resonance frequency on the sum of the Taft σ^* constants for the substituents in the compounds $R^1R^2R^3$ $SiCH_2Cl$ is given by

$$\nu^{77} = 34.053 + 0.31 \sum_i \sigma_i^*$$

compared with

$$\nu^{77} = 32.039 + 0.594 \sum_i \sigma_i^* \quad (28)$$

for the corresponding carbon compounds. Thus the effect in question is most marked when R^1, R^2 and R^3 are electron-releasing groups such as CH_3.

As for almost all 'anomalous' behaviour of second-row atoms, the conventional explanation of this behaviour is based upon the fact that the heteroatom has vacant d orbitals. In a manner analogous to the *donation* of the lone pair electrons of first-row substituents to the carbon–halogen bond, these orbitals are supposed to *accept* the electrons of the carbon–chlorine bond and thus, by decreasing the 3p population of the chlorine atom, increase the ^{35}Cl resonance frequency. This hypothesis has the merit of explaining the reduced ability to transmit substituent effects for, since almost all the substituents have lone pair electrons, these electrons can also interact with the vacant d orbitals, restore electrons to the C—Cl bond and reduce the ^{35}Cl resonance frequency. This effect opposes the purely inductive electron-withdrawing power of the substituent.

It seems difficult to verify this hypothesis. In the first place the idea that the observed frequencies are in any way anomalous depends on our faith in the oversimplified concept of electronegativity. In the second, although the hypothesis explains two separate observations, it is difficult to use it to predict anything. The multiplicity of d orbitals makes it unfortunately unlikely that there would be the marked stereo chemical dependence which is so revealing for first-row substituents. However, at the present it is the best we can do and, after all, may well be correct.

E. Condensed Aromatic Hydrocarbons

The studies of substituted benzene derivatives find a natural extension in the study of halo-derivatives of polycyclic hydrocarbons. From the theoretical point of view the main differences that one would expect to find among the monohalo derivatives would be a difference in the partial double bond character of the carbon–halogen bond, but it has been seen above that the frequencies of the substituted halobenzenes seem to be mainly determined by inductive effects. It is therefore perhaps not surprising that the ^{35}Cl resonance frequencies of the compounds in Table 20 are so similar, even though the partial double bond characters of the chlorine–carbon bond, particularly for 9-chloroanthracene, are expected to be very different from each other.

F. Heterocyclic Halides

1. *Pyridine and the polyazines*

Of all the heterocyclic systems this is the one which has been the most extensively studied by halogen nuclear quadrupole resonance. As usual it is the ^{35}Cl resonances which provide the most data and attention will be focused on these.

TABLE 20. ^{35}Cl resonance frequencies (in megahertz at 77 K) of chloro derivatives of condensed polycyclic hydrocarbons

Molecule	ν^{77}	Ref.
Chlorobenzene	34.62	3
1-Chloronaphthalene	34.606	65
2-Chloronaphthalene	34.69	115
9-Chloroanthracene	34.51	117
9-Chlorophenanthrene	34.837	117
	35.167	
1,4-Dichlorobenzene	34.780 (phase I)	5
	34.765 (phase II)	
	35.208 (phase III)	
1,4-Dichloronaphthalene	35.125	116
9,10-Dichloroanthracene	34.899	65

Table 21 shows the resonance frequencies for a variety of chloroazines, together with those of analogous chlorobenzenes for reference. The most striking thing about these frequencies is the low values for chlorine atoms adjacent to nitrogen. This is most clearly shown in the chloropyridines where, on the basis of the inductive effect of the electronegative nitrogen atom, one would expect the resonance frequencies to be in the order 2-chloropyridine > 3-chloropyridine > 4-chloropyridine ≃ chlorobenzene. The last three items are correctly predicted but the frequency of 2-chloropyridine instead of being the highest is the lowest of all. This point is even more strikingly illustrated by the frequencies of 2-chloroquinoline. We are also fortunate in the present case that most of the monohalopyridines have been studied by microwave spectroscopy where, because of the small mass difference between a nitrogen atom

TABLE 21. ^{35}Cl quadrupole resonance frequencies (in megahertz at 77 K) of chloroazines and the analogous chlorobenzenes

Molecule	ν^{77}	Ref.
Chlorobenzene	34.622	32
o-Dichlorobenzene	35.580, 35.755, 35.824	62
m-Dichlorobenzene	34.809, 34.875, 35.030	62
p-Dichlorobenzene	34.780	62
sym-Trichlorobenzene	35.545, 35.884, 36.115	118
2-Chloropyridine	34.194	119, 120
3-Chloropyridine	35.234	119
4-Chloropyridine	34.739	119
2,6-Dichloropyridine	34.654	121, 122
3,5-Dichloropyridine	35.62	121, 122
2-Chloroquinoline	33.271	119, 120
3-Chloropyridazine	34.52, 35.33	123
3,6-Dichloropyridazine	35.906, 36.015, 36.427, 36.433	124
2-Chloropyrimidine	34.42, 34.52	123
2-Chloropyrazine	35.02, 35.09	124
2,3-Dichloropyrazine	36.10	123
2,6-Dichloropyrazine	35.219, 35.694	124
2-Chloroquinoxaline	34.22	123
2,3-Dichloroquinoxaline	35.66	123
Cyanuric chloride	36.738, 36.771	120, 125

and a CH group compared to the large mass of a halogen, the inertial axis essentially coincides with the C—X bond even for the 2- and 3-halopyridines. These results are shown in Table 21 and it can be clearly seen that the low frequencies of the 2-halopyridines is not an artefact of the solid state. For polychloro derivatives or for the polyazines the frequencies behave in a fairly systematic additive manner with the notable exception of the effect of two nitrogen atoms adjacent to a halogen as in 2-chloropyrimidine where the effect of the *ortho* nitrogen atoms is not cumulative.

It was originally suggested[120] that the low frequency of 2-chloropyridine was due to conjugation with the nitrogen atom,

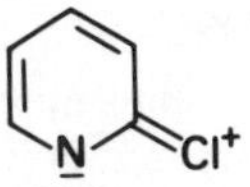

facilitated by the high electronegativity of nitrogen. If it is assumed that in 3-chloropyridine the only effect of the nitrogen atom is an inductive one, increasing the resonance frequency over that of chlorobenzene by 0.6 MHz, and that this inductive effect will be about three times greater in the 2-position, then we may use equation (17) to calculate the additional double bonding required to produce the observed frequency reduction as 5.8%. Added to the 3.1% double-bond character of *p*-dichlorobenzene this implies a total double-bond character of 8.9% and an asymmetry parameter of 0.21. Unfortunately, however, although the asymmetry parameters of 3,5-dichloropyridine[121,122] (0.086) is very similar to that of *p*-dichlorobenzene, in 2,6-dichloropyridine it is only 0.118[121,122]. These values are similar to the asymmetry parameters for 2- and 3-chloropyridine measured in the gas phase (Table 22). Even worse, in cyanuric chloride, where the lowering effect of the adjacent nitrogen atoms is by no means apparent, the asymmetry parameter is quite high (0.23, 0.21)[5]. We are forced to conclude either that the asymmetry parameters are not a good measure of conjugation – and there is other evidence to support this point of view (Section III.E) – or that the explanation of the low frequency of 2-chloropyridine is incorrect.

An alternative explanation may be the same as that proposed for the low frequency of the α-chloro ethers, namely partial donation of the nitrogen lone pair electrons to the C—Cl bond:

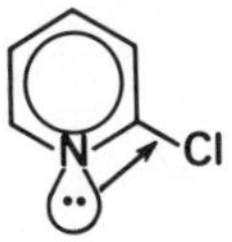

Certainly the lone pair orbitals and the C—Cl bond are now held in the most favourable position for such interaction to occur. In the absence of further information

TABLE 22. Gas-phase halogen quadrupole coupling constants (in megahertz) for monohalopyridines

	2-Halopyridine			3-Halopyridine			4-Halopyridine		
Nucleus	e^2Qq	η	Ref.	e^2Qq	η	Ref.	e^2Qq	η	Ref.
^{35}Cl	−71.10	0.10	126	−72.15	0.07	127	−72.0	0.12	128
^{79}Br	552.2	0.10	129	–	–	–	557.5	0.08	129
^{127}I	−1881	0.08	130	–	–	–	−1938	0.04	130

it is not possible to decide which of the two mechanisms is responsible for the low frequency of 2-chloropyridine, and indeed both mechanisms may be contributing to some extent. CNDO calculation[131] indicate that both interactions occur but that the lone pair interaction is more important. INDO calculations on these systems have also been reported[132] and, like the CNDO calculations, reproduce the experimental trend in 2-, 3- and 4-chloropyridine.

Finally, substituent effects in these systems have been discussed by Bray and his coworkers[133].

2. *Other nitrogen-containing heterocycles*

Several studies have been made of other nitrogen-containing heterocycles but the data are limited[134–139]. The chloroimidazoles have been covered the most thoroughly[140] and some representative results are shown in Table 23. The difference between 4-chloro- and 5-chloroimidazole is noteworthy: the lower frequency of the 4-chloro derivative is an immediate consequence of the lone pair donation mechanism since in the 4-chloro derivative the lone pair on nitrogen and the C—Cl bond are in the same plane whereas in 5-chloroimidazole the lone pair axis and the C—Cl bond are perpendicular.

TABLE 23. ^{35}Cl resonance frequencies (in megahertz at 77 K) of chloro-*N*-methylimidazoles[140]

Compound	ν^{77}
N=, N—CH_3, Cl	36.720 36.924
N=, N—CH_3, Cl	35.034
N=, N—CH_3, Cl, Cl	35.840 37.639 37.758
Cl, N=, N—CH_3, Cl, Cl	36.420 36.912 37.260 37.860

TABLE 24. ^{35}Cl resonance frequencies (in megahertz at 77 K) of chlorothiophenes[141]

Substituent	ν^{77}
2-Cl	36.920, 37.044
2,5-Dichloro	37.018, 37.071
2,3,4,5-Tetrachloro	37.517, 38.500
2-Cl, 5-CN	37.530, 37.416
2-Cl, 5-CHO	37.123, 37.200
2-Cl, 5-CH_3	36.68
2-Cl, 5-$Si(CH_3)_3$	36.321
2-Cl, 5-$SiCl_3$	37.252

3. *Halothiophenes*

Apart from the polyazines the only other heterocyclic system to have been studied in depth is thiophene. Unusually, too, in this case there is as much, if not more, data for the bromo- and iodothiophenes as for the chlorothiophenes. In Table 24 are shown the ^{35}Cl resonance frequencies for several substituted chlorothiophenes. For the 5-substituted 2-chlorothiophenes the ^{35}Cl resonance frequency is related to the Hammett σ parameter of the substituent in the *para* position[141] by the equation

$$\nu^{77} = 36.872 + 0.88\,\sigma_p. \tag{29}$$

This should be compared with the correlation for substituted chlorobenzenes[63]

$$\nu^{77} = 34.826 + 1.024 \sum_i \sigma_i. \tag{30}$$

For the bromo derivatives (Table 25), where a much wider variety of compounds with various substitution patterns are available, the correlation for the ^{79}Br resonance of 2-bromothiophenes[142] is

$$\nu^{77} = 290.9 + 6.7 \sum_i \sigma_i \tag{31}$$

and for 3-bromothiophenes[142]

$$\nu^{77} = 278.6 + 7.0 \sum_i \sigma_i. \tag{32}$$

In these equations the positions of the substituent constants are taken as *ortho, meta* or *para* according to whether the positions of the bromo group and the substituent are separated by zero, one or two carbon atoms. There are thus no '*para*-substituted'

TABLE 25. ^{79}Br resonance frequencies (in megahertz at 77 K) of bromothiophenes[142]

Substituent	ν^{77}
2-Br	290.93
3-Br	277.40, 279.8
2,5-Br_2	291.68, 291.98, 292.73, 293.07
	293.86, 294.47, 295.09
3,4-Br_2	285.15, 288.40, 289.46, 292.44
2,3,4,5-Br_4	296.62, 296.89, 305.51, 305.94

3-bromothiophenes. These correlations should be compared with the correlation for the ^{79}Br resonances of bromobenzenes[65]:

$$\nu^{77} = 271.6 + 9.14 \sum_i \sigma_i. \tag{33}$$

Thus both for ^{35}Cl and ^{79}Br resonances the general correlation seems to indicate that the thiophene ring transmits substituent effects slightly less efficiently than the benzene ring.

For the iodothiophenes the existence of two resonance frequencies provides at once the value of the asymmetry parameter. As for the chloro and bromo derivatives the frequencies of the 2-iodothiophenes are higher than those of the corresponding iodobenzenes. The asymmetry parameters are usually similar, with one or two notable exceptions (Table 26)

A limited study of the effect of substituents in the iodothiophenes yields the correlation for the 5-substituted 2-iodothiophenes[67,143]:

$$e^2Qq^{298} = 1986 + 179\,\sigma_p. \tag{34}$$

The corresponding equation for the *para*-substituted iodobenzenes is

$$e^2Qq^{298} = 1837 + 112\,\sigma_p. \tag{35}$$

In this case the thiophene ring appears to transmit electronic effects *more* effectively than a benzene ring. It must be realized, however, that the correlations for the thiophene derivatives are based on rather few compounds and the slopes and intercepts in the equation are subject to considerable errors and in the author's opinion the results show that the thiophene and benzene rings transmit electronic effects to about the same extent. This conclusion agrees with that based on purely chemical evidence[144].

G. Carbonyl Chlorides

As befits an important class of organochlorine compounds, the carboxylic acid chlorides have been extensively studied. The low resonance frequency of this group (Table 27) is usually ascribed to conjugation and partial double bond formation (see Section IV.F for the analogous 2-chloropyridines), but may well be due to the interaction between the lone pair electrons on the oxygen atom and the carbon–chlorine bond (Section IV.D). The only measured asymmetry parameter, that

TABLE 26. ^{127}I quadrupole coupling constants (in megahertz at 298 K) of 5-substituted 2-iodothiophenes and the corresponding *para*-substituted iodobenzenes[67,143]

	2-Iodothiophenes		*p*-Iodobenzenes	
Substituent	e^2Qq/h	η	e^2Qq/h	η
—I	1997	0.0360	1841	0.0424
	2009	0.0608		
—COOH	1940	0.0565	1841	0.053
	1968	0.0355		
$-COOCH_3$	2052	0.0159	1867	0.0787
$-COCH_3$	2052	0.0155	1884	0.0913
$-NO_2$	2084	0.0529	1889	0.0661

TABLE 27. ^{35}Cl resonance frequencies (in megahertz at 77 K) of substituted carbonyl chlorides, RCOCl

Substituent	ν^{77}
$-CH_3$	28.855, 29.070
$-C_6H_5$	29.918
$-CH_2Cl$	30.437
$-CHCl_2$	32.147
$-CCl_3$	33.721
$-CF_3$	34.432
CH_3O-	33.224
$(C_2H_5)_2N-$	31.877
$-Cl$	35.081, 36.225

of carbonyl chloride ($\eta = 0.25$)[32] lends support to the double-bonding hypothesis, but a more extensive study of the asymmetry parameters in these systems would be welcome.

The effect of substituents[145] and of chain length[146] on the ^{35}Cl resonance frequency has been studied. The resonance frequency for RCOCl compounds is given by the relationship

$$\nu^{77} = 28.972 + (1.544 \pm 0.111)\sigma^*, \tag{36}$$

where σ^* is the Taft inductive parameter of R. This equation implies that the $\rangle C{=}O$ group transmits the electronic effect of the substituent somewhat more effectively than does the $\rangle CH_2$ group (Section IV.A). There is also an important qualitative difference between the two series inasmuch as the effect of the methoxy group is now more or less in accordance with the electronegativity of the oxygen atom and of course the corresponding amino compounds no longer ionize spontaneously. This is no doubt mainly due to the fact that in the chloroformates the basic system is planar because the lone pair p-electrons of the alkoxy group now conjugate with the carbonyl group and thus are completely misaligned for interaction with electrons of the C—Cl bond (Section IV.D).

Analogous to the carbonyl chlorides are the chloroimines and the chloroiminium salts (Table 28). The frequencies of *N*-phenyldichloroimine are not too dissimilar from those of carbonyl chloride, although the relative electronegativities of oxygen and nitrogen would have led one to anticipate that the chloroimines would have had the lower frequencies. The chloroiminium salts have notably higher frequencies, as would have been expected from the positive charge on the nitrogen atom.

H. Acetylenes

Substituent effects in haloacetylenes have been studied in only a few compounds (Table 29). Their interpretation is hindered by the fact that both σ and π shifts may occur and these affect the resonance frequencies in opposite directions (equation 16). This is the probable reason for the small effect on going from chloroacetylene to cyanogen chloride[36].

In view of what we have already seen for the chlorobenzenes it is not surprising

TABLE 28. ^{35}Cl resonance frequencies (in megahertz at 77 K) of chloroimines and chloroiminium salts

Compound	ν^{77}	Ref.
$C_6H_5N{=}CCl_2$	36.128 37.047	147
$[(CH_3)_2N{=}C(H)Cl]^+ Cl^-$	35.786	148
$[(CH_3)_2N{=}C(Ph)Cl]^+ Cl^-$	37.050	106, 148
$[(CH_3)_2N{=}CCl_2]^+ Cl^-$	39.52 39.60 39.68 39.99	148

that the ^{35}Cl resonance frequencies of substituted *p*-chlorophenylacetylenes, $ClC_6H_4C{\equiv}C{-}R$, are but little affected by the nature of R[149].

I. Carboranes

Several studies of the carboranes show that the carboranyl group is highly electron withdrawing. Thus the ^{35}Cl resonance frequencies of a chlorine atom directly bonded to the carbon atom of the carborane system is in the neighbourhood of 41 MHz while that of the chloromethyl carboranes is around 37 MHz and of the carborane carboxylic acid chlorides around 33 MHz. (Table 30). The electron-withdrawing power of the three carboranes is similar and in the order *ortho* > *meta* > *para*. Studies of the transmission of substituent effects through the two carbon atoms of the *o*-carborane system have been made[150].

J. Polycoordinated Organic Halides

The halogen resonances of $C_6H_5ICl_2$[154], of $C_6H_5IO_2$[67], and of various diphenylhalonium compounds, $(C_6H_5)_2X^+Y^-$ [155–178], have been measured. The low

TABLE 29. ^{35}Cl resonance frequencies (in megahertz at 77 K) of chloroacetylenes[36]

Compound	ν^{77}
$HC{\equiv}CCl$	39.127, 39.294
$BrC{\equiv}CCl$	39.448, 39.573
$IC{\equiv}CCl$	39.439, 39.629
$NCC{\equiv}CCl$	39.614, 37.701
$N{\equiv}CCl$	41.736
$C_6H_5C{\equiv}CCl$	38.939, 39.144
$HOCH_2C{\equiv}CCl$	39.378, 39.384

TABLE 30. ^{35}Cl resonance frequencies (in megahertz at 77 K) of chlorine containing carboranes

Compound	ν^{77}	Ref.
1-Phenyl-2-chloro-*o*-carborane	41.51	151
1-Methyl-2-chloro-*o*-carborane	41.124	151
1-Methyl-7-chloro-*m*-carborane	40.98	151
1-Methyl-12-chloro-*p*-carborane	40.47	151
1-Phenyl-2-chloromethyl-*o*-carborane	36.33, 36.95	150
1-Methyl-2-chloromethyl-*o*-carborane	37.02	152
o-Carborane carbonyl chloride	33.018, 33.276	153
m-Carborane carbonyl chloride	32.091	153
p-Carborane carbonyl chloride	32.400, 32.430	153

frequencies are in accord with the higher coordination number of the halogen nucleus[4].

K. Molecular Complexes

1. Donor–acceptor complexes

The effect of certain catalysts is to form a complex with the substrate which then reacts more readily than the uncomplexed molecule. The variations in electronic structure which are responsible for the reactivity change produced by complex formation can be studied by NQR spectroscopy in model systems and a number of such studies involving the carbon–halogen bond have been made.

Most of the systems studied can be represented by the general formula DMX, where D is an electron-donor site, M represents the rest of the molecule and X the halogen nucleus. This molecule is studied both in the free state and in the form of a complex with Lewis acid, ADMX.

Several different factors may be varied. For a given molecule the effect of different donors may be studied and for a given molecular framework the halogen nucleus, considered as a probe, can be placed at different distances from the reactive site.

A fairly complete study of this kind has been carried out on the chloropyridines[159]. Chloro substituents can be placed in the 2-, 3- or 4-positions and stable complexes are formed with a variety of Lewis acids, including H^+. Table 31 shows typical results. It transpires that the frequency shift – naturally enough to higher frequencies – is but

TABLE 31. ^{35}Cl resonance frequencies (in megahertz at 77 K) of complexed chloropyridines[159]

	ν^{77}		
Complex	2-Cl	3-Cl	4-Cl
Free base, DMX	34.194	35.238	34.89*
DMX HCl	37.555	37.05	35.94
DMX BCl_3	38.572	37.305	36.585
DMX $SbCl_5$	37.872	37.076	–
DMX $HgCl_2$	38.190	–	36.32*
DMX $BiCl_3$	37.413	–	36.399

little influenced by the nature of the acceptor and, not surprisingly, the magnitude of the shift is greater the closer the halogen is to the site of complexation. The frequency shift in the 2-chloropyridines corresponds to an increase in covalent character of about 7%.

In Table 32 are shown the frequencies of complexes where the organochlorine atom is attached to a carbon atom directly bonded to the donor atom. The greatest shift occurs for the BCl_3 complex of chloromethyl methyl ether. This may reflect not only the partial charge on the oxygen atom but also the fact that the lone pair electrons can no longer interact with the electrons of the C—Cl bond. The negligible shift for cyanogen chloride is no doubt a consequence of the fact that shifts occur both in the π and σ populations of the chlorine atom and that these have opposing effects on the resonance frequency.

2. *Sandwich compounds*

A few studies have been made of the effect of complexation by transition metal carbonyls and related compounds. The most extensive of these[162] concerns complexes of the general formula $[RC_6H_4Cl\cdots Fe\cdots C_5H_5]^+PF_6^-$. The ^{35}Cl NQR frequency of the chlorobenzene fragment is related to the inductive parameter of

TABLE 32. ^{35}Cl resonance frequencies (in megahertz at 77 K) of organochlorine nuclei in donor–acceptor complexes

Complex	ν^{77} Complex	ν^{77} Free donor	Ref.	$\Delta\nu$
2-chloropyridine $N \rightarrow SbCl_5$ (Cl on ring)	38.572	34.194	159	4.38
$C_6H_5C(Cl)=O \rightarrow SbCl_5$	29.918	34.164	159	4.25
$ClC\equiv N \rightarrow SbCl_5$	42.005	41.736	160	0.26
$CH_3(ClCH_2)O \rightarrow BCl_3$	35.772	30.180	159	5.59
$[CH_3(ClCH_2)O] \rightarrow SnCl_4$	34.320	30.180	161	4.14
$CH_3(ClCH_2)S \rightarrow BCl_3$	37.963 38.031	33.104	159	4.89

R, σ_I, by the equations

$$\nu^{77} = 37.13 + 0.82\,\sigma_I \quad \text{for } para \text{ substituents.} \tag{37}$$

$$\nu^{77} = 37.13 + 0.72\,\sigma_I \quad \text{for } meta \text{ substituents.} \tag{38}$$

Comparison with the results for uncomplexed chlorobenzenes shows that the basic resonance frequency has been increased by 2.5 MHz while the substituent effect seems slightly less marked.

Although it is possible that it is the positive charge which is responsible for the increase in resonance frequency, a similar increase has been observed for the neutral complex $C_6H_5Cl—Cr(CO)_3$ (36.83 MHz, chlorobenzene 34.62 MHz)[163].

3. *Van der Waals' complexes*

Apart from the above complexes, where there is a specific donor–acceptor interaction, there have been extensive studies of systems for which the molecular interaction is more diffuse, as for example, with the complexes of trichloracetic acid[164], of chloroform[165], of chloranil[166,167] or of picryl chloride[168]. The frequency shifts are here much less marked and NQR does not in general throw very much light on the nature of the interaction. Other complexes are discussed by Weiss[19].

L. Conformation and Configuration

1. *Introduction*

Differences in chemical reactivity brought about by differences in conformation or configuration are well known, but it is not always clear that these differences stem from differences in the electronic structure of the ground state. NQR seems to afford a way of answering this question and accordingly a number of studies have been devoted to this point. It is, however, in only one system – the chloro ethers – that a large and unambiguous effect has been observed.

This is not to say that conformational and configurational effects in other systems do not exist, but as we have seen on several previous occasions, the fluctuations introduced by solid state effects and even by local intramolecular field gradients are often of a magnitude comparable to the effect we are trying to study. Thus, to take a typical example, in 1,3,5-trichlorobenzene, where all three chlorine atoms are identical in the free molecule, three ^{35}Cl NQR frequencies are observed with a spread of 0.57 MHz and, in perhaps an extreme case, two resonances are observed for $COCl_2$ separated by 1.14 MHz. As usual therefore we can draw no firm conclusion from isolated measurements unless the effect is very large. The difficulty is often, however, to have a sufficiently large number of similar systems for often the only strictly comparable molecules are just two geometrical isomers.

2. Cis/trans *isomers about a carbon–carbon double bond*

The final remark of the above paragraph can be immediately illustrated with this system. As Table 33 shows, a relatively large number of *cis/trans* isomers have been studied, but there is no *a priori* reason for supposing that the difference between two *cis/trans* isomers should be either of similar magnitude or of similar sign for two different substituents. In accordance with this both positive and negative differences can be seen and, with two or three exceptions, the differences are small.

In addition to these studies a number of compounds of the general formula

TABLE 33. ^{35}Cl NQR frequencies (in megahertz at 77 K) of *cis*- and *trans*-substituted vinyl chlorides

X	Y	X(Cl)C=C(H)Y	X(Cl)C=C(Y)H	$\Delta\nu$	Ref.
H	Cl	34.837, 34.894	34.497	0.36	169
H	COOH	35.261	34.315, 34.337	0.94	169
CH_3	COOH	34.250[a]	34.980[a]	−0.73	169
Ph	SO_2Ph	35.647, 35.765 35.920, 36.076	34.964	1.11	169
Ph	Cl	35.692 37.796	35.68 35.81	1.98	169
H	OCH_3	34.219	35.033	−0.81	170
H	HgCl	33.132	32.806	0.32	5
H	HgBr	33.078	33.246	−0.17	5
H	HgBr	32.95	33.77	−0.82	5

[a]These two frequencies were incorrectly interchanged in Ref. 169.

$R^1R^2C{=}CCl_2$ have been investigated and, once again, the frequency difference between the resonances of the two chlorine atoms is small[170–175].

3. Cis/trans *isomers about a carbon–carbon single bond*

Although in this case there are possibilities of a large number of comparable systems there are nevertheless at least two different types of isomeric pairs. For cyclohexane-like systems, where the substituents on adjacent carbon atoms have a staggered conformation, *trans* isomers are usually axial–axial and *cis* isomers axial–equatorial. There also exist, however, systems where adjacent substituents eclipse each other and in this case we might expect the interactions to be more marked.

For cyclohexane-like systems the NQR spectra of the various isomers of 1,2,3,4,5,6-hexachlorocyclohexane show a tendency for axial chloro substituents to have frequencies around 0.5 MHz lower than those in the equatorial position, although the assignments are not always clear. The frequency spread in those compounds is, however, over 1.5 MHz, and this must be due to a combination of intra- and intermolecular interactions[176–180]. In 1,2-dichlorocycloheptane and 1,2-dichlorocyclooctane the frequency difference between *cis* and *trans* isomers is only about 0.1 MHz and a similar difference occurs in two isomeric dichlorocholresteryl benzoates[169]. In agreement with the results for the hexachlorocyclohexanes, for several 5-chloro-1,3-dioxanes and 5-chlorodioxathianes[181] the resonance frequency for a chlorine atom in the axial position is about 0.5 MHz lower than that of a chlorine atom in the equatorial position.

Systems containing the CCl_3 group also are expected to show stereochemical effects

Cl, H, H, Cl, Cl, R

(1)

Cl, R, R, Cl, Cl, H

(2)

TABLE 34. ^{35}Cl NQR frequencies (in megahertz at 77 K) of $R^1R^2_2CCCl_3$

R^1	R^2	ν^{77}	$\Delta\nu$	Ref.
Cl	H	39.02	0	93
CH_2Cl	H	38.150 38.304 38.825	0.67	93
$(CH_2)_8Cl$	H	37.310 38.308 38.528	1.22	93
H	OH	38.189 39.433 39.519	1.33	182
H	p-ClC_6H_4	38.487 38.814 39.039	0.55	183

since in only exceptional cases is it possible for the three chlorine atoms to be equivalent. Thus for a compound RCH_2CCl_3 (**1**) or R_2CHCCl_3 (**2**) we have at least two different chlorine atoms.

Such an effect was noted early in the history of NQR and, as the data in Table 34 show, the splitting is in the region 0.5–1.3 MHz. Compounds where the CCl_3 group is attached to a conjugated system are discussed in Section IV.K.5.

Table 35 shows results for eclipsed systems. As expected the differences are large

TABLE 35. ^{35}Cl NQR frequencies (in megahertz at 77 K) of eclipsed vicinal dichloro derivatives

	cis	*trans*	$\Delta\nu$	Ref.
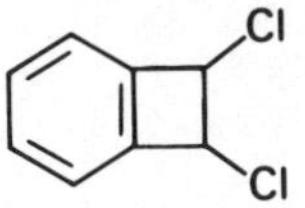	34.603 34.730	33.915 33.958	0.73	184
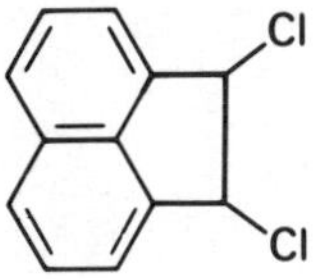	35.652 35.836	33.232	2.42	184
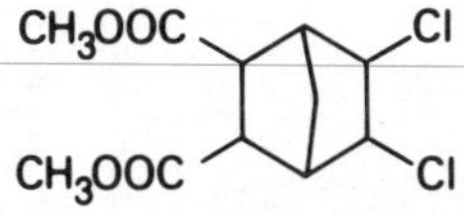	35.234[a] 35.335	34.403[a] 34.689	0.69	184
	35.446 35.748	33.996 35.184	1.01	185

[a] *Exo*, *exo*-dicarbomethoxy derivative.

here, that for the dichloroacenaphtenes being particularly noticeable. In all four pairs the *cis* isomer has the greatest frequency.

4. Syn/anti and endo/exo isomers

Table 36 shows a few representative values for pairs of isomers of polycyclic systems where the difference resides in the *syn/anti* or *exo/endo* position of the chlorine atom. From what has gone before we would expect the differences to be small, as indeed they are.

TABLE 36. Stereochemical effects in polycyclic systems (ν in megahertz at 77 K)

Compounds	ν^{77}	Ref.
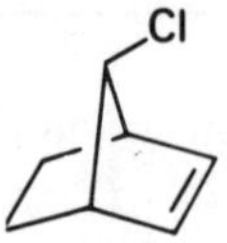 Cl	33.021	186, 187
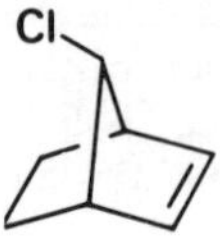 Cl	32.158	186, 187
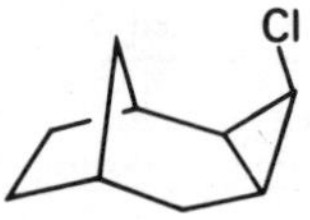 Cl	34.750	188
 Cl	34.050	188
 Cl, Cl	36.620 36.850	188
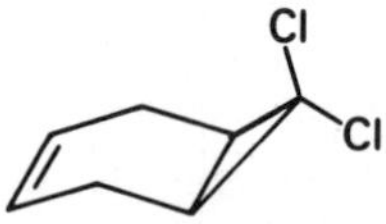 Cl, Cl	36.420 36.751	188
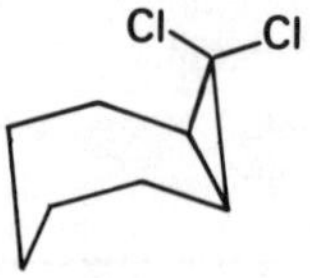 Cl, Cl	36.342 36.446 36.536 36.618	188

5. The effect of orientation with respect to a conjugated system: hyperconjugation

The interaction between a single σ-bond and a conjugated system is usually referred to as hyperconjugation and, although originally applied to interactions involving C—H bonds, can in principle be studied through the halogen NQR spectroscopy of carbon–halogen bonds. Such hyperconjugation must in fact be orientation dependent for the C—X bond, attaining a maximum in **3** and zero in **4**. Dewar and Herr[189] invoked

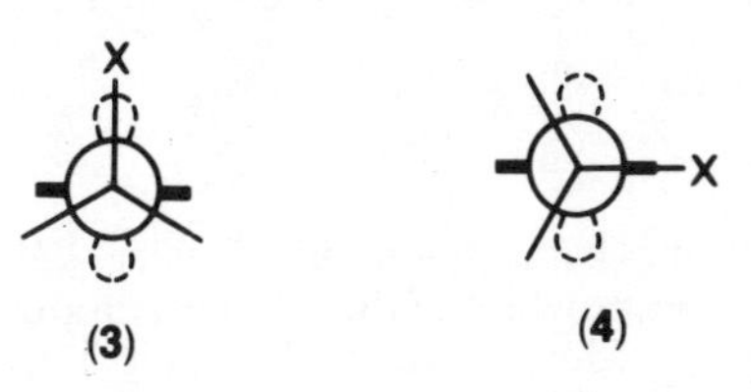

this effect to explain the effects of substituents in substituted benzyl chlorides where, in the absence of *ortho* substituents or in the presence of two *ortho* substituents, the benzyl chlorides have form **3** while one *ortho* substituent will favour form **4**. It was also used to explain the difference (0.84 MHz) between the resonance frequencies of the CCl_3 group in *p*-chlorobenzotrichloride[190]. The frequencies of trichloromethyl groups attached to a conjugated system have been discussed by Ardalan and Lucken[191] and although the frequency spreads are frequently more than 1 MHz the explanation of this is certainly more complex than the one given above. The splittings observed here are comparable in magnitude with those observed when the CCl_3 group is attached to a saturated system so that in fact there is no compelling reason to ascribe the effect to hyperconjugation.

6. α-Chloro ethers

As mentioned in Section IV.D the mechanism proposed to explain the low NQR frequency of α-chloro ethers has definite stereochemical consequences and, since the lowering of the frequency of α-chloro ethers is several megahertz, these should be easily discernible. Evidence from structural studies[192] and conformational equilibria[193], where the anomalous preference of an α-chloro substituent to adopt the axial position in cyclic ethers was baptised the 'anomeric affect', also indicated that this would be a fruitful field for study.

The first clear evidence of the effect was noted in 1970 and substantiated by extensive studies of several different systems in the following years[194–198]. These studies show that, as predicted by the lone pair/C—Cl bond interaction hypothesis, two otherwise equivalent chlorine atoms have resonance frequencies 2.42 ± 0.13 MHz lower when the C—Cl bond is in the axial position than when it is in the equatorial position. This large and regular frequency difference is absolutely unique in ^{35}Cl NQR spectroscopy. For comparison the total range of frequencies for RCH_2Cl compounds is 9 MHz (Table 8).

Finally a variety of molecular orbital calculations of various degrees of sophistication confirm that the flow of electrons from the lone pair oxygen orbital to the C—Cl bond does indeed occur and is of the correct magnitude to explain the NQR frequency differences[199–201].

M. Analysis and Structure Determination

1. Introduction

Analytical applications of NQR are by no means numerous, in particular for organic chemistry the powerful tool provided by NMR is usually more than adequate for the solution of most analytical and structural investigations. In some circumstances, however, particularly for heavily chlorinated compounds, the method can be used to advantage.

2. Qualitative analysis

One of the practical applications – as usual using ^{35}Cl NQR – arose from the problem of determining the structure of the different products, C_8Cl_8, arising from the exhaustive chlorination of cyclooctatetraene[202]. Table 37 shows the structures of the three isomers and their ^{35}Cl NQR frequencies. NQR could not of course arrive at these structures alone. It is possible, however, to differentiate between several chemically reasonable structures and this is the technique applied in this case. Thus the frequencies of the γ isomer are all relatively low, thus indicating the absence of any

TABLE 37. ^{35}Cl resonance frequencies (in megahertz at 77 K) of isomers C_8Cl_8

Isomer	Structure	ν^{77}
α	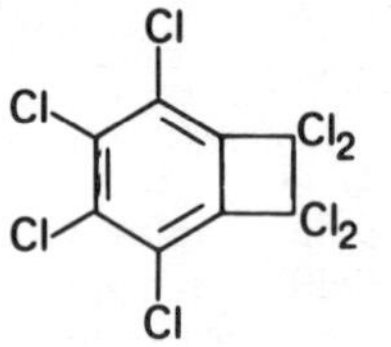	35.53 35.95 36.99(2)[a] 37.91 38.18 39.63 39.91
β	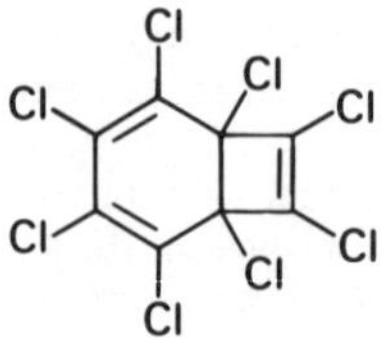	37.32 37.49 37.60(2)[a] 37.94(2)[a] 38.12 38.51
γ	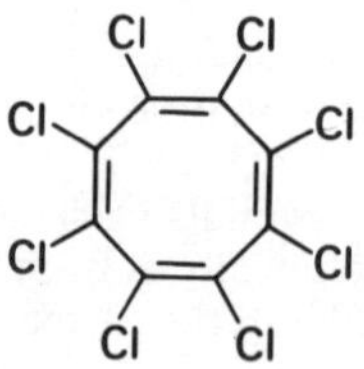	36.60 37.13(2)[a] 37.27

[a]The figures in parentheses after the frequencies indicate the relative intensity of this resonance in cases where this is different from unity.

TABLE 38. ^{35}Cl resonance frequencies (in megahertz at 77 K) of complexes of boron trichloride with organic nitriles[204]

Compound	ν^{77}
BCl_3	21.580
$CH_3CN.BCl_3$	22.300
CCl_3CN	41.553
	41.666
	41.730
$CCl_3CN.BCl_3$	23.084
	23.156
	37.821
	40.263
	41.163
	41.679

gem-dichloro groups or small rings, and sufficiently close together to indicate that all C—Cl bonds are basically similar. Octachlorocyclooctatetraene is the only reasonable candidate for this. For the other two likely structures the α isomer having the highest frequencies must contain the CCl_2 groups, leaving the remaining structure for the β isomer.

The addition of BCl_3 to organic nitriles usually yields donor–acceptor complexes $RCN—BCl_3$. For nitriles with very electronegative substituents there was infrared evidence that addition takes place across the C≡N triple bond[203], and this was confirmed by NQR[204] through the appearance of a new resonance typical of a chlorine–carbon bond (Table 38). The final product probably consists of a dimer having a four-membered boron–nitrogen ring:

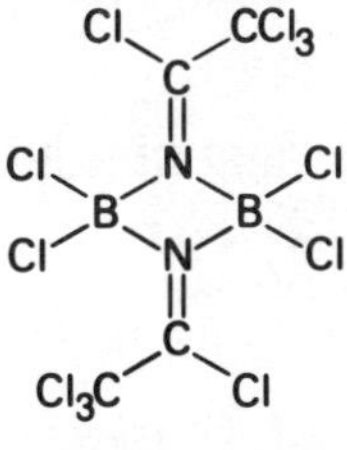

Cl CCl3
C
N
Cl B B Cl
Cl N Cl
C
Cl3C Cl

A related example is the reaction between chlorodimethyl ether and certain Lewis acids. With BCl_3 or $SnCl_4$ it reacts to give a donor–acceptor complex with the oxygen atom as the donor site. This is clearly revealed by the increase in frequency of the organochlorine nucleus (Section IV.J). With $SbCl_5$ on the other hand, the NQR spectrum shows the disappearance of the typical organochlorine resonance and here the product is the ionic complex $CH_2{=}\overset{+}{O}CH_3.SbCl_6^-$ [205]. Two reviews of analytical applications of halogen NQR have been published in the last decade[206,207].

3. *Structure determination*

Several different types of crystalline structural information are available from NQR. For example the three resonances in 1,3,5-trichlorobenzene show that the threefold symmetry of the isolated molecule is lost in the crystal (to be strictly correct this could also mean that there are three inequivalent molecules, each with threefold symmetry, in the unit cell). Likewise the two resonances for *p*-chlorophenol show that the unit cell contains two non-equivalent phenol molecules. Such information is of use to the crystallographer wishing to make an X-ray diffraction study. In the same vein discontinuities in the frequency/temperature relationship are diagnostic of a phase change.

Zeeman studies of single crystals yield more detailed information. An early example of this was 1,2,3,4,5-tetrachlorobenzene[208], where the three principal directions of the ^{35}Cl field gradient tensors define the orientation of the whole molecule, again providing a starting point for the X-ray crystallographer.

When heavily chlorinated molecules are studied the principal directions of the various chlorine atoms are often sufficient to give a fairly complete structural picture, although of course bond lengths are not obtained. Examples of this are perchloro(1,2-dimethylene cyclobutane)[209], perchloro-5-methylcyclopentadiene[210] and perchloro-3-cyclopentanone[211], where the rings have been shown to be almost planar and a number of 'bond angles' determined. Whether this technique presents any advantages over modern conventional crystallography is, however, doubtful.

N. NQR in Excited Triplet States

A variety of experiments combining the techniques of optical spectroscopy and magnetic resonance have yielded information about quadrupole coupling constants in excited triplet states. The measurements are, however, time-consuming and can only be applied to molecules having appropriate triplet lifetimes and spectra so that data are by no means extensive. The subject was reviewed in 1975 by Harris and Buckley[212] and we complete this review by including a selection of more recent references.[213–219]. The interpretation of differences in quadrupole coupling constants between ground and excited states is complicated by the fact that the molecular geometry in the excited state may be very different from that in the ground state and is, in addition, unknown. In fact a large change in quadrupole coupling constant has been taken to *imply* a major change in molecular geometry.

V. REFERENCES

1. H. Schüler and T. Schmidt, *Z. Phys.*, **94**, 457 (1935).
2. H. G. Dehmelt and H. Krüger, *Naturwissenschaften*, **37**, 111 (1950).
3. T. P. Das and E. L. Hahn, *Nuclear Quadrupole Resonance Spectroscopy*, Academic Press, London and New York (1958).
4. E. A. C. Lucken, *Nuclear Quadrupole Coupling Constants*, Academic Press, London and New York (1969).
5. G. K. Semin, T. A. Babushkina and G. G. Yakobson, *Nuclear Quadrupole Resonance in Chemistry*, Wiley, New York (1975).
6. *Nuclear Quadrupole Resonance* (Ed. F. Boschke), *Topics in Current Chemistry*, **30**, Springer Verlag, Berlin (1972).
7. H. Chihara and N. Nakamura in *MTP International Review of Science*, *Physical Chemistry*, *Series 1*, Vol. 4 (Ed. C. A. McDowell), Butterworths, London (1972), pp. 125.
8. M. G. Voronkov and V. P. Feshin in *Determination of Organic Structures by Physical Methods*, Vol. 5, Academic Press, London and New York (1973), pp. 169–234 and 323–340.
9. E. A. C. Lucken in *Physical Methods in Heterocyclic Chemistry*, Vol. IV, Academic Press, London and New York (1971), pp. 21–53.
10. M. G. Voronkov and V. P. Feshin, *Org. Mag. Resonance*, **9**, 665 (1977).
11. *Advances in Nuclear Quadrupole Resonance*, Vols I–III (Ed. J. A. S. Smith), Heyden, London (1974, 1975, 1978).
12. H. Haas and D. A. Shirley, *J. Chem. Phys.*, **58**, 3339 (1973).

13. H. Ackerman, D. Dubbers and H. J. Stöckmann in *Advances in Nuclear Quadrupole Resonance*, Vol. III (Ed. J. A. S. Smith, Heyden, London (1978), pp. 1–66.
14. Y. Morino and M. Toyama, *J. Chem. Phys.*, **35**, 1285 (1961).
15. H. R. Brooker and R. B. Creel, *J. Chem. Phys.*, **61**, 3658 (1974).
16. J. Darville, A. Gerard and M. T. Calende, *J. Mag. Resonance*, **16**, 205 (1974).
17. J. Darville and A. Gerard, *Comput. Phys. Commun.*, **9**, 173 (1975).
18. N. E. Ainbinder and G. Svetlov, *Zh. Strukt. Khim.*, **14**, 766 (1973).
19. A. Weiss in *Nuclear Quadrupole Resonance* (Ed. F. Boschke), *Topics in Current Chemistry*, **30**, Springer Verlag, Berlin (1972), pp. 1–76.
20. H. Jones, Ph.D. thesis, University of Wales (1969).
21. H. Jones, J. Sheridan and O. I. Stiefvater, *Z. Naturforsch.*, **32a**, 866 (1977).
22. S. G. Kukolich and A. C. Nelson, *J. Chem. Phys.*, **57**, 869 (1972).
23. Y. Morino and C. Hirose, *J. Mol. Spect.*, **24**, 204 (1967).
24. Y. Morino and C. Hirose, *J. Mol. Spect.*, **22**, 99 (1964).
25. T. Oka and H. Hirakawa, *J. Phys. Soc. Japan*, **12**, 820 (1957).
26. S. J. Tetenbaum, *Phys. Rev.*, **86**, 440 (1952).
27. C. H. Townes, A. N. Holden and F. R. Merritt, *Phys. Rev.*, **74**, 926 (1948).
28. T. Oka, H. Hirakawa and A. Miyahara, *J. Phys. Soc. Japan*, **12**, 39 (1957).
29. A. Javan, *Phys. Rev.*, **99**, 1302 (1955).
30. M. Imachi, T. Tanaka and E. Hirota, *J. Mol. Spect.*, **63**, 265 (1976).
31. A. K. Garrison, J. W. Simmons and C. Alexander, *J. Chem. Phys.*, **45**, 413 (1966).
32. R. Livingstone, *J. Phys. Chem.*, **57**, 496 (1953).
33. R. H. Schwendemann and G. D. Jacobs, *J. Chem. Phys.*, **36**, 1245 (1962).
34. D. Kivelson, E. B. Wilson and D. R. Lide, *J. Chem. Phys.*, **32**, 205 (1960).
35. R. Bersohn, *J. Chem. Phys.*, **22**, 2078 (1954).
36. P. Gerber, H. Labhart and E. Kloster-Jensen, *Helv. Chim. Acta*, **54**, 2030 (1971).
37. W. J. Lafferty, D. R. Lide and R. A. Toth, *J. Chem. Phys.*, **43**, 2063 (1965).
38. W. Caminati and A. M. Mirri, *Chem. Phys. Lett.*, **12**, 127 (1971).
39. H. Zeldes and R. Livingston, *J. Chem. Phys.*, **21**, 1418 (1953).
40. L. Flanagan and L. Pierce, *J. Chem. Phys.*, **38**, 2963 (1963).
41. A. Schawlow, *J. Chem. Phys.*, **22**, 1211 (1954).
42. S. Kojima, K. Tsukada, S. Ogawa and A. Shimauchi, *J. Chem. Phys.*, **21**, 1415 (1953).
43. T. Kasuya and T. Oka, *J. Phys. Soc. Japan*, **15**, 296 (1960).
44. H. W. Morgan and J. H. Goldstein, *J. Chem. Phys.*, **22**, 1427 (1954).
45. M. Moloney, *J. Chem. Phys.*, **50**, 1981 (1969).
46. A. M. Mirri and W. Caminati, *Chem. Phys. Lett.*, **8**, 409 (1971).
47. J. Hatton and B. Rollin, *Trans. Faraday Soc.*, **50**, 538 (1954).
48. C. H. Townes and B. P. Dailey, *J. Chem. Phys.*, **17**, 782 (1949).
49. R. Bersohn, *J. Chem. Phys.*, **22**, 2078 (1954).
50. K. P. R. Nair, J. Hoeft and E. Tiemann, *J. Mol. Spect.*, **78**, 506 (1979).
51. D. de Kerkhove Varent, *Ann. Soc. Sci. Bruxelles*, **84**, 277 (1970).
52. E. W. Kaiser, *J. Chem. Phys.*, **53**, 1686 (1970).
53. F. A. Van Dijk and A. Dymanus, *Chem. Phys. Lett.*, **4**, 170 (1969).
54. F. A. Van Dijk and A. Dymanus, *Chem. Phys. Lett.*, **2**, 235 (1968).
55. G. Cazzoli, D. G. Lister and P. G. Favero, *J. Mol. Spect.*, **42**, 286 (1972).
56. L. C. Krisher, R. A. Gsell and J. M. Bellama, *J. Chem. Phys.*, **54**, 2287 (1971).
57. R. Kewley, P. M. McKinney and A. G. Robrette, *J. Mol. Spect.*, **34**, 390 (1970).
58. R. W. Taft Jr in *Steric Effects in Organic Chemistry* (Ed. M. S. Newman), Wiley, New York, (1956), p. 619.
59. I. P. Biryukov and M. G. Voronkov, *Coll. Czech. Chem. Commun.*, **32**, 830 (1967).
60. D. Biedenkapp and A. Weiss, *J. Chem. Phys.*, **49**, 3933 (1968).
61. W. Pies, H. Rager and A. Weiss, *Org. Mag. Resonance*, **3**, 147 (1971).
62. W. Pies and A. Weiss, *Z. Naturforsch.*, **26b**, 555 (1971).
63. W. Pies and A. Weiss in *Advances in Nuclear Quadrupole Resonance*, Vol. I (Ed. J. A. S. Smith), Heyden, London (1974), pp. 57–70.
64. N. Nowak, W. Pies and A. Weiss in *Proceedings of the 2nd International Symposium on NQR Spectroscopy* (Ed. A. Colliginiani), Vallerini, Pisa (1975), p. 165.
65. W. Pies and A. Weiss, *J. Mag. Resonance*, **30**, 469 (1978).

66. W. Pies and A. Weiss, *Chem. Ber.*, **111**, 335 (1978).
67. D. V. Ramanamurti, P. Venkatacharyulu and D. Premaswarup, *Org. Mag. Resonance*, **12**, 655 (1979).
68. P. R. Wells, *Progr. Phys. Org. Chem.*, **6**, 111 (1968).
69. H. C. Meal, *J. Amer. Chem. Soc.*, **74**, 6121 (1952).
70. P. J. Bray and R. G. Barnes, *J. Chem. Phys.*, **27**, 551 (1957).
71. E. N. Tsvetkov, G. K. Semin, D. I. Lobanov and M. I. Kabachnik, *Tetrahedron Lett.*, 2521 (1967).
72. H. O. Hooper and P. J. Bray, *J. Chem. Phys.*, **33**, 334 (1960).
73. R. V. S. Rama Rao and C. R. K. Murty, *Indian J. Appl. Sci.*, **4**, 334 (1966).
74. G. W. Ludwig, *J. Chem. Phys.*, **25**, 159 (1956).
75. G. Litzistorf, S. Sengupta and E. A. C. Lucken, *J. Mag. Resonance*, **42**, 307 (1981).
76. P. Venkatacharyulu and D. Premaswarup, *J. Phys. Soc. Japan*, **37**, 1594 (1974).
77. T. Kiichi, N. Nakamura and H. Chihara, *J. Mag. Resonance*, **6**, 516 (1972).
78. S. Kojima, K. Tsukada, S. Ogawa and A. Shimanchi, *J. Chem. Phys.*, **21**, 2237 (1953).
79. H. Robinson, H. G. Dehmelt and W. Gordy, *J. Chem. Phys.*, **22**, 511 (1954).
80. C. D. Ritchie and W. F. Sager, *Progr. Phys. Org. Chem.*, **2**, 234 (1964).
81. G. Chakrapani, V. V. S. Sarma and C. R. K. Murty, *J. Phys. Soc. Japan,* **34**, 994 (1973).
82. P. Bucci, P. Cecchi and E. Scrocco, *Ric. Sci.*, **4**, 129 (1964).
83. P. Bucci, P. Cecchi and A. Colligiani, *J. Chem. Phys.*, **50**, 530 (1969).
84. R. Ambrosetti, A. Colligiani, P. Grigolini and F. Salvetti, *J. Chem. Phys.*, **60**, 459 (1974).
85. R. Rao and M. T. Rogers, *J. Mag. Resonance*, **8**, 392 (1972).
86. R. Angelone, P. Cecchi and A. Colligiani, *J. Chem. Phys.*, **53**, 4096 (1970).
87. P. Bucci, P. Cecchi and A. Colligiani, *J. Amer. Chem. Soc.*, **86**, 2513 (1964).
88. R. Ambrossetti, R. Angelone, P. Cecchi and A. Colligiani, *J. Chem. Phys.*, **54**, 2915 (1971).
89. P. Bucci, P. Cecchi and A. Colligiani, *J. Chem. Phys.*, **50**, 530 (1969).
90. A. Colligiani, R. Ambrosetti and P. Cecchi, *J. Chem. Phys.*, **55**, 4400 (1971).
91. M. G. Voronkov, V. P. Feshin and E. P. Popova, *Latv. P.S.R. Zinat. Akad. Vestis*, **2**, 33 (1970).
92. G. K. Semin, *Dokl. Akad. Nauk SSSR*, **158**, 1169 (1964).
93. G. K. Semin and V. I. Robas, *Zh. Strukt. Chim.*, **7**, 117 (1966).
94. V. P. Feshin, M. G. Voronkov and N. I. Berestennikov, *Isż. Sib. Old. Akad. Nauk SSSR Ser. Khim. Nauk*, **6**, 167 (1974).
95. W. H. Flygare, A. Narath and W. D. Gwinn, *J. Chem. Phys.*, **36**, 200 (1962).
96. R. J. Myers and W. D. Gwinn, *J. Chem. Phys.*, **20**, 1420 (1952).
97. T. Kawaguchi, K. Tanaka, T. Takeuchi and T. Watanabe, *Bull. Chem. Soc. Japan*, **46**, 62 (1973).
98. J. E. Todd, M. A. Whitehead and K. E. Weber, *J. Chem. Phys.*, **39**, 404 (1963).
99. V. S. Grechiskin, I. V. Murin, V. P. Sirkov and M. Z. Yusupov, *Zh. Fiz. Khim.*, **45**, 2891 (1971).
100. A. N. Murin, I. V. Murin, V. P. Kosakov and V. P. Sirkov, *Zh. Strukt. Khim.*, **14**, 158 (1973).
101. F. Delay, M. Geoffroy, E. A. C. Lucken and P. Müller, *JCS Faraday II*, **71**, 463 (1975).
102. V. P. Feshin, T. K. Voropaeva and A. A. Retinskii, *Dokl. Akad. Nauk SSSR*, **220**, 1380 (1975).
103. V. P. Feshin, M. G. Voronkov, S. M. Shostakovskii, N. S. Nikolskii, P. A. Nikitin and A. A. Retinskii, *Zh. Org. Khim.*, **12**, 2302 (1976).
104. M. G. Voronkov, V. P. Feshin and J. Polis, *Teor. Eksp. Khim.*, **7**, 555 (1971).
105. S. A. Petukhov, M. I. Vakhrin and M. P. Sivkova, *Zh. Fiz. Khim. SUN*, **53**, 122 (1979).
106. V. P. Kukhar, V. I. Pasternak, M. I. Povolotskii and N. G. Pavlenko, *Zh. Org. Khim.*, **10**, 449 (1974).
107. E. A. C. Lucken, *J. Chem. Soc.*, 2954 (1959).
108. V. P. Feshin, P. A. Nikitin and M. G. Voronkov, *Izv. Sibirskogo Otoelini Akad. Nauk SSSR*, **5**, 159 (1979).
109. V. P. Feshin, P. A. Nikitin and M. G. Voronkov, *Dokl. Akad. Nauk SSSR*, **238**, 1404 (1977).
110. E. Scrocco, *Adv. Chem. Phys.*, **5**, 319 (1963).

111. Z. Ardalan and E. A. C. Lucken, *Helv. Chim. Acta*, **56**, 1715 (1973).
112. M. G. Voronkov and V. P. Feshin, *Dokl. Akad. Nauk SSSR*, **209**, 400 (1973).
113. M. G. Voronkov, V. P. Feshin, L. S. Romanenko, J. Pola and V. Chvalovsky, *Coll. Czech. Chem. Commun.*, **41**, 2718 (1976).
114. M. G. Voronkov, V. P. Feshin, V. F. Mironov, S. A. Mikhaiyauts and T. K. Gar, *Zh. Obschch. Khim.*, **41**, 22ii (1971).
115. R. M. Myasnikov, V. I. Robas and G. K. Semin, *Zh. Strukt. Khim.*, **6**, 474 (1965).
116. V. S. Grechishkin and G. B. Seifer, *Zh. Strukt. Khim.*, **5**, 914 (1964).
117. L. V. Galishevskya, Yu. I. Manzhura and G. B. Seifer, *Isvest. Vyssh. Uchebn. Zaved. Khim. Tekhnol. SSSR*, **19**, 1622 (1976).
118. P. J. Bray, R. G. Barnes and R. Bensohn, *J. Chem. Phys.*, **25**, 813 (1956).
119. P. J. Bray, S. Moscowitz, H. O. Hooper, R. G. Barnes and S. C. Segel, *J. Chem. Phys.*, **28**, 99 (1958).
120. M. J. S. Dewar and E. A. C. Lucken, *J. Chem. Soc.*, 2653 (1958).
121. E. A. C. Lucken and C. Mazeline, in *Proceedings of XIII. Colloque Ampère Leuven*, North Holland Publishing Co., Amsterdam (1964), p. 235.
122. P. Bucci, P. Cecchi and A. Colligiani, *J. Amer. Chem. Soc.*, **87**, 3027 (1965).
123. C. J. Turner, *JCS Perkin II*, 1250 (1975).
124. H. D. Stidham and H. H. Farrell, *J. Chem. Phys.*, **49**, 2463 (1968).
125. H. Negita and I. Satou, *J. Chem. Phys.*, **27**, 602 (1957).
126. F. Scappini and A. Guarneri, *Z. Naturforsch.*, **27a**, 1011 (1972).
127. R. D. Brown and J. Matouskova, *J. Mol. Struct.*, **29**, 33 (1975).
128. W. Caminati and P. Forti, *Chem. Phys. Lett.*, **38**, 222 (1976).
129. W. Caminati and P. Forti, *Chem. Phys. Lett.*, **15**, 343 (1972).
130. W. Caminati and P. Forti, *Chem. Phys. Lett.*, **29**, 239 (1974).
131. S. Ardjomande, thesis, University of Geneva, (1974).
132. M. Redshaw, *JCS Perkin II*, 1821 (1976).
133. H. O. Hooper and P. J. Bray, *J. Chem. Phys.*, **30**, 957 (1959).
134. Z. Troutelj, J. Pirnat and L. Ehrenberg, in *Advances in Nuclear Quadrupole Resonance*, Vol. I (Ed. J. A. S. Smith), Heyden, London (1974), p. 71.
135. V. P. Feshin, S. A. Hiller, L. Y. Avota and M. G. Voronkov, *Khim. Geterosikl. Soediu SSSR*, **3**, 392 (1976).
136. V. P. Feshin, S. A. Hiller, L. Y. Avota and M. G. Voronkov, *Khim. Geterosikl. Soediu SSSR*, **3**, 392 (1976).
137. T. A. Babushkina, T. S. Leonova, A. M. Alymov and T. L. Knotsyanova, *Zh. Fiz. Khim. SUN*, **53**, 140 (1978).
138. V. V. Makarskii, M. G. Vronkov, V. P. Feshin, V. A. Lopyrev, N. I. Berestennikov, E. F. Shibanova and L. I. Volkova, *Dokl. Akad. Nauk SSSR*, **220**, 101 (1975).
139. T. A. Babushkina, G. K. Semin, S. D. Sokolov and I. M. Yudintseva, *Izv. Akad. Nauk SSSR, Ser. Khim.*, **10**, 2376 (1970).
140. A. N. Nesmyanov, D. N. Kravtsov, A. P. Zhukov, P. M. Kochgerin and G. K. Semin, *Dokl. Akad. Nauk SSSR*, **179**, 102 (1968).
141. M. G. Voronkov, V. P. Feshin, E. A. Chernyshev, V. I. Savushkina, P. A. Nikitin and V. A. Kotikov, *Dokl. Akad. Nauk SSSR*, **212**, 395 (1973).
142. Yu. P. Dormidontov, V. S. Grechishkin and S. I. Gushchin, *Org. Mag. Resonance*, **4**, 599 (1972).
143. M. P. Bogaard, thesis, University of Sydney (1967).
144. H. H. Jaffé and H. L. Jones, *Adv. Heterocyclic Chem.*, **3**, 209 (1964).
145. S. A. Petukhov, G. S. Posyagin, A. L. Fridman and N. A. Kolobov, *Reakts. Sposobnost. Org. Soedim*, **10**, 657 (1973).
146. V. P. Feshin, M. G. Vronkov and N. I. Berestennikov, *Izv. Sib. Old. Akad. Nauk SSSR, Ser. Khim.*, **6**, 167 (1974).
147. R. M. Hart, M. A. Whitehead and L. Krause, *J. Chem. Phys.*, **56**, 3038 (1972).
148. G. Jugie, J. A. S. Smith and G. J. Martin, *JCS Perkin II*, 925 (1975).
149. E. V. Bryukhova, I. R. Gol'ding and A. M. Sladkov, *Izv. Akad. Nauk SSSR, Ser. Khim.*, **5**, 1160 (1973).
150. V. I. Stanko, E. V. Bryukhova, T. A. Babushkina, T. V. Klimova, N. S. Titova and G. K. Semin, *Zh. Strukt. Khim.*, **14**, 377 (1973).

151. T. A. Babushkina, G. A. Anorova, E. V. Bryukhova and V. I. Stanko, *Zh. Strukt. Khim.*, **15**, 708 (1974).
152. M. G. Voronkov, V. P. Feshin, A. P. Suyakin, V. N. Kalinin and L. J. Zakharkin, *Geterosikl. Soediu*, **4**, 565 (1970).
153. E. V. Bryukhova, T. A. Babushkina, A. I. Klimova, V. J. Stanko, *Izv. Akad. Nauk SSSR, Ser. Khim.*, **7**, 1676 (1973).
154. J. S. Evans and G. Y.-S. Lo, *J. Phys. Chem.*, **71**, 2730 (1967).
155. T. L. Khotsyanova, T. A. Babushkina, V. V. Saatsazov, T. P. Tolstoia and G. K. Semin, *Dokl. Akad. Nauk SSSR*, **222**, 403 (1975).
156. T. L. Khotsyanova, T. A. Babushkina and V. V. Saatsazov, *Izv. Akad. Nauk SSSR, Ser. Fiz.*, **39**, 2530 (1975).
157. T. L. Khotsyanova, T. A. Babushkina, V. V. Saatsazov, T. P. Tolstoia, I. N. Lisichkina and G. K. Semin, *Koord. Khim. SSSR*, **2**, 1567 (1976).
158. V. V. Saatsazov, T. A. Babushkina, T. L. Khotsyanova and G. K. Semin, *Zh. Fiz. Khim. SUN*, **53**, 18 (1979).
159. S. Ardjomande and E. A. C. Lucken, *JCS Perkin II*, 453 (1975).
160. M. Burgard and E. A. C. Lucken, *J. Mol. Struct.*, **14**, 397 (1972).
161. Y. K. Maksyutin, V. P. Makridin, E. N. Gur'yanova and G. K. Semin, *Izv. Akad. Nauk SSSR, Ser. Khim.*, **7**, 1634 (1970).
162. A. N. Nesmeyanov, G. K. Semin, T. L. Khotsyanov, E. V. Bryukhova, N. A. Volkenau and E. I. Sirotkina, *Dokl. Akad. Nauk SSSR*, **202**, 854 (1972).
163. D. A. Brown and E. A. C. Lucken, unpublished observations.
164. D. Biedenkapp and A. Weiss, *Ber. Bunsenges. Phys. Chem.*, **70**, 788 (1966).
165. V. P. Anterov, V. S. Grechishkin and M. Z. Yusupov, *Zh. Fiz. Khim.*, **47**, 1267 (1973).
166. H. Chihara and N. N. Nakamura, *Bull. Chem. Soc. Japan*, **44**, 2676 (1971).
167. R. Basu and M. Gosh, *J. Chem. Phys.*, **68**, 2510 (1978).
168. Yu. K. Maksyutin, T. A. Babushkina, Ye. N. Guryanova and G. K. Semin, *Theoret. Chim. Acta*, **14**, 48 (1969).
169. Z. Ardalan-de Weck, F. Delay, M. Geoffroy and E. A. C. Lucken, *JCS Perkin II*, 1867 (1976).
170. M. G. Voronkov, V. P. Feshin, A. N. Mirskova, N. I. Berestennikov and B. A. Shainyan, *Dokl. Akad. Nauk SSSR*, **224**, 1103 (1975).
171. A. S. Atavin, A. N. Mirskova and E. F. Zorina, *Zh. Org. Khim.*, **9**, 1389 (1973).
172. V. P. Feshin, M. G. Voronkov, V. D. Simonov, V. Z. Estrina and J. B. Iasman, *Dokl. Akad. Nauk SSSR*, **218**, 1400 (1974).
173. G. G. Levkovskaya, A. N. Mirskova, E. V. Bryukhova, V. P. Kazakov and A. S. Atavin, *Izv. Akad. Nauk SSSR, Ser. Khim.*, **4**, 793 (1975).
174. V. P. Feshin, M. G. Voronkov, L. S. Mironov, L. S. Romanenko and T. K. Gar, *Teor. Eksp. Khim.*, **12**, 260 (1976).
175. V. P. Feshin, M. G. Voronkov, V. D. Simonov, V. Z. Estrina, Yu. B. Yasman and E. N. Shitova, *Dokl. Akad. Nauk SSSR*, **220**, 414 (1975).
176. Y. Morino, I. Miyagawa, T. Chiba and T. Shimozawa, *J. Chem. Phys.*, **25**, 185 (1956).
177. Y. Morino, T. Chiba, T. Shinozawa and M. Toyama, *J. Phys. Soc. Japan*, **13**, 869 (1958).
178. Y. Morino, T. Chiba, T. Shinozawa and M. Toyama, *Rev. Universelle Mines, 9ème Sér.*, **15**, 591 (1959).
179. Y. Morino, M. Toyama and K. Itoh, *Acta Cryst.*, **16**, 129 (1963).
180. G. Soda, M. Toyama and Y. Morino, *Bull. Chem. Soc. Japan*, **38**, 1965 (1965).
181. L. Cazaux and G. Jugie, *J. Mol. Struct.*, **39**, 219 (1977).
182. H. Chihara and N. Nakamura, *Bull. Chem. Soc. Japan*, **45**, 3530 (1972).
183. P. J. Bray, *J. Chem. Phys.*, **23**, 703 (1955).
184. Z. Ardalan, E. A. C. Lucken and S. Masson, *Helv. Chim. Acta*, **56**, 1720 (1973).
185. B. A. Arbusov, S. G. Vulfson, I. A. Safin, I. P. Biryukov and A. N. Vereshchagin, *Izv. Akad. Nauk SSSR, Ser. Khim.*, 1243 (1970).
186. H. Chihara, N. Nakamura and T. Irie, *Bull. Chem. Soc. Japan*, **42**, 3034 (1969).
187. M. J. S. Dewar, M. L. Herr and A. P. Marchand, *Tetrahedron*, **27**, 2371 (1971).
188. F. Delay, M. Geoffroy, E. A. C. Lucken and P. Müller, *JCS Faraday II*, **71**, 463 (1975).
189. M. J. S. Dewar and M. Herr, *Tetrahedron*, **27**, 2377 (1971).

190. T. Kiichi, N. Nakamura and H. Chihara, *J. Mag. Resonance*, **6**, 516 (1972).
191. Z. Ardalan and E. A. C. Lucken, *Helv. Chim. Acta*, **56**, 1724 (1973).
192. C. Altona, thesis, Leiden (1964).
193. E. Eliel, *Angew. Chem. Int. Edn*, **11**, 739 (1972).
194. P. Linscheid and E. A. C. Lucken, *Chem. Commun.*, 425 (1970).
195. Z. Ardalan and E. A. C. Lucken, *Helv. Chim. Acta*, **56**, 1715 (1973).
196. S. David and L. Guibé, *Carbohydrate Res.*, **20**, 440 (1971).
197. L. Guibé, J. Augé, S. David and O. Eisenstein, *J. Chem. Phys.*, **58**, 5579 (1973).
198. M. Sabir, J. A. S. Smith, O. Riobé, A. Lebouc, J. Delaunay and C. Cousseau, *Chem. Commun.*, 19 (1977).
199. S. Wolfe, A. Rauk, L. M. Tel and I. G. Csizmadia, *J. Chem. Soc. B*, 136 (1971).
200. S. Wolfe, *Acc. Chem. Res.*, 102 (1972).
201. Z. de Weck-Ardalan, E. A. C. Lucken and J. Weber, *J. Mol. Struct.*, **32**, 101 (1976).
202. A. Roedig, R. Helm, R. West and R. M. Smith, *Tetrahedron Lett.*, 2137 (1969).
203. A. Meller and W. Maringgele, *Monats. Chem.*, **99**, 2504 (1968).
204. S. Ardjomande and E. A. C. Lucken, *Helv. Chim. Acta*, **54**, 176 (1971).
205. B. A. Komarov, B. A. Rozenburg and N. S. Enikolopyan, *Izv. Akad. Nauk SSSR, Ser. Khim.*, 1874 (1974).
206. H. Fitzky, in *Advances in Nuclear Quadrupole Resonance*, Vol. I (Ed. J. A. S. Smith) Heyden, London (1974) pp. 79–113.
207. E. Brame, *Anal. Chem.*, **43**, 135 (1971).
208. C. Dean, M. Pollak, B. M. Craven and G. A. Jeffrey, *Acta Cryst.*, **11**, 710 (1958).
209. K. Mano, *J. Mag. Resonance*, **26**, 393 (1975).
210. K. Mano, *J. Mag. Resonance*, **29**, 403 (1978).
211. K. Mano, D. Giezendanner, S. Sengupta and E. A. C. Lucken, *J. Mol. Struct.*, **58**, 221 (1980).
212. C. B. Harris and M. J. Buckley in *Advances in Nuclear Quadrupole Resonance*, Vol. II (Ed. J. A. S. Smith), Heyden, London (1975), pp. 15–70.
213. G. Kothandraman, D. W. Pratt and D. S. Tinti, *J. Chem. Phys.*, **63**, 3337 (1975).
214. J. A. Mucha and D. W. Pratt, *J. Chem. Phys.*, **60**, 5339 (1977).
215. C. von Borczyskowski and E. Boroske, *Chem. Phys.*, **35**, 367 (1978).
216. K. P. Dinse and C. von Borczyskowski, *Chem. Phys.*, **44**, 93 (1979).
217. M. Deimling and K. P. Dinse, *Chem. Phys. Lett.*, **69**, 587 (1980).
218. J. Wonig, G. P. M. van der Velden and W. S. Weeman, *Chem. Phys.*, **51**, 97 (1980).
219. M. J. Buckley and C. B. Harris, *J. Chem. Phys.*, **56**, 137 (1972).

The Chemistry of Functional Groups, Supplement D
Edited by S. Patai and Z. Rappoport
© 1983 John Wiley & Sons Ltd

CHAPTER **5**

1,2-Dehalogenations and related reactions

ENRICO BACIOCCHI
Dipartimento di Chimica, Università di Perugia, Perugia, Italy

I. INTRODUCTION

The term 1,2-dehalogenation indicates a particular type of β-elimination which involves the loss of two halogen atoms from 1,2-dihaloalkanes and 1,2-dihaloalkenes with formation of alkenes and alkynes, respectively. Generally, both reactions are reduction processes which can be conveniently represented as in equations (1) and (2) (X, Y = halogens)[1]. Reactions (1) and (2) are promoted by a great variety of species

$$\begin{matrix} R \\ R' \end{matrix}\!\!>\!CX\!-\!CY\!<\!\!\begin{matrix} R'' \\ R''' \end{matrix} + 2\,e \longrightarrow \begin{matrix} R \\ R' \end{matrix}\!\!>\!C\!=\!C\!<\!\!\begin{matrix} R'' \\ R''' \end{matrix} + X^- + Y^- \qquad (1)$$

$$RCX\!=\!CYR' + 2\,e \longrightarrow RC\!\equiv\!CR' + X^- + Y^- \qquad (2)$$

(reductants)[2], the majority of which can be usefully divided into three categories: (a) two-electron or nucleophilic reductants; (b) one-electron reductants (metal ions and radical ions); (c) metals. Furthermore, dehalogenations can be promoted by electrochemical means[3], by high energy radiation[4] and, in some cases, simply by heating[5].

The range of reactivity covered by the possible reductants is enormous and the dehalogenation of all dihalides can be performed when the appropriate reductant and experimental conditions are used. Recently, a 1,4-elimination of fluorine from a perfluoroalkene has been reported[6].

Another important consequence of the great structural differences between the reductants is that different reaction mechanisms can operate not only when reductants belonging to different categories are considered but, sometimes, also when they belong to the same category. Generalizations are therefore extremely difficult and should be made with caution.

Even though we will discuss the mechanistic aspects of these reactions later on, some general considerations are appropriate here.

In most cases the dehalogenation process involves the attack of the reductant on one of the halogen atoms, whereas the other halogen atom has the role played by the leaving group in the more common β-elimination of HX from alkyl halides. However, the reaction of dihalides with mercurials[7] is probably a four-centre reaction and the two halogens could play exactly the same role. In general the tendency of the various halogens to be attacked by either an ionic or a radical reductant decreases in the order $I > Br > Cl \gg F$, and the same order holds for the leaving group ability of halides. The two facts account for the extremely low reactivity of difluorides (poor group with respect to the interaction with the nucleophile and poor leaving group) and allow a general reactivity order of different dihalides to be predicted. Thus, a 1,2-dibromide should be more reactive than the corresponding 1-bromo-2-chloro derivative and the latter more reactive than the corresponding 1,2-dichloride. This has been observed experimentally, for example, in the iodide ion-promoted dehalogenations[8,9]. Because of their reactivity and the easy preparation, dibromides are frequently the substrates of choice in dehalogenation processes.

The above considerations also explain why elimination is possible even when one of the two halogen atoms in the vicinal dihalide is replaced by another group, e.g. OH, OTs, OR, OCOR, which can act as the leaving group. However, since such groups are poorer leaving groups than halogens the elimination is slower than in the case of vicinal dihalides, and requires the use of the strongest reductants[10]. Moreover, in this case the simple reduction of the carbon–halogen bond has a greater chance to compete with the elimination reaction.

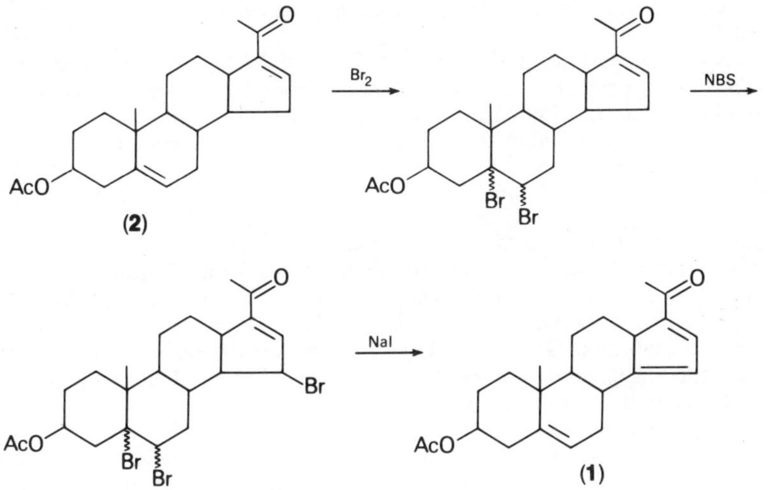

SCHEME 1

Dehalogenations (mostly debrominations) have been used as a means for the purification of olefins for many years[11]. Olefins are converted into dibromides which can be easily separated, having a higher boiling point. The pure olefin is then obtained from the dibromide by dehalogenation. Another frequent use of the bromination–debromination procedure is in the protection of double bonds in reactions which can affect them. An example is the synthesis of the 3-β-acetoxy-20-keto-5,14,16-triene (**1**) from **2**, reported in Scheme 1[12]. By addition of bromine to the 5,6 double bond a dibromide is obtained, which then undergoes allylic bromination by *N*-bromo-succinimide (NBS). The 5,6 double bond is regenerated by addition of sodium iodide. Under the experimental conditions of the debromination reaction hydrogen bromide is also lost.

Dehalogenation reactions have also been successfully employed as a part of synthetic sequences which generate a structurally new double bond; this point is illustrated by the synthesis of bicyclo[4.2.0]octa-2,7-diene (**3**)[13] from cyclooctatetraene (Scheme 2). In this case the starting material is not the dihalide of the resulting alkene. Moreover, dehalogenation reactions have been shown to provide a

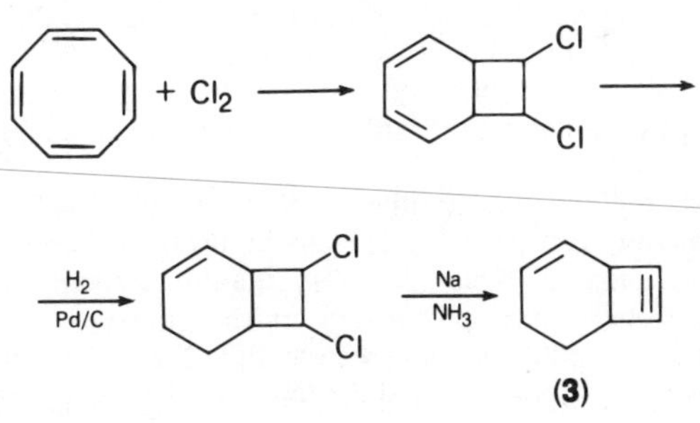

SCHEME 2

simple route to adamantene which, in principle, is extendable to other anti-Bredt alkenes[15–18].

Thus, the frequent statement[14] that 1,2-dehalogenations have scarce synthetic utility since dihalides are generally prepared from alkenes and alkynes is not warranted[13]. This is all the more true if we also consider the preparative value of the closely related elimination reactions of halohydrins, haloethers, etc. For example, the sequence

$$-COCHR- \longrightarrow -COCBrR- \longrightarrow -CH(OH)CBrR- \longrightarrow -CH{=}CR-$$

is a general method for introducing unsaturation in a steroid nucleus[19].

Of the two dehalogenations described in equations (1) and (2), process (1) is by far the more important as regards generality and scope. This chapter will therefore be devoted mainly to a discussion of the most significant features of 1,2-dehalogenations from vicinal dihaloalkanes and related reactions, with particular reference to problems concerning the mechanism, stereochemistry and relationships between reactivity and structure. The subject has been reviewed several times in the past[20–24], the most recent and comprehensive review being that due to Saunders and Cockerill[24].

Finally, it should be pointed out that, in addition to those mentioned above, other dehalogenation reactions from 1,3- and 1,4-dihalides are possible. In the 1,3-dehalogenations cyclopropane derivatives are formed and in the 1,4-dehalogenations cyclobutanes or dienes are obtained. Even though these reactions are outside the scope of this chapter, reference to them will be made whenever necessary.

II. DEHALOGENATIONS PROMOTED BY TWO-ELECTRON REDUCTANTS

A. General

Dehalogenation from vicinal dihalides can be promoted by a great variety of nucleophiles or two-electron reductants. These include halide ions[8,25–27] except fluorides, sulphur nucleophiles (e.g. thiolates[2,28,29] sodium hydrogen sulphide[30], thiourea[31,32], sodium thiosulphate[33] and sodium benzenesulphinate[2]), phosphorus nucleophiles (e.g. phosphines[34,35], alkylphosphites[36] and phosphides[16,37]), arsenides[37], cyanide ions[38], sodium hydrogen telluride[30], sodium selenide[39], alkyl and aryl selenides[40], hydrides (e.g. lithium aluminium hydride[21], sodium hydride[41] and sodium trimethoxyborohydride[42]), organometallic compounds (e.g. butyl-[43] and phenyllithium[44] and organocuprates[45]) and nitrogen and oxygen nucleophiles (e.g. amines[2,46,47], alkoxides[48] and hydroxide ions[49]). Dehalogenation can be also promoted by the solvent, e.g. dimethylformamide (DMF)[26,50] or ethanol[2,51], without the addition of external reductants. For many of these nucleophiles only some of the available references are indicated; these should be regarded as the sources for other literature data. Other lists of nucleophiles have been reported[2,21,24]. It should also be noted that for most of these species the classification as two-electron reductants is deduced from their general behaviour and not from evidence concerning dehalogenation reactions specifically.

Iodide ion is certainly the nucleophile most frequently used in dehalogenation reactions. Generally, the reaction is carried out by using an excess of sodium iodide in organic solvents, such as acetone or methanol. Recently, iodide-promoted dehalogenations have been also performed in the presence of a phase-transfer catalyst[52,53]. Catalytic amounts of sodium iodide and of a phase transfer agent (hexadecyltributylphosphonium bromide) in the presence of excess sodium thiosulphate have been used[52]. Iodide ion which is converted to I_3^- in the dehalogenation process (equation 3) is continuously regenerated by thiosulphate.

$$\text{RCHBrCHBrR}' + 3\text{I}^- \longrightarrow \text{RCH{=}CHR}' + \text{I}_3^- + 2\,\text{Br}^- \quad (3)$$

Among the more recently studied dehalogenating nucleophiles, sodium hydrogen telluride[30] appears promising. This reductant can be prepared *in situ* from tellurium powder and sodium borohydride and has the advantage that tellurium can be recovered quantitatively and re-employed. Organocopper reagents have also turned out to be effective reductants, especially useful in the debrominations of dibromo-esters[45]. Moreover, simple heating of a dibromide in DMF in the presence of thiourea oxide can provide a very convenient procedure for debromination[50].

A common feature of dehalogenations induced by nucleophiles, regardless of the detailed reaction mechanism, is the nucleophilic attack (two-electron transfer) of the reductant at one of the halogen atoms. The transition state involves the nucleophile and the dihalide; second-order kinetics, first-order in each reactant, are generally observed.

The most suitable substrates for these reactions are α,β-disubstituted vicinal dihalides, RCHXCHXR′ (X = halogen), especially dibromides. Further substitution normally favours dehalogenation; in contrast, with 1,2-dihaloethane or with primary dihalides of structure $RCHXCH_2X$, the competition between dehalogenation and an S_N2 nucleophilic substitution of X^- can become important.

Whereas we will deal later in more detail with this competition, as well as with the competition between dehalogenation and dehydrohalogenation, it is useful to mention here that dehalogenation can be the final outcome of the reaction between a dihalide and a nucleophile, even when the latter first attacks the carbon atom to give a substitution product. This occurs when such a product, unlike the starting material, can be converted into the final olefin by the nucleophile itself.

This pathway, also designated as the indirect or the substitution–elimination mechanism, can be of significant importance when primary dihalides react with iodide ions. S_N2 substitution leads in this case to an iodohalogeno derivative which is very easily dehalogenated by further reaction with iodide ions (equations 4 and 5).

$$\text{RCHXCH}_2\text{X} + \text{I}^- \longrightarrow \text{RCHXCH}_2\text{I} + \text{X}^- \quad (4)$$

$$\text{RCHXCH}_2\text{I} + 2\text{I}^- \longrightarrow \text{RCH{=}CH}_2 + \text{I}_3^- + \text{X}^- \quad (5)$$

Evidence supporting the operation of an indirect mechanism has been found for the iodide-promoted dehalogenations from *meso*-1,2-dibromo-1,2-dideuteroethane[54] and some vicinal dichloro- and bromochloroalkanes[55]. These reactions exhibit a *syn* stereochemistry, in agreement with an initial substitution step occurring with inversion of configuration followed by an *anti* elimination (see below). Interestingly, the stereochemical outcome of these reactions has been exploited to devise an efficient method for the interconversion of geometrical olefins via a two-step sequence: halogenation (with Cl_2 or BrCl) and iodide-induced dehalogenation[55].

Almost certainly primary dibromides too are dehalogenated via the indirect mechanism, as structural effects on the rate are those expected for a biomolecular nucleophilic substitution (see Section II.D.2). This suggests that the nucleophilic displacement is the rate-determining step[56]. A substitution–elimination mechanism has been also suggested for the dehalogenations of 2,3-dihalogenotetrahydrofuran promoted by iodide ions[57].

B. Mechanisms

The nucleophile-promoted dehalogenation in which two bonds are formed and two are broken is a very complex process, with a number of possible reaction pathways which is even greater than that available for the base-catalysed HX eliminations[58]. To simplify the discussion we will regroup these pathways into two main mechanistic schemes.

In the first scheme, dehalogenations are supposed, from the mechanistic point of view, to resemble closely the β-elimination of HX, as originally suggested by Winstein, Pressman and Young[59]. By analogy with the latter process, a concerted mechanism of dehalogenation (equation 6, where X = halogen and Nu = nucleophile), and two stepwise mechanisms involving the formation of a carbanion (equation 7) and a

$$\mathrm{XC{-}CX} \xrightarrow{\mathrm{Nu^-}} \left[\mathrm{Nu\cdots X\cdots C_\beta \cdots C_\alpha \cdots X}\right]^- \longrightarrow \mathrm{{>}C{=}C{<}} + \mathrm{NuX} + \mathrm{X^-} \tag{6}$$

$$\mathrm{XC{-}CX} \xrightarrow[-\mathrm{NuX}]{\mathrm{Nu^-}} \mathrm{{-}\bar{C}_\beta{-}C_\alpha(X){-}} \longrightarrow \mathrm{{>}C{=}C{<}} + \mathrm{X^-} \tag{7}$$

$$\mathrm{XC{-}CX} \xrightarrow{-\mathrm{X^-}} \mathrm{{-}C_\beta(X){-}\overset{+}{C}_\alpha{-}} \xrightarrow{\mathrm{Nu^-}} \mathrm{{>}C{=}C{<}} + \mathrm{X^-} \tag{8}$$

carbocation (equation 8) as reaction intermediates can be envisaged. In the above equations the carbon atom bearing the halogen which departs as an anion is the α-carbon (C_α), and the carbon bearing the halogen attacked by the reductant is the β-carbon (C_β). This notation will be used throughout the chapter.

The concerted mechanism of equation (6) is analogous to the $E2$ mechanism of dehydrohalogenation reactions and can be labelled either in the same way or, as suggested by Ko and Parker[60], by the notation E2Hal. Likewise, we can also speak of E1cB and $E1$ mechanisms of dehalogenation (equations 7 and 8, respectively). However, it must be recognized that the former notation is formally incorrect since the initially formed carbanion is not the conjugate base of the substrate.

As in the $E2$ dehalogenations, the rupture of the two carbon–halogen bonds might not be synchronous in the E2Hal mechanism. Thus, a spectrum of transition-state structures (Scheme 3), from the $E1$-like (**4**) to the E1cB-like (**6**), which differ in the relative extent of C_α—X and C_β—X bond breaking, should in principle be possible for concerted dehalogenations, according to the theory of the variable $E2$ transition state[61]. In the middle of the spectrum is the 'central' transition state structure (**5**), in which formation and rupture of the various bonds is synchronous, or nearly so, and no appreciable charge is developed on C_α and C_β. On the basis of the analogy

$$\mathrm{Nu\cdots X\cdots C_\beta\cdots \overset{\delta+}{C}_\alpha\cdots X} \quad (\mathbf{4}) \qquad \mathrm{Nu\cdots X\cdots C_\beta\cdots C_\alpha\cdots X} \quad (\mathbf{5}) \qquad \mathrm{Nu\cdots X\cdots \overset{\delta-}{C}_\beta\cdots C_\alpha\cdots X} \quad (\mathbf{6})$$

SCHEME 3

between dehalogenations and dehydrohalogenations it has also been suggested[62] that the former reactions might take place by a mechanism corresponding to Winstein and Parker's $E2C$ mechanism[63], where the nucleophile interacts simultaneously with both the β-halogen and C_α (transition state **7**).

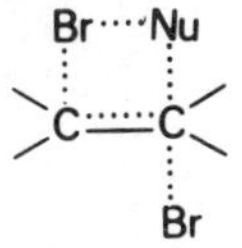

(7)

For steric and electronic factors similar to those at work in $E2$ dehydrohalogenations[64], dehalogenations occurring by a concerted mechanism are expected to follow preferentially an *anti* stereochemistry. However, strict stereospecificity is not expected for $E1$ and $E1cB$ dehalogenations.

The alternative mechanistic framework for 1,2-dehalogenations is based on the view that, by the principle of microscopic reversibility, the mechanism of these reactions should be closely related to that of the reverse reaction, i.e. the halogen addition to alkenes. According to this premise, Hine and Braider first suggested that the reversible iodide-catalysed deiodination of 1,2-diiodoethane takes place by a stepwise mechanism (equation 9), involving the same bridged intermediate thought to be implied in the iodine addition process[56]. The sequence of equation (9) has been

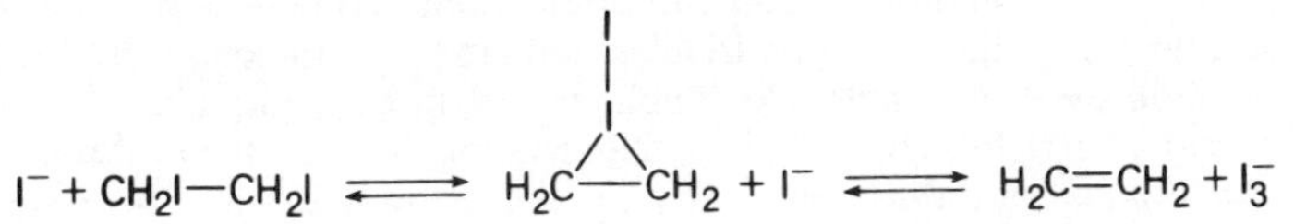

$$I^- + CH_2I{-}CH_2I \rightleftharpoons H_2C\overset{I{-}I^{+}}{\triangle}CH_2 + I^- \rightleftharpoons H_2C{=}CH_2 + I_3^- \qquad (9)$$

extended by Miller and coworkers[25] to include a number of interconvertible intermediate bridged species, halonium ions and open carbonium ions (Scheme 4).

Although Scheme 4 was originally devised to account for the iodide-promoted dehalogenations from *meso*- and D,L-1,2-dibromo-1,2-diphenylethane, there is no doubt that it can be considered a reasonable mechanistic sequence for any kind of nucleophile-promoted dehalogenation. Since the formation of the bridged intermediate requires that the two halogen atoms assume an *anti* arrangement, an *anti* stereochemistry can be also predicted for dehalogenations occurring by this 'halonium

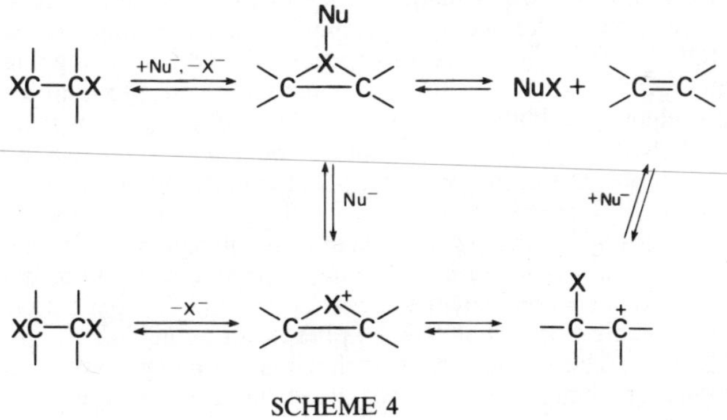

SCHEME 4

ion mechanism' provided that no interconversion between isomeric intermediates takes place before they decompose to the olefin.

In the following sections we will present the experimental results concerning stereochemistry and reactivity in nucleophile-promoted dehalogenation reactions in relation to the above mechanistic schemes. It is anticipated that very often dehalogenations can be reasonably described in terms of both the $E2$Hal and the halonium ion mechanisms. In these cases, as correctly pointed out by Saunders and Cockerill[24], the preference for one mechanism or the other is a question of personal choice.

C. Stereochemistry

Most of the studies on the stereochemistry of dehalogenation reactions deal with dibromoalkanes as substrates. However, even in this limited group of substrates, the structure of the dibromide can significantly influence the stereochemical course of the reaction.

With *meso* and *erythro* diasteroisomers of vicinal dibromides of the type RCHBrCHBrR′, a very marked *anti* stereochemistry is generally observed with a great variety of nucleophiles, independently of the substrate structure. For example, yields of ≥90% of *trans*-2-butene are obtained in the reactions of *meso*-2,3-dibromobutane with iodide ion[59], lithium aluminium hydride[21], selenides[39b,40], thiolates[28], $C_3H_5Fe(CO)_2P(C_6H_5)_3^-$ [65] and phenyllithium[44]. *Trans*-stilbene is the exclusive reaction product in the debromination from *meso*-1,2-dibromo-1,2-diphenylethane promoted by iodide[8,9,25,66,67], bromide[8,27,68] and chloride[8,26] ions, lithium aluminium hydride[21], sodium hydride[41], sodium borohydride[42], sodium trimethoxyborohydride[42], thiosulphate[33], thiourea[31], *p*-methylbenzenthiolate[2] and triphenylphosphine[34,35]. Likewise, only cinnamic acid is formed in the dehalogenation from *erythro*-dibromocinnamic acid by iodide ions[69] and thiourea[31]. *Anti* stereoselectivity has also been observed in the iodide ion-promoted dehalogenations of medium ring dibromoalkanes[70] and in some dehalogenations in DMF in the absence of nucleophiles[50].

A quite different situation arises with D,L or *threo* dibromides, as in this case the substrate structure can influence the stereochemical course of the reaction somewhat.

Simple aliphatic dibromides of the type RCHBrCHBrR′ (R, R′ = alkyl groups) again exhibit predominant *anti* stereoselectivity, at least with most of the nucleophiles tested (iodide ions[59,70], lithium aluminium hydride[21], thiolates[28], selenides[39b], $C_6H_5Fe(CO)_2P(C_6H_5)_3^-$ [65], phenyllithium[44]). In contrast, the stereoselectivity of α,β-disubstituted *threo* and D,L-dibromides where R and (or) R′ are phenyl groups or groups capable of stabilizing a negative charge, appears to depend on the nature of the nucleophile and the experimental conditions (e.g. solvent, temperature). Thus, whereas *cis*-stilbene is formed predominantly from D,L-1,2-dibromo-1,2-diphenylethane with iodide ions, *p*-methylbenzenethiolates and phenylsulphinates[2], prevalent or exclusive formation of *trans*-stilbene is obtained with bromide ions[27,68], *n*-butyllithium[43], lithium aluminium hydride[21], sodium trimethoxyborohydride[39], sodium borohydride[42], triphenylphosphine[2,35] and sodium hydride[41] (where *cis*-stilbene → *trans*-stilbene isomerization is, however, possible under the reaction conditions).

Since the latter nucleophiles give *trans*-stilbene when the substrate is *meso*-1,2-dibromo-1,2-diphenylethane, we are dealing in these cases with stereoconvergent processes where the *trans* olefin is the major product from both diastereoisomers. Another example of stereoconvergent reaction is the debromination of *meso*- and D,L-diethyl-1,2-dibromosuccinate with triphenylphosphine[35].

The conclusion from these results is that nucleophile-promoted dehalogenations from dibromides generally follow an *anti* stereochemistry. However, drastic deviations from this pattern are possible for reactions involving D,L or *threo* dibromoalkanes substituted by aryl or electron-withdrawing substituents.

The preference for an *anti* coplanar arrangement of the two bromine atoms in the transition state of the dehalogenation reaction is also supported by reactivity data. However, this is nearly exclusively limited to iodide ions as nucleophiles. Thus, a *meso* or *erythro* dibromide is generally debrominated faster than the corresponding D,L or *threo* isomer, in agreement with the fact that the former can more easily assume the conformation appropriate for an *anti* elimination (compare **8** with **9**). Moreover, in

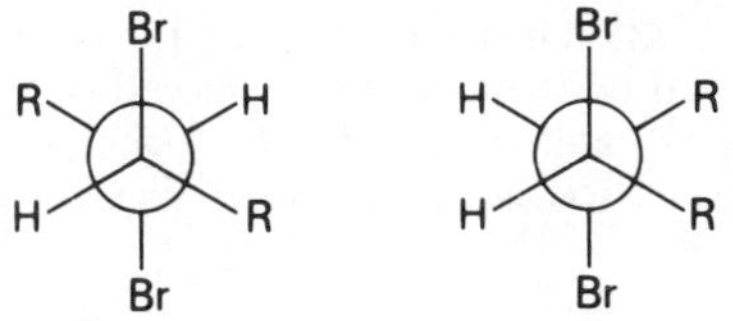

(**8**) (**9**)

going to the resulting olefin the interaction between the two R groups in **9** is expected to increase due to the eclipsing effect, with a consequent increase in the reactivity of the *meso* (*erythro*) form over that of the D,L (*threo*) one. A substantial difference in rate is observed when R = phenyl: the $k_{meso}/k_{\text{D,L}}$ ratio changes from 65 in methanol[25] to more than 300 in DMF[67] for the reactions of 1,2-dibromo-1,2-diphenylethanes with iodide ions. With bromide ions in DMF, $k_{meso}/k_{\text{D,L}}$ is 60[27]. Much smaller values are obtained when R = CH_3. In the reactions of 2,3-dibromobutane $k_{meso}/k_{\text{D,L}}$ is between 1.7 and 3, depending on the temperature and the solvent[66,71–73].

The strong preference for an *anti* diaxial conformation is confirmed by the observation that 5α,6β-dibromocholestan-3β-yl benzoate with two axial bromines undergoes fast dehalogenation by iodide ions in acetone (67% reaction in 225 min at 5°C) whereas the 5β,6α-isomer does not react under identical experimental conditions[74]. No reaction is also observed between iodide ions and a derivative of *cis*-2,3-dibromo-[2.2.1]bicycloheptane which has two *cis* coplanar bromine atoms[75].

Likewise, *trans*-1,2-dibromocyclohexane reacts with iodide ions *c*. 11 times faster than the *cis* isomer[76] and a much higher *trans*:*cis* reactivity ratio has been estimated for the reaction of the same nucleophile with 1,1,2-tribromocyclohexanes[77]. The former ratio is probably a lower limit since the *cis*-dibromide should react by the substitution–elimination mechanism.

Interestingly, an *anti* alignment of the halogen atoms is also favoured in eliminations from 1,2-dihaloalkenes; thus, *trans*-1,2-diiodoethylene reacts with iodide ions to give acetylene much more rapidly than the *cis* isomer[78].

If we now look at the results of the stereochemical studies in terms of possible reaction mechanisms, the first observation is that both the *E*2Hal and the halonium ion mechanism could be operating in the *anti* elimination from *meso* and *erythro* dibromides. In order to make the latter mechanism compatible with an *anti* stereochemistry, no isomerization between the bridged intermediates should take place. This is plausible since with *meso* and *erythro* dibromides both stereoelectronic and conformational factors are aligned and the most stable bridged species is formed. It follows that the *E*2Hal and the halonium ion mechanism cannot be distinguished for these reactions on stereochemical grounds alone.

The discussion of the significant deviations from an *anti* stereoselectivity which occur in some dehalogenations of D,L and *threo* dibromides is more complex and at least three possibilities have to be considered.

If we follow the view of the close analogy between dehydrohalogenations and dehalogenations (equations 4–6), these deviations could be attributed to the intervention of a concerted *syn* dehalogenation or of an *E*1cB mechanism of dehalogenation, operating in various combinations with the *E*2Hal mechanism. In the D,L and the *threo* dihalides the conformational and stereoelectronic factors are opposite and it is more difficult to achieve an *anti* coplanarity of the two halogen atoms than in the case of *meso* and *erythro* isomers. In some systems the opposition of these conformational factors to *anti* dehalogenation might become so important that the reactions of the D,L and the *threo* isomers are forced to utilize reaction pathways different from the concerted *anti* route which is preferred by the *meso* and *erythro* forms.

A concerted *syn* dehalogenation has been proposed to account for the predominant *syn* stereoselectivity observed in bromide ion-promoted dehalogenations of D,L-1,2-dibromo-1,2-diphenylethane[24] and in some dehalogenations induced by lithium aluminium hydride[21]. A transition state (**10**) involving an interaction between the

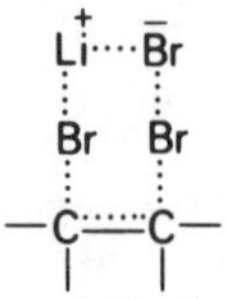

(10) 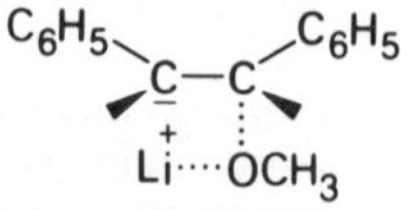 **(11)**

substrate and the ion-paired nucleophile, similar to that suggested for HX eliminations[79], has been considered possible for the former reaction[24]. However, the observation of a predominant *anti* stereochemistry in the iodide-promoted dehalogenations of medium ring dibromocycloalkanes[70] (a system very suitable to undergo *syn* dehydrohalogenations[79]), would indicate that, at least for the halide ion-induced reactions, a concerted *syn* mechanism is not highly probable.

The intermediacy of the coordinated carbanion **11** has been suggested to account for the *syn* elimination from *erythro*-1-bromo-2-methoxy-1,2-diphenylethane promoted by *n*-butyllithium in non-polar solvents[43,80]. To account for the *syn* stereospecificity the additional hypothesis that metal–halogen exchange proceeds with retention of configuration has been made. Interestingly, stereospecificity disappears when the reaction is carried out in diglyme, a very effective solvent for solvating the metal cation, where the carbanion is probably present as a solvent-separated ion pair.

However, it is not certain that this mechanism may operate also with dibromides. When *n*-butyllithium reacts with 1,2-dibromo-1,2-diphenylethane a stereoconvergent reaction is observed: *trans*-stilbene is predominantly formed from both the *meso* and the D,L isomer, as frequently observed with other nucleophiles. This may indicate that the carbanion is no longer coordinated with Li^+, since coordination with bromine is less effective than with oxygen, and can therefore undergo rotation. Alternatively, since bromine is a much better leaving group than methoxy, an intermediate carbanion is not formed and a different mechanism is operating.

The second possible explanation for the stereochemical results, which is limited to halide-promoted dehalogenations, is the intervention of a substitution–elimination mechanism. Accordingly, the overall stereochemical outcome of this mechanism is a *syn* elimination and it is reasonable that this mechanism would operate only in the reactions of substrates which exhibit the greater reluctance to undergo the normal *anti* elimination process. However, no evidence is available to support a substitution–elimination route for the reactions of α,β-disubstituted dibromides. In fact, data for iodide-promoted dehalogenations of D,L-1,2-dibromo-1,2-diphenylethane are contradictory to this possibility[27].

Finally, the deviations from *anti* stereoselectivity observed with some D,L and *threo* isomers can be also rationalized by the halonium ion mechanism. Unlike the reaction of the *meso* and *erythro* diasteroisomers, the less stable bridged species is formed from these substrates. It could therefore show a strong tendency to isomerize into the more stable isomer, presumably via the open carbonium ion (Scheme 4), before the decomposition to alkene takes place. Variable degrees of stereoselectivity should result depending on the role played by conformational factors. When complete isomerization occurs, the reaction is thermodynamically controlled.

Conformational factors are not expected to be very important in the 2,3-dibromobutane systems, and *anti* stereoselectivity is generally observed in the reaction of both diasteroisomers. In contrast, the conformational factors should be much more significant in the 1,2-dibromo-1,2-diphenylethane system, and the less stable bridged species formed from the D,L isomer can more easily isomerize. Thus, predominant or exclusive formation of *trans*-stilbene, which is the most stable olefin, is observed on some occasions.

The significant dependence of the stereoselectivity upon the nature of the solvent and the nucleophile which has been noted in these reactions[2] is in agreement with the above suggestion since both the solvent and the nucleophile might influence to some extent the interconversion rate between the bridged species. Moreover, the halonium ion mechanism also accounts nicely for the observation that the dehalogenation of *threo* and D,L dibromides is frequently accompanied by formation of solvolysis products[25,81]. These products could originate by reaction of the bridged species or the open carbonium ion with the solvent.

Summing up, even though clear-cut evidence is not available, the most reasonable possibility out of the three discussed above is that ascribing the observed deviations from the *anti* stereochemistry to the intervention of the halonium ion mechanism. Since this mechanism is also able to account for the *anti* stereochemistry exhibited by most of the nucleophile-promoted dehalogenations, it provides a mechanistic framework to which *all* the results concerning the dehalogenation stereochemistry can be accommodated.

However, even though this unifying interpretation is very attractive, the alternative hypothesis that the halonium ion mechanism intervenes *only* when the operation of the *anti* $E2$Hal mechanism becomes unlikely due to conformational factors is also plausible and can rationalize experimental facts as well. At this stage of our discussion a firm choice between these two hypotheses is impossible. We will, however, further discuss this point after having dealt with structural effects on the rate of dehalogenation reactions.

D. Reactivity and Structure

1. The halogen

The role played by the nature of the halogen with respect to the reactivity of dihalides is twofold. One of the halogen atoms serves as the electrophilic centre, and undergoes the nucleophilic attack (whatever is the detailed mechanism) whereas the other halogen serves as the leaving group and is displaced as an anion.

Most information on this subject refers to reactions where iodide ion is the nucleophile. Nevertheless, the conclusions should be substantially valid, in most cases, for reactions with other nucleophiles.

If we first consider the halogen as electrophilic centre, the reactivity generally decreases in the order $I > Br > Cl \gg F$, which parallels the order of the ease of electrochemical reduction of the carbon–halogen bonds[82]. The polarizability of the halogen, which decreases in the same order as the reactivity, probably plays an important role in this respect since it should diminish the energy of the interaction between the halogen and the nucleophile in the transition state. Additional factors contributing to the observed reactivity order could be the strength of the carbon–halogen bond, which increases in the order I < Br < Cl < F, and the neighbouring group ability of the halogens, which decreases in the same order[83]. The former factor should, however, be more important for dehalogenations occurring by the *E*2Hal mechanism, whereas the latter should play a role only when the halonium ion mechanism is operating.

The energy associated with the formation of the halogen–nucleophile bond appears, on the contrary, to play a secondary role. Accordingly, if this factor were important, deviation from the observed reactivity order would have been expected[84] since the strength of the halogen–nucleophile bond increases in going from iodine to fluorine. A reasonable explanation is that the formation of this bond has not progressed to a large extent in the transition state of dehalogenation reactions.

Quantitative information on the relative tendency of the various halogens to undergo a nucleophilic attack in a dehalogenation reaction are available only for iodine and bromine with iodide ions as nucleophiles. Thus, *erythro*-1-chloro-2-iodo-1,2-diphenylethane reacts with iodide ions, either in methanol or in DMF, *c*. 10^6 times faster than *erythro*-1-bromo-2-chloro-1,2-diphenylethane[85]. A smaller but still significant iodine:bromine reactivity ratio of *c*. 10^4 is observed when dehalogenations by iodide ions of 1,2-dibromoethane and 1-bromo-2-iodoethane are compared in acetone[86]. (Even though 1,2-dibromoethane reacts mainly by a substitution–elimination mechanism, a small part of the dehalogenation appears to proceed by a direct elimination. The rate of this direct elimination was measured by following the initial rate of iodine formation.) The difference in the iodine:bromine ratio can be justified, at least partially, by the higher overcrowding in the diphenylethane system than in the ethane system. Therefore, a steric relief of strain in the former system can facilitate the loss of ICl compared with that of BrCl.

Qualitatively, the greater reactivity of bromine with respect to chlorine and of chlorine with respect to fluorine is deduced, for example, from the observation that iodide ions dehalogenate 1,2-dibromodiphenylethane much faster than they dehalogenate the corresponding dichloro derivative[9], or from the fact that only Cl_2 is eliminated in the reaction of the chlorofluoro derivative **12** with Ph_3P (equation 10)[87].

$$CF_2ClCFClCF_2CFClCF_2CO_2Et \xrightarrow{Ph_3P} CF_2{=}CFCF_2CFClCF_2CO_2Et \quad (10)$$

When the nucleofugal halogen is considered, the expected reactivity order is again I > Br > Cl > F, which is the order of the leaving group abilities of the halogens in nucleophilic displacements[88].

Quantitative data for the bromine:chlorine leaving group effect (k_{Br}/k_{Cl}) have been obtained by comparing dehalogenation rates of *meso*-1,2-dibromo- and *erythro*-1-bromo-2-chloro-1,2-diphenylethane[8]. With iodide ion as a nucleophile the k_{Br}/k_{Cl} ratio is 87 in methanol and 260 in DMF, thus suggesting a substantial degree of C_α-halogen bond breaking in the transition state. A value of *c*. 30 for the k_{Br}/k_{Cl} ratio can be calculated from the data for iodide-promoted dehalogenations of *trans*-1,2-dibromo- and *trans*-1-bromo-2-chloro- cyclohexanes in methanol[76].

The iodine:bromine leaving group effect appears to be significantly smaller than the bromine:chlorine one: it is reported that iodide ion dehalogenates 1,2-diiodoethane only three times faster than 1-bromo-2-iodoethane[89].

The great reluctance of fluorine to depart as an anion is nicely illustrated by the observation that whereas when 1,2-dibromoethane is treated with iodide ions it gives ethylene, via 1-bromo-2-iodoethane, 1-bromo-2-fluoroethane forms 1-iodo-2-fluoroethane[86].

When the leaving group is not a halogen atom, the reactivity order again follows the order of leaving group ability observed in nucleophilic displacements[88]. Thus, in the reaction of a series of *trans*-1-bromo-2-X-cyclohexanes with iodide ion, the reactivity of different X groups has been found to decrease in the order OBs > OTs > ONO_2[90].

From the combined effect of the halogens as electrophilic centres and as nucleofugal groups, the reactivity order diiodides > iodobromides > dibromides > bromochlorides > dichlorides > chlorofluorides > difluorides can be predicted for the dehalogenation reaction.

2. The organic moiety

In studies of the effect of the organic moiety most quantitative studies again concern dibromides as substrates and iodide ion as a nucleophile. Beside the fact that iodide ion-promoted reactions have long been the most important ones among the dehalogenations induced by nucleophiles, it should also be noted that kinetic studies of these reactions can be conveniently carried out by measuring the concentration of the triiodide ion formed (equation 3).

As already anticipated, primary dibromides $RCHBrCH_2Br$ are generally dehalogenated by iodide ions via the substitution–elimination mechanism. For these substrates, structural effects are, therefore, those expected for S_N2 displacements on ethyl bromide[91]. As shown in Table 1, the replacement of hydrogen at the β-carbon by an alkyl group slows down the dehalogenation rate[92], in line with the operation of a steric effect which hinders the initial nucleophilic attack at carbon.

An opposite situation holds for the direct dehalogenation of α,β-disubstituted dibromoalkanes. Both in acetone and methanol (Table 2) dehalogenation rate increases with the branching in the alkyl moiety. The reactivity order is therefore secondary–secondary < secondary–tertiary < tertiary–tertiary. It has been suggested that this order is indicative of a transition state with significant S_N1 character[25]. However, the reactivity order of Table 2 also parallels the stability order of the resulting olefin and it is possible as well to suggest a transition state structure which incorporates many of the essential features of the final product.

The effect of aromatic substituents on the dehalogenation rate gives useful information on the transition state structure since steric effects are kept constant in this case.

TABLE 1. Reactivity data for the dehalogenations of $RCHBrCH_2Br$ promoted by iodide ions in methanol at 75°C[92]

R	$k_2 \times 10^4$, s^{-1} M^{-1}
H	4.27
CH_3	0.18
C_2H_5	0.25
n-C_3H_7	0.28
Br	0.16
C_6H_5	5.03
CO_2H	5.07

TABLE 2. Relative reactivity for the iodide-induced dehalogenations from 1,2-disubstituted dibromides in acetone and methanol

Substrates	Relative reactivity	
	Acetone (20°C)[a]	Methanol (59°C)[b]
$CH_3CHBrCHBrCH_3$ (*meso*)	1	1
trans-1,2-Dibromocyclohexane		0.45[c]
$CH_3CHBrCBr(CH_3)_2$	2.50	80.2
$(CH_3)_2CBrCBr(CH_3)_2$	109	711[d]
$Br_2CHCHBr_2$		0.23[e]
$C_6H_5CHBrCHBrC_6H_5$ (*meso*)	209	561[f]

[a]Ref. 66.
[b]Ref. 73.
[c]Ref. 76. Extrapolated from data at other temperatures.
[d]Ref. 92. Extrapolated from data at other temperatures and compared with *trans*-1,2-dibromocyclohexane.
[e]Ref. 93. Extrapolated from data at other temperatures.
[f]Ref. 25.

Benzalacetophenone, *p*-nitrobenzalacetophenone and *m*-chlorobenzalacetophenone dibromides are dehalogenated by iodide ion at a very similar rate[94], the substituent exerting only a small activating effect. Baciocchi and Schiroli have investigated the dehalogenation from *erythro* *p*-methoxy- and *p*-nitro-substituted 1,2-dibromo-1,2-diphenylethanes promoted by iodide ions in methanol and by iodide, bromide and chloride ions in DMF[8]. The results for the reactions with iodide and bromide ions (Table 3) show that in either solvent substituent effects are very small, with both electron-releasing (OCH_3) and electron-withdrawing (NO_2) substituents increasing the dehalogenation rate to a very small extent. Similar results have been also obtained in the reactions promoted by chloride ions.

This reactivity pattern has been substantially confirmed by a later study on the dehalogenations of *p,p'*-disubstituted 1,2-diphenyl-1,2-dibromoethanes by iodide ions in acetone[96].

TABLE 3. Kinetic data for the dehalogenation of *meso*- or *erythro*-1,2-dibromo-1-(*p*-R^1-phenyl)-2-(*p*-R^2-phenyl)ethanes promoted by iodide and bromide ions and triphenylphosphine in DMF

Substituent		k_2, s^{-1} M^{-1}		
R^1	R^2	I^- (25°C)[a]	Br^- (25°C)[a]	Ph_3P (50°C)[b]
H	H	5.20×10^{-3}	7.35×10^{-3}	1.07×10^{-2}
H	OCH_3	1.43×10^{-2}	18.20×10^{-5}	1.64×10^{-2}
H	NO_2	1.15×10^{-2}	25.50×10^{-5}	3.10×10^{-1}
OCH_3	OCH_3	3.58×10^{-2}	37.90×10^{-5}	8.75×10^{-2}
NO_2	NO_2	2.20×10^{-2}	167×10^{-5}	1.69
OCH_3	NO_2	1.14×10^{-1}	121×10^{-5}	3.4×10^{-1}

[a]Ref. 8.
[b]Ref. 95.

These data clearly suggest that no significant amount of charge is present at either the α or the β carbon atoms in the transition state of the halide ion-induced dehalogenation from 1,2-dibromo-1,2-diphenylethanes. Moreover, the observation that substituent effects appear to be nearly unaffected by the nature of the attacking halide is indicative of a weak interaction between the halogen and the nucleophile in the transition state. This conclusion is in agreement with previous considerations concerning the role of this interaction with respect to the effect of the halogen nature on the dehalogenation rate.

The iodide ion-promoted dehalogenations of aryl-substituted *erythro*-2,3-dibromo-3-phenylpropionic acid appear to exhibit a different substituent effect than the reactions of 1,2-dibromo-1,2-diphenylethanes[97]. It has been found in the former reactions that, compared with the unsubstituted compound, a *p*-OCH_3 group causes a rate enhancement of 500-fold, while a *p*-NO_2 group slightly retards the dehalogenation rate. A ρ value of -0.84 has been determined for this reaction from the dehalogenation rates of a series of *para* and *meta* aryl-substituted 2,3-dibromo-3-phenylpropionic acids, not including the *p*-methoxy-substituted compound[98]. Substituent effects in the iodide-promoted dehalogenation of the dibromides of styrene and α-naphthylethylene have been also determined[99,100]. However, in some cases these reactions could take place by the substitution–elimination mechanism[99].

Structural effects in the iodide ion-induced dehalogenations can be satisfactorily rationalized by both the *E*2Hal and the halonium ion mechanism. Thus, the small substituent effect observed in the dehalogenations from *meso*-1,2-dibromo-1,2-diphenylethanes, together with the significant leaving group effects previously discussed, are fully consistent with an *E*2Hal mechanism characterized by a central transition state with a well developed double bond. The substituent effect would arise from the conjugation between the substituent and the double bond, thereby accounting for the observation that both the NO_2 and the OCH_3 groups exert comparable effects on the rate. Small effects are expected, as actually found, since it is known that these groups contribute very little stabilization to a styrene-type system[101]. In line with this interpretation, the 4-methoxy-4′-nitro compound exhibits a reactivity larger than that expected on the basis of the reactivities of the *p*-nitro and *p*-methoxy compounds since there is an enhanced conjugation involving simultaneously the two substituents and the developing double bond.

The *E*2Hal mechanism with its spectrum of transition states (Scheme 3) can also account for the strong rate-enhancing effect of the *p*-OCH_3 group on the dehalogenation of 2,3-dibromo-3-phenylpropionic acid. It is sufficient to assume that in this case the transition-state structure is no longer 'central' but *E*1-like.

On the other hand, small substituent effects are also predicted by the halonium ion mechanism if the formation of the bridged intermediate is the slow step, provided that the formation of the bond between the β-halogen and C_α occurs nearly synchronously with the breaking of the C_α-halogen bond. If the breaking of this bond is, instead, more advanced than the formation of the former bond, positive charge should develop on C_α and strong favourable effects of electron-releasing substituents become possible. Thus, both the halonium ion and the *E*2Hal mechanisms can account for the substituent effects observed in the reactions of *meso*-1,2-dibromo-1,2-diphenylethanes and *erythro*-2,3-dibromo-3-phenylpropionic acids.

It is also possible that the *E*2Hal and the halonium ion mechanisms can operate side by side in the halide ion-promoted dehalogenations, the prevalence of one or the other depending on the substrate structure. For example, on the basis of the presence or the absence of a common ion effect, it has been suggested that *meso*-2,3-dibromobutane reacts with iodide in acetone by the halonium ion mechanism, whereas *meso*-1,2-dibromo-1,2-diphenylethane is debrominated by the *E*2Hal mechanism[66]. The preference for the halonium ion mechanism might be higher in the reactions of β-substituted alkyl iodides, due to the very great neighbouring group anchimeric ability of iodine, as well as in the solvolytic dehalogenations where, in the absence of a strong nucleophile, the departure of the nucleofugal group may require some assistance from the neighbouring halogen. For example, the formation of bridged intermediates has been suggested in the eliminations of iodine acetate from *trans*-diaxial-2,3-acetoxyiodo-5-cholestanes in acetic acid[102], in the debromination from vicinal dibromides in DMF[50] and in the alkyne-forming solvolysis of 1,2-disubstituted 2-iodovinyl-2,4,6-trinitrobenzenesulphonates[103].

Studies on the substituent effects in dehalogenations from D,L and *threo* dibromides might be very useful for the distinction between the *E*2Hal and the halonium ion mechanism since, as previously observed, the latter mechanism should be the more probable one for these substrates in several cases. It would be interesting, therefore, to know whether the substituent effects in the reactions of D,L and *threo* dibromides are similar or not to those exhibited by the *meso* and *erythro* isomers.

Substituent effects in the dehalogenation from *meso*-1,2-dibromo-1,2-diphenylethane promoted by triphenylphosphine[95] are also reported in Table 3. In this reaction a *p*-nitro group exerts a significant rate-enhancing effect, *c*. 60-fold, whereas the similar effect of the *p*-methoxy group is much smaller, being *c*. 1.5-fold. The strong sensitivity of the dehalogenation reactions which are promoted by phosphorus nucleophiles to the presence of electron-withdrawing groups is also confirmed by the observation that whereas 1,2-dibromoethane and 1,2-dibromopropane are not dehalogenated by triethyl phosphite, dehalogenation takes place when at least an electron-withdrawing substituent is present at one of the bromine-bearing carbon atoms of these compounds[36,104].

In contrast to the iodide-promoted dehalogenations, the substituent effects observed in the reactions with triphenylphosphine fit only the *E*2Hal mechanism. The observed strong rate-enhancing effect of the *p*-nitro group can be rationalized by an *E*2Hal transition state with a high carbanion character (Scheme 3). The neutral character of the nucleophile could play a role in this respect since the incipient positive charge on the phosphorus atom might contribute to stabilizing the incipient negative charge on the carbon (cf. structure **12**), thus favouring a carbanionic transition state.

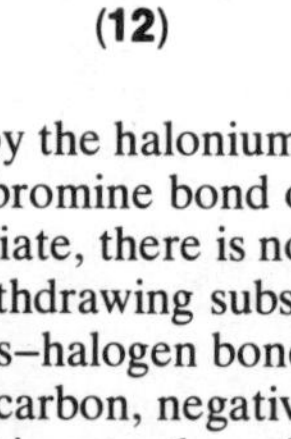

(12)

Rationalization of these findings by the halonium ion mechanism seems much more difficult. As no breaking of the C_β–bromine bond occurs in the transition state of the step leading to the bridged intermediate, there is no way to account for the significant rate-enhancing effect of electron-withdrawing substituents.

If the formation of the phosphorus–halogen bond were more advanced than that of the bond between β-bromine and α-carbon, negative charge would accumulate on the bromine atom. However, even in this case there is no possibility that this charge is stabilized by electron-withdrawing substituents via a conjugative mechanism.

The *E*2Hal mechanism, though, is not the sole route available for the triphenylphosphine-promoted dehalogenation. When a strong electron-attracting substituent

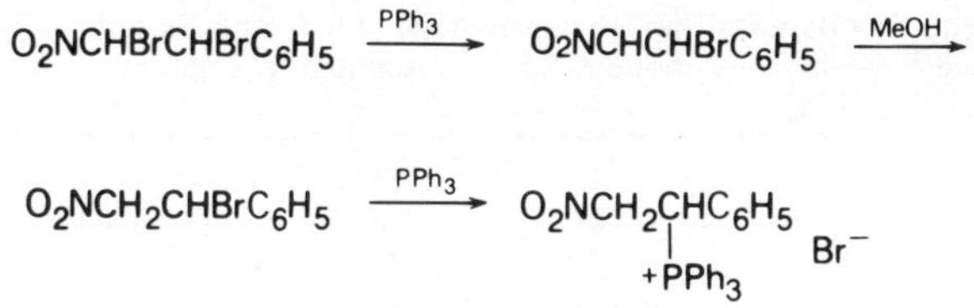

$$O_2NCHBrCHBrC_6H_5 \xrightarrow{PPh_3} O_2N\bar{C}HCHBrC_6H_5 \xrightarrow{MeOH}$$

$$O_2NCH_2CHBrC_6H_5 \xrightarrow{PPh_3} O_2NCH_2CH(\overset{+}{P}Ph_3)C_6H_5 \; Br^-$$

SCHEME 4a

(e.g. NO_2) is directly bonded to the β-carbon, the initial attack of triphenylphosphine at bromine can lead to a carbanionic intermediate, i.e. to the *E*1cB mechanism. Evidence in this respect has been found in the reaction of 1,2-dibromo-1-nitro-2-phenylethane with triphenylphosphine in methanol; 2-nitro-1-phenylethyltriphenylphosphonium bromide was obtained, presumably by protonation of the intermediate carbanion followed by quaternization (Scheme 4a)[35]. The intermediacy of carbanions has also been shown in the 1,3-debromination of $PhCH(Br)SO_2CH(Br)Ph$ by triphenylphosphine[105]. The observation of an *E*1cB mechanism for the dehalogenations with triphenylphosphine probably makes plausible the suggestion that these reactions can also proceed by the *E*2Hal mechanism when no particularly stable carbanion can be formed.

Finally, an interesting structural effect on dehalogenation rate is the strong rise in rate caused by a 3-hydroxyl group in BH_4^--promoted dehalogenations from $5\alpha,6\beta$-dichlorocholestane derivatives[106]. This acceleration has been explained by assuming an *E*2Hal mechanism and suggesting electrophilic assistance by the 3-hydroxyl group to the departure of chlorine as shown in structure **13**.

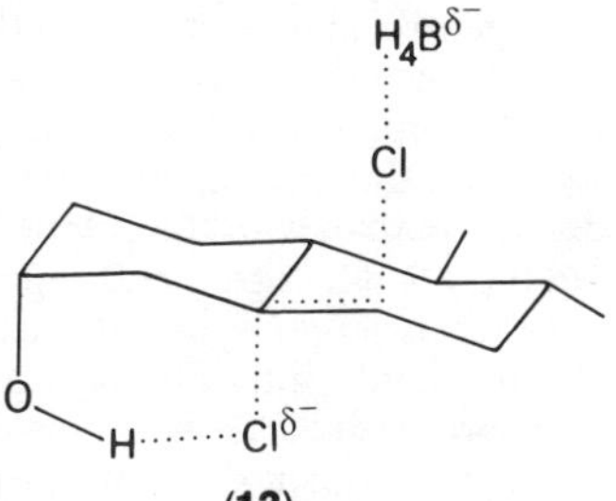

$H_4B^{\delta-}$ ··· Cl ; O–H ··· $Cl^{\delta-}$

(13)

3. The nucleophile

Bromine bonded to carbon is a relatively soft electrophilic centre[107]. The more suitable nucleophiles for debrominations are therefore soft nucleophiles, i.e. highly polarizable species of low basicity such as triphenylphosphine, thioalkoxides, selenides, iodides, carbanions and hydrides. Hard nucleophiles, e.g. hydroxide ion and amines, can be also effective as debrominating agents, but their scope is generally limited to very reactive substrates of particular structure.

This is all the more so when iodine is the electrophilic centre. With chlorine, however, which is a less soft centre than bromine, the softness of the nucleophile should be a less important factor. Unfortunately, quantitative data on the role of the nucleophile on dehalogenation reactions are available only with iodine and bromine as electrophilic centres.

When the nucleophilic attack occurs either on iodine or on bromine, as in the reactions of *erythro*-1-chloro-2-iodo-1,2-diphenylethane and *meso*-1,2-dibromo-1,2-

TABLE 4. Reactivity data for the dehalogenations of *meso*-1,2-dibromo-1,2-diphenylethane and *erythro*-1-chloro-2-iodo-1,2-diphenylethane promoted by some nucleophiles in methanol and DMF at 25°C

	k_2, s^{-1} M^{-1}			
	meso-1,2-Dibromo-1,2-diphenylethane[a]		*erythro*-1-Chloro-2-iodo-1,2-diphenylethane[b]	
Nucleophile	MeOH	DMF	MeOH	DMF
I^-	1.09×10^{-4}	5.20×10^{-3}	0.86	48
Br^-	n.r.[c,d]	7.35×10^{-5}	0.022	9.5
Cl^-	n.r.[c]	1×10^{-5}	0.0036	6.8
F^-	n.r.[c]	n.r.[c,e]		
PPh_3	1.14×10^{-3}[f]	1.70×10^{-3}		

[a]Data from Ref. 8 unless otherwise indicated.
[b]Data from Ref. 85.
[c]No dehalogenation was observed.
[d]A k_2 value of *c*. 10^{-6} M^{-1} s^{-1} can be estimated from Ref. 108.
[e]Ref. 29.
[f]Ref. 95. Data are in ethanol. Slightly larger values of k_2 are expected in methanol[109].

diphenylethane, respectively, the reactivity order of halide ions is $I^- > Br^- > Cl^- \gg F^-$ (with fluoride ion being ineffective), in both protic and dipolar aprotic solvents (Table 4)[8,25,26,85,95]. This order parallels the polarizability order of halogens.

As shown in Table 4, the reactivity spread is larger in protic solvents, in agreement with the role of specific hydrogen-bonding solvation of the nucleophile anion, which is more important the smaller is the nucleophile[110]. Thus, only iodide ions can induce debromination from *meso*-1,2-dibromo-1,2-diphenylethane in methanol. In DMF the differences in reactivity between the various halides are quite small, being slightly larger with the dibromo derivative than with the chloroiodo compound. This confirms previous suggestions that in the 1,2-dihalogeno-1,2-diphenylethane system there is no strong interaction between the nucleophile and the halogen in the transition state. However, the situation can be significantly different in other systems. Thus, in the bromine eliminations from *meso*-diethyl 2,3-dibromosuccinate, iodide ions are about 1000 times more reactive than bromide ions[68]. This reactivity ratio compares with that (70) observed in the dehalogenations from *meso*-1,2-dibromo-1,2-diphenylethane.

It is noteworthy that in dipolar aprotic solvents the reactivity order of halide ions in dehalogenation reactions is exactly the reverse of that found in the S_N2 nucleophilic substitutions, thus indicating that bromine and iodine are softer electrophilic centres than saturated carbon. This observation makes the *E*2C-like transition state **7** for dehalogenation reactions highly unlikely.

In Table 4 data for the reactions with triphenylphosphine are also reported. It is seen that in DMF triphenylphosphine is slightly less effective than iodide ions. In contrast, in a protic solvent, triphenylphosphine dehalogenates *c*. 10 times faster than iodide ions. Interestingly, when triphenylphosphine and iodide ions attack the saturated carbon of methyl iodide the reverse reactivity order is again observed[111].

Among phosphorus nucleophiles the reactivity order towards bromine is $Bu_3P > PPh_3 > Ph_2POC_2H_5 > PhP(OC_2H_5)_2 > P(OC_2H_5)_3$, which is also the basicity order for these nucleophiles[35]. Basicity can be indeed an important factor in determining the nucleophilic reactivity when structurally related nucleophiles having the same reacting atom are considered[112].

TABLE 5. Kinetic data for the dehalogenation of *erythro*-1-chloro-2-iodo-1,2-diphenylethane induced by a variety of nucleophiles in methanol at 25°C[85]

Nucleophile	k_2, s^{-1} M^{-1}
$SeCN^-$	1.46
I^-	0.86
CN^-	0.40
$SC(NH_2)_2$	0.32
Piperidine	0.061
Br^-	0.022
Morpholine	0.015
Cl^-	0.0036
Hydrazine	0.0025
Ammonia	0.0005

More detailed information on the factors playing a role with respect to the reactivity of the nucleophile towards halogen comes from a study of the dehalogenation rates of *erythro*-1-chloro-2-iodo-1,2-diphenylethane with a variety of nucleophiles[85]. Some results are reported in Table 5.

The reactivity of these nucleophiles is satisfactorily correlated by the Edwards equation[113]:

$$\log(k/k_0) = \alpha E + \beta H$$

where E and H are the oxidative dimerization potential and the basicity of the nucleophile, respectively. From the plot an α-value of 2.13 and a β-value of -0.12 are calculated, thus confirming that polarizability and desolvation factors play a much larger role than basicity in the nucleophilic reactivity towards iodine. The secondary role of basicity is also shown by the small Brönsted β-value (0.29) observed in the dehalogenation of *erythro*-1-chloro-2-iodo-1,2-diphenylethane promoted by substituted pyridines in 60% dioxan[114].

The reactivity of a series of aliphatic amines in the same reaction can be correlated with the stability constants of the charge transfer complexes between the amines and iodine[114]. Since the slope of this correlation is close to unity the charge distribution in the transition state of the dehalogenation reaction is probably similar to that of a charge transfer complex, with little nitrogen–iodine bond formation. A further observation is that the steric effects of the amines are less important than electronic effects as secondary amines are more reactive than primary ones.

4. The solvent

In protic solvents the rate of dehalogenation reactions is not particularly sensitive to the nature of the solvent. Thus, in the reaction of iodide with *trans*-1,2-dibromocyclohexane very similar rate constants are observed in methanol[76] and in *n*-propanol[90]. With thiophenoxide ion and the same substrate the dehalogenation rate is slightly faster in ethanol than in methanol at 75°C, but the reverse holds at 100°C[60].

More significant rate changes are observed on going from protic to dipolar aprotic solvents. As shown in Table 4, *meso*-1,2-dibromo-1,2-diphenylethane and *erythro*-1-chloro-2-iodo-1,2-diphenylethane react with iodide ions *c.* 50 times faster in DMF than in methanol. Larger rate enhancements are observed in the reactions promoted by bromide and chloride ions in agreement with the fact that solvation energy of halide ions in protic solvents decreases in the order $Cl^- > Br^- > I^-$.

With thiophenoxide ion the rate effect for the change from a protic to dipolar aprotic solvent is two to three orders of magnitude larger than with halide ions[60]. In the reaction of *trans*-1,2-dibromocyclohexane with thiophenoxide ion a very impressive rise in rate (40 000-fold) is observed as we move from methanol to DMF[60]. This rate increase has been attributed to a tight transition state which is a very weak hydrogen bond acceptor (small negative charge on sulphur and bromine, low degree of breaking of the C_α—Br bond and little developed double bond). If this interpretation is correct, the much smaller solvent effect observed in the dehalogenation of 1,2-dihalogeno-1,2-diphenylethanes by halide ions would suggest a loose transition state, which is a good hydrogen bond acceptor, for these reactions, in agreement with previous conclusions.

E. Competition with Other Reactions

As anticipated, substitution at saturated carbon and dehydrohalogenation can compete with dehalogenation of a dihalide. The importance of these two competing reactions depends mainly on the nature of the substrate as well as of the nucleophile.

Since the reactivity of the halogens as electrophilic centres decreases in the order $I > Br > Cl \gg F$, the probability of competition by substitution and HX elimination will *increase* in the order $I < Br < Cl \ll F$. Thus, whereas 1,2-diiodoethane reacts with thiophenoxide ions to give ethylene, only the substitution product is obtained from 1,2-dibromoethane[115]. With diphenylmethylcarbanion as the nucleophile, dehalogenation takes place with 1,2-dibromoethane and substitution with 1,2-dichloroethane[116].

The nature of the halogen leaving group is also important. For example, with lithium aluminium hydride *trans*-1,2-dibromocyclohexane forms cyclohexene, whereas *trans*-1-bromo-2-fluorocyclohexane is exclusively converted into 2-fluorocyclohexane[21].

S_N2 substitution can more effectively compete with dehalogenations in the reactions of primary dibromides. That it is indeed so, with iodide as a nucleophile, has been previously mentioned. With thiolates, substitution prevails over bromine elimination in the case of 1,2-dibromoethane, 1,2-dibromopropane and ethyl 3,4-dibromobutanoate[29]. With other primary dibromides of the type $RCHBrCH_2Br$ ($R = C_6H_5$, C_6H_{13}, C_2H_5, $C(CH_3)_3$) dehalogenation is the main reaction[29]. No olefin is obtained in the reaction of 1,2-dibromoethane and 1,2-dibromopropane with sodium thiosulphate[117].

In D,L and *threo* dibromides *anti* dehalogenation is disfavoured by the conformational factors discussed above. These substrates are therefore more prone to the competing processes than the *meso* or the *erythro* forms. Thus, *meso*-1,2-dibromo-1,2-diphenylethane reacts with lithium aluminium hydride to give *trans*-stilbene exclusively, whereas the D,L-diastereoisomer is reduced in part to 1,2-diphenylethane[21].

Conformational factors become still more important when competition between dehalogenation and dehydrohalogenation is considered. Since an *anti* stereochemistry is preferred for both eliminations, conformational factors will favour dehalogenation with *meso* and *erythro* dibromides, but dehydrohalogenation with D,L and *threo* isomers. As a consequence, chloride ion debrominates *meso*-1,2-dibromo-1,2-diphenylethane in DMF, but promotes loss of hydrogen bromide from the D,L isomer in the same solvent[26]. Similarly, it has been reported that in the reaction of thiolate ions with *meso*- and D,L-2,3-dibromobutane, the dehydrobromination is a more important side-reaction with the latter substrates[28].

Competition between dehalogenation and dehydrohalogenation is also strongly influenced by the nature of the nucleophile, since halogen is a much softer centre than hydrogen. Generally, the harder the nucleophile the more effectively dehydrohalogenation will compete with dehalogenation. For example, *meso*-1,2-dibromo-1,2-diphenylethane is dehalogenated by soft nucleophiles (e.g. iodide and bromide ions, triphenylphosphine) but undergoes a dehydrobromination reaction when reacted with ethanolic potassium hydroxide[118] and fluoride ions[26]. Actually, debromination from *meso*-1,2-dibromo-1,2-diphenylethane promoted by potassium or caesium fluoride in DMSO has been reported[119]; however, it has also been suggested that this reaction probably occurs via the intermediacy of the dimsyl anion ($^{-}CH_2SOCH_3$), a soft nucleophile.

Another example is the reaction of *erythro*-ethyl *p*-nitrocinnamate dibromide, which is dehalogenated by pyridine and triphenylphosphine, whereas it is dehydrohalogenated by acetate ions[49]. Similarly, 1,2-dibromooctane forms 1-octyne by elimination of 2 mol of hydrogen bromide in its reaction with strong bases such as potassium hydroxide, potassium *tert*-butoxide or $Et_3CO^-K^+$, but gives debromination to 1-octene when it reacts with quinoline[120].

Since the nucleophilic reactivity is strongly influenced by the solvent, the solvent too can play a role in determining the relative importance of dehalogenation versus dehydrohalogenation. This point is convincingly illustrated by the results of Parker and coworkers on the reaction of *erythro*-1,2-dibromo-1-(4-nitrophenyl)-2-phenylethane with cyanide ions[38] (equation 11). In ethanol, elimination of hydrogen bromide

$$O_2N-C_6H_4-CBr{=}CH-C_6H_5 \ (90\%) \xleftarrow[C_2H_5OH]{CN^-} O_2N-C_6H_4-CHBr-CHBr-C_6H_5$$

$$\xrightarrow[DMF]{CN^-} O_2N-C_6H_4-CH{=}CH-C_6H_5 \ (100\%) \qquad (11)$$

accounts for 90% of the reaction, whereas in DMF exclusive dehalogenation is observed. This phenomenon has been rationalized by suggesting that the transition state for the debromination reaction is tight, while that for the dehydrohalogenation reaction is loose. Thus, the former reaction takes greater advantage in the transfer from protic to dipolar aprotic solvents. This conclusion, however, should not be taken as a general one, since we have previously observed that loose transition states are also possible for dehalogenation reactions.

Considering the substrate structure, the presence of electron-attracting substituents (especially carbonyl groups) appears to favour dehalogenation over both dehydrohalogenation and substitution. We recall, in this respect, that primary dibromides substituted with electron-attracting β-substituents are dehalogenated by triethyl phosphite whereas the unsubstituted compound undergoes nucleophilic attack at carbon[36]. Furthermore, hydroxide ions give *cis*-α-bromostilbene from *meso*-1,2-dibromo-1,2-diphenylethane (equation 12), but they debrominate the *N,N*-diethylamide of *erythro*-2,3-dibromo-3-phenylpropanoic acid[49] (equation 13). The $CONR_2$ group appears to be more effective than the $CO_2C_2H_5$ group in favouring debromination since *erythro*-ethyl 2,3-dibromo-3-phenylpropanoate is dehybrominated by acetate ions[49] (equation 14).

The replacement of methyl groups by phenyl groups also appears to favour dehalogenation relative to dehydrohalogenation, as shown by the finding that *erythro*-2,3-dibromo-3-phenylpropanoic acid is debrominated by triphenylphosphine whereas

$$C_6H_5-CHBrCHBr-C_6H_5 \xrightarrow{OH^-} C_6H_5-CH{=}CBr-C_6H_5 \qquad (12)$$

(*meso*) (*cis*)

$$C_6H_5-CHBrCHBrCON(C_2H_5)_2 \xrightarrow{OH^-} C_6H_5-CH{=}CHCON(C_2H_5)_2 \qquad (13)$$

(*erythro*) (*trans*)

$$C_6H_5-CHBrCHBrCO_2C_2H_5 \xrightarrow{AcO^-} C_6H_5-CH{=}CBrCO_2C_2H_5 + C_6H_5-CHBr{=}CHCO_2C_2H_5 \qquad (14)$$

(*erythro*)

erythro-2,3-dibromobutanoic acid is dehydrobrominated[35]. As a further example, the reaction of bromide ion with *trans*-1,2-dibromo-1,2-cyclohexane in DMF yields both bromine and hydrogen bromide[68]; under the same experimental conditions, only debromination is observed with *meso*-1,2-dibromo-1,2-diphenylethane[8,27].

Steric effects too could explain, at least in part, these observations, since it is expected that both overcrowding in the substrate and large steric requirements of the base will favour dehalogenation with respect to dehydrohalogenation. In crowded systems, the halogen atom is more accessible to the nucleophile than the hydrogen atom and the loss of a halogen molecule provides more steric relief than that of a molecule of a hydrohalogenic acid. On the other hand, attack on halogen does not have large steric requirements, as shown in the reactions between α-halogenoacetophenones and diphenylphosphine, where the progressive substitution of the hydrogen atoms of the $-CH_2Br$ group with methyl groups exerts a relatively small rate-retarding effect[121].

In agreement with the above reasoning is the observation that isocoumarin dibromide and other vicinal dibromides undergo exclusive debromination when they react with 1,8-bis(dimethylamino)naphthalene (proton sponge), a strong base with very high steric requirements[46]. Likewise, in the reaction of 3,4-dibromoheptane with sodium alkoxides debromination competes more effectively with dehydrobromination when the alkoxide is sodium *tert*-butoxide than when it is sodium ethoxide[48].

III. DEHALOGENATIONS PROMOTED BY ONE-ELECTRON REDUCTANTS

A. General

With nucleophiles as dehalogenating agents the two electrons involved in the process are transferred from the reductant to the substrate in a single step. With one-electron reductants the two electrons are transferred, one at a time, in separate steps.

One-electron reductants include radical anions (e.g. sodium naphthalenide[122], sodium biphenyl[123] and disodium dihydrophenanthrenediide[124]), carbanions (e.g. $Ph_2\bar{C}CN$[125]), free or complexed metal ions (e.g. chromium(II)[1,126–130], titanium(II)[131,132], tin(II)[133,134], cobalt(II)[135,136], tungsten(II)[137], copper(I)[138], iron(II)[139] and vanadium(II)[140]), radicals (e.g. phenyl[141], tri-*n*-butyltin[142,143] and silyl[144] radicals, and $C_3H_5Fe(CO)_2P(C_6H_5)_3$[65]).

The great utility of alkali dihydroarylides as dehalogenating agents has been emphasized[145]. These radical anions react very rapidly at low temperature and are effective with dibromides and dichlorides even when other reductants fail. Thus, disodium dihydrophenanthrenediide has been used in the synthesis of 'hemi' Dewar naphthalene (benzobicyclo[2.2.0]hexa-1,5-diene) (**14**), starting from the *cis*-dichloride

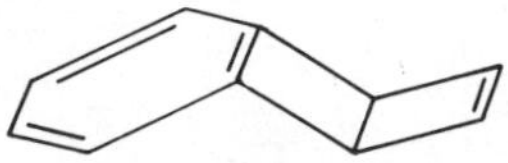

(**14**) 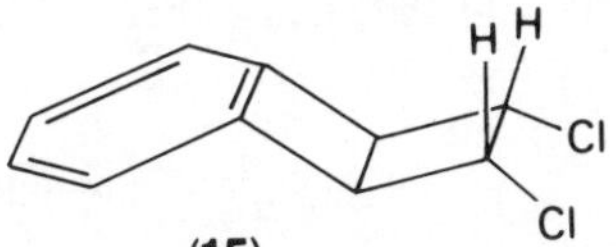 (**15**)

15[124]. The cyclobutene derivative **16** has been obtained by dehalogenation of **17** promoted by sodium naphthalenide[145].

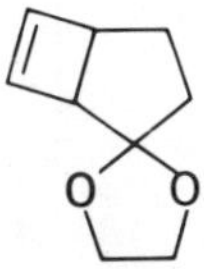

(**16**) 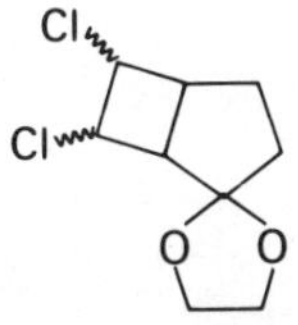 (**17**)

Among metal ions, chromium(II) has been found to be very effective for dehalogenations in the steroid series[126,146]. Better results have been obtained than with more conventional reductants such as zinc and iodide ions. Thus, the smooth conversion of Δ^5-3-hydroxysteroids to Δ^4-3-ketones by bromine addition, oxidation to ketone and dehalogenation by chromous chloride has been accomplished[147]. The reducing power of chromous salts can be greatly enhanced by complexation with ethylenediamine or related ligands[148], so making it possible to perform β-elimination reactions not only from dibromides and dichlorides but also from β-halohydrins, β-haloethers and β-haloamines[1]. Vicinal bromofluorides are also dehalogenated[149,150].

Selective dehalogenations have been carried out with bis(η-cyclopentadienyl)-dinitrosyl chromium[130]. With this reductant it is possible to remove vicinal halogens while leaving other non-benzylic halogen atoms unaffected, as shown by the equation (15).

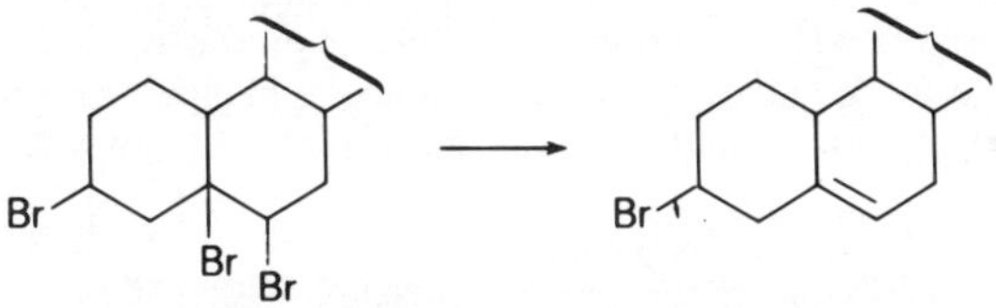

(15)

Recently, the use of titanium(II)[132] and vanadium(II)[140] has been recommended. In both cases a mixture of titanium(III) or titanium(IV) or of vanadium(III) chlorides with lithium aluminium hydride is used and the reactions are carried out in tetrahydrofuran. Titanium(II) is also extremely effective in the reduction of bromohydrins; however, when the dehalogenation can be conducted either with titanium(II) or vanadium(II) the latter reductant appears more convenient from the point of view of the experimental conditions required. Among other things, vanadium(III) chloride is stable to air whereas titanium(III) or titanium(IV) chlorides are not.

B. Dehalogenations by Radical Anions and Carbanions

1. Mechanism

A one-electron transfer mechanism has been demonstrated for dehalogenations promoted by sodium naphthalenide[122], especially through a CIDNP investigation[122a]. On the basis of the results obtained, the mechanistic sequence reported in Scheme 5 has been proposed.

$$C_{10}H_8^{-\cdot} + \overset{|}{\underset{|}{C}}X-\overset{|}{\underset{|}{C}}X \longrightarrow \cdot\overset{|}{\underset{|}{C}}-\overset{|}{\underset{|}{C}}X + X^- + C_{10}H_8$$
(18)

$$\cdot\overset{|}{\underset{|}{C}}-\overset{|}{\underset{|}{C}}X \;(18) \xrightarrow{-X^\cdot \text{ (a)}} \;>C=C<$$

$$\cdot\overset{|}{\underset{|}{C}}-\overset{|}{\underset{|}{C}}X \;(18) \xrightarrow[+C_{10}H_8^{-\cdot}]{\text{(b)}} {}^-\overset{|}{\underset{|}{C}}-\overset{|}{\underset{|}{C}}X \;(19) \xrightarrow{-X^-} \;>C=C<$$

SCHEME 5

The initial transfer of one electron to the halogen atom results in the formation of the haloalkyl radical **18**, presumably through the formation of an alkyl halide radical anion which immediately loses an X^- anion. The overall process is a dissociative electron transfer[151]. The final olefinic product can be obtained from **18** via two competing reaction paths. Radical **18** either undergoes a fragmentation process and is directly converted to the alkene derivative (path (a)) or it can be further oxidized to the carbanion **19** which forms the olefin by loss of X^- (path (b)). With dibromo derivatives ($X = Br$) path (a) seems more important than path (b).

A substantially similar electron transfer pathway has been suggested for the dehalogenations promoted by some tertiary carbanions ($Me_2\bar{C}CO_2Et$ and $Ph_2\bar{C}CN$)[125]. It has been observed that dehalogenations by $Ph_2\bar{C}CN$ are completely inhibited in the presence of *p*-dinitrobenzene, a powerful electron acceptor[152]. Moreover, the conversion reported in equation (16), which is promoted by $(CH_3)_2\bar{C}NO_2$, is catalysed by light and inhibited by *p*-dinitrobenzene[153]. Hence, this process, closely related to a dehalogenation reaction, also takes place by an electron transfer mechanism.

$$p\text{-}NO_2C_6H_4CH(Cl)CNO_2(CH_3)_2 \xrightarrow{(CH_3)_2\bar{C}NO_2} p\text{-}NO_2C_6H_4CH=C(CH_3)_2 \qquad (16)$$

The possibility that other carbanions considered to act as two-electron reductants[116,154,155] behave instead as one-electron reductants has been suggested[125]. However, it should be noted that the relative tendency of a nucleophilic species to operate as a one- or a two-electron donor may be a very subtle function of the structure, as well as of the experimental conditions, thus making generalizations very difficult. For example, whereas $(CH_3)_2\bar{C}CO_2C_2H_5$ is a dehalogenating agent, presumably by a one-electron transfer mechanism[125], the diethyl malonate anion

$\bar{C}H(CO_2Et)_2$ reacts with *trans*-1,2-dibromocyclohexane via a nucleophilic substitution reaction followed by a dehydrohalogenation[156].

2. *Stereochemistry*

Data concerning the stereochemistry of one-electron transfer dehalogenations promoted by radical ions and carbanions are somewhat contradictory. With $Ph_2\bar{C}CN$ a stereoconvergent reaction results, both *meso*- and D,L-3,4-dibromohexane give exclusively *trans*-3-hexene[125]. With sodium naphthalenide in dimethoxyethane a predominant '*anti*' course is instead observed with *erythro*- and *threo*-2,3-dibromo-3-methylpentane[122b] and, even though to a less extent, with *meso*- and D,L-2,3-dibromobutanes[122a], whereas the same mixture of *trans*- and *cis*-2-butene is obtained from *meso*- and D,L-2,3-dichlorobutane[122a].

It has been rightly noted[122a] that this variety of stereochemical results is not surprising, as the steric course of these dehalogenations may depend on a number of factors whose relative weight depends upon the reactant's structure. The first step may be non-stereospecific or only partially stereospecific, and the haloalkyl radical may undergo rotameric relaxation, thus destroying any stereochemical integrity with which it might be formed. Moreover, the possible lack of stereospecificity of some process involving an intermediate subsequent to the haloalkyl radical, and the relative importance of the two competing pathways available to the radical, may also play a significant role in determining the stereochemical outcome of the reaction. Thus, the greater *anti* stereoselectivity of the dehalogenation from diastereoisomeric 2,3-dibromobutanes has been explained by suggesting that with this substrate the fragmentation step can compete with the reduction step. The rotameric relaxation of the 2-bromo-2-butyl radical should be retarded by bromine bridging[157], as shown in structure **20** for the radical derived from *meso*-2,3-dibromobutane. Consequently, it

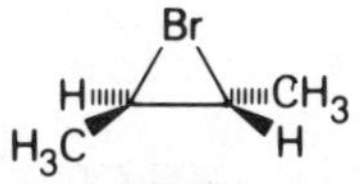

(20)

will become competitive or slower than the loss of bromine, and some degree of *anti* stereospecificity is allowed. With 2,3-dichlorobutane CIDNP measurements suggest that the exclusive path of the haloalkyl radical **18** is the reduction to anion. Moreover, the rotameric relaxation of the free radical has been calculated to be faster than its conversion to carbanion. This is in line with the fact that chlorine is less effective than bromine as a bridging group. It is thus possible to rationalize the fact that both the *meso* and the D,L isomers of this substrate give the same proportions of *cis*- and *trans*-2-butene.

C. Dehalogenations by Metal Ions and Radicals

1. *Mechanism*

Out of the dehalogenations promoted by free or complexed metal ions, those induced by chromium(II) have been the more intensively investigated. A variety of chromium(II) compounds, e.g. chromous sulphate, chloride and perchlorate, and chromium(II) complexed by ethylenediamine can be used in a number of solvents such as ethanol, DMF, DMSO and pyridine. Since the reaction takes place under homogeneous conditions, quantitative studies are possible.

$$\underset{|}{\overset{|}{C}}X-\underset{|}{\overset{|}{C}}X + 2\,Cr(II) \longrightarrow \;\rangle C{=}C\langle\; + 2\,Cr(III) + (2\,X^-) \qquad (17)$$

The stoichiometry of the reaction is given by equation (17), where X is a halogen[127,128] and X^- can be either free or bonded to chromium(III). The rate expression is given by equation (18).

$$\text{Rate} = k[\text{Dihalide}][\text{Cr(II)}] \qquad (18)$$

As the stoichiometry of the rate-determining step is of lower order than that of the overall reaction, the reductive elimination by chromium(II) has to be a multistep process.

Chromium(II) is also a very effective reducing agent able to convert alkyl monohalides to the corresponding alkanes[148,158–160]. This reaction, for which several lines of evidence, including kinetics, stereochemistry and reactivity, strongly support the mechanism reported in Scheme 6[159–160] (SH is a proton donor), exhibits character-

$$RX + Cr(II) \longrightarrow R^{\bullet} + Cr(III)X$$

$$R^{\bullet} + Cr(II) \longrightarrow RCr(III)$$

$$RCr(III) + SH \longrightarrow RH + Cr(III)S$$

SCHEME 6

istics very similar to those of chromium(II)-promoted dehalogenations. Both reactions display the same order of halogen reactivity I > Br > Cl and second-order kinetics. Moreover, the chromium(III) species produced by alkyl monohalides is Cr(III)X and at least one halogen from the dihalide is converted to this species. Thus, it seemed very reasonable to suggest that chromium(II)-promoted dehalogenations also involve a halogen atom transfer to form a *β*-haloalkyl radical in the first step of the reaction (equation 19). With dibromides the removal of one of the halogens by Cr(II) is

$$X\underset{|}{\overset{|}{C}}-\underset{|}{\overset{|}{C}}X + Cr(II) \longrightarrow \cdot\underset{|}{\overset{|}{C}}-\underset{|}{\overset{|}{C}}X + Cr(III)X \qquad (19)$$

probably assisted by the neighbouring group effect of the other bromine atom, and the radical formed has the bridged structure **21** (Scheme 7, X = Br). This hypothesis is in agreement with information available on other reactions involving *β*-bromoalkyl radicals[157], and in the specific case under consideration is strongly supported by the observations that dehalogenative reduction of dibromides is much faster than reduction of the corresponding alkyl bromides. More significantly, *trans*-1,2-dibromocyclohexane is dehalogenated by chromium(II) *c.* 3000 times more rapidly than the *cis* isomer[128]. This finding is consistent with the fact that the *anti* periplanar conformation required for an optimum neighbouring group assistance is possible only with the *trans*-dibromide. Moreover, the intervention of a bridged 2-bromoalkyl radical is also in agreement with the high *anti* stereoselectivity exhibited by the chromium(II)-promoted debromination of the 2,3-dibromobutanes[129] (see below).

Replacement of one bromine atom by a group which is less capable of providing neighbouring group assistance causes a decrease in the elimination rate, as shown by the data reported in Table 6 for a series of *erythro*-2-bromo-3-X-butanes (X = Cl,

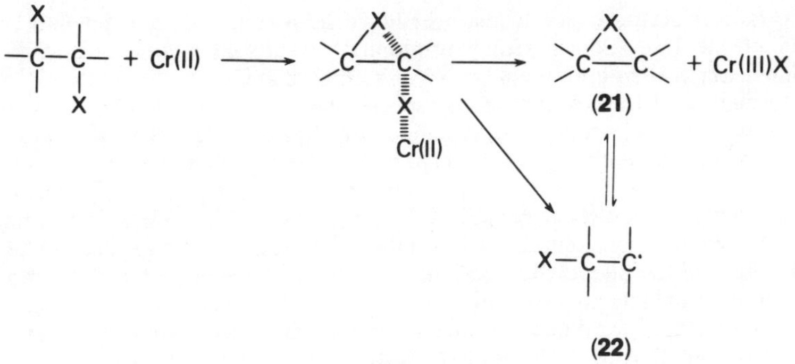

X–C–C–X + Cr(II) ⟶ [bridged X···Cr(II) complex] ⟶ (21) + Cr(III)X

(21) ⇌ X–C–C· (22)

SCHEME 7

TABLE 6. Reactivity data for Cr(II)-promoted dehalogenation[a] of *erythro*-3-X-2-bromobutanes at 0°C[128]

X	$k_2 \times 10^4$, s^{-1} M^{-1}	Relative rate
H[b]	0.14	1
OH	0.65	4.6
OAc	2.1	15
OTs	7.7	55
Cl	51	364
Br[c]	1600	11 428

[a]In 85% DMF containing 5% water, 10% ethanol, 0.90 M perchloric acid and 0.057 M Cr(II).
[b]Reduction to butane.
[c]*Meso* form.

OH, $OCOCH_3$, OTs)[128]. In these reactions the intermediate free radical is better formulated as an open species, even though the possibility that a first-formed bridged species is in a fast equilibrium with a more stable open radical (**22**) (Scheme 7) cannot be excluded.

It has been proposed that the bridged or open radical is converted to the olefin according to the pathways reported in Scheme 8[127–129]. In path (a) a fragmentation of

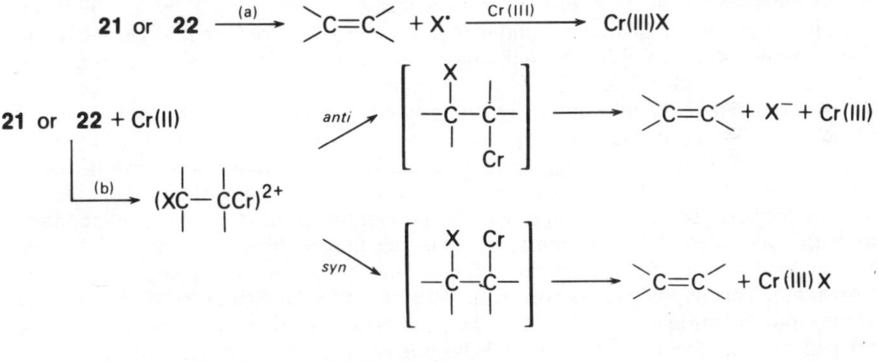

21 or **22** —(a)→ C=C + X· —Cr(III)→ Cr(III)X

21 or **22** + Cr(II) —(b)→ $(XC{-}CCr)^{2+}$

anti → [X–C–C–Cr] ⟶ C=C + X^- + Cr(III)

syn → [X Cr –C–C–] ⟶ C=C + Cr(III)X

SCHEME 8

the free radical occurs which leads directly to the alkene, whereas in path (b) the reaction of the free radical with chromium(II) results in a β-haloalkylchromium compound, which then undergoes *syn* or *anti* elimination.

The formation of the alkylchromium species has been suggested by analogy with the mechanism of reduction of alkyl halides by chromium(II) (Scheme 6), and the competition between path (a) and path (b) by the observation that, when X = Br, the amount of bromide ion formed increases with increasing chromium(II) concentration until it reaches a plateau characteristic of the alkyl moiety. Moreover, the fact that the amount of bromide ion generated is somewhat less than that expected from the decomposition of the alkylchromium intermediate, even when path (a) is completely suppressed by using an excess of chromium(II), suggests the two elimination modes of **21** reported in Scheme 8. Whereas in the *anti* elimination a free bromide ion is formed, an intramolecular transfer of bromine to chromium giving the species Cr(III)Br may represent the driving force of the *syn* process.

In the competition between path (a) and path (b) and between *anti* and *syn* eliminations from the alkylchromium intermediate, the nature of the leaving group, X, is also very important. When the leaving group is poorer than bromine (e.g. X = Cl, OH, OAc, OTs), path (a) in Scheme 8 is probably of little significance due to the relatively high heats of formation of the $X^{\bullet}$ radical. Moreover, in this case, the alkylchromium intermediates undergo *syn* elimination exclusively since the breaking of the carbon-leaving group bond requires intramolecular assistance by chromium.

An initial halogen atom transfer, similar to that suggested for the chromium(II)-promoted dehalogenations, has also been proposed for the reactions induced by [(η-cyclopentadienyl)dinitrosyl]chromium[130], pentacyanocobaltate[135,136], titanium(II)[132], tin(II)[134], $C_3H_5Fe(CO)_2P(C_6H_5)_3$[65], tri-*n*-butyltin radicals[142,143], silyl radicals[144] and phenyl radicals[141].

2. *Stereochemistry*

As anticipated, the chromium(II)-promoted debromination of *meso*- and D,L-2,3-dibromobutanes shows *anti* stereoselectivity. The degree of *anti* stereoselectivity is, however, variable, depending upon the nature of the solvent and upon the ratios Cr(II)/dibromide and ligand/Cr(II) (Table 7)[129]. The highest *anti* stereoselectivity is observed in DMSO or in pyridine as solvent or by using high ligand/Cr(II) ratios. With 2-bromo-3-X-butanes (X = Cl, OH, OAc, OTs) the situation is drastically different. The same mixture of *trans*- and *cis*-2-butenes is obtained in each case, from the two diastereoisomeric compounds.

The stereochemical data can be satisfactorily accommodated by the reaction sequences reported in Schemes 7 and 8[129]. Thus, the *anti* stereoselectivity observed in the debromination of 2,3-dibromobutanes is explained by three considerations: (a) The bridged radical (X = Br) is stereospecifically generated from the dibromide and is expected to have a sufficient stability to maintain stereochemical integrity before fragmentation or reaction with chromium(II). (b) The stereochemical integrity is also preserved in the subsequent formation of the alkylchromium intermediate since this step certainly involves a stereospecific *trans* opening of the bridged radical and the formed carbon–chromium bond is sufficiently strong. (c) An *anti* elimination from this intermediate is the most probable reaction as bromine is a good leaving group which does not need extensive neighbouring chromium assistance in order to be eliminated. This hypothesis, however, holds only when the difference in the stability of the isomeric bridged species **23** and **24** is not large, as certainly is the case with the bridged radicals formed from the 2,3-dibromobutanes. With other structures (e.g. R = phenyl, or a much bulkier group than methyl) this difference can become so large

TABLE 7. Stereoselectivity in the dehalogenation of *meso*- and D,L-2,3-dibromobutanes by chromium(II) at 25°C under different reaction conditions[129]

Substrate	[Cr(II)]/[Dibromide]	Solvent	Ethylene-diamine, M	2-butene, % *trans*	2-butene, % *cis*
meso-2,3-Dibromobutane	1.56	90% EtOH		80	20
	1.56	90% DMF		82	18
	1.56	90% DMSO		97	3
	1.56	88% pyridine		92	8
D,L-2,3-Dibromobutane	1.56	90% EtOH		64	36
	1.56	90% DMF		43	57
	1.56	90% DMSO		6	94
	4.88	Pyridine		18	82
	5	86% DMF[a]	0.18[b]	35	65
	5	65% DMF[a]	3.6[b]	14	86

[a] At −15°C.
[b] [Cr(II)] = 0.083 M.

as to make the **23** ⇌ **24** isomerization a much faster process, thus increasing the probability that it occurs before the reaction of the radical with chromium(II), and diminishing the *anti* stereoselectivity. Accordingly, it has been observed that chromium(II) exclusively converts D,L-1,2-dibromo-1,2-diphenylethane into *trans*-stilbene[2].

(**23**) (**24**)

The variations in the *anti* stereoselectivity observed as the solvent is changed or in the presence of ethylenediamine (Table 7) have been rationalized by suggesting that the relative importance of the *anti* and *syn* eliminations from the alkylchromium intermediate may depend on the solvent and the ligands coordinated to chromium(II), which influence the extent of the possible interaction between the metal and bromine in the elimination process[129]. Thus, the maximum *anti* stereoselectivity observed in DMSO, pyridine or in the presence of ethylenediamine might be due to the fact that chromium is so tightly coordinated by these solvents or by ethylenediamine as to become less available for bond formation with bromine, thus lowering the probability of a competing *syn* elimination.

Likewise, the increase in *anti* stereoselectivity as the Cr(II)/dibromide ratio increases has been explained by suggesting that as the chromium(II) concentration rises, its reaction with the radical has more chance to compete with other possible processes (e.g. conversion into a open free radical, isomerization) which diminish the stereoselectivity.

Finally, the lack of stereospecificity observed in the elimination of *erythro* and *threo*-2-bromo-3-X-butanes (X = Cl, OAc, OH, OTs) is expected, since the open radical thought to be involved in these cases, alone or in equilibrium with the bridged species, should undergo fast rotation and lose asymmetry around carbon. The same proportions of *trans*- and *cis*-2-butenes are thus obtained from each pair of the diastereoisomeric substrates.

Predominant *anti* elimination has been observed in the dehalogenation of *meso*- and D,L-2,3-dibromobutanes and of 1,2-dibromo-1-deuterohexane with tri-*n*-butyltin hydride[142]; interestingly, the *anti* stereoselectivity increases by increasing the hydride concentration, but disappears in the dehalogenations of the isomeric 2,3-dichlorobutanes (where the main reaction is reduction of the carbon–chlorine bond), and diethyl 2,3-dibromosuccinates. These results can be accounted for by the free radical chain mechanism shown in Scheme 7 by following the same reasoning as used to discuss the chromium(II)-promoted dehalogenations. Thus, in the case of the dibromobutanes a bromine-bridged radical is formed. It can lose a bromine atom by fragmentation or by reaction with an organotin hydride before isomerization (Scheme 9). However, when a less stable bridged radical is formed from the dichloro compound or when the relative stability of the two possible isomeric bridged radicals is much larger than in the dibromobutane system (e.g., with diethyl 2,3-dibromosuccinate) isomerization of the bridged radical is possible and stereospecificity disappears.

It is noteworthy that in these dehalogenations bromine and chlorine exert a very similar neighbouring assistance in the formation of the bridged radical as *meso*-2,3-dibromobutane is only *c.* 60% faster than *erythro*-2-bromo-3-chlorobutane[142].

A proclivity towards *anti* elimination, increasing with increasing reductant concentration, has been observed as well in the light-catalysed debrominations of *erythro*- and *threo*-2,3-dibromo-4-methylpentane promoted by hexabutylditin[143].

Anti stereochemistry has been also found in the eliminations of 2-bromo-3-sulphinylbutanes to butenes promoted by tributyltin radicals[161]. In contrast, β-phenylthioalkyl bromides react with trialkyltin radicals in a non-stereospecific way[162]. It has been suggested that the stereoselectivity results from rapid loss of phenylsulphinyl radical from the non-equilibrated alkyl radical.

The reaction stereochemistry has also been investigated for dehalogenations promoted by $C_3H_5Fe(CO)_2P(C_6H_5)_3$[65]. From both *meso*- and D,L-3,4-dibromohexane

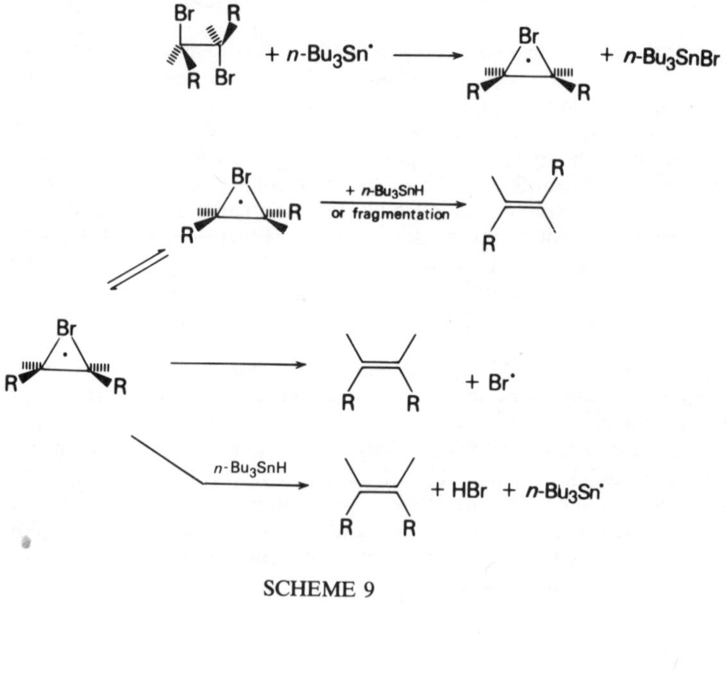

SCHEME 9

prevalent formation of *trans*-3-hexene is observed. Some isomerization of the bromo radical occurs in this case, but the equilibration is not complete since the two diastereoisomers give a different product distribution.

D. Dehalogenations by Other Reductants

A one-electron transfer mechanism has been also proposed to account for the facile *syn* dehalogenation of **25** promoted by hydroxide ions in diethyl ether–DMSO (equation 20)[163]. The observation that the *trans* isomer of **25** eliminates chlorine less

$$(C_6Cl_5)_2CH{-}C_6Cl_4{-}C(Cl){=}C(Cl){-}C_6Cl_4{-}CH(C_6Cl_5)_2 \xrightarrow{OH^-} (C_6Cl_5)_2\bar{C}{-}C_6Cl_4{-}C{\equiv}C{-}C_6Cl_4{-}\bar{C}(C_6Cl_5)_2 \qquad (20)$$

(25)

readily than the *cis* isomer has been rationalized by assuming that the formation of the radical anion is more favoured with the latter isomer since in this case the ethylene system is less shielded to the approach of the electron donor. However, beside the fact that electron transfer reactions should be little sensitive to steric effects, it should be also noted that the electron might first be transferred to the LUMO (lowest unoccupied molecular orbital) of the aromatic system.

Another one-electron reductant is probably sodium dithionite, which dehalogenates *meso*- and D,L-2,3-dibromobutane in a non-stereospecific way, giving mixtures of the same composition of *trans*- and *cis*-2-butene. Other mechanisms for this reaction have, however, been suggested[164].

IV. DEHALOGENATIONS PROMOTED BY METALS

A. General

Metals are certainly the most versatile and powerful dehalogenating agents with the broadest area of applications. Besides dihalides, β-substituted alkyl halides (halohydrins, haloethers, etc.) undergo elimination as well. The reduction of iodohydrins by zinc is a step in a sequence which converts epoxides into olefins[165]. Metals are also particularly effective in the dehalogenations of 1,2-dihalocyclobutane[166–168] and in the conversion of 2,3-dihalogenopropenes to allenes[169].

Among metals, zinc is by far the most common reductant, the method of Gladston and Tribe for the zinc-promoted dehalogenation of dibromides going back to 1874[170]. Zince dust is usually used in solvents like alcohols, dioxane, acetic anhydride, acetic acid and DMF. In addition to zinc, other metals, e.g. magnesium[2,70,171], sodium[2,10,13,171] and to a less extent copper[9], aluminium[2], lead[2], cadmium[2] and lithium[70] have been used. Recently, sodium in liquid ammonia has been recommended as one of the best dehalogenating systems[13].

A single metal can also be effectively replaced by metal couples, e.g. zinc–mercury[2], zinc–silver[172], copper–tin[173] and zinc–copper[174,175] which provide a means to enhance and (or) regulate the reactivity of the reducing system.

In spite of the widespread use of metals as dehalogenating reagents, relatively little work has dealt with the mechanistic aspects of these reactions. The main reason is probably that quantitative studies are very difficult in this area since these reactions generally occur under heterogeneous conditions. The surface of the metal is involved and its state (dust, granules, etc.) may be therefore of great importance, as also may be possible treatments for activation. As a consequence, no firm mechanistic conclusions are generally possible.

B. Stereochemistry

1. Dehalogenations promoted by zinc

These reactions are those for which most information is available. When the substrate is a 1,2-dibromoalkane of the type $RCHBrCHBrR^1$ (R, R^1 = alkyl groups), zinc-promoted dehalogenations follow an *anti* stereochemistry. Thus, from *meso*-2,3-dibromobutane predominant formation of *trans*-2-butene is observed. whereas *cis*-2-butene is the main product of the dehalogenation of D,L-2,3-dibromobutane[10,171,176,177]. In refluxing ethanol the yield of *trans*-2-butene from the *meso* form is 96–97%; that of *cis*-2-butene from the D,L isomer is 96%[10]. The stereoselectivity is smaller (*c*. 90%) when the reaction is carried out with the substrate in the vapour phase[178].

A significant *anti* stereoselectivity (between 90 and 95%) is also exhibited in the reactions of higher homologues of 2,3-dibromobutanes, such as 7,8-dibromooctadecanes[179] and 5,6-dibromodecanes[70], and in the debromination of 1,1,2-tribromocyclohexane[77]. Thus, the earlier suggestion[175] that stereoselectivity decreases steadily on increasing the molecular weight of the dibromoalkane does not seem to be substantiated. It is noteworthy that the *anti* stereochemistry exhibited by the zinc-promoted dehalogenations has been successfully exploited in a sequence devised to invert olefin geometry[179].

An exception to this behaviour is the *syn* stereoselectivity (80–95%) found in the zinc-promoted dehalogenations of *trans*-1,2-dibromocyclodecanes and *trans*-1,2-dibromocyclododecanes[70]. However, this phenomenon might be related to the particular conformational situation of medium rings[180].

In zinc-promoted dehalogenations from purely aliphatic dibromides the requirement of an *anti* periplanar geometry of the bromine atoms is less stringent than in halide ion-promoted reactions. For example, a derivative of *cis*-2,3-dibromo[2.2.1]-bicycloheptane can be debrominated by zinc with relative facility, whereas it is unreactive towards iodide ions[75].

Replacement of one or both bromine atoms of the dibromide by chlorine atoms significantly reduces the *anti* stereoselectivity of the dehalogenation process[179]. The *anti* stereoselectivity practically disappears when eliminations from β-bromoethers and β-bromoesters are considered. Thus, equimolar amounts of *cis*- and *trans*-2-butene are obtained from *erythro*-2-bromo-3-methoxybutane[10]. The phenomenon should probably be ascribed to the decrease in leaving group ability as we go from vicinal dibromides to vicinal dichlorides and to β-bromoethers or β-bromoesters. However, the elimination from *trans*-2-X-cyclohexyl bromides (X = OH, OTs, OAc) promoted by zinc–copper in ethanol is *c*. 10 times faster than from the corresponding *cis* isomers[181].

Anti stereochemistry is generally not observed in the reactions of 1,2-disubstituted dibromides, when the substituents are groups capable of interacting favourably with a negative charge. Debromination of these substrates leads to the predominant formation of the more stable olefin from both the diastereoisomeric substrates. This was shown for the reactions of *meso*- and D,L-1,2-dibromo-1,2-diphenylethanes[2,9,171] and *meso* and D,L sodium 2,3-dibromosuccinates with zinc[171]. Likewise, *threo*-1,2-dibromo-1-phenyl-2-triphenylsilylethane exclusively affords *trans*-triphenylsilylstyrene when dehalogenated by zinc in acetic acid[182].

2. Dehalogenations promoted by other metals

Magnesium exhibits a high degree of *anti* stereoselectivity in the reaction with 2,3-dibromobutanes[171]; its stereoselectivity is, however, significantly smaller than that of zinc in the dehalogenation from vicinal dibromodecanes[70]. Magnesium, like zinc, leads to prevalent formation of *trans*-stilbene when it reacts with both diastereoisomeric 1,2-dibromo-1,2-diphenylethanes[171].

Alkali metals, e.g. sodium and lithium, do not seem to display any significant stereoselectivity[10,171]. Thus, mixtures of *trans*- and *cis*-2-butenes of similar composition have been obtained in the debromination of *meso*- and D,L-2,3-dibromobutanes with sodium in liquid ammonia[171]. However, *syn* stereoselectivity is observed in dehalogenations from medium ring dibromides promoted by lithium[70].

Stereochemistry has been also investigated for the dehalogenations of D,L-1,2-dibromo-1,2-diphenylethane promoted by copper[9] and other metals or metallic couples (cadmium, lead, zinc–mercury, aluminium)[2]. In each case the predominant product was *trans*-stilbene, as observed with zinc, with the exception of aluminium in DMSO which afforded 53% of *cis*-stilbene.

C. Reactivity and Structure

1. The substrate

For the reasons stated above, quantitative data of reactivity are not available for metal-promoted dehalogenations. Nonetheless, some considerations of qualitative nature are possible in the few cases where product yields, obtained under similar experimental conditions, can be compared.

A reactivity order of dibromides > bromochlorides > dichlorides can be deduced from the reaction conditions required to perform dehalogenations from 7,8-dihalogenooctadecanes by zinc[179]. Attack on chlorine is certainly much easier than on fluorine. This has been exploited for the synthesis of olefins (equation 21) which are potential monomers in polymerization processes[183].

$$(F_2C(Cl)CF(Cl)CO_2CH_2)_2 \xrightarrow{Zn} (CF_2C{=}CFCO_2CH_2)_2 \qquad (21)$$

The presence of substituents can influence the reaction rate as well. Thus, it has been reported that 1,2-dichlorooctane is not dehalogenated by Zn after 44 days at 101°C, whereas 1,1,2-trichloroethane undergoes 98% dechlorination in 72 h[184]. The presence of fluorine substituents, however, has been found to retard the dehalogenation rate[185].

2. The metal

The reactivity of the reducing system has been strongly enhanced by passing from zinc in methanol to a metallic couple, e.g. zinc–copper, in a dipolar aprotic solvent[174,186].

The copper–tin couple appears to be less reactive and more selective than the zinc–copper couple. Thus, using the former couple, it is possible to accomplish the selective debromination of vicinal dibromides containing other reducible halogens[173], as shown in equation (22). With a zinc–copper couple the third bromine atom is reduced as well.

$$\underset{\displaystyle CH_2Br}{HO_2C\overset{|}{C}(Br)}CH_2CH(Br)CO_2H \xrightarrow{Sn-Cu,\,THF} \underset{\displaystyle CH_2}{HO_2C\overset{||}{C}}CH_2CH(Br)CO_2H \qquad (22)$$

As expected, zinc is more reactive than copper in the debromination of 1,2-dibromo-1,2-diphenylethanes[9].

Interestingly, dehalogenation by metals is not accompanied by competitive dehydrohalogenation. Metals used in dehalogenation reactions are soft electron donors (and have low electronegativity and are easy to oxidize) and interaction with the halogen rather than with the hydrogen is strongly preferred.

D. Mechanisms

The hypotheses concerning the mechanisms of metal-promoted dehalogenations are very uncertain since they are based on very little information, which is nearly exclusively of a stereochemical nature.

The prevalent *anti* stereochemistry of zinc-induced dehalogenations from purely aliphatic dibromides suggests that these reactions might occur by a concerted mechanism (cf. structure **26**) where the transfer of two electrons from the metal to the substrate and the breaking of the C_α—Br bond occur in the same step[10,171]. There is no doubt that an *anti* coplanar arrangement of the two bromine atoms should be the favoured one for such a mechanism.

Zn
⋮
Br
⋮
—C$_\beta$····C$_\alpha$—
⋮
Br

(26)

A stepwise mechanism involving the formation of a carbanion or of an organometallic compound as intermediate has instead been proposed for the elimination of those substrates where bromine has been replaced by a poorer leaving group or where substituents capable of interacting favourably with a negative charge are present on the α- and (or) the β-position of the dibromo compound[10,171]. If this intermediate (which is most probably a carbanion[20,165,187]) is sufficiently stable to equilibrate before the loss of the leaving group, the low stereospecificity noted above for the reaction of these dibromides is accounted for. However, the lifetime of the carbanion or the organometallic intermediate must be very short since products characteristic of the reactions of these species with the solvent have never been observed. A likely possibility is that the intermediate is closely associated with the metal surface[24].

In some cases, however, concerted and stepwise mechanisms should occur side by side, since significant amounts of *cis*-stilbene are formed in the dehalogenations of D,L-1,2-dibromo-1,2-diphenylethane promoted by zinc and other metals[2]. Exclusive formation of the much more stable *trans*-stilbene should be observed from an intermediate carbanion which is free to rotate.

It should be noted that a very short lived carbanion, which loses the leaving group before equilibration, might also be involved in the metal-promoted dehalogenations which occur with an *anti* stereochemistry[20]. The stereochemical outcome would be justified, in this case, on the ground that in the more favoured initial conformations of diastereoisomeric dibromo derivatives, **8** and **9**, the two bromine atoms are *anti* to each other.

To explain the predominant *syn* stereochemistry of the zinc-promoted dehalogenations of dibromocyclodecanes, it has been suggested that, owing to the particular conformational characteristics of medium rings, a steric interaction might cause preferential surface adsorption of the conformer leading to *syn* elimination[24]. However, if this interpretation is correct it must be assumed that the steric requirements of the zinc-promoted dehalogenations are quite different from those of the corresponding reactions induced by iodide ions. Indeed, an *anti* stereochemistry has been observed in the reactions of dibromocyclodecanes and dibromocyclododecanes with iodide ions[70].

It is probable that the mechanisms proposed for the dehalogenations by zinc hold also in the case of the reactions promoted by magnesium: predominant concerted mechanism for the purely aliphatic dibromides and intervention of a stepwise carbanionic mechanism for the other dibromo derivatives.

The stepwise mechanism should always hold for dehalogenations promoted by sodium and lithium since these reductants do not display any stereospecificity. However, owing to the ability of these metals (especially sodium in liquid ammonia) to act as strong electron donors there is also a possibility of one-electron transfer processes involving the formation of radical intermediates.

V. ELECTROCHEMICAL DEHALOGENATIONS

The reduction of a vicinal dihalide to olefin can also be carried out electrochemically. The two electrons in this case are provided by the cathode (generally mercury) and the process is expressed by equation (1).

Owing to the mildness of the reaction conditions, electrochemical dehalogenation offers an attractive alternative to chemical dehalogenation. A frequent application of this reduction method is found in the synthesis of perhalogenoethylenes. For example, 2-chloro-1,1,2-trifluoroethylene is obtained by electrochemical reduction of 1,2,2-trichloro-1,1,2-trifluoroethane[188,189]. Electroreduction has been usefully employed to obtain the dimer of benzocyclobutadiene from 1,2-dibromobenzocyclobutene[123], and the highly strained olefin $\Delta^{1,4}$-bicyclo[2.2.0]hexene (**27**), identified as the Diels–Alder reduction of vicinal dibromides upon the dihedral angle between the carbon–bromine

Recently, the selective monoprotection of the less alkylated double bond has been accomplished by controlled potential electrolysis[191].

Electrochemical dehalogenation has been reviewed several times[3,192,193], including a review in the present series[3]. A brief outline of the main characteristics of the reaction is therefore sufficient here.

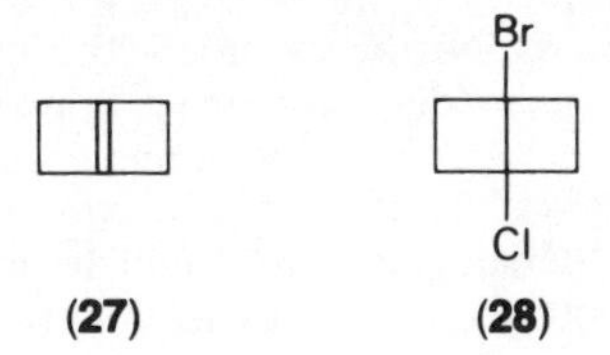

(27) **(28)**

Electroreduction of vicinal dibromoalkanes takes place with a rigorous *anti* stereochemistry. Thus, *meso*-2,3-dibromobutane is electrochemically dehalogenated to *trans*-2-butene, whereas exclusive formation of *cis*-2-butene is observed with the D,L-isomer[194]. This stringent stereospecificity agrees well with polarographic studies that have shown a marked dependence of the half-wave potential ($E_{1/2}$) for the reduction of vicinal dibromides upon the dihedral angle between the carbon–bromine bonds. The minimum value of $E_{1/2}$ was observed for the *anti* periplanar arrangement, i.e. for a dihedral angle of 180°[195].

Another striking piece of evidence for the preferred *anti* stereochemistry of electrochemical dehalogenations is given by the large difference in $E_{1/2}$ (0.81 V) for the reduction of the two isomeric dibromides **29** and **30**; **29** is reduced much more easily than **30** so that when a 60:40 mixture of the two bromides is subjected to electrolysis in DMF at a controlled potential (−1.3 V versus SCE) only **29** is reduced and **30** can

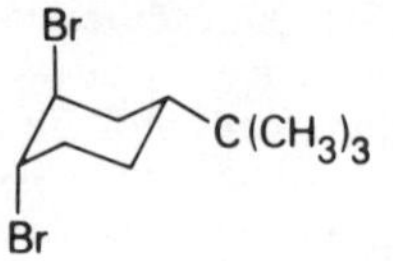

Br C(CH$_3$)$_3$ Br

(29)

Br Br C(CH$_3$)$_3$

(30)

be recovered[194]. A nice separation of the diequatorial isomer from the diaxial one is thus performed.

The observed *anti* stereochemistry has led to the suggestion of a concerted mechanism, involving the transfer of two electrons in a single step, for the electroreduction of vicinal dihalides. This suggestion appears to be supported by the shape of the chrono-amperometric (potentiostatic) curve[194]. The intervention of a carbanion intermediate which loses a bromide ion has been excluded on the ground that products of proton capture are absent when the reaction is carried out in the presence of proton donor. However, an intermediate carbanion which undergoes loss of bromide ion before reacting with the proton donor and before equilibration could still be consistent with the stereochemistry and the reaction products.

The more significant observation against the intervention of a carbanion intermediate is probably that the reduction of a monohalide, which involves a carbanion intermediate[196], is much more difficult than that of a vicinal dihalide. This is indicated by the $E_{1/2}$ values of ethyl bromide and 1,2-dibromoethane, −2.08 and −1.52 V, respectively. This difference is too high to be attributed to an inductive effect of the halogen on the carbanion formation, and therefore suggests that dihalides react by a pathway not available to monohalides[193].

In contrast to the simple dibromoalkane systems, evidence for the intervention of a carbanionic intermediate has been obtained in the electroreduction of 1,2-dibromo-1-chloro-1,2,2-trifluoroethane, a compound which can form a very stable carbanion[197]. Likewise, the stereoconvergent debromination of *meso*- and D,L-diethyl 2,3-dibromosuccinate to give diethyl fumarate[198] might be rationalized on the basis of a carbanion undergoing complete rotameric relaxation before losing the bromide ion. Thus, diethyl fumarate, which is more stable than diethyl maleate by *c.* 4 kcal mol^{-1} [199] is formed from both diastereoisomers. However, the possibility that the two diastereoisomers react by different mechanisms cannot be excluded: the *meso* form utilizes the concerted pathway whereas an intermediate carbanion is formed from the D,L isomer, where the conformation with the *anti* bromine atoms is disfavoured.

It should finally be noted that most of the results concerning reactivity and stereochemistry of electrochemical dehalogenations might be explained just as well by a mechanism which involves two one-electron transfer steps, similar to that previously suggested for the dehalogenations promoted by radical anions (Section III.B.1). Accordingly, if a radical is formed in the first step, dibromides are expected to be much more reactive than monobromides, as observed, since a bridged radical is obtained in the former case. Moreover, when such a bridged radical does not isomerize before undergoing further reduction and the carbanion formed in the second step loses the bromide ion before rotation, *anti* stereoselectivity is observed, as with the 2,3-dibromobutanes. In contrast, when complete equilibration of the bridged radical is possible, as with diethyl 2,3-dibromosuccinate, a stereoconvergent reaction results.

Probably, a study of the electrochemical reduction of β-substituted alkyl bromides, e.g. the tosylates of 2-bromo-3-butanol, would be useful to distinguish between the concerted two-electron transfer mechanism and the stepwise one-electron transfer mechanism. In the former case the bromotosylate should be reduced more easily than the dibromide as the tosyloxy group is a better leaving group than bromine. The reverse situation should be obtained if a radical is formed, as bromine is a much more effective neighbouring group than tosyloxy. Moreover, in this case the bromotosylate should no longer exhibit an *anti* stereochemistry (see Section III.B.2).

Acknowledgements. The financial support of the Italian National Council of Research (C.N.R.) is gratefully recognized.

VI. REFERENCES

1. J. K. Kochi, D. M. Singleton and L. J. Andrews, *Tetrahedron*, **24**, 3503 (1968).
2. I. M. Mathai, K. Schug and S. I. Miller, *J. Org. Chem*., **35**, 1733 (1970).
3. J. Casanova and L. Eberson, in *The Chemistry of Functional Groups: The Chemistry of the Carbon–Halogen Bond* (Ed. S. Patai), Wiley-Interscience, Chichester (1974), p. 1001.
4. S. P. Mishra and M. C. R. Symons, *Tetrahedron Lett*., 2597 (1975).
5. E. T. McBee, C. W. Roberts and J. D. Idol, Jr, *J. Amer. Chem. Soc*., **77**, 4942 (1955).
6. R. D. Chambers, A. A. Lindley and H. C. Fielding, *J. Fluorine Chem*., **13**, 85 (1979).
7. S. W. Bennett, C. Eaborn, R. A. Jackson and R. W. Walsingham, *J. Organometal. Chem*., **27**, 195 (1971).
8. E. Baciocchi and A. Schiroli, *J. Chem. Soc. B*, 554 (1969).
9. R. E. Buckles, J. M. Bader and R. J. Thurmaier, *J. Org. Chem*., **27**, 4523 (1962).
10. H. O. House and R. S. Ro, *J. Amer. Chem. Soc*., **80**, 182 (1958).
11. F. J. Soday and C. E. Boord, *J. Amer. Chem. Soc*., **55**, 3293 (1933).
12. A. J. Solo and B. Singh, *J. Org. Chem*., **30**, 1658 (1965).
13. E. L. Allred, B. R. Beck and K. J. Voorhes, *J. Org. Chem*., **39**, 1426 (1974).
14. See references cited in Ref. 13.
15. J. I. C. Cadogan and R. Leardini, *JCS Chem. Commun*., 783 (1979).
16. D. G. Gillespie and B. J. Walker, *Tetrahedron Lett*., 1673 (1977).
17. W. Burns, D. Grant, M. A. McKervey and G. Step, *JCS Perkin I*, 234 (1976).
18. D. Lenoir and J. Firl, *Annalen*, 1467 (1974).
19. D. R. James, R. W. Rees and C. W. Shoppee, *J. Chem. Soc*., 1370 (1955).
20. D. V. Banthorpe, *Elimination Reactions*, Elsevier, Amsterdam (1962), Chap. 6.
21. J. F. King and R. G. Pews, *Canad. J. Chem*., **42**, 1294 (1964).
22. A. F. Cockerill in *Comprehensive Chemical Kinetics*, Vol. 9 (Ed. C. H. Bamford and C. F. H. Tipper), Elsevier, Amsterdam (1973).
23. N. A. LeBel in *Advances in Alicyclic Chemistry*, Vol. 3 (Ed. H. Haat and G. J. Karabatsos), Academic Press, New York (1971).
24. W. H. Saunders, Jr and A. F. Cockerill, *Mechanisms of Elimination Reactions*, Wiley, New York (1973), pp. 332–376.
25. C. S. Tsai Lee, I. M. Mathai and S. I. Miller, *J. Amer. Chem. Soc*., **92**, 4602 (1970).
26. W. K. Kwok and S. I. Miller, *J. Org. Chem*., **35**, 4034 (1970).
27. W. K. Kwok, I. M. Mathai and S. I. Miller, *J. Org. Chem*., **35**, 3420 (1970).
28. P. Dorn, *Diss. Abstr*., **22**, 3396 (1962).
29. V. Janout, M. Prochàzka and M. Paleček, *Coll. Czech. Chem. Commun*., **41**, 617 (1976).

30. K. Ramasamy, S. K. Kalyanasundaram and P. Shanmugam, *Synthesis*, 311 (1978).
31. K. M. Ibne-Rasa, N. Muhammad and Hasibullah, *Chem. Ind.* (Lond.), 1418 (1966).
32. T. C. Sharma and M. M. Bokadia, *Indian J. Chem. B*, **14**, 65 (1976); N. J. Reddy and T. C. Sharma, *Indian J. Chem. B*, **17**, 645 (1979).
33. K. M. Ibne-Rasa, A. Rashid Tahir and A. Rahman, *Chem. Ind.* (Lond.), 232 (1973).
34. I. J. Borowitz, D. Weiss and R. K. Crouch, *J. Org. Chem.*, **36**, 2377 (1971).
35. C. J. Devlin and B. J. Walker, *JCS Perkin I*, 1249 (1972).
36. J. P. Schroeder, L. B. Tew and V. M. Peters, *J. Org. Chem.*, **35**, 3181 (1970).
37. D. G. Gillespie and B. J. Walker, *Tetrahedron Lett.*, 4709 (1975).
38. J. Avraamides and A. J. Parker, *Tetrahedron Lett.*, 4043 (1971).
39. (a) M. Prince, B. W. Bremer and W. Brenner, *J. Org. Chem.*, **31**, 4292 (1966).
(b) M. Prince and B. W. Brenner, *J. Org. Chem.*, **32**, 1655 (1967).
40. M. Sevrin, J. N. Denis and A. Krief, *Tetrahedron Lett.*, 1877 (1980).
41. P. Caubere and J. Moreau, *Tetrahedron*, **25**, 2469 (1969).
42. J. F. King, A. D. Allbutt and R. G. Pews, *Canad. J. Chem.*, **46**, 805 (1968).
43. T. Sugita, J. Nakagawa, K. Nishimoto, Y. Kasai and K. Ichikawa, *Bull. Chem. Soc. Japan*, **52**, 871 (1979).
44. H. J. S. Winkler and H. Winkler, *Ann. Chem.*, **705**, 76 (1967).
45. G. H. Posner and J. S. Ting, *Synthetic Commun.*, **3**, 281 (1973).
46. T.-L. Ho and C. M. Wong, *Synthetic Commun.*, **5**, 87 (1975).
47. A. Kasal, *Coll. Czech. Chem. Commun.*, **41**, 2040 (1976).
48. V. A. Cherkasova and L. A. Tamm, *Zh. Org. Khim.*, **4**, 2108 (1968).
49. A. J. Speziale and C. C. Tung, *J. Org. Chem.*, **28**, 1353 (1963).
50. R. Caputo, L. Mangoni, P. Monaco, G. Palumbo and L. Previtera, *Atti del X Convegno di Chimica Organica della S.C.I.* (Organic Chemistry Division), Ferrara, Italy (1977), CP64.
51. W. Tadros, R. R. Tadros and M. S. Ishak, *Helv. Chim. Acta*, **59**, 1925 (1976).
52. D. Landini, S. Quici and F. Rolla, *Synthesis*, 397 (1975).
53. S. L. Regen, *J. Org. Chem.*, **42**, 875 (1977).
54. W. M. Schubert, H. Steadly and B. S. Rabinovitch, *J. Amer. Chem. Soc.*, **77**, 5755 (1955).
55. P. E. Sonnet and J. E. Oliver, *J. Org. Chem.*, **41**, 3284 (1976).
56. J. Hine and W. H. Brader, Jr, *J. Amer. Chem. Soc.*, **77**, 361 (1955).
57. J. Pichler and I. Borkovcová, *Coll. Czech. Chem. Commun.*, **39**, 779 (1974).
58. F. G. Bordwell, *Acc. Chem. Res.*, **5**, 374 (1972).
59. S. Winstein, D. Pressman and W. G. Young, *J. Amer. Chem. Soc.*, **61**, 1645 (1939).
60. E. C. F. Ko and A. J. Parker, *J. Amer. Chem. Soc.*, **90**, 6447 (1968).
61. J. F. Bunnett, *Surv. Progr. Chem.*, **5**, 53 (1969).
62. J. Csapilla, *Chimia*, **18**, 37 (1964).
63. G. Biale, A. J. Parker, S. G. Smith, I. D. R. Stevens and S. Winstein, *J. Amer. Chem. Soc.*, **92**, 115 (1970); G. Biale, D. Cook, D. J. Lloyd, A. J. Parker, I. D. R. Stevens, J. Takahashi and S. Winstein, *J. Amer. Chem. Soc.*, **93**, 4735 (1971); P. Beltrame, G. Biale, D. J. Lloyd, A. J. Parker, M. Ruane and S. Winstein, *J. Amer. Chem. Soc.*, **94**, 2240 (1972).
64. W. H. Saunders and A. F. Cockerill, *Mechanisms of Elimination Reactions*, Wiley, New York (1973), Chap. 3.
65. P. A. Wegner and M. S. Delaney, *Inorg. Chem.*, **15**, 1918 (1976).
66. J. Mulders and J. Nasielski, *Bull. Soc. Chim. Belg.*, **72**, 322 (1963).
67. I. M. Mathai and S. I. Miller, *J. Org. Chem.*, **35**, 3416 (1970).
68. F. Badea, T. Constantinescu, A. Juvara and C. D. Nenitzescu, *Justus Liebigs Ann. Chem.*, **706**, 20 (1967).
69. E. R. Trumbull and K. M. Ibne-Rasa, *J. Org. Chem.*, **28**, 1907 (1963).
70. J. Sicher, M. Havel and M. Svoboda, *Tetrahedron Lett.*, 4269 (1968).
71. R. T. Dillon, W. G. Young and H. J. Lucas, *J. Amer. Chem. Soc.*, **52**, 1953 (1930).
72. R. T. Dillon, *J. Amer. Chem. Soc.*, **54**, 952 (1932).
73. W. G. Young, D. Pressman and C. D. Coryell, *J. Amer. Chem. Soc.*, **61**, 1640 (1939).
74. D. H. R. Barton and E. Miller, *J. Amer. Chem. Soc.*, **72**, 1066 (1950).
75. J. A. Berson and R. Swidler, *J. Amer. Chem. Soc.*, **76**, 4057 (1954).
76. H. L. Goering and H. H. Espy, *J. Amer. Chem. Soc.*, **77** 5023 (1955).
77. C. L. Stevens and J. A. Valicenti, *J. Amer. Chem. Soc.*, **87**, 838 (1965).

78. S. I. Miller and R. M. Noyes, *J. Amer. Chem. Soc.*, **74**, 3403 (1952).
79. J. Sicher, *Angew. Chem. Int. Edn*, **11**, 200 (1972).
80. T. Sugita, K. Nishimoto and K. Ichikawa, *Chem. Lett.*, 607 (1973).
81. C. F. Van Duin, *Rec. Trav. Chim. Pays-Bas*, **43**, 341 (1924); **45**, 345 (1926).
82. J. Casanova and L. Eberson in *The Chemistry of Functional Groups: The Chemistry of the Carbon–Halogen Bond* (Ed. S. Patai), Wiley-Interscience, Chichester (1974), p. 979.
83. A. Streitwieser, Jr, *Solvolytic Displacement Reactions*, McGraw-Hill, New York (1962), p. 120.
84. B. B. Jarvis and W. P. Tong, *J. Org. Chem.*, **41**, 1557 (1976) and references therein.
85. E. Baciocchi and C. Lillocci, *JCS Perkin II*, 38 (1973).
86. F. Declerck, J. Mulders and J. Nasielski, *Bull. Soc. Chim. Belg.*, **71**, 518 (1962).
87. P. Johncock, *Synthesis*, 551 (1977).
88. A. Streitwieser, Jr, *Solvolytic Displacement Reactions*, McGraw-Hill, New York (1962), p. 82.
89. A. Slator, *J. Chem. Soc.*, 1697 (1904).
90. S. J. Cristol, J. Q. Weber and M. C. Brindell, *J. Amer. Chem. Soc.*, **78**, 598 (1956).
91. A. Streitwieser, Jr, *Solvolytic Displacement Reactions*, McGraw-Hill, New York (1962), p. 6.
92. D. Pressman and W. G. Young, *J. Amer. Chem. Soc.*, **66**, 705 (1944).
93. (a) W. Hückel and H. Waiblinger, *Justus Liebigs Ann. Chem.*, **666**, 17 (1963).
(b) W. G. Lee and S. I. Miller, *J. Phys. Chem.*, **66**, 655 (1962).
94. T. L. Davis and R. Heggie, *J. Org. Chem.*, **2**, 470 (1938).
95. S. Alunni, E. Baciocchi and V. Mancini, *JCS Perkin II*, 140 (1977).
96. J. Nasielski and V. Guiette-Limbourg, *Bull. Soc. Chim. Belg.*, **81**, 351 (1972).
97. M. Youmas, K. M. Ibne-Rasa and M. A. Kashmiri, *Pak. J. Sci. Ind. Res.*, **17**, 105 (1975).
98. C. Tuzun and K. H. Malic, *Commun. Fac. Sci. Univ. Ankara, Ser. B*, **21**, 11 (1974); *Chem. Abstr.*, **81**, 49124 (1974).
99. L. H. Schwartzman and B. B. Corson, *J. Amer. Chem. Soc.*, **78**, 322 (1956).
100. A. M. Shur, R. I. Ischchenko, N. A. Barba, *Zh. Vses. Khim. O-va*, **19**, 471 (1974); *Chem. Abstr.*, **82**, 42529c (1975).
101. A. M. Easton, M. J. A. Habib, J. Park and W. E. Watts, *JCS Perkin II*, 2290 (1972); W. M. Schubert, *J. Amer. Chem. Soc.*, **94**, 559 (1972); S. C. Bernstein, *J. Org. Chem.*, **33**, 3486 (1968).
102. M. Adinolfi, M. Parrilli, G. Barone, G. Laonigro and L. Mangoni, *Gazz. Chim. It.*, **107**, 263 (1977); **105**, 1259 (1975).
103. P. Bassi and U. Tonellato, *JCS Perkin II*, 1283 (1974).
104. S. Dershowitz and S. Proskauer, *J. Org. Chem.*, **26**, 3595 (1961).
105. F. G. Bordwell and B. B. Jarvis, *J. Amer. Chem. Soc.*, **95**, 3585 (1973).
106. Y. Houminer, *JCS Perkin I*, 277 (1975).
107. R. G. Pearson and J. Songstad, *J. Amer. Chem. Soc.*, **89**, 1827 (1967).
108. E. Baciocchi and P. L. Bocca, *Ric. Sci.*, **37**, 1182 (1967).
109. W. A. Henderson, Jr, and S. A. Buckel, *J. Amer. Chem. Soc.*, **82**, 5794 (1960).
110. A. J. Parker, *Adv. Phys. Org. Chem.*, **5**, 173 (1967).
111. R. G. Pearson, H. Sobel and J. Songstad, *J. Amer. Chem. Soc.*, **90**, 319 (1968).
112. S. R. Hartshorn, *Aliphatic Nucleophilic Substitution*, Cambridge University Press, Cambridge (1973), p. 44.
113. J. O. Edwards, *J. Amer. Chem. Soc.*, **76**, 1540 (1954).
114. E. Baciocchi and C. Lillocci, *JCS Perkin II*, 802 (1975).
115. J. Hine and W. H. Brader, Jr, *J. Amer. Chem. Soc.*, **75**, 3964 (1953).
116. W. G. Kofron and C. R. Hauser, *J. Amer. Chem. Soc.*, **90**, 4126 (1968).
117. B. G. Gavrilov and V. E. Tischenko, *J. Gen. Chem. U.S.S.R.*, **18**, 1687 (1948); *Chem. Abstr.*, **43**, 2569h (1949).
118. P. Pfeiffer, *Z. Phys. Chem. (Leipzig)*, **48**, 40 (1904).
119. M. M. Kremlev and Yu. A. Fialkov, *Ukr. Khim. Zh.* (Russ. Ed.), **42**, 1058 (1976).
120. V. A. Cherkasova, N. Le Cam, L. P. Ivanova, N. L. Lebedeva, *Kratk. Tezicy-Vsev. Sovesch. Probl. Mekh. Geteroliticheskikh Reacts*, 136 (1974); *Chem. Abstr.*, **85**, 123214k (1976).

121. I. J. Borowitz, K. C. Kirby, P. E. Rusek and E. Lord, *J. Org. Chem.*, **34**, 2687 (1969).
122. (a) J. F. Garst, J. A. Pacifici, V. D. Singleton, M. F. Ezzel and J. I. Morris, *J. Amer. Chem. Soc.*, **97**, 5242 (1975).
(b) W. Adam and J. Arce, *J. Org. Chem.*, **37**, 507 (1972).
123. R. D. Rieke and P. M. Hudnall, *J. Amer. Chem. Soc.*, **95**, 2646 (1973).
124. R. N. McDonald and D. G. Frickey, *J. Amer. Chem. Soc.*, **90**, 5315 (1968).
125. D. G. Korzan, F. Chen and C. Ainsworth, *JCS Chem. Commun.*, 1053 (1971).
126. P. L. Julian, W. Cole, A. Magnani and E. W. Meyer, *J. Amer. Chem. Soc.*, **67**, 1728 (1945).
127. W. C. Kray, Jr and C. E. Castro, *J. Amer. Chem. Soc.*, **86**, 4603 (1964).
128. D. M. Singleton and J. K. Kochi, *J. Amer. Chem. Soc.*, **89**, 6547 (1967).
129. J. K. Kochi and D. M. Singleton, *J. Amer. Chem. Soc.*, **90**, 1582 (1968).
130. B. W. S. Kolthammer, P. Legzdins and D. T. Martin, *Tetrahedron Lett.*, 323 (1978).
131. J. E. McMurry and T. Hoz, *J. Org. Chem.*, **40**, 3797 (1975).
132. G. A. Olah and G. K. S. Prakash, *Synthesis*, 607 (1976).
133. W. K. Kwok and S. I. Miller, *Canad. J. Chem.*, **45**, 1161 (1967).
134. W. K. Kwok and S. I. Miller, *J. Amer. Chem. Soc.*, **92**, 4599 (1970).
135. J. Halpern and J. P. Maher, *J. Amer. Chem. Soc.*, **87**, 5361 (1965).
136. P. B. Chock and J. Halpern, *J. Amer. Chem. Soc.*, **91**, 582 (1969).
137. K. B. Sharpless, M. A. Umbreit, M. T. Nieh and T. C. Flood, *J. Amer. Chem. Soc.*, **94**, 6538 (1972).
138. H. Nozaki, T. Shirafaji and Y. Yamamoto, *Tetrahedron*, **25**, 3461 (1969).
139. H. Bretschneider and M. Ajtai, *Monats. Chem.*, **74**, 57 (1941).
140. T.-L. Ho and G. A. Olah, *Synthesis*, 170 (1977).
141. W. C. Danen and K. A. Rosa, *J. Org. Chem.*, **40**, 619 (1975).
142. R. J. Strunk, P. M. Di Giacomo, K. Aso and H. G. Kuivila, *J. Amer. Chem. Soc.*, **92**, 2849 (1970).
143. H. G. Kuivila and C. H.-C. Pian, *Tetrahedron Lett.*, 2561 (1973).
144. A. Hosomi and H. Sakurai, *J. Amer. Chem. Soc.*, **94**, 1384 (1972).
145. C. G. Scouten, F. E. Barton, Jr, J. R. Burgess, P. R. Story and J. F. Garst, *JCS Chem. Commun.*, 78 (1969).
146. J. R. Hanson, *Synthesis*, 1 (1974).
147. H. Hirschmann, F. B. Hirschmann and J. W. Corcorau, *J. Org. Chem.*, **20**, 572 (1955).
148. J. K. Kochi and P. Mocadlo, *J. Amer. Chem. Soc.*, **88**, 4049 (1966).
149. C. H. Robinson, L. Finckenor, E. P. Oliveto and D. Gould, *J. Amer. Chem. Soc.*, **81**, 2191 (1959).
150. A. Bowers, L. C. Ibáñez, E. Denot and R. Becerra, *J. Amer. Chem. Soc.*, **82**, 4001 (1960).
151. L. Miller, *J. Chem. Educ.*, **48**, 168 (1971); M. J. S. Dewar and R. C. Dougherty, *The PMO Theory of Organic Chemistry*, Plenum, New York (1975), p. 126.
152. R. C. Kerber, G. W. Urry and N. Kornblum, *J. Amer. Chem. Soc.*, **87**, 4520 (1965).
153. D. J. Girdler and R. K. Norris, *Tetrahedron Lett.*, **431**, 2375 (1975).
154. W. G. Kofron and C. R. Hauser, *J. Org. Chem.*, **35**, 2085 (1970).
155. F. Rash, S. Boatman and C. R. Hauser, *J. Org. Chem.*, **32**, 372 (1967).
156. M. Mousseron and F. Winternitz, *Bull. Soc. Chim. Fr.*, 604 (1946).
157. P. S. Skell and K. J. Shea in *Free Radicals*, Vol. II (Ed. J. K. Kochi), Wiley, New York (1973), Chap. 26, pp. 809–852.
158. C. E. Castro and W. C. Kray, Jr, *J. Amer. Chem. Soc.*, **85**, 2768 (1963).
159. J. K. Kochi and D. D. Davis, *J. Amer. Chem. Soc.*, **86**, 5264 (1964).
160. L. H. Slaugh and J. H. Raley, *Tetrahedron*, **20**, 1005 (1964).
161. T. E. Boothe, J. L. Green, Jr, P. B. Shevlin, M. R. Willcott, III, R. R. Inners and A. Cornelis, *J. Amer. Chem. Soc.*, **100**, 3874 (1978).
162. T. E. Boothe, J. L. Greene, Jr and P. B. Shevlin, *J. Amer. Chem. Soc.*, **98**, 951 (1976).
163. M. Ballaster, J. Castañer and A. Ibáñez, *Tetrahedron Lett.*, 2147 (1974).
164. T. Kempe, T. Norin and R. Caputo, *Acta Chem. Scand. B*, **30**, 366 (1976).
165. J. W. Cornforth, R. H. Cornforth and K. K. Mathew, *J. Chem. Soc.*, 112 (1959).
166. E. K. G. Schmidt, L. Brener and R. Pettit, *J. Amer. Chem. Soc.*, **92**, 3240 (1970).
167. M. P. Cava and M. J. Mitchell, *Cyclobutadiene and Related Compounds*, Academic Press, New York (1967).

168. C. E. Berkoff, R. C. Coockson, J. Hudec and R. O. Williams, *Proc. Chem. Soc.*, 312 (1961).
169. H. N. Crips and E. F. Kiefer, *Org. Synthesis*, **42**, 12 (1962).
170. J. H. Gladston and A. Tribe, *Ber.*, **7**, 364 (1874).
171. W. M. Schubert, B. S. Rabinovitch, N. R. Larson and V. A. Sims, *J. Amer. Chem. Soc.*, **74**, 4590 (1952).
172. M. P. Cava and K. T. Buch, *J. Amer. Chem. Soc.*, **95**, 5805 (1973).
173. P. Dowd and L. K. Marwaba, *J. Org. Chem.*, **41**, 4035 (1976).
174. H. Blancou, P. Moreau and A. Commeyras, *Tetrahedron*, **33**, 2061 (1977).
175. W. G. Young, S. J. Cristol and T. Skei, *J. Amer. Chem. Soc.*, **65**, 2099 (1943).
176. W. G. Young and H. J. Lucas, *J. Amer. Chem. Soc.*, **52**, 1964 (1930).
177. W. G. Young and S. Winstein, *J. Amer. Chem. Soc.*, **58**, 102 (1936).
178. M. Gordon and J. V. Hay, *J. Org. Chem.*, **33**, 427 (1968).
179. P. E. Sonnet and J. E. Oliver, *J. Org. Chem.*, **41**, 3279 (1976).
180. J. Sicher in *Progress in Stereochemistry*, Vol. 6 (Eds. P. B. D. de la Mare and W. Klyne), Butterworth, London (1962).
181. S. J. Cristol and L. E. Rademacher, *J. Amer. Chem. Soc.*, **81**, 1600 (1959).
182. A. Brook, J. M. Duff, P. Hitchcock and R. Mason, *J. Organomet. Chem.*, **113**, C11 (1976).
183. O. Paleta, A. Posta, F. Liska and J. Okrauhlik, *Czech.*, **158**, 839 (1976), *Chem. Abstr.*, **85**, P22243 (1976); O. Paleta, A. Posta and J. Okrauhlick, *Sb. Vys. Sk. Chem.-Technol. Praze, Org. Chem. Technol. C*, **23**, 5 (1976), *Chem. Abstr.*, **86**, 139369w (1977).
184. T. Alfrey, Jr, H. C. Haas and C. W. Lewis, *J. Amer. Chem. Soc.*, **74**, 2097 (1952).
185. H. Kimoto, T. Takahashi and H. Muramatsu, *Bull. Chem. Soc. Jpn*, **53**, 764 (1980).
186. H. Blancon and A. Commeyras, *J. Fluorine Chem.*, **9**, 309 (1977).
187. L. Crombie and S. Harper, *J. Chem. Soc.*, 1715 (1950).
188. S. Wawzonek and S. Willging, *J. Electrochem. Soc.*, **124**, 860 (1977).
189. H. Matschin, R. Voigtlaender and B. Hesse, *Z. Chem.*, **17**, 107 (1977).
190. K. B. Wiberg, W. F. Bailey and M. E. Jason, *J. Org. Chem.*, **39**, 3803 (1974); J. Casanova and H. R. Rogers, *J. Org. Chem.*, **39**, 3803 (1974).
191. U. Husstedt and H. J. Schafer, *Synthesis*, 964 (1979).
192. L. Perrin, *Progr. Phys. Org. Chem.*, **3**, 165 (1966).
193. A. Fry, *Synthetic Organic Electrochemistry*, Harper & Row, New York (1972), pp. 180–185.
194. J. Casanova and H. R. Rogers, *J. Org. Chem.*, **39**, 2408 (1974).
195. J. Závada, T. Krupiĉka and J. Sicher, *Coll. Czech. Chem. Commun.*, **28**, 1644 (1963).
196. A. Fry, *Synthetic Organic Electrochemistry*, Harper & Row, New York (1972), p. 170.
197. L. G. Feoktestov and M. M. Gol'din, *Electrokhimya*, **4**, 490 (1968).
198. P. J. Elving, I. Rosenthal and A. Martin, *J. Amer. Chem. Soc.*, **77**, 5218 (1955).
199. E. L. Eliel, N. L. Allinger, S. J. Angyal and G. A. Morrison, *Conformational Analysis*, Interscience, New York (1965), p. 24.

The Chemistry of Functional Groups, Supplement D
Edited by S. Patai and Z. Rappoport
© 1983 John Wiley & Sons Ltd

CHAPTER 6

Electrochemical oxidation, reduction and formation of the C—X bond – direct and indirect processes

JAMES Y. BECKER

Department of Chemistry, Ben-Gurion University of the Negev, Beer Sheva, Israel

I. INTRODUCTION

Halogen compounds occupy a central position in organic chemistry and, therefore, electrochemistry of the carbon–halogen bond is a subject of considerable importance. The present chapter covers mainly preparative and mechanistic aspects which are directly related to the C—X bond. It supplements the previous excellent review by Casanova and Eberson[1] on the electrochemistry of alkyl halides, and discusses new developments which have not been covered before. Since there is a vast literature on the cathodic reduction of alkyl halides[2–9], including a recent book by Hawley[10], this subject is covered only briefly (Section V) and the emphasis is on the following aspects:

(1) Many previous reviews on electrochemistry of organic compounds[2–10] deal with alkyl halides, and are concerned almost exclusively with electrochemical reduction. Electrochemical oxidation is scarcely mentioned and only in the past decade this aspect has begun to draw much attention. Therefore, special consideration will be given to the *anodic oxidation* of alkyl halides (Section II).

(2) The formation of the C—X bond by halogenation (Section III), and its cleavage by direct anodic and cathodic processes, as well as by indirect pathways through electrochemically generated electron transfer species, are discussed. Electron transfer mediators include aromatic hydrocarbons, inorganic species, transition metal complexes, etc. (Section IV).

(3) The present review also contains a coverage of the electrochemical properties of carbon–pseudohalogen bonds including carbon–azide bonds. The electrochemistry of these bonds has not previously been reviewed (except for polarographic data on isocyanates).

No attempt has been made either to chronicle or to categorize the voluminous literature, although this reviewer hopes that most of the work which appeared since Ref. 1 and up to February 1981 (according to *Current Contents*) is included. Electrochemical techniques and their application in organic chemistry, as well as principles of electrochemistry, will not be treated since many textbooks and reviews deal with these subjects. In the present discussion, electrochemical reactions of the carbon–halogen bond will first be divided according to the nature of the electrochemical process (direct oxidation–reduction, halogenation or indirect cleavage) and then subdivided in each section according to the nature of the halogen.

II. DIRECT ANODIC CLEAVAGE OF THE C—X BOND

In contrast to the vast literature on electrochemical reduction of alkyl halides, both from the point of view of mechanism and synthetic applications, relatively very little has been published on their anodic oxidation. Since the more electronegative halogens either cause the reactions of interest to take place in other parts of the molecule or make the appropriate compounds difficult to oxidize, most of the oxidation reactions were confined to iodoaryl and iodoalkyl compounds, which appear to be oxidized readily in the region of accessible potentials. However, in the past decade, some progress has been made and electrochemical oxidation under proper conditions has been extended to other halides, usually aliphatic compounds.

A. Carbon–Iodine Bond

Miller and Hoffman[11] were the first to investigate the anodic oxidation of aliphatic iodides in acetonitrile at a platinum electrode (Table 1). Their oxidation potential was sufficiently low with respect to solvent–electrolyte breakdown to permit oxidation to be observed. Whereas tertiary iodides solvolysed much too rapidly in acetonitrile–$LiClO_4$ medium to allow controlled potential electrolysis, primary and secondary iodides were found to be stable in that medium and form *N*-alkylacetamides by a Ritter-type reaction on a carbenium ion intermediate. Further evidence for the involvement of carbocations was deduced from the fact that only a

TABLE 1. Oxidation potentials of alkyl iodides measured on Pt anode in acetonitrile

Compound	$E_{1/2}$, V[a]	Reference
Methyl iodide	2.12	11
Neopentyl iodide	2.14	11
Isopropyl iodide	2.04	11
t-Butyl iodide	1.87	11
2-Octyl iodide	Not reported	13
n-Propyl iodide	Not reported	13
Cyclohexyl iodide	2.0[b]	14
Cyclobutyl iodide	2.0[b]	14
Cyclopropylmethyl iodide	2.0[b]	14
2-Adamantyl iodide	2.0[b]	15
2-Iodo-1-methyladamantane	2.0[b]	15
2-Iodo-1,3,5,7-tetramethyladamantane	2.0[b]	15

[a] Against Ag/0.01 M $AgNO_3$.
[b] Applied controlled potential for electrosynthesis.

rearranged product, namely *N-t*-pentylacetamide, was formed by the oxidation of neopentyl iodide.

Laurent and Tardivel[12] also studied the anodic oxidation of various alkyl iodides in acetonitrile and revealed the synthetic potentiality of this reaction which was previously reviewed in this series (Ref. 1, p. 1037).

The electrochemical oxidation of alkyl iodides was extended to alicyclic compounds among which iodocyclohexane[14] produced a mixture of *N*-cyclohexylacetamide and *N*-cyclohex-2-enylacetamide:

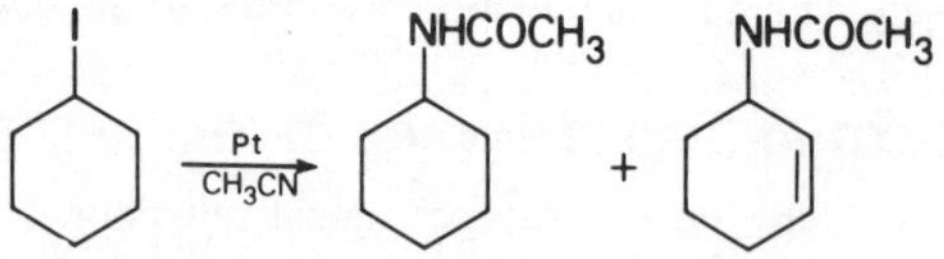

NHCOCH3 NHCOCH3
Pt
CH3CN
+

The latter product is attributed to the oxidation of cyclohexene which was previously formed by elimination of a proton from a carbocation intermediate.

Iodocyclobutane and iodomethylcyclopropane were oxidized at platinum in acetonitrile and both yielded the same mixture of two isomeric products (~80% total yield): *N*-cyclobutylacetamide and *N*-(cyclopropylmethyl)acetamide[16]. It was suggested that a common intermediate, $[C_4H_7]^+$, is responsible for the results obtained from both iodides:

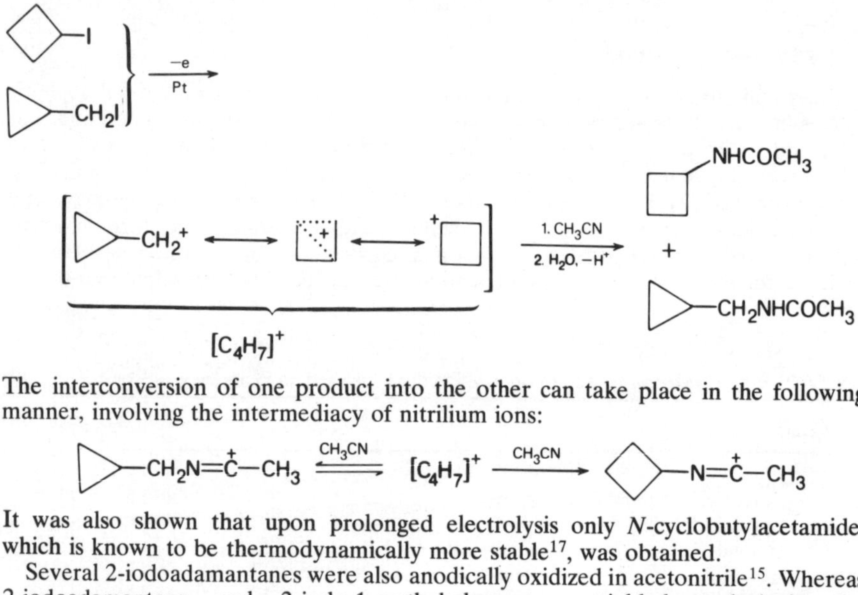

−e
Pt
CH2I

NHCOCH3
CH2+
1. CH3CN
2. H2O, −H+
+
CH2NHCOCH3
$[C_4H_7]^+$

The interconversion of one product into the other can take place in the following manner, involving the intermediacy of nitrilium ions:

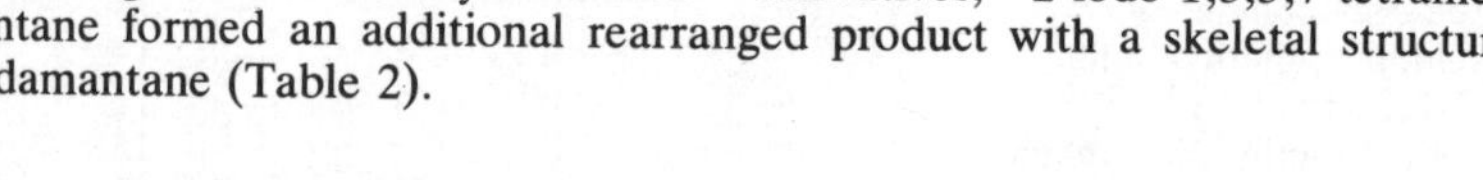

$$\text{c-C}_3\text{H}_5\text{—CH}_2\overset{+}{\text{N}}\text{=C—CH}_3 \xrightleftharpoons{CH_3CN} [C_4H_7]^+ \xrightarrow{CH_3CN} \text{c-C}_4\text{H}_7\text{—}\overset{+}{\text{N}}\text{=C—CH}_3$$

It was also shown that upon prolonged electrolysis only *N*-cyclobutylacetamide, which is known to be thermodynamically more stable[17], was obtained.

Several 2-iodoadamantanes were also anodically oxidized in acetonitrile[15]. Whereas 2-iodoadamantane and 2-iodo-1-methyladamantane yielded exclusively the corresponding 2-*N*-adamantylacetamido derivatives, 2-iodo-1,3,5,7-tetramethyladamantane formed an additional rearranged product with a skeletal structure of protoadamantane (Table 2).

B. Carbon–Bromine Bond

It has been shown that adamantane[18,19] undergoes $2e^-$ oxidation to produce *N*-(1-adamantyl)acetamide (75% isolated yield). The same product (89%) was formed

TABLE 2. Anodic oxidation of 2-iodoadamantanes in acetonitrile[a]

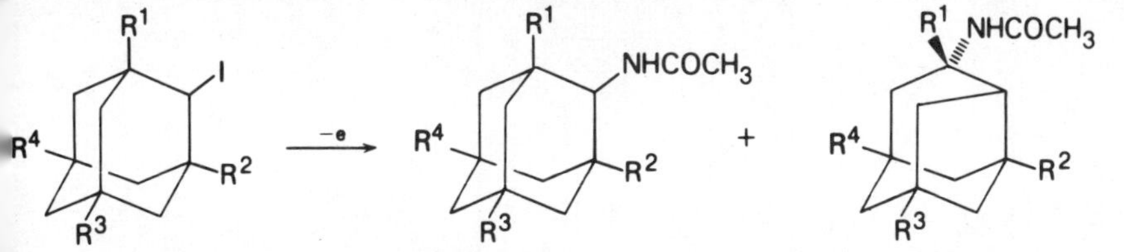

	2-iodoadamantane $\xrightarrow{-e}$ (R¹, NHCOCH₃ product)	+ (R¹ NHCOCH₃ rearranged product)	Total yield
$R^1 = R^2 = R^3 = R^4 = H$	>98	–	86
$R^2 = R^3 = R^4 = H, R^1 = Me$	>98	–	85
$R^1 = R^2 = R^3 = R^4 = Me$	~90	~10	93

[a]Data from Ref. 15. Platinum working anode against Ag/0.1 N $AgNO_3$ was employed. All derivatives were potentiostatically oxidized at 1.7 V.

also by $1e^-$ oxidation of 1-bromoadamantane[20]. This result implicates the involvement of 1-adamantyl carbenium ion in the reaction, since it is known that such an intermediate may lead to this product in acetonitrile[21]. It was postulated that in the oxidation of 1-bromoadamantane, a carbocation intermediate may arise either from direct anodic oxidation of the adamantyl moiety or from the non-bonding electrons of the bromine atom. When 2-bromoadamantane was potentiostatically oxidized[22] at 2.5 V (against Ag/Ag^+), consuming 2 F mol^{-1}, two acetamides were formed: *N*-(2-adamantyl)acetamide (derived from carbon–bromine bond cleavage) and a mixture of isomers of 2-bromoadamantylacetamides (derived from tertiary carbon–hydrogen bond fission), the latter being highly favourable. However, on introducing one methyl group at a bridgehead position the oxidation product predominantly undergoes one-electron cleavage to give a non-brominated acetamide due to C—Br bond fission. It has been pointed out that 2-bromo-1-methyladamantane is solvolysed 30 times faster than 2-bromoadamantane and the different electrochemical results have been interpreted in terms of solvent assistance to the halogen loss (see for details Section II.D). To complicate the matter, when four methyl groups are introduced in all four available bridgehead positions, the compound is not oxidized at $\leqslant 3.1$ V and the total product yield decreased from 80 to 50%. The breaking of the *sec*-C—H bond was found to be favoured over that of the C—Br bond. The latter cleavage yielded also a rearranged product (Table 3).

Unlike secondary adamantyl bromides which gave products due to a competitive cleavage between C—Br and C—H bonds, other simple bromoalkanes yielded products[23] derived from an exclusive C—Br bond breaking, indicating a selectivity towards bromine substitution (Table 4). Since all alkyl bromides studied have $E_p \geqslant 2.8$ V (except for *t*-bromobutane, 2.47 V), controlled potential electrolyses were carried out on the foot of the waves to avoid solvent interference. Thus, primary, secondary and tertiary bromides were found to be oxidizable under these conditions and yielded products with reasonably good yields. Furthermore, products due to Wagner–Meerwein rearrangement were occasionally observed. Anodic oxidation of 1-bromo-2-methylpropane, which contains a primary bromide and a tertiary hydrogen, formed the product derived only from C—Br bond fission. This result is quite surprising in light of those obtained from the electro-oxidation of 2-bromoadamantanes. The discrepancy may be explained qualitatively on the basis of photoelectron spectroscopy measurements compared with oxidation potentials of both bromoalkanes and their parent hydrocarbons (Table 5). As one can see, the $E_{p/2}$ for

TABLE 3. Anodic oxidation of 2-bromoadamantane derivatives in acetonitrile at platinum anode[a]

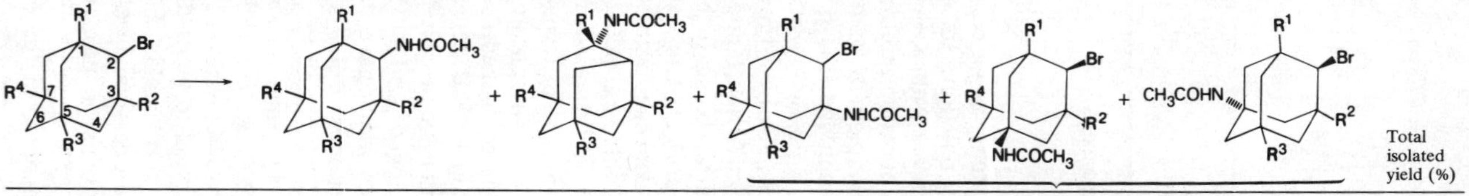

Substrate (2-bromoadamantane, positions 1–7; R^1, R^2, R^3, R^4)	→ 2-acetamido product	+ 1-acetamido product	+ bromo-acetamido products (three structures, bracketed)	Total isolated yield (%)
$R^1 = R^2 = R^3 = R^4 = H$	5	–	95 (2:4:3)	80
$R^2 = R^3 = R^4 = H$, $R^1 = CH_3$	85	Trace	15 (not separable)	85
$R^1 = R^2 = R^3 = R^4 = CH_3$	27	9	64[b]	52[c]

[a]Data from Ref. 15. Working potentials are 2.5–2.6 V against Ag/0.1 N $AgNO_3$.
[b]Two isomers were formed according to NMR and MS: 2-bromo-(*N*-4)- and (*N*-6)-acetamidoadamantanes.
[c]In addition, 15% of 1,3,5,7-tetramethyladamantanone was formed.

TABLE 4. Voltammetric data[a] and oxidation products[23]

Reactant (20 mmol)	Supporting electrolyte	Current yield, %[b]	Products
t-Butyl bromide	TEAF[c]	83	*N*-*t*-Butylacetamide (**1**)
2-Bromopropane	$LiClO_4$	70	*N*-2-Acetamidopropane (**2**)
	TEAF	40	
2-Bromobutane	$LiClO_4$	60	*N*-2-Acetamidobutane (**3**)
	TEAF	41	
1-Bromopentane	$LiClO_4$	40	*N*-1- (**4**), *N*-2- (**5**), *N*-3-Acetamido-pentane (**6**)
1-Bromo-2-methylpropane	$LiClO_4$	50	**1** + **3** + *N*-isobutylacetamide (**7**)

[a]All preparative oxidations have been carried out at 2.35 V against $Ag/AgNO_3$ 0.1 M in 10 ml of acetonitrile at room temperature. Cyclic voltammograms of all the listed compounds give E_p values in the range 2.4–2.8 V (against Ag/Ag^+) at 0.5 V s^{-1} scan rate.
[b]Based on 2 F mol^{-1}.
[c]Tetraethylammonium fluoroborate.

adamantane is lower than that for its corresponding bromide while their corresponding ionization potentials (IPs) are in reverse order (or close to each other). On examining other simple alkanes and their bromides it is clear that now the $E_{p/2}$ values for bromobutanes are lower than those corresponding to their parent hydrocarbons, along with the same trend observed for their IP values. On the grounds of the above observation it was suggested that initial removal of electron from bromoadamantane is from the HOMO which is primarily located on the adamantyl moiety[19]. However, in the case of other bromoalkanes, as shown in Table 5, it was postulated[23] that the HOMO consists mainly of the lone pair orbital of the bromine, and therefore these compounds are easier to oxidize than the corresponding hydrocarbons. This argument explains nicely why, in 2-bromoadamantane, C—H competes with the C—Br bond fission (both IP and $E_{p/2}$ values of the bromide are close to the corresponding values of its parent hydrocarbon). In other alkyl bromides the energy gap for oxidation is higher and thus the bond cleavage is more selective towards that of the C—Br bond.

Becker has also studied[27] the effect of supporting electrolyte and substrate concentration on the product yield from the anodic oxidation of simple aliphatic

TABLE 5. Oxidation and ionization potentials of alkyl bromides and their parent hydrocarbons

Substrate	$E_{p/2}$, V[a]	Reference	IP, eV	Reference
Adamantane	2.36	20	9.68	25
1-Adamantyl bromide	2.54	20	9.54	25
n-Butane	–[b]	24	10.63	26
1-Bromobutane	2.50	27	10.13	26
2-Bromobutane	2.40	27	9.98	26
2-Methylpropane	–[b]	24	10.56	26
t-Butyl bromide	2.25	27	9.89	26
Isobutyl bromide	2.50	27	10.09	26

[a]Against Ag/0.1 N $AgNO_3$. Sweep rate, 0.5 V s^{-1}.
[b]$E_{p/2}$ values of aliphatic hydrocarbons are in the range 3.0–3.6 V.

TABLE 6. Electrochemical data[a] and oxidation products of $Br(CH_2)_nCH_3$

$Br(CH_2)_nCH_3$		Total mF	Current	Relative percentage of oxidation products, %[c]			
n	Concn, M	consumed	yield, %[b]	1-$NHCOCH_3$-alkane	2-$NHCOCH_3$-alkane	3-$NHCOCH_3$-alkane	4-$NHCOCH_3$-alkane
1	1.6	4.4	79	100			
2	1.6	4.5	52	40	60		
3	1.6	4.6	40	32	68		
4	1.56	4.0	40	33	33	33	
5	1.54	5.0	50	27	41	32	
6	1.47	4.0	12	22	45	30	3
7	1.47[d]	5.2	10	5	24	16	44

[a]Data from Ref. 29. All oxidations were carried out at 2.35 V against Ag/0.1 N $AgNO_3$ in acetonitrile. The cyclic voltammogram of each of the alkyl bromides mentioned above gave an ill-defined wave in the range 2.6–2.8.
[b]Based on 2 F mol^{-1}.
[c]Percentages shown express product distributions of isolated materials determined by GLC. On this basis the current yield is treated as 100%.
[d]A small amount of acetamide and other impurities were observed.

bromides. Since the oxidation takes place on the foot of the wave, obviously the higher the concentration of the substrate the more efficient is the bromide oxidation, which competes favourably with the background oxidation. Regarding the effect of electrolytes, the results show better current yields with perchlorate anions than with fluoroborate ions since the latter were found[28] to be adsorbed more easily on platinum and consequently decrease the concentration of the depolarizer on the electrode surface.

The effect of chain length on the anodic oxidation of straight-chain 1-bromoalkanes, $Br(CH_2)_nCH_3$ (n = 1–7), was investigated[29]. It has been found that all substrates undergo an exclusive C—Br bond cleavage to produce acetamido derivatives in acetonitrile solvent (Table 6). The product distribution indicates isomerizations in the direction of the more stable carbocation, namely from primary alkyl products to secondary ones. However, it was realized that hydride migration steps occurred in competition with a nucleophilic attack by the solvent on the carbocation intermediate, because isomerization does not always end up completely with the most stable carbocation.

Current yields were strongly affected by the chain length and it was found that the longer the chain the lower the yield. The preference for a certain acetamido derivative obtained from the long-chain bromides led to the hypothesis that besides the possibility of a series of consecutive 1,2-hydride shifts, a remote rearrangement route consisting of a six-membered ring transition state may also be considered, although there is no clear-cut proof (such as using a labelled compound) to support such a hypothesis:

$$R(CH_2)_5Br \xrightarrow{-e/Pt} \left[\text{R–CH(H)} \cdots CH_2\text{–}Br^{+\cdot} \; (H_2C, CH_2, CH_2 \text{ ring}) \right] \xrightarrow{-Br^{\cdot}} R\overset{+}{C}H(CH_2)_3CH_3$$

It is noteworthy that none of the carbocation intermediates underwent deprotonation since no olefinic products were detected.

A study of the effect of ring size on the anodic oxidation of alicyclic bromides has been conducted by Becker and Zemach[30] and revealed that in addition to the formation of amides in acetonitrile, olefins were also formed in certain cases (Table 7).

TABLE 7. Products from anodic oxidation of alicyclic bromides in acetronitrile[a]

Substrate (0.5 M)	*N*-Cycloalkylacetamide, %	Cycloalkene, %[b]	1-Bromo-2-hydroxy-cycloalkane, %
Bromocycloheptane	26	1–2	–
Bromocyclohexane	60	14	Trace
Bromocyclopentane	45	4	Trace
Bromocyclobutane	60	–	–
Bromocyclopropane	CH_2=$CHCH_2NHCOCH_3$ (45)	–	–

[a]Data were taken from Ref. 30. Tetraethylammonium fluoroborate (0.2 M) was used as an electrolyte. Electrolyses were carried out at 2.5 V (against Ag/0.1 N $AgNO_3$) at 0 °C. Yields are based either on isolated products or calibration curves.
[b]Trapped as 1,2-dibromocycloalkane.

The same authors also studied the effects of concentration, temperature, electrolyte and electricity consumption of the anodic process. Only bromocyclopropane undergoes ring opening during the course of its anodic oxidation, due to the ring strain in the three-membered ring. Also, it is well established that a positive charge generated on a cyclopropyl ring causes ring opening to an allylic cation[31]. Therefore, this result is to be expected and the following mechanism, involving a skeletal rearrangement in the cyclopropyl ring, was suggested:

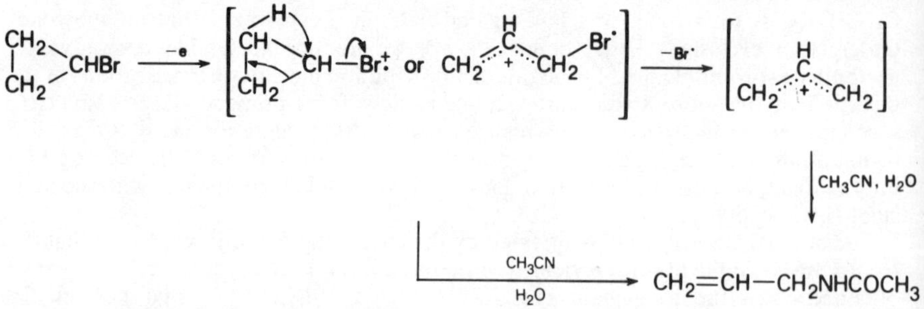

$$\text{cyclo-}(CH_2CH_2CHBr) \xrightarrow{-e} \left[\text{cyclo-}(CH_2CH\text{—}CH)\text{—}Br^{+\cdot} \text{ or } CH_2\overset{H}{\underset{+}{\cdots C \cdots}}CH_2 \; Br^{\cdot} \right] \xrightarrow{-Br^{\cdot}} \left[CH_2\overset{H}{\underset{+}{\cdots C \cdots}}CH_2 \right] \xrightarrow[H_2O]{CH_3CN} CH_2{=}CH{-}CH_2NHCOCH_3$$

Bromocyclobutane yielded only, *N*-cyclobutylacetamide and no *N*-methylcyclopropylacetamide was detected since the former is thermodynamically more stable[17]. However, both amides were obtained when the electrolysis duration was short, enabling also product formation from the cyclopropylmethyl cation[16].

Unlike the unstrained five-, six- and seven-membered rings which also produced cycloalkenes, no cyclobutene or propene was detected.

C. Carbon–Chlorine and Carbon–Fluorine Bonds

The only aliphatic chloro- and fluoroalkanes, the anodic oxidation of which is reported, are adamantyl derivatives. Miller and Koch[20] oxidized 1-haloadamantanes and concluded that they undergo C—H bond cleavage at the bridgehead exclusively, to form haloadamantyl acetamides (Table 8). Others[15] reported on the electrochemical oxidation of 2-chloro- and 2-fluoroadamantanes. Surprisingly, the latter gave a product derived from substitution of fluorine by an acetamido group. This result was rationalized by the following scheme, which suggests elimination of hydrogen fluoride from the electrochemically generated cation radical:

TABLE 8. Products of electro-oxidation of adamantyl chlorides and fluorides in acetonitrile on platinum anode

Substrate	Oxidation potential[a]	Isolated yield, %	Haloadamantyl acetamides (%)	Adamantyl acetamides(%)	Reference
1-Adamantyl chloride	2.5	91	100	–	20
1-Adamantyl fluoride	2.5	51	100	–	20
2-Adamantyl chloride	2.6	82	100[b]	–	15
2-Adamantyl fluoride	2.6	40	65[b]	35	15

[a]Against Ag/0.1 N $AgNO_3$.
[b]Three isomers.

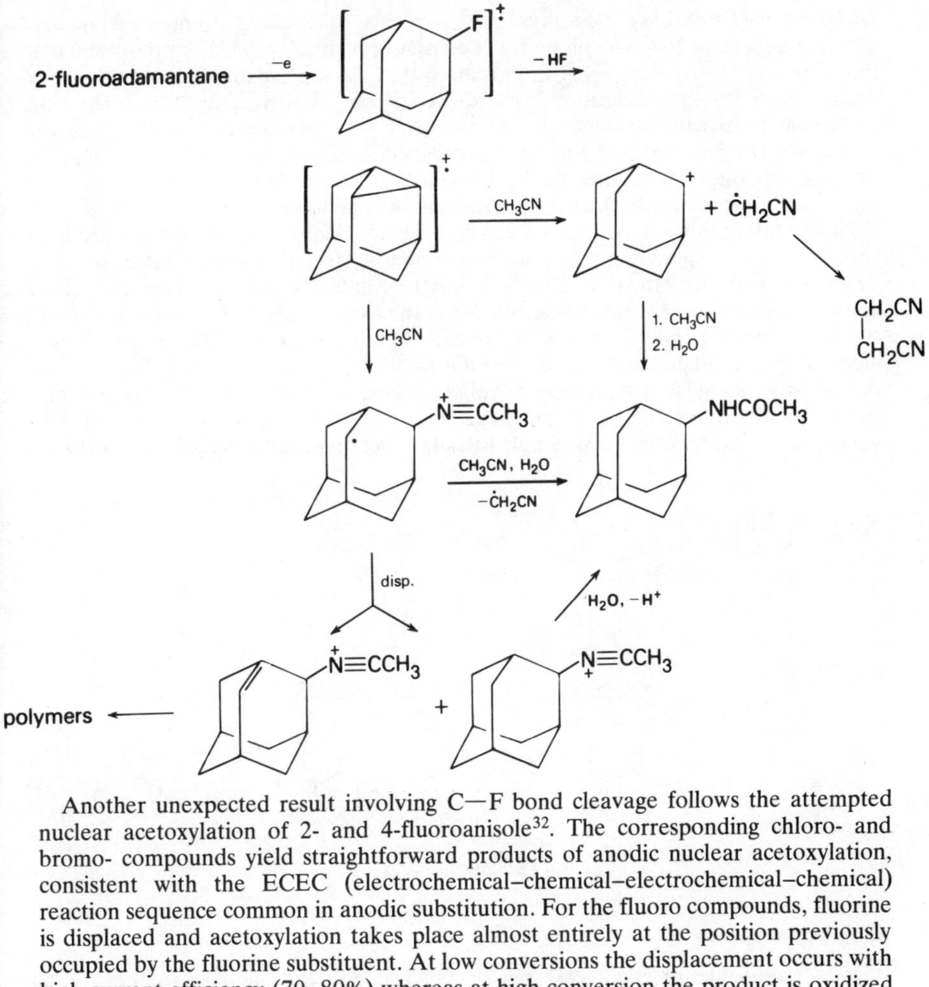

2-fluoroadamantane $\xrightarrow{-e}$ $[\text{C}_{10}\text{H}_{15}\text{F}]^{+\cdot}$ $\xrightarrow{-HF}$

$[\ ]^{+\cdot}$ $\xrightarrow{CH_3CN}$ (adamantyl cation)$^{+}$ + $\dot{C}H_2CN$ → CH_2CN–CH_2CN

$\downarrow CH_3CN$ ... $\downarrow$ 1. CH_3CN 2. H_2O

$\overset{+}{N}{\equiv}CCH_3$ $\xrightarrow[-\dot{C}H_2CN]{CH_3CN,\ H_2O}$ $NHCOCH_3$

disp.

$H_2O, -H^+$

polymers ← $\overset{+}{N}{\equiv}CCH_3$ + $\overset{+}{N}{\equiv}CCH_3$

Another unexpected result involving C—F bond cleavage follows the attempted nuclear acetoxylation of 2- and 4-fluoroanisole[32]. The corresponding chloro- and bromo- compounds yield straightforward products of anodic nuclear acetoxylation, consistent with the ECEC (electrochemical–chemical–electrochemical–chemical) reaction sequence common in anodic substitution. For the fluoro compounds, fluorine is displaced and acetoxylation takes place almost entirely at the position previously occupied by the fluorine substituent. At low conversions the displacement occurs with high current efficiency (70–80%) whereas at high conversion the product is oxidized further. The mechanism is not clear, but a plausible explanation is given in which a fluoride ion is lost at some intermediary stage, as follows:

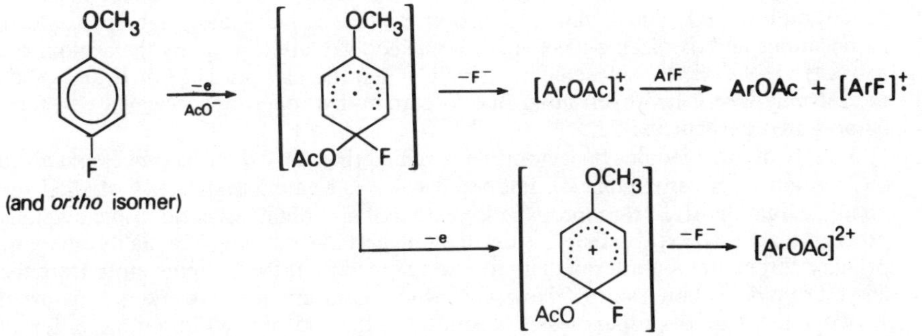

OCH_3 / F (and *ortho* isomer) $\xrightarrow[AcO^-]{-e}$ $[OCH_3,\ AcO,\ F]$ $\xrightarrow{-F^-}$ $[ArOAc]^{+\cdot}$ $\xrightarrow{ArF}$ ArOAc + $[ArF]^{+\cdot}$

$\xrightarrow{-e}$ $[OCH_3,\ +,\ AcO,\ F]$ $\xrightarrow{-F^-}$ $[ArOAc]^{2+}$

Different n-values have been reported for this reaction, 2 from voltammetric measurements and 1 from coulometry at constant potential. It has been proposed that this difference in n-values gives an indication that the starting material is oxidized by an intermediate, e.g. the dication in the above scheme. However, the authors have not been able to explain satisfactorily the reduction of either the radical cation or the dication to the final product. Interestingly, the same behaviour is observed[33] when the fluoro compounds are oxidized by Ag(II), and it is reasonable to assume that the same mechanism is valid for both chemical and anodic oxidation.

Unlike fluoroanisoles, neither 4-fluorotoluene nor fluorobenzene gave any products resulting from displacement of fluorine on anodic oxidation in the same medium. Therefore, it is of interest to know how polyfluorinated arenes would behave. These compounds contain only fluorine substituents and fluorine displacement or addition to the aromatic ring would be feasible reaction pathways. Blum and Nyberg[34] have studied the anodic oxidation of hexafluorobenzene, octafluoronaphthalene and decafluorobiphenyl in trifluoroacetic acid. After hydrolysis the former two compounds gave tetrafluorobenzoquinone and hexafluoronaphthoquinone in 75% and 60% yields, respectively, whereas attempts to isolate octafluorobiphenoquinone failed:

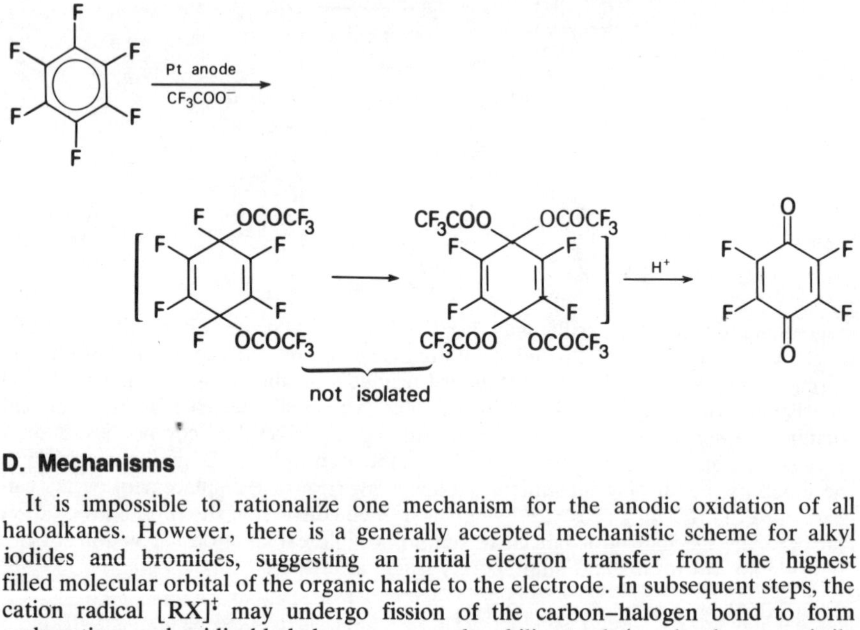

Pt anode, CF_3COO^-

H^+

not isolated

D. Mechanisms

It is impossible to rationalize one mechanism for the anodic oxidation of all haloalkanes. However, there is a generally accepted mechanistic scheme for alkyl iodides and bromides, suggesting an initial electron transfer from the highest filled molecular orbital of the organic halide to the electrode. In subsequent steps, the cation radical $[RX]^{\ddagger}$ may undergo fission of the carbon–halogen bond to form carbocations and oxidizable halogen, or a nucleophilic attack (e.g. by the acetonitrile solvent) via S_N2 type displacement. Usually, when the halogen is Cl or F, the cation radical may lose a hydrogen atom due to carbon–hydrogen bond cleavage to form halogenated derivatives.

A study of alkyl iodides in acetonitrile revealed that the oxidation process involved the formation of carbenium ion intermediates[11,12]. Keating and Skell[35] studied the anodic oxidation of n-bromopropane in methanol and obtained cyclopropane among other products. To explain their results, they suggested formation of a highly energetic primary carbenium ion intermediate following ejection of the bromine atom from the alkyl bromide cation radical. However, when Laurent and coworkers[36] oxidized 1-iodo-1,1-dideuteriopropane electrochemically, their results led to a quite different conclusion. By analysing the mass spectra of the isolated *N*-propylacetamides it was found that 98% of the deuterium atoms are localized on the α-carbon ($CH_3CH_2CD_2NHCOCH_3$). This points to an attack of the nucleophilic solvent on the radical cation via an S_N2 type mechanism:

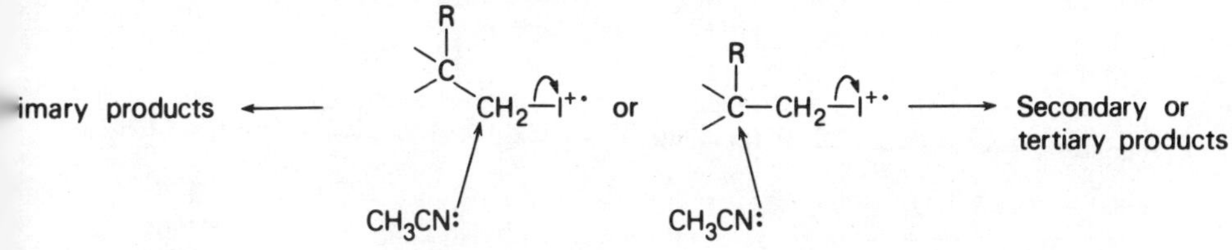

$$\text{Primary products} \longleftarrow \underset{\uparrow \; CH_3CN:}{\left(>C(R)-CH_2-I^{+\bullet}\right)} \quad \text{or} \quad \underset{\uparrow \; CH_3CN:}{\left(>C(R)-CH_2-I^{+\bullet}\right)} \longrightarrow \text{Secondary or tertiary products}$$

If carbenium ions are to be formed, then a mixture of three primary products is to be expected[37], according to the following scheme:

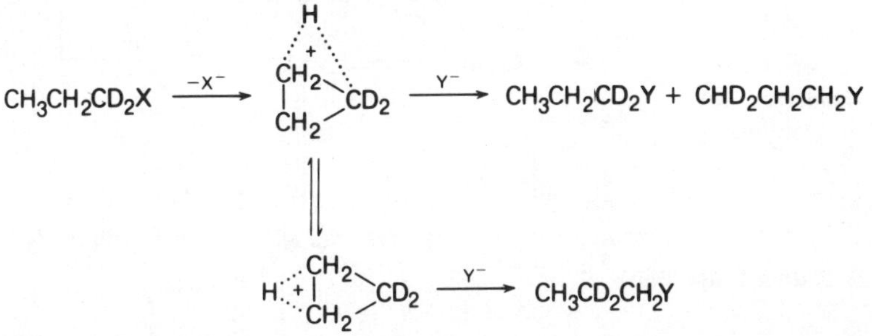

$$CH_3CH_2CD_2X \xrightarrow{-X^-} [\text{H-bridged } CH_2CH_2CD_2]^+ \xrightarrow{Y^-} CH_3CH_2CD_2Y + CHD_2CH_2CH_2Y$$

$$\Updownarrow$$

$$[\text{H-bridged } CH_2CD_2CH_2]^+ \xrightarrow{Y^-} CH_3CD_2CH_2Y$$

Further support for the favoured S_N2 type mechanism evolves from a comparison between the results obtained from anodic oxidation of alkyl iodides with those obtained from a Kolbe-type process (which is known to produce carbocations under proper conditions). Some of the results are described in Table 9. Obviously, the differences are indicative of more than one mechanism (at least for the formation of the primary substituted amide). The extent of rearrangement of carbenium ions produced by oxidation of alkyl iodides has been compared with that of cations produced by trifluoroacetolysis of tosylates in a comprehensive paper by Laurent and coworkers[12]. They conclude that, for electro-oxidation of iodide, the solvent (acetonitrile) assists cation formation. This conclusion is reinforced by the observation[39] of 20% inversion in the major product (*N*-(2-octyl)acetamide) of the anodic oxidation of optically active 2-iodooctane.

The complexity of the reaction of alkyl bromides suggests that more than one mechanism is operative in order to explain rearranged, as well as olefinic products.

TABLE 9. Products obtained from anodic oxidation of alkyl iodides and carboxylates, in acetonitrile on Pt anode

Substrate	*N*-(1-Alkyl)acetamide, %	*N*-(2-Alkyl)acetamide, %	Reference
$CH_3CH_2CH_2I$	60	40	36
$CH_3CH_2CH_2COO^-$	–	~100	38
$CH_3CH_2CH_2CH_2I$	58	42	36
$CH_3CH_2CH_2CH_2COO^-$	–	~100	38

The following scheme describes possible routes for obtaining products from the organic moieties. The fate of the halogen, which does not have a straightforward interpretation, will be discussed elsewhere in this chapter:

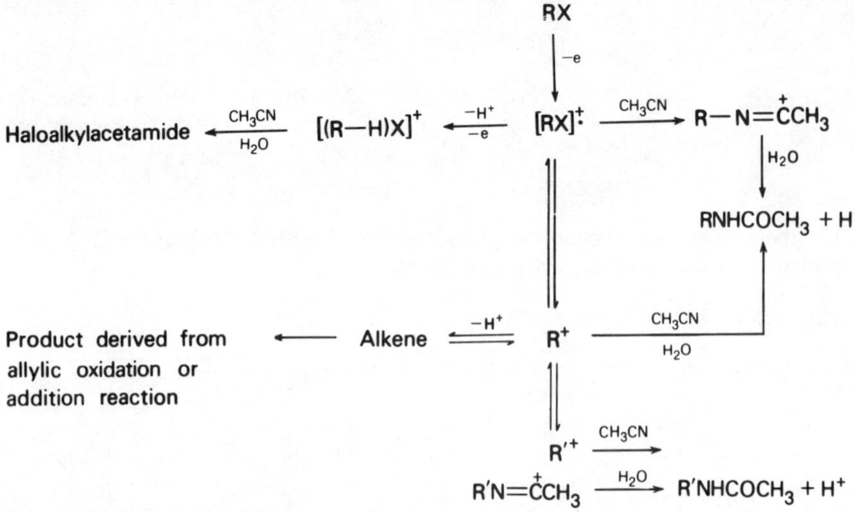

$$
\begin{array}{ccccccc}
 & & & & RX & & \\
 & & & & \downarrow{-e} & & \\
\text{Haloalkylacetamide} & \xleftarrow[H_2O]{CH_3CN} & [(R{-}H)X]^+ & \xleftarrow[-e]{-H^+} & [RX]^{+\cdot} & \xrightarrow{CH_3CN} & R{-}N{=}\overset{+}{C}CH_3 \\
 & & & & \Downarrow & & \downarrow{H_2O} \\
 & & & & & & RNHCOCH_3 + H^+ \\
 & & & & & & \uparrow \\
\text{Product derived from allylic oxidation or addition reaction} & \longleftarrow & \text{Alkene} & \underset{}{\overset{-H^+}{\rightleftharpoons}} & R^+ & \xrightarrow[H_2O]{CH_3CN} & \\
 & & & & \Downarrow & & \\
 & & & & R'^+ & \xrightarrow{CH_3CN} & \\
 & & & & R'N{=}\overset{+}{C}CH_3 & \xrightarrow{H_2O} & R'NHCOCH_3 + H^+
\end{array}
$$

See Addendum (page 286)

III. C—X BOND FORMATION BY DIRECT ANODIC OXIDATION OF 'X' SPECIES

Electrochemical halogenation of organic substrates has been mainly confined to anodic fluorination, for which numerous patents and publications exist[40]. However, considerable advances have been made in anodic bromination, chlorination and iodination, both from the synthetic and mechanistic viewpoints. The relative ease of oxidation is in the expected order $I^- > Br^- > Cl^- > F^-$.

For halogenations in I^-, Br^- or Cl^- electrolytes, Pt and carbon (in their various forms) are usually suitable, and there are no special requirements for anode materials other than that they be inert to electrochemical oxidation and to the electrogenerated halogen species. In contrast, fluorinations in liquid HF require nickel or carbon anodes for perfluorination to occur. The specific reasons will be discussed later in this section.

The present section illustrates C—X bond formation via direct anodic oxidation of halogen moieties (as well as pseudohalogens).

A. Carbon–Iodine Bond

In chemical iodination of organic compounds in aqueous media it has often been postulated that either iodonium ion (I^+) or protonated hypoiodous acid, $(H_2OI)^+$, are involved[41]. In the presence of oxidizing agents, evidence for IO^+, I_2^+, I_3^+ and I_5^+ has been reported[42]. The powerful electrophilicity of positive iodine species has been studied by several groups[42,43].

Popov and Geske[44] were the first to study the electrochemical oxidation of iodine in acetonitrile as a part of an investigation of the $I^-{-}I_2{-}I_3^-$ redox system. Their results indicated that only a portion of the number of coulombs consumed in iodine oxidation were assayable as positive iodine species. This did not permit elucidation of the oxidation pathway.

TABLE 10. Oxidation products from iodination of aromatic compounds in acetonitrile–$LiClO_4$ on Pt anode[a]

Substrate	Percentage yield of iodoaromatic isomers from a pre-formed agent, %[b]	Percentage yield of iodoaromatic from mixtures of iodine and aromatic substrates, %
Benzene	96	11
Toluene	47 *para*, 47 *ortho*	16 *para*, 16 *ortho*, 2 *meta*
p-Xylene	100	50 (12)[c]
Anisole	56 *para*, 24 *ortho*	13 *para*, 6 *ortho*, 1 *meta*
Nitrobenzene	0	0

[a]Data from Ref. 45. Iodination was conducted in a divided cell at 1.6–1.7 V against Ag/0.1 N $AgNO_3$.
[b]Based on 3.9 mmol of iodinating agent.
[c]Side-chain substitution product: *N*-α(*p*-xylylacetamide).

Miller and coworkers[45] studied the electro-oxidation of mixtures of iodine and benzenoid hydrocarbons and the results pointed to electrophilic iodination of the aromatic ring, competing with a side-chain substitution reaction in a particular case:

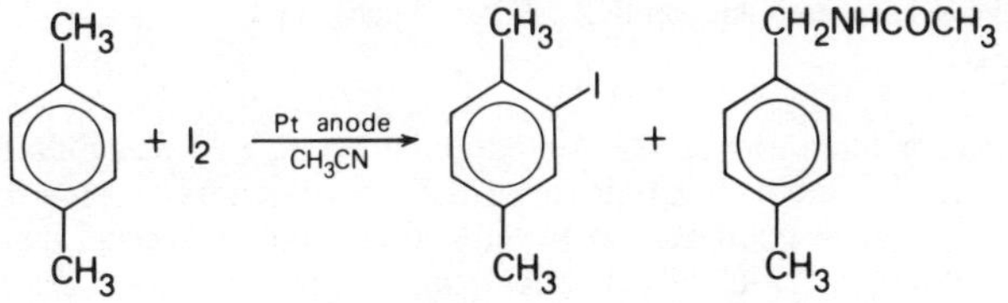

Pt anode
CH_3CN

These authors reported that a supression of the side-chain substitution, as well as increasing the total product yield (Table 10), was achieved when the iodinating agent was separately formed (upon electro-oxidation of iodine in acetonitrile–$LiClO_4$ solution) followed by addition of the aromatic compound, after the electrolysis was discontinued. Furthermore, it was shown that polyiodination can occur by varying either the mole ratio between 'I^+' and substrate or the amount of electricity consumption:

$$I_2 - 2e^- \longrightarrow 2\,'I^+' \text{ (could be in the form of } CH_3\overset{+}{C}{=}NI)$$

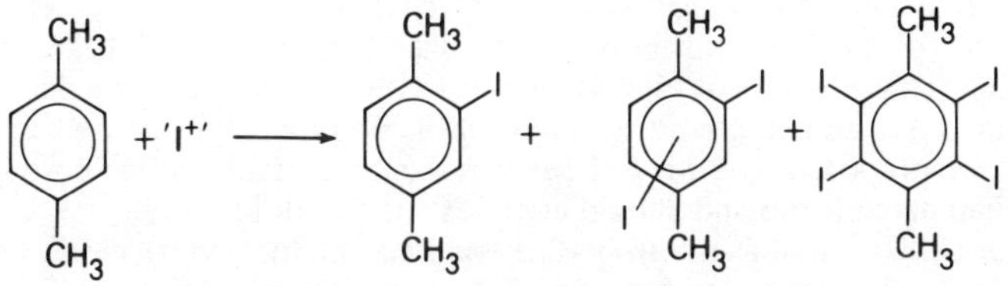

(a mixture of isomers)

Whereas nitrobenzene has not been iodinated in acetonitrile, it has been found by Parker and coworkers[46] that the anodic oxidation of iodine in trifluoroacetic acid containing solvents produces a highly reactive iodine species which iodinates even the most highly deactivated monosubstituted benzenes such as nitrobenzene and benzotrifluoride. Representative examples are summarized in Table 11.

TABLE 11. Products from iodination of aromatic compounds in the presence of I_2

Substrate	Solvent[a]	Product (%)
Iodobenzene	AN	1,4-Diiodobenzene (46) 1,2-Diiodobenzene (23)
Iodobenzene	DCE + 10% TFA	1,4-Diiodobenzene (77)
Benzonitrile	AN	No reaction
Benzonitrile	DCE + 10% TFA	3-Iodobenzonitrile (40)
Benzaldehyde	DCE + 10% TFA	3-Iodobenzaldehyde (74)
Benzotrifluoride	DCE + 10% TFA	3-Iodobenzotrifluoride (97)
4-Chloronitrobenzene	DCE + 10% TFA	3-Iodo-4-chloronitrobenzene (56)
Nitrobenzene	AN	No reaction
Nitrobenzene	DCM	No reaction
Nitrobenzene	DCE	No reaction
Nitrobenzene	AN + 10% TFA	No reaction
Nitrobenzene	DCM + 10% TFA	3-Iodonitrobenzene (47)
Nitrobenzene	DCE + 10% TFA	3-Iodonitrobenzene (78)

[a]All experiments in acetonitrile (AN) have been carried out at a controlled potential of 1.9 V against Ag/0.1 N Ag^+, on Pt anode using $LiClO_4$[45]. Data in other media, 1,2-dichloroethane (DCE), dichloromethane (DCM) and trifluoroacetic acid (TFA), are taken from Ref. 46, using tetrabutylammonium fluoroborate (TBAF) at a constant current of 150 mA. In all experiments 2 mmol of I_2 have been used and about 2.5 F mol^{-1} consumed.

Regarding the mechanism of the iodination process, two possibilities have been suggested so far. One involves an attack of the electrogenerated electrophile, I^+, on the aromatic ring, as has been demonstrated before, and the second suggests an ECE (electrochemical–chemical–electrochemical) mechanism in which the iodine cation radical reacts as an electrophile:

$$I_2 \xrightarrow{-e} I_2^{+\bullet}$$

$$I_2^{+\bullet} + ArH \longrightarrow ArI + H^+ + I^\bullet$$

$$I^\bullet \xrightarrow{-e} I^+$$

Both mechanisms satisfy the stoicheiometry of the iodinations, which requires the consumption of at least 2 F mol^{-1}.

A kinetic study by Miller and Watkins[47] includes measurements of relative rates towards 'I^+' in acetonitrile for substituted benzenes, deuterium isotope effects and a linear Hammett plot with σ^+ constants ($\rho^+ = -6.27$, Figure 1). The data are interpreted in terms of electrophilic substitution via a σ-complex with rate-limiting deprotonation of this complex for all compounds studied. In dichloromethane the ρ-value is -2.85 and it has been proposed that this difference is due to the fact that dichloromethane is less basic than acetonitrile and should complex the 'I^+' differently.

It has been shown[48] that a carbon–iodine bond may be formed by anodic coupling of iodobenzene to benzene to form a diphenyliodonium cation and by anodic self-coupling to form iodophenylphenyliodonium cation, as a potentially useful electrosynthesis of iodonium salts:

$$PhI + PhH \xrightarrow{-2e} Ph\overset{+}{I}Ph + H^+ \qquad (1)$$

$$2\ PhI \xrightarrow{-2e} IC_6H_4\overset{+}{I}Ph + H^+ \qquad (2)$$

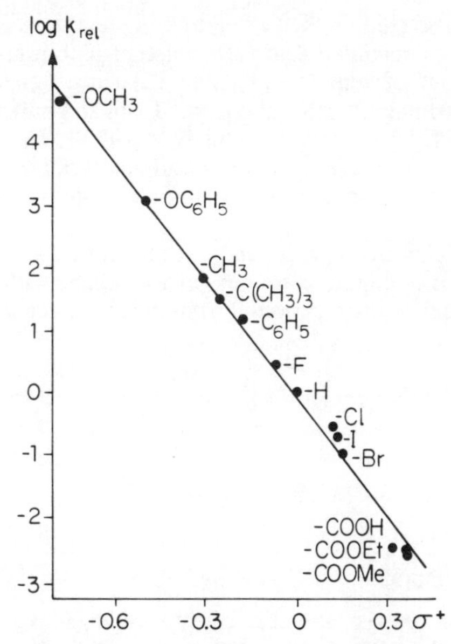

FIGURE 1. Hammett plot for iodination of substituted benzenes by electrochemically generated positive iodine species in CH_3CN. (Reproduced from Ref. 47.)

This coupling reaction may serve as an electrochemical method to produce phenol from benzene in two steps. Therefore, the reaction was reinvestigated by Wendt and coworkers[49] with the aim of optimizing the yield. In the dependence of the anodic $E_{1/2}$ of the iodoarenes (Table 12) on the electrode material one would expect an increase of coupling efficiencies in the order gold < platinum < lead dioxide < carbon since the adsorption of the substrate at the electrode surface increases in this direction. However, this prediction was fulfilled only for platinum and gold electrodes and failed for lead dioxide and carbon at which no iodonium ions are produced. The suggested reason for this behaviour is that polymerization reactions (for instance, with solvent molecules) prevail at carbon and PbO_2 anodes due to a catalytic effect.

TABLE 12. Half-wave potentials and coupling yields for the anodic iodoarene oxidation at different anode materials in acetonitrile–0.1 M $NaClO_4$[49]

Anode	$E_{1/2}$[a] of iodobenzene	Coupling yield, %[b]	$E_{1/2}$ of iodotoluene	Coupling yield, %[c]
Au	2.2	20–30	2.18	20
Pt	2.05	80	2.00	70
C	1.95	0	1.68	0
PbO_2	1.8 ± 0.05	0	–	–

[a]V, against SCE.
[b]For tenfold excess of benzene.
[c]For tenfold excess of toluene.

The same authors[49] also studied the two competing reactions (1) and (2). From their preparative results they concluded that the choice of platinum–acetonitrile gave the best selectivity in favour of reaction (1) over (2) since benzene was found to be adsorbed much more strongly than iodobenzene. Even at a bulk concentration ratio of 1 : 1 the formation of Ph_2I^+ was almost exclusive (>97%). The same trend was found for iodotoluene–toluene. However, at other anode-material/electrolyte interfaces the selectivity was worse and the competing anodic self-coupling reaction (2) was increased.

A continuous process has been proposed[50] for the synthesis of aromatic amines from benzene by the reaction of aliphatic amines (or ammonia) with the iodonium cations produced as a result of anodic oxidation of aryl iodides in acetonitrile in the presence of benzene:

$$\mathrm{PhH + ArI \xrightarrow[-H^+]{-2e} [ArIPh]^+ \xrightarrow{RNH_2} ArI + H^+ + PhNHR}$$

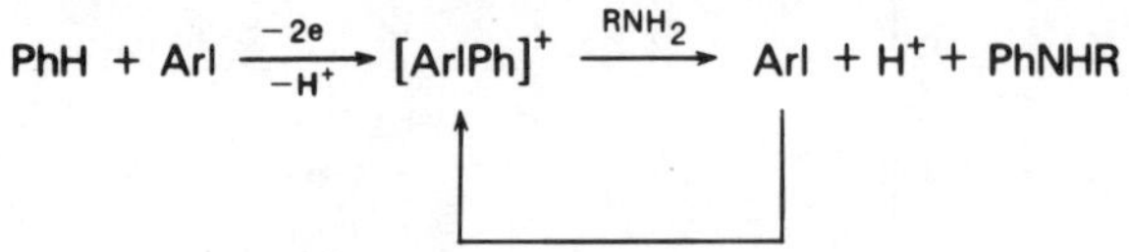

B. Carbon–Bromine Bond

Bromide ion, in acetonitrile solvent, can be oxidized to Br_3^- and Br_2 in two consecutive steps[51–53]. Popov and Geske[52] were the first to report the existence of a third anomalous anodic wave (observed at +1.5 V against Ag/0.1 M $AgNO_3$), but they gave no interpretation. The existence of this third step was confirmed by Magno and coworkers[54] and a mechanism was elucidated suggesting the electrochemical formation of Br^+ species. This hypothesis was supported by the fact that bromobenzene was detected when bromide ion was electrolysed in acetonitrile in the presence of benzene and at a potential corresponding to the third wave and below that of the oxidation of benzene. They suggested that bromobenzene is formed as in the following scheme:

$$\mathrm{[Br^+] + PhH \longrightarrow PhBr + H^+}$$

A radical reaction mechanism was excluded on the basis of their experimental results which showed no bromobenzene formation at working potential values *lower* than the third wave (at which bromine radicals are believed to be formed).

An extended study of anodic bromination of aromatic compounds in anhydrous acetic acid on Pt electrode was advanced[55] and, ring-, chain- and solvent-bromination took place (Table 13). Interestingly it was found that both electrochemical and photochemical steps are essential in the formation of aromatic bromo-substituted compounds. This conclusion was reached since no bromo aromatics were formed when electrolyses were carried out in the dark at a controlled potential of 1.0 V. On the other hand, illumination of the same solution (without electrolysis) showed that ring substitution was very slow and can be neglected as a side raction in the course of an electrolysis. Therefore, the following scheme was suggested:

$$\mathrm{2\,Br^- \longrightarrow Br_2 + 2e}$$

$$\mathrm{Br_2 + ArH \rightleftharpoons [ArH \cdot Br_2]}$$

where $[ArH \cdot Br_2]$ is a π-complex, non-oxidizable intermediate;

$$\mathrm{ArH \cdot Br_2 \xrightarrow{light} Intermediate\ product\ (detected\ by\ UV)}$$

TABLE 13. Electrolysis at 1.0 V (against SCE) at 40°C in acetic acid[a]

Aromatic substrate	Bromine species	n	Illumination	Percentage yield[b] of ring substitution, %	Percentage yield[b] of chain substitution, %	Percentage yield[b] of $CH_2BrCOOH$, %
p-Xylene	Br^-	2.05	Light	18	–	42
p-Xylene	Br^-	1.08	Dark	–	–	–
p-Xylene	Br_2	1.18	Light	25	–	44[c]
p-Xylene	Br_2	0.04	Dark	–	–	–
Toluene	Br^-	1.88	Light	14	3–4	52
Toluene	Br^-	1.12	Dark	–	–	–
Toluene	Br_2	1.2	Light	15.5	11	40
Toluene	Br_2	0.03	Dark	–	–	–
Naphthalene[d]	Br^-	1.8	Light	65	–	–
Benzene[d]	Br^-	1.0	Light	–	–	–
Benzene[d,e]	Br^-	2.2	Light	35	–	50
Benzene[d,e]	Br^-	1.0	Dark	–	–	–

[a] Most of the results were averaged from two to four runs and taken from Ref. 55.
[b] All yields are with respect to Br^- or Br_2 concentration.
[c] In one run the yield was 70%.
[d] Data are taken from Ref. 56. (Experimental conditions were slightly different.)
[e] The solution was oxidized at +1.65 V (SCE).

which is detected by ultraviolet light;

$$\text{Intermediate product} \longrightarrow \text{oxidizable species (possibly solvated ArH}\cdot\text{Br}_2\text{);}$$

$$\text{Oxidizable species} \xrightarrow{\text{HOAc}} \text{ArBr} + \text{BrCH}_2\text{COOH} + 2\,\text{H}^+ + 2\,\text{e}$$

The intermediacy of Br_2 in the anodic bromination reaction was confirmed by electrolysing mixtures of Br_2 and aromatic compounds, yielding quite similar results to those obtained upon starting from Br^- (Table 13).

The same authors[55–57] have also shown that bromination can take place even more rapidly and without the involvement of a photochemical step when higher potentials are applied (1.2–1.4 V correspond to the last oxidation wave of the system Br^-/ArH). The similarity in the results obtained by the two bromination routes points to similar mechanisms, except for the initial steps:

$$\text{ArH} \xrightarrow{-1\,\text{e}} \text{ArH}^+,$$

$$\text{ArH}^+ + 2\,\text{Br}^- \xrightarrow{-1\text{e}} \text{ArH}\cdot\text{Br}_2$$

followed by the same steps in acetic acid, as described in the former scheme.

A considerable influence of the substrate concentration on the limiting current of the third wave was found in acetonitrile and this fact was taken into account by postulating[57b] the following steps, after a non-oxidizable species, $[ArH \cdot Br_2]$ is formed:

$$[\text{ArH}\cdot\text{Br}_2] + \text{ArH} \overset{\text{slow}}{\rightleftharpoons} [2\,\text{ArHBr}^\cdot]\ (\sigma\text{-complex}),$$

$$[\text{ArHBr}^\cdot] \xrightarrow{-\text{e}} [\text{ArHBr}^+]\ (\sigma\text{-complex}),$$

$$[\text{ArHBr}^+] \xrightarrow{-\text{e}} \text{ArBr} + \text{H}^+$$

The first step of the three explains the dependence of the voltammetric wave and of the product yield on the substrate concentration. An alternative mechanism was suggested for the formation of aryl bromides, involving the intermediacy of Br^+:

$$\text{Br}^+ + \text{ArH} \longrightarrow [\text{ArHBr}^+]\ (\sigma\text{- complex}),$$

$$[\text{ArHBr}^+] \longrightarrow \text{ArBr} + \text{H}^+$$

In the case of naphthalene bromination[56], a thermal reaction between Br_2 and naphthalene was suggested, due to the relatively high rate of this reaction (4×10^{-3} l mol^{-1} s^{-1}, at 40°C). Since with benzene no bromination took place at 1.0 V (unlike the results obtained for toluene and *p*-xylene), it was concluded that no oxidizable intermediates were formed between Br_2 and benzene during the electrolysis. However, at potentials at which the benzene ring is oxidizable (1.65 V against SCE), bromination did take place (Table 13).

The advantage of using acetonitrile solvent instead of acetic acid was nicely demonstrated[57] in the electrochemical bromination of aromatic hydrocarbons on Pt, where the yields of the brominated products were near 100%, mainly because the solvent was not brominated (Table 14). Furthermore, bromination could occur either mostly on the chain or mostly on the ring, depending on illumination conditions. It was also established, by separate control experiments, that the chain-substitution reaction was mainly due to photo-activation of an oxidized intermediate, e.g. $[ArH \cdot Br_2]^+$, and not due to bromine radicals.

TABLE 14. Electrolysis of Br_2 and aromatic hydrocarbon mixtures at 1.1 V (against SCE), in acetonitrile at 40°C[a]

ArH	Illumination	Percentage ring substitution, %[b]	Percentage chain substitution, %
Toluene	Light	5.7	71.0
	Dark	59.0	40.2
p-Xylene	Light	6.2	66.0
	Dark	54.3	19.4
Benzene	Dark	No reaction	

[a]Data from Ref. 57a.
[b]With respect to charge exchange.

The synthetic usefulness of the anodic bromination of phenol was clearly demonstrated by Gileadi and Bejerano[58]. It is well known that chemical bromination of phenol with Br_2 in aqueous solutions yields tribromophenol quantitatively. Also, *para* and *ortho* bromophenols can be prepared separately in high yields[59]. These authors have shown that electrolysing a mixture of Br^- and phenol afforded either monobromophenol or dibromophenol, as desired, in high yields, avoiding the formation of tribromophenol completely. The results, obtained in highly acidic solutions on platinized platinum or titanium electrodes, showed that *p*-bromophenol was the main product when one equivalent of charge was passed. 2,4-Dibromophenol became the main product when two equivalents of charge were consumed. Other isomers, *o*-bromophenol and 2,6-dibromophenol, were formed in small quantities; *m*-bromophenol was not detected at all. Only 10% of tribromophenol were found at two equivalent charges and none at less than one equivalent charge. Product distribution as a function of charge is demonstrated in Figure 2.

C. Carbon–Chlorine Bond

Anodic chlorination of aromatic compounds in acetic acid has been conducted in the presence of concentrated HCl[60] or LiCl[61]. Table 15 shows that mono-substitution takes place with no side-chain chlorination, but with relatively low overall yield of isolated products. Chlorination of aromatic compounds in the same solvent, via anodic oxidation of Cl[62], has proved to give similar results with almost equal isomeric distributions, suggesting that the same intermediate may be responsible for the halogenation.

Recently, Mastragostino and coworkers[63] have investigated the anodic chlorination of an α,β-unsaturated ester (methyl butenoate) in HOAc–LiCl media. Both addition and substitution products have been formed, the latter being in a higher yield than that obtained from a chemical chlorination of the same ester. This difference is explained by assuming that a chlorocarbo-cation is formed at the electrode surface followed by proton elimination to form a monochloro-substituted product. In the non-electrochemical process, the chlorocarbocation could couple more easily with Cl^- of the nucleophilic solvent.

Regarding the mechanism, the generally accepted reaction scheme[64] is not

$$X^- \longrightarrow X^{\cdot} + e$$

$$RH + X^{\cdot} \longrightarrow RX + H^+ + e$$

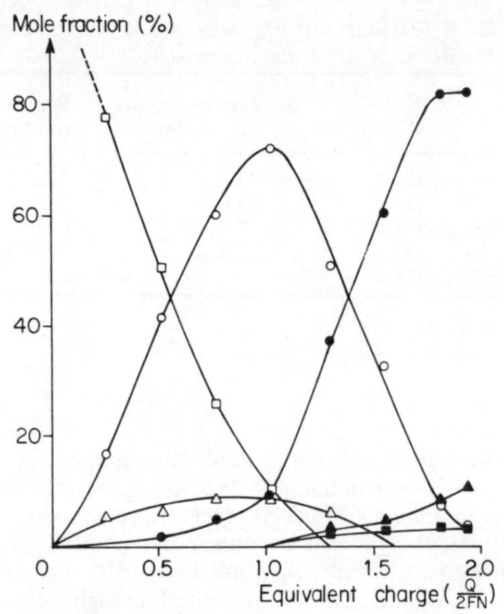

FIGURE 2. Product distribution of bromophenol derivatives as a function of charge equivalents consumed: □, phenol; ○, *p*-bromophenol; △, *o*-bromophenol; ●, 2,4-dibromophenol; ■, 2,6-dibromophenol; ▲, tribromophenol. (Reproduced from Ref. 58.)

supported by experimental results. Other suggested mechanisms[65], e.g. photochemical and thermal chlorination, do, however, explain some of these results.

With regard to the anodic chlorination mechanism, it has been suggested[66] that, under voltammetric conditions, the final product (starting from Cl^-) is Cl_3^-:

$$Cl^- \xrightarrow{-1e} Cl^\bullet$$

$$Cl^\bullet + Cl^- \longrightarrow Cl_2^{\bar{\bullet}}$$

TABLE 15. Anodic chlorination of aromatic compounds in acetic acid, at +1.4 V (against SCE), on Pt, using LiCl[a]

ArH	t, °C	n^b	Current yield, %	Chloro derivatives
Chlorobenzene	40	1.0	–	–
Benzene	40	1.1	15	Chlorobenzene
Benzene	30	1.1	15	Chlorobenzene
Toluene	40	1.9	90	*o*- and *p*-chlorotoluene
Toluene	30	1.7	65	*o*- and *p*-chlorotoluene
p-Xylene	40	2.0	100	Chloro-*p*-xylene
p-Xylene	30	1.9	90	Chloro-*p*-xylene

[a]Data from Ref. 61.
[b]Number of electrons used for chloro-derivative generation.

$$2\,Cl_2^{\cdot -} \longrightarrow Cl_3^- + Cl^-$$

or

$$Cl_2^{\cdot -} + Cl^{\cdot} \longrightarrow Cl_3^-$$

However, in electrolysis, in which a longer time is available, the final product becomes Cl_2:

$$Cl_3^- \xrightarrow{\text{Slow}} Cl_2 + Cl^-.$$

In anodic chlorination, addition of various organic substrates to the solutions containing Cl^- results in an increase in the limiting current of the voltammetric wave of Cl^-. Since none of the aromatic substrates mentioned in Table 15 is oxidized at the potential to which the oxidation wave of Cl^- is displayed, the increase of the oxidation wave is undoubtedly attributable to a reaction between one of the products of the electrodic process ($Cl^{\cdot}$, Cl_2, $Cl_2^{\cdot -}$, Cl_3^- etc.) and the organic substrate to form an oxidizable chemical species, at the potential corresponding to the Cl^- oxidation. Since no chlorinated compound in the side chain was detected (electrolysis has been performed in the dark) the authors exclude the involvement of $Cl^{\cdot}$, Cl_3^- and Cl_2 as responsible for the chlorination. They suggested an overall scheme of the process and singled out $Cl_2^{\cdot -}$ as the halogenating agent:

$$2\,Cl^- \xrightarrow{-2e} 2\,Cl^{\cdot} \xrightarrow{+2\,Cl^-} 2\,Cl_2^{\cdot -} \longrightarrow Cl_3^- + Cl^-$$

$$2\,Cl_2^{\cdot -} \xrightarrow{ArH} 2\,Cl^- + ArHCl_2 \qquad Cl_3^- + Cl^- \longrightarrow Cl_2 + Cl^-$$

$$2\,Cl^- + ArHCl_2 \xrightarrow{-e} \sigma\text{-}[ArHCl]^+ + 1/2\,Cl_2$$

$$\sigma\text{-}[ArHCl]^+ \xrightarrow{\text{Fast}} ArCl + H^+$$

It has also been shown that the chlorination is most probably of the second order with respect to $Cl_2^{\cdot -}$ and of the first order with respect to C_6H_6. In the case of electrochemical chlorination from Cl_2 and the aromatic compounds, it has been suggested[62] that the same intermediate, $ArHCl_2$, is initially formed by chemical reaction between ArH and Cl_2, followed by electrodic oxidation (to form $[ArHCl]^+$) and subsequent fast formation of ArCl. From kinetic results it has been concluded that the formation of the $ArHCl_2$ intermediate from $Cl_2^{\cdot -}$ is much faster than from Cl_2, and suggested that the latter reaction involves the prior formation of a π-complex.

D. Carbon–Fluorine Bond

Electrochemical fluorination of organic compounds has been studied quite extensively and reviewed by Burdon and Tatlow[67] in 1960, Nagase in 1967[68], Weinberg in 1974[69] and Rozhkov in 1976[70] and 1980[71]. The process is not only attractive from an industrial point of view, but also for being one of the mildest methods for obtaining perfluorinated organic compounds. In contrast, chemical fluorination with F_2 is often explosive. Only a brief account of the mechanism involved and new developments in the field will be presented in this section. For more comprehensive details the reader is referred to the above publications.

Electrochemical fluorinations are carried out mainly at a nickel anode or its alloy in anhydrous liquid HF in which the organic substrate is dissolved or suspended. With non-conducting substrate–HF solutions, conductivity additives such as NaF or KF are needed. Completely fluorinated products are obtained upon using nickel electrodes, whereas at a porous carbon anode, in a solution of HF–KF, high yields in partial fluorination have been reported[72]. Partial fluorination generally occurs most successfully with increasing concentration of the substrate[73]. The porous electrode keeps the insoluble feed confined to the pores and does not permit to break out into the bulk of the solution. Porous electrodes have been used for introducing a high concentration of gaseous or low boiling substances[68]. Porous and other carbons have been used extensively in molten inorganic fluorides. Here the anode generally serves as the source of carbon in the product while the inorganic melt provides the fluorine, e.g.[74]

$$\text{Molten salt (NaF:LiF)} \xrightarrow[\text{anode}]{\text{Porous carbon}} CF_4 + C_2F_6 + C_3F_8$$

Usually fluorination takes place with preservation of the original carbon skeleton and functional groups:

$$(CH_3)_2NCH{=}O \longrightarrow (CF_3)_2NCF{=}O$$

All hydrogens in alkanes, amines, ethers, carboxylic acids and their derivatives and in other groups of organic compounds are replaced by fluroine atoms. However, the process is sometimes accompanied by fragmentation[75] between a carbon and a functional group, e.g.

$$CH_3C(=O)Cl \longrightarrow CF_3C(=O)F + CF_4 + CHF_3$$

$$C_6H_5NH_2 \longrightarrow C_6F_{11}NF_2 + C_6F_{12}$$

I and Br are replaced by electrofluorination but chlorine is usually retained in the product, for example

$$Cl-CCl_2(Br)-CCl_2(I)... \quad Cl_2(Br)C{-}CCl_2(I)\ \text{i.e.}\ ClC(Cl)(Br)-C(Cl)(I)Cl \xrightarrow[\text{HF}]{\text{anode}} ClC(Cl)(F)-C(Cl)(F)Cl$$

Upon using a Pt anode in acetonitrile containing fluoride ions, it has been found that electrofluorination may take place by direct anodic oxidation of the organic substrate followed by reaction with the electrolyte. This has been demonstrated for various compounds by several groups[76–79]. In all these cases the reactions are conducted at potentials below that for the discharge of F^- but sufficiently positive for electro-oxidation of the organic compound.

1. Mechanism

Whereas it is fairly established that chemical fluorination (by F_2) is a radical reaction, the electrochemical process appears to be much more complicated. Various mechanisms proposed include the involvement of radicals, high-valent nickel fluorides generated at the anode or complex nickel fluorides. The implication is that there is no general mechanism for all electrofluorinations. let us consider the mechanisms proposed in the literature on electrochemical fluorination, mainly at nickel anodes.

a. Radical mechanism. This mechanism suggests that fluorine radicals are generated electrochemically at the anode as a consequence of the first oxidation step of F^-. Perfluorinated organic compounds are formed by subsequent homolytic substitution of hydrogen atoms. Others suggested that molecular fluorine which is adsorbed on the nickel anode covered with a film of difluoride is involved in the fluorination process. The latter suggestion has been disputed because free F_2 is found during the induction period (but not later) with new Ni anodes[69]. However, at carbon anodes it is likely that an adsorbed fluorine atom or free F_2 is the generated fluorinating agent[72]. At any rate, the radical mechanism has been proved to be oversimplified in explaining many results. The fact that fluorination has been affected by electrode material, its crystalline structure and other parameters[70] points to the possibility that other mechanisms may be involved.

b. Fluorination with high-valent nickel fluorides. According to this mechanism, species such as NiF_3 or NiF_4 are generated at the anode, following initial formation of nickel difluoride. Then an organic molecule reacts with the modified surface of the anode. This mechanism does not provide an acceptable explanation for many facts. For instance, there is no good correlation between the electrochemical results and those obtained from known chemical fluorinating agents[68,70]. Fluorination[81] by Co(III) fluoride involves detachment of functional groups from the organic molecule, whereas the electrochemical process is distinguished by preservation of the functional group of the initial organic material.

c. Fluorination by complex nickel fluorides. In this mechanism, highly conducting complexes of high-valent nickel fluorides and the organic compound, $(RH)_2NiF_6$ and $(RH)_3NiF_6$, are generated at the anode. Direct fluorination occurs on decomposition of these complexes with a resultant weakening of carbon–hydrogen bonds which are attacked by the fluorine radical. Indirect evidence for this scheme was achieved when coloured anions, $[NiF_6]^{3-}$ and $[NiF_6]^{2-}$ were detected[82] in the anode layer during the electrolysis of a solution of KF–HF. Furthermore, it has been previously shown[83] that, by the action of F_2 on mixtures of potassium and nickel chlorides, both K_3NiF_6 and K_2NiF_6 can be formed.

While Weinberg[69] states that the above complexes satisfy most objectives, Rozhkov[70] claims that, in spite of their great attractiveness, they cannot explain certain facts such as the favourable formation of incompletely fluorinated products upon increasing the organic substrate concentration, the achievement of complete perfluorinated products although electron-withdrawing substituents are attached to the organic substrate, the increase in fragmentation of the organic molecule upon lowering the current density, etc.

d. Carbon–fluorine bond formation as a result of ionic fluorination of positive organic species. This mechanism suggests that the organic substrate undergoes direct electron transfer (below the discharge potential of F^-) at the anode to give a cationic species which then reacts with the electrolyte medium.

Since this type of halogenation is out of the scope of this chapter and has already been well surveyed by Rozhkov[70,71], only a few illustrative examples are presented.

(All examples were carried out in acetonitrile on Pt, using R_4NF–HF. Potentials are against SCE except for the last entry.)

Benzene (and substituted benzenes) ⟶ C_6H_5F (40%)[84*]

Naphthalene $\xrightarrow{2.06}$ 1-fluoronaphthalene (27%) and 1,4-difluoronaphthalene (70%)[85*]

Diphenylanthracene $\xrightarrow{1.45}$ 9,10-difluoro-9,10-diphenyl-9,10-dihydroanthracene (75%)[87]

$Ph_2C{=}CH_2 \xrightarrow{2.26} Ph_2CFCH_2F$ (63%)[88]

$Ph_3CH \xrightarrow{1.8} Ph_3CF$ (80%)[89]

2,4,6-$R_3C_6H_2OH$ $\xrightarrow{1.1\ (Ag/Ag^+)}$ 4-fluoro-2,4,6-R_3-cyclohexa-2,5-dienone (65%)[90]

Other mechanisms, involving a loose complex of NiF_2 and F_2 formed at the anode or dissociative chemisorption of the organic substrate followed by reaction with solvent and further electron transfer, have been proposed but very little discussed in the literature[69,91].

Recently, the fluorination of nitrogen-containing compounds has been studied by several groups. Plashkin and Dolnakov[92] have studied the fluorination of *N*,*N*-dialkyl anilines:

$C_6H_5N(CH_3)_2$ ⟶ $C_6F_{11}N(CF_3)_2$ + $CF_3(CF_2)_5N(CF_3)_2$

Plashkin[93] has extended the work to heterocyclic *N*-isoalkylamines and reported that the fluorination is more effective when electrolytic nickel coating has been used than solid nickel anode. In addition to the usual perfluorinated products, other rearranged ones were formed, for example:

*See footnotes in Ref. 86.

$CH(CH_3)_2$-N-piperidine $\longrightarrow$ $CF(CF_3)_2$-N-(F) + $CF_2CF_2CF_3$-N-(F) + CF_2CF_3-N-(F) + C_3F_7-N-(F)-CF_3

(31–48%) (25–32%) (6–12%) (15–32%)

Electrochemical fluorination of alkyl-substituted pyridines affords[94] the corresponding perfluoro-(*N*-fluoroalkylpiperidines) in yields (2–26%) which depend to a large extent on the position and number of alkyl substituents. However, chloropyridines also undergo cleavage during fluorination with the production of NF_3 [95].

The electrochemical fluorination of benzene containing one or two trifluoromethyl groups has been carried out[96] and afforded perfluorocyclohexane derivatives with CF_3 groups in good yields (up to ~50%). The use of 2-trifluoromethylbenzonitriles yielded perfluorodimethylcyclohexanes: the nitrile group was converted into a trifluoromethyl group, and released NF_3.

Fluorination of aliphatic esters gives perfluoro products as well as degradation derivatives with an overall current efficiency of 8–22%[97]. Perfluorination of dithiols as well as cyclization have been performed[98]:

$$HS(CH_2)_4SH \longrightarrow SF_5(CF_2)_4SF_5 + (F_2)_4\text{-ring-}SF_4$$

Other examples of cyclization during fluorination include diols, $X(CH_2CH_2CH_2OH)_2$, where X is CH_2, O, S, NH, NMe or NEt. These are converted into perfluorinated heterocycles which in part undergo cleavage to form open-chain products[99].

E. Carbon–Thiocyanate and Carbon–Selenocyanate Bonds

Many examples of producing vicinal dithiocyanates by addition of thiocyanogen (generated by chemical methods[100]) to olefins have been reported[101]. Surprisingly, little attention has been paid to the electrochemical route to vicinal dithiocyanates, proceeding via the anodic preparation of thiocyanogen.

It has been found that pseudohalide anions (SCN^-, $SeCN^-$) can be converted to the corresponding pseudohalogens (via the intermediacy of X_3^-) by electrochemical oxidation on Pt anode in acetonitrile[102–103]. The synthetic utility of thiocyanation of both activated and non-activated olefins in acidic media[104] and in acetonitrile[105] is shown in Table 16.

The effect of solvent and irradiation was briefly outlined[104]. The addition of thiocyanogen to styrene was very slow in methanol and no dithiocyanate addition product could be detected. However, when the solution was irradiated during electrolysis in methanol, two main products were obtained, 1-phenyl-1,2-dithiocyanatoethane (81%) and 1-phenyl-1-methoxy-2-thiocyanatoethane (13%) as well as other unidentified products. It was established that the minor identified product was formed in a photochemical reaction from the major product. This was confirmed by irradiating a solution of the latter compound in methanol. The same phenomenon repeated itself in acetonitrile; no reaction took place without irradiation and only the dithiocyanate addition product (80%) was formed after irradiation. Unlike the former solvents, the addition of thiocyanogen to styrene was reasonably fast in glacial acetic acid and no

TABLE 16. Thiocyanation of olefins

Substrate	Solvent	Electrochemical conditions	Yield, %	Types of product, %: —C(SCN)—C(SCN)—	—C(NCS)—C(SCN)—	—C(OH)—C(SCN)—	Ref.
n-Octene	Acetic acid	*a*	80[c]	100			104
Butadiene	Acetic acid	*a*	66[c]	100			104
Butadiene	EtOH–H_2O	*b*	42[c]	1,4-Adduct			112
Cyclohexene	Acetic acid	*a*	57[c]	84	16		104
Oleic acid	Acetic acid	*a*	60[c]	100			104
Styrene	Acetic acid	*a*	67[c]	100			104
Styrene	Acetic acid	*b*	92[c]	100			112
Styrene	Acetic acid + HCl	*a*	61.5[c]	100			104
trans-2-Butene	Acetonitrile	*b*	25[d]	85	15		105b
cis-2-Butene	Acetonitrile	*b*	40[d]	70	30		105b
Ethoxyethylene	Acetonitrile	*b*	75[d]		100		105b
n-Butoxyethylene	Acetonitrile	*b*	80[d]		100		105b
Tetramethylethylene	Acetonitrile	*b*	80[d]	70	30		105b
Tetramethylethylene	Acetonitrile + H_2O	*b*	80[d]	40	23	37	105b
2,4,6-Trimethylstyrene	Acetonitrile	*b*	80[d]	65	55		105b
2,4,6-Trimethylstyrene	Acetonitrile + H_2O	*b*	80[d]	65	28	7	105b

[a]Rotating platinum anode, undivided cell, constant current density.
[b]Three-compartment cell with Pt anode, controlled potential electrolysis (0.7 V against Ag/0.01 M Ag^+), 0.1 M $LiClO_4$ as electrolyte.
[c]Current yield.
[d]Total product yield. Similar product mixture obtained for selenocyanation.

photoinitiation was necessary. However, it was found that thiocyanation of double bonds with electron-withdrawing groups, e.g. vinyl chloride, was effective in acetic acid only upon irradiation.

Electrochemical thiocyanation of aromatic compounds has been known for several decades[105a]. Recently Cauquis and Pierre[105b] have studied the thiocyanation of activated olefins in acetonitrile, which all produced a mixture of two isomers as demonstrated in Table 16. These results show the ambident character of the ion SCN^- in aprotic medium and the strong nucleophilic nature of the nitrogen in the ion[106], leading at least partially also to isothiocyanation.

On addition of water to the electrolysis mixture it was found that hydroxylation competes with thiocyanation and products containing SCN and OH groups were also formed.

The first electrochemical selenocyanation of organic compounds was conducted on substituted anilines[107] and the reported yields were rather poor (10–20%). Cauquis and Pierre[108] attributed these low yields to the presence of water, which is known to react with $(SCN)_2$ and $(SeCN)_2$[102b,103a]. These authors extended the study of both thiocyanation and selenocyanation of substituted anilines and phenol under improved conditions (pseudohalide anions were oxidized in the presence of the aromatic substrate at Pt anode in acetonitrile at −10°C) and the yields went up to 70–80% (Table 17).

1. Mechanism

Among the various literature reports on thiocyanation of double bonds two possible mechanisms have been suggested.

(i) A radical one which involves the electrochemically generated $SCN^\bullet$ and/or photodissociation of the pseudohalogen $(SCN)_2$ [109]. This mechanism is supported by the fact that it has been shown in several cases and by different authors that thiocyanation in benzene, acetonitrile and methanol could be accelerated by light[101a,104,111].

TABLE 17. Thiocyanation and selenocyanation of aromatic compounds in acetonitrile[a]

Substrate	Applied anodic potential, V[b]	Yield of substitution products, %	
		4-SCN	4-SeCN
Aniline	0.40	78	
	0.25		70
N-Methylaniline	0.35	65	
	0.25		62
N,N-Dimethylaniline	0.20		70
N-Ethylaniline	0.35	55	
	0.25		72
N,N-Diethylaniline	0.20		65
o-Toluidine	0.35	70	
	0.25		75
m-Toluidine	0.35	62	
	0.25		68
Phenol[c]	0.60	70	

[a]Data from Ref. 108. Potentials are against Ag/0.01 M Ag^+. 0.1 M $LiClO_4$ as electrolyte.
[b]All working potentials were 200–300 mV below the $E_{1/2}$ of the organic substrate.
[c]This compound reacts very slowly with selenocyanogen.

(ii) An ionic mechanism was also suggested since not all results could be explained by the radical mechanism. For instance, electrolysing SCN^- in acetonitrile in the dark affords $(SCN)_2$ as the initial reactant towards olefins. It was previously shown[102b] that SCN^- oxidizes at lower potential than the olefins studied, and therefore any radical mechanism is excluded under these circumstances. Furthermore, addition reactions are generally slow and their duration is incompatible with the life-time existence of SCN radicals in solution due to their known instability[102b]. More support for the ionic mechanism lies in the fact that thiocyanation can occur in acidic media in good yield and without any effect of irradiation[104].

The supporters of the ionic mechanism suggest a complex formation between thiocyanogen and the double bond of the alkene as the first step in this electrophilic addition, followed by a formation of a cyclic cyanosulphonium ion (in analogy to halonium ions formation[111]) and then opening by SCN^- or other anions present to yield products:

$$SCN^- \xrightarrow{-e} SCN^\bullet \longrightarrow (SCN)_2$$

$$>C{=}C< + (SCN)_2 \longrightarrow \text{[}C{\cdots}C\text{]}{\cdots}\overset{\delta^+}{S}(CN){-}\overset{\delta^-}{SCN}$$

$$\text{[}C{\cdots}C\text{]}{\cdots}\overset{\delta^+}{S}(CN){-}\overset{\delta^-}{SCN} \longrightarrow \text{[}C{-}C\text{]}{>}\overset{+}{S}CN + [S{=}C{=}N^- \longleftrightarrow {}^-S{-}C{\equiv}N]$$

$$\text{[}C{-}C\text{]}{>}\overset{+}{S}CN + [SCN]^- \longrightarrow {-}\underset{NCS}{C}{-}\underset{SCN}{C}{-} + {-}\underset{NCS}{C}{-}\underset{NCS}{C}{-}$$

Table 16 shows that in acetic acid or in acetonitrile + water, the amount of products containing C–NCS vanished or went down. It has been suggested[104,105,112,113] that in protic solvents hydrogen bonds are formed with the nitrogen atom of the $[SCN]^-$ derived from the heterolytic fission, which decrease both its nucleophilic character and, consequently, the formation of the isothiocyanate isomers.

de Klein[104] had also studied the acid effect on the thiocyanation reaction and suggested that the acid catalyses the heterolytic fission of the S—S bond in the transition state:

$$\text{[}C{\cdots}C\text{]}{\cdots}S(CN){\cdots}SC\overset{+}{N}H$$

The main product in methanol–HCl was found to be 1-phenyl-1-methoxy-2-thiocyanatoethane along with 1,2-dithiocyanato-1-phenylethane as a side product. This result can be accounted for by the above hypothesis. Further support for the latter can be derived from the experimental result that the yield of addition product of thiocyanogen in acetic acid decreases markedly (from 67% to

15%) when potassium acetate is added as a base, decreasing the acidity of the solution and hence the catalytic effect.

F. Carbon–Azide Bond

The anodic discharge of the azide ion to nitrogen molecules occurs according to the reaction in the following scheme,

$$2\,N_3^- \xrightarrow{-e} 2\,N_3^{\bullet} \longrightarrow N_6 \longrightarrow 3\,N_2$$

has been studied in aqueous solutions on Pt anode by several authors[114] and was reported as the basis of an analytical method for the determination of the azide ion[115].

Azide ion has also been oxidized in aqueous alkaline solution in the presence of nitroalkanes at platinum on nickel anode. It has been postulated[116] that *gem*-azidonitroalkanes can be prepared in good yields via initially formed azide radicals, according to the following scheme:

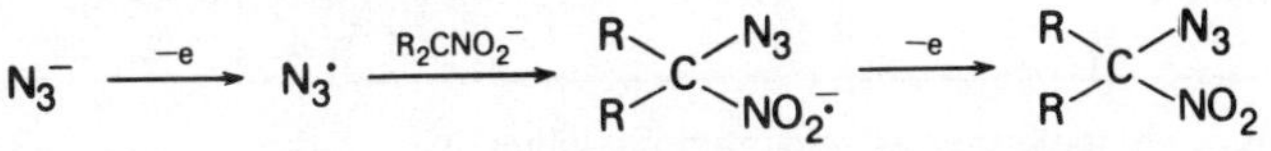

$$N_3^- \xrightarrow{-e} N_3^{\bullet} \xrightarrow{R_2CNO_2^-} R_2C(N_3)NO_2^{\bullet -} \xrightarrow{-e} R_2C(N_3)NO_2$$

The generation of azide radicals in organic solvents and investigation of their addition to olefins for the syntheses of substituted hydrocarbons has been advanced by Schäfer and coworkers[117,118]. Intermediacy of azide radicals has been postulated since the gas evolution at the anode is strongly decreased in the presence of olefins:

$$N_3^{\bullet} + {>}C{=}C{<} \longrightarrow {\bullet}\overset{|}{\underset{|}{C}}-\overset{|}{\underset{|}{C}}-N_3$$

$$N_3-\overset{|}{\underset{|}{C}}-\overset{|}{\underset{|}{C}}-\overset{|}{\underset{|}{C}}-\overset{|}{\underset{|}{C}}-N_3 \quad \text{(dimer)}$$

$${\bullet}\overset{|}{\underset{|}{C}}-\overset{|}{\underset{|}{C}}-N_3 \longrightarrow {+}\overset{|}{\underset{|}{C}}-\overset{|}{\underset{|}{C}}-N_3 \xrightarrow{N_3^-} N_3-\overset{|}{\underset{|}{C}}-\overset{|}{\underset{|}{C}}-N_3 \quad \text{(monomer)}$$

$$N_3-\overset{|}{\underset{|}{C}}-\overset{|}{\underset{|}{C}}-H + {>}C{=}C{<}_{CH_2N_3} \quad \text{(allylic substitution)}$$

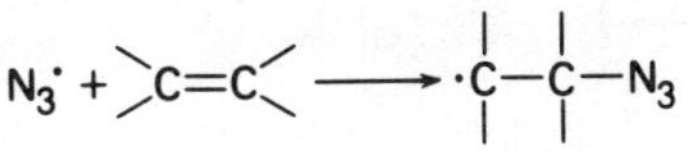

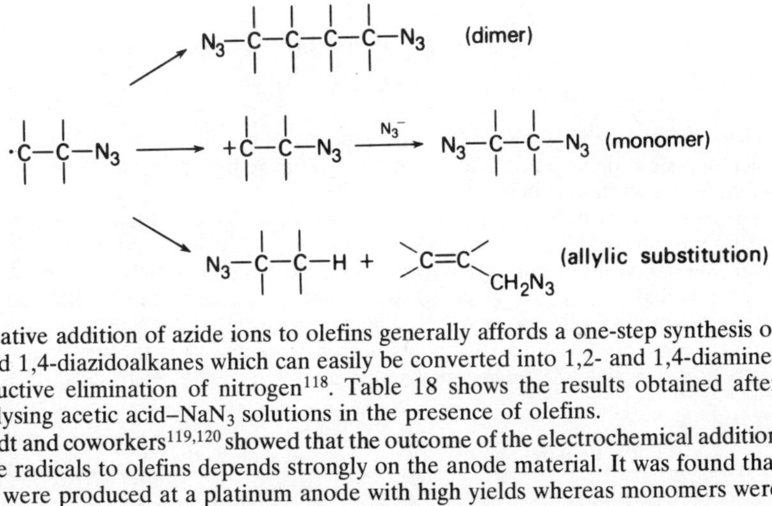

Oxidative addition of azide ions to olefins generally affords a one-step synthesis of 1,2- and 1,4-diazidoalkanes which can easily be converted into 1,2- and 1,4-diamines by reductive elimination of nitrogen[118]. Table 18 shows the results obtained after electrolysing acetic acid–NaN_3 solutions in the presence of olefins.

Wendt and coworkers[119,120] showed that the outcome of the electrochemical addition of azide radicals to olefins depends strongly on the anode material. It was found that dimers were produced at a platinum anode with high yields whereas monomers were especially produced on a carbon anode. This behaviour was attributed to the poor adsorbtion of the radicals at platinum, which gives rise to formation of dimers, relative to the strong radical adsorption at carbon anodes at which the formation of monomers is preferred. In the latter case, the radicals which are produced are generally

TABLE 18. Azidoalkanes obtained on electrolysing N_3^- in glacial acetic acid on Pt anode at +1.6 V (against Ag/AgCl) in a non-divided cell, at 40°C[118]

Substrate	Yield, %[a]	Products
Styrene	57	$[N_3CH_2CH(Ph)]_2$
α-Methylstyrene	45	$[N_3CH_2C(Ph)(Me)]_2$
Cyclooctene	24	1,2-Diazidocyclooctane Azidocyclooctane 3-Azidocyclooctene
1-Octene	17	$CH_3(CH_2)_5CH(N_3)CH_2N_3$ $CH_3(CH_2)_6CH_2N_3$ $CH_3(CH_2)_4CH{=}CHCH_2N_3$ $CH_3(CH_2)_4CH(N_3)CH{=}CH_2$
1,1-Diphenylethylene	15	$Ph_2C(N_3)CH_2N_3$
	33	$Ph_2C(OAc)CH_2N_3$
Cyclohexene	18	1,2-Diazocyclohexane Azidocyclohexane 3-Azidocyclohexene

[a]Yield of products referred to current consumption.

immediately oxidized to carbocations at the applied anodic potential:

$$>C{=}C< + N_3^{\bullet} \longrightarrow \left(\cdot\overset{|}{\underset{|}{C}}-\overset{|}{\underset{|}{C}}-N_3\right)_{ads}$$

$$\left(\cdot\overset{|}{\underset{|}{C}}-\overset{|}{\underset{|}{C}}-N_3\right)_{ads} \xrightarrow{-e} +\overset{|}{\underset{|}{C}}-\overset{|}{\underset{|}{C}}-N_3$$

A nice example of the effect of electrode material on the anodic $N_3^{\bullet}$ addition to styrene in acetonitrile is demonstrated in Table 19. Yields are high (70–100%) provided that styrene concentration is high enough (> 0.1 M).

The electrode kinetics and the kinetics of the consecutive addition to olefins as well as the conversion of $N_3^{\bullet}$ to N_6 has been studied thoroughly by Wendt and coworkers[119]. Their investigation was conducted at different electrode materials, namely platinum, graphite and glassy carbon. Table 20 shows that the rates of addition of $N_3^{\bullet}$ to olefins are slowest on platinum anode and are equal or greater than the homogeneous kinetic diffusion-limited value at carbon. This is clear evidence that the scavenger radical reaction is essentially heterogeneous and is catalysed by the electrode surface.

TABLE 19. Products obtained from addition of azide to styrene[120]

	Dimers	Monomers
Pt	$[CH(Ph)CH_2N_3]_2$	$PhCH(N_3)CH_2N_3$
C	$PhCH(N_3)CH_2CH(Ph)CH_2N_3$ (trace)	$PhCH(N_3)CH_2N_3$ $PhCH{=}CHN_3$ $PhCH(NHCOCH_3)CH_2N_3$

TABLE 20. Rate constants for the addition of $N_3^{\bullet}$ to alkenes at various anodes[a]

	Rates (M^{-1} s^{-1}) measured in acetonitrile		
Alkene	Pt	GC[b]	C
1-Octene	2.2×10^5	3.0×10^5	1.1×10^9
Cyclohexene	6.0×10^5	1.5×10^6	4.9×10^9
cis-Stilbene	1.0×10^5	7.5×10^8	8.4×10^9
trans-Stilbene	4.0×10^7	4.5×10^8	2.1×10^{11}
Styrene	2.0×10^7	1.5×10^8	7.0×10^{10}
α-Methylstyrene	4.0×10^7	7.5×10^8	1.4×10^{11}
1,1-Diphenylethylene	6.0×10^7	6.0×10^8	2.1×10^{11}
1,4-Diphenyl-1,3-butadiene	4.0×10^8	1.5×10^{10}	6.4×10^{12}

[a]Data from Ref. 119.
[b]GC = glassy carbon.

G. Carbon–Isocyanate Bond

To date very little has been published on the electrochemistry of isocyanates, and even this is mostly confined to polarographic data[121]. Cauquis and Pierre[122] have studied the anodic oxidation of isocyanate anion and suggested, along with another group[123], that Pt catalyses the decomposition of the unstable $(OCN)_2$ to oxygen and cyanogen[124]. according to the following scheme:

$$OCN^- \xrightarrow{-e} OCN^{\bullet}$$

$$2\ OCN^{\bullet} \longrightarrow (OCN)_2$$

$$(OCN)_2 \longrightarrow (CN)_2 + O_2$$

Electrochemical isocyanation of organic compounds is also a rare reaction in the electrochemical literature. In fact, there is one example[125] on the formation of 1-naphthylisocyanate by anodic oxidation of naphthalene in molten ammonium cyanate.

IV. INDIRECT CLEAVAGE OF THE C—X BOND BY ELECTROCHEMICALLY GENERATED MEDIATORS

The field of homogeneous electron transfer to organic substrates by electrochemically generated mono- and doubly-charged species has received increasing attention in recent years. The major importance of such an indirect electrolysis lies in the capability of oxidizing or reducing a substrate at a potential below the one required for the direct electrochemical process. In certain cases, the direct process is not feasible at potentials accessible in organic solvents. For instance, the reductive cleavage of alkyl fluorides can take place only through homogeneous electron transfer (see Chapter 5 in Ref. 5).

There are numerous examples in which an electron carrier between the electrode and the substrate acts as a catalyst[126–136]. These reported examples include both homogeneous reductions and oxidations of organic substrates, usually catalysed by organic redox couples, e.g. an aromatic hydrocarbon and its corresponding anion or cation radical. Other types of electron transfer mediators have been reported too, and include, for instance, transition metal ions[137–140]. Detailed kinetic analyses of homogeneous redox catalyses of electrochemical reactions by various types of mediators have been published quite recently[140,141]. As will be shown later in this section, not all electrochemically generated species react in a catalytic manner and some of the products are due to S_N2 type displacement or nucleophilic addition. The discussion will be confined exclusively to reactions involving the cleavage or formation of a carbon–halogen bond.

A. Reactions of Alkyl Halides with Electrogenerated Organic Species

1. Anion radicals and dianions

The reaction between aromatic anion radicals (formed electrochemically or from an alkali metal and aromatic hydrocarbon) and an alkyl halide has been the subject of many studies[142–148]. Naphthalene anion radical has been a favourite reactant and several types of alkyl halides, including primary alkyl fluorides[148], have been included in these studies. The results are interpreted in terms of electron transfer from the aromatic anion radical to the alkyl halide:

$$ArH^{\overline{\cdot}} + RX \longrightarrow ArH + R^{\cdot} + X^{-} \quad (ArH = \text{naphthalene})$$

The aliphatic radical may react with an aromatic anion radical, either by coupling or electron transfer or both:

$$R^{\cdot} + ArH^{\overline{\cdot}} \xrightarrow{e\ \text{transfer}} ArH + R^{-} \xrightarrow{H^{+}} RH$$

$$R^{\cdot} + ArH^{\overline{\cdot}} \xrightarrow{\text{Coupling}} \text{[1-R-1-H-dihydronaphthalenyl anion]}$$

These competing reactions depend on the structure and nature of the aliphatic halide.

The results obtained by Garst and coworkers[146] ruled out the possibility that alkylated products are to be formed by a nucleophilic displacement mechanism, since they were found not to be affected much by extreme variation in the nature of the halogen (iodine through fluorine):

$$ArH^{\overline{\cdot}} + RX \not\longrightarrow [ArHR]^{\cdot} + X^{-}$$

They suggested that alkylated products may be formed by reacting $R^{\cdot}$ or R^{-} with an aromatic hydrocarbon, in addition to the reaction between $R^{\cdot}$ and $ArH^{\overline{\cdot}}$. A further study by Sargent and Lux[147] threw light on this point and it is now generally agreed that regardless of the nature of either the alkyl fragments or the halogen atom, all alkyl halides react with $ArH^{\overline{\cdot}}$ via initial electron transfer and yield alkylation products only as a result of radical–radical anion combination:

$$ArH^{\overline{\cdot}} + RX \longrightarrow ArH + [RX]^{\overline{\cdot}}$$

$$[RX]^{\overline{\cdot}} \longrightarrow R^{\cdot} + X^{-}$$

$$ArH^{\overline{\cdot}} + R^{\cdot} \longrightarrow \text{[1-R-1-H-dihydronaphthalenyl anion]}$$

The carbanion-coupled product may undergo a protonation and/or another coupling reaction with an alkyl halide to form dialkylated aromatic hydrocarbon:

H^+

RX

The alkyl free radicals, generated after transferring electron from naphthalene anion radical to alkyl iodides, undergo combination to dimers, and this reaction was found to compete favourably with hydrogen abstraction, disproportionation and further electron transfer to form a carbanion:

$$RCH_2CH_2^{\bullet} \longrightarrow \text{A dimer}$$

$$RCH_2CH_2^{\bullet} \xrightarrow{\text{Disp.}} RCH{=}CH_2 + RCH_2CH_3$$

$$RCH_2CH_2^{\bullet} \xrightarrow{\text{SH}} RCH_2CH_3$$

$$RCH_2CH_2^{\bullet} \xrightarrow{\text{ArH}^{\bullet-}} \text{ArH} + RCH_2CH_2^-$$

Unlike alkyl iodides, bromides and chlorides[143] yielded substantial quantities of hydrocarbon products which could not arise from radical–radical or radical–solvent reactions. It has been suggested that carbanions were the immediate precursors for the major hydrocarbon products.

Lund, Michel and Simonet[128] have studied the electron-transfer reactions between electrochemically generated anion radicals of various aromatic and heterocyclic compounds and less easily reducible compounds (halobenzenes, azides, a sulphonate, and sulphonamide) by means of classical polarography and cyclic voltammetry[132]. Typical examples of 'catalytic currents' in polarography and in cyclic voltammetry are demonstrated in Figures 3 and 4, respectively. These authors calculated the rate of homogeneous electron transfer between a variety of mediators and substrates and the results are shown in Table 21. Clearly, there is a dependence between these rate constants and ΔE (the difference between half-wave potentials of a mediator and a substrate) and the smaller this difference, the higher the rate constant for a given substrate. On the basis of their available results they concluded that the rate of the electron exchange between the mediator and the substrate is the limiting factor when the latter decomposes fast. However, the poorer the leaving group attached to the substrate, the more significant becomes the rate of the decomposition of the substrate.

Since an anion radical transfers only one electron at a time to a molecule outside the electrical sphere of the double layer, at which the electrode has no influence on the stereochemistry of the product, whereas an electrode can transfer two electrons at a

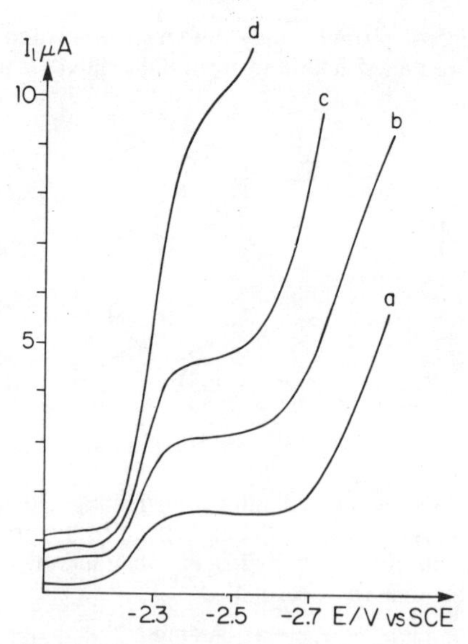

FIGURE 3. Polarograms of chrysene (4×10^{-4} M) in DMF containing tetrabutylammonium iodide and various concentrations of bromobenzene. (a) 0 M; (b) 5×10^{-4} M; (c) 1.5×10^{-3} M; (d) 4×10^{-3} M. (Reproduced from Ref. 128.)

time, a stereochemical influence may be observed in the latter case. An attempt[128] to test such a possibility was carried out for the reduction (Table 21) of *d,l*- and *meso*- α,α'-dichlorostilbenes by means of both anion radicals and directly at the electrode. The results from both cases gave *trans*-stilbene preferentially, suggesting that two one-electron steps are involved. However, direct electrolysis of *d,l*- α,α'-dibromostilbene gave 40–60% *cis*-stilbene and indirect electrolysis yielded only 5% *cis*-stilbene. Probably in the latter case, the intermediate has more time to rotate about the central bond due to a larger time gap between the two electron transfers.

Sease and Reed[130] studied the catalytic electrochemical reduction of alkyl halides via anion radicals of aromatic hydrocarbons and found it to be limited only to alkyl chlorides. They explain this unusual behaviour by the lower polarity of the C—Br bond and the more positive electrode potential (relative to alkyl chlorides), which is closer to the electrocapillary maximum. A failure to reduce alkyl fluorides by electrochemically generated organic mediators is attributed to the presence of tetraalkylammonium cation, which was previously found, by Garst and Barton[148], to decrease strongly the rate of electron transfer. They reported that rates of electron transfer between aromatic radical anions and alkyl halides are highly dependent upon the nature of the cation associated as counterion with the radical anion and the extent of ion pairing in the system[151]. Electron transfer from a tight ion pair to a molecule is slower than from a loose ion pair or a free radical anion[152]. For instance, one can see from Table 21 that the rate of the reaction between sodium naphthalide and *n*-hexyl

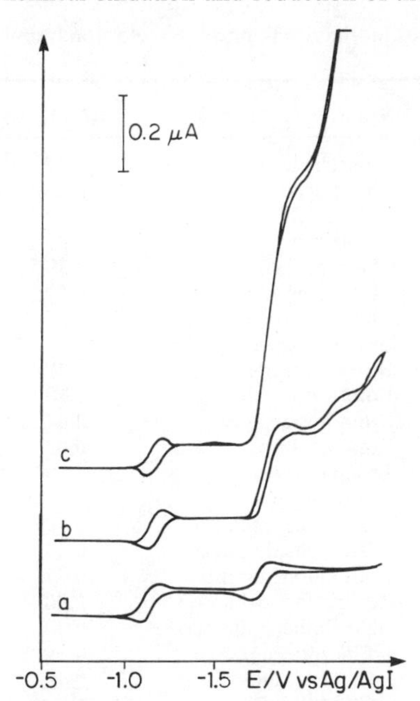

FIGURE 4. Cyclic voltammograms of perylene (3.3×10^{-3} M) in DMF/0.1 M tetrabutylammonium iodide at 10 mV/s^{-1} sweep rate in the presence of different concentrations of 1,4-dichlorobenzene. (a) 0 M; (b) 4.5×10^{-3} M; (c) 3.5×10^{-2} M. (Reproduced in part from the first article in Ref. 132.)

chloride in dimethoxyethane (in which the sodium and naphthalene radical anions exist as loose ion pairs), is higher than that observed for the same chloride upon electrogeneration of napththalene anion radical, in the presence of Et_4N^+ as cation. The reduction of the supporting electrolyte (amalgam formation) sometimes hinders the reduction of a difficulty reducible substrate. Contrastingly, it was found[129] that anthracene anion radical transfers an electron to 2-chloropyridine in the presence of Li^+, although the polarographic wave of this cation masks the reduction of 2-chloropyridine.

In a different case[129], when tetrabutylammonium iodide (TBAI) was used as supporting electrolyte, an electron was transferred from the anion radical of 2-methylnaphthalene to the cation to form the tetrabutylammonium radical. The latter decomposes slowly into tributylamine and butyl radical, which may couple with the anion radical to form butylated methyltetrahydronaphthalene.

Lund and Simonet[131] have studied the indirect electroreduction process also by cyclic voltammetry and controlled potential electrolysis. Their results are in agreement with the findings by chemical methods[142–148], that the coupiing between an anion radical from an aromatic hydrocarbon and an alkyl halide occurs between the anion radical and the aliphatic radical. From a preparative point of view, there are two

TABLE 21. Rate constants for the reaction between electron transfer mediators (M) and alkyl and aryl halides (S)

M	S	ΔE, V	k, $M^{-1} s^{-1}$ [a]	Ref.[b]
Biphenyl	Chlorobenzene	0.27	2.3×10^4	128
1-Methylnaphthalene	Chlorobenzene	0.28	1.2×10^4	128
Naphthalene	Chlorobenzene	0.32	5.0×10^3	128
2-Methylphenanthrene	Chlorobenzene	0.37	1.6×10^3	128
Phenanthrene	Chlorobenzene	0.39	1.2×10^3	128
Terphenyl	Chlorobenzene	0.52	1.7×10^1	128
Benzonitrile	Chlorobenzene	0.55	0.8×10^1	128
Chrysene	Bromobenzene	0.36	4.0×10^3	128
Methyl benzoate	Bromobenzene	0.39	1.3×10^2	128
Diethyl fumarate	Bromobenzene	1.10	0.0	149
Ethyl cinnamate	Bromobenzene	0.64	0.46	149
Anthracene	Bromobenzene	0.42	0.45×10^1	149
trans-Stilbene	Bromobenzene	0.27	1.8×10^3	149
Biphenyl	Fluorobenzene	0.32	9.0×10^1	128
Quinoxaline	1-Bromonaphthalene	0.39	1.5×10^1	128
Acridine	1-Bromonaphthalene	0.45	0.5×10^1	128
Quinoxaline	9-Bromophenanthrene	0.32	2.7×10^1	128
Benzophenone	*d,l*-Stilbene dichloride	0.30	3.0×10^3	128
Benzophenone	*meso*-Stilbene dichloride	0.28	9.0×10^3	128
Naphthalene	1-Chlorohexane	>0.39	3.2×10^2	130
Naphthalene	1-Chlorohexane	>0.40	1.08×10^3	151[c]
Anthracene	1-Chlorohexane	>1.0	6.6×10^{-2}	151[c]
Biphenyl	1-Chlorohexane	>0.3	0.32×10^1	151[c]
Biphenyl	1-Fluorohexane	–[e]	1.4×10^{-4}	148[d]
Naphthalene	1-Fluorohexane	–[e]	2.2×10^{-4}	148[d]
Phenanthrene	1-Chlorohexane	>0.43	4.5×10^2	130, 150
Phenanthrene	1-Chlorooctane	>0.43	4.9×10^2	130, 150
Diethyl fumarate	Butyl chloride[f]	1.50	0.0	149
Ethyl cinnamate	Butyl chloride[f]	1.04	0.15	149
Anthracene	Butyl chloride[f]	0.92	0.25	149
trans-Stilbene	Butyl chloride[f]	0.67	3.0×10^1	149
Diethyl fumarate	Butyl bromide[f]	1.01	0.47	149
Ethyl cinnamate	Butyl bromide[f]	0.55	3.0×10^2	149
Anthracene	Butyl bromide[f]	0.43	1.5×10^3	149
trans-Stilbene	Butyl bromide[f]	0.17	1.9×10^4	149
Diethyl fumarate	Benzyl chloride	0.86	3.0×10^1	149
Ethyl cinnamate	Benzyl chloride	0.40	5.5×10^3	149
Anthracene	Benzyl chloride	0.28	2.2×10^4	149
Diethyl fumarate	Allyl chloride	0.84	1.2×10^1	149
Ethyl cinnamate	Allyl chloride	0.38	1.5×10^3	149
Anthracene	Allyl chloride	0.26	5.8×10^3	149
Diethyl fumarate	Allyl bromide	0.18	3.7×10^3	149

[a]Error in the range 5–15%.
[b]All measurement in Ref. 149 were carried out in DMSO; DMF was used in other experiments.
[c]In dimethoxyethane.
[d]In THF.
[e]Not known.
[f]Presumably they are all primary halides since there is no specification in Ref. 149.

aspects: catalytic reduction and reductive coupling. The electrocatalytic reduction of alkyl halides has the advantage that it takes place with much less overvoltage than is required for the direct reduction at the electrode. The coupling reaction usually gives a mixture of alkylated, partly hydrogenated products, and competes with other reactions which form aliphatic hydrocarbons. In spite of the disadvantage of forming a complex mixture (see below), inaccessible compounds may sometimes be formed, e.g. 1-*t*-butylpyrene is the main product from electrolytic reduction of pyrene in DMF–TBAI in the presence of *t*-butyl chloride[153] (other conventional ways of preparing it were unsuccessful). Some preparative examples are demonstrated below:

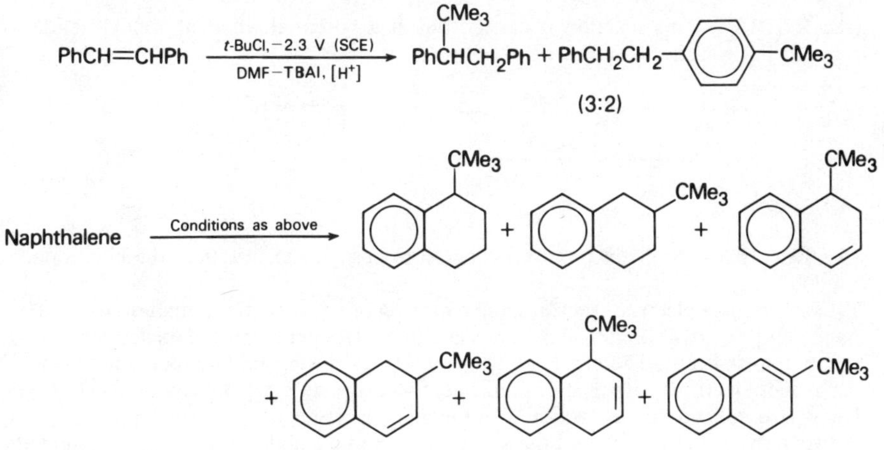

$$\text{PhCH=CHPh} \xrightarrow[\text{DMF-TBAI, [H}^+]]{t\text{-BuCl, }-2.3\text{ V (SCE)}} \text{PhCH(CMe}_3\text{)CH}_2\text{Ph} + \text{PhCH}_2\text{CH}_2\text{-C}_6\text{H}_4\text{-CMe}_3 \quad (3:2)$$

Naphthalene $\xrightarrow{\text{Conditions as above}}$ (*t*-butyl-substituted tetrahydro- and dihydronaphthalenes, CMe₃)

(see Ref. 155);

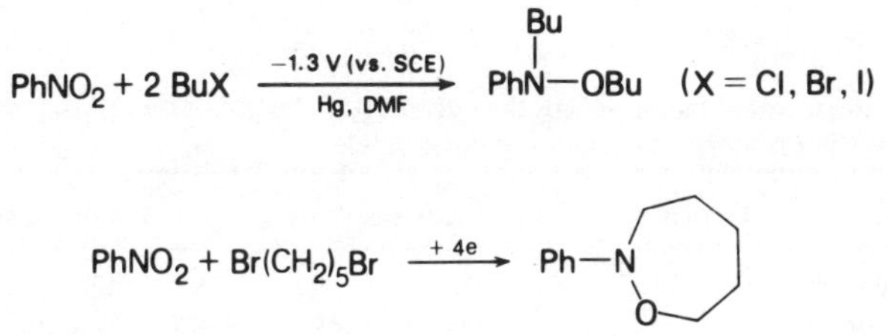

$$\text{PhNO}_2 + 2\text{ BuX} \xrightarrow[\text{Hg, DMF}]{-1.3\text{ V (vs. SCE)}} \text{PhN(Bu)—OBu} \quad (\text{X} = \text{Cl, Br, I})$$

$$\text{PhNO}_2 + \text{Br(CH}_2)_5\text{Br} \xrightarrow{+4e} \text{Ph—N}\langle\text{(CH}_2)_5\text{O}\rangle$$

(see Ref. 156);

$$\text{O=C—C=O} \xrightarrow[\text{DMF}]{2e^-,\text{ RX}} \text{RO—C=C—OR} + \text{RO—C=C—OR} + \text{O=C—C(R)—OR} + \text{O=C—C(R)—OH}$$

$$\text{PhN=C—C=NPh} \xrightarrow[\text{DMF}]{2e^-,\text{ RX}} \text{PhNR—C=C—RNPh} + \text{PhNR—C=C—RNPh} +$$

$$\text{PhNR—C=C—NHPh} + \text{PhNH—C=C—PhNH}$$

cis and *trans*

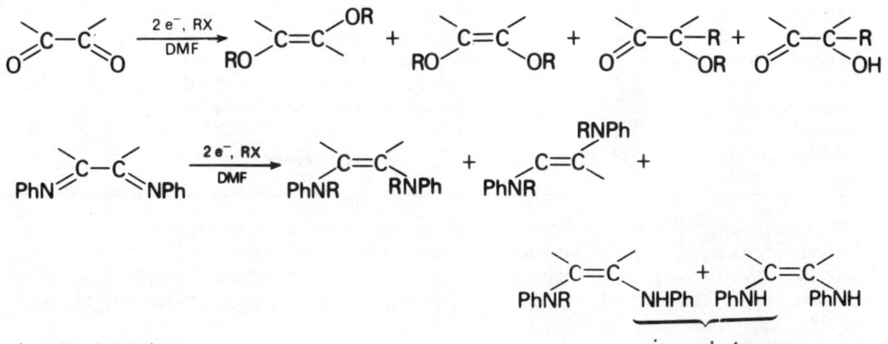

(see Ref. 154);

$$\text{PhC}\equiv\text{CH} \xrightarrow[\text{HMPA-TBAI}]{\text{Pt cathode, RX}} \text{PhC}\equiv\text{CR} \quad (\text{R} = \text{Me, Et, } n\text{-Bu})$$

(see Ref. 157 – this reaction is limited only to alkyl halides which are more easily reducible than phenylacetylene);

$$\text{PhCH}_2\text{N}{=}\text{C(Me)}{-}\text{COOR}' \xrightarrow[\text{Bu}_4\text{N}^+,\ \text{DMF}]{2\,e^-,\ \text{R}''\text{X}} \text{PhCH}_2\text{NHC(Me)(R}'')\text{COOR}' \xrightarrow[\text{2. H}_2/\text{Pd-C}]{\text{1. H}^+\text{ hydrolysis}} \text{R}''{-}\text{C(Me)(NH}_2){-}\text{COOH}$$

(see Ref. 158 – this reaction is carried out at a potential where the alkyl halide is reduced but not the Schiff base);

$$\text{R—S—S—R} \xrightarrow[\text{DMF}]{2\,e^-,\ \text{LiCl}} 2\,\text{RS}^- \begin{cases} \xrightarrow{\text{H}^+} \text{RSH} \\ \xrightarrow{\text{R}'\text{X}} \text{R—S—R}' + \text{X}^- \end{cases}$$

(see Ref. 159 – the S—S bond must be more easily reducible than the electrophile itself).

In some cases electron transfer from a dianion of an aromatic compound to an alkyl halide may be possible[154]. The dianions are much stronger bases and nucleophiles than the anion radicals. Therefore, it was found, for instance, that perylene dianion exchanges electrons with dichlorobenzene, but abstracts a proton from butyl chloride. It was also noticed that a reaction with the dianion is only observed when the reaction between the anion radical and the substrate is infinitesimally slow. The enhancement

TABLE 22. Enhancement factor, R^*, of the current peak height of the *second* reduction peak of the mediator in the presence of certain aromatic halides[132]

Aromatic halide	Concn, M × 10^2	Mediator	Concn, M × 10^4	R^{*a}
1,4-Dichlorobenzene	3.5	Perylene	33.0	9.5
1,4-Dichlorobenzene	0.6	Dimethyl terephthalate	42.0	2.2
1,4-Dichlorobenzene	2.3	4-Benzoylpyridine	45.0	5.3
1,4-Dichlorobenzene	0.57	3-Benzoylpyridine	45.0	4.5
Benzotrifluoride	3.4	Perylene	33.0	6.5
Benzotrifluoride	2.3	Dimethyl terephthalate	85.0	4.0
Benzotrifluoride	2.3	4-Benzoylpyridine	45.0	4.5
Benzotrifluoride	2.3	2-Benzoylpyridine	45.0	10.0
Bromobenzene	3.2	Dimethyl terephthalate	42.0	5.3
Bromobenzene	3.2	4-Benzoylpyridine	45.0	3.4
p-Bromoanisole	4.0	3,6-Diphenylpyridazine	70.0	6.5
Butyl chloride[b]	–[c]	Anthracene	–[c]	–[c]

[a] $R^* = i_p/i_p^0$, where i_p^0 is the current peak height in the absence of halo-substrate.
[b] Alkyl halides may react as proton donors by elimination of hydrogen halide. The observed increase in the current peak height on addition of butyl chloride to anthracene dianion is probably due to a catalytic reduction of the butyl chloride, but it could also be caused by further reduction of the coupled product, *n*-butyldihydroanthracene.
[c] Not reported.

effect during the course of reduction of substituted halobenzenes in the presence of dianions as electron transfer mediators is shown in Table 22.

When the tendency to proton abstraction is diminished for compounds forming less basic dianions, an S_N2 reaction between the dianion and the substrate (two electrons are transferred simultaneously) or an electron exchange forming two anion radicals may take place.

A schematic and qualitative scale of a possible distribution of the relative half-wave potentials of aromatic electron transfer mediators (M), alkyl radicals and alkyl halides was suggested by Lund and Simonet[131], as follows:

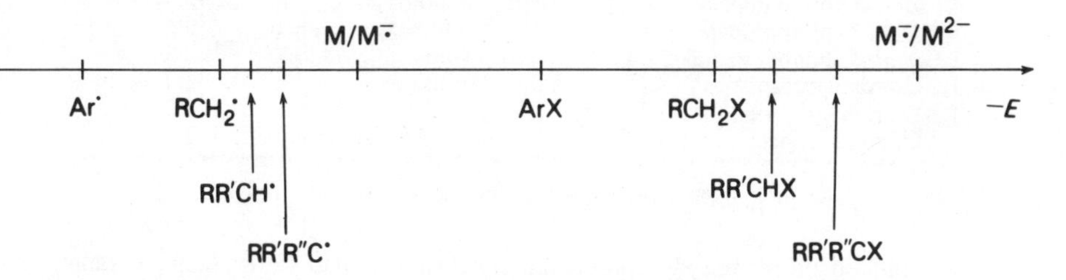

On the basis of electron affinity considerations of the components participating in the reaction it is reasonable to assume that electron affinity increases in the order RR′R″C· < RR′CH· < $RCH_2^{\cdot}$ < Ar·. On examining the schematic representation one may postulate that the rate of electron transfer from an aromatic anion radical to a primary alkyl radical is faster than the transfer to a tertiary alkyl radical. It is also assumed that coupling between aliphatic radicals and electron transfer mediators would be favoured for tertiary alkyl halides compared with primary ones, since the rate of the coupling reaction is insensitive to the radical structure.

The catalytic regeneration of substrates by an electron transfer mediator has been extended[132] to 1,2- and 1,3-dihalides, employing anion radicals of aromatic hydrocarbons, heteroaromatic compounds, ketones, esters, nitriles and olefins, as well as dianions of some of these groups of compounds. An interesting result has been reported upon reacting anthracene with 1,3-dibromopropane. A ring-closure product due to coupling in the 1- and 2-positions of anthracene was formed:

$$\text{anthracene} + Br(CH_2)_3Br \longrightarrow \text{(ring-closure product)} + \text{9-}(CH_2)_3Br\text{-anthracene}$$

The catalytic effect of the homogeneous electron transfer observed for dihalides is shown in Table 23 and represented by the enhancement factor, R^*, of the current peak height of the first peak emerged from the formation of the radical anion of the mediator.

2. Cation radicals

As has been demonstrated in the preceding subsection, extensive work has been done on the properties of aromatic anion radicals as electron transfer carriers. In contrast, relatively little has been published on cation radicals as electron transfer mediators. Among the few known examples, the reaction[161,162] of 9,10-diphenylanthracene (DPA) cation radical with halides and with SCN^- has been

TABLE 23. Enhancement factor R^* of the current peak height of the *first* reduction peak of the mediator in the presence of dihalides and halo-substituted benzenes[132]

Dihalide	Concn, M × 10^2	Mediator	Concn, M × 10^4	R^{*a}
1,2-Dichloroethane	1.4	Phenanthrene	9.4	11.6
1,2-Dichloroethane	4.2	*p*-Methoxyacetophenone	11.0	4.5
cis-1,2-Dichloroethylene	4.4	Pyrene	8.2	4.9
cis-1,2-Dichloroethylene	4.4	Methyl nicotinate	3.6	7.6
cis-1,2-Dichloroethylene	4.4	*p*-Methoxyacetophenone	3.3	11.4
1-Bromo-3-chloropropane	3.4	Anthracene	93.0	2.1
1-Bromo-3-chloropropane	2.1	*p*-Methoxyacetophenone	3.3	11.5
1-Bromo-3-chloropropane	3.4	Pyrene	8.2	4.0
1,3-Dibromopropane	3.1	Anthracene	93.0	2.9
1,3-Dibromopropane	3.3	Pyrene	8.2	4.7
1,3-Dibromopropane	0.69	*p*-Methoxyacetophenone	2.2	11.9

[a]$R^* = i_p/i_p^0$, where i_p^0 is the current peak height in the absence of dihalide.

investigated more thoroughly and the conclusion is that both electron transfer and addition processes are feasible (Table 24):

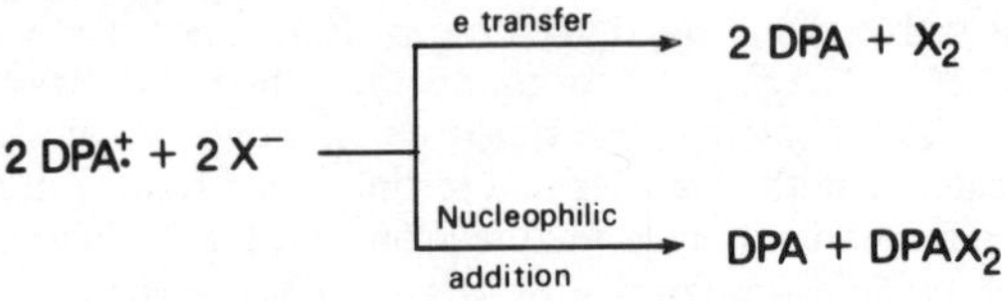

$$2\,DPA^{+\cdot} + 2\,X^- \begin{cases} \xrightarrow{\text{e transfer}} 2\,DPA + X_2 \\ \xrightarrow{\text{Nucleophilic addition}} DPA + DPAX_2 \end{cases}$$

The mechanism has been discussed by Evans and Blount[163] and supported by their previous kinetic work with $DPA^{+\cdot}$ and Cl^- [164]. The following scheme is suggested:

$$DPA^{+\cdot} + X^- \rightleftharpoons [DPAX]^{\cdot}$$

$$DPA^{+\cdot} + [DPAX]^{\cdot} \rightleftharpoons [DPAX]^{+} + DPA$$

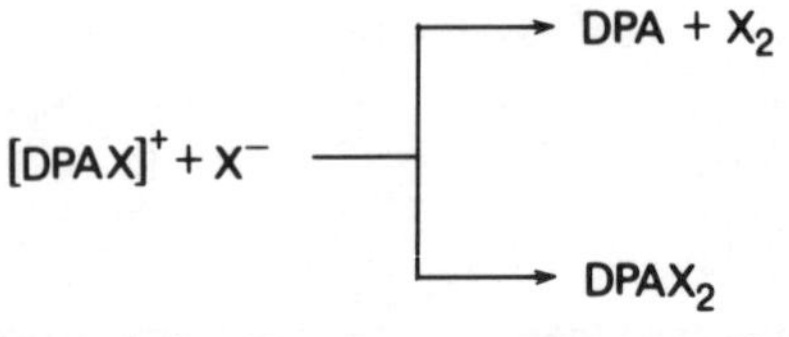

$$[DPAX]^{+} + X^- \begin{cases} \longrightarrow DPA + X_2 \\ \longrightarrow DPAX_2 \end{cases}$$

However, not all cation radicals react according to this scheme. For example, thianthrene cation radical has been shown to undergo an initial disproportionation step followed by a reaction of its dication with a substrate[165].

The reactivity of radical cations toward nucleophiles presents[166] some puzzling features and the reactions of the radical cations of perylene[167,168], thianthrene[165,169], phenothiazine[169] and dibenzodioxin[170] are representative of the complexities involved (Table 25). No systematic reactivity pattern is discernable when one tries to relate their reactivities to oxidation potentials of the nucleophiles; although iodide ion is a good nucleophile, it is too easily oxidized by all cation radicals studied to permit nucleophilic substitution in the ring. Recently, Eberson and coworkers[171,172] have proposed an interesting approach by employing the Dewar–Zimmerman rules[173]. They suggested that these rules may give testable predictions about the reactions between radical cations and nucleophiles. The proposal is based on the basic assump-

TABLE 24. Electrochemical, kinetic and products data on the reaction between $DPA^{+\cdot}$ and various nucleophiles[a]

Nucleophile	$E_{1/2}$, V[b]	Percentage of DPA regenerated, %	Reaction type	Products	k_{obs}, M^{-1} s^{-1}[e]
I^-	0.50	98 ± 3	e transfer	I_2, I_3^-[c]	1.49 (±0.55) × 10^7
SCN^-	0.69	98 ± 3	e transfer	$(SCN)_2$, $(SCN)_x$[c]	3.35 (±0.23) × 10^6
Br^-	0.85	95 ± 4	e transfer	Br_2	6.91 (±0.85) × 10^5
Cl^-	1.11	53 ± 3	Nucleophilic addition	$DPA(Cl)_2$[d]	9.26 (±0.62) × 10^9[f]

[a]Data from Ref. 161–164.
[b]Voltammetric half-wave potentials for oxidation at a Pt anode in CH_3CN against SCE.
[c]I_3^- and parathiocyanogen, $(SCN)_x$, are formed respectively when an excess of the nucleophile is present.
[d]9,10-Dichloro-9,10-diphenyl-9,10-dihydroanthracene.
[e]Weighted average values, measured by stopped-flow kinetic spectrophotometry.
[f]M^{-2} s^{-1}.

TABLE 25. Reactivity patterns of radical cations versus nucleophiles in homogeneous solutions

Nucleophile (X^-)	E°, V[a] ($X^{\cdot}/X^-$)	Reaction type[b] [Perylene]‡	[Thianthrene]‡	[Phenothiazine]‡	[Dibenzodioxin]‡
F^-	3.0	No reaction	Not known	No reaction[c]	Not known
Cl^-	2.1	ET	NA	NA[d]	Not known
Br^-	1.7	ET	Not known	NA[d]	Not known
I^-	1.2	ET	ET	ET	ET
NO_2^-	0.7	NA	Not known	NA	NA
CN^-	1.6	NA	ET	ET	ET
CH_3COO^-	1.8	NA	Not known	Not known	Not known
$PhCOO^-$	1.1	NA	Not known	Not known	Not known
H_2O	1.1	NA	NA	NA	NA
Pyridine	1.8	NA	NA	NA	NA
Amines	1.2	Not known	NA	Not known	ET

[a] Data from Refs 166, 168–170 and 172, in acetonitrile. The last three entries correspond to $E_{1/2}$.
[b] Against Ag/Ag^+ (0.1 M) reference electrode; ET, electron transfer; NA, nucleophilic addition.
[c] F^- acts as a base toward the phenothiazine radical cation, producing a dimer, 3,10′-biphenothiazine, and unidentified green product.
[d] The corresponding 3-halo and 3,7-dihalo phenothiazines are formed.

tion that the transition state of the electron transfer reaction does not require any interaction between the two reacting components. Nucleophilic attack does require such interaction to form a new bond. They conclude that the nature of the transition state (aromatic or antiaromatic) influences the pathway which takes place. The same idea has been applied to the reaction between radical anions and electrophiles and accordingly, it has been clearly demonstrated that the protonation of radical anions derived from $4n + 2$ aromatic systems is considerably slower than that of the analogous carbanion[172].

In contradiction to Eberson's postulate, Rozhkov and coworkers[174] have presented evidence that fluoride ion can react as a nucleophile towards radical cations. Evans and Hurysz[175] also arrived at conclusions contrary to those of Eberson on examining the rates of reaction of perylene radical cation with various nucleophiles.* According to their results, the rate constants for Br^- and Cl^- are of the same order of magnitude despite the fact that these anions differ in oxidation potential by at least 250 mV, and therefore their reactivity should be different. This behaviour indicates that either bromide is reacting more slowly than expected or chloride much faster. However, a good correlation is obtained for various nucleophiles, excepting I^-, when a nucleophilic reactivity scale has been used. Consequently, these authors concluded that iodide is the only electron donor which clearly reduces perylene radical cation by electron transfer, whereas Cl^- and Br^- reduce by a mechanism involving nucleophilic addition, as follows:

$$[\text{PeryH}]^{+\cdot} + X^- \longrightarrow \left[\text{Pery}\begin{matrix}\diagup H\\ \diagdown X\end{matrix}\right]^{\cdot} \rightleftharpoons \text{PeryH} + X^{\cdot}.$$

The above mechanism is well known[176] for reversible addition of a free radical to aromatics whenever the radical is more stable than the aromatic radical.

*It should be noted that the rates measured by Evans and Hurysz do not take into account the background reaction, e.g., the reaction between perylene radical cation and MeOH. This missing point is important and its absence may lead to different conclusions.

B. Electrochemical Generation of Inorganic Species as Electron Transfer Mediators

The previous section demonstrated various possibilities of reductive carbon–halogen bond cleavage by indirect electrolysis through a homogeneous electron transfer by aromatic anion radicals and dianions, or carbon–halogen bond formation via a displacement process.

The present section describes the oxidative breaking of the carbon–halogen bond indirectly by electrochemically generated positive halogen species. It is shown that alkyl halides can be oxidized at potentials with lower absolute values than that required for the direct process at the electrode. A prerequisite for such a 'catalytic' process is a fast and irreversible chemical follow-up reaction of the substrate after the homogeneous electron transfer with the mediator. Here, the cleavage of the carbon–halogen bond is the irreversible step.

1. Positive iodine species

The results obtained from coulometry and other electrochemical data on the anodic oxidation of alkyl iodides are inconsistent with a simple S_N1 or S_N2 type mechanism[11,36,177]. Miller and coworkers[178] discovered that a stable I^+ species could be formed by iodine oxidation at potentials necessary for alkyl iodide oxidation. They demonstrated that an indirect process involving I^+ and alkyl iodides could contribute to carbon–iodine cleavage to form the same product obtained from direct anodic alkyl iodide chemistry, and postulated the following scheme:

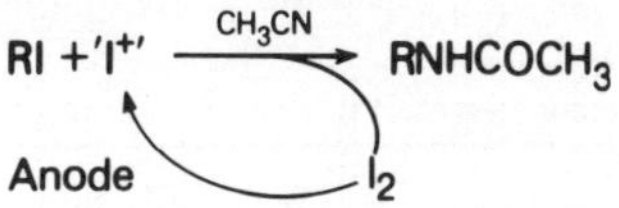

$$RI + 'I^{+}' \xrightarrow{CH_3CN} RNHCOCH_3$$

Anode ← I_2

Several alkyl iodides were reacted with electrochemically generated positive iodine species from anodic oxidation of I_2 in acetonitrile and the results are summarized in Table 26.

An attempt to determine the percentage of *N*-alkyl acetamides coming from direct anodic cleavage of alkyl iodides and that formed by the indirect cleavage upon reaction with iodonium species was advanced[13] by examining the products obtained

TABLE 26. Reaction products of electrochemically generated I^+ and alkyl iodides in acetonitrile at 1.7 V (against Ag/0.1 N Ag^+) on platinum anode

Iodides	Ref.	Products and yield, %
2-Octyl	12	*N*-(2-Octyl)acetamide (54) + *N*-(3-octyl)acetamide (22)
1-Adamantyl	13	*N*-(1-Adamantyl)acetamide (60)
Cyclohexyl	13	*N*-(Cyclohexyl)acetamide (45)
n-Propyl	13	*N*-(*n*-Propyl)acetamide (19) + *N*-(isopropyl)acetamide (12)
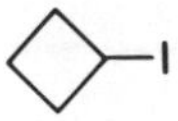 ◇—I	16	*N*-Cyclobutylacetamide (61) + *N*-(cyclopropylmethyl)acetamide (15)
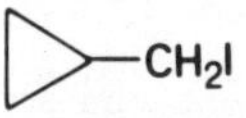 ▷—CH_2I	16	*N*-Cyclobutylacetamide (72)[a] + *N*-(cyclopropylmethyl)acetamide (8)

[a]Under appropriate conditions this was the only product formed.

from *S*-2-octyl iodide by the two methods. On the basis of results which indicate great similarity between the two processes, the problem remained unresolved.

The nature of the positive iodine species has been postulated to be $[CH_3\overset{+}{C}NI]$ or *N*-iodoacetamide[13,178]. In protic media, other species have been suggested[41,42], such as $(H_2O)I^+$, I^+, IO^+, etc., which are perhaps also present in 'wet' aprotic media.

The electrophilicity of electrochemically generated positive iodine species was discussed elsewhere (Section III.A). However, its potentiality to cleave other types of carbon–halogen bonds has not been reported.

2. *Positive bromine species*

Recently Becker and Zemach have reported that electrochemically generated positive bromine species could partake in the cleavage of both C–Br and C–Cl bonds[179]. On anodic oxidation of Br_2 in acetonitrile at platinum anode, a new absorption band appears in the ultraviolet ($\lambda_{max} = 269$ nm) which does not correspond either to the solvent or to Br_2. Based on some literature reports on the UV spectra of halide and polyhalide ions in acetonitrile[180] it has been suggested that a complex is formed between positive bromine species and acetonitrile which may have a structure of $[CH_3CNBr]^+$ or a 'sandwich', $[(CH_3CN)_2Br]^+$, with various counterions, e.g. Br^-, Br_3^- and the electrolyte employed (ClO_4^-).

The reactivity of 'Br^+' species in the fission of carbon–bromine bonds has been studied[179] for a variety of alkyl bromides, primary through tertiary, cyclic and non-cyclic compounds (Table 27). The type of products was the same as that obtained by direct anodic oxidation of alkyl bromides, namely, *N*-alkylacetamides and olefins.

TABLE 27. The effect of electrogenerated positive bromine species on various alkyl bromides[a]

Substrate (M)	Br_2, M	Total mF consumed	F mol^{-1} (Br_2)	Reaction time, h	Products (%)[b]
Cyclohexyl bromide (0.5)	0.3	4.93	0.51	5.0	*N*-Cyclohexylacetamide (10) 1,2-Dibromocyclohexane (10) 3-Acetamidocyclohexene (2)
Cyclopentyl bromide (0.53)	0.27	2.05	0.51	4.0	*N*-Cyclopentylacetamide (14) 1,2-Dibromocyclopentane (1)
Cycloheptyl bromide (0.53)	0.27	2.76	0.69	2.0	*N*-Cyclohepthylacetamide (13) 1,2-Dibromocycloheptane (1)
Isopropyl bromide (0.53)	0.27	1.61	0.41	7.0	Isopropylacetamide (8)
sec-Butyl bromide (0.53)	0.27	2.05	0.51	5.0	2-Acetamidobutane (10) 1-Acetamidobutane (trace)
1-Bromo-2-methylpropane (0.53)	0.27	4.18	0.91	3.5	*t*-Butylacetamide (32) 1-Acetamido-2-methylpropane (trace)
1-Bromobutane (0.53)	0.27	2.97	0.37	8.0	1-Acetamidobutane (trace)
2-Bromopentane (0.53)	0.53	3.10	0.38	5.0	2-Acetamidopentane (10)

[a]Data from Ref. 179. Experimental conditions: 15 ml acetonitrile–0.1 M $LiClO_4$; nitrogen atmosphere-controlled potential at 1.6 V against Ag/0.1 N $AgNO_3$ and at 38 ± 1°C, using Pt anode.

[b]Controlled experiments, under the same conditions but without electrolysis, show only trace amounts of *N*-alkylacetamides.

However, the yield of isolated products was usually ~15% (in one case, up to 32%), significantly lower than the product yield obtained from the direct process at the electrode, 40–75%[23,27,30]. Although the indirect carbon–bromine bond cleavage seems to be less efficient, its advantage is that it takes place at lower potentials (~1.6 V) compared with the required potentials for the direct process (~2.5 V). The difference in potentials at which the two processes take place enables one to estimate the relative contribution of each of the processes to the carbon–bromine bond breaking. It is noteworthy that such an estimation was not feasible in the case of alkyl iodides, since both direct and indirect electrolyses occur at about the same potentials.

The potentiality of electrochemically generated 'Br^{+}' species towards carbon–chlorine bond cleavage in alkyl chlorides was examined[179]. It was found that such cleavage does occur in secondary and tertiary compounds (Table 28), but it seems to be less efficient than for alkyl bromides. This result is not surprising due to the fact that the C–Cl bond is stronger than the C–Br bond. The main importance of indirect C–Cl bond fission arises from the fact that attempts at direct anodic cleavage have been unsuccessful[20,179].

$$RCl + \text{'}Br^{+}\text{'} \xrightarrow{-BrCl} [\dot{R^{+}}] \xrightarrow[2.\ H_2O]{1.\ CH_3CN} RNHCOCH_3 + H^{+}$$

An interesting feature of using 'Br^{+}' in a double mediatory system has been reported recently[181]. Secondary alcohols were found to be oxidized to ketones in reasonable yields in the presence of *n*-octyl methyl sulphide and tetraethylammonium bromide as mediators, while the same thioether in the presence of tetraethylammonium *p*-toluenesulphonate gave a rather poor yield. This result indicates that this sulphide works as a better mediator in the presence of bromide ion. In a typical example, 2-octanol yields 25% of 2-octanone without the presence of Br^{-} but 85% of the ketone in its presence. A reaction mechanism involving an initial oxidation of Br^{-} to Br^{+}, following the formation of sulphonium ion intermediate by reacting bromonium ion with sulphide, has been postulated; this is similar to the

TABLE 28. The effect of electrogenerated positive bromine species (at 1.6 V against Ag/Ag^{+}) relative to non-electrolysed solutions of Br_2 and acetonitrile on C—Cl bond cleavage of various chloroalkanes[a]

Substrate (M)	Br_2 (M)	Reaction time, h	Total mF consumed	F mol^{-1} (Br_2)	Products (%)
Cyclohexyl chloride (0.2)	0.8	24	–	–	No products
Cyclohexyl chloride (0.33)	1.33	4	10	0.5	*N*-Cyclohexylacetamide (8.6) 1,2-Dibromocyclohexane (trace) 3-Acetamidocyclohexene (1)
sec-Butyl chloride (0.32)	0.16	24	–	–	No products
sec-Butyl chloride (0.33)	1.33	7	1.65	0.4	2-Acetamidobutane (3)
					2-Acetamidobutane (7)
t-Butyl chloride (0.32)	0.10	24	–	–	*N*-(*t*-butyl)acetamide (2)
t-Butyl chloride (0.53)	0.27	0.5	2.41	0.6	*N*-(*t*-butyl)acetamide (22.5)

[a]Data from Ref. 179. All experiments were carried out at 38 ± 1°C using $LiClO_4$ as an electrolyte.

chemical oxidation of alcohols by means of dimethyl sulphide and halogens[182].

$$Br^- \xrightarrow{-2e^-} Br^+$$

$$Br^+ + RSCH_3 \longrightarrow RS^+(Br)CH_3$$

$$RS^+(Br)CH_3 + R^1R^2CHOH \xrightarrow{-H^+} RS(Br)(CH_3)\text{–O–CH}R^1R^2 \rightleftharpoons [RS^+(CH_3)\text{–O–CH}R^1R^2]\,Br^-$$

$$[RS^+(CH_3)\text{–O–CH}R^1R^2]\,Br^- \xrightarrow{-H^+} Br^-\,RS^+(CH_2^-)\text{–O–C(H)}R^1R^2 \longrightarrow RSCH_3 + R^1COR^2 + Br^-$$

Recently, the elctrolytic transformation of olefins into oxyselenides has been reported[183]. The reaction is promoted with a trace amount of halide ion (Cl^-, Br^-, I^-), of which positive species are electrochemically generated. The reaction has shown high regioselectivity and gave high product yield.

3. *Superoxide ion*

Alkali metal superoxides are well known but their insolubility makes them of little preparative use, unless incorporated into a crown ether[184]. However, superoxide ion may be generated electrochemically from oxygen[185]. The discovery[186] that $O_2^{\cdot -}$ is a respiratory intermediate of aerobic organisms has prompted widespread interest in the chemical properties and reactivity of $O_2^{\cdot -}$ with organic functional groups (carbonyl systems[187], tosylates[188], alkyl halides[188–192], etc.) as well as with metal ions as a complexation ligand[184,193].

Peover and coworkers[189] have studied the nucleophilic reactions of electrogenerated superoxide ion with alkyl bromides. Their study, by cyclic voltammetry at mercury-covered platinum electrode in DMF–Bu_4NClO_4, is consistent with a nucleophilic substitution mechanism. The displacement of bromide by superoxide gives an alkyl hydroperoxy radical which has an electron affinity at least as large as that of oxygen:

$$O_2 + e \rightleftharpoons O_2^{\cdot -} \quad (E^{\circ\prime} = -0.50 \text{ against NHE})$$

$$n\text{-BuBr} + O_2^{\cdot -} \rightleftharpoons n\text{-BuO}_2^{\cdot} + Br^-$$

This radical can be further reduced, either by the electrode or by bulk reaction with superoxide:

$$n\text{-BuO}_2^{\cdot} \overset{+e}{\rightleftharpoons} n\text{-BuO}_2^-$$

$$n\text{-BuO}_2^{\cdot} + O_2^{\cdot -} \rightleftharpoons n\text{-BuO}_2^- + O_2$$

On polarography at the dropping mercury electrode (DME), the cathodic current corresponding to reduction of oxygen was approximately doubled by the addition of excess of n-BuBr, as required by the overall stoicheiometry of the substitution.

$$n\text{-BuBr} + O_2 + 2\,e \longrightarrow n\text{-BuO}_2^- + Br^-$$

Table 29 demonstrates some representative results obtained upon reacting alkyl bromides with electrogenerated $O_2^{\cdot-}$ and with KO_2-crown ether. The solvent, as well as the nature of the alkyl bromide, have tremendous effects on the products: primary bromides yield mainly peroxides and small amounts of alcohols in CH_3CN, DMF or benzene, whereas tertiary bromides form alcohols and alkenes preferentially in DMF and DMSO. In the latter solvent, no peroxides are formed by any type of bromide. It seems that with secondary and tertiary halides competing elimination reactions account for olefin formation.

A polarographic study of oxygen with added acetyl chloride has revealed essentially a four-electron wave, as is required by the formation of diacetyl peroxide and its subsequent reduction:

$$2\,CH_3COCl + O_2 \xrightarrow{+2e} (CH_3CO)_2O_2 + 2\,Cl^-$$

$$(CH_3CO)_2O_2 \xrightarrow{+2e} 2\,CH_3COO^-$$

Stable solutions of tetraethylammonium superoxide in aprotic media have provided a basis for a chronopotentiometric study of the kinetics for the nucleophilic displacement of alkyl chlorides to give peroxide radicals ($RCl + O_2^{\cdot-} \xrightarrow{k} RO_2^{\cdot} + Cl^-$). The reaction is found to be first order in superoxide and occurs with 1:1 stoicheiometry. The rates of reactions of butyl bromides and iodides are too rapid for quantitative evaluation by this method. However, Danen and Warner[192] succeeded in measuring the rate constants for alkyl bromides reaction by using the stopped-flow technique. Kinetic results by several groups[189,191,194] are consistent with an S_N2 mechanism; the reaction rates follow the order $1° > 2° \gg 3°$ for alkyl halides and tosylates and the attack by $O_2^{\cdot-}$ results in an inversion of configuration[188,190]. Some of the kinetic results in DMSO and pyridine are summarized in Table 30.

4. Sulphur dioxide radical anion

An interesting synthesis of sulphones consists of the cathodic reduction of SO_2 in aprotic media, containing an alkyl halide[195,196]:

$$SO_2 \xrightarrow{+e} SO_2^{\cdot-} \xrightarrow[+e]{2\,RX} RSO_2R + 2\,X^-$$

Dihalides usually give polymeric or cyclic sulphones, for example:

$$C_6H_4(CH_2Br)_2\ (ortho) + SO_2 \xrightarrow{+2\,e} 2\,Br^- + C_6H_4(CH_2)_2SO_2\ (\text{cyclic sulphone})$$

Under appropriate conditions, non-symmetric sulphones can be formed:

$$CH_2{=}CHCH_2Br + CH_3Br + SO_2 \xrightarrow{+2\,e} CH_2{=}CHCH_2SO_2CH_3 + 2\,Br^-$$

C. Electrochemical Generation of Transition Metal Complexes as Electron Transfer Mediators

Organometallic species are commonly used as reagents for organic synthesis. Surprisingly, there have been few attempts to couple the electrochemical generation of

TABLE 29. Product composition after the generation of one equivalent of $O_2^{\cdot-}$ in the presence of alkyl bromides (supporting electrolyte–Bu_4NClO_4) compared with the results obtained from KO_2-crown ether

Reactant	Solvent	Electrode	Yield, %					Ref.
			Peroxide (ROOR)	Alcohol	Alkenes	Aldehyde	Reactant	
n-BuBr	CH_3CN	Hg	79	12	2	5	0	189
n-BuBr	DMF	Hg	70	<20	–	–	1	189
n-BuBr	CH_3CN	Graphite	~40	20	–	–	8	189
n-BuBr	DMF	Vitreous carbon	57	–	–	–	3	189
t-BuBr	DMF	Hg	–	31	53	–	15	189
n-$C_{18}H_{37}$Br	Benzene	KO_2-18-crown-6 ether	61	18	–	–	–[c]	190
$C_6H_{13}CH(Me)Br$	Benzene	KO_2-18-crown-6 ether	55	–	37[a]	–	–[c]	190
Bromocyclohexane	Benzene	KO_2-18-crown-6 ether	–	–	67	–	–[c]	190
Bromocyclopentane	Benzene	KO_2-18-crown-6 ether	42	–	24	–	–[c]	190
1-Bromooctane	DMSO	KO_2-18-crown-6 ether	–	63	<1	12	–[c]	188
2-Bromooctane	DMSO	KO_2-18-crown-6 ether	–	51	34[a]	<1[b]	–[c]	188
$Me(CH_2)_2C(Me)_2Br$	DMSO	KO_2-18-crown-6 ether	–	20	30	–	–[c]	188

[a] A mixture of isomers.
[b] 2-Octanone.
[c] Not reported.

TABLE 30. Second-order rate constants for the reaction of $O_2^{\cdot-}$ with primary, secondary and tertiary alkyl halides in DMSO and pyridine[a]

Substrate	Solvent	k_2, M^{-1} s^{-1} (25°C)		
		1°	2°	3°
A. Butyl halides				
BuCl[b]	$(CH_3)_2SO$	3.2	0.6	0.4
BuCl	Pyridine	2.3	0.5	<0.003
BuBr	Pyridine	90	54	<0.01
B. Alkyl bromides				
CH_3Br	$(CH_3)_2SO$	670	–	–
CH_3CH_2Br	$(CH_3)_2SO$	350	–	–
$CH_3(CH_2)_3Br$	$(CH_3)_2SO$	150	–	–
$(CH_3)_2CHBr$	$(CH_3)_2SO$	–	65	–

[a]Data from Ref. 184.
[b]The reactions for butyl bromides and iodides in DMSO were found to be too rapid for quantitative evaluations of their rate constants.

reactive intermediates with synthetic reactions[196]. Only in the past few years, has the role of odd-electron and zero-valent transition metal complexes in organic chemistry become a subject of current interest[197–201], for example, oxidative addition reactions, radical chain processes and the activation of carbon–hydrogen bonds. The advantage of employing electrochemical methods for generating odd-electron complexes is quite obvious, since no additional reagents are needed.

Quite recently, there have been several reports[202–204] on the electrocatalytic reduction of alkyl halides through electrochemically generated transition metal complexes as electron transfer mediators. Electrochemical reduction[202] of a d^8 Rh complex, $[Rh(diphos)_2]Cl$, to the corresponding d^8 Rh complex, $[Rh(diphos)_2]^0$, in the presence of cyclohexyl halides yielded, catalytically, cyclohexyl radicals according to the following scheme:

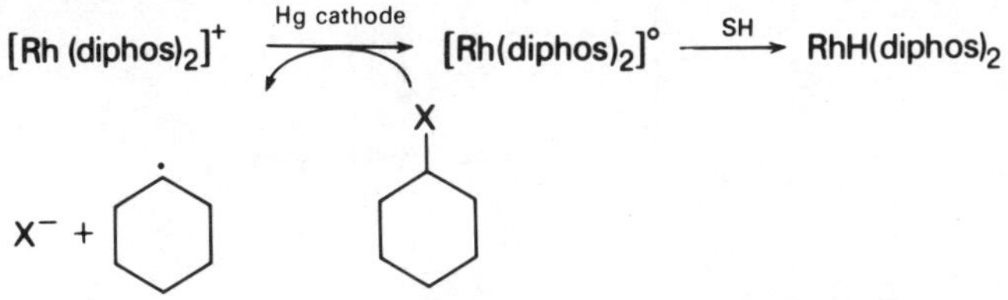

$$[Rh(diphos)_2]^+ \xrightarrow{\text{Hg cathode}} [Rh(diphos)_2]^\circ \xrightarrow{SH} RhH(diphos)_2$$

(catalytic cycle: $[Rh(diphos)_2]^\circ$ + cyclohexyl–X → $[Rh(diphos)_2]^+$ + X^- + cyclohexyl radical)

On changing various conditions (solvent, electrode material and substrate), several types of products may be formed from the cyclohexyl radical:

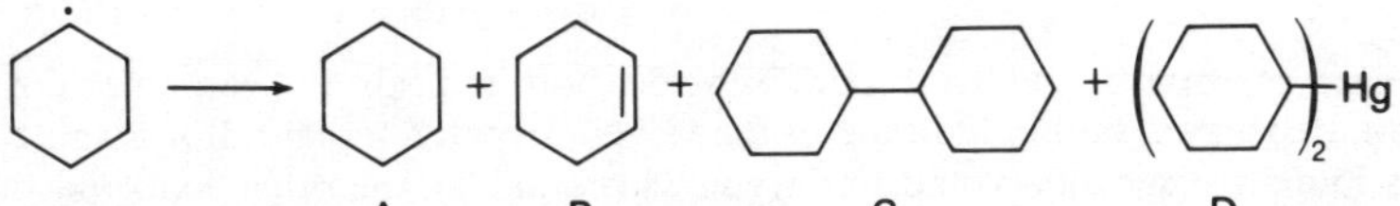

$$C_6H_{11}^{\cdot} \longrightarrow \underset{A}{C_6H_{12}} + \underset{B}{C_6H_{10}} + \underset{C}{C_6H_{11}\text{–}C_6H_{11}} + \underset{D}{(C_6H_{11})_2Hg}$$

The results shown in Table 31 demonstrate a pronounced catalytic effect, especially for the reduction of cyclohexyl iodide at mercury in acetonitrile and cyclohexyl chloride at platinum in dimethyl sulphoxide. The formation of radical termination products points to radical nature of the propagating species, $[Rh(diphos)_2]^0$.

TABLE 31. Electrochemical reduction of cyclohexyl halides via electrochemically generated $[Rh(diphos)_2]^{\circ}$ complex[202]

Solvent	Electrode	Halide	% Halide[a]	n^b	Types of product
Benzonitrile	Hg	Cl	33	4.1	A + B + C + D
Acetonitrile	Hg	Cl	50	3.4	D
Acetonitrile	Graphite	Cl	33	4.5	A + B
Acetonitrile	Hg	Br	10	5.2	A + B + D
Acetonitrile	Hg	I	10	23.0	A + B + C
Dimethylacetamide	Pt	Cl	10	3.7	A + B
Dimethylsulphoxide	Pt	Cl	33	16.0	Unidentified

[a]Percentage of cyclohexyl halide by volume, in the solution.
[b]The coulometric n value is based on the equivalents of the $[Rh(diphos)_2]Cl$ present.

The fact that cyclohexyl chloride could not be reduced directly under the same conditions but without the Rh(0) complex exemplifies the utility of the electrocatalytic Rh(I)/Rh(0) system.

Pletcher and coworkers[203] have studied the catalytic nature of Ni(I) species, derived electrochemically from Ni(II) square planar complexes, towards the reduction of alkyl halides in acetonitrile (Figure 5). It was found that hydrocarbon products are formed and a mechanism was postulated, involving not only a simple electron transfer mechanism, but also organonickel species as intermediate in the catalytic process:

$$\text{Ni(II)} + e \rightleftharpoons \text{Ni(I)}$$

$$\text{Ni(I)} + \text{RX} \longrightarrow \text{R–Ni(III)–X}$$

$$\text{R–Ni(III)–X} \longrightarrow \text{R}^{\bullet} + \text{Ni(II)} + \text{X}^-$$

$$\text{R–Ni(III)–X} \xrightarrow{+e} \text{R–Ni(II)–X}$$

$$\text{R–Ni(III)–X} \xrightarrow{+\text{Ni(I)}} \text{Ni(II)} + \text{R–Ni(II)–X}$$

$$\text{R–Ni(II)–X} \xrightarrow{\text{Y}^-} \text{R}^- + \text{Y–Ni(II)–X}$$

As one can see from the above general scheme, both radicals and carbanions may be involved, depending on the life-time of the C—Ni(II) intermediate. If it is short-lived (as was found for secondary and tertiary alkyl bromides), then this bond will rapidly decompose to form $R^{\bullet}$, and regenerate the electroactive complex, as was found for secondary and tertiary alkyl bromides. However, with primary bromides it was suggested that a 'Ni—C' intermediate would have sufficient life-time to undergo reduction at the electrode or in solution before the spontaneous cleavage to a radical might occur. Only the spontaneous cleavage process forming radicals was found to be

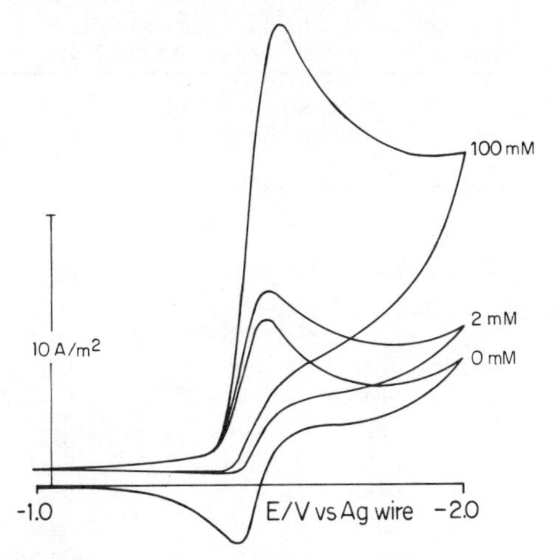

FIGURE 5. Cyclic voltammograms of 1 mM Ni(salen) (see footnote *a* to Table 32) in CH_3CN–Bu_4NBF_4 before and after the addition of 1-bromooctane (the numbers on the curves indicate the concentration of the bromide). Potential sweep rate 0.3 V s^{-1}. (Reproduced from Ref. 203.)

catalytic, whereas the other pathway is an ECE mechanism, leading to an electroinactive complex.

Some representative results are described in Table 32 and it is obvious that most of the products are consistent with a free-radical intermediate which determines the product distribution by the following reactions:

$$2\,R^{\bullet} \longrightarrow R{-}R$$

$$2\,R^{\bullet} \xrightarrow{\text{Disp.}} RH + \text{olefin}$$

$$R^{\bullet} + CH_3CN \longrightarrow RH + \dot{C}H_2CN$$

It was found that in the case of primary halides, dimerization was the major route, while with tertiary radicals disproportionation predominates. For secondary radicals, all the three reactions listed above were of similar importance. In certain cases, products derived from a carbanion intermediate were observed:

$$R^{-} + CH_3CN \longrightarrow RH + \bar{C}H_2CN$$

$$\bar{C}H_2CN + RBr \longrightarrow RCH_2CN + Br^{-}$$

The results with *n*-octyl iodide are different from those with the bromide and seem to represent the intermediate situation where both mechanisms, the catalytic and ECE are competing. Probably it is due to a weaker Ni—C bond in the iodo complex than in the bromo one.

In conclusion, the reductions of alkyl iodides and bromides in the presence of Ni(I) intermediates occur at considerably less negative potentials (e.g. −1.20 V) than in solutions containing alkyl halide alone (~ −2.2 V). It was also demonstrated by

TABLE 32. Products from controlled potential reduction of the nickel(II) complexes in the presence of a tenfold excess of alkyl halide. Medium AN/Bu_4NBF_4 [203]

	Ni(salen)		$Ni(teta)^{2+}$	
Substrate	Product, %[a]	n[b]	Product, %[a]	n[b]
n-$C_8H_{17}I$	n-$C_{16}H_{34}$ (72) n-C_8H_{18} (5) $C_6H_{13}CH{=}CH_2$ (2)	11	n-$C_{16}H_{34}$ (61) n-C_8H_{18} (29) $C_6H_{13}CH{=}CH_2$ (5) $C_9H_{19}CN$ (6)	5
n-$C_8H_{17}Br$	$C_{16}H_{34}$ (77) n-C_8H_{18} (4) $C_6H_{13}CH{=}CH_2$ (2)	12	n-$C_{16}H_{34}$ (8) n-C_8H_{18} (66) n-$C_9H_{19}CN$ (6)	2
$C_6H_{13}CH(Br){-}CH_3$	$C_{16}H_{34}$ (27)[c] n-C_8H_{18} (32) $C_6H_{13}CH{=}CH_2$ (7)	11	$C_{16}H_{34}$ (46)[c] n-C_8H_{18} (33) $C_6H_{13}CH{=}CH_2$ (11)	11
$C_5H_{11}{-}C(CH_3)(Br){-}CH_3$	$C_{16}H_{34}$ (9)[c] $C_5H_{11}CH(CH_3)_2$ (28) $C_5H_{11}C(CH_3){=}CH_2$ (32) $C_4H_9CH{=}C(CH_3)_2$ (6)	11	$C_{16}H_{34}$ (9)[c] $C_5H_{11}CH(CH_3)_2$ (31) $C_5H_{11}C(CH_3){=}CH_2$ (32) $C_4H_9CH{=}C(CH_3)_2$ (9)	11

[a]Organic yields based on octyl halide consumed. Ni(salen) = [*N,N'*-ethylene-*bis*(salicylideneiminato]nickel(II); $Ni(teta)^{2+}$ = (5,5,7,12,12,14-hexamethyl-1,4,8,11-tetraazacyclotetradecane)nickel(II).
[b]Faradays per mole when current had dropped to 5% of initial value.
[c]Non-straight-chain hexadecanes (two or three isomers).

employing two different nickel (II) complexes, that the weaker the Ni—C bond in the intermediate, the more significant is the catalytic effect.

The effect of activated olefins on the indirect reduction of alkyl bromides in the presence of Ni(I) intermediate was also studied and showed that an insertion reaction takes place, forming a new Ni—C bond which undergoes cleavage by further reduction. This hypothesis was found to be valid only when an electron-withdrawing group is attached to the double bond. The following overall process was suggested[204] in the presence of such olefins and alkyl bromides:

$$\mathrm{Ni(II)} \xrightleftharpoons{+e} \mathrm{Ni(I)}$$

$$\mathrm{Ni(I)} + \mathrm{RBr} \longrightarrow \mathrm{R{-}Ni(III){-}Br}$$

$$\mathrm{R{-}Ni(III){-}Br} + \mathrm{CH_2{=}CHCN} \longrightarrow \mathrm{RCH_2CH(CN){-}Ni(III){-}Br}$$

$$\mathrm{RCH_2CH(CN){-}Ni(III){-}Br} \xrightarrow[HX]{+e} \mathrm{RCH_2CH_2CN} + \mathrm{X{-}Ni(II){-}Br}$$

According to the suggested scheme, 2e per molecule of Ni(II) are transferred, and the reaction was not found to be catalytic because, in the final stage, the nickel species is not electroactive and the reaction does not continue further. The reaction was studied with primary, secondary and tertiary butyl bromides in the presence of acrylonitrile and the same type of product was dectected in good yield. The reaction was extended to a range of olefins in the presence of 2-bromobutane and it was found that both coulometry and yield are dependent on the olefin structure. The reaction depicted in the above scheme seems to be general for several types of alkyl bromides and with olefins conjugated or attached to electron-withdrawing groups. However, wherever the yield is low, C_4 and C_8 hydrocarbons were also formed, indicating that the first Ni—C species mentioned in the scheme decomposes rapidly to generate radicals and does not survive long enough to react with an olefin to form the second Ni—C species.

Recently it has been shown that other[205] nickel complexes, e.g. triphenylphosphine nickel(II), can be electrochemically reduced in ethereal solvents as well as in ethanol to zero-valent nickel, at various cathode materials (gold, platinum, nickel). The kinetics of the reaction of Ni(0)L_4 with PhX (X = I, Br, Cl) have been studied and the rates follow the order I > Br > Cl:

$$\text{Ni(0)L}_4 + \text{PhX} \longrightarrow \text{PhNiXL}_2 + 2\,\text{L}.$$

Electrolysis of PhX in excess at the reduction potential of $PhNiXL_2$ allows the catalytic reduction of PhX.

Other groups[206] have demonstrated the catalytic electrochemical reduction of alkyl halides by vitamin B_{12} derivatives and vitamin B_{12} model compounds. Recently, Scheffold and coworkers[207a] have reported a novel 1,4 addition of alkyl halides to Michael olefins employing controlled potential electrolysis. Bromoalkylcyclohexenones, bearing *all* the functional groups required in one molecule, yielded two types of product, bicyclic ketones and open-chain derivatives, as follows:

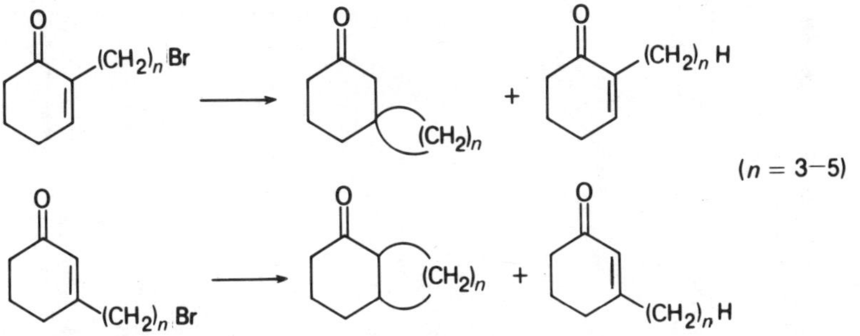

O (CH2)n Br → O (CH2)n + O (CH2)n H

(n = 3–5)

O (CH2)n Br → O (CH2)n + O (CH2)n H

The suggested mechanism involves the initial formation of a catalytically active Co(I) species by electrochemical reduction of a Co(III) complex, followed by a rapid reaction with alkylating agent to form octahedral alkyl–Co(III) complex. The latter is electrochemically reduced to a Co(II) intermediate which reacts with a Michael olefin to form the type of products mentioned above and regenerates the Co(I) complex. This chemically catalysed reductive cleavage provides a versatile tool for the formation of C—C bonds.

Unlike former examples, and others[207b] in which the formation of an organometallic intermediate is proposed, a recent work by Costa and coworkers[208] describes a catalytic reaction between electrogenerated Co(I) complex and tertiary organic halides which does not afford the organocobalt derivative. Instead, the reaction gives an unstable intermediate which in turn decomposes, regenerating Co(II) and alkenes.

D. Miscellaneous

1. Templates

An interesting example of intramolecular electron transfer has been reported by Breslow and coworkers[209], who used templates to direct the free-radical chlorination of steroids to specific positions. An interesting application of such template methods to electrochemical functionalization of a steroid has been found successful[210], as is demonstrated in the following scheme:

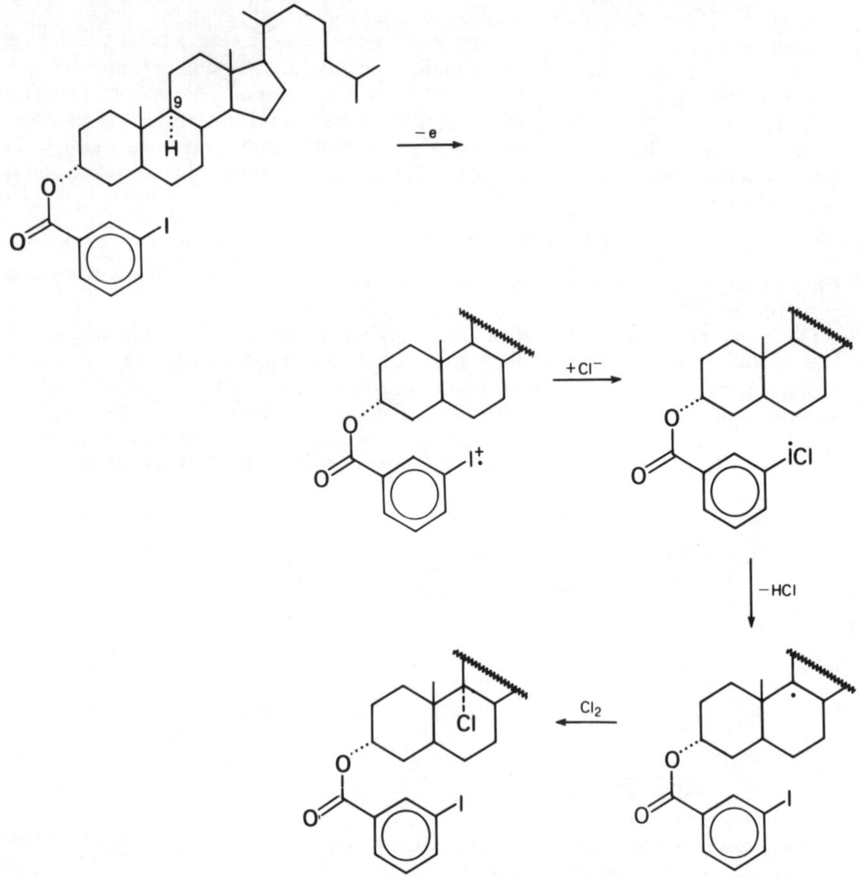

In this intramolecular electron transfer process, the esteric steroid in the scheme shows an oxidation peak at +2.6 V (against Ag wire on cyclic voltammetry). When it was oxidized in acetonitrile in the presence of Cl^-, in the dark, at +1.8 V, the steroid molecule was not chlorinated and only chlorine was evolved. However, oxidation at +2.7 V, followed by saponification, resulted in the corresponding chlorinated cholestan-3-α-ol, in 35–71% isolated yield. The results are interpreted in terms of an electrochemically initiated chain reaction sustained by a remote abstraction of hydrogen from the 9 position to form a radical intermediate which reacts with electrochemically generated chlorine to produce the chlorinated steroid.

2. Induction

Another interesting example involves catalysis of chemical reactions by electrodes. Pinson and Saveant[211] were the first to demonstrate that the nucleophilic substitution reaction of 4-bromobenzophenone (ArX) with thiophenolate (Nu^-) can be triggered electrochemically. The catalytic nature of the reaction has been ascribed to the 'exchange' current, since only 0.2 F/mol^{-1} of substrate has been required to convert 95% of ArX to ArNu:

$$ArX + e \rightleftharpoons ArX^{-\cdot} \qquad (X = Br)$$

$$ArX^{-\cdot} \rightleftharpoons Ar^{\cdot} + X^-$$

$$Ar^{\cdot} + Nu^- \longrightarrow ArNu^{-\cdot} \qquad (Nu^- = PhS^-)$$

$$ArNu^{-\cdot} - e \longrightarrow ArNu$$

Later experiments conducted by Tilborg and coworkers[212] on the same system have shown inconsistency with the above mechanism and suggested the possible occurrence of a bimolecular radical chain, as follows:

$$ArNu^{-\cdot} + ArX \rightleftharpoons ArNu + ArX^{-\cdot}$$

The reaction may be terminated by, for example, recombination of the chain carrying the radicals to form a dimer:

$$2\left[PhS{-}C_6H_4{-}COR\right]^{-\cdot} \xrightarrow{+2H^+} HO{-}C(R)(C_6H_4SPh){-}C(R)(C_6H_4SPh){-}OH$$

Indeed, a pinacol was obtained as the main by-product. According to this mechanism the catalytic nature of the reaction is attributed to disproportionation of the radical anion, $ArNu^{-\cdot}$, with starting material, ArX. Which mechanism is the true one may be of importance with respect to the applicability of the reaction, although both may operate simultaneously.

A comprehensive account of the electron-induced catalytic nucleophilic aromatic substitution with mechanistic features has been accomplished quite recently by Saveant[213]. It is suggested that due to the fact that the substitution reaction is triggered by setting the potential at the reduction wave of the substrate, the best candidate to be attacked by a nucleophile is the neutral radical ($Ar^{\cdot}$) formed upon cleavage of the initial radical anion. This mechanism is similar to the $S_{RN}1$ mechanism previously established by Bunnett[214], and which is based on the electrophilic reactivity of aryl radicals towards nucleophiles.

3. Modified electrodes

Since the first publications on chemical modification of SnO_2 [215] and graphite[216] electrode surface, there has been a tremendous interest in this field. The use of surface modified electrodes is widespread in areas such as photoelectrodes, synthesis, catalysis and basic research on the properties of immobilized molecules. The most interesting molecules to attach to a functionalized electrode surface are obviously the electrochemically reactive ones which can be applied on electrocatalysis. In such a catalytic process, the attached molecule acts as a fast electron transfer mediator for a substrate dissolved in the contacting solution. The basic idea of the process is expressed in the following scheme:

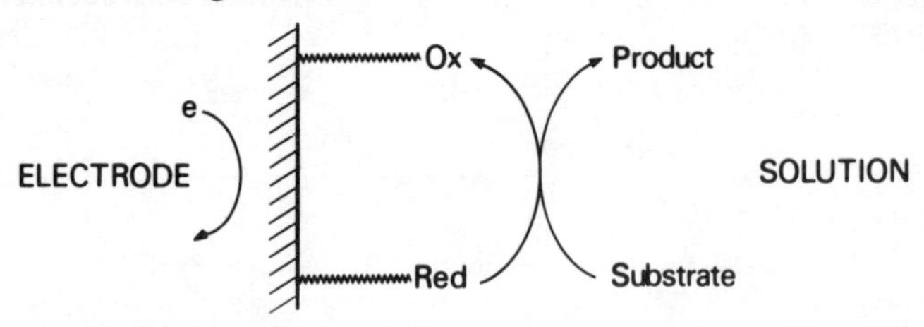

Two excellent reviews have been published recently by Murray[217] and by Snell and Keenan[218], which cover the literature to date on the field of chemically modified electrodes. Nowadays there is a variety of methods for immobilizing reagents on electrodes such as metal oxide surfaces (SnO_2, RuO_2, TiO_2 and oxides of Pt, Au, Si and more), oxidized sites on carbon surfaces as well as oxide-free carbon surfaces. Other approaches of surface modification which do not involve chemical binding, include chemisorption (irreversible adsorption) of π-systems on pyrolytic graphite and attaching a polymer (in which the reagent of interest is incorporated) to the electrode surface by various coating methods. Polymer coating seems to be appealing because it is technically simple, affords a 'stable' surface and provides a relatively large electrochemical response since multiple layers (10–1000) of redox sites react. However, in spite of the vast literature in this growing field, the information available so far does not give a straightforward answer on whether reactivity is promoted or diminished by immobilization of the electroactive reagent and seems to be highly dependent upon experimental conditions.

Examples of electrocatalysis include oxidation of ascorbic acid[219] by attached benzidine, oxidation of dihydronicotinamide adenosine diphosphate (NADH) by the use of a quinone-modified graphite as well as polymer-modified surface[219,220]. Electrocatalytic reductions involve the reduction of organic dihalides by using a polymer-modified electrode[221] (the turnover number for catalyst sites was estimated to be 10^4 for this reduction[221]) and reduction of molecular oxygen by metalloporphyrin

$$\text{PhCHBrCHBrPh} \xrightarrow[+2e]{\text{Modified electrode}} \text{PhCH=CHPh} + 2\,\text{Br}^-$$

derivatives[222–224] chemisorbed on pyrolytic graphite or adsorbed on glassy carbon[225], and by poly(p-nitrostyrene) dip coated on Pt[226]. So far, the most successful electrocatalysis on a modified electrode has been the 4e reduction of O_2 to H_2O without significant production of H_2O_2 by an adsorbed dicobalt face-to-face porphyrin dimers[224].

V. DIRECT CATHODIC REDUCTION OF THE C—X BOND

Electrochemical reduction of organic halogen-containing compounds in aqueous and non-aqueous media has been studied quite extensively, from mechanistic and synthetic viewpoints. The field has been reviewed by several authors and fairly recently by Hawley[10]. Since this author does not feel the need to re-review the current literature, this section is mainly confined to results which supplement the former review in this series[1] and which shed light on the complex mechanism of the reduction process.

With regard to the electrochemical reduction of carbon–pseudohalogen bonds, very little has been published. The second part of this section treats the reduction of the carbon–azide bond.

A. Organic Halides

The reductive cleavage of organic halides is irreversible[227] and is in general independent of the pH of the medium. The ease of the reduction is in the order RI > RBr > RCl > RF and tertiary RX > secondary RX > primary RX. No direct reduction of an unactivated C—F bond in a monofluoroalkane has been reported. Aromatic and vinyl halides are more difficult to reduce than alkyl halides. Vicinal dihalides and *gem*-dihalides are reduced more easily than the corresponding monohalides.

Macroscale reduction of monohalides leads to the corresponding hydrocarbon and some coupling products. Vicinal dihalides afford olefins and/or saturated hydrocarbons. α,ω-Dihalides form cyclic products and the corresponding hydrocarbons, depending on the proximity of the halides. *gem*-Halides generally afford hydrocarbons, although carbene intermediates have been detected.

1. Reduction pathways

In spite of the voluminous data on the reductive cleavage of the carbon–halogen bond, it is impossible to force all of the data into a strictly ionic or free-radical mechanism, because of the essentially ambivalent nature of the process[228]. In principle, there are several possible mechanisms of electron addition and C—X bond rupture and these are illustrated by the following schemes.

Scheme 1. A cleavage of the carbon–halogen bond in a single two-electron process to give a carbanion:

$$\mathrm{RX + 2e \longrightarrow R^- + X^-}$$

Scheme 2. Two one-electron transfers, the first concerted with bond rupture. When only one reduction wave is observed, the second step is assumed to be fast, suggesting that the standard potential for the reduction of the radical is anodic of the potential required for the first electron transfer step:

$$\mathrm{RX + e \xrightarrow{r.d.s.} R^{\bullet} + X^-}$$

$$\mathrm{R^{\bullet} + e \xrightarrow{fast} R^-}$$

Scheme 3. The first step is a composite one and consists of generation of a radical anion, which subsequently decomposes as follows:

$$\mathrm{RX + e \longrightarrow [RX]^{\overline{\bullet}}}$$

$$\mathrm{[RX]^{\overline{\bullet}} \longrightarrow R^{\bullet} + X^-}$$

$$\mathrm{R^{\bullet} \xrightarrow{SH} RH}$$

$$\mathrm{2\,R^{\bullet} \longrightarrow R{-}R}$$

or

$$\mathrm{R^{\bullet} + e \longrightarrow R^-}$$

Scheme 4. The addition of a second electron to the radical anion occurs synchronously with C—X bond cleavage:

$$RX + e \longrightarrow [RX]^{\cdot-}$$

$$[RX]^{\cdot-} + e \longrightarrow R^- + X^-$$

Scheme 5. Formation of a dianion from the radical anion:

$$[RX]^{\cdot-} + e \longrightarrow [RX]^{2-}$$

Scheme 6. A reduction which involves organomercury intermediates:

$$RX + Hg \longrightarrow RHgX$$

$$RHgX + e \longrightarrow RHg^{\cdot}_{ads} + X^-$$

$$\text{or} \quad RX + e \longrightarrow RHg^{\cdot}_{ads} + X^-$$

$$2\,RHg^{\cdot}_{ads} \longrightarrow R_2Hg + Hg^0$$

$$R_2Hg + 2\,e \longrightarrow 2\,R^- + Hg^0$$

The mechanism outlined in Scheme 3, involving a radical anion and free-radical intermediates, is the generally accepted one, although it has some limitations. Some halogenated aromatic ketones, nitro and nitrile compounds, and halogenated polynuclear aromatics can be reduced to fairly stable radical anions which could be characterized by electrochemical and ESR methods[229]. In these cases, cleavage into free radicals and halide ions occurs in the bulk solution and the final products result, at least partially, from free-radical reaction pathways[230]. It is assumed that the decay of the radical anion is very fast (there is theoretical[231] and kinetic[232] evidence for this) or even concerted[172] with the first electron transfer to prevent escape of $RX^{\cdot}$ from the electrode surface by diffusion. Recently, Saveant and coworkers[233] have also proposed essentially a simultaneous electron transfer bond-breaking process for the reduction of aliphatic halides.

There is no direct evidence for the existence of free radicals, and attempts to detect them by ESR measurements and spin trapping techniques have failed[234–236]. However, the intermediacy of radicals has been established indirectly through isolation of dimeric products, organometallic compounds at mercury or lead cathodes[237,238], and rearrangement products[239]. Some authors have pointed out that dimerisation in the electrochemical reduction of organic halides may well be the result of a nucleophilic Wurtz–Fittig type attack of the carbanion intermediate on the starting halide, which would be even more likely with the highly reactive benzylic or allylic species[235,240].

For halo-compounds which show two reduction waves in d.c. polarography, the second electron transfer was attributed[236] to the reduction of $R^{\cdot}$ to the corresponding R^-, thus suggesting that this process occurs at a more negative potential than that of the first electron transfer. The formation of carbanions in the second step can be demonstrated, for example, by trapping the carbanion with carbon dioxide[241] or by isolating Hofmann-elimination products (when reduction is carried out in the presence of quaternary ammonium salt as electrolyte)[228,238]:

$$R^- + Et_4N^+ \longrightarrow RH + CH_2{=}CH_2 + Et_3N$$

2. Acyclic monohaloalkanes

Although most simple monohaloalkanes are electrochemically reduced in a single step[242], some compounds exhibit two polarographic waves. Consequently, new mechanisms have emerged, involving radical anions, radicals and organomercury intermediates, as illustrated in the schemes above (Section V.A.1). Just to present a simple example, *t*-butyl bromide gives a single 2e wave in DMSO, but *t*-butyl iodide affords two waves. Fry[243] has suggested that both the ease of reduction of the C—I bond and the difficulty of reducing the *t*-butyl radical to the corresponding carbanion give rise to separate one-electron waves from butyl iodide.

Recently, interesting results have been reported on the electrochemistry of primary haloalkanes. Peters and coworkers[244] have studied the reduction of 1-bromo- and 1-iododecane in DMF at Hg cathode and originally stated that the process involves one electron and no carbanions. More recently they reported[245] that both radicals and carbanions lie on the product-forming pathways. For 1-iododecane this group has observed two well resolved waves ($E_{1/2}{}^1 = -1.61$, $E_{1/2}{}^2 \approx -2.1$ against SCE) by d.c. polarography study and suggested that the decyl radical arising from this reduction is stabilized sufficiently to allow the two waves to be resolved. This stabilization is mainly due to the strong interaction of the decyl radical with the mercury electrode. Evidently, the isolation of didecylmercury (at potentials more positive than the second reduction wave) in nearly quantitative yield supports this hypothesis. Contrastingly, 1-bromodecane does not show two waves and does not afford a detectable amount of mercury compound at any reduction potential. They further studied the reduction of 1-iododecane by means of chronocoulometry, polarography, cyclic voltammetry and controlled potential coulometry. It was reported that the relative time scale of the techniques is of great importance (when potentials negative of the polarographic maximum (*ca.* −2.05 V) have been employed) and strongly affects the *n*-value. Electrolysis with the latter technique in DMF + 50 mM H_2O for 10 min affords $n = 1$. On the other hand, with the other three methods, where electrolysis can be done with much shorter times, an *n*-value of 2 is obtained (at potentials more negative than −2.1 V). At these negative potentials, the decyl carbanion is the predominant intermediate, and this was confirmed by deuterium-labelled trapping experiments. The amount of water present also has a strong effect on the results. When the concentration of H_2O is high (50 mM), then $n = 1$ results from the protonation of electrolytically generated decyl anions by H_2O, followed by producing 1-decene and 1-decanol, as follows:

$$n\text{-decyl}^- + H_2O \longrightarrow n\text{-decane} + OH^-$$

$$n\text{-decyl I} + OH^- \begin{cases} \xrightarrow{E_2} \text{1-decene} + H_2O + I^- \\ \xrightarrow{S_N2} \text{1-decanol} + I^- \end{cases}$$

However, when the concentration of H_2O is low (2 mM), the *n*-value approaches 2 and decyl carbanions are presumably protonated by DMF to produce mostly decane:

$$n\text{-decyl}^- + HCON(Me)_2 \longrightarrow n\text{-decane} + {}^-CON(Me)_2$$

At more negative electrolysis potentials, products of high molecular weight were formed, presumably from 1-decane and radicals derived from DMF and *N*-methylformamide as an impurity.

Wagenknecht[246] has studied the reduction of *n*-BuBr in a CO_2-saturated solution of DMF, at a Hg cathode. Approximately equal amounts of three products were obtained, one from direct reduction of butyl bromide and the other two from indirect reduction by $CO_2^{\cdot-}$. Apparently, the reduction on Hg was found to be a one-electron transfer:

$$2\,BuBr \xrightarrow{+2e} Bu_2Hg + Hg + 2\,Br^-$$

$$2\,CO_2 \xrightarrow{+2e} 2\,CO_2^{\cdot-} \longrightarrow {}^-OC(=O)\text{—}C(=O)O^- \xrightarrow[-2\,Br^-]{BuBr} BuO_2CCO_2Bu \quad \text{(dibutyl oxalate)}$$

$$CO_2^{\cdot-} + BuBr \longrightarrow Br^- + [BuCO_2^{\cdot}] \xrightarrow{+1e} BuCOO^-$$

$$CO_2^{\cdot-} + Bu^{\cdot} \longrightarrow BuCO_2^- \xrightarrow[-Br^-]{BuBr} BuCO_2Bu \quad \text{(butyl valerate)}$$

The same reduction on graphite appears to be a 2e process leading to butyl carbanion which reacts chemically to afford the following products:[246]

$$BuBr \xrightarrow[\text{graphite}]{2e} Br^- + Bu^-$$

$$Bu^- + BuBr \longrightarrow \text{Butane + Butene}$$

$$Bu^- + BuBr \longrightarrow \text{Octane}$$

$$Bu^- + CO_2 \longrightarrow BuCO_2^- \xrightarrow[-Br^-]{BuBr} BuCO_2Bu$$

$$Bu^- + HCONMe_2 \longrightarrow \text{Butane} + {}^-CONMe_2$$

$${}^-CONMe_2 \xrightarrow[BuBr]{CO_2} BuOC(=O)C(=O)NMe_2 \quad \text{(butyl-}N,N\text{-dimethyloxamate)}$$

$$Bu^- + Et_4N^+ + 2\,CO_2 + 2\,BuBr \xrightarrow{+2e} Et_3N + 2\,Br^- + BuO_2CCH(CH_3)CO_2Bu$$

(dibutyl α-methylmalonate)

When butyl iodide was reduced at a graphite cathode, both butyl valerate and butyl *N*,*N*-dimethyloxamate were formed. Due to the fact that butyl-*N*,*N*-dimethyloxamate is not obtained when either BuBr or BuI is reduced on Hg, Wagenknecht suggests the absence of Bu^- as an intermediate in the reduction of butyl halide at a Hg cathode. Consequently, two distinct pathways were described above for the formation of butyl valerate, via butyl radical (on Hg) and via butyl carbanion (on graphite).

As has been demonstrated above, the electrode material is known to play an important role in the electroreduction of alkyl halides. Other examples reveal that the final products of reduction of alkyl halides on Hg are generally alkanes; on lead, tetraalkyllead is formed[237], and reduction on a tin[247] cathode gives a product distribution which is dependent upon the size of the alkyl group. The major product from the reduction of methyl iodide in DMF is Me_4Sn, but from higher alkyl iodides dimers as well as monomeric hydrocarbons are formed. It is assumed that steric factors, as well as the strength of metal–alkyl bonds, dictate the product distribution.

3. *Acyclic polyhaloalkanes*

It has been found[248] that the ease of reduction of *gem*-polyhaloalkanes at a glassy carbon electrode, in anhydrous DMF proceeds in the expected order, $CX_4 > RCX_3 > RCHX_2 > RCH_2X$. However, the reduction of CH_3Cl at Hg is reported to occur more readily than that of CH_2Cl_2. This behaviour clearly indicates that, at mercury, polychloroalkanes may not be reduced by simple stepwise loss of Cl^- as on glassy carbon, due to chemical involvement of the mercury before or during the electroreduction step.

Semmelhack and Heinsohn[249] have shown that 2-haloethylchloroformates yield 2-haloethoxycarbonyl derivatives when they react with amino, hydroxy and thiol groups. Removal of the protecting group from these derivatives occurs in nearly quantitative yield with a minor side-reaction, as follows:

$$PhCO_2CH_2CCl_3 \xrightarrow[-1.65\ V]{MeOH} PhCO_2CH_2\bar{C}Cl_2 \longrightarrow \begin{cases} CH_2{=}CCl_2 + PhCO_2^- \quad (87\%) \\ PhCO_2CH_2CHCl_2 \quad \text{(by-product)} \end{cases}$$

Merz[250] reduced 2,2,2-trichloroethanol and trichlorophenylethyl derivatives electrochemically. The major product is reported to be 1,3-dichloroalkene and its formation is affected by both the acidity of the solution as well as by the nature of the substituent adjacent to the trichloromethyl group.

The predominant reduction pathway of vicinal polyhaloalkanes involves β-elimination of two halide ions in an overall 2e step and olefin formation, although stepwise loss of halogen from 1,2-dihalides has also been claimed.

Several groups[251,252] have recently studied the electrochemical reduction of various *meso*-vicinal dibromides in order to try to determine the stereochemistry of the reaction. It has been found that all exhibit high specificity for *antiperiplanar* elimination of halide ion to form the (*E*)-olefin:

$$\text{Newman projection (Br, R, H / Br, R, H)} \xrightarrow{+2\,e} (H)(R)C{=}C(R)(H) + 2\,Br^-$$

In contrast, the behaviour of *d,l*-diastereomers is more complicated and does not appear to be predictable[251,252]. For example:

$$\text{Newman projection (X, H, R / X, H, R)} \xrightarrow{+2\,e} \begin{cases} \xrightarrow{R = Ph,\ X = Cl} (H)(Ph)C{=}C(Ph)(H) + 2\,Cl^- \quad (E)\text{-stilbene} \\ \xrightarrow{R = Me,\ X = Br} (H_3C)(H)C{=}C(CH_3)(H) + 2\,Br^- \quad (Z)\text{-2-butene} \end{cases}$$

There is evidence with the reduction of 1,1,2,2-tetrabromoethane that potential does not affect the stereochemistry of the reduction product[253]. It has been suggested that the (*E*)-1,2-dibromoethene is formed from the antiplanar rotamer, whereas the

(*Z*)-isomer is formed from the predominant (85%) synclinal rotamer. Other factors have been suggested to affect the (*Z*)/(*E*) product ratio, such as steric effects[254], an anionic intermediate and differences in ion pair formation[252].

The mechanism of the reduction of vicinal dihalides to olefinic products, by a concerted 2e process or via a discrete carbanion intermediate, is still debatable:

$$X{-}\overset{|}{\underset{|}{C}}{-}\overset{|}{\underset{|}{C}}{-}X \xrightarrow{+2e} [X\cdots\overset{|}{\underset{|}{C}}\cdots\overset{|}{\underset{|}{C}}\cdots X]^{2-} \xrightarrow{-2X^-} \text{Alkene}$$

or

$$X{-}\overset{|}{\underset{|}{C}}{-}\overset{|}{\underset{|}{C}}{-}X \xrightarrow[-X^-]{+e} X{-}\overset{|}{\underset{|}{C}}{-}\overset{|}{\underset{|}{C}}^{\bullet} \xrightarrow{+e} X{-}\overset{|}{\underset{|}{C}}{-}\overset{|}{\underset{|}{C}}^{-} \xrightarrow{-X^-} \text{Alkene}$$

Some investigators are in favour of a concerted process due to the remarkable ease with which dihalides are reduced[255], or due to the absence of a carbanion intermediate (to form any proton capture or Hofmann degradation products)[251]. Garst and coworkers[256] have studied the reduction of 2,3-dihalobutanes by sodium naphthalide and with the assistance of CIDNP experiments suggested that the mechanism for 2,3-dibromobutane may involve a loss of a bromine atom from either the initial radical anion or the bromoalkyl radical:

$$[Br{-}\overset{|}{\underset{|}{C}}{-}\overset{|}{\underset{|}{C}}{-}Br]^{\overline{\bullet}} \xrightarrow{-Br^-} Br{-}\overset{|}{\underset{|}{C}}{-}\overset{|}{\underset{|}{C}}^{\bullet} \longrightarrow \text{Alkene} + Br^{\bullet}$$

$$[Br{-}\overset{|}{\underset{|}{C}}{-}\overset{|}{\underset{|}{C}}{-}Br]^{\overline{\bullet}} \longrightarrow \text{Alkene} + Br^- + Br^{\bullet}$$

Mishra and Symons[257] have studied the reduction of 1,2-dihaloethanes and with the assistance of ESR and ^{60}Co γ-ray techniques reached the conclusion that the reaction pathway involves a loss of $X_2^{\overline{\bullet}}$ from initially formed radical anion, as follows:

$$[X{-}\overset{|}{\underset{|}{C}}{-}\overset{|}{\underset{|}{C}}{-}X]^{\overline{\bullet}} \longrightarrow \text{Alkene} + X_2^{\overline{\bullet}}$$

Electroreduction of 1,3-dihalides at a Hg cathode affords cyclization as the principal reaction pathway, whereas higher homologues yield substantial quantities of organomercury derivatives accompanied by little or no cycloalkane. However, on Pt cathode in THF, moderate yields of 3–7 ring cycloalkanes can be formed[258]. The generally accepted concerted 2e pathway for the formation of cycloalkane requires stereospecific reduction. Fry and Britton[259], have shown that reduction of both *meso*- and *d,l*-2,4-dibromopentane is non-stereospecific and affords almost equal amounts of products from either diastereomer:

$$CH_3\underset{Br}{\underset{|}{C}}HCH_2\underset{Br}{\underset{|}{C}}HCH_3 \xrightarrow[-2\,Br^-]{+2\,e} \text{trans-1,2-dimethylcyclopropane} + \text{cis-1,2-dimethylcyclopropane}$$

	trans (CH_3 / CH_3)	cis (CH_3 CH_3)
meso–	41%	45%
dl–	44%	39.5%

Moreover, they also obtained pentane and 1- and 2-pentene, but no dimers or organomercurial products. In the light of their results they are in favour of a stepwise reduction pathway of 1,3-dihalides.

The product distribution from the reduction of 1,3-dibromopropane at Pt in DMF has been shown by Wiberg and Epling[260] to be potential dependent; cyclopropane is the major product at −2.65 V (against Hg) whereas propane becomes the main product at −1.45 V. By performing labelling experiments, they concluded that propane is formed predominantly via hydrogen atom abstraction by the propyl radical and not a carbanion intermediate. Similar product-distribution potential dependence has been observed for 1,4-dibromobutane at Pt, but on Hg mainly dibutylmercury along with minor by-products were detected[261].

In view of the fact that cyclopropanes are formed in good yields upon reduction of 1,3-dihalopropanes, attempts have been made to synthesize cyclopropanones electrochemically from α,α′-dihaloketones[262–264]. However, non-cyclized products (involving the intermediacy of an enol allylic halide and 2-hydroxyallylic cation) are formed with the exception of the successful trapping of several cyclopropanones as their hemiketals. With regard to the formation of a cyclic product, an interesting result was obtained by cathodic reduction of α,α′-dibromophosphinate on Hg in DMSO[265]. The products formed confirm strongly the intermediacy of a cyclic derivative, oxyphosphirane:

$$\mathrm{PhCHBrP(O)(OMe)CHBrPh} \xrightarrow[-\mathrm{Br}^-]{+2e} \mathrm{Ph\bar{C}HP(O)(OMe)CHBrPh} \xrightarrow[-\mathrm{Br}^-\ +2\mathrm{H}^+]{+2e} \mathrm{(PhCH_2)_2P(O)OMe}\ (24\%)$$

$$\mathrm{Ph\bar{C}HP(O)(OMe)CHBrPh} \xrightarrow{-\mathrm{Br}^-} \text{oxyphosphirane (Ph, H / Ph, H; P(=O)OMe)} \xrightarrow{-\mathrm{P(O)OMe}} \mathrm{Ph(H)C{=}C(H)Ph}\ (43\%) + \mathrm{Ph(H)C{=}C(Ph)H}\ (33\%)$$

4. *Allyl halides*

Lately, Bard and Merz have reinvestigated[235] the electrochemical reduction of allyl halides in non-aqueous media, by means of cyclic voltammetry and coulometry. They reduced allyl iodide, allyl bromide, (*E*)-3-bromo-1-phenyl-1-propene and (*E*)-5-bromo-2,2,6,6-tetramethyl-3-heptene on vitreous carbon and each compound gave a single 2e reduction wave, with half-peak potentials of −1.38, −1.64, −1.11 and −1.89 V (against SCE), respectively. However, on Pt and Hg, multiple waves were formed because of halide surface effects (which have been noticed previously[226] in the electrochemical reduction of 9,10-dichloro-9,10-dihydro-9,10-diphenylanthracene at Pt and Au electrodes in DMF or CH_3CN). These authors suggest that the reduction of allyl halides on Hg involves organomercury intermediates which are reduced at potentials corresponding to the second polarographic wave, according to the mechanism outlined in Scheme 6 (Section V.A.1). Their conclusion is that the allyl halides are reduced via a two-electron electrode reaction and the reduction of the allyl radical to the allyl anion cannot be seen as a separate step. Moreover, the reduction of the allyl radicals occurs at potentials substantially more positive than the potential required for the first electron transfer to the allyl halide. This was confirmed by electrochemical oxidation of allyl anion, previously formed by reduction of allyl halide, with solvated electrons in liquid ammonia. These authors' principal conclusion is in contrast to previous statements in the literature since it leads to the mechanism first suggested by Von Stackelberg and Stracke[267], which is outlined in Scheme 1 (Section V.A.1).

Polyhaloallyl compounds, of the general formulae $R_2CXCH{=}CHX$ and $R_2C{=}CHCHX_2$, have been electrochemically reduced by Doupeux and Simonet[268]. 3,3-Dibromo-1,1-diphenylpropene has been found to undergo prior chemical reaction with the mercury cathode to give a readily reducible organomercury derivative, followed by two consecutive 2e waves to form 1,1-diphenylpropane, as follows:

$$Ph_2C{=}CHCHBr_2 \xrightarrow{2\,Hg} Ph_2C{=}CHCH(HgBr)_2 \xrightarrow[-2\,Br^-]{2e} Ph_2C{=}CHCH\begin{cases}Hg^{\bullet}\\Hg^{\bullet}\end{cases} \xrightarrow[2\,H^+]{2e}$$

$$Ph_2C{=}CHCH_3 \xrightarrow[+2\,H^+]{2e} Ph_2CHCH_2CH_3$$

However, unlike the above compound, the reduction of 1,1-dichloro- or 1-bromo-1-chloro- derivatives does involve a carbanion intermediate, which undergoes either protonation to give 1-chloro-3,3-diphenylpropene or intramolecular cyclization with concomitant loss of Cl^- to yield 1,1-diphenylcyclopropene. The former product could be reduced further at more negative potentials to form 1,1-diphenylpropene and then to 1,1-diphenylpropane.

Brillas and Costa[269] have reported that the electroreductions of (E)-1,4-dibromo- or (E)-1,4-dichloro-2-butene each consisted of two waves. The second was attributed to the reduction of the product, 1,3-butadiene, and the first to the addition of 2e to give a halide ion and anionic intermediate:

$$XCH_2CH{=}CHCH_2X \xrightarrow{+2e} X^- + XCH_2CH{=}CHCH_2^- \xrightarrow{-X^-} CH_2{=}CHCH{=}CH_2$$

Halogenated acetylenes have been investigated by Peters and coworkers[270–272]. Since the isolated triple bond in 1-bromo-5-decyne is electrochemically inert, only a single (n = 1) polarographic wave ($E_{1/2} = -2.45$ V, against SCE) has been observed and attributed to the cleavage of the C—Br bond, to form the following products:

$$CH_3(CH_2)_3C{\equiv}C(CH_2)_4Br \longrightarrow CH_3(CH_2)_3C{\equiv}C(CH_2)_3CH_3 + \text{1-}n\text{-Bu-cyclohexene}$$

59% 38%

Since the cyclic product does not contain bromine atom, these workers postulated that it arises from either adsorbed radical $CH_3(CH_2)_3C{\equiv}C(CH_2)_4^{\bullet}$ or an organomercury intermediate, $CH_3(CH_2)_3C{\equiv}C(CH_2)_4HgBr$ or $CH_3(CH_2)_3C{\equiv}C(CH_2)_4]_2Hg$. The nature of the products from the electroreduction of 6-bromo-1-phenyl-1-hexyne, as well as their distribution, have been found to be potential dependent[270]. At −2.45 V (against SCE) and n = 1, 1-phenylhexyne (58%), 1-phenylcyclohexene (14%), benzylidenecyclopentane (12%) and 1-phenyl-1-hexyne-5-ene (9%) have been formed, presumably all due to the involvement of the radical, $PhC{\equiv}C(CH_2)_4^{\bullet}$. Electrolysis at more negative potential (−2.60 V) afforded n = 3 and higher yields of the carbocyclic products: cyclohexylbenzene (23%), benzylidenecyclopentane (20%), 1-phenylcyclohexene (11%), 1-benzylcyclopentene (7%), 1-phenyl-1-hexyne (17%) and 1-phenyl-1-hexene (13%). At this potential, a reduction of the triple bond was postulated too:

$$PhC{\equiv}C(CH_2)_4Br \xrightarrow{+e} Ph\dot{C}{=}\bar{C}(CH_2)_4Br \longrightarrow Br^- + (\text{Ph}\dot{C}H{=}\text{cyclopentylidene}) \longrightarrow PhCH{=}\text{cyclopentylidene}$$

The redox behaviour of 6-chloro-1-phenyl-1-hexyne has been found[271] to be more complex. Here, the major product was benzylidenecyclopentane (81%) along with various hexenes, one hexyne and one hexane. The following mechanistic scheme has been proposed:

$$\text{Ph}\dot{\text{C}}{=}\bar{\text{C}}(\text{CH}_2)_4\text{Cl} \xrightarrow{-\text{Cl}^-} \text{PhC}{\equiv}\text{C}(\text{CH}_2)_3\text{CH}_2^{\bullet} \longrightarrow \text{PhC}{\equiv}\text{C}(\text{CH}_2)_3\text{CH}_3$$

$$\text{Ph}\dot{\text{C}}{=}\bar{\text{C}}(\text{CH}_2)_4\text{Cl} \xrightarrow{\text{H}^+} \text{Ph}\dot{\text{C}}{=}\text{CH}(\text{CH}_2)_3\text{CH}_2\text{Cl} \longrightarrow (E)\text{—PhCH}{=}\text{CH}(\text{CH}_2)_3\text{CH}_2\text{Cl} \xrightarrow[-\text{Cl}^-,+\text{H}^+]{\text{Cathode}} \text{PhCH}_2\text{—cyclopentane}$$

At higher concentration of the substrate, an allenic derivative, 6-chloro-1-phenyl-1,2-hexadiene is formed[272], which is further reduced to a radical anion. This intermediate may undergo either intramolecular cyclization or protonation, as follows:

$$\text{PhCH}{=}\text{C}{=}\text{CH}(\text{CH}_2)_3\text{Cl} \xrightarrow{+\text{e}} \text{Ph}\dot{\text{C}}\text{H}\bar{\text{C}}{=}\text{CH}(\text{CH}_2)_3\text{Cl} \xrightarrow{-\text{Cl}^-}$$

$$\left[\text{PhCH}^{\bullet}\text{—(cyclopent-1-enyl)} \longleftrightarrow \text{PhCH}{=}\text{(cyclopentan-1-ylidene-2-yl radical)}\right] \xrightarrow[+\text{H}^+]{+\text{e}} \text{1-(CH}_2\text{Ph)cyclopentene} + \text{CHPh}{=}\text{cyclopentane}$$

5. *Benzyl halides and halomethyl arenes*

It is accepted that benzyl iodide reduction at Hg involves the intermediacy of benzylmercury radical, and a similar but slower reaction is assumed for benzyl bromide and chloride. The following mechanism has been suggested[240,241,273] (which is slightly different from that in Scheme 6 (Section V.A.1)):

Initiation
$$\text{PhCH}_2\text{I} + \text{e} \longrightarrow [\text{PhCH}_2\text{I}]^{\bar{\bullet}} \longrightarrow \text{PhCH}_2^{\bullet} + \text{I}^-$$

Propagation
$$\text{PhCH}_2^{\bullet} + \text{Hg} \longrightarrow \text{PhCH}_2\text{Hg}^{\bullet}$$
$$\text{PhCH}_2\text{Hg}^{\bullet} + \text{e} \longrightarrow \text{PhCH}_2^{-} + \text{Hg}$$

or

$$\text{PhCH}_2\text{Hg}^{\bullet} + \text{PhCH}_2\text{I} \longrightarrow \text{PhCH}_2\text{Hg}^{+} + \text{PhCH}_2^{\bullet} + \text{I}^-$$
$$\text{PhCH}_2\text{Hg}^{+} + \text{I}^- \longrightarrow \text{PhCH}_2\text{HgI} \rightleftharpoons 1/2\,(\text{PhCH}_2)_2\text{Hg} + 1/2\,\text{HgI}_2$$

Termination
$$\text{PhCH}_2^{\bullet} + \text{SH} \longrightarrow \text{PhCH}_3 + \text{S}^{\bullet}$$
$$\text{PhCH}_2^{\bullet} + \text{PhCH}_2\text{Hg}^{\bullet} \longrightarrow (\text{PhCH}_2)_2\text{Hg}$$
$$2\,\text{PhCH}_2^{\bullet} \longrightarrow \text{PhCH}_2\text{CH}_2\text{Ph}$$

The presence of a carbanion intermediate has been established[240,241] in the reduction of benzyl chloride, and more recently in 1-bromo-1-phenylethane derivatives[274].

$$PhCH_2Cl + 2\,e \longrightarrow PhCH_2^- + Cl^-$$

$$PhCH_2^- \xrightarrow{CO_2} PhCH_2COO^-$$

$$PhCH_2CO_2^- + PhCH_2Cl \longrightarrow PhCH_2COOCH_2Ph + Cl^-$$

The reduction of *o*- and *p*-nitrobenzyl bromide, chloride and thiocyanate at Hg or Pt affords mainly bibenzyl in high yield (≥85%) and only a small amount of nitrotoluene[275]. The suggested mechanism is in line with the radical mechanism outlined in Scheme 3 (Section V.A.1), excluding the intermediacy of a carbanion[276a]. In contrast to the above results, *m*-nitrobenzyl halides form *m*-nitrotoluene as the major product in an overall 1e step[276b], and it has been suggested that the radical anion of the *meta* isomer is more stable than its *ortho* or *para* counterparts[276a].

Other related compounds investigated by various groups include α,α'-dibromoxylenes[277], benzhydryl halides[278,307], and trichloro- and trifluoromethylbenzenes[279]. Generally, the reduction involves cleavage of the carbon–halogen bond followed by subsequent substitution by hydrogen:

$$X\text{-}C_6H_4\text{-}CF_3 \xrightarrow[\text{MeOH}]{\text{Pb cathode}} X\text{-}C_6H_4\text{-}CH_3 \quad (X = CN, COOMe)$$

An interesting case is the cathodic reduction of $\alpha,\alpha,\alpha',\alpha'$-tetrabromo-*o*-xylene, which has been investigated by Rampazzo and coworkers[280]. This compound exhibits two major voltammetric peaks at a Hg cathode in DMF. Electrolysis at the first wave affords 1,2-dibromobenzocyclobutene and polymers, whereas at the second wave a tetracyclic product is formed, resulting from a rearrangement of a Diels–Alder benzocyclobutadiene dimer, as follows:

$$o\text{-}C_6H_4(CHBr_2)_2 \xrightarrow[\text{2nd wave}]{+4\,e,\ -4\,Br^-} [\text{benzocyclobutadiene}] \longrightarrow \text{tetracyclic product}$$

6. Alicyclic halides

This topic has been well treated in the previous review by Casanova and Eberson in this series[1] and has been reviewed recently by Hawley (Chapter 2 in Ref. 10). Therefore, only recent developments in this field are outlined below.

Casanova and Rogers[281] and Wiberg and coworkers[282] have reported on a practical and convenient method for the synthesis of highly strained olefins from the electrolytic reduction of 1-bromo-4-chlorobicyclo[2.2.0]hexane. $\Delta^{1,4}$-bicyclo[2.2.0]hexene is formed in quantitative yield and trapped with cyclopentadiene:

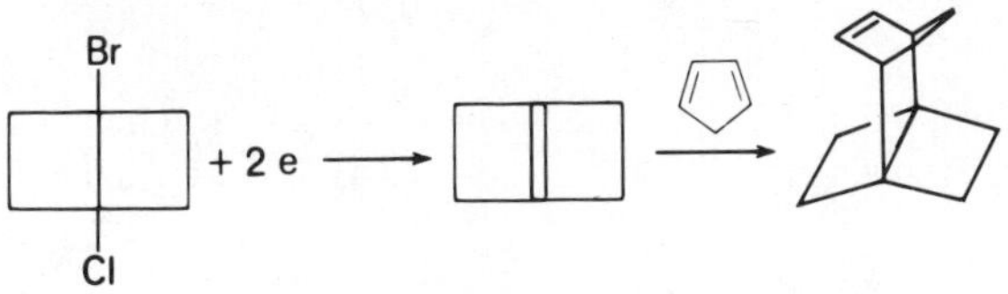

Br

+ 2 e

Cl

Electroreduction of 1,4-dibromobicyclo[2.2.2]octane afforded [2.2.2]propellane (12%), which was trapped by chlorine to obtain 1,4-dichlorobicyclo[2.2.2]octane[283]:

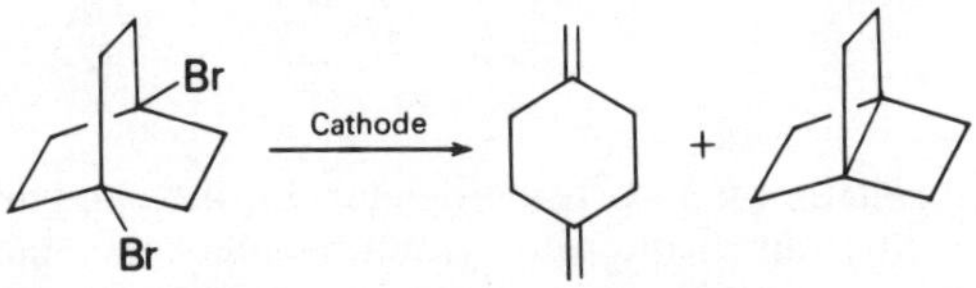

Br

Cathode

+

Br

Recently, Peters and coworkers[284] have found evidence for the intermediacy of [2.2.1]propellane upon reducing 1,4-dihalonorbornanes electrochemically at low temperature, although attempts to isolate it were unsuccessful.

In contrast to a previous report[285] that anionic intermediates or a concerted process must be involved in the electroreduction of dihalobornanes, Azizullah and Grimshaw[286] have reported that the intermediacy of radicals is favoured in the reduction of certain dibromobornanes. The following mechanism has been suggested for the reduction of *endo*-2,*endo*-6-dibromobornane in DMF:

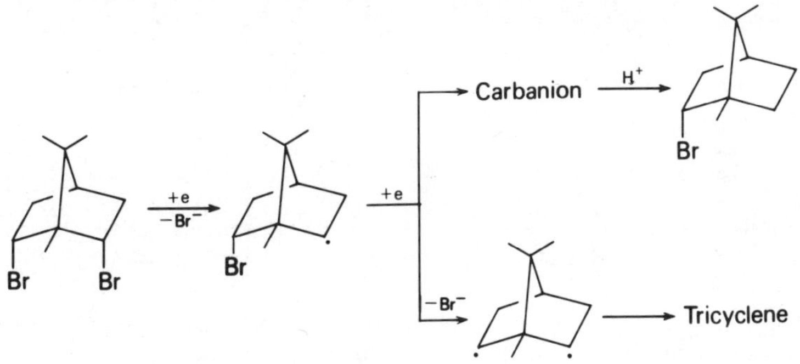

Carbanion $\xrightarrow{H^+}$

Br

$\xrightarrow[-Br^-]{+e}$ $\xrightarrow{+e}$

Br Br Br

$-Br^-$

Tricyclene

7. *Aryl halides*

Most published works on halobenzenes are limited to polarographic studies. The generally accepted mechanism is the one outlined in Scheme 1 (Section V.A.1). However, radical intermediates have been suggested for dimerization of certain halogenated aryl compounds[287] and for hydrogen atom abstraction from the solvent[288] as well as for intramolecular cyclization[289]. The reductions of *meta* and *para* halobenzenes usually proceed by the stepwise removal of halogen[290]. However, *o*-dibromobenzene behaves differently and Wawzonek and Wagenknecht[291] have suggested the intermediacy of benzyne.

In the reduction of halogenated nitrobenzenes, the intermediacy of the nitrophenyl radical has been demonstrated by anion capture[292]:

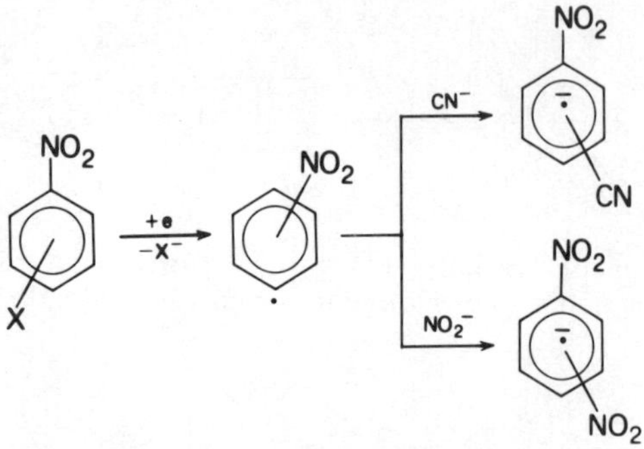

$$\text{NO}_2\text{C}_6\text{H}_4\text{X} \xrightarrow[-X^-]{+e} \cdot\text{C}_6\text{H}_4\text{NO}_2 \begin{cases} \xrightarrow{CN^-} [\text{NO}_2\text{C}_6\text{H}_4\text{CN}]^{\overline{\cdot}} \\ \xrightarrow{NO_2^-} [\text{NO}_2\text{C}_6\text{H}_4\text{NO}_2]^{\overline{\cdot}} \end{cases}$$

The ease of dehalogenation follows the order I > Br > Cl and *ortho* ≫ *para* > *meta*. Alkyl substituents adjacent to the nitro group decrease the stabilities of the corresponding radical anions and enhance the rate of halide ion loss, as was found by Hawley and coworkers[293]. Other kinetic studies regarding the effect of temperature, electrode potential, etc., on halide ion loss have also been conducted[294,295].

It has been shown previously[296], that certain substituted benzonitriles afforded stable radical anions whereas others, e.g. *p*-amino- and *p*-fluoro-, decomposed rapidly to form cyanophenyl radicals followed by dimerization:

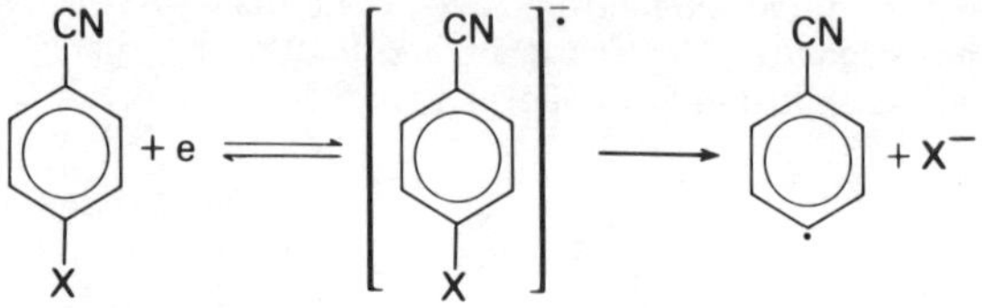

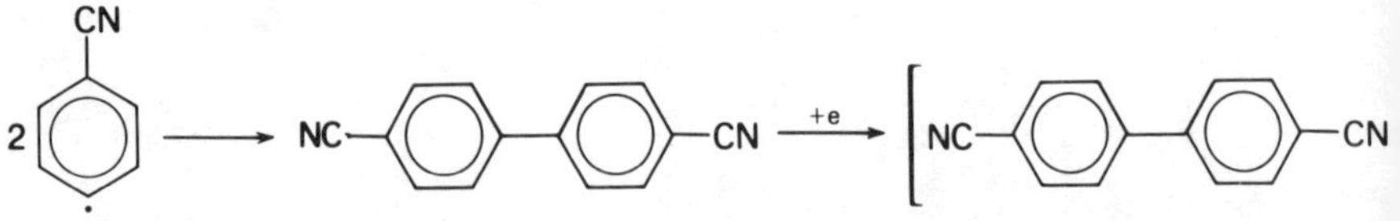

$$p\text{-NCC}_6\text{H}_4\text{X} + e \rightleftharpoons [p\text{-NCC}_6\text{H}_4\text{X}]^{\overline{\cdot}} \longrightarrow p\text{-NCC}_6\text{H}_4^{\cdot} + X^-$$

$$2\, p\text{-NCC}_6\text{H}_4^{\cdot} \longrightarrow \text{NC}-\text{C}_6\text{H}_4-\text{C}_6\text{H}_4-\text{CN} \xrightarrow{+e} [\text{NC}-\text{C}_6\text{H}_4-\text{C}_6\text{H}_4-\text{CN}]$$

Kemp and coworkers[297] studied the reduction of fluorobenzonitriles by solvated electrons in ammonia and found no dimeric products, but only benzonitrile. Later, Hawley and coworkers[287,298] reported that electrochemical reductive dehalogenation of halobenzonitriles occurs by at least five different pathways in DMF.

Electrochemical reduction of halobenzophenones have been studied by Saveant and coworkers[126,299] and the following mechanism has been suggested:

$$\text{PhCOC}_6\text{H}_4\text{X} + e \rightleftharpoons [\text{PhCOC}_6\text{H}_4\text{X}]^{\overline{\cdot}} \xrightarrow{-X^-} \text{PhCOC}_6\text{H}_4^{\cdot}$$

$$\text{PhCOC}_6\text{H}_4^{\cdot} + \text{SH} \longrightarrow \text{PhCOPh} + \text{S}^{\cdot}$$

The above scheme has been supported by the results obtained by Grimshaw and coworkers[300], precluding the intermediacy of carbanion intermediates (no incorporation of deuterium into benzophenone resulted when reductions were carried out in DMF–1% D_2O). However, the intermediacy of both aryl radicals and carbanions has been suggested in the electroreduction of halobenzamide derivatives[300,301]. In certain cases, deuterium did incorporate into the site previously occupied by the halogen.

A vast literature on polarographic studies of substituted mono- and polyhalobenzenes is available, and, for many of them, 2e reductive dehalogenations have been observed. Cockrell and Murray[302] have managed to prepare dehalogenated products with excellent isotopic purity, via aryl carbanion intermediates.

8. Halogenated polycyclic arenes

In contrast to polychlorobenzenes[290], the polychlorobiphenyls give a large number of reduction products[303,304]. Some of the polychlorobiphenyls also undergo reductive dechlorination, losing two chlorines simultaneously. The intermediacy of benzyne is excluded since the chlorines lost are usually *meta* to one another.

Campbell[305] has studied the cyclic voltammetry of fluorobiphenyl derivatives and it seems that substitution of chlorine with fluorine greatly enhances the stability of the corresponding fluorobiphenyl radical anion. Campbell also studied the effect of proton donors on the reduction of fluorinated biphenyls and naphthalene and suggested a similar mechanism for all systems studied, in which the fluorinated radical anion is first protonated and then further reduced to the corresponding anion, followed by loss of F^- to give the final aromatic hydrocarbon:

$$\mathrm{ArF} + e \rightleftharpoons [\mathrm{ArF}]^{\overline{\cdot}}$$

$$[\mathrm{ArF}]^{\overline{\cdot}} + \mathrm{H}^+ \longrightarrow [\mathrm{ArHF}]^{\cdot}$$

$$[\mathrm{ArHF}]^{\cdot} + e \longrightarrow [\mathrm{ArHF}]^-$$

$$[\mathrm{ArHF}]^- \longrightarrow \mathrm{ArH} + \mathrm{F}^-$$

Halonaphthalenes have been also studied by Renaud[306] in the presence of D_2O and relatively high isotope purity (>88%) of deuterated naphthalene was obtained from the iodo- or bromonaphthalene, whereas only ~50% was obtained from chloronaphthalene.

Electrochemical reduction of 9-halo- and 9,9-dihalofluorenes has been conducted by Hawley and coworkers[307,308]. Usually, the process involves an overall 2e reductive cleavage of a halide ion and formation of the corresponding ion. The final products, monomers and dimers (see below) and their distribution are dependent (at least) upon reduction potential, proton availability, nucleophilic attack by the carbanion intermediate and on the parent compound.

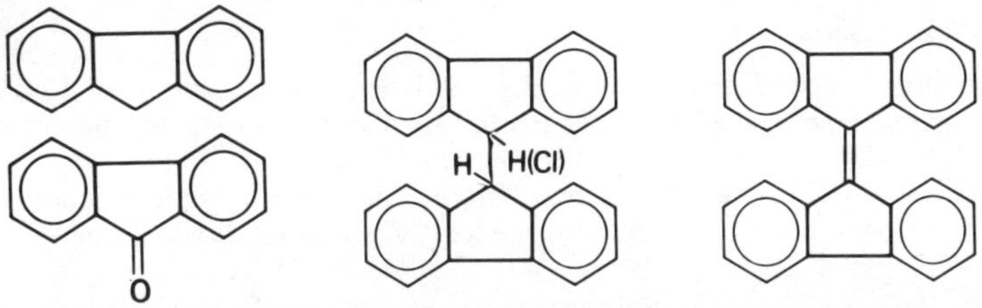

O

H H(Cl)

Hawley and coworkers[308] have also studied the electrochemical reduction of mono- and dihalobifluorenyls. The controlled potential electrolysis of 9-chlorobifluorenyl yielded bifluorenyl as the major product (77%) and bifluorenylidene as the minor one (21%). However, the reduction of 9,9′-dichlorobifluorenyl affords bifluorenylidene in high yield. 9,9′-Dibromobifluorenyl behaved in a more complicated manner since its reduction is electrocatalysed by both bifluorenylidene radical anion and dianion. If the reduction occurs either directly at the electrode surface (Pt) or indirectly by the bifluorenyl radical anion, bifluorenyl is formed in high yield, whereas little or none of this product is observed when the reduction is mediated by the dianion.

The electroreduction of bromofluorenones yielded fluorenone as the product, whereas no decomposition was observed of the corresponding chlorofluorenone radical anions[309].

9-Bromo-, 9,10-dibromo- and 9,10-dichloroanthracenes[289,310] have been studied in various solvents. A complex mixture of products, including anthracene, di-9-anthrylmercury, anthraquinone and others has been reported[289]. 9,10-Dihalo-, 9,10-dihydro- and 9,10-diphenylanthracenes have been studied at Pt and Au cathodes and their redox behaviour has been shown to be affected by the presence of halide ions and water[266,311]. The isolated reduction product is diphenylanthracene. The dichloro derivative ($DPACl_2$) has been studied also in connection with the electrogenerated chemiluminescence phenomenon which accompanies the reduction of this compound concomitantly with diphenyl anthracene (DPA)[311,312]. It has been suggested that direct excited singlet formation occurs by one of the following pathways:

$$DPA^{\cdot -} + DPACl_2 \longrightarrow [DPACl]^{\cdot} + Cl^- + DPA$$

$$[DPACl]^{\cdot} + DPA^{\cdot -} \rightarrow \rightarrow + {}^1DPA^* + Cl^-$$

or

$$DPA^{\cdot +} + DPA^{\cdot -} \rightarrow \rightarrow DPA + {}^1DPA^*$$

9. *Halogenated heterocycles*

Among all heterocycles, the polarographic reductions of halopyridines have been studied most extensively. The ease of reduction follows the expected order, I > Br > Cl > F. Regarding the position of the substituent, the ease of reduction usually follows the order 4-X > 2-X > 3-X. Protonation or *N*-alkylation shifts the reduction potential to more positive values. Recently, the reduction of dihalopyridines on the mercury dropping electrode has been reported[313]. Most of the studies have been carried out in aqueous or *mixed* aqueous–organic solutions and very few in pure organic solvents[314]. In the few controlled potential electrolysis studies, it has been found that dehalogenation takes place in the reduction of halopyridines.

Other halogenated heterocycles, 6-chloro- and 6-fluoroquinolines[315], monohalo derivatives of quinazoline, quinoxaline and phenazine[315–317], 1,3-thiazoles[317,318] and thiophenes[319] have been studied polarographically.

B. Organic Azides

The electroreduction of organic azides has been studied very little. Lund[320] has found a strong dependence of $E_{1/2}$ on pH for alkyl azides and no dependence for acyl azides (Table 33).

Azides undergo 2e reduction in two different pathways[321]. One reaction involves nitrogen–nitrogen bond cleavage with the loss of N_2, from all alkyl, phenyl and acyl azides:

$$RN_3 \xrightarrow[+2\,H^+]{+2\,e} RNH_2 + N_2$$

$$RCON_3 \xrightarrow[+2\,H^+]{+2e} RCONH_2 + N_2$$

The second pathway involves carbon–nitrogen bond cleavage and takes place whenever the azide possesses an active methylene group, such as phenacyl azide

$$PhCOCH_2N_3 \xrightarrow[+H^+]{+2\,e} PhCOCH_3 + N_3^-$$

TABLE 33. Polarographic half-wave potentials ($-V$ versus SCE) of some azides RN_3 in aqueous solution at room temperature[320]

Substituent	Concentration, M	pH 0.9	2.6	4.3	6.2	9.1	11.9	Product
Allyl	4.8×10^{-4}	(0.88)	(0.91)	0.98	1.07	1.16[a]	–[b]	Unknown
Phenyl	4.1×10^{-4}	0.82	0.90	0.92	0.93	0.98	0.94	Aniline
						1.48	1.47	
Benzyl	1.5×10^{-3}	0.86	(0.85)	0.95[c]	1.04	1.08	1.06	Benzylamine
4-Pyridylmethyl	3.0×10^{-4}	0.82	0.82	0.86	0.97	1.13	1.10	4-Picoline
Phenacyl	2.5×10^{-4}	0.68	0.76	0.85	0.98	1.07	0.99	Acetophenone
		1.00	1.13	1.29	1.44	1.52	1.33	
Benzoyl	2.7×10^{-4}	0.22	0.22	0.21	0.20	0.21	–[b]	Benzamide
Isonicotinoyl	2.7×10^{-4}	0.15	0.15	0.16	0.15	0.15[a]	–[b]	Isonicotinamide
		0.65	0.73	0.88	1.05	1.20		

[a]pH 8.3.
[b]Not measured.
[c]pH 4.8.
Values in parenthesis are uncertain. Products have been isolated and identified[322].

Armand and Souchay have studied[323] the polarographic reduction of carbamoyl azide, as the basis for nitrite determination. The reaction was reported to consume 3Fmol^{-1} in acidic media (pH = 1–4), in contrast to Lund's results[321] which found $n = 2$:

$$H_2N{-}C(=O){-}N_3 + 3e + 3H^+ \xrightarrow{-0.75\text{ V against SCE}} H_2N{-}C(=O){-}NHNH_2 + 1/2\,N_2$$

$$H_2N{-}C(=O){-}N_3 + 2e + 2H^+ \longrightarrow H_2N{-}C(=O){-}NH_2 + N_2$$

Later, in 1970, Kononenko and coworkers[324] studied the electroreduction of substituted azidobenzenes (Table 34) in DMF and observed one 2e reduction wave for all compounds, leading to the formation of the corresponding amines. They also observed a linear correlation (correlation coefficient 0.997) of $E_{1/2}$ with *para* and *meta* substituent constants resulting in $\rho = 0.33 \pm 0.007$. This indicates the existence of an inductive effect of the substituents on the reduction process. The same group has also investigated conjugated diazides (Table 34) of the type $N_3C_6H_4{-}CH{=}Y{=}CH{-}C_6H_4N_3$ (where Y = a conjugated system) and stated that the change from mono- to diazides containing bridge groupings that lengthen the chain of conjugation leads to the facilitation of this reduction ($-E_{1/2} = 0.9$–1.1 V).

Malyugina and coworkers[325,326] have extended the electroreduction to various isomers of halo derivatives of azidobenzene as well as to azidoquinolines and azidonaphthalenes, in ethanol–water solutions. All compounds have shown at least two irreversible reduction waves and some of them three (Table 35). This time, it has been suggested that the reduction of the azido group takes place stepwise in two waves. The

TABLE 34. $E_{1/2}$ of azidobenzenes and diazides in DMF–Et_4NI (against SCE)[324]

Compound	$E_{1/2}$, V
1-Azido-4-nitrobenzene	−1.206[a]
1-Azido-3-nitrobenzene	−1.201[b]
1-Azido-4-methylbenzene	−1.519
1-Azido-3-methylbenzene	−1.499
1-Azido-4-methoxybenzene	−1.544
1-Azido-3-methoxybenzene	−1.408
Azidobenzene	−1.46
Y in $N_3C_6H_4CH{=}Y{=}CHC_6H_4N_3$	
$=CH-C(=O)-CH=$	−1.000
O	−0.935
O	−1.010
O … CH_3	−1.098
O	−1.115
$=N-N=$	−1.163

[a]Two additional polarographic waves, at −0.862 V and −1.777 V, have been observed and correspond to the nitro group.
[b]As in the former footnote, but at −0.707 and −1.907 V.

third wave has been attributed to the carbon–halogen bond fission. The data presented in Table 35 show that the introduction of halogen in the azidobenzene molecule facilitates the reduction of the azido group in comparison with unsubstituted azidobenzene. In these two-step reductions, a linear correlation has been found only with $E_{1/2}$ of the second reduction wave and the Taft σ^* constants for *meta* halo-substituted azidobenzenes. Their results for other azidoaromatic compounds indicate that

TABLE 35. $E_{1/2}$ (against SCE) of azidobenzenes, azidonaphthalenes and azidoquinolines in 25% EtOH–H_2O, 0.1 M $LiClO_4$ [325,326]

	$E_{1/2}$, V			
Compound	1st wave	2nd wave	3rd wave	4th wave
Azidobenzene	−1.15	−1.58	–	–
1-Azido-2-chlorobenzene	−0.97	−1.53	–	–
1-Azido-3-chlorobenzene	−0.94	−1.47	–	–
1-Azido-4-chlorobenzene	−0.97	−1.49	–	–
1-Azido-2-bromobenzene	−0.92	−1.51	–	–
1-Azido-3-bromobenzene	−0.92	−1.48	–	–
1-Azido-4-bromobenzene	−0.92	−1.47	–	–
1-Azido-2-iodobenzene	−0.91	−1.48	–	–
1-Azido-3-iodobenzene	−0.91	−1.36	−1.64	–
1-Azido-4-iodobenzene	−0.90	−1.33	−1.70	–
1-Azido-2,6-dichlorobenzene	−0.91	−1.20	−1.65	–
1-Azido-2,4,6-tribromobenzene	−0.80	−1.22	−1.71	–
1-Azidonaphthalene	−1.03	−1.56	–	–
2-Azidonaphthalene	−0.93	−1.48	–	–
7-Azidoquinoline	−0.67	−1.30	−1.57	−1.86
3-Azidoquinoline	−0.72	−1.35	−1.62	–
6-Azidoquinoline	−0.81	−1.17	−1.53	−1.82
5-Azidoquinoline	−0.90	−1.34	−1.52	–
8-Azidoquinoline	−1.04	−1.63	−1.83	–
8-Azido-2-methylquinoline	−1.04	–	–	–

azidonaphthalenes exhibit two 1e reduction waves whereas azidoquinolines are reduced by one 2e step.

The different electrochemical behaviour obtained for various azidobenzene derivatives by different groups may be accounted for by the influence of media (aqueous alcoholic solution and $LiClO_4$ versus DMF and tetraethylammonium iodide). Both solvent and electrolyte could stabilize the first 1e reduction intermediate and consequently give two distinctive polarographic waves.

VI. ACKNOWLEDGEMENTS

The final portion of this review was written during my tenure as a Visiting Professor at the University of British Columbia. I wish to thank the Chemistry Department of this university for its warm hospitality and to express my sincere gratitude and appreciation to S. F. B. Pickett for proofing a preliminary draft of this chapter and to the able assistance of Mrs Joycie Miura for her tireless typing. Thanks are also due to Mrs E. Solomon and Miss R. Stein, from the Ben-Gurion University of the Negev, for their technical assistance at the earlier stages of the writing. Last but not least I am grateful to all colleagues with whom I have corresponded throughout the writing of this chapter and especially to Profs. H. Lund and L. Eberson.

VII. REFERENCES

1. J. Casanova and L. Eberson in *The Chemistry of the Carbon–Halogen Bond* (Ed. S. Patai), Wiley, New York and London (1973), Part 2, Chap. 15.
2. *Technique of Electro-organic Synthesis*, Vol. V (Eds A. Weisberger and N. L. Weinberg), Wiley, New York (1975).
3. *Organic Electrochemistry* (Ed. M. M. Baizer), Marcel Dekker, New York (1973).

4. C. K. Mann and K. K. Barnes, *Electrochemical Reactions in Nonaqueous System*, Marcel Dekker, New York (1970).
5. A. J. Fry, *Synthetic Organic Electrochemistry*, Harper and Row, New York (1972).
6. M. R. Rifi and F. H. Covitz, *Introduction to Organic Electrochemistry*, Marcel Dekker, New York (1974).
7. L. Eberson and H. Schäfer, *Topics Curr. Chem.* **21**, (1971).
8. L. Eberson and K. Nyberg, *Adv. Phys. Org. Chem.*, **12**, 1 (1976).
9. S. D. Ross, M. Finkelstein and E. J. Rudd, *Anodic Oxidation*, Academic Press, New York, (1975).
10. M. D. Hawley in *Encyclopedia of the Electrochemistry of the Elements*, Vol. XIV (Eds A. J. Bard and H. Lund), Marcel Dekker, New York (1980).
11. L. L. Miller and A. K. Hoffman, *J. Amer. Chem. Soc.*, **89**, 593 (1967).
12. A. Laurent, E. Laurent and R. Tardivel, *Tetrahedron*, **30**, 3423 (1974).
13. L. L. Miller and B. F. Watkins, *Tetrahedron Lett.*, 4495 (1974).
14. A. Laurent and R. Tardivel, *Compt. Rend., C*, **271**, 324 (1970).
15. F. Vincent, R. Tardivel and P. Mison, *Tetrahedron*, **32**, 1681 (1976).
16. E. Laurent and R. Tardivel, *Tetrahedron Lett.*, 2779 (1976).
17. M. Hanack and M. J. Schneider, *Angew. Chem. Int. Ed.*, **6**, 666 (1967).
18. M. Fleischmann and D. Pletcher, *Tetrahedron Lett.*, 6255 (1968).
19. L. L. Miller and V. R. Koch, *J. Amer. Chem. Soc.*, **95**, 8631 (1973).
20. V. R. Koch and L. L. Miller, *Tetrahedron Lett.*, 693 (1973).
21. H. Stetter, M. Schwarz and A. Hirschorn, *Chem. Ber.*, **92**, 1629 (1959); W. Haaf, *Angew. Chem.*, **73**, 144 (1961); T. Sasaki, S. Eguchi and T. Toru, *Bull. Chem. Soc. Japan*, **41**, 236 (1968).
22. F. Vincent, R. Tardivel and P. Mison, *Tetrahedron Lett.*, 603 (1975).
23. J. Y. Becker and M. Münster, *Tetrahedron Lett.*, 455 (1977).
24. D. Clark, M. Fleischmann and D. Pletcher, *JCS Perkin II*, 1578 (1973); *J. Electroanal. Chem.*, **43**, 133 (1973).
25. G. D. Mateesu and S. D. Worley, *Tetrahedron Lett*, 5285 (1972).
26. R. W. Kiser, *Introduction to Mass Spectroscopy and its Applications*, Prentice-Hall, Englewood Cliffs, N.J. (1965), pp. A-16.
27. J. Y. Becker, *J. Org. Chem.*, **42**, 3997 (1977).
28. E. A. Mayeda and L. L. Miller, *Tetrahedron,* **28**, 3375 (1972).
29. J. Y. Becker, *Tetrahedron Lett.*, 1331 (1978).
30. J. Y. Becker and D. Zemach, *JCS Perkin II*, 914 (1979).
31. C. H. DePuy, L. G. Schnack and J. W. Hausser, *J. Amer. Chem. Soc.*, **88**, 3343 (1966); P. S. Skell and S. R. Sandler, *J. Amer. Chem. Soc.*, **80**, 2024 (1958); H. M. Frey, *Adv. Phys. Org. Chem.* **4**, 147 (1966).
32. K. Nyberg and L. G. Wistrand, *JCS Chem. Commun.*, 898 (1976).
33. K. Nyberg and L. G. Wistrand, *J. Org. Chem.*, **43**, 2613 (1978).
34. Z. Blum and K. Nyberg, *Acta Chem. Scand. B*, **33**, 73 (1979); L. Eberson and L. Jönsson, *JCS Chem. Commun.*, 1187 (1980) and 133 (1981).
35. J. T. Keating and P. S. Skell, *J. Org. Chem.*, **34**, 1479 (1969).
36. A. Laurent, E. Laurent and R. Tardivel, *Tetrahedron Lett.*, 4861 (1973).
37. G. A. Olah and P. v. R. Schleyer, *Carbonium Ions*, Wiley, New York (1970), pp. 530; C. C. Lee, S. Vassie and E. C. F. Ko, *J. Amer. Chem. Soc.*, **94**, 8931 (1972).
38. J. M. Kornprobst, A. Laurent and E. Laurent, *Bull. Soc. Chim. Fr.*, 3657 (1968).
39. A. Laurent, E. Laurent and R. Tardivel, *Tetrahedron*, **30**, 3431 (1974).
40. N. L. Weinberg in Ref. 2, Vol. V, Part II, Chap. 1.
41. E. Berliner, *J. Amer. Chem. Soc.,* **78**, 3632 (1956) and earlier references in this series.
42. J. Arotsky and M. C. R. Symons, *Quart. Rev.,* **16**, 282 (1962); R. A. Garrett, R. J. Gillespie and J. B. Senior, *Inorg. Chem.*, **4**, 563 (1965); R. J. Gillespie and J. B. Milne, *Inorg. Chem.*, **5**, 1577 (1966).
43. I. Masson, *J. Chem. Soc.*, 1708 (1938).
44. A. I. Popov and D. H. Geske, *J. Amer. Chem. Soc.*, **80**, 1340 (1958).
45. L. L. Miller, E. P. Kujawa and C. B. Campbell, *J. Amer. Chem. Soc.*, **92**, 2821 (1970).
46. R. Lines and V. D. Parker, *Acta Chem. Scand. B*, **34**, 47 (1980).
47. L. L. Miller and B. F. Watkins, *J. Amer. Chem. Soc.,* **98**, 1515 (1976).
48. L. L. Miller and A. K. Hoffman, *J. Amer. Chem. Soc.,* **89**, 593 (1967).
49. H. Hoffelner, H. W. Lorch and H. Wendt, *J. Electroanal. Chem.*, **66**, 183 (1975).

50. H. Behret, Ger. Offen. 2 154 348 (1973); *Chem. Abst.*, **79**, 18341 (1973).
51. I. M. Kolthoff and J. F. Coetzee, *J. Amer. Chem. Soc.*, **79**, 1852 (1957).
52. A. I. Popov and D. H. Geske, *J. Amer. Chem. Soc.*, **80**, 5346 (1958).
53. T. Iwasita and M. C. Giordano, *Electrochim. Acta*, **14**, 1045 (1969).
54. F. Magno, G. A. Mazzocchin and G. Bontempelli, *J. Electroanal. Chem.*, **47**, 461 (1973).
55. G. Casalbore, M. Mastragostino and S. Valcher, *J. Electroanal. Chem.*, **61**, 33 (1975).
56. G. Casalbore, M. Mastragostino and S. Valcher, *J. Electroanal. Chem.*, **68**, 123 (1976).
57. (a) G. Casalbore, M. Mastragostino and S. Valcher, *J. Electroanal. Chem.*, **77**, 373 (1977).
    (b) G. Casalbore, M. Mastragostino and S. Valcher, *J. Electroanal. Chem.*, **87**, 411 (1978).
58. E. Gileadi and T. Bejarano, *Electrochim. Acta*, **21**, 231 (1976).
59. A. I. Vogel, *Practical Organic Chemistry*, 3rd edn, Longmans, London, on Phenols in Chap. IV; O. Rabe, *Chem. Abst.* **75**, 129523a (1971).
60. F. Fitcher and L. Glanzstein, *Chem. Ber.*, **49**, 2473 (1916).
61. M. Mastragostino, G. Casalbore and S. Valcher, *J. Electroanal. Chem.*, **56**, 117 (1974).
62. M. Mastragostino, G. Casalbore, S. Valcher and L. Pastorelli, *J. Electroanal. Chem.*, **90**, 439 (1978).
63. M. Mastragostino, G. Casalbore, S. Valcher and C. Zucchi, *Annali di Chimica*, **69**, 307 (1979).
64. A. P. Tomilov, S. G. Hairanovskii, H. Ya. Fioshin and V. A. Smirnov, *The Electrochemistry of Organic Compounds*, Halsted Press, New York (1972), p. 447.
65. F. Acquah, A. T. Kuhn and C. J. Mortimer, *J. Appl. Chem. Biotechnol.*, **22**, 1195 (1972).
66. M. Mastragostino, G. Casalbore and S. Valcher, *J. Electroanal. Chem.*, **44**, 37 (1973).
67. J. Burdon and J. C. Tatlow in *Advances in Fluorine Chemistry*, Vol. 1 (Eds M. Stacey, J. C. Tatlow and A. G. Sharp), Academic Press, New York (1960), p. 129.
68. S. Nagase in *Fluorine Chemistry Review*, Vol. 1 (Ed. P. Tarrant), Marcel Dekker, New York (1967), p. 77.
69. N. L. Weinberg in Ref. 2, Vol. V., Part II, Chap. 7, pp. 4–78.
70. I. N. Rozhkov, *Russ. Chem. Revs. Engl. Edn*, **45**, 615 (1976).
71. I. N. Rozhkov in the 2nd ed of Ref. 3 (Eds M. M. Baizer and H. Lund); private communication from H. Lund.
72. H. M. Fox, F. N. Ruehlen and W. V. Childs, *J. Electrochem. Soc.*, **118**, 1246 (1971); see also US Patents in *Chem. Abstr.*, **73**, 51709c (1970); **74**, 70920b (1971).
73. H. Schmidt and H. D. Schmidt, *J. Prakt. Chem.*, **2**, 105 (1955); G. A. Sokol'skii and M. A. Dmitriev, *J. Gen. Chem. USSR*, **31**, 1026 (1961).
74. F. Olstowski and J. J. Newport, US Patent 3,033,767 (1962); *Chem. Abstr.*, **57**, 4472b (1962).
75. V. S. Plashkin, G. P. Tataurov and S. V. Sokolov, *J. Gen. Chem. USSR*, **36**, 1705 (1966).
76. H. Schmidt and H. Meinert, *Angew. Chem.*, **72**, 109 (1960).
77. J. D. Domijan, C. J. Ludman, E. M. McCarbon, R. F. O'Malley and V. J. Roman, *Inorg. Chem.*, **8**, 1534 (1969).
78. I. L. Knunyants, I. N. Rozhkov, A. V. Bukhtiarov, M. M. Gol'din and R. V. Kudryavstev, *Izv. Akad. Nauk SSSR Ser. Khim.*, 1207 (1970); *Chem. Abstr.*, **73**, 65752 (1970).
79. I. N. Rozhkov, A. V. Bukhtiarov, N. D. Kuleshova and I. L. Knunyants, *Dokl. Akad. Nauk SSSR*, **193**, 1322 (1970); *Chem. Abstr.*, **74**, 70878u (1971).
80. B. Chang, H. Yanase, K. Nakanishi and N. Watanabe, *Electrochim. Acta*, **16**, 1179 (1971).
81. J. Burdon, I. W. Parsons and J. C. Tatlow, *Tetrahedron*, **28**, 43 (1972).
82. L. Stein, J. M. Nail and G. R. Alms, *Inorg. Chem.*, **8**, 2472 (1969).
83. J. W. Mellor, *A Comprehensive Treatise on Inorganic and Theoretical Chemistry*, Vol. 15, Longmans, Green, London (1936), p. 406.
84. I. N. Rozhkov and I. Ya. Alyev, *Tetrahedron*, **31**, 977 (1975); I. N. Rozhkov, A. V. Bukhtiarov and I. L. Knunyants, *Izv. Akad. Nauk SSSR, Ser. Khim.*, 1130 (1972).
85. I. N. Rozhkov, A. V. Bukhtiarov, N. D. Kuleshova and I. L. Knunyants, *Dokl. Akad. Nauk SSSR*, **193**, 1322 (1970); *Chem. Abstr.*, **74**, 70878u (1971); I. L. Knunyants, I. N. Rozhkov, A. V. Bukhtiarov, M. M. Gol'din and R. V. Kudryavstev, *Izv. Akad. Nauk SSSR Ser. Khim.*, 1207 (1970); *Chem. Abstr.*, **73**, 65752 (1970).
86. It has been claimed by several authors that Rozhkov's fluorination procedure could not be repeated (see a note in Ref. 172 by Eberson and coworkers and a recent publication by Laurent and coworkers: *Nouveau J. de Chimie*, **4**, 453 (1980)). Recently Eberson has tried to react a pre-formed salt of [Naphthalene]‡ with F^- and found no fluorination product, although the radical cation disappears (a private communication).

87. I. N. Rozhkov, N. F. Gambaryan and E. G. Galpern, *Tetrahedron Lett.* 4819 (1976); C. J. Lundman, E. M. McCarron and R. F. O'Malley, *J. Electrochem. Soc.*, **119**, 874 (1972).
88. A. Bensadat, G. Bodennec, E. Laurent and R. Tardivel, *Tetrahedron Lett.*, 3799 (1977); I. N. Rozhkov, I. Ya. Aliev and I. L. Knunyants, *Bull. Acad. Sci. USSR, Div. Chem. Sci.*, **25**, 1361 (1976).
89. I. N. Rozhkov, I. L. Knunyants and I. Ya. Aliev, *Izv. Akad. Nauk SSSR, Ser. Khim.*, 1227 (1972); 2390 (1974).
90. I. Ya. Aliev, I. N. Rozhkov and I. L. Knunyants, *Bull. Akad. Sci., USSR*, **24**, 2309 (1975).
91. G. N. Kokhanov and S. A. Per'kova, *Sov. Electrochem.*, **3**, 867 (1967).
92. V. S. Plashkin and Yu. P. Dolnakov, *J. Appl. Chem. USSR*, **48**, 734 (1975).
93. V. S. Plashkin and Yu, P. Dolnakov, *J. Appl. Chem. USSR*, **48**, 738 (1975).
94. V. J. Davis, R. N. Haszeldine and A. E. Tipping, *JCS Perkin I*, 1263 (1975).
95. Y. Inoue, S. Nagase, K. Kodura, H. Baba and T. Abe, *Bull. Chem. Soc. Japan*, **46**, 2204 (1973).
96. M. Yonekura, S. Nagase, H. Baba, K. Kodaira and T. Abe, *Bull. Chem. Soc. Japan*, **49**, 1113 (1976).
97. V. V. Berenblit, B. A. Byzov, V. I. Gracher, I. M. Dolgopol'skii and Yu. P. Dolnakov, *J. Appl. Chem. USSR*, **48**, 742 (1975).
98. T. Abe, S. Nagase and H. Baba, *Bull. Chem. Soc. Japan*, **46**, 3845 (1973).
99. T. Abe, S. Nagase and H. Baba, *Bull. Chem. Soc. Japan*, **46**, 2524 (1973).
100. E. Soderback, *Ann. Chim.*, **443**, 142 (1925); H. P. Kaufman and J. Liepe, *Ber. Dt. Pharm. Ges.*, **33**, 139 (1923).
101. (a) R. G. R. Bacon and N. Kharasch, *Organic Sulfur Compounds*, Vol. 1, Pergamon Press, New York (1961), p. 318.
     (b) J. L. Wood in *Organic Reactions*, Vol. 3, Wiley, New York (1946), p. 240.
102. (a) G. Cauquis and G. Pierre, *Bull. Soc. Chim. Fr.*, 2244 (1972).
     (b) G. Cauquis and G. Pierre, *Compt. Rend. C*, **266**, 883 (1968).
103. (a) G. Cauquis and G. Pierre, *Compt. Rend. C*, **269**, 740 (1969).
     (b) G. Cauquis and G. Pierre, *Bull. Soc. Chim. Fr.*, 1225 (1972).
104. W. J. de Klein, *Electrochim. Acta*, **18**, 413 (1973).
105. (a) N. N. Melnikov and E. M. Cherkasova, *J. Gen. Chem. USSR*, **14**, 113 (1944) (*Chem. Abstr.*, **39**, 934_5 (1945); and S. I. Sklyarenko, **9**, 1819 (1939) (*Chem. Abstr.*, **34**, 3699_4 (1940)).
     (b) G. Cauquis and G. Pierre, *Tetrahedron*, **34**, 1475 (1978).
106. E. L. Wagner, *J. Chem. Phys.*, **43**, 2728 (1965).
107. N. N. Melnikov and F. M. Cherkasova, *J. Gen. Chem. USSR*, **16**, 1025 (1946) (*Chem. Abstr.*, **41**, 2697_d (1947).
108. G. Cauquis and G. Pierre, *Compt. Rend. C*, **272**, 609 (1971).
109. (a) R. G. R. Bacon, R. G. Guy, R. S. Irwin and T. A. Robinson, *Proc. Chem. Soc.*, 304 (1959).
     (b) R. G. Guy and J. J. Thompson, *Chem. Ind.*, **14**, 99 (1970).
110. L. S. Silbert, J. R. Russel and J. S. Showell, *J. Am. Oil Chem. Soc.*, **50**, 415 (1973).
111. P. B. D. de la Mare and R. Bolton in *Electrophilic Addition to Unsaturated Systems*, Elsevier, New York (1966), p. 113.
112. G. Smith, U.S. Patent, 3,472,747 (1969); *Chem. Abstr.*, **72**, P38326k (1970).
113. J. F. McGhie, W. A. Ross, F. J. Julietti and B. F. Grimwood, *J. Chem. Soc.*, 4638 (1962).
114. H. P. Stout, *Trans. Faraday Soc.*, **41**, 64 (1945); E. Briner and P. Winkler, *J. Chim. Phys.*, **20**, 201 (1923); J. G. N. Thomas, *Trans. Faraday Soc.*, **58**, 1412 (1962); E. H. Riesenfeld and F. Müller, *Z. Electrochem.*, **41**, 87 (1935); L. I. Krishtalik and G. E. Titova, *Soviet Electrochem.*, 249 (1968).
115. G. A. Ward and C. M. Wright, *J. Electroanal. Chem.*, **8**, 302 (1964).
116. P. E. Iverson in *The Encyclopedia of the Electrochemistry of the Elements*, Vol. XIII (eds A. J. Bard and H. Lund), Marcel Dekker, New York (1979), Chap. 5 and references therein.
117. H. Schäfer and A. Alazrak, *Angew. Chem. Int. Edn.*, **7**, 474 (1968); H. Schäfer, *Chemie Ing. Techn.*, **41**, 179 (1969).
118. H. Schäfer, *Angew. Chem. Int. Edn.*, **9**, 158 (1970).
119. V. Plzak and H. Wendt, *Ber. Bunsenges. Phys. Chem.*, **83**, 481 (1979).

120. K. Köster, P. Riemenschneider and H. Wendt, *Israel J. Chem.*, **18**, 141 (1979).
121. O. Hammerich and V. D. Parker in *The Chemistry of the cyanates and Their Thio-Derivatives* (Ed. S. Patai), Wiley, New York and Chichester (1977), Chap. 9.
122. G. Cauquis and G. Pierre, *Bull. Soc. Chim. Fr.*, 997 (1975).
123. D. T. Sawyer and R. J. Day, *J. Electroanal. Chem.*, **5**, 195 (1963).
124. A similar mechanistic scheme has been suggested for the anodic discharge of $TeCN^-$. See G. Cauquis and G. Pierre, *Bull. Soc. Chim. Fr.*, 736 (1976).
125. English Patent No. 1 141 638 (1966) as cited by L. Eberson and K. Niberg in *Tetrahedron*, **32**, 2185 (1976).
126. J. Pinson and J. M. Saveant, *JCS Chem. Commun.*, 933 (1974).
127. S. Margel and M. Levy, *J. Electroanal. Chem.*, **56**, 259 (1974).
128. H. Lund, M. A. Nichel and J. Simonet, *Acta Chem. Scand.*, *B*, **28**, 901 (1974).
129. J. Simonet, M. A. Michel and H. Lund, *Acta Chem. Scand. B*, **29**, 217 (1975).
130. J. W. Sease and C. R. Reed, *Tetrahedron Lett.*, 393 (1975).
131. H. Lund, M. A. Michel and J. Simonet, *Acta Chem. Scand. B*, **29**, 489 (1975).
132. H. Lund and J. Simonet, *J. Electroanal. Chem.*, **65**, 205 (1975); E. Hobolth and H. Lund, *Acta Chem. Scand. B*, **31**, 395 (1977); C. Degrand and H. Lund, *Acta Chem. Scand. B*, **31**, 593 (1977).
133. V. G. Mairanovsky and N. F. Loginova, *Zh. Obshch. Khim.*, **45**, 2112 (1975).
134. V. G. Mairanovsky, N. F. Loginova and I. A. Titova, *Dokl. Akad. Nauk SSSR*, **223**, 643 (1975).
135. S. G. Mairanovsky, L. I. Kosichenko and S. Z. Taits, *Elektrokhimiya*, **8**, 1250 (1977).
136. A. M. Martre and J. Simonet, *J. Electroanal. Chem.*, **97**, 287 (1979).
137. J. Wellmann and E. Steckhan, *Chem. Ber.*, **110**, 3561 (1977).
138. K. Nyberg and L. G. Wistrand, *Acta Chem. Scand. B*, **29**, 629 (1975).
139. T. Chiba and Y. Takata, *Bull. Chem. Soc. Japan*, **51**, 1418 (1978).
140. G. Farnia, G. Sansona and E. Vianello, *J. Electroanal. Chem.*, **108**, 245 (1980).
141. C. P. Andrieux, J. M. Dumas-Bouchiat and J. M. Saveant, *J. Electroanal. Chem.*, **87**, 39 (1978); **87**, 55 (1978); **88**, 43 (1978).
142. G. D. Sargent, J. N. Cron and S. Bank, *J. Amer. Chem. Soc.*, **88**, 5363 (1966).
143. J. F. Garst, P. W. Ayers and R. C. Lamb, *J. Amer. Chem. Soc.*, **88**, 4260 (1966).
144. S. J. Cristol and R. V. Barbour, *J. Amer. Chem. Soc.*, **88**, 4262 (1966); **90**, 2832 (1968).
145. J. F. Garst, *Acc. Chem. Res.*, **4**, 400 (1971).
146. J. F. Garst, J. T. Barbas and F. E. Barton, II, *J. Amer. Chem. Soc.*, **90**, 7159 (1968).
147. G. D. Sargent and G. A. Lux, *J. Amer. Chem. Soc.*, **90**, 7160 (1968).
148. J. F. Garst and F. E. Barton, II, *J. Amer. Chem. Soc.*, **96**, 523 (1974) and B. N. Abels, *J. Amer. Chem. Soc.*, **97**, 4926 (1975).
149. S. Margel and M. Levy, *J. Electroanal. Chem.*, **56**, 259 (1974).
150. W. E. Britton and A. J. Fry, *Anal. Chem.*, **47**, 95 (1975).
151. S. Bank and D. A. Jackett, *J. Amer. Chem. Soc.*, **98**, 7742 (1976); **97**, 567 (1975).
152. B. Bockrath and L. M. Dorfman, *J. Phys. Chem.*, **77**, 2618 (1973).
153. P. E. Hansen, A. Berg and H. Lund, *Acta Chem. Scand. B*, **30**, 267 (1976).
154. J. Simonet and H. Lund, *Bull. Soc. Chim. Fr.*, 2547 (1975).
155. W. Schmidt and E. Steckhan, *J. Electroanal. Chem.*, **89**, 215 (1978).
156. J. H. Wagenknecht, *J. Org. Chem.*, **42**, 1836 (1977).
157. M. Tokuda, T. Taguchi, O. Nishio and M. Itoh, *JCS Chem. Commun.*, 606 (1976).
158. T. Iwasaki and K. Harada, *JCS Chem. Commun.*, 338 (1974).
159. P. E. Iversen and H. Lund, *Acta Chem. Scand. B*, **28**, 827 (1974).
160. W. Schmidt and E. Steckhan, *Angew. Chem. Int. Edn. Engl.*, **17**, 673 (1978).
161. G. Cauquis and G. Pierre, *Bull Soc. Chim. Fr.*, 736 (1976).
162. R. E. Sioda, *J. Phys. Chem.*, **72**, 2322 (1968).
163. J. F. Evans and H. N. Blount, *J. Amer. Chem. Soc.*, **100**, 4191 (1978).
164. J. F. Evans and H. N. Blount, *J. Org. Chem.*, **41**, 516 (1976).
165. J. J. Silber and H. J. Shine, *J. Org. Chem.*, **36**, 2923 (1971); K. Kim, V. J. Hull and H. J. Shine, *J. Org. Chem.*, **39**, 2534 (1974); K. Kim and H. J. Shine, *J. Org. Chem.*, **39**, 2537 (1974); H. J. Shine and J. J. Silber, *J. Amer. Chem. Soc.*, **94**, 1026 (1972).
166. A. J. Bard, A. Ledwith and H. J. Shine, *Adv. Phys. Org. Chem.*, **13**, 155 (1976).
167. C. V. Ristagno and H. J. Shine, *J. Org. Chem.*, **36**, 4050 (1971).

168. H. J. Shine and C. V. Ristagno, *J. Org. Chem.*, **37**, 3424 (1972).
169. H. J. Shine, J. J. Silber, R. J. Boussey and T. Okuyama, *J. Org. Chem.*, **37**, 2691 (1972).
170. H. J. Shine and L. R. Shade, *J. Heterocyclic Chem.*, **11**, 139 (1974).
171. L. Eberson, *JCS Chem. Commun.*, 826 (1975); L. Eberson and K. Nyberg, *Acta Chem. Scand. B,* **32**, 235 (1978); L. Eberson, L. Jönsson and L. G. Wistrand, *Acta Chem. Scand. B*, **32**, 520 (1978).
172. L. Eberson, Z. Blum, B. Helgée and K. Nyberg, *Tetrahedron*, **34**, 731 (1978); L. Eberson, *Adv. Phys. Org. Chem*., **18**, 79 (1982).
173. M. J. S. Dewar, *Angew. Chem.*, **83**, 859 (1971); H. E. Zimmerman, *Acc. Chem. Res.*, **4**, 272 (1971).
174. I. N. Rozhkov, N. P. Gambaryan and E. G. Galpern, *Tetrahedron Lett.*, 4819 (1976).
175. T. R. Evans and L. P. Hurysz, *Tetrahedron Lett.*, 3103 (1977).
176. A. L. J. Beckwith in *MTP International Review of Science, Organic Chemistry, Series 1*, Vol. 10, Butterworth, London (1973), p. 1; R. A. Jackson in *MTP International Review of Science, Organic Chemistry, Series 1*, Vol. 10, Butterworth (1973), p. 205 and JCS Chem. Commun., 573 (1974).
177. A. Laurent and R. Tardivel, *Compt. Rend. C*, **272**, 8 (1971).
178. L. L. Miller, E. P. Kujawa and C. B. Campbell, *J. Amer. Chem. Soc.*, **92**, 2821 (1970).
179. J. Y. Becker and D. Zemach, *JCS Perkin II*, 336 (1981).
180. A. I. Popov and R. F. Swensen, *J. Amer. Chem. Soc.*, **77**, 3724 (1955); J. Jortner and A. Treinin, *Trans. Faraday Soc.*, **58**, 1503 (1962).
181. T. Shono, Y. Matsumura, J. Hayashi and M. Mizoguchi, *Tetrahedron Lett.*, 1867 (1980).
182. E. J. Corey and C. V. Kim, *J. Amer. Chem. Soc.,* **94**, 7586 (1972).
183. S. Torii, K. Uneyama and M. Ono, *Tetrahedron Lett.*, 2741 (1980).
184. D. T. Sawyer and M. J. Gibian, *Tetrahedron*, **35**, 1471 (1979) and references therein.
185. A. D. McElroy and J. S. Hashman, *Inorg. Chem.*, **3**, 1798 (1964); D. L. Maricle and W. G. Hodgson, *Anal. Chem.*, **37**, 1562 (1965); M. E. Peover and B. S. White, *JCS Chem. Commun.*, 183 (1965) and *Electrochimica Acta* **11**, 1061 (1966); E. J. Johnson, K. H. Pool and R. E. Hamm, *Anal. Chem.*, **38**, 183 (1966); D. T. Sawyer and J. L. Roberts, *J. Electroanal. Chem*., **12**, 90 (1966).
186. J. M. McCord and I. Fridovich, *J. Biol. Chem.*, **244**, 604 (1969); I. Fridovich, *Acc. Chem. Res.*, **5**, 321 (1972).
187. M. J. Gibian, D. T. Sawyer, T. Ungermann, R. Tangpoonpholvivat and M. M. Morrison, *J. Amer. Chem. Soc.*, **101**, 640 (1979); G. W. Gokel, H. M. Gerdes and N. W. Rebert, *Tetrahedron Lett*., 653 (1976).
188. J. San Fillipo, Jr, C.-I. Chern and J. S. Valentine, *J. Org. Chem.*, **40**, 1678 (1975).
189. R. Dietz, A. E. J. Forno, B. E. Larcombe and M. E. Peover, *J. Chem. Soc. B*, 816 (1970).
190. R. A. Johnson and E. G. Nidy, *J. Org. Chem.*, **40**, 1680 (1975).
191. V. Merritt and D. T. Sawyer, *J. Org. Chem.*, **35**, 2157 (1970).
192. W. C. Danen and R. J. Warner, *Tetrahedron Lett.*, 989 (1977).
193. E. J. Johnson, K. H. Pool and R. E. Hamm, *Anal. Chem.*, **39**, 888 (1967); B. H. J. Bielski and P. C. Chan, *J. Amer. Chem. Soc.*, **100**, 1920 (1978).
194. F. Magno, R. Seeber and S. Valcher, *J. Electroanal. Chem.*, **83**, 131 (1977).
195. D. Knittel and B. Kastening, *J. Appl. Electrochem.*, **3**, 291 (1973); D. Knittel and B. Kastening, *Ber. Bunsenges. Phys. Chem.*, **77**, 833 (1973).
196. H. Lehmkuhl in *Organic Electrochemistry* (Ed. M. M. Baizer), Marcel Dekker, New York (1973), p. 621.
197. A. V. Kramer, J. A. Labinger, J. S. Bradley and J. A. Osborn, *J. Amer. Chem. Soc.*, **96**, 7145 (1974).
198. R. L. Sweany and J. Halpern, *J. Amer. Chem. Soc.*, **99**, 8335 (1977).
199. J. Y. Chen and J. K. Kochi, *J. Amer. Chem. Soc.*, **99**, 1450 (1977).
200. D. R. Kidd, C. P. Cheng and T. L. Brown, *J. Amer. Chem. Soc.*, **100**, 4103 (1978).
201. D. E. Webster, *Adv. Organometal. Chem*., **15**, 147 (1977).
202. J. A. Sofranko, R. Eisenberg and J. A. Kampmeier, *J. Amer. Chem. Soc*., **101**, 1042 (1979).
203. C. Gosden, K. P. Healy and D. Pletcher, *JCS Dalton*, 972 (1978); J. Y. Becker, J. B. Kerr, D. Pletcher and R. Rosas, *J. Electroanal. Chem*., **117**, 87 (1981).
204. C. Gosden and D. Pletcher, *J. Organometal. Chem.*, **186**, 401 (1980) and K. P. Healy and D. Pletcher, *J. Organometal. Chem.*, **161**, 109 (1978).

205. M. Troupel, Y. Rollin, C. Chevrot, F. Pfluger and J.-F. Fauvarque, *J. Chem. Res. (S)*, 50 (1979); M. Troupel, Y. Rollin, S. Sibille, J.-F. Fauvarque and J. Périchon, *J. Chem. Res. (S)*, 24 and 147 (1980); S. Sibille, J. C. Folest, J. Coulombeix, M. Troupel, J.-F. Fauvarque and J. Périchon, *J. Chem. Res. (S)*, 268 (1980).
206. L. Walder, G. Rytz, K. Meier and R. Scheffold, *Helv. Chim. Acta*, **61**, 3013 (1978); G. Rytz, L. Walder and R. Scheffold in *Vitamin* B_{12} (Eds B. Zagalak and W. Friedrich), Walter de Gruyter, Berlin and New York (1979), p. 173; D. Lexa, J. M. Saveant and J. P. Soufflet, *J. Electroanal. Chem.*, **100**, 159 (1979); H. A. O. Hill, J. M. Pratt, M. P. O'Riordan, F. R. Williams and R. J. P. Williams, *J. Chem. Soc. A*, 1859 (1971).
207. (a) R. Scheffold, M. Dike, S. Dike, T. Herold and L. Walder, *J. Amer. Chem. Soc.*, **102**, 3644 (1980).
(b) S. Margel and F. C. Anson, *J. Electrochem. Soc.*, **125**, 1232 (1978).
208. A. Puxeddu, G. Costa and N. Marsich, *JCS Dalton*, 1489 (1980).
209. R. Breslow, R. J. Corcoran and B. B. Snider, *J. Amer. Chem. Soc.*, **96**, 6791, 6792 (1974); **97**, 6580 (1975); R. Breslow, *Acc. Chem. Res.*, **13**, 170 (1980).
210. R. Breslow and R. Goodin, *Tetrahedron Lett.*, 2675 (1976).
211. J. Pinson and J.-M. Savéant, *JCS Chem. Commun.*, 933 (1974).
212. W. J. M. van Tilborg, C. J. Smit and J. J. Scheele, *Tetrahedron Lett.*, 2113 (1977); 776 (1978).
213. J.-M. Savéant, *Acc. Chem. Res.*, **13**, 323 (1980) and references therein.
214. J. F. Bunnett, *Acc. Chem. Res.*, **5**, 139 (1975).
215. P. R. Moses, L. Wier, and R. W. Murray, *Anal. Chem.*, **47**, 1882 (1975).
216. B. F. Watkins, J. R. Behling, E. Kariv and L. L. Miller, *J. Amer. Chem. Soc.*, **97**, 3549 (1975).
217. R. W. Murray, *Acc. Chem. Res.*, **13**, 135 (1980).
218. K. D. Snell and A. G. Keenan, *Chem. Soc. Revs*, **8**, 259 (1979).
219. J. F. Evans, T. Kuwana, M. T. Henne and G. P. Royer, *J. Electroanal. Chem.*, **80**, 409 (1977).
220. C. Degrand and L. L. Miller, *J. Amer. Chem. Soc.*, **102**, 5728 (1980).
221. J. B. Kerr and L. L. Miller, *J. Electroanal. Chem.*, **101**, 263 (1979).
222. D. C. S. Tse and T. Kuwana, *Anal. Chem.*, **50**, 1315 (1978).
223. A. P. Brown, C. Koval and F. C. Anson, *J. Electroanal. Chem.*, **72**, 379 (1976); J. Zagal, R. K. Sen and E. Yeager, *J. Electroanal. Chem.*, **83**, 207 (1977).
224. J. P. Collman, M. Marrocco, P. Denisevich, C. Koval and F. C. Anson, *J. Electroanal. Chem.*, **101**, 117 (1979); J. P. Collman, P. Denisevich, Y. Konai, M. Marrocco, C. Koval and F. C. Anson, *J. Amer. Chem. Soc.*, **102**, 6027 (1980); J. Y. Becker, D. Dolphin and S. F. B. Pickett, submitted for publication.
225. A. Bettelheim, R. J. H. Chan and T. Kuwana, *J. Electroanal. Chem.*, **99**, 391 (1979).
226. M. R. Van De Mark and L. L. Miller, *J. Amer. Chem. Soc.*, **100**, 3223 (1978).
227. The first reversible electroreduction of organic halogen compounds has been reported recently by A. Merz and R. Tomahogh, *Angew. Chem. Int. Edn. Engl.*, **18**, 938 (1979).
228. P. J. Elving, *Canad. J. Chem.*, **55**, 3392 (1977).
229. L. Nadjo and J. M. Saveant, *J. Electroanal. Chem.*, **30**, 41 (1971); J. G. Lawless and M. D. Hawley, *J. Electroanal. Chem.*, **21**, 365 (1969); D. E. Bartak, K. J. Houser and M. D. Hawley, *J. Amer. Chem. Soc.*, **95**, 6033 (1973).
230. J. Grimshaw, R. J. Haslett and J. Trocha-Grimshaw, *JCS Perkin I*, 2448 (1977); W. J. M. van Tilborg, C. J. Smit and J. J. Scheele, *Tetrahedron Lett.*, 2113, 3651 (1977).
231. E. Canadell, P. Karafiloglon and L. Salem, *J. Amer. Chem. Soc.*, **102**, 855 (1980).
232. J. F. Garst, R. D. Roberts and J. A. Pacifici, *J. Amer. Chem. Soc.*, **99**, 3528 (1977).
233. C. P. Andrieux, J. M. Duman-Bouchiant and J. M. Saveant, *J. Electroanal. Chem.*, **87**, 55 (1978).
234. A. K. Hoffmann, W. G. Hodgson, D. L. Maricle and W. J. Jura, *J. Amer. Chem. Soc.*, **86**, 631 (1964); G. J. Gores, C. E. Koeppe and D. E. Bartak, *J. Org. Chem.*, **44**, 380 (1979).
235. A. J. Bard and A. Merz, *J. Amer. Chem. Soc.*, **101**, 2959 (1979).
236. R. Breslow and J. L. Grant, *J. Amer. Chem. Soc.*, **99**, 7745 (1977); R. Breslow and R. Goodin, *J. Amer. Chem. Soc.*, **98**, 6077 (1976).
237. M. Fleischmann, D. Pletcher and C. J. Vance, *J. Electroanal. Chem.*, **29**, 325 (1971); D. Britz and H. Luft, *Ber. Bunsenges. Phys. Chem.*, **77**, 836 (1973).
238. J. W. Sease and R. C. Reed, *Tetrahedron Lett.*, 393 (1975).

239. L. Eberson, *Acta Chem. Scand.*, **22**, 3045 (1968); E. Grovenstein, Jr, and Y.-M. Cheng, *JCS Chem. Commun.*, 101 (1970).
240. M. M. Baizer and J. L. Chruma, *J. Org. Chem.*, **37**, 1951 (1972).
241. S. Wawzonek, R. C. Duty and J. H. Wagenknecht, *J. Electrochem. Soc.*, **111**, 74 (1964).
242. O. R. Brown and K. Taylor, *J. Electroanal. Chem.*, **50**, 211 (1974).
243. A. J. Fry and R. L. Krieger, *J. Org. Chem.*, **41**, 54 (1976).
244. G. M. McNamee, B. C. Willett, D. M. La Perriere, D. G. Peters, *J. Amer. Chem. Soc.*, **99**, 1831 (1977).
245. D. M. La Perriere, B. C. Willett, W. F. Carroll, Jr, E. C. Torp and D. G. Peters, *J. Amer. Chem. Soc.*, **100**, 6293 (1978).
246. J. H. Wagenknecht, *J. Electroanal. Chem.*, **52**, 489 (1974).
247. M. Fleischman, G. Mengoli and D. Pletcher, *Electrochim. Acta*, **18**, 231 (1973).
248. F. L. Lambert, B. L. Hasslinger and R. N. Franz, III, *J. Electrochem. Soc.*, **122**, 737 (1975).
249. M. F. Semmelhack and G. E. Heinsohn, *J. Amer. Chem. Soc.*, **94**, 5139 (1972).
250. A. Merz, *Angew. Chem. Int. Edn Engl.*, **16**, 57 (1977); *Electrochim. Acta*, **22**, 1271 (1977).
251. A. Inesi and L. Rampazzo, *J. Electroanal. Chem.*, **54**, 289 (1974); J. Casanova and H. R. Rogers, *J. Org. Chem.*, **39**, 2408 (1974).
252. H. Lund and E. Hobolth, *Acta Chem. Scand. B*, **30**, 895 (1976).
253. M. M. Gol'din, L. G. Feoktistov, V. R. Polishchuk and L. S. German, *Electrokhimiya*, **7**, 916 (1971); **9**, 67 (1973).
254. L. G. Feoktistov and M. M. Gol'din, *Zh. Obshch. Khim.*, **43**, 515 (1973); L. G. Feoktistov and M. M. Gol'din, *Elektrokhimiya*, **4**, 490 (1968).
255. F. L. Lambert and G. B. Ingall, *Tetrahedron Lett.*, 3231 (1974); A. J. Fry, *Fortschr. Chem. Forsch.*, **34**, 1 (1972).
256. J. F. Garst, J. A. Pacifici, V. D. Singleton, M. F. Ezzel and J. I. Morris, *J. Amer. Chem. Soc.*, **97**, 5242 (1975).
257. S. P. Mishra and M. C. R. Symons, *Tetrahedron Lett.*, 2597 (1975).
258. S. Satoh, M. Itoh and M. Tokuda, *JCS Chem. Commun.*, 481 (1978).
259. A. J. Fry and W. E. Britton, *Tetrahedron Lett.*, 4363 (1971); *J. Org. Chem.*, **38**, 4016 (1973).
260. K. B. Wiberg and G. A. Epling, *Tetrahedron Lett.*, 1119 (1974).
261. J. Casanova and H. R. Rogers, *J. Amer. Chem. Soc.*, **96**, 1942 (1974); O. R. Brown and E. R. Gonzalez, *J. Electroanal. Chem.*, **43**, 215 (1973).
262. J. P. Dirlam, L. Eberson and J. Casanova, *J. Amer. Chem. Soc.*, **94**, 240 (1972).
263. A. J. Fry and R. Scoggins, *Tetrahedron Lett.*, 4079 (1972); A. J. Fry and J. J. O'Dea, *J. Org. Chem.*, **40**, 3625 (1975); A. J. Fry and G. S. Ginsburg, *J. Amer. Chem. Soc.*, **101**, 7439 (1979).
264. L. Rampazzo, A. Inesi and A. Zeppa, *J. Electroanal. Chem.*, **76**, 175 (1977).
265. A. J. Fry and L. L. Chung, *Tetrahedron Lett.*, 645 (1976).
266. K. G. Boto and A. J. Bard, *J. Electroanal. Chem.*, **65**, 945 (1975).
267. M. von Stackelberg and W. Stracke, *Z. Electrochem.*, **53**, 118 (1949).
268. M. Doupeux and J. Simonet, *Bull. Soc. Chim. Fr.*, 1219 (1972).
269. E. Brillas and J. M. Costa, *J. Electroanal. Chem.*, **69**, 435 (1976).
270. W. M. Moore and D. G. Peters, *Tetrahedron Lett.*, 453 (1972).
271. W. M. Moore, A. Salajegheh and D. G. Peters, *J. Amer. Chem. Soc.*, **97**, 4954 (1975).
272. W. M. Moore and D. G. Peters, *J. Amer. Chem. Soc.*, **97**, 139 (1975).
273. L. W. Marple, L. E. Hummelstedt and L. B. Rogers, *J. Electrochem. Soc.*, **107**, 437 (1960); L. B. Rogers and A. J. Diefenderfer, *J. Electrochem. Soc.*, **114**, 942 (1967); O. R. Brown, H. R. Thirsk and B. Thornton, *Electrochim. Acta*, **16**, 495 (1971); J. Grimshaw and J. S. Ramsay, *J. Chem. Soc. B*, 60 (1968).
274. R. B. Yamasaki, M. Tarle and J. Casanova, *J. Org. Chem.*, **44**, 4519 (1979).
275. D. E. Bartak, T. M. Shields and M. D. Hawley, *J. Electroanal. Chem.*, **30**, 289 (1971); D. E. Bartak and M. D. Hawley, *J. Amer. Chem. Soc.*, **94**, 640 (1972); P. Peterson, A. K. Carpenter and R. F. Nelson, *J. Electroanal. Chem.*, **27**, 1 (1970).
276. (a) C. M. Pak and W. M. Gulick, Jr, *Electrochim. Acta*, **18**, 1025 (1973); M. Mohammad, *Anal. Chem.*, **49**, 60 (1977).
(b) J. G. Lawless, D. E. Bartak and M. D. Hawley, *J. Amer. Chem. Soc.*, **91**, 7121 (1969).
277. F. H. Covitz, *J. Amer. Chem. Soc.*, **89**, 5403 (1967).

278. Y. Matsui, T. Soga and Y. Date, *Bull. Chem. Soc. Japan*, **44**, 513 (1971).
279. H. Lund and N. J. Jensen, *Acta Chem. Scand. B*, **28**, 263 (1974); J. P. Coleman, Naser-ud-din, H. G. Gilde, J. H. P. Utley, B. C. L. Weedon and L. Eberson, *JCS Perkin II*, 1903 (1973); H. Lund, *Acta Chem. Scand.*, **13**, 192 (1959); J. P. Coleman, H. G. Gilde, J. H. P. Utley and B. C. L. Weedon, *JCS Chem. Commun.*, 738 (1970).
280. L. Rampazzo, A. Inesi and R. M. Bettolo, *J. Electroanal. Chem.*, **83**, 341 (1977).
281. J. Casanova and H. R. Rogers, *J. Org. Chem.*, **39**, 3803 (1974).
282. K. B. Wiberg, W. F. Bailey and M. E. Jason, *J. Org. Chem.*, **41**, 2711 (1976).
283. K. B. Wiberg, G. A. Epling and M. Jason, *J. Amer. Chem. Soc.*, **96**, 912 (1974).
284. W. F. Carroll, Jr, and D. G. Peters, *J. Amer. Chem. Soc.*, **102**, 4127 (1980).
285. M. R. Rifi, *J. Amer. Chem. Soc.*, **89**, 4442 (1967).
286. Azizullah and J. Grimshaw, *JCS Perkin I*, 425 (1973).
287. M. R. Asirvatham and M. D. Hawley, *J. Amer. Chem. Soc.*, **97**, 5204 (1975).
288. F. M'Halla, J. Pinson and J. M. Saveant, *J. Amer. Chem. Soc.*, **102**, 4120 (1980).
289. J. Grimshaw and R. J. Haslett, *JCS Perkin I*, 657 (1980).
290. S. Wawzonek and S. M. Heilman, *J. Electrochem. Soc.*, **121**, 516 (1974); A. J. Fry, M. A. Nitnick and R. G. Reed, *J. Org. Chem.*, **35**, 1232 (1970); S. O. Farwell, F. A. Beland and R. D. Geer, *J. Electroanal. Chem.*, **61**, 303 (1975).
291. S. Wawzonek and J. H. Wagenknecht, *J. Electrochem. Soc.*, **110**, 420 (1963).
292. D. E. Bartak, W. C. Danen and M. D. Hawley, *J. Org. Chem.*, **35**, 1206 (1970); G. A. Russell and W. C. Danen, *J. Amer. Chem. Soc.*, **88**, 5663 (1966); J. F. Bunnett and B. F. Gloor, *J. Org. Chem.*, **38**, 4156 (1973); J. K. Kim and J. F. Bunnett, *J. Amer. Chem. Soc.*, **92**, 7463 (1970).
293. W. C. Danen, T. T. Kensler, J. G. Lawless, M. F. Marcus and M. D. Hawley, *J. Phys. Chem.*, **73**, 4389 (1969).
294. R. F. Nelson, A. K. Carpenter and E. T. Seo, *J. Electrochem. Soc.*, **120**, 206 (1973).
295. R. P. Van Duyne and C. N. Reilley, *Anal. Chem.*, **44**, 158 (1972); J. C. Lawless and M. D. Hawley, *J. Electroanal. Chem.*, **23**, App. 1 (1969).
296. P. H. Rieger, I. Bernal, W. H. Reinmuth and G. K. Fraenkel, *J. Amer. Chem. Soc.*, **85**, 683 (1963).
297. A. R. Buick, T. J. Kemp, G. T. Neal and T. J. Stone, *J. Chem. Soc. A*, 666 (1969).
298. K. J. Houser, D. E. Bartak and M. D. Hawley, *J. Amer. Chem. Soc.*, **95**, 6033 (1973); D. E. Bartak, K. J. Houser, B. C. Rudy and M. D. Hawley, *J. Amer. Chem. Soc.*, **94**, 7526 (1972).
299. J. M. Saveant, *J. Electroanal. Chem.*, **29**, 87 (1971).
300. J. Grimshaw and J. Trocha-Grimshaw, *JCS Perkin II*, 215 (1975).
301. J. Grimshaw and J. Trocha-Grimshaw, *Tetrahedron Lett.*, 993 (1974).
302. J. R. Cockrell and R. W. Murray, *J. Electrochem. Soc.*, **119**, 849 (1972).
303. S. O. Farwell, F. A. Beland and R. D. Geer, *Anal. Chem.*, **47**, 895 (1975).
304. S. O. Farwell, F. A. Beland and R. D. Geer, *J. Electroanal. Chem.*, **61**, 315 (1975).
305. B. H. Campbell, *Anal. Chem.*, **44**, 1659 (1972).
306. R. N. Renaud, *Canad. J. Chem.*, **52**, 376 (1974).
307. F. M. Triebe, K. J. Borhani and M. D. Hawley, *J. Amer. Chem. Soc.*, **101**, 4637 (1979).
308. F. M. Triebe, K. J. Borhani and M. D. Hawley, *J. Electroanal. Chem.*, **107**, 375 (1980).
309. J. Grimshaw and J. Trocha-Grimshaw, *J. Electroanal. Chem.*, **56**, 443 (1974).
310. T. Matsumoto, M. Sato and S. Hirayama, *Chem. Lett.*, 603 (1972).
311. T. M. Siegel and H. B. Mark, Jr, *J. Amer. Chem. Soc.*, **94**, 9020 (1972); C. P. Keszthelyi and A. J. Bard, *J. Org. Chem.*, **39**, 2936 (1974).
312. T. M. Siegel and H. B. Mark, Jr, *J. Amer. Chem. Soc.*, **93**, 6281 (1971); N. E. Tockel, C. P. Keszthelyi and A. J. Bard, *J. Amer. Chem. Soc.*, **94**, 4872 (1972).
313. S. Kashti-Kaplan and E. Kirowa-Eisner, *Israel J. Chem.*, **18**, 75 (1979); *J. Electroanal. Chem.*, **103**, 119 (1979).
314. E. Farina, L. Nucci, G. Biggi, F. D. Cima and F. Pietra, *Tetrahedron Lett.*, 3305 (1974): P. T. Cottrell and P. H. Rieger. *Mol. Phys.*, **12**, 149 (1967).
315. K. Alwair and J. Grimshaw, *JCS Perkin II*, 1811 (1973).
316. E. Laviron, *Bull. Chim. Soc. Fr.*, 2840 (1963); K. Sugino, K. Shirai, T. Sekine and K. Odo, *J. Electrochem. Soc.*, **104**, 667 (1957); H. Lund, *Acta Chem. Scand.*, **18**, 1984 (1964).
317. P. E. Iversen, *Synthesis*, 484 (1972).
318. E. Laviron, *Bull. Chim. Soc. Fr.*, 2350 (1961).

319. M. Person and R. Mora, *Bull. Soc. Chim. Fr.*, **521**, 528 (1973); S. G. Mairanovskii and A. D. Filonova, *Elektrokhimiya*, **1**, 1044 (1965).
320. H. Lund, unpublished results (1965), cited by P. E. Iversen in *The Encyclopedia of the Electrochemistry of the Elements*, Vol. XIII (Eds A. J. Bard and H. Lund), Marcel Dekker, New York (1979), Chap. 5.
321. H. Lund, *Osterr. Chem. Ztg*, **68**, 43 (1967).
322. H. Lund, *Acta Chem. Scand.*, **14**, 1927 (1960); **17**, 2325 (1963).
323. J. Armand and P. Souchay, *Chim. Anal.* (Paris), **44**, 239 (1962).
324. L. V. Kononenko, T. A. Yurre, V. N. Dmitrieva, L. S. Éfros and V. D. Bezuglyi, *J. Gen. Chem. USSR*, **40**, 1345 (1970).
325. N. I. Malyugina, N. L. Gusarskaya and A. V. Oleinik, *J. Gen. Chem. USSR* **45**, 1801 (1975).
326. N. I. Malyugina, L. N. Vertyulina, K. Z. Koryttsev and A. V. Oleinik, *J. Gen. Chem. USSR*, **44**, 478 (1974).
327. J. Y. Becker and B. Zinger, *J. Amer. Chem. Soc.*, **104**, 2327 (1982).

ADDENDUM

E. Carbon–Isocyanate and Carbon–Isothiocyanate Bonds

Recently, Becker and Zinger have studied[327] the direct anodic oxidation of organic isocyanates and isothiocyanates in acetonitrile. It has been found that phenyl- and 1-naphthylisocyanate undergo polymerization whereas alkyl isocyanates undergo α-cleavage. For instance:

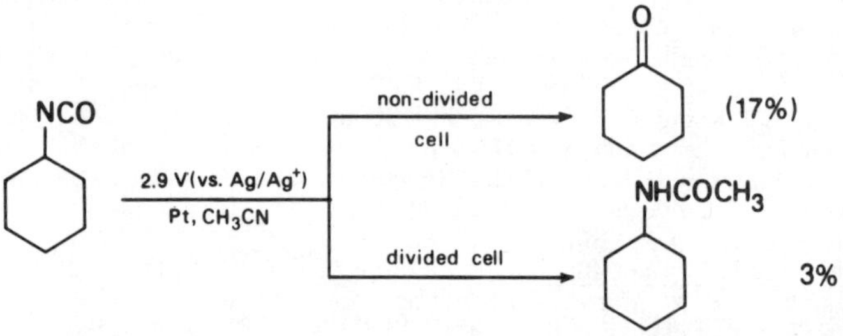

NCO
2.9 V(vs. Ag/Ag$^+$)
Pt, CH_3CN
non-divided cell
(17%)
divided cell
$NHCOCH_3$
3%

Alkyl isothiocyanates are oxidized at lower anodic potentials than the corresponding isocyanates and afford products not only due to α-cleavage process but also due to oxidation of the C═S functional group. Tertiary and secondary compounds yielded isocyanates, and primary alkylisothiocyanates formed mostly cyclic dimers containing S—S bond. It is assumed that the HOMO in isothiocyanates involves also the non-bonding electrons of the sulphur atom. For example:

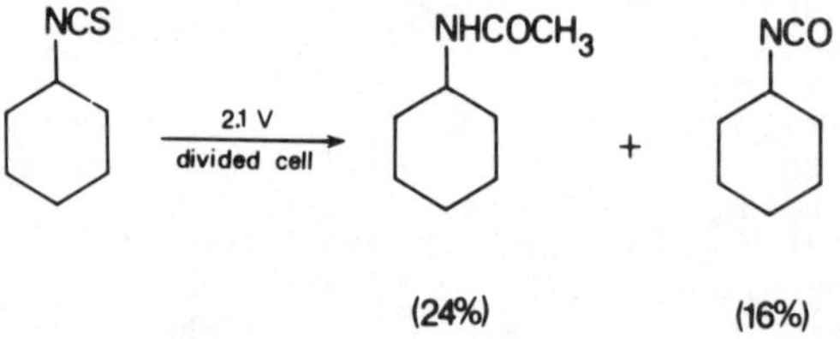

NCS
2.1 V
divided cell
$NHCOCH_3$
+
NCO
(24%)
(16%)

The Chemistry of Functional Groups, Supplement D
Edited by S. Patai and Z. Rappoport
© 1983 John Wiley & Sons Ltd

CHAPTER 7

Pyrolysis of aryl azides

L. K. DYALL

Department of Chemistry, University of Newcastle, New South Wales, Australia 2308

I. INTRODUCTION

The pyrolysis of organic azides was last given a comprehensive review by Abramovitch and Kyba[1], using literature references to the end of 1969. Substantial interest in this field has continued, and there have been exciting new developments. Major research efforts have been made on understanding how singlet and triplet nitrenes interconvert and give rise to products, on the high-temperature nitrene–carbene interconversion, the nature of neighbouring group assistance to azide pyrolysis, and to pyrolyses which involve ring opening.

This review concentrates on aryl azides, with some attention to sulphonyl azides. Little new information on alkyl azide pyrolysis has been published in recent years. Unsaturated azides have received considerable attention, and this work is reviewed by Professor H. W. Moore in another chapter in this volume.

II. SOLUTION PYROLYSES INVOLVING ARYL NITRENES

A. Experimental Conditions

Unless there are special effects operating, a temperature of about 150°C is needed to obtain a half-life of 2 h in the decomposition of a phenyl azide. The pyrolysis generally proceeds as in equation (1), via a singlet nitrene which can interconvert with the lower-energy triplet[1]. The final products form by further reactions of these transient species.

$$ArN_3 \xrightarrow[-N_2]{\Delta} Ar\ddot{\underset{\cdot\cdot}{N}} \rightleftharpoons Ar\dot{\underset{\cdot\cdot}{N}}\cdot \qquad (1)$$

Singlet Triplet

There have been important advances in methods of controlling the singlet–triplet system to maximize the yields of desired products (see Sections II.B and II.C). Suschitzky and his colleagues[2] advocate use of the highest reaction temperature commensurate with stabilities of reactants and products, so that the nitrenes are energetic enough to give useful reactions. Singlet nitrene reactions are favoured by use of tetrachlorothiophen, which not only provides a high reflux temperature (240°C) but also simplifies the reactions by having no abstractable hydrogen. For triplet nitrene reactions, deactivation of the initially formed singlet is required, and can be achieved via the 'heavy atom' effect[3] of bromobenzene. Excellent deactivation to the triplet species, in combination with a higher reaction temperature, is provided by

2,3,5-tribromothiophen[2]. It is always recommended that triplet reactions be performed under nitrogen because oxygen is a triplet quencher[4].

B. Nitrene Traps

1. *Trapping of singlet aryl nitrenes from azide pyrolysis*

There is at this time no universal trapping agent for singlet aryl nitrenes under pyrolysis conditions. While many can be trapped with aniline to make azepines[1], this method has limitations. Nitroazepines are unstable[5,6], certain halogenoazepines react further with aniline[6], and the trapping experiment is often complicated by an alternative reaction between the azide and the amine to yield a 'crossed' azo compound[7]. However, most singlet aryl nitrenes can be effectively removed from a reaction system by promoting singlet-to-triplet crossing (see Section II.C).

2. *Trapping of triplet aryl nitrenes*

The selective trapping of triplet arylnitrenes from the singlet–triplet equilibrium can usually be achieved with a nitrosoarene (equation 2)[8,9,10]. The preferred trap is 4-nitroso-*N,N*-dimethylaniline, known[11] to be a radical quencher and spin trap. Both isomers of the azoxy compound have been reported. Another very effective trap is the masked 1,2-dinitroso compound, benzofuroxan (equation 3)[12].

$$ArN_3 + p\text{-}ONC_6H_4NMe_2 \longrightarrow Ar{-}\overset{+}{N}(O^-){=}N{-}C_6H_4{-}NMe_2 \qquad (2)$$

$$2\,ArN_3 + \text{benzofuroxan} \xrightarrow[155\,^\circ C]{PhBr} 1,2\text{-}C_6H_4(\overset{+}{N}(O^-){=}NAr)_2 + 1,2\text{-}C_6H_4(N{=}NAr)_2 \qquad (3)$$

(Ar = 4-methoxyphenyl) (40%) (5%)

C. Roles of Singlet and Triplet Nitrenes in Product Formation

There have been very important advances in our understanding of which products arise from singlet and which from triplet nitrenes. Most of these contributions have come from the research group of Hans Suschitzky at Salford. The evidence from thermal experiments is summarized below; readers should consult the original papers for the photochemical evidence, which is also relevant.

Interest in the reactions of aryl nitrenes centres on the formation of cyclized products by reaction with another part of the same molecule. Nitrenes are also known to enter into intermolecular reactions (to yield amine, azo compound, and various polymers)[1] but these reactions have not attracted much recent interest and will be given very little attention in this review.

The formation of cyclic products in azide pyrolyses does not necessarily involve 'free' nitrenes. Cases of neighbouring group participation by *ortho* substituents in the loss of nitrogen are well known, and do not appear to proceed via nitrenes (see Section

III). The present section is devoted to those cases where nitrenes are actually produced in the rate-determining step, and the reaction products arise in subsequent steps.

1. Use of 'heavy atom effects' and triplet traps

The use of both these effects to identify origins of products was first reported in 1974 by Suschitzky and his colleagues, and has proved a powerful method. Data for pyrolysis of 1-naphthyl azide are given in Table 1 and demonstrate quite clearly where the various products arise[13].

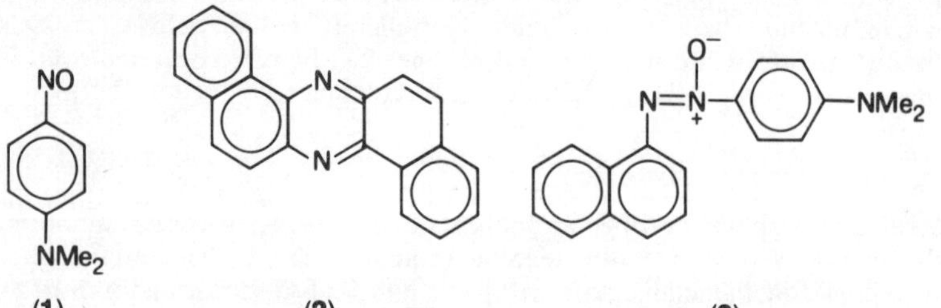

(1) (2) (3)

The phenazine **2** must be derived from the triplet nitrene: its yield is increased by changing solvent from cumene to bromobenzene (the heavy atom effect) while its formation is totally inhibited by the nitroso trap **1** which removes the triplet nitrene in the form of the adduct **3**. The azo compound and amine are likewise seen to be triplet-derived products; there are in fact no tractable products from singlet nitrene in this particular reaction. Similar work has now been reported for other aryl azides[4].

Further light on the behaviour of singlet and triplet nitrenes is shed by Suschitzky's work with 2-azidobiphenyls[2] (see equation 5).

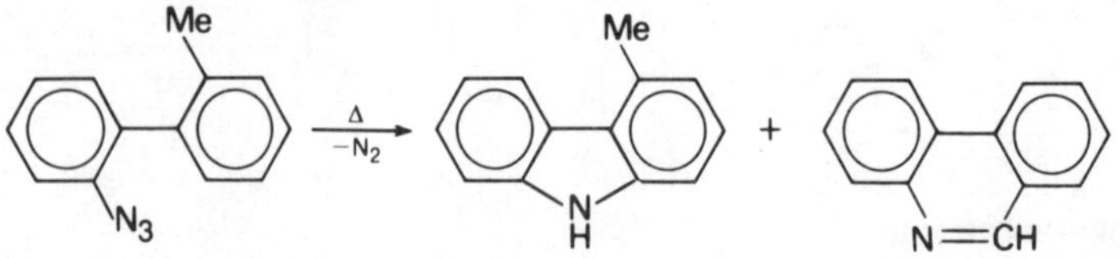

(4) $\xrightarrow[-N_2]{\Delta}$ (5) + (6) **(5)**

The data in Table 2 identify the carbazole **5** as a singlet-derived product. Triplet converters of increasing efficiency make some inroads into its yield. These inroads are

TABLE 1 Use of heavy atom and spin trap effects to identify origins of pyrolysis products from 1-naphthyl azide[13]

Product (R = 1-naphthyl)	PhBr	Yields (%) PhBr + **1**	Cumene	Cumene + **1**
R—N=N—R	20	0	11	0
$R—NH_2$	10	Trace	28	Trace
Phenazine **2**	12	0	5	0
Adduct **3**	—	92	—	61

TABLE 2. Heavy atom effects on products of pyrolysis of 2′-methyl-2-azidobiphenyl (**4**)[2]

Solvent	Temperature, °C	Carbazole **5** yield, %	Phenanthridine **6** yield, %
Paraffin	360	100	0
PhBr	155	98	1.5
2,3,5-Tribromothiophen	156	95.5	4.5

compensated by the appearance of phenanthridine, which must therefore arise from the triplet nitrene. In both this and further work[16], these workers propose that phenyl azide → azepine and 2-azidobiphenyl → carbazole are best viewed as concerted cyclizations of a singlet nitrene intermediate. Triplet nitrenes are seen as the precursors of amines, azo compounds, and phenazines with the azo compounds being dimers of 'lazy' triplets which lack the thermal energy to enter into other reactions.

2. *Substituent effects on product formation from aryl nitrenes*

Ring substituents can exert a considerable influence on the balance of products derived from singlet and triplet precursors. 4-Methoxyphenyl azide[4] gives triplet-derived products (phenazine, azo compound, amine) while other aryl azides can give singlet-derived products.

The most clear-cut set of results on substituents effects operating in competitive situations between singlet and triplet nitrenes was reported by Suschitzky and his group[4,15] (see equation 6 and Table 3). Here we see the singlet-derived product **8** favoured by electron-attracting substituents (5-chloro, 4-trifluoromethyl) in the 2-azidoaryl ring of **7**. Triplet-derived products (**9** and **10**) dominate for a 5-dimethylamino group and, of course, when bromobenzene is used as solvent.

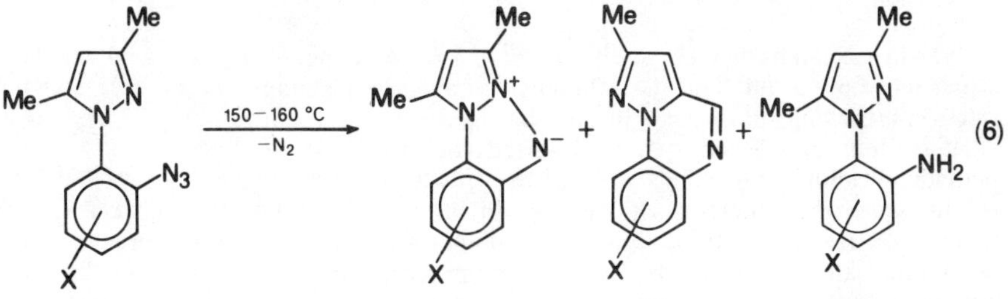

(6)

TABLE 3. Substituent effects on the products of pyrolysis of azides **7**[15]

Solvent	Substituent X	Yields, % **8**	Yields, % **9**	Yields, % **10**
Cumene	5-NMe_2	0	44	41
Cumene	5-Cl	39	5	2
PhBr	5-Cl	8	6	21
Decalin	4-CF_3	67	0	0

3. *Solvent effects*

Apart from the heavy atom effects, there has been little exploration of solvent effects on the ratio of products derived from singlet and triplet sources. Suschitzky[4] reports that diglyme stabilizes (and thus deactivates) the singlet nitrene derived from the azide **7** (X = 5-chloro). Thus, only a trace of the singlet-derived product **8** is obtained. Both the triplet-derived products **9** and **10** are observed, **10** in 83.5% yield. This remarkably high yield of amine is attributed to the plentiful supply of nearby hydrogen atoms when the triplet forms within the coordination shell of its singlet predecessor.

D. Substitution by Nitrenes into Adjacent Aryl Rings

1. *Formation of carbazoles*

This reaction (reviewed in Ref. 1) is now believed to proceed via a concerted cyclization of the singlet nitrene[2] (equation 7).

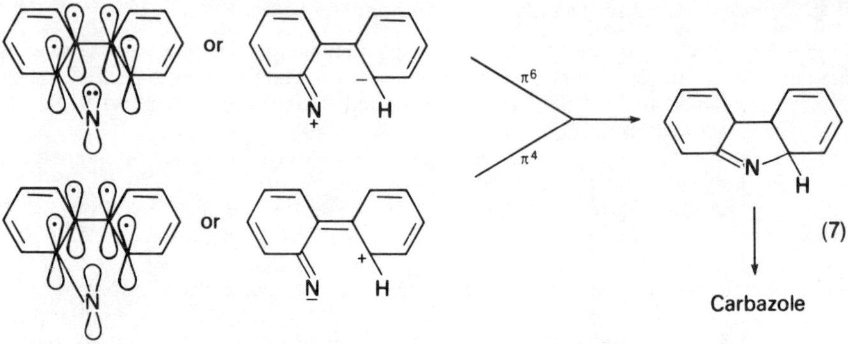

or

or

π^6

π^4

Carbazole

(7)

The nitrene can participate as either a π^2 or a π^0 component (depending on whether it uses its empty or filled orbital), thereby accommodating either electron-releasing or electron-attracting substituents in the rings.

Hall, Behr and Reed[17] have reported that pyrolysis of 2-azido-2′,3′,4′,5′,6′-pentadeuteriobiphenyl yields carbazole with (33 ± 5)% N—H and (67 ± 5)% N—D, which they interpreted in terms of hydrogen abstraction by a nitrene (from solvent or the 2′-position) prior to ring closure. Further studies of exchange possibilities are needed before one attempts to reconcile the result with Suschitzky's scheme (equation 7).

2. *Substitution by nitrenes into adjacent heteroaromatic rings*

A bridgehead nitrogen (as in the azide **11**) leads to poor yields of cyclized product, presumably because it impairs the concertedness of the ring closure described above[2].

However, the cyclization goes reasonably well (via the singlet nitrene) when there is

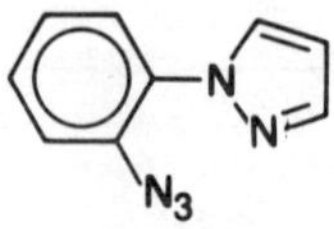

N_3

(11)

a nitrogen atom at the *ortho* position (see equation 6 and Table 3 above). When there is competition between cyclization of a singlet nitrene to sulphur or nitrogen, the cyclization goes exclusively to nitrogen (equation 8)[18,19].

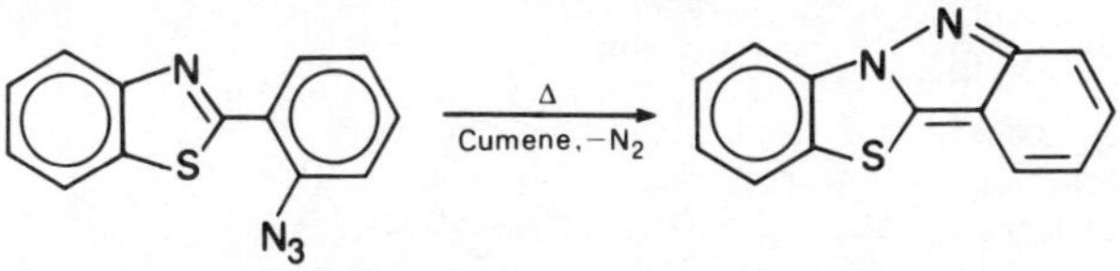

N S N_3 $\xrightarrow[\text{Cumene}, -N_2]{\Delta}$ N N S (79%) (8)

3. *Reactions via spirodienes and azanorcaradienes*

There have been extensive studies of pyrolysis in which an azide of type **12** yields a cyclic product **13**. Substituted examples for X = S reveal that the cyclization proceeds with rearrangement (equation 9)[20,21,22].

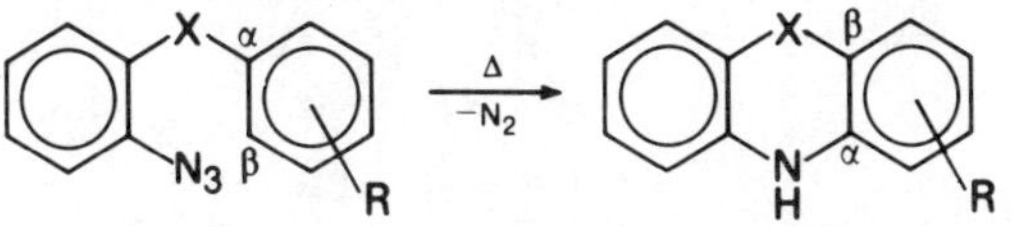

X α N_3 β R $\xrightarrow[-N_2]{\Delta}$ X β N H α R (9)

(12) **(13)**

(X = O or S)

Cadogan has reviewed the field[23] and points out that all the observed cyclic products can be rationalized by assuming that a singlet nitrene forms either a spirodiene intermediate **14** (preferred by Cadogan[23]) or the highly strained azanorcaradienes **15** (preferred by Gurnos Jones[33,34]). No evidence exists to distinguish between them.

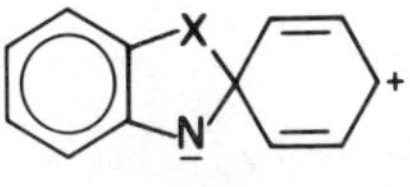

X N + X N

(14) 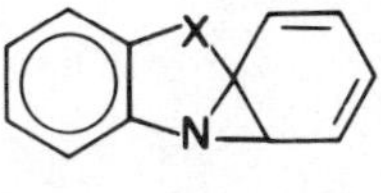 **(15)**

($X = O, S, SO_2, CH_2$)

The methylene-bridged azides (**12**; $X = CH_2$) differ from the others in that ring expansion is a major pathway, and will be treated separately.

a. Sulphur and oxygen bridges. The general mechanism[24] for phenothiazine formation is given in equation (10).

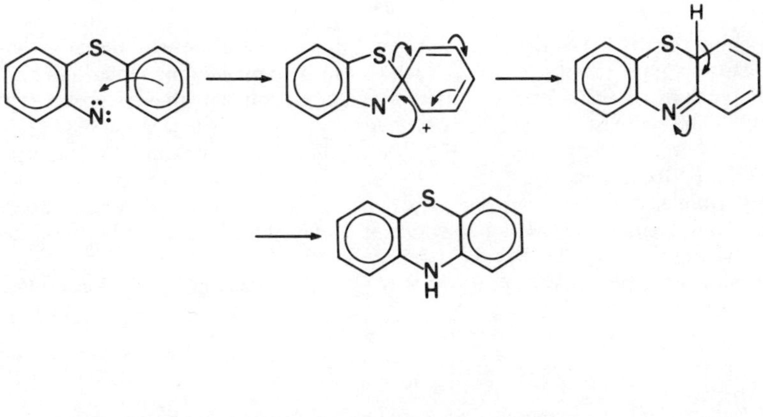

S N: ⟶ S N + ⟶ H S N (10)

⟶ S N H

When the *ortho*-positions in the phenyl group are blocked with methyl groups, one might expect the nitrene to abstract hydrogen. In practice, it is the sulphur atom which attaches to the methyl group, a result which Cadogan and Kulik[24] rationalize with equation (11).

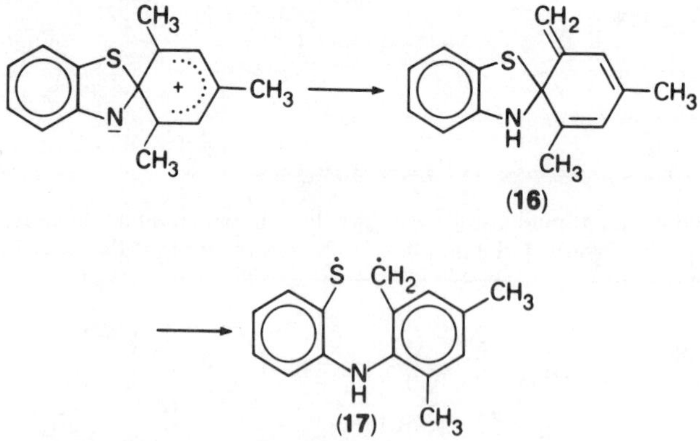

(16)

(11)

(17)

The diradical **17** was invoked as the precursor to the thiazepine **18**, because the suprafacial 1,3-sigmatropic shift in **16** is disallowed. Cadogan[23] has since suggested that the shift could proceed if sulphur uses its d orbitals.

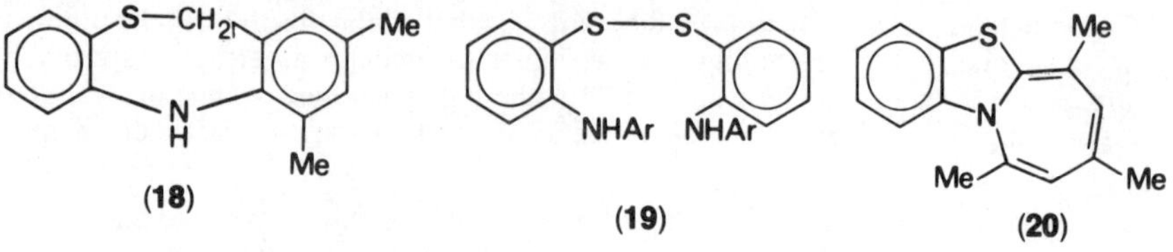

(18) (19) (20)

(Ar = 2,4,6-trimethylphenyl)

The reaction also yields the disulphide **19**, and a trace of the azepine **20** has been reported[25]. Suschitzky has established an unusual feature of the pyrolysis shown in equation (11): all three products form from both singlet and triplet precursors. A diradical spirodiene (**21**; X = S) must be proposed.

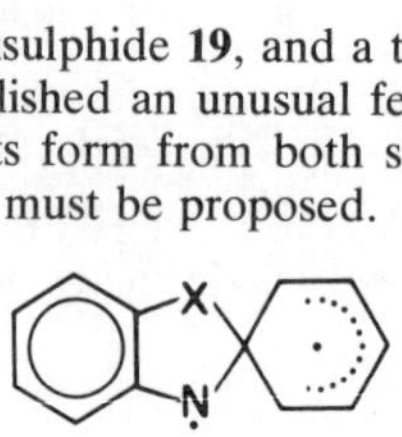

(21)

Blocking the *ortho*-positions of the phenyl ring of the azide with methoxyl groups (**22**) leads to quite curious reactions[24] which can be explained by equation (12).

This pattern of behaviour is, however, changed when a thienyl group replaces phenyl[26]. The simplest thienyl azide **23a** gave only polymers but the dimethyl azide **23b** gave **24**, an example of the hitherto unknown heterocycle, pyrrolo[2,1-*b*]benzothiazole.

This product (and others) arose via the benzothiazole thioketal **27**, which apparently forms as in equation (13). The behaviour of azides of type **23** is further modified by benzo fusions[27].

An interesting suggestion to explain why pyrolysis of **23a** gives polymers is that the

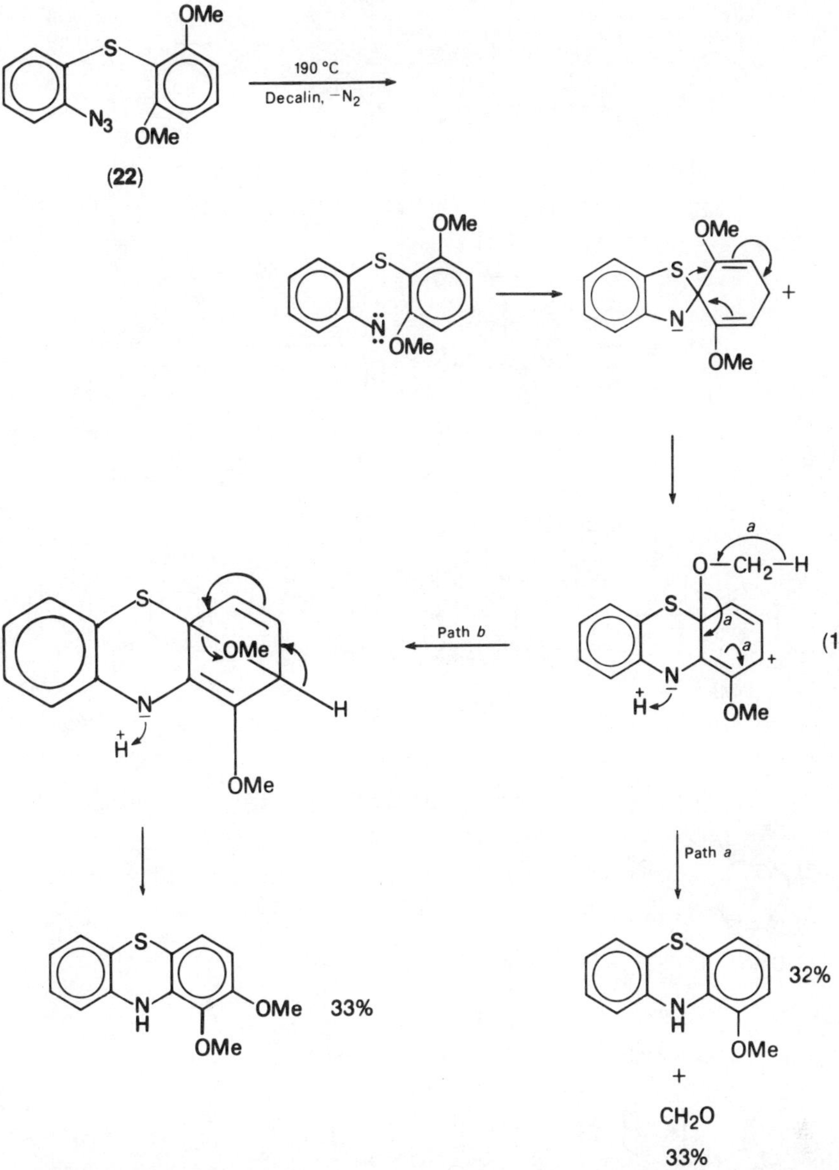

(22)

190 °C

Decalin, $-N_2$

Path b

(12)

Path a

33%

32%

+

CH_2O

33%

spirodiene intermediate **26** (R = H) opens out to give **25** whereas methyl groups in **26** prevent this reaction[26].

The 2-azidobiphenyl ethers (**12**; X = 0) pyrolyse in a way similar to the sulphides (**12**; X = S), but there are important differences. 2-Azidobiphenyl ether does not yield a phenoxazine (**13**; X = O) on pyrolysis[28] but the cyclization does occur if both *ortho* positions are blocked by methoxyl groups[29]. The apparent pathway (equation 14) resembles equation (12) except that the spirodiene intermediate can rearrange with

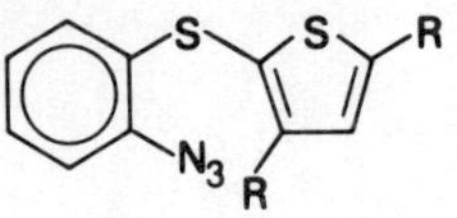

(23a) R = H

(23b) R = Me

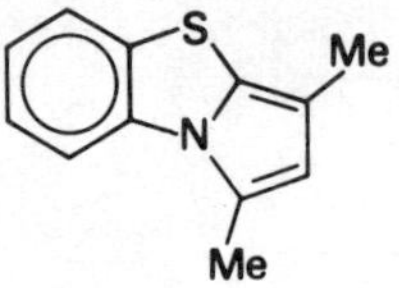

(24)

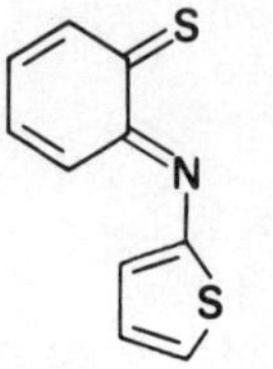

(25)

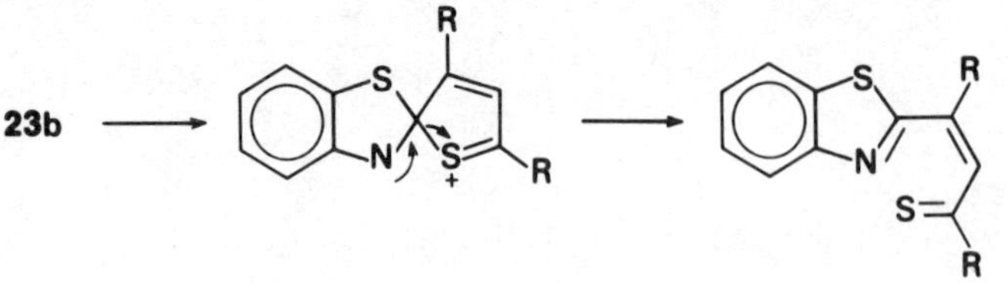

23b → (26) (R = Me) → (27) (R = Me) (13)

15%

35%

+ CH_2O 15%

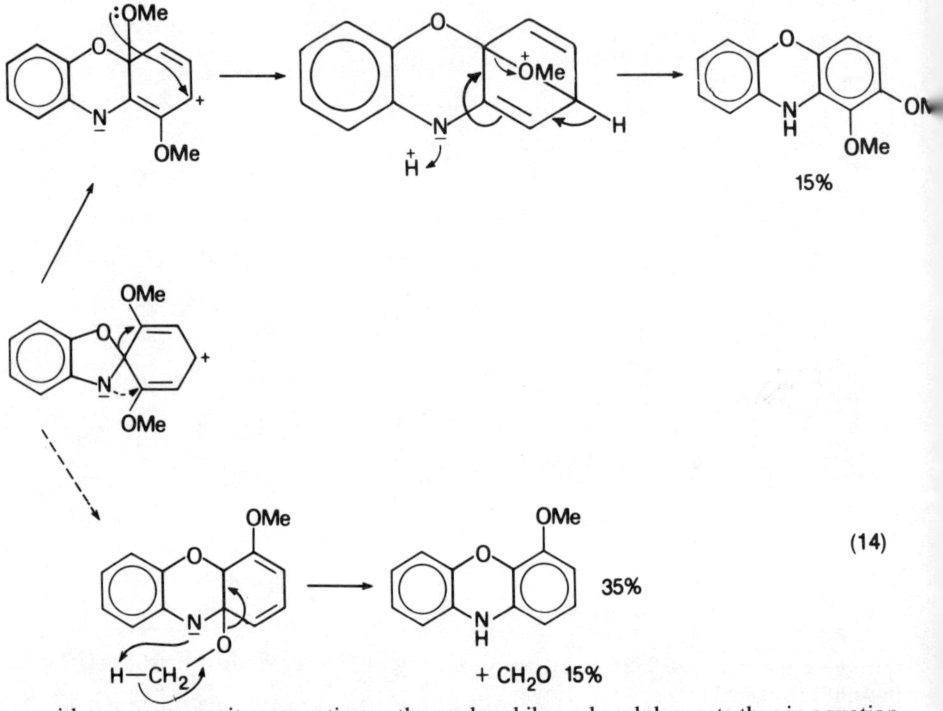

(14)

either oxygen or nitrogen acting as the nucleophile; only sulphur acts thus in equation (12). The existing data do not make it clear why the azidobiphenyl ether **12** (X = O) cyclizes only if the *ortho* positions are blocked, though the explanation offered for the thienyl cases **23** may be relevant. When the *ortho* positions are blocked with methyls, the oxygen analogue of **18** is the major product and no phenoxazine is obtained[29].

The sulphone bridge (**12**; X = SO_2) leads to more complicated mixtures of pyrolysis products. As well as rearrangement via a spirodiene intermediate, it is necessary to invoke the (singlet) nitrene as an electrophile substituting into the adjacent aryl ring. The ratios of substitution isomers in the phenothiazine-5,5-dioxide products are controlled not only by electronic effects, but also by proximity ones[30].

b. Methylene bridges. Decompositions of azides of type **12** (X = CH_2) have been extensively studied by Gurnos Jones and his group at Keele. The products (which are often quite complex) include the ring-expanded 10*H*-azepino[1,2-*a*]indoles (e.g. **28**), as well as acridans (**29**) and acridines (**30**) (equation 15)[31]. The acridine is believed to arise from the acridan by oxidation[32].

(15)

The three products form in about equal proportions in the example shown, but in bromobenzene solvent there is very little **28** in a product mixture dominated by **29** and **30**. Naphthalene (a solvent of low triplet energy) had a similar effect on product distribution. Thus, **29** and **30** are identified as triplet-derived products, with the azepinoindole **28** derived from singlet nitrene. Jones and his coworkers[33,34] have accounted for products in the thermolysis of azides **12** (X = CH_2) in terms of ring openings and H-shifts from an azanorcaradiene intermediate **15** but their subsequent discovery of triplet nitrene involvement requires that a species like **21** (X = CH_2) be added to the reaction intermediates. The ratios of azepine to acridan/acridine are strikingly affected by substituents in the benzyl group of **12**, and the pyrolysis products from the 2′-methoxy derivative include an acridan which has lost methoxyl in a fashion similar to the sulphides and ethers (see Section II.C.3.a above)[32,33].

c. Ester bridges. The azides **31** (R = H, Me, Cl, Br or COOEt) give no cyclic products when pyrolysed in solution, but flash vacuum pyrolysis gives a carbazole. The mechanism as shown in equation (16) was proposed[35].

Appropriate activation of the electrophilic nitrene (by the carbonyl group), and of the aryl ring to be substituted, is apparently essential: no cyclization occurs if the ester function in **31** is reversed.

E. Rates of Aryl Azide Pyrolysis

There is now a considerable body of data on the rates of pyrolysis of aryl azides[36–40]. The decompositions are first-order in azide, though Dyall[40] has shown that the order

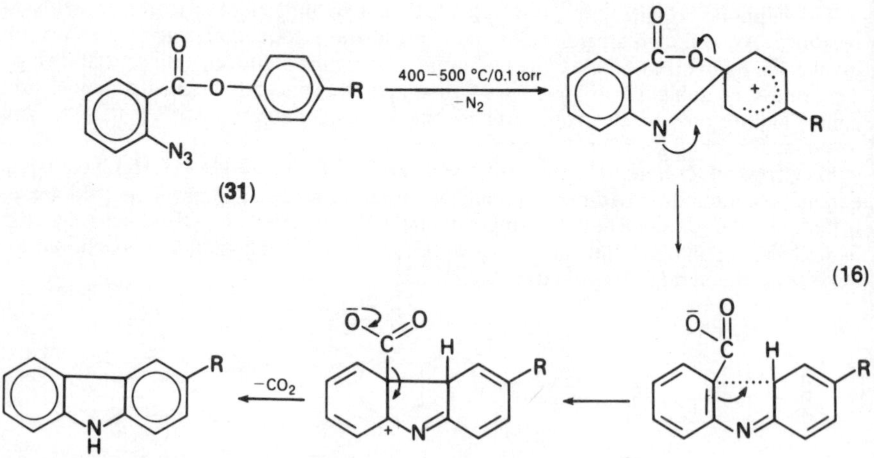

400–500 °C/0.1 torr, $-N_2$; $-CO_2$

(31)

(16)

increases above 0.01 M concentrations, presumably due to radical-induced decomposition which is known to operate in certain azide pyrolyses[36,41].

Rates have usually been measured by nitrogen evolution[42,43] but in our experience it is more convenient to monitor the azido band near 2150 cm^{-1} in the infrared spectrum[44].

For phenyl azides, the rate of decomposition is increased by *all* substituents (see Table 4), though the total range of rates is only fivefold if we exclude the special cases of neighbouring group assistance (Section III.D). The largest rate increases shown in Table 4 are exerted by *ortho* substituents with lone electron pairs and/or π-bonds, and these presumably stabilize the nitrene-like transition state by exerting a through-space interaction with the vacant nitrogen orbital[45]. Solvent has little effect on rate unless triazoline formation between azide and solvent is possible or radical chains can develop[36,38]. Isokinetic relationships are common: the variations of E_{act} and ΔS_{act} bear a linear relationship both for substituent change[37,40] and for solvent change[36].

To most organic chemists, the interest in these rate data lies in the estimation of half-lives under selected pyrolysis conditions. The rates of these simple pyrolyses are especially useful for comparison with those in which neighbouring group assistance for

TABLE 4. Kinetic data for pyrolysis of aryl azides in decalin solution[40,46]

Azide	k_{rel}(393 K)	$t_{1/2}$(393 K), h	$\log k = -a/T + b$	
			$a =$	$b =$
Phenyl	1	35.5	7130	12.867
2-Methylphenyl	1.27	28.0	6607	11.641
2-Chlorophenyl	5.03	7.1	6763	12.635
4-Chlorophenyl	1.55	22.9	7218	13.281
2-Cyanophenyl	4.92	7.2	6080	10.914
4-Cyanophenyl	2.28	15.6	6960	12.793
1-Naphthyl	4.31	8.2	7074	13.361
2-Naphthyl	2.82	12.6	7735	14.856
2-Biphenylyl[37]	1.54	23.2	7009	12.747

extrusion of nitrogen is believed to occur (see Section III). Some useful data for these purposes are collected in Table 4.

III. SOLUTION PYROLYSES WITH NEIGHBOURING GROUP ASSISTANCE

A. Cyclization Reactions Leading to Five-membered Rings

The general reaction shown in equation (17) has been widely used as a method of synthesis of five-membered heterocycles; the group —X=Y is most commonly nitro, carbonyl or arylazo[1].

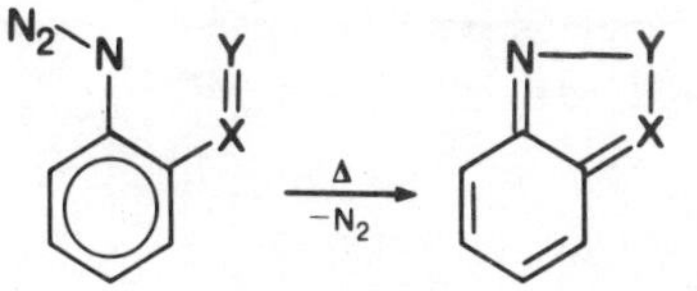

(17)

It has been known for nearly a century that the loss of nitrogen accompanying these cyclizations is often appreciably faster than from pyrolyses of aryl azides in which there is no cyclization. The rate increase can be very large, e.g. 2-azidoazobenzene pyrolyses 21,180 times faster than phenyl azide at 393 K in decalin solution[46]. (2-Cyanophenyl azide pyrolyses only 4.92 times faster.) Some kind of neighbouring group assistance is clearly indicated, and a number of studies of reaction kinetics have been made in an attempt to understand how it operates. Two schools of thought have emerged.

B. Possible Mechanisms for Thermal Cyclizations

1. *Single-stage cyclization and nitrogen extrusion*

Some elements of this mechanism have been in the literature[47–52] since 1956, but it was first given a formal description by Dyall and Kemp[53] in 1968. The π-electron reorganization can be represented in various ways, one of which is shown in equation (18).

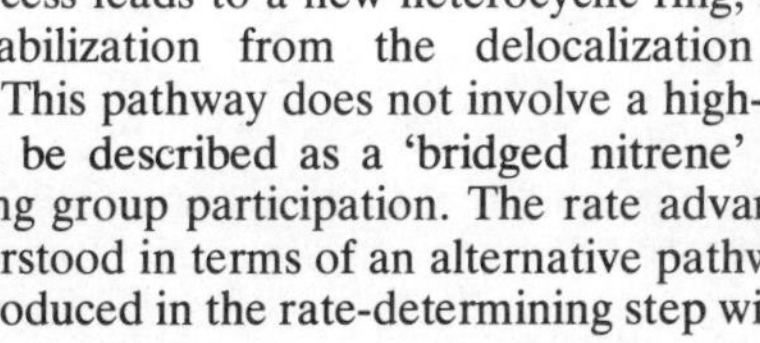

(18)

This electrocyclic process leads to a new heterocyclic ring, and the transition state must derive some stabilization from the delocalization energy of the new heteroaromatic system. This pathway does not involve a high-energy nitrene species; the transition state can be described as a 'bridged nitrene' if one uses the formal language of neighbouring group participation. The rate advantage conferred by this mechanism can be understood in terms of an alternative pathway, in which a 'free' or 'unbridged' nitrene is produced in the rate-determining step with cyclization occurring subsequently.

Effective π-electron reorganization of the type shown in equation (18) requires coplanarity of the participating orbitals. One would expect substituents at the 3- or 6-position in the phenyl azide to twist the system out of coplanarity and thereby slow the reaction rate. Very substantial steric effects have in fact been observed (see Table

TABLE 5. Steric effects on relative rates of aryl azide pyrolysis (decalin, 393 K)[40,46]

Substituent in phenyl azide	k_{rel}	Reference
None	1^a	40
2-Nitro	738	40
3-Methyl-2-nitro	65.9	40
6-Methyl-2-nitro	11.8	40
2-Acetyl	287	40
3-Methyl-2-acetyl	5.1	46
6-Methyl-2-acetyl	10.5	46

^{a}The rate constant is 5.43×10^{-6} s^{-1}.

5) and in extreme cases the cyclic products must form via a nitrene and its subsequent intramolecular capture.

2. Two-stage pathway of 1,3-dipolar addition and nitrogen extrusion

A different view of cyclization mechanisms has been advanced by Hall, Behr and Reed[17], who drew upon the 1,3-dipolar properties of the azido group[1] to form an adduct **32** from which nitrogen is subsequently extruded (equation 19).

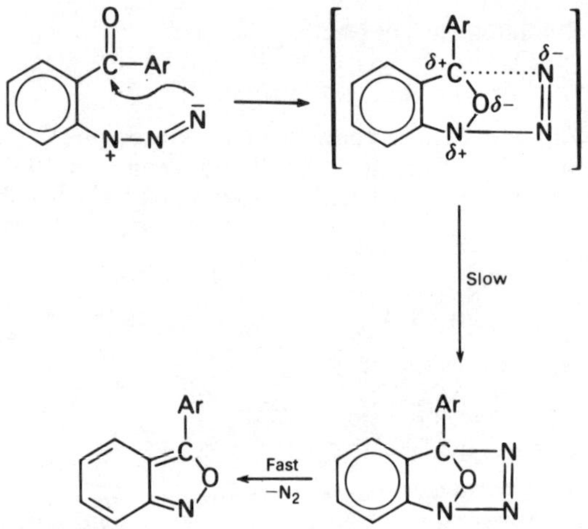

Slow

Fast
$-N_2$

(19)

(32)

Hall and his coworkers suggested that there is a spectrum of possible mechanisms for cyclization of *ortho*-substituted phenyl azides, represented by the transition states **33–36**.

The first of these, **33**, is an unassisted cyclization proceeding via a nitrene, as in the formation of carbazole from 2-azidobiphenyl (see Section II.D). The second is the 'bridged' nitrene of the Dyall–Kemp mechanism, and was accepted for —X=Y being nitro by Hall. (The data in Table 5 for acetyl was not published at that time.) The 1,3-dipolar addition pathway is represented by the third component, **35**, of the

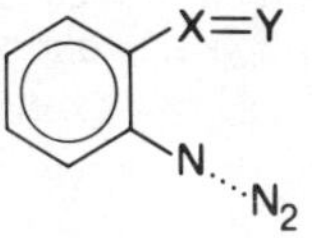

(33) 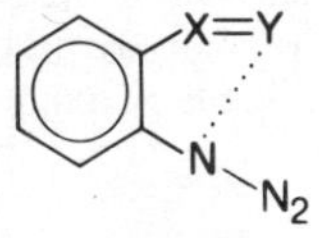 (34) 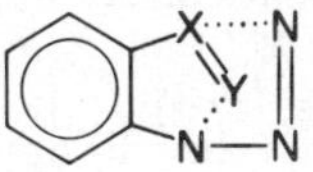 (35) 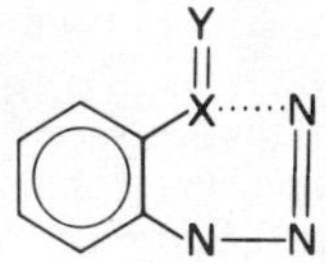 (36)

spectrum, or by **36**, which differs from **35** only in the extent to which N···X bond formation precedes Y···N bond formation. Hall regarded the 1,3-dipolar addition pathway as the normal one for assisted pyrolyses of azides.

C. Comparison of the Two Mechanisms for Rate-enhanced Cyclization

1. Data consistent with Hall's 1,3-dipolar addition mechanism

Studies of the mechanism of assisted pyrolysis have centred on kinetic measurements. Hall, Behr and Reed[17] have presumed that equation (18) implies a nucleophilic attack by the —X=Y group upon N-1 of the azido group, and looked for effects of a substituent R (in **37**) upon the nucleophilicity of the carbonyl group.

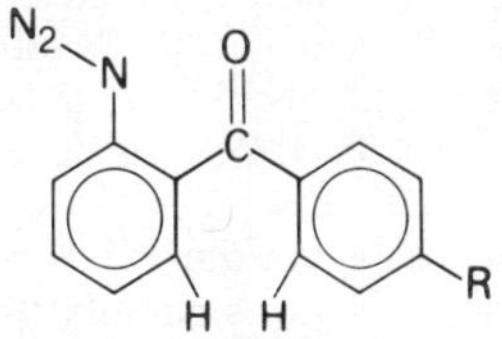

(37) 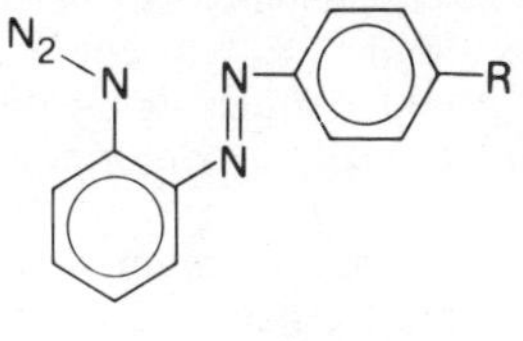 (38)

The system is complicated by the lack of coplanarity of the two aryl rings in **37**, but it was assumed that the angle of twist is so great that the principal conjugative effects of R are exerted on the carbonyl group and do not extend to the other ring. It was thus argued that an electron-releasing R group should (on the basis of equation 18) increase the rate of an electrocyclic process, while an electron-attracting group should retard it.

The substituent effects on rate are quite small (see Table 6) but are quite definitely the reverse of those expected by Hall and coworkers on the basis of the Dyall–Kemp mechanism. The effects do, however, fit the 1,3-dipolar mechanism (equation 19), provided it is assumed that N···C bond formation progresses ahead of O···N bond

TABLE 6. Substituent effects on relative rates of pyrolysis of 2-azidobenzophenones (decalin, 393 K)[17] and 2-azidoazobenzenes (anisole, 333 K)[54]

Substituent (4′-R in **37**)	OMe	Me	H	Br	NO_2
k_{rel}	0.64	0.92	1[a]	1.01	1.60
Substituent (4′-R in **38**)	NMe_2	Me	H	COMe	NO_2
k_{rel}	0.432	0.720	1[b]	1.33	1.45

[a]The actual rate constant at 393 K is $3.57 \times 10^{-4}\ s^{-1}$.
[b]The actual rate constant at 333 K is $7.63 \times 10^{-4}\ s^{-1}$.

formation. In further work, Hall and Dolan[54] studied the effects of substituents 4′-R on the rates of pyrolysis of 2-azidoazobenzenes, **38**. Again, the substituent effects on the rates are very small (Table 6) but do appear to support the 1,3-dipolar addition mechanism. It should be noted that the results in Table 6 rule out involvement of a nitrene intermediate, because *all* substituents would then enhance the rate (see Section II.E).

Additional arguments brought forward by Hall, Behr and Reed point out that, for equation (18), the various *ortho* substituents —X=Y should exert rate effects related to their nucleophilicities, whereas they do not. The most effective neighbouring group in azide pyrolysis, namely phenylazo, is not known as a very effective nucleophile.

Counter-arguments to support the Dyall–Kemp mechanism in the face of Hall's evidence are readily produced. First, it is too simplistic to focus attention on just one component (the nucleophilic bridging of the *ortho* substituent to N-1 of azido in equation 18) of an electrocyclic process. Substituent effects will operate on *all* components and it is difficult to say what the overall effect will be. Hall's argument that the substituent effects from 4′-R in the azidobenzophenone **37** do not reach the azidoaryl ring is improbable, it being known that these effects reach the remote ring by a π-bond polarization mechanism[55]. Thus, as 4′-R is varied there will be a subtle interplay between azido-carbonyl conjugation and the angle of twist between the ring systems[40]. Angle-of-twist effects on pyrolysis rates can be quite large (see Table 5) and it is therefore not at all clear how to interpret the small rate variations reported (see Table 6). Similar arguments can be applied to the data for azidoazobenzenes. In addition, the attempt to correlate nucleophilicities of *ortho* substituents with neighbouring group abilities in azide pyrolysis quite overlooks the stabilization produced in the transition state of equation (18) by the delocalization energy of the new heterocycle[53].

Although these counter-arguments weaken the case for the 1,3-dipolar addition mechanism, it still remains an attractive possibility. A number of relevant intramolecular 1,3-dipolar additions have been reported[56–59] and examples[57] are shown in equation (20). The adduct **32** in equation (19) involves very substantial angle strains but it nevertheless might provide a pathway of lower activation energy than the alternative route *via* a free nitrene.

(20)

The direct detection of the 1,3-dipolar adduct **32** is unfavourable in view of its relative rates of formation and decomposition. Thus, the definitive experiment to distinguish the mechanisms of equations (18) and (19) is the steric effect of a flanking substituent, as already described for nitro and acetyl neighbouring groups (Table 5). There is no apparent reason why a 6-substituent (as in **39**) should appreciably inhibit the formation of the dipolar adduct **32**; indeed its presence should force the two groups into the conformation **40** which is favourable for addition.

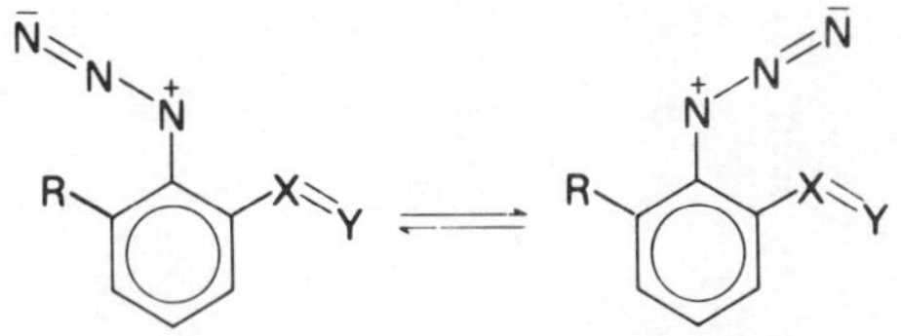

(39) **(40)**

The most general way of introducing such flanking substituents into a series of *ortho*-substituted phenyl azides was to dibrominate in the 4- and 6-positions. The azide pyrolysis rate comparisons[60] are presented in Table 7.

To interpret these data, one must remember that introducing the two bromines into phenyl azide will increase the pyrolysis rate at 393 K (by 5.84-fold). In addition, those azides whose *ortho* substituent (—X=Y) is totally inhibited from exerting a neighbouring group effect will derive a rate advantage along the 'free nitrene' route from other substituent effects of the —X=Y group. A rough measure of this influence is provided by an *ortho* cyano group which exerts the same sort of electronic effects as nitro, benzoyl, etc., but cannot participate in the building of a new heterocycle. Thus, in assessing the Table 7 data, the evidence for neighbouring group assistance will be a pyrolysis rate greater than that of 4,6-dibromo-2-cyanophenyl azide.

It is seen that the introduction of a flanking bromine provides support for the electrocyclic Dyall–Kemp mechanism, and the results cannot be reconciled with equation (19). The flanking bromine reduces the neighbouring group assistance by nitro and benzoyl to an undetectable level, and that by phenylazo to a very low level.

TABLE 7. Steric effects of flanking bromine on rates of assisted pyrolysis (decalin solutions 393 K)[60]

Substituent in phenyl azide	k_{rel}
None	1[a]
2-Nitro	738
2-Phenylazo	21,180
2-Benzoyl	70.0
2-Formyl	22.8
2,4-Dibromo	1[b]
4,6-Dibromo-2-cyano	10.0
4.6-Dibromo-2-nitro	9.43
4,6-Dibromo-2-phenylazo	169.3
4,6-Dibromo-2-benzoyl	8.56
4,6-Dibromo-2-formyl	10.7

[a]The rate constant is 5.43×10^{-6} s^{-1}.
[b]The rate constant is 3.17×10^{-5} s^{-1}.

The assistance given by formyl is anyway small so that the steric effect has little to work on, but the result does suggest that formyl belongs to the same mechanistic group as the other three *ortho* substituents. It should be noted that there is no evidence for the bromines complicating the experiments by introducing triplet character into the transition states: the result obtained here for nitro is the same as with a flanking methyl (Table 5).

Further support for the Dyall–Kemp mechanism is provided by the azides **41** and **42**.

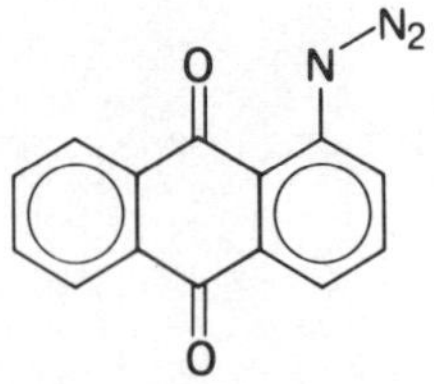

(41) 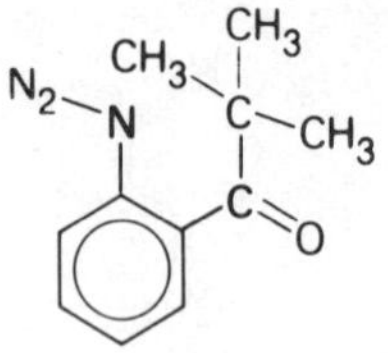 **(42)**

In the azidoanthraquinone **41**, the stereochemistry is ideal for the electrocyclic mechanism, but according to Cenco–Peterson models, the 1,3-dipolar addition involves especially severe strain, and very little rate enhancement could be expected from this pathway. In practice, this azide pyrolyses (with formation of anthra[9,1-*cd*]isoxazol-6-one) extremely rapidly[46]. With **42**, the 1,3-dipolar addition is blocked off by the *t*-butyl group but nevertheless the pivaloyl group provides a substantial increase (61.7-fold) in rate over phenyl azide[46].

Work has been done[44] on the pyrolysis of 2-substituted 1-naphthyl azides and 1-substituted 2-naphthyl azides. Again, the results can be understood in terms of the electrocyclic mechanism, with steric effects being exerted by the 8-hydrogen.

At this point in time, all the assisting *ortho*-substituents which have been studied in detail can be fitted to the Dyall–Kemp mechanism. No convincing example of pyrolysis via the 1,3-dipolar addition mechanism has yet been established, though it is not improbable that some will be discovered (see Section III.F).

D. Magnitudes of Neighbouring Group Effects in Azide Pyrolysis

1. *Quantitative data*

The neighbouring group effect is defined (following the description given for nucleophilic substitution)[61] as k_{obs}/k_n, where k_{obs} is the measured rate constant for the assisted pyrolysis and k_n is the corresponding rate via the alternative nitrene pathway (equation 21).

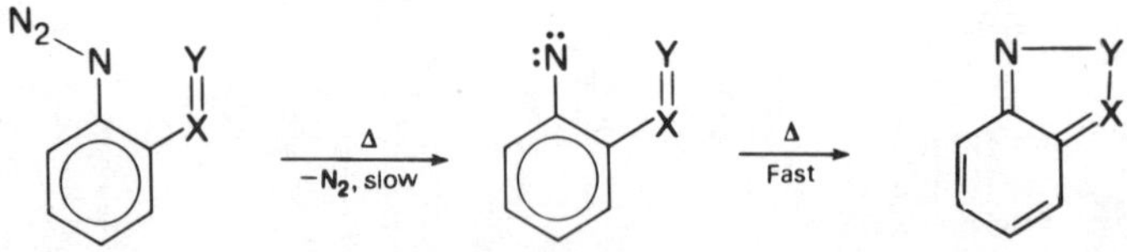

(21)

Of course, k_n cannot be experimentally measured. We can assume (on the basis of data given in Section II.E) that the *ortho* substituent would affect k_n via its normal stereoelectronic effects. For —X=Y type *ortho* substituents, these can be approximated by the *ortho* cyano group, which obviously cannot get effectively involved in an electrocyclic transition state. In Table 8, k_{rel} data are presented on the

TABLE 8. Magnitudes of neighbouring group effects, in pyrolysis of 2-substituted phenyl azides (decalin, 393 K)

2-Substituent	k_{rel}[a]
Phenylazo	4080
Nitro	142
Acetyl	55.3
Benzoyl	13.5
Pivaloyl	11.9
Formyl	4.4
Methoxycarbonyl	0.94

[a]Defined as k_{obs}/k_n, where k_n is the rate constant for 2-cyanophenyl azide (see text). The k_n value, 2.67×10^{-5} s^{-1} at 393 K, is 4.92 times that for phenyl azide.

basis that k_n is the rate constant for 2-cyanophenyl azide. A temperature of 393 K is chosen because it falls close to the experimental ones used in measuring both k_{obs} and k_n. Choice of other temperatures has some effect on the magnitude of k_{rel} but does not alter the general picture.

There is a considerable range of neighbouring group ability. One entry (methoxycarbonyl) obviously provides no assistance to azide pyrolysis (and yields no cyclic product). 1-Azidoanthraquinone does not readily fit into this table but (if compared to 1-azidonaphthalene) the k_{rel} value[46] is 1710. These k_{rel} values represent a composite of delocalization energies in initial and final states, as well as involving conformational possibilities (not always favourable) in the initial azide. It would be of considerable interest to see molecular orbital calculations on the energy surfaces involved in the Dyall–Kemp electrocyclic mechanism.

These neighbouring group abilities have turned out to be useful in the interpretation of oxidative cyclizations of *ortho*-substituted anilines[62].

While values of E_{act} and ΔS_{act} are available for a large number of assisted pyrolyses of azides, they have proved of little value in diagnosis of mechanism. Negative values of ΔS_{act} have been obtained and have been seen to be appropriate in magnitude for each of the rival mechanisms[53,54]. One of the most outstanding examples of assisted pyrolysis, 1-azidoanthraquinone, has a quite small negative entropy of activation[46]. Dyall[40] notes that rather large negative ΔS_{act} values are measured for 2-cyanophenyl azide and 2-methylphenyl azide where no cyclic transition state is indicated by other evidence. He suggested that ΔS_{act} values may reflect not only the cyclic nature of the transition state but also the degree of unloosening of the nitrogen molecule; these factors would partly cancel. For 2-azidobenzophenones, 2-azidoazobenzenes and benzylidene-2-azidoanilines, Hall and coworkers[17,54] report a linear (isokinetic) relationship between E_{act} and ΔS_{act}.

2. *Qualitative data*

There are a number of examples of cyclization via azide pyrolysis which do, or might, involve neighbouring group participation, but for which precise kinetic data are not available. The outstanding example is 2-thiobenzoylphenyl azide, whose crystals decomposed explosively at room temperature (or smoothly in solution at 0°C) to yield the expected cyclized product[63]. Both 3-chloro-2-nitrosophenyl azide[64] and the sodium salt of 2-azidobenzaldehyde oxime[65] decompose below 100°C and must involve very considerable neighbouring group assistance. Other possible participating substituents are thioacetyl[63], azomethine groups[66,67,68] and vinyl groups[69] though some of these azide decompositions occur at temperatures high enough to generate a free nitrene which could subsequently cyclize to the *ortho* substituent (equation 21). A quite unusual example is reported by Mosby and Silva[70] (equation 22).

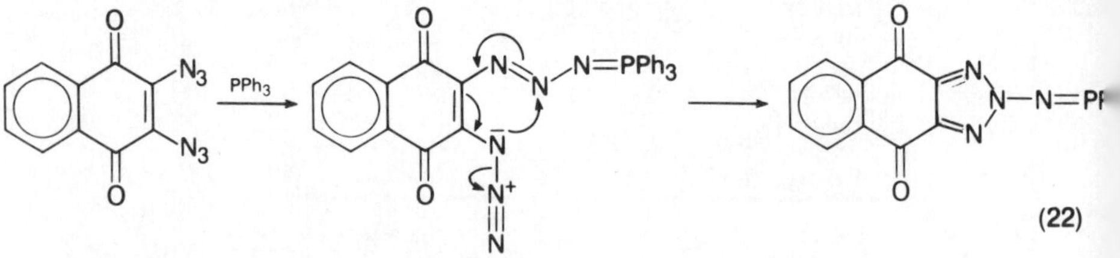

(22)

Curiously, however, 1,2-diazidobenzene reacts with triphenylphosphine without yielding such a cyclic product.

2-Cyclopropylphenyl azide apparently does not pyrolyse with neighbouring group assistance[71]. The product (quinoline) can be explained in terms of nitrene insertion into the cyclopropyl ring and subsequent aromatization.

E. Anomalous Neighbouring Group Effects in Azide Pyrolysis

A number of *ortho*-substituted aryl azides fail to yield the expected cyclized products on pyrolysis. Our understanding of these failures is only partial at this time.

1. Cases of unstable furoxan and isoxazole products

The azides **43** and **45** do not yield the expected cyclic products (**44** and **46**) on pyrolysis[40,53,72,73]. The reaction of **43** shows no kinetic evidence of neighbouring group participation, which has been ascribed to the unstable **44** not being able to contribute any stabilization to the cyclic transition state in the Dyall–Kemp mechanism[40,53]. **43** must form a nitrene which, being unable to cyclize to form a stable product with the *ortho* substituent, reacts in other ways[73].

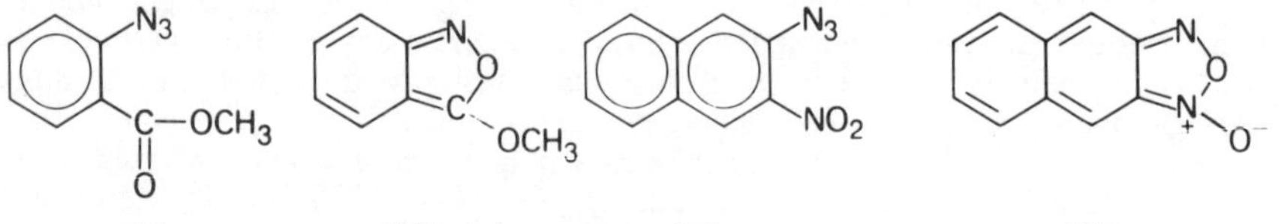

(43) **(44)** **(45)** **(46)**

The behaviour of **45** is most interesting. If fails to yield **46** on pyrolysis, but Selvarajan and Boyer[72] note that the reaction is three-quarters complete in only 8 h at 100°C in octane solution and claim that there must be anchimeric assistance. The approximate half-life of 4 h is to be compared with 146 h for 2-naphthyl azide; an *ortho* substituent would reduce that by a factor of about five[46] to 30 h. It would appear that we have with **45** the only known example in azide pyrolysis of neighbouring group assistance without formation of a cyclic product. More precise rate data would throw light on the question of whether the naphthofuroxan **46** might have a transitory existence. 6-Azido-7-nitroquinoline also fails to yield the expected quinofuroxan[74], but no estimate of its pyrolysis rate is available.

2. *Azidonitropyridines and azidonitroquinolines*

2-Azido-3-nitropyridine (which exists in its tetrazolo form **47**) pyrolyses smoothly to the pyridofuroxan **48** at 170°C in various solvents[75,76] but surprisingly the azide **49** fails[77] to yield the furoxan **50**. Likewise, the failure[77,78] of **51** to cyclize to **52** is difficult to explain in view of the successful thermal cyclization[74] of **53** to **54**.

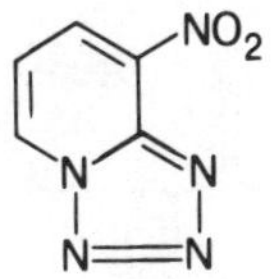

(47) 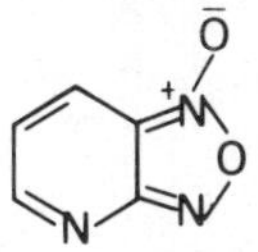 **(48)** 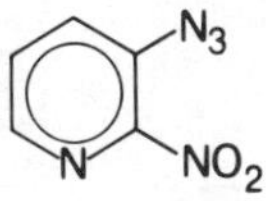 **(49)** 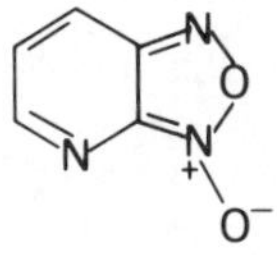 **(50)**

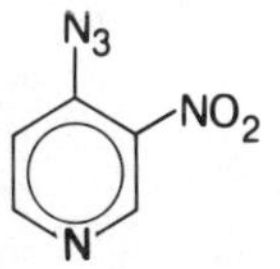

(51) 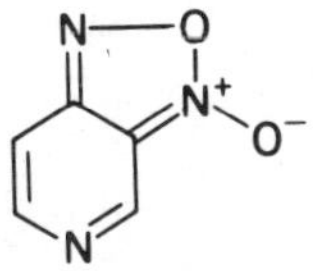 **(52)** 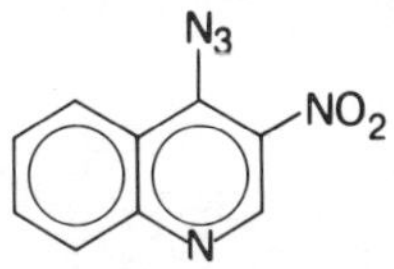 **(53)** 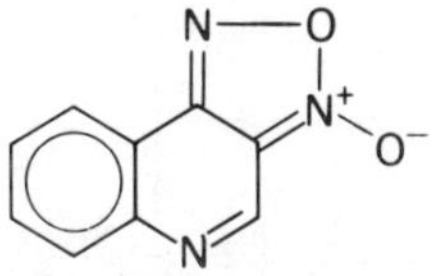 **(54)**

The azidopyridine-1-oxides behave like the azidopyridines: neither 3-azido-4-nitropyridine-1-oxide nor 4-azido-3-nitropyridine-1-oxide yields a furoxan[78]. However, whereas the (masked) azidopyridine **47** did yield a furoxan, its *N*-oxide yields tars[79], possibly because a lower-energy path of ring opening (see Section IV.A) is available.

Little is known about the stabilities of most of the expected pyrido- and quinofuroxan products, but it has been established[77] that **48** is 4.6 kJ mol^{-1} more stable than **50**. This rather small energy difference can hardly account for the total failure of the pyrolysis **49** → **50**.

It is surprising that the azide **55** pyrolyses smoothly[78] to the pyridofuroxan **56** in benzene at 80°C when **51** requires a temperature of 179°C for decomposition (and yields no cyclic product). There are many unexplained features of these pyrolyses of azidonitroheterocycles which await further study.

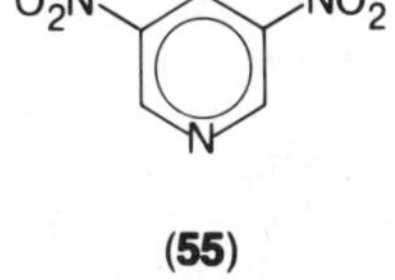

(55) 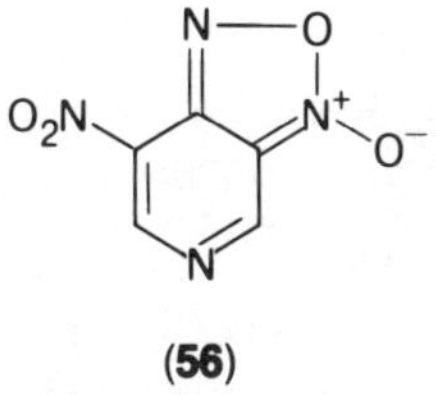 **(56)**

F. Other Examples of Possible Bridging to Incipient Aryl Nitrenes

The assisted pyrolyses discussed above have all involved an *ortho* substituent and the azido group in a five-membered cyclic transition state. A number of workers have attempted to build five- or six-membered heterocyclic rings across the 2,2′-positions of biphenyl or the 1,8-positions of naphthalene. These reactions have not received anything like the attention given to *ortho*-substituted aryl azides, but some success has been achieved.

1. The 8-substituted-1-naphthyl azide system

Three successful cyclizations have now been reported. In the first of these, Bradbury, Rees and Storr[80] heated the amide **57** in boiling 1,2,3-trichlorobenzene (218°C) to obtain a mixture of the perimidine **58** and the oxazole **59** (R = Me or Ph) (equation 23). The products were thought to arise from competitive attack by the

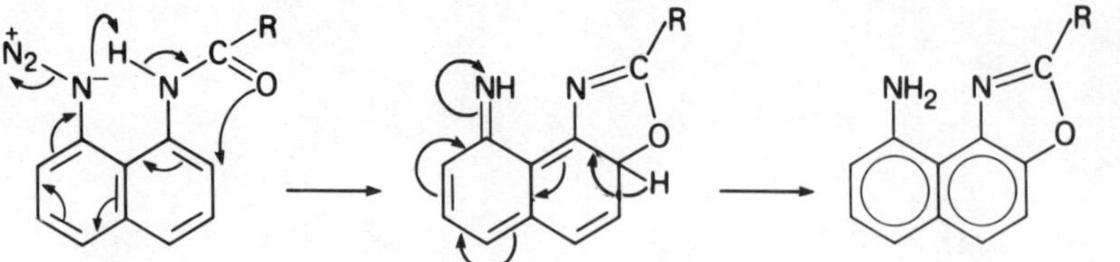

(57) (59)

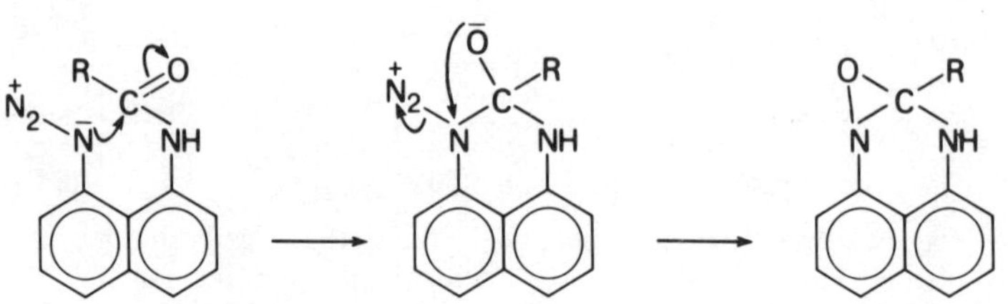

(23)

(60)

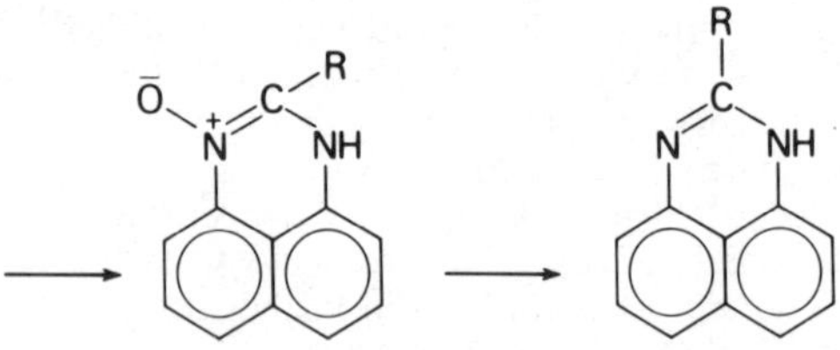

(60) (58)

nucleophilic N-1 of the azido group on either the adjacent amido N—H, or on the amide carbonyl. Assisted expulsion of nitrogen was suggested by the fact that **57** could be decomposed smoothly at about 140°C, whereas other 8-substitued-1-naphthyl azides required 180°C. As would be expected, *N*-methylation of **57** slowed up the azide decomposition considerably, and different products were obtained.

8-Nitro-1-naphthyl azide does not decompose to useful products in hot solvents, but cyclic products are obtained[81] in the gas phase at 280°C (equation 24). It was suggested that a free nitrene is involved, and the success of the intramolecular reaction

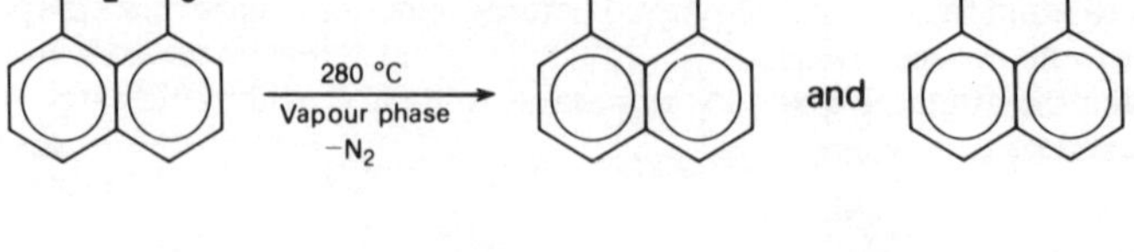

NO_2 N_3 —(280 °C, Vapour phase, $-N_2$)→ and (24)

in the gas phase is presumably due to the absence of opportunity for reactions with other molecules.

Spagnolo, Tundo and Zanirato[82] attempted to make the naphtho[1,8-*de*]triazene **63** from the azide **62** (equation 25). The pyrolysis occurred readily (2–3 h in toluene solution at 110°C) which suggested neighbouring group assistance. (For comparison, 1-naphthyl azide[46] has $t_{1/2}$ 24.8 h at 110°C in decalin.) Only the inner salt **61** was obtained, which suggested addition of the azide function to the azo group without involvement of the nitrene which might be expected to give rise to other products as well.

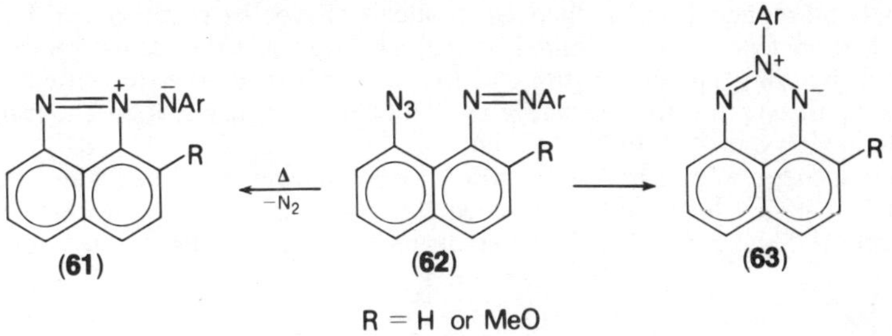

(61) ← Δ, $-N_2$ — (62) → (63) (25)

R = H or MeO

2. *The 2′-substituted-2-azidobiphenyl system*

Pyrolysis of 2-azido-2′-nitrosobiphenyl in solution yields benzo[*c*]cinnoline-*N*-oxide (equation 26), and 2-nitreno-2′-nitrosobiphenyl was suggested as an intermediate[83].

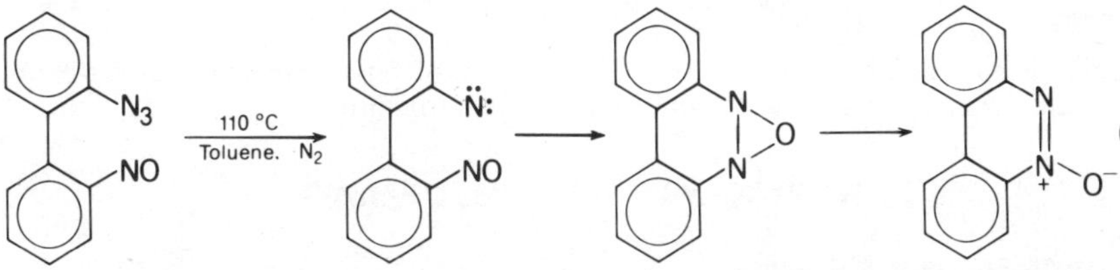

(26) [110 °C, Toluene, N_2]

While one could view this cyclization as the intramolecular version of triplet nitrene trapping, one must note the quite low temperature of decomposition. Some kind of neighbouring group assistance is indicated, and one can suggest a variant of the Hall 1,3-dipolar addition mechanism (equation 27).

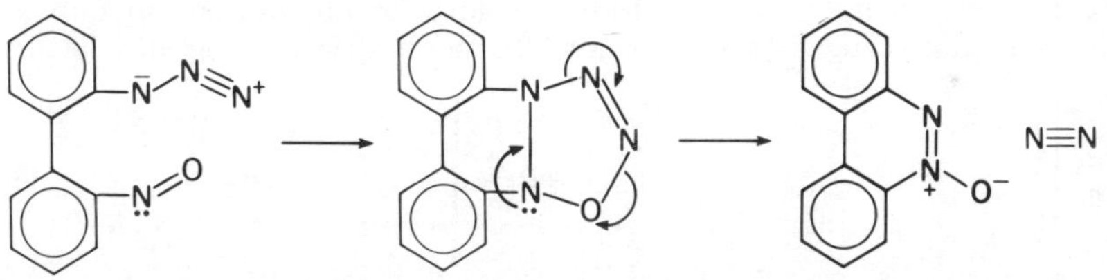

(27)

2′-Arylazo-2-azidobiphenyls also yield a cyclic product on pyrolysis in boiling *o*-dichlorobenzene (179°C)[84]. The expected inner salts **64** were not isolated, but the benzocinnolines **65** (obtained in yields between 34 and 43%) are presumed to be decomposition products from them (equation 28). At a lower temperature (refluxing bromobenzene, 156°C) the product **64** was detected by thin-layer chromatography in the initial stages of reaction, but did not survive into the later stages. When

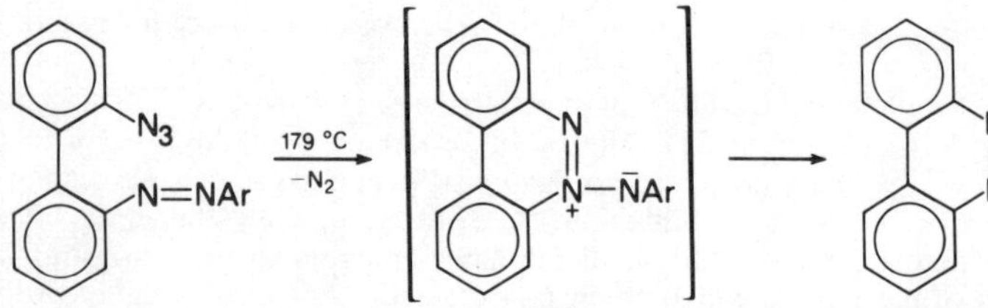

(28)

(64) (65)

Ar = 4-chlorophenyl or 2-biphenylyl, products of type **64** were isolated from partly completed reactions and identified. It was suggested that the reaction proceeds with some degree of neighbouring group assistance rather than via a free nitrene. It would be useful to establish this assistance by kinetic measurements. A mechanism such as equation (27) could of course operate.

Not all the pyrolyses of 2′-substituted-2-azidobiphenyls proceed with extrusion of nitrogen. Smith, Clegg and Hall[85] heated 2-azido-2′-cyanobiphenyl at 180°C to obtain a tetrazene (equation 29), and the result is significant in view of the equation (27) suggestion.

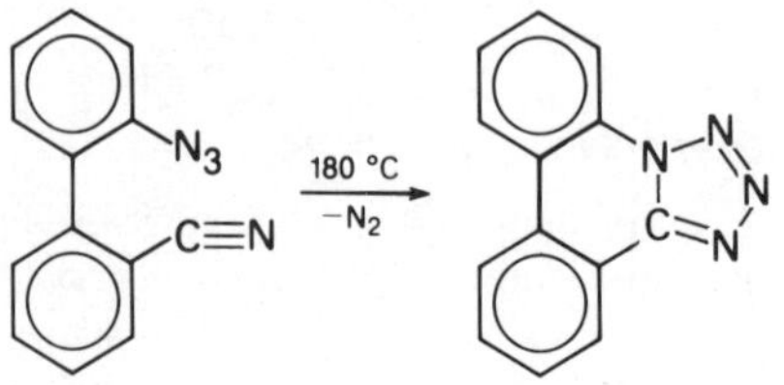

(29)

There is no product of this type from the pyrolysis of 2-cyanophenyl azide[40], where of course the angle strains in the adduct would be prohibitive.

IV. PYROLYSIS WITH RING OPENING

A. Heteroaromatic *N*-Oxides

An unusual reaction was reported in 1973 by Abramovitch and Cue[86,87] in which azidoheteroaromatic *N*-oxides yielded ring-contracted products. The temperatures of pyrolysis were quite low (e.g. benzene at 80°C), which indicates that the nitrogen loss does not involve generation of a nitrene. Both the ease of nitrogen loss, and the nature of the products obtained, were explained in terms of ring opening to form an unsaturated nitroso compound **66** (equation 30). The mechanism is closely related to

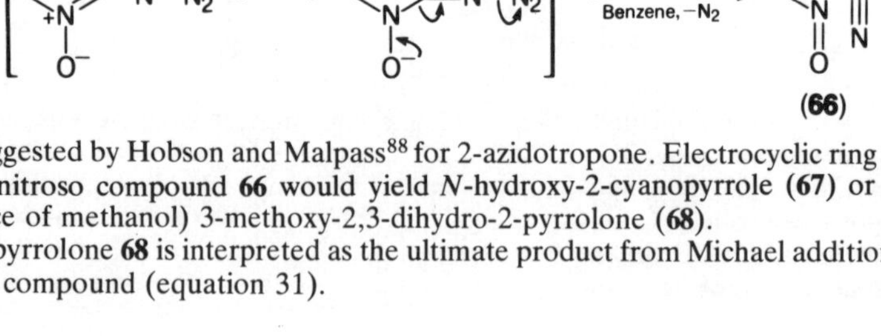

(30)

(66)

that suggested by Hobson and Malpass[88] for 2-azidotropone. Electrocyclic ring closure of the nitroso compound **66** would yield *N*-hydroxy-2-cyanopyrrole (**67**) or (in the presence of methanol) 3-methoxy-2,3-dihydro-2-pyrrolone (**68**).

The pyrrolone **68** is interpreted as the ultimate product from Michael addition to the nitroso compound (equation 31).

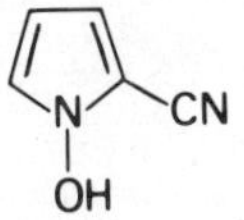

(31)

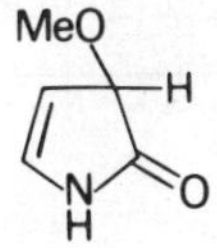

The deoxygenation **69** → **70** is a likely thermal process. In further work, Abramovitch and Cue[79,89] used a blocking methyl group to prevent H-shift in the initial cyclization product, **71**. In this *N*-oxide form, its nitrone character was demonstrated by formation of the 1 : 1 adduct, **72**, with phenylisocyanate (equation 32).

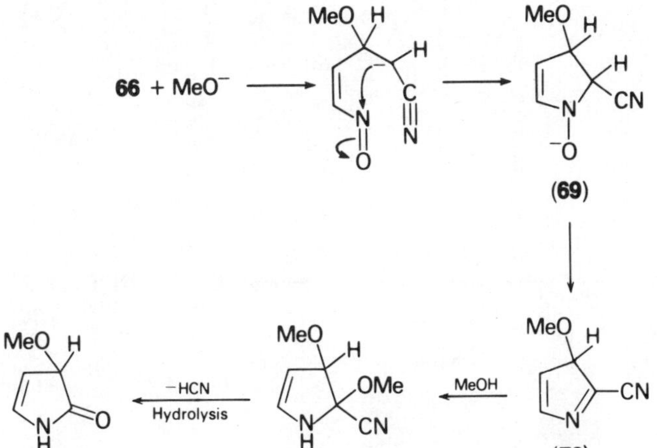

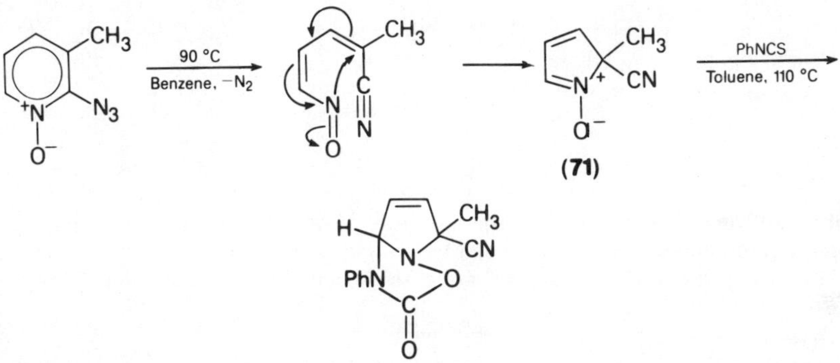

(32)

The reaction gives access to a whole family of these unsaturated nitroso intermediates. The 3-azidopyridazine-2-oxides pyrolyse to *N*-nitroso intermediates whose chemistry is quite complex (equation 33)[90].

An additional mode of re-cyclization is displayed by ring-halogenated 2-azidopyridine-1-oxides (equation 34). Provided that the 6-position is substituted, a

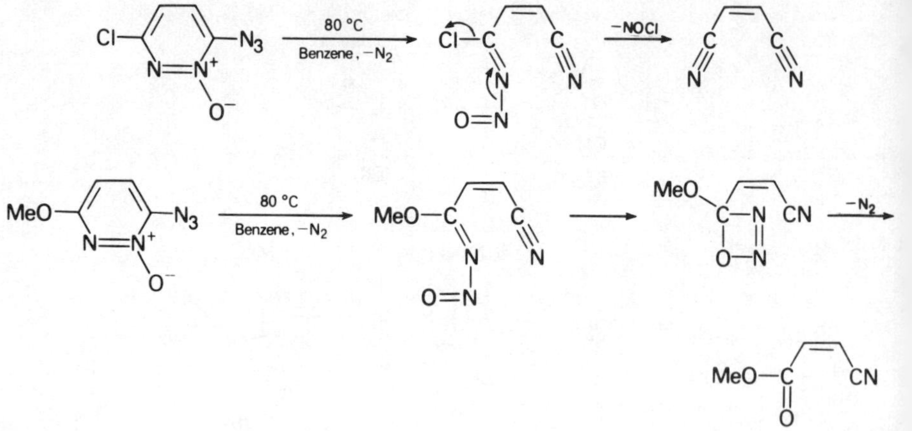

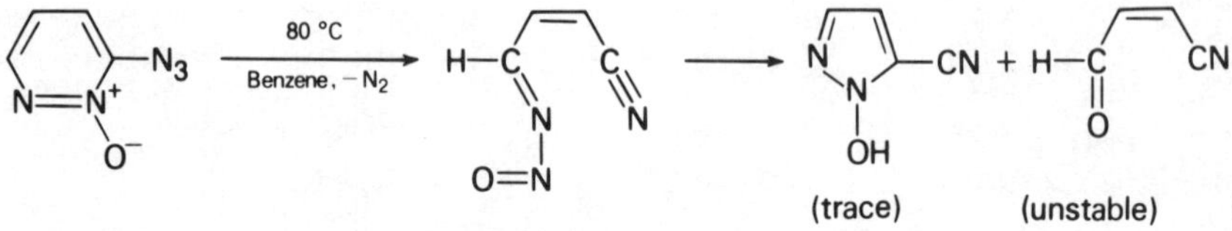

(33)

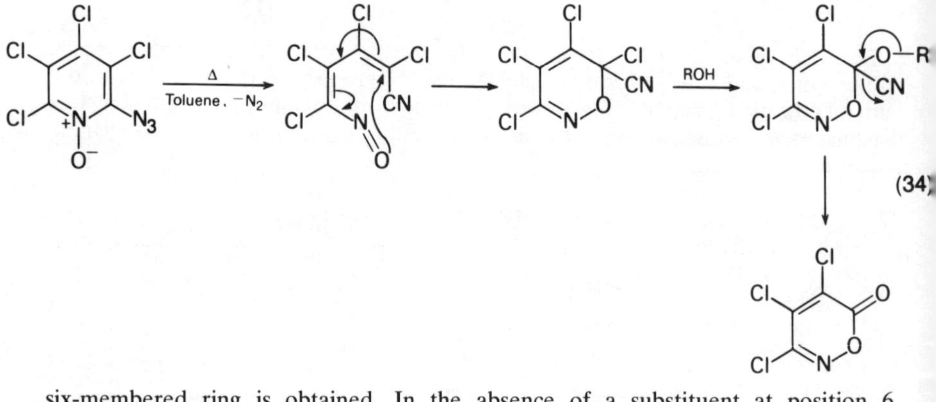

(34)

six-membered ring is obtained. In the absence of a substituent at position 6, nucleophiles are able to open the new ring[91].

An interesting variation was shown when the azidoquinazoline dioxide **73** was pyrolysed[92].

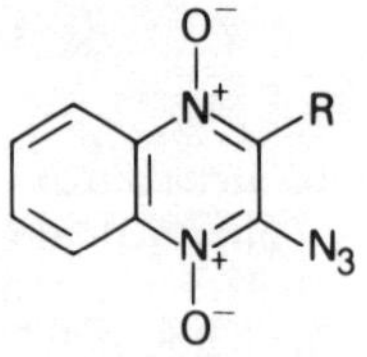

(73) 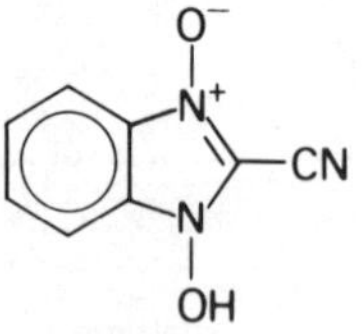 **(74)** 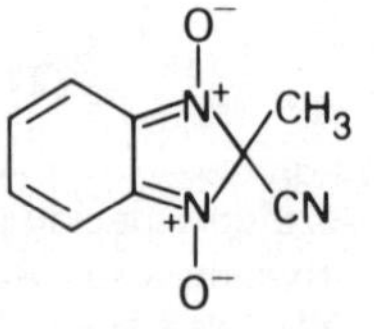 **(75)** 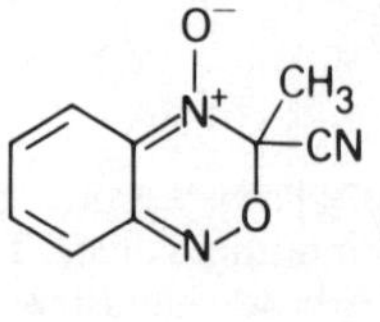 **(76)**

From **73** (R = H) the ring-contracted product **74** was isolated. However, for R = Me the expected product **75** was not obtained; the six-membered ring product **76** was isolated instead. When this pyrolysis (in benzene) was halted in the early stages, both **75** and **76** were isolated, and it was shown that **75** isomerizes into **76**.

B. Azidoquinones

The 2-azidoquinones undergo pyrolysis at 80–110°C to produce a five-membered ring, the yields being excellent if the *ortho* position is blocked[93]. The mechanism shown in equation (35) has been suggested[94]. Direct formation of the azirine

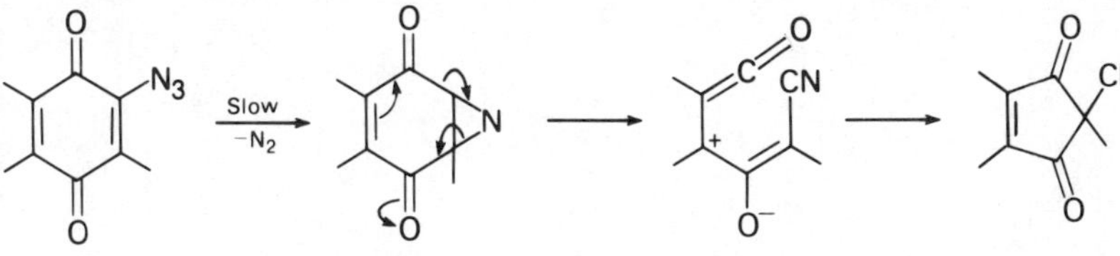

Slow, $-N_2$ (35)

(**77**)

intermediate **77** was preferred, though nitrene involvement was not ruled out. There is a useful analogy for direct formation of the azirine from the thermal behaviour of vinyl azides[95] (equation 36).

$$>C{=}C<\overset{+}{N}{-}N_2 \longrightarrow \text{azirine} + N_2 \quad (36)$$

The thermal reactions of azidoquinones are discussed in detail by H. W. Moore in another chapter in this volume.

C. *ortho*-Diazido Compounds

There have been two interesting examples of the well known ring opening of *ortho* diazides to yield 1,4-dicyano-1,3-butadienes[96]. In the first of these (equation 37), the products arise by two different routes[97]. One of these routes invokes the carbene–nitrene interconversion (see Section V). The other example (equation 38) demonstrates that flash vacuum pyrolysis has been a sadly neglected technique in synthesis via azide pyrolysis. The pyrolysis produced a polymer when conducted in diglyme solution[98].

D. Azidopyrazoles

Smith and Dounchis[99] have reported two types of ring opening for azidopyrazoles (equation 39).

The authors noted the presence of amines and azo compounds among the products and therefore suggested nitrene intermediates. In the absence of any kinetic data for pyrolysis of azidopyrazoles, one cannot guess whether a nitrene intermediate would be accessible at such modest temperatures.

V. NITRENE–CARBENE REARRANGEMENTS

The interconversion of pyridylcarbene (**78**) and phenylnitrene (**82**) at high temperatures is now a well known reaction[1]. When phenyl azide is pyrolysed in the gas phase, the nitrene **82** yields the triplet-derived product (aniline) under comparatively gentle conditions, but at higher temperatures there is extensive rearrangement and evidence for the presence of **78** is obtained[1,100]. The reactions have (until recently) been interpreted in terms of the equilibria shown in equation (40).

The high temperature reaction has been studied in detail by the research groups of W. D. Crow and C. Wentrup. In order to understand the formation of the ring-contracted product, 2-cyanocyclopentadiene (**83**), phenyl azide with a ^{14}C label at C-1 was pyrolysed (600°C/0.05 torr). The nitrile **83** which was isolated had only 27% ^{14}C in the cyano group. This randomization of carbon indicates that either

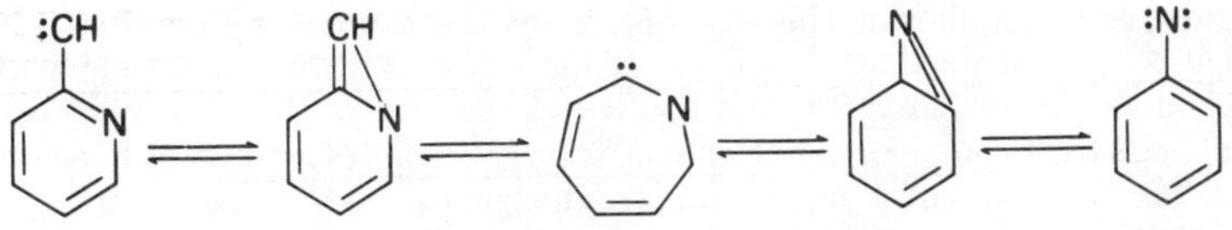

(40)

(78) (79) (80) (81) (82)

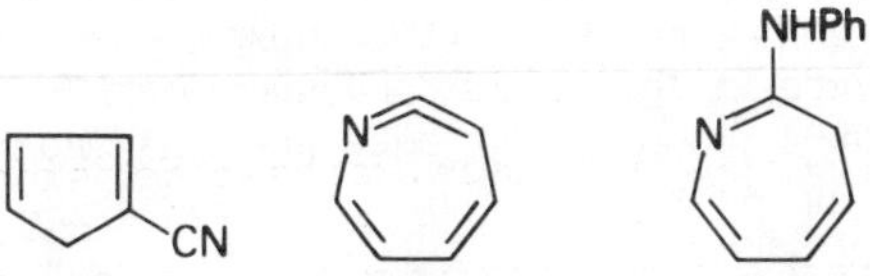

(83) (84) (85)

cycloperambulation of nitrogen, or cycles of ring expansion and contraction, have occurred, in combination with H-shift[101]. Much elegant work has been done with isotopic labels and calculations of the energy changes associated with equation (40). This work has been reviewed by Sharp[102] and Wentrup[103,104]. However, very important recent developments necessitate a re-interpretation of much of the data.

Photochemical studies reported in 1978 by Chapman and his colleagues[105,106] demonstrated that the intermediates lying between **78** and **82** are not the bicyclic species **79** and **81**, but the heterocumulene **84**. They did not find it necessary to invoke the carbene **80**. In the usual trapping experiment with aniline (to yield the azepine **85**), the species **84** was found to be involved, rather than the azirine **81** which has hitherto been invoked[5,107].

Evidence for the intermediacy of carbodiimide types of heterocumulene intermediates in pyrolyses of azido heterocycles has soon followed. At about 100°C/10^{-4} torr, the flash vacuum pyrolysis (FVP) of the tetrazolo[1,5-*c*]quinazoline (**86**) gave 4-azido-2-phenylquinazoline, which ring-contracted to **90** at temperatures above 300°C. The tetrazolo[1,5-*a*]quinoxaline (**88**) likewise flash pyrolysed first to 2-azido-3-phenylquinoxaline and then to the same product **90**. As well as this nitrile, **86** gave the amine derived from **87**, while **88** gave only the amine derived from **89**. That different amines are obtained provides an argument that nitrenes **87** and **89** do not interconvert, while the common product, **90**, from **86** and **88** argues for the intermediate carbodiimide **91**. Labelling of **86** with ^{15}N as indicated gives the product

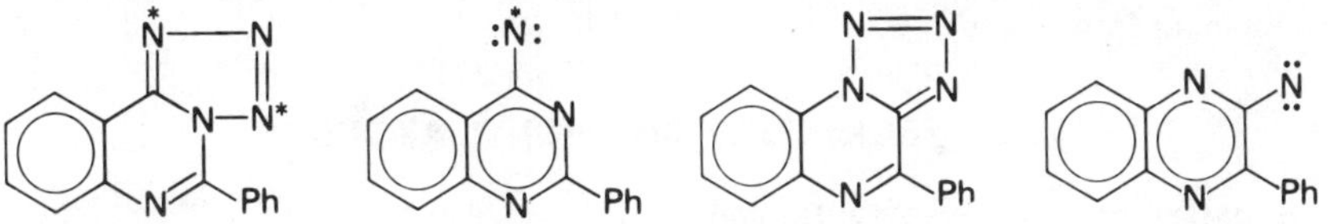

(86) (87) (88) (89)

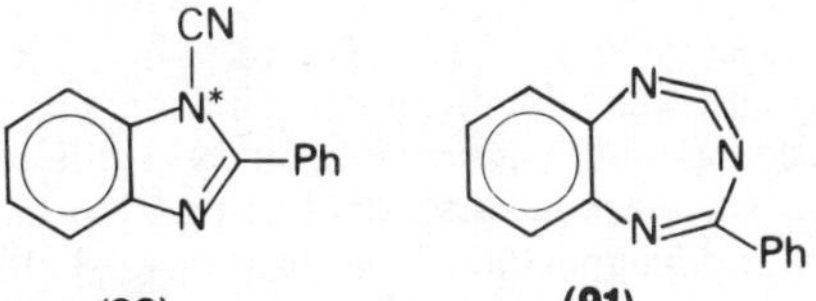

(90) (91)

90 labelled only where shown. This result is also understandable in terms of **91**, but is not explained by a direct ring contraction of the nitrene **87**, which would give 100% ^{15}N label in the cyano group[108]. The species **91** has not yet been isolated from a pyrolysis, but heating **86** or **88** in benzene at 160–180°C gives a dimer of it, as well as the nitrile **90**. Photolysis of 4-azido-2-phenylquinazoline in argon matrix at 10 K has yielded **91**, which was identified from its infrared spectrum.

Several examples of isolation of carbodiimides from flash vacuum pyrolyses have been reported by Wentrup and Winter[109] who used a low-temperature trap and infrared spectroscopy to isolate and identify transitory species. The first effect of pyrolysis of **92** was to produce the hitherto unknown azide **93**, which at higher temperatures produced the carbodiimide **94** (equation 41). When **94** was allowed to

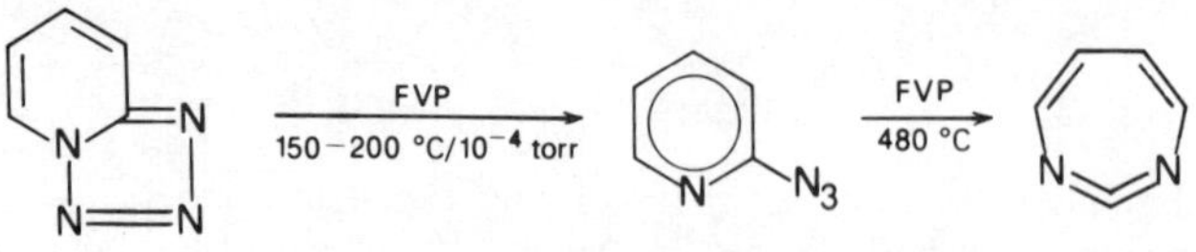

FVP 150–200 °C/10^{-4} torr; FVP 480 °C (41)

(92) **(93)** **(94)**

warm up from −196°C, it decomposed into 2-cyanopyrrole and a trace of glutacononitrile NCCH=CHCH$_2$CN. 2-Cyanopyrrole is the usual pyrolysis product obtained from **92** and its formation via **94** could be observed in FVP experiments at temperatures in the 480°C range. In further experiments, it was shown that the two tetrazolo compounds **95** and **97** both gave the same carbodiimide **96**, and in each case the tetrazolo species first isomerized to its corresponding azide.

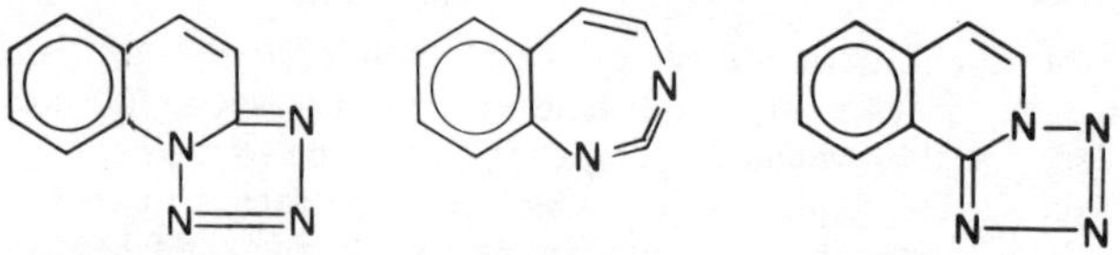

(95) **(96)** **(97)**

It appears that all the data on interconversions of nitrenes and carbenes, and of rearrangements of isotopically labelled nitrenopyridines[110] prior to forming the ring-contracted cyanopyrroles, can be accommodated in terms of carbodiimide intermediates. Wentrup points out that carbodiimides prepared by FVP can be trapped in quantity at −196°C, so that we can expect a great deal of synthetic development from them[109].

VI. PYROLYSIS OF SULPHONYL AZIDES

A. Substitution into Aromatic Solvents

The pyrolysis of sulphonyl azides RSO_2N_3 at temperatures above 120°C has long been known to give sulphonamides, RSO_2NHAr, when conducted in aromatic solvents[1]. A definitive study of the reaction system (equation 42) has been reported by Abramovitch and coworkers[111,112].

When tetracyanoethylene was added, the yield of the sulphonamide RSO_2NHPh was reduced and a 1:1 adduct of the sulphonylazepine **99** (hitherto not detected in these reactions) was isolated. This azepine was identified as the kinetic product, and the sulphonamide as the thermodynamic one[111]. In further work, the sulphonyl azide was decomposed in a series of mono-substituted benzene solvents. The isomer ratios in the

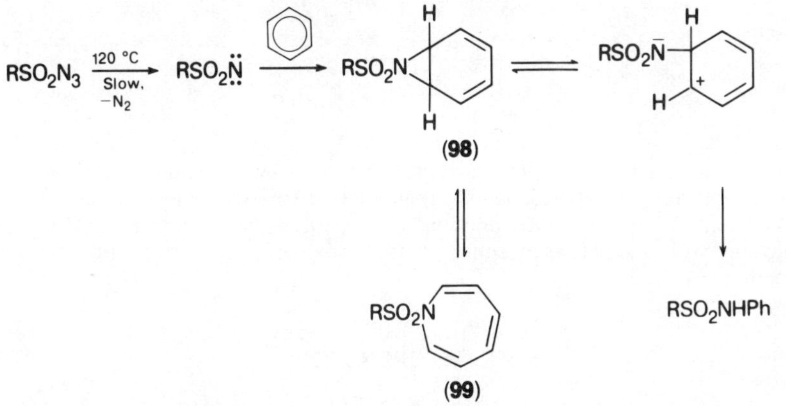

$$RSO_2N_3 \xrightarrow[\text{Slow, } -N_2]{120\,^\circ C} RSO_2\ddot{N}: \longrightarrow (98) \rightleftharpoons RSO_2\bar{N}\text{-cation} \longrightarrow RSO_2NHPh; \quad (98) \rightleftharpoons (99) \qquad (42)$$

products were interpreted in terms of the singlet nitrene adding to form a benzaziridine **98** which ring-opened to (mostly) a *meta*-substituted sulphonamide. Some singlet crossed to triplet, which substituted into all positions of the substrate ring, but mostly into the *ortho* position in the usual fashion of a highly electrophilic diradical. When nitrobenzene was the substrate, none of the electrophilic singlet nitrene was able to substitute before singlet-to-triplet crossing occurred. The observed products were those of radical substitution by the triplet[112].

B. Cyclizations

The possibility of obtaining cyclizations of sulphonyl nitrenes, to parallel those observed for aryl nitrenes (see Section II.D), has been extensively explored. In general, the attempts have had little success, most of the sulphonyl azide pyrolysis products arising from the alternative pathways of hydrogen abstraction, substitution into solvent, or Curtius rearrangement ($ArSO_2N \rightarrow ArN{=}SO_2$)[1]. While 2-biphenylsulphonyl azide pyrolyses in dodecane solution to give a fair yield of the cyclic sulphonamide, azides **100a–100c** give poor yields of **101** (15, 0 and 1% respectively)[113,114]. Substitution into *ortho* alkyl groups is also not very successful,

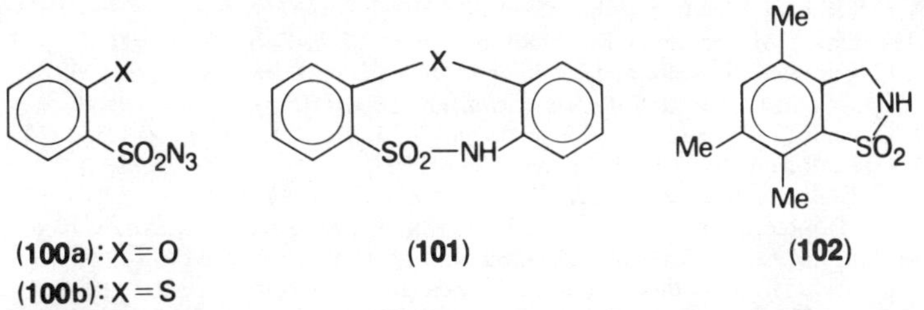

(**100a**): X=O
(**100b**): X=S
(**100c**): X=CO

(**101**)

(**102**)

even the favourable case of 2,3,5,6-tetramethylphenylsulphonyl azide giving only 15% of the cyclic product **102**[115]. In an alternative approach to attempted cyclization, 2-phenylethylsulphonyl azides were subjected to flash vacuum pyrolysis, but only low yields of the expected cyclic sulphonamides were obtained[116].

The most useful reaction to emerge from these pyrolyses is the insertion of an arylsulphonyl nitrene into an *ortho* dialkylamino group (equation 43)[117].

These mesoionic thiadiazole dioxides are formed in 60–80% yields. However, the

NMe_2 SO_2N_3 $\xrightarrow[\text{PhCl}, -N_2]{130\ ^\circ\text{C}}$ $Me_2\overset{+}{N}$—$\overset{-}{N}$ SO_2 **(43)**

reaction is successful only if the *ortho* amino group is dimethylamino or a six-membered ring. Other dialkylamino groups either undergo Cope elimination of an alkyl group from the mesoionic compound, or react as if this cyclized product were in equilibrium with its parent nitrene, so that other nitrene-derived products are obtained.

VII. REFERENCES

1. R. A. Abramovitch and E. P. Kyba in *The Chemistry of the Azido Group* (Ed. S. Patai), Interscience Publishers, New York (1971), pp. 221–329.
2. J. M. Lindley, I. M. McRobbie, O. Meth-Cohn and H. Suschitzky, *JCS Perkin I*, 2194 (1977).
3. A. G. Anastassiou, *J. Amer. Chem. Soc.*, **89**, 3184 (1967).
4. J. M. Lindley, I. M. McRobbie, O. Meth-Cohn and H. Suschitzky, *JCS Perkin I*, 982 (1980).
5. R. Huisgen, D. Vossius and M. Appl, *Chem. Ber.*, **91**, 1 (1958).
6. R. K. Smalley and H. Suschitzky, *J. Chem. Soc.*, 5922 (1964).
7. E. F. V. Scriven, H. Suschitzky and G. V. Garner, *Tetrahedron Lett.*, 103 (1973).
8. G. V. Garner, K. B. Niewiadomski and H. Suschitzky, *Chem. Ind. (London)*, 462 (1972).
9. R. De Luca and G. Renzi, *Chem. Ind (London)*, 923 (1975).
10. A. B. Bulacinski, B. Nay, E. F. V. Scriven and H. Suschitzky, *Chem. Ind. (London)*, 746 (1975).
11. E. G. Janzen, *Acc. Chem. Res.*, **4**, 31 (1971).
12. A. B. Bulacinski, E. F. V. Scriven and H. Suschitzky, *Tetrahedron Lett.*, 3577 (1975).
13. S. E. Hilton, E. F. V. Scriven and H. Suschitzky, *JCS Chem. Commun.*, 853 (1974).
14. B. Nay, E. F. V. Scriven, H. Suschitzky and D. R. Thomas, *JCS Perkin I*, 611 (1980).
15. I. M. McRobbie, O. Meth-Cohn and H. Suschitzky, *Tetrahedron Lett.*, 925 (1976).
16. J. M. Lindley, I. M. McRobbie, O. Meth-Cohn and H. Suschitzky, *Tetrahedron Lett.*, 4513 (1976).
17. J. H. Hall, F. E. Behr and R. L. Reed, *J. Amer. Chem. Soc.*, **94**, 4952 (1972).
18. I. M. McRobbie, O. Meth-Cohn, and H. Suschitzky, *Tetrahedron Lett.*, 929 (1976).
19. D. Hawkins, J. M. Lindley, I. M. McRobbie and O. Meth-Cohn, *JCS Perkin I*, 2387 (1980).
20. J. I. G. Cadogan, S. Kulik and M. J. Todd, *JCS Chem. Commun.*, 736 (1968).
21. M. Messer and D. Farge, *Bull. Soc. Chim. Fr.*, 2832 (1968).
22. J. I. G. Cadogan, S. Kulik, C. Thomson and M. J. Todd, *J. Chem. Soc. C*, 2437 (1970).
23. J. I. G. Cadogan, *Acc. Chem. Res.*, **5**, 303 (1972).
24. J. I. G. Cadogan and S. Kulik, *J. Chem. Soc. C*, 2621 (1971).
25. I. M. McRobbie, O. Meth-Cohn and H. Suschitzky, *J. Chem. Res.*, Synop, 17 (1977).
26. J. M. Lindley, O. Meth-Cohn and H. Suschitzky, *JCS Perkin I*, 1198 (1978).
27. D. G. Hawkins, O. Meth-Cohn and H. Suschitzky, *JCS Perkin I*, 3207 (1979).
28. P. A. S. Smith, B. B. Brown, R. K. Putney and R. R. Reinisch, *J. Amer. Chem. Soc.*, **75**, 6335 (1953).
29. J. I. G. Cadogan and P. K. K. Lim, *JCS Chem. Commun.*, 1431 (1971).
30. J. I. G. Cadogan, J. N. Done, G. Lunn and P. K. K. Lim, *JCS Perkin I*, 1749 (1976).
31. R. N. Carde, G. Jones, W. H. McKinley and C. Price, *JCS Perkin I*, 1211 (1978).
32. G. R. Cliff and G. Jones, *JCS Chem. Commun.*, 1705 (1970).
33. G. Jones and G. R. Cliff, *J. Chem. Soc. C*, 3418 (1971).
34. R. N. Carde and G. Jones, *JCS Perkin I*, 2066 (1974).
35. M. G. Clancy, M. M. Hesabi and O. Meth-Cohn, *JCS Chem. Commun.*, 1112 (1980).
36. M. Takabayashi and T. Shingaki, *Kogyo Kagaku Zasshi*, **64**, 469 (1961).

37. P. A. S. Smith and J. H. Hall, *J. Amer. Chem. Soc.*, **84**, 480 (1962).
38. P. Walker and W. A. Waters, *J. Chem. Soc.*, 1632 (1962).
39. R. Huisgen and M. Appl, *Chem. Ber.*, **92**, 2961 (1959).
40. L. K. Dyall, *Aust. J. Chem.*, **28**, 2147 (1975).
41. J. E. Leffler and H. H. Gibson, *J. Amer. Chem. Soc.*, **90**, 4117 (1968).
42. K. E. Russell, *J. Amer. Chem. Soc.*, **77**, 3487 (1955).
43. E. A. Birkhimer, B. Norup and T. A. Bak, *Acta Chem. Scand.*, **14**, 1894 (1960).
44. G. Boshev, L. K. Dyall and P. R. Sadler, *Aust. J. Chem.*, **25**, 599 (1972).
45. R. Gleiter and R. Hoffmann, *Tetrahedron*, **24**, 5899 (1968).
46. L. K. Dyall, *Aust. J. Chem.*, **30**, 2669 (1977).
47. T. F. Fagley, J. R. Sutter and R. L. Oglukian, *J. Amer. Chem. Soc.*, **78**, 5567 (1956).
48. M. S. Gibson, *Tetrahedron*, **18**, 1377 (1962).
49. R. A. Abramovitch and B. A. Davis, *Chem. Rev.*, **64**, 149 (1964).
50. A. C. G. Gray and A. R. Katritzky, *J. Chem. Soc. B*, 1958 (1965).
51. S. Patai and Y. Gotshal, *J. Chem. Soc. B*, 489 (1966).
52. G. L'abbé, *Ind. Chim. Belge*, **T33**, 543 (1968).
53. L. K. Dyall and J. E. Kemp, *J. Chem. Soc. B*, 976 (1968).
54. J. H. Hall and F. W. Dolan, *J. Org. Chem.*, **43**, 4608 (1978).
55. M. T. Shapiro, *Tetrahedron*, **33**, 1091 (1977).
56. A. Padwa, *Angew. Chem. Int. Edn Engl.*, **15**, 123 (1976).
57. R. Fusco, L. Garanti, and G. Zecchi, *J. Org. Chem.*, **40**, 1906 (1975).
58. L. Garanti and G. Zecchi, *JCS Perkin II*, 1176 (1979).
59. O. Tsuge, K. Ueno and A. Inaba, *Heterocycles*, **4**, 1 (1976).
60. N. J. Dickson and L. K. Dyall, *Aust. J. Chem.*, **33**, 91 (1980).
61. B. Capon, *Quart. Rev.*, **18**, 45 (1964).
62. L. K. Dyall and J. E. Kemp, *Aust. J. Chem.*, **20**, 1625 (1967).
63. J. Ashby and H. Suschitzky, *Tetrahedron Lett.*, 1315 (1971).
64. A. J. Boulton, P. B. Ghosh and A. R. Katritzky, *Tetrahedron Lett.*, 2887 (1966).
65. E. Bamberger, *Chem. Ber.*, **35**, 1885 (1902).
66. L. Krbechek and H. Takimoto, *J. Org. Chem.*, **29**, 1150 (1964).
67. L. Krbechek and H. Takimoto, *J. Org. Chem.*, **29**, 3630 (1964).
68. J. H. Hall and D. R. Kamm, *J. Org. Chem.*, **30**, 2092 (1965).
69. R. J. Sundberg, H. F. Russell, W. V. Ligon and L.-S. Lin, *J. Org. Chem.*, **37**, 719 (1972).
70. W. L. Mosby and M. L. Silva, *J. Chem. Soc.*, 1003 (1965).
71. S. S. Mocholov, A. N. Fedotov, A. I. Sizov and Yu. S. Shabarov, *Zh. Org. Khim.*, **15**, 1425 (1979).
72. R. Selvarajan and J. H. Boyer, *J. Org. Chem.*, **36**, 3464 (1971).
73. R. Purvis, R. K. Smalley, W. A. Strachan and H. Suschitzky, *JCS Perkin I*, 191 (1978).
74. Altaf-ur-Rahman, A. J. Boulton, D. P. Clifford and G. J. T. Tiddy, *J. Chem. Soc. B*, 1516 (1968).
75. J. H. Boyer, D. I. McCane, W. J. McCarville and A. T. Tweedie, *J. Amer. Chem. Soc.*, **75**, 5298 (1953).
76. B. Stanovic and M. Tisler, *Chimia*, **25**, 272 (1971).
77. J. J. Eatough, L. S. Fuller, R. H. Good and R. K. Smalley, *J. Chem. Soc. C*, 1874 (1970).
78. A. S. Bailey, M. W. Heaton and J. I. Murphy, *J. Chem. Soc. C*, 1211 (1971).
79. R. A. Abramovitch and B. W. Cue, *J. Amer. Chem. Soc.*, **98**, 1478 (1976).
80. S. Bradbury, C. W. Rees and R. C. Storr, *JCS Perkin I*, 72 (1972).
81. R. W. Alder, G. A. Niazi and M. C. Whiting, *J. Chem. Soc. C*, 1693 (1970).
82. P. Spagnolo, A. Tundo and P. Zanirato, *J. Org. Chem.*, **43**, 2508 (1978).
83. L. A. Neiman, V. I. Maimind and M. M. Shemyakin, *Izv. Akad. Nauk SSSR, Ser. Khim.*, 1357 (1964).
84. A. Spagnolo, A. Tundo and P. Zanirato, *J. Org. Chem.*, **42**, 292 (1977).
85. P. A. S. Smith, J. M. Clegg and J. H. Hall, *J. Org. Chem.* **23**, 524 (1958).
86. R. A. Abramovitch and B. W. Cue, *Heterocycles*, **1**, 227 (1973).
87. R. A. Abramovitch and B. W. Cue, *J. Org. Chem.*, **38**, 73 (1973).
88. J. D. Hobson and J. R. Malpass, *J. Chem. Soc. C*, 1645 (1967).
89. R. A. Abramovitch and B. W. Cue, *Heterocycles*, **2**, 297 (1974).
90. R. A. Abramovitch and I. Shinkai, *JCS Chem. Commun.*, 703 (1975).

91. R. A. Abramovitch, I. Shinkai, B. W. Cue, F. A. Ragan and J. L. Atwood, *J. Heterocycl. Chem.*, **13**, 415 (1976).
92. J. P. Dirlam, B. W. Cue, and K. J. Gambatz, *J. Org. Chem.*, **43**, 76 (1978).
93. H. W. Moore, *Chem. Soc. Rev.*, **2**, 415 (1973).
94. W. Weyler, D. S. Pearce, and H. W. Moore, *J. Amer. Chem. Soc.*, **95**, 2603 (1973).
95. G. L'abbé, *Angew. Chem., Int. Edn Engl.*, **14**, 775 (1975).
96. J. H. Hall and E. Patterson, *J. Amer. Chem. Soc.*, **89**, 5856 (1967).
97. D. S. Pearce, M-S. Lee and H. W. Moore, *J. Org. Chem.*, **39**, 1362 (1974).
98. M. E. Peek, C. W. Rees and R. C. Storr, *JCS Perkin I*, 1260 (1974).
99. P. A. S. Smith and H. Dounchis, *J. Org. Chem.*, **38**, 2958 (1973).
100. W. D. Crow and M. N. Paddon-Row, *Aust. J. Chem.*, **28**, 1755 (1975).
101. W. D. Crow and M. N. Paddon-Row, *Tetrahedron Lett.*, 2231 (1972).
102. J. T. Sharp, *Ann. Reports Chem. Soc.*, 229 (1972).
103. C. Wentrup in *Reactive Intermediates*, Vol. 1 (Ed. R. A. Abramovitch), Plenum Press, London (1980), p. 263.
104. C. Wentrup, *Topics Current Chem.*, **62**, 173 (1976).
105. O. L. Chapman and J.-P. LeRoux, *J. Amer. Chem. Soc.*, **100**, 282 (1978).
106. O. L. Chapman, R. S. Sheridan and J-P. LeRoux, *J. Amer. Chem. Soc.*, **100**, 6245 (1978).
107. R. Huisgen and M. Appl, *Chem. Ber.*, **91**, 12 (1958).
108. C. Wentrup, C. Thétaz, E. Tagliaferri, H. J. Lindner, B. Kitschke, H.-W. Winter and H. P. Reisenauer, *Angew. Chem., Int. Edn Engl.*, **19**, 566 (1980).
109. C. Wentrup and H.-W. Winter, *J. Amer. Chem. Soc.*, **102**, 6159 (1980).
110. R. Harden and C. Wentrup, *J. Amer. Chem. Soc.*, **98**, 1259 (1976).
111. R. A. Abramovitch, T. D. Bailey, T. Takaya and V. Uma, *J. Org. Chem.*, **39**, 340 (1974).
112. R. A. Abramovitch, G. N. Knaus and V. Uma, *J. Org. Chem.*, **39**, 1101 (1974).
113. R. A. Abramovitch, C. I. Azogu and I. T. McMaster, *J. Amer. Chem. Soc.*, **91**, 1219 (1969).
114. R. A. Abramovitch and D. P. Vanderpool, *JCS Chem. Commun.*, 18 (1977).
115. R. A. Abramovitch, T. Chellathurai, W. D. Holcomb, I. T. McMaster and D. P. Vanderpool, *J. Org. Chem.*, **42**, 2920 (1977).
116. R. A. Abramovitch and W. D. Holcomb, *J. Amer. Chem. Soc.*, **97**, 676 (1975).
117. J. Martin, O. Meth-Cohn and H. Suschitzky, *JCS Perkin I*, 2451 (1974).

The Chemistry of Functional Groups, Supplement D
Edited by S. Patai and Z. Rappoport
© 1983 John Wiley & Sons Ltd

CHAPTER **8**

Vinyl, aryl and acyl azides

HAROLD W. MOORE

Department of Chemistry, University of California, Irvine, California 92717, U.S.A.

DOROTHY M. GOLDISH

Department of Chemistry, California State University Long Beach, Long Beach, California 90840, U.S.A.

I. INTRODUCTION

This chapter discusses the chemistry of those compounds in which an azide group is bonded to an sp^2 carbon atom. The emphasis is placed upon vinyl azides, aryl azides, acyl azides and related heterocyclic analogues. The chemistry of such unsaturated azides continues to gain interest viewed from both synthetic and mechanistic aspects. However, this discussion will focus primarily upon the variety of chemical transformations rather than the mechanistic details. Since recent review articles have appeared on reactions involving nitrenes[1], azidoquinones[2] and the chemistry of azirines[3], these topics will not be presented in detail here. Rather, this chapter will discuss the reactions reported since the last review in this series[4], and will not attempt to be a comprehensive compendium of the transformations of unsaturated azides.

II. PREPARATION OF VINYL AZIDES

Since the last review in this series[4], only a few new methods for the synthesis of vinyl azides have appeared. The standard routes still remain the addition of IN_3 to alkynes, the elimination of HI from vicinal iodoazides, and the nucleophilic displacement (addition–elimination) of halide by azide in β-haloenones and related compounds. Two new methods of potential synthetic importance have appeared and are outlined below.

A. Conversion of Triflones to Vinyl Azides

1,2-Diphenylethyl triflone, upon treatment with sodium hydride followed by *p*-toluenesulphonyl azide, gave a high yield of 1-azido-1,2-diphenylethene[5]. The procedure has been used to prepare a number of cyclic and acyclic vinyl azides. This, along with the report that the starting triflones are easily prepared, suggests a potentially versatile route to vinyl azides[5].

B. Conversion of Nitrosooxazolidones to Vinyl Azides

A novel synthesis of terminal vinyl azides was observed when the 3-nitrosooxazolidone **1** was treated with base in the presence of azide ion. This

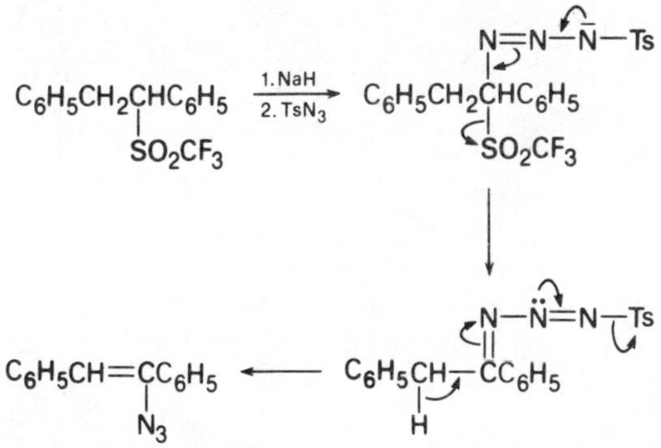

$$C_6H_5CH_2CH(SO_2CF_3)C_6H_5 \xrightarrow[2.\ TsN_3]{1.\ NaH} C_6H_5CH_2C(N{=}N{-}\bar{N}{-}Ts)(SO_2CF_3)C_6H_5$$

$$\downarrow$$

$$C_6H_5CH{=}C(N_3)C_6H_5 \longleftarrow C_6H_5CH(H){-}C(={N}{-}\ddot{N}{=}N{-}Ts)C_6H_5$$

resulted in a 50% yield of the vinyl azide **2**[6]. A detailed mechanistic study of this reaction has shown it to involve the intermediacy of a vinyl diazonium ion[7]. However, the synthetic scope has not been elaborated.

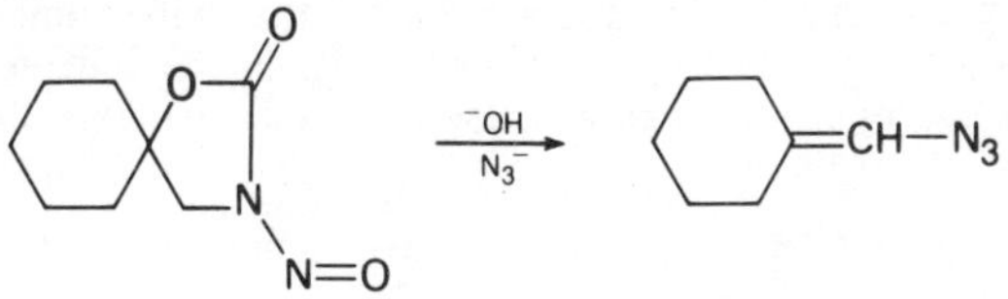

$$(\mathbf{1}) \xrightarrow[N_3^-]{^-OH} (\mathbf{2})\ \text{cyclohexylidene}{=}CH{-}N_3$$

(**1**) (**2**)

III. REACTIONS OF VINYL AZIDES

The most general reaction of vinyl azides is their thermal of photochemical conversion to azirines, i.e. **3** → **4**. Even though this transformation is a well known reaction, its

(**3**) (**4**)

mechanism still remains unresolved. Recent theoretical studies suggest an initial electrocyclic ring closure to a triazole which gives the azirine upon loss of molecular nitrogen[8]. This cannot be differentiated from a concerted process, but a nitrene intermediate to **4** is rejected on the basis of kinetic data. In any regard, the majority of the reactions of vinyl azides in which nitrogen is lost involve azirine intermediates. These in turn may proceed to zwitterionic species or vinyl nitrenes, or undergo concerted rearrangements. Even though the mechanistic details behind the reactions outlined in this review will not be elaborated, one can envisage the above reaction sequences as the most usual pathways available.

A. Fragmentation of Vinyl Azides to Nitriles

An area gaining extensive study in recent years is the pyrolytic conversion of appropriately substituted vinyl azides to nitriles. In many examples the punultimate precursor to the observed products has been established as a zwitterion. In a general sense, cyclic vinyl azides of structure **5** cleave to zwitterions **6** when X is a substituent capable of stabilizing a positive charge and Y and/or Z are anion stabilizing groups

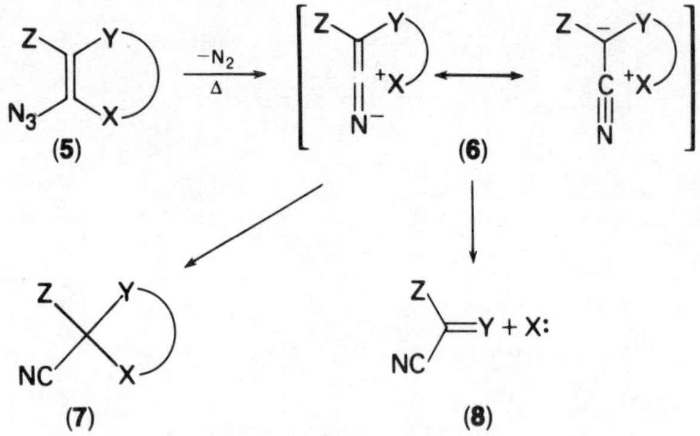

Z Y N3 X (5) $-N_2$ Δ [Z Y +X N− ⟷ Z − Y C +X N] (6)

Z Y NC X (7) Z NC =Y + X: (8)

SCHEME 1

(Scheme 1). The zwitterionic intermediate **6** can then ring close to **7** (ring contraction) or cleave to **8** (fragmentation)[9]. Nearly all of the examples given in this section can be explained by such a mechanism, and in some cases the zwitterionic intermediate has been unambiguously established. Triazoles, azirines or nitrenes may precede zwitterion formation, but in most cases these have not been established.

1. *Synthesis of cyanoketenes*

The most efficient method known for the generation of cyanoketenes is the pyrolysis of appropriately substituted vinyl azides of the type shown in Scheme 1. For example, 2,5-diazido-1,4-benzoquinones are ideal and give two molecules of the corresponding cyanoketene, **11**, upon thermolysis in refluxing benzene[10,11]. The reaction has been shown to proceed with initial ring contraction to **9**, which subsequently cleaves to the cumulenes via zwitterion **10**.

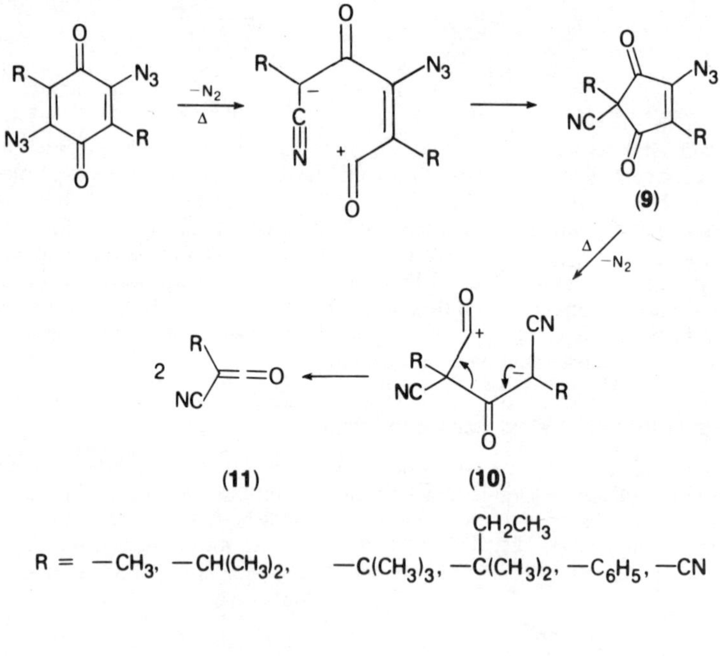

O R N3 N3 R O $-N_2$ Δ O R N3 C N + R O → O N3 R NC R O (9)

Δ $-N_2$

2 R NC =C=O (11) ← O + CN R NC R O (10)

R = $-CH_3$, $-CH(CH_3)_2$, $-C(CH_3)_3$, $-C(CH_3)_2CH_2CH_3$, $-C_6H_5$, $-CN$

Attempts to extend the scope of this reaction to include halocyanoketenes failed due to the insolubility of the diazidoquinone precursors. However, 4-azido-3-halo-5-methoxy-2(5*H*)-furanones, **12**, were found to cleave readily to methyl formate and the halocyanoketenes **13** in refluxing benzene[12,13].

(12) $\xrightarrow[-CH_3OCHO]{-N_2}$ **(13)**

R = Cl, Br, I

These thermolysis routes to cyanoketenes are superior to classical methods of ketene synthesis such as dehydrohalogenation of acid halides. For example, *t*-butylcyanoketene (TBCK) can be prepared in nearly quantitative yield from 2,5-diazido-3,6-di-*t*-butyl-1,4-benzoquinone. However, attempts to prepare TBCK from 2-cyano-3,3-dimethylbutanoyl chloride by treatment with triethylamine gave only 3,5-dicyano-2,2,6,6-tetramethyl-3,4-heptadiene[14]. This was shown to arise from a rapid amine-catalysed dimerization of the initially formed ketene.

2. *Synthesis of cyclobutane-1,3-diones and 2-oxetanones*

As noted earlier, 4-azido-2-cyanocyclopentene-1,3-diones ring open to zwitterions upon thermolysis and these subsequently cleave to two molecules of cyanoketenes, e.g. **9** → **10** → **11**[10]. The course of this reaction has been shown to depend upon the substituents at position 2 of the cyclopentenedione. In **9** the 2-cyano group apparently destabilizes the zwitterion and facilitates fragmentation to cumulenes. On the other hand, if the cyano group is replaced by an alkyl group, the thermolysis gives high yields of the corresponding cyclobutane-1,3-diones, **16**. A zwitterionic intermediate, **15**, was established by trapping experiments[15].

(14) $\xrightarrow[\Delta]{-N_2}$ **(15)** ⟶ **(16)**

	R^1	R^2	Yield, %
(a)	$-CH_3$	$-CH_3$	80
(b)	$-CH_3$	$-CH_2CH_3$	90
(c)	$-CH_3$	$-CH_2C_6H_5$	86
(d)	$-CH_3$	$-(CH_2)_4-CH=CH_2$	80

The ring closures of zwitterions such as **15** were also shown to depend upon steric factors. For example, if position 2 of the starting azide was monosubstituted, then the exclusive product was a 2-oxetanone rather than a cyclobutanedione, e.g. **17** → **18**. It is also noteworthy that **16** and **18** were the exclusive products formed when TBCK was

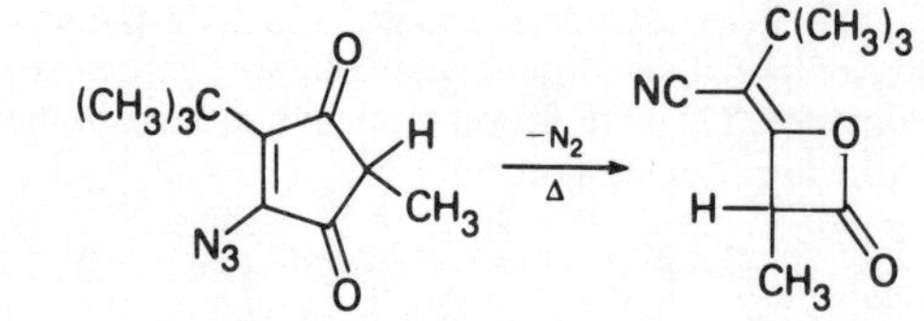

(17) (18)

treated with the corresponding aldo- or ketoketenes. Thus, these cycloadditions must also involve zwitterion intermediates.

3. *Synthesis of 3-cyano-2-azetidinones*

Inspection of Scheme 1 reveals a new synthetic route to 3-cyano-2-azetidinones (β-lactams). That is, thermolysis of 4-azido-2-pyrrolinones in refluxing benzene results in their conversion to zwitterions **19** which subsequently ring close to the azetidinones **20**[16].

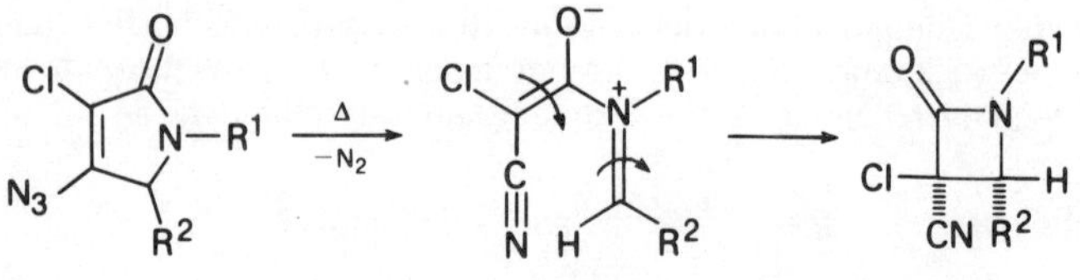

(19) (20)

	R^1	R^2	Yield, %
(a)	$-CH_3$	$-OCH_3$	55
(b)	$-CH_3$	$-OCH_2CH_3$	60
(c)	$-CH_2CH_3$	$-OCH_2CH_3$	88
(d)	$-CH(CH_3)_2$	$-OCH_2C_6H_5$	90
(e)	$-C_6H_{11}$	$-OCH_2CH_3$	62
(f)	$-C_6H_{11}$	$-C_6H_5$	49

Attempts to extend this methodology to include bicyclic examples have met with only partial success. For example, **21** (n = 3,4) could be prepared, but attempts to ring contract **22** gave only the acyclic amide **23**.

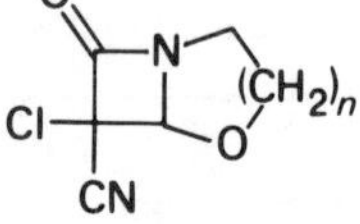

(21)

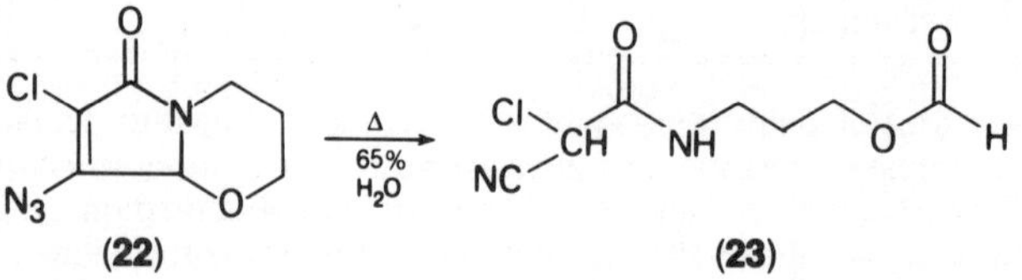

(22) (23)

The mechanism of these ring contractions to give β-lactams was again established as involving zwitterionic intermediates such as **19**[17]. Evidence for such comes from a series of trapping experiments as well as their independent generation upon treatment of formimidates and imines with chlorocyanoketene.

4. *Synthesis of β-lactones*

In analogy to the above ring contractions of azidopyrrolinones, 4-azido-3-chloro-5-(4-methoxyphenyl)-2(5*H*)-furanone, **24**, rearranges to the β-lactone **26**. This decarboxylates under the reaction conditions to give 1-chloro-1-cyano-2-(4-methoxyphenyl)ethene **27**[18]. The intermediacy of the zwitterion, **25**, was established by its independent generation from chlorocyanoketene and 4-methoxybenzaldehyde. The scope of the ring contraction was not explored, but the cycloaddition mode was shown to be quite general as a synthetic route to 1-chloro-1-cyano-2-arylethenes, **27**.

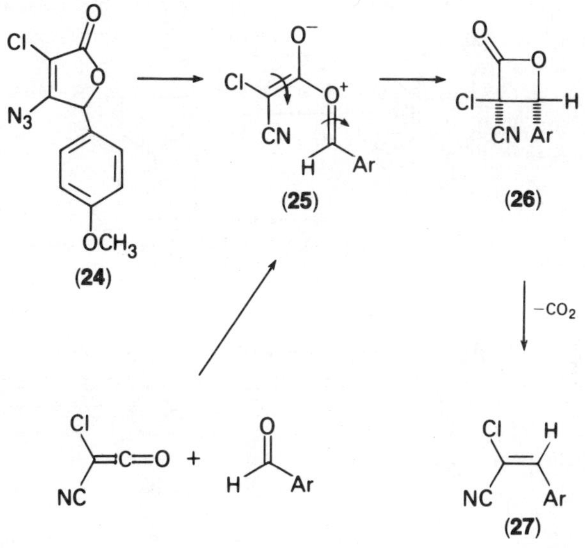

5. *Synthesis of* N-*cyano compounds*

An interesting transformation which is viewed as involving an intermediate unsaturated azide capable of zwitterion formation is the pyrolytic conversion of the geminal diazide **28** to **30**[19]. This is suggested as arising from an initial ring expansion to **29** followed by ring contraction via a zwitterionic intermediate as outlined below.

Still another transformation involving the formation of an *N*-cyano compound was observed when 9-dinitromethylenefluorene and 9-bromonitromethylenefluorene, **31**, were treated with sodium azide in DMSO[20]. Here, a rapid evolution of nitrogen and a high yield of *N*-cyano-9-iminofluorene were observed. This transformation is particularly interesting when compared to the reaction of 9-dichloromethylenefluorene, **32**, with sodium azide to give 9-cyano-9-azidofluorene[21]. No azide intermediates were isolated in any of these reactions. However, it is reasonable to assume that azide ion initiated the reaction of **31** by attack at position 9 followed by rearrangement, while initial attack on **32** takes place at the methylidene carbon.

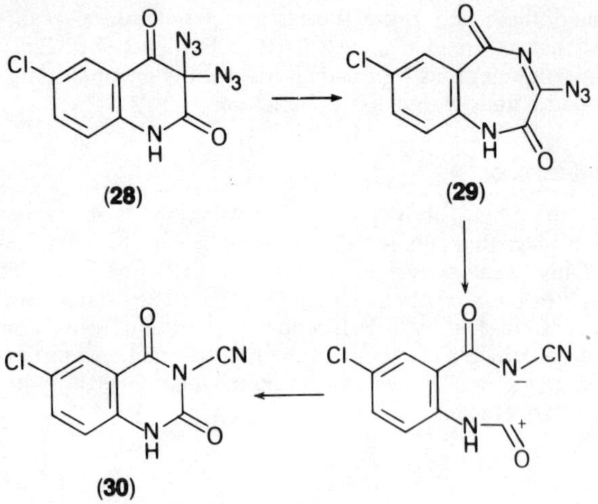

O N$_3$ Cl N$_3$ N O H

O Cl N N$_3$ N O H

(28) (29)

O Cl N CN N O H

O Cl N CN N H O

(30)

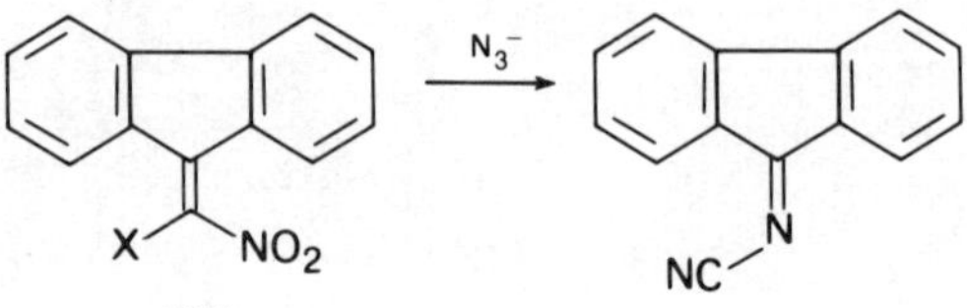

N_3^-

X NO_2

N NC

(31)

X = Br, NO_2

N_3^-

Cl Cl

N≡C N_3

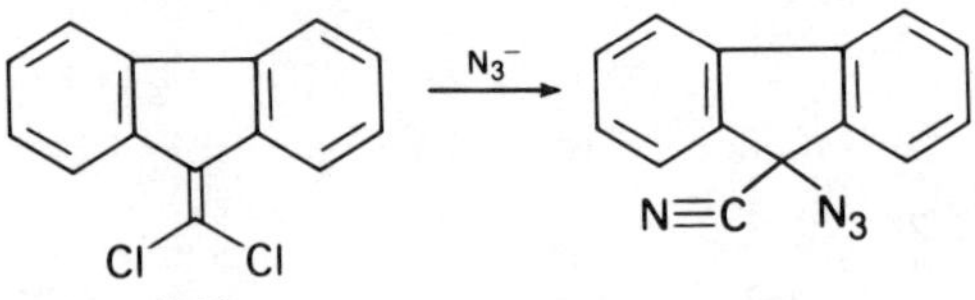

(32)

6. Other examples

Other examples of ring contractions of cyclic vinyl azides that can be viewed as arising via the general zwitterionic mechanism outlined in Scheme 1 are provided below.

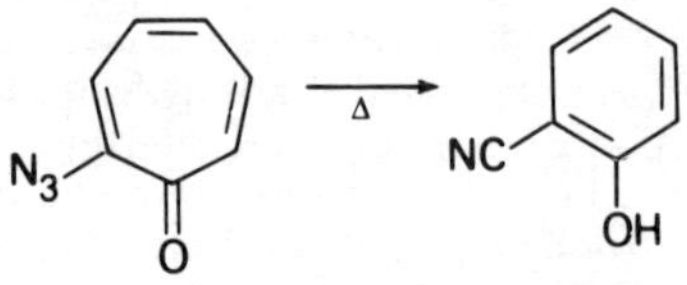

N_3 O Δ NC OH (Ref. 22)

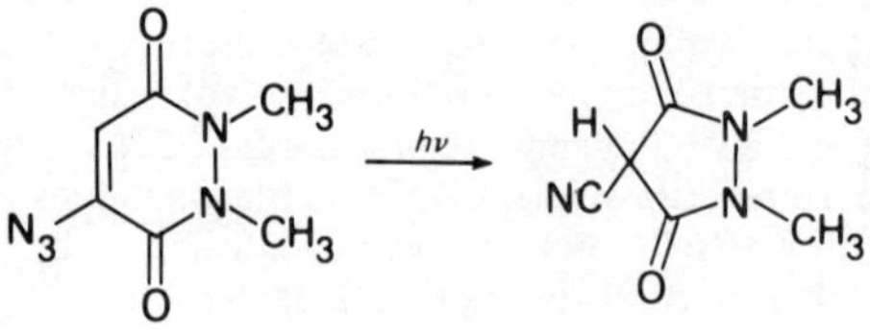

O N CH_3 N CH_3 N_3 O $h\nu$ O H N CH_3 NC N CH_3 O (Ref. 23)

CH_3, N_3, N, O^- —Δ→ CH_3, NC, $\overset{+}{N}$, O^- (Ref. 24)

$\overset{+}{N}$, N_3, O^- —Δ→ O, CN, $\overset{+}{N}$, O^- (Ref. 25)

O, O, N_3 —$h\nu$→ O, O, CN (Ref. 26)

C_6H_5, O, N_3, O —$-N_2$, $-CO$, Δ→ C_6H_5, C=O, NC (Ref. 27)

O, CH_3, N_3, CH_3, O —Δ→ O, CH_3, NC, CH_3, O (Ref. 28)

$SO_2C_6H_5$, N, CH_3, N_3, CH_3, O —Δ→ $SO_2C_6H_5$, N, CH_3, NC, CH_3, O (Ref. 29)

O, H, CH_3, N_3, O, CH_3, OCH_3 —Δ→ O, H, CH_3, NC, O, CH_3, OCH_3 (Ref. 29)

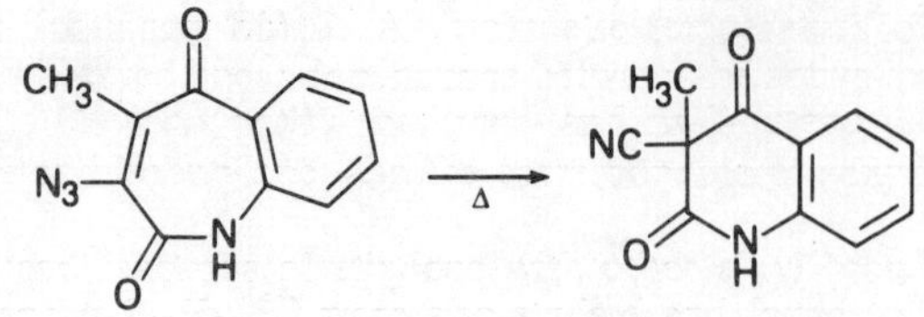

(Ref. 29)

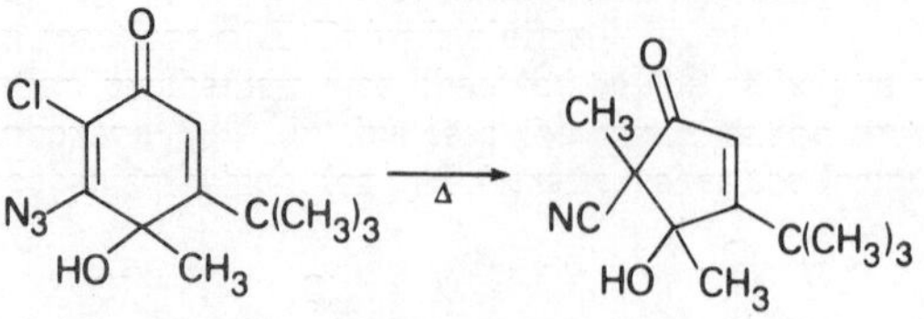

(Ref. 29)

The conversion of vinyl azides to nitriles is not limited to cyclic compounds. For example, a series of acyclic enazidocarbonyl compounds, **33**, were shown to rearrange thermally to nitriles **34**[30,31]. The mechanism of these transformations has not been

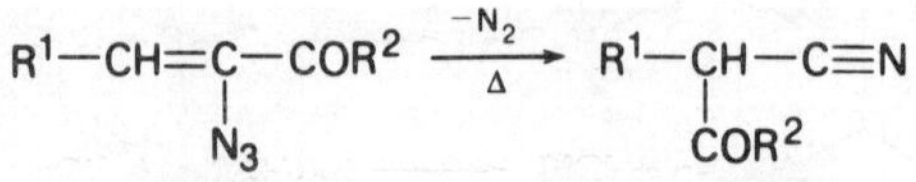

$$R^1-CH=C(N_3)-COR^2 \xrightarrow[\Delta]{-N_2} R^1-CH(COR^2)-C\equiv N$$

(33) **(34)**

R^1	R^2	Yield, %
C_6H_5	C_6H_5	73
C_6H_5	$4\text{-}CH_3C_6H_4$	65
$4\text{-}CH_3C_6H_4$	$4\text{-}ClC_6H_4$	74
$4\text{-}ClC_6H_4$	$4\text{-}NO_2C_6H_4$	80
$4\text{-}CH_3C_6H_4$	CH_3	75
$4\text{-}ClC_6H_4$	CH_3	70
2,6-Di-ClC_6H_3	$4\text{-}ClC_6H_4$	45
2,6-Di-ClC_6H_3	OCH_2CH_3	100

established. However, the intermediacy of an azirine has been suggested. Nitrile formation was observed only for those cases in which R^1 is an aryl group. The alkyl series gave the corresponding azirines as stable products. It is, of course, possible to envisage a mechanism analogous to the zwitterion process outlined in Scheme 1. That is, cleavage of the azirine to the ion pair is facilitated when R^1 is an aryl group, which lends more stabilization to the anion via resonance delocalization.

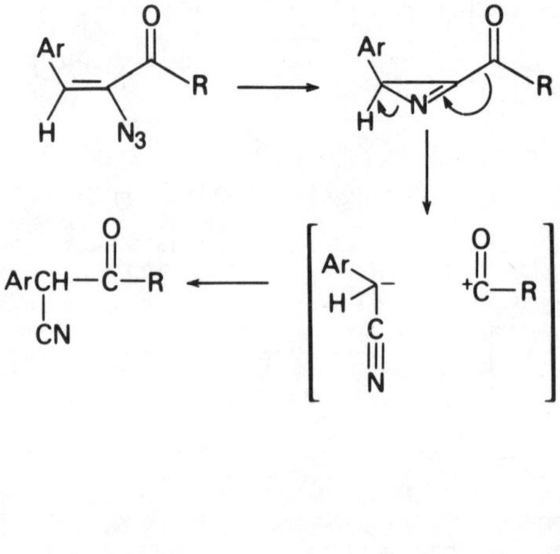

7. Synthesis of cinnamoyl cyanides

One last example of a fragmentation that could conceivably proceed via a zwitterion mechanism (Scheme 1) is that reported for the thermolysis of azido isoxazoles[32]. Specifically, the azido isoxazoles **35** ($n = 1,2$) were observed to give acetonitrile and greater than 90% yield of the correponding cinnamoyl cyanides **37** ($n = 1,2$) when subjected to thermolysis in refluxing decalin. The mechanism of this transformation was not established. However, it can be rationalized via the zwitterionic intermediate **36**.

$C_6H_5-(CH{=}CH)_n$ (35) $\xrightarrow[-N_2]{\Delta}$ $C_6H_5-(CH{=}CH)_n$ (36) $\xrightarrow{-CH_3C\equiv N}$ $C_6H_5-(CH{=}CH)_n-C(=O)-CN$ (37)

If the zwitterion **36** does function as the intermediate in this reaction, its formation should depend upon anion stabilization. Thus, azido isoxazoles in which the styryl group is replaced by an alkyl substituent should be less likely to undergo this reaction.

8. Fragmentation of 2-azidooxiranes

The previously unknown 2-azidooxiranes have been prepared by epoxidation of vinyl azides using *N*-benzoylperoxycarbamic acid[33]. These epoxides are relatively stable, having half-lives of 2 days in refluxing benzene. The products of thermolysis are the corresponding ketones and hydrogen cyanide, which are formed in high yields.

$$\xrightarrow{\Delta} \longrightarrow R^1R^2C{=}O + R^3-C\equiv N$$

R^1	R^2	R^3
$C(CH_3)_3$	H	H
$C(CH_3)_3$	CH_3	H
C_6H_5	C_6H_5	H

An analogous fragmentation was observed when 1-azidocyclohexene was treated with *m*-chloroperbenzoic acid. This resulted in its spontaneous conversion to 5-cyanopentanal; the azidooxirane was proposed as the intermediate[34].

$$\xrightarrow{m\text{-}ClC_6H_4CO_3H} [\ ] \longrightarrow NC-(CH_2)_4-CH{=}O$$

B. Intramolecular Cyclizations

A considerable variety of heterocyclic compounds has been prepared by intramolecular cyclization reactions involving the vinyl azide moiety and appropriately situated sites of unsaturation or alkyl groups capable of nitrene insertions. In most cases, these transformations are viewed as proceeding via nitrene or azirine intermediates. The mechanistic details will not be elaborated here since detailed discussions have appeared[2,35,36]. Rather, selected examples will be used to illustrate the general types of products that can be obtained.

1. Synthesis of indolequinones

Thermolysis of 2-azido-3-alkenyl-1,4-quinones in refluxing benzene results in conversion to indolequinones in high yield[37]. The scope of this reaction has been examined and it has recently been utilized for the synthesis of 7-chloro-6-methyl-1,2,5,8-tetrahydro-3*H*-pyrrolo[1,2-*a*]indole-5,8-diones, **38**, a compound having the basic ring system of the mitomycin antineoplastic antibiotics. Since the starting azidoquinones are easily prepared[2], this ring closure constitutes one of the simplest and most general route to indolequinones.

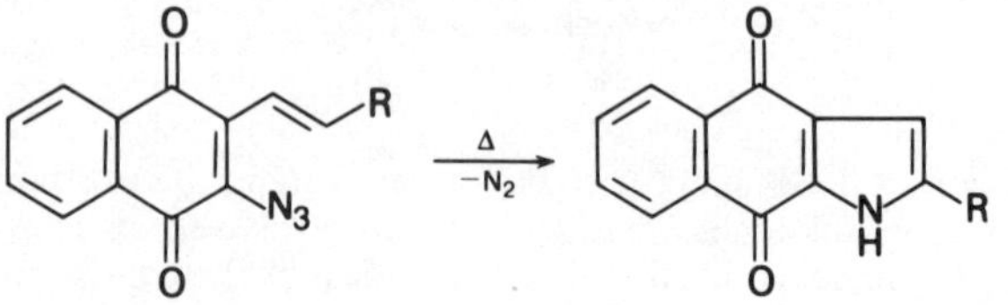

Δ, $-N_2$

	R	Yield, %
(a)	CH_3	90
(b)	$CH_2CH_2CH_3$	81
(c)	$CH_2(CH_2)_8CH_3$	87
(d)	C_6H_5	92
(e)	$CH_2(CH_2)_2OCOCH_3$	88

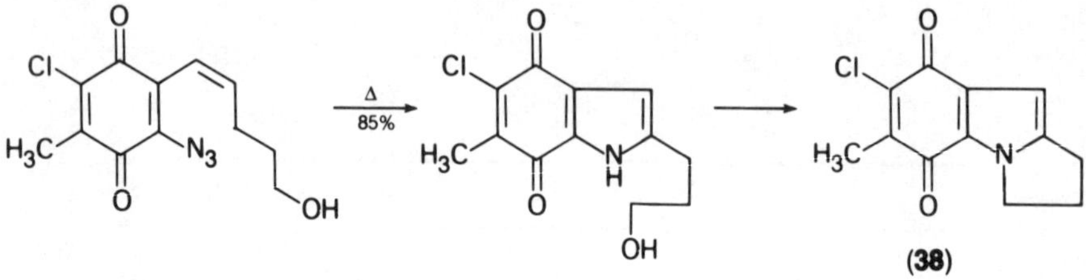

Δ, 85%

(38)

2. Synthesis of indoles and pyridines

The formation of indoles upon thermolysis of β-azidostyrenes is a well known reaction of synthetic importance. For example, thermolysis of α-azidocinnamate esters **39** in xylene gives greater than 90% yields of the indoles **40**[39]. At lower temperatures the azirines can be isolated; these have been shown to function as precursors to the indoles. Analogous results were obtained for the benzofuran **41**.

If the aryl group bears alkyl substituents in both *ortho* positions, nitrene insertions

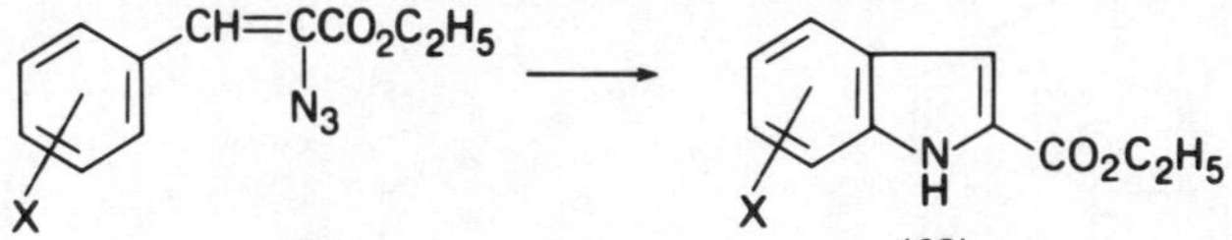

(39) (40)

X = p-CH_3, o-CH_3, p-Cl, o-Cl, p-Br, o-Br,
p-OCH_3, m-OCH_3, o-OCH_3, p-F

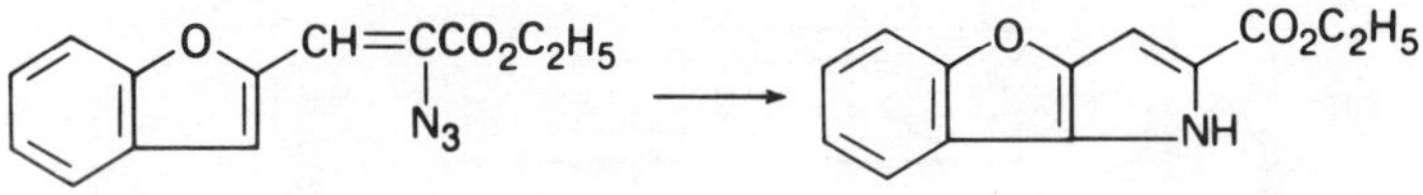

(41)

take place to give *c*-fused pyridines[41]. These reactions are accomplished in refluxing toluene or bromobenzene in the presence of air which oxidatively converts the initially formed dihydro derivative to the pyridine. Most other routes to such heterocyclic compounds requires acidic conditions. Thus this ring closure has potential for the synthesis of acid-sensitive pyridine derivatives.

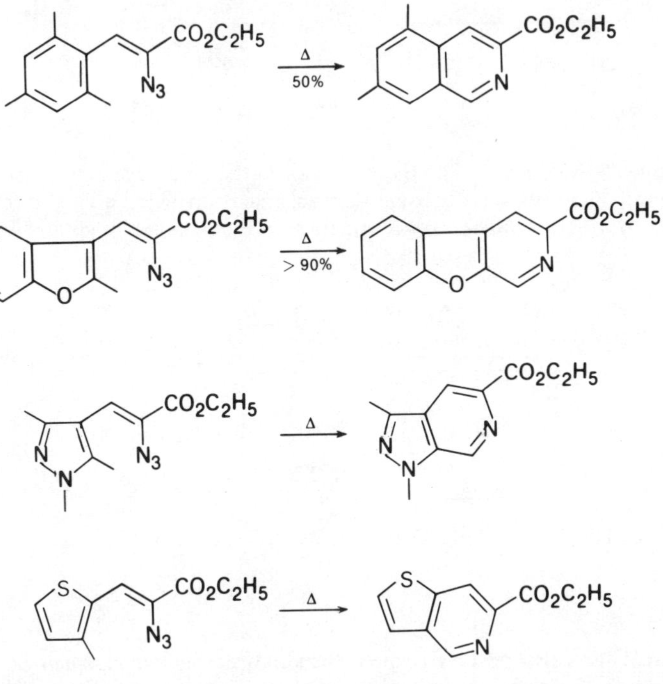

In a related study **42** was thermolysed to give the 3*H*-azepine, **43**, which rearranged to the 1*H* isomer at higher temperatures[41].

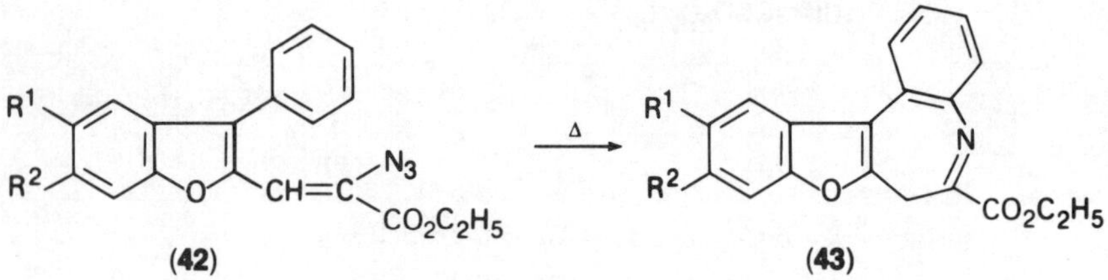

(42) → (43) [Δ]

R^1	R^2
C_6H_5	H
H	CH_3

3. Synthesis of oxazoles and isoxazoles

Ring closures involving intramolecular cyclizations to carbonyl groups have also been reported. These result in oxazole and isoxazole formation. For example, when the azide **44** was thermolysed in refluxing toluene, the products were **45**, **46** and **47**[42].

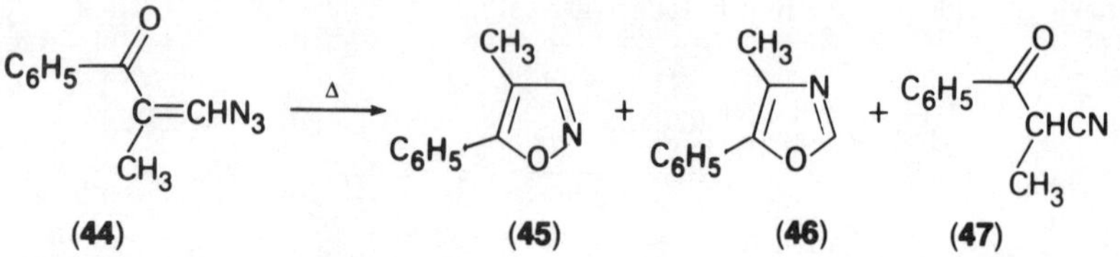

(44) → (45) + (46) + (47) [Δ]

The azirine **48** was shown to be the precursor to the heterocyclic products **45** and **46**. The former, **45**, arises via a thermal rearrangement involving a vinyl nitrene **49**, and the latter by a base-catalysed transformation to the isocyanide anion, **50**.

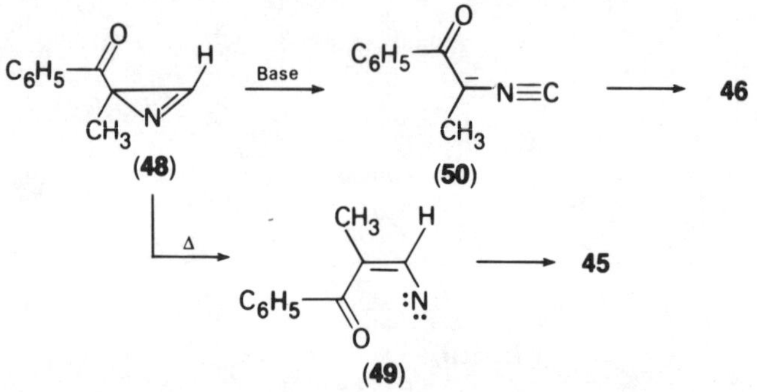

(48) → (50) → 46 [Base]; (48) → (49) → 45 [Δ]

Isoxazoles are also observed when 2-azido-3-carboalkoxy-1,4-benzoquinones are thermolysed[43,44], a reaction similar to that observed in the aromatic series. These ring closures take place in preference to the previously described ring contraction or fragmentation reaction of azidoquinones to cyclopentendiones or cyanoketenes.

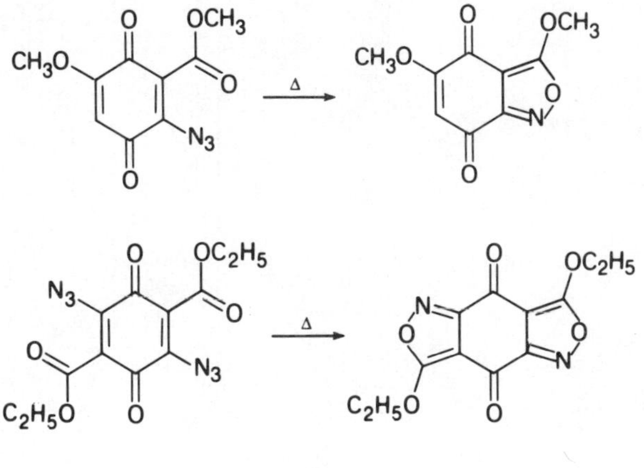

4. *Synthesis of 5-dialkylamino-1,2,3-triazoles*

1-Azido-1-dialkylaminoalkenes (α-azidoenamines) have not been isolated. However, their intermediacy has been suggested when α-chloroenamines are treated with sodium azide in acetonitrile at room temperature[45]. The proposed azide intermediate spontaneously undergoes electrocyclic ring closure followed by proton transfer to give 5-dialkylamino-1,2,3-triazoles.

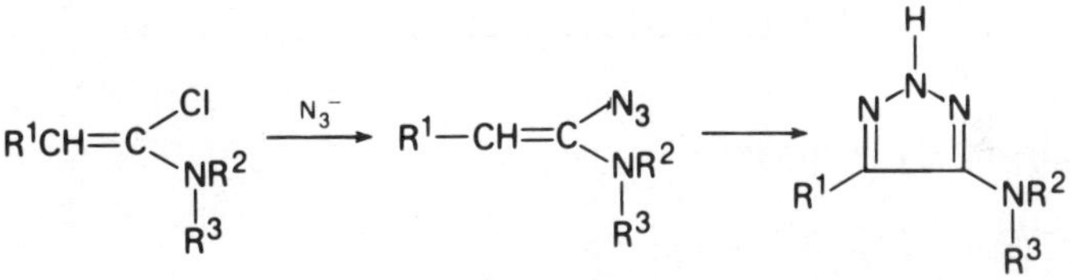

R^1	R^2	R^3	Yield, %
CH_3	$-(CH_2)_4-$		80
CH_3	CH_3	CH_3	60
CH_3	$-(CH_2)_2-O-(CH_2)_2-$		58
C_6H_5	CH_3	CH_3	84
C_6H_5	$-(CH_2)_2-O-(CH_2)_2-$		62

5. *Synthesis of benzoxazines*

An interesting case of an intramolecular cyclization is that reported for the 1-azidohydrazones **51**[46]. Thermolysis of these compounds in refluxing benzene for 2–3 h gives 63–79% yields of 3, 4, 4a, 5-tetrahydro-[1,2,4]triazino[6.1-*c*][1.4]-benzoxazines **53**. The authors propose formation of the intermediate **52** which then undergoes an intramolecular Diels–Alder cyclization to generate **53**.

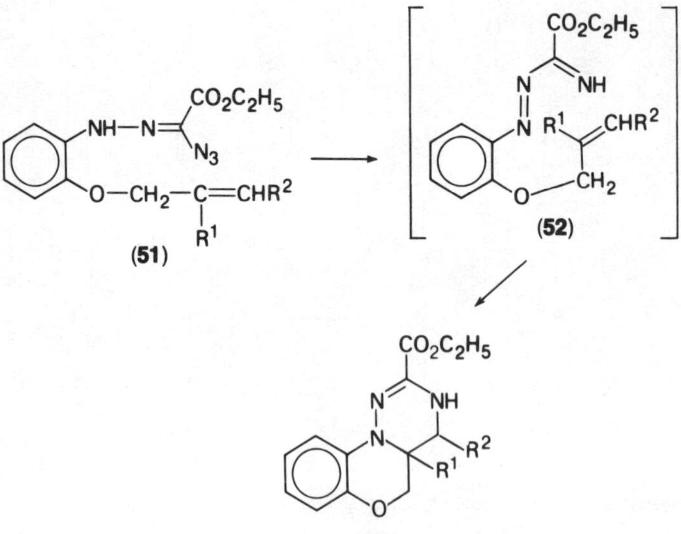

(51) → [(52)] → (53)

R^1	R^2
H	H
CH_3	H
H	C_6H_5

C. 1,3-Dipolar Cycloadditions

1. *Intermolecular cycloadditions*

It is well established that organic azides readily undergo 1,3-dipolar cycloadditions. Thus it is somewhat surprising that only a few reports have appeared for such reactions of vinyl azides. Acetylenedicarboxylate and simple vinyl azides readily undergo cycloaddition to give 1-alkenyl-1,2,3-triazoles **54**[47]. If a monosubstituted alkyne is used, both regioisomers are obtained[48]. Still another route to vinyltriazoles, and one which is regiospecific, involves the cycloaddition of vinyl azides to 2-oxoalkylidene-phosphoranes to give the betaine. Subsequent loss of triphenylphosphine oxide results in only regioisomer **55**[49]. Condensation of active methylene compounds with simple vinyl azides also gives 1-alkenyl-1,2,3-triazoles[48]. However, this method failed with α-azido vinyl ketones, for which only complex mixtures were obtained.

α-Azidostyrene adds to 2,5-dimethyl-3,4-diphenylcyclopentadienone in benzene at ambient temperature to give an 88% yield of **56**[50]. This cycloaddition is very sensitive

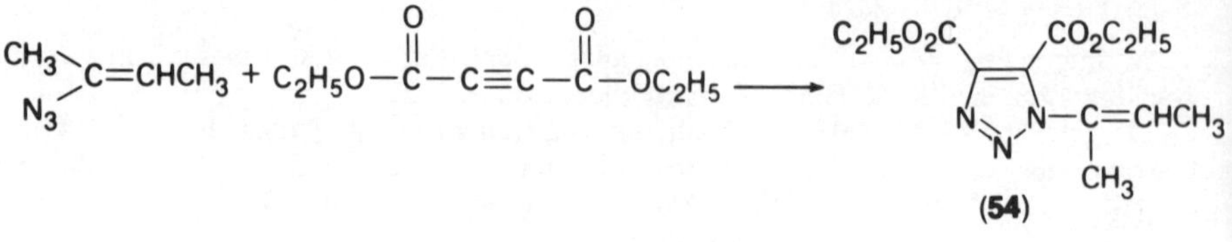

$CH_3(N_3)C{=}CHCH_3 + C_2H_5O{-}C(O){-}C{\equiv}C{-}C(O){-}OC_2H_5 \longrightarrow$ (54)

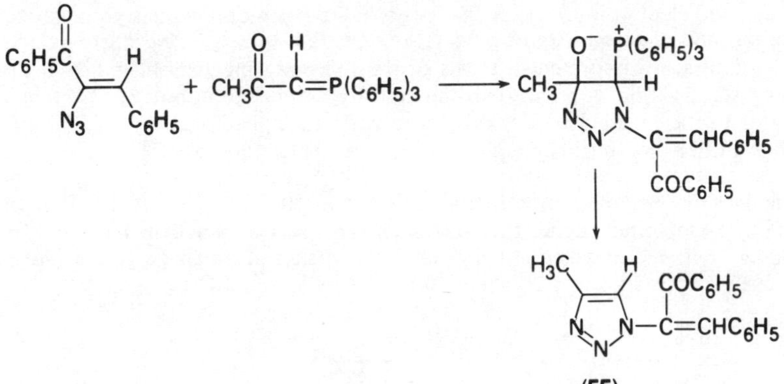

O
C_6H_5C H
N_3 C_6H_5
+ O H
$CH_3C—C{=}P(C_6H_5)_3$
O^- $\overset{+}{P}(C_6H_5)_3$
CH_3 H
N N N
$C{=}CHC_6H_5$
COC_6H_5

H_3C H COC_6H_5
N N N
$N—C{=}CHC_6H_5$

(55)

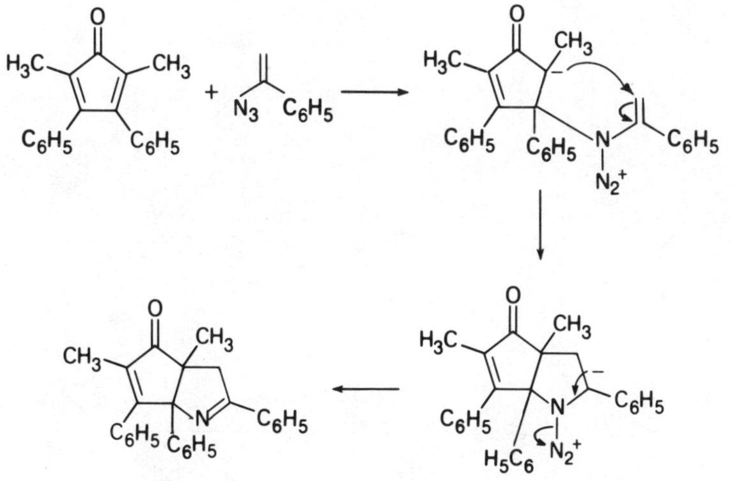

O
H_3C CH_3
C_6H_5 C_6H_5
+
N_3 C_6H_5
O
H_3C CH_3
C_6H_5 C_6H_5 N C_6H_5
N_2^+

O
H_3C CH_3
C_6H_5 N C_6H_5
H_5C_6 N_2^+

O
CH_3 CH_3
C_6H_5 C_6H_5 N C_6H_5

(56)

O
C_6H_5N N N N_3
O
+
N C_6H_5
O
C_6H_5N N N
O N C_6H_5
N_2^+

O
$C_6H_5—N$ N N N C_6H_5
O

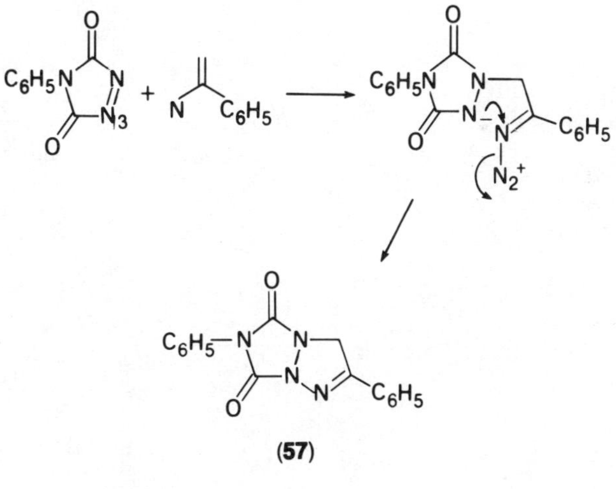

(57)

to steric and electronic effects. Substituents in the β-position of the styrene prevent the reaction, and electron-withdrawing groups on the aromatic ring decrease the rate. These data are interpreted in terms of the following mechanism in which negative charge stabilization is realized in the transition state. It is noteworthy that this is counter to the reaction of α-azidostyrene with triazolinedione to give **57**[51], in which carbon rather than nitrogen appears to be the nucleophilic site.

β-Azidostyrene reacts with cyclones by a different pathway. For example, when the azide was decomposed in refluxing toluene in the presence of 2,5-dimethyl-3,4-diphenylcyclopentadienone, the 3*H*-azepine **59** was obtained in high yield[52]. This product is viewed as arising by an initial conversion of the azide to the 1-azirine **58** which then cycloadds to the cyclone followed by loss of CO to give **59**.

C_6H_5 H H N_3 → C_6H_5 H H N **(58)** → (C_6H_5, C_6H_5, H_3C, CH_3, O; −CO) → C_6H_5 H C_6H_5 CH_3 H_3C N C_6H_5 **(59)**

C_6H_5(N_3)C=CH_2 + $(C_6H_5)_2$=C=O → O H_2C $(C_6H_5)_2$ C_6H_5 N N_2^+ **(60)** → O C_6H_5 C_6H_5 N C_6H_5 → O C_6H_5 C_6H_5 N H C_6H_5 **(61)**

$CH_3-(CH_2)_3-C(N_3)=CH_2$ —$(C_6H_5)_2C=C=O$→ N_3 C_4H_9 C_6H_5 O C_6H_5 **(62)**

Cycloadditions of vinyl azides to diphenylketene have also been reported. For example α-azidostyrene and the ketene react slowly to give the enamino ketone **61** with loss of nitrogen[53]. The reaction is viewed as involving nucleophilic attack of the β-carbon of the vinyl azide upon the ketene to give the zwitterion **60**. Subsequent ring closure results in **61**. When 2-azido-1-hexene was employed, the reaction took a different course and gave a 30% yield of cyclobutanone **62**, a 2 + 2 cycloaddition product.

Vinyl azides or their respective azirine thermolysis products have been shown to react with nitriles in the presence of BF_3 to give imidazoles in 16–82% yield[54].

$$C_6H_5(N_3)C{=}CHC_6H_5 + R{-}CN \xrightarrow{BF_3} \text{4,5-}(C_6H_5)_2\text{-2-R-imidazole}$$

$$R = -CH_3,\ -CH_2C_6H_5,\ -C_6H_5,\ -C(CH_3)_3,\ -CH_2{+}CH_2{+}_8CH_3$$

$$-CH_2-C(=O)-OC_2H_5,\quad -(CH_2)_4-CN$$

α- and β-Azidostyrenes as well as 2-azido-*trans*-2-butene react with the electron-deficient alkenes acrylonitrile and methyl acrylate at reflux temperatures to give 1-vinylaziridines in yields ranging from 35 to 83%[55]. This provides a useful synthesis of these unusual enamines, which are not available by conventional methodology.

$$R^1R^2C{=}C(R^3)N_3 + CH_2{=}CH_2{-}R^4 \longrightarrow R^1R^2C{=}C(R^3){-}N\text{(2-}R^4\text{-aziridine)}$$

R^1	R^2	R^3	R^4	Yield, %
C_6H_5	H	H	CN	35
C_6H_5	H	H	CO_2CH_3	56
H	H	C_6H_5	CN	45
H	H	C_6H_5	CO_2CH_3	68
H	CH_3	CH_3	CN	73
H	CH_3	CH_3	CO_2CH_3	83

The electron-rich double bond of vinyl azides allows their utilization as dipolarophiles in cycloadditions with nitrile imines[56] and nitrile oxides[57]. Like enamines, the azide group controls the regiochemistry and adducts **63** and **64** are the observed products. However, these readily lose HN_3 to give the corresponding aromatic pyrazoles **65** and isoxazoles **66**.

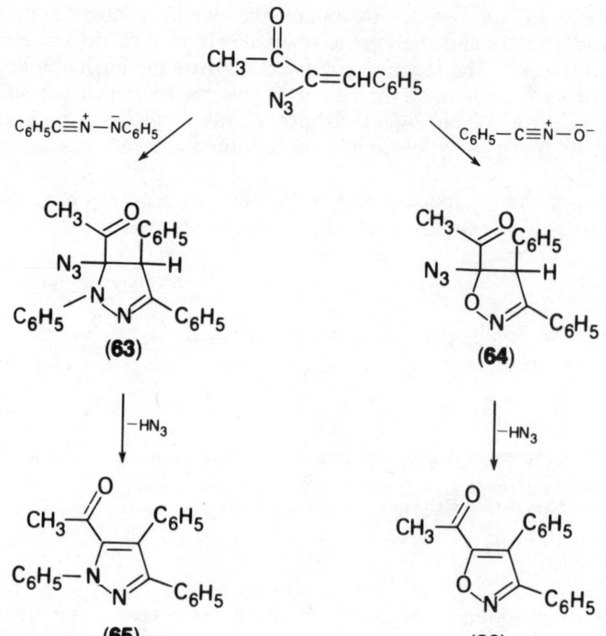

2. *Intramolecular cycloadditions*

The synthetic potential of intramolecular 1,3-dipolar cycloaddition of vinyl azides to appropriately situated alkenes and alkynes has not yet been realized in that only one study has been reported[58]. When the azide **67** was allowed to stand at 0°C for 3 days, it cyclized quantitatively to give the Δ^2-1,2,3-triazoline **68**. Further heating caused nitrogen loss to form the azepine **69**. In an analogous manner the alkynyl analogue was converted to **70** and 1-(1-azidoethenyl)-2-(2-propenyl)benzene to **71**.

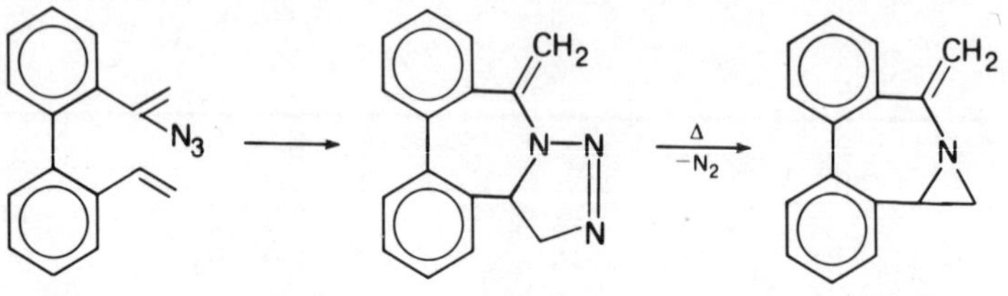

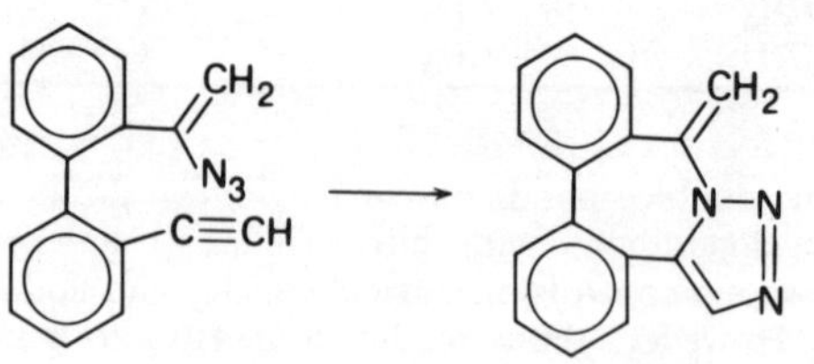

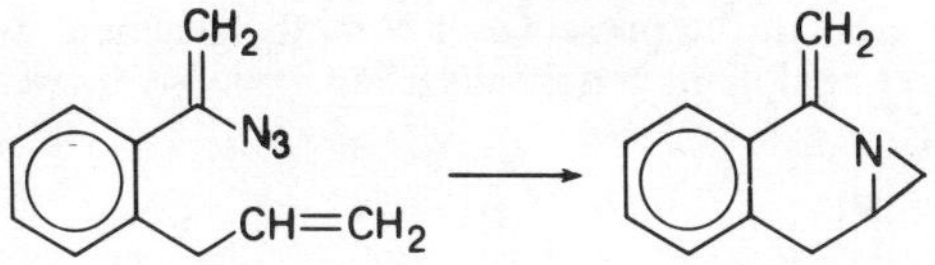

(71)

D. Ring Expansion of Cyclic Vinyl Azides

A series of papers has reported ring expansion of cyclic β-azidoenones under thermal or photolytic conditions. These reactions show many similarities to reactions of aryl azides, discussed in a later section.

When 3-azido-2-cyclohexenones were thermally decomposed in refluxing methanol, 25–36% yields of the 4, 5, 6, 7-tetrahydro-2-methoxy-1*H*-azepin-4-ones **73** were isolated[59]. This rearrangement is viewed as proceeding via the cyclic keteneimines **72**

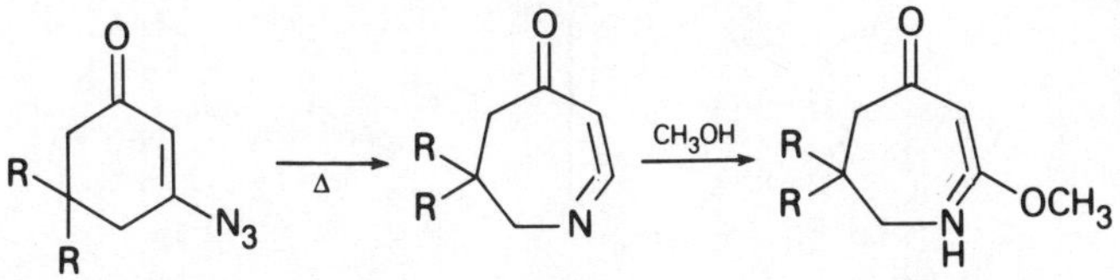

(71) (72) (73)

R = H, CH_3

which proceed to the azepines upon addition of methanol. When the rearrangement was carried out in ethanol or aniline, 2-ethoxy- or 2-anilino- derivatives of **73** were obtained. In a related study, the thermolysis and photolysis of 2-alkyl-3-azido-2-cyclohexen-1-ones **74** were found to give 2-amino-2-alkyl-3,3-dimethoxycyclohexanones **76**, which were proposed to arise via the azirine **75**[60].

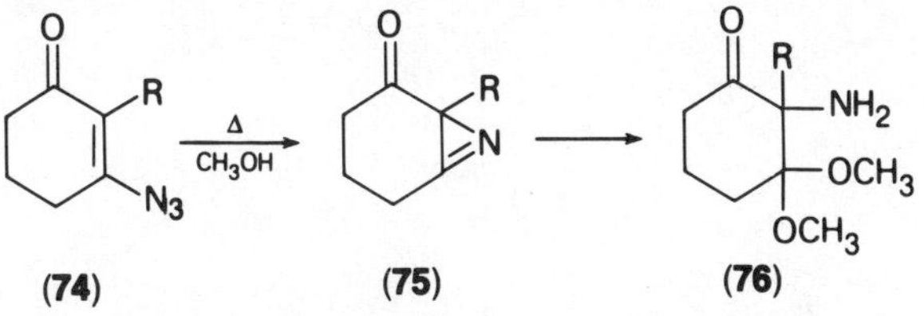

(74) (75) (76)

R = $-CH_3$, $-CH_2CH{=}CH_2$, $-CH_2C_6H_5$

It is most interesting that an opposite trend was observed for the photolytic decomposition of 6-azido-1,3-dimethylpyrimidines in the presence of amines. That is, ring expansion was observed for 5-substituted derivatives and not for the unsubstituted compounds. Specifically, photolysis of 6-azido-1,3-dimethylthymine **77** (R = CH_3) in the presence of methylamine gave a 40% yield of **78**, while, under analogous conditions, 6-azido-1,3-dimethyluracil **77** (R = H) gave a 42% yield of **79**. Photolysis of **77** (R = CH_3) in methanol gave **80** (43%), a product analogous to **76**. An initially formed azirine intermediate is suggested as common to all of these products.

The formation of 6-alkylamino-5-amino-1,3-dimethyluracils, **79**, from **77** has been employed in a novel synthesis of lumazines and fervenulinus[63]. For example, photolysis of **77** (R = H) in the presence of *N*-methylglycine ethyl ester gave a 73% yield of **81**, and, under analogous conditions using formylhydrazine, **82** was isolated in 55% yield.

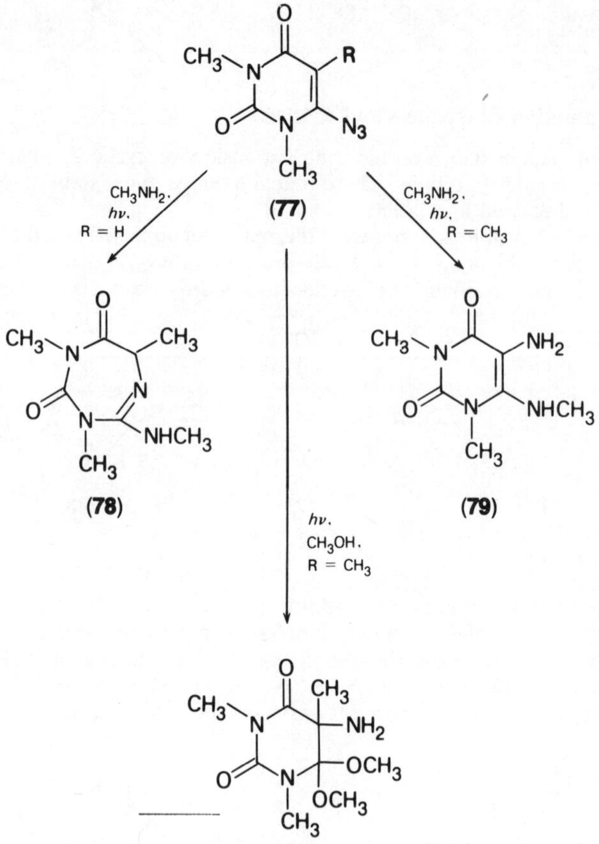

E. Reactions of Vinyl Azides with Electrophiles

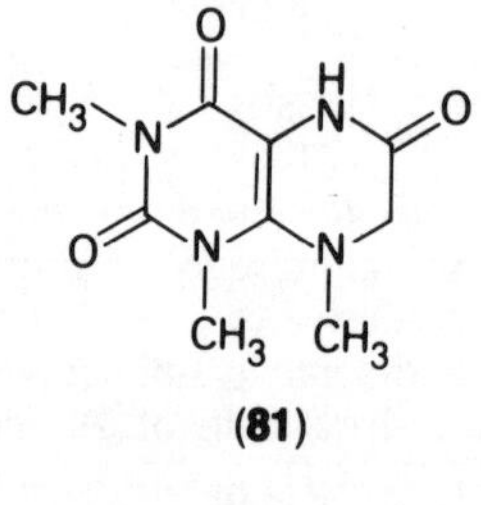

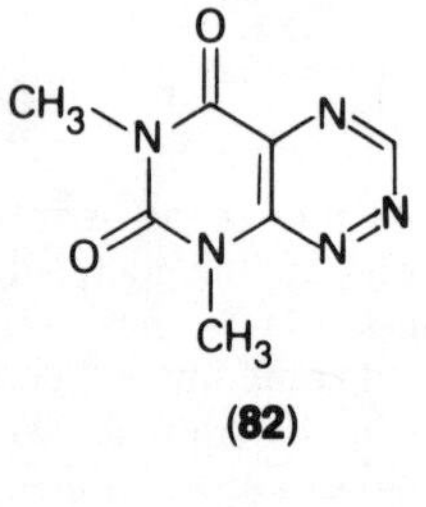

The electron-releasing property of the azide group imparts some nucleophilic character to the alkene π-bond. Therefore, vinyl azides, like enamines, are susceptible to electrophilic attack at the β-position. A number of studies have appeared describing the reactions of vinyl azides with electrophilic reagents such as protic acids, halogens, nitrosyl tetrafluoroborate, nitryl tetrafluoroborate, acyl chlorides, ketenes and dichlorocarbene. These reactions will not be discussed since they have been the subject of a review article[64]. Selected examples which illustrate the reactions are given below, using 1-azido-1-phenyl-1-propene for illustrative purposes.

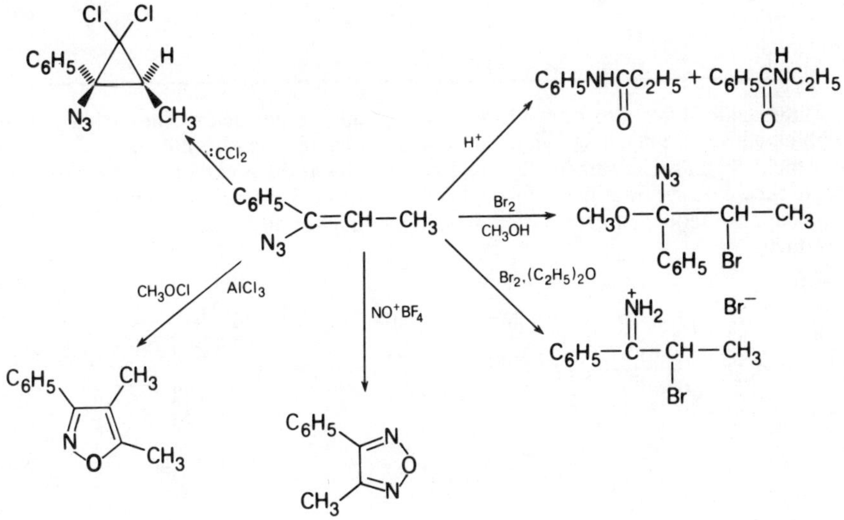

F. Conversion of Vinyl Azides to Aldehydes and Ketones

Among the more interesting examples of simple functional group transformations of vinyl azides is their utility as precursors to aldehydes and ketones by reduction to the enamines. A mild and convenient method involves treatment of vinyl azides with sodium sulphide in methanol in the presence of a catalytic amount of triethylamine, followed by work-up with aqueous acid[65]. Analogous transformations have been reported when the reducing agent was zinc–acetic acid[66] or triethyl phosphite[67]. On the other hand, reduction with lithium aluminum hydride gives the corresponding saturated amines[68].

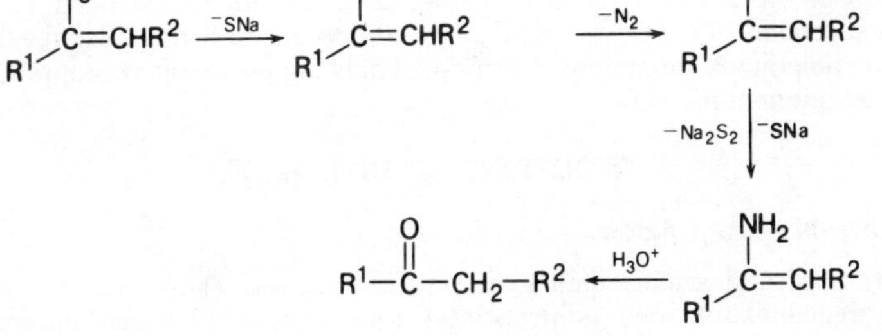

R^1		R^2	Yield, %
$-C_6H_5$		$-H$	91
$-C_6H_5$		$-CH_3$	92
$-n\text{-}C_4H_9$		$-H$	90
$-H$		$-t\text{-}C_4H_9$	82
	$-(CH_2)_6-$		85
$-C(=O)-C_6H_5$		$-C_6H_5$	74

Vinyl azides have also been utilized as acyl anion equivalents. They react readily with alkyllithium reagents to yield, after acidic work-up, mainly ketones or aldehydes in which alkylation has taken place regiospecifically at the β-carbon[69]. Triazines have been shown to be intermediates since they can be isolated by omitting the acid work-up. Subsequent treatment of these with aqueous acid then gives the alkylated products.

$$R^1C(N_3)=CHR^2 \xrightarrow{1.\ R^3Li;\ 2.\ H^+} R^1C(\overset{+}{N}H_2-N=N-R^3)=CHR^2 \longrightarrow R^1C(\ddot{N}H_2)=CHR^2 + R^3-\overset{+}{N}\equiv N \longrightarrow R^1C(=NH)-CH(R^2)R^3 \xrightarrow{H_3O^+} R^1-\overset{O}{\overset{\|}{C}}-HC(R^2)(R^3)$$

R^1	R^2	R^3	Yield, %
$-C_6H_5$	$-H$	$-CH_3$	82
$-C_6H_5$	$-CH_3$	$-CH_3$	32
$-n\text{-}C_4H_9$	$-H$	$-CH_3$	82
$-H$	$-t\text{-}C_4H_9$	$-CH_3$	71

IV. PREPARATION OF ARYL AZIDES

No new general methods of preparing aryl azides have been developed since the last review in this series[4] Aryl azides are generally prepared from diazonium salts upon treatment with sodium azide. Some perhalogenated aryl azides have been prepared by azide displacement of a halo[70], nitro[71] or hydrazino group[72], but these reactions are limited to compounds containing electron-withdrawing groups at appropriate sites to enhance addition–elimination reactions.

V. REACTIONS OF ARYL AZIDES

A. Photolysis of Aryl Azides

Photolysis of aryl azides often gives rise to nitrenes. There are few reports of insertion reactions of such intermediates with solvent, although intramolecular insertion reactions occur when an aromatic ring is *ortho* to the azide group. In addition, ring expansions to azepines and nucleophilic trapping of azirine intermediates have been reported.

1. *Formation of carbazoles*

Photolysis (or thermolysis) of the azide **83** gives mainly the carbazole, **84**, rather than the phenanthridine, **85**[73]. Formation of the carbazole is attributed to a singlet nitrene intermediate while the phenanthridine is a triplet nitrene product. Evidence for this comes from an investigation of the photolysis of **83** in the presence of acetophenone, a triplet sensitizer. Under these conditions, the yield of carbazole decreases markedly and the triplet-derived products, **85**, **86** and **87**, are formed in substantial yields. If there is not an obvious site for singlet nitrene attack, as in the

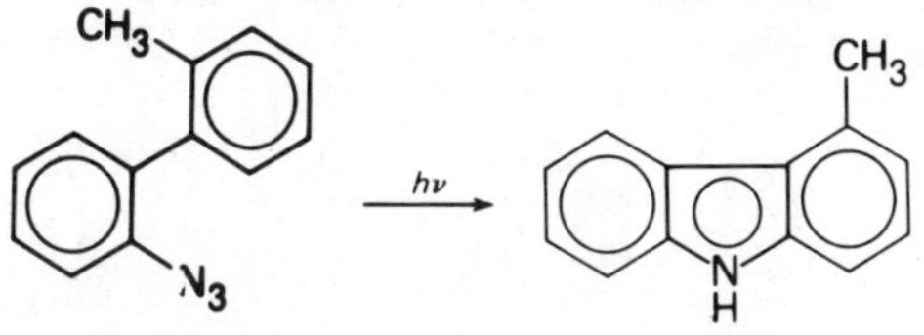

CH_3 N_3 $h\nu$ CH_3 N H

(83) **(84)**

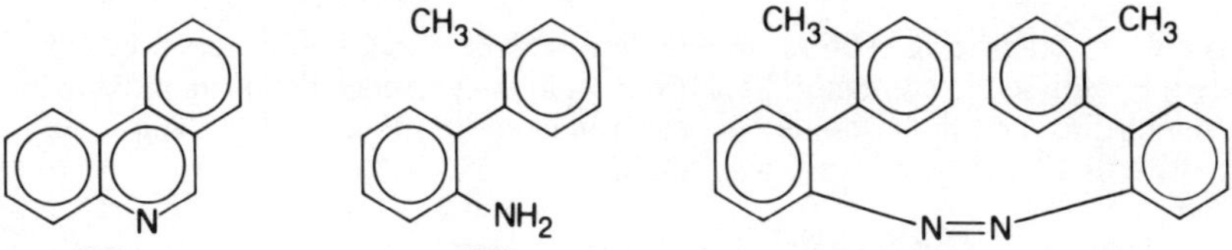

N CH_3 NH_2 CH_3 CH_3 N=N

(85) **(86)** **(87)**

thiophene derivative, **88**, only the amine and azo compounds are formed when the photolysis is carried out in cold acetophenone. However, at higher temperatures (107°C) the product of insertion into the methyl groups, **89**, is formed in as high as

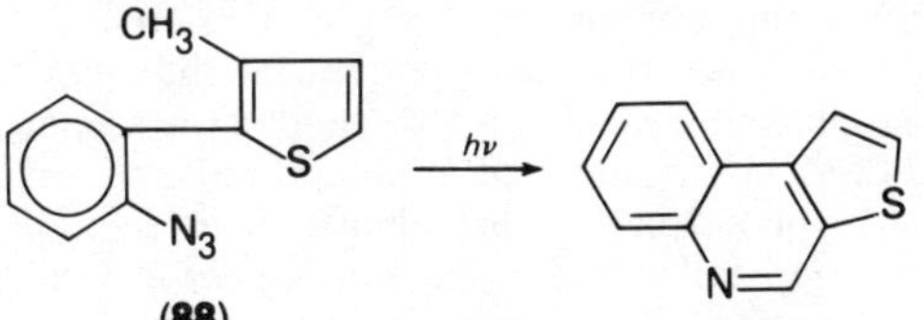

CH_3 S N_3 $h\nu$ S N

(88) **(89)**

30% yield[73,74]. Temperature also affects the conversion of **83** to **84** in that the yield of **84** can be increased to 90% by accomplishing the photolysis in chlorobenzene at 107°C, a temperature below that at which the thermolytic conversion of **83** to **84** occurs.

2. *Formation of 3*H*-azepines*

Photolysis of aromatic azides in the presence of nucleophiles, especially amines, produces the righ-expanded 3*H*-azepines, **90**[75]. However, when an aryl ring is *ortho* to the azide group, the previously mentioned carbazole formation is competitive with ring expansion[76,77]. Kinetic evidence indicates that there are at least two intermediates involved in these competitive pathways[77]. The relative amounts of azepine and carbazole are affected by the concentration of the amine and by the wavelength of light. When X in **91** was CF_3, the yield of carbazole **92** was 80% in the absence of diethylamine, but dropped to 20% if as little as 1% of amine was present. When X was H or CH_3, up to 50% amine was needed to maximize the yield of the azepine, **93**. In addition, more azepine was produced using light of 2537 Å than if the wavelength exceeded 2900 Å[78].

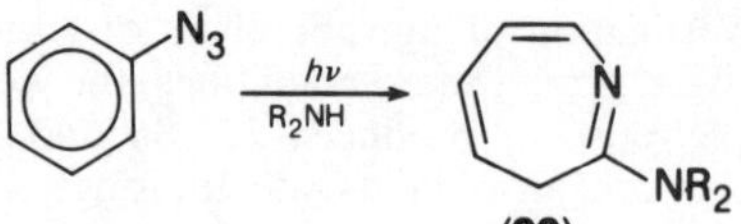

N_3 — $h\nu$, R_2NH → NR_2

(90)

X, X, N_3 — $h\nu$, $(C_2H_5)_2NH$ → X, N H, X + X, X, N H, $N(C_2H_5)_2$

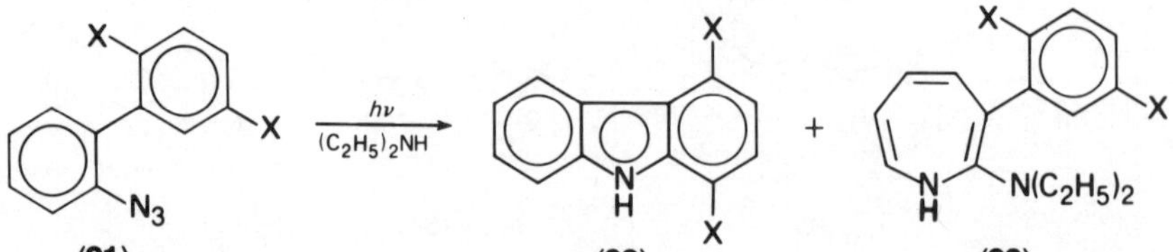

(91) **(92)** **(93)**

There are conflicting reports on whether azepine formation occurs when phenyl azide is photolysed in methanol[79,80]. However, in the presence of 3 M methoxide ion in methanol/dioxane, the azepine **94** was obtained in 35% yield[79]. This could be increased to 48% if 18-crown-6 was added.

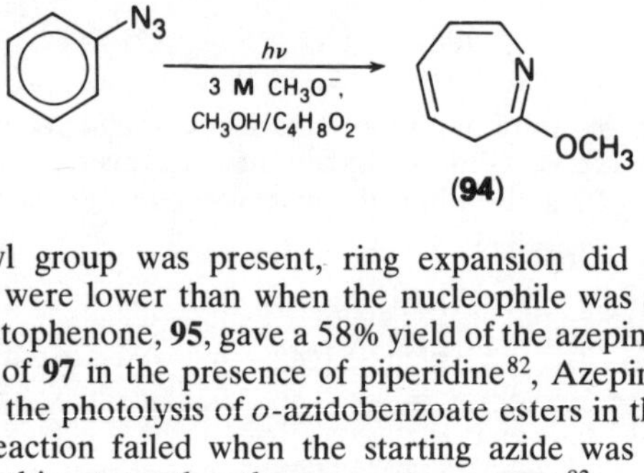

N_3 — $h\nu$, 3 M CH_3O^-, $CH_3OH/C_4H_8O_2$ → N, OCH_3

(94)

When an *o*-acetyl group was present, ring expansion did occur in methanol, although yields still were lower than when the nucleophile was an amine[81,82,83]. For example, *o*-azidoacetophenone, **95**, gave a 58% yield of the azepine **96** in methanol but a quantitative yield of **97** in the presence of piperidine[82], Azepine yields of 58–72% were obtained from the photolysis of *o*-azidobenzoate esters in the presence of other alcohols, but the reaction failed when the starting azide was substituted with an *o*-carboxylic acid or thioester rather than an acetoxy group[83].

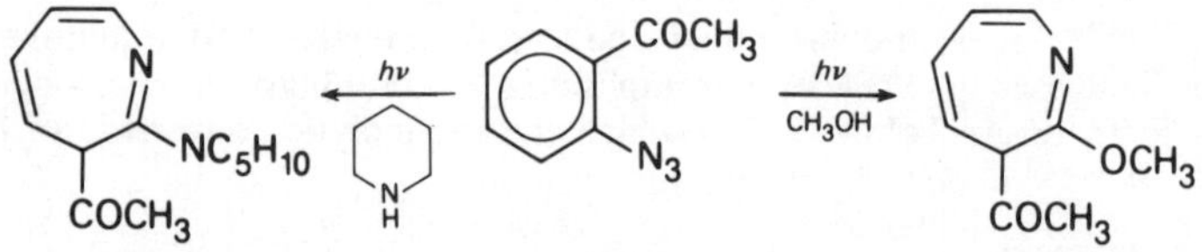

N, NC_5H_{10}, $COCH_3$ ← $h\nu$, piperidine (N H) — $COCH_3$, N_3 — $h\nu$, CH_3OH → N, OCH_3, $COCH_3$

(97) **(95)** **(96)**

In order to gain mechanistic information on the intermediates formed in the photolysis of aryl azides, azidobenzene was photolytically decomposed in a matrix at 8°K[84]. Under these conditions the product detected was the cyclic ketenimine **98**. This

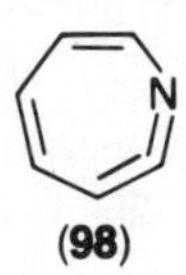

(98)

can be viewed as a possible intermediate to azepine products, which could result from addition of the amine followed by proton transfer.

3. *Nitrogen migration*

Aryl azides containing fused aromatic rings undergo a different reaction when photolysed in the presence of nucleophiles. For example, 2-azidonaphthalene **99** gives **101**, a product in which the nitrogen has migrated and the nucleophile is bonded to the position originally occupied by the azide group[85,86]. The reaction always occurs between positions 1 and 2 rather than 2 and 3, i.e. **99** → **101**. It appears to involve the azirine intermediate **100**[87]; such can be detected if the photolysis is carried out in a matrix at 12°K[88]. The fact that the product is **101** rather than **103** can be explained by a preference for the aromatic azirine **100** over the non-aromatic analogue, **102**.

(102) (99) (100)

(103) (101)

With few exceptions, the azide group must be β to the ring junction in order to observe efficient rearrangement. Usually, if the azide group is in the α-position, as in **104**, a complex mixture is obtained which includes the amine, **105**, and the azo compound, **106**, and no detectable rearranged products[85].

(104) (105) (106)

The difference between α- and β-azides has been attributed to at least two factors: (1) there would be less favourable resonance stabilization of the nitrene formed from the α-azide, thus enhancing triplet nitrene formation and products therefrom; (2) nucleophilic attack at the α-position on an azirine intermediate would be substantially hindered over β-attack due to *peri* interaction.

Anomalous behaviour of some α-aryl azides has been reported. 5-Azidoquinoline,

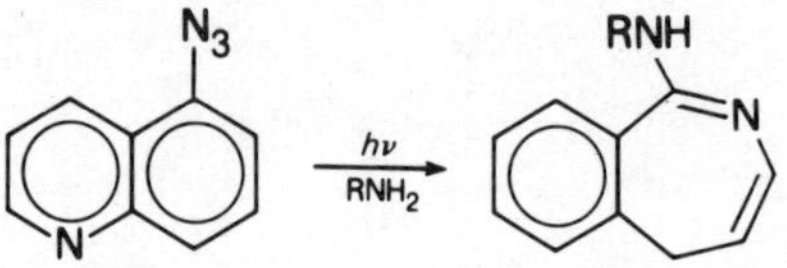

(107) (108)

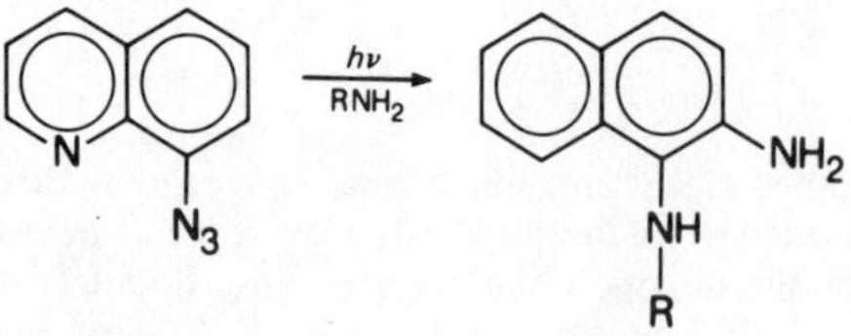

(109) (110)

107, forms the azepine 108 when photolysed in the presence of amines[90], while the 8-azido isomer, 109, gives the diamine 110[89]. In contrast, the α-azidoindoles 111 and 112 gave complex mixtures while the β-isomer 113 gave the expected diamines 114[91]. In piperidine, 1-azidonaphthalene, 115, gave both diamine regioisomers, 116 and 117; this appears to be the only reported case in which both isomers are formed. When the photolysis was carried out in the presence of tetramethylenediamine (TMEDA), a singlet sensitizer, only 117, the unusual isomer, was formed, in 75% yield.

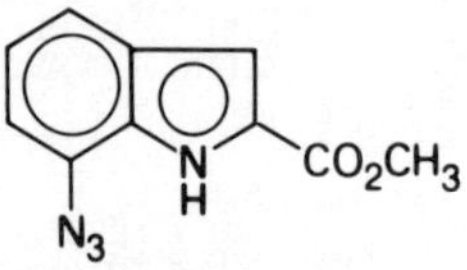

(111) 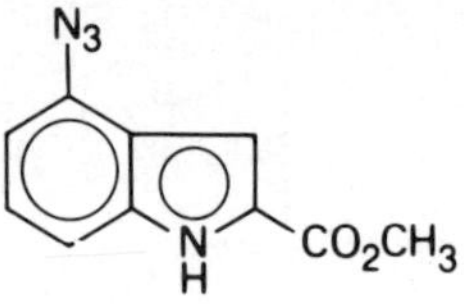 (112)

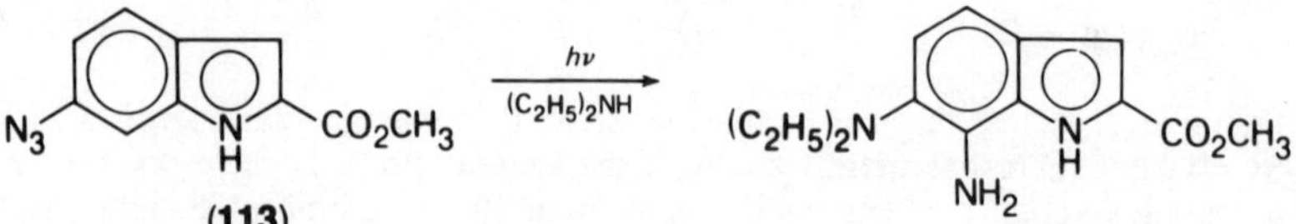

(113) (114)

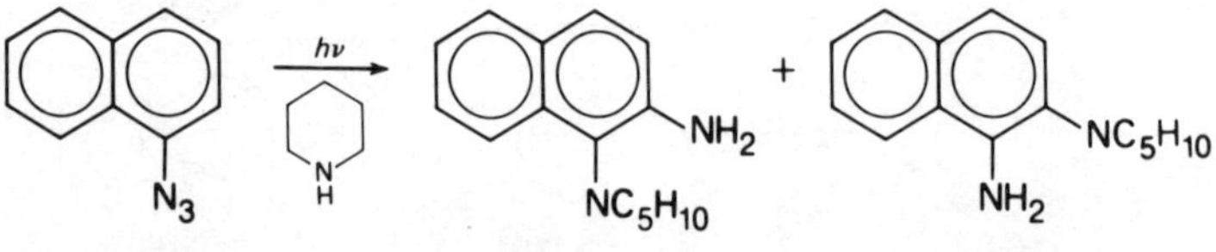

(115) (116) (117)

Anomolous behaviour of some β-azides also occurs. Some 6-azidobenzo[*b*]-thiophenes form azepines under certain conditions. For example, when 118 is irradiated for no more than 9 h or when pyrene is used as a cosolvent for the irradiation, the product is 119, in 12% or 22% yield, respectively. Irradiation for 18 h in the absence of pyrene gives, instead, a 12% yield of the expected diamine 120[93].

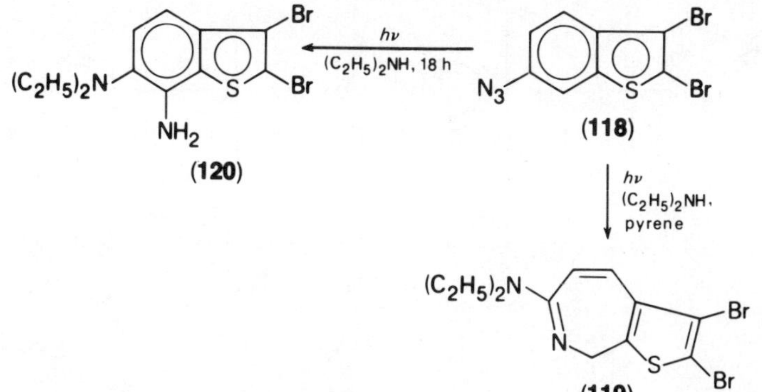

(120) (118) (119)

Very few reactions involving an azide group on a heterocyclic ring have been reported. In contrast to the behaviour of azide on carbocyclic rings, 4-azido-7-chloroquinoline, **121**, an α-azide, gives ring expansion and, in the presence of some amines, the rearranged diamine **125**[94]. The β-azide **126** appears to be very

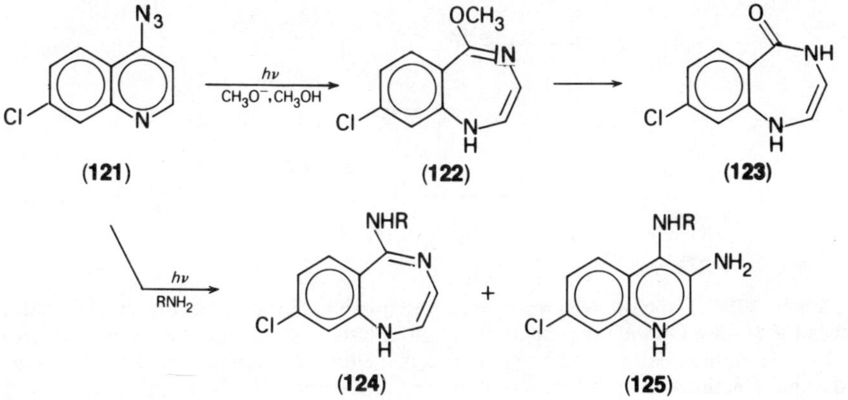

(121) (122) (123) (124) (125)

Reagent	Azepine, %	Diamine, %
CH_3O^-	40–70% (**122** + **123**)	—
$C_6H_{11}NH_2$	40	25
n-$C_4H_9NH_2$	35	—
$PhCH_2NH_2$	55	25

sensitive to the reagents used. As with **121**, the use of methoxide seems to favour ring expansion[94]. Surprisingly, no such ring expansion occurs in the photolysis of 3-azidopyridine, **127**, in the presence of methoxide[79].

Secondary amines are the most commonly used nucleophiles to intercept reactive intermediates formed in the photolysis of fused aryl azides. In addition, methoxide ion has been employed, but the course of the reaction is greatly influenced by the base concentration and reaction conditions. For example, photolysis of 2-azidoanthracene, **128**, in the presence of 3 M potassium methoxide in methanol–dioxane, followed by brief refluxing or by overnight standing, gave nearly a quantitative yield of the

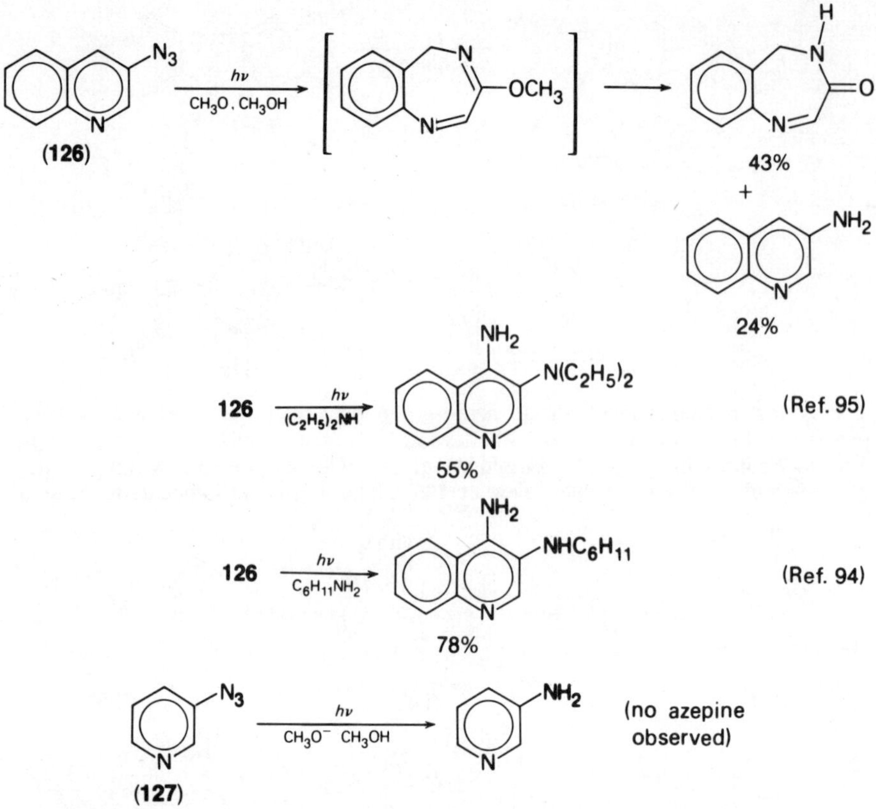

(126) $\xrightarrow[CH_3O^-, CH_3OH]{h\nu}$ [─OCH$_3$] ⟶ 43% + 24%

126 $\xrightarrow[(C_2H_5)_2NH]{h\nu}$ 55% (Ref. 95)

126 $\xrightarrow[C_6H_{11}NH_2]{h\nu}$ 78% (Ref. 94)

(127) $\xrightarrow[CH_3O^- \; CH_3OH]{h\nu}$ (no azepine observed)

azepine, **129**[96]. If the solution was neutralized immediately after photolysis, the product was **130** (60%). On the other hand, if the photolysis was accomplished in the presence of 0.5 M methoxide at room temperature or below, to minimize base-catalysed and thermal reactions, the product was a mixture of **130** and **131**. These differences can be

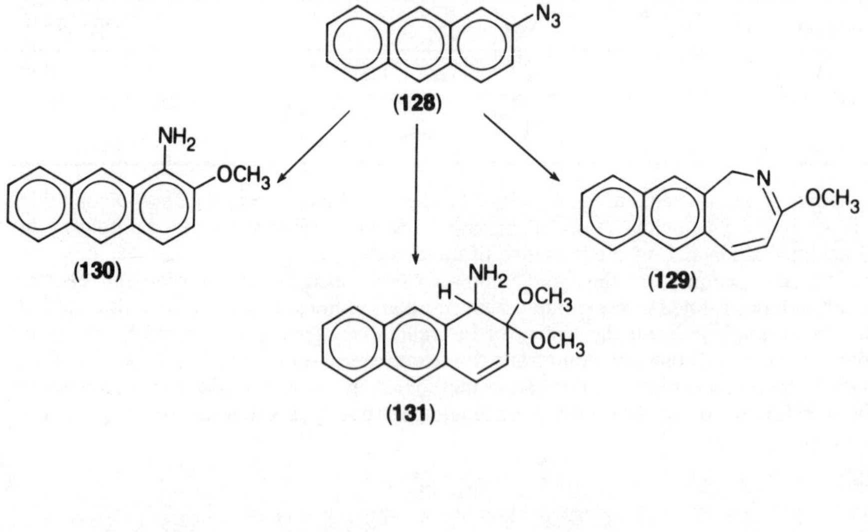

(128)

(130) (131) (129)

explained by considering the aziridine intermediate **132**. Immediate neutralization allows acid-catalysed ring opening to give **130**. Electrocyclic ring opening of **132** to a 1*H*-azepine is a slower process, since it would involve an initial loss of aromatic stabilization energy. In the low temperature experiment, ring opening by methoxide attack would give **131**, which could form **130** by loss of methanol.

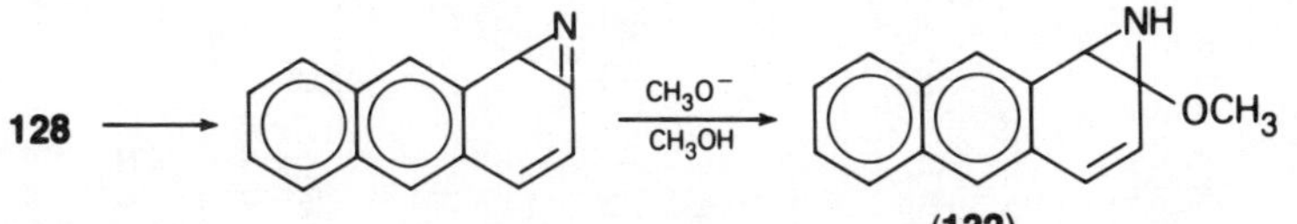

(132)

Photolysis of aryl azides in the presence of ethanethiol, which provides both a nucleophile and a weak proton source, appears to give only rearranged amines. This is true even for those azides which one might expect to give azepine, such as 4-azidotoluene **133**[97].

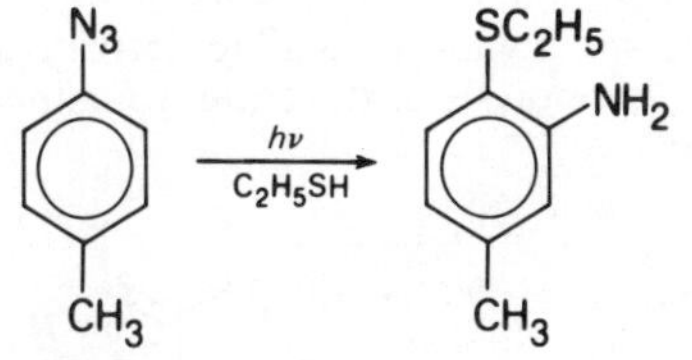

(133)

B. Thermolysis of Aryl Azides

Thermal decomposition of aromatic azides produces products of both intra- and intermolecular reactions. However, the former have been most extensively studied, and these transformations will be emphasized here.

1. Formation of carbazoles and related compounds

Thermolysis of appropriately substituted aryl azides gives carbazoles, often in better yields than those obtained by photolysis[73]. The reaction can be used to make not only carbazole derivatives but also related heterocyclic analogues such as **134**[98]. As with photolysis, the nature and yields of the products are affected by changes which influence the singlet nitrene, the intermediate leading to carbazoles[99,100].

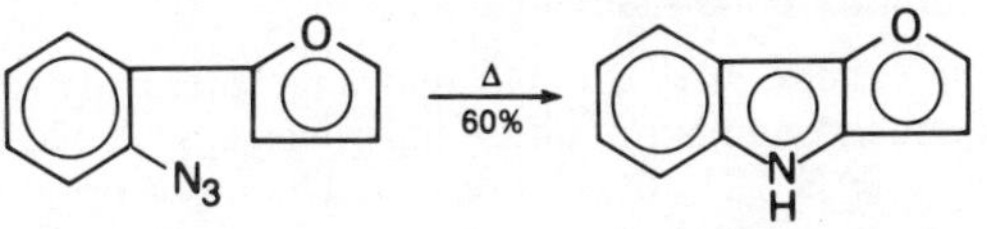

(134)

In the thermolysis of **135** the yield of the singlet-derived product, **136**, is increased when X is an electron-withdrawing group, thus increasing the electrophilic character of the nitrene intermediate[100]. Electron-releasing substituents enhanced the triplet-derived product, **137**, when thermolysis was carried out in the presence of acetophenone.

When **138** was decomposed in either tetralin or acetonitrile, a quantitative yield of

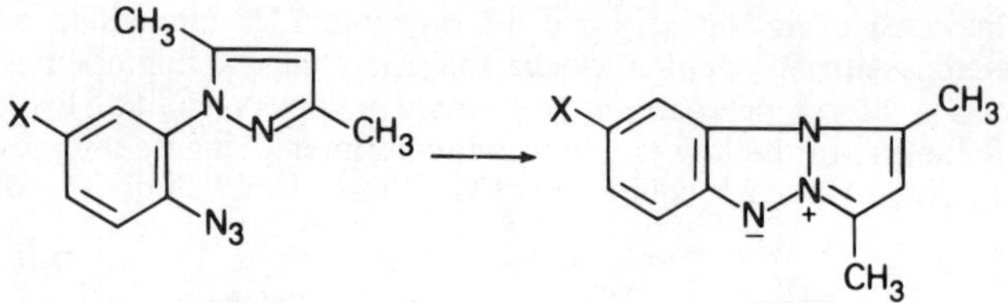

(135) (136)

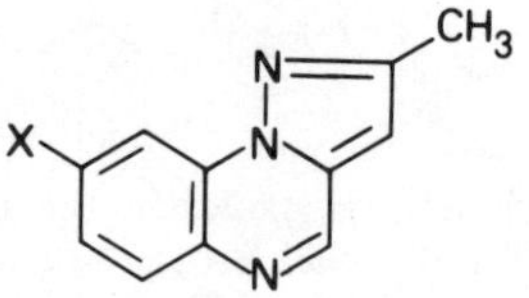

(137)

139 was reported[101]. However, if this azide was heated in dibutylamine, **139** was obtained in only 50% yield along with the aniline derivative **140** (40%). A carboline, **142**, was produced by the thermolysis of the trifluoromethyl derivative **141**[99].

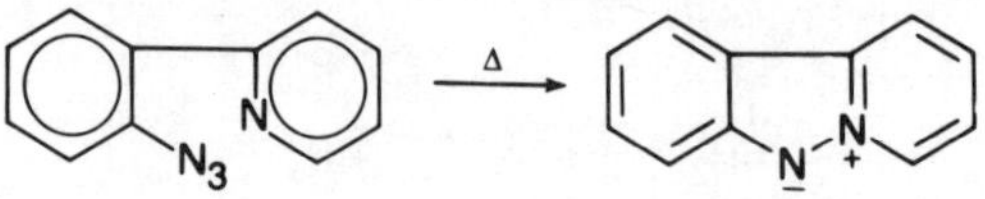

(138) (139)

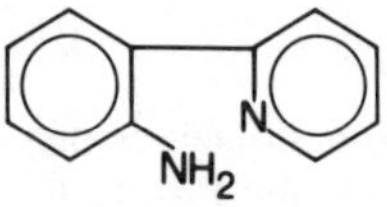

(140)

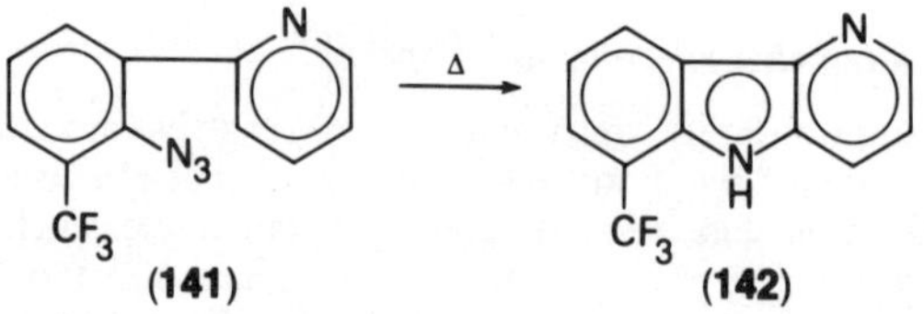

(141) (142)

2. *Ortho-substitution with rearrangement*

Aryl azides of the general structure **143** rearrange thermally to give a variety of heterocyclic compounds. A number of different aryl groups (Ar) have been employed; the X group is commonly S, CH_2, SO_2, O or CO. These reactions have recently been reviewed[102,103], therefore only selected examples will be outlined here.

The 'typical' reaction on an unsubstituted example appears deceivingly to involve simply a nitrene insertion. However, in those cases in which the aryl ring (Ar) bears

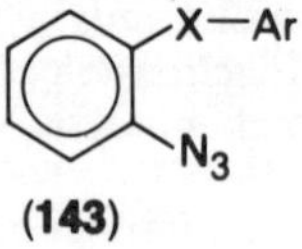

(143)

substituents, a more complex rearrangement is revealed. For example, thermolysis of **144** gives **145**[104], and **146** gives the azepine, **147**[105].

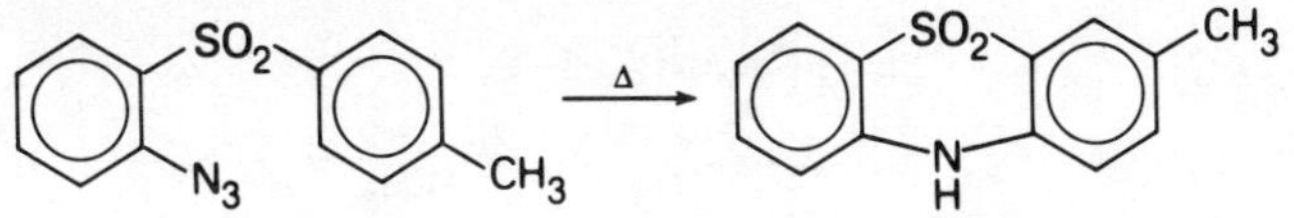

(**144**) (**145**)

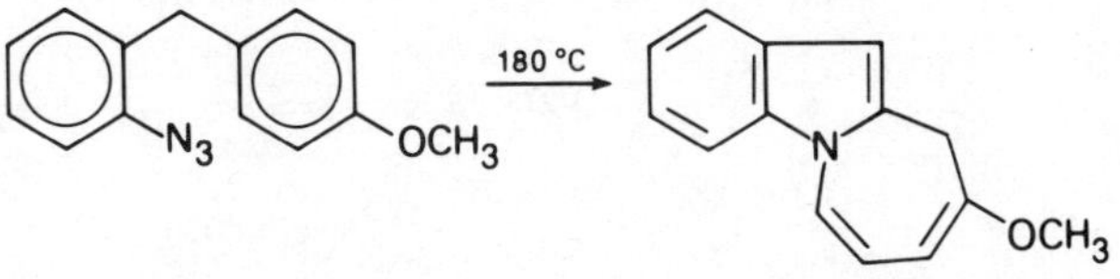

(**146**) (**147**)

The mechanism of these reactions is generally accepted to involve the spiro intermediate **148**. When the aryl ring of **143** is an aromatic heterocycle, still other reaction pathways from the spiro intermediate are observed, e.g. **149** → **150**[106,107].

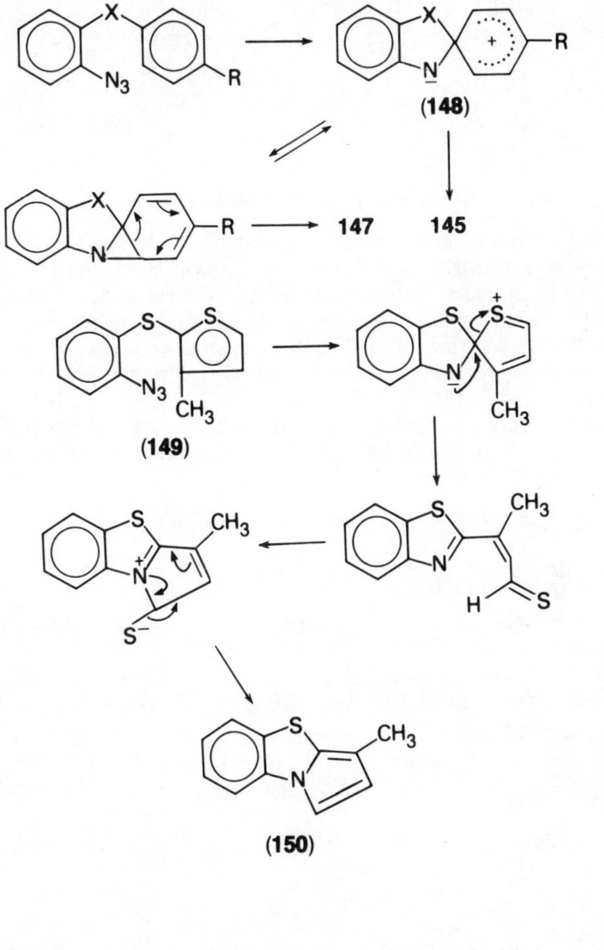

(**148**)

147 **145**

(**149**)

(**150**)

Thermolysis of **151** gave the carbazole **152**, while photolysis of **151** in the presence of acetophenone gave the phenothiazine **153**[103].

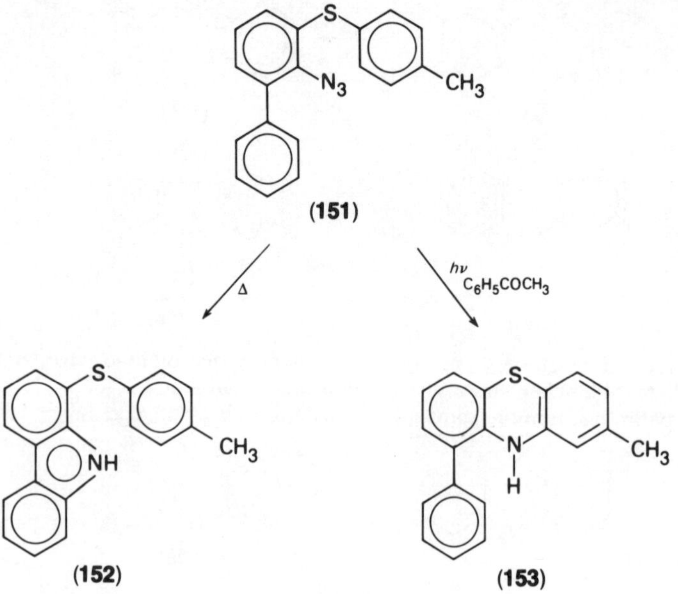

(**151**)

Δ $h\nu$ $C_6H_5COCH_3$

(**152**) (**153**)

3. *Intramolecular cyclizations with polar substituents*

Aryl azides having a polar unsaturated substituent in the position *ortho* to the azide group readily lose nitrogen and form cyclic products. A variety of heterocyclic compounds can be prepared by this methodology. The reaction occurs at temperatures below that required for the generation of nitrenes and is generally free of products associated with such intermediates. The mechanisms proposed are either a concerted process or one involving an initial 1,3-dipolar cycloaddition followed, in most cases, by nitrogen loss, i.e. **154** → **155**[108,109,110] or **156** → **157** → **158**[111,112]. It is reasonable that both mechanistic pathways are available, and the choice depends upon the specific case in question. In any regard, the following examples illustrate the variety of products

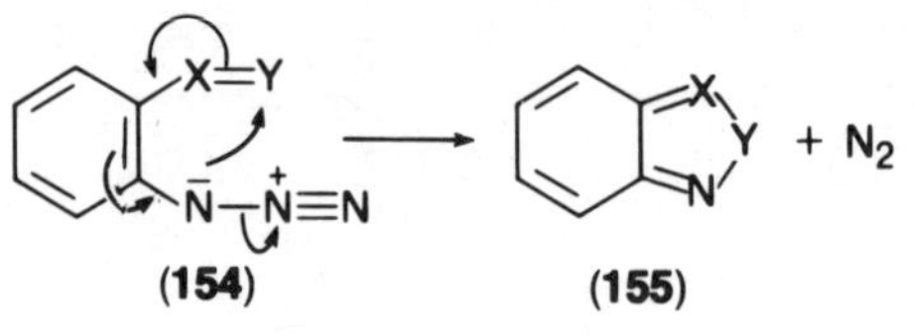

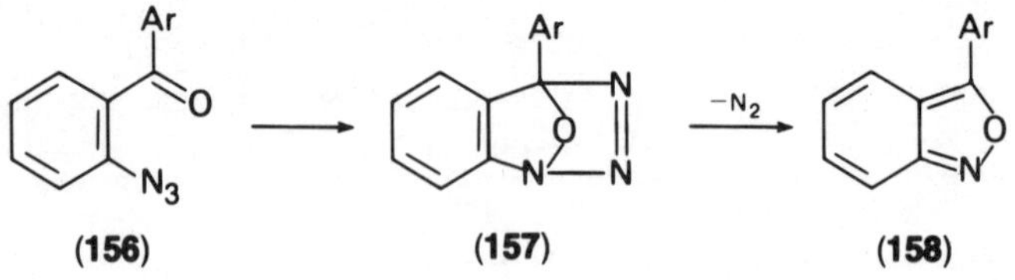

(**154**) (**155**) + N_2

(**156**) (**157**) $-N_2$ (**158**)

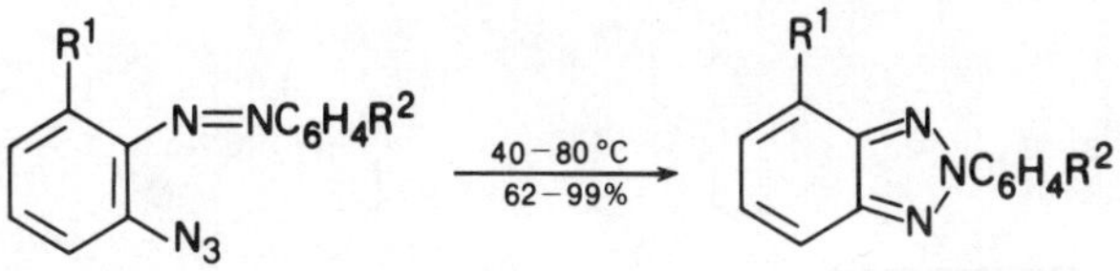

(Refs. 109,111)

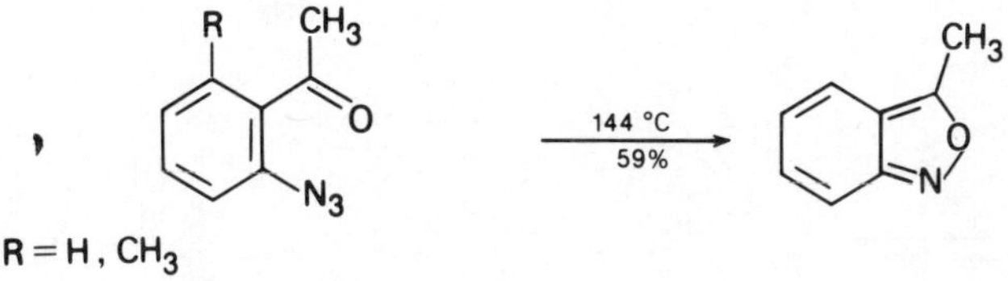

40–80 °C, 62–99%

R[1], $N{=}NC_6H_4R^2$, N_3 → R[1], $NC_6H_4R^2$

(Ref. 110)

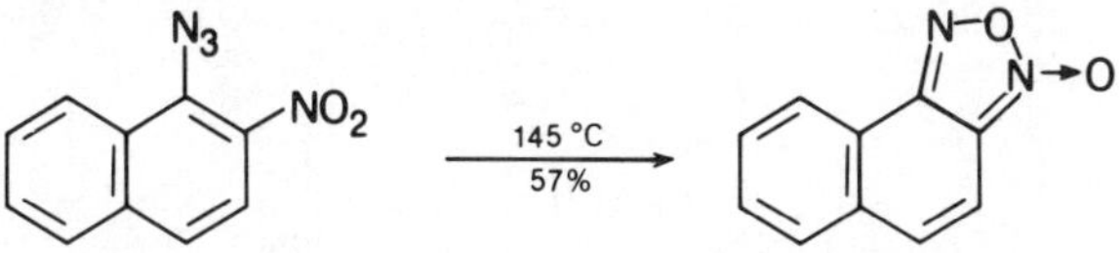

144 °C, 59%

R, CH_3, O, N_3 → CH_3, O, N

R = H, CH_3

(Ref. 108)

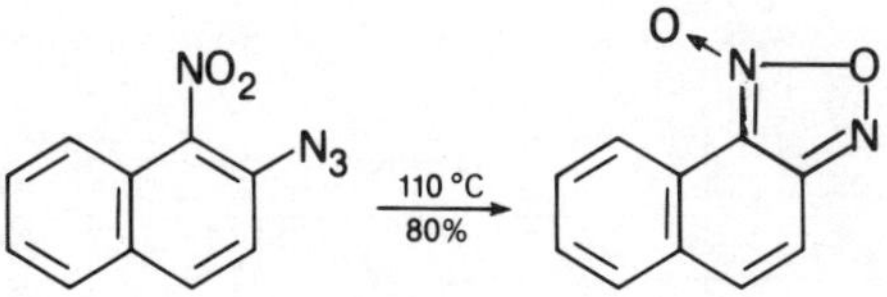

145 °C, 57%

N_3, NO_2 → N–O, N→O

(Ref. 108)

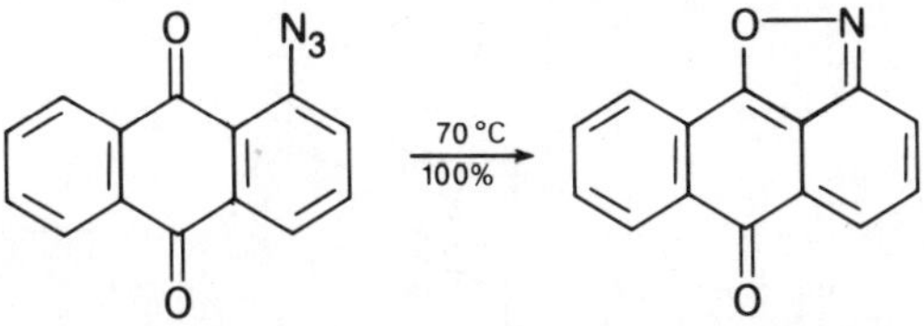

110 °C, 80%

NO_2, N_3 → O←N, O, N

(Ref. 110)

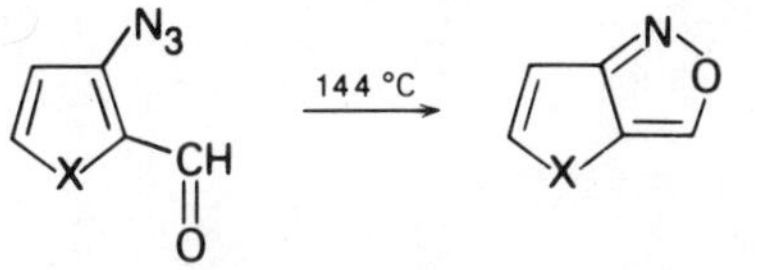

70 °C, 100%

O, N_3, O → O–N, O

(Ref. 113)

144 °C

N_3, X, CH=O → N, O, X

X	Yield, %
O	3
S	22
Se	40

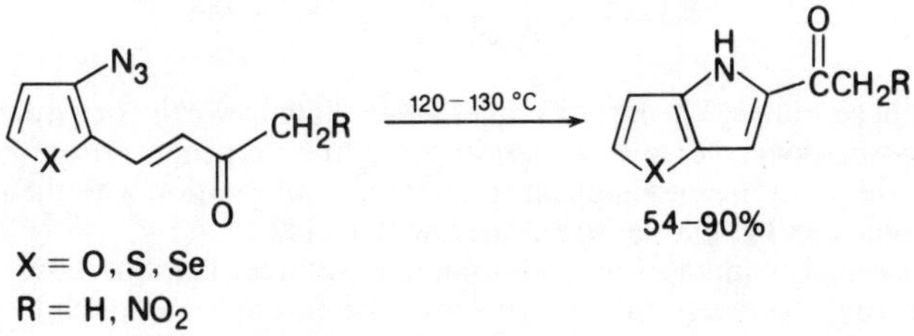

(Ref. 115)

120–130 °C

N_3, X, CH_2R, O → H, N, O, $\overset{\|}{C}CH_2R$, X

54–90%

X = O, S, Se
R = H, NO_2

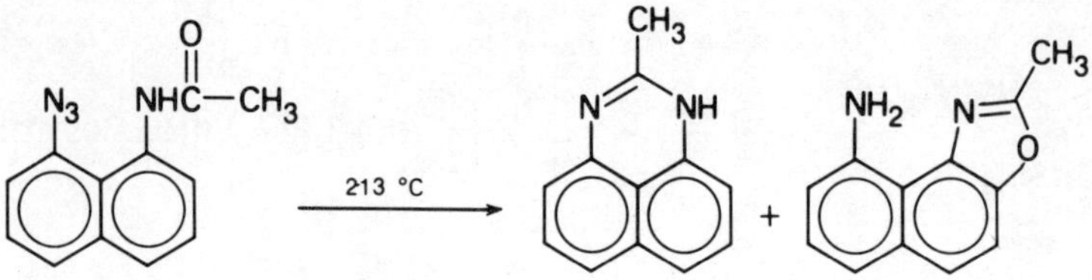

(Ref. 115)

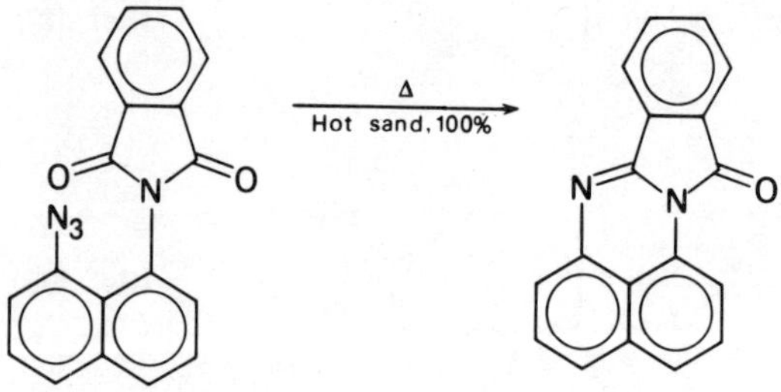

(Ref. 115)

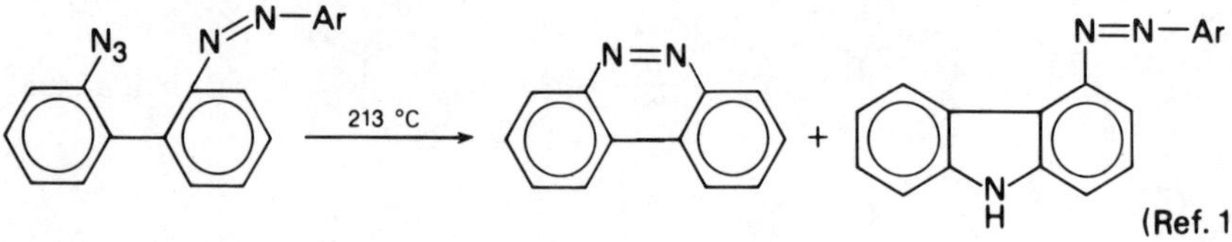

(Ref. 116)

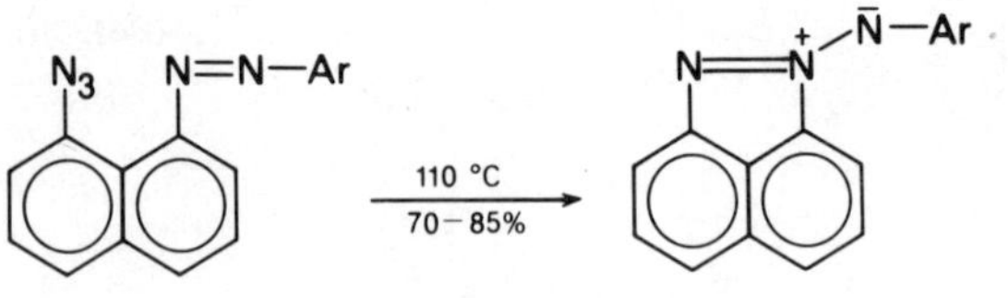

(Ref. 117)

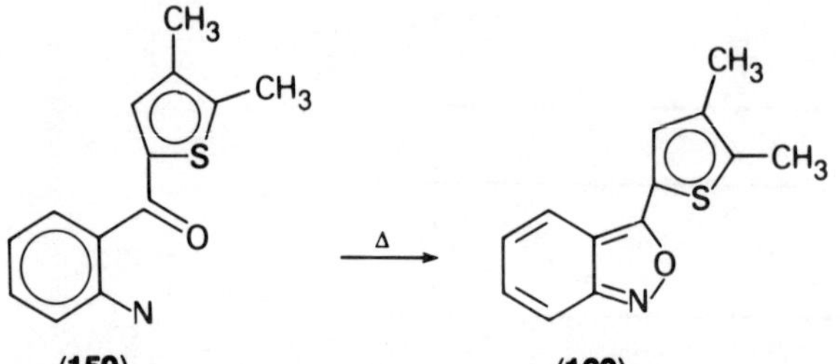

(Ref. 103)

(159) (160)

available by these routes. The last example, **159** → **160**, is worthy of further comment. The formation of isoxazoles such as **160** appears to be a general reaction. However, upon further thermolysis, these appear to undergo equilibration with the nitrene, **161**, and this proceeds to **163** via the spiro intermediate **162**[103].

Aryl azides can also undergo cycloaddition to unsaturated groups more remote from the aromatic ring, forming tricyclic systems. For example, a series of compounds

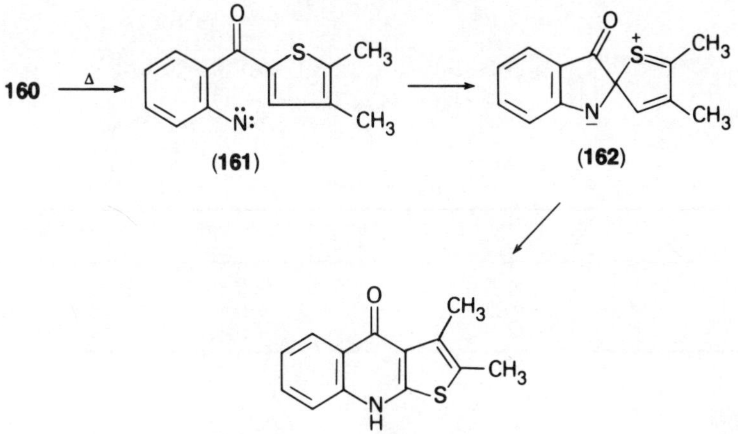

160 —Δ→ (161) → (162) → (163)

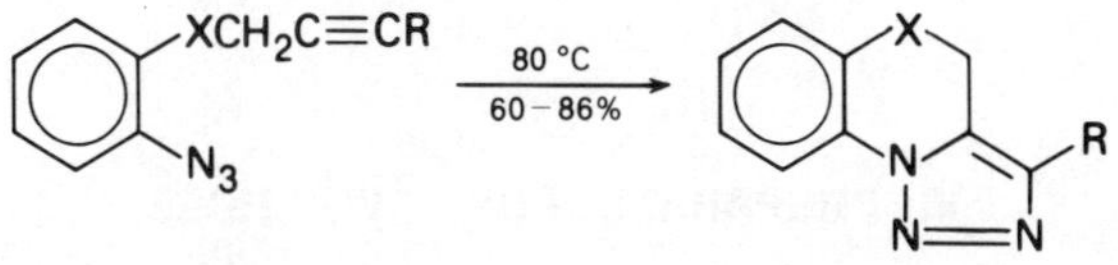

O, S, CH_3, CH_3, N: (161)

O, $\overset{+}{S}$, CH_3, CH_3, N (162)

O, CH_3, CH_3, S, N, H (163)

$XCH_2C{\equiv}CR$, N_3 (164) —80 °C, 60–86%→ X, N, R, N=N (165)

X = O, S; R = H, C_6H_5

containing a triple bond, **164**, gave the fused triazoles, **165**. When a similar addition occurred with a side-chain double bond, the triazole (presumably formed initially) lost nitrogen to give aziridines, e.g. **166** → **167** and **168** → **169**[119].

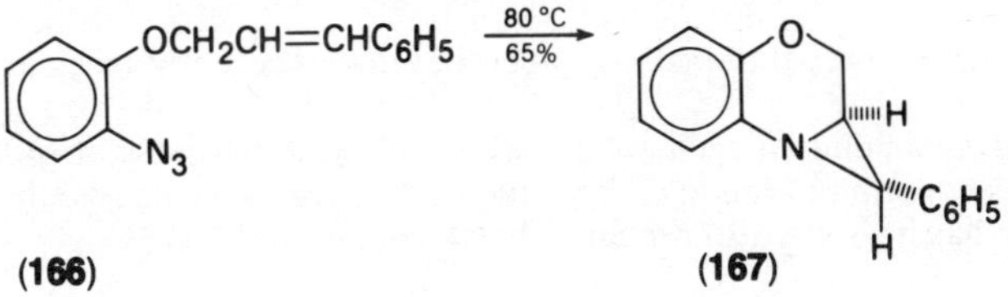

$OCH_2CH{=}CHC_6H_5$, N_3 (166) —80 °C, 65%→ O, N, H, C_6H_5, H (167)

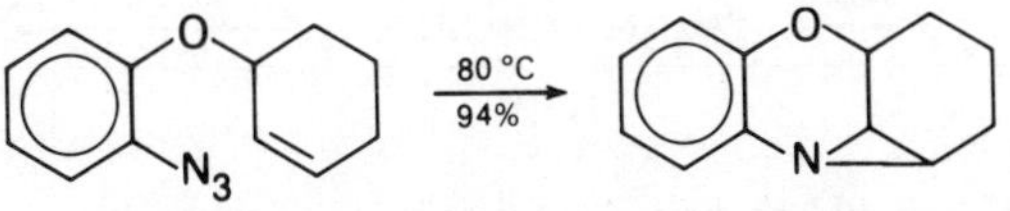

O, N_3 (168) —80 °C, 94%→ O, N (169)

The stereochemistry of the side-chain affect the course of the reaction. In the *cis* nitrile **170** the azide group adds to the C≡N triple bond. In contrast, the *trans* nitrile **171** reacts at the carbon–carbon bond[120].

$$\text{(170)} \xrightarrow[\text{Toluene}]{110\,°\text{C}} \text{tetrazolo[1,5-a]quinoline}$$

R^1	R^2	R^3	Yield, %
—H	—H	—H	28
—Cl	$-C_6H_5$	—H	78
—H	—H	$-CH_3$	85

$$\text{(171)} \xrightarrow[\text{DMSO, 85\%}]{140\,°\text{C}} \text{indole-2-carbonitrile}$$

VI. PREPARATION OF ACYL AZIDES

Acyl azides have most commonly been prepared by treating acid chlorides with sodium azide. This method is not entirely satisfactory since sodium azide has low solubility in organic solvents and thus must often be used in aqueous solution. Reactive acid chlorides may be hydrolysed under these conditions, while unreactive examples may be inert. If the mixture is heated, then the possibility of the Curtius rearrangement exists, i.e.

$$R-\overset{O}{\overset{\|}{C}}-Cl \xrightarrow{NaN_3} R-\overset{O}{\overset{\|}{C}}N_3 \xrightarrow{\Delta} RN{=}C{=}O$$

Several alternative preparations have been developed recently to circumvent these difficulties.

Tetramethylguanidinium azide, **172**, reacts with acid chlorides at sufficiently low temperatures to permit isolation of the azide with no rearrangement to the isocyanate. This reagent has been used to prepare *t*-butyl azidoformate, **173**[121].

$$(CH_3)_2N-\overset{NH}{\overset{\|}{C}}-N(CH_3)_2 \xrightarrow{HN_3} \underset{\textbf{(172)}}{(CH_3)_2N-\overset{\overset{+}{N}H_2N_3^-}{\overset{\|}{C}}-N(CH_3)_2} \xrightarrow[0\,°\text{C}]{ClCO_2C(CH_3)_3} \underset{\textbf{(173)}}{N_3CO_2C(CH_3)_3}$$

A mixture of pyridine and hydrazoic acid in toluene serves as a soluble source of azide ion. At 0°C treatment of a solution of this reagent with an acyl chloride gives an immediate precipitate of pyridinium chloride, leaving the acyl azide in solution. Quantitative yields of the acyl azides are reported[122]. This method has been used successfully for conversion of the highly hindered acid chloride **174** to the corresponding acyl azide[123].

$$\mathrm{RC(=O)Cl + HN_3 + C_5H_5N \longrightarrow RC(=O)N_3 + C_5H_5NH^+Cl^-}$$

$$\textbf{(174)} \xrightarrow[0\ ^\circ\mathrm{C}]{\mathrm{HN_3-pyridine}}$$

A method that avoids handling hazardous hydrazoic acid is the use of tetrabutylammonium azide. This salt can be extracted from aqueous solution by dichloromethane, and can be isolated in pure crystalline form and used in a number of different organic solvents. Yields of the acyl azides were 84–97%, based on evolution of nitrogen upon heating, or 52–89%, based on isolation of the product of the reaction of the isocyanate with aniline[124].

$$\mathrm{(C_4H_9)_4N^+OH \xrightarrow{NaN_3} (C_4H_9)_4N^+\ N_3^- \xrightarrow{RC(=O)Cl} RC(=O)N_3}$$

$$\mathrm{RC(=O)N_3 \xrightarrow[-N_2]{\Delta} RN{=}C{=}O \xrightarrow{C_6H_5NH_2} RNHC(=O)NHC_6H_5}$$

Diphenyl phosphorazidate, **175**, can be used to convert a carboxylic acid directly to the azide, without requiring the intermediate acid chloride[125]. The mechanism appears to involve a cyclic transition state, **176**[126]. The same reagent can be used to produce peptide coupling directly, with no need for isolation of the acyl azide, using amino acids containing a wide variety of functional groups[127].

$$\underset{\textbf{(175)}}{\mathrm{(C_6H_5O)_2P(=O)N_3}} \xrightarrow{\mathrm{RCO_2H}} \mathrm{RC(=O)N_3}$$

(176)

A reagent that appears to have wide applicability, when the acyl azide is to be converted to the isocyanate, is trimethylsilyl azide (TMSA). This reagent reacts not only with acid chlorides and anhydrides[128,129], but also with especially reactive esters and lactones[130]. The reaction is carried out without a solvent or in refluxing carbon tetrachloride or toluene, and the acyl azide is immediately converted to the isocyanate.

$$\mathrm{RC(=O)Cl + (CH_3)_3SiN_3 \longrightarrow RC(=O)N_3 + (CH_3)_3SiCl}$$

If an acid chloride is unreactive under these conditions, addition of trace amounts of powdered potassium azide and 18-crown-6, along with the trimethylsilyl azide, effects the reaction in good yield[131].

TMSA, KN_3, 18-crown-6; $-N{=}C{=}O$; H_2O, HCl; $-NH_2 \cdot HCl$; 88%

TMSA, KN_3, 18-crown-6; N_3; OC_2H_5; $N{=}C{=}O$; C_2H_5OH; 61%

The reaction of TMSA with anhydrides produces, in addition to the isocyanate, the trimethylsilyl ester. This may be hydrolysed readily to recover the acid. When a cyclic anhydride is used, the product, **177**, contains two functional groups that can react with

$$C_3H_7COCC_3H_7 \xrightarrow{TMSA} C_3H_7COSi(CH_3)_3 + C_3H_7N{=}C{=}O$$

$$\xrightarrow{TMSA} (CH_3)_3SiOCCH_2CH_2N{=}C{=}O$$

(177)

each other. Although **177** is found entirely in the open chain form, **178** appears to give an equilibrium mixture containing 20% of the cyclized product, **179**. Hydrolysis of the mixture gives isatoic anhydride, **180**.

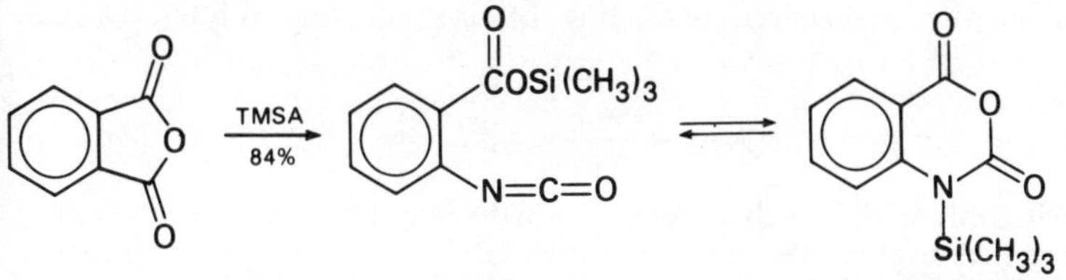

(178) 80%. (179) 20%

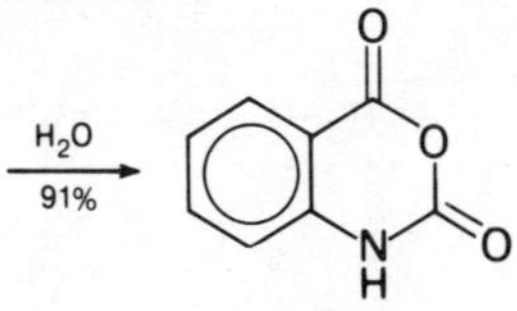

(180)

The reaction of maleic anhydride with TMSA to give **181** is notable in a number of respects. The reaction occurs so vigorously that it must be modified by dilution with benzene[132]. Reaction occurs at the carbonyl, rather than at the double bond, and the product, **181**, exists entirely in the cyclic form. Substituted maleic anhydrides could

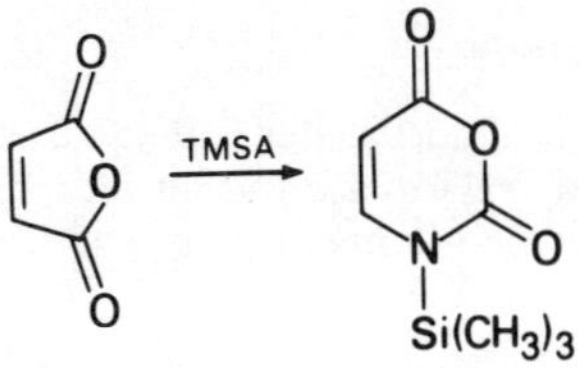

(181)

give two possible products, **182** and **183**. When R is methyl or a halogen, the product is **182**. When R is a phenyl group, **182** still predominates, but 10–45% of **183** is also formed. The product ratios do not appear to be very predictable. The two products, **182** and **183** are separable, giving access to both systems, which are of potential biological interest[133].

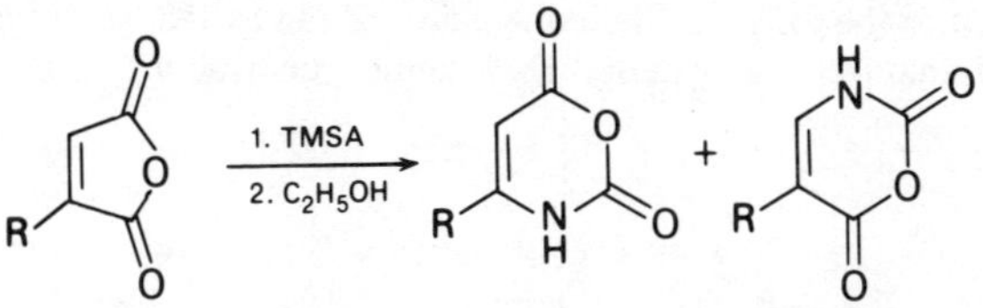

(182) (183)

R	Yield, %	**182**, %	**183**, %
C_6H_5—	36	90	10
p-FC_6H_4—	66	80	20
p-ClC_6H_4—	26	70	30
o-ClC_6H_4—	30	55	45
p-$CH_3OC_6H_4$—	58	70	30

Acyl azides can also be prepared by treating the acyl hydrazide with nitrous acid.

$$\mathrm{RC(=O)NHNH_2} \xrightarrow[\mathrm{HCl}]{\mathrm{NaNo_2}} \mathrm{RC(=O)N_3}$$

An attempt to use this method with a γ-keto acyl hydrazide, **184**, led, not to the azide, but to a pyridazine derivative, **185**, formed by internal condensation. The preparation of the desired azide from the corresponding acid chloride was successful[134].

$$\mathrm{C_6H_5{-}C(=O){-}CH_2{-}C(=CH{-}C_6H_4{-}R){-}C(=O){-}NHNH_2} \xrightarrow[\mathrm{HCl}]{\mathrm{NaNO_2}} \text{(185)}$$

(184) (185)

VII. REACTIONS OF ACYL AZIDES

A. Thermolysis of Acyl Azides

The most common reaction of acyl azides is the Curtius rearrangement to isocyanates. This rearrangement may take place during the preparation of the azide if the reaction is carried out at elevated temperatures. The isocyanate can then add protic reagents.

$$\mathrm{RC(=O)N_3} \xrightarrow{\Delta} \mathrm{R{-}N{=}C{=}O} \xrightarrow{\mathrm{H_2O}} \mathrm{RNH_2}$$

$$\mathrm{R{-}N{=}C{=}O} \xrightarrow{\mathrm{R'OH}} \mathrm{RNH{-}C(=O){-}OR'}$$

Intramolecular cycloaddition reactions of the isocyanate often occur when an aryl group is adjacent to the azide. In the conversion of **186** to **188**, such cyclization occurs rapidly enough that the isocyanate, **187**, cannot be isolated when R^1 is phenyl,

$$\text{(186)} \xrightarrow{\Delta} \text{(187)} \xrightarrow{\Delta\Delta} \text{(188)}$$

although it is isolable when R^1 = H and R^2 = methyl[135]. When **189** is heated in dry dimethylacetamide at 60–70°C, both azide groups rearrange, but cyclization occurs with only one, giving **190**. Heating in ethanol produces **191** rather than **192**, indicating that cyclization is rapid relative to alcoholysis of the isocyanate[136]. For **193**, both acyl azide groups rearrange before further reaction occurs[137].

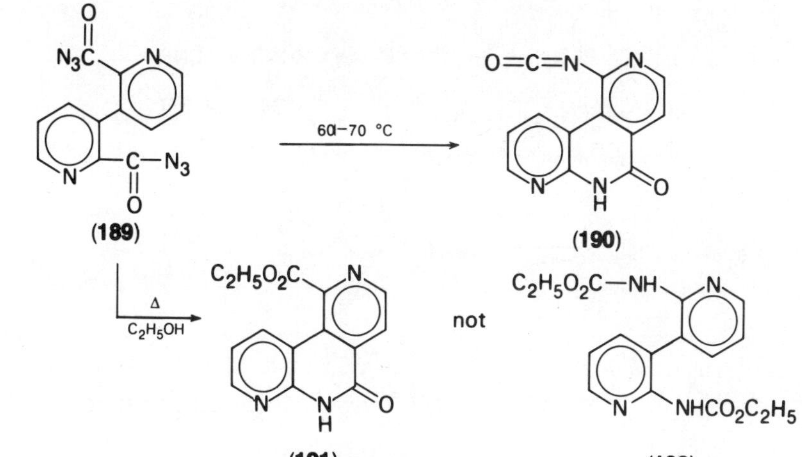

(189) (190) (191) (192) 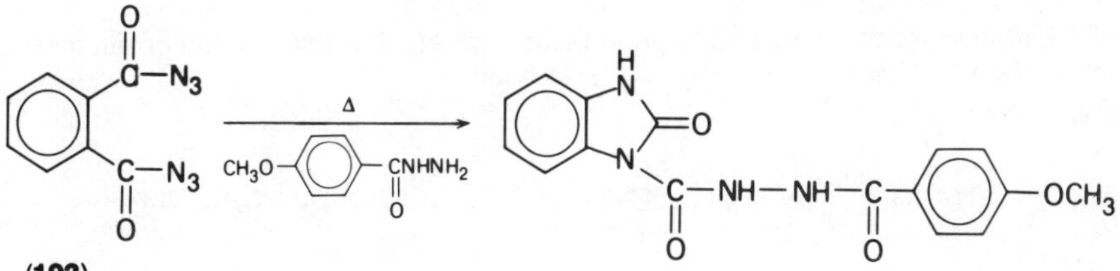 (193)

B. Photolysis of Acyl Azides

Photolysis of acyl azides gives both nitrene formation and rearrangement to the isocyanate. Photolysis of **194** gives, among other products, **195** that is 98% optically pure[138], a result consistent with formation of the singlet nitrene.

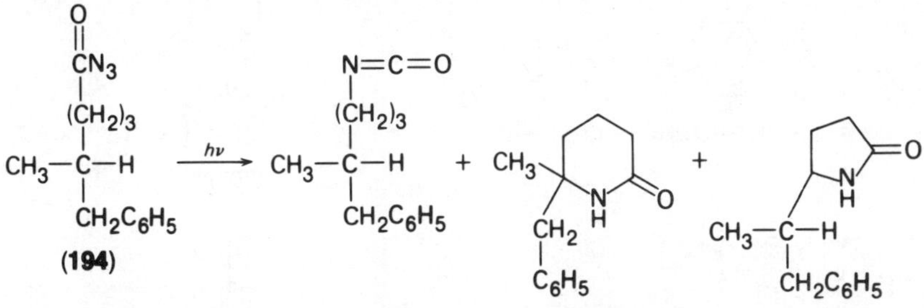

(194) (195)

The relative amounts of isocyanate and nitrene depend markedly on the nature of the solvent[139]. In halogen-free solvents (except acetone) the yield of isocyanate is 40–50%. In halogenated methanes, the yield of isocyanate is higher. In acetone, only 5% isocyanate was formed along with 55% of **198**. An analogous product, **199**, was formed in acetonitrile, along with 47% of the isocyanate. The usual nitrene products in hydrocarbon solvents were the amide resulting from insertion in a C—H bond, **196**,

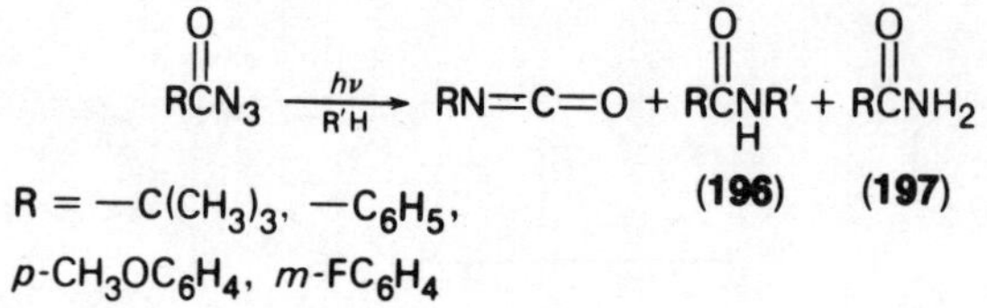

$$RC(=O)N_3 \xrightarrow[R'H]{h\nu} RN{=}C{=}O + RC(=O)NHR' \ (\mathbf{196}) + RC(=O)NH_2 \ (\mathbf{197})$$

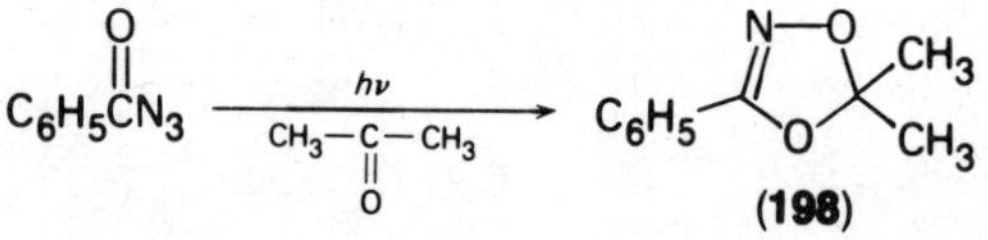

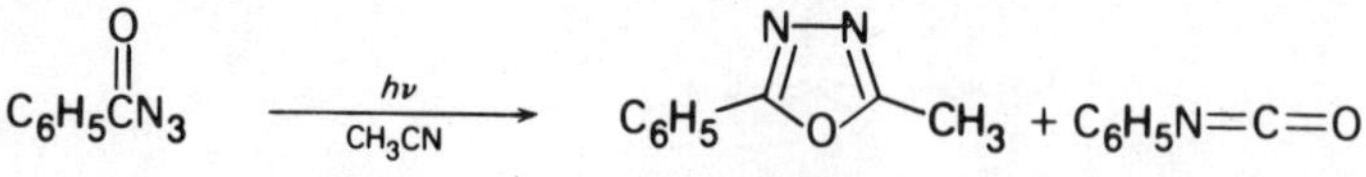

$R = -C(CH_3)_3$, $-C_6H_5$, p-$CH_3OC_6H_4$, m-FC_6H_4

$$C_6H_5C(=O)N_3 \xrightarrow[CH_3-C(=O)-CH_3]{h\nu} \mathbf{(198)}$$

$$C_6H_5C(=O)N_3 \xrightarrow[CH_3CN]{h\nu} \mathbf{(199)} + C_6H_5N{=}C{=}O$$

and 0–7% of **197**. In methanol, 27% of **197** was formed at 10–17°C, but only 1.5% at −60°C; the other product in addition to the isocyanate was **200**, resulting from insertion in the O—H bond rather than a C—H bond.

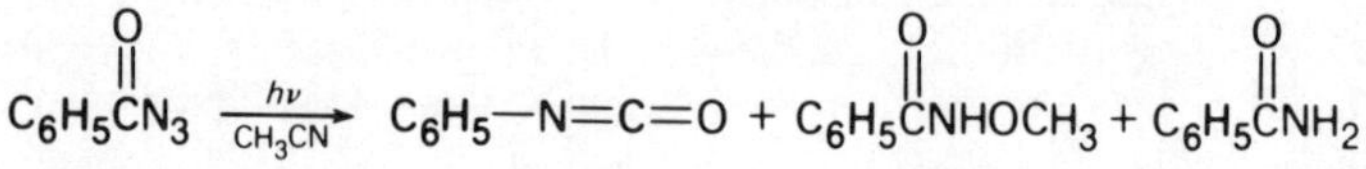

$$C_6H_5C(=O)N_3 \xrightarrow[CH_3CN]{h\nu} C_6H_5{-}N{=}C{=}O + C_6H_5C(=O)NHOCH_3 \ (\mathbf{200}) + C_6H_5C(=O)NH_2$$

Photolysis of acyl azides in the presence of diketene gives approximately 17% of **201**, in addition to the photo-Curtius rearrangement[140]. The mechanism is thought to involve addition of a nitrene to the carbon–carbon double bond to form **202**, followed by rearrangement to **201**.

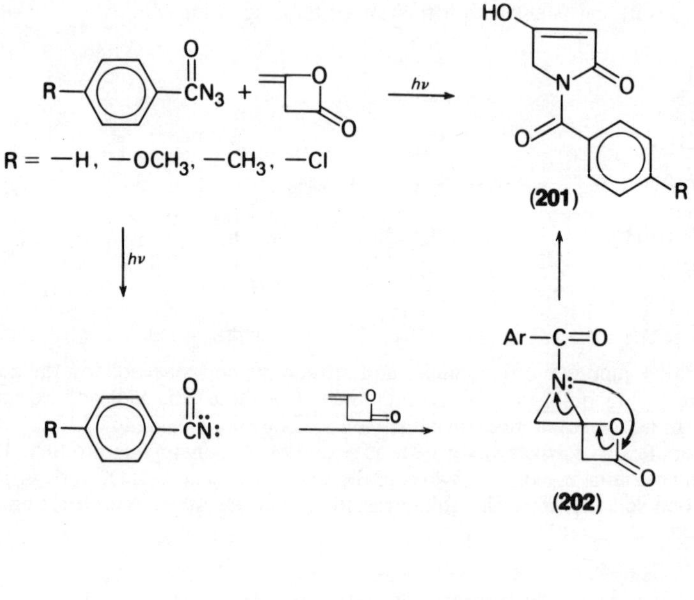

$R = -H$, $-OCH_3$, $-CH_3$, $-Cl$

(201)

(202)

C. Addition of Acyl Azides to Double Bonds

A number of examples have been reported of addition of acyl azides to double bonds[141,142,143] to form triazoles or, with loss of nitrogen, aziridines.

When an acyl azide is added to an α-keto or α-carbethoxy phosphorus ylide, **203**, the initially formed N-1 substituted triazoles **204** isomerized to the N-2 substituted triazole **205** under the basic reaction conditions[144]. Working up the reaction mixture prior to

$$R^1-\overset{O}{\overset{\|}{C}}-\overset{R^2}{\overset{|}{C}}=P(C_6H_5)_3 + R^3-\overset{O}{\overset{\|}{C}}N_3 \longrightarrow (204) \longrightarrow (205)$$

(203) **(204)** **(205)**

$R^1 = CH_3-,\ C_6H_5-$

$R^2 = H-,\ CH_3-$

$R^3 = CH_3-$, *m*- and *p*-$CH_3OC_6H_4-$

C_2H_5O-, *m*- and *p*-$NO_2C_6H_4-$

completion allowed isolation of **204**. No isomerization occurred when a solution of **204** was heated, with or without the addition of triphenyl phosphine oxide. Aryl migration did occur in the presence of the ylide **203** or other bases such as triethylamine. When a mixture of two different compounds (**204** with different R^3 groups) was used, the cross-product was formed, indicating that the reaction is not an intramolecular process. This isomerization was found to be general for 1-acyl triazoles. As a result of this facile rearrangement compounds identified previously as 1-acyl triazoles were in fact the 2-acyl isomers.

VIII. REFERENCES

1. C. Wentrup, *Topics Curr. Chem.*, **62**, 173 (1976); B. Iddon, O. Meth-Cohn, E. F. V. Scriven, H. Suschitzky and P. T. Gallagher, *Angew. Chem. Int. Edn Engl.*, **18**, 900 (1979).
2. H. W. Moore, *Chem. Soc. Rev.*, **2**, 415 (1974).
3. D. J. Anderson and A. Hassner, *Synthesis*, 483 (1975); A. Hassner, *Heterocycles*, **14**, 1517 (1980).
4. *The Chemistry of the Azide Group* (Ed. S. Patai), Wiley, New York (1971).
5. J. B. Hendrickson and P. Skipper, *Tetrahedron*, **32**, 1627 (1976).
6. M. S. Newman and W. C. Liang, *J. Org. Chem.*, **38**, 2438 (1973).
7. W. Kirmse and O. Schnurr, *J. Amer. Chem. Soc.*, **99**, 3539 (1977).
8. L. A. Burke, G. Leroy, M. T. Nguyen and M. Sana, *J. Amer. Chem. Soc.*, **100**, 3668 (1978).
9. H. W. Moore, *Acc. Chem. Res.*, **12**, 125 (1979).
10. W. Weyler, W. G. Duncan and H. W. Moore, *J. Amer. Chem. Soc.*, **97**, 6187 (1975).
11. R. Neidlein and E. Bernhard, *Angew. Chem. Int. Edn Engl.*, **17**, 369 (1978).
12. H. W. Moore, L. Hernandez and A. Sing, *J. Amer. Chem. Soc.*, **98**, 3728 (1976).
13. D. M. Kunert, R. Chambers, F. Mercer, L. Hernandez and H. W. Moore, *Tetrahedron Lett.*, 929 (1978).
14. H. W. Moore and W. G. Duncan, *J. Org. Chem.*, **38**, 156 (1973).
15. H. W. Moore and D. Scott Wilbur, *J. Org. Chem.*, **45**, 4483 (1980).
16. H. W. Moore, L. Hernandez, D. M. Kunert, F. Mercer and A. Sing, *J. Amer. Chem. Soc.*, **103**, 1769 (1981).
17. H. W. Moore, L. Hernandez and R. Chambers, *J. Amer. Chem. Soc.*, **100**, 2245 (1978).

18. H. W. Moore, F. Mercer, D. Kunert and P. Albaugh, *J. Amer. Chem. Soc.*, **101**, 5435 (1979).
19. G. Landen and H. W. Moore, *Tetrahedron Lett*., 2513 (1976).
20. S. Hoz and D. Speizman, *Tetrahedron Lett*., 4855 (1979).
21. G. Smolinsky and C. A. Pryde, *J. Org. Chem.*, **33**, 2411 (1968).
22. J. D. Hobson and J. R. Malpass, *J. Chem. Soc. C*, 1645 (1967).
23. T. Sasaki, K. Kenematsu and M. Murata, *Tetrahedron*, **29**, 529 (1973).
24. R. A. Abramovitch and B. W. Cue, *J. Org. Chem.*, **38**, 173 (1973).
25. R. A. Abramovitch and B. W. Cue, *Heterocycles*, **1**, 227 (1973).
26. E. M. Smith, E. L. Shapiro, G. Teutsch, L. Weber, H. L. Herzog, A. T. McPhail, P. W. Tschang and J. Meinwald, *Tetrahedron Lett*., 3519 (1974).
27. R. D. DeSelms, *Tetrahedron Lett*., 1179 (1969).
28. W. Weyler, D. Pearce and H. W. Moore, *J. Amer. Chem. Soc.*, **95**, 2603 (1973).
29. D. Rutolo, D. Stevens, H. R. Shelden, G. Landen, R. Chambers and H. W. Moore, unpublished results.
30. D. Knittel, H. Hemetsberger, R. Leipert, and H. Weidmann, *Tetrahedron Lett.*, 1459 (1970).
31. T. L. Gilchrist, C. W. Rees and J. A. R. Rodrigues, *JCS Chem. Commun*., 627 (1979).
32. G. Kumar, K. Rajagopalan, S. Swaminathan and K. K. Balasubramanian, *Tetrahedron Lett*., 4685 (1979).
33. E. P. Kyba and D. C. Alexander, *Tetrahedron Lett.*, 4563 (1976).
34. E. Zbiral, *Synthesis*, 285 (1972).
35. G. L'abbé, *Angew. Chem. Int. Edn Engl.*, **14**, 775 (1975).
36. A. Padwa, J. Smolanoff and A. Tremper, *J. Amer. Chem. Soc.*, **97**, 4682 (1975).
37. P. Germeraad and H. W. Moore, *J. Org. Chem.*, **39**, 774 (1974).
38. H. W. Moore, Y. L. Sing and R. S. Sidhu, *J. Org. Chem*., **45**, 5057 (1980).
39. H. Hemetsberger, D. Knittel and H. Weidmann, *Monats. Chem.*, **101**, 161 (1970); H. Hemetsberger, D. Knittel and H. Weidmann, *Monats. Chem.*, **100**, 1599 (1969).
40. K. Isomura, H. Taguchi, T. Tanaka and H. Taniguchi, *Chem. Lett.*, 401 (1977).
41. K. Yakushiin, S. Yoshina and A. Tanaka, *Heterocycles*, **6**, 721 (1977).
42. K. Isomura, Y. Hirose, H. Shuyama, S. Abe, G. Ayabe and H. Taniguchi, *Heterocycles*, **9**, 1207 (1978).
43. W. Shäfer, H. W. Moore and A. Aguado, *Synthesis*, 30 (1974).
44. R. Neidlein, G. Humberg, A. Gieren and C. Hahn, *Chem. Ber.*, **111**, 3346 (1978).
45. M. Henriet, M. Houtekie, B. Techy, R. Touillaux and L. Ghosez, *Tetrahedron Lett*., 223 (1980).
46. L. Garanti and G. Zechhi, *Tetrahedron Lett*., 559 (1980).
47. A. N. Thakore, J. Buchshriber and A. C. Oehlschlager, *Canad. J. Chem.*, **51**, 2406 (1973).
48. G. L'abbé and A. Hassner, *Bull. Soc. Chim. Belg.*, **80**. 209 (1971); G. L'abbé, J. E. Galle and A. Hassner, *Tetrahedron Lett*., 303 (1970).
49. P. Ykman, G. Mathys, G. L'abbé and G. Smets, *J. Org. Chem.*, **37**, 3213 (1972).
50. A. Hassner, D. J. Anderson and R. H. Reuss, *Tetrahedron Lett*., 2463 (1977).
51. A. Hassner, D. Tang and J. Keogh, *J. Org. Chem.*, **41**, 2102 (1976).
52. D. J. Anderson and A. Hassner, *J. Org. Chem.*, **38**, 2565 (1973).
53. H. Hassner, A. S. Miller and M. J. Haddadin, *J. Org. Chem.*,. **37**, 2682 (1972).
54. H. Bader and H. J. Hansen, *Chimia*, **29**, 264 (1975).
55. Y. Nomura, N. Hatanaka and Y. Takeuchi, *Chem. Lett.*, 901 (1976).
56. G. L'abbé and G. Mathys, *J. Heterocyclic Chem.*, **11**, 613 (1974).
57. G. L'abbé and G. Mathys, *J. Org. Chem.*, **39**, 1221 (1974).
58. A. Padwa, A. Ku, H. Ku and A. Mazzu, *Tetrahedron Lett.*, 551 (1977); A. Padwa, A. Ku, H. Ku and A. Mazzu, *J. Org. Chem.*, **43**, 66 (1978).
59. Y. Tamura, Y. Yoshimura and Y. Kita, *Chem. Pharm. Bull*., **20**, 871 (1972).
60. Y. Tamura, Y. Yoshimura, T. Nishimura, S. Kato and Y. Kita, *Tetrahedron Lett.*, 351 (1973).
61. S. Senda, K. Hirota, T. Asao, K. Muruhashi and N. Kitawara, *Tetrahedron Lett.*, 1531 (1978).
62. S. Senda, K. Hirota, M. Suzuki, T. Asao and K. Maruhashi, *JCS Chem. Comm.*, 731 (1976).

63. S. Senda, K. Hirota, T. Asao and K. Murahashi, *J. Amer. Chem. Soc.*, **99**, 7358 (1977).
64. G. L'abbé, *Angew. Chem. Int. Edn Engl.*, **14**, 775 (1975).
65. B. A. Belinka, Jr and A. Hassner, *J. Org. Chem.*, **44**, 4712 (1979).
66. A. Hassner, R. J. Isbister and A. Friederang, *Tetrahedron Lett.*, 2939 (1969).
67. J. B. Hendrickson, K. W. Bair and P. M. Keehn, *J. Org. Chem.*, **42**, 2935 (1977).
68. J. B. Hendrickson and P. L. Skipper, *Tetrahedron*, **32**, 1627 (1976).
69. A. Hassner and B. A. Belinka, Jr, *J. Amer. Chem. Soc.*, **102**, 6185 (1980).
70. R. E. Banks and G. R. Sparkes, *JCS Perkin I*, 2964 (1972).
71. I. R. A. Bernard, G. E. Chivers, R. J. W. Cremlyn and K. G. Mootoosany, *Aust. J. Chem.*, **27**, 171 (1974).
72. R. E. Banks and A. Prakash, *JCS Perkin I*, 1365 (1974).
73. J. M. Lindley, I. M. McRobbie, O. Meth-Cohn and H. Suschitzky, *Tetrahedron Lett.*, 4513 (1976); *JCS Perkin I*, 2194 (1977).
74. I. M. McRobbie, O. Meth-Cohn and H. Suschitzky, *Tetrahedron Lett.*, 929 (1976).
75. W. von E. Doering and R. A. Odum, *Tetrahedron*, **22**, 81 (1966).
76. R. J. Sundberg, M. Brenner, S. R. Suter and B. D. Das, *Tetrahedron Lett.*, 2715 (1970).
77. R. J. Sundberg, D. W. Gillespie and B. A. DeGraff, *J. Amer. Chem. Soc.*, **97**, 6193 (1975).
78. R. J. Sundberg and R. W. Heintzleman, *J. Org. Chem.*, **39**, 2546 (1974).
79. E. F. V. Scriven and D. R. Thomas, *Chem. Ind.*, 385 (1978).
80. R. J. Sundberg and R. H. Smith, Jr, *J. Org. Chem.*, **36**, 295 (1971).
81. A. C. Mair and M. F. G. Stevens, *J. Chem. Soc. C*, 2317 (1971).
82. M. A. Berwick, *J. Amer. Chem. Soc.*, **93**, 5780 (1971).
83. R. K. Smalley, W. A. Strachan and H. Suschitzky, *Synthesis*, 503 (1974).
84. O. L. Chapman and J.-P. LeRoux, *J. Amer. Chem. Soc.*, **100**, 282 (1978).
85. B. Iddon, H. Suschitzky and D. S. Taylor, *JCS Chem. Comm.*, 879 (1972); *JCS Perkin I.*, 579 (1974).
86. S. E. Hilton, E. F. V. Scriven and H. Suschitzky, *JCS Chem. Comm.*, 853 (1974).
87. J. Rigaudy, C. Igier and J. Barcelo, *Tetrahedron Lett.*, 1837 (1979).
88. I. R. Dunkin and P. C. P. Thomson *JCS Chem. Comm.*, 499 (1980).
89. S. E. Carroll, B. Nay, E. F. V. Scriven and H. Suschitzky, *Synthesis*, 710 (1975).
90. B. Nay, E. F. V. Scriven, H. Suschitzky and Z. U. Khan, *Synthesis*, 757 (1977).
91. E. F. V. Scriven, H. Suschitzky, D. R. Thomas and R. F. Newton, *JCS Perkin I*, 53 (1979).
92. S. E. Carroll, B. Nay, E. F. V. Scriven and H. Suschitzky, *Tetrahedron Lett.*, 943 (1977).
93. B. Iddon, M. W. Pickering and H. Suschitzky, *JCS Chem. Comm.*, 759 (1974).
94. F. Hollywood, E. F. V. Scriven, H. Suschitzky, D. R. Thomas and R. Hull, *JCS Chem. Comm.*, 806 (1978).
95. B. Nay, E. F. V. Scriven, H. Suschitzky, D. R. Thomas and S. E. Carroll, *Tetrahedron Lett.*, 1811 (1977).
96. J. Rigaudy, C. Igier and J. Barcelo, *Tetrahedron Lett.*, 3845 (1975).
97. S. E. Carroll, B. Nay, E. F. V. Scriven, H. Suschitzky and D. R. Thomas, *Tetrahedron Lett.*, 3175 (1977).
98. A. Tanaka, K. Yakushijin and S. Yoshina, *J. Heterocycl. Chem.*, **14**, 975 (1977).
99. R. A. Abramovitch and J. Kalinowski, *J. Heterocycl. Chem.*, **11**, 857 (1974).
100. I. M. McRobbie, O. Meth-Cohn and H. Suschitzky, *Tetrahedron Lett.*, 925 (1976).
101. J. H. Boyer and C.-C. Lai, *JCS Perkin I*, 74 (1977).
102. J. I. G. Cadogan, *Acc. Chem. Res.*, **5**, 303 (1972).
103. O. Meth-Cohn, *Heterocycles*, **14**, 1479 (1980).
104. J. I. G. Cadogan, J. N. Done, G. Lunn and P. K. K. Lim, *JCS Perkin I*, 1749 (1976).
105. G. R. Cliff and G. Jones, *J. Chem. Soc. C*, 3418 (1971).
106. J. M. Lindley, O. Meth-Cohn and H. Suschitzky, *JCS Perkin I*, 1198 (1978).
107. D. G. Hawkins, O. Meth-Cohn and H. Suschitzky, *JCS Perkin I*, 3207 (1979).
108. G. Boshev, L. K. Dyall and P. R. Sadler, *Aust. J. Chem.*, **25** 599 (1972).
109. L. K. Dyall, *Aust. J. Chem.*, **28**, 2147 (1975).
110. L. K. Dyall, *Aust. J. Chem.*, **30**, 2669 (1977).
111. J. H. Hall, F. E. Behr and R. L. Reed, *J. Amer. Chem. Soc.*, **94**, 4952 (1972).
112. J. H. Hall and F. W. Dolan, *J. Org. Chem.*, **43**, 4608 (1978).
113. S. Gronowitz, C. Westerlund and A.-B. Hörnfeldt, *Chem. Ser.*, **10**, 165 (1976).
114. M. C. Paulmier, *Compt. Rend. C*, **281**, 317 (1975); *Chem. Abstr.*, **84**, 43919h (1976).

115. S. Gronowitz, C. Westerlund and A.-B. Hörnfeldt, *Acta Chem. Scand. B*, **30**, 391 (1976).
116. S. Bradbury, C. W. Rees and R. C. Storr, *JCS Perkin I*, 72 (1972).
117. P. Spagnolo, A. Tundo and P. Zanirato, *J. Org. Chem.*, **42**, 292 (1977).
118. P. Spagnolo, A. Tundo and P. Zanirato, *J. Org. Chem.*, **43**, 2508 (1978).
119. O. Tsuge, K. Ueno and A. Inaba, *Heterocycles*, **4**, 1 (1976).
120. L. Garanti and G. Zecchi, *J. Org. Chem.*, **45**, 4767 (1980).
121. K. Sakai and J.-P. Anselme, *J. Org. Chem.*, **36**, 2387 (1971).
122. J. W. van Reijendam and F. Baardman, *Synthesis*, 413 (1973).
123. A. K. Banerjee, P. C. Caraballo, H. E. Hurtado and M. C. Carrasco, *Heterocycles*, **14**, 315 (1980).
124. A. Brändstrom, B. Lamm and I. Palmertz, *Acta Chem. Scand. B,* **28**, 699, (1974).
125. C. T. Shioiri and S. Yamada, *Chem. Pharm. Bull.*, **22**, 849 (1974).
126. T. Shiori, K. Ninomiya and S. Yamada, *J. Amer. Chem. Soc.*, **94**, 6203 (1973); H. Saikachi and T. Kitagawa, *Chem. Pharm. Bull.*, **25**, 1651 (1977).
127. T. Shiori and S. Yamada, *Chem. Pharm. Bull*, **22**, 855, 859 (1974).
128. S. S. Washburne and W. R. Peterson, Jr, *Synth. Commun.*, **2**, 227 (1972).
129. H. R. Kricheldorf, *Synthesis*, 551 (1972).
130. H. R. Kricheldorf, *Chem. Ber.*, **106**, 3765 (1973).
131. J. D. Warren and J. B. Press, *Synth. Commun.*, **10**, 107 (1980).
132. S. S. Washburne, W. R. Peterson, Jr and K. A. Berman, *J. Org. Chem.*, **37**, 1738 (1972).
133. J. D. Warren, J. H. MacMillan and S. S. Washburne, *J. Org. Chem.*, **40**, 743 (1975); J. H. MacMillan and S. S. Washburne, *J. Heterocycl. Chem.*, **12**, 1215 (1975).
134. W. I. Awad, A. I. Hashem and K. El-Badry, *Indian J. Chem.*, **13**, 1139 (1975); *Chem. Abstr.*, **84**, 43568t (1976).
135. K. Ito, K. Yakushijin, S. Yoshina, A. Tanaka and K. Yamamoto, *J. Heterocycl. Chem.*, **15**, 301 (1978).
136. R. L. Williams and M. G. El Fayoumy, *J. Heterocycl. Chem.*, **9**, 1021 (1972).
137. A. F. M. Fahmy and S. A. Esawy, *Indian J. Chem.*, **11**, 871 (1973); *Chem. Abstr.*, **80**, 36810 (1974).
138. S. Terashima, M. Nara and S. Yamada, *Chem. Pharm. Bull*, **18**, 1124 (1970).
139. E. Eibler and J. Sauer, *Tetrahedron Lett.*, 2569 (1974).
140. T. Kato, Y. Suzuki and M. Sato, *Chem. Pharm. Bull*, **27**, 1181 (1979).
141. T. Sasaki and Y. Suzuki, *Yuki Gosei Kagaku Kyokai Shi*, **29**, 880 (1971); *Chem. Abstr.*, **76**, 99184s (1972).
142. V. P. Semenov, S. V. Ozernaya, I. M. Stroiman and K. A. Ogloblin, *Khim. Geterotsikl. Soedin*, 1613 (1976); *Chem. Abstr.*, **86**, 171301d (1977).
143. V. P. Semenov, A. P. Prosypkina, O. F. Gavrilova and K. A. Ogloblin, *Khim. Geterotsikl. Soedin*, 464 (1977); *Chem. Abstr.*, **87**, 68211w (1977).
144. P. Ykman, G. L'abbé and G. Smets, *Tetrahedron Lett.*, 5225 (1970).

The Chemistry of Functional Groups, Supplement D
Edited by S. Patai and Z. Rappoport
© 1983 John Wiley & Sons Ltd

CHAPTER **9**

Recent advances in the radiation chemistry of halocarbons

ABRAHAM HOROWITZ
Soreq Nuclear Research Centre, Yavne, Israel

I. INTRODUCTION

Extensive studies of the radiation chemistry of halocarbons have been carried out since the previous review of this topic appeared in this series[1]. In that review, by Bühler, emphasis was placed on the fundamental radiolytic processes such as electron attachment and electron scavenging, reactions of excited molecules and reactions of other short-lived intermediates. Many of the recent studies contribute to the understanding of basic radiolytic behaviour by providing more detailed information that supports existing models and concepts. The basic picture of the radiation chemistry of halocarbons, as it appeared in the earlier review, was not altered by these studies. This situation is not limited to halocarbons, but reflects developments and the current status of radiation chemistry in general.

The present review of the radiation chemistry of halocarbons covers the literature to the end of 1980 and deals with the main topics covered by Bühler[1]. A relatively large amount of space has been devoted to radiation-induced radical reactions, a subject that has not been extensively discussed previously. In the current survey an effort was made to preserve the continuity of the review without presenting material already covered by Bühler. On the whole this goal was achieved by organizing the material along the lines established in the previous work. In some cases, however, a certain degree of overlap between the two texts could not be avoided.

II. OUTLINE OF BASIC FEATURES

As a basis for a survey of recent work it is necessary to first recall the known and well established general features of the radiation chemistry of halocarbons, which are as follows:

(1) In the initial steps of interaction of ionizing radiation with halocarbons (RX), molecular cations and excited halocarbon molecules are formed.

$$RX \longrightarrow RX^{+\cdot} + e^- \quad (1)$$

$$RX \longrightarrow RX^{**} \quad (2)$$

(2) These species carry excess energy either by virtue of being in the highly excited electronic states of both RX and RX^+ or as kinetic energy in the case of the electron. Under most conditions (with the exception of a few cases such as at very low RX pressures) most of this energy is dissipated in the medium. Consequently, almost all the chemical change induced by radiation can be described in terms of reactions in which RX and RX^+ are in one or a few low-lying excited states and the electrons are thermalized.

(3) Initially formed excited molecules as well as additional excited molecules generated in the neutralization reaction between RX^+ and the thermal electrons

$$RX^+ + e^- \longrightarrow RX^* \quad (3)$$

either decay by radiative and non-radiative processes or decompose into stable products (P^1, P^2) or radicals ($R^{1\cdot}$ and $R^\cdot$):

$$RX^* \longrightarrow P^1 + P^2 \quad (4)$$

$$RX^* \longrightarrow R^{1\cdot} + P^2X^\cdot \quad (5)$$

$$RX^* \longrightarrow R^\cdot + X^\cdot \quad (6)$$

In halocarbons, with the exception of fluoro compounds, the C—X bond is weaker than the C—C or C—H bond. Therefore pathways involving the rupture of the C—X bond represent the most favourable route of decomposition of the excited RX molecules. On the other hand, in fluorocarbons, where the C—F bond is stronger than the C—C bond, C—C scission is favoured.

(4) Because of the high electron affinity of the halogen atom, RX compounds

readily undergo dissociative and non-dissociative electron capture reactions in which anions and radicals are formed:

$$RX + e^- \longrightarrow (RX^-) \xrightarrow{\text{Fast}} R^{\bullet} + X^- \quad (7)$$

$$RX + e^- \longrightarrow RX^- \quad (8)$$

Again the behaviour of fluorocarbons is different from that of the other halocarbons. While in the latter compounds dissociative electron capture is usually observed, the former compounds in their reactions with thermal electrons often generate long-lived molecular anions.

(5) Depending on experimental conditions such as phase and temperature, most of the product formation in irradiated halocarbons can be accounted for by unimolecular ionic and radical reactions and by bimolecular reactions between the ions and radicals and with RX. These reactions, and therefore the nature of the final products formed, strongly depend not only on the properties of the halogen substituent but also on the structure and composition of R.

Studies aimed at the elucidation of the radiation chemistry of different systems can be divided into two complementary categories: studies concerned mainly with the detection and reactions of intermediates and studies in which the primary object is to determine the overall chemical change. Both of these aspects will be dealt with in the present review.

III. FORMATION AND REACTIONS OF CHARGED INTERMEDIATES

A. The Reactions of Electrons with Haloalkanes

1. ESR studies

Most ESR studies of intermediates formed as a result of electron attachment to halocarbons have been carried out in rigid glassy or solid matrices at temperatures around 77 K. These conditions allow detection and identification of radical anion species and radicals which have very short lifetimes in the liquid and gas phases at considerably higher temperatures. In some cases phase and temperature also determine the nature of species formed. Therefore, ESR findings under a given set of conditions cannot always be extrapolated straightforwardly to other systems and different conditions.

Mishra and Symons[2,3] have carried out extensive ESR studies of electron capture processes. They have shown[4] that decomposition pathways following electron capture in compounds having the general formula $RCCl_3$ depend on the properties of R as well as on the environment in which the process takes place. For R = H, CH_3, Cl and CCl_3CO the main species formed at 77 K in pure compounds and in methanol glassy solutions were the radicals $^{\bullet}CHCl_2$, $^{\bullet}CCl_2CH_3$, $^{\bullet}CCl_3$ and $^{\bullet}CCl_2COCCl_3$, respectively. These compounds thus follow the generally accepted and well established[5] pattern of dissociative electron capture in simple chloro-, bromo- and iodoalkanes.

On the other hand, in pure trichloronitromethane ($R = NO_2$)[6] the dissociative electron capture takes the route

$$CCl_3NO_2 + e^- \longrightarrow CCl_3^- + {}^{\bullet}NO_2 \quad (9)$$

while in CD_3OD the main process appears to be

$$CCl_3NO_2 + e^- \longrightarrow {}^{\bullet}CCl_3 + NO_2^- \quad (10)$$

with a minor contribution of the pathway

$$CCl_3NO_2 + e^- \longrightarrow Cl_2^{-\bullet} + ClCNO_2 \tag{11}$$

This difference between the dissociation in polar media and inert media was ascribed to preferential solvation of small anionic species in CD_3OD which promotes the dissociation into NO_2^-. Solvent effect was observed also for the dissociative electron capture by $CCl_3P(O)(OH)_2$. In this case the two modes of dissociation are as follows:

$$CCl_3P(O)(OH)_2 + e^- \longrightarrow {}^{\bullet}CCl_3 + H_2PO_3^- \tag{12}$$

$$CCl_3P(O)(OH)_2 + e^- \longrightarrow Cl_2\dot{C}P(O)(OH)_2 + Cl^- \tag{13}$$

The first process (equation 12) was observed in the neat compound while in CD_3OD, $Cl_2\dot{C}P(O)(OH)_2$ was the predominant radical species.

Preferential ejection of small anionic species in polar medium is revealed in the electron capture reactions of chlorobromomethanes[7]. In methanol $BrCH_2Cl$, $BrCHCl_2$ and $BrCCl_3$ eject Cl^- and not Br^-, as in the pure compounds. Solvation in this case is assumed to take place by hydrogen bonding at the more electronegative Cl atom:

$$BrCH_2Cl + e^- \xrightarrow{MeOH} BrH_2C{-}Cl^{-}\cdot\cdot HOMe \longrightarrow BrH_2\dot{C} + Cl^- + HOMe \tag{14}$$

A different aspect of the effect of the medium on the fate of the species formed by electron capture is revealed in solid solutions of acetonitrile. In this solvent alkyl radical–halide ion weak interaction results in the formation of an adduct having the structure $R^{\bullet}\cdots X^-$. This type of behaviour was reported by Sprague and Williams[8] for MeBr and by Mishra and Symons[9] for MeBr, MeI and EtI. For the decay of the $CH_3^{\bullet}\cdots Br^-$ pair a rate constant of 0.012 min^{-1} at 77 K and an activation energy of 1.56 kcal mol^{-1} was recently estimated by Sprague[10].

Ejection of dihalide radical anion $X_2^{-\bullet}$ in gamma-irradiated 1,2-dihalides such as $ClCH_2CH_2Cl$, $BrCH_2CH_2Br$ and ICH_2CH_2I was rationalized by Mishra and Symons[11] in terms of the interesting single-step electron capture process:

$$XCH_2CH_2X + e^- \longrightarrow CH_2{=}CH_2 + X_2^{-\bullet} \tag{15}$$

However, these authors did not exclude the possibility that the dihalide radical anion is formed in an alternative two-stage process:

$$XCH_2CH_2X + e^- \longrightarrow XCH_2CH_2^{\bullet} + X^- \tag{16}$$

$$XCH_2CH_2 + X^- \longrightarrow CH_2{=}CH_2 + X_2^{-\bullet} \tag{17}$$

One would expect that in a similar fashion $ClCH_2CH_2CN$ would give the $ClCN^{-\bullet}$ anion. However, this compound, as well as $ClCH_2CN$ and Br_2CHCN, undergoes a simple one-stage reaction in which a halide ion is eliminated and a cyano radical is formed[12].

Non-dissociative electron capture, as evidence by the formation of parent radical anions, has been reported for CBr_4[13]. In general, stable radical anions are generated only from fluorocarbons or from substituted halo compounds in which the strongly electronegative substituent can accommodate the additional electron. Williams and colleagues[14,15] have clearly identified the radical anions c-$C_3F_6^{-\bullet}$, c-$C_4F_8^{-\bullet}$ and c $C_5F_{10}^{-\bullet}$ in ESR studies of gamma-irradiated solid solutions of parent perfluorocycloalkanes in neopentane and tetramethylsilane at 77 K. Under these conditions compounds such as CF_3Cl, CF_3Br, CF_3I and CF_2Cl_2 also yield parent radical anions. These radical anions are not as stable as those derived from perfluorocycloalkanes and dissociate readily at temperatures around 100 K. As evidenced by the ESR spectra, fluoride anion is not released in this dissociation reaction. Only $^{\bullet}CF_3$ radical is formed in irradiated CF_3Cl, CF_3Br and CF_3I.

ESR spectra assigned to radical anionic species[2] have been observed in irradiated α-iodoacetamide[16,17], bromomalonamide[18] and monochloroacetic acid[19]. For the iodo compound it has been suggested[2,19] that, following electron capture, the initially formed anion radical **1** undergoes rearrangement and the ESR spectrum is of the rearranged species **2**:

$$ICH_2CONH_2 + e^- \longrightarrow ICH_2C(NH_2)O^{-\bullet} \qquad (18)$$

$$\underset{(\mathbf{1})}{ICH_2C(NH_2)O^{-\bullet}} \longrightarrow \underset{(\mathbf{2})}{H_2\dot{C}{-}C(I)(NH_2){-}O^-} \longrightarrow H_2\dot{C}CONH_2 + I^- \qquad (19)$$

In monochloroacetic acid the anionic species is presumably the β-chloroacetic acid anion radical $ClCH_2C(OH)O^{-\bullet}$ [2,19], which undergoes dissociation followed by hydrogen abstraction:

$$ClCH_2C(OH)O^{-\bullet} \longrightarrow Cl^- + {}^{\bullet}CH_2CO_2H \qquad (20)$$

$${}^{\bullet}CH_2CO_2H + ClCH_2CO_2H \longrightarrow CH_3CO_2H + Cl\dot{C}HCO_2H \qquad (21)$$

2. *Electron scavenging in aqueous solutions*

Absolute rate constants for the reaction of hydrated electrons (e_{aq}^-) determined mainly by the combination of pulse radiolysis with fast measurement of conductivity build-up, are summarized in Table 1. In addition to providing kinetic information, studies of this type in which $G(X^-)$ is determined also permit examination of the fate of the species formed in the electron capture process. (G here and elsewhere stands for yield in molecules per 100 eV of absorbed energy.)

Based on the observation that $G(X^-)$ in pulse radiolytic studies, particularly in acidic solution, was considerably lower than in continuous irradiation studies and lower than $G(e_{aq}^-)$, Koester and Asmus[23] suggested a non-dissociative route for electron capture by fluorobenzene. In subsequent studies by Lichtscheidl and Getoff[22,24] it was shown, however, that the low $G(X^-)$ values were due to incomplete scavenging of e_{aq}^- because of the competition with the fast ($k = 10^{10}\ \mathrm{M}^{-1}\,\mathrm{s}^{-1}$) neutralization reaction[25]:

$$e_{aq}^- + H^+ \longrightarrow H^{\bullet} \qquad (22)$$

It was also shown that the part of e_{aq}^- scavenged by RX was quantitatively converted into X^-. These authors thus concluded that for all the aromatic compounds listed in Table 1 the expected $G(X^-)$ values, at low dose rate and in basic solution, should be equal to $G(e_{aq}^-)$, i.e. to 2.7, provided of course that the RX was present at the necessary high concentration.

A rather interesting phenomenon was observed in photolytic and radiolytic steady state studies in which the rate constants for the reaction of solvated electrons with organic halides were determined competitively versus electron scavenging by N_2O:

$$e_{aq}^- + N_2O \longrightarrow N_2 + O^- \qquad (23)$$

TABLE 1. Rate constants (in $M^{-1} s^{-1}$) for the reaction between hydrated electrons and halocarbons

Chloroethylenes[20,21]	k	Halobenzenes	k	Other haloaromatics	k
$CH_2{=}CHCl$	2.5×10^9	$PhCl$[22]	5×10^8	$PhCH_2Cl$[24]	4.5×10^9
$CHCl{=}CHCl$	7.5×10^9	$PhBr$[22]	1×10^{10}	$PhCH_2CH_2Cl$[24]	5.3×10^8
$CH_2{=}CCl_2$	2.3×10^{10}	PhF[22]	6×10^7	o-$ClC_6H_4NH_2$[22]	5.4×10^8
$CHCl{=}CCl_2$	1.9×10^{10}	PhF[21,23]	7×10^7	m-$ClC_6H_4NH_2$[22]	5.3×10^8
$CCl_2{=}CCl_2$	1.3×10^{10}	o-$C_6H_4F_2$[23]	1.2×10^9	p-$ClC_6H_4NH_2$[22]	5.2×10^8
		p-$C_6H_4F_2$[23]	2×10^9		
		1,2,3,4-$C_6H_2F_4$[23]	2.6×10^{10}		
		C_6HF_5[21,23]	1.6×10^{10}		
		C_6F_6[21,23]	2×10^{10}		

For $CHCl_3$, CCl_4, PrBr and other alkyl halides known to undergo dissociative electron capture the rate constant ratio $k(e_{aq}^- + RX)/k(e_{aq}^- + N_2O)$ determined by product analysis, was found to be considerably lower than the value estimated from the directly determined individual rate constants. This anomaly was rationalized by Logan and Wilmot[26] by the assumption that, in those cases in which it was observed, electron capture was reversible. As these authors noticed, the main argument against this assumption is that it requires a rather long-lived $RX^{-\cdot}$. It is then not clear why during its long lifetime this species escapes reaction with other solutes. It should be stressed that, apart from the complications reported by Logan and Wilmot, their results clearly indicate that direct pulse radiolytic determination of $k(e_{aq}^- + RX)$ is a more reliable method than the competitive determination of this rate constant based on product analysis in steady state studies.

Nitro-[27,28], cyano-[29] and aceto-[30] substituted aromatic halogen-containing compounds yield relatively stable and optically detectable anion radicals in their reactions with e_{aq}^-:

$$XC_6H_4Y + e_{aq}^- \longrightarrow XC_6H_4\dot{Y}^- \tag{24}$$

$$XAC_6H_4Y + e_{aq}^- \longrightarrow XAC_6H_4\dot{Y}^- \tag{25}$$

where Y is the substituent at which the anion radical centre is initially formed and X denotes a halogen atom bound to the benzene ring directly or via a bridging group A. Using mainly kinetic spectrophotometric pulse radiolytic techniques Behar and Neta[27–30] studied the formation and subsequent reactions of these anion radicals. In some cases the reduction of the parent compounds was carried out by $(CH_3)_2\dot{C}OH$ radicals and $(CH_3)_2\dot{C}O^-$ anion radicals formed upon radiolysis of neutral and alkaline solutions of isopropyl alcohol, respectively. The decay of the halosubstituted anion radicals was monitored and found to obey either second order or first order kinetics. In the latter case, intramolecular electron transfer followed by unimolecular cleavage of the C—X bond took place,

$$XC_6H_4Y^- \text{ or } XAC_6H_4Y^- \longrightarrow X^- + {}^{\cdot}C_6H_4Y \text{ or } {}^{\cdot}AC_6H_4Y \tag{26}$$

and was verified by direct determination of the halide anions formed upon gamma-radiolysis of the same solutions.

The dehalogenation rate constants determined by Behar and Neta are summarized in Table 2. Inspection of these rate constants reveals some characteristic reactivity patterns. Thus in the series of the nitro-centred anion radicals[27,28] the strongly bound halogen atoms attached directly to the benzene ring do not undergo cleavage. On the other hand, when the C—X bond is relatively weak, as it is in the nitrobenzyl halides, dehalogenation occurs readily. Dehalogenation is also inhibited by insertion of various groups between the halogen atom and the benzene ring. Presumably, as has been suggested by Neta and Behar, this behaviour reflects the interference of the inserted group with the intramolecular electron transfer from the initially formed radical anion to the halogen atom. The limited data did not permit the authors to arrive at a quantitative structure–reactivity correlation. They were, however, able to show a qualitative correlation between reactivity in dehalogenation and spin density distribution in the anion radical as estimated from ESR parameters[30].

It is worth mentioning that in contrast to the previously described behaviour halogenated aliphatic peroxyl anion radicals ${}^{\cdot}O_2CCl_2O^-$, ${}^{\cdot}O_2CHClO_2^-$, ${}^{\cdot}O_2CF_2CO_2^-$ and ${}^{\cdot}O_2CH_2CO_2^-$ do not undergo intramolecular dehalogenation[31]. Instead, as has been shown in pulse radiolytic studies, these anion radicals can transfer the electron to

TABLE 2. Rate constants (in s^{-1}) of dehalogenation of anion radicals $XAC_6H_4Y^{\bullet -} \rightarrow X^- + {}^{\bullet}AC_6H_4Y$

XA	Y: $\dot{N}O_2$	Y: CN^- [29]	Y: $CH_3\dot{C}O^-$ [30]
o-Cl	—	9×10^6	1.5×10^3
o-Br	—	—	5×10^5
o-I	<1 [28]	—	—
p-Cl	<1 [28]	5×10^6	~10^2
p-Br	—	>3×10^7	5×10^3
p-I	<1 [28]	—	1.4×10^5
m-Cl	<1 [28]	4×10^4	—
m-Br	—	8×10^6	~10^2
o-$ClCH_2$	1×10^4 [27]	—	—
o-$BrCH_2$	4×10^5 [27]	—	—
p-$ClCH_2$	4×10^3 [27]	>3×10^7	—
p-$BrCH_2$	1.7×10^5 [27]	>6×10^7	—
p-ICH_2	5.7×10^5 [27]	—	—
m-$ClCH_2$	<5 [27]	—	1.5×10^4
m-$BrCH_2$	6×10 [27]	1.3×10^7	—
p-ICH_2	3×10^3 [27]	—	—
p-$BrCH_2CH_2$	0.2 [28]	—	—
p-$BrCH_2CO$	4.1×10^4 [28]	—	—
m-$BrCH_2CO$	1.5×10^2 [28]	—	—
p-$ClCH_2CO_2$	<1 [28]	—	—
p-ICH_2CO_2	30	—	—
o-$ClCH_2CONH$	<3	—	—
p-$Cl(CH_2CH_2O)_2$	<1	—	—

compounds such as ascorbic acid. The ascorbic acid anion radical is formed with a G value of 2.7, indicating that the electron transfer reaction of the peroxyl anion radical is the only pathway by which these species react.

3. Electron scavenging in non-polar liquids

In non-polar liquids only a small fraction of the electrons formed succeed in escaping the coulombic field of the parent cation. These electrons are known as free electrons and their yield is denoted as G_{fi}. Unless an electron scavenger is added to the system the major part of the initially formed electrons recombine with the parent cations. These 'paired' electrons are non-homogeneously distributed throughout the solution and their yield is denoted as G_{gi}.

Practically all rate constants of electron scavenging in non-polar liquids, mainly alkanes, were determined from product analysis in continuous irradiation studies. Quite obviously the rate constants as well as their exact meaning depend strongly on the model used for the electron scavenging process. The most successful and widely applicable model of electron scavenging in non-polar liquids was proposed by Warman, Asmus and Schuler[32]. Most of the recent studies[33–44] of the electron capture reactions of halocarbons are based on this model and therefore its main aspects will be briefly presented here. A more detailed discussion can be found in the original work[32] as well as in the earlier review in this series.

The model assumes that in a pure non-polar liquid such as an alkane (RH) the main processes are as follows:

$$RH \longrightarrow \text{Molecular products} \quad (27)$$

$$RH \longrightarrow [RH^+ + e^-] \longrightarrow \text{Products } (P_{RH}) \quad (28)$$

$$RH \longrightarrow RH^+ + e^- \longrightarrow \text{Products } (P_{RH}) \quad (29)$$

Electron capture by an electron scavenger (S),

$$e^- + S \longrightarrow S^- \longrightarrow \text{Products } (P_S) \quad (30)$$

changes the original product distribution. The yields of products formed as a result of the neutralization reactions of 'paired' and free electrons (reactions 28 and 29, respectively) is reduced. This reduction in the yield of products derived from the solvent (P_{RH}) is accompanied by concurrent formation of products from the scavenger (P_S). The semi-empirical correlation of Warman and coworkers, based on the approach of Hummel[45], predicts that the relationship between P_S and the electron scavenging reactivity of the solute should be given by the following expression:

$$G(P_S) = G_{fi} + \frac{(G_{gi}\alpha_S[S])^{1/2}}{1 + (\alpha_S[S])^{1/2}}, \quad (30a)$$

where α_S is (by definition) the reactivity of the scavenger (S) towards electrons; α_S is proportional to the rate constant $k(e^- + S)$.

The applicability of this equation was tested in a number of solvents containing halocarbons as well as other electron scavengers. It was found to have very good predictive power, particularly at low scavenger concentrations when equation (30a) reduces to the more simple form

$$G(P_S) = G_{fi} + G_{gi}[\alpha_S[S]]^{1/2}. \quad (30b)$$

The effect of external electric field on electron scavenging in the n-hexane–MeBr system was also found[46] to obey equations (30a) and (30b). Furthermore, it has been shown[32,35] that the decrease in P_{RH}, particularly when it is H_2, is proportional to the increase in P_S, i.e. $\Delta[P_{RH}] = -f\Delta[P_S]$. When this condition is observed and f is known, α_S can be estimated from the variation of $[P_{RH}]$ with $[S]$. As has been stated before, the best agreement between the experimentally determined $G(P_S)$ values and those predicted on the basis of the correlation given by equations (30a) and (30b) is observed in systems containing a single electron scavenger at low concentration. At high solute concentrations and in systems in which two scavengers compete for the electrons, departure from the predicted behaviour is observed. Typical examples include systems in which HCl formation from solutions of PhCl and $PhCH_2Cl$ in cyclohexane[38] and solutions of $CHCl_3$ and CCl_4 in n-hexane[40,41] were determined. In these systems G(HCl) was found to exceed the values expected from equation (30a). A similar observation was reported for G(MeH) in solutions of MeCl and MeBr in cyclohexane. In competitive studies of electron scavenging by EtBr and MeCl in cyclohexane, G(Me) was lower than calculated from the separately determined α_{MeCl} and α_{EtBr} values[33], and the same pattern was revealed in concentrated solutions of MeCl with other electron scavengers[36].

The causes of the complications occurring at high scavenger concentrations and in competitive systems can be traced back to the assumptions and approximations made in the derivation of equation (30a). Thus, as has been noted by Davids and coworkers[39], reactions of species other than electrons with the solute may lead to the same products as those formed as a result of electron capture. The probability that such reactions will occur increases at high solute concentrations. Furthermore, at these high concentrations the added solute may alter the initial separation distribution between the electrons and the molecular cations. According to Yakovlev and coworkers[47], who used a non-homogeneous model for the kinetics of electron–ion recombination, serious deviation from equation (30a) can be expected when the solute concentration is in excess of 1 M.

Complications in competitive studies can be ascribed to electron transfer reactions between the solutes. These reactions can be very fast, as has been shown by Bockrath and Dorfman[48], who found rate constants close to the diffusion-controlled limit for the electron transfer between BuI and biphenylide (9.4×10^9 M^{-1} s^{-1}) and naphthalenide (7.4×10^9 M^{-1} s^{-1}) radical anions. A lifetime of 30 ns has been estimated[33,36] of $MeCl^{-\bullet}$ in cyclohexane. This lifetime appears to be sufficiently long to allow electron transfer to another solute present at high concentration to compete with the unimolecular dissociation:

$$MeCl^{-\bullet} \longrightarrow Me^{\bullet} + Cl^{-} \quad (31)$$

The estimation of absolute α_S values from $G_{gi}\alpha^{1/2}$ values obtained from slopes of plots of $G(P_S)$ against $[S]^{1/2}$ requires knowledge of G_{gi} which is estimated from experiments at high [S]. G_{gi} cannot be determined as accurately as the product $G_{gi}\alpha_S^{1/2}$ and therefore, as has been rightly pointed out by Davids and coworkers[39], relative α_S values in the same solvent are more reliable than their absolute values. Inspection of

TABLE 3. Electron scavenging α_S values (in M^{-1}) and relative rate constants

Solvent	S	α_S	$k(e^- + S)/k(e^- + MeBr)$
Cyclohexane	MeCl	5[36], 3.42[39]	0.31[36], 0.38[39]
	MeBr	16[36], 9[39]	1.00[36], 1.00[39]
	MeI	22[36]	1.38[36]
	EtCl	0.24[36]	0.02[36]
	EtBr	10[36]	0.63[36]
	CCl_4	12[36], 8.1[39]	0.75[36], 0.90[39]
	Ph_2	15[36], 2.7[39], 6.3[39]	0.94[36], 0.3[39], 0.7[39]
	c-C_4F_8	14[36]	—
	c-C_6F_{12}	21[36]	—
	N_2O	8[36]	—
	SF_6	16[36]	0.9
	PhBr	10.6[43]	—
n-Hexane	MeBr	10.2[40]	1.00
	CCl_4	13.3[41]	1.30
	$CHCl_3$	7.3[42]	0.72
	N_2O	3.2[40]	0.31
	SF_6	11.2[40]	1.10
	PhBr	12.1[43]	—
Isooctane	MeBr	24.7[34], 23.6[40]	1.00[34], 100[40]
	EtBr	5[34]	0.20
	SF_6	65[34], 65[40]	2.63[34], 2.75[40]
	N_2O	4.0[40]	0.17
	PhBr	13.3[43]	—
Neopentane[50]	MeBr	—	1.00
	EtBr	—	0.12
	MeCl	—	~0
	CCl_4	—	6.2
	SF_6	—	6.2
	N_2O	—	0.2

TABLE 4. Typical rate constants (in $M^{-1}\ s^{-1} \times 10^{12}$) for the reaction of the quasi-free electrons determined from electron mobilities[19]

Solvent	Reactant						
	SF_6	N_2O	MeCl	CCl_4	EtBr	EtCl	$ClCH{=}CCl_2$
Cyclohexane	4.0	2.4	1.40	2.7	2.0	—	—
Isooctane	58	96	—	—	5.1	<0.009	35
n-Hexane	1.9	1.09	—	1.32	—	2.6	2.6

recently determined absolute and relative α_S values summarized in Table 3 shows that this is indeed the case.

The α_S values of Table 3 fall within a narrow range as compared to the large spread observed between the gas-phase electron capture rate constants for the same compounds. Despite the narrow spread in α_S values, they still show a solvent dependence. Ito and Hatano[49,50] suggested that the comparatively large spread in relative $k(e^- + S)$ values in neopentane can be considered to be an indication of the gas-like character of electron scavenging in this liquid to which diffusion kinetics, assumed in other solvents, does not apply. The relative $k(e^- + S)$ values in neopentane were derived from competitive studies which were based on the assumption that conventional competitive kinetics could be applied to this system because only free electrons were captured. This assumption does not seem to be completely justified under all the experimental conditions used by Ito and Hatano[49,50]. Thus the relative reactivities obtained may not be too accurate. Despite these inaccuracies the qualitative conclusions of Ito and Hatano are supported by the findings that the rate constants of the reactions of quasi-free electrons with various scavengers show a larger spread in neopentane than in *n*-hexane and cyclohexane[51,52]. In this context it is worth mentioning that it has been suggested by Boriyev and Yakovlev[53] that in neopentane. which has the highest electron mobility of the hydrocarbons listed in Table 3, the use of diffusion concepts leads to overestimated α-values.

In theory, $\alpha_S = f \times k(e^- + S)$, where the coefficient f is essentially a function of mobility or, which is the same, the diffusion coefficients of the ionic species in the liquid. Various models have been suggested[45,47,52–56] to estimate the effect of mobility and spatial distribution of ion pairs on the rates of electron scavenging. In general, all these models yield, for the compounds listed in Table 3, $k(e^- + S)$ values of the order of magnitude of 10^{11}–$10^{13}\ M^{-1}\ s^{-1}$. It should be emphasized though that the rate constants thus obtained represent mean values derived by averaging procedures that depend on the model used. Typical $k(e^- + S)$ values estimated by Allen and coworkers[52] from electron mobility determinations are presented in Table 4.

B. Cationic Species

1. *Identity of cationic species*

The identity of the cationic species formed upon exposure of halocarbons to ionizing radiation was the subject of several recent investigations which were conducted mainly in condensed systems. Evidence obtained in these studies indicates that, because of various secondary reactions, primary molecular cations generated by electron ejection from the irradiated halocarbon molecule are not the only positively charged species present in the irradiated system.

In the pulse radiolysis of *n*-BuCl at 133 K, as well as in matrix isolation studies of gamma-irradiated solid *n*-BuCl at 77 K, Arai and coworkers[57] found evidence for the formation of two different cationic species having a common precursor. These species are long-lived butene cations and short-lived butyl chloride cations with absorption maxima at 450 and 550 nm, respectively.

$$n\text{-BuCl} \longrightarrow (n\text{-BuCl}^{+\cdot})^* \begin{cases} \nearrow n\text{-BuCl}^{+\cdot} \\ \searrow C_4H_8^{+\cdot} + HCl \end{cases} \qquad (32)$$

The common precursor $(n\text{-BuCl}^{+\cdot})^*$ is a vibrationally or electronically excited n-BuCl cation which undergoes a very fast charge transfer reaction with added biphenyl:

$$(n\text{-BuCl}^{+\cdot})^* + Ph_2 \longrightarrow n\text{-BuCl} + Ph_2^{+\cdot} \qquad (33)$$

Similar behaviour was also observed in s-BuCl.

In the pulse radiolysis of 1,2-dichloroethane (DCE) at 24°C at least two different cationic species are formed. Kinetic evidence for the existence of these two distinguishable positively charged species was obtained by Wang and coworkers[58], who monitored the formation of phenylcarbenium ions from various solutes in their studies of charge transfer reactions. These authors have suggested that the two cationic species are: (a) the chloroethyl cation $ClCH_2CH_2^+$ whose yield has been estimated as 0.2 and (b) the parent radical cation $ClCH_2CH_2Cl^{\cdot+}$ and/or $(ClCH{=}CH_2)^{+\cdot}$, which is formed by HCl elimination from the parent radical cation with a total yield of 0.68. The former cation presumably may be present in equilibrium with the chloronium ion **3**:

$$ClCH_2CH_2^+ \rightleftharpoons CH_2\text{—}CH_2 \text{ (bridged by } Cl^+) \qquad (34)$$

(3)

In liquid carbon tetrachloride, the absorption band centred around 500 nm has been attributed to the $CCl_4^{\cdot+}$ radical cation or to the $CCl_4^{\cdot+}Cl^-$ ion pair[59]. Recently Bühler and Hurni[60] carried out a rather detailed study of the effect of cation scavengers on the 500 nm absorption in pulse-irradiated, pure liquid CCl_4. Their results seem to indicate that the CCl_3^+ cation within the ion pair $CCl_3^+Cl^-$ is responsible for the 500 nm absorption. This ion pair is presumably formed by the following mechanism:

$$CCl_4^{+\cdot} \longrightarrow CCl_3^+ + Cl^\cdot \qquad (35)$$

$$e^- + CCl_4 \longrightarrow {}^\cdot CCl_3 + Cl^- \qquad (36)$$

$$CCl_3^+ + Cl^- \longrightarrow (CCl_3^+\ Cl^-) \qquad (37)$$

or

$$CCl_4^{+\cdot} + Cl^- \longrightarrow Cl^\cdot + CCl_3^+\ Cl^- \qquad (38)$$

A lifetime of 33 ± 3 ns at −22°C and an activation energy of 10.9 ± 2.1 kJ has been determined for the recombination of the $CCl_3^+\ Cl^-$ ion pair. The CCl_3^+ characteristic absorption with λ_{max} at 500 nm was also detected in Freon-113 ($CFCl_2CFCl_2$)[61]. In this case CCl_3^+ is paired with the Freon$^-$ parent radical anion and decays with a lifetime of 144 ns at −22°C and an activation energy of 8.8 ± 3.0 kJ. In pure $CFCl_2CFCl_2$, a band with λ_{max} at 380 nm was observed and assigned to the $CFCl_2CFCl_2^{+\cdot}$ radical cation. This radical cation decays mainly in a second order neutralization reaction with the molecular radical anion $CFCl_2CFCl_2^{-\cdot}$ (λ_{max} = 440 nm).

Studies by Sukhov and coworkers[62,63] indicate that ion pairs are also formed in solid and glassy *i*-PrCl. They assigned the IR absorption at 2545 cm^{-1} in irradiated crystalline isopropyl chloride to the complex $[C_3H_7]^+[ClHCl]^- \cdot C_3H_6$.

On the whole, with the exception of radical cation dimers $(RX—XR)^{+\cdot}$ formed in the reaction of parent radical cations $RX^{+\cdot}$ with RX[11], cationic species were not detected in low temperature ESR studies. However, ESR studies provide indirect evidence for reactions of parent cations such as the formation of radicals $R_2\dot{C}X$ (X = Cl, Br and I) by H^+ loss from initially formed $R_2CHX^{+\cdot}$ [64–66].

2. Charge transfer reactions

Studies of the mechanism and kinetics of charge transfer reactions in chloroalkanes have been conducted by a number of investigators[57–59,67–75]. In most of these works, the formation and decay of solute cations with known absorption spectra were monitored and conclusions about the reaction mechanism were inferred from the kinetic observations. This method yields reliable information on rates and mechanisms of the reactions of cations produced from the solute. However, it provides only indirect, and therefore speculative, evidence on the cationic species formed from the solvent which are involved in charge transfer reactions.

It was suggested that mobile hole capture[57,67,68] and long-range electron transfer reactions are the mechanisms of positive charge transfer in butyl chloride glasses. Only the latter mechanism is consistent with the results of recent pulse radiolysis studies by Nosaka and coworkers[70] in solutions of both pyrene and diphenyl in *s*-BuCl. Kira and coworkers[71] reached the same conclusion on the basis of their studies of positive charge scavenging by amines in pulse-irradiated *s*-BuCl[71]. In their study the rates of charge transfer to amines were found to increase with decreasing ionization potential of the amine. On the other hand, a model based on the assumption that the cross-section for the hole scavenging is a function of the hole velocity was proposed by Arai and Imamura[69]. By means of this model, the authors were able to come forward with a quantitative explanation of the dependence of the cation yield on solute concentration in BuCl. Most probably, the hole or holes used in these models represent the previously mentioned vibrationally excited butyl chloride cations (see preceding section).

Pulse radiolysis studies of arylcarbenium ions in the chloroalkanes MeCl[75], 1,2-dichloroethane[72–75] and 1,1,2-trichloroethane[75] at 24°C were carried out by Dorfman and coworkers. Absolute rate constants for the reaction of these cations with halide ions[72–74], tertiary alkyl amines[72–75], aliphatic alcohols[72,74] and various nucleophilic species[74] were determined in these studies. The arylcarbenium ions were formed from solute cations $RCl^{+\cdot}$ by dissociative charge transfer reactions such as[72]

$$RCl^{+\cdot} + (PhCH_2)_2Hg \longrightarrow RCl + PhCH_2^+ + PhCH_2^{\cdot} + Hg \quad (39)$$

$$RCl^{+\cdot} + Ph_2CHBr \longrightarrow RCl + Ph_2CH^+ + Br^{\cdot} \quad (40)$$

$$RCl^{+\cdot} + Ph_3COH \longrightarrow RCl + Ph_3C^+ + OH^{\cdot} \quad (41)$$

In 1,2-dichloroethane rate constants of 1.3×10^{10} and 1.6×10^{10} $M^{-1}\,s^{-1}$ which approach the diffusion-controlled limit were observed[74] for the formation of benzyl and benzhydryl cations from $(PhCH_2)_2Hg$ and Ph_2CHBr, respectively. Lower rate constants were determined for the formation of trityl cations, namely 8.4×10^9, 5.7×10^8 and 4×10^8 $M^{-1}\,s^{-1}$, when the solutes were Ph_3CBr, Ph_3COH and Ph_3CCl, respectively[74]. As previously mentioned, it was shown by Wang and coworkers[58] that the two cationic species derived from 1,2-dichloroethane may participate in this type

TABLE 5. Rate constants (in $M^{-1} s^{-1}$) for the charge transfer reactions of cationic species derived from 1,2-dichloroethane at 24°C[58]

Cation[a]	Substrate	k
$S_1^{+\cdot}$	Ph_2	1.3×10^{10}
$S_1^{+\cdot}$	NH_3	6.8×10^{9}
S_2^{+}	NH_3	1.3×10^{10}
$S_1^{+\cdot}$	Ph_2CHBr	1.2×10^{10}
$S_1^{+\cdot}$	Ph_2CHBr	4×10^{9}

[a] $S_1^{+\cdot} = ClCH_2CH_2Cl^{+\cdot}$ and/or $ClCHCH_2^{+\cdot}$; $S_2^{+} = ClCH_2CH_2^{+}$.

of charge transfer reaction. Typical rate data for the reactions of the cationic species in 1,2-dichloroethane are summarized in Table 5.

Rate constants at 20°C for the charge transfer reaction of carbon tetrachloride radical cation to alkyl chlorides, normal and cyclic alkanes, alkenes and aromatic compounds were determined by Mehnert and coworkers[59]. In their pulse radiolysis study the effect of these compounds on the 360 nm absorption in CCl_4, assigned to the $CCl_4^{+\cdot}$ radical cation, was monitored. They showed that the rate constants thus obtained could be reasonably well correlated with the difference between the gas-phase ionization potential of the solvent and the solutes.

C. Excited States and Energy Transfer

The formation of excited halocarbon molecules has been postulated in radiolysed pure halocarbons as well as in their solutions in various hydrocarbons. Undisputed direct spectroscopic evidence in support of these assumptions is not available as yet. In this respect it is quite obvious that little progress has been achieved since the earlier report in the radiation chemistry of halocarbons appeared in this series[1]. However, with the recent development of the extremely fast sub-nanosecond pulse radiolytic techniques, additional information supporting the occurrence of energy transfer reactions of halocarbons has been obtained. In studies of dilute solutions of aromatic solutes in hydrocarbons in which this technique was employed, the effect of added halocarbon, mainly CCl_4, on the emission from aromatic solutes was determined.

In cyclohexane, Beck and Thomas[76,77] observed a reduction in the fluorescence yields from 9,10-diphenylanthracene (DPA-S_1) and an increase in the rate of formation of this species when CCl_4 was added. They rationalized this observation by assuming that the following competing reactions take place:

$$c\text{-}C_6H_{12}^* + DPA \longrightarrow c\text{-}C_6H_{12} + DPA^*\ (S_1) \longrightarrow DPA + h\nu \tag{42}$$

$$c\text{-}C_6H_{12}^* + CCl_4 \xrightarrow{k_{43}} c\text{-}C_6H_{12} + CCl_4 \tag{43}$$

This assumption does not take into account the possibility of direct DPA* fluorescence quenching by CCl_4. It was verified by the Stern–Volmer treatment of the fluorescence yields in the presence of CCl_4 which gave a k_{43} value of $2.5 \times 10^{11}\ M^{-1} s^{-1}$. This compares well with $k_{43} = 3 \times 10^{11}\ M^{-1} s^{-1}$ derived from the rates of formation of DPA* in the $c\text{-}C_6H_{12}$–DPA–CCl_4 system. A similar study of cyclohexane solutions of paraterphenyl (PT) was carried out by Jonah and coworkers[78]. The reduction of

$PT(S_1)$ fluorescence by CCl_4 was attributed in this case both to interference with the formation of the emitting state and to its quenching:

$$PT(S_1) + CCl_4 \longrightarrow PT + CCl_4 \qquad (44)$$

A rate constant of $2.9 \times 10^9\ M^{-1}\ s^{-1}$ was estimated for this reaction.

Tabata and coworkers used 2,5-diphenyloxazole (PPO) as a scintillator in cyclohexane[79,80]. They explained the formation of PPO* in terms of a slow and a fast process. The slow process was attributed to energy transfer from excited cyclohexane to PPO. Various alternatives were considered for the fast process including direct excitation by sub-excitation electrons, the rapid recombination of solute ions and direct excitation by Cerenkov light. A rate constant of $(3 \pm 1) \times 10^{11}\ M^{-1}\ s^{-1}$ was estimated for the interference of CCl_4 with the formation of PPO* by the first of these fast processes, while for its direct quenching reaction with PPO*,

$$PPO^* + CCl_4 \longrightarrow PPO + CCl_4 \qquad (45)$$

a rate constant of $(2 \pm 1) \times 10^9\ M^{-1}\ s^{-1}$ was determined. Fast and slow PPO* formation was also reported for PPO solutions in toluene[81]. As in the case of cyclohexane solutions, CCl_4 appears to be much more efficient in inhibiting the slow process than the fast process.

In steady state irradiation studies perfluorocarbons were employed by Walter and coworkers[82] to quench fluorescence induced by both β-ray radiolysis and 184.9 nm photolysis of bicyclohexyl. Perfluorodecalin and perfluoromethylcyclohexane were found to have almost the same suppressing effect on radiation-induced and photo-initiated fluorescence.

It should be noted that in none of the radiolytic studies cited here has conclusive irrefutable evidence been presented for the reaction between excited solvent molecules and CCl_4 as being the cause of the observed quenching effects. A large body of spectroscopic evidence exists for the formation of these excited solvent molecules in alkanes and in aromatic liquids[83–91]. However, as has been pointed out by some authors[78,87], these excited hydrocarbon solvent molecules could be formed by geminate ion recombination. If this indeed is the case then the interference of CCl_4 with these recombination processes and the charge transfer reactions to the aromatic solutes could be conceived as an alternative explanation for its effect on the emission from scintillators.

On the other hand, the occurrence of energy transfer from excited solvent molecules to halocarbon solutes is strongly supported by the findings of photochemical studies such as those of Walter and coworkers[82] and Hatano and coworkers[92,93]. In the latter work photolysis of liquid cyclohexane was carried out at 163 nm and the effect of CCl_4, MeI, MeBr and EtBr on hydrogen formation was determined. Kinetic analysis of the reduction in hydrogen formation in terms of energy transfer reactions between the excited cyclohexane and these compounds gave an energy transfer rate constant of $8.6 \times 10^{10}\ M^{-1}\ s^{-1}$ for CCl_4. This value compares reasonably well with the $(3 \pm 1) \times 10^{11}\ M^{-1}\ s^{-1}$ obtained in the aforementioned work of Beck and Thomas[77]. Also worth noting is the fact that these two rate constant values markedly exceed the diffusion controlled limit – a situation that is quite common in other energy transfer reactions in solutions.

Finally, attention should be drawn to the fact that electronically excited halocarbon molecules may not necessarily be formed in radiolytic energy transfer reactions such as (43)–(45). However, if formed, their detection would certainly be considered as important evidence in favour of the energy transfer mechanism of these reactions.

D. Charge Transfer Complexes

The formation of transient absorption spectra assigned to charge transfer (CT) complexes of bromocarbons (RBr · Br) has been observed in both the pulse radiolysis of liquid bromobenzene (λ_{max} = 560 nm)[94] and in the gamma-radiolysis of CH_3Br (λ_{max} = 400 nm), CH_2Br_2 (λ_{max} = 385 nm), $CHBr_3$ (λ_{max} = 435 nm) and CBr_4 (λ_{max} = 490 nm) in 3-methylpentane glass at 77 K[95]. The formation of the same absorption bands was also observed in analogous photochemical experiments. Based on this observation and the decrease in the absorption intensity caused by the addition of charge scavengers, it has been suggested by both Bossy and Bühler[94] and Bajaj and Iyer[95] that charge neutralization and radical–molecule reactions are responsible for the formation of the charge transfer complexes:

$$RBr^+ + Br^- \longrightarrow (RBr \cdot Br) \tag{46a}$$

$$RBr + Br^\bullet \longrightarrow (RBr \cdot Br) \tag{46b}$$

For the unimolecular decay of PhBr · Br, presumably to bromocyclohexadienyl radical, a rate constant of $5.4 \times 10^4\ s^{-1}$ was estimated.

CT complex formation was also postulated for irradiation of alkyl iodides in both 3-methylpentane at 77 K[96] and in aqueous solutions[97]. In 3-methylpentane the reaction sequence involved in the formation and destruction of the complex is presumably

$$CH_3I^{+\bullet} + I^- \longrightarrow (CH_3I \cdot I) \longrightarrow CH_3^* + I^\bullet + I^\bullet \tag{47}$$

In water, absorption spectra assigned to inner and outer CT complexes were observed in the reaction of hydroxyl radical with alkyl iodides (R = Me, Et, *n*-Pr, *i*-Pr and *n*-Bu)[97]:

$$^\bullet OH + RI \longrightarrow \underset{\text{Outer CT}}{(HO \cdot RI)} \longrightarrow \underset{\text{Inner CT}}{(HO^- R^+)} \tag{48}$$

For EtI at pH 4.23, the inner CT complex decays with a rate constant of $5.8 \times 10^9\ M^{-1}\ s^{-1}$.

IV. RADICAL REACTIONS

Both free radicals and charged species are formed in the initial steps of radiolysis. However, under most conditions, and in the absence of charge stabilization by solvation, neutralization reactions are faster than radical combination reactions. Consequently, the lifetime of radicals is in general longer than the lifetime of charged intermediates. In addition, the energy released in the neutralization process may be sufficient to create excited molecules that are able to undergo homolytic dissociation into radicals. The net effect of such a reaction sequence is the partial conversion of charged species into radicals.

In theory, since the energy of ionizing radiation by far exceeds the energy needed to break any bond in the irradiated compound, one would expect a large number of different radicals to be formed upon radiolysis. The actual situation is much less complicated. In most compounds selective bond rupture results in the predominant formation of only a few different radicals. Furthermore, in all irradiated systems a large degree of control of the nature of the radicals formed can be achieved by the appropriate choice of experimental conditions such as pH, temperature and composition.

Continuous, low dose rate radiolytic methods are particularly suited for kinetic studies of radical chain reactions. The advantages of these methods, when employed in dilute liquid solutions, include radical generation which, for all practical purposes, can be considered as uniform in time and space and which remains constant over a wide range of temperatures and reactant concentrations. The chemistry of radiation-induced free radical chain reactions and their synthetic applications has been the subject of a number of reviews[98–101] in which a more detailed discussion of the general aspects of this method can be found.

A. Reactions of Hydrogen Atoms in Aqueous Solutions

Hydrogen atom reactions in aqueous solutions of haloaromatic compounds were studied by Lichtscheidl and Getoff[102]. Rate constant data derived in that study are summarized in Table 6. Three different routes for the reaction with hydrogen atoms were observed in the haloaromatic compounds studied. Addition of hydrogen atoms to the benzene ring resulting in the formation of cyclohexadienyl radicals was the only reaction path detected for PhF[102], PhCl[102], PhBr[102] and *p*-FC_6H_4CN[103]. Chlorine abstraction, accompanied by the formation of benzyl radical, was the only observed route of hydrogen removal in its reaction with $PhCH_2Cl$. In the case of phenethyl chloride, hydrogen atoms add to the benzene ring and abstract hydrogen from the side chain. The respective rate constants[102] are 2×10^9 and 1.5×10^8 $\text{M}^{-1}\,\text{s}^{-1}$.

TABLE 6. Rate constants for the reaction of hydrogen atoms with haloaromatic compounds in water

Substrate	k, $\text{M}^{-1}\,\text{s}^{-1} \times 10^{-9}$	Mode	Ref.
PhF	1.5[a]	Addition	102
PhCl	1.2[b]	Addition	102
PhBr	1.3[a]	Addition	102
BzCl	0.95[c]	Cl-abstraction	102
p-EtC_6H_4Cl	2.0[b]	Addition	102
p-EtC_6H_4Cl	0.15[b]	H-abstraction	102
p-FC_6H_4CN	0.67[a]	Addition	103

[a]Determined from the kinetics of radical build-up.
[b]Average value determined from radical build-up and competitively, versus $H^{\bullet} + CH_3OH$.
[c]Determined competitively versus $H^{\bullet} + CH_3OH$.

Rate constants for the reaction between hydrogen atoms and 5-halouracils were determined by Neta and Schuler[104] and found to be 1.8, 1.6 and 2.2×10^8 $\text{M}^{-1}\,\text{s}^{-1}$ when the halo substituent was F, Cl and Br, respectively.

B. Reactions of Hydroxyl Radicals

The reactions between halosubstituted benzene derivatives and hydroxyl radicals result in oxidative replacement of the halogen by OH:

$$\text{Y-C}_6\text{H}_4\text{-X} + \text{HO}^{\bullet} \longrightarrow \text{Y-C}_6\text{H}_4(\text{HO})(\text{X})^{\bullet} \longrightarrow \text{Y-C}_6\text{H}_4\text{-O}^{\bullet} + \text{HX} \quad (49)$$

This type of reaction, studied mainly by pulse radiolysis combined with conductivity detection, was observed in substituted fluorobenzenes[23], 4-fluorobenzonitrile[103] and 4-haloanisoles (Y = MeO; X = F, Cl and Br)[105]. Incidentally, the occurrence of OH oxidative replacement reactions is not limited to halosubstituted benzenes. Similar reactions were observed for X = OCH_3[106,107], CO_2H[106,107], NO_2[108] and NH_2[109]. Two routes of hydrogen chloride formation from haloanisoles were observed by Latif and coworkers[105]. When the substituents F, Cl and Br were located in the *ortho* and *para* positions to the halogen, fast first order formation of HX was observed. On the other hand, when these substituents were in the *meta* position, HX was formed by a second order process for which the rate constants 1.7×10^{10}, 1.5×10^{9} and 1.3×10^{9} $M^{-1} s^{-1}$ were estimated when X = F, Cl and Br, respectively. The occurrence of the fast HX formation in *ortho* and *para*-substituted anisoles was explained as an indication that in these compounds preferential addition to the X-substituted position took place. The difference in the reactivity between the *ortho*- and *para*-substituted anisoles and the *meta*-substituted anisoles was ascribed to the *ortho*- and *para*-directing activity of the methoxy group. It is worth mentioning that the decay of the hydroxy-adduct by a second order process ($k = 4 \times 10^{8}$ $M^{-1} s^{-1}$) was also observed in the case of *p*-fluorobenzonitrile[103].

The reaction of 5-halosubstituted uracils (X = F, Cl and Br) with hydroxyl radicals results in release of hydrogen halide[110,111]. The reaction presumably proceeds by addition to the site of the halogen atom, followed by rapid dehydrohalogenation[110]:

$$HO^{\cdot} + \text{5-X-uracil (H–N–C(=O)–C–X=C–H–N(H)–C(=O))} \longrightarrow \text{(H–N–C(=O)–C=O, }{}^{\cdot}\text{C–H, N–H, O=C)} + H^{+} + X^{-} \qquad (50)$$

Hydroxyl addition followed by HCl elimination was observed in the reaction of hydroxyl radicals with the chloroethylenes 1,2-$C_2H_2Cl_2$, C_2HCl_3 and C_2Cl_4. In all

TABLE 7. Rate constants for the reaction of hydroxyl radicals with various compounds in aqueous solution

Substrate	k, $M^{-1} s^{-1} \times 10^{9}$	Ref.
C_6H_5F	8	20
o-$C_6H_4F_2$	4.5	20
p-$C_6H_4F_2$	6	20
1,2,3,4-$C_6H_2F_4$	5	20
C_6HF_5	4	20
C_6F_6	2	20
p-FC_6H_4CN	3.5	103
5-F-Uracyl	5.4	103
5-Cl-Uracyl	5.4	103
5-Br-Uracyl	5.4	103
$CH_2{=}CHCl$	7.1	103
$CHCl{=}CHCl$	7.5	103
$CH_2{=}CCl_2$	23	103
$CHCl{=}CCl_2$	19	103
$CCl_2{=}CCl_2$	13	103

three compounds the hydroxyl adduct is formed at a chlorine-substituted carbon. These adducts are unstable and lose HCl:

$$\rangle C{=}C\langle^{\ }_{Cl} + HO^{\bullet} \longrightarrow -\dot{C}-\underset{Cl}{\underset{|}{C}}-OH + \rangle\dot{C}-C\langle^{H}_{O} + H^{+} + Cl^{-} \quad (51)$$

Adducts formed by hydroxyl attack on vinyl chloride and vinylidene chloride do not eliminate hydrogen chloride. In these compounds addition takes place at the carbon that does not carry a chlorine atom and the radical thus formed is stable.

The rate constants for hydroxyl addition to halobenzenes, uracils and chloroethylenes are presented in Table 7.

C. Dehalogenation of Halomethanes

Radiation-induced dehalogenation of halomethanes (RX) has been studied in alkane[41,42,112–125], silane[126–128] and alcohols[129] solutions, mostly under conditions where the dehalogenation proceeds by a free radical chain mechanism. In alkanes (R^1H) the propagation step is composed of two reactions, X-transfer followed by hydrogen atom abstraction:

$$R'^{\bullet} + RX \xrightarrow{k_X} R'X + R^{\bullet} \qquad (52)$$

$$R^{\bullet} + R'H \xrightarrow{k_H} RH + R'^{\bullet} \qquad (53)$$

and therefore $G(R'X) = G(RH) = G(-RX) = G(-R'H)$.

In cyclohexane, the dehalogenation of CH_2Cl_2[112], $CHCl_3$[112], CCl_4[112,113,119], CCl_3CN[115], $CHCl_2CN$[116], $CH_2(Cl)CN$[116], $CH_2(Br)CN$[117], CHI_3[118] and $CFCl_3$[120] was studied. Typical $G(R'X)$ values in these systems ranged from 10^2 to 10^3 depending on experimental conditions such as dose rate, temperature and RX concentration. Arrhenius parameters of the X-transfer and H-transfer reactions were determined in the c-C_6H_{12}–RX systems and are presented in Table 8. The H-transfer rate constants k_H were determined utilizing equations (53a) and (53b).

$$G(RH) = \frac{k_H}{k_t^{1/2}} \alpha^{1/2} G(R_2)^{1/2} [R'H] \qquad (53a)$$

$$G(RH) = \frac{k_H}{(2k_t)^{1/2}} \alpha^{1/2} G(R'_0)^{1/2} [R'H] \qquad (53b)$$

TABLE 8. Arrhenius parameters in the dehalogenation reactions of halomethanes (RX) in liquid cyclohexane

RX	$\log A_X$, $M^{-1}\ s^{-1}$	E_X, kcal mol^{-1}	Ref.	$\log A_H/A_t^{1/2}$	$E_H - \frac{1}{2}E_t$, kcal mol^{-1}	Ref.
CCl_4	9.40	5.88	112	3.28	8.81	113
$CHCl_3$	9.45	10.16	112	—	—	—
CH_2Cl_2	9.24	13.67	112	—	—	—
$CH_2(Cl)CN$	8.58	9.99	116	—	—	—
$CHCl_2CN$	8.80	6.55	116	—	—	—
CCl_3CN	8.20	2.09	115	3.95	13.70	115
$CH_2(Br)CN$	8.70	8.57	117	4.22	11.96	117
$CFCl_3$	8.95	7.79	120	—	—	—

where k_t is the termination rate constant, α is a known coefficient that converts G values into rates of formation and $G(R_0')$ stands for the yield of radicals in cyclohexane. The Cl-transfer Arrhenius parameters were determined in competitive studies, mainly versus addition to chloroethylenes (see below).

The chain length of dehalogenation, at a given dose rate and RX concentration, is determined by the relative magnitudes of k_H and k_{Cl}. With the exception of CH_2Cl_2, $k_{Cl} > k_H$ both in the chloromethanes and in the chloroacetonitriles. Hence, hydrogen abstraction from the solvent is the rate-determining step in the chain dehalogenation of chloromethanes in cyclohexane.

Dehalogenation of CCl_4 was also studied in n-hexane[42,114]. Tuan and Gaumann[42] determined the following Arrhenius expressions for k_{Cl} and k_H for 3-hexyl radical:

$$\log k_{Cl}\,(\text{M}^{-1}\,\text{s}^{-1}) = 8.26 - 5.02/\theta,$$

$$\log k_H\,(\text{M}^{-1}\,\text{s}^{-1}) = 7.15 - 9.18/\theta,$$

where $\theta = 2.303RT$ in kcal mol^{-1}. Katz and coworkers[114] found the overall k_H in n-hexane to be given by

$$\log k_H/k_t^{1/2}\,(\text{M}^{-1/2}\,\text{s}^{-1/2}) = 3.69 - 9.62/\theta.$$

The addition of a small amount of alkane to CCl_4 is sufficient to cause dechlorination, which otherwise does not occur readily. Under these conditions, which necessarily result in short chains, Alfassi and Feldman[121] determined the Arrhenius parameters for the abstraction from 2,3-dimethylbutane and found k_H to be given by

$$\log k_H/k_t^{1/2}\,(\text{M}^{-1/2}\,\text{s}^{-1/2}) = 2.40 - 6.96/\theta.$$

Alfassi and coworkers also investigated the dehalogenation of CCl_4 in the presence of cholosteric and cholostanic esters[122–125]. For these compounds the Arrhenius parameters fall within a narrow range: 7.5 ± 0.3 for $\log A_H(\text{M}^{-1}\,\text{s}^{-1})$ and 6.8 ± 0.4 kcal mol^{-1} for E_H[122].

The radiation-induced dechlorination of CCl_4, $CHCl_3$, CH_2Cl_2 and CH_3Cl in liquid Cl_3SiH[126] and of CH_2Cl_2, $CHCl_3$, CCl_4 and CCl_3CN in liquid Et_3SiH[127] was studied by Aloni and coworkers. In addition, the radiolytic method was used for the competitive determination of the Arrhenius H-abstraction parameters from silanes in solutions of CCl_4 in mixtures of cyclohexane with triethylsilane and trimethylsilane[128]. The Arrhenius parameters determined in those studies are presented in Tables 9 and 10.

TABLE 9. Arrhenius parameters for chlorine transfer in the dechlorination reactions of chloromethanes (RCl) in silanes[a]

RCl	Radical	$\log A_{Cl}/A_{Br}$	$E_{Cl} - E_{Br}$[b]	Ref.
CH_3Cl	Cl_3Si	0.85	2.53	126
CH_2Cl_2		0.05	3.40	126
$CHCl_3$		0.20	4.60	126
CCl_4		−0.63	5.18	126
CH_2Cl_2	Et_3Si	0.40	1.11	127
$CHCl_3$		0.28	0.19	127
CCl_4		1.25	−0.17	127
CCl_3CN		1.86	0.24	127

[a]Competitively determined versus Br abstraction from c-$C_6H_{11}Br$ in Cl_3SiH and from n-$C_5H_{11}Br$ in Et_3SiH.
[b]In kcal mol^{-1}.

TABLE 10. Arrhenius parameters for hydrogen abstraction in the dechlorination reactions of carbon tetrachloride in silanes

Substrate	log A_H, M^{-1} s^{-1}	E_H, kcal mol^{-1}	Ref.	D(Si—H), kcal mol^{-1}	Ref.
Cl_3SiH(g)	8.21	9.02	130	91.2	132a
Me_3SiH	8.49	8.70	128	90.0	132b
Et_3SiH	8.62	8.06	128	—	—
c-C_6H_{12}	8.79	11.08	113	95.5	133

The dehalogenation mechanism in silanes is similar to the mechanism in alkanes:

$$\gtrdot Si^{\bullet} + RX \longrightarrow \gtrdot SiX + R^{\bullet} \quad (54)$$

$$R^{\bullet} + \gtrdot SiH \longrightarrow RH + \gtrdot Si^{\bullet} \quad (55)$$

The hydrogen abstraction is the rate-determining step both in Et_3SiH and Me_3SiH. The data of Table 9 show that, for the CCl_3 radical in the silanes studied, E_H is lower by about 2 kcal mol^{-1} than for the analogous reaction of $^{\bullet}CCl_3$ radicals in cyclohexane. The same conclusion can be reached on the basis of gas-phase photolytic studies in Cl_3SiH[130] and can be expected to be generally true for other chloromethyl radicals. Dechlorination of chloromethanes is considerably faster in silanes than in cyclohexane because of the lower E_H. This is also shown by the experimentally determined G values of dechlorination, which in some cases were as high as 10^4.

The higher reactivity of silanes in dechlorination reactions simply reflects the fact that the overall ΔH of dechlorination is higher in these compounds than in alkanes because the Si—Cl bond is stronger than the C—Cl bond while the reverse order of bond dissociation energies is found for the Si—H and C—H[131,132] bonds. In so far as X-transfer reactions are concerned, gas-phase[134] studies indicate that the activation energies of these reactions are lower for silyl radicals than for alkyl radicals, in agreement with the Si—Cl and C—Cl bond strength difference.

Dehalogenation in alcohols differs in only one respect from dehalogenation in alkanes and silanes. The abstraction of hydrogen atom occurs only at the α position to the hydroxyl and therefore, in the Cl-transfer products, the OH and Cl are located on the same carbon atom. These products are unstable and eliminate hydrogen chloride. Such behaviour was observed by Radlowski and Sherman[129] in the i-PrOH–CCl_4 system at room temperature. In this system the three main products, acetone, chloroform and hydrogen chloride, were formed in equal yields according to the following reaction scheme:

$$(CH_3)_2\dot{C}OH + CCl_4 \longrightarrow [(CH_3)_2C(OH)Cl] + {}^{\bullet}CCl_3 \quad (56)$$

$$[(CH_3)_2C(OH)Cl] \xrightarrow{\text{Fast}} (CH_3)_2CO + HCl \quad (57)$$

$${}^{\bullet}CCl_3 + (CH_3)_2CHOH \longrightarrow CHCl_3 + (CH_3)_2\dot{C}OH \quad (58)$$

The rate-determining step in this system, as in silanes and alkanes, is the hydrogen abstraction reaction, for which a rate constant of 3 M^{-1} s^{-1} was determined at room temperature.

D. Dehalogenation of Haloethanes

The radiation-induced chain dehalogenation of C_2Cl_6[135,136], C_2HCl_5[136,137], *sym*-$C_2H_2Cl_4$[137], *asym*-$C_2H_2Cl_4$[136,138], CH_3CCl_3[139], 1,1,2-$C_2H_3Cl_3$[139,140], 1,1-$C_2H_4Cl_2$[139], CF_3CCl_3[136], CCl_2BrCH_2Cl[138] and *sym*-$C_2Br_2F_4$[141] has been studied in cyclohexane. The dehalogenation mechanism for compounds such as CF_3CCl_3 is identical to the dehalogenation mechanism for halomethanes:

$$R'^{\cdot} + CF_3CCl_3 \longrightarrow R'Cl + CF_3\dot{C}Cl_2 \tag{59}$$

$$CF_3\dot{C}Cl_2 + R'H \longrightarrow CF_3CCl_2H + R'^{\cdot} \tag{60}$$

However, in the majority of the haloethanes ($EX_2 = C_2H_{3-n}X_{3+n}$ (n = 1–3)) listed, each of the two carbons carries a halogen atom and therefore the dehalogenation mechanism is different.

$$R^{\cdot} + EX_2 \xrightarrow{k_X} RX + EX^{\cdot} \tag{61}$$

$$EX^{\cdot} + RH \xrightarrow{k_H} EXH + R^{\cdot} \tag{62}$$

$$EX^{\cdot} \xrightarrow{k_{el}} E + X^{\cdot} \tag{63}$$

$$X^{\cdot} + RH \longrightarrow HX + R^{\cdot} \tag{64}$$

In these compounds the haloethyl radicals (EX) can eliminate a halogen atom and therefore, on the average, a reacting haloethane molecule loses more than one halogen atom.

In the c-C_6H_{12}–EX_2 system the following relationship holds:

$$\frac{G(\mathrm{E})}{G(\mathrm{EHX})} = \frac{G(\mathrm{HCl})}{G(\mathrm{EHX})} = \frac{G(\mathrm{RX}) - G(\mathrm{EHX})}{G(\mathrm{EHX})} = \frac{k_{el}}{k_H}\frac{1}{[\mathrm{RH}]}. \tag{64a}$$

Since $E_{el} > E_H$, the elimination reaction in which equal amounts of HX and haloethylene–E are formed becomes increasingly important at high temperatures. However, because A_{el} is higher by several orders of magnitude than A_H, the elimination reaction is the rate-determining step of dehalogenation only at low temperatures at which $k_H > k_{el}$. Quite obviously, at temperatures at which this inequality is reversed the hydrogen-abstraction reaction becomes the rate-determining step. This behaviour is illustrated by the C_2Cl_6 data given in Table 11.

TABLE 11. Kinetic data for the radiation induced dechlorination of hexachloroethane in various liquids

Solvent	$G(C_2Cl_4)/G(C_2HCl_5)$ (50°C)	$G(C_2Cl_4)/G(C_2HCl_5)$ (100°C)	$\log A_{el}/A_H$, M	$E_{el} - E_H$, kcal mol^{-1}	k_{el}/k_H (25°C)
c-C_6H_{12}[135]	1.7	6.25	5.52	6.49	0.76
Et_3SiH[142]	0.35	1.68	5.50	9.08	0.0073
MeOH[143]	4.6	—	—	—	50
EtOH[143]	0.11 (25°C)	—	—	—	1.69
i-PrOH[143]	4.5	—	—	—	0.32

Equation (64a) can be utilized for the determination of the relative Arrhenius parameters of k_{el}/k_H from which the Arrhenius parameters of the elimination reaction can be estimated, provided that A_H and E_H are known. The procedure adopted in the derivation of the data given in Table 11 was to assume that all abstraction reactions of chloroethyl radicals have equal Arrhenius parameters and that these parameters are the same as those determined for the $^{\bullet}CCl_3$ radical[113]. The E_{Cl} values thus obtained are generally in good agreement with gas-phase chlorine elimination data for the same chloroethyl radicals[144].

In Et_3SiH the dechlorination of C_2Cl_6, C_2HCl_5 and *sym*-$C_2H_2Cl_4$ was studied[143]. The mechanism in these systems is identical to the dechlorination mechanism in alkanes. However, because the $Et_3Si—H$ bond is weaker than the c-$C_6H_{11}—H$ bond, the activation energy of hydrogen abstraction is lower in Et_3SiH than in cyclohexane. Consequently the E/EX product ratio is smaller in the silane than in cyclohexane. This effect can be seen from the C_2Cl_6 data presented in Table 11.

Table 11 also includes data for the dechlorination of C_2Cl_6 in the alcohols MeOH, EtOH and *i*-PrOH, which were determined by Sawai and coworkers[143]. As in the case of the reaction with CCl_4, following chlorine transfer all three alcohols undergo rapid oxidation, forming formaldehyde, acetaldehyde and acetone, respectively.

Finally it is worth mentioning that in Et_3SiH[145] and in cyclohexane[136] the chlorine transfer rate constants for chloromethanes and chloroethanes having the structure CCl_3X correlate well with the Taft σ^* parameters. For E_{Cl} in cyclohexane a more precise correlation has been proposed by Katz and Rajbenbach[146]:

$$E_{Cl} = \alpha D + \rho^*\sigma^* + \delta E_s + \text{constant}, \tag{64b}$$

where D is the C—Cl bond dissociation energy, σ^* and E_s are the Taft polar and steric substituent constants of X and α, ρ^* and δ are the respective sensitivity parameters.

E. Chlorination

The exposure to chlorine is one of the most widely used methods for the chlorination of alkanes. Since this reaction proceeds by a free radical chain mechanism, it can also be initiated by ionizing radiation. This type of study of the gas-phase gamma-radiation-induced chlorination of methane was recently carried out by Nel and Van der Linde[147], who determined the effect of dose, dose rate, chlorine concentration and mixture density on the yields of products at room temperature. The radiation-induced reaction was found to be very fast, as indicated by G(MeCl), which in some cases exceeded 10^5. Since exhaustive chlorination was carried out, the MeCl was accompanied by higher chlorination products, namely CH_2Cl_2, $CHCl_3$ and CCl_4. A kinetic simulation model based on rate data for 21 different radical reactions was developed and found to account for the product distribution in the system and its time (dose) dependence.

Radiolytic chlorination of liquid propionitrile by chlorine at 55°C was studied by Konnecke and coworkers[148]. At the experimental conditions used, the principal product, α,α-dichloropropionitrile, was formed with G values ranging from 10^3 to 10^4 and conversion as high as 65%. This product was accompanied by small amounts of the α,β-dichloro isomer and by β-chloropropionitrile. Presumably the chlorination in this system is a two-stage process of consecutive chlorination.

$$Cl^{\bullet} + CH_3CH_2CN \longrightarrow CH_3\dot{C}HCN + HCl \tag{65}$$

$$CH_3\dot{C}HCN + Cl_2 \longrightarrow CH_3CH(Cl)CN + Cl^{\bullet} \tag{66}$$

$$Cl^{\bullet} + CH_3CH(Cl)CN \longrightarrow CH_3\dot{C}(Cl)CN + \dot{C}H_2CH(Cl)CN + HCl \quad (67)$$

$$CH_3\dot{C}(Cl)CN + Cl_2 \longrightarrow CH_3CCl_2CN + Cl^{\bullet} \quad (68)$$

$$\dot{C}H_2CH(Cl)CN + Cl_2 \longrightarrow ClCH_2CH(Cl)CN \quad (69)$$

The predominance of the α,α isomer indicates that Cl atoms attack CH_3CH_2CN primarily at the α-position.

It is also worth noting that the previously discussed dechlorination reactions of chloromethanes and chloroethanes can also be considered as chlorination routes for alkanes and silanes. In this sense chlorine transfer from alkylsulphonyl chlorides ($R'SO_2Cl$) could *a priori* be utilized for the chlorination of alkanes by the following mechanism established in radiolytic studies of the decomposition of $MeSO_2Cl$, n-$PrSO_2Cl$ and n-$BuSO_2Cl$ in cyclohexane[149–151]:

$$R^{\bullet} + R'SO_2Cl \longrightarrow RCl + R'SO_2^{\bullet} \quad (70)$$

$$R'SO_2^{\bullet} \longrightarrow R'^{\bullet} + SO_2 \quad (71)$$

$$R'^{\bullet} + RH \longrightarrow R'H + R^{\bullet} \quad (72)$$

$$R^{\bullet} + SO_2 \longrightarrow RSO_2^{\bullet} \quad (73)$$

However, alkylsulphonyl chlorides are not good chlorinating agents because the liberated SO_2 acts as a chain inhibitor, even if present in only very small amounts.

F. Reactions of Olefins

Radiation-induced radical reactions of olefins with various substrates result in the formation of 1:1 adducts which, in some cases, particularly at high olefin concentrations, are accompanied by telomers. In subsequent discussion these 1:1 adducts will be referred to as condensation products: while not strictly accurate, this nomenclature is common usage among radiation chemists. The exact mechanism of condensation depends on the properties of the substrate and the olefin. Simple olefins form 1:1 saturated adducts in a two-stage process in which the addition is followed by hydrogen abstraction if the solvent is an alkane and by halogen transfer if it is a halogenated solvent with a weak C—X bond.

$$R^{\bullet} + >C{=}C< \longrightarrow R{-}\overset{|}{\underset{|}{C}}{-}\overset{|}{\underset{|}{C}}^{\bullet} \quad (74)$$

$$R{-}\overset{|}{\underset{|}{C}}{-}\overset{|}{\underset{|}{C}}^{\bullet} + RH(X) \longrightarrow R{-}\overset{|}{\underset{|}{C}}{-}\overset{|}{\underset{|}{C}}{-}H(X) + R^{\bullet} \quad (75)$$

The radiolytic synthesis of 1-cyclohexyl-2*H*-hexafluoropropane by condensation of hexafluoropropene with cyclohexane was studied by Podkhalyuzin and Nazarova[152] and it represents a typical example of this type of condensation reaction.

$$c\text{-}C_6H_{11}^{\bullet} + CF_2{=}CF_3 \longrightarrow c\text{-}C_6H_{11}CF_2\dot{C}FCF_3 \quad (76)$$

$$c\text{-}C_6H_{11}CF_2\dot{C}FCF_3 + c\text{-}C_6H_{12} \longrightarrow c\text{-}C_6H_{11}CF_2CHFCF_3 + c\text{-}C_6H_{11}^{\bullet} \quad (77)$$

G values of product formation as high as 8×10^3 at 200°C were observed in this system and the overall activation energy of the condensation process was estimated to be 3.4 ± 0.7 kcal mol^{-1} [152].

The condensation of propene and 2-methylpropene with CCl_4[153] is an example of a system in which the chain transfer occurs via chlorine abstraction. Accordingly, the propagation step in this system is described by the following reactions:

$$CCl_3^{\bullet} + \gt C{=}C\lt \longrightarrow CCl_3{-}\overset{|}{\underset{|}{C}}{-}\overset{|}{\underset{|}{C}}^{\bullet} \qquad (78)$$

$$CCl_3{-}\overset{|}{\underset{|}{C}}{-}\overset{|}{\underset{|}{C}}^{\bullet} + CCl_4 \longrightarrow CCl_3{-}\overset{|}{\underset{|}{C}}{-}\overset{|}{\underset{|}{C}}{-}Cl + CCl_3^{\bullet} \qquad (79)$$

Kim and coworkers[153] studied this reaction and determined the Arrhenius parameters for $^{\bullet}CCl_3$ addition (Table 12), assuming that the overall activation energy of product formation was determined by the addition reaction.

The condensation between alkenes and alkyl bromides, which was extensively studied by Myshkin and coworkers[154–162], provides an additional example of a system in which halogen abstraction is the chain transfer step:

$$R^{\bullet} + \gt C{=}C\lt \longrightarrow R{-}\overset{|}{\underset{|}{C}}{-}\overset{|}{\underset{|}{C}}^{\bullet} \qquad (80)$$

$$R{-}\overset{|}{\underset{|}{C}}{-}\overset{|}{\underset{|}{C}}^{\bullet} + RBr \longrightarrow R{-}\overset{|}{\underset{|}{C}}{-}\overset{|}{\underset{|}{C}}{-}Br + R^{\bullet} \qquad (81)$$

In addition to the 1:1 adducts, the formation of telomers was observed in some of the systems studied, such as EtBr–ethylene and EtBr–propylene. In the growing telomer radical, evidence was found for the occurrence of isomerization by a 1,5 shift of a hydrogen atom[154]. Typical activation energies for the addition of 5.5[155], 2.8[156] and 3.8 kcal mol^{-1} [156] were determined for the reaction of ethyl radicals with ethylene, propylene and 1-hexene, respectively. E_{Br} was found to be 3.7 kcal mol^{-1} for the chain

TABLE 12. Radiolytically determined Arrhenius parameters for the addition (ad) of c-C_6H_{11} and CCl_3 radicals to olefins

	$CCl_3^{\bullet}$			c-$C_6H_{11}^{\bullet}$		
Olefin[a]	log A_{ad}, M^{-1} s^{-1}	E_{ad}, kcal mol^{-1}	Ref.	log A_{ad}, M^{-1} s^{-1}	E_{ad}, kcal mol^{-1}	Ref.
cis-CHCl=CHCl	7.76	8.53	166	8.72	7.30	163
trans-CHCl=CHCl	8.16	8.49	166	9.13	7.25	163
CHCl=CCl_2	8.04	8.05	166	9.10	6.75	163
C_2Cl_4	7.58	9.49	165	8.68	7.30	163
CHF=CCl_2	—	—	—	8.64	6.38	164
CClF=CCl_2	—	—	—	8.72	8.49	164
$CH_3CH{=}CH_2$	5.1	2.70	153	—	—	—
$(CH_3)_2C{=}CH_2$	4.6	2.10	153	—	—	—

[a]The site of addition is underlined.

transfer to EtBr (equation 81)[156]. An isokinetic temperature of 220 ± 15 K was derived from the observed linear dependence between the activation energies of ethyl radical addition to ethylene, propylene, *trans*- and *cis*-2-butene, 1-hexene, 1-octene, trimethylethylene and isobutylene, and the corresponding pre-exponential Arrhenius coefficients[158].

Radiation-induced reactions of ethylene with other alkyl bromides (R = *n*-Pr[159], *n*-Bu[160], *i*-Pr[162] and *n*-$C_{10}H_{21}$[161]) were also studied and found to follow the pattern observed in the addition reactions of EtBr with olefins.

A different condensation mechanism was observed in the radiolytic reactions of chloroethylenes in cyclohexane[163,164]. In this case the condensation proceeds by an addition–elimination mechanism which, for cyclohexane, takes the form:

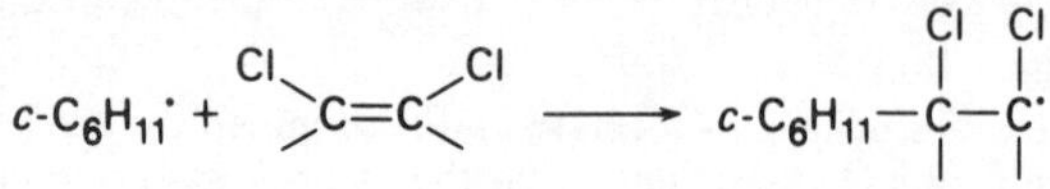

$$c\text{-}C_6H_{11}^{\bullet} + \text{ClCH=CHCl} \longrightarrow c\text{-}C_6H_{11}\text{—CHCl—CHCl}^{\bullet} \quad (82)$$

$$c\text{-}C_6H_{11}\text{—CHCl—CHCl}^{\bullet} \longrightarrow c\text{-}C_6H_{11}\text{CH=CHCl} + Cl^{\bullet} \quad (83)$$

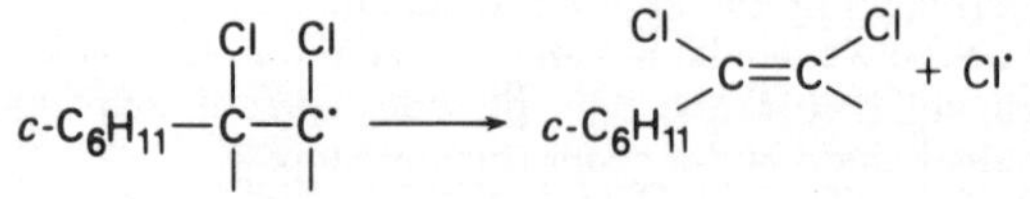

$$Cl^{\bullet} + c\text{-}C_6H_{12} \longrightarrow HCl + c\text{-}C_6H_{11}^{\bullet} \quad (84)$$

This mechanism applies to chloroethylenes with at least two vicinal chlorine atoms such as *cis*-$C_2H_2Cl_2$, *trans*-$C_2H_2Cl_2$, C_2HCl_3, C_2Cl_4 and C_2Cl_3F.

Because the reaction between $Cl^{\bullet}$ and CCl_4 is slow, condensation between CCl_4 and chloroethylenes has to be carried out in the presence of cyclohexane[165,166] as a chain transfer agent. The overall chain propagation mechanism in the *c*-C_6H_{12}–CCl_4–chloroethylene system is given by the following reaction scheme:

$$c\text{-}C_6H_{11}^{\bullet} + CCl_4 \longrightarrow c\text{-}C_6H_{11}Cl + {}^{\bullet}CCl_3 \quad (85)$$

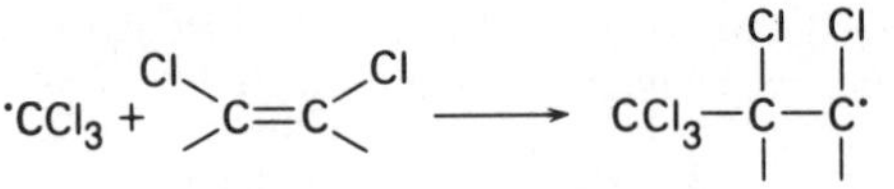

$$^{\bullet}CCl_3 + \text{ClCH=CHCl} \longrightarrow CCl_3\text{—CHCl—CHCl}^{\bullet} \quad (86)$$

$$CCl_3\text{—CHCl—CHCl}^{\bullet} \longrightarrow CCl_3\text{CH=CHCl} + Cl^{\bullet} \quad (87)$$

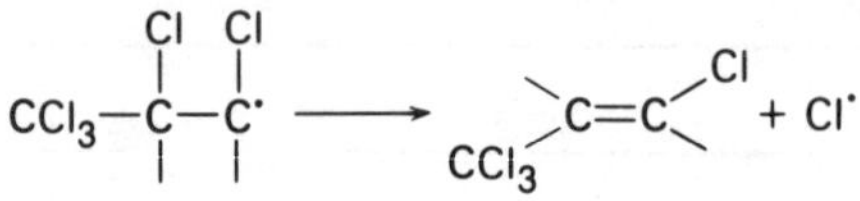

$$Cl^{\bullet} + c\text{-}C_6H_{12} \longrightarrow HCl + c\text{-}C_6H_{11}^{\bullet} \quad (88)$$

$$^{\bullet}CCl_3 + c\text{-}C_6H_{12} \longrightarrow CHCl_3 + c\text{-}C_6H_{11}^{\bullet} \quad (89)$$

The concentration of CCl_4 should be considerably higher than that of the chloroethylene, so that the addition of cyclohexyl radicals to the chloroethylene is negligible.

In the condensation reaction of $^{\bullet}CCl_3$ with chloroethylenes only 1:1 chlorovinylated adducts are formed. The chlorovinylation reactions of cyclohexane and CCl_4 were studied over a wide temperature range. The Arrhenius parameters for the addition determined in those studies are presented in Table 12. Comparison of the $^{\bullet}CCl_3$ data for chloroethylenes with the data for the addition of these radicals to propene and 2-methylpropene reveals a large difference in both the activation energies and the pre-exponential coefficients, that cannot be ascribed to the effect of chlorine substitution. It appears that the unusually low A-factors and activation energies determined by Kim and coworkers[153] are the consequence of the method used in their derivation, in which a zero activation energy was assigned to the combination reaction of CCl_3 radicals. This cannot be valid since even if this reaction is diffusion controlled it should have an activation energy equal to the diffusion activation energy of the medium in which it was studied.

Radiolytic chlorovinylation of Et_3SiH proceeds by a similar mechanism[167,168]. However, the $Et_3Si—Cl$ bond is considerably stronger than the $CCl_3—Cl$ or $c\text{-}C_6H_{11}—Cl$ bonds. Consequently, the reaction between Et_3SiH and C_2Cl_4 results in the simultaneous formation of both chlorinated and chlorovinylated silane by the following reaction sequence:

$$Et_3Si^{\cdot} + C_2Cl_4 \xrightarrow{k_{ad}} Et_3SiC_2Cl_4^{\cdot} \qquad (90)$$

$$Et_3Si^{\cdot} + C_2Cl_4 \xrightarrow{k_{Cl}} Et_3SiCl + C_2Cl_3^{\cdot} \qquad (91)$$

$$Et_3SiC_2Cl_4^{\cdot} \longrightarrow Et_3SiC_2Cl_3 + Cl^{\cdot} \qquad (92)$$

$$Cl^{\cdot} + Et_3SiH \longrightarrow HCl + Et_3Si^{\cdot} \qquad (93)$$

$$C_2Cl_3^{\cdot} + Et_3SiH \longrightarrow C_2HCl_3 + Et_3Si^{\cdot} \qquad (94)$$

At 65°C k_{ad}/k_{Cl} = 109[167] and 28.7[168] for C_2Cl_3H and C_2Cl_4, respectively.

V. PRODUCTS AND OVERALL MECHANISM IN PURE COMPOUNDS

In this section we will concentrate on studies aimed at the elucidation of the overall mechanism of radiolysis in pure compounds. In such investigations an attempt is made to determine all the radiolytic products from a given compound and to establish the mechanism of formation of each of them by examining the effect of various parameters and scavengers on their distribution and yields. A system with an added scavenger cannot be described as a pure system and, therefore, in this sense the title of this section is slightly misleading. To avoid misunderstanding it should be noted that this title refers primarily to the stated goal of the investigation rather than to the method used.

A total of 29 products were detected by Frank and Hanrahan[169] in gamma-irradiated gaseous ethyl bromide. The major products are listed in Table 13. From the variation of the yields of these products with total dose it is evident that secondary reactions take place. The effect of oxygen on the product formation was

TABLE 13. Major products of the radiolysis of gaseous ethyl bromide[169]

Product	HBr	H_2	C_2H_6	C_2H_4	C_2H_2	CH_4	CH_3Br	1,1-$C_2H_4Br_2$	1,2-$C_2H_4Br_2$	C_2H_3Br
G^a	6.02	1.39	2.70	0.78	0.16	0.080	0.080	0.88	0.12	0.32
G^b	0	1.39	2.70	0.86	0.13	0.26	0.24	1.66	0.31	0.75

^{a}Total dose (0–5) × 10^{20} eV g^{-1}.
^{b}Total dose (4.5–5) × 10^{20} eV g^{-1}.

also determined and the results in both pure EtBr and its mixture with O_2 were compared to photolytic decomposition under the same conditions. In the mechanism they proposed for EtBr radiolysis the authors also relied heavily on mass spectrometric information and made the simplifying assumption that primary ionic processes in the radiolysis of EtBr were identical to the fragmentation processes in the mass spectrometer at considerably lower pressures. Still their mechanism includes a total of 37 reactions comprised of 10 primary processes.

Neutral

$$C_2H_5Br \longrightarrow C_2H_5^{\bullet} + Br^{\bullet} \tag{95}$$

$$\longrightarrow C_2H_4 + HBr \tag{96}$$

$$\longrightarrow CH_3^{\bullet} + {}^{\bullet}CH_2Br \tag{97}$$

Ionic

$$C_2H_5Br \longrightarrow C_2H_5Br^{+} + e^{-} \quad (31.9\%) \tag{98}$$

$$\longrightarrow C_2H_5^{+} + Br^{\bullet} + e^{-} \quad (25.3\%) \tag{99}$$

$$\longrightarrow C_2H_3^{+} + H_2 + Br^{\bullet} + e^{-} \quad (2.2\%) \tag{100}$$

$$\longrightarrow C_2H_4^{+} + HBr + e^{-} \quad (5\%) \tag{101}$$

$$\longrightarrow C_2H_2^{\bullet+} + H_2 + HBr + e^{-} \quad (7.3\%) \tag{102}$$

$$\longrightarrow Br^{+} + C_2H_5^{\bullet} + e^{-} \quad (2.2\%) \tag{103}$$

$$e^{-} + C_2H_5Br \longrightarrow C_2H_5^{\bullet} + Br^{-} \tag{104}$$

The radiolysis of 1-chloropropane vapour[170–172] and liquid[173], both at 300 K, and of the solid at 70 K[173,174], was investigated by Ceulemans and coworkers. The main results are summarized in Table 14.

By detailed investigation of the formation of C_4 and C_5 products in the gas phase and the effect of added cyclohexane on their distribution, Geurts and Ceulmans were able to come forward with a mechanism of the ionic C–C bond cleavage reactions in 1-chloropropane[170,171].

$$C_3H_7^{\bullet+*} \longrightarrow C_2H_5^{+} + {}^{\bullet}CH_2Cl \tag{105}$$

$$C_2H_5^{+} \longrightarrow C_2H_3^{+} + H_2 \text{ or } 2H^{\bullet} \tag{106}$$

$$C_2H_5^{+} + C_3H_7Cl \longrightarrow C_2H_6 + C_3H_7Cl^{+} \tag{107}$$

$$C_3H_7Cl^{+*} \longrightarrow C_2H_5^{+} + {}^{\bullet}CH_2Cl \tag{108}$$

Presumably the C_4 and C_5 products are formed in subsequent neutralization and radical combination reactions.

TABLE 14. Zero dose extrapolated G values of the main products of solid and liquid 1-chloropropane radiolysis at 77 and 300 K[172]

Product	G (77 K)	G (300 K)
HCl	3.46	5.20
H_2	1.09	1.90
C_3H_6	1.85	0.86
C_3H_8	1.18	2.33
$Cl_2CHCH_2CH_3$	0.54	0.31
$ClCH_2CHClCH_3$	0.63	0.55
$ClCH_2CH_2CH_2Cl$	0.29	0.04

To a large extent this mechanism was based on the observations that C_4 and C_5 products were not formed upon photolysis of 1-chloropropane and that addition of cyclohexane reduced the yield of these products, mainly C_5. However, it is worth noting that although ion–molecule reaction of alkyl chlorides are known to take place in the gas phase they do not represent the only route leading to C_4 and C_5 products. This fact was recognized by the authors, who suggested an alternative mechanism of C—C bond scission based on the fast dissociation of an excited $C_3H_7Cl^{**}$ molecule formed by the neutralization of 1-chloropropane ions by electrons:

$$C_3H_7^+ + e^- \longrightarrow C_3H_7Cl^{**} \longrightarrow {}^{\bullet}C_2H_5 + {}^{\bullet}CH_2Cl \qquad (109)$$

Examination of the data given in Table 14 shows that the main radiolytic products of 1-chloropropane in both the liquid and solid phases are H_2, HCl, C_3H_8, C_3H_6 and isomers of $C_3H_6Cl_2$. Small amounts of hexane and chlorohexanes are also formed. The yields of some of these products are higher in the liquid phase. The observation seems to indicate that, at least in part, these products are formed via free radical reactions that require considerable activation energies. The following reaction sequence was suggested for the formation of the main product, hydrogen chloride:

$$C_3H_7Cl \longrightarrow C_3H_7Cl^{+\bullet} + e^- \qquad (110)$$

$$C_3H_7Cl^* \longrightarrow C_3H_7Cl \qquad (111)$$

$$C_3H_7Cl^{+\bullet} + e^- \longrightarrow C_3H_7^{\bullet} + Cl^{\bullet} \qquad (112)$$

$$C_3H_7Cl^* \longrightarrow C_3H_7^{\bullet} + Cl^{\bullet} \qquad (112a)$$

$$C_3H_7Cl^{+\bullet} + e^- \longrightarrow C_3H_6 + HCl \qquad (113)$$

$$C_3H_7Cl^{+\bullet} + Cl^- \longrightarrow C_3H_7Cl + Cl^{\bullet} \qquad (114)$$

$$Cl^{\bullet} + C_3H_7Cl \longrightarrow HCl + C_3H_6Cl^{\bullet} \qquad (115)$$

This sequence includes reactions of both excited 1-chloropropane molecules and ions. Incidentally, the occurrence of radical reactions is also supported by the observation that in the gas phase cyclohexane did not appreciably reduce the HCl yield. Evidently, chlorine abstracts the allylic hydrogen atom from cyclohexane.

The influence of aggregation state on the decomposition of 1-chloropropane was demonstrated by Junge and Ceulemans[174,175] by comparing dose effects on product formation in polycrystalline 1-chloropropane at 70 K with those observed in glassy 1-chloropropane at the same temperature. The observed differences between the two systems were rationalized by assuming that chloride ion migration has a greater range in the glassy state.

The influence of dose[174] and additives[175] on the formation of the main products in the gamma-radiolysis of solid 1-chloropentane was investigated by Ceulemans and Claes. The main products in pure 1-chloropropane were formed with the following, zero dose extrapolated yields: $G(H_2) = 1.65$, $G(\text{pentane}) = 1.2$, $\Sigma G(\text{pentenes}) = 3$, $G(C_{10}) \sim 0.3$, $\Sigma G(C_5H_{10}Cl_2) = 1.6$. The yields of H_2, HCl, C_5H_{10} and $C_{10}H_{22}$ decreased with dose while the reverse trend was observed in the variation of $G(C_5H_{12})$ and $G(C_5H_{10}Cl_2)$ with dose, indicating that at high irradiation doses, secondary reactions took place. The overall mechanism of radiolytic decomposition suggested by the authors included both ionic and radical reactions and was quite similar to the mechanism of chloropropane radiolysis discussed earlier.

It should be noticed that Ceulemans and coworkers, in their studies of the radiation chemistry of both 1-chloropropane and 1-chloropentane, were concerned mainly with zero dose extrapolated yields and primary products. Their mechanism does not

TABLE 15. Zero dose extrapolated *G*-values of the main products formed in the radiolysis of CF_3I vapour[177]

Product	*G*, pure	*G* with 6.5% HI
I_2	0.50	3.05
CF_4	0.55	0.26
C_2F_4	0.006	0.002
C_2F_6	0.11	0.004
CF_2I_2	0.016	0.008
C_3F_8	0.012	0.003
C_2F_5I	0.001	0.0002
CHF_3	—	0.76
CH_2FI	—	1.12
H_2	—	1.67

account for the formation of all primary products. Had this been done the mechanism would be no less complicated than the mechanism suggested for ethyl bromide.

The C—F bond is considerably stronger than the C—Cl, C—Br or C—I bonds. Hence the radiation chemistry of fluorocarbons can be expected to be less complicated than the radiation chemistry of other halocarbons. This should be particularly true in so far as radical reactions are concerned. To some extent this expectation is borne out by the experimental findings in 1,1,2,2-tetrafluorocyclobutane[176], and trifluoromethyl iodide[177], both in the gas phase, as well as in the radiolysis of liquid perfluorocyclohexane[178].

In 1,1,2,2-tetrafluorocyclobutane[176] 25 different products are formed. However, only six of them are formed with *G* values higher than 0.1. The main products, with *G* values in parentheses, are HF (2.3), H_2 (0.52), 1,1-$C_2H_2F_2$ (0.47), C_2F_4 (0.17), C_2H_2 (0.20) and *c*-$C_4H_3F_3$ (0.20). The last two are mainly secondary products.

A smaller number of products is formed in CF_3I[177] (Table 15). Still, based on experiments with added HI and on the mass spectral fragmentation pattern of CF_3I, 21 possible reactions were considered by Hsieh and Hanraham[177] as being responsible for the products formed.

Radiolysis of liquid perfluorocyclohexane leads to the formation of the products listed in Table 16. Fluorine is probably also formed, but its yield was not determined. The small number of C_1–C_5 products, as well as the effect of added iodine, indicate that in the liquid phase ion fragmentation reactions are considerably less important

TABLE 16. *G*-values of the main products formed in the radiolysis of liquid *c*-C_6F_{12}[178]

Product	*G*
(*c*-$C_6F_{11})_2$	0.95
c-C_6F_{11}–C_6F_{11}	0.43
c-C_6F_{11}–C_6F_{13}	0.11
Σ C_1–C_6	0.3
Σ C_7–C_{12}	0.4
c-$C_6F_{11}I$[a]	3.1
$C_6F_{11}I$[a]	0.68
$C_6F_{13}I$[a]	0.13
$I(CF_2)_6I$[a]	0.11

[a] In the presence of I_2.

than in the gas phase. Consequently, fewer radicals and ions are formed and therefore the overall mechanism of c-C_6F_{12} decomposition can be described by a relatively small number of reactions:

$$c\text{-}C_6F_{12} \longrightarrow C_6F_{12}^{\cdot+} + e^- \quad (116)$$

$$\longrightarrow C_6F_{11}^+ + F^\cdot + e^- \text{ (or } F^-) \quad (117)$$

$$\longrightarrow c\text{-}C_6F_{12}^* \quad (118)$$

$$e^- + C_6F_{12} \longrightarrow c\text{-}C_6F_{12}^{\cdot-} \quad (119)$$

$$\longrightarrow c\text{-}C_6F_{11}^\cdot + F^- \quad (120)$$

$$c\text{-}C_6F_{12}^* \longrightarrow c\text{-}C_6F_{11}^\cdot + F^\cdot \quad (121)$$

$$F^\cdot + c\text{-}C_6F_{12} \longrightarrow n\text{-}C_6F_{13}^\cdot \quad (122)$$

$$2c\text{-}C_6F_{11}^\cdot \longrightarrow (c\text{-}C_6F_{11})_2 \quad (123)$$

$$c\text{-}C_6F_{11}^\cdot + C_6F_{11}^\cdot \longrightarrow c\text{-}C_6F_{11}\text{—}C_6F_{11} \quad (124)$$

$$c\text{-}C_6F_{11}^\cdot + C_6F_{13}^\cdot \longrightarrow c\text{-}C_6F_{11}\text{—}C_6F_{13} \quad (125)$$

In this scheme neutralization reactions leading to the formation of excited c-C_6F_{12} molecules were not taken into consideration.

On the whole, information obtained by product analysis studies does not provide a sufficient basis for a comprehensive mechanism of radiolysis even when supplemented by information from more direct methods such as ESR, pulse radiolysis and mass spectrometry. A major drawback of product analysis is the need to expose the compound studied to relatively large radiation doses in order to form detectable amounts of products. Under these conditions, the analysis is complicated by the occurrence of secondary reactions. On the other hand, from the practical point of view, the information obtained is highly relevant since it addresses itself directly to the problem of the chemical changes induced by ionizing radiation.

VI. REFERENCES

1. R. E. Bühler in *Chemistry of the Carbon–Halogen Bond* (Ed. S. Patai), Wiley, London (1973), pp. 795–864.
2. M. C. R. Symons, *Radiat. Phys. Chem.*, **15**, 453 (1980).
3. S. P. Mishra and M. C. R. Symons, *Discuss. Faraday Soc.*, **63**, 175 (1977).
4. S. P. Mishra and M. C. R. Symons, *Int. J. Radiat. Phys. Chem.*, **7**, 617 (1975).
5. P. B. Ayscongh, *Electron Spin Resonance in Chemistry*, Methuen, London (1967).
6. S. P. Mishra, M. C. R. Symons and B. W. Tattershall, *JCS Faraday I*, **71**, 1772 (1975).
7. S. P. Mishra and M. C. R. Symons, *J. Chem. Res.*, 1660 (1977).
8. E. D. Sprague and F. Williams, *J. Chem. Phys.*, **54**, 5425 (1971).
9. S. P. Mishra and M. C. R. Symons, *JCS Perkin II*, 391 (1973).
10. E. D. Sprague, *J. Phys. Chem.*, **83**, 849 (1979).
11. S. P. Mishra and M. C. R. Symons, *JCS Perkin II*, 1492 (1975).
12. S. P. Mishra, G. W. Neilson and M. C. R. Symons, *JCS Faraday II*, **70**, 1165 (1974).
13. S. P. Mishra and M. S. R. Symons, *JCS Chem. Commun.*, 577 (1973).
14. M. Shiotani and F. Williams, *J. Amer. Chem. Soc.*, **98**, 4006 (1976).
15. A. Hasegawa, M. Shiotani, and F. Williams, *Discuss. Faraday Soc.*, **63**, 157 (1977).
16. G. W. Neilson and M. C. R. Symons, *JCS Chem. Commun.*, 717 (1973).
17. R. F. Picone and M. T. Rogers, *J. Magn. Res.*, **14**, 279 (1974).
18. R. F. Picone and M. T. Rogers, *J. Chem. Phys.*, **61**, 4814 (1974).

19. P. O. Samskog, A. Lund, G. Nilsson and M. C. R. Symons, *Chem. Phys.*, **42**, 369 (1979).
20. R. Koester and K. D. Asmus, *Z. Naturforsch.*, **B26**, 1108 (1971).
21. R. Koester and K. D. Asmus in *Proceedings of the 3rd Tihany Symposium on Radiation Chemistry*, Akademiai Kiado, Budapest (1972), p. 1195.
22. J. Lichtscheidl and N. Getoff, *Int. J. Radiat. Phys. Chem.*, **8**, 661 (1976).
23. R. Koester and K. D. Asmus, *J. Phys. Chem.*, **77**, 749 (1973).
24. J. Lichtscheidl and N. Getoff, *Monats. Chem.*, **110**, 1367 (1979).
25. L. M. Dorfman and I. A. Taub, *J. Amer. Chem. Soc.*, **85**, 2370 (1963).
26. S. R. Logan and P. B. Wilmot, *Int. J. Radiat. Phys. Chem.*, **6**, 1 (1974).
27. P. Neta and D. Behar, *J. Amer. Chem. Soc.*, **102**, 4978 (1980).
28. D. Behar and P. Neta, *J. Phys. Chem.*, **85**, 690 (1981).
29. P. Neta and D. Behar, *J. Amer. Chem. Soc.*, **103**, 103 (1981).
30. D. Behar and P. Neta, *J. Amer. Chem. Soc.*, **103**, 2280 (1981).
31. J. Packer, R. L. Wilson, D. Bahnemann and K. D. Asmus, *JCS Perkin II*, 296 (1980).
32. J. M. Warman, K. D. Asmus and R. H. Schuler, *Adv. Chem. Ser.*, **82**, 25 (1968).
33. P. P. Infelta and R. H. Schuler, *J. Phys. Chem.*, **76**, 987 (1972).
34. S. J. Rzad and K. M. Bansal, *J. Phys. Chem.*, **76**, 2734 (1972).
35. K. M. Bansal and S. J. Rzad, *J. Phys. Chem.*, **76**, 2381 (1972).
36. G. W. Klein and R. H. Schuler, *J. Phys. Chem.*, **77**, 979 (1973).
37. S. J. Rzad, G. W. Klein and P. P. Infelta, *Chem. Phys. Lett.* **24**, 33 (1974).
38. A. Horowitz and L. A. Rajbenbach, *Int. J. Radiat. Phys. Chem.*, **5**, 163 (1973).
39. E. L. Davids, J. M. Warman and A. Hummel, *JCS Faraday I*, **71**, 1252 (1975).
40. J. Kroch, E. Hankiewicz and J. Piekarska, *Int. J. Radiat. Phys. Chem.*, **8**, 289 (1976).
41. N. Q. Tuan and T. Gaumann, *Radiat. Phys. Chem.*, **10**, 263 (1977).
42. N. Q. Tuan and T. Gaumann, *Radiat. Phys. Chem.*, **11**, 183 (1978).
43. M. Tanaka and K. Fueki, *J. Phys. Chem.*, **77**, 2524 (1973).
44. R. Grob and J. Casanovas, *Radiat. Phys. Chem.*, **15**, 325 (1980).
45. A. Hummel, *J. Chem. Phys.*, **49**, 4840 (1968).
46. S. J. Rzad and G. Bakale, *Chem. Phys.*, **59**, 2768 (1973).
47. B. S. Yakovlev, K. K. Ametov and G. F. Novikov, *Radiat. Phys. Chem.*, **11**, 219 (1978).
48. B. Bockrath and L. M. Dorfman, *J. Phys. Chem.*, **77**, 2618 (1973).
49. K. Ito and and Y. Hatano, *J. Phys. Chem.*, **78**, 853 (1974).
50. K. Mori, K. Ito and Y. Hatano, *J. Phys. Chem.*, **79**, 2093 (1975).
51. A. O. Allen and R. A. Holroyd, *J. Phys. Chem.*, **78**, 796 (1974).
52. A. O. Allen, T. Gangwer and R. A. Holroyd, *J. Phys. Chem.*, **79**, 25 (1975).
53. I. A. Boriyev and B. S. Yakovlev, *Int. J. Radiat. Phys. Chem.*, **8**, 511 (1976).
54. A. Hummel in *Advances in Radiation Chemistry*, Vol. 4 (Eds M. Burton and J. L. Magee), Wiley, New York (1974), Chap. 1.
55. M. Tachiya, *Chem. Phys.*, **66**, 2282 (1977).
56. H. Watanabe, *Radiat. Phys. Chem.*, **10**, 251 (1977).
57. S. Arai, A. Kira and M. Imamura, *J. Phys. Chem.*, **83**, 1946 (1979).
58. Y. Wang, J. J. Tria and L. M. Dorfman, *J. Phys. Chem.*, **83**, 1946 (1979).
59. R. Mehnert, O. Brede, J. Bos and W. Naumann, *Ber. Bunsenges. Phys. Chem.*, **83**, 992 (1979).
60. R. E. Bühler and B. Hurni, *Helv. Chim. Acta*, **61**, 90 (1978).
61. B. Hurni and R. E. Bühler, *Radiat. Phys. Chem.*, **15**, 231 (1980).
62. F. F. Sukhov, N. A. Slovokhotova, L. S. Borshagovskaya, A. A. Karatun and A. N. Pankratov, *Dokl. Akad. Nauk SSSR*, **219**, 667 (1974).
63. F. F. Sukhov, A. A. Karatun, L. S. Borshagovskaya, A. N. Pankratov and N. A. Slovokhotova, *Khim. Vys. Energ.*, **13**, 122 (1979).
64. S. P. Mishra, G. W. Neilson and M. C. R. Symons, *J. Amer. Chem. Soc.*, **95**, 605 (1973).
65. S. P. Mishra, G. W. Neilson and M. C. R. Symons, *JCS Faraday II*, **69**, 1425 (1973).
66. S. P. Mishra, G. W. Neilson and M. C. R. Symons, *JCS Faraday II*, **70**, 1165 (1974).
67. A. Kira, T. Nakamura and M. Imamura, *J. Phys. Chem.*, **81**, 591 (1977).
68. A. Kira, T. Nakamura and M. Imamura, *J. Phys. Chem.*, **82**, 1961 (1978).
69. S. Arai and M. Imamura, *J. Phys. Chem.*, **83**, 337 (1979).
70. Y. Nosaka, A. Kira and M. Imamura, *J. Phys. Chem.*, **83**, 2273 (1979).
71. A. Kira, Y. Nosaka and M. Imamura, *J. Phys. Chem.*, **84**, 1882 (1980).

72. R. L. Jones and L. M. Dorfman, *J. Amer. Chem. Soc.*, **96**, 5715 (1974).
73. R. J. Sujdak, R. L. Jones and L. M. Dorfman, *J. Amer. Chem. Soc.*, **98**, 4875 (1976).
74. L. M. Dorfman, Y. Wang, H. Y. Wang and R. J. Sujdak, *Faraday Discuss. Chem. Soc.*, **63**, 149 (1977).
75. V. M. De Palma, Y. Wang and L. M. Dorfman, *J. Amer. Chem. Soc.*, **100**, 5416 (1976).
76. G. Beck and J. K. Thomas, *J. Phys. Chem.*, **76**, 3856 (1972).
77. G. Beck and J. K. Thomas, *Chem. Phys. Lett.*, **16**, 318 (1972).
78. C. D. Jonah, M. C. Sauer Jr, R. Cooper and A. D. Trifunac, *Chem. Phys. Lett.*, **63**, 535 (1979).
79. S. Tagawa, Y. Katsumura and Y. Tabata, *Chem. Phys. Lett.*, **64**, 258 (1969).
80. Y. Katsumura, S. Tagawa and Y. Tabata, *J. Phys. Chem.*, **84**, 833 (1980).
81. S. Tagawa, Y. Katsumura, T. Ueda and Y. Tabata, *Radiat. Phys. Chem.*, **15**, 287 (1980).
82. L. Walter, F. Hirayama and S. Lipsky, *Int. J. Radiat. Phys. Chem.*, **8**, 237 (1976).
83. G. A. Salmon, *Int. J. Radiat. Phys. Chem.*, **8**, 13 (1976).
84. T. Miyazaki, *Int. J. Radiat. Phys. Chem.*, **8**, 57 (1976).
85. M. S. Henry and W. P. Helman, *J. Chem. Phys.*, **56**, 5734 (1972).
86. F. S. Dainton, T. Morrow, G. A. Salmon, and G. F. Thompson, *Proc. Roy. Soc.* (London), *A*, **328**, 457 (1972).
87. J. K. Thomas, *Int. J. Radiat. Phys. Chem.*, **8**, 1 (1976).
88. J. H. Baxendale and M. Fiti, *JCS Faraday II*, **68**, 218 (1972).
89. R. Bensasson, J. T. Richards, T. Gangwer and J. K. Thomas, *Chem. Phys. Lett.*, **14**, 430 (1972).
90. T. Gangwer and J. K. Thomas, *Radiat. Res.* **54**, 192 (1973).
91. R. Bensasson and E. J. Land, *Trans. Faraday Soc.*, **67**, 1904 (1971).
92. J. Nafisi-Movaghar and Y. Hatano, *J. Phys. Chem.*, **78**, 1899 (1974).
93. T. Wada and Y. Hatano, *J. Phys. Chem.*, **81**, 1057 (1977).
94. J. M. Bossy and R. E. Bühler, *Int. J. Radiat. Phys. Chem.*, **6**, 85 (1974).
95. P. N. Bajaj and R. M. Iyer, *Radiat. Phys. Chem.*, **16**, 21 (1980).
96. S. K. Saha and R. M. Iyer, *Radiat. Eff.*, **36**, 219 (1980).
97. U. Bruhlmann, H. Bühler, F. Marchetti and R. E. Bühler, *Chem. Phys. Lett*, **21**, 412 (1973).
98. C. D. Wagner, *Adv. Radiat. Chem.*, **1**, 199 (1969).
99. I. V. Vereshchinski, *Adv. Radiat. Chem.*, **3**, 75 (1972).
100. B. Chutny and J. Kucera, *Radiat. Res. Rev.*, **5**, 1, 55, 93, 135 (1974).
101. A. Horowitz, *Amer. Chem. Soc. Symp. Ser.*, **69**, 161 (1978).
102. J. Lichtscheidl and N. Getoff, *Montas. Chem.*, **110**, 1377 (1979).
103. H. Klever and D. Schulte-Frohlinde, *Ber. Bunsenges. Phys. Chem.*, **80**, 1259 (1978).
104. P. Neta and R. H. Schuler, *Radiat. Res.*, **47**, 612 (1971).
105. N. Latif, P. O'Neill, D. Schulte-Frohlinde and S. Steenken, *Ber. Bunsenges. Phys. Chem.*, **82**, 468 (1978).
106. P. O'Neill and S. Steenken, *Ber. Bunsenges. Phys. Chem.*, **81**, 550 (1977).
107. S. Steenken and P. O'Neill, *J. Phys. Chem.*, **81**, 505 (1977).
108. K. Eiben, D. Schulte-Frohlinde, C. Savrez and H. Zorn, *Int. J. Radiat. Phys. Chem.*, **3**, 409 (1971).
109. P. Neta and R. W. Fessenden, *J. Phys. Chem.*, **78**, 523 (1974).
110. K. M. Bansal, L. K. Patterson and R. H. Schuler, *J. Phys. Chem.*, **76**, 2386 (1972).
111. L. K. Patterson and K. M. Bansal, *J. Phys. Chem.*, **76**, 2392 (1972).
112. M. G. Katz, A. Horowitz and L. A. Rajbenbach, *Int. J. Chem. Kinet.*, **7**, 183 (1975).
113. M. G. Katz, G. Baruch and L. A. Rajbenbach, *Int. J. Chem. Kinet.*, **8**, 131 (1976).
114. M. G. Katz, G. Baruch and L. A. Rajbenbach, *Int. J. Chem. Kinet.*, **8**, 599 (1976).
115. Y. Gonen (Geliebter), A. Horowitz and L. A. Rajbenbach, *JCS Faraday I*, **72**, 901 (1976).
116. Y. Gonen (Geliebter), A. Horowitz and L. A. Rajbenbach, *JCS Faraday I*, **73**, 886 (1977).
117. Y. Gonen (Geliebter), L. A. Rajbenbach and A. Horowitz, *Int. J. Chem. Kinet.*, **9**, 361 (1977).
118. A. D. Belapurkar and R. M. Iyer, *Radiat. Eff. lett.*, **43**, 25 (1979).
119. T. Q. Nguyen, T. M. Daud and T. Gaumann, *Radiat. Phys. Chem.*, **15**, 223 (1980).
120. G. Baruch, L. A. Rajbenbach and A. Horowitz, *Int. J. Chem. Kinet.*, **13**, 473 (1981).
121. Z. B. Alfassi and L. Feldman, *Int. J. Chem. Kinet.*, **12**, 379 (1980).

122. L. Feldman and Z. B. Alfassi, *Radiat. Phys. Chem.*, **15**, 687 (1980).
123. Z. B. Alfassi, L. Feldman and A. P. Kushelevsky, *Radiat. Eff.*, **32**, 67 (1977).
124. Z. B. Alfassi, A. P. Kushelevsky and L. Feldman, *Mol. Cryst. Liquid Cryst.*, **34**, 33 (1979).
125. L. Feldman, A. P. Kushelevsky and Z. B. Alfassi, *Radiat. Eff.*, **40**, 151 (1979).
126. R. Aloni, A. Horowitz and L. A. Rajbenbach, *Int. J. Chem. Kinet.*, **8**, 673 (1976).
127. R. Aloni, L. A. Rajbenbach and A. Horowitz, *Int. J. Chem. Kinet.*, **11**, 899 (1976).
128. G. Baruch and A. Horowitz, *J. Phys. Chem.*, **84**, 2535 (1980).
129. C. Radlowski and W. V. Sherman, *J. Phys. Chem.*, **74**, 3043 (1970).
130. B. N. Ormson, W. D. Perrymore and M. L. White, *Int. J. Chem. Kinet.*, **9**, 663 (1977).
131. H. Sakurai in *Free Radicals*, Vol. II (Ed. J. K. Kochi), Wiley, New York (1973), Chap. 25, pp. 741–808.
132a. A. M. Doncaster and R. M. Walsh, *JCS Faraday I*, **72**, 1216 (1976).
132b. R. M. Walsh and J. M. Wells, *JCS Faraday I*, **72**, 100 (1976).
133. K. C. Ferguson and E. Whittle, *Int. J. Chem. Kinet.*, **2**, 479 (1971).
134. P. Cadman, G. M. Tisley and A. F. Trotman-Dickenson, *J. Chem. Soc. A*, 1370 (1969).
135. A. Horowitz and L. A. Rajbenbach, *J. Phys. Chem.*, **74**, 678 (1970).
136. M. G. Katz and L. A. Rajbenbach, *Int. J. Chem. Kinet.*, **7**, 785 (1975).
137. M. G. Katz, A. Horowitz and L. A. Rajbenbach, *Trans. Faraday Soc.*, **67**, 2354 (1971).
138. R. Aloni, M. G. Katz and L. A. Rajbenbach, *Int. J. Chem. Kinet.*, **7**, 699 (1975).
139. M. G. Katz, G.Baruch and L. A. Rajbenbach, *Int. J. Chem. Kinet.*, **9**, 55 (1977).
140. M. G. Katz, G. Baruch and L. A. Rajbenbach, *JCS Faraday I*, **72**, 1903 (1976).
141. M. Ben-Yehuda, M. G. Katz and L. A. Rajbenbach, *JCS Faraday I*, **70**, 908 (1974).
142. R. Aloni, L. A. Rajbenbach and A. Horowitz, *Int. J. Chem. Kinet.*, **13**, 23 (1981).
143. T. Sawai, N. Ohara and T. Shimokawa, *Bull. Chem. Soc. Japan*, **51**, 1300 (1978).
144. J. A. Franklin and G. H. Huybrechts, *Int. J. Chem. Kinet.*, **1**, 3 (1969).
145. Y. Nagai, K. Yamazaki, I. Shiojima, N. Kabori and M. Hayashi, *J. Organometal. Chem.*, **9**, 21 (1967).
146. M. G. Katz and L. A. Rajbenbach, *Int. J. Chem. Kinet.*, **10**, 955 (1978).
147. P. B. Nel and H. J. Van der Linde, *Z. Phys. Chem. Neue Folge*, **97**, 155 (1975).
148. H. G. Konnecke, M. Bar and H. Langguth, *Radiochem. Radioanal. Lett.*, **37**, 319 (1979).
149. A. Horowitz and L. A. Rajbenbach, *J. Amer. Chem. Soc.*, **97**, 10 (1975).
150. A. Horowitz, *Int. J. Chem. Kinet.*, **7**, 927 (1975).
151. A. Horowitz, *Int. J. Chem. Kinet.*, **8**, 709 (1976).
152. A. T. Podkhalyuzin and M. P. Nazarova, *Khim. Vys. Energ.*, **13**, 130 (1979).
153. V. Kim, A. G. Shostenko and M. D. Gasparyan, *React. Kinet. Catal. Lett.*, **12**, 479 (1979).
154. A. G. Shostenko, V. E. Myshkin and V. Kim, *React. Kinet. Catal. Lett.*, **10**, 311 (1979).
155. V. E. Myshkin, A. G. Shostenko, P. A. Zagorets, N. M. Sharapova and A. I. Pchelkin, *Khim. Vys. Energ.*, **13**, 127 (1970).
156. V. E. Myshkin, A. G. Shostenko, P. A. Zagorets, A. I. Pchelkin and K. G. Markova, *Khim. Vys. Energ.*, **12**, 11 (1978).
157. A. G. Shostenko, V. E. Myshkin and V. Kim, *React. Kinet. Catal. Lett.*, **11**, 343 (1979).
158. V. E. Myshkin, A. G. Shostenko, P. A. Zagorets and A. I. Pchelkin, *Zh. Org. Khim.*, **13**, 696 (1977).
159. V. E. Myshkin, A. G. Shostenko, P. A. Zagorets and V. Kim, *Teor. Eksp. Khim.*, **14**, 277 (1978).
160. V. E. Myshkin, A. G. Shostenko, P. A. Zagorets and N. M. Sharapova, *Izv. Vys. Uchebn. Zaved. Khim. Khim. Tekhnol.*, **20**, 1446 (1977).
161. A. G. Shostenko, V. E. Myshkin and V. Kim, *Vysokomolekul. Soedin. Ser. A*, **20**, 1560 (1978).
162. A. G. Shostenko, V. E. Myshkin and V. Kim, *Khim. Vys. Energ.*, **13**, 475 (1978).
163. A. Horowitz and L. A. Rajbenbach, *J. Amer. Chem. Soc.*, **95**, 6308 (1973).
164. A. Horowitz, A. Mey-Marom and L. A. Rajbenbach, *Int. J. Chem. Kinet.*, **6**, 2651 (1974).
165. A. Horowitz and G. Baruch, *Int. J. Chem. Kinet.*, **11**, 1263 (1979).
166. A. Horowitz and G. Baruch, *Int. J. Chem. Kinet.*, **12**, 883 (1980).
167. R. Aloni, L. A. Rajbenbach and A. Horowitz, *J. Organometal. Chem.*, **171**, 155 (1979).
168. R. Aloni, L. A. Rajbenbach and A. Horowitz, unpublished results.
169. A. J. Frank and R. J. Hanrahan, *J. Phys. Chem.*, **80**, 1533 (1976).
170. A. Guerts and J. Ceulemans, *Int. J. Radiat. Phys. Chem.*, **8**, 353 (1976).

171. A. Guerts and J. Ceulemans, *Radiat. Phys. Chem.*, **11**, 285 (1978)
172. D. De Jonge and J. Ceulemans, *Bull. Soc. Chim. Belg.*, **84**, 871 (1975).
173. D. De Jonge and J. Ceulemans, *Bull. Soc. Chim. Belg.*, **83**, 397 (1974).
174. J. Ceulemans and P. Claes, *Bull. Soc. Chim. Belg.*, **81**, 405 (1972).
175. J. Ceulemans and P. Claes, *Bull. Soc. Chim. Belg.*, **81**, 417 (1972).
176. A. R. Ravinshakara and R. J. Hanrahan, *Radiat. Phys. Chem.*, **10**, 183 (1977).
177. T. Hsieh and R. J. Hanrahan, *Radiat. Phys. Chem.*, **12**, 153 (1978).
178. G. A. Kennedy and R. J. Hanrahan, *J. Phys. Chem.*, **78**, 360 (1974).

The Chemistry of Functional Groups, Supplement D
Edited by S. Patai and Z. Rappoport
© 1983 John Wiley & Sons Ltd

CHAPTER **10**

Organic chemistry of astatine

KLARA BEREI and LÁSZLÓ VASÁROS
Central Research Institute for Physics, Budapest, Hungary

I. INTRODUCTION

The peculiar feature of the fifth element of the halogen group is that so far not one of its stable isotopes has been found in Nature; hence the origin of its name: $\alpha\sigma\tau\alpha\tau\omega\sigma$ = unstable. The longest-lived of the four astatine descendents of the natural radioactive decay chains, ^{219}At, has a half-life of less than 1 min[1] and the total amount of all four natural astatine isotopes in the Earth's crust does not exceed 50 mg, which means that astatine is by far the rarest element[2].

Isotopes that can be generally used for chemical studies, i.e. those with a half-life of several hours, can be produced only via nuclear reactions in cyclotrons or heavy ion accelerators. This may well be the reason why research work on astatine chemistry has been limited to a select few nuclear centres.

On the other hand, all investigations in this field have to be performed on tracer scale utilizing techniques which allow astatine to be measured by its radioactivity. The highest concentration of astatine that has been obtained[3] is 10^{-8} M; that needed for typical chemical experiments is in the region of 10^{-13}–10^{-15} M. The accompanying radiation sets a limit to the concentration applicable to chemical studies. (For example, because of its specific activity of 1.5×10^{16} α-particles min^{-1} cm^{-3}, a 1 M solution of ^{211}At would inevitably be a source of intense radiation and heat effects, thereby making chemical investigations impossible.) Working with tracer amounts often results in poor reproducibility due to the masking effects of impurities which are sometimes present in much higher concentrations, even in high quality commercial reagents, than the astatine compound being investigated.

Although immediately after the discovery[4] of element 85 its biological behaviour captured the interest of scientists[5], systematic studies on the chemistry of organic astatine compounds did not follow for quite a long time. One particular obstacle may have been that astatine has a more pronounced positive character than the lighter members of the halogen group. It was originally described as a metal[4] and this view remained prevalent for some time. Inorganic systems were predominantly studied and this is well reflected in numerous reviews (see for example Refs 6–10) dealing almost exclusively with the chemical behaviour of astatine in aqueous solutions.

The main aim of this chapter is to survey the results of mostly very recent investigations on the organic chemistry of astatine. These studies clearly demonstrate that the reactions of astatine in organic systems, and also the properties of its organic compounds, characterize this element as the fifth halogen.

II. PREPARATION AND MEASUREMENT OF ASTATINE

Only three out of the 24 known astatine isotopes, viz. ^{209}At, ^{210}At and ^{211}At, are regularly used for chemical studies on account of their relatively long half-lives and also of the favourable conditions for their production by means of nuclear facilities. Experiments carried out with HAt beams using ^{217}At ($T_{1/2}$ = 0.032 s) represent a valuable exception to this rule. Here, in contrast, the advantages of short-lived nuclides for radioactive monitoring are utilized to gain information on some chemical properties of astatine compounds[11].

Except for the direct measurement of its atomic absorption spectrum[12], all experimental information concerning the physical and chemical properties of astatine has been obtained by detecting the radioactivity of its isotopes. It is, therefore, of basic importance for any study in this field to prepare well defined astatine isotopes and to measure their radioactivity without interference by radiation from other isotopes present. Only a brief survey of the methods applied for these purposes can be given here.

A. Preparation of Suitable Isotopes

Nuclear processes used to produce astatine isotopes most suitable for chemical studies are summarized in Table 1. (It is worth mentioning that irradiation of metallic bismuth by 32 MeV α-particles in the 60 inch (150 cm) cyclotron of the Crocker Radiation Laboratory in Berkeley, California, led to the discovery of astatine in 1940.) According to the threshold energies of the (α, xn) reactions[13], only ^{211}At can be

TABLE 1. Nuclear processes used to produce the longest-lived astatine isotopes and their decay

Nuclear process	Isotope	Particle energy[13], MeV	$T_{1/2}$ h	Decay
^{209}Bi (α, 4n)	^{209}At	>34	5.5	EC (95.9%) α (4.1%)
^{209}Bi (α, 3n)	^{210}At	>28	8.1	EC (99.8%) α (0.2%)
^{209}Bi (α, 2n) ^{211}Rn (EC)	^{211}At	>20	7.2	EC (58.1%) α (41.9%)
^{232}Th (p, sp) ^{238}U (p, sp)	^{209}At, ^{210}At, ^{211}At	660		

sp = spallation; EC = electron capture.

obtained in a reasonable purity from the other two isotopes by irradiation of bismuth using α-particles with energy up to 28 MeV. ^{211}At is also the most widely favoured astatine isotope for chemical studies since the somewhat longer-lived ^{210}At is a health hazard, decaying into the radiotoxic ^{210}Po with a half-life of 138 days.

In routine procedures bismuth is irradiated either in metallic form[3,4,14,15], fused or vapourized on aluminium or copper backing plates, or as bismuth oxide pellets[16] pressed into holes in an aluminium plate. The target is water-cooled during the irradiation to avoid the melting of the bismuth (m.p. 271 °C) and evaporation of the astatine. The radioactive halogen itself can be removed from the irradiated target by distillation at high temperaures[14,15,17,18] (dry methods) or by extraction into organic solvents after dissolving the target in strong inorganic acid[16], occasionally combined with distillation[14,15] (wet methods).

A mixture of ^{209}At, ^{210}At and ^{211}At isotopes can be obtained (among numerous other spallation products) by bombarding thorium or uranium with 660 MeV protons in a synchrocyclotron[19,20] (Table 1). The separation of astatine isotopes in this case is more complicated due to the wide product spectrum forming in spallation reactions (sp). Nevertheless, a number of wet separation techniques have been developed[19,21,22] based essentially on selective adsorption of astatide on metallic tellurium from hydrochloric acid solution. More recently, the introduction of gas thermochromatography[20,23] has provided a simple and elegant technique for fast and selective separation of astatine from other spallation products – including isotopes of other halogens.

The same procedures can be utilized to separate neutron-deficient noble gas isotopes (decaying into halogens by electron capture (EC)) from the spallation products. Isolation of the radon isotopes from this mixture is easily performed by gas chromatography in a column packed with molecular sieves[24]. The longest-lived ^{211}Rn ($T_{1/2}$ = 14.6 h) can then be used as a source of ^{211}At (Table 1). Introducing ^{211}Rn into an organic substance enables us to study the reactions of recoil ^{211}At atoms *in situ*[26] (see Section IV.B.6).

Radon and astatine isotopes can also be obtained by heavy ion-induced nuclear reactions[14,26,27] or by photospallation[28]. Especially promising is a recently reported procedure[27] for producing ^{211}Rn and hence ^{211}At by irradiating bismuth:

$$^{209}\mathrm{Bi}(^{7}\mathrm{Li},5n)^{211}\mathrm{Rn}. \quad (1)$$

This is carried out using 60 MeV ^{7}Li ions (equation 1) in larger cyclotrons and Van de Graaff accelerators. However, this has not yet become a routine technique.

B. Nuclear Properties and Measurement

^{211}At decays[29] partly by emitting α-particles to the long-lived ^{207}Bi and partly by electron capture to ^{211}Po which, in turn, is an α-emitting isotope, see Figure 1. Due to the very short half-life of ^{211}Po (0.5 s) an equilibrium between the two isotopes is reached very rapidly, with a controlling half-life of the longer-lived astatine. This means that for each decaying ^{211}At nucleus one α-particle is emitted either by itself or by ^{211}Po with energies of 5.9 and 7.45 MeV, respectively. The characteristic α-spectrum, therefore, serves as a distinctive signature of ^{211}At. The other two long-lived astatine isotopes do not interfere since they decay by α-emission only to an insignificant extent (see Table 1).

The requirement of virtually weightless samples to avoid self-absorption, however, severely restricts α-counting as a means of assaying ^{211}At. Measurement of the 80 keV X-rays originating from the electron capture branch of its decay is preferable. This can be carried out with simple NaI(T1) scintillation counters. The X-radiation of ^{207}Bi, always present as a daughter element of ^{211}At and also decaying by electron capture, can be neglected: due to its long half-life it contributes less than 0.002% of the astatine activity.

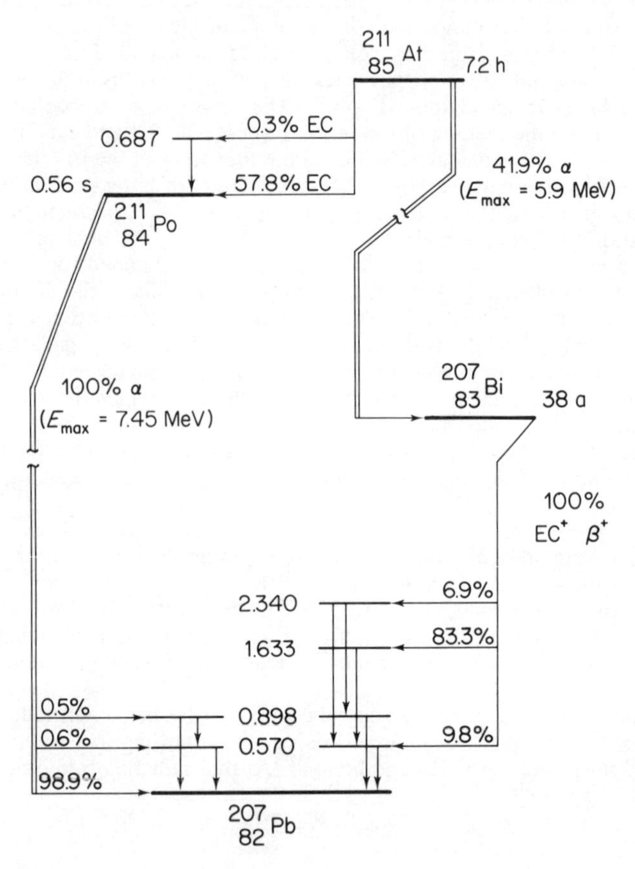

FIGURE 1. Simplified decay scheme of ^{211}At (Ref. 29). Energy values in MeV.

Similar techniques apply generally to the measurement of the other astatine isotopes, with specific modifications arising from their decay schemes.

In a number of studies, such as, for example, in the identification of newly synthesized organic astatine compounds[30–32], it does not matter very much if the activity observed originates from the mixture of the three long-lived isotopes. In these cases an average half-life is measured and a comparison of the activities in different samples can be made with satisfactory precision.

If the reactions of ^{211}At originating from ^{211}Rn by electron capture are studied, the measurement is complicated by the presence of ^{207}Po, arising from ^{211}Rn by α-decay; a further complication is due to ^{211}Rn itself being in equilibrium with its ^{211}At daughter. Adsorption of polonium on metallic tellurium at higher pH values is generally used[22,25] for its removal from aqueous solutions where it usually concentrates. Radon dissolves mainly in organic solutions and its evaporation cannot always be achieved completely. It can, however, easily be separated from organic astatine compounds by means of gas chromatographic techniques, and its radioactivity can thus be taken into account[25].

Since sufficient activity is usually available, normal radiometric equipment can be used to measure astatine without special requirements of high sensitivity or low background.

III. PHYSICAL AND CHEMICAL PROPERTIES OF ASTATINE

The physical properties of astatine are generally estimated by extrapolation from the data available for neighbouring elements and for the other members of the halogen group. Some data of this kind are shown in Table 2. Probably the only exception so far is the direct measurement of the atomic absorption spectrum[12], performed with a gaseous sample containing 10^{-9}–10^{-10} g of astatine.

TABLE 2. Physical properties of astatine

Property	Value	Reference
At		
Covalent radius	158 pm	31
Ionic radius, r_{At^-}		
Gas	197 pm	33
Crystal	230 pm	33
Ionization potentials		
IP_1	9.5 eV	34
IP_2	18.2 eV	34
Electron affinity	2.8 eV	35
Electronegativity	2.2	36
Atomic refraction	19.3 cm^3	36
$\Delta H_f^\circ[At^-/g]$ at 298 K	-195 kJ mol^{-1}	9
At_2[a]		
Internuclear distance	316 pm	31
Dissociation energy		
D_{At_2}	116 kJ mol^{-1}	37
$D_{At_2^+}$	232 kJ mol^{-1}	37
Ionization potential, IP_{At_2}	8.3 eV	37
Melting point	244°C	10
Boiling point	309°C	10

[a]Existence of molecular astatine, more precisely of At_2^+, has been established recently in the plasma ion source of a mass separator[38].

TABLE 3. Halogen redox potentials (in volts) in 0.1 M acid

X	X^-–X_2(aq.)	X_2(aq.)–HOX	HOX–HXO_2	HXO_3–HXO_4
Cl	−1.40	−1.53	−1.35	−1.13
Br	−1.09	−1.51	−1.42	—
I	−0.62	−1.31	−1.07	(−1.6)
At	(−0.3)	(−1.0)	(−1.5)	<−1.6

Reproduced from E. H. Appelman, *J. Amer. Chem. Soc.*, **83**, 805 (1961) by permission of the American Chemical Society.

It is difficult to describe the chemical nature of astatine without ambiguity. The first investigators[4] considered it to show closer resemblance to polonium than to iodine by virtue of its ability to precipitate with hydrogen sulphide and its very reduced tendency to do so with silver nitrate. Somewhat later even the existence of astatine cations in acidic aqueous solutions could be proved[39,40]. On the other hand, immediately after its discovery astatine was shown to behave in biological systems in a very similar way to iodine.[5] The volatility of astatine, presumably present in the elementary state, as well as its extractability with a variety of solvents[41], also seems to be consistent with its halogen character.

At this point it is interesting to mention that, according to computer predictions, the next (hypothetical) member of the halogen family, i.e. the heavier homologue of astatine (element 117), would be a typical metal with prevalent +1 and +3 oxidation states[42]. The observed amphoteric character of astatine[6–10] is, therefore, not surprising. It is assumed to exist in five different oxidation states in aqueous systems. Appelman[43] has made an estimation of the corresponding redox potentials in acidic media compared with those for other halogens (see Table 3).

In organic media astatine is most probably present in a relatively volatile, elementary state, generally designated as At(0). Its exact appearance has not yet been clarified. The existence of the At_2 form is excluded by the very low concentrations of astatine. So far as the $At^{\bullet}$ radical is concerned, it is very unlikely to survive due to its reactivity. Most probably At(0) is bound in some way or other to the organic species present.

Astatine is capable of both electrophilic and nucleophilic reactions in the presence of oxidizing or reducing reagents, respectively. The more pronounced positive character of astatine as compared with that of iodine is reflected in the milder oxidizing conditions necessary to perform electrophilic substitution (see Section IV.B.5).

IV. SYNTHESIS AND IDENTIFICATION OF ORGANIC COMPOUNDS

Early reports concerning the preparation of organic astatine compounds of mainly biological importance[44] were not followed by many others for more than a decade due to often contradictory results[6,45], leading to the myth that astatine exhibited a capricious character when reacting in organic systems. Besides the interference of impurities, as already mentioned, these contradictions and poor reproducibility of results may well have been caused by the use of experimental techniques like coprecipitation, distillation, etc., which were inadequate for the separation of tracer amounts of astatine compounds from the macro-scale components present. The considerable progress achieved lately in organic astatine chemistry implies the application of chromatographic methods capable of separating and identifying tracer as well as macro-scale amounts[46,47].

In the syntheses described below macro-scale amounts of iodine are often used to act as a 'non-isotopic' carrier for astatine present in tracer amounts. Sometimes, especially in earlier investigations, the analogous compounds of the two halogens obtained were also identified together. The presence of astatine in the same chemical form as iodine could in these cases be proved by measuring its α-radiation. There is a growing tendency to prepare and use organic compounds of carrier-free astatine. Therefore, in the following, mention will be made whenever iodine carrier was used to prepare and/or identify astatine compounds.

A. Compounds of Multivalent Astatine

Because of the more pronounced positive character of astatine compared with that of other halogens, one of the first attempts in this field was aimed at preparing organic derivatives of astatine in +3 and +5 valency state. Norseyev and coworkers[48,49] obtained the following types of compounds: $ArAtCl_2$ (**1**), Ar_2AtCl (**2**) and $ArAtO_2$ (**3**), where Ar = C_6H_5 or p-$C_6H_4CH_3$, via the reaction schemes seen below. Macro-scale amounts of iodine carrier labelled with ^{131}I isotope were always added, resulting

$$Ar_2ICl + At^- \longrightarrow Cl^- + Ar_2I\,.\,At \xrightarrow{170\,^\circ C} ArI + ArAt \xrightarrow[0\,^\circ C]{Cl_2} \underset{(\mathbf{1})}{ArAtCl_2} \qquad (2)$$

$$\underset{(\mathbf{1})}{ArAtCl_2} + Ar_2Hg \longrightarrow ArHgCl + \underset{(\mathbf{2})}{Ar_2AtCl} \qquad (3a)$$

$$\underset{(\mathbf{1})}{ArAtCl_2} + ArHgCl \longrightarrow HgCl_2 + \underset{(\mathbf{2})}{Ar_2AtCl} \qquad (3b)$$

$$\underset{(\mathbf{1})}{ArAtCl_2} + OCl^- + 2\,OH^- \xrightarrow{70-100\,^\circ C} \underset{(\mathbf{3})}{ArAtO_2} + 3\,Cl^- + H_2O \qquad (4)$$

in the formation of analogous iodine compounds along with those of astatine at each stage of the syntheses. (Although the mixture of the two compounds is referred to in the text, for the sake of clarity it has been omitted when writing the reaction schemes. The mode of writing KI(At) and ArI(At), etc., signifies a mixture of the major component (KI, ArI, etc.) carrying the minor component (KAt, ArAt, etc.).)

To prepare $ArI(At)Cl_2$ (**1**), first KI(At) is added to the aqueous solution of Ar_2ICl. The crystalline $Ar_2I.I(At)$ formed is centrifuged and washed with small quantities of ethyl alcohol, then sealed in glass ampoules and heated for a few minutes at 170–190 °C. The product of the thermal decomposition, ArI(At), is dissolved in chloroform, cooled to 0 °C and chlorinated into the end product (**1**) (see equation 2). It is a yellow precipitate which can be recrystallized from chloroform.

This substance is used as starting material for synthesizing $Ar_2I(At)Cl$ (**2**) by slowly adding Ar_2Hg to its hot chloroform solution (equations 3a, b). After cooling, $HgCl_2$ precipitates, leaving a chloroform solution which contains a mixture of **1** and **2**. The latter can be extracted into the aqueous phase, as has been proved by paper chromatographic analysis of the two phases[48].

$ArI(At)O_2$ (**3**) is formed if sodium hydroxide solution and acetic acid are added to the crystals of substance **1** and the mixture is then chlorinated until the yellow crystals of **1** completely transform into the white amorphous precipitate of **3** (equation 4).

The carrier iodine compounds were also used for identifying the corresponding astatine derivatives by means of paper chromatography[48] and thin layer chromatography (TLC)[49]. β- and α-activities for iodine and astatine products were measured, respectively.

B. Compounds of Monovalent Astatine

Quite a few organic compounds containing stable C–At bonds have been prepared and unambiguously identified using a variety of chemical procedures – principally during the last decade – in spite of the doubts and difficulties previously mentioned. Since progress in this field is likely to continue during and after the appearance of this volume, it seems justifiable to review the results on the basis of the methods most often applied for the synthesis rather than to concentrate on the groups of compounds obtained.

However, Table 4 has been included in order to summarize the main groups of organic astatine derivatives successfully synthesized.

TABLE 4. Preparation and identification of organic compounds of monovalent astatine

Compound	Identification[a]	Preparation[b]	Reference
$AtCH_2COOH$	IEC	At^- for I (hom.)	50, 51
n-$C_nH_{2n+1}At$ (n = 2–6)	GLC	At^- for I (het.) EC recoil At for H	50, 52 56, 86
i-$C_nH_{2n+1}At$ (n = 3–5)	GLC	At^- for I (het.) EC recoil At for H	56
At At	GLC	EC recoil At for H	86
C_6H_5At	GLC	At^- for I (hom., het.) At^- for Br (het.) At^+ for H (hom.) (α, 2n) recoil in $(C_6H_5)_3Bi$ EC recoil At for H or Cl (X = F, Cl, Br, I) Diazonium decomp. $(C_6H_5)_2I.At$ decomp. AtCl, AtBr for Cl or Br	50, 53 30, 32 82 50, 53 25, 85–87 50, 53 50, 53 14, 83
AtC_6H_4X (o-, m-, p-) (X = F, Cl, Br, I)	GLC	At^- for Br (het.) At^+ for H (hom.) EC recoil At for H or Cl Diazonium decomp. AtCl, AtBr for H	32 82 14, 25, 87 14, 60, 62 14, 83
$AtC_6H_4CH_3$ (o-, m-, p-)	GLC	Diazonium decomp.	14, 62
$AtC_6H_4NH_2$ (o-, m-, p-) (o-, m-, p-) (o-, p-)	HPLC TLC	EC recoil At for H AtCl, AtBr for H Mercury compound	14, 87 14, 83 65, 68

TABLE 4. *continued*

Compound		Identification[a]	Preparation[b]	Reference
$AtC_6H_4NO_2$	(*o*-, *m*-, *p*-)	HPLC	At^- for Br (het.)	58
	(*o*-, *m*-, *p*-)	TLC	EC recoil At for H	58
	(*m*-)		Mercury compound	65
AtC_6H_4OH	(*o*-, *p*-)	HPLC	AtCl, AtBr for H	14, 83
		TLC	Mercury compound	65, 68
AtC_6H_4COOH	(*o*-, *m*-, *p*-)	Extr.		44
		TLC	Diazonium decomp.	65
		CC		68
4-At-*N*,*N*-dimethylaniline		TLC	Mercury compound	70
4-At-anisole		TLC	Mercury compound	70
4-At-phenylalanine		PEP		
3-At-4-methoxyphenylalanine		PEP		
3-At-tyrosine		IEC	At^+ for H (hom.)	75
		PEP	Mercury compound	59, 70
3-At-5-I-tyrosine		PEP	Mercury compound	59
4-At-imidazole		TLC	Mercury compound	69
4-At-2-I-imidazole				
5-At-4-methylimidazole				
5-At-2-I-4-methylimidazole		PEP		
5-At-histidine				
5-At-uracil		HPLC	Diazonium decomp.	14, 46, 61, 64
		TLC	AtCl, AtBr for H	14
			Mercury compound	65, 69
5-At-deoxyuridine		HPLC	Diazonium decomp.	14, 46, 64
			AtCl, AtBr for Br or I	

[a]GLC = gas liquid chromatography; HPLC = high pressure liquid chromatography; IEC = ion exchange chromatography; PEP = paper electrophoresis; CC = column chromatography; Extr. = extraction.
[b](hom.) = homogeneous; (het.) = heterogeneous.

1. *Homogeneous halogen exchange*

Halogen atoms of the haloacetic acids are readily replaced by another halogen in aqueous solutions. Samson and Aten[50,51] have taken advantage of this phenomenon to prepare astatoacetic acid. Astatide ion (with iodide as a carrier) was allowed to react with an aqueous solution of iodoacetic acid at 40 °C according to equation (5):

$$ICH_2COOH + At^- \longrightarrow AtCH_2COOH + I^- \tag{5}$$

The product was extracted with ethyl ether, then the solvent evaporated to dryness and the residue recrystallized from carbon tetrachloride. The presence of astatine in the form of astatoacetic acid was proved by ion exchange chromatography. The entire

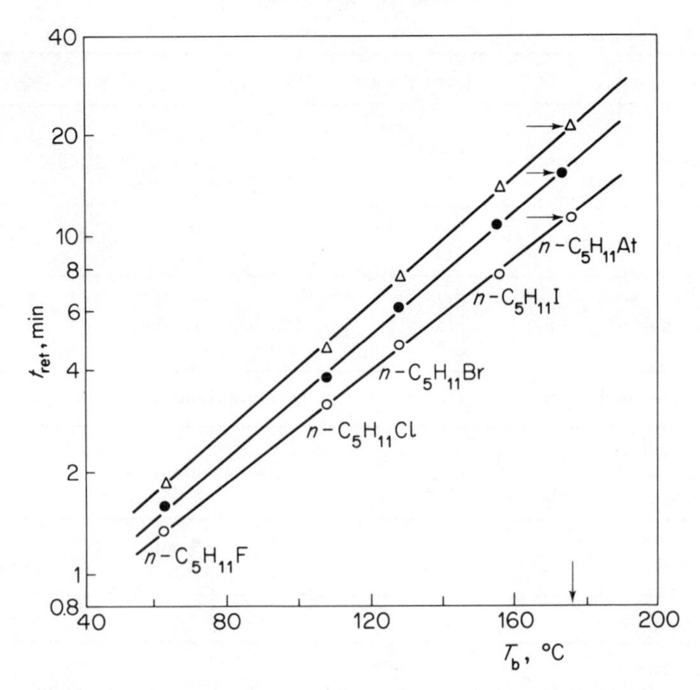

FIGURE 2. Logarithmic retention time for *n*-pentyl halogenides, measured using dinonyl phtalate stationary phase at 140 °C (△), 150 °C (●) and 160 °C (○), versus the respective boiling points of the compounds. (Reproduced from G. Samson, *Organic Compounds of Astatine*, dissertation, Universität Amsterdam (1971), by permission of the author.)

astatine activity could be eluted from the column as a single peak closely following the peak of ICH_2COOH labelled with ^{131}I.

Essentially the same procedure can be used to synthesize a number of *n*-alkyl astatides[50,52] as well as astatobenzene[50,53] at room temperature. An intense field of ionizing radiation increases the rate of the exchange reaction leading to formation of astatobenzene.

Halogen exchange between $(C_6H_5)_2I.I$ and At^- in hot ethyl alcohol solution (very similar to that described by equation 2 for Ar_2ICl and At^-) gives rise to formation of diphenyliodonium astatide ($(C_6H_5)_2I.At$). Decomposition of this compound at 175 °C has also been utilized to prepare astatobenzene[50,53].

Samson and Aten[50,52,53] were the first to use gas–liquid chromatography (GLC) to isolate organic products of astatine. This technique not only allows their separation from the corresponding products of iodine but also serves to identify them by means of sequential analysis of the analogous halogen compounds. An example of this for *n*-pentyl halogenides is shown in Figure 2. Furthermore, the difference in the GLC retention times (t_{ret}) for analogous halogen derivatives was made use of when establishing the boiling points of the corresponding astatine compounds[50,52,53]. The method was later developed and extended to determine several physicochemical properties of these compounds, as discussed in Section V.B.

2. *Heterogeneous halogen exchange*

n-Alkyl astatides[50,52] and astatobenzene[50,53] have also been prepared by means of gas chromatographic halogen exchange, as had been described earlier[54,55] for radioactive labelling of volatile organic compounds. In this case At^- is adsorbed on the solid phase and the iodine compound flows through the GLC column in an inert gas stream.

$$RI_{vapour} + At^-_{solid} \longrightarrow RAt + I^- \qquad (6)$$

where R = n-C_nH_{2n+1} (n = 2–6) or R = i-C_nH_{2n+1} (n = 3–5) or R = C_6H_5.

The exchange reaction (equation 6) was performed at a temperature of 130–200 °C in a short (15 cm) column packed with Kieselguhr and connected to a longer one for separation and analysis of the products formed. In a simplified procedure Norseyev and coworkers[56] used only the analysing column, with At^- adsorbed at its inlet, to obtain *n*- and *i*-alkyl astatides. Each iodide gave rise to the corresponding astatide – as could be established from the GLC behaviour[50,52,56,57]. Figure 3 shows a plot of the logarithmic retention time values versus boiling points for analogous alkyl iodides and alkyl astatides.

Kolachkovsky and Khalkin[30] obtained astatobenzene by an exchange reaction between At^- adsorbed on sodium iodide and bromobenzene at the boiling temperature of the latter. This technique was further developed by using sealed ampoules which enabled the temperature of the reacting systems to be increased[31,32]. A detailed study

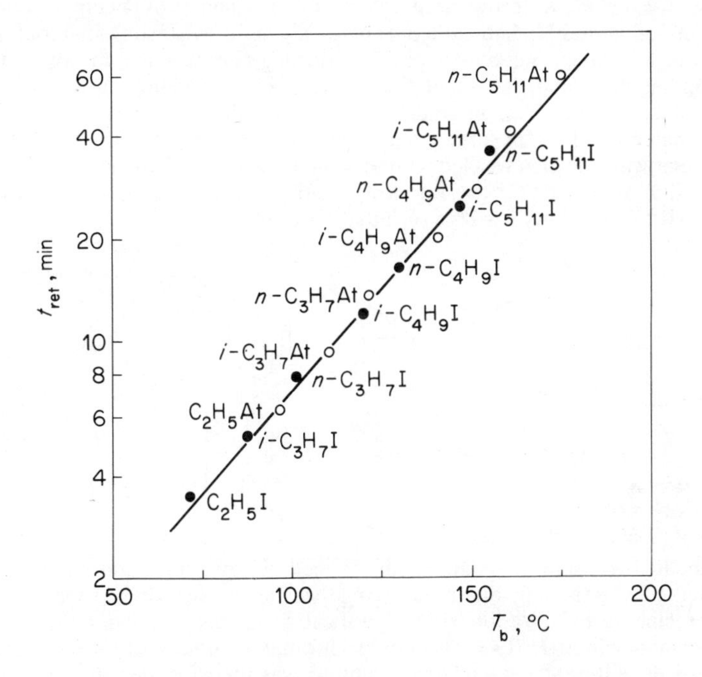

FIGURE 3. Dependence of logarithmic retention time for alkyl iodides (●) and alkyl astatides (○), measured using dionyl phtalate, on boiling points of the compounds[56].

concerning the influence of reaction time, temperature and some other factors on the synthesis yields has been carried out. As a result, the following conditions have been found to be optimal for preparing astatobenzene and the isomers of astatohalobenzenes from the corresponding bromine or iodine compounds[32]. Aqueous solution of astatine containing sodium hydroxide is evaporated to dryness, then a small amount of water is added, the ampoule sealed off and heated for about an hour at 250 °C. The isomers of astatonitrobenzene were prepared[58] at lower temperatures (50–60 °C). GLC and high pressure liquid chromatography (HPLC) served to identify the compounds formed. Yields of 50–70% and fairly good reproducibility could be obtained with these methods, thereby making the use of iodine carrier unnecessary.

Visser and colleagues[59] reported lower yields (1–5%) for At → I exchange in the solid phase when astatine and iodotyrosine or 3,5-diiodotyrosine reacted at 120 °C in a vacuum.

3. *Decomposition of diazonium salts*

In early attempts to produce benzoic acid and hence serum albumin labelled with astatine, Hughes and coworkers[44] utilized decomposition of the corresponding diazonium salts and later so did Samson and Aten in preparing astatobenzene[50,53].

More recently Meyer, Rössler and Stöcklin carried out systematic studies on the application of these reactions for synthesizing astatohalobenzene, astatotoluene and astatoaniline isomers[14,60], as well as 5-astatouracil[14,61] and 5-astatodeoxyuridine[14,64]. A comparison with analogous processes of carrier-free ^{125}I or ^{131}I under similar conditions was also made to throw more light on the mechanism of decomposition of the diazonium compounds[62]. Substituted anilines are used as starting materials for the synthesis of astatobenzene derivatives. The starting materials are converted into the corresponding diazonium salts (equation 7). The excess of sodium nitrite is destroyed by urea. Subsequently, At^- in sodium sulphite solution is added and the reaction mixture heated to 80 °C for some minutes (equation 8). The product is extracted with diethyl ether which is then washed with sodium hydroxide, dried and the diethyl ether evaporated at about 30 °C. The resulting astatine compound, identified by GLC or partition HPLC, contains 12–26% of initial radioactivity[14,60].

$$\text{X-C}_6\text{H}_4\text{-NH}_2 \xrightarrow[-5\,^\circ\text{C}]{\text{HCl, NaNO}_2} \text{X-C}_6\text{H}_4\text{-}\overset{+}{\text{N}}\equiv\text{N Cl}^- \tag{7}$$

$$\text{X-C}_6\text{H}_4\text{-}\overset{+}{\text{N}}\equiv\text{N Cl}^- \xrightarrow[-5\,^\circ\text{C}]{\text{At}^-} \left[\text{X-C}_6\text{H}_4\text{-}\overset{+}{\text{N}}\equiv\text{N At}^-\right] \xrightarrow[80\,^\circ\text{C}]{-\text{N}_2} \text{X-C}_6\text{H}_4\text{-At} \tag{8}$$

Essentially the same procedure was applied to obtain 5-astatouracil[14,61] from 5-aminouracil with the only difference being that the product can be separated from a yellow precipitate by filtration. The identification in this case can be made by ion exchange or partition HPLC. The liquid chromatographic sequence of uracil and halouracils on different ion exchange columns was used for identification of astato derivatives, as is shown in Figure 4. Iodouracil is analysed both in macro-scale concentrations and as a carrier-free compound of ^{125}I ($\sim 10^{-13}$ mol ml^{-1}) prepared by the same method as astatouracil. The chromatographic pattern indicates that the major

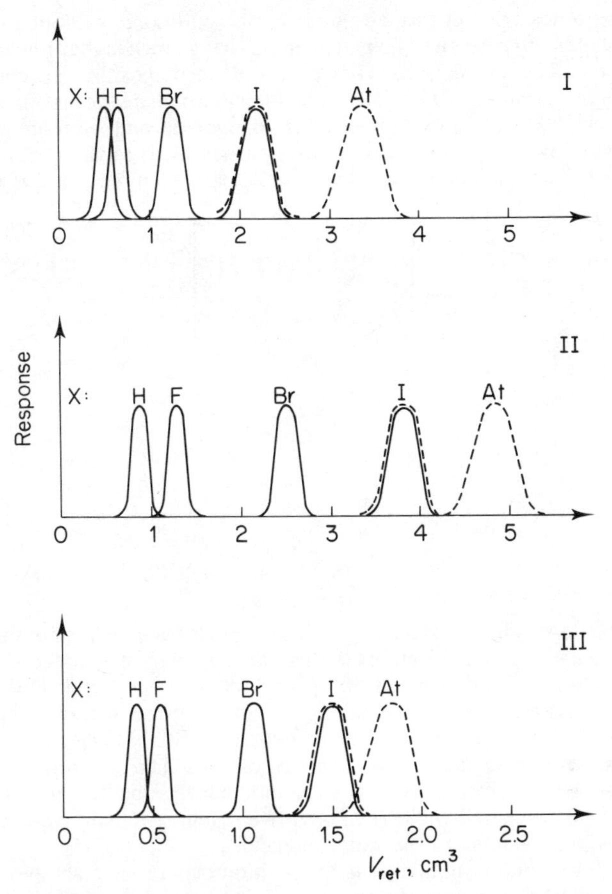

FIGURE 4. Sequential analysis of uracil, 5-X-uracils (X = F, Br, I) and carrier-free ^{125}I-uracil and ^{211}At-uracil using ion exchange HPLC. I. Aminex A7 at 25 °C, 180 bar (1.8×10^7 Pa), with 0.05 M $KHSO_4$ eluent; II. Aminex A25 at 25 °C, 50 bar (5×10^6 Pa), with 2 M HCOOH eluent; III. Aminex A27 at 60 °C, 30 bar (3×10^6 Pa), with 10^{-2} M $NaNO_3$ eluent. (Reproduced from G.-J. Meyer, K. Rössler and G. Stöcklin, *J. Labelled Compd. Radiopharm.*, **12**, 449 (1976), by permission of John Wiley and Sons, Ltd.)

radioactive product (~30% yield) formed by decomposition of the diazonium salt in the presence of At^-, is really 5-astatouracil (see Figure 4). Reinjection of this fraction gives only one and the same peak again even after keeping it at 80 °C for 30 min, which shows the stability of the C–At bond in the compound.

Data obtained for astatohalobenzene formation and for the products of carrier-free iodine might provide additional information on the widely discussed mechanism of the diazonium ion decomposition[63]. Since water is always present in much higher concentration than the trace amounts of carrier-free iodide or astatide, phenol is the product of diazonium ion decomposition to be expected. The fact that iodohalobenzenes and astatohalobenzenes are still formed under these conditions, without catalyst, giving reasonable yields, suggests that the halide anions have much higher reactivity than the hydroxide ion. The peculiar selectivity of the decomposition reaction could be explained by the formation of a relatively stable intermediate: according to Meyer and his colleagues[14,62], a complex between the halogenide and diazonium ion (**4**). Its formation is followed by an electron transfer which leads to the release of nitrogen while the phenyl and halogen radicals recombine, as is demonstrated in equation (9).

$$\text{X-}C_6H_4\text{-}\overset{+}{N}\equiv N + At^- \longleftrightarrow \left[\text{X-}C_6H_4\text{-}\bar{N}=\bar{N}^+\ At^-\right] \ (\mathbf{4}) \xrightarrow{e^-\ \text{transfer}} \text{X-}C_6H_4^{\bullet}\ (\mathbf{5}) + N_2 + At^{\bullet} \longrightarrow \text{X-}C_6H_4\text{-At} \qquad (9)$$

The heavier halogens have a tendency to form complexes which is not expected from hydroxide ion, and the high selectivity of this step may suggest an especially favourable interaction between the diazonium group and the I^- or At^- ion. Due to its lower electronegativity and higher polarizability, astatine is the better complex-forming agent. This is well in line with the observed two- to threefold higher yields for astatine than for iodine products under comparable conditions. The electron transfer (**4** → **5**) in equation (9) may proceed at lower excitation levels in the presence of I^- and especially At^- as compared with the competitive reaction of hydrolysis because of the relatively low polarizability of the water molecules.

The isomer distribution obtained in competition experiments with equimolar mixtures of *ortho*, *meta* and *para* diazonium salts[14,62] seems to confirm the suggested mechanism of intermediary complex formation. As it can be seen from Table 5, both ^{131}I and ^{211}At react preferentially with the *ortho* isomers of haloanilines, and this preference decreases in the series from fluorobenzene to iodobenzene. Though the

TABLE 5. Relative isomer distribution of astato- and iodobenzene derivatives in equimolar mixtures of *ortho*, *meta*, and *para* diazonium salts

Initial substrates	Astatohalobenzenes			Iodohalobenzenes		
	ortho	*meta*	*para*	*ortho*	*meta*	*para*
	(*ortho* + *meta* + *para* = 100)					
o-, *m*-, *p*-$FC_6H_4NH_2$	65 ± 3	25 ± 3	10 ± 3	82 ± 3	11 ± 3	7 ± 4
o-, *m*-, *p*-$ClC_6H_4NH_2$	50 ± 3	22 ± 3	28 ± 3	79 ± 2	3 ± 1	18 ± 1
o-, *m*-, *p*-$BrC_6H_4NH_2$	44 ± 3	26 ± 3	30 ± 3	59 ± 3	24 ± 1	17 ± 1
o-, *m*-, *p*-$IC_6H_4NH_2$	34 ± 4	34 ± 4	32 ± 4	64 ± 6	27 ± 6	9 ± 1

Reproduced from G.-J. Meyer, K. Rössler and G. Stöcklin, *J. Amer. Chem. Soc.*, **101**, 3121 (1979), by permission of the American Chemical Society.

ortho selectivity is stronger for the I^- ion, its dependence on the electronegativity of the halogen substituent already present in the molecule is even more clearly expressed for the At^- ion.

This phenomenon was explained by the dependence of halogenide–diazonium ion complex stability on the extent of covalency of the participating bonds. Thus, complex **4** might be further stabilized by additional charge delocalization brought about by the substituent present in the position *ortho* to the diazonium group. Accordingly, the differences in isomer distribution observed in the competition experiments could be attributed to different rates of complex formation, depending on the electronegativity of the *ortho* substituent. On the other hand, the somewhat lower *ortho* selectivity of astatide compared with that of iodide was explained by the higher steric hindrance for the bulkier halogen.

5-Astatodeoxyuridine (**7**), which is likely to be of special interest for biological studies (see Section VI), can be prepared from the corresponding amino derivative with a yield of only 2–3%, the main product (20–25%) being 5-astatouracil (**8**)[14,64]. The same is true for the iodination of 5-aminodeoxyuridine. Hydrolysis of the *N*-glycosyl bond of the starting substance (**6**) in the course of diazotation is assumed to be responsible for this phenomenon[14].

Diazotation

(**6**) (**7**) (**8**) (10)

Other laboratories[65–67] have obtained high yields (~90%) of astatobenzoic acid isomers by decomposition of diazonium salts. The products can be identified[65] by TLC. The *para* isomer is then used to prepare a biologically stable astatinated protein (bovine serum albumin) by a condensation reaction between the carboxylic group and the amine function of the protein, as has been reported by Friedman, Zalutsky and colleagues[66,67]. The benzoic acid derivative was chosen as an intermediate because aromatic halogen compounds are more stable against halogen displacement than the aliphatic ones. Astatinated protein is separated from the unreacted *p*-At-benzoic acid by column chromatography, overall yield of labelling being 12%. The labelled protein was found to be stable *in vivo* over a 20 h period.

A somewhat modified procedure has been also applied for synthesizing ^{211}At-labelled antibody proteins[68] which showed no loss of immunological specificity.

4. *Astatination via mercury compounds*

Astatine can be built into aromatic and heterocyclic molecules with high yields under relatively mild conditions using the method generally known for converting chloromercury compounds into iodides[71], as it was shown by Visser and colleagues[59,69,70]. The sequence of the reactions leading to astatinated benzene derivatives is shown schematically by equation (11):

$$X\text{-}C_6H_5 \xrightarrow[H_2SO_4]{Hg^{2+}} X\text{-}C_6H_4\text{-}Hg^+ \xrightarrow{Cl^-} X\text{-}C_6H_4\text{-}HgCl \xrightarrow[KI_3]{At^-/I_2} X\text{-}C_6H_4\text{-}At \quad (11)$$

The aromatic or heterocyclic substrate is dissolved or suspended in sulphuric acid and a somewhat less than stoichiometric amount of mercury sulphate is added. The mixture is stirred for several hours at room temperature or at 60 °C, depending on the substrate. Thereafter, twice the stoichiometric amount of sodium chloride is added at room temperature, followed after 5 min by diluted sodium hydroxide–sodium sulphite solution of astatide containing an iodide carrier and by KI_3. This mixture is stirred for an additional 5–30 min. The mercury iodide precipitate is filtered or dissolved by adding an excess of potassium iodide to the system. The astatinated products, except the amino acids, are extracted with organic solvents and identified by TLC. Astatoamino acids are analysed in aqueous solution using paper electrophoresis. Astatine derivatives of phenol, aniline, dimethylaniline, anisol, phenylalanine, uracil[70] and tyrosine[69,70], as well as of imidazoles and histidine[69], can be prepared by this method with 50–95% yields.

Compared with the decomposition of diazonium salts, astatination via chloromercury derivatives – besides the usually higher yields – has the advantage that side reactions can be avoided. This means that after removing the inorganic fraction the required product is generally present in more than 95% purity[70]. On the other hand, it should be kept in mind that the substitution pattern (isomer distribution) is determined by the mercuration reaction. Therefore, only *ortho* and *para* astatophenol or astatoaniline can be obtained by this technique, while the *meta* isomers originate if nitrobenzene is the starting material.

For mercuration of substances such as phenol that possess highly activated substitution sites, it is not necessary to use strong acidic media. Thus, for example, *ortho*- and *para*-chloromercuryphenol can be prepared using $Hg(OAc)_2$ followed by the reaction with sodium chloride as described earlier[72]. The corresponding astatophenols are then produced smoothly by interaction with At^- at room temperature, with 95% yield.

The higher reactivity of astatine compared with iodine in the reactions with chloromercury compounds has been established[70]. This is reflected in the higher yields of some astatinated as compared to those of iodinated products. Furthermore, astatinated products, though with lower yields, are also obtainable without iodine carrier present while carrier-free ^{131}I fails to react, as is shown, for example, with tyrosine, aniline and nitrobenzene. Both ionic electrophilic and radical mechanisms had earlier been proposed for the halogenation of chloromercury compounds. For astatine reacting in the absence of an iodine carrier, however, a strong indication of the radical mechanism has been found[70] which can be explained by easy oxidation of At^- to At(0) at low pH values.

5. *Electrophilic substitution*

Electrophilic substitution is one of the most characteristic features of halogen atoms. It is surprising, therefore, that not very much is known about this aspect of astatine chemistry. One of the reasons for this area not yet having been clarified is that most investigations related to the substitution of a hydrogen atom by a positive astatine species were directed towards labelling complicated molecules such as proteins and lymphocytes[73–76]. Such systems are, of course, not best suited for studying chemical reaction mechanisms. Their investigation is, however, justified by the potential importance of ^{211}At-labelled biomolecules in medical applications, as will be discussed in Section VI.

Neirinckx and coworkers[73] could label lymphocytes by electrolysis in isotonic solution leading to the formation of At^+ ions on the platinum gauze anode. Varying the electrode potential, the highest labelling yields can be obtained if the potential difference is 3–7 V. However, a rapid decomposition of the product is also observed in these cases.

Further information concerning the factors affecting the electrolytic astatination can be obtained from the work of Aaij and colleagues[74]. They investigated different techniques for electrophilic labelling of keyhole limpet haemocyanin (KLH). Electro-oxidation at pH 7.4 with 1 V potential difference for about 30 min led to coupling of about 30% of astatine present to the protein. The value of the electrode potential seems to be crucial as far as the protein denaturation is concerned: samples obtained under conditions described above do not show any significant denaturation whereas if the voltage is increased to 4–5 V the latter process becomes very fast.

It was proved in the same study that the chloramine-T technique, successfully used to oxidize species to form I^+ for labelling proteins with radioiodine, is very inefficient in the case of astatine. A probable explanation is that, due to the difference in oxidation potentials between the two halogens, chloramine-T may oxidize astatine to a higher valency state, which is not capable of electrophilic substitution in the molecules investigated.

On the other hand, oxidation by hydrogen peroxide at pH 7.4 in the presence of a small amount of potassium iodide resulted in 60% astatination of KLH. Both electrolytic oxidation and that with hydrogen peroxide were also applied for labelling human gamma-globulin and tuberculin. However, despite successful incorporation of astatine into the proteins under conditions suitable for electrophilic substitution, neither the real mechanism of labelling nor the type or site of the At bond could be determined. Thus, it remained a question whether the astatine built into the KLH by oxidation with hydrogen peroxide is bound in the tyrosine group as a positive ion or forms a complex, as AtI, with the protein molecule[74]. Though several studies have attempted to clarify these questions, we feel that no reliable answer has yet been found.

Thus, according to Vaughan and Fremlin[75] reaction of astatine with L-tyrosine in the presence of hydrogen peroxide and potassium iodide results in formation of astatotyrosine at pH $\geqslant$ 9. The product is identified by ion exchange chromatography; however, a loss of astatine with a chemical half-life of 310 min is observed. Similar instability is observed[76] if proteins (rabbit IgG immunoglobulin and the light chain fragment) are labelled by the same technique at pH 7–7.4. The authors assume that astatine bound originally to the tyrosyl residue of the protein is readily released due to the very unstable nature of the C—At bond and reacts non-specifically with other groups, finally being trapped by the tertiary structure of the protein.

Investigations carried out by Visser and colleagues[59] with astatotyrosine have shown, however, that although astatotyrosine is fairly stable in acidic solutions, it decomposes rapidly at pH $\geqslant$ 7, especially in the presence of oxidizing agents, similarly to astatoiodotyrosine, as is demonstrated in Table 6. This behaviour was then attributed to the generally known sensitivity of *o*-halophenols to oxidation[77] rather than to the weakness of the C—At bond. Fast deastatination at higher pH values can be explained by the formation of the reactive phenolate ion, as was also observed for deiodination of iodotyrosine[78,79]. Consequently, astatotyrosine is very unlikely to sur-

TABLE 6. Stability of astatotyrosine and astatoiodotyrosine at room temperature[59]

Compound	pH	Additive	Chemical half-life
3-Astatotyrosine	<1	0.2 M H_2SO_4	No decomposition in 20 h
	4	—	No decomposition in 20 h
	7	—	14–17 h
	7	H_2O_2	0.5 h
	10	—	45 min
	11.5	—	3–4 min
3-Astato-5-iodotyrosine	<1	0.2 M H_2SO_4	No decomposition in 20 h
	7	—	0.5–1.5 h

vive the conditions described for electrophilic astatination (hydrogen peroxide in neutral or alkaline media) of proteins[75,76]. Instead, a complex formation between oxidized astatine and protein was suggested without specifying its exact structure[80].

The controversy between different research groups[80,81] shows that the chemistry of these processes is not wholly understood yet. Further systematic studies and also more unequivocal techniques identifying the products are necessary prior to using electrophilic substitution to produce astatine-labelled proteins stable under physiological conditions.

More straightforward work concerning the electrophilic substitution reactions of astatine has been carried out recently with benzene and its monosubstituted derivatives, C_6H_5X, where X = H, F, Cl, Br[82]. The reactions were performed in homogeneous mixtures of the aromatic compound and acetic acid containing $H_2Cr_2O_7$ as oxidizing agent. The redox potential of the media being 1.0 V, the astatine is presumably present in the At^+ form[8] (see Table 3).

Under these conditions no significant hydrogen substitution in benzene can be detected below 80 °C. At higher temperatures, however, substitution yields of up to 50% are observed in short time periods (see Figure 5). The isomer distribution of

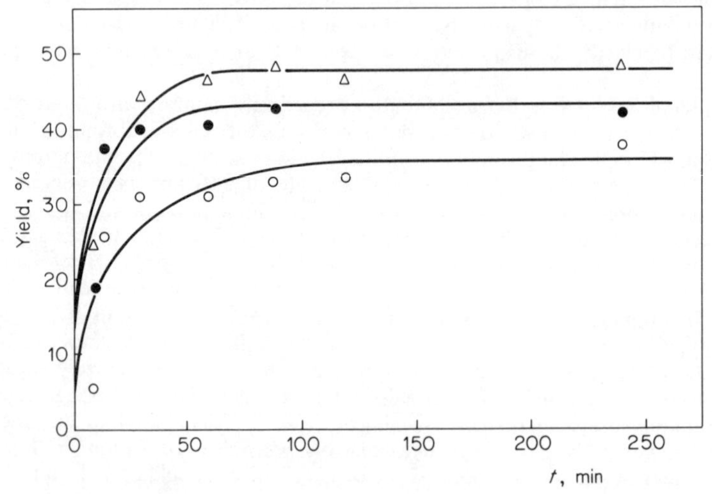

FIGURE 5. Kinetic curves of electrophilic At^+ for H substitution in benzene[82] at 80 °C (○) 100 °C (●) and 120 °C (△).

TABLE 7. Yield and isomer distribution as a result of electrophilic At^+ for hydrogen substitution in halobenzenes[82]

Halobenzenes	Yield, %	Isomer distribution (*ortho* + *meta* + *para* = 100)		
		ortho	*meta*	*para*
C_6H_5F	3.3	7 ± 1		93 ± 6
C_6H_5Cl	0.2	15.4 ± 0.8	1.6 ± 0.2	83 ± 5
C_6H_5Br	0.1	20 ± 1	2.0 ± 0.2	78 ± 5

hydrogen substitution products in monohalobenzenes clearly demonstrates the electrophilic character of the reacting astatine (see Table 7).

Reactions of AtCl and AtBr (prepared by interaction of ^{211}At with Cl_2 and Br_2 at room temperature[14,15]) in monosubstituted benzene derivatives, C_6H_5X, where X = F, Cl, Br, NH_2, OH, CH_3, were also considered to be mainly electrophilic processes as reported by Meyer and colleagues[14,64,83], by analogy with those of carrier-free ^{125}ICl[84]. This seems to be confirmed by enhanced hydrogen substitution in the activated aromatic compounds such as aniline and phenol as compared with that in halobenzenes and also by the isomer distribution of the products. In contrast to the well established mechanism of the iodine chloride reaction with aromatics, astatine chloride and astatine bromide should react in a different way, as shown by the significant extent of halogen exchange (30–40%) with halobenzenes. This phenomenon, together with the high *ortho* selectivity of hydrogen substitution in halobenzenes and aniline, has been interpreted[14,83] as being an attack of the polarized interhalogen at the electronegative site of the aromatic substrate (i.e. at the halogen atom) followed by a complex formation. This complex should then react in two different ways: either by normal aromatic substitution (proton removal) or by electrophilic halogen replacement reaction. It has also been assumed that both reactions are assisted by a Lewis base always being present in the reaction mixture.

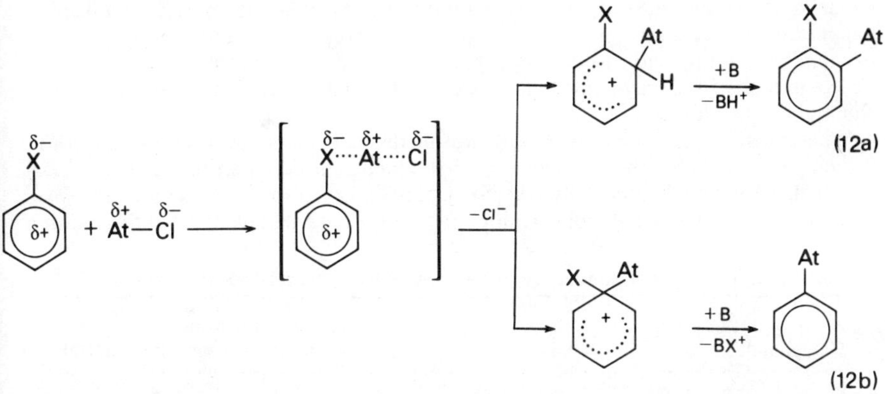

δ− X, δ+ + δ+ At—Cl δ− → [δ− X····δ+ At····Cl δ−, δ+] —(−Cl⁻)→

→ +B / −BH⁺ (12a)

→ +B / −BX⁺ (12b)

where X = F, Cl, Br; B = Lewis base.

It should be emphasized, however, that the mechanism proposed in equations (12) needs further study and more detailed information, especially on the ratio of the observed two directions as a function of reaction conditions. Better statistics of the experimental data would also be necessary to prove the assumption described above[14,83].

The high halogen replacement yields observed with halobenzenes initiated investigations with $^{211}AtCl$ in order to prepare 5-^{211}At-deoxyuridine from the corresponding iodine derivative. In this case, however, only yields of 3–4% could be obtained[14,64].

6. *Reactions of recoil astatine*

The fact that astatine can be obtained only via nuclear transformations offers a good opportunity to synthesize its compounds by immediate reactions of recoil astatine. Atoms originated in nuclear processes generally have an excess of kinetic, excitation and also ionization energy which increases their reactivity. This can give rise to chemical reactions prohibited for thermal species due to the activation energy needed.

Samson and Aten[50,53] were the first to use recoil astatination to synthesize astatobenzene by irradiating triphenylbismuth with α-particles in a synchrocyclotron. Astatobenzene, as one of the products of recoil astatine formed in nuclear reaction is separated and identified by GLC.

Norseyev and colleagues applied reactions of ^{211}At formed by electron capture from ^{211}Rn (see Table 1) in benzene and aliphatic hydrocarbons to obtain astatobenzene[85], *n*- and *i*-alkyl astatides[56] as well as cyclopentyl and cyclohexyl astatide[86].

More systematic studies were carried out on the replacement reactions of EC-produced astatine in gaseous, liquid and crystalline benzene and halobenzenes[25,82] as well as in liquid nitrobenzene[59] and aniline[87].

After its separation from the other spallation products and its subsequent purification as described in Section II.B, carrier-free ^{211}Rn is introduced into thoroughly evacuated glass ampoules containing the organic substrate. The ampoules are sealed and ^{211}Rn is allowed to decay for 14 h, until the equilibrium with ^{211}At is reached. Organic and inorganic fractions are separated by extraction of the substrate with carbon tetrachloride and aqueous sodium hydroxide solution containing a small amount of sodium sulphite as reducing agent. Identification and determination of the yields of individual organic products are performed by GLC and HPLC.

Considerable amounts of replacement products were obtained for benzene and halobenzenes[87] with the highest yields for liquid systems. The hydrogen replacement yields in aniline and nitrobenzene[59] do not differ significantly from those obtained in halobenzenes (as is shown in Table 8). This finding, together with the nearly statistical isomer distribution, confirms the assumption that the hydrogen replacement in aromatic compounds by decay-activated astatine is a hot homolytic process rather than thermal electrophilic one.

Whereas the extent of hydrogen substitution decreases in the series fluoro-, chloro-, bromo-, iodobenzene, the replacement of the halogens shows an opposite tendency (see Table 9). This is especially true for yields observed in the presence of a small amount (0.5–1.0 mol%) of iodine commonly used as radical scavenger for thermalized

TABLE 8. Hydrogen replacement of recoil ^{211}At in liquid benzene and its derivatives[59,87]

Compound	Yield, %	Isomer distribution (*ortho* + *meta* + *para* = 100)
C_6H_6	22.8 ± 2.5	
C_6H_5F	14.4 ± 4.0	38:40:22
C_6H_5Cl	10.7 ± 1.7	40:40:20
C_6H_5Br	7.8 ± 0.9	56:30:14
C_6H_5I	3.7 ± 0.2	48:36:16
$C_6H_5NH_2$	5.2 ± 1.2	52:34:14
$C_6H_5NO_2$	6.5 ± 0.4	36:44:20

TABLE 9. Halogen replacement by recoil ^{211}At in liquid halobenzenes[87]

Compound	Yield, %	
	Neat	+0.5 mol% I_2
C_6H_5F	4.9 ± 0.9	3.6 ± 0.4
C_6H_5Cl	35.3 ± 5.2	18.8 ± 0.3
C_6H_5Br	41.0 ± 5.0	27.6 ± 3.3
C_6H_5I	44.0 ± 2.0	32.8 ± 1.8

halogen atoms, i.e. to distinguish between the products of hot and thermal reactions of recoil halogens.

Competition between the halogen and hydrogen replacement seems to be responsible for the opposite tendency in their product yields through the series of halobenzenes. This may imply a common activated state for both reactions, e.g. some kind of short-lived excited intermediate complex formed as a result of a highly inelastic atom–molecule collision of astatine with the aromatic molecule, similar to that postulated earlier[88,89] for the hot replacement reactions of the other halogens in analogous systems:

$$At_{hot} + C_6H_5X \longrightarrow [C_6H_5X\cdots At]_{exc.} \begin{cases} \longrightarrow C_6H_5At + X^{\bullet} & (13a) \\ \longrightarrow C_6H_4XAt + H^{\bullet} & (13b) \end{cases}$$

where X = F, Cl, Br, I.

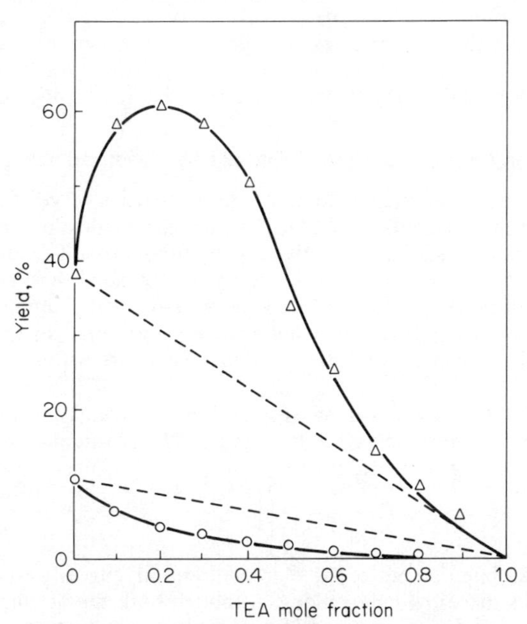

FIGURE 6. Effect of dilution with triethylamine (TEA) on ^{211}At for Cl (Δ) and ^{211}At for H (○) replacement in chlorobenzene[92]. Dashed lines represent the theoretical dilution curve.

Moreover, the halogen replacement yields both in liquid and gas phase show a linear dependence on the reciprocal bond strength of the halogen to be replaced. This again is consistent with the bond energy dependence of the hot halogen replacement established for other recoil halogens formed in different nuclear reactions[88,89] and decays[90,91]. Other factors, such as the increasing polarizability of the substituents in the same series of halobenzenes or steric effects may, however, also be of importance.

On the other hand, investigations on systems diluted with solvents of different ionization potentials (IPs) suggest a more complicated pattern of recoil astatine reactions[92]. Dilution of chlorobenzene with triethylamine (TEA) having an IP lower than astatine increases considerably the replacement of chlorine atoms while decreasing that of hydrogen atoms (Figure 6). The opposite tendency was observed when diluting chlorobenzene with carbon tetrachloride or hexafluorobenzene – both having higher IPs than astatine. Since astatine is formed in the electron capture originally in the charged state, its neutralization before taking part in the chemical reactions depends on the IPs of the surrounding molecules. Therefore, the phenomena observed in different media, as described above, may indicate a significant participation of At^+ in replacing hydrogen while neutral astatine atoms seem to prevail in halogen replacement.

V. PHYSICOCHEMICAL PROPERTIES OF ORGANIC COMPOUNDS

Even though the number of organic astatine compounds prepared and unequivocally identified has increased rapidly over the past few years, not too many of their properties are known precisely; this is because of the obvious difficulties in measuring microscale concentrations. Most of the data concerning the physicochemical properties of organic astatine compounds have been obtained by making use partly or entirely of extrapolation from the properties of analogous halogen derivatives. Along with the development of techniques for synthesizing and identifying astatine compounds some direct methods for establishing their characteristics have recently come in sight.

A. Extrapolation from Properties of Other Halo Compounds

In a study aimed at predicting some of the properties of volatile compounds of superheavy elements, Bächmann and Hoffmann[93] also made a number of estimations for the corresponding astatine derivatives. The authors found relationships between physicochemical constants giving monotonic plots for alkyl derivatives of elements which belong to the same group of the periodic system. Extrapolation of the properties according to these plots is possible because the molecular structure of the alkyl derivatives does not change essentially for the elements within one and the same group.

Thus, since both the atomic volume (v_A) and the electronegativity (χ) of the heavy atoms exhibit a definite influence on the Van der Waals' interactions of their organic derivatives, the boiling temperature (T_b) was plotted against a relationship of the former quantities:

$$T_b = f(Zv_A/\chi), \tag{14}$$

where Z is the atomic number of the element. Smooth curves were obtained for the methyl as well as the ethyl halogenides. Although both the atomic volume and the electronegativity of astatine were established – likewise by extrapolation – from corresponding values measured for the other halogens, the T_b values determined on the basis of equation (14) agree reasonably well with those obtained using other methods of extrapolation (see Table 10).

TABLE 10. Some physicochemical properties obtained by extrapolation

Property	Quantity used for extrapolation[a]	Value	Reference
CH_3At			
T_b, °C	Zv_A/χ	73 ± 5	93
	W_M	72 ± 2	94
	p_v	77 ± 5	95
IP, eV	Zr_A^2/χ	8.8	95
$D_{C\text{-}At}$, kJ mol^{-1}	Z/χ	205	95
C_2H_5At			
T_b, °C	Zv_A/χ	103 ± 5	93
	t_{ret} (GLC)	98 ± 2	50, 52
IP, eV	Zr_A^2/χ	8.65	95

[a]W_M = molecular volume; p_v = vapour pressure.

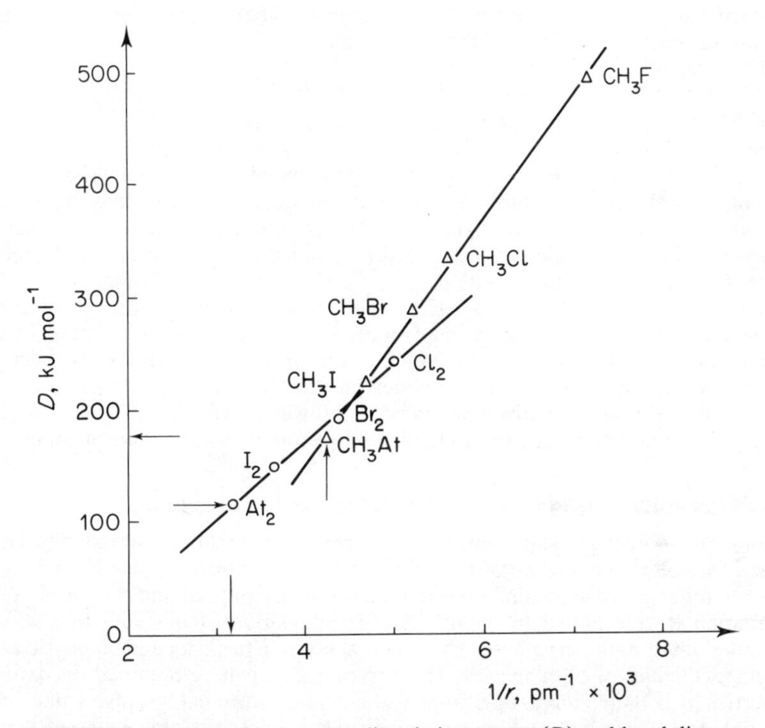

FIGURE 7. Relationship between dissociation energy (D) and bond distance (r) for halogen molecules (○) and methyl halogenides (△). (Reproduced from L. Vasáros, K. Berei, Yu. V. Norseyev and V. A. Khalkin, *Magy. Kém. Folyóirat*, **80**, 487 (1974), by permission of the Hungarian Chemical Society.)

TABLE 11. Dissociation energy values for some organic astatine and iodine compounds

	$D_{C\text{-}X}$, kJ mol^{-1}			
	X = At		X = I	
Compound	Calculated[31]	Experimental[108]	Calculated[31]	Experimental[96]
CH_3X	176		226	226
C_2H_5X	167		218	213
$n\text{-}C_3H_7X$	163	162 ± 10	213	209
$i\text{-}C_3H_7X$	159	152 ± 10	209	192–218
$n\text{-}C_4H_9X$	163		213	205
C_6H_5X	205	187 ± 20	255	252 ± 24[108]
FC_6H_4X	146		197	
ClC_6H_4X	125		175	

The dissociation energy (D_{C-At}) of methyl astatide has been determined by extrapolation from other methyl halogenides on the basis of the relationship

$$D_{C\text{-Hal}} = f(Z/\chi). \tag{15}$$

Furthermore, the ionization potentials for methyl and ethyl astatide have also been estimated using the dependence of this former constant on the covalent atomic radius (r_A) and the electronegativity of the heavy atom[95]:

$$\text{IP} = f(Zr_A^2/\chi). \tag{16}$$

The values for both latter quantities are also given in Table 10.

Another method proposed as a means of estimating the dissociation energy for some aliphatic and aromatic astatine compounds is also based primarily on the assumption of identical molecular structure of analogous derivatives[31]. According to the linear relationship found between D values and the reciprocal covalent radii for halogen molecules[96], first the covalent radius of At_2 (r_{At_2}) was estimated by extrapolation (see Figure 7) using a theoretical value for D_{At_2}[37]. Hence the bond distance (r_{C-At}) can easily be calculated and the D_{C-At} can again be determined by extrapolation from corresponding values of analogous halogen compounds (as is shown in Figure 7 for the methyl halogenide series). D_{C-At} values for other compounds can also be calculated using Szabo's method[97] of bond energy decrements. D_{C-At} values for some aliphatic and aromatic astatine derivatives estimated in this way are shown in Table 11 together with calculated and measured values for analogous iodine compounds, for comparison.

B. Determination Based on Gas Chromatographic Behaviour

Besides the separation and identification of carrier-free astatine compounds, GLC is applied for determining some of their features. Gas chromatographic behaviour of a substrate reflects its distribution between the stationary (liquid) and the gas phase; the distribution is determined by intermolecular interaction of this substrate with the molecules of the stationary phase. These interactions, in turn, depend on physicochemical characteristics of both species. Thus, information can be obtained on particular properties of volatile compounds from systematic gas chromatographic studies using different stationary phases of known characteristics. Actually, this is one of the very few techniques suitable for studying the physicochemical properties of astatine compounds due to its equal ability to separate species present in micro-scale as well as in macro-scale amounts.

It was first utilized by Samson and Aten[52] to establish the boiling points of n-alkyl astatides (n-$C_nH_{2n+1}At$, where $n = 2$–6) by extrapolation from the T_b values of other alkyl halogenides. In this case the boiling points were plotted simply against the logarithmic values of retention time obtained under identical experimental conditions for the analogous alkyl derivatives of the five halogens (see Figure 2).

Norseyev and coworkers[56,57] used the same method to establish the boiling points of n- and i-alkyl astatides. They could also show that the T_b dependence on the logarithmic retention time is linear for these compounds, similarly to that for the corresponding iodine derivatives (see Figure 3). The boiling points of astatobenzene[14,50,53,98], astatohalobenzenes and astatotoluenes[14,61] can be likewise estimated. T_b values established based on gas chromatographic behaviour are summarized in Table 12.

The chromatographic behaviour of astatine compounds in relation to that of other halogen derivatives has also provided a means of calculating the 'effective' atomic number of astatine. This latter quantity has allowed a rough estimation of physicochemical constants, such as T_b, heat of vaporization (ΔH_v), IP, D and bond distance for a number of simple aliphatic compounds of astatine and also for astatobenzene[99].

One of the factors limiting the accuracy of estimations described above is that the absolute retention time values depend on the given experimental conditions of the chromatographic separation. Therefore, if we introduce retention index[100] (I_x), which represents a relative value, i.e. retention time of the measured compound com-

TABLE 12. Boiling points of some organic astatine compounds based on gas chromatographic behaviour

Compound	T_b, °C	Ref.	Compound		T_b, °C	Ref.
CH_3At	66 ± 3	99	C_6H_5At		212 ± 2	50, 53
					219 ± 3	14
					216 ± 2	102
C_2H_5At	98 ± 2	50, 52	$AtC_6H_4CH_3$	o	237 ± 2	102
				m	237 ± 2	
				p	236 ± 2	
n-C_3H_7At	123 ± 2	50, 52	AtC_6H_4F	o	213 ± 2	102
				m	206 ± 2	
				p	209 ± 2	
i-C_3H_7At	112 ± 2	56	AtC_6H_4Cl	o	258 ± 2	102
				m	255 ± 3	
				p	253 ± 2	
n-C_4H_9At	152 ± 3	50, 52	AtC_6H_4Br	o	303 ± 3	14
				m	304 ± 3	
				p	305 ± 3	
i-C_4H_9At	142 ± 3	56	AtC_6H_4I	o	336 ± 4	14
				m	337 ± 4	
				p	337 ± 4	
n-$C_5H_{11}At$	176 ± 3	50, 52	$AtC_6H_4NO_2$	o	303 ± 3	58
				m	297 ± 3	
				p	303 ± 3	
i-$C_5H_{11}At$	163 ± 3	56				
n-$C_6H_{13}At$	201 ± 2	50, 52				

pared with that of a standard compound (usually an n-hydrocarbon) under the same conditions, we are able to improve considerably the precision of determination.

I_x can be calculated as follows[100]:

$$I_x = 100\,\frac{\log t(x) - \log t(n)}{\log t(n+1) - \log t(n)} + 100n, \tag{17}$$

where $t(x)$ is the retention time of the component x, $t(n)$ is the retention time of the n-alkane with n C-atoms, $t(n + 1)$ is the retention time of the n-alkane with $n + 1$ C-atoms, all measured under the same conditions; $t(n) \leqslant t(x) \leqslant t(n + 1)$.

An extensive study has been carried out to establish retention indices for aromatic halo compounds, including those of astatine, with a variety of stationary phases[31,32,101]. Comparison of I_x values obtained with stationary phases of different polarities allows a more reliable estimation of physicochemical constants such as T_b, ΔH_v, bond refraction (R_{C-X}) and dipole moment (μ), for At derivatives of benzene, halobenzenes, toluene and nitrobenzene[31,32,102,103]. T_b and ΔH_v can also be directly related to the gas chromatographic parameters.

Known ΔH_v values for halogenated benzene derivatives can be used to construct I_x versus ΔH_v plots from which the corresponding heat of vaporization values for analogous astatine compounds can be determined. Since the heat of vaporization is closely related to the boiling temperature (Trouton's rule), similar correlation is to be expected for the boiling points of these compounds. Figure 8 shows the linear dependence of retention indices for monosubstituted benzene derivatives, measured using Squalane stationary phase, on their normal boiling temperature as an example (the

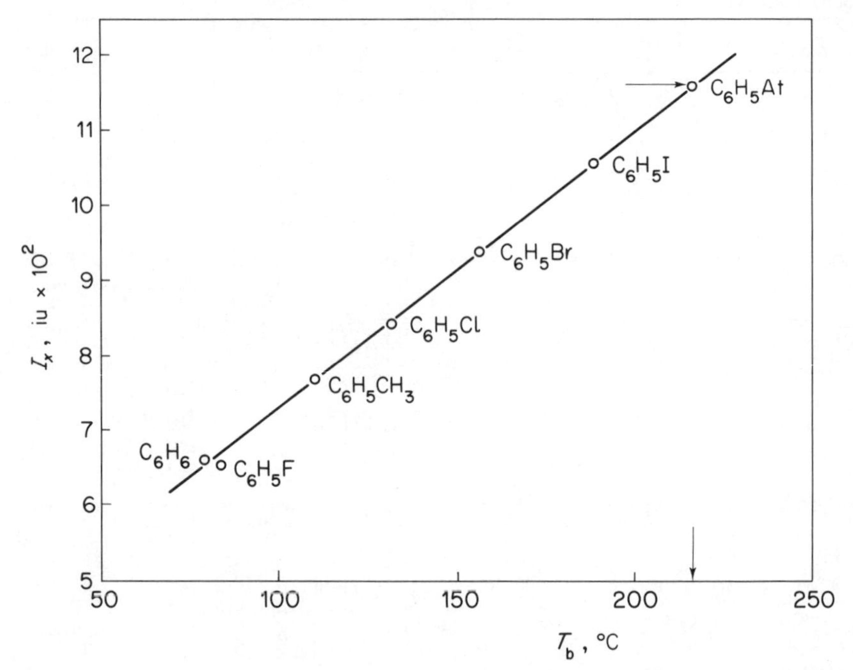

FIGURE 8. Retention index values (I_x) measured using Squalane stationary phase versus boiling temperature of benzene, toluene and halobenzenes[103] (iu = index units).

experimental value for fluorobenzene does not fit the straight line and has, therefore, been omitted in the calculations). Similar linear dependence has been obtained for T_b and ΔH_v in the cases of dihalobenzenes, halotoluenes and halonitrobenzenes. This allows one to extrapolate the values of these two quantities for the corresponding astatine derivatives.

It should be stressed at this point that physicochemical properties governed by dispersion forces, such as T_b, ΔH_v, R_{C-X}, etc., can be established with the greatest accuracy using non-polar stationary phases, such as for example Squalane or Apiezon. Polar phases, frequently used in earlier experiments for determining boiling points of astatine compounds by extrapolation, induce inaccuracy due to the polar interactions between the solute and the solvent (stationary phase) involved.

ΔH_v can also be established directly from the absolute retention volumes (V_g) measured at different column temperatures (T_c) near to the boiling point of the investigated compound by means of the equation[104]

$$\ln V_g = \frac{\Delta H_s}{RT_c} + K, \tag{18}$$

where ΔH_s is the heat of solution ($\simeq \Delta H_v$ for non-polar solvents[104]), R is the gas constant and K is a constant. Heat of vaporization values for astatobenzene and astatotoluenes calculated by this method do not differ significantly from those obtained by extrapolation. The average values[102] are given in Table 13.

Boiling temperatures for the same compounds have also been established by direct calculation using the empirical relationship of Kistiakowsky[105]:

$$\Delta H_v = \left(1 + \frac{2\mu}{100}\right) T_b(8.75 + R \ln T_b). \tag{19}$$

(μ values of the corresponding iodine compounds were used for these calculations.) The T_b values determined by the two different methods[102] are presented in Table 12.

For a series of halobenzenes and substituted halobenzenes a linear relationship has been found[103] between the retention index increments (δI_x) observed with non-polar stationary phases and dispersity factors of the corresponding halogens, as shown in

TABLE 13. Physicochemical constants based on gas chromatographic behaviour for some aromatic astatine compounds[102,103]

Compound		ΔH_v, kJ mol^{-1}	R_{C-At}, cm^3 mol^{-1}	μ_{C-At}, D
C_6H_5At		42.8	20.8	1.06 1.60[a]
$AtC_6H_4CH_3$	*ortho*	46.3		
	meta	46.6	20.7	0.90
	para	46.7		
AtC_6H_4F	*ortho*	44.6		
	meta	43.4	22.0	
	para	42.5		
AtC_6H_4Cl	*ortho*	50.8		
	meta	49.0	21.5	
	para	47.5		
		Average:	21.3	

[a]Ref. 14.

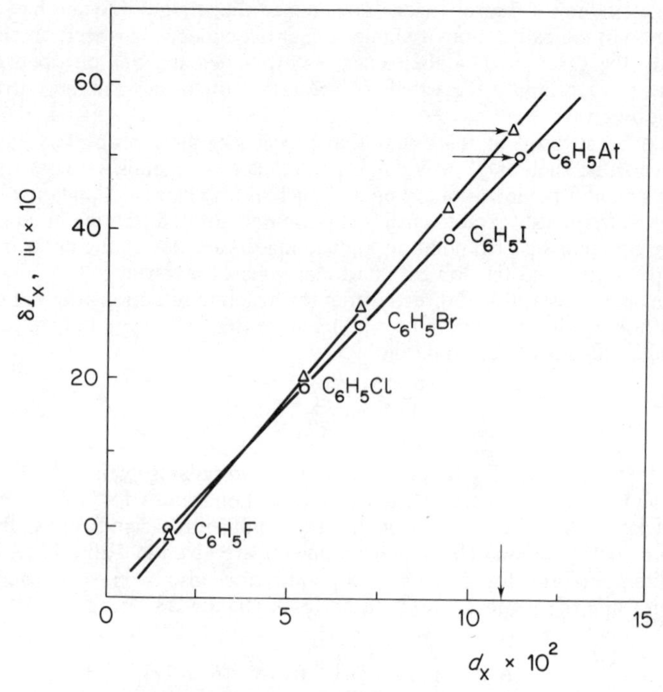

FIGURE 9. Retention index increments for halobenzenes (δI_X) measured using Squalane (○) and Apiezon (△) stationary phases versus dispersity factors of corresponding halogens[103] (iu = index units).

Figure 9. δI_x can be defined as the change in the retention index of a benzene derivative caused by the introduction of an additional halogen X into the aromatic ring:

$$\delta I_X = I_{ArX} - I_{ArH}. \tag{20}$$

The dispersity factor of corresponding halogens (d_X) can be calculated from the dispersion energy relationship[106] determined by the interactions between the functional groups of the solute and solvent:

$$\delta U_d \simeq A_d \frac{\alpha_X}{(r_0 + r_X)^3} = A_d d_X, \tag{21}$$

where δU_d is the dispersion energy increment, α_X is the polarizability of X, r_0 is the Van der Waals' radius of the solvent functional group, r_X is the Van der Waals' radius of X and A_d is a constant.

From the linear plots between δI_X and d_X the latter value for astatine can be estimated and the α_{At} can be calculated according to the equation (21). R_{C-At} is then determined using the relationship between R and α (see for example Ref. 107):

$$R = \frac{4\pi N}{3}\alpha, \tag{22}$$

where N is Avogadro's constant. The values are given in Table 13.

In order to estimate dipole moments of the C—At group (μ_{C-At}) for some aromatic astatine compounds, the differences between the retention indices observed with polar (e.g. polyethylene glycol (PEG)) and non-polar (e.g. Squalane) stationary phases have been related to the polarity factors of the halogens (p_X). This latter quantity is determined by the equation of orientation interaction energy[106] determined by the interactions between the functional groups of the solute and solvent:

$$U_{or} \approx A_{or} \frac{\mu_X^2}{T_c(r_0 + r_X)^3} = A_{or} p_X, \tag{23}$$

where U_{or} is the orientation energy, μ_X is the dipole moment of the C—X group, T_c is the absolute temperature of the column (in kelvins), r_0 is the Van der Waals' radius of solvent functional groups, r_X is the Van der Waals' radius of X and A_{or} is a constant. Figure 10 shows the $\Delta I_{ArX}^{PEG} = I_{ArX}^{PEG} - I_{ArX}^{Squalane}$ versus p_X plots[103] for halobenzenes and *p*-halotoluenes from which the μ_{C-At} values could be determined using equation (23). (This treatment involves the assumption that the difference between the retention indices observed on polar and non-polar stationary phases is controlled mainly by the orientation interactions between the solute and the solvent.) Dipole moments obtained this way for astatobenzene and *p*-astatotoluene are also given in Table 13. The value of 1.06 D for astatobenzene is lower than that reported earlier by Meyer[14], viz. 1.60 D. This latter value was derived from the difference in retention indices observed with silicon oil (non-polar) and silicon oil containing Bentone 34 (polar) stationary phases for halogenated fluorobenzenes.

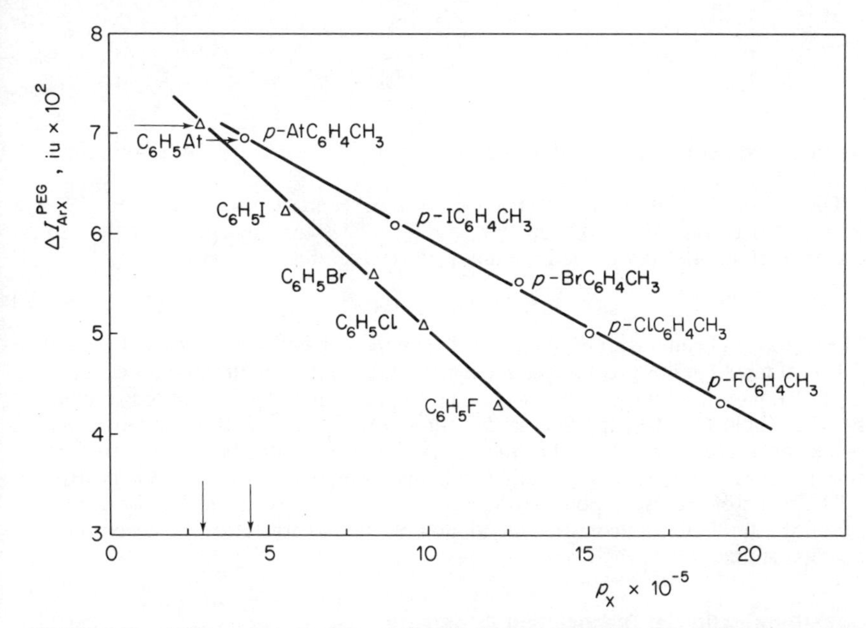

FIGURE 10. Differences in retention indices for halobenzenes (Δ) and *p*-halotoluenes (O) measured using polyethylene glycol (PEG) compared with Squalane stationary phases (ΔI_{ArX}^{PEG}) as a function of polarity factors for the corresponding halogens[103].

In contrast to the majority of procedures discussed up to this point, in the following two sections techniques are described where the conclusions are drawn from measurements of definite properties (thermal decomposition, solubility) of the astatine compounds themselves. The application of such direct methods is a very significant step forward in the still obscure field of astatine chemistry.

C. Kinetic Determination of C—At Dissociation Energy

The dissociation energy of the C—At bond for astatobenzene and *n*- and *i*-propyl astatides has been established experimentally[108] using pyrolytic decomposition of these compounds. The generally used method, well known as the toluene carrier gas technique[109,110], was slightly modified by connecting the pyrolytic oven – a Pyrex tube – directly to the gas chromatograph. This ensures continuous removal of non-dissociated original compound from the reaction zone and also its instantaneous separation from the products of pyrolysis as well as measurement. The temperature of the GLC column was kept low enough ($\leqslant 140\,^{\circ}C$) to avoid additional decomposition during the analysis. (Absence of such decomposition was proved by showing that $98 \pm 2\%$ of injected compound was eluted from the column in the same chemical form.)

The reaction rate of the monomolecular decomposition described in equation (24) follows the first order law and can be calculated according to equations (25) and (26).

$$RAt \longrightarrow R^{\bullet} + At^{\bullet} \qquad (24)$$

where $R = C_6H_5$, *n*-Pr, *i*-Pr;

$$\frac{dc}{dt} = kc; \qquad (25)$$

$$k = \frac{\ln(c/c_0)}{t}; \qquad (26)$$

where c_0 is the concentration of RAt at $t = 0$, c is the concentration of RAt at time t, k is the rate constant.

Dissociation energy was established using the Arrhenius equation (equation 27) and taking into consideration that in this case $E_a \simeq D$ since the energy of the radical recombination does not exceed the limits of the experimental error:

$$k = A\,e^{-E_a/RT} \simeq A\,e^{-D/RT}. \qquad (27)$$

The values of D could then be determined from the slopes of $\ln k$ versus $1/T$ plots. The values obtained in this way for astatine compounds together with those for carrier-free $C_6H_5{}^{131}I$, measured by the same technique to prove the reliability of the results, are listed in Table 11. The experimental D values are very close to those obtained earlier by extrapolation[31] and show that the C—At bond in astatobenzene is considerably stronger than in aliphatic, especially secondary, compounds, as is to be expected.

The value of the pre-exponential factor $A = 3 \times 10^{13}$, obtained for the decomposition of astatobenzene, confirms the monomolecular character of the decomposition reaction studied.

D. Determination of Dissociation Constants

Distribution of acids and bases between organic and aqueous phases at various acidities has been used to establish the dissociation constants (K_a) for astatoacetic acid[50,51] and for the isomers of astatobenzoic acid, astatophenol and astatoaniline as well as for astatouracil[65,70]. According to equations (28) and (29), pK_a values for acids and bases can be evaluated from $1/S$ versus $1/[H^+]$ or versus $[H^+]$ plots, respectively:

$$\frac{1}{S} = \frac{1}{S_0} + \frac{1}{S_0}\frac{K_a}{[H^+]}, \tag{28}$$

$$\frac{1}{S} = \frac{1}{S_0} + \frac{1}{S_0}\frac{[H^+]}{K_a}, \tag{29}$$

where S is the distribution coefficient for dissociated acid or base and S_0 is the distribution coefficient for undissociated acid or base.

Samson and Aten[50,51] used diisopropyl ether and water with buffer-adjusted acidities to determine the distribution of astatoacetic acid and ion exchange chromatography for the analysis. Visser and coworkers[65,70] chose heptane as the organic extractant for halogenated benzoic acids, phenols and anilines; benzene was chosen for halouracils. The analysis in this case was performed by TLC. The pK_a values for astato compounds and also for the corresponding iodine derivatives determined in these investigations are shown in Table 14.

An estimation of Hammett σ-constants and hence of the field and resonance effects was made for halophenols and haloanilines, and among them for the astatine derivatives, based on the acidity constants. A considerably weaker field effect was found for astatine than for the other halogens. The resonance effect is about the same as for iodine but again much weaker than that obtained for the other members of the halogen family[65].

TABLE 14. Dissociation constants for some astato and iodo compounds in aqueous solution at 0°C

Compound		pK_a		Reference
		X = At	X = I	
XCH_2COOH		3.78	3.14	50, 51
		3.70[a]	3.12[a]	
XC_6H_4COOH	*ortho*	2.71 ± 0.02	2.70 ± 0.02	65
	meta	3.77 ± 0.02	3.70 ± 0.02	
	para	4.03 ± 0.02	3.94 ± 0.03	
$XC_6H_4NH_2$	*ortho*	3.03 ± 0.03	2.65 ± 0.01	
	meta	3.90 ± 0.03	3.65 ± 0.02	
	para	4.04 ± 0.02	3.80 ± 0.02	
XC_6H_4OH	*ortho*	8.92 ± 0.03	8.50 ± 0.01	
	meta	9.33 ± 0.03	9.07 ± 0.02	
	para	9.53 ± 0.03	9.29 ± 0.01	
5-X-Uracil		8.97 ± 0.01	8.25 ± 0.01	

[a]At 22°C.

VI. ^{211}At IN NUCLEAR MEDICINE

The attention paid to organic astatine compounds, and leading to the recent progress in this field, stems mainly from their potential applications in medicine. This is explained by the nuclear characteristics of the ^{211}At isotope and also by its halogenous nature. By virtue of its decay by α-emission (see Section II.B), it provides intensely ionizing radiation in a short range (60 μm in water). Thus, 1 g of tissue containing 10 kBq of ^{211}At will receive a dose equivalent of 6500 μSv min^{-1} whereas the same activity of ^{125}I will give only 20 μSv min^{-}. If attached to an appropriate biomolecule, it allows, in principle, a maximal destruction of some target cells with relatively small damage to neighbouring healthy tissues.

Inorganic astatine, similarly to iodine, concentrates in the thyroid gland and can thus provide a unique means for treating hyperthyroidism. It has been shown by Hamilton and colleagues[111,112] in animal studies that the parathyroid glands remain unaltered even when the thyroid is totally destroyed.

Selective cytotoxicity might be applied in tumour therapy, for destroying malignant cells, if tumour-specific antibodies are labelled with ^{211}At. Using astatinated DNA predecessors like astatouracil and astatodeoxyuridine as carriers of α-activity into the centres of cell proliferation seems to have potential importance[14,46], as has been indicated by radiation therapeutic studies on animals with ^{125}I-labelled deoxyuridine[113]. Rössler and coworkers[64] have found the concentration of ^{211}At in the tumour tissue to be three times higher than that of the corresponding radioiodine-labelled compounds.

The labelling of red blood cells with ^{211}At may lead to selective α-irradiation of the spleen[114] since damaged erythrocytes are rapidly captured by this organ.

The success of organ transplantation can also be influenced by ^{211}At-labelled surface membrane antigens (lymphocytes). The defence mechanism of the recipient based upon specific recognition and destruction of non-self antigens by host lymphocytes results in the rejection of the transplanted organ. This often lethal side-effect can be suppressed or avoided by specific elimination of immunocompetent lymphocytes, which is, therefore, of paramount importance in transplantation attempts. Neirinckx and coworkers[73] were able to suppress to a considerable extent the rejection process by labelling baboon lymphocytes with ^{211}At. The mean survival time for the baboon group receiving astatine-labelled lymphocytes has been found to be three times longer than that for the untreated group.

Organic molecules tagged with ^{211}At are, therefore, in considerable demand for biological studies aimed at practical applications in nuclear medicine. To produce controllable circumstances the two main problems chemists face in this field are: (1) building ^{211}At selectively into certain positions of appropriate biomolecules; (2) ensuring strong enough binding of astatine in these molecules so that they remain stable under *in vivo* conditions – the loss of the ^{211}At label leads most probably to the spreading of α-radioactivity throughout the body, thereby damaging healthy organs too.

It is not surprising, therefore, that, parallel with the studies on potential immunological and therapeutic applications of ^{211}At, quite a few reports have been published on the hazards involved. It has been found, for example, that exposure of rats to sublethal amounts of astatine results in the appearance of numerous mammary tumours – including malignant ones[112]. An extensive study[115] of the metabolic effects and embryotoxicity of ^{211}At has shown increased embryo lethality, intrauterine growth retardation and induced malformations. The first two effects were found to be comparable to that produced by external X-ray exposure of similar dose range. More recently, measurable loss of reproductive capacity of cultured mammalian cells in the presence of ^{211}At, even in extremely low concentrations, has been reported[116].

Clearly, a great deal more work on the selective labelling of biomolecules with ^{211}At, to yield stable products and to provide knowledge on side-effects, is needed prior to the wide application of organic astatine compounds in human therapy.

VII. ACKNOWLEDGEMENTS

We wish to express our sincere gratitude to Professor V. A. Khalkin and Dr Yu. V. Norseyev for the opportunity to work in an international group dealing with astatine chemistry in JINR, Dubna, USSR. Without their hospitality and cooperation, this work would not have been possible. The many useful suggestions and the constructive criticism of Drs J. Márton, H. Illy and Zs. Kardos during the preparation of this manuscript are also deeply appreciated.

VIII. REFERENCES

1. E. K. Hyde and A. Ghiorso, *Phys. Rev.*, **90**, 267 (1953).
2. I. Asimov, *J. Chem. Educ.*, **30**, 616 (1953).
3. E. H. Appelman, *Chemical Properties of Astatine*, thesis, University of California, Berkeley, UCRL-9025 (1960).
4. D. R. Corson, K. R. MacKenzie and E. Segrè, *Phys. Rev.*, **57**, 1087 (1940); D. R. Corson, K. R. MacKenzie and E. Segrè, *Phys. Rev.*, **58**, 672 (1940).
5. J. G. Hamilton and M. H. Soley, *Proc. Nat. Acad. Sci.*, **26**, 483 (1940); J. G. Hamilton and M. H. Soley, *J. Appl. Phys.*,**12**, 314 (1941).
6. A. H. W. Aten, Jr, *Adv. Inorg. Chem. Radiochem.*, **6**, 207 (1964).
7. V. D. Nefedov, Yu. V. Norseyev, M. A. Toropova and V. A. Khalkin, *Usp. Khim.*, **37**, 193 (1968).
8. E. H. Appelman in *MTP International Review of Science, Inorganic Chemistry, Series I*, Vol. 3 (Ed. V. Gutman), Butterworth, London (1972), p. 183.
9. A. J. Downs and C. J. Adams, *The Chemistry of Chlorine, Bromine, Iodine and Astatine*, Pergamon Press, Oxford (1975).
10. W. A. Chalkin, E. Hermann, J. W. Norseev and I. Dreyer, *Chemiker Z.*, **101**, 470 (1977).
11. J. R. Grover, F. M. Kiely, E. Lebowitz and E. Baker, *Rev. Sci. Instr.*, **42**, 293 (1971); J. R. Grover and C. R. Iden, *J. Chem. Phys.*, **61**, 2157 (1974); J. R. Grover, C. R. Iden and H. V. Lilienfeld, *J. Chem. Phys.*, **64**, 4657 (1976); J. R. Grover, D. E. Malloy and J. B. A. Mitchell, *VIIth Int. Symp. Mol. Beams* (*Abstr.*) (Trento, Italy), May 28, 1979.
12. R. MacLaughlin, *J. Opt. Soc. Amer.*, **54**, 965 (1964).
13. K. J. Hofstetter and J. D. Stickler, *Phys. Rev.*, **9**, 1072 (1974).
14. G.-J. Meyer, *Zur Reaktivität und Selektivität anorganischer Formen des Radioelementes Astat bei Substitutionsreaktionen an aromatischen Systemen*, thesis, Jül-1418 (1977).
15. G.-J. Meyer and K. Rössler, *Radiochem. Radioanal. Lett.*, **25**, 377 (1976).
16. A. H. W. Aten, Jr, T. Doorgeest, U. Hollstein and P. H. Moeken, *Analyst*, **77**, 774 (1952).
17. G. Barton, A. Ghiorso and I. Perlman, *Phys. Rev.*, **82**, 13 (1951).
18. E. H. Appelman, E.-N. Sloth and M. H. Studier, *Inorg. Chem.*, **5**, 766 (1966).
19. B. N. Belyaev, Wang Yun-Yui, E. N. Sinotova, L. Németh and V. A. Khalkin, *Radiokhimiya*, **2**, 603 (1960).
20. V. M. Vachtel, G. V. Vinel, C. Vilov, I. I. Gromova, A. F. Novgorodov, Yu. V. Norseyev, V. A. Khalkin and V. G. Tsumin, *Isotopenpraxis*, **12**, 441 (1976).
21. Yu. V. Norseyev and V. A. Khalkin, *J. Inorg. Nucl. Chem.*, **30**, 3239 (1968).
22. M. Bochvarova, D. K. Tyung, I. Dudova, Yu. V. Norseyev and V. A. Khalkin, *Radiokhimiya*, **14**, 858 (1972).
23. J. Merinis and G. Bouissieres, *Radiochim. Acta*, **12**, 140 (1968).
24. A. Kolachkovsky and Yu. V. Norseyev, JINR P6-6923, Dubna, USSR (1969); *Nucl. Sci. Abstr.*, **27**, 22317.
25. K. Berei, L. Vasáros, Yu. V. Norseyev and V. A. Khalkin, *Radiochem. Radioanal. Lett.*, **26**, 177 (1976).
26. H. Freiesleben, H. C. Britt, J. R. Birkelund and J. R. Huizenga, University of Rochester, COO-3496-44, 94 (1974); *Nucl. Sci. Abstr.*, **30**, 22833.
27. G.-J. Meyer and R. M. Lambrecht, *Int. J. Appl. Radiat. Isotopes*, **31**, 351 (1980).
28. J. Visser, G. A. Brinkman and C. N. M. Bakker, *Int. J. Appl. Radiat. Isotopes*, **30**, 745 (1979).
29. L. J. Jardine, *Phys. Rev. C*, **11**, 1385 (1975).
30. A. Kolachkovsky and V. A. Khalkin, JINR 12-9473, Dubna, USSR (1976); *Chem. Abstr.*, **88**, 89231 p.
31. L. Vasáros, K. Berei, Yu. V. Norseyev and V. A. Khalkin, *Magy. Kém. Folyóirat* **80**, 487 (1974).

32. L. Vasáros, K. Berei, Yu. V. Norseyev and V. A. Khalkin, *Radiochem. Radioanal. Lett.*, **27**, 329 (1976).
33. G. A. Krestov, *Radiokhimiya*, **4**, 690 (1962).
34. W. Finkelnburg and F. Stern, *Phys. Rev.*, **77**, 303 (1950).
35. R. J. Zollweg, *J. Chem. Phys.*, **50**, 4251 (1969).
36. W. Gardy and W. Thomas, *J. Chem. Phys.*, **24**, 439 (1956).
37. R. W. Kiser, *J. Chem. Phys.*, **33**, 1265 (1960).
38. N. A. Golovkov, I. I. Gromova, Yu. V. Norseyev, V. G. Sandukovsky, L. Vasáros and M. Yanicky, *Radiochem. Radioanal. Lett.*, **44**, 67 (1980).
39. G. I. Johnson, R. F. Leninger and E. Segrè, *J. Chem. Phys.*, **17**, 1 (1949).
40. I. Dreyer, R. Dreyer and V. A. Khalkin, *Radiochem. Radioanal. Lett.*, **36**, 389 (1978).
41. H. M. Neumann, *J. Inorg. Nucl. Chem.*, **4**, 349 (1957).
42. B. Fricke and J. T. Waber, *Actinides Rev.*, **1**, 433 (1971).
43. E. H. Appelman, *J. Amer. Chem. Soc.*, **83**, 805 (1961).
44. W. L. Hughes and D. Gitlin, US AECD, BNL-314 (1954); W. L. Hughes and J. Klinenberg, US AECD, BNL-367 (1955); W. L. Hughes, E. Smith and J. Klinenberg, US AECD, BNL-406 (1956); W. L. Hughes and D. Gitlin, *Fed. Proc.*, **14**, 229 (1955).
45. J. J. C. Schats and A. H. W. Aten, Jr, *J. Inorg. Nucl. Chem.*, **15**, 197 (1960).
46. K. Rössler, W. Tornau and G. Stöcklin, *J. Radioanal. Chem.*, **21**, 199 (1974).
47. K. Berei, L. Vasáros and Zs. Kardos, *J. Radioanal. Chem.*, **21**, 419 (1974).
48. V. D. Nefedov, Yu. V. Norseyev, H. Savlevich, E. N. Sinotova, M. A. Toropova and V. A. Khalkin, *Dokl. Akad. Nauk SSSR*, **144**, 806 (1962).
49. Yu. V. Norseyev and V. A. Khalkin, *Chem. Zvesti*, **21**, 602 (1967).
50. G. Samson, *Organic Compounds of Astatine*, thesis, Universität Amsterdam (1971).
51. G. Samson and A. H. W. Aten, Jr, *Radiochim. Acta*, **9**, 53 (1968).
52. G. Samson and A. H. W. Aten, Jr., *Radiochim. Acta*, **12**, 55 (1969).
53. G. Samson and A. H. W. Aten, Jr., *Radiochim. Acta*, **13**, 220 (1970).
54. F. Schmidt-Bleek, G. Stöcklin and W. Herr, *Angew. Chem.*, **72**, 778 (1960); G. Stöcklin in *Proceedings of the International Symposium on the Preparation and Biomedical Application of Labelled Molecules*, Euratom, Brussels (1964), p. 481.
55. S. Krutzik and H. Elias, *Radiochim. Acta*, **7**, 26 and 33 (1967).
56. M. Gesheva, A. Kolachkovsky and Yu. V. Norseyev, *J. Chromatogr.*, **60**, 414 (1971).
57. A. Kolachkovsky and Yu. V. Norseyev, *J. Chromatogr.*, **84**, 175 (1973).
58. L. Vasáros, Yu. V. Norseyev, M. Perez, V. I. Fominikh and V. A. Khalkin, to be published in JINR communications.
59. G. W. M. Visser, E. L. Diemer and F. M. Kaspersen, *Int. J. Appl. Radiat. Isotopes*, **30**, 749 (1979).
60. G.-J. Meyer, K. Rössler and G. Stöcklin, *Radiochem. Radioanal. Lett.*, **21**, 247 (1975).
61. G.-J. Meyer, K. Rössler and G. Stöcklin, *J. Labelled Compd. Radiopharm.*, **12**, 449 (1976).
62. G.-J. Meyer, K. Rössler and G. Stöcklin, *J. Amer. Chem. Soc.*, **101**, 3121 (1979).
63. H. H. Hodgson, *Chem. Rev.*, **40**, 251 (1947); J. G. Carey, G. Jones and I. T. Millar, *Chem. Ind. (London)*, 1018 (1959); D. H. Hey, S. H. Jones and M. J. Perkins, *Chem. Commun.*, 1438 (1970); P. R. Singh and R. Kumar, *Aust. J. Chem.*, **25**, 2133 (1972).
64. K. Rössler, G.-J. Meyer and G. Stöcklin, *J. Labelled Compd. Radiopharm.*, **13**, 271 (1977).
65. G. W. M. Visser, E. L. Diemer and F. M. Kaspersen, *Rec. Trav. Chim.*, **99**, 93 (1980).
66. M. R. Zalutsky, A. M. Friedman, F. C. Buckingham, W. Wung, F. P. Stuart and S. J. Simonian, *J. Labelled Compd. Radiopharm.*, **13**, 181 (1977).
67. A. M. Friedman, M. R. Zalutsky, W. Wung, F. Buckingham, P. V. Halpern, Jr., G. H. Scherr, B. Wainer, R. L. Hunter, E. H. Appelman, R. M. Rothberg, F. W. Fitch, F. P. Stuart and S. J. Simonian, *Int. J. Nucl. Med. Biol.*, **4**, 219 (1977).
68. A. T. M. Vaughan, *Int. J. Appl. Radiat. Isotopes*, **30**, 576 (1979).
69. G. W. M Visser, E. L. Diemer and F. M. Kaspersen, *Int. J. Appl. Radiat. Isotopes*, **31**, 275 (1980).
70. G. W. M. Visser, E. L. Diemer and F. M. Kaspersen, *J. Labelled Compd. Radipharm.*, **17**, 657 (1980).
71. Houben-Weyl, *Methoden der organischen Chemie*, Vol. V/4, Thieme Verlag, Stuttgart (1960), p. 589.

72. O. Dimroth, *Chem. Ber*., **31**, 2154 (1898).
73. R. D. Neirinckx, J. A. Myburgh and J. A. Smit in *Proceedings of the Symposium on the Development and Radiopharmacology of Labelled Compounds*, Vol. II, IAEA, Vienna (1973), p. 171; *Chem. Abstr*., **82**, 14941n; J. A. Smit, J. A. Myburgh, R. D. Neirinckx, *Clin. Exp. Immunol*., **14**, 107 (1973); *Chem. Abstr*., **79**, 30327k.
74. C. Aaij, W. R. J. M. Tschroots, L. Lindner and T. E. W. Feltkamp, *Int. J. Appl. Radiat. Isotopes*, **26**, 25 (1975).
75. A. T. M. Vaughan and J. H. Fremlin, *Int. J. Appl. Radiat. Isotopes*, **28**, 595 (1977).
76. A. T. M. Vaughan and J. H. Fremlin, *Int. J. Nucl. Med. Biol*., **5**, 229 (1978).
77. J. M. Lj. Mihailović and Z. Čekovićin, *The Chemistry of the Hydroxyl Group*, Part I (Ed. S. Patai), Wiley Interscience, New York (1971).
78. J. R. Tata, *Biochem. J*., **72**, 214 (1959).
79. A. Taurog, *Endocrinology*, **73**, 45 (1963).
80. G. W. M. Visser and F. M. Kaspersen, *Int. J. Nucl. Med. Biol*., **7**, 79 (1980).
81. A. T. M. Vaughan, *Int. J. Nucl. Med. Biol*., **7**, 80 (1980).
82. L. Vasáros, Yu. V. Norseyev and V. A. Khalkin, JINR P-12-80-439, Dubna, USSR (1980).
83. G.-J. Meyer, K. Rössler and G. Stöcklin, *Radiochim. Acta*, **24**, 81 (1977).
84. R. M. Lambrecht, C. Mantescu, C. Redvanly and A. P. Wolf, *J. Nucl. Med*., **13**, 266 (1972).
85. V. D. Nefedov, M. A. Toropova, V. A. Khalkin, Yu. V. Norseyev and V. I. Kuzin, *Radiokhimiya*, **12**, 194 (1970).
86. V. I. Kuzin, V. D. Nefedov, Yu. V. Norseyev, M. A. Toropova, V. A. Khalkin and E. S. Filatov, *Khim. Vys. Energ*., **6**, 181 (1972); *Chem. Abstr*., **77**, 27357y.
87. L. Vasáros, Yu. V. Norseyev, G.-J. Meyer, K. Berei and V. A. Khalkin, *Radiochim. Acta*, **26**, 171 (1979).
88. K. Berei and G. Stöcklin, *Radiochim. Acta*, **15**, 39 (1971).
89. K. Berei and L. Vasáros, *Radiochim. Acta*, **21**, 75 (1974).
90. A. Shanshal, *Jülich Report*, Jül-1402 (1977); *Chem. Abstr*., 87, 200248v.
91. H. H. Coenen, H.-J. Machulla, A. Shanshal and G. Stöcklin, *Abstr. 9th Int. Hot Atom Chem. Symp*., Blacksburg, VA, USA (1977), p. 78.
92. L. Vasáros, Yu. V. Norseyev, V. A. Khalkin and K. Berei, *Radiochim. Acta*, in press.
93. K. Bächmann and P. Hoffmann, *Radiochim. Acta*, **15**, 154 (1971).
94. P. Hoffmann, *Radiochim. Acta*, **17**, 169 (1972).
95. P. Hoffmann, *Radiochim. Acta*, **19**, 69 (1973).
96. V. I. Vedeneyev, L. V. Gurvich, V. N. Kondratyev, V. A. Medvedyev and V. L. Frankevich, *Dissociation Energy of the Chemical Bond* (in Russian), Izdatyelstvo AN SSSR, Moscow (1962); M. J. S. Dewar, *The Electronic Theory of Organic Chemistry*, Oxford University Press, Oxford (1952).
97. Z. G. Szabó, *Z. Elektrochem*., **61**, 1083 (1957); Z. G. Szabó and T. Bérces, *Acta Chim. Hung*., **22**, 461 (1960).
98. V. I. Kuzin, V. D. Nefedov, Yu. V. Norseyev, M. A. Toropova and V. A. Khalkin, *Radiokhimiya*, **12**, 414 (1970).
99. Yu. V. Norseyev and V. D. Nefedov, *Investigations on Chemistry, Technology and Application of Radioactive Species* (in Russian), Interuniversity Compilation, Technological University of Leningrad, Leningrad (1977).
100. E. Kováts, *Helv. Chim. Acta*, **41**, 1915 (1958); A. Wehrli and E. Kováts, *Helv. Chim. Acta*, **42**, 2709 (1959); E. Kováts, *Adv. Chromatogr*., **1**, 229 (1965).
101. L. Vasáros, Yu. V. Norseyev and V. A. Khalkin, JINR 12-12188, Dubna, USSR (1979).
102. L. Vasáros, Yu. V. Norseyev and V. A. Khalkin, JINR P 6-80-158, Dubna, USSR (1980).
103. L. Vasáros, Yu, V. Norseyev and V. A. Khalkin, to be published in JINR communications.
104. J. R. Condor in *Progress in Gas Chromatography*, Vol. 6 (Ed. J. H. Purnell), Wiley, New York (1968), p. 209.
105. W. Kistiakowsky, *Z. Phys. Chem*., **107**, 65 (1923).
106. N. V. Volkenstein, *Stroyeniye i fizicheskiye svoistva molekul*, AN SSSR, Moscow (1955), p. 129.
107. G. M. Barrow, *Physical Chemistry*, McGraw-Hill, New York (1961), p. 298.
108. L. Vasáros, Yu. V. Norseyev and V. A. Khalkin, to be published in JINR communications.
109. T. L. Cottrell, *The Strength of Chemical Bonds*, Butterworth, London (1959), p. 59.

110. M. Szwarc, *Proc. Roy. Soc.* (London) *A*, **207**, 5 (1951).
111. J. G. Hamilton, P. W. Durbin, C. W. Asling and M. E. Johnston in *Proceedings of the International Conference on the Peaceful Uses of Atomic Energy*, Vol. 10, Geneva (1956), p. 176.
112. P. W. Durbin, C. W. Asling, M. E. Johnston, M. W. Parrott, N. Jeung, M. H. Williams and J. G. Hamilton, *Radiat. Res.*, **9**, 378 (1956).
113. W. D. Bloomer and S. J. Adelstein, *Nature*, **265**, 620 (1977).
114. G. Samson, *11. Jahrestagung Ges. Nuclearmed., Athens 1973*, Schattauer Verlag, Stuttgart (1974), p. 506; *Nucl. Sci. Abstr.*, **32**, 19939.
115. C. Borras, R. O. Gorson and R. L. Brent, *Phys. Med. Biol.*, **22**, 118 (1977); C. Borras, R. L. Brent, R. O. Gorson and J. F. Lamb, Jefferson University, Philadelphia, COO-3268-5 (1974); *Chem. Abstr.*, **82**, 166749u.
116. C. R. Harris, S. J. Adelstein, T. J. Ruth and A. P. Wolf, *Radiat. Res.*, **74**, 590 (1978).

The Chemistry of Functional Groups, Supplement D
Edited by S. Patai and Z. Rappoport
© 1983 John Wiley & Sons Ltd

CHAPTER **11**

Positive halogen compounds

A. FOUCAUD

Département de Physique Cristalline et Chimie Structurale, Université de Rennes, France

I. INTRODUCTION

A halogen compound can be considered as a 'positive halogen' compound when, opposed to some nucleophilic reagents, the result of the first steps of the reaction is the electron transfer represented in equation (1): the halogen atom X is finally transferred with only a sextet of electrons. Frequently, XNu is an intermediate.

$$RX + Nu^- \longrightarrow R^- + XNu \qquad (1)$$

This reaction may be a nucleophilic substitution on the halogen atom, or the consequence of the dissociation of RX (equation 2) or, perhaps, in some cases, the result of a discrete electron transfer (equations 3 and 4).

$$RX \rightleftharpoons R^- + X^+ \qquad (2)$$

$$RX + Nu^- \longrightarrow (RX^{\dot{-}}\ Nu^{\bullet}) \qquad (3)$$

$$(RX^{\dot{-}}\ Nu^{\bullet}) \longrightarrow R^- + XNu \qquad (4)$$

The reaction expressed by equation (1) is determined by several factors:

(1) The nucleophilic reactivity of Nu. The nucleophile must be highly polarizable[1,2]. For instance, the reactions of phenacyl bromide, in protic solvent, proceed via initial attack of the carbon atom with pyridine[3] (equation 5) and of bromine atom with triphenylphosphine[4] (equation 6). The relative reactivities of the halogens toward

$$PhCOCH_2Br + C_5H_5N \xrightarrow{MeOH} PhCOCH_2{-}\overset{+}{N}C_5H_5\ \ Br^- \qquad (5)$$

$$PhCOCH_2Br + PPh_3 \xrightarrow{MeOH} PhCOCH_2^-\ \ Br\overset{+}{P}Ph_3 \xrightarrow{MeOH} PhCOCH_3 + MeBr + OPPh_3 \qquad (6)$$

nucleophilic attack are in the sequence in which the soft acceptor character is increasing: $F < Cl < Br < I$.

(2) The electron density at halogen and the stabilization of the incipient ion which results from the removal of positive halogen by nucleophile. Hence, the attack on halogen should be enhanced by further substitution at the α-carbon atom by electron-withdrawing groups.

(3) The bulk of groups on the carbon atom bonded to the halogen, which may forbid the normal S_N2 displacement of halide ion for steric reasons.

A large diversity of reaction types proceed through nucleophilic attack on halogen. For instance, halogenation of ketones by halogen and reduction of halo compounds by thiol anions or phosphines belong to this class of reactions. The typical sources of positive halogens are halogen molecules, hypohalites, *N*-haloamides, polyhalo compounds and generally R—X compounds (X = halogen) were R is a good anionic leaving group.

It is important to note that these halo compounds may also be, in certain conditions, halogenating agents by a free radical mechanism. For instance, hydrocarbons can be halogenated with *t*-butyl hypochlorite, *N*-chloroamines[5] or *t*-butyl hypobromite[5,6] by a free radical chain reaction, e.g. *N*-haloamides are precursors of amidyl radicals[7] and *N*-halosuccinimides are precursors of succinimidyl radicals[8].

This review is not exhaustive, but it does include examples representative of the reactions of organic positive halogen compounds and of their synthetic uses. The emphasis in this chapter is on more recent contributions to the subject. Mention should be made of previous reviews covering certain aspects of the chemistry of positive halogen[9,10].

II. COMPOUNDS WITH HALOGEN–OXYGEN BONDS

A. Alkyl Hypohalites

Alkyl hypochlorites[11,12], hypobromites[13] and hypoiodites[14] have been known for many years, and their preparation and homolytic cleavages have been reviewed. Fluorinated alkyl hypochlorites have been prepared more recently. The reaction of chlorine monofluoride with alcohols gives the corresponding fluoroalkyl hypochlorites[15]:

$$R_FOH + ClF \longrightarrow R_FOCl + HF$$

Chlorine monofluoride adds to fluorinated ketones[16,17] or carbonyl fluoride[16] in the presence of CsF, and yields fluorinated hypochlorites:

$$R_FCOR_F + ClF \xrightarrow[-20\,°C]{MF} R_FCF(OCl)R_F \quad (M = K, Rb, Cs)$$

$$F_2CO + ClF \xrightarrow[-20\,°C]{CsF} CF_3OCl$$

Pentafluoro sulphur hypochlorite (**1**) was prepared from chlorine monofluoride and F_4SO[17,18]:

$$F_4SO + ClF \xrightarrow[-20\,°C]{CsF} \underset{(\mathbf{1})}{SF_5OCl}$$

The conversion of fluorinated alkoxide salts to the corresponding hypochlorites by reaction with chlorine monofluoride has been reported[18,19]:

$$R_FOM + ClF \longrightarrow R_FOCl + MF$$

The fluorinated hypochlorites appear to be much more stable than the corresponding alkyl hypochlorites, many of which are difficult to isolate[19]. The hypochlorites may react violently with certain reducing agents and the fluorinated alkyl hypochlorites are expected to be quite toxic.

Hypofluorites, recently synthesized, can act as a source of positive fluorine. Trifluoromethyl hypofluorite has been widely employed in organic synthesis[20]. Pentafluoroethyl hypofluorite (**2**) was prepared by the reaction of fluorine with sodium trifluoroacetate, at −75 °C[21,22]:

$$CF_3CO_2Na + 2\,F_2 \longrightarrow \underset{(\mathbf{2})}{CF_3CF_2OF} + NaF + \tfrac{1}{2}O_2$$

1. *Reactions with sulphides*

The oxidation of a number of sulphides with *t*-butyl hypochlorite gives an alkoxysulphonium salt (**3**), according to an anionic mechanism: oxygen did not inhibit the formation of **3** and no significant change in product conversion was observed in the presence of cyclohexane; in particular, cyclohexyl chloride could not be detected[23].

$$R_2S + t\text{-BuOCl} \longrightarrow R_2\overset{+}{S}Cl\ t\text{-BuO}^- \longrightarrow \underset{(\mathbf{3})}{(R_2\overset{+}{S}\ OBu\text{-}t)Cl^-} \longrightarrow \underset{(\mathbf{4})}{R_2\overset{+}{S}(O\ Bu\text{-}t)SbCl_6^-}$$

The alkoxysulphonium chlorides (**3**) are too unstable to be isolated, but can be transformed into the readily isolable hexachloroantimonates (**4**) by addition of antimony pentachloride to the solution of **3**[24].

Johnson and Rigau[25] have reported evidence of the formation of a tetravalent sulphur species (**5**) in such oxidations.

Indeed, treatment of thiane **6** with *t*-butyl hypochlorite gives an ion pair (**7**) which reacts quickly to form the sulphurane (**5**).

$$\text{(6) thiane} \xrightarrow{t\text{-BuOCl}} \text{(7) } \overset{+}{S}\text{—Cl} + t\text{-BuO}^- \longrightarrow \text{(5) } S(\text{OBu-}t)(\text{Cl})$$

The treatment of **5** with mercuric chloride, even in the presence of ethanol, gives *t*-butoxysulphonium trichloromercurate (**8**). However, in the same conditions, **7** gives rise to ethoxysulphonium trichloromercurate (**9**). These results differentiate the salt **7** from the sulphurane **5**.

$$\overset{+}{S}\text{—OBu-}t\ \ HgCl_3^- \quad \textbf{(8)} \qquad \overset{+}{S}\text{—OEt}\ \ HgCl_3^- \quad \textbf{(9)}$$

The oxidation of sulphides having at least one hydrogen at the α-carbon by hypochlorites gives α-substitution products **10**.

Chlorosulphonium salts (**11**) react with the alcohol (solvent) R^2OH to give alkoxysulphonium chlorides (**12**). This salt can either lose a proton, yielding the α-substitution product **10** or form the corresponding sulphoxide (**13**) by elimination of alkyl halide. This reaction could be useful for the synthesis of certain hemithioacetals (**10**) or for the oxidation of sulphides to sulphoxides (**13**) without concomitant formation of sulphones[24].

$$RSCH_2R^1 \xrightarrow{t\text{-BuOCl}} R\text{—}\overset{+}{S}(Cl)\text{—}CH_2R^1\ \ t\text{-BuO}^- \ \textbf{(11)} \xrightarrow{R^2OH} R\text{—}\overset{+}{S}(OR^2)\text{—}CH_2R^1\ \ Cl^- \ \textbf{(12)}$$

$$\textbf{(12)} \longrightarrow R\text{—}\overset{+}{S}(OR^2)\text{—}\overset{-}{C}H\text{—}R^1 \longrightarrow R\text{—}S\text{—}CH(OR^2)\text{—}R^1 \ \textbf{(10)}$$

$$\textbf{(12)} \longrightarrow R\text{—}S(=O)\text{—}CH_2R^1 + R^2Cl \ \textbf{(13)}$$

Walling and Mintz found that the oxidation of diphenyl sulphide or dimethyl sulphide into sulphoxides requires approximately 2 mol of *t*-butyl hypochlorite; a part of the hypochlorite is decomposed[26].

$$Ph_2S + t\text{-BuOCl} \longrightarrow Ph_2SO + i\text{-}C_4H_8 + HCl$$

$$t\text{-BuOCl} + HCl \longrightarrow t\text{-BuOH} + Cl_2$$

$$Ph_2S + Cl_2 \longrightarrow Ph_2SCl_2 \xrightarrow{t\text{-BuOH}} Ph_2SO + t\text{-BuCl} + HCl$$

$$Ph_2SCl_2 + i\text{-}C_4H_8 \longrightarrow Ph_2S + HCl + CH_2{=}C(CH_3)CH_2Cl$$

According to Swern and coworkers, oxidation of sulphides with *t*-butyl hypochlorite, immediately followed by reaction with amide anions R^1NH^- or malononitrile anion, leads to sulphimides (**14**) or ylides (**15**), respectively[27].

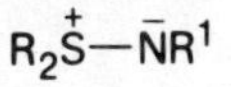

(**14**), R^1 = CN, $PhSO_2$, R^2CO

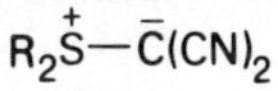

(**15**)

A stable tetracoordinate sulphur compound (**16**) was prepared by treatment of hypochlorite **17** with sulphide **18**, followed by addition of the potassium alkoxide (**20**) to the intermediate compound **19**[28].

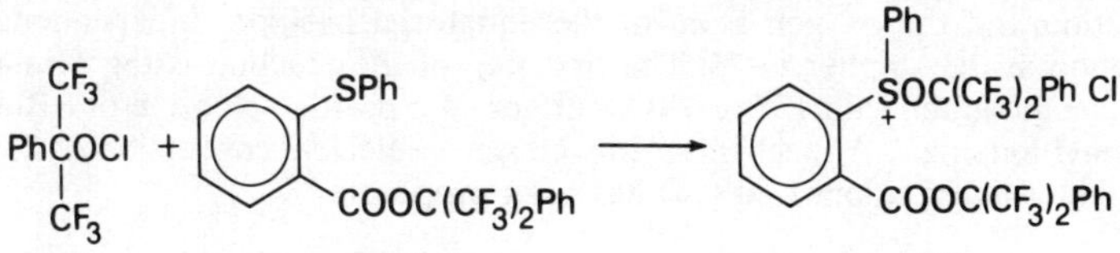

(**17**) (**18**) (**19**)

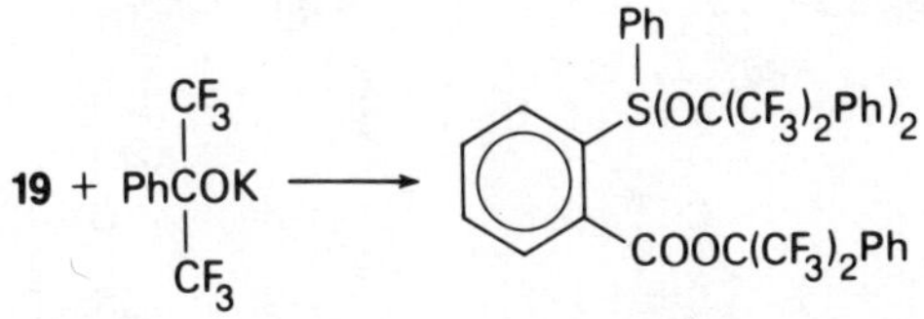

(**20**) (**16**)

t-Butyl hypochlorite has been used for the synthesis of optically active chlorosulphuranes **22** from sulphides **21** by Martin and Balthazor[29,30] and for the preparation of stable spirosulphuranes (**24**) from sulphides (**23**)[31].

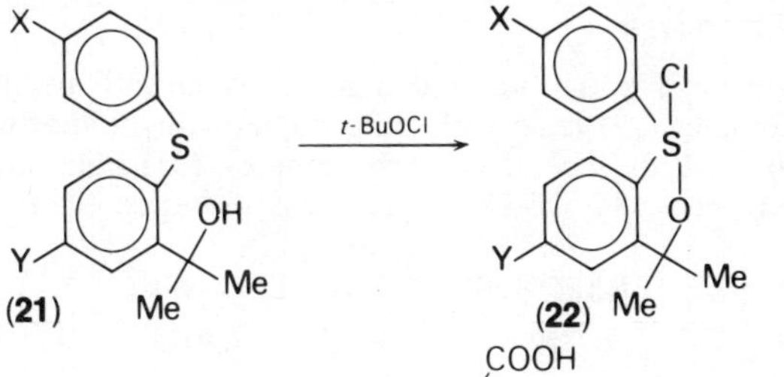

(**21**) (**22**)

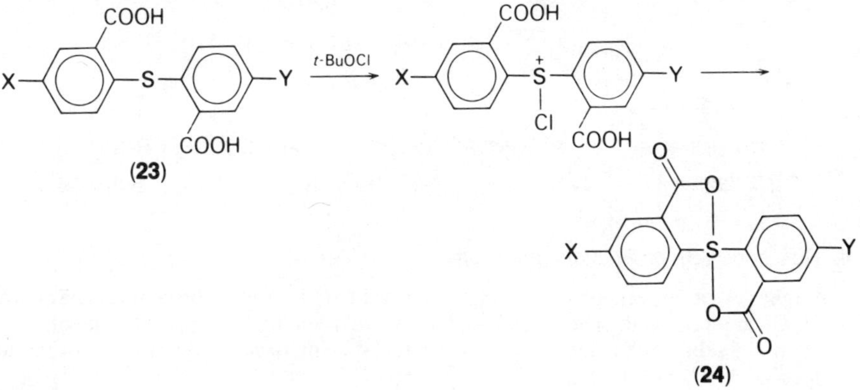

(**23**) (**24**)

2. Reactions with sulphoxides

The halogenation of sulphoxides with *t*-butyl hypohalites has been described by Iriuchijima and Tsuchihashi. This reaction seems to be ionic in the presence of pyridine[32].

$$\text{RSOCH}_2\text{Ph} \xrightarrow[\text{X = Cl, Br}]{t\text{-BuOX}} \text{RSOCHXPh}$$

The stereochemistry of the chlorination of cyclic sulphoxides with hypochlorite has been studied. The two isomers of thiane 1-oxide (**25**) have been chlorinated with *t*-BuOCl in the presence of pyridine. The chlorine atom of the sulphoxide **26** is in the axial position and the oxygen atom in the equatorial position, independent of the configuration at the sulphur in **25**. The first step of the reaction is the formation of chloro-oxosulphonium salts (**27** or **28**) by attack of a positive chlorine of *t*-BuOCl on the sulphinyl lone pair. A mechanism for the stereoselective conversion of **27** and **28** into an α-chlorinated sulphoxide (**26**) has been proposed[32–35].

(**25**) 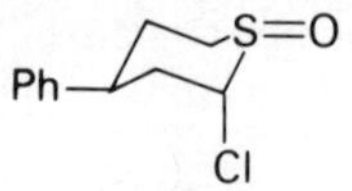 (**26**)

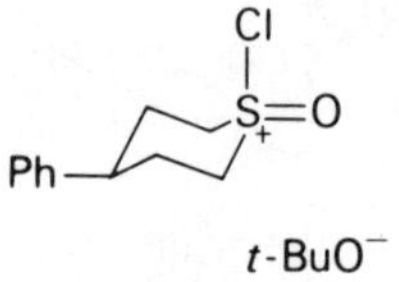

t-BuO$^-$ t-BuO$^-$

(**27**) 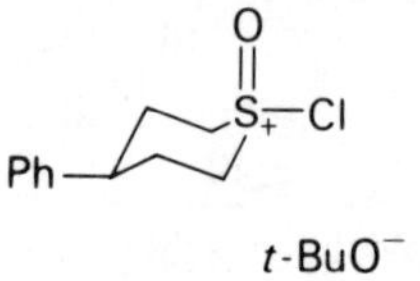 (**28**)

3. Reactions with amides

Reactions involving hypofluorites and amides have recently been investigated. Secondary sulphonamides (**29**) react with trifluoromethyl hypofluorite to give *N*-fluoro derivatives (**30**)[20]. With secondary carboxamides (**31**), the initially formed *N*-fluoroamides (**32**) react with CF_3OF to yield *N,N*-difluoroamines (**33**)[36].

$$\underset{\mathbf{(29)}}{\text{R}^1\text{SO}_2\text{NHR}^2} \xrightarrow{\text{CF}_3\text{OF}} \underset{\mathbf{(30)}}{\text{R}^1\text{SO}_2\text{NFR}^2}$$

$$\underset{\mathbf{(31)}}{\text{R}^1\text{CONHR}^2} \xrightarrow{\text{CF}_3\text{OF}} \underset{\mathbf{(32)}}{\text{R}^1\text{CONFR}^2} + \text{CF}_3\text{O}^- + \text{H}^+$$

$$\underset{\mathbf{(32)}}{\text{R}^1\text{CONFR}^2} \xrightarrow{\text{CF}_3\text{OF}} \text{R}^1\text{CO}\overset{+}{\text{N}}\text{F}_2\text{R}^2\ \ \text{CF}_3\text{O}^- \longrightarrow \text{R}^1\text{COOCF}_3 + \underset{\mathbf{(33)}}{\text{R}^2\text{NF}_2}$$

4. Reactions with amines and imidates

A number of conversions of aromatic amines (**34**) into *N*-chloroamines (**35**) with *t*-BuOCl are reported. The *N*-chloroamines are used for the preparation of azasulphonium salts, useful intermediates for the specific *ortho* alkylation of aromatic amines[37].

$$X-C_6H_4-NH_2 \xrightarrow{t\text{-BuOCl}} X-C_6H_4-NHCl + t\text{-BuOH}$$

(34) (35)

The *N*-fluorination of aziridines[38] and the ring opening of *N*-substituted aziridines[39] by reaction with CF_3OF have been observed.

t-Butyl hypochlorite has been used to convert imidates into their *N*-chloro derivatives[40].

$$R^1C(=NH)OR^2 \xrightarrow{t\text{-BuOCl}} R^1C(=NCl)OR^2$$

5. *Reactions with phosphites and phosphines*

Oxidation of phosphites and phosphines by alkyl hypochlorites has been studied. It appears that the reaction is likely to be an anionic process, the first stage of which is the formation of a chlorophosphonium salt (**36**). In the presence of an excess of hydrolytic solvent, such as alcohol R^1OH, a rapid proton transfer between alkoxide ion and alcohol leads to the formation of an alkoxyphosphonium chloride (**37**), then alkyl chloride **38** is formed. Optically active tetrahydrolinalyl hypochlorite reacted with triphenylphosphite to give optically active tetrahydrolinalyl chloride of inverted configuration.

$$R^1OCl + R_3P \longrightarrow \underset{(36)}{R_3\overset{+}{P}Cl\ R^1O^-} \longrightarrow R_3\overset{+}{P}OR^1\ Cl^- \longrightarrow R_3P{=}O + R^1Cl$$

$$(36) \xrightarrow{R^2OH} \underset{(37)}{R_3\overset{+}{P}OR^2\ Cl^-} \longrightarrow \underset{(38)}{R_3P{=}O + R^2Cl}$$

When the reaction was conducted under ultraviolet irradiation, the yield of chloride **38** was markedly reduced. If the reaction involves radicals, irradiation should improve the yields[41].

The reactions of *cis*- and *trans*-2-methoxy-4-methyl-1,3-dioxaphosphorinane (**39**) with neopentyl hypochlorite to give phosphates (**40**) have been investigated. Only one isomer of **39** undergoes stereospecific conversion to **40**. These stereochemical results are explained by formation of pentasubstituted phosphorus intermediates (**41** and **42**),

POMe Me (39) O O P OMe Me (40)

Cl P—OCH$_2$Bu-*t* OMe Me (41) ⇌ OMe P—OCH$_2$Bu-*t* Cl Me (42)

one of which is more stable. These intermediates can interconvert by pseudo-rotation[42].

The reaction of the hypochlorite **43** with triphenylphosphine gives the bridgehead chloride **44**[41].

$$\text{(43) bicyclo-OCl} \xrightarrow{PPh_3} \text{bicyclo-}\overset{+}{O}PPh_3Cl^- \xrightarrow{\Delta} \text{(44) bicyclo-Cl} + Ph_3P{=}O$$

6. *Reactions with alkenes and arenes*

The reactions of alkyl hypochlorites with alkenes are either ionic or radical additions. The ionic additions occur essentially in protic solvents (alcohols, carboxylic acids or water) and the radical additions occur in aprotic solvents, except, perhaps, in the presence of boron trifluoride[43]. *t*-Butyl hypobromite[44] and especially fluorinated alkyl hypochlorites have attracted interest for their electrophilic reactions with olefins.

Polyfluoroalkyl hypochlorites and pentafluoro sulphur hypochlorite (**1**) add to both unsubstituted and halogen-substituted terminal olefins to produce ethers. The direction of addition is such that when hypochlorites are added to asymmetrical olefins the main product is an ether in which the chlorine atom of the hypochlorite has become bonded to the most nucleophilic olefinic carbon[45,46].

$$SF_5OCl + CF_2{=}CF_2 \longrightarrow SF_5OCF_2CF_2Cl$$

$$CF_3OCl + CF_2{=}CH_2 \longrightarrow CF_3OCF_2CH_2Cl$$

This result argues for an ionic mechanism involving a positive halogen rather than a radical process[46].

Fluoroalkyl hypofluorites react with olefinic linkages to afford, mainly by *cis* addition, products bearing fluorine linked to the carbon constituting the most nucleophilic terminus of the olefinic substrate[21,22,48–52]. The mechanism of this reaction is suggested to involve an addition of positive fluorine, for instance on a nucleophilic olefin **45**, to give a cationic intermediate (**46**). This intermediate can be stabilized by fragmentation to a fluoroketone (**47**). A radical process is excluded because the reaction is not affected by the presence of radical scavengers[48].

$$\underset{(45)}{R^1(H)C{=}C(R^2)OMe} \xrightarrow{CF_3OF} \underset{(46)}{R^1(F)CH{-}\overset{+}{C}(R^2)OMe} \longrightarrow \underset{(47)}{R^1(F)CH{-}C(R^2){=}O}$$

The α-fluorination of carbonyl compound with trifluoromethyl hypofluorite has been reported. The carbonyl compounds are first converted into their trimethylsilyl-enol ethers. The reaction of these ethers, in an inert solvent, with trifluoromethyl hypofluorite gives the corresponding α-fluorocarbonyl compounds (**48**). No hydrolysis step is required. This method is general. Ketones, aldehydes, esters, amides and carboxylic acids can be fluorinated on the carbon adjacent to the carbonyl group[50].

$$R^1COCH_2R^2 \longrightarrow R^1C(OSiMe_3){=}CHR^2 \xrightarrow[-70\,^\circ C]{CF_3OF} \underset{(48)}{R^1COCHFR^2} + COF_2 + Me_3SiF$$

The *cis* addition of pentafluoroethane hypofluorite (**2**) to *cis* and *trans* stilbenes (**49**) has been attributed to the involvement of a tight ion pair such as (**51**). *Cis*-**49** gives mainly *erythro*-**50** and *trans*-**49** gives mainly *threo*-**50**[47,51].

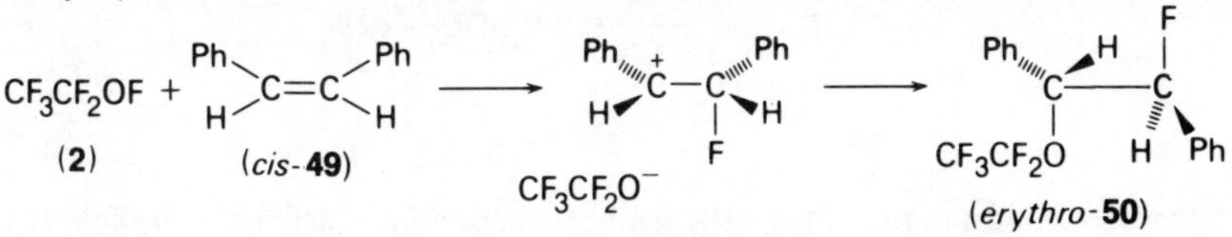

CF_3CF_2OF + PhHC=CHPh (*cis*-**49**) ⟶ [PhHC⁺—CHFPh] $CF_3CF_2O^-$ (**51**) ⟶ PhHC(OCF_2CF_3)—CHFPh (*erythro*-**50**)

(**2**)

Electrophilic addition to an aromatic ring may be achieved with alkyl hypofluorites. For instance, 2,3-benzofuran reacts with trifluoromethylhypofluorite to give the addition compound **52**[49].

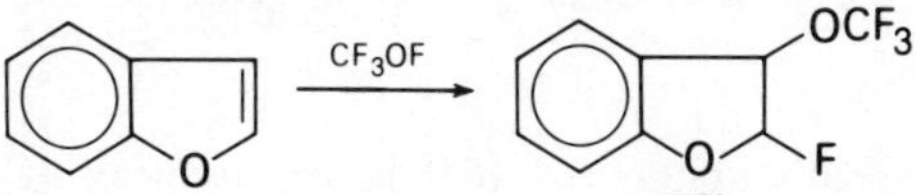

2,3-benzofuran $\xrightarrow{CF_3OF}$ 3-OCF_3-2-F-2,3-dihydrobenzofuran (**52**)

With *trans*-4-methoxystilbene (**53**), aromatic electrophilic fluorination takes place on the activated ring, to give **54**[52].

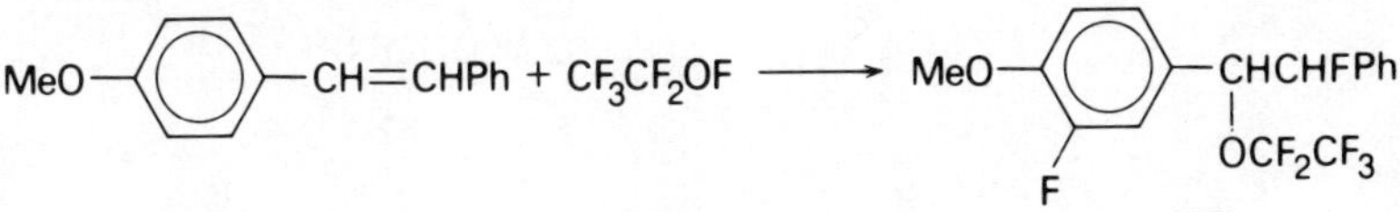

$$\text{MeO}-C_6H_4-\text{CH=CHPh} + CF_3CF_2OF \longrightarrow \text{MeO}-C_6H_3(\text{F})-\text{CHCHFPh}(OCF_2CF_3)$$

(**53**) (**2**) (**54**)

7. *Reactions with alkanes*

Fluorination at saturated carbon is a very interesting reaction involving direct electrophilic attack on the electrons of the C—H σ-bond, with retention of configuration. Thus, adamantane (**55**) reacts with CF_3OF to give 1-fluoroadamantane (**56**)[53].

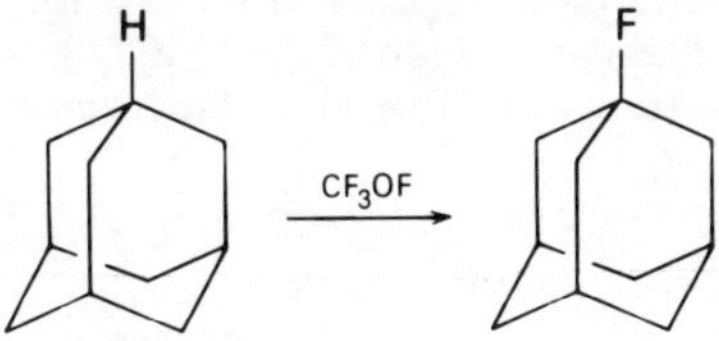

(**55**) $\xrightarrow{CF_3OF}$ (**56**)

Recently, Rozen and coworkers have shown that selective fluorination of a tertiary carbon–hydrogen bond, with high electron density, can be performed with elemental fluorine which acts as an electrophile[54].

B. Chloroperoxyalkanes

The reaction of chlorine monofluoride with trifluoromethyl hydroperoxide (**57**) produces chloroperoxytrifluoromethane (**58**)[55].

$$CF_3OOH + ClF \xrightarrow{-110\ °C} CF_3OOCl + HF$$

(**57**) (**58**)

An alternative method of synthesis is the reaction of fluorocarbonyl trifluoromethyl peroxide (**59**) with ClF in the presence of CsF[56].

$$CF_3OOC(=O)F + ClF \xrightarrow{CsF} CF_3OOCl + COF_2$$

(**59**) (**58**)

Chloroperoxytrifluoromethane (**58**) undergoes addition reactions with olefins, to yield trifluoromethylperoxy derivatives. The reactions proceed by an electrophilic mechanism, in which the positive chlorine of **58** adds to the most nucleophilic carbon[57].

$$CF_3OOCl + CF_2{=}CFCl \longrightarrow CF_3OOCF_2CFCl_2$$

(**58**)

The addition to 1,2-difluoroethylene (**64**) is stereospecific. Pentafluorosulphurperoxy hypochlorite (**60**) undergoes addition reactions with alkenes, forming pentafluorosulphurperoxy derivatives: 1,1-difluoroethylene (**61**) leads to peroxy compounds **62** and ether **63**[57].

$$SF_5OOCl + CF_2{=}CH_2 \longrightarrow SF_5OOCF_2CH_2Cl\ (\mathbf{62}) + SF_5OCF_2CH_2Cl\ (\mathbf{63})$$

(**60**) (**61**)

The mechanism for the reaction appears to be that for electrophilic addition of positive chlorine, followed by the addition of the SF_5OO group. The stereospecific *cis* addition observed with **64** rules out a free radical mechanism[57].

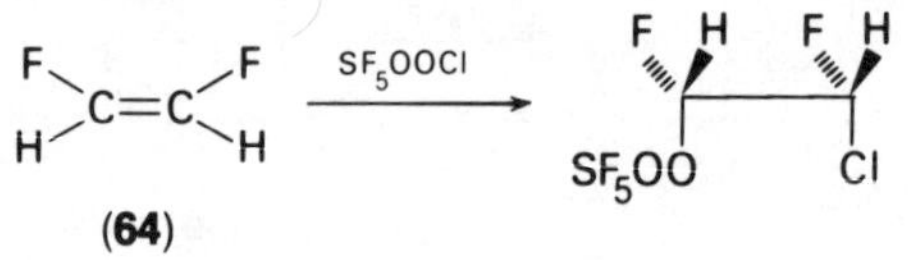

$$FHC{=}CHF\ (\mathbf{64}) \xrightarrow{SF_5OOCl} SF_5OO{-}CHF{-}CHF{-}Cl$$

(**64**)

C. Halogen Derivatives of Acids

Halogen derivatives of oxyacids are powerful electrophilic halogen sources[58–60]. The kinetics and stereochemistry of the reactions of electrophilic chlorine acetate (**65a**)[61–62] and bromine acetate (**65b**)[63] with some unsaturated compounds are described.

MeCOOX

(**65a**, X = Cl)

(**65b**, X = Br)

$$CF_3SO_3H + ClF \longrightarrow \underset{(\mathbf{66})}{CF_3SO_2OCl} + HF$$

$$CF_3CO_2H + ClF \longrightarrow \underset{(\mathbf{67})}{CF_3COOCl} + HF$$

In 1978, Des Marteau reported a procedure for the preparation of chlorine derivatives **66** and **67** of trifluoromethanesulphonic acid and trifluoroacetic acid[64].

The compound **66** is probably the most electrophilic chlorine compound known. It readily undergoes reactions with covalent chlorides to form chlorine and trifluoromethanesulphonates (**68**)[64].

$$\underset{(\mathbf{66})}{CF_3SO_2OCl} + RCl \longrightarrow \underset{(\mathbf{68})}{CF_3SO_2OR} + Cl_2$$

Additions of **66** and CF_3SO_2OBr to simple alkenes occur readily at low temperature to give the esters **69** and **70** in high yield[65].

$$CF_3SO_2OX + CH_2{=}CH_2 \longrightarrow \underset{(\mathbf{69})}{CF_3SO_2OCH_2CH_2X}$$

$$\mathbf{69} + CF_3SO_2OX \longrightarrow \underset{(\mathbf{70})}{CF_3SO_2OCH_2CH_2OSO_2CF_3} + X_2$$

Addition to halogenated alkenes which are normally resistant to attack by electrophiles occurs. It is of interest to note that the stereospecificity of the reaction with alkenes implies, as does the reaction of hypofluorite (**2**) to olefins, a *cis* addition. For instance, the olefin **64** gives the *threo* isomer **71**[65].

$$\underset{(\mathbf{64})}{FHC{=}CHF} \xrightarrow[X = Cl, Br]{CF_3SO_2OX} \underset{threo\text{-}\mathbf{71}}{CF_3SO_2O{-}CHF{-}CHFX}$$

The displacement of the chlorine atom from **71**, X = Cl, proceeds with retention of configuration: an S_Ei type mechanism has been suggested[66].

$$CF_3S(O)_2O{-}Cl + Cl{-}CHFCHFOSO_2CF_3 \longrightarrow \underset{(\mathbf{72})}{CF_3SO_2OCHFCHFOSO_2CF_3} + Cl_2$$

It has been found that electrophilic aromatic substitution is performed with *bis*(trifluoroacetoxy)phenyliodine (**73**), in the presence of iodine. The iodine trifluoroacetate (**74**) formed when **73** is treated with iodine, is probably the reagent responsible for the iodination[67].

$$\underset{(\mathbf{73})}{PhI(OCOCF_3)_2} + I_2 \longrightarrow PhI + 2\ \underset{(\mathbf{74})}{CF_3COOI}$$

$$ArH + \underset{(\mathbf{74})}{CF_3COOI} \longrightarrow ArI + CF_3COOH$$

III. COMPOUNDS WITH HALOGEN–NITROGEN BONDS

Two types of mechanism can be envisaged for the first step of the ionic reactions of *N*-halo compounds with nucleophiles: (1) nucleophilic attack of Nu on the *N*-halo compounds,

$$R_2NX + Nu \longrightarrow R_2\overset{+}{N}{-}Nu\ X^- \quad (7)$$

(2) nucleophilic attack of Nu on the halogen atom

$$R_2NX + Nu \longrightarrow R_2N^- + \overset{+}{Nu}X \quad (8)$$

An interesting study of the thermodynamic aspects of the reactions of these *N*-halo compounds has been published[68,69]. Pitman and coworkers have compared the equilibrium constants for competitive reactions (9) and (10)[69].

$$\underset{(\mathbf{76})}{R_2NH} + HOCl \rightleftharpoons \underset{(\mathbf{75})}{R_2NCl} + H_2O \quad (9)$$

$$\underset{(\mathbf{76})}{R_2NH} + HOCl \rightleftharpoons \underset{(\mathbf{77})}{R_2\overset{+}{N}H_2} + ClO^- \quad (10)$$

The results are explained in terms of polarization of the halogen–nitrogen bond for **75** and of the basic properties of **76**. The polarization of the N—Cl bond involves a positive chlorine when **75** derives from weak bases **76** (sulphonamides, succinimide), and a negative chlorine when **75** derives from strong bases **76** (amines).

A. *N*-Haloimides

While the reaction of disulphides, weak nucleophiles, with *N*-bromosuccinimide is a radical process, initiated by benzoyl peroxide[70], the nucleophilic sulphides R^1SR^2 are oxidized by *N*-halosuccinimides according to an ionic process. It has been suggested that the first step of the reaction of sulphides with **78** is the attack on the halogen atom by the sulphur, to lead to a chlorosulphonium salt (**79**), then a fast nucleophilic substitution by succinimidate ion gives the unstable sulphonium chloride **80**[74]. However, **79** has never been isolated[75].

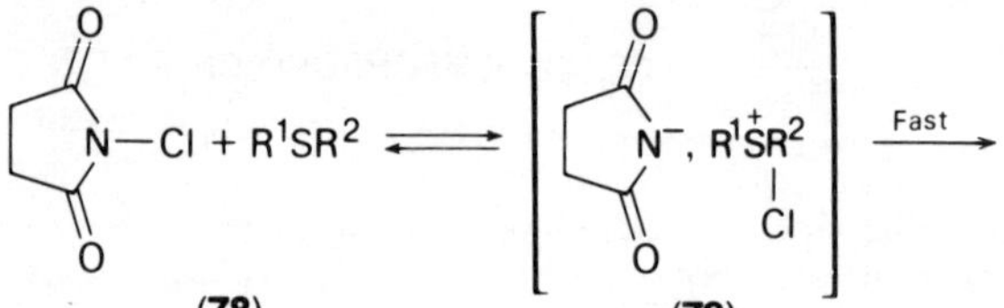

(succinimide)N—Cl + R^1SR^2 ⇌ [(succinimide)N^-, $R^1\overset{+}{S}R^2$(—Cl)] —Fast→

(**78**) (**79**)

(succinimide)N—$\overset{+}{S}R^1R^2$ Cl^- —$SbCl_5$→ (succinimide)N—$\overset{+}{S}R^1R^2$ $SbCl_6^-$

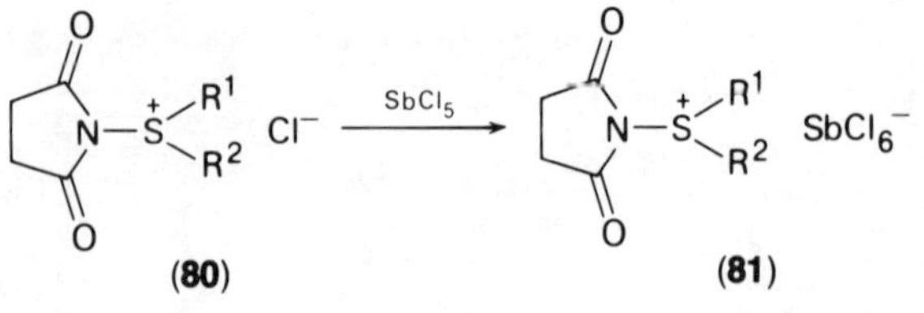

(**80**) (**81**)

The first step being probably reversible, the succinimidosulphonium salts (**81**) are isolated by the addition of antimony pentachloride[71]. The reactivity of *N*-chloro compounds with sulphides increases in the series *N*-chloroamines < *N*-chloroamides < *N*-chloroimides. Failure of di-*t*-butyl sulphide to react with *N*-chlorosuccinimide (**78**) has been considered as a result of steric hindrance to attack on the bulky thioether. Diphenyl sulphide and thiophene do not react with **78**. This result is attributed to the reduction in nucleophilicity of the sulphur atom by the aromatic rings[72].

Considerable interest has been devoted to the formation of succinimidosulphonium salts and their utilizations in synthesis. Rearrangement of **80**, $R^1 = R^2 = Me$, by thermolysis gives monochlorinated thioether **82**[73,75]. Triethylamine reacts with **80** to give the rearranged products **83** and **84**[75]. The azasulphonium salts (**80**) are reagents for the synthesis of *ortho* alkylated phenols from phenols[76,77], for selective conversion of allylic and benzylic alcohols to halides under neutral conditions[78], for oxidation of primary and secondary alcohols to carbonyl compounds[79–80], for the synthesis of sulphimides[81,82] and for the chlorination of enamines[83].

(**80**) ($R^1 = R^2 = Me$) → (**85**) NH + $ClCH_2SMe$ (**82**)

(**80**) —NEt_3→ (**83**) N, OCH_2SMe + (**84**) NCH_2SMe

Tsuchihashi and Ogura[84] have reported the chlorination of sulphoxide **86** by *N*-chlorosuccinimide (**78**). In the presence of pyridine, two chlorinated sulphoxides **87** and **88** were isolated. When an insoluble base such as potassium acetate was used, only α-chlorobenzyl methyl sulphoxide (**87**) was obtained. The addition of hydroquinone changes the amount of **87** produced by the chlorination of **86**, but the yield of **88** remains unchanged.

$$\underset{(\mathbf{87})}{PhCH(Cl)SOMe} \longleftarrow \underset{(\mathbf{86})}{PhCH_2SOMe} \longrightarrow \underset{(\mathbf{88})}{PhCH_2SOCH_2Cl}$$

The authors[84] suggest two mechanisms: a radical reaction in the presence of insoluble base, which gives the product **87**, and an ionic process, which gives the chlorosulphoxides **87** and **88** and which is mainly operative in the presence of pyridine. Kinetic isotope effects have been examined by using the deutero derivatives **89** and **90**. The authors found that proton transfer is the rate-determining step for the ionic process.

$$\underset{(\mathbf{89})}{PhCD_2SOMe} \qquad \underset{(\mathbf{90})}{PhCH_2SOCD_3}$$

Although not suggested by the original workers, presumably the ionic reaction proceeds in fact via a chlorooxosulphonium salt, as has been shown for the chlorination of sulphoxides **25** with *t*-butyl hypochlorite[35].

Trialkyl phosphites were found to react similarly to sulphides with *N*-halosuccinimides. The rearrangement of the halophosphonium succinimidate **(92)** which results from the attack by phosphite on the halogen gives the phosphoramide **91**[85,86].

$$\text{succinimide-N–X} \xrightarrow{(EtO)_3P} (EtO)_3\overset{+}{P}X \ {}^{-}N\text{(succinimide)}\ \textbf{(92)} \longrightarrow X^{-}\ (EtO)_3\overset{+}{P}\text{–N(succinimide)}\ \textbf{(93)} \longrightarrow EtX + (EtO)_2P(=O)\text{–N(succinimide)}\ \textbf{(91)}$$

B. *N*-Haloamides

A polar mechanism has been suggested for the reaction of *N*-bromoacetamide **(94)** with sulphides.

$$\underset{\textbf{(94)}}{MeCONHBr} \xrightarrow[\text{Slow}]{R^1SR^2} \underset{\textbf{(95)}}{MeCONH^{-}\ Br\overset{+}{S}R^1R^2} \xrightarrow{\text{Fast}} \underset{\textbf{(96)}}{MeCONH\overset{+}{S}R^1R^2\ Br^{-}}$$

The reaction of **94** with di-*t*-butyl sulphide failed to yield sulphonium bromide, presumably due to steric hindrance by the *t*-butyl groups. Methyl phenyl sulphide and diphenyl sulphide do not react with **94**. *N*-Acetyldialkylsulphimides **(97)** were obtained by treatment of sulphonium bromides **96** with NEt_3[87].

$$\underset{\textbf{(97)}}{MeCON{=}SR^1R^2}$$

The secondary *N*-chloroamides **(98)** are transformed by dimethyl sulphide into sulphonium chlorides **(99)** which are rearranged in basic media to give iminoethers **(100)**[88].

$$\underset{\textbf{(98)}}{R^1CON(Cl)R^2} \xrightarrow{Me_2S} \underset{\textbf{(99)}}{R^1CON(\overset{+}{S}Me_2)R^2\ Cl^{-}} \xrightarrow{NEt_3} \underset{\textbf{(100)}}{R^1C(OCH_2SMe){=}NR^2}$$

When the *N*-chloroacetanilides **(101)**, prepared through the reaction of the appropriate acetanilide with calcium hypochlorite, were allowed to react with dimethyl sulphide, *N*-acyl-*N*-arylazasulphonium chloride **(102)** was formed. The treatment of **102** with NEt_3 gave the sulphur ylide **103**, which rearranged to give a mixture of acetanilides **(104)** and iminoethers **(105)**[89].

$$\text{X-C}_6\text{H}_4\text{-N(Cl)COMe}\ (\mathbf{101}) \xrightarrow[-15\,^\circ\text{C}]{\text{Me}_2\text{S}} \text{X-C}_6\text{H}_4\text{-N(}^+\text{SMe}_2\text{)COMe Cl}^-\ (\mathbf{102}) \xrightarrow{\text{NEt}_3} \text{X-C}_6\text{H}_4\text{-N(}\overset{+}{\text{S}}\text{(Me)}^-\text{CH}_2\text{)COMe}\ (\mathbf{103})$$

$$\text{X-C}_6\text{H}_3\text{(CH}_2\text{SMe)-NHCOMe}\ (\mathbf{104}) \longleftarrow \mathbf{103} \longrightarrow \text{X-C}_6\text{H}_4\text{-N=C(Me)OCH}_2\text{SMe}\ (\mathbf{105})$$

Trialkyl phosphites react with *N*-haloamides (**106**) and with *N*-halo-*N*-alkylamides (**98**) to give nitriles (**107**) and *N*-haloimines (**108**), respectively[90].

$$\text{R}^1\text{CONHX}\ (\mathbf{106}) \xrightarrow{(\text{RO})_3\text{P}} (\text{RO})_3\overset{+}{\text{P}}\text{OC(R}^1\text{)=NH X}^- \longrightarrow (\text{RO})_3\text{P=O} + \text{X(R}^1\text{)C=NH}$$

$$\text{X(R}^1\text{)C=NH} \longrightarrow \text{R}^1\text{CN} + \text{XH}\ (\mathbf{107})$$

$$\text{R}^1\text{CON(Cl)R}^2\ (\mathbf{98}) \xrightarrow{(\text{RO})_3\text{P}} (\text{RO})_3\overset{+}{\text{P}}\text{OC(R}^1\text{)=NR}^2\ \text{Cl}^- \longrightarrow (\text{RO})_3\text{P=O} + \text{R}^1\text{(Cl)C=NR}^2\ (\mathbf{108})$$

C. *N*-Halosulphonamides

In their study of the kinetics of the reaction of sulphides with *N*-chloroarenesulphonamides (**109**), Ruff and Kucsman show that the chloronium ion transfers from **109** to the sulphides, yielding the chlorosulphonium salt (**110**). The rearrangement of **110** produces sulphimides (**111**), while its hydrolysis will give sulphoxides (**112**)[91].

$$\text{R}^1\text{SR}^2 + \text{ArSO}_2\text{NHCl}\ (\mathbf{109}) \xrightarrow{\text{Slow}} \text{ArSO}_2\text{NH}^-\ \text{R}^1\overset{+}{\text{S}}\text{(Cl)R}^2\ (\mathbf{110}) \xrightarrow{\text{Fast}} \text{ArSO}_2\text{=S(R}^1\text{)R}^2 + \text{HCl}\ (\mathbf{111})$$

$$\mathbf{110} \xrightarrow{\text{H}_2\text{O}} \text{ArSO}_2\text{NH}_2 + \text{R}^1\text{SOR}^2 + \text{HCl}\ (\mathbf{112})$$

The steric effects in the reaction of sulphides with chloramine-T has been studied[92]. The reaction of thioxanthene (**113**) with chloramine-T in methanol–methylene chloride, in the presence of small amounts of acetic acid, gives thioxanthene *N*-*p*-toluenesulphonylsulphimide (**114**). It is suggested that **113** is first converted to a chlorosulphonium salt (**115**). A direct attack of the anion on the sulphur atom of **115** may lead to the sulphimide **114**, and a competitive elimination of hydrogen chloride from **115** followed by an attack of *p*-toluenesulphonamide anion at the 9 position of the resulting ion **116** may account for the formation of 9-(*N*-*p*-toluenesulphonamido)thioxanthene (**117**)[93].

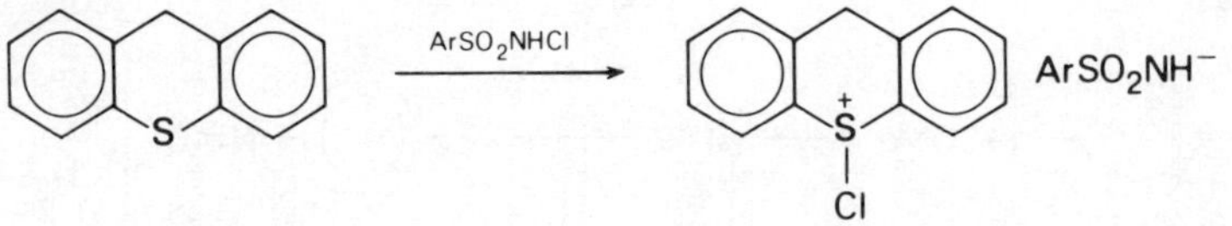

ArSO$_2$NHCl; ArSO$_2$NH$^-$

(**113**) (**115**)

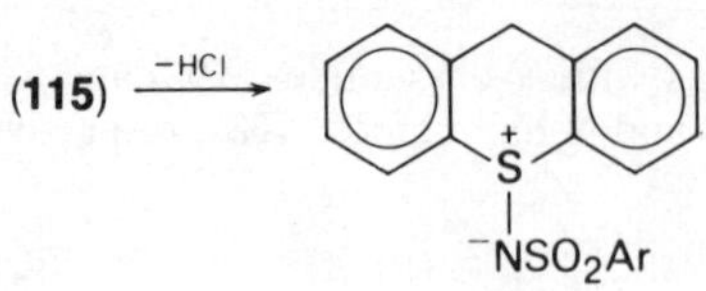

(**115**) $\xrightarrow{-HCl}$ $^-$NSO$_2$Ar

(**114**)

(**115**) → + ArSO$_2$NH$^-$ → NHSO$_2$Ar

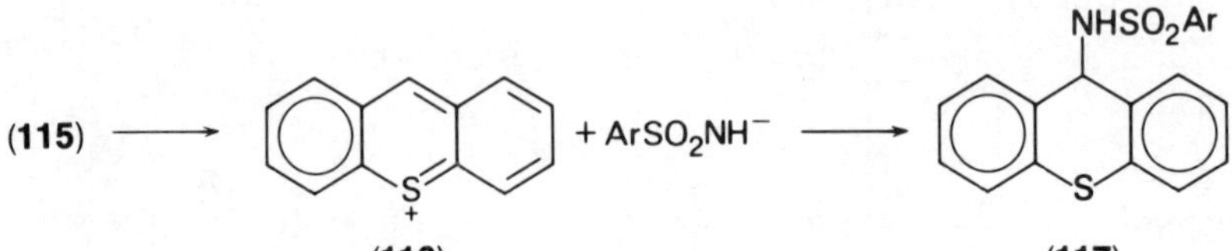

(**116**) (**117**)

The syntheses of sulphuranes **24** have been carried out starting from sulphides **23** and chloramine-T[32,94].

D. Iodine Isocyanate and Iodine Azide

The reactions of iodine isocyanate or iodine azide with alkenes are consistent with transfer of a positive iodine. The addition of INCO to cyclic olefins generally occurs in a stereospecific manner: the iodine and isocyanate groups are introduced *trans* to each other. Cyclohexene gives the product **118**. With styrene, the addition is regiospecific, suggesting involvement of a benzylic carbonium ion (**119**). The reactions of INCO with alkenes have synthetic interest for the synthesis of carbamates, aziridines, oxazolidones, aminoalcohols, diamines and azepines[95].

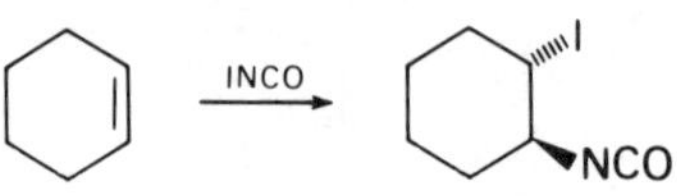

INCO; I; NCO

(**118**)

PhCH=CH$_2$ ⟶ PhCH—CH$_2$ (I$^+$ bridged) $\bar{N}$CO ⇌ Ph$\overset{+}{C}$H—CH$_2$I + $^-$NCO ⟶ PhCH(NCO)CH$_2$I

(**119**)

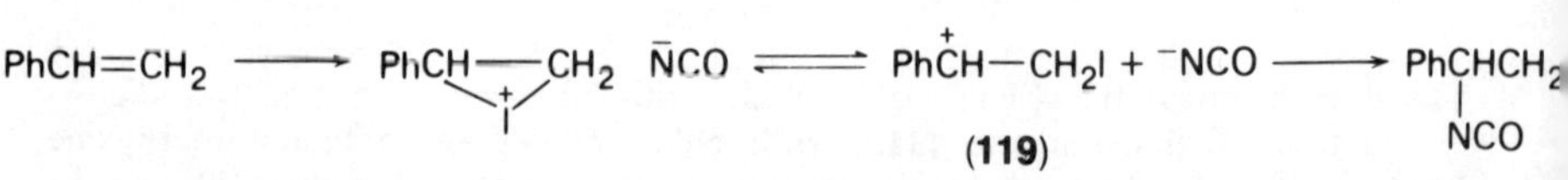

Iodine azide adds in a stereospecific manner to alkenes. Thus, cyclohexene gives the *trans* adduct **120**[96].

(120) (121)

Cambie and coworkers have reinvestigated the reaction of iodine azide with alkenes. The stereochemistry depends on the polarity of the solvent. Addition of iodine azide to cyclohexene in chloroform–water gives *cis*- and *trans*-iodo azides (**120** and **121**). Intermediates in these reactions are suggested: cyclic iodonium ions give a *trans* adduct, while carbocationic intermediates allow *cis* adduct formation[97].

E. *N*-Chlorobenzotriazole

The oxidation of sulphides by 1-chlorobenzotriazole (**122**) is a useful method for the preparation of sulphoxides without concomitant formation of sulphones, aminosulphonium salts (**123c**) and alkoxysulphonium salts (**123b**)[98]. Apparently, the nature of the oxidation reaction is like that of *t*-butyl hypochlorite with sulphides[24,100], the chlorosulphonium salt (**123a**) being an intermediate.

(122) $\xrightarrow{R_2S}$ (123a) ClS^+R_2 $\xrightarrow{MeOH}$ (124) + R_2SO + MeCl

(123a) ⟶ $^+SR_2Cl^-$

$\xrightarrow{R^1OH,\ AgBF_4}$ $R_2\overset{+}{S}—OR^1,\ BF_4^-$ + AgCl + **124** (123b)

$\xrightarrow{R^1_2NH,\ AgBF_4}$ $R_2\overset{+}{S}—NR^1_2,\ BF_4^-$ + AgCl + **124** (123c)

The reaction of sulphoxides with **122**, in the presence of pyridine, affords the corresponding α-chlorosulphoxides[101].

$$\mathbf{122} + R^1SOCH_2R^2 \longrightarrow R^1—\overset{Cl}{\underset{O}{S^+}}—CH_2R^2 \longrightarrow R^1SOCHClR^2$$

N-Halopyrazoles have been prepared and can act as sources of positive halogen[99].

F. *N*-Haloamidines

It has been shown that the reactions of *N*-chloroamidines (**125**) with enamines, involving transfer of positive chlorine, lead to an ion pair. Further reaction of the latter, depending on the nature of the substituents, leads to imidazoles (**126a**) or amidines (**126b**)[102].

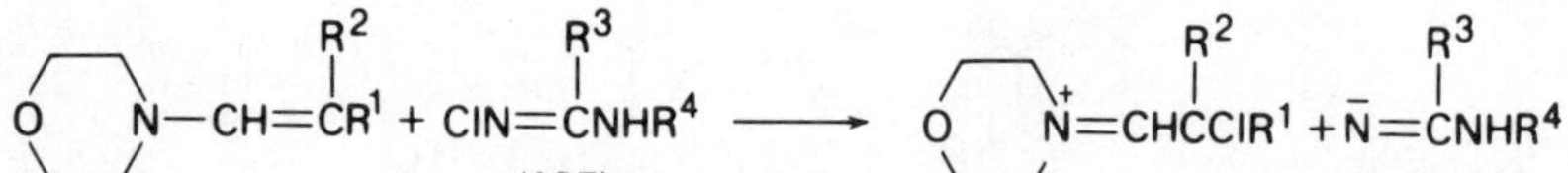

$$\text{O}\langle\rangle\text{N}-\text{CH}=\text{C}(\text{R}^2)\text{R}^1 + \text{ClN}=\text{C}(\text{R}^3)\text{NHR}^4 \longrightarrow \text{O}\langle\rangle\overset{+}{\text{N}}=\text{CHC}(\text{R}^2)\text{ClR}^1 + \bar{\text{N}}=\text{C}(\text{R}^3)\text{NHR}^4$$

(**125**)

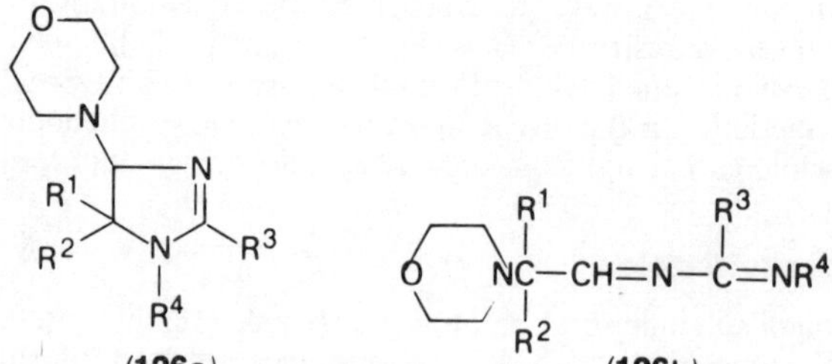

(**126a**) (**126b**)

G. *N*-Haloamines

Treatment of anilines with calcium hypochlorite resulted in the formation of *N*-chloroanilines (**127**). The conversion of **127** into azasulphonium salts through reaction with methyl alkyl sulphides and the treatment of these salts with base yields an azasulphonium ylide **128**. This ylide undergoes Sommelet–Hauser type rearrangement, useful in the synthesis of *ortho* substituted anilines[103–104].

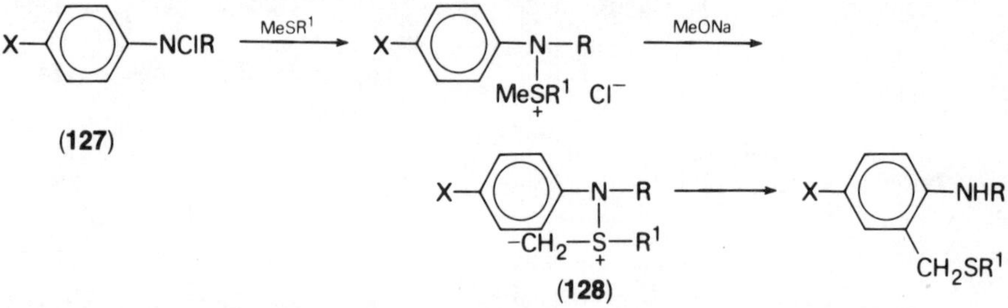

$$\text{X}-\text{C}_6\text{H}_4-\text{NClR} \xrightarrow{\text{MeSR}^1} \text{X}-\text{C}_6\text{H}_4-\text{N(R)}-\overset{+}{\text{S}}\text{MeR}^1\ \text{Cl}^- \xrightarrow{\text{MeONa}}$$

(**127**)

$$\text{X}-\text{C}_6\text{H}_4-\text{N(R)}-\overset{+}{\text{S}}(\text{R}^1)-\text{CH}_2^- \longrightarrow \text{X}-\text{C}_6\text{H}_3(\text{CH}_2\text{SR}^1)-\text{NHR}$$

(**128**)

However, the mechanism of the reactions of *N*-chloroanilines with sulphides does not, in all cases, involve a transfer of positive halogen. Study of the thermal rearrangement of **127** into *o*-chloroanilines (**129**) showed that the mechanism involved the generation of an electron-deficient nitrogen species (**130**) (nitrenium ion[105]) and chloride anion[106], as could be expected from polarization of the N—Cl bond in amines[69].

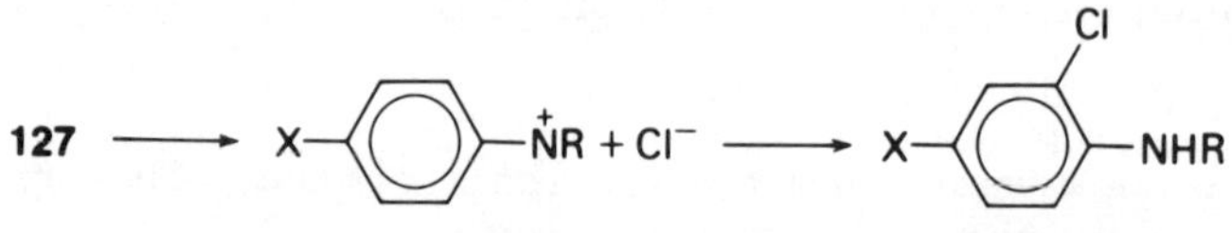

$$\textbf{127} \longrightarrow \text{X}-\text{C}_6\text{H}_4-\overset{+}{\text{N}}\text{R} + \text{Cl}^- \longrightarrow \text{X}-\text{C}_6\text{H}_3(\text{Cl})-\text{NHR}$$

(**130**) (**129**)

However, in the presence of acid catalysts, two competing mechanisms seem to be involved in the rearrangement of *N*-chloroanilines (**127**) to *ortho*-chloroanilines (**129**): one mechanism is favoured by electron-donating substituents and involves nitrenium ion **130**; the other mechanism is promoted by electron-withdrawing substituents, which can involve positive chlorine, via the protonated *N*-haloamines (**127a**)[106].

$$\mathbf{127} \xrightarrow{H^+} X\text{-}C_6H_4\text{-}\overset{+}{N}HR(Cl) \rightleftharpoons X\text{-}C_6H_4\text{-}NHR + Cl^+ \longrightarrow \mathbf{129} + H^+$$

(**127a**)

IV. COMPOUNDS WITH HALOGEN–CARBON BONDS

Nucleophiles usually react with alkyl halides via displacement of a halide ion from the carbon atom. However, if the normal S_N2 reaction on the carbon atom is made difficult, displacement on the halogen atom may occur to release a carbanion, chiefly in the presence of carbanion-stabilizing groups and if the central carbon atom is made less accessible through steric effects.

A. Polyhaloalkanes

1. Reactions with trivalent phosphorus compounds

The reactions of phosphines with tetrahalomethanes, extensively developed in preparative chemistry for halogenation, dehydration and P—N bond formation, have been reviewed[107,108]. More recent developments in this field are discussed here. These reactions lead to heterolytic bond cleavage by a direct attack of the phosphorus atom on the halogen atom.

$$R_3P + CX_4 \longrightarrow R_3\overset{+}{P}X\ CX_3^- \longrightarrow R_3P\langle^{X}_{CX_3} \longrightarrow R_3\overset{+}{P}\text{—}CX_3\ X^-$$

Addition of radical trapping reagents and ultraviolet irradiation exert no effect on the course of the reaction[109]. An enormous solvent dependence is observed. The reaction is extremely fast in acetonitrile and slow in CCl_4[108]. These findings indicate that charge separation in the transition state of the rate-determining step and the solvation capacity of the solvent are important. The trichloromethylphosphonium chloride (**131**), which can be isolated[110], reacts with remaining phosphine very fast to give, via the short lived dichloromethylenephosphorane (**132**)[111], the chloride **133**[112,113].

$$\underset{(\mathbf{131})}{Ph_3\overset{+}{P}\text{—}CCl_3\ Cl^-} \xrightarrow{PPh_3} \underset{(\mathbf{132})}{Ph_3P{=}CCl_2} + Cl_2PPh_3 \longrightarrow$$

$$Ph_3\overset{+}{P}\text{—}CCl_2\text{—}\overset{+}{P}Ph_3\ \ 2\,Cl^- \xrightarrow{PPh_3} \underset{(\mathbf{133})}{Ph_3P{=}CCl\text{—}\overset{+}{P}Ph_3\ Cl^-} + Cl_2PPh_3$$

Tertiary aliphatic phosphines react with CCl_4 to give phosphonium salts **134** directly[114].

$$\underset{(\mathbf{134})}{R_3P{=}CCl\text{—}\overset{+}{P}R_3\ Cl^-} \qquad \underset{(\mathbf{135})}{Ph_3P{=}C{=}PPh_3}$$

The reaction of trisdimethylaminophosphine with **133** leads to bis(triphenylphosphoranylidene)methane (**135**)[115]. and the reaction with **131** leads to the crystalline ylide **132**[112], which has been also prepared by the dechlorination of **131** by **135**[111].

$$\mathrm{Ph_3\overset{+}{P}{-}CCl_3Cl^-} \xrightarrow[\text{(or 135)}]{\mathrm{P(NMe_2)_3}} \mathrm{Ph_3P{=}CCl_2}$$
(131) **(132)**

The reactions of phosphines with tetrahalomethanes are widely used in organic chemistry.

The reaction of $Ph_3P–CCl_4$ with alcohols gives rise to an elegant method for the synthesis of primary and secondary alkyl chlorides[116]. Recent mechanistic studies show that this reaction generally proceeds in two steps (equations 11 and 12)[117].

$$\mathrm{CCl_4 + PPh_3} \xrightarrow{\mathrm{ROH}} \underset{\textbf{(136)}}{\mathrm{Ph_3P(OR)Cl}} + \mathrm{CHCl_3} \quad (11)$$

$$\textbf{136} \longrightarrow \mathrm{RCl + Ph_3P{=}O} \quad (12)$$

The structure of the intermediate **136**, R = *t*-$BuCH_2$, was elucidated by NMR spectroscopy. In the case of sterically unhindered primary alcohols, reaction (11) is probably slower than (12), but for secondary alcohols, (11) is faster than (12). The synthetic utility of mild bromination of primary alcohols with $Ph_3P–CBr_4$ is suggested by the work of Kocienski, Cernigliaro and Feldstein[118]. Barstow and Hruby have shown that $Ph_3P–CCl_4$ or $Ph_3P–BrCCl_3$ are good reagents for the formation of amide bonds[119].

$$\mathrm{R^1COOH} \xrightarrow[\mathrm{PPh_3}]{\mathrm{CCl_4}} \mathrm{R^1COCl} \xrightarrow{\mathrm{R^2NHR^3}} \mathrm{R^1CONR^2R^3}$$

The reaction of trisdimethylaminophosphine with CCl_4 in the presence of alcohols gives stable salts **137**, useful for the synthesis of many derivatives **138** from alcohols ROH[120–124].

$$\mathrm{(Me_2N)_3P + CCl_4} \xrightarrow{\text{Fast}} \mathrm{(Me_2N)_3\overset{+}{P}Cl\ \overset{-}{C}Cl_3} \xrightarrow{\mathrm{ROH}}$$

$$\mathrm{(Me_2N)_3\overset{+}{P}OR\ Cl^-} \xrightarrow{\mathrm{KPF_6}} \underset{\textbf{(137)}}{\mathrm{(Me_2N)_3\overset{+}{P}OR\ PF_6^-}} \xrightarrow{\mathrm{A^-}} \underset{\textbf{(138)}}{\mathrm{RA}}$$

(A^- = Br^-, I^-, CN^-, PhS^-, SCN^-, MeO^-, N_3^-).

The joint action of trisdimethylaminophosphine and carbon tetrachloride on vicinal diols affords either epoxide or spirophosphorane[125]. The reaction of secondary alcohols with $Ph_3P–CCl_4$ is dependent on the solvent used. In excess of carbon tetrachloride, the formation of alkyl chloride is the main reaction, whereas in acetonitrile, Ph_3PCl_2 formed with ylide **132** is responsible for the formation of olefins **(139)**[126].

$$\mathrm{Ph_3PCl_2 + R^1CH_2CH(OH)R_2} \longrightarrow \mathrm{Ph_3P{=}O + 2\ HCl} + \underset{\textbf{(139)}}{\mathrm{R^1CH{=}CHR^2}}$$

Zhmurova and Yurchenko have found that the reaction of $PPh_3–CCl_4$ with diphenylketimine **(140)** may have two pathways. Besides the normal Appel reaction that gives phosphonium chloride **(141a)**, the formation of an iminophosphorane **(141b)** is observed[127].

$$\underset{\textbf{(140)}}{\mathrm{Ph_2C{=}NH}} \qquad \underset{\textbf{(141a)}}{\mathrm{Ph_2C{=}N\overset{+}{P}Ph_3\ Cl^-}} \qquad \underset{\textbf{(141b)}}{\mathrm{Ph_2C(CCl_3)N{=}PPh_3}}$$

It has been demonstrated that the position of the equilibrium shown in equation (13) depends upon the nature of the halogen X.

$$2\,Ph_3P \xrightarrow{BrCX_3} Ph_3\overset{+}{P}CX_3\ Br^- + Ph_3P \rightleftharpoons \underset{(\mathbf{142})}{Ph_3P{=}CX_2} + Ph_3PXBr \qquad (13)$$

When X = Cl or F, the equilibrium lies predominantly to the left, and for X = Br, it lies to the right[128,129]. Ylide **142**, X = Br, can be alkylated to give **143**. The treatment of **143** with butyllithium gives rise to α-bromoalkylides (**144**), useful reagents for the synthesis of trisubstituted bromoolefins by Wittig reactions[130].

$$Ph_3P{=}CBr_2 \xrightarrow{RBr} \underset{(\mathbf{143})}{Ph_3\overset{+}{P}{-}CBr_2R\ Br^-} \xrightarrow{n\text{-BuLi}} \underset{(\mathbf{144})}{Ph_3P{=}CBrR}$$

The substitution of bromotrichloromethane for carbon tetrachloride for the preparation of **132** gives higher yields and no by-product formation[131].

R_3P–CCl_4 or R_3P–CBr_4 are versatile synthetic reagents for the dihalomethylenation of carbonyl groups[132–134]. Salmond has reported the preparation of dichloromethylenetrisdimethylaminophosphorane (**145**) and its use for the conversion of aldehydes to the corresponding olefins by Wittig reaction[135].

$$2(Me_2N)_3P \xrightarrow{BrCCl_3} \underset{(\mathbf{145})}{(Me_2N)_3P{=}CCl_2} + (Me_2N)_3PClBr$$

$$\mathbf{145} + RCHO \longrightarrow RCH{=}CCl_2 + (Me_2N)_3P{=}O$$

The reaction of di-*t*-butyl alkyl phosphines **146** with tetrahalomethanes yields new P-haloalkylides (**147**); the attack of X_3C^- on the phosphorus atom of the intermediate salt **148** being at a disadvantage for steric reasons, the abstraction of a proton is favoured[136].

$$\underset{(\mathbf{146},\ R{=}H,\ Me,\ Pr)}{t\text{-}Bu_2PCH_2R} \xrightarrow[X = Br,\ Cl]{CX_4} \underset{(\mathbf{148})}{t\text{-}Bu_2\overset{+}{P}(X)CH_2R\ X_3C^-} \longrightarrow \underset{(\mathbf{147})}{t\text{-}Bu_2P(X){=}CHR} + CHX_3$$

As has been shown previously, the reaction of triphenylphosphine and carbon tetrachloride with active hydrogen-containing substrates has synthetic interest. However, the separation of the products from the triphenylphosphine oxide can present problems, if the products are not volatile. These separation problems can be overcome by using polymer-supported phosphine.

Carboxylic acids can be easily converted into acid chlorides by treatment with phosphine resin in CCl_4[138]. The insoluble cross-linked polystyryldiphenylphosphine has been used for the formation of the amide bond in peptide synthesis[137]. Appel and Willms have reported an interesting modification of this technique for peptide synthesis. The condensation reaction can be carried out in homogeneous solution with a linear soluble polymer-supported phosphine, instead of insoluble polymer[139].

Cross-linked polymer-supported phosphines and CCl_4 have been advantageously used for the conversion of primary amides (**149**) and oximes (**150**) to the corres-

$$\underset{(\mathbf{149})}{RCONH_2} \longrightarrow RCN \longleftarrow \underset{(\mathbf{150})}{RCH{=}NOH}$$

ponding nitriles and for the conversion of secondary amines (**151**) to chloroimines (**152**)[140].

$$\underset{(\mathbf{151})}{R^1CONHR^2} \longrightarrow \underset{(\mathbf{152})}{R^1CCl{=}NR^2}$$

It is also probable that the mechanism of the elimination of halogen from 1,2-dihalo compounds involves attack on halogen. In the presence of alcohol, monodehalogenation takes place with phosphites, the condition for this reaction is the presence of electronegative substituents attached to a carbon bearing one of the halogen atoms[141,142].

$$ClCH_2CCl_2CN \xrightarrow{P(OR)_3} (RO)_3\overset{+}{P}Cl\ ClCH_2\overset{-}{C}ClCN \xrightarrow{EtOH} ClCH_2CHClCN$$

Borowitz and coworkers have studied the debromination of vicinal dibromides by trivalent phosphorus compounds. They have found that triphenylphosphine is more reactive towards bromine than triethylphosphite[143]. Phosphines react readily with hexachloroethane to give tetrachloroethylene (**153**)[144].

$$CCl_3CCl_3 \xrightarrow{Ph_3P} Ph_3PCl_2 + \underset{(\mathbf{153})}{CCl_2{=}CCl_2}$$

2. *Reactions with carbanions*

The reactions of carbanions with polyhalomethanes, 1,2-dihaloethanes, 1,1,2,2-tetrabromoethane or hexachloroethane have been studied for many years[145–148]. For instance, the reaction of **154** in liquid ammonia with tetrahalomethane gives 1,1,2,2-tetraphenylethane (**155**).

$$\underset{(\mathbf{154})}{2\ Ph_2CHK} \xrightarrow{XCCl_3} \underset{(\mathbf{155})}{Ph_2CHCHPh_2}$$

Kofron and Hauser have suggested that these reactions comprise a displacement on halogen by carbanion leading to the halogenated compound **156**. Then the S_N2 reaction of **156** with **154** gives the dimer **155**[145].

$$\mathbf{154} \xrightarrow{XCCl_3} \underset{(\mathbf{156})}{Ph_2CHX} + K^+\ ^-CCl_3$$

The trichloromethyl anion $^-CCl_3$ decomposes into dichlorocarbene which is trapped by cyclohexene, to yield dichloronorcarane. In the same way, a nucleophilic displacement on halogen has been proposed for the reaction of **154** with 1,2-dibromoethane or 1,2-diiodoethane[147].

However, more recent studies have shown that a number of reactions with a formal two-electron transfer can be more accurately described as two one-electron steps[149]. Evidence of free radicals during the exchange reaction between alkyllithium compounds and alkyl or aryl iodide (equation 14) has been reported.

$$R^1I + R^2Li \rightleftharpoons R^1Li + R^2I \qquad (14)$$

Chemically induced dynamic nuclear polarization in the protons of alkyl halides has been observed[150,151] and free radicals have been detected by ESR[152]. The reaction of the anion of diphenylacetonitrile (**157**) with 1,2-dibromoethane was entirely quenched by *p*-dinitrobenzene[153]. The stereochemistry of the reaction of **157** with 3,4-dibromohexane (**158**) suggests an electron transfer mechanism rather than an *E*2 pathway[154].

$$Ph_2\bar{C}CN \quad (\mathbf{157}) \qquad EtCHBrCHBrEt \quad (\mathbf{158})$$

Ainsworth and coworkers[154] have suggested that the reaction of **154** with CCl_4, which has been described as a displacement at halogen, followed by a S_N2 reaction[145], may, however, be rationalized in terms of an electron transfer process.

Reactions between phenylacetonitrile or α-alkylphenylacetonitriles (**159**) and tetrahalomethane on solid potassium hydroxide in *t*-butyl alcohol have been studied. α-Halo-α-alkylphenylacetonitriles (**160**) and succinonitriles (**161**) are formed. It is suggested that α-alkylphenylacetonitriles can be halogenated via an ionic process and that α-halo-α-alkylphenylacetonitriles (**160**) are converted into **161** via a radical process[155].

$$PhCH(R)CN \ (\mathbf{159}) \xrightarrow{CX_4} PhC(R)(X)CN \ (\mathbf{160}) \qquad PhC(R)(CN){-}C(R)(CN)Ph \ (\mathbf{161})$$

However, many reactions of carbanions with tetrahalomethanes give halogenated products without coupling. Carbon tetrachloride reacts with ketones, alcohols and sulphones in the presence of potassium hydroxide and *t*-butyl alcohol, leading to a variety of products.

Chlorinated products are sometimes isolated, such as α-chlorosulphones **163** and **165**[157].

$$PhCH(Me)SO_2Ph \ (\mathbf{162}) \xrightarrow[KOH,\ t\text{-}BuOH]{CCl_4} PhCCl(Me)SO_2Ph \ (\mathbf{163})$$

$$Ph_2CHSO_2Ph \ (\mathbf{164}) \xrightarrow[KOH,\ t\text{-}BuOH]{CCl_4} Ph_2CClSO_2Ph \ (\mathbf{165})$$

In most cases, the chlorinated products which are formed in the first step of the reaction, react further in alkaline media. The α-chloroketone formed from ketone **166** gives the carboxylic acid **167** via the Favorski reaction[156].

$$Me_2CHCOMe \ (\mathbf{166}) \xrightarrow[KOH,\ t\text{-}BuOH]{CCl_4} Me_2CHCOCH_2Cl \xrightarrow{KOH} Me_3CCOOH \ (\mathbf{167})$$

Dichlorosulphones formed from sulphones (**168**) give alkenesulfinic acids (**169**) via the Ramberg–Backlund reaction[157–159]. The conversion of benzhydrylsulphones (**170**) into 1,1-diaryl alkenes (**171**) on treatment with CCl_4–KOH has been reported[157].

$$RCH_2SO_2CH_2R \ (\mathbf{168}) \rightarrow RCCl_2SO_2CH_2R \rightarrow RCH{=}C(R)SO_3H \ (\mathbf{169})$$

$$RCH_2SO_2CH_2R \ (\mathbf{168}) \rightarrow RCHClSO_2CHClR \rightarrow RCH{=}C(R)SO_3H \ (\mathbf{169})$$

$$Ph_2CHSO_2CH(R^1)(R^2) \ (\mathbf{170}) \xrightarrow[KOH]{CCl_4} Ph_2C{=}C(R^1)(R^2) \ (\mathbf{171})$$

The chlorination of succinimides (**172** and **174**) gives the expected chlorinated succinimides (**173** and **175**); however, **175** is converted by KOH into a β-lactam (**176**)[160].

$$(\mathbf{172}, R = H), (\mathbf{174}, R = Me) \xrightarrow[KOH]{CCl_4} (\mathbf{173}, R = H), (\mathbf{175}, R = Me) \xrightarrow[KOH]{R\,=\,Me} (\mathbf{176})$$

Amides, esters[160], phosphonates[161] and succinonitriles[162] have been chlorinated in the same way, with CCl_4–KOH. Trichloroethylene (**177**) has been converted into tetrachloroethylene (**178**) by CCl_4 in a catalytic two-phase system[163].

$$CCl_2{=}CHCl \ (\mathbf{177}) \xrightarrow[NaOH]{CCl_4} CCl_2{=}CCl_2 \ (\mathbf{178})$$

Although there was no positive proof, Meyers and Kolb[164] have suggested that α-halogenation of ketones involves the reaction of enolates (**180**) with CCl_4 in a discrete electron transfer, which gives a radical–anion radical pair (**181**), which does not require initiation by oxygen. This mechanism would not be a radical chain process. The net result is a transfer of a halogen atom with a sextet of electrons, the leaving group being a carbanion.

$$R^1_2CHCOR^2 \ (\mathbf{179}) \xrightarrow{KOH} [R^1_2C{\cdots}C(R^2){\cdots}O]^- \ (\mathbf{180}) \xrightarrow{CCl_4} \left[R^1_2C{\cdots}C(R^2){\cdots}O^{\bullet} \ \ CCl_4^{\overset{\bullet}{-}}\right] \ (\mathbf{181}) \longrightarrow R^1_2CClCOR^2 + {}^-CCl_3$$

The reactions of carbanions which derive from esters[165] (**182**) and sulphones[166] (**183**) with tetrahalomethanes CX_4, X = Cl, Br, to give halo compounds have been recently described. Lithiated cyclic sulphones have been chlorinated by hexachloroethane[167].

It is difficult, for all these reactions, to ascertain the exact mechanism: it may be either a mechanism with a radical–anion radical pair intermediate or a one-step displacement on halogen atom.

$$R^1_2CHCOOR^2 \xrightarrow[CX_4]{LiN(i\text{-}Pr)_2} R^1_2CXCOOR^2$$

(**182**) (**184**) $R^1 = H, Me$; $R^2 = Me, CH_2{=}CHCH_2$

$$RCH_2SO_2C_6H_4CH_3\text{-}p \xrightarrow[CX_4]{BuLi} RCX_2SO_2C_6H_4CH_3\text{-}p$$

(**183**) (**185**)

B. Aryl Halides

The base-induced dehalogenation of polyhalobenzenes (**186**) has been reported. Halogens located *ortho* to another halogen were preferentially removed; iodine is more reactive than bromine and dechlorination was not observed. Bunnett has suggested a nucleophilic displacement by a carbanion on halogen; aryl anions (**187**) might be intermediates, the tests for aryl radical intermediates being negative. The reagent can be the carbanion **189** formed by the deprotonation of dimethylsulphoxide with potassium *t*-butylate[168–171].

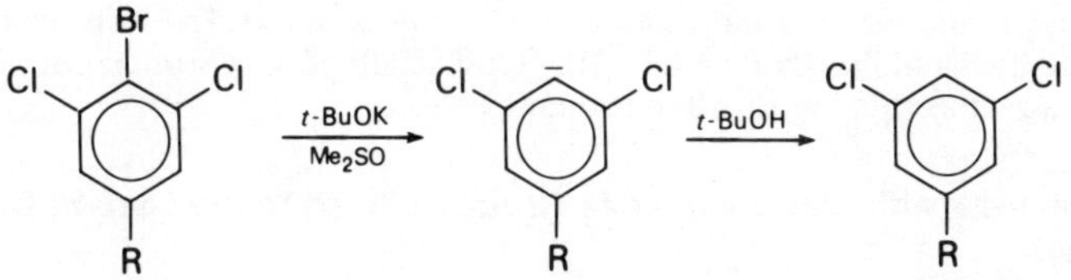

Br, Cl, Cl, R (**186**) $\xrightarrow[Me_2SO]{t\text{-BuOK}}$ Cl, Cl, R, − (**187**) $\xrightarrow{t\text{-BuOH}}$ Cl, Cl, R (**188**)

$$Me_2SO + t\text{-BuOK} \rightleftharpoons MeSOCH_2^- + t\text{-BuOH}$$

(**189**)

Dehalogenation of *o*-haloiodobenzenes (**190**) may be performed in the dark, with diethyl phosphite ion in liquid ammonia. The probable mechanism is a nucleophilic attack of iodine to displace an *o*-halophenyl anion (**191**).

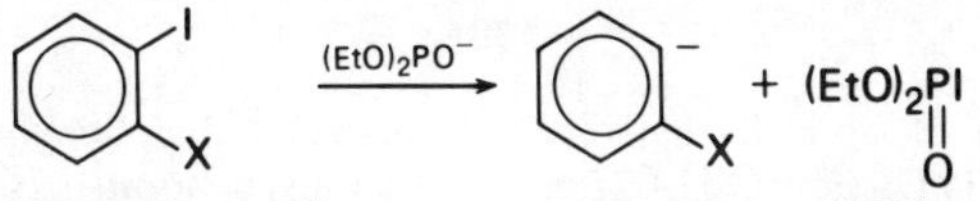

I, X (**190**, X = F, Cl, Br, I) $\xrightarrow{(EtO)_2PO^-}$ −, X (**191**) + $(EtO)_2P(O)I$ (**192**)

The anion **191** takes a proton from the solvent to give halobenzene (**193**). The formation of aniline from **190**, X = Br or I appears to derive from benzyne (**194**). It is important to note that the photo-stimulated reaction of **190** with diethyl phosphite anion occurs by the $S_{RN}1$ mechanism and gives essentially biphosphonate (**195**)[172].

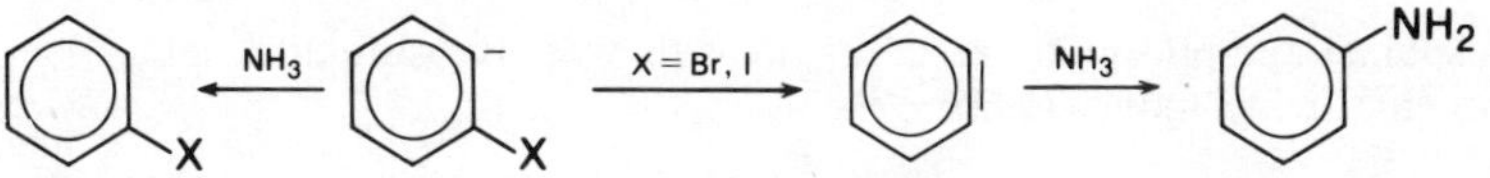

X (**193**) $\xleftarrow{NH_3}$ −, X (**191**) $\xrightarrow{X = Br, I}$ (**194**) $\xrightarrow{NH_3}$ NH_2

$PO(OEt)_2$, $PO(OEt)_2$ (**195**)

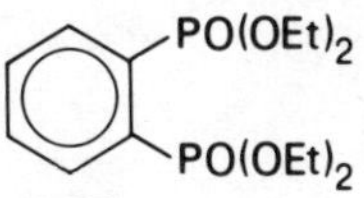

Halogen migrations in reactions of haloisothiazoles or halothiophenes with bases have also been observed[173,174].

C. Haloalkynes

Three mechanisms can explain the reactions of haloacetylenes (**196**) with nucleophiles Nu^-: addition of nucleophile on acetylenic carbons (equations 15 and 16) and nucleophilic attack on halogen atom (equation 17). A kinetic study of the reaction and

$$RC{\equiv}CX + Nu^- \;(\mathbf{196}) \longrightarrow R\bar{C}{=}C(X)(Nu) \quad (15)$$

$$RC{\equiv}CX + Nu^- \longrightarrow (R)(Nu)C{=}\bar{C}X \quad (16)$$

$$RC{\equiv}CX + Nu^- \longrightarrow RC{\equiv}C^- + NuX \quad (17)$$

the product formation indicate that equation (17) is the preferred mechanism with polarizable nucleophiles ($Nu^- = S^{2-}, RS^-$), especially if X is bromine or iodine. The iodides react more readily than the bromides[175,176].

D. Compounds with Electronegative Groups Linked to the Carbon Bearing Halogen

1. Reactions with phosphines and phosphites

The kinetics of the reaction of α-halobenzyl phenyl sulphones (**197**) with triarylphosphines in aqueous dimethylformamide have been investigated in detail. The results are consistent with a nucleophilic displacement on the halogen atom by the phosphine[177–181].

$$\underset{(\mathbf{197}),\ X = ,\ Cl\ Br,\ I)}{ArCHXSO_2Ph} + Ar_3P \xrightarrow{\text{Slow}} \underset{(\mathbf{198})}{(Ar\bar{C}HSO_2Ph\ Ar_3\overset{+}{P}X)} \xrightarrow[H_2O]{\text{Fast}} ArCH_2SO_2Ph + Ar_3PO + HX$$

The reaction of optically active *R*-(+)-benzylmethylphenylphosphine (**200**) with halosulphone (**197**, Ar = C_6H_4CN-*m*) in aqueous acetonitrile gives phosphine oxide **199**. The observed inversion of stereochemistry in oxidation of **200** is in agreement with a direct displacement on the halogen atom. A racemic phosphine oxide (**199**) would result if a radical cation **201** were the intermediate[182].

$$\underset{(\mathbf{199})}{PhCH_2(Me)(Ph)P{=}O} \qquad \underset{(\mathbf{200})}{PhCH_2(Me)(Ph)P} \qquad \underset{(\mathbf{201})}{PhCH_2(Me)(Ph)P^{+\cdot}}$$

Phosphines and phosphites react readily with *N,N*-disubstituted α-haloamides (**202**) to give chlorovinylamines (**203**)[183–185].

$$\underset{(\mathbf{202})}{R^1CCl_2CONR^2_2} \xrightarrow{Ph_3P} Ph_3\overset{+}{P}Cl\ R^1CCl{=}C(O^-)(NR^2_2) \longrightarrow R^1CCl{=}C(\overset{+}{O}PPh_3)(NR^2_2)\ Cl^- \longrightarrow \underset{(\mathbf{203})}{R^1CCl{=}C(Cl)(NR^2_2)} + Ph_3P{=}O$$

The Perkow reaction of trialkylphosphites with α-halocarbonyl compounds proceeds by a mechanism involving initial attack of phosphorus atom on the carbonyl group[186–189].

$$\text{BrCH}_2\text{COR}^1 \xrightarrow{(\text{RO})_3\text{P}} \text{BrCH}_2\overset{\text{R}^1}{\underset{^+\text{P(OR)}_3}{\text{C}}}\text{—O}^- \longrightarrow (\text{RO})_3\overset{+}{\text{P}}\text{O}\overset{\text{R}^1}{\text{C}}\text{=CH}_2\ \ \text{Br}^- \longrightarrow (\text{RO})_2\text{P(O)O}\underset{\text{R}^1}{\text{C}}\text{=CH}_2 + \text{RBr}$$

The monodebromination of α,α-dibromoketones (**204**) by triethylphosphite in the presence of acetic acid is in contrast with the reactions of other α-haloketones and α,α-dichloroketones, wherein enol phosphates are formed. For α,α-dibromoketones, attack on bromine by phosphite is likely, at least in the presence of acid, since carbonyl addition might be unfavourable for steric reasons[190].

$$\underset{(\mathbf{204})}{\text{PhCOCBr}_2\text{Me}} \xrightarrow[\text{MeCOOH}]{(\text{EtO})_3\text{P}} \text{PhCOCHBrMe}$$

The reaction of α-haloketones with tertiary phosphines gives either keto-phosphonium halides (**205**) or enol phosphonium halides (**206**). The formation of **205**, in aprotic media, has been shown to proceed via S_N2 displacement of halide ion[191–193].

$$\text{ArCOCHXR} \xrightarrow{\text{PPh}_3} \underset{(\mathbf{205})}{\text{ArCO}\underset{\text{R}}{\text{CH}}\overset{+}{\text{P}}\text{Ph}_3\ \text{X}^-}$$

The evidence for this conclusion has included kinetic studies of the reactions of α-haloketones with triphenylphosphine[194] and with optically active phosphine[192].

Enol phosphonium halides (**206**) are formed under anhydrous conditions via enolate halophosphonium intermediates (**207**), which occur by attack of the phosphine on halogen.

$$\text{PhCO}\underset{\text{X}}{\text{C}}\text{R}^1_2 \xrightarrow{\text{R}_3\text{P}} \underset{(\mathbf{207})}{\text{Ph}\underset{\text{O}^-}{\text{C}}\text{=CR}^1_2\ \text{R}_3\overset{+}{\text{P}}\text{X}} \longrightarrow \underset{(\mathbf{206})}{\text{Ph}\underset{\text{O}\overset{+}{\text{P}}\text{R}_3\text{X}^-}{\text{C}}\text{=CR}^1_2}$$

The attack on halogen by a phosphine should be enhanced by further substitution at the α-carbon by bulky and electron-withdrawing groups, such as a phenyl substituent or bromine. The S_N2 displacement on carbon is subject to steric factors. However, the displacement of phosphine on bromine should be less dependent on steric factors[186,192]. The rate of rearrangement of **208** is dependent on the phosphine. This fact appears with the reaction of phosphine with methyl trichloroacetate (**209**):

$$\underset{(\mathbf{209})}{\text{Cl}_3\text{CCOOMe}} \xrightarrow{\text{R}_3\text{P}} \underset{(\mathbf{208})}{\text{R}_3\overset{+}{\text{P}}\text{Cl}\ \bar{\text{C}}\text{Cl}_2\text{—COOMe}}$$

When R = Ph, this rate of rearrangement is fast. The ylide **210** is formed.

$$\mathbf{208},\ \text{R} = \text{Ph} \longrightarrow \text{Ph}_3\overset{+}{\text{P}}\text{CCl}_2\text{COOMe Cl}^- \xrightarrow{\text{Ph}_3\text{P}} \underset{(\mathbf{210})}{\text{Ph}_3\text{P=CClCOOMe}}$$

When R = NMe_2, the rate of the rearrangement is slow. In this case, the anion of the ion pair **208** can be trapped by an aldehyde: the epoxide **211** was isolated[195].

$$\mathbf{208},\ R{=}NMe_2 \xrightarrow{i\text{-PrCHO}} i\text{-Pr}\underset{\underset{O^-}{|}}{C}HCCl_2COOMe + (Me_2N)_3\overset{+}{P}Cl \longrightarrow$$

$$i\text{-PrCH}\overset{O}{\frown}CClCOOMe + (Me_2N)_3PCl_2$$

(211)

The debromination of α-bromoketones with phosphines is catalysed by acids or by protic solvents. The kinetic data suggest pathways involving rate-determining attack of phosphine on the bromine of protonated bromoketones[196]. The presence of triethylamine slows down the acid-catalysed debromination enough that ketophosphonium bromide formation predominates[193,196].

$$ArCOCH_2Br \overset{H^+}{\rightleftharpoons} Ar\overset{\overset{+}{O}H}{\overset{||}{C}}CH_2Br \xrightarrow{Ph_3P} Ar\overset{OH}{\overset{|}{C}}{=}CH_2 \longrightarrow ArCOCH_3$$

The corresponding α-chloroketones gives ketophosphonium chlorides (via S_N2 displacement) even in protic media. This result is in agreement with the fact that attack on bromine by Ph_3P is a more facile process than attack on chlorine[191,193]. Perchlorinated ketones undergo α,β-dechlorination readily when treated with phosphines or phosphites[197].

$$CCl_3CCl_2COCCl_2CCl_3 \xrightarrow{2\ Ph_3P} CCl_2{=}CClCOCCl{=}CCl_2 + 2\ Ph_3PCl_2$$

The reaction of triphenylphosphine with hexachloroacetone is easier than with CCl_4. The rate of the determining step, which appears to be initial abstraction of positive chlorine by PPh_3, is increased because $^-CCl_2{-}CO{-}CCl_3$ is a better leaving group than $^-CCl_3$. Allylic alcohols **(212)** react with Ph_3P–hexachloroacetone to produce excellent yields of the corresponding chlorides **(213)**, with preservation of double bond geometry and with inversion of configuration for optically active alcohols. Isolation is accomplished simply by flash distillation, hexachloroacetone and pentachloroacetone not being volatile[198,199].

$$\underset{\mathbf{(212)}}{(H)(Me)C{=}C(H)CH(OH)R} \xrightarrow[Ph_3P]{CCl_3COCCl_3} \underset{\mathbf{(213)}}{(H)(Me)C{=}C(H)CH(Cl)R}$$

Generally, α-monohalonitriles do not have a very positive halogen atom. Chloroacetonitrile **215** reacts with triethylphosphite to give the normal Arbuzov product[200] and PPh_3 reacts with bromoacetonitrile to give phosphonium salt, although attack at bromine is possible when MeOH is added[201]. Ethyl di-*t*-butylphosphinite **(214)** appears to be an exceptionally halophilic phosphinite. It gives the normal Arbuzov reaction with simple alkyl halides. However, when the alkyl halide is chloroacetonitrile **(215)**, the attack of **214** takes place at both carbon (path *a*) and chlorine (path *b*) with preference for the latter[202].

Phosphine oxide **(216)**, di-*t*-butylphosphinyl chloride **(217)** and succinonitrile **(218)** were isolated.

It is suggested that the important halophilicity of **214** is a consequence of the steric

$$t\text{-Bu}_2\text{POEt} + \text{ClCH}_2\text{CN} \xrightarrow{a} t\text{-Bu}_2\overset{+}{\text{P}}(\text{OEt})-\text{CH}_2\text{CN}\ \text{Cl}^- \longrightarrow t\text{-Bu}_2\text{P}(=\text{O})-\text{CH}_2\text{CN}$$

(214) (215) (216)

$$t\text{-Bu}_2\text{POEt} + \text{ClCH}_2\text{CN} \xrightarrow{b} t\text{-Bu}_2\overset{+}{\text{P}}(\text{OEt})-\text{Cl}\ \ \bar{\text{C}}\text{H}_2\text{CN}$$

$$t\text{-Bu}_2\overset{+}{\text{P}}(\text{OEt})\text{Cl} + \text{Cl}^- \longrightarrow t\text{-Bu}_2\text{P}(=\text{O})\text{Cl} + \text{EtCl}$$

(217)

$$\mathbf{215} + \bar{\text{C}}\text{H}_2\text{CN} \longrightarrow \text{NCCH}_2\text{CH}_2\text{CN} + \text{Cl}^-$$

(218)

requirements of the *t*-butyl groups. For the same reason, *i*-Pr_2POEt presents only a slight tendency to attack at halogen atom of **215**[203]. Di-iodomethane and 1,2-dibromoethane are also attacked at the halogen atoms by **214**.

Phosphines and phosphites react readily with compounds bearing very positive halogen, such as α-bromo α-cyano esters (**219**), α-bromo α-cyano nitriles (**223**) and α-bromo α-cyano imides (**231**).

Trialkyl[204–206] and triaryl phosphites[207] react with **219** to give, in the first step, ion pair **220**. The rearrangement of **220**, as a consequence of steric hindrance of S_N2 displacement on the phosphorus atom by halophosphonium ion and of electronic effects[208], gives essentially keteniminophosphonium salts (**221**), then *N*-phosphorylated ketenimines (**222**), which are useful for the synthesis of a number of heterocyclic systems[205–207,209].

$$\text{R}^1\text{CBr}(\text{COOMe})(\text{CN}) \xrightarrow{(\text{RO})_3\text{P}} \left[\text{R}^1\text{C}^-(\text{COOMe})(\text{CN})\quad (\text{RO})_3\overset{+}{\text{P}}\text{Br}\right] \longrightarrow (\text{MeOCO})(\text{R}^1)\text{C}=\text{C}=\text{N}-\overset{+}{\text{P}}(\text{OR})_3\ \text{Br}^-$$

(219) (220) (221)

$$\longrightarrow (\text{MeOCO})(\text{R}^1)\text{C}=\text{C}=\text{N}\sim\text{P(O)(OR)}_2 + \text{RBr}$$

(222)

Iminophosphoranes (**225**) are generated via (**224**) upon addition of phosphites to bromonitriles (**223**)[210–211].

$$\text{R}^1\text{CBr(CN)}_2 \xrightarrow{(\text{RO})_3\text{P}} (\text{NC})(\text{R}^1)\text{C}=\text{C}=\text{N}-\overset{+}{\text{P}}(\text{OR})_3 + \text{Br}^- \longrightarrow (\text{NC})(\text{R}^1)\text{C}=\text{CBr}-\text{N}=\text{P(OR)}_3$$

(223) (224) (225)

Dihalomalononitriles (**226**) react with phosphite to give the Arbuzov rearrangement product (**227**) and with Ph_3P to give **228**[212].

$$(\text{NC})_2\text{C}(\text{PO(OEt)}_2)_2 \xleftarrow{(\text{EtO})_3\text{P}} \text{X}_2\text{C(CN)}_2 \xrightarrow{\text{PPh}_3} \text{X(CN)}=\text{CX}-\text{N}=\text{PPh}_3$$

(227) (226, X = Cl, Br) (228)

The reaction of triethyl phosphite with **229** gives a ketenimine **(230)** and, to a certain extent, tetraphenylsuccinonitrile (**161**, R = Ph). The formation of **230** is an ionic process, involving attack of the phosphite at halogen, but the formation of **161**, R = Ph, is likely to be a radical process.

$$\underset{\textbf{(229)}}{Ph_2CClCN} \xrightarrow[(EtO)_3P]{} \underset{\textbf{(230)}}{Ph_2C{=}C{=}N{-}PO(OEt)_2}$$

Enol phosphates **(233)**, iminophosphoranes **(234)** and phosphoranes **(235)** are the products of the reactions of imides **(231)** with phosphites and phosphines[214–217].

R = OMe; R = Ph; R = OPh; R_3P; $Br\overset{+}{P}R_3$

(231) **(232)** **(233)** **(234)** **(235)**

The reaction of nitriles **(223)** with the complex trimethyl phosphite–silver nitrate **(236)** is a convenient method for the selective replacement of a positive halogen by a nitro group, via an intermediate ion pair **(237)**[218]. The same reaction is observed with esters **(219)** and imides **(231)**.

$$\textbf{223} + \underset{\textbf{(236)}}{(MeO)_3P{-}Ag{-}ONO_2} \longrightarrow \underset{\textbf{(237)}}{[R^1\bar{C}(CN)_2\ Br\overset{+}{P}(OMe)_3 + AgNO_3]} \longrightarrow$$

$$[R^1\bar{C}(CN)_2,\ NO_2O\overset{+}{P}(OMe)_3] \longrightarrow \underset{\textbf{(238)}}{R^1C(CN)_2NO_2}$$

Dihaloimides **(239)** react with phosphines to yield the intermediates **240** which rearrange by a reversible process into the kinetically controlled phosphobetaïne **241**, and slowly, by an irreversible process, into the thermodynamically stable phosphobetaïne **242**. The conversion of **241** into **242**, via **240**, is catalysed by halide ions[219].

$$\text{(239)} \xrightarrow{2\,Ph_3P} \text{(240)} + 2\,Cl\overset{+}{P}Ph_3$$

$$Cl_2PPh_3 + \text{(241)} \rightleftharpoons \mathbf{240} \longrightarrow \text{(242)} + Cl_2PPh_3$$

The reactions of 1-chloro-1-nitroalkanes[220], 1-chloro-1-nitrocycloalkanes[221] and 1-halo-1-nitroalkanes (**243**)[222–224] with phosphines are very complex, depending on the nature of substituents. Various mechanisms are suggested and discussed, including attack of phosphine on the halogen or on the oxygen of the nitro group. In certain cases, the first step is likely to be the attack of the halogen atom by phosphine to give an ion pair (**244**) which has been trapped by reaction with a protic solvent[224].

$$\underset{\mathbf{(243)}}{RCH{=}CBr{-}NO_2} \xrightarrow{Ph_3P} \underset{\mathbf{(244)}}{RCH{=}CNO_2^-\ Ph_3\overset{+}{P}Br} \xrightarrow{MeOH} \underset{\mathbf{(245)}}{RCH{=}CHNO_2} + MeBr + Ph_3PO$$

α-Bromo α-nitro esters (**246**), which have a very positive halogen atom, react with triphenyl phosphine to give an ion pair (**247**) that can be trapped with methanol. The rearrangement of **247** has been studied[225].

$$\underset{\mathbf{(246)}}{R^1CH_2CBr(NO_2)COOR^2} \xrightarrow{Ph_3P} \underset{\mathbf{(247)}}{(R^1CH_2\overset{-}{C}(NO_2)COOR^2\ \ Ph_3\overset{+}{P}Br)} \xrightarrow{MeOH} \underset{\mathbf{(248)}}{R^1CH_2CH(NO_2)COOR^2}$$

2. *Reactions with arene sulphinates*

The reaction of α-halosulphones (**249**) with arene sulphinates (**250**) involves nucleophilic displacement on halogen by the nucleophile[226]:

$$\underset{\mathbf{(249)}}{ArCHXSO_2Ph} + \underset{\mathbf{(250)}}{Ar^1SO_2^-} \xrightarrow{Slow} \underset{\mathbf{(251)}}{Ar\overset{-}{C}HSO_2Ph} + \underset{\mathbf{(252)}}{Ar^1SO_2X} \xrightarrow[H_2O]{Fast} \underset{\mathbf{(253)}}{ArCH_2SO_2Ph} + \underset{\mathbf{(254)}}{Ar^1SO_3^-} + HX$$

Moreover, a nucleophilic displacement on the halogen atom of α-haloacylmalonates (**255**) by arene sulphinate ion has been observed. In wet *t*-butanol, the acylmalonate **257** is formed. When these reactions are carried out in dry *t*-butanol, enol sulphonate **258** is isolated[227].

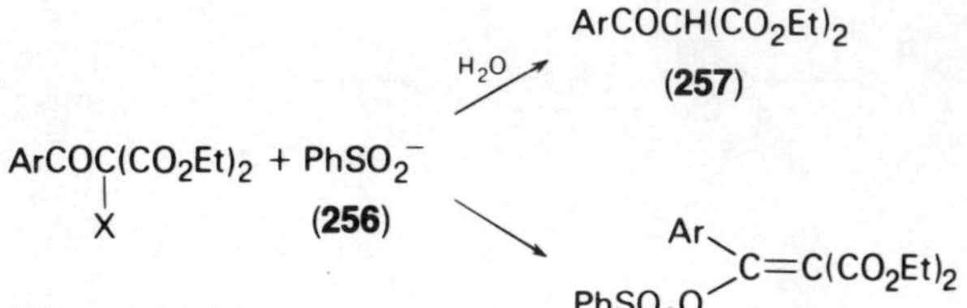

$$\mathrm{ArCOC(CO_2Et)_2(X) + PhSO_2^-} \xrightarrow{H_2O} \mathrm{ArCOCH(CO_2Et)_2}$$

(255, X = Cl, Br) (256) (257)

$$\mathrm{ArCOC(CO_2Et)_2(X) + PhSO_2^-} \longrightarrow \mathrm{Ar(PhSO_2O)C{=}C(CO_2Et)_2}$$

(258)

The debromination of α-cyano α-bromo esters (**219**) and α-cyano α-bromosuccinimides (**231**) by sodium arene sulphinates in dry *t*-butanol gives unstable ketenimines (**259**) and enol sulphonates (**260**) respectively[228].

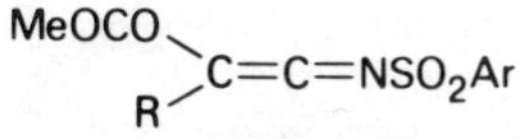

$\mathrm{MeOCO(R)C{=}C{=}NSO_2Ar}$

(**259**)

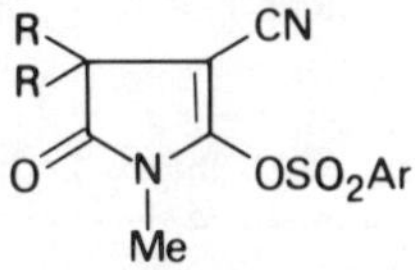

(**260**)

3. Reactions with enamines

Laskovics and Schulman have reported the preparation of α-chloroketones (**261**) by the reaction of enamines (**262**) with hexachloroacetone. They suggest that the reaction involves a nucleophilic displacement on chlorine, the pentachloroacetonide ion (**264**) being a good leaving group. This is an interesting method for mild and regioselective chlorination[229,230].

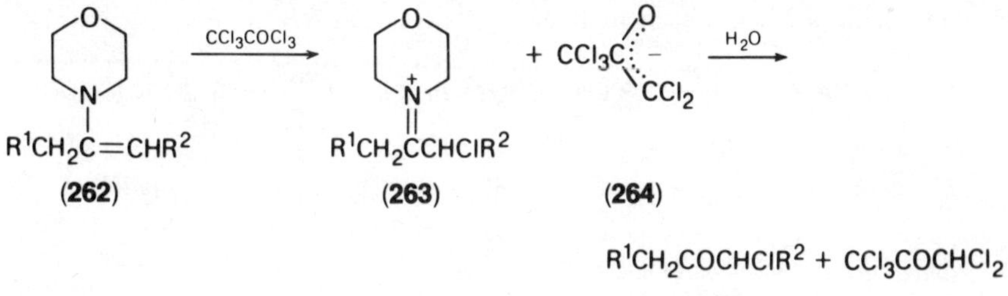

$$\mathrm{R^1CH_2C{=}CHR^2}\ (\mathbf{262}) \xrightarrow{CCl_3COCl_3} \mathrm{R^1CH_2C(=N^+)CHClR^2}\ (\mathbf{263}) + \mathrm{CCl_3C(O)CCl_2^-}\ (\mathbf{264}) \xrightarrow{H_2O} \mathrm{R^1CH_2COCHClR^2}\ (\mathbf{261}) + \mathrm{CCl_3COCHCl_2}$$

4. Reactions with halide ions

The key step of the reduction of the unsaturated diketones **265** to the saturated diketones **266** by treatment with sodium iodide in acidic media is the attack on the iodine atom of the intermediate **267** by the iodide ion[231].

$$\mathrm{R^1COCH{=}CHCOR^2}\ (\mathbf{265}) \xrightarrow{HI} \mathrm{R^1COCHI{-}CH_2COR^2}\ (\mathbf{267}) \xrightarrow[H^+]{I^-} \mathrm{R^1COCH_2CH_2COR^2}\ (\mathbf{266}) + I_2$$

The reduction of α-haloketones (**268**, X = Br, Cl) with sodium iodide and trimethylsilyl chloride has been reported. Probably the iodide ion attacks the halogen and the ketone is converted to silyl ether (**269**)[232].

$$\underset{(268)}{RCOCH_2X} \xrightarrow[Me_3SiCl]{I^-} \underset{(269)}{R{-}\overset{\overset{\displaystyle OSiMe_3}{|}}{C}{=}CH_2} + IX + Cl^-$$

The reduction of α-haloketones by sodium iodide was also achieved by the use of complexing reagents of the carbonyl group, such as triethylamine–sulphur dioxide or pyridine–sulphur trioxide[233].

A closely related method of reduction of haloketones has been reported by Gemal and Luche. Aryl bromo- and iodoalkyl ketones (**270**) are easily converted to the corresponding ketones by NaI in the presence of acid. Aliphatic ketones are more resistant to reduction[234].

$$\underset{(270,\ X = Br,\ I)}{ArCOCH_2X} \overset{H^+}{\rightleftharpoons} Ar{-}\overset{\overset{\displaystyle ^+OH}{||}}{C}{-}CH_2X \xrightarrow{I^-} Ar{-}\overset{\overset{\displaystyle OH}{|}}{C}{=}CH_2 + IX$$

V. COMPOUNDS WITH HALOGEN–SULPHUR BONDS

The many mechanistic investigations of the reactions of sulphenyl chlorides, reviewed by Marino[235], show a polarization of the halogen–sulphur bond that promotes ionic reactions involving a positive sulphur atom and a negative chloride ion.

$$RSCl \longrightarrow RS^+ + Cl^-$$

Among the few compounds which show an opposite polarization of the halogen sulphur bond are the halodialkylsulphonium ions (**271**), which can act as a source of positive halogen.

$$\underset{(271)}{R_2\overset{+}{S}{-}Cl} + Cl^- \rightleftharpoons R_2S + Cl_2$$

Enamines (**272**) have been brominated by bromodimethylsulphonium bromide. The hydrolysis of the products gives α-bromo ketones (**273**) in high yields[236].

$$(272) \xrightarrow{Me_2\overset{+}{S}{-}Br\ Br^-} \text{iminium salt (2-bromo, } Br^-) \xrightarrow{H_2O} (273)$$

The reduction of sulphoxides with phosphorus pentachloride and *N,N*-diethylaniline (**274**) or 1-morpholino-1-cyclohexane (**272**) probably involves an intermediate sulphonium salt (**275**), which may react with **272** or **274** to give chlorinated products **276** and **277**[237].

In the absence of amine or enamine, **275** can undergo a Pummerer rearrangement[237].

When sulphoxides react with activating reagents, such as triphenylphosphine and

$$\mathrm{RSOR} \xrightarrow{\mathrm{PCl_5}} \mathrm{R\overset{+}{S}(OPCl_4)R}\ \mathrm{Cl^-} \longrightarrow \mathrm{R\overset{+}{S}(Cl)R}\ \mathrm{Cl^-} + \mathrm{Cl_3P{=}O}$$

(275)

$$\mathbf{275} + \mathbf{272} \longrightarrow \text{(276)}\cdot\mathrm{HCl} + \mathrm{RSR}$$

(276)

$$\mathbf{275} + \mathrm{PhNEt_2} \longrightarrow \mathrm{Cl{-}C_6H_4{-}NEt_2\cdot HCl} + \mathrm{RSR}$$

(274) (277)

iodine[238] or trimethylsilyl iodide and iodide ion or other bases such as CN^-, MeS^-, S^{2-}, NCS^-, SO_3^{2-}, $S_2O_3^{2-}$, sulphides are obtained, generally in good yields. The mechanism suggested involves a sulphurane **278**[239,240].

$$\mathrm{R^1SOR^2} \xrightarrow{\mathrm{Me_3SiI}} \left[\mathrm{R^1S(OSiMe_3)(I)R^2}\right] \xrightarrow{\mathrm{I^-}} \mathrm{R^1SR^2} + \mathrm{I_2} + \mathrm{Me_3SiO^-}$$

(278)

The reductions of sulphonyl halides with trimethylsilyl iodide and sodium iodide[241] and of arenesulphonic acids with triphenylphosphine and iodine[242] to sulphides have been reported. These reactions probably take place in the same way as the reduction of **278**. Corson and Pews have shown that the first step of the reaction of *p*-toluensulphonyl chloride (**279**) with sodium cyanide is an attack of the nucleophile at the chlorine atom to give cyanogen chloride[243].

$$p\text{-}\mathrm{MeC_6H_4SO_2Cl} + \mathrm{CN^-} \longrightarrow p\text{-}\mathrm{MeC_6H_4SO_2^-} + \mathrm{ClCN}$$

(279)

The reaction of benzenesulphonyl chloride with sodium iodide in acetone has been recently reinvestigated. The stoichiometry of the reaction is consistent with the formation of a complex (**280**), which in turn reacts with the sodium phenylsulphinate to yield benzenesulphonyl iodide (**281**)[244].

$$\mathrm{PhSO_2Cl} + 3\,\mathrm{NaI} \xrightarrow{\mathrm{Me_2CO}} \mathrm{(NaI)_2ICl(MeCOMe)_6} + \mathrm{PhSO_2Na} \longrightarrow \mathrm{PhSO_2I}$$

(280) (281)

Trifluoromethanesulphonyl chloride (**280**) is a very efficient chlorinating agent of stabilized carbanions. The treatment of acidic compounds **283** with **280** in the presence of triethylamine or 1,5-diazabicyclo[5,4,0]undec-5-one gives chlorinated compounds (**282**) in high yields[245].

Aryl chlorosulphates (**284**) react with nucleophiles Nu^- such as $S_2O_3^{2-}$, CN^-, I^- or SO_3^{2-}. The mechanism suggested is a nucleophilic displacement on chlorine atom[246].

$$XCH_2Y + CF_3SO_2Cl \xrightarrow{NEt_3} XCCl_2Y$$
(**283**) (**280**) (**282**)

X = Y = MeCO
X = MeCO, Y = CO_2Et
X = Y = CO_2Me
X = CN, Y = CO_2Me

$$ArOSO_2Cl + Nu^- \longrightarrow ArOSO_2^- + ClNu$$
(**284**)

VI. REFERENCES

1. R. G. Pearson, *J. Chem. Educ.*, **45**, 581, 643 (1968).
2. G. Klopman and R. F. Hudson, *Chemical Reactivity and Reaction Paths*, Wiley, New York (1974).
3. R. G. Pearson, S. H. Langer, F. V. Williams and W. J. McGuire, *J. Amer. Chem. Soc.*, **74**, 5130 (1952).
4. P. A. Chopard, R. F. Hudson and G. Klopman, *J. Chem. Soc.*, 1379 (1965).
5. E. S. Huyser in *The Chemistry of the Carbon–Halogen Bond*, Part 1, (Ed. S. Patai), Wiley, Chichester (1973), Chap. 8, pp. 549–607.
6. P. Brun and B. Waegell, *Tetrahedron*, **32**, 517 (1976).
7. P. Mackiewicz and R. Furstoss, *Tetrahedron*, **34**, 3241 (1978).
8. P. S. Skell and J. C. Day, *Acc. Chem. Res.*, **11**, 381 (1978).
9. B. Miller in *Topics in Phosphorus Chemistry*, Vol. 2 (Eds M. Grayson and E. J. Griffith), Interscience, New York (1965), pp. 134–199.
10. J. I. G. Cadogan and R. K. Mackie, *Chem. Soc. Rev.*, **3**, 87 (1974).
11. M. Anbar and D. Ginsburg, *Chem. Rev.*, **54**, 925 (1954).
12. C. Walling, *Bull. Soc. Chim. Fr.*, 1609 (1968).
13. R. A. Sneen and N. P. Matheny, *J. Amer. Chem. Soc.*, **86**, 5503 (1964).
14. K. Heusler and J. Kalvoda, *Angew. Chem. Int. Edn*, **3**, 525 (1964).
15. D. E. Young, L. R. Anderson, D. E. Gould and W. B. Fox, *J. Amer. Chem. Soc.*, **92**, 2313 (1970).
16. D. E. Gould, L. R. Anderson, D. E. Young and W. B. Fox, *J. Amer. Chem. Soc.*, **91**, 1310 (1969).
17. C. J. Schack and W. Maya, *J. Amer. Chem. Soc.*, **91**, 2902 (1969).
18. C. J. Schack, R. D. Wilson, J. S. Muirhead and S. N. Cohn, *J. Amer. Chem. Soc.*, **91**, 2907 (1969).
19. D. E. Gould, L. R. Anderson, D. E. Young and B. W. Fox, *Chem. Commun.*, 1564 (1968).
20. R. H. Hesse, *Israel J. Chem.*, **17**, 60 (1978).
21. S. Rozen and Y. Menahem, *Tetrahedron Lett.*, 725 (1979).
22. O. Lerman and S. Rozen, *J. Org. Chem.*, **45**, 4122 (1980).
23. L. Skattbol, B. Boulette and S. Solomon, *J. Org. Chem.*, **32**, 3111 (1967).
24. C. R. Johnson and M. P. Jones, *J. Org. Chem.*, **32**, 2014 (1967).
25. C. R. Johnson and J. J. Rigau, *J. Amer. Chem. Soc.*, **91**, 5398 (1969).
26. C. Walling and M. J. Mintz, *J. Org. Chem.*, **32**, 1286 (1967).
27. D. Swern, I. Ikeda and G. F. Whitfield, *Tetrahedron Lett.*, 2635 (1972).
28. J. C. Martin and R. J. Arhart, *J. Amer. Chem. Soc.*, **93**, 2339 (1971).
29. T. M. Balthazor and J. C. Martin, *J. Amer. Chem. Soc.*, **97**, 5634 (1975).
30. J. C. Martin and T. M. Balthazor, *J. Amer. Chem. Soc.*, **99**, 152 (1977).
31. I. Kapovits, J. Rabai, F. Ruff and A. Kucsman, *Tetrahedron*, **35**, 1869 (1979).
32. S. Iriuchijima and G. Tsuchihashi, *Tetrahedron Lett.*, 5259 (1969).
33. S. Iriuchijima, M. Ishibashi and G. Tsuchihashi, *Bull. Chem. Soc. Japan*, **46**, 921 (1973).
34. J. Klein and H. Stollar, *J. Amer. Chem. Soc.*, **95**, 7437 (1973).
35. M. Cinquini, S. Collona and D. Landini, *JCS Perkin II*, 296 (1972).

36. D. H. R. Barton, R. H. Hesse, M. M. Pechet and H. T. Toh, *JCS Perkin I*, 732 (1974).
37. P. G. Gassman and T. J. Van Bergen, *J. Amer. Chem. Soc.*, **95**, 590, 591 (1973).
38. M. Seguin, J. C. Adenis, C. Michaud and J. J. Basselier, *J. Fluorine Chem.*, **15**, 37 (1980).
39. M. Seguin, J. C. Adenis, C. Michaud and J. J. Basselier, *J. Fluorine Chem.*, **15**, 201 (1980).
40. H. E. Baumgarten, J. E. Dirks, J. M. Petersen and R. L. Zey, *J. Org. Chem.*, **31**, 3708 (1966).
41. D. B. Denney and R. R. Di Leone, *J. Amer. Chem. Soc.*, **84**, 4737 (1962).
42. J. H. Finley and D. B. Denney, *J. Amer. Chem. Soc.*, **92**, 362 (1970).
43. G. E. Heasley, V. M. McCully, R. T. Wiegman, V. L. Heasley and R. A. Skidgel, *J. Org. Chem.*, **41**, 644 (1976); V. L. Heasley, D. F. Shellhamer, R. K. Gipe, H. C. Wiese, M. L. Oakes and G. E. Heasley, *Tetrahedron Lett.*, **21**, 4133 (1980).
44. A. Bresson, G. Dauphin, J. M. Geneste, A. Kergomard and A. Lacourt, *Bull. Soc. Chim. Fr.*, 2432 (1970).
45. W. Maya, C. J. Schack, R. D. Wilson and J. S. Muirhead, *Tetrahedron Lett.*, 3247 (1969).
46. L. R. Anderson, D. E. Young, D. E. Gould, R. Juurik-Hogan, D. NuechTerlein and W. B. Fox, *J. Org. Chem.*, **35**, 3730 (1970).
47. D. H. R. Barton, R. H. Hesse, G. P. Jackman, L. Ogunkoya and M. M. Pechet, *JCS Perkin I*, 739 (1974).
48. D. H. R. Barton, L. J. Danks, A. K. Ganguly, R. H. Hesse, G. Tarzia and M. M. Pechet, *Chem. Commun.*, 227 (1969); D. H. R. Barton, L. S. Godinho, R. H. Hesse and M. M. Pechet, *Chem. Commun.*, 804 (1968).
49. D. H. R. Barton, A. K. Ganguly, R. H. Hesse, S. N. Loo and M. M. Pechet, *Chem. Commun.*, 806 (1968).
50. W. S. Middleton and E. M. Bingham, *J. Amer. Chem. Soc.*, **102**, 4845 (1980).
51. S. Rozen and O. Lerman, *J. Amer. Chem. Soc.*, **101**, 2782 (1979).
52. S. Rozen and O. Lerman, *J. Org. Chem.*, **45**, 672 (1980).
53. D. H. R. Barton, R. H. Hesse, R. E. Markwell, M. M. Pechet and H. T. Toh, *J. Amer. Chem. Soc.*, **98**, 3034 (1976); D. H. R. Barton, R. H. Hesse, R. E. Markwell, M. M. Pechet and S. Rozen, *J. Amer. Chem. Soc.*, **98**, 3036 (1976).
54. C. Gal, G. Ben Shoshan and S. Rozen, *Tetrahedron Lett.*, **21**, 5067 (1980).
55. C. T. Ratcliffe, C. V. Hardin, L. R. Anderson and W. B. Fox, *J. Amer. Chem. Soc.*, **93**, 3886 (1971).
56. N. Walker and D. D. Des Marteau, *J. Amer. Chem. Soc.*, **97**, 13 (1975).
57. M. J. Hopkinson, N. S. Walker and D. D. Des Marteau, *J. Org. Chem.*, **41**, 1407 (1976).
58. D. D. Des Marteau, *Inorg. Chem.*, **7**, 434 (1968).
59. K. Seppelt, *Chem. Ber.*, **106**, 157 (1973).
60. F. Aulike and D. D. Des Marteau, *Fluorine Chem. Rev.*, **8**, 73 (1977).
61. P. B. D. de la Mare, C. J. O'Connor and M. A. Wilson, *JCS Perkin II*, 1150 (1975).
62. P. B. D. de la Mare, M. A. Wilson and J. M. Rosser, *JCS Perkin II*, 1480 (1973).
63. M. A. Wilson and P. D. Woodgate, *JCS Perkin II*, 141 (1976).
64. D. D. Des Marteau, *J. Amer. Chem. Soc.*, **100**, 340 (1978).
65. Y. Katsuhara and D. D. Des Marteau, *J. Org. Chem.*, **45**, 2441 (1980).
66. Y. Katsuhara and D. D. Des Marteau, *J. Amer. Chem. Soc.*, **101**, 1039 (1979).
67. E. B. Merkushev, N. D. Simakhina and G. M. Koveshinikova, *Synthesis*, 486 (1980).
68. T. Higuchi, A. A. Hussain and I. H. Pitman, *J. Chem. Soc. B*, 626 (1969).
69. I. H. Pitman, H. Dawn, T. Higuchi and A. A. Hussain, *J. Chem. Soc. B*, 1230 (1969).
70. W. Groebel, *Chem. Ber.*, **93**, 284 (1960).
71. E. Vilsmaier, J. Schütz and S. Zimmerer, *Chem. Ber.*, **112**, 2231 (1979).
72. A. D. Dawson and D. Swern, *J. Org. Chem.*, **42**, 592 (1977).
73. E. Vilsmaier and W. Sprügel, *Tetrahedron Lett.*, 625 (1972).
74. D. L. Tuleen and T. B. Stephens, *J. Org. Chem.*, **34**, 31 (1969).
75. E. Vilsmaier, K. H. Dittrich and W. Sprügel, *Tetrahedron Lett.*, 3601 (1974); E. Vilsmaier and W. Sprügel, *Ann. Chem.*, **747**, 151 (1971).
76. P. Claus and W. Rieder, *Tetrahedron Lett.*, 3879 (1972).
77. P. G. Gassman and D. R. Amick, *Tetrahedron Lett.*, 889 (1974).
78. E. J. Corey, C. U. Kim and M. Takeda, *Tetrahedron Lett.*, 4339 (1972).
79. E. J. Corey and C. U. Kim, *J. Amer. Chem. Soc.*, **94**, 7586 (1972).
80. J. P. McCormick, *Tetrahedron Lett.*, 1701 (1974).

81. P. K. Claus, W. Rieder, P. Hofbauer and E. Vilsmaier, *Tetrahedron*, **31**, 505 (1975).
82. A. D. Dawson and D. Swern, *J. Org. Chem.*, **42**, 592 (1977).
83. E. Vilsmaier, W. Sprügel and K. Gagel, *Tetrahedron Lett.*, 2475 (1974).
84. G. Tsuchihashi and K. Ogura, *Bull. Chem. Soc. Japan*, **44**, 1726 (1971).
85. A. K. Tsolis, W. E. McEwen and C. A. Vanderwerf, *Tetrahedron Lett.*, 3217 (1964).
86. E. Gaydou, G. Peiffer, A. Guillemonat and J. C. Traynard, *Compt. Rend. C*, **275**, 547 (1972); J. M. Desmarchelier and T. R. Fukuto, *J. Org. Chem.*, **37**, 4218 (1972); D. J. Scharf, *J. Org. Chem.*, **39**, 922 (1974).
87. H. Kise, G. F. Whitfield and D. Swern, *J. Org. Chem.*, **37**, 1121 (1972).
88. E. Vilsmaier and R. Bayer, *Synthesis*, 46 (1976).
89. P. G. Gassman and R. J. Balchunis, *Tetrahedron Lett.*, 2235 (1977).
90. J. M. Desmarchelier and T. R. Fukuto, *J. Org. Chem.*, **37**, 4218 (1972).
91. F. Ruff and A. Kucsman, *JCS Perkin II*, 509 (1975); F. Ruff, I. Kapovits, J. Rabai and A. Kucsman, *Tetrahedron*, **34**, 2767 (1978); F. Ruff, G. Szabo, J. Vajda, I. Kövesdi and A. Kucsman, *Tetrahedron*, **36**, 1631 (1980).
92. F. Ruff, K. Komoto, N. Furukawa and S. Oae, *Tetrahedron*, **32**, 2763 (1976).
93. Y. Tamura, Y. Nishikawa, K. Sumoto, M. Ikeda, M. Murase and M. Kise, *J. Org. Chem.*, **42**, 3226 (1977).
94. I. Kapovits and A. Kalman, *Chem. Commun.*, 649 (1971).
95. A. Hassner, M. E. Lorber and C. Heathcock, *J. Org. Chem.*, **32**, 540 (1967); A. Hassner, R. P. Hoblitt, C. Heathcock, J. E. Kropp and M. Lorber, *J. Amer. Chem. Soc.*, **92**, 1326 (1970); G. Swift and D. Swern, *J. Org. Chem.*, **32**, 511 (1967).
96. F. W. Fowler, A. Hassner and L. A. Levy, *J. Amer. Chem. Soc.*, **89**, 2077 (1967); A. Hassner, *Acc. Chem. Res.*, **4**, 9 (1971).
97. R. C. Cambie, P. S. Rutlege, T. Smith-Palmer and P. D. Woodgate, *JCS Perkin I*, 2250 (1977); R. C. Cambie, R. C. Hayward, P. S. Rutledge, T. Smith-Palmer, B. E. Swedlund and P. D. Woodgate, *JCS Perkin I*, 180 (1979).
98. W. D. Kingsbury and C. R. Johnson, *Chem. Commun.*, 365 (1969); C. R. Johnson, C. C. Bacon and W. D. Kingsbury, *Tetrahedron Lett.*, 501 (1972).
99. R. Hüttel, H. Wagner and P. Jochum, *Ann. Chem.*, **593**, 13 (1955); R. Hüttel, O. Schäfer and G. Welzel, *Ann. Chem.*, **598**, 186 (1956).
100. R. Harville and S. P. Reed, *J. Org. Chem.*, **33**, 3976 (1968).
101. M. Cinquini and S. Colonna, *Synthesis*, 259 (1972).
102. L. Citerio, D. Poca and R. Stradi, *JCS Perkin I*, 309 (1978); L. Citerio, D. Pocar, M. Saccarello and R. Stradi, *Tetrahedron*, **35**, 2453 (1979).
103. P. G. Gassman and G. Gruetzmacher, *J. Amer. Chem. Soc.*, **96**, 5487 (1974); P. G. Gassman and C. T. Huang, *J. Amer. Chem. Soc.*, **95**, 4454 (1973).
104. P. G. Gassman and T. J. Van Bergen, *J. Amer. Chem. Soc.*, **95**, 2718 (1973).
105. P. G. Gassman, *Acc. Chem. Res.*, **3**, 26 (1970).
106. P. G. Gassman and G. A. Campbell, *J. Amer. Chem. Soc.*, **94**, 3891 (1972).
107. J. I. G. Cadogan and R. K. Mackie, *Chem. Soc. Rev.*, **3**, 87 (1974).
108. R. Appel, *Angew. Chem. Int. Edn*, **14**, 801 (1975).
109. R. Rabinowitz and R. Marcus, *J. Amer. Chem. Soc.*, **84**, 1312 (1962).
110. R. Appel, F. Knoll, W. Michel, W. Morbach, H. D. Wihler and H. Veltmann, *Chem. Ber.*, **109**, 58 (1976).
111. R. Appel, F. Knoll and H. Veltmann, *Angew. Chem. Int. Edn*, **15**, 315 (1976).
112. R. Appel and H. Veltmann, *Tetrahedron Lett.*, 399 (1977).
113. R. Appel and W. Morbach, *Synthesis*, 699 (1977).
114. R. Appel and H. F. Schöler, *Chem. Ber.*, **111**, 2056 (1978).
115. R. Appel, F. Knoll, H. Schöler and H. D. Wihler, *Angew. Chem. Int. Edn*, **15**, 702 (1976).
116. I. M. Downie, J. B. Holms and J. B. Lee, *Chem. Ind.*, 950 (1966).
117. M. A. Jones, C. E. Summer Jr, B. Franzus, T. T.-S. Huang and E. I. Snyder, *J. Org. Chem.*, **43**, 2821 (1978).
118. P. J. Kocienski, G. Cernigliaro and G. Feldstein, *J. Org. Chem.*, **42**, 353 (1977).
119. L. E. Barstow and V. J. Hruby, *J. Org. Chem.*, **36**, 1305 (1971).
120. R. Boigegrain, B. Castro and C. Selve, *Tetrahedron Lett.*, 2529 (1975).
121. B. Castro, M. Ly and C. Selve, *Tetrahedron Lett.*, 4455 (1973).
122. B. Castro and C. Selve, *Bull. Soc. Chim. Fr.*, 3009 (1974).

123. B. Castro and C. Selve, *Bull. Soc. Chim. Fr.*, 2296 (1971).
124. I. M. Downie, J. B. Lee and M. F. S. Matough, *Chem. Commun.*, 1350 (1968).
125. R. Boigegrain and B. Castro, *Tetrahedron*, **32**, 1283 (1976).
126. R. Appel and H. D. Wihler, *Chem. Ber.*, **109**, 3446 (1976).
127. I. N. Zhmurova and V. G. Yurchenko, *Zh. Obshch. Khim.*, **50**, 52 (1980); *Chem. Abstr.*, **93**, 46774 (1980).
128. R. H. Smithers, *J. Org. Chem.*, **45**, 173 (1980).
129. D. G. Naae, H. S. Kesling and D. J. Burton, *Tetrahedron Lett.*, 3789 (1975).
130. R. H. Smithers, *J. Org. Chem.*, **43**, 2833 (1978).
131. B. A. Clement and R. L. Soulen, *J. Org. Chem.*, **41**, 556 (1976).
132. E. J. Corey and P. L. Fuchs, *Tetrahedron Lett.*, 3769 (1972).
133. C. Gadreau and A. Foucaud, *Bull. Soc. Chim. Fr.*, 2068 (1976).
134. G. H. Posner, G. L. Loomis and H. S. Sawaya, *Tetrahedron Lett.*, 1373 (1975).
135. W. G. Salmond, *Tetrahedron Lett.*, 1239 (1977).
136. O. I. Kolodiazhnyi, *Tetrahedron Lett.*, **21**, 3983 (1980).
137. R. Appel, W. Strüver and L. Willms, *Tetrahedron Lett.*, 905 (1976).
138. P. Hodge and G. Richardson, *JCS Chem. Commun.*, 622 (1975).
139. R. Appel and L. Willms, *J. Chem. Res. (S)*, **84** (1977).
140. C. R. Harrison, P. Hödge and W. Rogers, *Synthesis*, 41 (1977).
141. J. P. Schroeder, L. B. Tew and V. M. Peters, *J. Org. Chem.*, **35**, 3181 (1970).
142. K. C. Pande and G. Trampe, *J. Org. Chem.*, **35**, 1169 (1970).
143. I. J. Borowitz, D. Weiss and R. K. Crouch, *J. Org. Chem.*, **36**, 2377 (1971).
144. R. Appel and H. Schöler, *Chem. Ber.*, **110**, 2382 (1977).
145. W. G. Kofron and C. R. Hauser, *J. Org. Chem.*, **28**, 577 (1963).
146. C. R. Hauser, W. G. Kofron, W. R. Dunnavant and W. F. Owens, *J. Org. Chem.*, **26**, 2627 (1961).
147. W. G. Kofron and C. R. Hauser, *J. Amer. Chem. Soc.*, **90**, 4126 (1968).
148. W. G. Kofron and C. R. Hauser, *J. Org. Chem.*, **35**, 2085 (1970).
149. K. A. Bilevitch, N. N. Bubnov and O. Y. Okhlobystin, *Tetrahedron Lett.*, 3465 (1968).
150. H. R. Ward, R. G. Lawier and R. A. Cooper, *J. Amer. Chem. Soc.*, **91**, 746 (1969).
151. A. R. Lepley and R. L. Landau, *J. Amer. Chem. Soc.*, **91**, 748, 749 (1969).
152. G. A. Russel and D. W. Lamson, *J. Amer. Chem. Soc.*, **91**, 3967 (1969).
153. K. C. Kerber, G. W. Urry and N. Kornblum, *J. Amer. Chem. Soc.*, **87**, 4520 (1965).
154. D. G. Korzan, F. Chen and C. Ainsworth, *Chem. Commun.*, 1053 (1971).
155. R. Seux, G. Morel and A. Foucaud, *Tetrahedron*, **31**, 1335 (1975).
156. C. Y. Meyers, A. M. Malte and W. S. Matthews, *J. Amer. Chem. Soc.*, **91**, 7510 (1969).
157. C. Y. Meyers, W. S. Matthews, G. J. McCollum and J. C. Branca, *Tetrahedron Lett.*, 1105 (1974).
158. C. Y. Meyers and L. L. Ho, *Tetrahedron Lett.*, 4319 (1972).
159. C. Y. Meyers, L. L. Ho, G. J. McCollum and J. C. Branca, *Tetrahedron Lett.*, 1843 (1973).
160. G. Morel, R. Seux and A. Foucaud, *Bull. Soc. Chim. Fr.*, 1865 (1975).
161. P. Coutrot, C. Laurenco, J. F. Normant, P. Perriot, P. Savignac and J. Villieras, *Synthesis*, 615 (1977).
162. G. Morel, R. Seux and A. Foucaud, *Tetrahedron Lett.*, 1031 (1971).
163. A. Jonczyk, A. Kwast and M. Makosza, *J. Org. Chem.*, **44**, 1192 (1979).
164. C. Y. Meyers and V. M. Kolb, *J. Org. Chem.*, **43**, 1985 (1978).
165. R. T. Arnold and S. T. Kulenovic, *J. Org. Chem.*, **43**, 3687 (1978).
166. H. Kotake, K. Inomata, H. Kinoshita, Y. Sakamoto and Y. Kaneto, *Bull. Chem. Soc. Japan*, **53**, 3027 (1980).
167. J. Kattenberg, E. R. De Waard and H. O. Huisman, *Tetrahedron*, **29**, 4149 (1973).
168. J. F. Bunnett, *Acc. Chem. Res.*, **5**, 139 (1972).
169. J. F. Bunnett and R. R. Victor, *J. Amer. Chem. Soc.*, **90**, 810 (1968).
170. J. F. Bunnett and C. E. Moyer, Jr, *J. Amer. Chem. Soc.*, **93**, 1183 (1971).
171. J. F. Bunnett and G. Scorrano, *J. Amer. Chem. Soc.*, **93**, 1190 (1971).
172. R. R. Bard, J. F. Bunnett and R. P. Traber, *J. Org. Chem.*, **44**, 4918 (1979).
173. M. G. Reinecke, H. W. Adickes, *J. Amer. Chem. Soc.*, **90**, 511 (1968).
174. D. A. De Bie and H. C. Van Der Plas, *Tetrahedron Lett.*, 3905 (1968).
175. M. C. Ver Ploegh, L. Donk, H. J. T. Bos and W. Drenth, *Rec. Trav. Chim.*, **90**, 765 (1971).

176. J. F. Arenc, *Rec. Trav. Chim.*, **82**, 183 (1963).
177. B. B. Jarvis and J. C. Saukaitis, *J. Amer. Chem. Soc.*, **95**, 7708 (1973).
178. B. B. Jarvis, R. L. Halper, Jr, and W. P. Tong, *J. Org. Chem.*, **40**, 3778 (1975).
179. B. B. Jarvis and B. A. Marien, *J. Org. Chem.*, **42**, 2676 (1977).
180. B. B. Jarvis and J. C. Saukaitis, *Tetrahedron Lett.*, 709 (1973).
181. B. B. Jarvis and B. A. Marien, *J. Org. Chem.*, **40**, 2587 (1975).
182. B. B. Jarvis and B. A. Marien, *J. Org. Chem.*, **41**, 2182 (1976).
183. A. J. Speziale and R. C. Freeman, *J. Amer. Chem. Soc.*, **82**, 903 (1960).
184. A. J. Speziale and L. J. Taylor, *J. Org. Chem.*, **31**, 2450 (1966).
185. A. J. Speziale and L. R. Smith, *J. Amer. Chem. Soc.*, **84**, 1868 (1962).
186. I. J. Borowitz, M. Anschel and S. Firstenberg, *J. Org. Chem.*, **32**, 1723 (1967).
187. I. J. Borowitz, S. Firstenberg, G. B. Borowitz and D. Schuessler, *J. Amer. Chem. Soc.*, **94**, 1623 (1972).
188. F. Fukui, R. Sudo, M. Masaki and M. Ohta, *J. Org. Chem.*, **33**, 5504 (1968).
189. M. F. Chasle and A. Foucaud, *Bull. Soc. Chim. Fr.*, 1535 (1972).
190. I. J. Borowitz, S. Firstenberg, E. W. R. Casper and R. K. Crouch, *Phosphorus*, **1**, 301 (1972).
191. I. J. Borowitz, K. Kirby and R. Virkaus, *J. Org. Chem.*, **31**, 4031 (1966).
192. I. J. Borowitz, K. Kirby, P. E. Rusek and E. W. R. Casper, *J. Org. Chem.*, **36**, 88 (1971).
193. I. J. Borowitz, P. E. Rusek and R. Virkaus, *J. Org. Chem.*, **34**, 1595 (1969).
194. I. J. Borowitz and H. Parnes, *J. Org. Chem.*, **32**, 3560 (1967).
195. J. Villieras, G. Lavielle, R. Burgada and B. Castro, *Compt. Rend. C*, **268**, 1164 (1969).
196. I. J. Borowitz, H. Parnes, E. Lord and K. C. Yee, *J. Amer. Chem. Soc.*, **94**, 6817 (1972).
197. K. Pilgram and H. Ohse, *J. Org. Chem.*, **34**, 1592 (1969).
198. R. M. Magid, O. S. Fruchey and W. L. Johnson, *Tetrahedron Lett.*, 2999 (1977).
199. R. M. Magid, O. S. Fruchey, W. L. Johnson and T. G. Allen, *J. Org. Chem.*, **44**, 359 (1979).
200. B. Fiszer and J. Michalski, *Roczniki. Chem.*, **28**, 185 (1954).
201. G. P. Schiemenz and H. Engelhard, *Chem. Ber.*, **94**, 578 (1961).
202. O. Dahl, *JCS Perkin I*, 947 (1978).
203. O. Dahl and F. K. Jensen, *Acta Chem. Scand. B*, **29**, 863 (1975).
204. A. Foucaud and R. Leblanc, *Tetrahedron Lett.*, 509 (1969).
205. E. Corre, M. F. Chasle and A. Foucaud, *Tetrahedron*, **28**, 5055 (1972).
206. R. Leblanc, E. Corre, M. Soenen-Svilarich, M. F. Chasle and A. Foucaud, *Tetrahedron*, **28**, 4431 (1972).
207. M. Svilarich-Soenen and A. Foucaud, *Tetrahedron*, **28**, 5149 (1972).
208. M. F. Pommeret-Chasle, F. Tonnard, M. Hassairi and A. Foucaud, *Tetrahedron*, **29**, 4219 (1973).
209. E. Corre and A. Foucaud, *Chem. Commun.*, 10 (1971).
210. R. Leblanc and A. Foucaud, *Tetrahedron Lett.*, 2441 (1969).
211. R. Leblanc, E. Corre and A. Foucaud, *Tetrahedron*, **28**, 4039 (1972).
212. V. P. Kukhar, E. I. Sagina and N. G. Pavlenko, *Zh. Obshch. Khim.*, **49**, 2217 (1979).
213. R. D. Partos and A. J. Speziale, *J. Amer. Chem. Soc.*, **87**, 5068 (1965).
214. M. F. Chasle-Pommeret, M. Leduc, A. Foucaud, M. Hassairi and E. Marchand, *Tetrahedron*, **29**, 1419 (1973).
215. M. F. Pommeret-Chasle, A. Foucaud, M. Leduc and M. Hassairi, *Tetrahedron*, **31**, 2775 (1975).
216. M. F. Pommeret-Chasle, A. Foucaud and M. Hassairi, *Tetrahedron*, **30**, 4181 (1974).
217. M. Leduc, M. F. Chasle and A. Foucaud, *Tetahedron Lett.*, 1513 (1970).
218. R. Ketari and A. Foucaud, *Tetrahedron Lett.*, **21**, 2237 (1980).
219. D. Leguern, M. A. Le Moing, G. Morel and A. Foucaud, *Tetrahedron*, **33**, 27 (1977).
220. M. Ohno and N. Kawake, *Tetrahedron Lett.*, 3935 (1966).
221. I. Sakai, N. Kawabe and M. Ohno, *Bull. Chem. Soc. Japan*, **52**, 3381 (1979).
222. C. J. Devlin and B. J. Walker, *Tetrahedron Lett.*, 1593 (1971).
223. C. J. Devlin and B. J. Walker, *JCS Perkin I*, 1428 (1973).
224. C. J. Devlin and B. J. Walker, *JCS Perkin I*, 453 (1974).
225. D. Leguern, G. Morel and A. Foucaud, *Bull. Soc. Chim. Fr.*, 252 (1975).
226. B. B. Jarvis and W. P. Tong, *J. Org. Chem.*, **41**, 1557 (1976).
227. I. Fleming and C. R. Owen, *JCS Chem. Commun.*, 1402 (1970).

228. M. Hassairi, M. F. Chasle-Pommeret and A. Foucaud, *Compt. Rend. C*, **275**, 1309 (1972).
229. F. M. Laskovics and E. M. Schulman, *Tetrahedron Lett.*, 759 (1977).
230. F. M. Laskovics and E. M. Schulman, *J. Amer. Chem. Soc.*, **99**, 6672 (1977).
231. M. D'Auria, G. Piancalli and A. Scettri, *Synthesis*, 245 (1980).
232. G. A. Olah, M. Arvanghi and Y. D. Vankar, *J. Org. Chem.*, **80**, 3531 (1980).
233. G. A. Olah, Y. D. Vankar and A. P. Fung, *Synthesis*, 59 (1979).
234. A. L. Gemal and J. L. Luche, *Tetrahedron Lett.*, **21**, 3195 (1980).
235. J. P. Marino in *Topics in Sulfur Chemistry*, Vol. 1 (Ed. A. Senning), G. Thieme, Stuttgart (1976), pp. 1–102.
236. G. A. Olah, Y. D. Vankar and M. Arvanaghi, *Tetrahedron Lett.*, 3653 (1979).
237. M. Waliska, M. Hatanaka, H. Nitta, M. Hatamura and T. Ishimaru, *Synthesis*, 67 (1980).
238. G. A. Olah, B. G. B. Gupta and S. C. Narang, *Synthesis*, 137 (1978).
239. G. A. Olah, B. G. B. Gupta and S. C. Narang, *Synthesis*, 583 (1977).
240. G. A. Olah, S. C. Narang, B. G. B. Gupta and R. Malhotra, *Synthesis*, 61 (1979).
241. G. A. Olah, S. C. Narang, L. D. Field and G. F. Salem, *J. Org. Chem.*, **45**, 4792 (1980).
242. K. Fujimari, H. Togo and S. Oae, *Tetrahedron Lett.*, **21**, 4921 (1980).
243. F. D. Corson and R. G. Pews, *J. Org. Chem.*, **36**, 1654 (1971).
244. L. M. Harwood, M. Julia and G. Le Thuillier, *Tetrahedron*, **36**, 2483 (1980).
245. G. H. Hakimelahi and G. Just, *Tetrahedron Lett.*, 3643 (1979).
246. E. Buncel, A. Raoult and L. A. Lancaster, *J. Amer. Chem. Soc.*, **95**, 5964 (1973).

The Chemistry of Functional Groups, Supplement D
Edited by S. Patai and Z. Rappoport
© 1983 John Wiley & Sons Ltd

CHAPTER **12**

Aspects of the chemistry of halophenols and halodienones

JUDITH M. BRITTAIN and PETER B. D. DE LA MARE
Chemistry Department, University of Auckland, Private Bag, Auckland, New Zealand

I. INTRODUCTION

The general properties and reactions of the phenols have been reviewed in many textbooks and comprehensive treatises[1], and aspects of their chemistry are dealt with in earlier volumes of the present series[2]. Resonance involving delocalization of the lone pair electrons on oxygen modifies the properties of the hydroxyl group, making phenol a stronger acid than ethanol; and the same electron delocalization results in phenol being more reactive than benzene with electrophiles. Phenol itself is potentially a mixture of the mobile tautomers **1**, **2** and **3**, all related by loss of a proton to the resonance-stabilized phenoxide ion **4**. The aromatic form **1** predominates over its

(1) ⇌ (2) ⇌ (3) $\underset{+H^+}{\overset{-H^+}{\rightleftharpoons}}$ (4)

tautomers, and the rate of proton exchange between the hydroxyl group and other hydroxylic compounds is rapid, being much faster than that of exchange with the *ortho* and *para* protons of the aromatic ring. The predominance of the hydroxylic tautomer is easily interpretable in terms of the resonance energy of the aromatic ring; it is perhaps not so apparent intuitively that further protonation would occur at the aromatic ring-carbon atoms in competition with protonation on oxygen. Experimentally[3], it is found that protonation of phenol in very strong acids occurs at both types of site, to give a mixture of the tautomers **5**, **6** and **7**. At temperatures above 0°C, exchange between the protons shown and such acidic media as trifluoromethanesulphonic acid is rapid; lower temperatures are required to enable the spectra of the individual forms to be recognized.

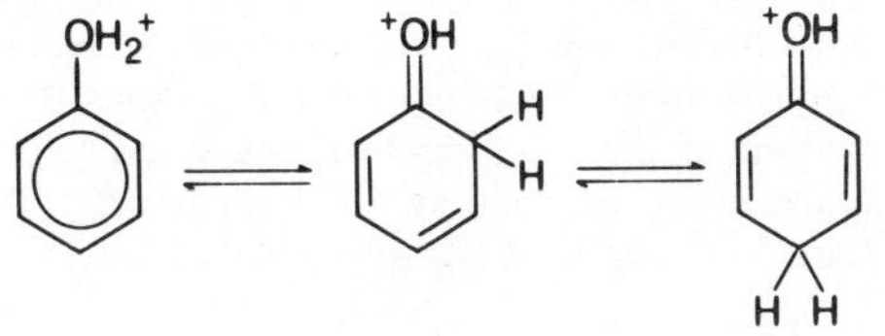

(**5**) (**6**) (**7**) 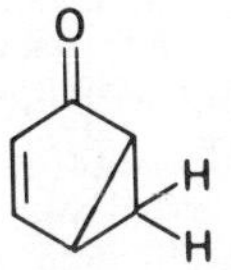 (**8**)

It is possible to introduce any of the halogens into a phenol by an electrophilic substitution, and normally these reactions result in the formation of compounds having halogen *ortho* and/or *para* to the hydroxyl group; electrons from the lone pair of the hydroxyl group cannot be readily localized at the *meta* position, for the same reason that the tautomeric form **8** is thermodynamically unstable relative to **1**, **2** and **3**.

The resulting halogenophenols are of considerable technical importance, particularly as germicides and fungicides, but for other purposes also. Intermediate stages in their formation are often dienones, which are sometimes easily isolated and can serve as sources of electrophilic halogen, but in other cases are unstable and subject to a variety of different modes of rearrangement. An understanding of the pathways available in these reactions is valuable, not only because it helps towards a detailed knowledge of the mechanisms involved in these reactions, but also because it allows us to formulate practical methods for synthesis of compounds which otherwise would be accessible only by circuitous methods.

In the present article, attention is focused particularly on the heterolytic sequences by which halogenophenols are obtained, and on the properties of these compounds and of the intermediate dienones which are so often concerned in their formation. Some homolytic reactions will also be considered, particularly where these are related to the processes of halogenation. Among the most important earlier relevant reviews, special mention should be made of articles by Waring[4], by Ershov and colleagues[5] and by Thomson[6].

II. PROPERTIES OF THE HYDROXYL AND HALOGEN SUBSTITUENTS; ACID STRENGTHS OF SUBSTITUTED PHENOLS

Before dealing with the combined and mutual effects of the hydroxyl and halogen substituents, a brief account is desirable of their individual properties. Inductively, the hydroxyl group is electron-withdrawing, because of the electronegativity of oxygen, as is shown for example by the fact that hydroxyacetic acid ($pK_a = 3.83$) is stronger acid than acetic acid ($pK_a = 4.76$)[7]. Conjugatively, however, it is powerfully electron-releasing, as is shown by the strength of benzoic acid ($pK_a = 4.19$) in comparison with *p*-hydroxybenzoic acid ($pK_a = 4.48$). Many other studies of equilibria, rates and physical properties can be used to exemplify these facts.

The group is relatively small; the barrier for free rotation of the OH group about its bond to the benzene ring[8] is only about 14 kJ mol^{-1}, and the most stable conformation has the O—H bond in the plane of the aromatic ring. In principle, large *ortho*-substituents could alter this situation, and so diminish the potential conjugation of the hydroxyl group with the benzene ring. Evidence has been adduced, however, that even in 2,6-di-*t*-butylphenol the hydrogen atom of the hydroxyl group lies in the plane of the benzene ring[9]. Theory requires, too, that in the least favoured conformation some conjugation of the lone pair electrons with the ring is possible.

To a first approximation, therefore, the mode of action of the hydroxyl group in releasing electrons to an attached unsaturated system is similar to that of a methoxyl group. Canonical forms like **9** and **10**, for example, contribute to the ground states of phenol and anisole respectively. This description is, however, not complete when reactions with electrophiles are being considered, for two related reasons. It is now generally accepted, despite earlier doubts, that the electrons of suitably oriented *sigma*-bonds can engage in hyperconjugative interaction with attached unsaturated systems. Structures like **11** and **12**, therefore, augment electron release from H—C and

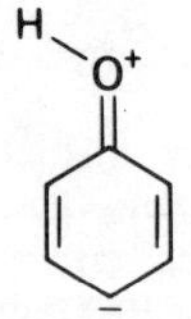

(**9**) 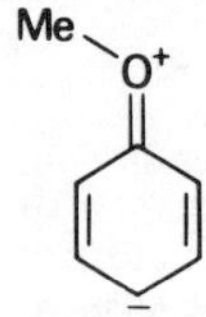 (**10**) 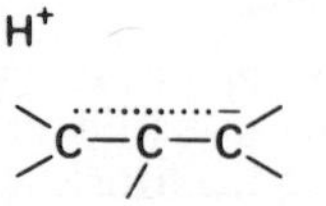 (**11**) 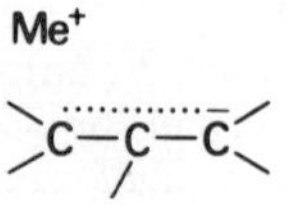 (**12**)

C—C bonds under conditions of suitably located electron demand. In general, H—C and C—C bonds contribute electron release by this process to similar extents, their relative effectiveness depending on the solvent. In most circumstances, the H—C bond is not sufficiently loosened that the proton is released to the reaction medium. Exactly the same physical situation can exist with H—O and Me—O bonds[10]. But proton loss from oxygen to solvent is usually very rapid, and the easiest result of H—O hyperconjugation can be the physical loss of a proton. The result is the incursion of a new reaction, which in the case of an electrophilic attack on phenol is a substitution with rearrangement. Proton loss from oxygen as compared with proton loss from carbon is more likely to become a concerted part of the rate-determining step.

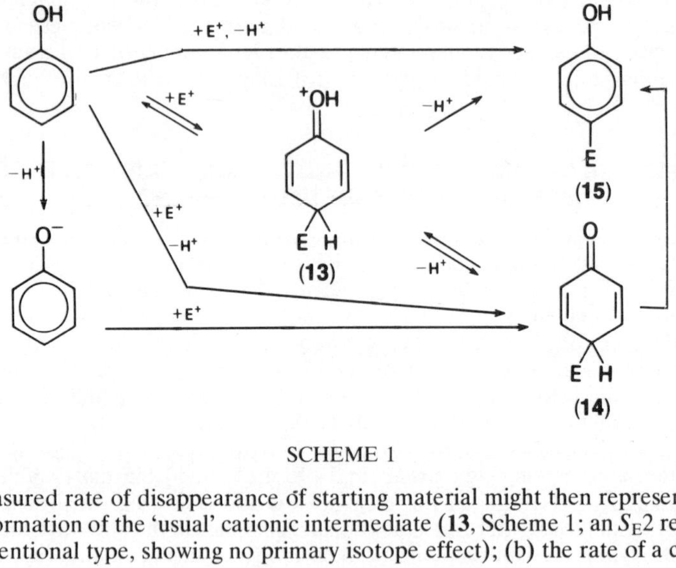

(**13**) (**15**) (**14**)

SCHEME 1

A measured rate of disappearance of starting material might then represent (a) the rate of formation of the 'usual' cationic intermediate (**13**, Scheme 1; an S_E2 reaction of the conventional type, showing no primary isotope effect); (b) the rate of a concerted formation either of the ketonic product (**14**, an S_E2' reaction) or of the enolic product (**15**, an S_E2 reaction showing a primary isotope effect); (c) the rate of electrophilic attack on a small pre-equilibrium concentration of phenoxide ion.

Whether the dienone **14** or the phenol **15** is the observed product depends on the rate and mechanism of conversion of **14** into **15**. Interpretation of the comparison of the overall rate of disappearance of starting material with the corresponding rate for anisole, or for benzene, depends on which of these routes is adopted. Effects of substituents on rates of reaction promoted by the hydroxyl group need, therefore, to be considered with particularly careful attention to the mechanism. They are likely also to be much more dependent on solvent than is the case for most groups, because hydrogen-bonding involving the OH group promotes O—H hyperconjugation and hence enhances the electron-releasing power of the substituent.

The effects of halogens as substituents have been discussed in an earlier volume[11]. If these elements are treated as a series, several sequences involving changes in properties are found to be combined. Fluorine is the most electron-withdrawing by virtue of the inductive effect; but it is also the most electron-releasing conjugatively, it forms hydrogen bonds the most strongly, it is the smallest, and it is the least polarizable. Each of these properties has an effect on other physical and chemical properties of a molecule containing a carbon–halogen bond, and the various effects are to a first approximation exerted independently. So if a series of halogen-substituted compounds is examined, such other properties as are modified by these effects can change along the series in a complicated way within which the individual effects are difficult to disentangle. One way of comparing these influences is through the language of linear free-energy relationships, using equation (1).

$$\log_{10}(K_R/K_H) \quad \text{or} \quad \log_{10}(k_R/k_H) = \sigma\rho. \qquad (1)$$

It is usual to start with a standard reaction for which the reaction constant ρ is defined as 1, so that substituent constants σ can be derived by determining the effects of substituents R on the relative rate (k_R/k_H) or equilibrium constant (K_R/K_H) for the reaction. For any other reaction, a logarithmic plot of the new value of K'_R/K'_H enables a value of ρ for that reaction to be derived; perfect correlation tells us that substituents affect the two reactions in a way which differs only through a scale factor.

The first scale of substituent constants applicable to a range of organic reactions was derived from the strengths of benzoic acids. Substituent constants derived in this way represent a combination of effects specific to this set of equilibria, and a selection of values[7,12–14] is given in Table 1. The values of $\sigma_{m\text{-R}}$ show that the hydroxyl and

TABLE 1. Effects of selected substituents on ionization of substituted benzoic acids

Substituent (R)	Reaction: ionization of substituted benzoic acids, $RC_6H_4CO_2H$			
	$\sigma_{p\text{-R}}$	$\sigma_{m\text{-R}}$	$\sigma_{o\text{-R}}$	$\sigma_{p\text{-R}} - \sigma_{m\text{-R}}$
OH	−0.37	0.10	1.22	−0.47
OMe	−0.27	0.08	0.12	−0.35
Me	−0.17	−0.07	0.29	−0.10
H	0	0	0	0
F	0.06	0.34	0.93	−0.28
Cl	0.23	0.37	1.28	−0.14
Br	0.27	0.39	1.35	−0.12
I	0.30	0.35	1.34	−0.05
NO_2	0.78	0.71	1.99	0.07

methoxyl group are electron-withdrawing; and so are all the halogens, there being only minor differences between them. The values of $\sigma_{p\text{-R}}$ show that this effect, which results in an increase in acid strength, is opposed by an effect in the opposite direction. This effect is considerable for the methoxyl group, and among the halogens it is most marked for fluorine. It can be interpreted as conjugative electron release **(16)** because

(16)

it shows up differentially between the values for the *para* and *meta* positions. In using these values for quantitative comparisons, it should be remembered that tabulations of values of σ sometimes obscure their origin. The experimental basis may be a single reaction series, or the statistical average from a number of series. In considering the related substituent effects for the ionization of phenols as acids, it is useful first to make as direct a comparison as possible with the ionization constants of substituted benzoic acids. For this purpose we have used the values of σ adopted by Barlin and Perrin[14], together with the corresponding values[15] for the ionization of substituted phenols in water at 25°C. The resulting plot is shown in Figure 1; the data are in Table 2.

The correlation between the effects of *meta* and *para* halogen substituents on the two ionizations is quite good. The situation becomes a little different if substituents

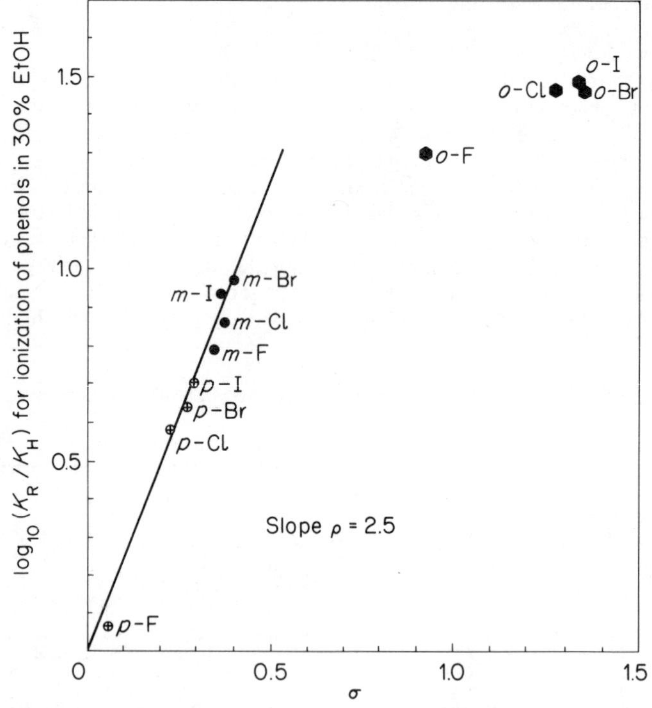

FIGURE 1. Linear free-energy plot for dissociation of halogen-substituted phenols.

TABLE 2. Data for linear free-energy comparison of dissociations of halogen-substituted benzoic acids and phenols

Substituent	Substituted benzoic acids: σ or σ_{app} (Ref. 14)	Substituted phenols	
		pK_a	$\log_{10} K_R/K_H$
H	0	10.00	0
o-F	0.93	8.70	1.30
o-Cl	1.28	8.53	1.47
o-Br	1.35	8.44	1.46
o-I	1.34	8.51	1.49
m-F	0.34	9.21	0.79
m-Cl	0.37	9.13	0.87
m-Br	0.39	9.03	0.97
m-I	0.35	9.06	0.94
p-F	0.06	9.91	0.09
p-Cl	0.23	9.42	0.58
p-Br	0.27	9.36	0.64
p-I	0.30	9.30	0.70

are compared over a wider range[15]. The dissociation of phenols is one of several reactions that have been used to define a scale of substituent constants, σ^-, to which electron-withdrawing conjugative effects make a larger contribution than they make to values of σ. The nitro group, for example, appears to engage in conjugation with O^- (**17**) more effectively than with CO_2^-, and consequently the value of $\sigma^-_{p\text{-}NO_2}$ (1.21) is larger than the value of $\sigma_{p\text{-}NO_2}$ (0.78).

(**17**)

Implicit behind the adoption of any single set of substituent constants is the assumption that inductive and resonance contributions are combined in a ratio which is defined for each substituent. Various approaches[16,17] have been used in which the proportionate contributions have been allowed to vary, so that a more general equation can be used to describe substituent effects over a wider range of reactions. Perhaps the most sophisticated treatment uses equation (2), in which the ratio ρ_I/ρ_R is treated as an additional variable parameter[18].

$$\log_{10} K_R/K_H = \rho_I\sigma_I + \rho_R\sigma_R \qquad (2)$$

Table 3 presents values of the resonance component of the substituent effect dissected in this way from the inductive contribution for three reactions relevant to the present discussion. The values of σ_I reflect the order of inductive electron withdrawal by the halogens. The values of σ_R illustrate similarly that as a conjugative electron donor the fluorine substituent is the most electron-releasing of these groups. These are the same conclusions as were drawn from the results given in Table 1. Since the resonance parameters σ_R and σ_R^- for the halogens derived from the three different ionization processes noted in Table 3 are nearly the same for each halogen, it can be concluded that the effects of these substituents on the ionization of 4-substituted phenols are determined mainly by the usual electronic effects of these groups.

Fischer and coworkers[19] have obtained related results for 4-substituted 2,6-dichlorophenols and 2,6-dimethylphenols in a number of solvents, including hydrox-

TABLE 3. Inductive and resonance contributions in ionization of some aromatic acids; values from Refs 12 or 17

Substituent (R)	$\sigma_I{}^a$	$\sigma_{R(BA)}{}^b$	$\sigma_{\bar{R}(A)}{}^c$	$\sigma_{\bar{R}(P)}{}^d$
OMe	0.29	−0.61	−0.45	−0.36
Me	−0.04	−0.11	−0.11	−0.06
H	0	0	0	0
F	0.54	−0.45	−0.45	−0.45
Cl	0.47	−0.23	−0.23	−0.23
Br	0.46	−0.19	−0.19	−0.19
I	0.39	−0.16	−0.11	−0.11
NO_2	0.76	−0.15	0.46	0.46

[a]Inductive substituent constants: from values of pK_a for substituted acetic acids in water.
[b]Resonance contributions to the ionization of *para*-substituted benzoic acids: $4\text{-}RC_6H_4CO_2H \rightleftharpoons 4\text{-}RC_6H_4CO_2^- + H^+$.
[c]Resonance contributions to the ionization of *para*-substituted anilinium ions: $4\text{-}RC_6H_4NH_3^+ \rightleftharpoons 4\text{-}RC_6H_4NH_2 + H^+$.
[d]Resonance contributions to ionization of *para*-substituted phenols: $4\text{-}RC_6H_4OH \rightleftharpoons 4\text{-}RC_6H_4O^- + H^+$.

ylic and dipolar aprotic media. It was found that the response of the dissociation constant to change in structure (i.e. the value of ρ^-) varied somewhat with the solvent and with the substitution pattern in the substrate. This very probably is the result of changes in the differential solvation of the phenol and its anion. No very marked anomalies were noted for halogen substituents.

Dissociation constants of 4-substituted 2,6-di-*t*-butylphenols, including a 4-bromo derivative, have been measured also in 50% ethanol as solvent[20]. The results were found to correlate quite well with those for other 4-substituted phenols, but again the response of dissociation constant to change in structure was significantly different from that for the ionization of phenols. For most substituents, the two *t*-butyl substituents are acid-weakening; but the Hammett plots intersect, and 2,6-di-*t*-butyl-4-nitrophenol is a stronger acid than 4-nitrophenol. It was concluded that two bulky *ortho*-*t*-butyl substituents may provide either steric hindrance or steric enhancement to ionization.

It is clear from these studies, and from comparisons of spectral data[9,21], that the effects of *para*- and *meta*-halogeno substituents on ionization of an aromatic hydroxyl group are not particularly anomalous. Results for *ortho*-halogenophenols deserve separate discussion; they are presented in Table 2 and Figure 1. As the substituent is placed closer to the centre of ionization, its inductive effect should have a more important influence on that centre, and this is reflected in the strengths of the phenols as acids. It is evident also that a value of ρ, which for *meta* and *para* substituents is approximately 2.5, would not represent the results for *ortho*-halogen substituents; for these, a value of about 1 would be more appropriate. Several groups[22,23] have used Taft's[24] approach, in which steric effects of *ortho* substituents are taken into account through equation (3).

$$\log(k/k_H) = \rho_E\sigma_E + \rho_s\sigma_s \tag{3}$$

In this equation, σ_s is a constant which, though primarily steric in origin, has in it an admixture of other influences specific to the defining reaction. It was found that the behaviour of halogenophenols was fairly well represented by using Taft's steric constants.

Barlin and Perrin[14] have surveyed the accuracy with which strengths of poly-

TABLE 4. Observed and calculated dissociation constants of di- and polychlorophenols as acids in water at 25°C

Positions of chlorine substituents	pK_a (found)	pK_a (calc)
2, 3	7.70[a]	7.16
2, 4	7.85[a]	7.95
2, 5	7.51[a]	7.66
2, 6	6.79[a]	7.06
3, 4	8.59[a]	8.55
3, 5	8.19[a]	8.26
2, 3, 4	7.59[b]	7.08
2, 3, 5	7.23[b]	6.79
2, 3, 6	6.12[b]	6.19
2, 4, 5	7.33[b]	7.08
2, 4, 6	6.42[b]	6.48
3, 4, 5	7.74[b]	7.68
2, 3, 4, 5	6.96[b]	6.21
2, 3, 4, 6	5.22[b]	5.61
2, 3, 5, 6	5.44[b]	5.32
2, 3, 4, 5, 6	5.26[b]	4.74

[a]Ref. 14.
[b]Ref. 25.

substituted phenols can be predicted from values of σ, using supplementary data for *ortho*-substituted phenols where necessary. Reasonably good agreement between observed and calculated values is obtained from treatment based on the principles of additivity of substituent effects. To exemplify the utility of this approach, we have calculated (Table 4) the strengths of di- and polychlorophenols as acids, using the values (Table 2) for monosubstituted phenols in water at 25°C. It should be emphasized that, for accurate use of this principle of additivity, a value of ρ appropriate to the reaction and solvent system under investigation is needed.

Gas-phase acidities of some phenols have been determined also, and have been correlated with gas-phase acidities of benzoic acids, with substituent constants, and with the results of theoretical calculations[26].

III. PHENOLS AS BASES

A. General Considerations

Phenols can react with acids, and a variety of reactions can be observed, depending on the conditions. Exchange between the hydroxyl group and other hydroxylic solvents is, of course, rapid. The rates of reaction of substituted phenols with methanol, for example, can be measured only at very low temperatures[27] (e.g. −78°C), when in the absence of acidic or basic catalysts the rates lie below those expected for a diffusion-controlled process. At temperatures from 25°C upwards, exchange at carbon occurs at rates convenient for study, and involves hydrogen atoms only in the positions activated conjugatively by the hydroxyl group[28,29]. It is clear from the effect of structure on the rate of reaction that the behaviour is that which would be expected for an electrophilic aromatic replacement. The range of acidity over which the reaction involves the phenol molecule rather than the phenoxide ion has been defined by careful studies of the dependence of rate upon the pH of the solution[30].

In order to define the nature of the transition state for exchange more closely, the detailed kinetic forms at high acidities have been studied. Gold and coworkers[31] examined the reactions of several substituted phenols, including 4-chlorophenol, comparing rates of dedeuteration and detritiation and examining also the solvent isotope effects. Substituted methoxybenzenes were examined also[32]; the work has been extended by Kresge and his coworkers[33–35] and an excellent summary is given by Taylor[36a]. It is now generally accepted that the reaction pathway involves a transition state in which proton-transfer to carbon has been partly effected, and that a single protonated intermediate lies between transition states of comparable but not quite equal free energy. The pathways for phenols do not seem to be substantially different from those for anisoles. Kresge and coworkers[37] have shown that the substrates [^{3}H]-2,4,6-trihydroxybenzene and [^{3}H]-2,4,6-trimethoxybenzene show different slopes of $\log_{10}k_{exch}$ against $-H_0$, but that the C-protonation of these substrates also have different acidity dependences. Allowance having been made for the latter difference, it was concluded that the extents of proton transfer in the transition states for isotope exchange were about the same. It would seem, therefore, that proton loss from the oxygen atom of C-protonated phenol is not concerned in the transition state leading to exchange at carbon. The fact that the hydroxyl group promotes more rapid exchange than the methoxyl group when exchanges at the 2-position in 4-methylphenol and its methyl ether are compared[32] can, therefore, perhaps be attributed to H—O hyperconjugation[10] (**18**).

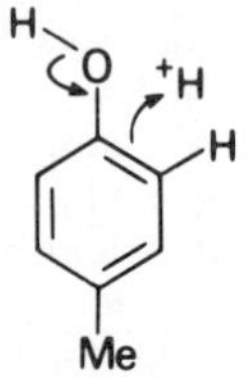

(**18**) 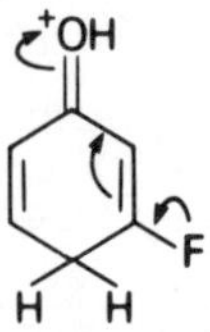 (**19**)

At acidities towards the top end of the range used for kinetic measurements, phenols become protonated to give conjugate acids in bulk concentration. Examination of solutions in perchloric acid and in sulphuric acid by using ultraviolet and NMR spectroscopy[33] showed that with the latter acid, rapid sulphonation could sometimes obscure the observation of protonation. For examples where this could be avoided, however, it was concluded that protonation often occurred both on carbon and on oxygen in competition. It was found that the acidity function for carbon protonation had a higher slope than that for oxygen protonation; hence for any given phenol, oxygen may be the more basic site in aqueous solution, but carbon becomes relatively more protonated at higher acidities. It appeared that, for very basic solutes, the rate of deprotonation was sufficiently reduced that reactions with other electrophiles could become slow; so that sulphonation by SO_3H^+, for example, was no longer competitive with C-protonation.

In other studies, solutions in superacid solvents have been examined at low temperatures, and a complex situation has been revealed[39–42]. The balance between carbon and oxygen protonation depends not only on the substrate and the acidity, but also on the temperature and the nature of the superacid system. Signals resulting from C-protonated species can be identified, so that proton exchange at carbon is slow on the ^{1}H NMR time-scale at temperatures of -40°C or less; at higher temperatures, on the other hand, broad signals characteristic of species undergoing slow exchange become apparent[43,44].

B. Protonation of Halophenols

Olah and Mo[45] have studied in detail the protonation of halophenols and haloanisoles in a range of superacid solvents at low temperatures, by using ^{1}H and ^{13}C NMR spectroscopy. The results have a number of features of interest; they are summarized in Table 5. The left-hand column gives the results for the strongest superacid, and the right-hand for the weakest.

For 4-halophenols, O-protonation only was observed, though the decomposition of the iodo compounds with liberation of iodine indicates that *ipso*-attachment of a proton is not completely inhibited. For 2- and 3-halophenols, C-protonation was noted only in the 4-position; no evidence for the presence of species protonated in the

TABLE 5. Positions (C = on carbon; O = on oxygen) of protonation of halophenols in 'superacid' solvents at −40°C

Compound	Hal	Solvent HF SbF_5 SO_2ClF	Solvent FSO_3H (1 mol) SbF_5 (1 mol) SO_2ClF	Solvent FSO_3H (1 mol) SbF_5 ($\frac{1}{4}$ mol) SO_2ClF	Solvent FSO_3H SO_2ClF
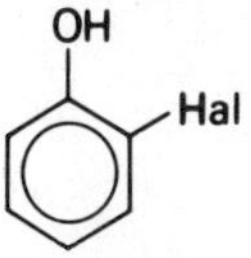 OH, Hal (2-halophenol)	F	C[a]	C[a]	C[a], O[b,c]	O[b,c]
	Cl	C[a]	C[a]	C[a], O[b,c]	O[b,c]
	Br	C[a]	C[a]	C[a]	O[b,c]
	I	Dec[g]	Dec[g]	C[a]	O[b,c]
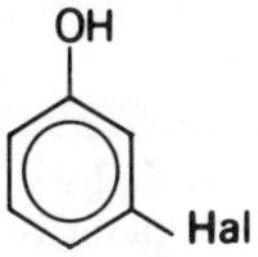 OH, Hal (3-halophenol)	F	C[d]	C[d]	C[d]	C[d]
	Cl	C[d]	C[d]	C[d], O[e,c]	C[d], O[e,c]
	Br	C[d]	C[d]	C[d], O[e,c]	C[d], O[e,c]
	I	Dec[g]	Dec[g]	C[d]	C[d], O[e,c]
 OH, Hal (4-halophenol)	F	O[f]	O[f]	O[f,c]	O[f,c]
	Cl	O[f]	O[f]	O[f,c]	O[f,c]
	Br	O[f]	O[f]	O[f,c]	O[f,c]
	I	Dec[g]	Dec[g]	Dec[g]	O[f,c]

[a] OH, Hal, +, H, H

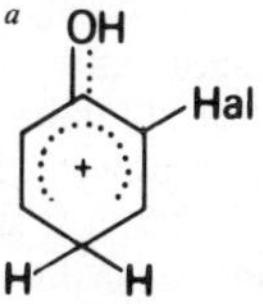

[b] $^{+}OH_2$, Hal

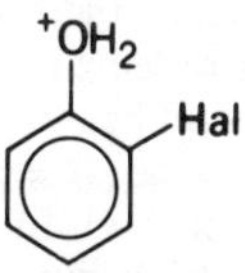

[c] Exchange with starting material.

[d] OH, +, Hal, H, H

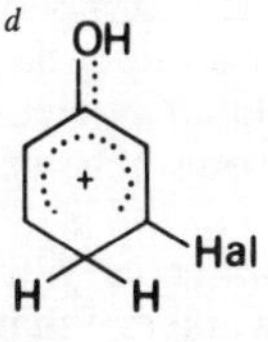

[e] $^{+}OH_2$, Hal

[f] $^{+}OH_2$, Hal

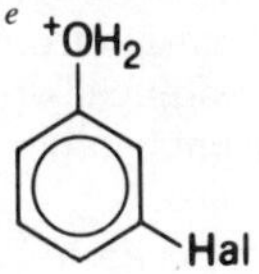

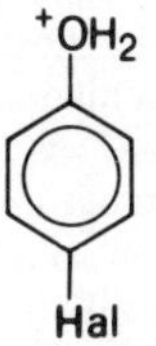

[g] Decomposed.

6-position was found. Protonation on carbon in competition with protonation on oxygen is favoured for the stronger superacids. Comparing the 2-halophenols, fluorine and chlorine allow some O-protonation, whereas bromine and iodine allow only C-protonation except in the weakest superacid; this was attributed to the greater inductive effect of the smaller halogens. For the 3-halophenols, on the other hand, both fluorine and iodine favoured C-protonation. This result, which is in line with the irregular sequence of reactivity of halobenzenes with electrophilic reagents[46], can be interpreted in terms of a conjugative electron-releasing effect (greatest for fluorine; **19**) and an effect of polarizability (greatest for iodine).

C. Protodehalogenation and Acid-catalysed Rearrangements of Halophenols

Although thermodynamically the positions in the benzene ring *ipso* to halogen are not the most favoured for protonation, reactions can occur through intermediates and transition states in which such protonation must occur. It was shown in 1924 that the removal of halogen from halophenols on treatment with hydrogen iodide in acetic acid followed the rate sequence *p*-I > *o*-I > *p*-Br > *o*-Br > *p*-Cl ≫ *m*-I, and that 4-iodoresorcinol was much more reactive than 4-iodophenol[47]. These results suggest strongly that the reaction is an electrophilic protodeiodination (equation 4).

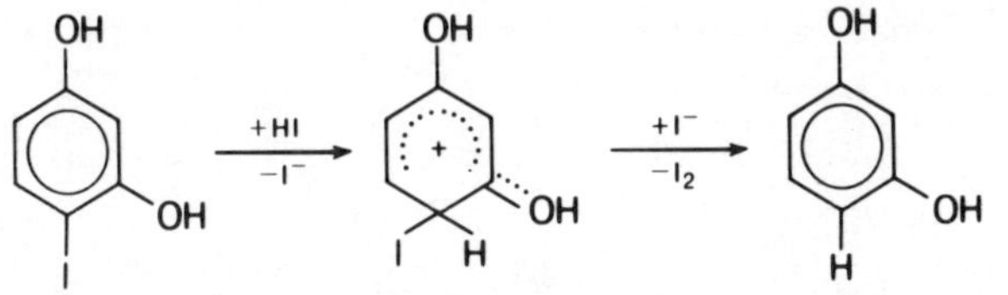

OH OH OH
+HI +I⁻
−I⁻ −I₂ (4)
OH OH OH
I H H
I

Gold and Whittaker[48] made a more detailed kinetic investigation of the reactions of 4-iodophenol and some of its derivatives with hydrogen iodide in acetic acid and in slightly aqueous acetic acid at a number of temperatures. The results established that the effects of alkyl and halogen substituents on the rate of protodeiodination were quantitatively very similar to the effects of these substituents on rates of nitration of hydrocarbons. The kinetic equation (5) described the results within the concentration ranges [HI] = 0.4–1 M, H_2O = 0–10 M.

$$v = k_0[\mathrm{RI}][\mathrm{HI}]^3/[\mathrm{H_2O}]^{3.5} \quad (5)$$

Dependence of rate on a high power of the concentration of hydrogen iodide is consistent with the view that the removal of I^+ can be assisted nucleophilically by a source of iodide ions. The complex dependence of rate on concentration of water is difficult to discuss quantitatively because it reflects change not only of the acidity of the medium but also of other ion-forming equilibria possible in the solution. Qualitatively, a reduction in rate might be expected, because the proton-donating power of the medium as measured by $-H_0$ is greatly affected by concentration of water[49]. It is shown in Figure 2 that this reduction is only partly reflected in the rate of proton transfer to carbon from the medium containing hydrogen iodide; a still more complex situation is evident for the rate of proton transfer from hydrogen chloride to cyclohexene[50], as is evident also in the figure. The original writers[48] concluded that molecular HI or the ion H_2I^+ were possible reagents; other processes involving both an electrophile and a nucleophile are possible also.

Protodehalogenation is involved in the reactions of some bromophenols with hydrogen bromide in aprotic solvents. It was shown by O'Bara and coworkers[51] that treatment of 4-bromophenol or its alkyl-substituted derivatives with hydrogen

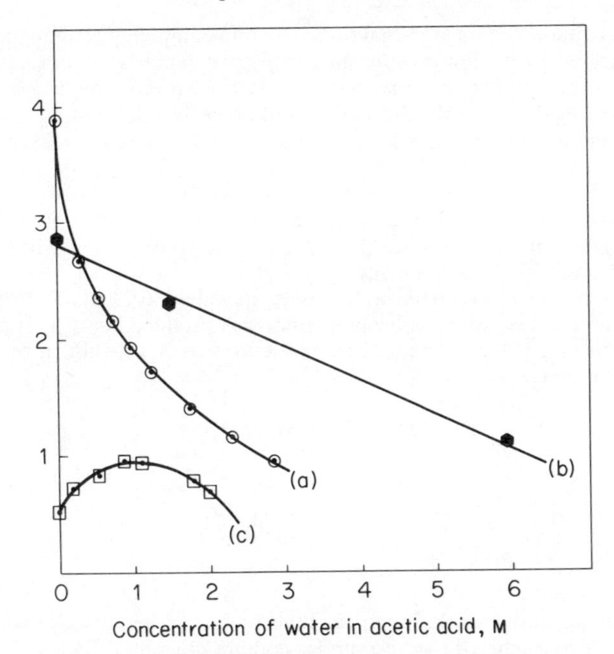

FIGURE 2. Effect of added water in solvent acetic acid on rates and equilibria. Plots against concentrations of water in acetic acid of (a) $3 - H_0$ (indicator measurements)[49]; (b) $7 + \log_{10}k_1$ (rate of protodeiodination of 2-chloro-4-iodophenol)[48]; (c) $8 + \log_{10}k_1$ (rate of addition of HCl to cyclohexene)[50].

bromide in chloroform or in carbon tetrachloride resulted in isomerization and disproportionation. Some of the results are given in Table 6.

The rate of the reaction was found to depend on the positions of the substituents; and in all cases mixtures rich in 2- and 4-substituted phenols were produced, with no 3-substituted phenols. The corresponding chloro compounds did not react. Positive bromine produced as an intermediate could be trapped not only by the starting material (whence the disproportionation) but also by added compounds. Thus the treatment of 4-bromo-3-methylphenol with hydrogen bromide in acetone resulted in the formation of much 3-methylphenol.

TABLE 6. Products of reaction of 4-bromophenol with HBr in $CHCl_3$ at 25°C

	Time, h				
	0	6	24	144	942
Component	Composition of reaction mixture, %				
Phenol	—	2	4	7	9
4-Bromophenol	100	95	89	66	41
2-Bromophenol	—	2	5	21	43
Dibromophenols	—	1	2	6	7

This isomerization results in partial rather than in complete thermodynamic control of the product ratio for bromination, since 3-bromophenols were not detected in the products. In a case studied more recently in which 2-bromo-4-methylphenol was the substrate, rearrangement to the 3-position could be effected by using the very strong acid trifluoromethanesulphonic acid (CF_3SO_3H)[44]. Under both of these conditions, the isomerization can be explained as resulting from reversible protodebromination. A different mechanism must, however, be concerned in the reaction of the complex between aluminium chloride and a deficiency of 4-bromophenol with hydrogen chloride at high temperatures, since this can be carried out in a way which gives 3- but no 2-bromophenols[52]. Results are shown in Table 7.

When the reaction was carried out by using equivalent amounts of 4-bromophenol and aluminium chloride, some 2-bromophenol was produced also. It is possible that 3-bromophenol is produced under these conditions by a 1,2-shift in a co-ordinated dienone (equation 6).

OH, Br (4-bromophenol) $\xrightarrow{+AlCl_3}$ $^{+}O—Al\bar{C}l_3$ (Br, H) $\longrightarrow$ $O—Al\bar{C}l_3$ (Br, H, +) $\xrightarrow{-AlCl_3}$ OH, Br (3-bromophenol) (6)

A close balance must exist between the relative ease of loss of H^+, loss of Br^+, and 1,2-shift of halogen under these and similar conditions.

It was suggested[51] that, because of the power of modern methods of separation of mixtures, isomerization with hydrogen bromide could be useful as a practical method of obtaining some *ortho*-substituted bromophenols. Another practical use for protodehalogenation in synthesis is in preparing *meta*-substituted compounds. It is often easy to polybrominate phenols by using excess of bromine, a reaction which can involve acid-catalysed rearrangement of a dienone. Protodebromination of the resulting compounds can then often be effected by heating them under reflux with aqueous (55%) hydrogen iodide, nitrogen being passed through to remove iodine as it is formed. Bromine is thus removed specifically from the *ortho* and *para* positions, *meta*-bromine being removed much more slowly.

The reaction can be described by equation (7), and examples of application of the method are given in sequences (8)–(11)[53,54].

$$ArBr + 2\,HI \longrightarrow ArH + HBr + I_2\uparrow \qquad (7)$$

This procedure is similar in principle to that used in a method developed earlier by Kohn and coworkers[55–57], in which a bromophenol was heated with aluminium tri-

TABLE 7. Products of reaction of 4-bromophenol (0.1 mol) with aluminium chloride (0.2 mol) and hydrogen chloride at 130°C

	Time, h					
	0	1	2	4	8	9
Component	Composition of reaction mixtures, %					
4-Bromophenol	100	92	84	67	49	46
3-Bromophenol	—	8	16	33	51	54

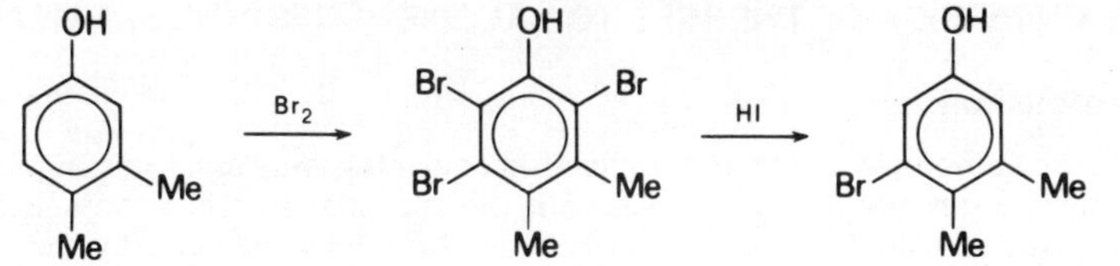

(8)

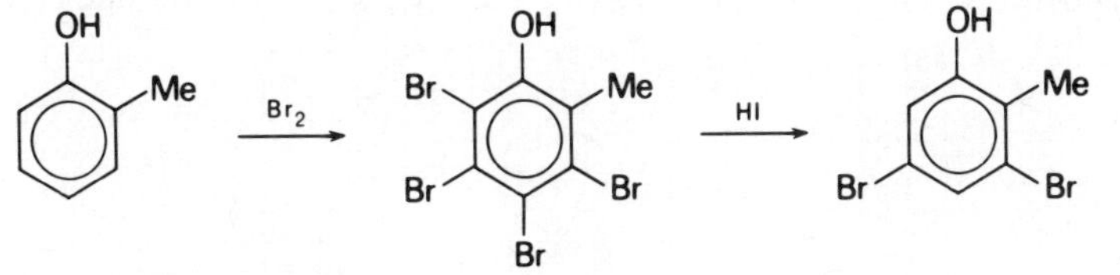

(9)

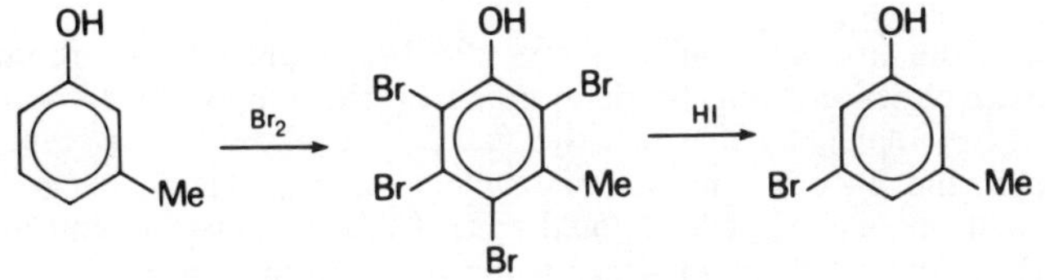

(10)

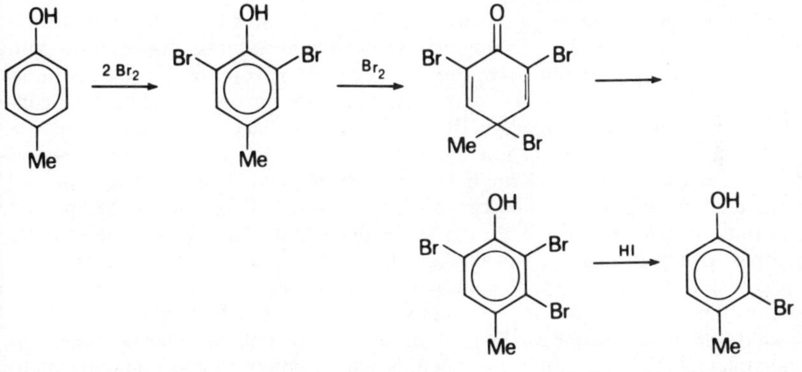

(11)

chloride in benzene to give the product of protodebromination at positions activated by the hydroxyl group. The exact nature of the acid providing the proton under these conditions cannot be regarded as certain; one possibility for the conversion of 4-bromophenol to phenol is shown in equations (12)–(15).

In general, reported yields were not as good as those found for protodebromination using aqueous hydrogen iodide as the reagent.

$$C_6H_4(OH)Br + AlCl_3 \rightleftharpoons H^+[AlCl_3OC_6H_4Br]^- \quad (12)$$

$$H^+[AlCl_3OC_6H_4Br]^- \longrightarrow Br^+[AlCl_3OC_6H_5]^- \quad (13)$$

$$Br^+[AlCl_3OC_6H_5]^- + C_6H_6 \longrightarrow C_6H_5Br + H^+[AlCl_3OC_6H_5]^- \quad (14)$$

$$H^+[AlCl_3OC_6H_5]^- \rightleftharpoons C_6H_5OH + AlCl_3 \quad (15)$$

IV. CONVERSION OF PHENOLS TO HALODIENONES; THE S_E2' REACTION

A. Introduction

The most commonly encountered mechanism of electrophilic replacement in aromatic compounds is usually designated S_E2, but the reaction proceeds through an intermediate (the so-called Wheland intermediate, **20**) which then can lose a proton or another positive species to give the product. The simplest possible representation of this sequence is shown in equation (16). The first stage may be rate-determining;

R, X + E^+ ⇌ R, X, E **(20)** ⇌ R, E + X^+ (16)

alternatively, the first and second stages can be jointly rate-determining. Product ratios in such reactions may be determined by the relative rates of formation of isomeric intermediates, by their relative rates of decomposition to give products of replacement, and by other modes of reaction open to them. We neglect for the moment any complications which could arise because of involvement of the nucleophile, which must have been provided with the electrophile.

For replacement in phenols, a possible new phenomenon must immediately be recognized, that proton loss from oxygen is likely to be much faster than proton loss from carbon. Scheme 1 showed the minimum degree of elaboration of the reaction path needed to describe electrophilic substitution at a single position in these substrates. The Wheland intermediate (**13**) still may exist as an intermediate, and may be concerned not only in the original reaction of the phenol with the electrophile, but also in the conversion of the dienone **14** into the substituted phenol **15**. Intermediates and dienones isomeric with **13** and **14** can likewise be formulated. The reaction of phenol to give the dienone **14**, and likewise of the dienone **14** to give the substituted phenol **15**, are both to be categorized as S_E2', where the prime indicates that the replacement has occurred with rearrangement of the double-bond system. Of these, the first is a non-isomeric and the second is an isomeric rearrangement.

The importance of dienones in the replacement reactions of phenols is relatively easy to assess when the former can be recognized independently as intermediates, and many examples of the halogenations of phenols and naphthols to give dienones are to be found in the early literature[58,59]. Lapworth[60] discussed the possibility that electrophilic substitutions of phenols generally went through the intermediacy of dienones; but many textbooks, even when this possibility is recognized, treat the hydroxyl group as if its influences could be compared directly with those of other substituents[61,62]. This approach is perhaps valid when attack by the electrophile is unambiguously rate-determining and when proton loss from oxygen is not in the rate-determining step. Remarkably few investigations bear directly on these points. In the following discussion, we deal first with situations in which dienones of various kinds have been prepared by halogenation of phenols, and secondly with situations in which they have been recognized as fleeting intermediates by physical methods. We then deal with cases in which the intermediacy of dienones is presumed, but has not been established.

B. Dienones Obtained by Fluorination

Fluorodienones prepared by electrophilic fluorination of phenols have been used in several synthetic investigations. Mills and coworkers[63] studied the reactions of a

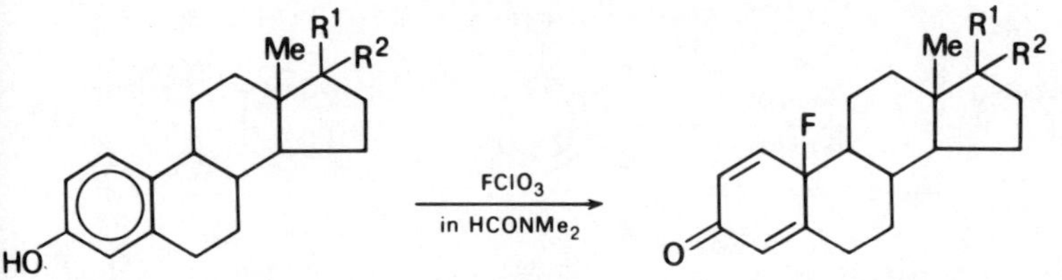

$$\xrightarrow[\text{in } HCONMe_2]{FClO_3} \qquad (17)$$

number of steroids analogous to oestrone with perchloryl fluoride in dimethylformamide; products were obtained in accordance with equation (17). Reactions of simpler substrates were studied by Taub[64] whose work established the results shown in Scheme 2. The two dienones **23** and **24** were separated by fractional crystallization, and were presumed to have been formed from the fluorophenols **21** and **22** respectively. It is interesting that, although the first stage of the fluorination gave substitution both *ortho* and *para* to the hydroxy group, the second stage apparently gave no product other than that of *ipso* attack.

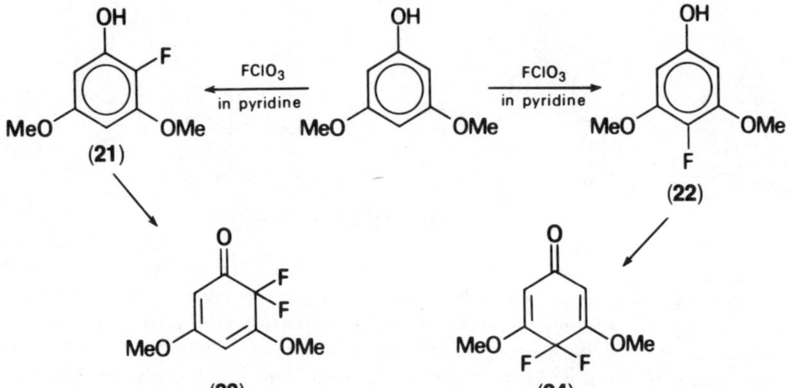

(**21**) (**22**) (**23**) (**24**)

Ultraviolet spectra:
(MeOH)

	(23)	(24)
λ_{max}	243 nm	240 nm
ε_{max}	7100	16 900
λ_{max}	355 nm	296 nm
ε_{max}	4900	4000

SCHEME 2. Products of fluorination of 3,5-dimethoxyphenol with perchloryl fluoride in pyridine.

The ultraviolet spectra recorded for the dienones and shown in Scheme 2 illustrate one of the important ways in which the fully conjugated 2,4-dienones produced by attack *ortho* to the hydroxyl group can be differentiated from 2,5-dienones, produced by the corresponding attack at the *para* position. Generally, the former have absorption of reasonable intensity with a maximum at considerably longer wavelength. The existence for **24** of subsidiary absorption with a maximum at 296 nm is, however, a feature which needs to be recognized. This band is usually regarded as an $n-\pi^*$ transition of the carbonyl group[65], such transitions usually being of much less intensity[66].

Barton and coworkers[67] have extended these observations by using other fluorinating agents, including fluoroxytrifluoromethane ($FOCF_3$). Examples of the various

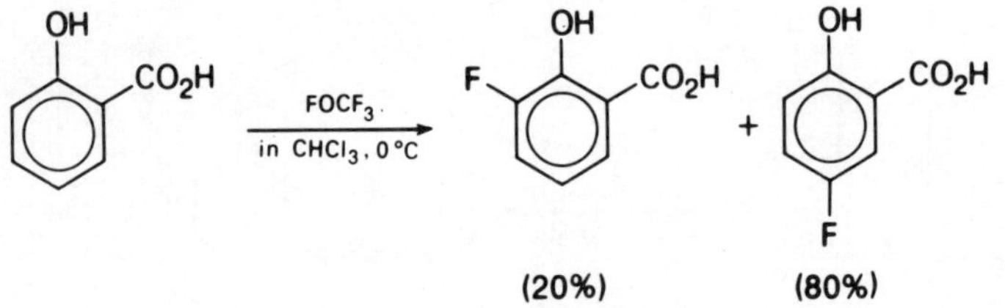

(18)

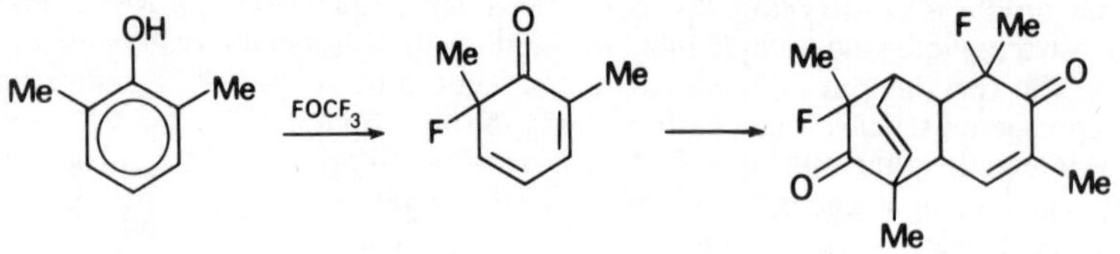

(19)

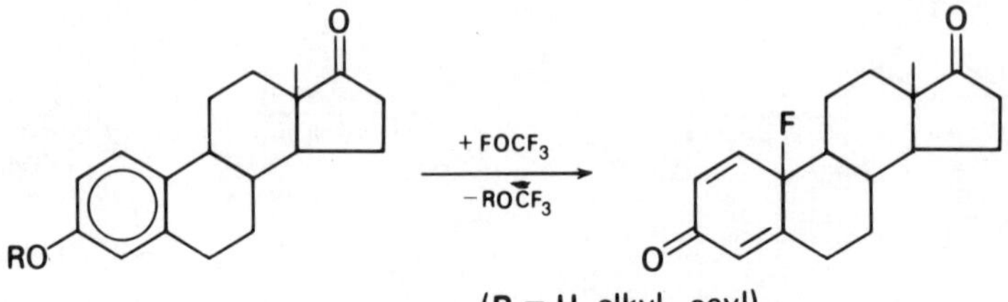

(20)

(R = H, alkyl, acyl)

classes of reaction reported are given in sequences (18)–(20). It is clear that fluorination under these conditions can result in substitution, in formation of a 2,4-dienone, or in formation of a 2,5-dienone, depending on the structure of the substrate.

Perfluorodienones have been prepared by reaction of a fluorinated hydrocarbon with chromium trioxide in hydrogen fluoride as solvent[68]. An addition–elimination pathway has been proposed for this conversion (sequence 21). The products can be protonated to give the corresponding cations by dissolving them in superacid solvents.

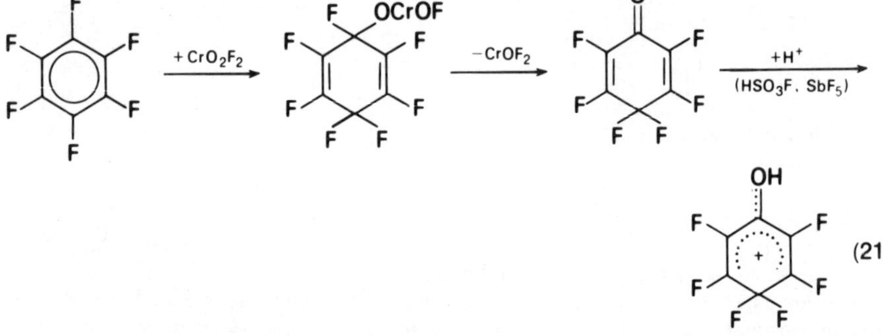

(21)

C. Dienones Obtained by Chlorination

The polychlorination of phenols results in the formation of compounds which liberate iodine from acidified potassium iodide, and in some earlier literature were regarded as aryl hypochlorites. Of recent years, however, it has become recognized through the use of spectroscopic techniques that such hypochlorites, if they are ever formed, are unstable and rearrange so rapidly to give isomeric dienones that no authentic aryl hypochlorites have yet been described. Thus when chlorine reacts with pentachlorophenol in non-polar solvents, the 3,5-dienone **25** is first produced, and can be isolated in good yield if reaction is carried out in the presence of pyridine or of other compounds which can remove hydrogen chloride[69]. The sodium salt of the phenol can be used, as can other chlorinating agents such as *t*-butyl hypochlorite. The dienone **25** rearranged into the 2,5-dienone **26** on being treated with iodine; and nearly complete conversion is effected when the chlorination is carried out with this or with aluminium trichloride as catalyst. When the reaction is carried out in acetic acid as solvent, with excess of chlorine and no removal of the formed hydrogen chloride, the 3,5-dienone **25** is partly captured by molecular chlorine with formation of an enone, 'octachlorophenol', probably having structure **27**. Adducts derived from **26** are formed also. Scheme 3 summarizes these findings.

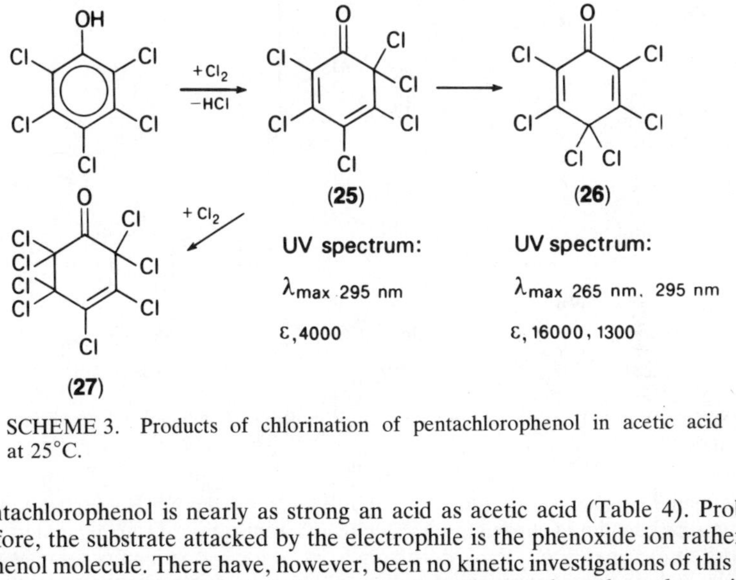

$+Cl_2$, $-HCl$; $+Cl_2$

(25) UV spectrum: λ_{max} 295 nm; ε, 4000

(26) UV spectrum: λ_{max} 265 nm, 295 nm; ε, 16000, 1300

(27)

SCHEME 3. Products of chlorination of pentachlorophenol in acetic acid at 25°C.

Pentachlorophenol is nearly as strong an acid as acetic acid (Table 4). Probably, therefore, the substrate attacked by the electrophile is the phenoxide ion rather than the phenol molecule. There have, however, been no kinetic investigations of this point. Noteworthy is the fact that the 3,5-dienone produced by attack on the *ortho* position is formed the more rapidly, but the 2,5-dienone is thermodynamically the more stable of the two isomers. Denivelle's papers[69] include much information concerning the infrared and Raman spectra of the products. Pentafluorophenol reacts similarly.

This work has been extended by study of the course of chlorination of phenol and of a number of alkyl-substituted phenols by Vollbracht and coworkers[70] and by Morita and Dietrich[71]. The results and general conclusions are similar; products of substitution at vacant *ortho* and *para* positions are formed, and these undergo further chlorination to give 2,5-dienones, 3,5-dienones and products of further addition of chlorine. Reactions under conditions of relatively high acidity seem to favour the formation of 2,5-dienones by attack at the position *para* to the hydroxy group. It is possible that attack on the phenol molecule favours this orientation, whilst attack on the phenolate ion favours the formation of the 3,5-isomer; but again it would require kinetic investigations to establish this point.

Recent studies have focused attention on the more labile dienones which can be formed by attack on an already substituted *para* position whilst retaining an unsubstituted *ortho* position. When 3,4-dimethylphenol is treated with chlorine in acetic acid, the products contain not only 6- and 2-chloro-3,4-dimethylphenol but also an enone **29** and two dienones **28** and **30**, which were assigned the structures shown in Scheme 4 from their ^{1}H NMR spectra[72].

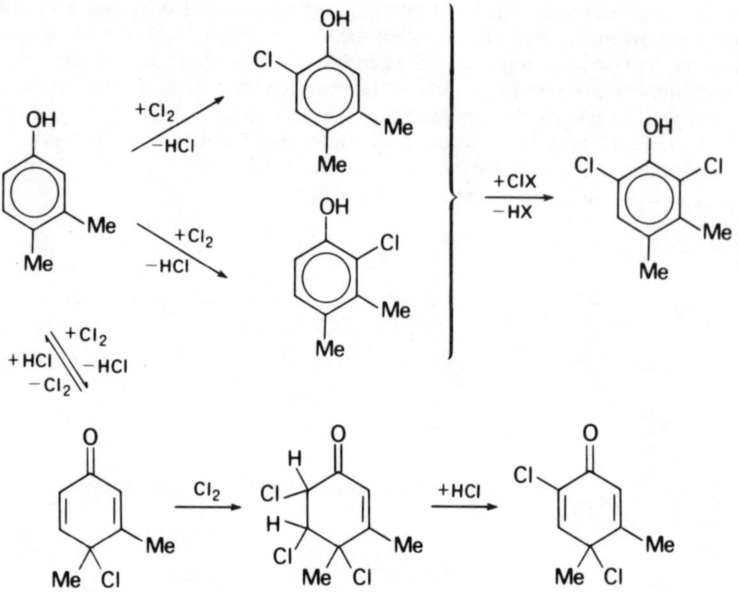

SCHEME 4. Products of chlorination of 3,4-dimethylphenol in acetic acid at 25°C.

The results of analysis of the reaction mixtures by ^{1}H NMR spectroscopy and separately by gas–liquid chromatography with different ratios of starting materials and after different times show that the dienone **28** is formed in competition with the products of 2- and 6-substitution; this dienone then slowly undergoes rearrangement to the mixture of chlorophenols. It also can react with chlorine, to give the trichloroenone **29**, which is formed in major proportion if excess of chlorine is used, and slowly loses hydrogen chloride to give the dienone **30**.

These results show that attack at an alkyl-substituted position *para* to a hydroxy group is under these conditions competitive with attack at an unsubstituted *ortho* position. Fischer and Henderson[73] have extended these results by studying the chlorination of a number of 4-alkyl-substituted phenols. 4-Chloro-2,5-dienones were obtained in particularly good yield by reaction in acetic anhydride; they found that the dienone **28** when purified is stable in acetic acid over a prolonged period. No 2-chloro-3,5-dienones were encountered in their chlorinations, even when 2,4-di- or 2,4,6-trimethylphenol was used. Although it is possible that *ortho* isomers are formed rapidly and rearrange to give their *para* isomers, it seems to the writers to be more likely that electron release from the hydroxy group of the phenol molecule is more activating for chlorination in the *para* than in the *ortho* position, in contrast with the results found when activation is by the O^- group.

A method of a different kind which has been used for the preparation of cyclohexadienones involves the action of gaseous chlorine on phenols in the solid state[74]. The reactions were very rapid, and generally led to mixtures, some of which were separated by crystallization and some by manual sorting of the crystals. Thus the chlorination of 4-*t*-butylphenol gave 32% of 2,4,6-trichloro-4-*t*-butylcyclohexa-2,5-dienone; similarly 2,5-di-*t*-butylphenol gave 60% of a mixture of 4,4,6-trichloro-2,5-di-*t*-butylcyclohexa-2,5-dienone, 4,6,6-trichloro-2,5-di-*t*-butylcyclohexa-2,4-dienone, and 2,4,6-trichloro-2,5-di-*t*-butylcyclohexa-3,5-dienone:

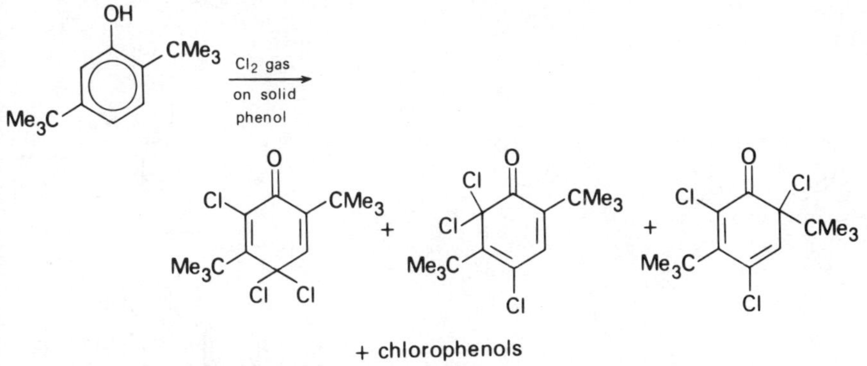

OH, CMe3, Me3C — Cl2 gas on solid phenol →

Cl, CMe3, Me3C, Cl Cl + Cl, Cl, CMe3, Me3C, Cl + Cl, Cl, CMe3, Me3C, Cl

+ chlorophenols

Trimethyl phosphate has been recommended as a medium for the formation of chloro- and bromodienones by halogenation[75]. It acts as an acceptor of hydrogen halide, so that the formation of the dienone is not reversed and acid is consumed as it is liberated.

Results similar to those described for alkyl- and halogen-substituted phenols have for many years been established also in the chemistry of the naphthols[76], and the availability of modern physical methods has led to renewed interest in the sequences possible in their reactions[77]. Thus the chlorination of 2,4-dichloro-1-naphthol in chloroform follows sequence (22). The intervention of addition–elimination sequences is shown in sequences (23), (24) and (25), for which all intermediates shown have been isolated and characterized[77,78].

The factors which determine whether attack occurs *ortho* or *para* to the hydroxy group in 1-naphthol are no better defined than they are for phenols.

Interesting sequences of reaction are known also for 2-naphthol. The compounds formed by successive uptake of three molecular equivalents of chlorine are shown in sequence (26)[79]. The conversion of the ketodichloride (**31**) into 6-bromo-1-chloro-2-naphthol by hydrogen bromide in acetic acid must follow the pathway (27), which is mechanistically analogous with that taken in the Orton rearrangement. Since the corresponding reaction with hydrogen chloride gives 1,4-dichloro-2-

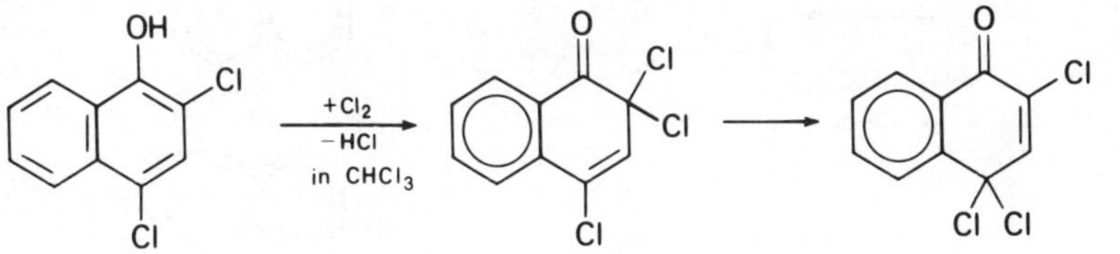

OH, Cl, Cl — $+Cl_2$, $-HCl$, in $CHCl_3$ → O, Cl, Cl, Cl → O, Cl, Cl Cl (22)

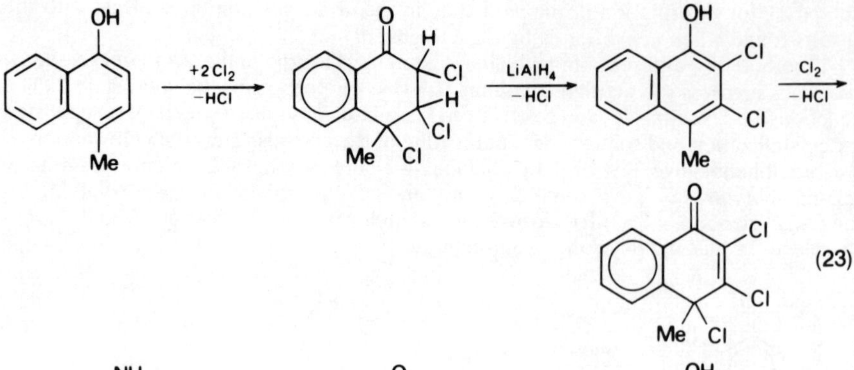

$\xrightarrow[-HCl]{+2Cl_2}$ $\xrightarrow[-HCl]{LiAlH_4}$ $\xrightarrow[-HCl]{Cl_2}$ (23)

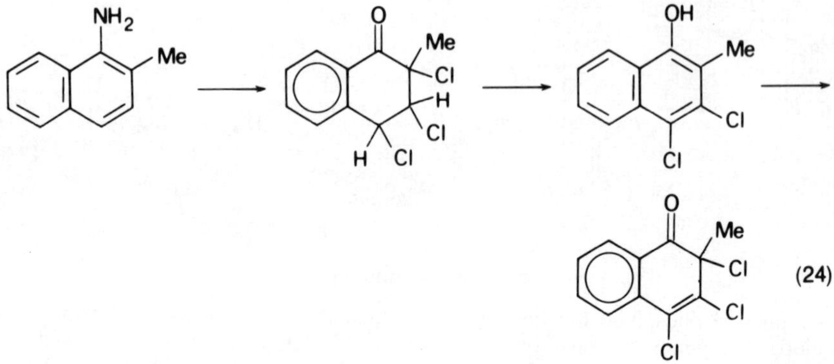

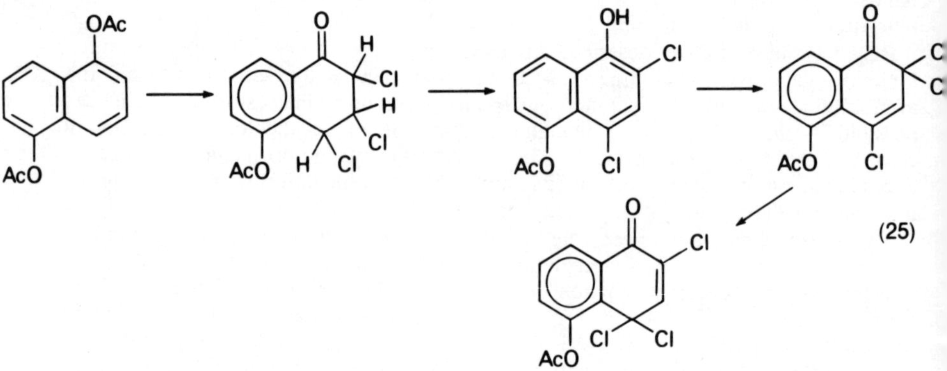

(24)

(25)

$\xrightarrow[-HCl]{+Cl_2}$ $\xrightarrow[-HCl]{+Cl_2}$ $\xrightarrow{+Cl_2}$ (26)

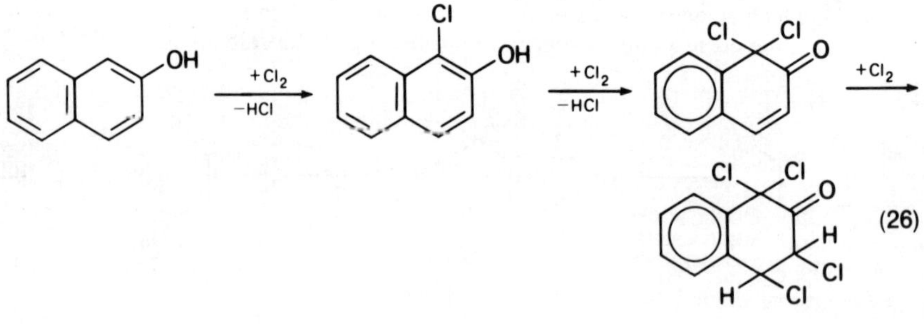

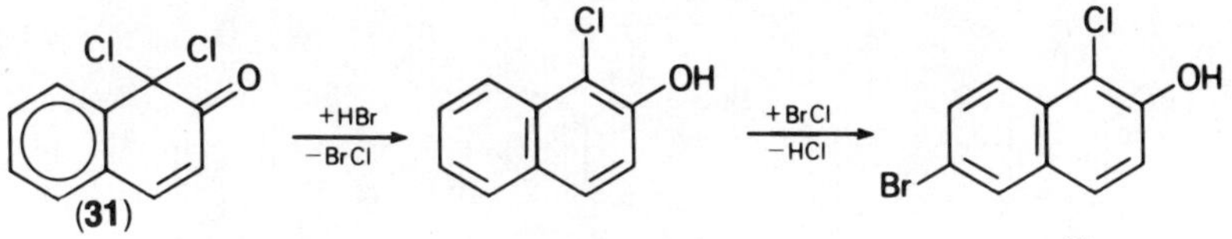

(27)

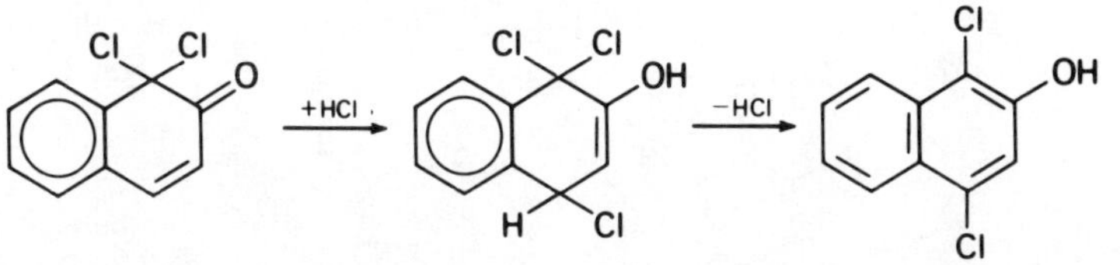

(28)

naphthol, it has been suggested that the addition–elimination pathway (sequence 28)[80,81] is taken in the latter reaction.

D. Dienones Obtained by Bromination

Many examples are known of dienones formed by the bromination of highly substituted phenols; but, as for their chloro analogues, differentiation between 4-bromo-2,5-dienones, 2-bromo-3,4-dienones and the as yet unknown aryl hypobromites proved difficult until modern spectroscopic techniques became available. Reviews by Ershov and coauthors[5] and by Waring[4] summarize the earlier literature. Compounds having the 2-bromo-3,5-dienone structure were prepared by Ershov and Volod'kin[82] from a number of 4-substituted 2,6-dialkylphenols by reaction with bromine in ether containing water and pyridine at 0°C or below (equation 29).

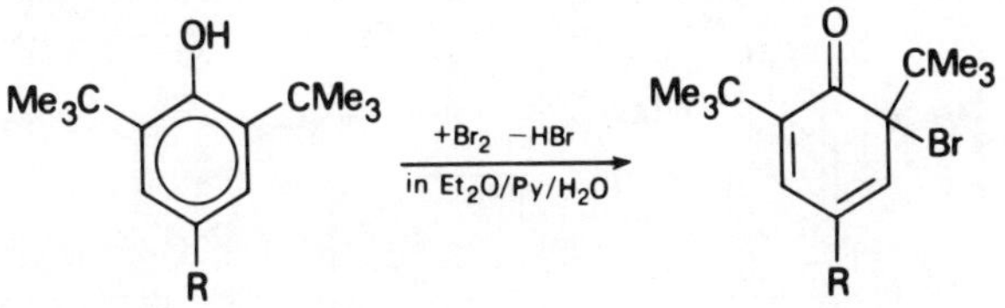

(29)

(R = NO_2, CO_2H, CN, CHO, Cl, Br, CH_2CN)

Many of these compounds were readily isomerized to give the corresponding 4-bromo-2,5-dienones.

Two factors may contribute to the formation of 2,4- (rather than 2,5-) dienones under these conditions. One is that the presence of a base favours the formation of the phenolate anion, which may intrinsically be more reactive at the position adjacent to the negative charge. The other is that in most of the compounds the 4-substituent would deactivate the *ipso* position for electrophilic attack, so that despite the bulkiness of the *ortho* substituent the 2-position might be the most activated.

Kinetic investigations of the formation of bromodienones have been made in a few cases, in most of which it was necessary to make allowance for subsequent decomposition or rearrangement. Grovenstein[83] interpreted the kinetics of bromination of 3,5-dibromo-4-hydroxybenzoic acid in terms of reaction through the dienone **32**. The effect of acidity on the rate showed that the substrate undergoing electrophilic attack was mainly the phenolate anion. Similarly, the intermediates **33** and **34** were considered to be involved in the brominations of 3,5-dibromo-2-hydroxybenzoic acid and 2,6-dibromophenol respectively. These results confirm the expected mechanistic similarity between electrophilic attack on substituted and unsubstituted positions *ortho* and *para* to a hydroxy group.

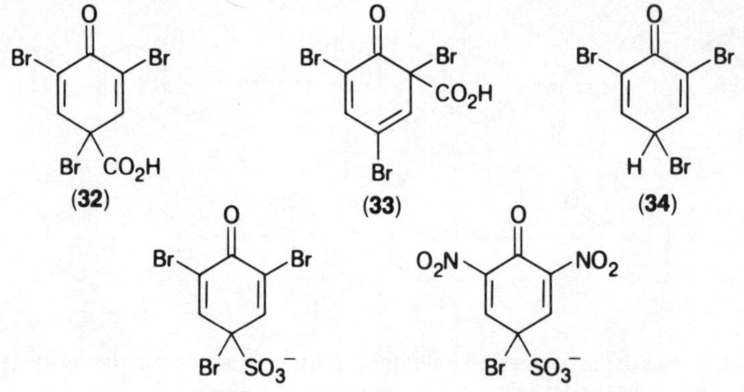

(32) (33) (34)

(35) (36)

In a similar investigation of bromodesulphonation[84], Cannell showed kinetically and spectroscopically that the dienones **35** and **36** were intermediates in the reactions of the appropriately substituted phenolsulphonic acids. Effects of substituents were as expected for a reaction in which the reagent was electrophilic bromine.

In all these experiments, carried out as they were in rather aqueous solvents, the aromatic species under attack was shown to be largely the phenolate ion. Study of the kinetics of bromination of 2,6-di-*t*-butylphenol in acetic acid and in slightly aqueous acetic acid, on the other hand, showed that the substrate under attack was the phenol molecule, yielding the dienone **37**, the main reaction being that shown in equation (30)[85]. Baciocchi and Illuminati[86] extended this work by comparing the rates of

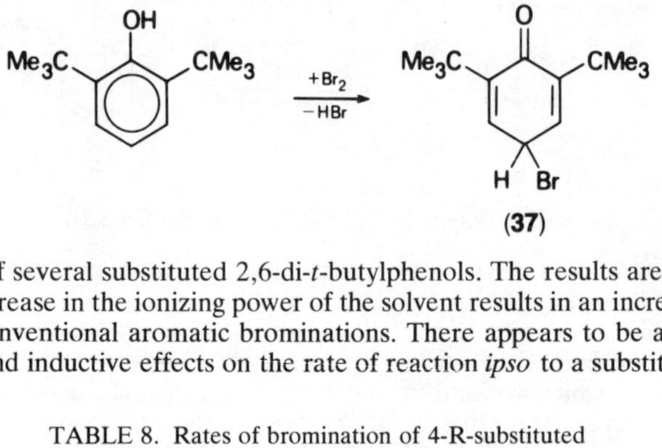

$$\xrightarrow[-HBr]{+Br_2} \qquad (30)$$

(37)

bromination of several substituted 2,6-di-*t*-butylphenols. The results are summarized in Table 8. Increase in the ionizing power of the solvent results in an increased rate, as is usual for conventional aromatic brominations. There appears to be a balance between steric and inductive effects on the rate of reaction *ipso* to a substituent, so that

TABLE 8. Rates of bromination of 4-R-substituted 2,6-di-*t*-butylphenols in acetic acid and in aqueous acetic acid at 25°C

	k_2, l mol^{-1} s^{-1}, in solvent	
R	HOAc	98% HOAc
H	4.8	25
Me	1.3	7
t-Bu	1.3	6
Br	$<10^{-4}$	$<10^{-5}$

alkyl groups inhibit slightly the rate of attack, whilst the effect of bromine is very much larger.

Neither of these investigations throws light on the question of whether the loss of the phenolic proton can be concerted with electrophilic attack by bromine. We will return to this question when dealing with the rates of bromination of phenols to give bromophenols.

Related polycyclic systems for which electrophilic substitution with rearrangement is a consequence of bromination include the reactions of the 2-naphthol-6,8-disulphonate anion to give **38** (equation 31)[87] and of 9-hydroxyanthracene (equation 32)[88].

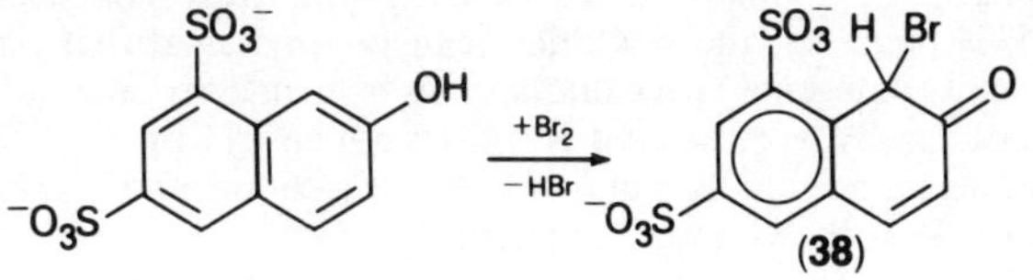

(31)

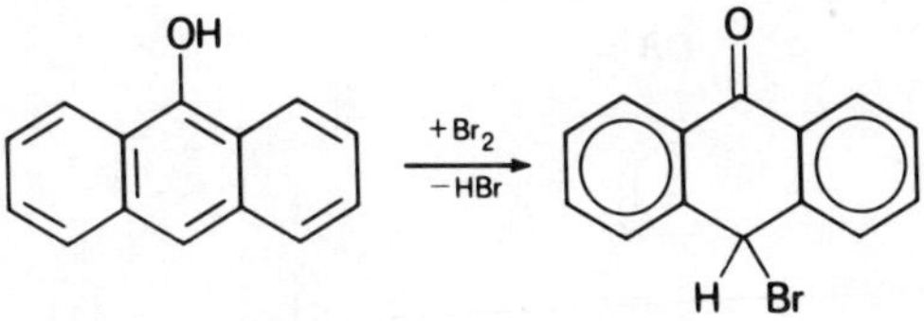

(**38**)

(32)

E. Intermediates Involved in Iodination

In comparison with fluoro-, chloro- and bromodienones, iododienones are very little known. Zollinger and coworkers[87] obtained evidence for the formation of a complex between I^+ and the 2-naphthol-6,8-disulphonate ion having stoicheiometry equivalent to that involved in the reaction of equation (31). Because of its ^{1}H NMR spectrum, however, the authors formulated it as having a different type of structure; they regarded it either as a π-complex (**39**) or as a naphthyl hypoiodite (**40**). Even if this is the case, it is to be noted that the 1-position in the 2-naphthol-6,8-disulphonate ion is very crowded, and the iodine cation is relatively large. Less hindered analogues might be found having the dienone structure, but as far as we are aware none has been examined by definitive physical techniques.

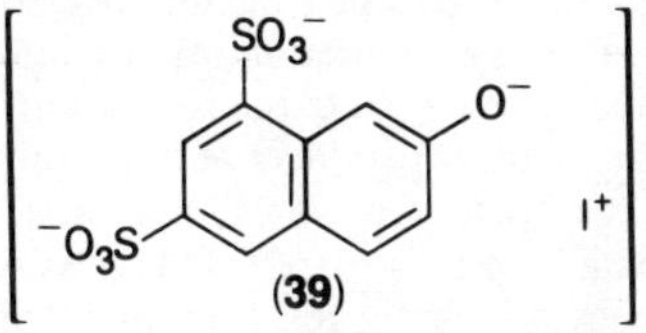

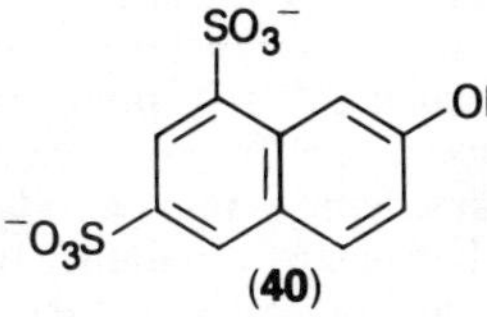

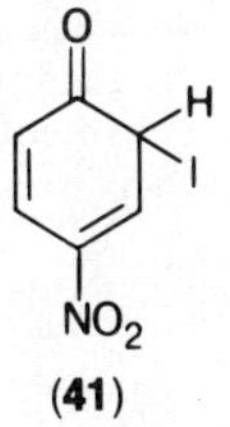

(**39**) (**40**)

(**41**)

Kinetic methods have been used[89] to show that the course of iodination of 4-nitrophenol involves an intermediate. The decomposition of this intermediate was found to be subject to a primary isotope effect, $k_H/k_D = 5.4$ at high concentrations of iodide ions but less when the concentration of iodide ions was reduced and the formation of the intermediate approached being rate-determining. The results were held to suggest that the intermediate should probably be formulated as in structure **41**.

F. Halodienones Formed by Halogenation of Derivatives of Phenols

The most usual electrophilic substitutions involve replacement of hydrogen as a cation by an electrophile; but many cases are known where groups other than a proton are replaced. This being so, the possibility can be envisaged that phenol ethers or phenol esters might sometimes react similarly, as is formalized in equation (33). Such processes will more easily be realized if R^1 (**42**) is not easily displaced, if other reactive positions in the nucleus are blocked, if R is stable as a cation, and if a further reagent is present to remove R as the cationic fragment.

OR, R^1 —$+E^+$→ O, E R^1 + R^+ (33)

(**42**)

(R^+ = Alkyl, Acyl, etc.; E^+ = an electrophile; R^1 = H, Alkyl, Acyl, CO_2H, SO_3H, etc.)

Reactions of this kind have been reviewed[90]; they have been studied only desultorily from a preparative point of view, and almost negligibly from a mechanistic standpoint. It is becoming clear, however, that they are of more importance than has been generally realized in determining the course taken in the reactions of some phenol ethers and esters with reagents providing electrophilic halogen. Some examples have already been quoted; thus equation (20) (R = Me) represents a fluorodealkylation with rearrangement, and the first stage of sequence (25) must include a chlorodeacylation with rearrangement. Others are discussed in papers cited in other connections[72]. The chemical consequences can be quite varied. They include the formation of unexpected substitution products, as in sequence (25), and of products of addition. A number of very general uncertainties remain to be resolved, however. One is whether concerted processes involving an external nucleophile can be realized. This possibility is formulated for halogenation in **43**, and evidence which suggests the availability of such a route has been presented[91]. The second is whether cyclic processes can occur (**44**). The third is whether the reaction can involve the formation of an adduct (e.g. **45** or **46**) and its decomposition. All of these points are equally relevant to the chemistry of enol ethers and enol esters, and merit further investigation. The relative ease of displacement of different groups from oxygen is also not known. Qualitatively, it can be deduced[91] that the acetyl cation is more easily displaced than the benzoyl cation, but other comparisons are lacking.

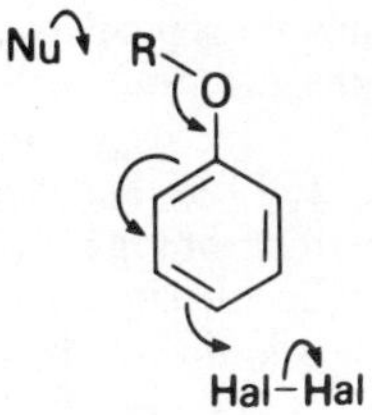

(43) 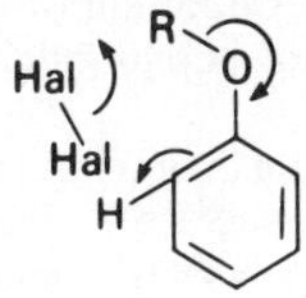 (44) 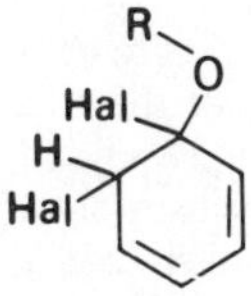 (45) 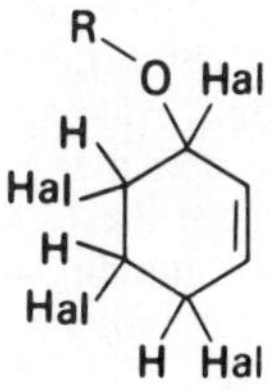 (46)

V. REACTIONS OF HALODIENONES

A. Introduction

Halodienones derived from phenols are reactive species which can undergo a variety of transformations depending on their structures and the conditions in which they find themselves. In the present review, we are concerned particularly with two types of process: first, those which could lead to the normal product of electrophilic replacement; and secondly, those which could divert the reaction towards products of other kinds.

B. Conversion of Halodienones to Phenols; Loss of the *ipso* Substituent

1. ipso-*Hydrogen*

When the *ipso* substituent (e.g. R^1 in **42**) is hydrogen, decomposition of a halodienone with loss of a proton is usually a facile reaction. In a few cases[83,85,92,93] the course of bromination of a phenol has been followed spectroscopically or kinetically, and the build-up of such a dienone and its decomposition have been observed separately. The detailed mechanistic pathway for proton loss from the dienone has, however, been studied only for the dienone **37** with acetic acid or slightly aqueous acetic acid as the solvent. The results reveal a complicated situation, which we discuss in terms of Scheme 5[85,93].

Under conditions of catalysis by acid, the rate in general follows the value of $-H_0$ as measured by the extent of protonation of suitable indicator bases, is faster in deuterioacetic acid than in acetic acid, and is subject to a moderate primary deuterium isotope effect ($k_{4-H}/k_{4-D} \simeq 4$) which varies somewhat with the solvent. These results indicate that the first stage of the reaction path under these conditions involves a pre-equilibrium protonation of the dienone **47** to give protonated species, which will include the free ion **51** and ion pairs (e.g. **48**, **49** or **50**) derived from it. These protonated forms then suffer rate-determining proton loss to give the bromophenol **53**.

Evidence for intervention of ion pairs as chemically significant intermediates comes from the nature and consequences of the additional catalysis provided by bromide ions. This catalysis reaches a limiting situation with sufficient increase in the concentration of bromide ions. Further addition of bromide ions then produces no further increase in rate, but addition of perchlorate ions reduces the rate of the catalysed reaction.

These results are interpreted as meaning that the limiting rate had been reached when all the protonated intermediate had been converted through the equilibria shown in Scheme 5 to the ion pair **49**. Reaction through this intermediate is appreciably different from reaction through other ion pairs. This is shown by a somewhat different primary deuterium isotope effect, $k_{4-H}/k_{4-D} \simeq 3$; and also by the fact that under these conditions the intermediate reverts reversibly to the unbrominated phenol **52** and bromine to the extent of about 9% of the total reaction.

The rearrangement of the dienone **47** is catalysed also by bases such as sodium acetate. Under these conditions the primary isotope effect, k_{4-H}/k_{4-D} is very large (*c.* 8). Kinetic arguments were adduced to establish that reaction under these conditions is not of the ion pair **50**, but instead of the free dienone reacting with the complex ion $[Na(OAc)_2]^-$.

These results throw light not only on the rearrangement of the dienone **47**, but also on the course of bromination of the phenol **52**. Three significantly different pathways can be recognized. The first is prominent in acetic acid at low concentrations of bromine and with high concentrations of perchloric acid. The phenol is converted in more than one stage but essentially irreversibly to the ion pair **48**, which then partitions to give mostly the dienone **47** with initially a small proportion of the bromophenol **53**. Further reaction then results in rearrangement of **47** to **53** via the same intermediate **48**.

The second pathway becomes important in the absence of added anions, and involves hydrogen bromide produced in the course of the reaction. The intermediate is now **49**, and the first stage of the reaction becomes partly reversed, but at the same time the formation of the bromophenol from **49** is faster than that from **48** so that the rate of conversion of the original phenol to the substituted phenol is catalysed.

The third pathway is dominant when bromination is carried out in the presence of excess of sodium acetate. The dienone **47**, the formation of which is followed by its base-catalysed rearrangement, becomes a real intermediate in the overall conversion of the phenol **52** to the bromophenol **53**.

The decomposition of the related dienone derived from 2,6-dimethylphenol (**47**, R = Me) was investigated only in outline, and was much more rapid than that of its more hindered analogues[85]. Other dienones which behave similarly, and have been shown to be formed in the course of bromination of the corresponding phenols, include those derived from 2,6-di-*s*-butyl, 2,6-di-isopropyl, 2-methyl-6-*t*-butyl, and 2-*t*-butylphenol, the last being the most unstable of those investigated[92].

2. *Other* ipso-*substituents*

There are many qualitative reports of the displacement of substituents other than hydrogen accompanying halogenation of substituted phenols, and most of these can be formulated as involving the intermediacy of a halodienone. Sequences (34), (35), (36) and (37), illustrate respectively the displacement of the substituents *t*-Bu, CH_2OH, CHO and Br by routes of this kind[94–97]. Complex pathways in the course of which a methyl group can become replaced have been reported also[98]; it is clear, too, that in rather special cases a substituent can be replaced *ortho* to the hydroxyl group (sequence 38)[99].

In two other investigations, mechanistic details relating to the group displacement have been provided. For decarboxylation, Grovenstein and coworkers[83] showed that, under conditions in which the loss of the group was rate determining, it was subject to

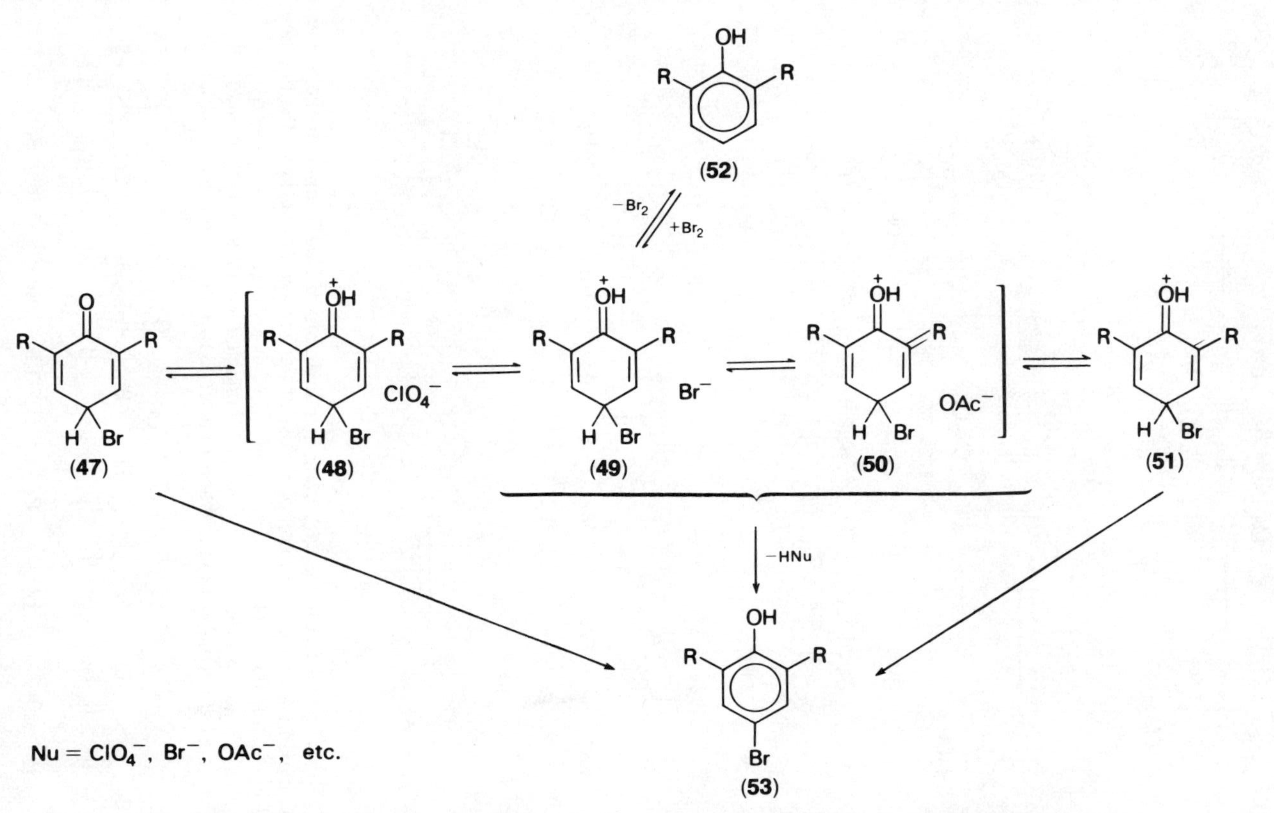

SCHEME 5. Pathways in the rearrangement of 4-bromo-2,6-di-*t*-butylcyclohexa-2,5-dienone (**47**; R = CMe_3).

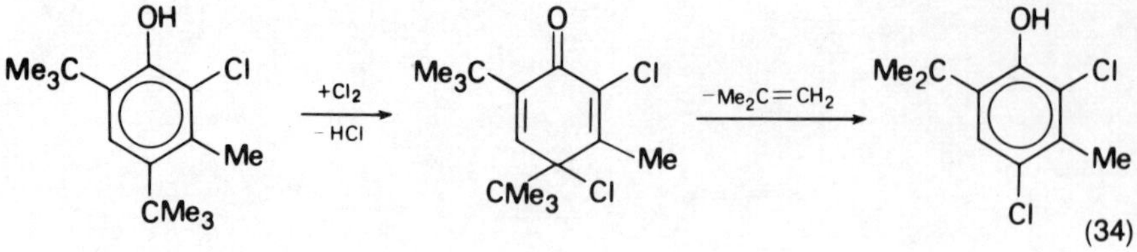

(34)

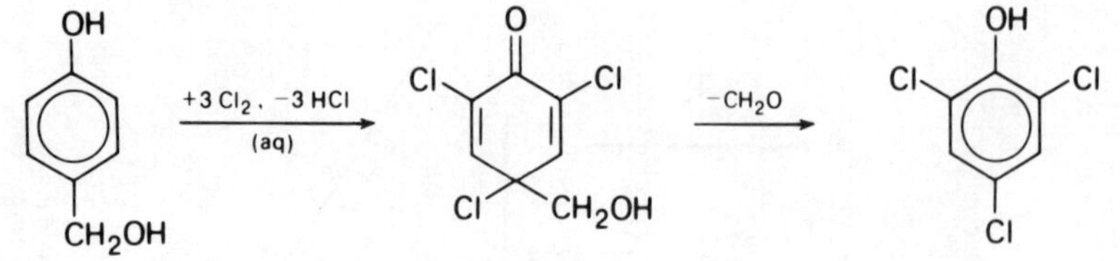

(35)

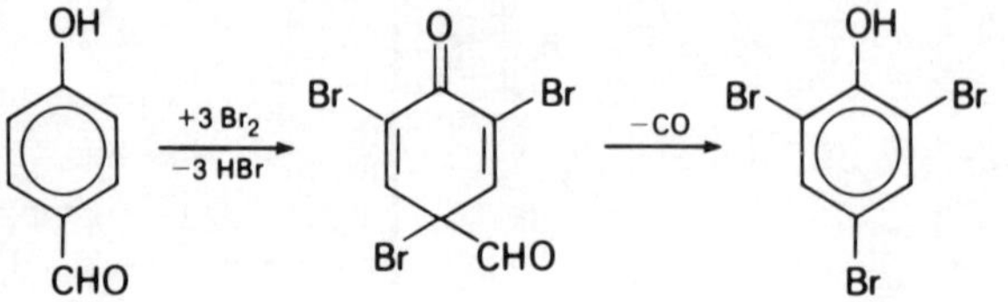

(36)

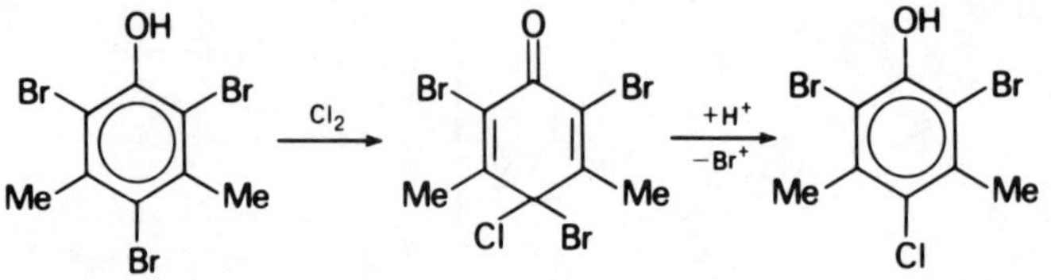

(37)

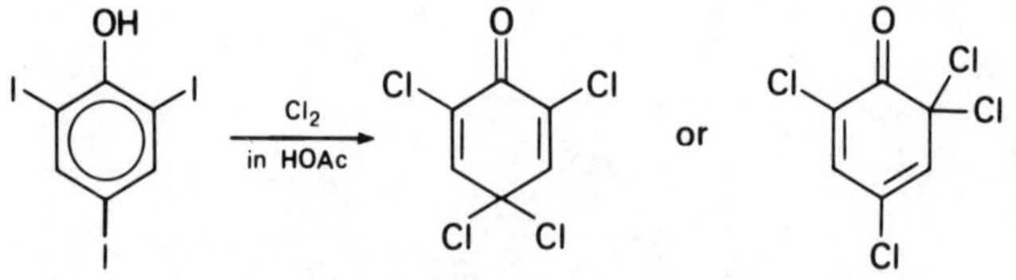

(38)

a $^{12}C/^{13}C$ isotope effect in the carbon dioxide liberated of 1.045, this value being near to the maximum expected for complete loss of the zero-point energy of the breaking bond in the transition state.

Cannell's study of bromodesulphonation[84] enabled a comparison to be made of the relative rates of decomposition in water of a series of related bromodienones, namely

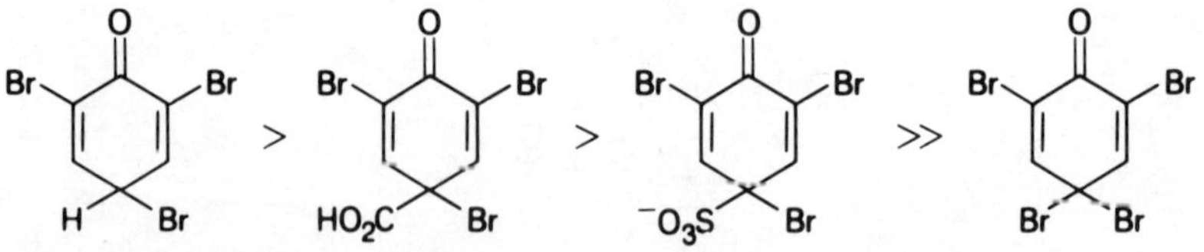

This is a useful qualitative comparison, but it is not as yet clear whether the dependence of rate on acidity is the same for all these compounds.

C. Conversion of Halodienones to Phenols; Halodienones as Sources of 'Positive Halogen'; 1,3-Migration

Chloro-, bromo- and iododienones are potentially sources of positive halogen. The reactions which can occur are of the form given for a 2,5-dienone in equation (39); they give back the phenol from which the dienone is formally derived by halogenation,

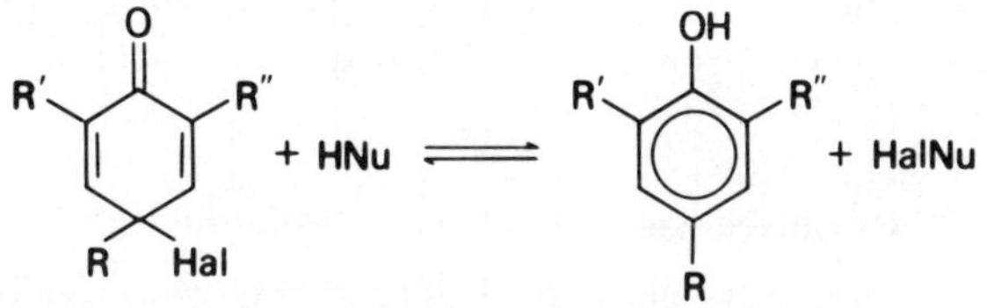

$$\text{(2,5-dienone: R', R'', R, Hal)} + \text{HNu} \rightleftharpoons \text{(phenol: OH, R', R'', R)} + \text{HalNu} \qquad (39)$$

and liberate an electrophile. The reactions are reversible; the position of equilibrium depends on the solvent, the halogen, and on any subsidiary equilibria possible in the system. The rate of attainment of equilibrium is often measurably slow, but is subject to catalysis by acids. Since it is also catalysed by reagents having nucleophilic power for halogen, the hydrogen halides are particularly good reagents for establishing the equilibrium. They help also to drive the reaction towards the phenol through formation of the trihalide anion.

The availability of processes of this kind leads in practice to instability of many dienones, particularly those in which one *ortho* substituent is hydrogen. In some discussions of the methods of preparation of dienones, there appears to have been confusion[5,84] between the effect of conditions on the rate and on the position of equilibrium.

The more stable dienones can be used as heterolytic halogenating agents analogous in their behaviour to the *N*-halogenoamides, anilides and imides, and to the *t*-butyl hypohalites. Thus 2,4,4,6-tetrabromocyclohexa-2,5-dienone has been recommended as a reagent for the monobromination of aromatic amines[100], for selective bromination of α,β-unsaturated ketones[100], and for the addition of BrOMe to olefinic compounds[101]. By the use of this type of method, chlorodienones by treatment with a hydrogen halide can become sources of electrophilic bromine through formation of bromine chloride, and of electrophilic iodine through formation of iodine chloride.

Probably analogous is the 1,3-rearrangement which can lead to the formation of 2,5- from 2,4-dienones, as in the conversion of **25** to **26** (Scheme 3), or in the isomerization of 3,5-dienones. Waring[4] has summarized the extensive work by Denivelle and his coworkers[69,102] and by Ershov and Volod'kin[103] on these reactions, which can occur spontaneously or under the influence of a variety of catalysts. It is apparent that intermolecular cationotropic pathways would be available; but other routes involving nucleophilic, homolytic or intramolecular migration of bromine are also possible. It has been shown, for example, that 2-bromo-2,6-di-*t*-butyl-4-cyanocyclohexa-3,5-dienone undergoes isomerization in chloroform at 0°C (equation 40), since the ^{1}H NMR signals, attributable to the *t*-butyl groups and separately observable at lower temperatures, become identical[104]. It was suggested that this

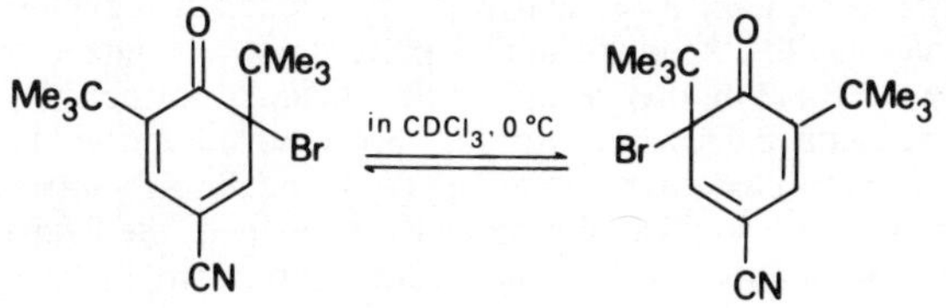

$$\text{(dienone: O, Me}_3\text{C, CMe}_3\text{, Br, CN)} \xrightleftharpoons{\text{in CDCl}_3,\ 0\ ^\circ\text{C}} \text{(dienone: O, Me}_3\text{C, Br, CMe}_3\text{, CN)} \qquad (40)$$

reaction adopts the pathway involving free radicals. Little is known, however, concerning the availability of these processes, some of which have bimolecular variants. Related also, but subject to equal possible mechanistic variability, are the exchanges of halogens and halide ions with halophenols and halodienones[105]. A summary of this work is given by Ershov's group[5]. The general conclusion is that all three atoms of bromine in 2,4,6-tribromophenol can be brought into equilibrium with labelled hydrogen bromide, but that one exchanges more rapidly than the other two. The exchange is slow in neutral solution, but becomes more rapid in the presence of hydrogen bromide and of pyridine.

D. Conversion of Halodienones to Phenols; 1,2-Migration

The cationotropic processes leading to 1,3-shifts or displacement of positive halogen discussed in the previous subsection are characteristically catalysed by acids; the conditions which lead instead to 1,2-shifts of halogen generally, however, require much higher acidities, and then are quite regioselective.

It has been known for many years that 4-bromo-2,5-dienones having a vacant 3- or 5-position when treated with sulphuric acid at quite low temperatures undergo a 1,2-shift of bromine to give ring-substituted phenols in nearly quantitative yield. Examples are given in equations (41)–(43)[53,54,106,107].

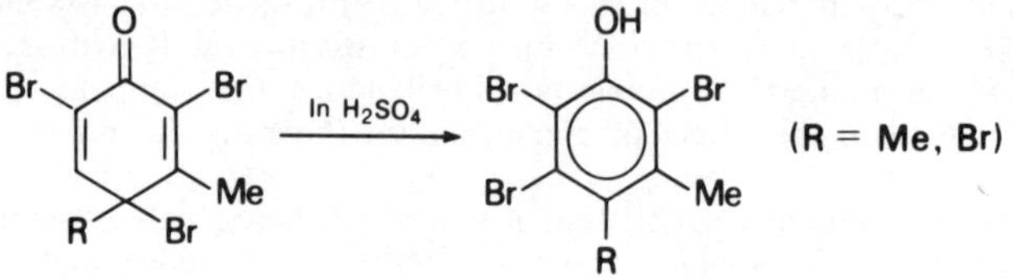

In H_2SO_4 (R = Me, Br) (41)

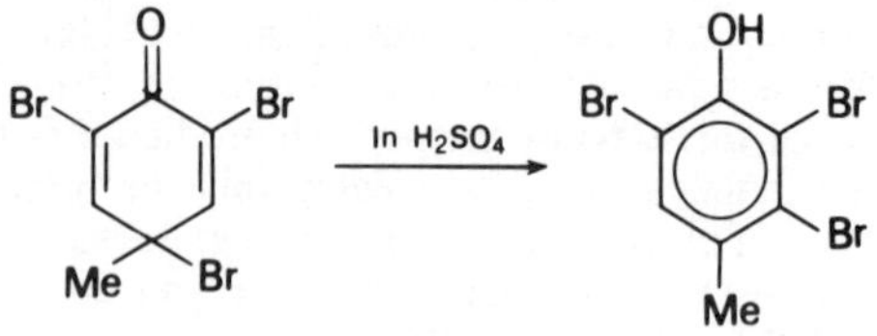

In H_2SO_4 (42)

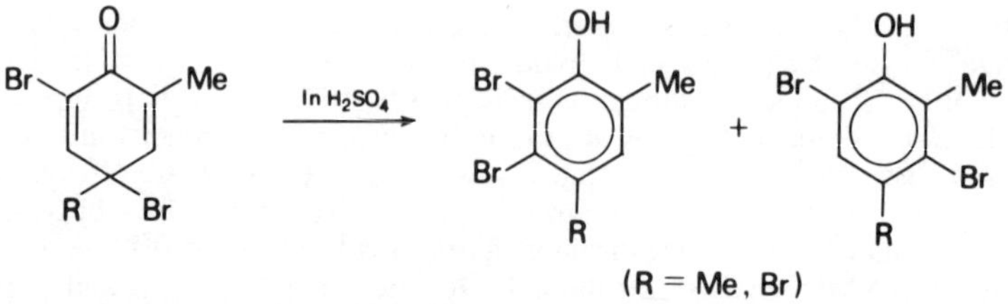

In H_2SO_4 (43)

(R = Me, Br)

Related chlorodienones also undergo this reaction, for which trifluoromethanesulphonic acid has been used as a catalyst[108]. The reaction is regioselective; in the examples of equation (43), bromine moves preferentially adjacent to bromine and consequently *para* to the 2-methyl group[53,54]. It is evident, furthermore, that bromine and chlorine move to the exclusion of the accompanying migration of an alkyl group.

The acidity-dependence of the reaction has not been closely studied; but a reagent at least similar in strength to 96% sulphuric acid is needed; the formation of chloro- or bromodienones under the usually quite acidic conditions of bromination is not

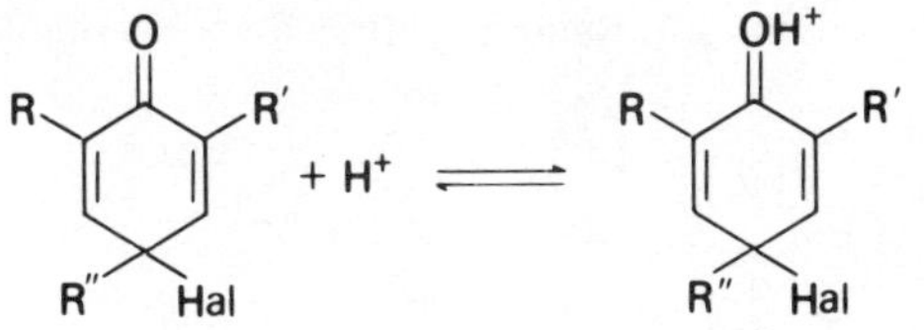

(44)

(**54**)

normally accompanied by this type of rearrangement. It can be presumed that the intermediate undergoing 1,2-shift is the conjugate acid of the dienone (**54**, sequence 44).

Little is known about the details of the stages of reaction which follow protonation. Structure **54** will, of course, be seen to be the Wheland intermediate expected to be formed by electrophilic attack *para* to a hydroxy group in a phenol, and there are various acid-catalysed rearrangements which appear to follow a similar course[4,109]. The most extensively discussed example is the so-called 'dienone–phenol rearrangement', important in the aromatization of steroidal ketones[110]. This also requires strongly acidic conditions, acetic anhydride and sulphuric acid being commonly used. Because of the observed relative migratory aptitudes of substituents[111], it is regarded as a rearrangement involving a carbocation produced by protonation of a carbonyl group. Its intramolecularity has been demonstrated; and regioselectivity of group migration is known[112].

Other reactions which appear to be analogous include those of 4-hydroxy- and 4-acetoxy-2,5-dienones; and those of the related Wheland intermediates involved in nitration[113,114]. Characteristic of a number of these processes is that they have been documented for 2,2-substituted 3,5-dienones just as they have for the isomeric 2,5-dienones (equation 45)[115,116]. It is fully to be expected that the 2-chloro and

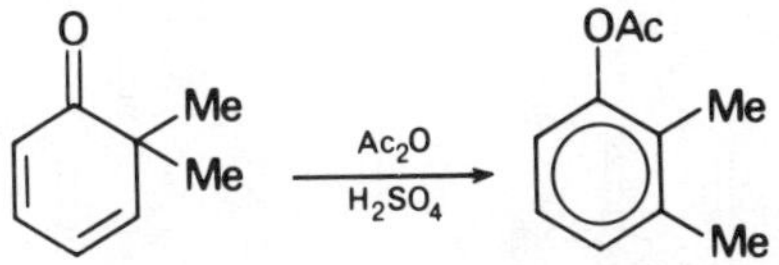

$$\xrightarrow[H_2SO_4]{Ac_2O}$$

(45)

2-bromo-3,5-dienones would similarly undergo this type of transformation. As far as the writers are aware, there is as yet no direct experimental illustration of this point, though indirect evidence has been adduced[44].

The sequence of migratory power of groups, $Br^+ > Cl^+ > Me^+$, derivable from the results for 4-substituted 2,5-dienones, refers apparently to groups competing directly for a 1,2-shift to a carbocationic centre. The same sequence has been derived from comparisons of the products of acid-catalysed reactions of 2-keto-1,2-dihydronaphthalenes[80,116,117], where inter- rather than intramolecular movement of the group is involved, and further complications could arise through involvement of a nucleophile[117,119].

It is useful to note that the corresponding reactions of nitrodienones can lead to 1,2-rearrangement in competition with 1,3-rearrangement; but that for these compounds, radical-cation processes are much more important[108,117].

E. Conversion of Halodienones to Phenols; the 'Quinobenzylic Rearrangement'

One of the most characteristic reactions of 4-methyl-4-bromocyclohexa-2,5-dienones is that they undergo the so-called 'quinobenzylic rearrangement', in which the dienone is converted into a (bromomethyl)phenol[4,5]. A simple example is shown in equation (46).

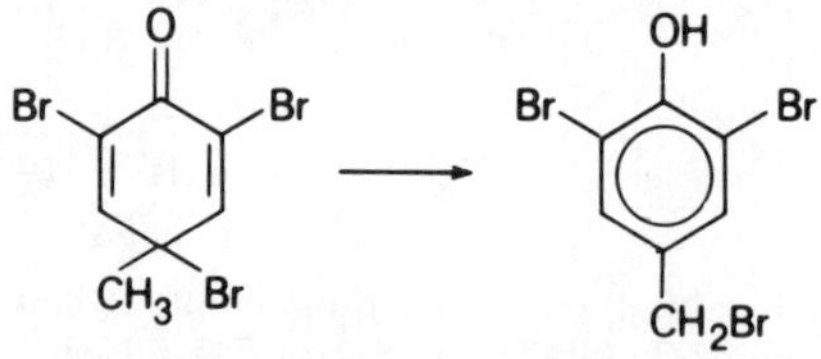

(46)

The reaction can occur spontaneously in the solid phase, when it appears to be autocatalytic, the rate often being dependent on the degree of purity of the sample. Rapid conversion can be effected by heat or by illumination of the sample. In favourable cases almost quantitative yield is obtained, and the reaction shows remarkable regiospecificity. Thus no 2-bromomethyl derivative accompanies the rearrangement of 4,6-dibromo-2,4-dimethylcyclohexa-2,5-dienone (equation 47)[120]. Similarly the

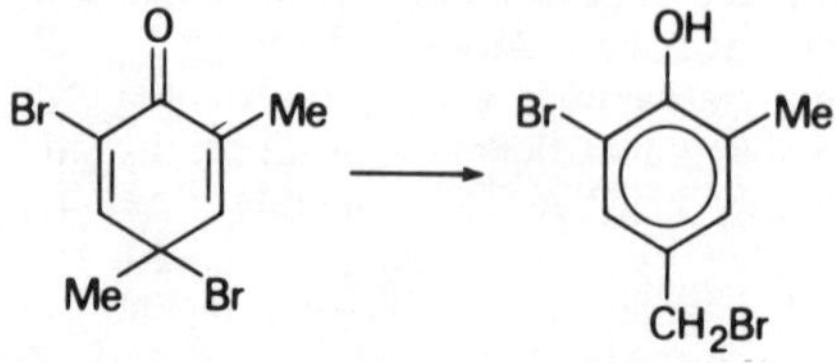

(47)

rearrangement of 2,4,6-tribromo-3,4-dimethylcyclohexa-2,5-dienone (equation 48) occurs with no accompanying formation of the 3-bromomethyl isomer[53]. A preliminary study[121] of this reaction in carbon tetrachloride showed that it was slow at

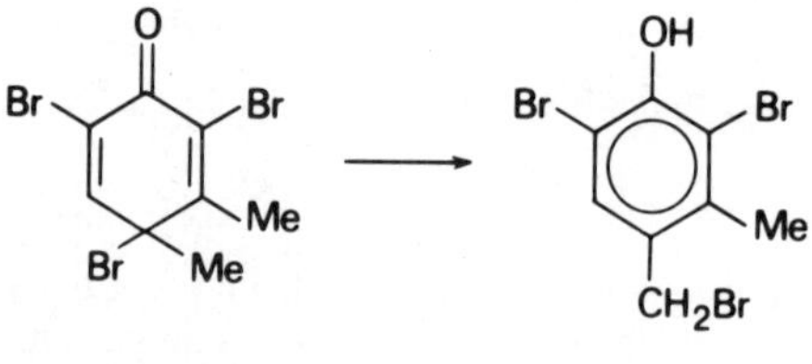

(48)

room temperature, and was only modestly autocatalytic. Dissolved oxygen had little influence on the progress of the reaction, which was also not affected dramatically by ordinary laboratory illumination.

In an earlier investigation of the kinetics of this type of reaction[122], the quinobenzylic rearrangement of 4-bromo-2,6-di-*t*-butyl-4-methylcyclohexa-2,5-dienone (**56** → **59**, Scheme 6) was shown to proceed with first-order kinetics, but more rapidly at higher initial concentration of substrate. It was suggested that this resulted from the solvent effect of the dienone in providing its own polar atmosphere, since the reaction was faster in bromobenzene than in benzene, and faster in benzene than in cyclohexane. Catalysis by traces of hydrogen bromide and of bases was noted, and it was recorded that during the rearrangement the compound is very sensitive to oxygen, becoming highly coloured and difficult to purify.

It has become generally accepted[5,122,123] that the rearrangement follows an elimination–addition pathway, with a quinone methide (e.g. **57**) as a key intermediate formed by dehydrobromination of the dienone. Quinone methides are normally difficult to isolate, since they polymerize very rapidly; but Filar and Winstein[124] showed that they are reasonably stable at low concentrations in inert media. More

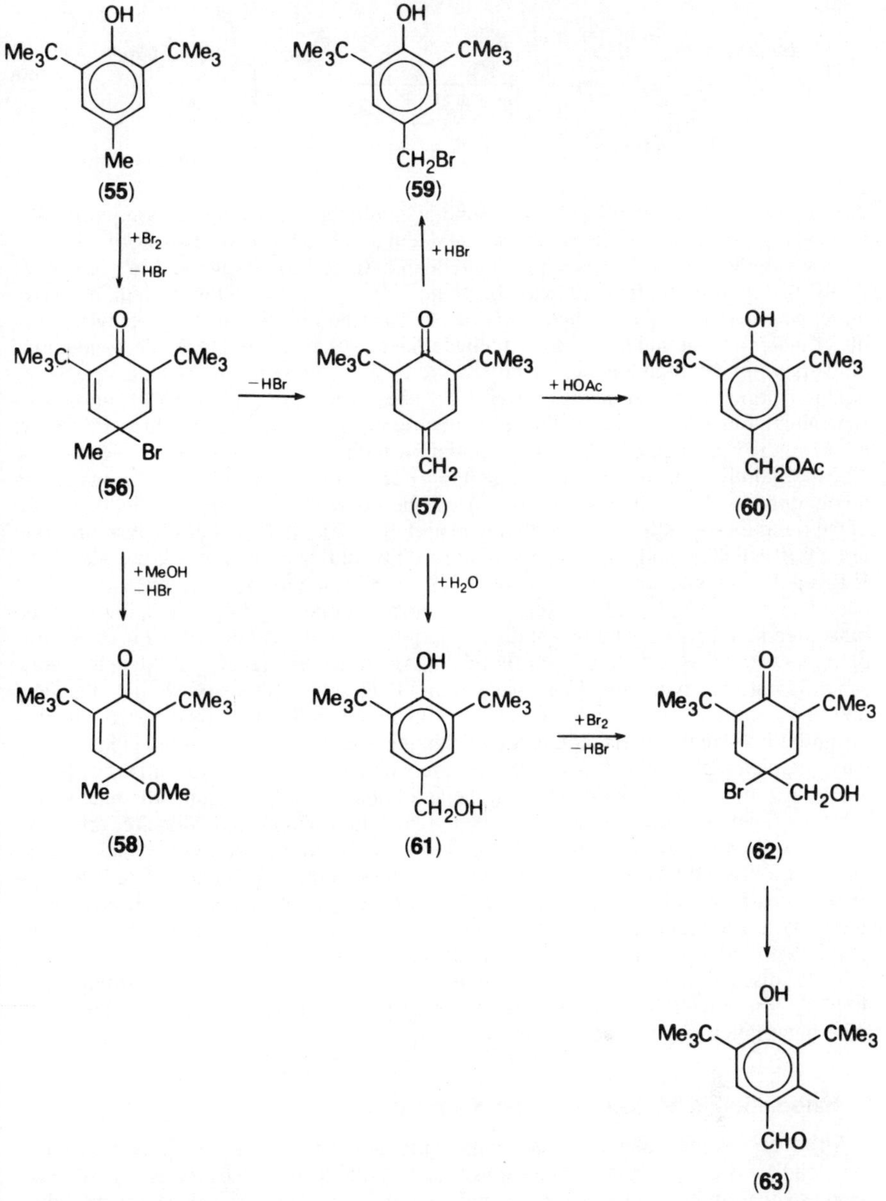

SCHEME 6. Pathways presumed to be available for bromination of 2,6-di-*t*-butyl-4-methylphenol.

recently, Volod'kin and coworkers[125] established that 4-bromo-2,6-di-*t*-butyl-4-cyanomethyl-2,5-dienone when left as a solid for 20–30 days at room temperature loses hydrogen bromide to give the quinone methide (equation 49). In this example, the

O, Me_3C, CMe_3, Br, CH_2CN $\xrightarrow[\text{(solid phase, 20 °C, 20–30 d)}]{-HBr}$ O, Me_3C, CMe_3, CHCN (49)

cyano group helps to make the compound less subject to further reaction. Normally, however, 1,6-addition to the unsaturated system leads to facile aromatization. Scheme 6 shows the varied pathways which have been assumed to be involved in reaction of 2,6-di-*t*-butyl-4-methylphenol with bromine[122,123] to give the bromodienone **56** in aqueous acetic acid, the methoxydienone **58** in methanol, the (acetoxymethyl)phenol **60** in glacial acetic acid, and the aldehyde **63** in aqueous *t*-butanol. The analogous conversion of **59** to **63** has also been reported[126].

Description of the bromination step (**55** → **56**) as an electrophilic substitution with rearrangement (S_E2'), and of the aromatizations of the quinone methide (**57** → **59**, **60**, **61**) as 1,6-conjugate additions can be supported by many good chemical analogies. The methanolysis of the bromodienone (**56** → **58**) can for the moment be regarded as a conventional S_N1 solvolysis, though it will be discussed separately in Section V.G. Since the dienone **56** is a doubly allylic tertiary bromide, it should easily lose bromide ion heterolytically, and then by loss of a proton would give the quinone methide (**57**). If this pathway is taken, the conversion of **56** to **57** is a heterolytic $E1$ reaction. Until recently, it seems to have been accepted by most writers[5,122,123] that the quinobenzylic rearrangement proceeds by this pathway, despite some indications of the involvement of free radicals. Ershov and coauthors[127] have, however, reviewed extensive work (mainly, but not exclusively, by Russian authors) which leads these authors to the view that the spontaneous rearrangements of 4-bromo-2,5-dienones to *p*-hydroxybenzyl bromides involve the intermediate formation of phenoxy radicals. It is not clear exactly what sequence of stages is envisaged, or even whether this mechanism is thought to apply both to spontaneous reaction in the solid phase and to spontaneous reaction in solution. Two negative points seem clear: first, atom chains involving free bromine atoms are not involved in the rearrangements in solution, otherwise attack on the methyl groups would be less discriminating. Presumably a sequence of radical processes occurring within a solvent cage[128] would explain the marked regiospecificity. Secondly, these reactions are not analogous to the free-radical dehydrobrominations of β-bromoketones[129], in which the α-hydrogen atom plays a vital role.

Some discussions[5] of quinobenzylic rearrangements suggest that 4-bromo- and 4-chloro-4-methyl-2,5-dienones behave similarly; but the latter reactions appear to have been much less extensively investigated.

F. Halodienones as Sources of Halogen Atoms

Although it would seem that free bromine atoms are not concerned in the sequences that lead to the quinobenzylic rearrangement, suitable bromodienones can become sources of bromine atoms. Kennedy and Ingold[130] examined the reactions of 2,4,6-tribromo-4-chlorocyclohexa-2,5-dienone (**64**) with a series of alkyl-substituted aromatic hydrocarbons in methylene dichloride at 40°C. The rates relative to that for ethylbenzene were shown to resemble closely those determined for other known sources of bromine atoms such as *N*-bromosuccinimide. It was concluded that the sequences involved must be similar, and that chains were sustained by the abstraction of hydrogen by free bromine atoms and not by the phenoxy radical which is concurrently produced. This conclusion was supported by a study[131] of structural effects on the bromination of $XC_6H_4CH_2OMe$ by 4-bromo-2,4,6-tri-*t*-butylcyclohexa-2,5-dienone (**65**) in carbon tetrachloride. Variation of the substituent X gave a ρ^+ value of -0.43, similar to that obtained in other reactions involving known sources of bromine atoms.

The irradiation of the dienone **65** in various solvents is known[132] to give the corresponding phenoxy radical **66** and hence 2,4,6-tri-*t*-butylphenol in good yield. In the presence of oxygen, the radical is attacked, and one of the final products is the quinone **67**[132,133].

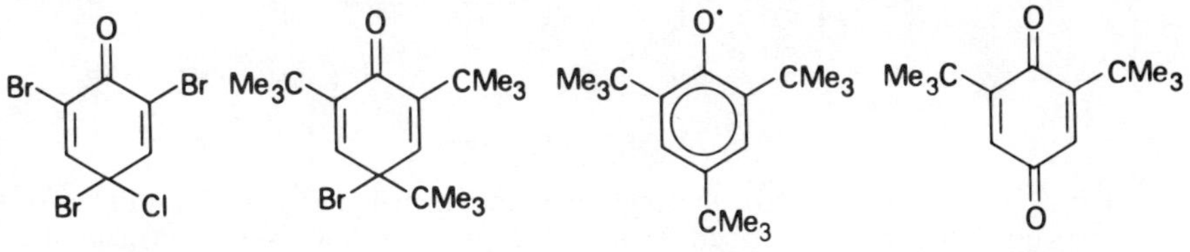

(**64**) (**65**) (**66**) (**67**)

G. Further Additions

When dienones are formed in the course of halogenation of phenols, it may happen that they will consume some of the reacting halogen to form products of further addition, particularly when excess of halogen is used. Examples have been given already (sequences 23–25). These products of addition may then be unstable, when products of substitution will result[72]. To understand the conditions likely to lead to these side-reactions, the mechanisms of addition to α,β-unsaturated carbonyl compounds need to be considered[134]. The carbonyl group deactivates the double bond for electrophilic addition; but hydrogen halides are powerful catalysts for another mechanism which has been described as 'nucleophilic', but almost certainly involves a pre-equilibrium protonation of the carbonyl group. The sequences of reactions which may follow have not yet been fully elucidated, but halide ions clearly play a very specific role. No 4-bromo-2,5-dienones have been investigated with this specifically in mind; but the availability of this 'nucleophilic' mechanism has been established for a range of α,β-unsaturated aldehydes, ketones and quinones[135].

The result of the availability of this mechanism is that intermediate dienones may become consumed by further addition when excess of molecular halogen is used, particularly in aprotic solvents. Anhydrous acetic acid is conducive to the mechanism of addition catalysed by hydrogen halides, but the addition of small proportions of water to the solvent profoundly reduces the effective acidity of the catalyst, and so disfavours reaction by this pathway. Halogenating reagents such as *N*-haloamides, alkyl hypohalites and acyl hypohalites are not expected to give reaction by this mechanism; it has been confirmed for *p*-benzoquinone, for example, that chlorine acetate in acetic acid is unreactive, whereas addition of molecular chlorine proceeds rapidly with catalysis by the hydrogen chloride formed[136].

Additions of hydrogen chloride to ketodienones have also been reported, as in the example given in sequence (28).

H. Nucleophilic Replacement of Halogen

4-Bromo- and 4-chloro-2,5-dienones are relatively stable at room temperature in aprotic solvents and solvents of low nucleophilic power, as is witnessed for example by the ease of their isolation from reactions in acetic acid and in aqueous acetic acid.

Their allylic character should, however, allow moderately rapid heterolysis of halogen, so permitting both the $E1$ and the S_N1 processes already referred to in relation to the conversion of the dienone **56** to **57** and **58** (Scheme 6). Other nucleophilic displacements of such dienones have been effected by using weak acids in the presence of silver ions[137], as is represented in sequence (50). Replacement reactions carried out

$$\text{4-MeC}_6\text{H}_4\text{OH} \xrightarrow[-50\,^\circ\text{C}]{\text{With SbCl}_5} \text{4-Me-4-Cl-cyclohexa-2,5-dienone} \xrightarrow[-\text{AgCl},\ -\text{H}^+]{+\text{HNu},\ +\text{Ag}^+} \text{4-Me-4-Nu-cyclohexa-2,5-dienone} \tag{50}$$

(Nu = OH, OMe, OAc)

under basic conditions can have the same type of stoicheiometry, and a number of these processes have been described. Typical examples, which for want of mechanistic investigation would be assumed to be S_N reactions, are given in equations (51)[138], (52)[139] and (53)[140]. Analogous reactions could be expected for 2-halo-2,5-dienones.

$$\text{2,6-(Me}_3\text{C)}_2\text{-4-Br-4-CMe}_3\text{-cyclohexa-2,5-dienone} \xrightarrow[-\text{Br}^-]{+\text{MeO}^-} \text{2,6-(Me}_3\text{C)}_2\text{-4-MeO-4-CMe}_3\text{-cyclohexa-2,5-dienone} \tag{51}$$

$$\text{2,6-(Me}_3\text{C)}_2\text{-4-Br-4-CMe}_3\text{-cyclohexa-2,5-dienone} \xrightarrow[-\text{HBr}]{+\text{C}_6\text{H}_{11}\text{NH}_2} \text{2,6-(Me}_3\text{C)}_2\text{-4-C}_6\text{H}_{11}\text{NH-4-CMe}_3\text{-cyclohexa-2,5-dienone} \tag{52}$$

$$\text{2,6-Me}_2\text{-3,5-Br}_2\text{-4-Me-4-Br-cyclohexa-2,5-dienone} \xrightarrow[\text{(with NaOAc in EtOH)}]{+\text{PhNH}_2,\ -\text{HBr}} \text{2,6-Me}_2\text{-3,5-Br}_2\text{-4-Me-4-NHPh-cyclohexa-2,5-dienone} \xrightarrow[\text{HOAc}]{\text{HCl}} \text{1-OC}_6\text{H}_4\text{NH}_2\text{-2,6-Me}_2\text{-3,5-Br}_2\text{-4-Me-benzene} \tag{53}$$

Allylic rearrangements accompanying nucleophilic replacement of halogen should also be possible. A reaction which could be a unimolecular example of this kind was reported by Crozier and Hewitt[141]. 3,5-Di-*t*-butyl-2,2-dichlorocyclohexa-3,5-dienone reacted with sodium methoxide in methanol to give 3,5-di-*t*-butyl-2,4-dichlorophenol, and it was suggested that the pathway was as shown in sequence (54). These authors reported two other base-catalysed rearrangements, one of which could involve a S_N2' rearrangement (sequence 55) and one a rearrangement of the Favorskii type (sequence 56).

Doubt has been thrown on conventional descriptions of these reactions through studies which suggest that radical anions and radicals can be involved[127]. The observation[138] that a blue colour develops during the reaction of the bromodienone **65** with

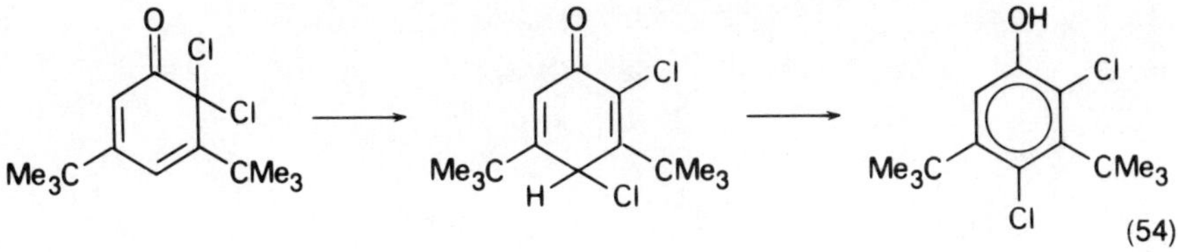

(54)

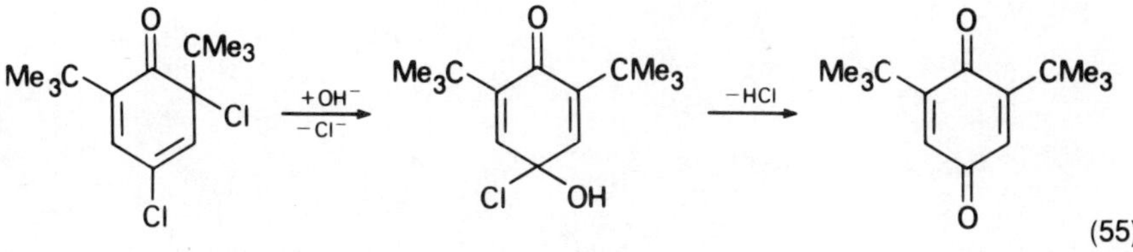

(55)

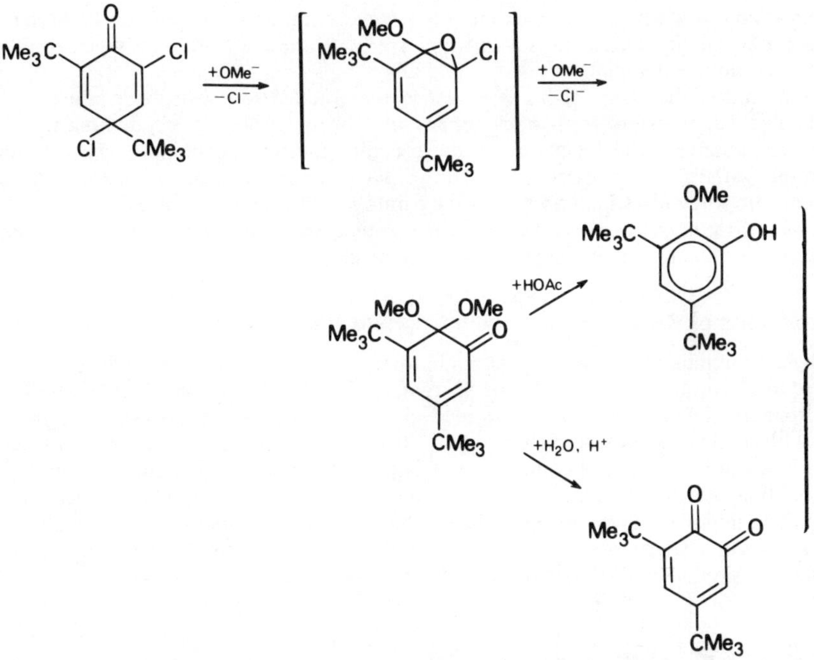

(56)

methoxide ions in methanol (equation 51) provided a hint in this direction. Prokof'ev and coworkers[142] showed that this dienone reacts with 2,4,6-tri-*t*-butyl phenol to give hydrogen bromide and two molecular equivalents of the phenoxide radical (equation 57) at a rate which was dependent on the concentration of base but nearly independent of the solvent, over the range of benzene, acetic acid, tetrahydrofuran, acetone, nitrobenzene, diethyl ether and ethanol. In other studies, the effect of change of structure of the aryloxide anion was investigated, and the rate was shown to be enhanced by electron release from a *para* substituent. The related reactions of the dienone **56** (Scheme 6) with morpholine, piperidine or other secondary and tertiary

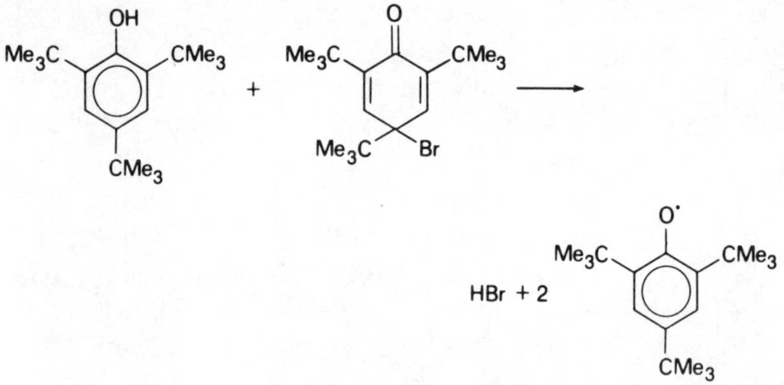

(57)

amines in benzene have similarly been interpreted as involving single-electron transfers, in this case from nitrogen to the carbonyl oxygen atom to give first a biradical complex which then decomposes to give ultimately the product of nucleophilic replacement[143]. Reactions with triphenylphosphine and with dimethyl sulphide have been interpreted similarly[139].

In summary, therefore, there is evidence that radical ions can be produced by single-electron transfers from nucleophiles to 4-bromocyclohexa-2,5-dienones, and that, accompanying the formation of these compounds, nucleophilic displacements with and without rearrangement can occur. There appears to be room for further mechanistic study to establish whether the formation of radical ions is essential to all, or some, of these reactions, and whether the displacements are produced through the same type of chain as is involved in the $S_{RN}1$ mechanism[144–146].

I. Reactions of Bromodienones with Grignard Reagents

These reactions often give complex mixtures of products, some of which can be regarded as formed by conventional processes. Volod'kin and coworkers[147] studied the reaction of 2,6-di-*t*-butyl-2-bromocyclohexa-3,5-dien-1-one-4-carboxylic acid (**68**) with methyl magnesium bromide at −70°C. The formation of **69** can be formulated as a normal Grignard coupling reaction; and the phenol **70** could be produced by coupling with allylic rearrangement followed by decarboxylation with rearrangement. The major product, however, is not that expected for addition of MeMgBr to the carbonyl group, or of similar conjugate addition; instead it is the MgBr salt of the phenol **71**, produced with loss of methyl bromide. Reactions of this kind have also

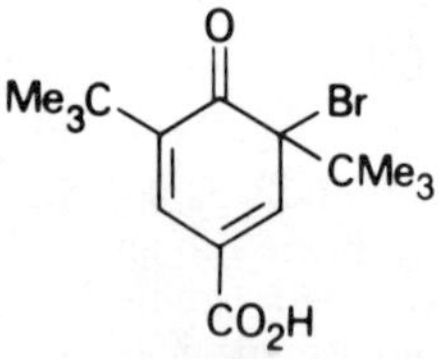

(**68**) 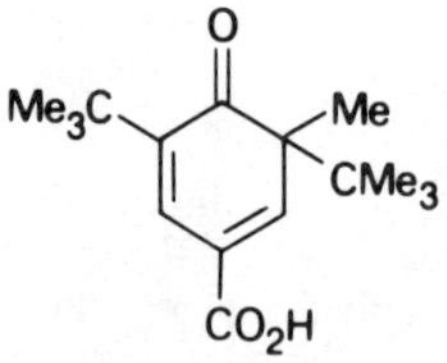 (**69**) 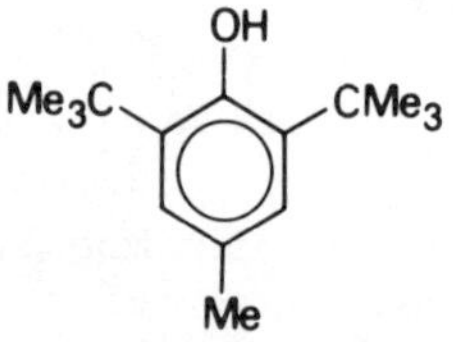 (**70**) 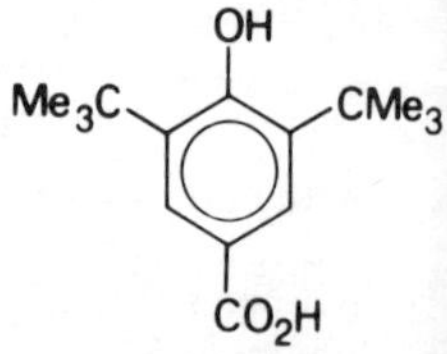 (**71**)

been reviewed extensively by Ershov and collaborators[127] and are believed to involve radical anions produced by one-electron transfers; these anions are believed to react further, often with other radicals formed within a solvent cage.

J. Diels–Alder Additions of Halodienones

We have already (in sequence 19) given an example of the easy dimerization of a halogeno-3,5-dienone by a Diels–Alder reaction. Somekawa and coworkers[148] have made an important study of the reactions typified by equation (58). The reactions were

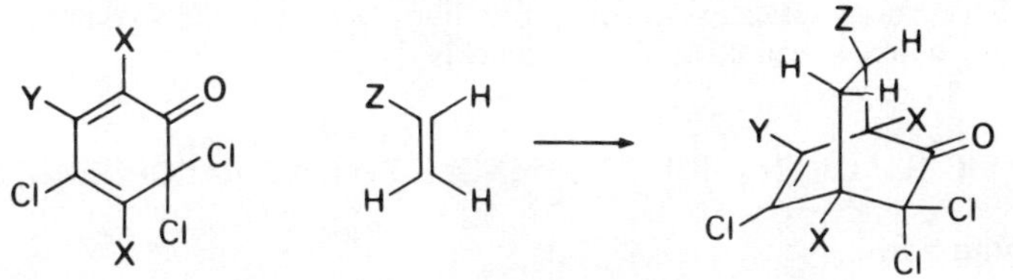

(58)

($X = Y = Cl$; $X = H$, $Y = Cl$; $X = Y = H$; and $Z = CO_2H$, CO_2Et, CN, $COCH_3$, OEt)

stereospecific, and gave exclusively the product shown, in which the α-carbon atom has attached itself *ortho* to the carbonyl group in the diene, and the electron-withdrawing group (Z) in the original dienophile is *endo* to the formed olefinic bond. The possible isomeric products **72**, **73** and **74** were not formed in detectable amount

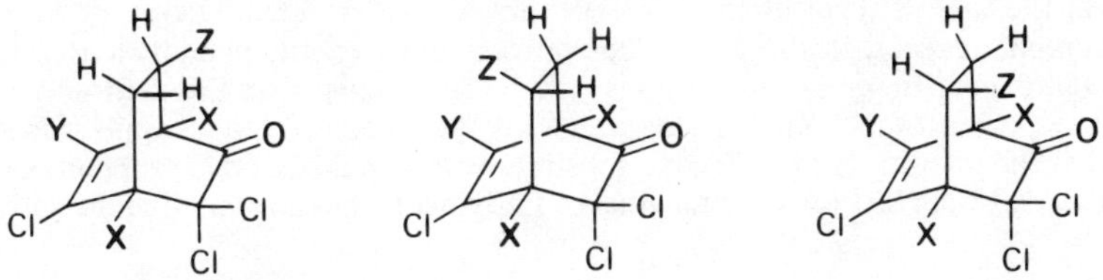

(72) **(73)** **(74)**

from any of the reactions studied. Simple molecular orbital and electrostatic theories were applied to consideration of the transition states expected for the formation of these isomers; possibility **74** was not considered explicitly, but would be highly disfavoured by steric interaction between Z and a chlorine atom from the CCl_2 group. It was concluded that the stereochemical outcome of reaction (58) cannot be explained in this way, and must be determined either by steric factors or by unknown electronic factors.

VI. SOME STRUCTURAL INFORMATION ON CYCLOHEXADIENONES

Waring's review[4] provides a great deal of information on the spectroscopic and other physical properties of cyclohexadienones, and here we refer only briefly to some pertinent later material.

Conjugative influences would encourage both 3,5-dienones and their 2,5-isomers to adopt planar conformations, but torsional and steric forces make it possible for both types of compound to exist in puckered forms. Studies by X-ray crystallography have shown that 2,4,4,6-tetrachlorocyclohexa-2,5-dienone[149] and 2,3,4,4,5,6-hexachlorocyclohexa-2,5-dienone[150] have the oxygen atom and the six carbon atoms of the ring approximately in one plane. In the solid phase, of course, crystal forces may impose planarity on an intrinsically non-planar molecule, but the rate of interconversion of hypothetical puckered forms would probably be very rapid. It has been shown[104] by using ^{1}H NMR spectroscopy at low temperatures that 4-bromo-2,4,6-tri-*t*-butylcyclohexa-2,5-dienone can exist in more than one form[104], but the preferred interpretation of this is that the *t*-butyl group becomes 'frozen' into a preferred conformation having non-equivalent methyl groups.

There have been desultory reports[53,54,73,104,151,152] of chemical shifts and coupling constants in the ^{1}H and ^{13}C NMR spectra of halogen-substituted cyclohexadienones, and most of the information can be interpreted in conventional ways. In general, it can be commented that the spectra remove any doubt as to the identity of the isolated compounds, which (particularly from the ^{13}C spectra) can be assigned readily to the 2,5 and 3,5-dienone series respectively. Equilibria between the two types of compound should also be recognizable easily and quickly.

VII. HALOGENATION OF PHENOLS TO GIVE HALOPHENOLS

A. Introduction

In Section 4, we dealt with the conversion of phenols to halodienones, and with the mechanistic aspects of reactions which have been shown to proceed in this way. We now turn to halogenations of phenols which result in the formation of halophenols. It is presumed that most, if not all, of these reactions involve dienones as intermediates which rapidly rearrange to give the described products.

For most aromatic substrates undergoing halogenation under conventional conditions, the kinetic form of the reaction can be used to identify the nature of the electrophilic reagent. With hydroxy derivatives, however, the phenoxide ion derived from the phenol by rapid proton transfer is more reactive than the neutral molecule by many powers of ten. Kinetic measurements therefore need to be applied in such a way that the identity of the effective substrate is resolved also; this problem becomes particularly important for the reactions of fairly acidic phenols in aqueous solvents.

B. Kinetic Forms and Structural Effects in Chlorination

The chlorination of phenol by molecular chlorine is very rapid; the few kinetic measurements that have been made relate to less reactive electrophiles. Brown and Soper[153] examined chlorination by diethylchloramine (Et_2NCl) in water. The rate of reaction was found to be independent of pH over the range 3.5–7.5, and it was concluded that the kinetic form should be interpreted as reflecting reaction between the phenoxide ion and the diethylchloroammonium cation (equation 59).

$$-d[Et_2NCl]/dt = k[ArO^-][Et_2N^+HCl] \qquad (59)$$

The relative rates of reaction of the phenols can then be calculated from their dissociation constants to follow the sequence

4-MeC_6H_4OH (9.0), C_6H_5OH (3.3), 2-ClC_6H_4OH (0.6), 4-ClC_6H_4OH (0.5).

This sequence accords qualitatively with that recorded for the corresponding reactions with hypochlorous acid.

Carr and England[154] examined the chlorination of phenol with *N*-chloromorpholine (**75**) in more acidic solution. With this *N*-chloroamine, unlike what has been found

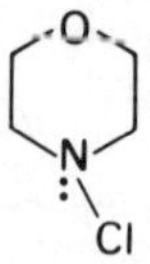

(**75**)

with other related compounds, catalysis by added chloride ion through the formation of free chlorine was not observed. The kinetic form was that of equation (60).

$$-d[R_2NCl]/dt = k_2[ArOH][H^+][R_2NCl] \quad (60)$$

It was concluded that the effective reagent is the protonated *N*-chloroamine, which at values of pH less than 2 or so attacks the phenol molecule.

The chlorination of phenol by very dilute hypochlorous acid in aqueous solution with silver perchlorate present to remove chloride ions has the kinetic form shown in equation (61)[155].

$$-d[ClOH]/dt = k_1[ClOH] + k_2[ClOH][H^+] + k_3[ClOH][H^+][ArH] \quad (61)$$

The interpretation of the first two terms has been the subject of controversy which has been discussed elsewhere[134b]. These terms make only a minor contribution and will be neglected in the present context. The final term requires a transition state involving the neutral aromatic molecule and the elements of $ClOH_2^+$. The products of reaction under these conditions have not been studied, but results obtained by us and by Professor L. Melander[156] suggest that the dienone probably formed as an intermediate may become oxidized to *p*-benzoquinone. Structural effects on the final term in the rate equation (61) put the value for the reactivity of phenol under these conditions as some four times that of anisole. This result is consistent with our view that the loss of a proton from phenolic oxygen is occurring in the transition state for chlorination, thus providing an extreme result of H—O hyperconjugation.

C. Kinetic Forms and Structural Effects in Bromination

Studies of the kinetics of the bromination of substituted phenols by bromine in water show that the main reaction involves bromine and the aryloxide ion. Thus the rate of bromination of 3-nitrophenol in water is reduced by increasing the concentration of hydrogen ion and of bromide ion[157]. A significant but small proportion of the reaction could occur by way of the tribromide ion, despite its negative charge. Rate coefficients reported for a number of compounds are summarized in Table 9[158].

It appears that for the more reactive aryloxide ions, the rate of reaction approaches that expected for a reaction occurring at every collision[159].

TABLE 9. Second-order rate coefficients (k_2, l mol^{-1} s^{-1}) for reactions of phenols, their anions and their O-methyl derivatives with bromine and with the tribromide ion

	k_2, l mol^{-1} s^{-1}, for reaction		
Aromatic compound	ArOH or ArOMe + Br_2	$ArO^- + Br_2$	$ArO^- + Br_3^-$
Phenol	1.8×10^5		
4-Bromophenol	3.2×10^3 [a]	7.8×10^9	
2,4-Dibromophenol	5.5×10^2	1.5×10^9	5.0×10^7
3-Nitrophenol	1.0×10^2	1.3×10^9	2.8×10^7
2,6-Dinitrophenol		5.4×10^6	1.0×10^5
2,4-Dinitrophenol		1.0×10^6	1.3×10^4
Anisole	4×10^4	—	—
4-Bromoanisole	5.0	—	—

[a] The value 3.2×10^9 given in the original paper[158] and repeated in some reviews[36b,159] is a misprint; we show the correct value. The authors are indebted to Professor R. P. Bell for correspondence clarifying this point.

Christén and Zollinger[87] extended these findings by examining the bromination of the 2-naphthol-6,8-disulphonate ion in aqueous solution (equation 31). For this compound, the rate was found to be independent of the acidity of the solution, and catalysis by acetate ion and by pyridine was observed. With hypobromous acid as the reagent, a primary isotope effect, ($k_{1-H}/k_{1-D} = 2.08$) was found. When molecular bromine was the reagent, the isotope effect varied with the concentration of bromine, and was increased by increase in concentration of bromide ion. It was concluded that the formation of the intermediate bromodienone was followed by rate-determining removal of the aromatic proton. Changes in the isotope effect reflected changes in the relative rate of loss of a proton and of bromide ion from the intermediate.

Reaction in dilute aqueous solution would not normally provide conditions of choice for controlled preparative bromination of phenols. There have been a number of studies of the rather rapid bromination of alkylphenols in solvent acetic acid. Under these conditions, even in very dilute solution, the rate is not reduced by the developing hydrogen ion or by added sulphuric acid[10,85], so the neutral molecule is the effective substrate for molecular bromine. Relative rates of bromination of a number of substituted phenols and anisoles are summarized in Table 10. From these results it was deduced that, whereas steric inhibition of resonance through interaction between the OMe group and an *ortho*-methyl substituent has an important influence in reducing the rate of bromination, the interaction between the OMe group and an *ortho*-hydrogen atom can have only a small effect, and does not provide a satisfactory explanation of the lesser reactivity of anisole than of phenol with molecular bromine. Significant contribution to the reactivity of the phenol molecule with electrophiles was consequently attributed to H—O hyperconjugation. That this has the expected limiting result of proton loss from oxygen seems to be confirmed by the finding of a significant isotope effect ($k_{PhOH,HOAc}/k_{PhOD,DOAc} = 1.8$–$1.9$) on the rate of bromination of phenol[160].

These results were carried out under conditions in which the kinetic form would be expected to be first order in molecular bromine. For most aromatic compounds, increase in the concentration of bromine results in the incursion of a kinetic term of higher order in bromine[161]. This term probably results from the assistance which can be given by an extra molecule of bromine to the breaking of the Br–Br bond. Yeddanapalli and Gnanapragasam[162] have reported that this term does not contribute

TABLE 10. Relative rates of bromination of substituted phenols and anisoles in acetic acid at 25°C

Substituents (as drawn)	Relative rate
OH	106
OH, Me	343
OH, Me, Me	471
OH, CMe_3	412
OMe, Me_3C, CMe_3	88
OMe	1.0
OMe, Me	6.5
OMe, Me, Me	0.126

significantly to the bromination of phenol in slightly aqueous acetic acid, though it is significant for anisole. This result is consistent with the view that the rate-determining step in the bromination of phenol is different from that of anisole in that proton loss has become significant. The intermediate $[HOC_6H_5 \cdot Br_2]$ can therefore lose a proton from oxygen rapidly rather than require the assistance of a further molecule of bromine to break the Br–Br bond. Significant differences between phenol and anisole were noted also in the iodine-catalysed brominations[162], which were studied both in acetic acid and in carbon tetrachloride as solvents. The kinetic form for reaction in acetic acid was interpreted in terms of equation (62), in which IBr is the catalyst for breaking of the Br—Br bond.

$$-d[Br_2]/dt = k_3[PhOR][Br_2][IBr] \quad (62)$$

In carbon tetrachloride, the same kinetic form prevailed for phenol, but for anisole the kinetic form was that of equation (63).

$$-d[Br_2]/dt = k_3[PhOR][Br_2][IBr]^2 \quad (63)$$

A negative activation energy was noted for the bromination of phenol in carbon tetrachloride, interpretable as resulting from the dissociation of one or more of the intermediate complexes concerned on the reaction path.

Related results have been reported by Karpinskii and Lyashenko[163] through comparison of the bromination of phenols and anisoles by 'dioxan dibromide' in benzene. Here the kinetic form was that of equation (64), and the temperature coefficient of reactivity (i.e. the apparent activation energy) was positive for anisoles, but negative for phenols.

$$-d[Br_2]/dt = [ArOH][\text{dioxan dibromide}]^2 \quad (64)$$

The significance of O—H bond-breaking in the reactions of phenols under these highly aprotic conditions is, however, not known, though at 25°C phenol was ten times more reactive than anisole.

D. Kinetic Forms and Structural Effects in Iodination

Most of the investigations of the iodination of phenols have been concerned with reactions in aqueous solution, when the reacting substrate is normally the aryloxide ion rather than the neutral molecule. The literature has been surveyed in detail by Taylor[36b], and some of the kinetic aspects have been discussed already (Section IV.E). Berliner[164] established the reactivity sequence PhO^- $(9.2 \times 10^9) > PhNH_2$ $(3.7 \times 10^5) > PhOH$ (1), and thus illustrated that iodination under these conditions has a large negative Hammett ρ value, in accordance with its formulation as a process in which electron-release to the reacting centre makes an important contribution to the rate. More recent studies[89,165,166] have been concerned with further comparisons of the kinetic behaviour of aryloxide ions with those of neutral, highly activated, aromatic substrates. It has been established through the study of primary isotope effects that the stage of proton loss from an intermediate can become rate determining. Thus Grovenstein and coworkers[89,165] showed by using the 2,4,6-trideuteriophenoxide ion that iodination can be subject to a kinetic isotope effect, $k_H/k_D = 4$; and by using the 4-nitrophenoxide ion, that the corresponding isotope effect decreased with the concentration of iodide ion. When proton loss is part of the rate-determining stage, it is usually not kinetically important whether the source of electrophilic iodine is H_2OI^+ or I^+, I_2, or some other iodinating species (ICl, IOAc, etc.); distinction does, however, become possible in principle when electrophilic attack becomes exclusively rate-determining.

E. Synthetic Aspects of the Formation of Halophenols by Halogenation of Phenols and Their Derivatives

1. *Introduction*

The mechanistic considerations described in the foregoing sections provide a guide to the methods which may be adopted towards the preparation of individual halophenols. They also indicate some of the types of condition that should be avoided if particular halophenols are required.

Modern methods of spectroscopic examination, including particularly ^{1}H and ^{13}C NMR spectroscopy, make the identification and establishment of purity of phenols much easier than it was when classical methods only were available. In particular, in the writers' experience, full 'single-resonance' ^{13}C NMR spectra, with appropriate single-frequency decoupling where necessary, has proved to be very valuable.

Comment is desirable on the separation of isomeric halophenols. This is usually possible by using column chromatography, but the separations are often very imperfect.

The writers' personal experience covers only a limited range, and therein has been concerned more extensively with bromo- and to a lesser extent with chlorophenols. The points which are discussed below are in part derived from our own and cognate observations.

2. *Successive introduction of bromine and chlorine into the 'reactive' positions in substituted phenols*

Since halogens are deactivating for electrophilic substitution, there is no great difficult in introducing halogen successively into any vacant positions *ortho* and *para* to the hydroxyl group. One proviso is necessary; if the acidity through the course of the reaction is such that the starting material is not dissociated to allow reaction through the phenoxide ion, but the product of the reaction is sufficiently dissociated to compete with the starting material for further halogen, polyhalogenation may occur even when a deficiency of halogen is used. This is evident from the early work of Francis and coworkers[167], who recorded the rates of the successive stages of bromination of phenol and a number of its derivatives in water by using competition methods. Selected results are given in Table 11. Under the standard conditions used by the investigators, it would seem that all the recorded stages of bromination of phenol and 4-methylphenol involve the neutral molecule; but that the second stage of bromination of 4-nitrophenol and the second and third stages of bromination of 3-nitrophenol are composite, involving in part the anion.

Many solvents, including carbon tetrachloride, chloroform, carbon disulphide and acetic acid, have been used successfully for preparative halogenations, which for best results should be carried out with careful control of the relative amounts of phenol and halogen. To minimize side-reactions of dienone intermediates, it is generally better to add a solution of the halogen slowly to a solution of the phenol. We have found that

TABLE 11. Relative rates (k_1, k_2, k_3) of successive stages of bromination of phenols in water at 25°C

Compound	k_1	k_2	k_3
Phenol	34	0.12	0.0083
4-Methylphenol	28	0.036	—
4-Nitrophenol	0.11	0.047	—
3-Nitrophenol	0.006	0.006	0.02

TABLE 12. Product proportions for chlorination of 3,4-dimethylphenol in acetic acid at 25°C

Initial concentrations				Proportions of phenolic products[b]			
[ArH], M	$[Cl_2]$, M	Reaction time prior to work-up (t), min	Total enone and dienone adducts[a]	Starting material	2-Cl	6-Cl	2,6-Di-Cl
0.1	0.2	5	0.66	0.01	0.02	0.43	0.54
0.1	0.2	5000	0.60	—	0.05	0.51	0.44
0.15	0.15	2	0.28	0.38	0.17	0.43	0.02
0.5	0.5	2	0.33	0.38	0.15	0.35	0.12
0.5	0.5	10 000	0.10	0.29	0.11	0.46	0.14
0.25	0.11	4000	0.14	0.44	0.17	0.38	0.01
0.25	0.11	10 000	0.09	0.44	0.17	0.38	0.01

[a]Determined by integration of appropriate signals in the ^{1}H NMR spectrum of the crude product.
[b]Determined by gas–liquid chromatography of the total product.

the product of monobromination of 4-methylphenol is isolated with particular ease when reaction is carried out in carbon tetrachloride as solvent. When several reactive positions are available, and a deficiency of halogen is used, mixtures of isomeric products are obtained, and the usefulness of the method in providing pure specimens of individual isomers depends on the ease of their separation from each other. In bromination, *para* normally predominates over *ortho* substitution sufficiently that the product of the former reaction can usually be obtained quite easily. Steam-distillation can sometimes be used[168], especially to separate mono- from di- and polychlorinated phenols.

In halogenation of phenols with bulky substituents in the 2- and 6-positions, dienones may be obtained as intermediates which do not rearrange to the phenol instantaneously. If this is suspected, it is as well to give the kinetic product time to react further, since an attempt to force the reaction to its natural conclusion by working up the reaction mixture may result in partial reversal of the original halogenation. Care is needed also in considering the results of gas–liquid chromatography of reaction mixtures, since intermediate dienones and any adducts which have been formed from them will decompose in complex ways and give an unsatisfactory account of the actual composition of the product.

To illustrate some of the problems, some results of chlorination of 3,4-dimethylphenol in acetic acid[72] are recorded in Table 12. The experiments show that, under the conditions used for these chlorinations, adducts were formed and were only partly stable after completion of the reaction, which was very rapid. These adducts decomposed when the crude reaction mixture was submitted to analysis by gas–liquid chromatography, and reverted in part to the starting material. Since it is unlikely that the initial course of the reaction is much changed by the change in initial concentration, the variation in the product ratios appears to reflect complex changes in the behaviour of intermediates. Less marked changes have been recorded for bromination, but this may reflect greater instability of intermediates and adducts rather than less complexity of the pathways taken.

3. *Halogenation carried out by reaction of halogen with the phenol without solvent or catalyst*

Positions *meta* to the hydroxyl group in phenol and even more so in halophenols should be heavily deactivated for further substitution; and so the facile reactions which lead to *ortho–para* substitution and to the formation of dienones easily take their expected precedence. Yet there are many known examples of the preparation of heavily brominated phenols by treatment of the phenol with the theoretical amount, or with excess of bromine without solvent. These reactions proceed at relatively low temperature. Where two isomers are possible, regioselectivity is found, as in the formation of different amounts of the two possibilities when 2,4-dimethylphenol is brominated (sequence 65)[120].

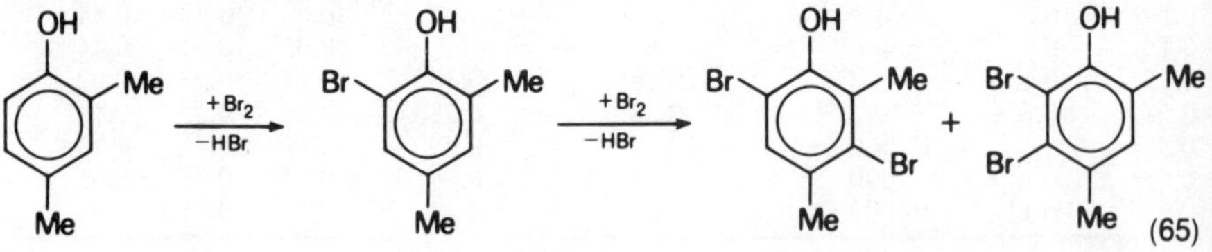

(65)

Tiessens[25] has used a similar method for the preparation of 2,3,4,6-tetrachlorophenol from 2,4,6-trichlorophenol (equation 66). The mechanisms of these halogenations are not known, but it can be speculated that a dienone is formed and rearranges under the influence of the hydrogen halide formed. Addition–elimination sequences are also possible.

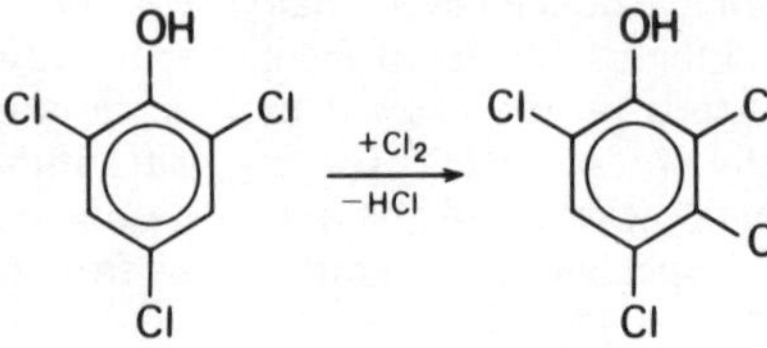

(66)

4. *Protodehalogenation*

The fact that bromine can be introduced into positions *meta* to a hydroxyl group in the above way is of additional synthetic importance because bromine or iodine in *ortho* and *para* positions can then be removed. Two reagents have been recommended for this purpose: benzene with aluminium chloride, and aqueous hydrogen iodide. In the writers' experience, better yields can usually be obtained by the latter method than have been reported for the former. Sequence (67) gives an example[54]. The principle

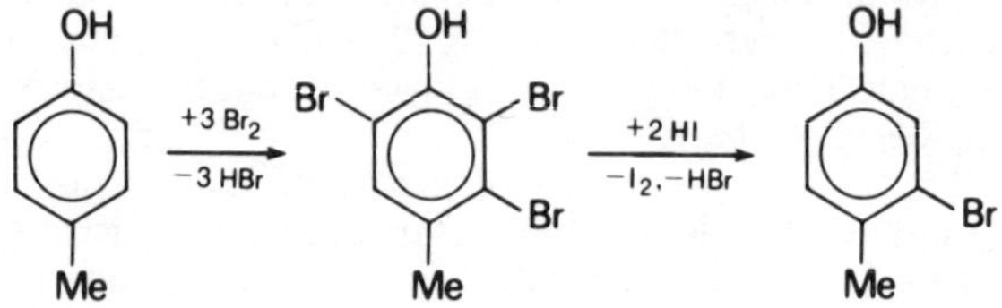

(67)

of the method could be applied in many ways that have not so far been tried; for example, by introducing iodine as a blocking group which would easily be removed, or chlorine as a group which would not.

5. *Bromination and chlorination by rearrangement of intermediate dienones*

Synthetic sequences similar to those discussed in the last two subsections can be developed by the isolation of intermediate dienones which then are rearranged by treating them with acid. A number of examples have already been referred to, and it has been noted that the regioselectivity of this type of bromination sometimes, though not always, is different from that which prevails when bromine is introduced into the *meta* position by reaction of the phenol with excess of bromine (sequence 68)[54]. Use of this method to introduce halogen into a position *meta* to a hydroxyl group has been recorded for chlorine also[108].

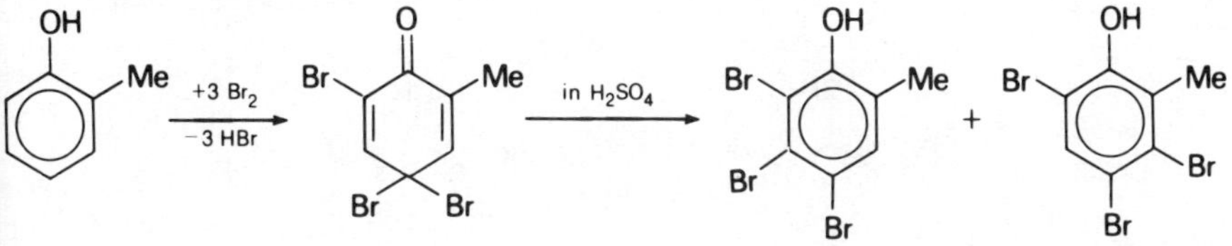

(68)

Some phenols when treated with bromine in superacid solvents give the product of *meta* substitution directly[44,169]. One interpretation of this result is that it involves reaction of the product of O-protonation (e.g. $4\text{-MeC}_6H_4OH_2^+$, formed from 4-methylphenol). Alternatively, reaction through a protonated dienone may be the pathway adopted[44]. The rather specific methods of acid-catalysed rearrangements of bromophenols (Section III.D) are related, and also have synthetic possibilities.

6. *Selective* ortho-*halogenation*

Pearson and coworkers[170] have developed a method for nearly specific halogenation *ortho* to a hydroxyl group. This involves the use of bromine, *t*-butylamine, and toluene at low temperatures. Their procedure has been applied successfully by other workers[54,171], modification being possible in the choice of amine. Examples are given in sequences (69) and (70). It seems likely that the function of the base is to ensure

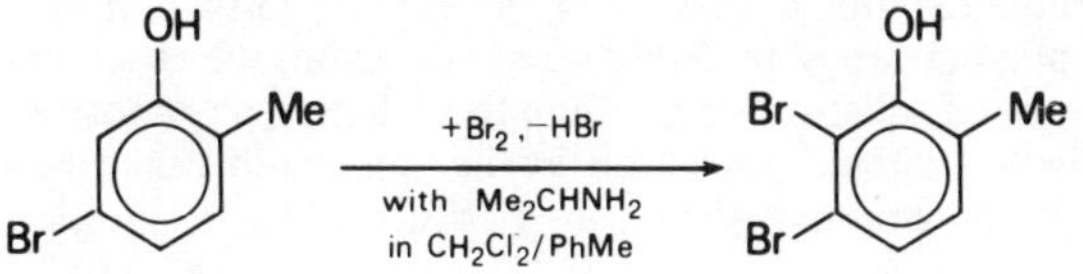

(69)

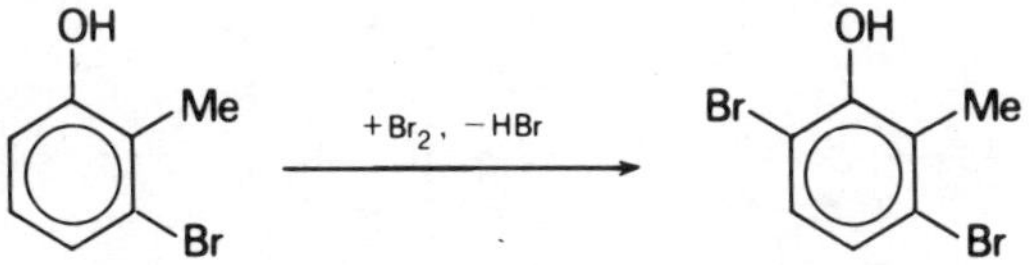

(70)

the formation of aryloxide ion, which then gives an *ortho*-dienone as a kinetic intermediate formed at a temperature low enough that it aromatizes more rapidly than it rearranges otherwise. The O^- substituent would favour attachment of an electrophile at the adjacent position. The involvement of an unstable aryl hypohalite has been suggested but has not been established[170]. The principle of the method has been applied also in chlorination[170] and in iodination[172].

7. *Side-chain halogenation*

In Section V.E we discussed the 'quinobenzylic' rearrangement, which provides a route for the conversion of 4-methylphenols via 4-bromo-4-methyl-2,5-dienones to

4-bromomethylphenols, and for the analogous reactions of phenols appropriately branched or substituted in a 4-alkyl side-chain. The corresponding reactions of chlorodienones are probably available. Conversion of a 4-methyl group to a 4-aldehyde substituent in good yield is also possible by using this type of pathway.

There are many reports of the bromination of alkylphenols formed by heating the phenol with excess of bromine, sometimes in a sealed tube. Examples are given in equations (71) and (72)[173,174].

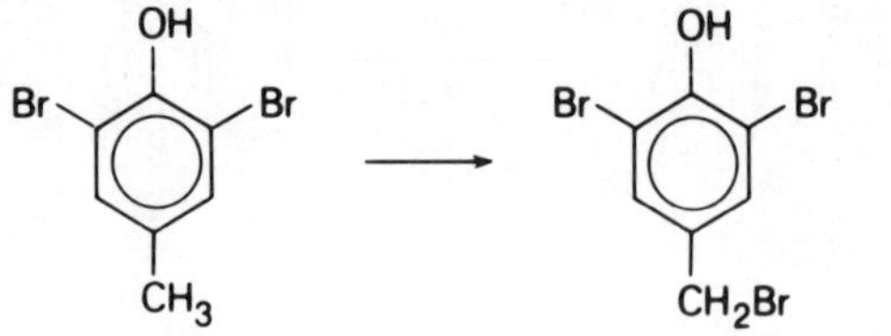

(71)

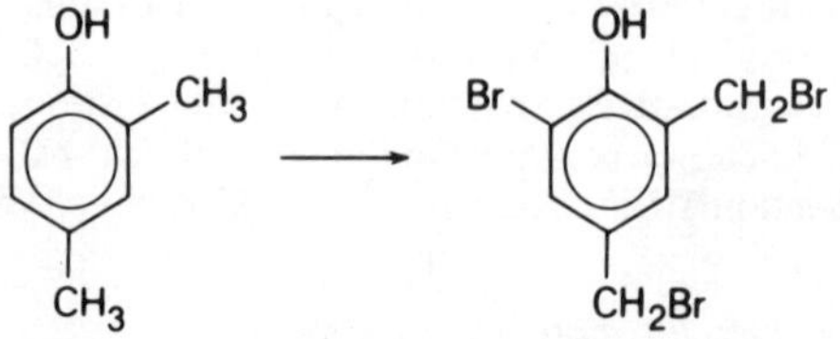

(72)

Reactions of this kind have generally been treated as mechanistically analogous to the 'quinobenzylic' brominations, but our own studies suggest that this is probably not the case. Side-chain bromination of alkylphenols will occur under much milder conditions than has been supposed generally; ordinary laboratory daylight at room temperature allows the reaction to proceed slowly in solvents like carbon tetrachloride. Very rapid reaction occurs when the reaction mixture is illuminated strongly. When only one alkyl group is available for substitution, the reaction can be carried to the required degree of substitution by using the calculated proportion of bromine. For modestly deactivated phenols, reaction in the side-chain can compete with nuclear substitution, as in the example shown in sequence (73). When more than one alkyl

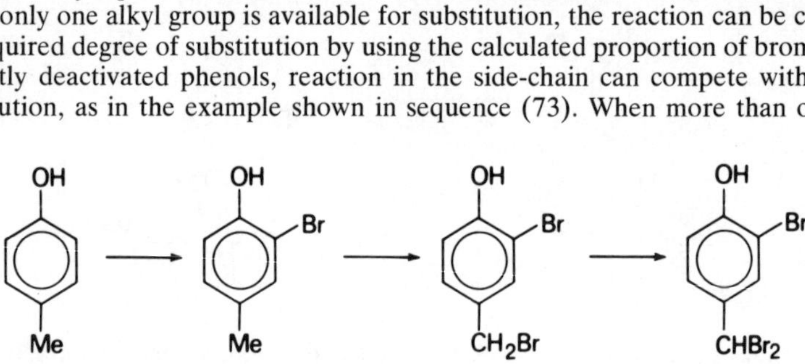

(73)

group is available, as with 2,4- or 3,4-dimethylphenol, reaction under these conditions is non-specific; substitution in both alkyl groups, and disubstitution, is evident from the ^{1}H NMR spectra of the products of attempted monosubstitution. Where bromination by this type of method has been preparatively successful, therefore, this has been the result of the easy isolation of a particular product because of its specific properties. Thus it has been confirmed[120] that 6-bromo-2,4-di(bromomethyl)phenol can be obtained from the complex reaction mixture resulting from bromination of 2,4-dimethylphenol in carbon tetrachloride under powerful illumination. Crystallization from *n*-hexane gave the required product.

8. *Miscellaneous reagents and conditions for chlorination*

In many investigations, molecular chlorine has been the reagent chosen for chlorination of phenols, and classical modifications of the conditions have been used when the position to be chlorinated has been unreactive. Thus iodine[175] has been used as a catalyst for the chlorination of 2,4,6-trichlorophenol at 40°C, and aluminium chloride[176] has been similarly recommended for the preparation of pentachlorophenol from phenol. Other special conditions already referred to include the use of antimony pentachloride as a chlorinating agent for the preparation of dienones[137]. The influence of solvent on the ratio of *ortho* to *para* chlorination has been discussed with particular reference to the influence of hydrogen-bonding[177,178]. For phenol, the *ortho*:*para* ratio is highest for reaction in non-polar aprotic solvents at high dilution, but becomes relatively low in dipolar aprotic and in hydroxylic solvents. A similar effect has been noted in bromination[179]. It seems likely, too, that hydrogen-bonding is concerned in determining the *ortho*:*para* ratio in chlorination in the solid phase, when it varies with the crystal face presented to the gas[180,181]. Polychlorination accompanying monochlorination is also a feature of chlorination of phenols in the solid phase[182].

Sulphuryl chloride can be a homolytic chlorinating agent, but it can also react electrophilically, especially with reactive aromatic compounds. It has been used for the chlorination of phenols[177], when it tends to give relatively low *ortho*:*para* ratios. A particularly good yield of 4-chloro-2-methylphenol was obtained from 2-methylphenol and sulphuryl chloride with aluminium chloride and diphenyl sulphide as cocatalysts[183].

Kosower and coworkers[184] included phenol in their study of halogenation using copper(II) halides as source of halogen. Rather low *ortho*:*para* ratios were obtained both with cupric chloride and with cupric bromide.

9. *Classical ('indirect') procedures for obtaining halophenols*

This review has concerned itself chiefly with methods of halogenation in which the hydroxyl function is already present in the molecule, and halogen is introduced into a vacant position by using a source of electrophilic halogen. Variants on this procedure include protection of the alkoxy group by the formation of ethers and esters. By this means the reactivity with electrophiles at positions *ortho* or *para* to the oxygen function is reduced, and *ortho* substitution is usually somewhat disfavoured with respect to *para* substitution. Procedures based on this principle are useful in some special applications (see also Section IV.F), but limitation of space prevents our considering them in detail.

The other type of use of protecting groups, in which an otherwise reactive aromatic position is first blocked by a substituent which can later be removed, has on the face of it been used quite extensively in the synthesis of particular phenols. Thus a procedure for preparing *o*-chlorophenol free from its isomers involves sulphonation, chlorination and protodesulphonation (sequence 74)[185].

It should be noted, however, that some early descriptions of syntheses carried out by using procedures of this kind are incorrect[53]. Much remains to be done towards elucidating pathways in brominations and chlorinations of phenols in sulphuric acid of different strengths, where halodeprotonation, sulphodeprotonation and halodesulphonation may be in competition.

Use of the different orientational patterns promoted by interconvertible groups, one or both of which can be removed or replaced when necessary (e.g. NO_2, NH_2), provides practicable routes for the laboratory syntheses of many aromatic compounds. Consultation of Beilstein's Handbook, or of Huntress's compendium of organic

OH → ($2 H_2SO_4$, $-2 H_2O$) → OH, SO_3H, SO_3H → ($+Cl_2$, $-HCl$) → OH, Cl, SO_3H, SO_3H → ($+2 H_2O$, $-2 H_2SO_4$, (steam distillation)) → OH, Cl (74)

chlorine compounds[186] provides a plethora of examples in the field of halophenols, one of which is set out in sequence (75)[25]. In another method, a particular halogen in a polyhalogeno compound is removed by nucleophilic displacement. One example, structurally favourable and industrially important, involves first the preparation of 1,2,4,5-tetrachlorobenzene and then its treatment with base (sequence 76).

Cl, Cl → Cl, Cl, NO_2 → Cl, Cl, NH_2 → Cl, Cl, NHAc →

NO_2, Cl, Cl, NHAc → NO_2, Cl, Cl, Cl, NHAc → NO_2, Cl, Cl, Cl, NH_2 → NO_2, Cl, Cl, Cl, Cl →

NH_2, Cl, Cl, Cl, Cl → OH, Cl, Cl, Cl, Cl (75)

→ (Cl_2, Fe catalyst) → Cl, Cl, Cl, Cl → (KOH) → OH, Cl, Cl, Cl (76)

VIII. INDUSTRIAL APPLICATIONS OF HALOPHENOLS AND HALODIENONES

Phenols, halophenols and their derivatives have found application in many diverse areas[187] but their main exploitation has been as aids to agriculture, and in the fields of polymers and drugs.

A. Applications to Agriculture

The utility of phenols in agriculture has been extensively studied. In the early 1900s phenols and alkylphenols (primarily cresols) were first employed as agents for soil sterilization. The discovery of the increase in herbicidal and fungicidal activity, when the phenols are substituted with nitro and (or) chloro groups, was somewhat later[188]. The halogen may also be incorporated in the form of a trifluoromethyl group[189]. Examples of a number of the simple chlorophenols, which commonly have been used as pesticides, are illustrated (Table 13).

The annual world production of chlorinated phenols alone has been estimated at 150 000 tons[190]. The method most widely used for their preparation is the direct chlorination of the appropriate chlorobenzene[191], followed by alkaline hydrolysis of the product. The selection of the method, however, depends upon the isomer required; a number of technically important compounds, including 2,4-dichloro-, 2,4,6-trichloro-, 2,3,4,6-tetrachloro- and pentachlorophenol, may also be manufactured by the direct chlorination of phenol.

Pentachlorophenol (PCP) continues to receive widespread application in diverse fields as a wood preservative[192], agricultural fungicide, defoliant and general herbicide. Until 1971 it constituted more than half the amount of herbicides used in Japan. Restrictions on its use were imposed at that time because its application to paddy fields led to adverse effects in other sectors. It has been used also as an aquatic weed-killer, and thus finds its way into the aquatic environment, especially in run-off waters and wood-treatment plant effluents. Studies on the photolysis of PCP by sunlight in aqueous solution were consequently undertaken to investigate the decomposition products formed. Sequences to explain the products formed have been presented (Scheme 7)[193]. The scheme provides a general illustration of pathways available for decomposition of chlorophenols.

Aromatic chlorine atoms are replaced by hydroxyl groups, and this process is followed by the well known oxidation by air to yield quinones, in a manner which is analogous to the formation of humic acids from 2,4-dichlorophenoxyacetic acid.

The levels of pentachlorophenol and other organochlorine insecticides and their decomposition products have been studied in relation to a number of waterways[194], and the toxicity to marine life has been quantified. Lethal threshold amounts of pentachlorophenol and of 2,4- and 2,5-dichlorophenols in shrimps and clams have recently been reported[195].

Not only the chlorophenols themselves but also the contaminants of the commercially available products give rise to environmental concern. Besides other chlorophenols, they may contain several percent of polychlorinated phenoxylphenols (predoxins). Polychlorinated dibenzo-*p*-dioxins (PCDDS), dibenzofurans (PCDFs) and diphenyl ethers often are present in the range of tens to hundreds of parts per million[190]. Some of the isomers of these compounds have biological properties which make them very hazardous[196]. Their 'concentration', and consequently their dangers as pollutants, may also be increased by the burning of vegetation which has been treated with the chlorophenols or their derivatives[197].

Various phenolic compounds occurring in soil, either as the direct pesticide or a product therefrom, have been found to be incorporated into humus complexes by microbial activity. In a recent study[198] experiments were carried out in which a phenol oxidase isolated from a fungus was incubated with humus constituents and in the presence of 2,4-dichlorophenol. Oligomeric products, dimers to pentamers, were formed by oxidative coupling. The combined incubation of naturally occurring phenols and 2,4-dichlorophenol, a main product of various herbicides, resulted in the formation of cross-coupling products. It was proposed that enzymic processes lead to the

TABLE 13. Structures of some phenols and their derivatives commonly employed in agricultural chemistry

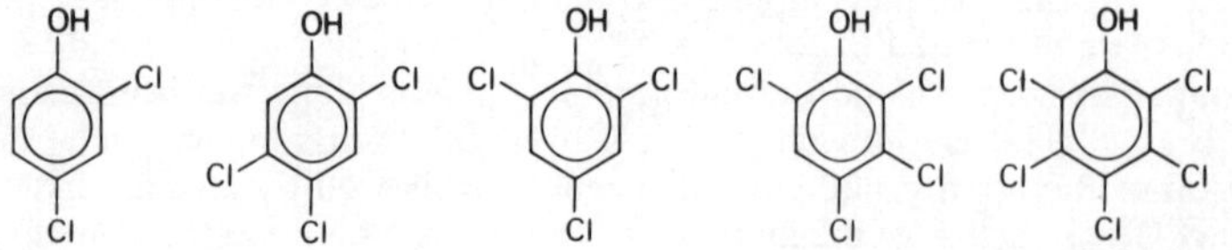

PCP

and the corresponding phenoxyacetic acids e.g.

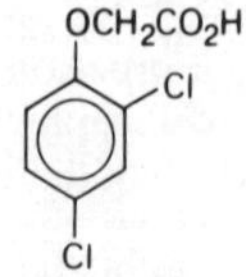

OCH_2CO_2H

2,4–D

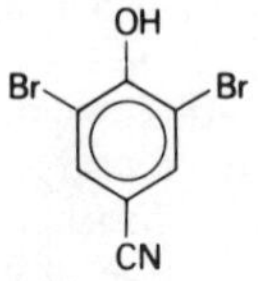

Bromoxynil

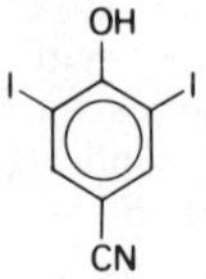

Ioxynil

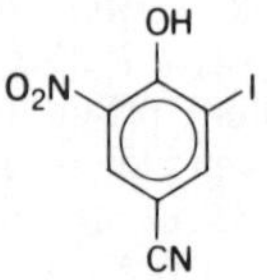

Nitroxynil

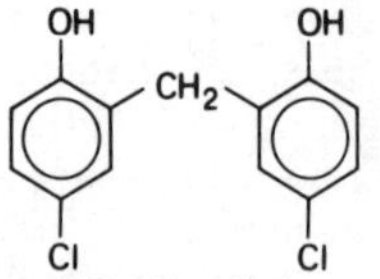

Dichlorophen

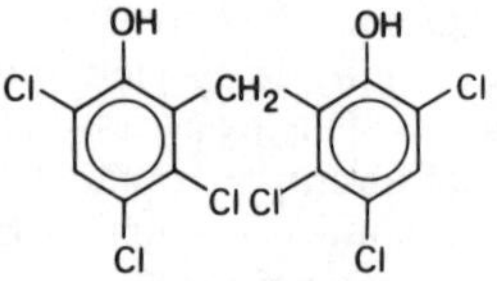

Hexachlorophene

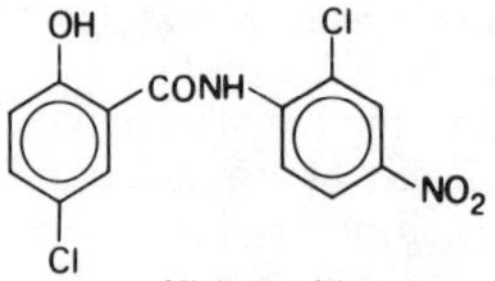

Niclosamide

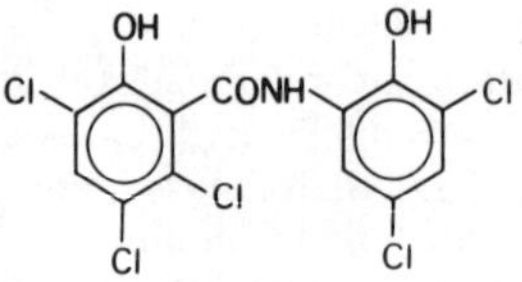

Oxyclozanide or Zanil

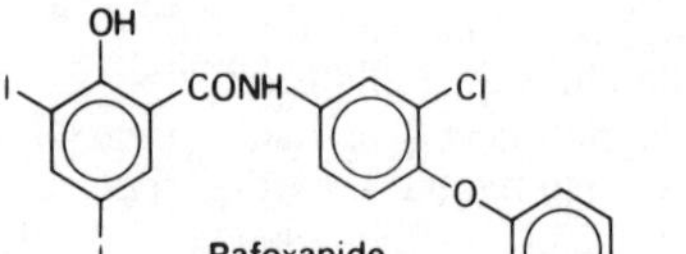

Rafoxanide

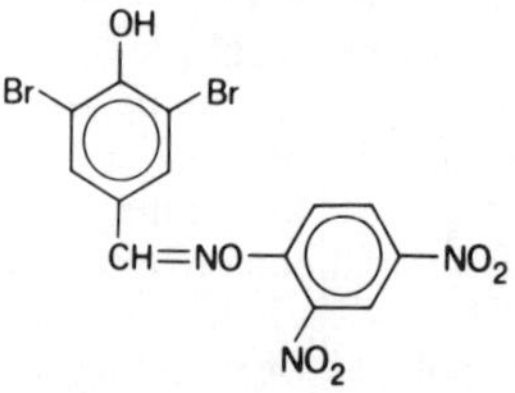

Bromofenoxim

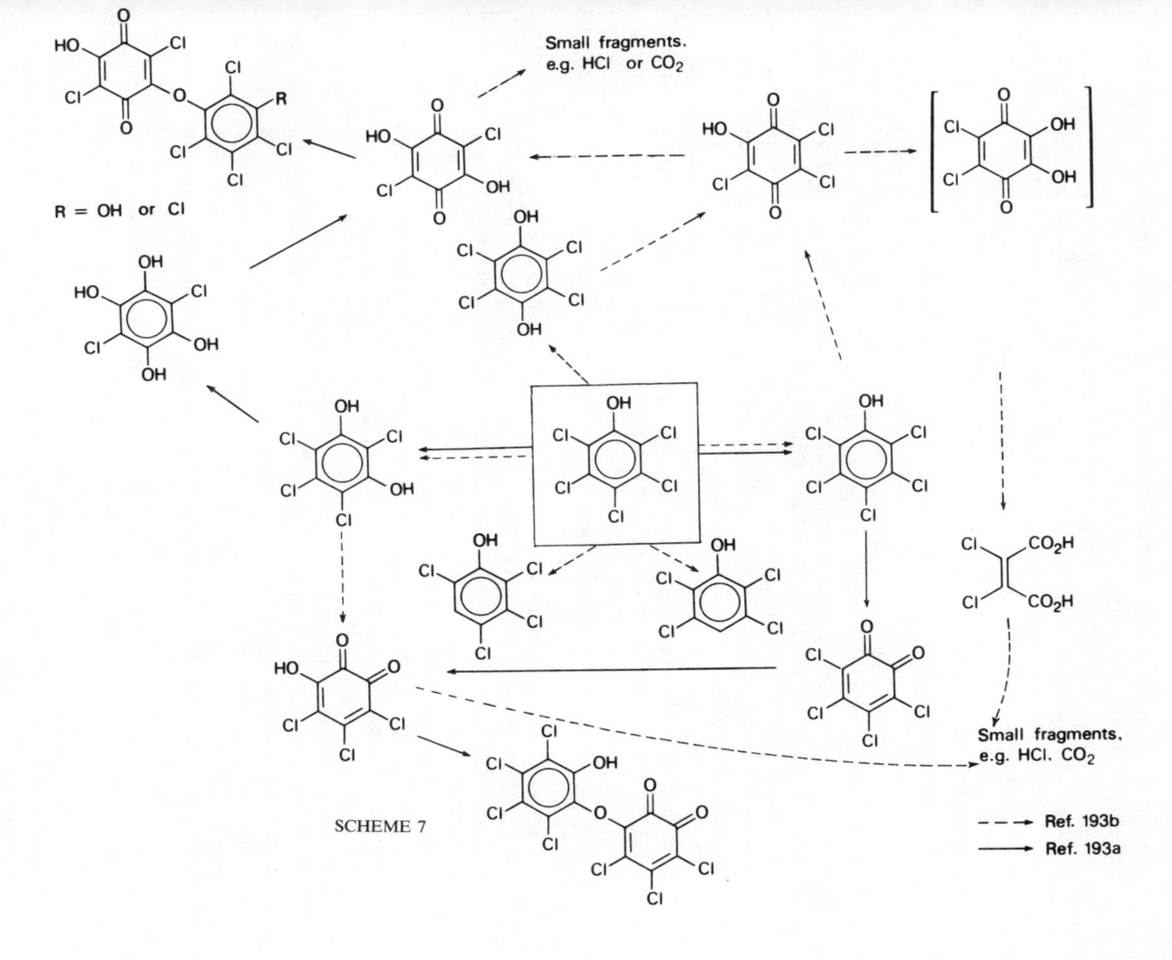

Small fragments. e.g. HCl or CO_2
R = OH or Cl
Small fragments. e.g. HCl. CO_2
Ref. 193b
Ref. 193a

SCHEME 7

incorporation of these compounds into soil organic matter and thus prolong their persistence in soil.

The mechanism of action of most of the currently used pesticides incorporating phenolic and halogen substituents has been thought to involve interference with reversible oxidative phosphorylation, which is an essential part of the biological degradation of foodstuffs. Several ways in which this interference could occur have been suggested, and these are not mutually exclusive since different functions might operate for different types of compound. Wilson and coworkers[199] have suggested that compounds such as halosalicylanilides may act by general acid or general base catalysis of the hydrolytic destruction of intermediates concerned in phosphorylation. Many of the pesticides in current use, including 2,4-dichlorophenoxyacetic acid and bromoxynil, disrupt photosynthetic oxidative phosphorylation in another way, however, namely by inhibiting the initial photosynthetic electron transfer[200]. The pesticides were themselves rapidly photo-oxidized, and the results obtained provided additional support for the hypothesis that depletion of the source of reducing potential is responsible for chloroplast photo-oxidation and plant death, following treatment with photosynthesis-inhibiting herbicides.

Yet another way in which metabolic processes could be interfered with can be described as a chemiosmotic hypothesis, in which it is considered that lipid-soluble acids will readily carry protons and will uncouple mitochondria by conducting protons across the lipid-containing mitochondrial membrane, therefore destroying the proton differential which is produced by electron transfer, and is required for the formation of ATP[201].

Various of the halophenol derivatives have been used internally to control infections of sheep and cattle by liver flukes. The use of these compounds to eradicate the parasites without damage to the host requires careful attention to the dosage; the therapeutic dosage in sheep may be one-third or more of the maximum tolerated dose[202]. A recent study of the efficacy of rafoxanide against *Fasciola gigantica* adult flukes in the liver of cattle showed reductions of the fluke numbers to less than 1% of the original value at a treatment level of 1.5 mg kg^{-1} [203]. Toxicity and efficacy studies for the use of nitroxynil in dromedaries showed good tolerance and successful treatment of various infestations at a dosage of 20% of that which is lethal[204].

The mode of action of these compounds has been a subject for considerable discussion because the adult flukes live in virtually anaerobic conditions in the bile duct, so that the possibility of normal oxidative phosphorylation is eliminated. It has been found[205], however, that fumarate may act as a terminal electron-acceptor in liver flukes, and phosphorylation occurs with the formation of succinate rather than water. In treatment of flukes with bromoxynil or with a number of other compounds, this process of electron transport is interfered with, and determination of the concentrations required to kill the isolated flukes suggests that interference with phosphorylation in the fluke is the cause of death[206].

2,4-Dichlorophenoxyacetic acid and its analogues have been found to be particularly beneficial for the control of infections of plants by mites and by powdery mildew. These infections occur superficially on the plant, and it is presumed that the aryloxyacetic acids may be more lipophilic than the phenols and hence may concentrate in the external fatty layers of the leaf on which mites and powdery mildew thrive. Once the compound has been taken up by the pest, it is presumed to generate the phenol. Some of the established pathways of metabolism of 2,4-dichlorophenoxyacetic acid in living organisms are summarized in Scheme 8[207].

The design and development of pesticides has been reviewed[208]. In the area of phenols, interest has focused on a number of their derivatives, including the diphenyl ethers (**76**) which are prepared by the reaction of the appropriate nitrobenzene and the

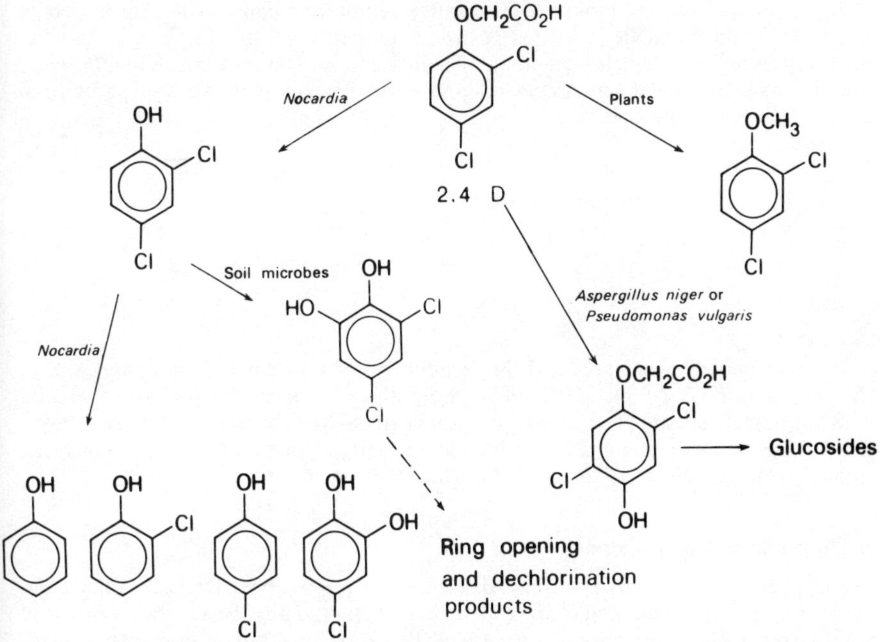

SCHEME 8

halophenol[209]. Halogenated phenol-4-sulphonamides (e.g. **77**) are reported to be useful for the complete control of wild oats in fields of grains, sugar beet or flax with little or no damage to the crop[210]. Heterocycles with halogenated phenolic substitu-

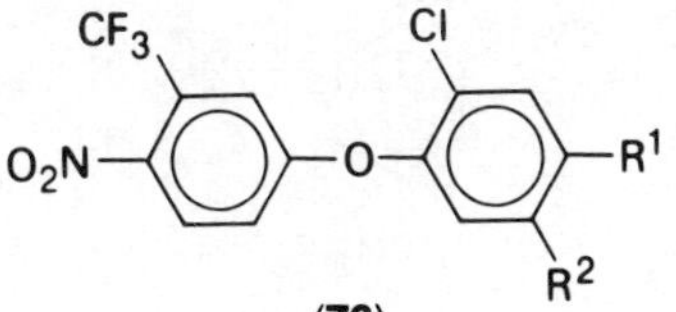

(**76**)

R^1, R^2, for example, = H, OMe respectively

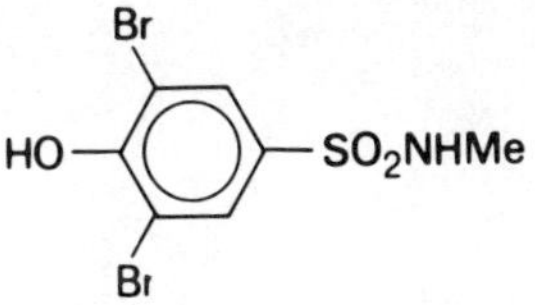

(**77**)

ents are also prominent in the current literature concerning pesticides. These include the isoxazolyl phenols (**78**)[211], all of which show fungicidal properties and improve the germination percentage and seedling vigour of many crops in soil infested with *Pythium* and *Fusarium*. Similar potential is exhibited by a range of chlorophenoxymethyloxadiazoles and triazoles[212].

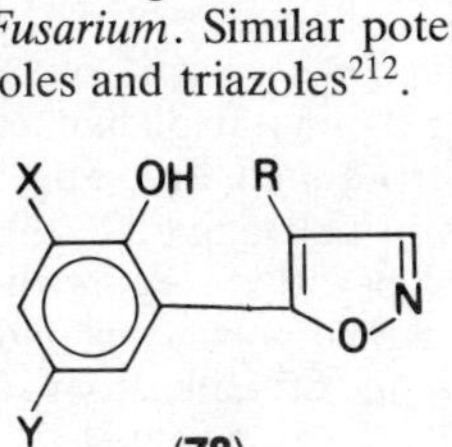

X = H, C_{1-3} alkyl, Cl or Br

Y = H, Cl or Br

R = H or C_{1-3} alkyl

(**78**)

Dienones have been employed more often as intermediates in the formation of insecticides and fungicides, than as pesticides themselves. An exception to this has been reported[213], in which a number of halocyclohexadienones, including **79**, were found to have fungicidal and insecticidal properties. The dienones were prepared from cyclohexenones or alkylphenols.

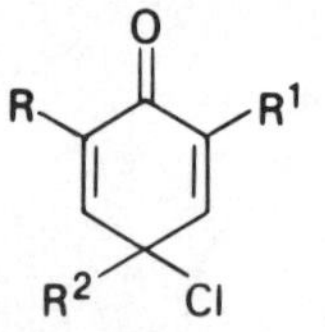

R, R^1 = Br, Cl

R^2 = C_{1-4} alkyl

(**79**)

The fluorination of phenols and the formation of fluorodienones was employed in the preparation of 7-fluoro-7-dechlorogriseofulvin[214], which is a mould metabolite with antifungal activity. Herbicides have also been derived from hexachlorocyclohexa-2,4- or 2,5-dienone, irradiation of which in an inert solvent yields dimers, which are stable herbicides and have the advantage that they are harmless to fish[215].

B. Applications to Polymers

The phenol–formaldehyde resins (Bakelite, and related products) are among the oldest synthetic polymers, and are still very important. Thus the reaction of phenol with formaldehyde, in the presence of alkali or acid, affords a polymer of high molecular weight in which the many phenol rings are linked by $—CH_2—$ groups (**80**). The formation of these materials involves electrophilic polycondensation in which the steps involved in building up and branching the chains are of similar rates[216].

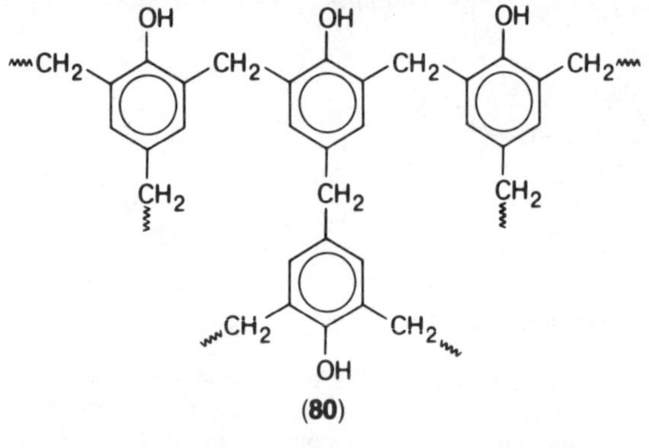

(**80**)

Halophenols have been utilized in the formation of another type of polymer, which does not involve the incorporation of an additional carbon unit, but instead the formation of ether linkages. Oxidation of 4-bromo-2,6-dimethylphenol, converted to the phenolate ion, with an oxidizing agent such as ferricyanide or iodine[217] results in the formation of polymers with a molecular weight in the range 2000–10 000 daltons. A radical mechanism is proposed, involving a propagation step in which the aryloxy radical displaces bromide from the phenolate ion. The polymer is found to be of the linear type (**81**); analysis indicates the presence of one bromine atom and one hydroxy group per chain.

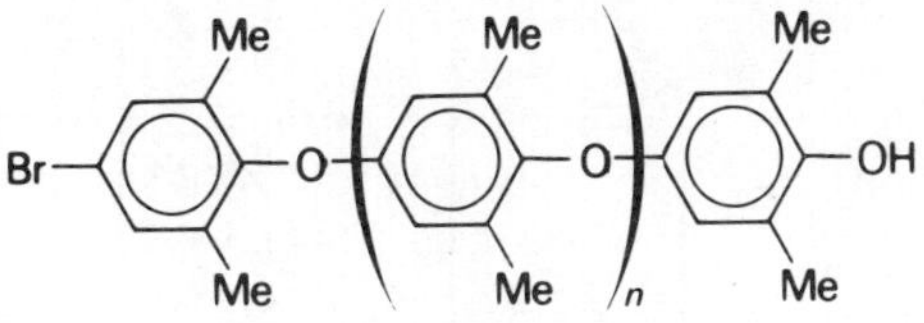

(81)

Formation of similar polymers from other halogenophenols (e.g. from 2,6-dimethyl-4-chlorophenol) has been exploited using 2,4,4,6-tetrabromocyclohexa-2,5-dienone as the catalyst for polymerization. The polymers obtained are reported to be of high molecular weight and exhibit good thermal and mechanical resistance in the presence of oxygen[218].

Some 2-halogenosubstituted phenols (e.g. 2-chloro-6-methylphenol), however, have been polymerized by using oxygen and an amine complex of Cu(I), and yield polymers of higher molecular weight (up to 71 000 daltons) which from studies of their intrinsic viscosities are apparently somewhat branched[219]. Elemental analyses confirm that up to 15% of the chlorine present in the original phenol is lost during oxidation. Thus 2-chloro-6-methylphenol is acting in part as a trifunctional monomer by reaction through the 2-position with elimination of chlorine. This would give rise to a unit of type **82** being incorporated into the structure of the polymer.

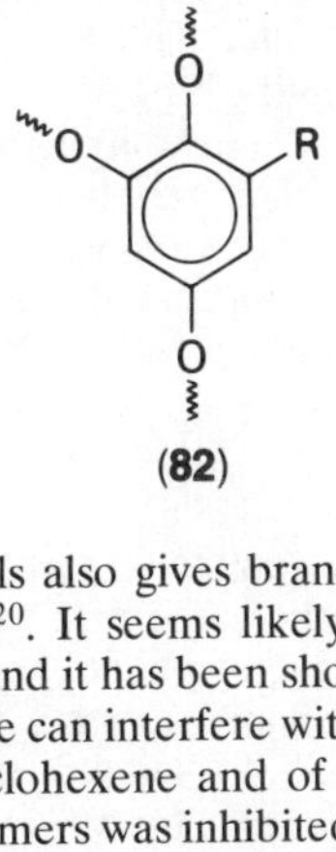

(82)

Oxidation of 2,6-dihalophenols also gives branched polymers, but these polymers are of lower molecular weight[220]. It seems likely that a number of these reactions involve radicals or radical ions, and it has been shown that as a consequence the intermediates produced in their course can interfere with other radical chains. For example, the auto-oxidation of 4-vinylcyclohexene and of cumene under catalysis by salts of cobalt or manganese to give polymers was inhibited by phenols containing one or more chlorine atoms *ortho* or *para* to the hydroxyl group[221]. The chlorophenols probably acted as free-radical chain-breakers, thus giving chlorocyclohexadienones which themselves would interfere with catalysis by metal ions.

A linear poly(phenylene oxide) in which the R groups are bromine (**81c**, $n = 50$) has been reported. This polymer is obtained quantitatively from 2,4,4,6-tetrabromo-2,5-cyclohexadienone and mercury bis(phenylacetylide), and evidence was produced to suggest that free-radicals are not likely to be intermediates. Instead, it was noted that polymerization starts only when reagents capable of providing acceptors for both positive and negative bromine are present. The proposed pathway involves a carbene as the polymerizing entity (Scheme 9)[222].

The halogenated phenol polymers have found many diverse uses in recent years. For example, bromo-substituted polyoxyphenylene polymers (particularly the polymer from 2,4,6-tribromophenol) have been incorporated into nylons to the extent of up to

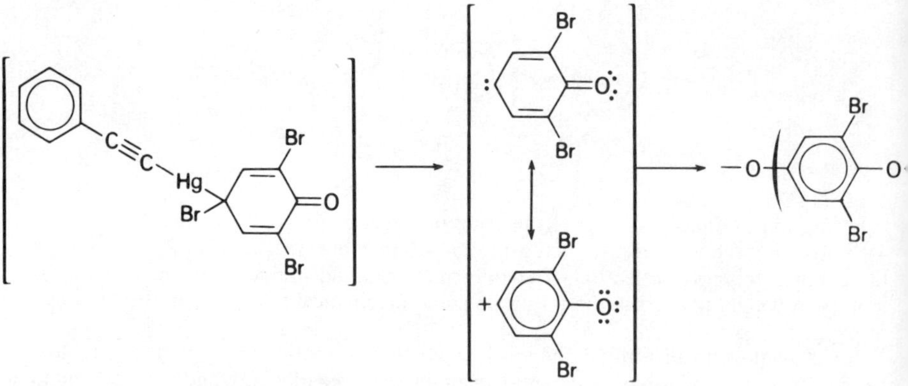

SCHEME 9

20% by weight by dry blending and injection moulding, and the resulting products have flame-retardant properties[223].

The property of flame retardation has also been applied to PVC where the addition of either bromo-poly(alkenylphenol) or -poly(*p*-hydroxystyrene) is found to increase fire resistance[224,225]. Polyester foams, comprising a polyester resin, a halogenated bisphenol polycarbonate, a fibrous filler, and a minor amount of a blowing agent, exhibit similar resistance to flammability[226].

Brominated phenol monopolymer exhibits good properties as an electrical insulator[227], and such polymers, which contain cross-linking potential, may be employed as curing accelerators for fast-drying epoxy-varnishes[228]. The gel time of epoxy-resins which incorporate a halogenated poly(alkenylphenol) and triphenylphosphite is extended by approximately a factor of seven at 140°C[229].

C. Applications to Drugs

Bacteriology and the use of disinfectants has developed rapidly, since Robert Koch in 1876 described a living germ as the cause of a human disease. Even prior to this time, Lister had recommended the use of phenol as a disinfectant.

A great number of phenols and derivatives, mostly halogenated (e.g. **83–88**), are still in current use. The destructive effect of phenol on the human skin was found to be moderated by the introduction of lipophilic (alkyl or chloro) groups into the molecule; a change initiated by Bechhold and Ehrlich in 1906[230]. Because these groups are lipophilic, it is held that these phenols, which are least harmful to the skin, are the most strongly antagonized by serum, because, having higher partition coefficients between oil and water, they enter the albumin core.

Studies to find the best conditions in relation to selective toxicity of the phenols as disinfectants have involved the investigation of equilibria between monomers and micelles in aqueous solution. The use of soaps to solubilize phenols in water, for use as disinfectants, depends on the formation of mixed micelles of the soap and phenol. Variation of the proportion of soap to phenol can lead to three different situations[231].

The first zone exists up to a figure which is identical with the critical micelle concentration for the soap, where the maximum bactericidal effect is obtained. It was concluded that the bactericidal action in this first zone is a combined attack of the phenol (mainly) and the soap on the protoplasmic membrane. As the soap concen-

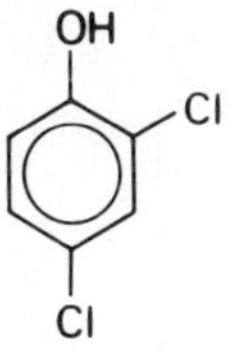

(**83**)

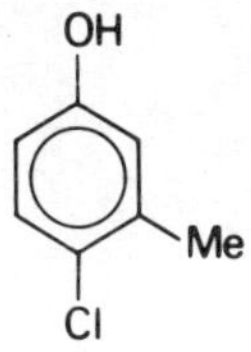

(**84**)

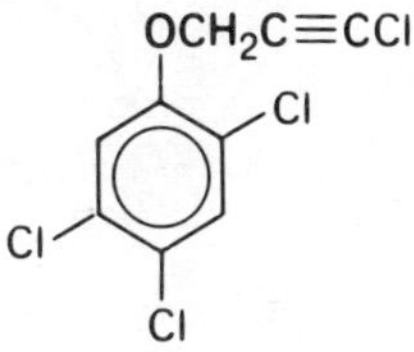

Haloprogin
(**85**)

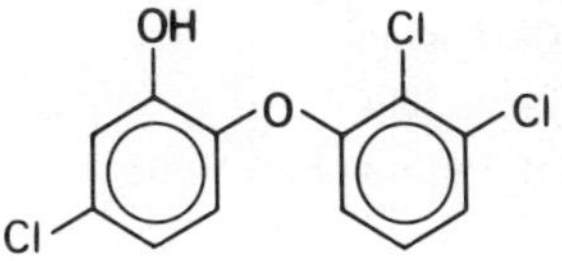

Triclosan
(**86**)

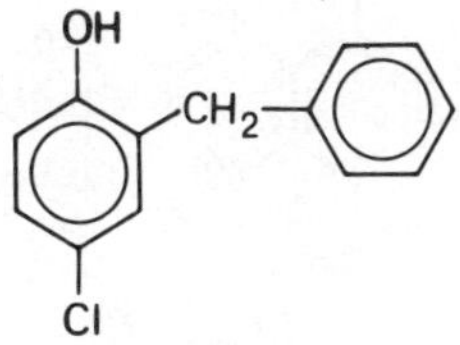

Chlorophene
(**87**)

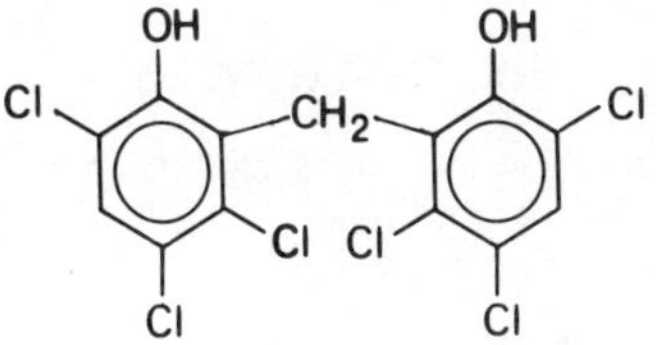

Hexachlorophene
(**88**)

tration increases beyond that of the critical micelle concentration a second zone of greatly diminished bactericidal effect is observed. This has been interpreted in terms of the phenol having entered the micelles, many more of which must have formed, and thus little of it is available for disinfection. Eventually, as the soap concentration is increased still further, a third zone is observed in which vigorous disinfection occurs, due to the toxicity of the soap itself. All phenols commonly used as disinfectants, including *p*-chloro-*m*-xylenol, exhibit these zones.

Haloprogin, which is important in drug therapy, was shown by Harrison and coworkers[232] to have activity against dermatophytes *Candida albicans* and related yeasts, and certain gram-positive bacteria. Its fungicidal activity has been attributed to disruption of the cell membrane, as shown by the loss of intracellular potassium ions.

As mentioned in Section VIII.A many phenols and derivatives can be employed as anthelmintics. Although these have in the past been used in treatment of hookworm infestations in man, they have been superseded by compounds somewhat less toxic to the host.

Compounds containing phenolic and halo substituents find extensive utility in treatment of hypofunction of the thyroid gland[233]. The compounds most commonly used are thyroxine (**89a**) and tri-iodothyronine (**89b**).

Other drugs containing both phenolic and halo substituents are the quinoline

(**89**a) X = I
(**89**b) X = H

(**90a**): X = SO_3H, chinifon
(**90b**): X = Cl, clioquinol
(**90c**): X = I, di-iodohydroxyquine

derivatives (**90a–c**). Chinifon (**90a**), previously called Yatren, was, on discovery in 1892, prescribed as a cure against all illness. However, its efficacy is now recognized along with (**90b**) and (**90c**) for treatment of amoebiasis[234].

IX. REFERENCES

1. D. A. Whiting in *Comprehensive Organic Chemistry*, Vol. 1 (Ed. Sir Derek Barton and W. D. Ollis), Pergamon, Oxford (1979), pp. 707–798; A. R. Forrester and J. L. Wardell in *Rodd's Chemistry of Carbon Compounds*, 2nd edn, Vol. IIIA (Ed. S. Coffey), Elsevier, Amsterdam (1971), pp. 289 ff., 334 ff.
2. D. A. R. Happer and J. Vaughan in *The Chemistry of the Hydroxyl Group*, Part 1 (Ed. S. Patai), Wiley, Chichester (1971), pp. 393–452; H.-D. Becker in *The Chemistry of the Hydroxyl Group*, Part 2 (Ed. S. Patai), Wiley, Chichester (1971), pp. 835–936; S. Forsén and M. Nilsson in *The Chemistry of the Carbonyl Group*, Vol. 2 (Ed. J. Zabicky), Wiley, Chichester (1970), pp. 157–240 (esp. pp. 168 ff).
3. T. Birchall, A. N. Bourns, R. J. Gillespie and P. J. Smith, *Canad. J. Chem.*, **42**, 1433 (1964); G. A. Olah and Y. K. Mo, *J. Org. Chem.*, **38**, 353 (1973).
4. A. J. Waring, *Adv. Alicyclic Chem.*, **1**, 129 (1966).
5. V. V. Ershov, A. A. Volodkin, and G. N. Bogdanov, *Russ. Chem. Rev.*, **32**, 75 (1963); V. V. Ershov and G. A. Nikiforov, *Russ. Chem. Rev.*, **35**, 817 (1966).
6. R. H. Thomson, *Quart. Rev.*, **10**, 27 (1956).
7. A. Albert and E. P. Serjeant, *The Determination of Ionization Constants*, Chapman and Hall, London (1971).
8. T. Kojima, *J. Phys. Chem. Japan*, **15**, 284 (1960).
9. K. U. Ingold, *Canad. J. Chem.*, **38**, 1092 (1960); **40**, 111 (1962).
10. P. B. D. de la Mare, *Tetrahedron*, **5**, 107 (1959).
11. G. Modena and G. Scorrano in *The Chemistry of the Carbon–Halogen Bond* (Ed. S. Patai), Wiley, Chichester (1973), p. 301 ff.
12. O. Exner in *Correlation Analysis in Chemistry* (Ed. N. B. Chapman and J. Shorter), Plenum, New York (1978).
13. C. K. Ingold, *Structure and Mechanism in Organic Chemistry*, 2nd edn, G. Bell, London (1969), p. 1208.
14. G. B. Barlin and D. D. Perrin, *Quart. Rev. Chem. Soc.*, **20**, 75 (1966).
15. A. I. Biggs and R. A. Robinson, *J. Chem. Soc.*, 388 (1961).
16. H. Van Bekkum, P. E. Verkade and B. M. Wepster, *Rec. Trav. Chim.*, **78**, 815 (1959).
17. Y. Yukawa and Y. Tsuno, *Bull. Chem. Soc. Japan*, **32**, 971 (1959).

18. S. Ehrenson, R. T. C. Brownlee and R. W. Taft, *Progr. Phys. Org. Chem.*, **10**, 1 (1973).
19. A. Fischer, G. J. Leary, R. D. Topsom and J. Vaughan, *J. Chem. Soc. B*, 782 (1966); 686, 846 (1967).
20. L. A. Cohen and W. M. Jones, *J. Amer. Chem. Soc.*, **85**, 3397 (1963).
21. L. A. Cohen and W. M. Jones, *J. Amer. Chem. Soc.*, **85**, 3402 (1963).
22. P. D. Bolton, F. M. Hall and I. H. Reece, *J. Chem. Soc. B*, 717 (1966); 1047 (1969).
23. M. T. Tribble and J. G. Traynham, *J. Amer. Chem. Soc.*, **91**, 379 (1969).
24. R. W. Taft in *Steric Effects in Organic Chemistry* (Ed. M. S. Newman), Wiley, New York (1956), p. 556.
25. G. J. Tiessens, *Rec. Trav. Chim.*, **50**, 112 (1931).
26. T. B. McMahon and P. Kebarle, *J. Amer. Chem. Soc.*, **99**, 2222 (1977); J. Catalán and A. Macias, *JCS Perkin II*, 1632 (1979).
27. M. S. Puar and E. Grunwald, *Tetrahedron*, **24**, 2603 (1968).
28. A. P. Best and C. L. Wilson, *J. Chem. Soc.*, 28 (1938).
29. P. F. Tryon, W. G. Brown and M. S. Kharasch, *J. Amer. Chem. Soc.*, **70**, 2003 (1948).
30. P. Bellingham, C. D. Johnson and A. R. Katritzky, *J. Chem. Soc. B*, 1226 (1967).
31. V. Gold and D. P. N. Satchell, *J. Chem. Soc.*, 3609 (1955); V. Gold, R. W. Lambert and D. P. N. Satchell, *J. Chem. Soc.*, 2461 (1960); V. Gold, J. R. Lee and A. Gitter, *J. Chem. Soc. B*, 32 (1971).
32. D. P. N. Satchell, *J. Chem. Soc.*, 3911 (1956).
33. A. J. Kresge, Y. Chiang and L. E. Hakka, *J. Amer. Chem. Soc.*, **93**, 6167 (1971).
34. A. J. Kresge, H. J. Chen, L. E. Hakka and J. E. Kouba, *J. Amer. Chem. Soc.*, **93**, 6174 (1971).
35. A. J. Kresge, S. G. Mylonakis, Y. Sato and V. P. Vitullo, *J. Amer. Chem. Soc.*, **93**, 6181 (1971).
36. R. Taylor in *Comprehensive Chemical Kinetics*, Vol. 13 (Ed. C. H. Bamford and C. F. H. Tipper), Elsevier, Amsterdam (1972), (a) pp. 194–266; (b) pp. 83–139.
37. A. J. Kresge, R. A. More O'Ferrall, L. E. Hakka and V. P. Vitullo, *Chem. Commun.*, 46 (1965).
38. J. W. Larsen and M. Eckert-Maksić, *Croat. Chem. Acta*, **45**, 503 (1973).
39. J. W. Larsen and M. Eckert-Maksić, *J. Amer. Chem. Soc.*, **96**, 4311 (1974).
40. G. A. Olah and Y. K. Mo, *J. Org. Chem.*, **38**, 353 (1973).
41. M. P. Hartshorn, K. E. Richards, J. Vaughan and G. J. Wright, *J. Chem. Soc. B*, 1624 (1971).
42. T. Birchall and R. J. Gillespie, *Canad. J. Chem.*, **42**, 502 (1964); T. Birchall, A. N. Bourns, R. J. Gillespie and P. J. Smith, *Canad. J. Chem.*, **42**, 1433 (1964).
43. D. M. Brouwer, E. L. Mackor and C. MacLean, *Rec. Trav. Chim.*, **85**, 109, 114 (1966).
44. J. M. Brittain, P. B. D. de la Mare and P. A. Newman, *Tetrahedron Lett.*, **21**, 4111 (1980).
45. G. A. Olah and Y. K. Mo, *J. Org. Chem.*, **38**, 2212 (1973).
46. P. B. D. de la Mare and P. W. Robertson, *J. Chem. Soc.*, 100 (1948).
47. J. B. Shoesmith, A. C. Hetherington and R. H. Slater, *J. Chem. Soc.*, **125**, 1312 (1924).
48. V. Gold and M. Whittaker, *J. Chem. Soc.*, 1184 (1951).
49. F. J. Ludwig and K. H. Adams, *J. Amer. Chem. Soc.*, **76**, 3853 (1954).
50. R. C. Fahey, M. W. Monahan and C. A. McPherson, *J. Amer. Chem. Soc.*, **92**, 2810 (1970).
51. E. J. O'Bara, R. B. Balsey and I. Starer, *J. Org. Chem.*, **35**, 16 (1970).
52. L. A. Fury Jr and D. E. Pearson, *J. Org. Chem.*, **30**, 2301 (1965).
53. J. M. Brittain, P. B. D. de la Mare, N. S. Isaacs and P. D. McIntyre, *JCS Perkin II*, 933 (1979).
54. J. M. Brittain, P. B. D. de la Mare and P. A. Newman, *JCS Perkin II*, 32 (1981).
55. M. Kohn and M. Jawetz, *Monats. Chem.*, **44**, 199, 204 (1923).
56. M. Kohn and M. Weissberg, *Monats. Chem.*, **45**, 295 (1924).
57. M. Kohn and A. Rosenfeld, *Monats. Chem.*, **46**, 122 (1926).
58. T. Zincke and C. Schaum, *Ber. dtsch. chem. Ges.*, **27**, 537 (1894).
59. J. Thiele and H. Eichwede, *Ber. dtsch. chem. Ges.*, **33**, 673 (1900).
60. A. Lapworth, *J. Chem. Soc.*, **79**, 1265 (1901).
61. J. March, *Advanced Organic Chemistry*, 2nd edn, McGraw-Hill, New York (1977), esp. pp. 482–485.

62. C. R. Noller, *Chemistry of Organic Compounds*, 3rd edn, Saunders, Philadelphia (1965).
63. J. S. Mills, J. Barrera, E. Olivares and H. Garcia, *J. Amer. Chem. Soc*., **82**, 5882 (1960).
64. D. Taub, *Chem. Ind*., 558 (1962).
65. C. Rao, *Ultraviolet and Visible Spectroscopy*, 3rd edn, Butterworth, London (1975), pp. 42–48.
66. H. Jaffe and M. Orchin, *Theory and Applications of Ultraviolet Spectroscopy*, Wiley, New York (1962), pp. 204–217.
67. D. H. R. Barton, A. K. Ganguly, R. H. Hesse, S. N. Loo and M. M. Pechet, *Chem. Commun*., 806 (1968).
68. N. G. Kostina and V. D. Shteingarts, *J. Org. Chem. USSR*, **9**, 578 (1973); **10**, 1721 (1974).
69. L. Denivelle and R. Fort, *Bull. Soc. Chim. Fr*., 1834 (1956); 724 (1957); 459 (1958); 392 (1959); L. Denivelle and Huynh Anh Hoa, *Bull. Soc. Chim. Fr*., 2171 (1974).
70. L. Vollbracht, W. G. B. Huysmans, W. J. Mijs and H. J. Hageman, *Tetrahedron*, **24**, 6265 (1968).
71. E. Morita and M. W. Dietrich, *Canad. J. Chem*., **47**, 1944 (1969).
72. P. B. D. de la Mare and B. N. B. Hannan, *Chem. Commun*., 1324 (1971); P. B. D. de la Mare, B. N. B. Hannan and N. S. Isaacs, *JCS Perkin II*, 1389 (1976).
73. A. Fischer and G. N. Henderson, *Canad. J. Chem*., **57**, 552 (1979).
74. R. Lamartine and R. Perrin, *J. Org. Chem*., **39**, 1744 (1974).
75. D. E. Pearson, S. D. Venkataramu and W. E. Childers Jr, *Synthetic Commun*., **9**, 5 (1979).
76. T. Zincke and O. Kegel, *Ber. dtsch. chem. Ges*., **21**, 1030, 3540 (1888).
77. P. B. D. de la Mare and H. Suzuki, *J. Chem. Soc. C*, 648 (1968).
78. P. B. D. de la Mare, S. de la Mare and H. Suzuki, *J. Chem. Soc. B*, 429 (1969).
79. T. Zincke and O. Kegel, *Ber. dtsch. chem. Ges*., **21**, 3378 (1888).
80. K. Fries and K. Schimmelschmidt, *Liebig's Ann*., **484**, 245 (1930) (see esp. p. 295 ff).
81. P. W. Robertson, *J. Chem. Soc*., 1883 (1956).
82. A. A. Volod'kin and V. V. Ershov, *Isv. Akad. Nauk SSSR, Otd. Khim. Nauk*, 152 (1963); V. V. Ershov and A. A. Volod'kin, *Isv. Akad. Nauk SSSR, Otd. Khim. Nauk*, 893 (1963).
83. E. Grovenstein Jr and U. V. Henderson Jr, *J. Amer. Chem. Soc*., **78**, 569 (1956); E. Grovenstein Jr and G. A. Ropp, *J. Amer. Chem. Soc*., **78**, 2560 (1956).
84. L. G. Cannell, *J. Amer. Chem. Soc*., **79**, 2927, 2932 (1957).
85. P. B. D. de la Mare, O. M. H. El Dusouqui, J. G. Tillett and M. Zeltner, *J. Chem. Soc*., 5306 (1964).
86. E. Baciocchi and G. Illuminati, *J. Amer. Chem. Soc*., **89**, 4017 (1967).
87. M. Christen and H. Zollinger, *Helv. Chim. Acta*, **45**, 2057, 2066 (1962); M. Christen, W. Koch, W. Simon and H. Zollinger, *Helv. Chim. Acta*, **45**, 2077 (1962).
88. W. Koch and H. Zollinger, *Helv. Chim. Acta*, **48**, 554 (1965).
89. E. Grovenstein Jr and N. S. Aprahamian, *J. Amer. Chem. Soc*., **84**, 212 (1962).
90. P. B. D. de la Mare, *Acc. Chem. Res*., **7**, 361 (1974).
91. P. B. D. de la Mare and B. N. B. Hannan, *Chem. Commun*., 156 (1970).
92. C. A. Fyfe and L. Van Veen Jr, *J. Amer. Chem. Soc*., **99**, 3366 (1977).
93. P. B. D. de la Mare, A. Singh, J. G. Tillett and M. Zeltner, *J. Chem. Soc. B*, 1122 (1971); P. B. D. de la Mare and A. Singh, *JCS Perkin II*, 1801 (1972); 59 (1973).
94. L. E. Forman and W. C. Sears, *J. Amer. Chem. Soc*., **76**, 4977 (1954).
95. K. V. Sarkanen and C. W. Dence, *J. Org. Chem*., **25**, 715 (1960).
96. A. W. Francis and A. J. Hill, *J. Amer. Chem. Soc*., **46**, 2498 (1924).
97. C. H. R. Elston, A. T. Peters and F. H. Rowe, *J. Chem. Soc*., 367 (1948).
98. N. V. Lyukshova, A. P. Krysin and V. A. Koptyug, *Isv. Sib. Otd. Akad. Nauk SSSR, Ser. Khim. Nauk*, 73 (1975), through *Chem. Abstr*., **84**, 43511u (1976).
99. S. A. Brazier and H. McCombie, *J. Chem. Soc*., **101**, 968 (1912); G. King and H. McCombie, *J. Chem. Soc*., **103**, 220 (1913).
100. V. Calö, L. Lopez, G. Pesce and P. E. Todesco, *Tetrahedron*, **29**, 1625 (1973); *JCS Perkin II*, 1192 (1974); G. Hallas and J. D. Hepworth, *Educ. Chem*., **11**, 25 (1974).
101. M. Tsubota, M. Iso and K. Suzuki, *Bull. Chem. Soc. Japan*, **45**, 1252 (1972).
102. L. Denivelle and R. Fort, *Compt. Rend*., **235**, 1698 (1952); **240**, 1550 (1955); *Bull. Soc. Chim. Fr*., 724 (1957); L. Denivelle and M. Hedayatullah, *Compt. Rend*., **253**, 2711 (1961).

103. V. V. Ershov and A. A. Volod'kin, *Isv. Akad. Nauk SSSR, Otd. Khim. Nauk*, 152 (1963).
104. A. Reiker, N. Zeller and H. Kessler, *J. Amer. Chem. Soc.*, **90**, 6566 (1968).
105. A. A. Yasnikov and E. A. Shilov, *Dokl. Akad. Nauk SSSR*, **78**, 925 (1951), through *Chem. Abstr.*, **46**, 5009h (1952); E. Schulek and K. Burger, *Acad. Chem. Acad. Sci. Hung.*, **17**, 211 (1958); E. Schulek, K. Burger and E. Körös, *Acad. Chem. Acad. Sci. Hung.*, **21**, 67 (1959).
106. H. Goldhahn, *Osterr. Chem.-Z.*, **44**, 244 (1941).
107. N. W. Janney, *Liebig's Ann.*, **398**, 354 (1913).
108. A. Fischer and G. N. Henderson, *JCS Chem. Commun.*, 279 (1979).
109. N. L. Wendler in *Molecular Rearrangements*, 2nd edn, Vol. 2 (Ed. P. De Mayo), McGraw-Hill Kogkusha, Tokyo (1977), esp. pp. 482–485.
110. D. N. Kirk and M. P. Hartshorn, *Steroid Reaction Mechanisms*, Elsevier, London (1968).
111. R. T. Arnold and J. S. Buckley Jr, *J. Amer. Chem. Soc.*, **71**, 1781 (1949).
112. F. G. Bordwell and K. M. Wellman, *J. Org. Chem.*, **29**, 509 (1964).
113. H. Suzuki, *Synthesis*, 217 (1977).
114. R. Moodie and K. Schofield, *Acc. Chem. Res.*, **9**, 287 (1976).
115. E. N. Marvell and E. Magoon, *J. Amer. Chem. Soc.*, **77**, 2542 (1955).
116. C. E. Barnes and P. C. Myrhe, *J. Amer. Chem. Soc.*, **100**, 973, 975 (1978).
117. C. L. Perrin, *J. Org. Chem.*, **36**, 420 (1971); C. L. Perrin and G. A. Skinner, *J. Amer. Chem. Soc.*, **93**, 3389 (1971).
118. S. R. Hartshorn, *Chem. Soc. Rev.*, **3**, 167 (1974).
119. P. B. Fischer and H. Zollinger, *Helv. Chim. Acta*, **55**, 2139 (1972).
120. S. C. Wong, MSc thesis, University of Auckland (1981).
121. P. D. McIntyre and P. B. D. de la Mare, unpublished results.
122. C. D. Cook, N. G. Nash and H. R. Flanagan, *J. Amer. Chem. Soc.*, **77**, 1783 (1955).
123. G. M. Coppinger and T. W. Campbell, *J. Amer. Chem. Soc.*, **75**, 734 (1953).
124. L. J. Filar and S. Winstein, *Tetrahedron Lett.*, **25**, 9 (1960).
125. A. A. Volod'kin, V. V. Ershov, L. I. Kudinova and T. I. Prokof'eva, *Isv. Akad. Nauk SSSR, Ser. Khim.*, 1680 (1975).
126. V. V. Ershov, A. A. Volod'kin, G. A. Nikiforov and K. M. Dyumaev, *Isv. Akad. Nauk SSSR, Otd. Khim. Nauk*, 1839 (1962).
127. V. V. Ershov, A. A. Volod'kin, A. I. Prokof'ev and S. P. Solodovnikov, *Russ. Chem. Rev.*, **42**, 740 (1973).
128. J. Franck and E. Rabinowitsch, *Trans. Faraday Soc.*, **30**, 120 (1934); L. Herk, M. Field and M. Szwarc, *J. Amer. Chem. Soc.*, **83**, 2998 (1961).
129. P. B. D. de la Mare and R. D. Wilson, *Tetrahedron Lett.*, 3247 (1975); *JCS Perkin II*, 2048 (1977).
130. B. R. Kennedy and K. U. Ingold, *Canad. J. Chem.*, **45**, 2632 (1967).
131. K. H. Lee, *Tetrahedron*, **25**, 4357 (1969).
132. K. Ogura and T. Matsuura, *Bull. Chem. Soc. Japan*, **43**, 3181 (1970).
133. E. Müller, K. Ley and G. Schlechte, *Angew. Chem.*, **69**, 204 (1957).
134. P. B. D. de la Mare, (a) *Quart. Rev.*, **3**, 126 (1949); (b) *Electrophilic Halogenation*, Cambridge University Press, Cambridge (1976).
135. P. B. D. de la Mare and P. W. Robertson, *J. Chem. Soc.*, 888 (1945); H. P. Rothbaum, I. Ting and P. W. Robertson, *J. Chem. Soc.*, 980 (1948).
136. R. C. Given, MSc thesis, University of Auckland (1976).
137. A. Nilsson, A. Ronlán and V. D. Parker, *Tetrahedron Lett.*, 1107 (1975).
138. E. Müller, K. Ley and W. Kiedaisch, *Chem. Ber.*, **87**, 1605 (1954).
139. M. Tsubota, *Nippon Kagaku Zasshi*, **89**, 602 (1968).
140. K. Fries and E. Brandes, *Liebig's Ann. Chem.*, **542**, 48 (1939); K. Fries and G. Oehmke, *Liebig's Ann. Chem.*, **462**, 1 (1928).
141. R. F. Crozier and D. G. Hewitt, *Aust. J. Chem.*, **25**, 183 (1972).
142. A. I. Prokof'ev, S. P. Solodovnikov, A. A. Volod'kin and V. V. Ershov, *Isv. Akad. Nauk SSSR, Ser. Khim.*, 174 (1971).
143. A. A. Volod'kin, D. Kh. Rasuleva, V. V. Ershov, A. I. Prokof'ev, S. P. Solodovnikov, N. N. Budnov and S. G. Kukes, *Isv. Akad. Nauk SSSR, Ser. Khim*, 1212 (1972).
144. N. Kornblum, *Angew. Chem. Int. Edn Engl.*, **14**, 734 (1975).

145. J. F. Bunnett, *Acc. Chem. Res.*, **11**, 413 (1978).
146. I. P. Beletskaya and V. N. Drozd, *Russ. Chem. Rev.*, **48**, 431 (1979).
147. A. A. Volod'kin, M. V. Tarkhanova, V. V. Ershov and A. L. Buchachenko, *Isv. Akad. Nauk SSSR, Ser. Khim.*, 1583 (1969).
148. K. Somekawa, T. Matsuo and S. Kumamoto, *Bull. Chem. Soc. Japan*, **42**, 3499 (1969).
149. A. M. Sørenson and B. Jerslev, *Acta Chem. Scand.*, **22**, 319 (1968).
150. S. Gali, C. Miravitlles and M. Font-Altaba, *Acta Cryst. B*, **31**, 2510 (1975).
151. R. Hollenstein and W. von Phillipsborn, *Helv. Chim. Acta*, **55**, 2030 (1972).
152. A. Rieker and S. Berger, *Org. Magn. Reson.*, **4**, 857 (1972).
153. L. O. Brown and F. G. Soper, *J. Chem. Soc.*, 3576 (1953).
154. M. D. Carr and B. D. England, *Proc. Chem. Soc.*, 350 (1958).
155. P. B. D. de la Mare, A. D. Ketley and C. A. Vernon, *J. Chem. Soc.*, 1290 (1954).
156. L. Melander, results personally communicated.
157. R. P. Bell and T. Spencer, *J. Chem. Soc.*, 1156 (1959).
158. R. P. Bell and D. J. Rawlinson, *J. Chem. Soc.*, 63 (1961).
159. J. H. Ridd, *Adv. Phys. Org. Chem.*, **16**, 1 (1978).
160. P. B. D. de la Mare and O. M. H. El Dusouqui, *J. Chem. Soc. B*, 251 (1967).
161. P. W. Robertson, P. B. D. de la Mare and W. T. G. Johnston, *J. Chem. Soc.*, 276 (1943).
162. L. M. Yeddanapalli and N. S. Gnanapragasam, *J. Chem. Soc.*, 4934 (1956); *J. Indian Chem. Soc.*, **36**, 745 (1959); N. S. Gnanapragasam, N. V. Rao and Y. N. Yeddanapalli, *J. Indian Chem. Soc.*, **36**, 777 (1959).
163. V. S. Karpinskii and V. D. Lyashenko, *J. Gen. Chem. USSR*, **30**, 164 (1960); **32**, 3922 (1962); **33**, 599 (1963).
164. E. Berliner, *J. Amer. Chem. Soc.*, **72**, 4003 (1950); **73**, 4307 (1951).
165. E. Grovenstein Jr and D. C. Kilby, *J. Amer. Chem. Soc.*, **79**, 2972 (1957).
166. E. Berliner, *J. Amer. Chem. Soc.*, **78**, 3632 (1956); **80**, 856 (1958); **82**, 5435 (1960).
167. A. W. Francis, A. J. Hill and J. Johnston, *J. Amer. Chem. Soc.*, **47**, 2211 (1925).
168. S. Takagi and M. Tanaka, *J. Pharm. Soc. Japan*, **517**, 247 (1925); S. Takagi and S. Ishimana, *J. Pharm. Soc. Japan*, **517**, 253 (1925).
169. J.-C. Jacquesy, M. P. Jouannetaud and S. Makani, *JCS Chem. Commun.*, 110 (1980).
170. D. E. Pearson, R. D. Wysong and C. V. Breder, *J. Org. Chem.*, **32**, 2358 (1967).
171. T. M. Cresp, R. C. F. Giles, M. V. Sargent, C. Brown and D. O'N. Smith, *JCS Perkin I*, 2435 (1974).
172. W. H. Hunter and T. T. Budrow, *J. Amer. Chem. Soc.*, **55**, 2122 (1933).
173. K. Auwers, *Ber. Dtsch. Chem. Ges.*, **36**, 1878 (1903).
174. K. Auwers and G. F. von Campenhausen, *Ber. Dtsch. Chem. Ges.*, **29**, 1129 (1896).
175. M. Kohn and J. Pfeifer, *Monats. Chem.*, **48**, 231 (1927).
176. W. C. Stoesser, US Patent 2,131,259 (1938), through *Chem. Abstr.*, **32**, 9102e (1938).
177. A. Campbell and D. J. Shields, *Tetrahedron*, **21**, 211 (1965).
178. M.-F. Vincent-Falquet-Berny and R. Lamartine, *Bull. Soc. Chim. Fr.*, 47 (1975).
179. G. Dutruc-Rosset, Doctoral thesis, Université Claude-Bernard, Lyon, France (1979).
180. R. Lamartine, *Compt. Rend. C*, **279**, 429 (1974); R. Lamartine and R. Perrin, *Compt. Rend. C*, **279**, 477 (1974).
181. R. Lamartine, R. Perrin, G. Bertholon and M. F. Vincent-Falquet, *J. Amer. Chem. Soc.*, **99**, 5436 (1977).
182. R. Perrin, R. Lamartine and G. Bertholon, *Mol. Cryst. Liq. Cryst.*, **52**, 303 (1979).
183. W. D. Watson, *Tetrahedron Lett.*, 2591 (1976).
184. E. M. Kosower, W. J. Cole, G.-S. Wu, D. E. Cardy and G. Meisters, *J. Org. Chem.*, **28**, 630 (1963).
185. S. Takagi and K. Kutani, *J. Pharm. Soc. Japan*, **517**, 260 (1925).
186. E. H. Huntress, *Organic Chlorine Compounds*, Wiley, New York (1948).
187. G. Kalopissis, A. Bugaut and F. Estradier, Ger. Offen. 2,237,270 through *Chem. Abstr.*, **78**, 149221r (1973).
188. A. Matthews, *J. Agric. Sci.*, **14**, 1 (1924).
189. R. L. Wain and D. P. Harper, *Nature*, **213**, 1155 (1967).
190. C.-A. Nilsson, A. Norstrom, K. Andersson and C. Rappe in *Pentachlorophenol: Chemistry, Pharmacology and Environmental Toxicology* (Ed. K. R. Rao) R. Env. Sci. Res. Series, Vol. 12, Plenum Press, New York (1978), pp. 313–324.
191. N. N. Melnikov, *Chemistry of Pesticides*, Springer Verlag, New York (1971).

192. D. G. Crosby and Ming-Yu Li in *Degradation of Herbicides* (Ed. P. C. Kearney and D. D. Kaufmann), Marcel Dekker, New York (1969), pp. 348–363.
193. (a) M. Kuwahara, N. Kato and K. Munakata, *Agric. Biol. Chem. Japan*, **30**, 232, 239 (1966); (b) A. S. Wong and D. G. Crosby in *Environmental Science Research*, Vol. 12 (Ed. K. R. Rao), Plenum Press, New York (1978), pp. 19–25.
194. For example, R. Frank, R. L. Thomas, M. Holdrinet, A. L. W. Kemp, H. E. Braun and R. Dawson, *Sci. Total Environ*., **13**, 101 (1979).
195. D. W. McLeese, V. Zitko and M. R. Peterson, *Chemosphere*, **8**, 53 (1979).
196. A. Poland, E. Glover and A. S. Kende, *J. Biol. Chem*., **251**, 4936 (1976).
197. R. Stehl and L. L. Lampaski, *Science*, **197**, 1008 (1977); B. Ahling, A. Lindskog, B. Jansson and G. Sundström, *Chemosphere*, **6**, 461 (1977).
198. J. M. Bollag, S.-Y. Liu and R. D. Minard, *Soil Sci. Soc. Amer. J*., **44**, 52 (1980).
199. D. F. Wilson, H. P. Ting and M. S. Koppelman, *Biochemistry*, **10**, 2897 (1971).
200. C. N. Giannopolitis and G. S. Ayers, *Weed Sci*., **26**, 440 (1978).
201. P. Mitchell, *Biol. Rev*., **41**, 445 (1966).
202. J. C. Boray, *Adv. Parasitol*., **7**, 95 (1969).
203. C. Razafindrakoto, A. Ranaivoson and J. P. Megard, *Rev. Elev. Med. Vet. Pays. Trop*., 165 (1978), through *Chem. Abstr*., **90**, 80991v (1979).
204. P. P. Delaveney and M. Guebre Negus, *Rev. Elev. Med. Vet. Pays. Trop*., 171 (1978), through *Chem. Abstr*., **90**, 80992w (1979).
205. C. Bryant, *Adv. Parasitol*., **8**, 139 (1970).
206. J. R. Corbett and J. Goose, *Pestic. Sci*., **2**, 119 (1971).
207. M. A. Q. Khan, M. L. Gassman and S. H. Ashrafi in *Environmental Dynamics of Pesticides* (Ed. R. Hague and V. H. Freed), Plenum Press, New York (1975), pp. 289–329.
208. A. Verloop and J. Tipker, *Pharmochem. Libr*., **2** (Biol. Act. Chem. Struct.), 63 (1977); E. Knuesli, *Pontif. Acad. Sci. Scr. Varia*, **41**, 755 (1976), through *Chem. Abstr*., **89**, 37873a (1978).
209. L. Schneider and D. E. Graham, US Patent 4,112,002 (1978), through *Chem. Abstr*., **90**, 121197m (1979).
210. American Cyanamid Co., Neth. Appl. 76,09,289 (1978), through *Chem. Abstr*., **89**, 101893m (1978).
211. M. W. Moon, US Patent 4,153,707 (1979), through *Chem. Abstr*., **91**, 103717x (1979).
212. A. K. Sen Gupta, O. P. Bajaj and Md. Mushtaq, *Indian J. Chem. B*, **16**, 629 (1978).
213. A. P. Leftwick and E. W. Parnell, Ger. Offen., 2,603,864 (1976), through *Chem. Abstr*., **86**, 29436h (1977).
214. D. Taub, C. H. Kuo and N. L. Wendler, *Chem. Ind*., 557 (1962); D. Taub, *Chem. Ind*., 558 (1962).
215. N. Kumamoto, *Kokai Tokkyo Koho* 70,14,763 (1970), through *Chem. Abstr*., **73**, 120184s (1970); J. E. Franz, US Patent 3,444,207 (1969), through *Chem. Abstr*., **71**, 38448f (1969).
216. R. T. Morrison and R. N. Boyd, *Organic Chemistry*, Allyn & Bacon, Boston, 1969.
217. G. D. Staffin and C. C. Price, *J. Amer. Chem. Soc*., **82**, 3632 (1960).
218. F. Feichtmayr and H. Naarmann, *Ger. Offen*. 2,046,353, through *Chem. Abstr*., **77**, 102418z (1972).
219. A. S. Hay, *J. Polymer Sci*., **58**, 581 (1962).
220. J. Peht and F. Laborie-Gardaix, *Compt. Rend*., **257**, 3407 (1963).
221. W. H. Starnes Jr. and H. J. Tarski, US Patent 3,541,171 (1970), through *Chem. Abstr*., **74**, 46093p (1971).
222. W. Reid and W. Merkel, *Angew. Chem. Eng. Edn*, **8**, 379 (1979).
223. R. C. Nametz and P. H. Burleigh, US Patent 4,141,880 (1979), through *Chem. Abstr*., **90**, 169547c (1979).
224. M. Kashiki, O. Miki and I. Arai, *Kokai Tokkyo Koho*, 78,142,458 (1978), through *Chem. Abstr*., **90**, 187944j (1979).
225. M. Kashiki, O. Miki and Y. Urata, *Kokai Tokkyo Koho*, 78,139,658 (1978), through *Chem. Abstr*., **90**, 1225956b (1979).
226. D. W. Fox, Canadian Patent 1,039,898 (1978), through *Chem. Abstr*., **90**, 104848g (1979).
227. A. Takahashi, M. Shimizu and T. Horio, *Proc. Electr./Electron. Insul. Conf*., **13**, 252 (1977), through *Chem. Abstr*., **90**, 7202j (1979).

228. H. Fujiwara, A. Takahashi and K. Fujii, *Kokai Tokkyo Koho*, 78,121,825 (1978), through *Chem. Abstr*., **90**, 88457w (1979).
229. A. Takahashi, O. Miki and H. Sasaki, *Kokai Tokkyo Koho*, 78,133,258 (1978), through *Chem. Abstr*., **90**, 122556q (1979).
230. H. Bechhold and P. Ehrlich, *Z. Physiol. Chem*., **47**, 173 (1906).
231. H. Berry, A. Cook and B. Wills, *J. Pharm. Pharmacol*., **8**, 425 (1956).
232. E. F. Harrison, P. Zwadyk Jr, R. J. Bequette, E. Hamlow, P. A. Tavormina and W. A. Zygmunt, *Appl. Microbiol*., **19**, 746 (1970); E. F. Harrison and W. A. Zygmunt, *Canad. J. Microbiol*., **20**, 1241 (1974).
233. *The Thyroid* (Ed. S. C. Werner and S. H. Ingbar), Harper & Row, New York (1971).
234. O. Schier and A. Marxer in *Progress in Drug Research*, Vol. 22 (Ed. E. Jucker), Birkhäuser Verlag, Stuttgart (1978).

The Chemistry of Functional Groups, Supplement D
Edited by S. Patai and Z. Rappoport
© 1983 John Wiley & Sons Ltd

CHAPTER **13**

α-Halogenated imines

NORBERT DE KIMPE† and ROLAND VERHÉ

Laboratory of Organic Chemistry§, Faculty of Agricultural Sciences, State University of Gent, Coupure 533, B-9000 Gent, Belgium

I. INTRODUCTION

α-Halogenated imino compounds (**1**) are the nitrogen homologues of α-halogenated ketones (**2**). While the latter class of compounds has been studied extensively in the

†'Bevoegdverklaard Navorser' of the Belgian 'Nationaal Fonds voor Wetenschappelijk Onderzoek'.

§Director: Prof. N. Schamp.

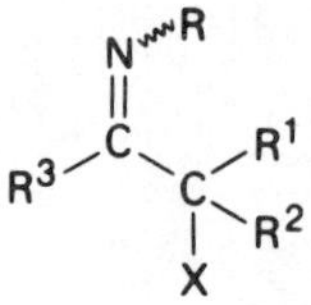

(1) X = halogen

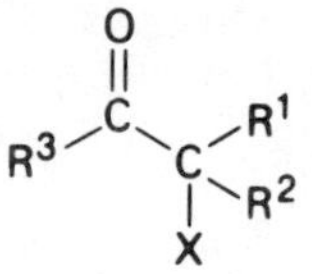

(2) X = halogen

literature, α-halo imines have only recently come to be used regularly in organic synthesis despite the fact that they were proven earlier to be valuable synthetic reagents.

Information regarding the use of α-halogenated imino compounds (**1**) in organic synthesis remained scattered until very recently when this matter was compiled in a review covering the synthesis[1] and reactivity[2] of these compounds. It is to focus attention on the versatility and potential of these reagents that this information has been brought together. The literature has been reviewed up to early 1980. Only α-halogenated imino compounds having a structural similarity with α-halogenated carbonyl compounds will be treated in this review. However, when necessary, some leading references to the chemistry of α-halo imidates, α-halo amidines and α-halo imidoyl cyanides will be provided. In general, α-halogenated imino compounds (**1**), having at least one α-hydrogen, do not tautomerize into the corresponding enamines, except when conjugation in the molecule with such substituents as CN, COOR, NO_2, etc. is possible. Such β-halo enamines are not subject to discussion here.

Some novel aspects of the chemistry of α-halo ketones will be discussed in another chapter in this volume. Throughout the text some comparisons will be made between the reactivities of α-halo carbonyl compounds and α-halo imines.

II. SYNTHESIS OF α-HALOGENATED IMINES

Comprehensive studies in the area of the synthesis of α-halo imines are of rather recent origin. The growing success of the use of these reagents in synthetic organic chemistry is predominantly due to the development of readily available syntheses of the title compounds. Indeed, early investigations in the field of α-halo imines met with difficulties, since no suitable conditions could be found for halogenation of imino compounds. Additionally, several α-halo imines were found to be unstable, especially towards hydrolytic and thermal reaction conditions. Two main strategies for the synthesis of α-halo imines (**1**) may be considered. The first one is the condensation of an α-halogenated carbonyl compound (**2**) with a primary amine under suitable reaction conditions, similar to the usual synthesis of imines starting from carbonyl compounds and primary amines[3,4]. The second approach involves the halogenation of imines (**4**). In both cases, carbonyl compounds (**3**) are the starting materials for such syntheses. The first method gives rise to the desired α-halo imines (**1**) only in special cases. Most often, a variety of side reactions is encountered, among others nucleophilic α-substitution[5,6], elimination of hydrogen halide[7], haloform-type reactions[8,36], Favorskii rearrangement[9–12] and rearrangement via intermediate epoxides[13–16,207]. In many cases, intermediately formed α-halo imino compounds were further transformed under the given reaction conditions to various final products[17–35].

The second approach to α-halo imines via halogenation of imines also met with major difficulties, especially in the older literature, because unstable immonium-type compounds resulted from this reaction. The latter were usually transformed into

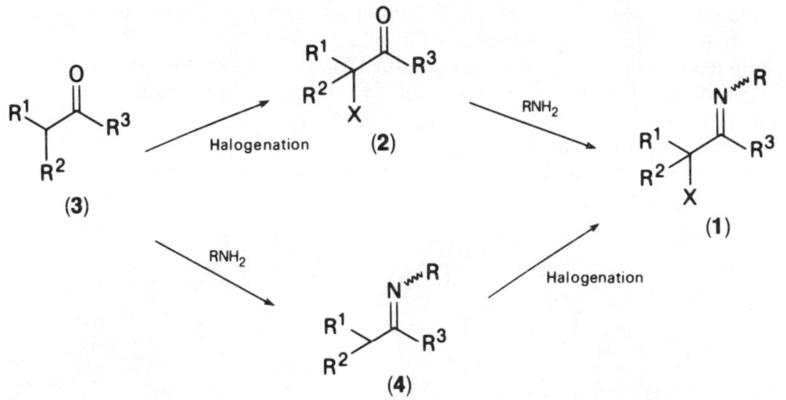

α-halo carbonyl compounds by aqueous work-up. The medium in which the imine is halogenated plays a predominant role, as will be demonstrated in the following sections.

Attention will be given now to the two aforementioned synthetic methods leading to α-halo imines, while the halogenation of enamines giving rise to the title compounds will also be discussed. Additionally, some miscellaneous methods for the synthesis of α-halogenated imines will be treated.

A. Condensation of α-Halogenated Carbonyl Compounds with Primary Amines

The reaction of α-fluoro carbonyl compounds with primary amines usually gives no difficulties in synthesizing the α-fluoro imines. 1,1,1-Trifluoroacetone (**7**) reacted with aniline in benzene for 2 days to give *N*-(1,1,1-trifluoro-2-propylidene)aniline (**9**) in 25% yield[37], while 2,2-difluorononanal (**5**) condensed smoothly with *t*-butylamine at room temperature to give the α,α-difluoro aldimine (**6**)[105]. In similar way, aromatic

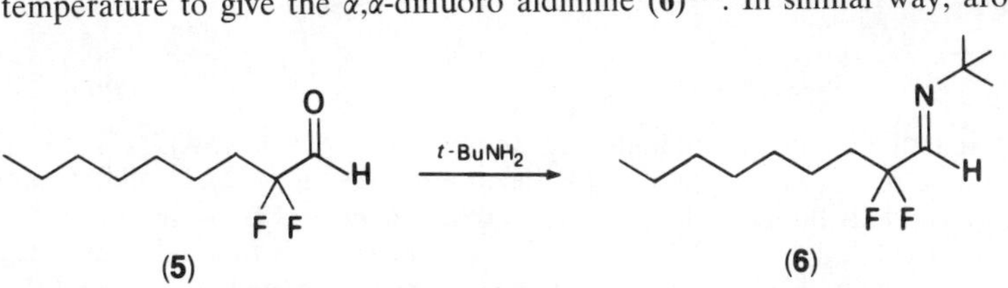

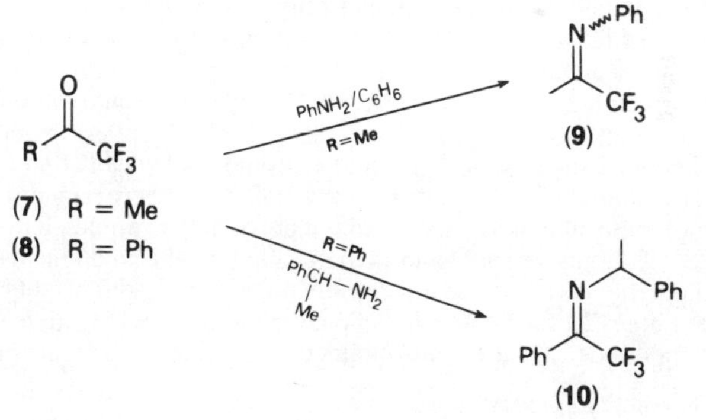

α-fluorinated ketones such as 2,2,2-trifluoroacetophenone (**8**) condensed with α-methylbenzylamine to afford α,α,α-trifluoroketimine (**10**)[38].

Iminophosphoranes can also be used in such iminations of α-fluorinated ketones[105], as exemplified by the synthesis of **9** from **7** by using **11**[37]. When *N*-trialkylstannyltriphenylphosphonimines (**13**) were used as reagents, hexafluoroacetone (**12**) was converted into the *N*-trialkylstannyl-substituted perfluoro ketimine (**14**)[39]. When applied to trimethylsilyltriphenylphosphonimine, the

$F_3C-CO-R$ (**7**) R = Me; (**12**) R = CF_3

(**7**) + $(C_6H_5)_3P{=}NC_6H_5$ (**11**) $\xrightarrow{\text{Ether}}$ $F_3C(CH_3)C{=}N\text{-}CH(CH_3)Ph$ (**9**)

(**12**) + $(C_6H_5)_3P{=}NSnR'_3$ (**13**) $\longrightarrow$ $(F_3C)_2C{=}N\text{-}SnR'_3$ (**14**)

corresponding *N*-trimethylsilylimine could only be isolated in 1% yield[39]. Oximation of α-fluoroketones[441] with hydroxylamine in ethanol in the presence of sodium acetate gives no side reactions, as exemplified by the synthesis of 5α-fluoro-6-oximinocholestane-3β-ol acetate (**16**)[428]. The same is true for the synthesis of α-fluoro hydrazones[412].

(**15**) $\xrightarrow[\text{NaOAc, EtOH}]{\Delta,\ NH_2OH\cdot HCl}$ (**16**)

Less reactive amino compounds, e.g. thiobenzamide derivatives (**17**), can also be used for direct condensation with hexafluoroacetone (**12**). Initially, however, 2,2,6,6-tetrakis(trifluoromethyl)-6*H*-1,3,5-oxathiazines (**18**) were formed, which could be pyrolysed into 2*H*-1,3-thiazetes (**19**), existing in thermal equilibrium with *N*-(perfluoroisopropylidene)thiocarboxamides (**20**)[40–42]. It will be demonstrated (*vide infra*) that the activated perfluoroketimines (**20**) show a fascinating reactive behaviour toward a variety of reagents with which they can undergo cycloadditions. Besides the direct imination of α-fluorinated carbonyl compounds, the direct condensation of α-halogenated ketones with aliphatic or aromatic primary amines to give α-halogenated ketimines has never been described*. Only less sterically hindered carbonyl compounds such as α-chloro- and α-bromoaldehydes (**21** and **22**) react in a straightforward manner with aliphatic primary amines in ethereal medium at $-30°C$ in the presence of molecular sieves to afford α-chloro- and α-bromoaldimines (**23**,**24**) in 27–73% yield[43,44]. Compounds **23** and **24** are rather unstable and sensitive to moisture, the *N*-*t*-butyl derivatives being the most stable ones. In the presence of excess primary amine, α-halo imines **23** and **24** are slowly converted into α-alkylamino aldimines (**25**). The latter compounds (**25**) are also obtained in a more rapid reaction from α-halo

*Note added in proof (page 601)

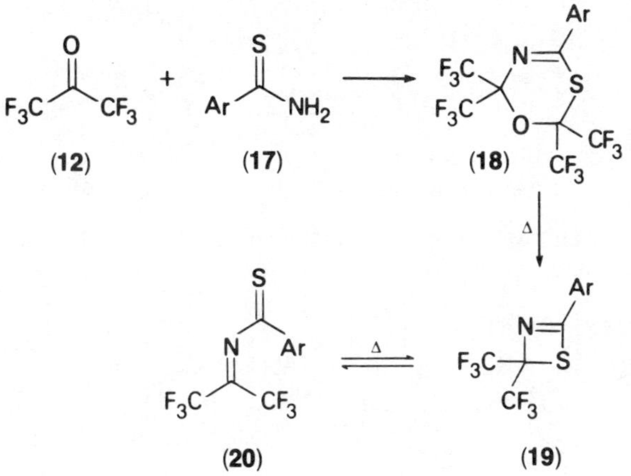

(12) (17) (18) (19) (20)

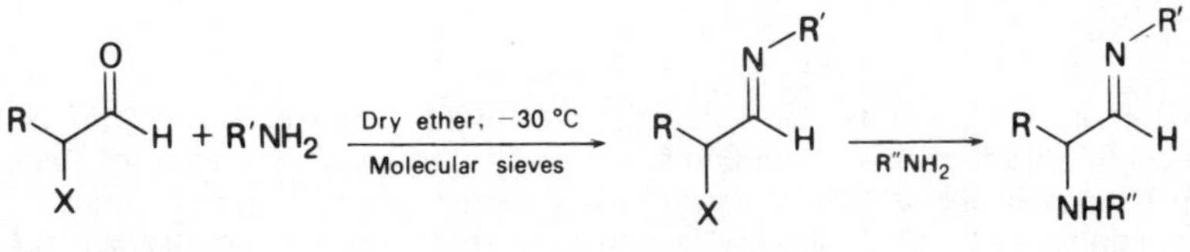

(21) X = Cl
(22) X = Br
(23) X = Cl
(24) X = Br
(25)

aldehydes and primary amines, indicating that the α-halo aldimines **23** and **24** are not intermediates in these reactions[43]. In more drastic conditions, chloral condensed with primary amines under catalytic influence of zinc chloride and under azeotropic water removal to give trichloroethylideneamines[331–333].

The usual carbonyl identification reagents such as 2,4-dinitrophenylhydrazine, hydroxylamine, semicarbazide, etc., also react with α-halo carbonyl compounds to afford the corresponding α-halo imino derivatives. Care should be taken, however, as regards the reaction conditions employed, since side reactions such as elimination, nitrosoölefin formation, etc., may take place.

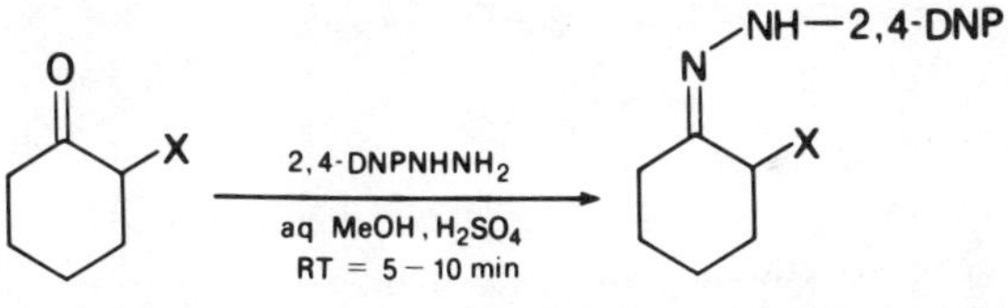

(26) X = Cl, Br
(27) X = Cl, Br

The Brady reagent[45], i.e. an aqueous methanolic solution of 2,4-dinitrophenylhydrazine sulphate containing excess sulphuric acid[23,46] could be successfully applied for the synthesis of α-halocyclohexanone 2,4-dinitrophenylhydrazones (**27**). The reaction of α-bromocycloalkanones (**28**; n = 3,5,9) with tosylhydrazine in ether produced crystalline α-bromo tosylhydrazones (**29**; n = 3,5,9)[47,48,50]. In a similar way, α-halogenated aldehydes gave the corresponding α-halogenated dinitrophenylhydrazones[199–202]. This method was also applied for the synthesis of aliphatic α-bromo tosylhydrazones[49,50,212]. The conversion into α-bromo

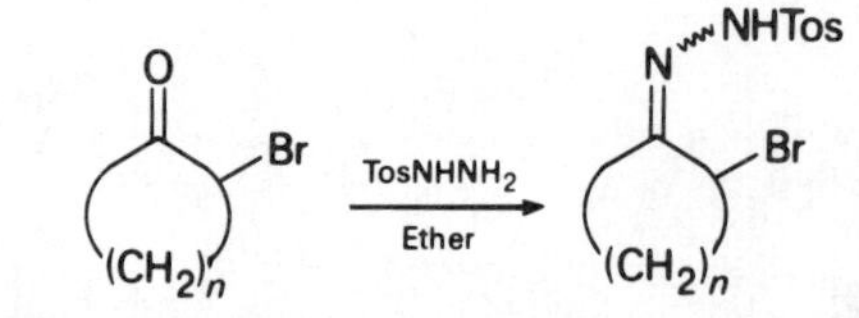

(28; n = 3, 5, 9) (29; n = 3, 5, 9)

tosylhydrazones seems to be a general reaction, occurring also with complex molecules like 14-bromodaunomycine (**30**)[51].

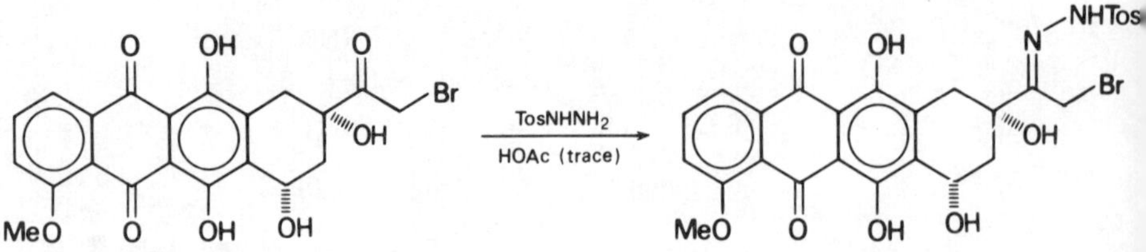

(**30**) (**31**)

In a similar way again, α-halo semicarbazones and related compounds were isolated under appropriate reaction conditions[52–59,109] but these compounds were subject to further transformations[110] into heterocyclic compounds[1].

Oximation of α-halo carbonyl compounds (**32**) requires controlled reaction conditions because of the possibility of side reactions of the initially formed α-halo oxime. Base-induced 1,4-elimination of hydrogen halide from α-halo oximes (**33**) yields nitrosoölefins (**34**) which are apt to undergo a variety of transformations[60–62].

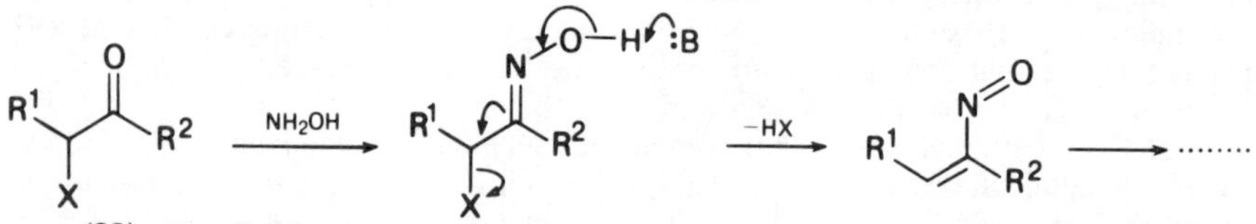

(**32**) X = Cl, Br (**33**) (**34**)

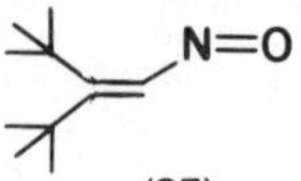

(**35**)

The intermediacy of these nitrosoölefins (**34**) was demonstrated by their isolation under appropriate reaction conditions (see for example compound **35**[63–65,218]). In order to avoid nitrosoölefin formation, it is recommended that oximations be performed in a slightly acid medium, such as in an aqueous calcium chloride solution[66,67] or with equimolecular amounts of sodium acetate in acetic acid[68,69]. Oximations of α-halo ketones can also be performed under milder conditions (such as NH_2OH/methanol/THF/room temperature, 18 h)[156,157,206,220,266].

B. Halogenation of Imino Compounds

Several halogenating agents have been found to convert imino compounds (**4**) into α-halo imines (**1**). However, most of them were not proven to be of general synthetic

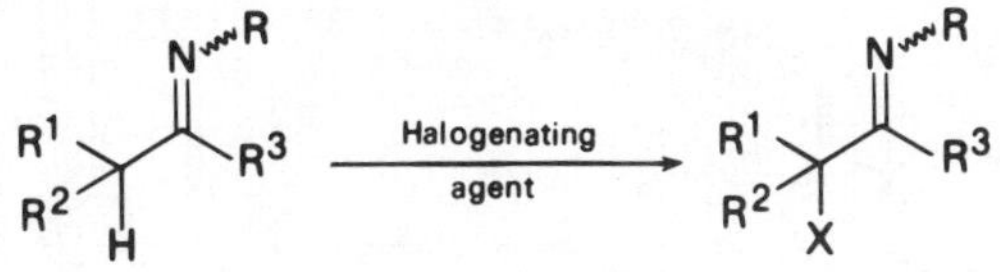

(4) (1)

interest and in many cases the α-halo imine formed could not be isolated, making hydrolysis to the more stable α-halo carbonyl compounds necessary. For example, halogenation with chlorine or bromine met with major difficulties by virtue of the instability of transient α-halogenated immonium halides[70–72].

Brominated acetophenone azines (**37**; X = Br) and (**38**; X = H) could be synthesized from the parent azine (**36**) with bromine in dichloromethane[73] or

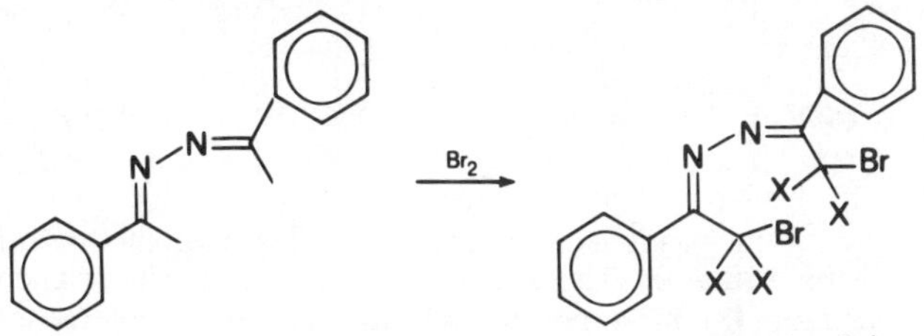

(36) (37; X = Br) (38; X = H)

methanol (0–5°C)[74], respectively. 2-Alkyloxazolines were halogenated with chlorine or bromine[75], while 2-pyrazolin-5-ones and 2-isoxazolin-5-ones (and related compounds) were also reported to be chlorinated at the active methylene function at the 4-position[76,77,208,286].

Probably due to their unstable nature, α-iodo ketimines have not been isolated hitherto. Recently, it was reported that a transient α-iodo ketimine (**40**) was used to transform methylketimines (**39**) into symmetric 1,4-diones (**41**) via lithiation (using lithium diisopropylamide = LDA), iodination, coupling of α-iodo ketimine (**40**) with the lithiated ketimine (anion of **39**) and acidic hydrolysis of the 1,4-diimine thus formed[78].

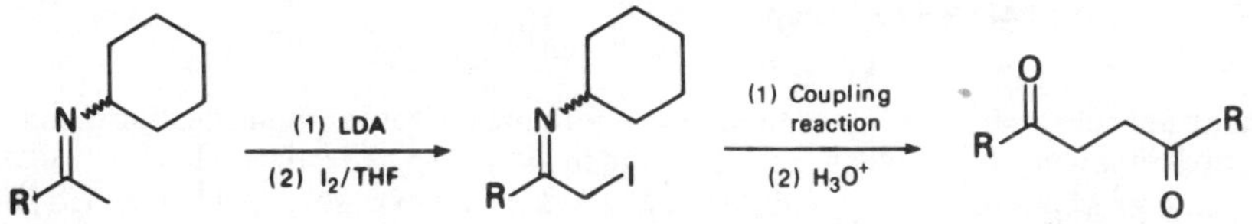

(39) (40) (41)

Among the halogenating agents of imines, *N*-halosuccinimide has been found to be the superior reagent for the synthesis of aliphatic and aromatic α-halogenated imino compounds. Ketimines are chlorinated at the less substituted α-position with *N*-chlorosuccinimide (NCS). The reaction is more regiospecific in carbon tetrachloride[79–83] than in ether[84]. The steroidal *N*-(2-hydroxyethyl)ketimines (**42**) were conveniently monochlorinated in ether, the resulting chloromethylketimine being hydrolysed in acidic medium to the corresponding α-chloroketone (**43**)[84]. *N*-Cyclohexyl and *N*-aryl methylketimines (**44** and **45**) were regiospecifically dichlorinated at the methyl function to produce dichloromethylketimines (**46** and

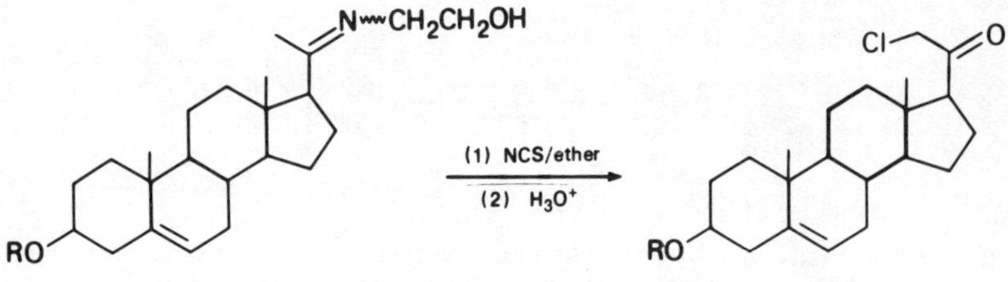

(42) R = H, Ac

(1) NCS/ether
(2) H_3O^+

(43)

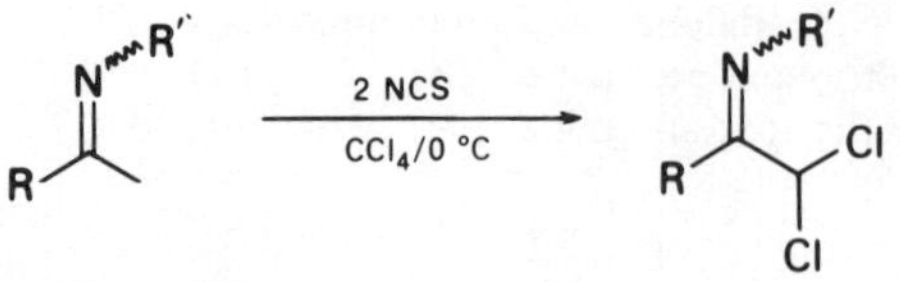

2 NCS
CCl_4/0 °C

(44) R′ = cyclohexyl
(45) R′ = aryl

(46) R′ = cyclohexyl
(47) R′ = aryl

47)[79–83], together with negligible amounts of 1,3-dichloromethylketimines and 1,1,1-trichloromethylketimines. The mechanism proceeds via chlorination of the less substituted enamine (**48**) in a non-radical manner. Steric interactions play an

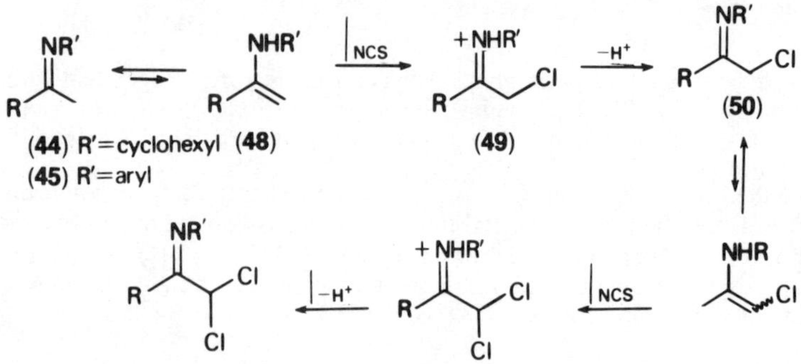

(44) R′=cyclohexyl (48) (49) (50)
(45) R′=aryl

(46) R′=cyclohexyl (52) (51)
(47) R′=aryl

important role in these halogenations and determine the regiospecific dichlorination of methylketimines. Even when the R-group in **44** or **45** is tertiary (R = *t*-Bu), the reaction proceeds to dihalogenation, but *N*-alkyl imines of diisopropyl ketone (**53**) did not suffer chlorination with *N*-chlorosuccinimide in carbon tetrachloride. α,α,α-Trichlorination of acetophenone imines (**45**; R = Ar) proceeds rapidly and quantitatively at reflux with NCS in carbon tetrachloride (5 min), while imines derived from dichloropinacolone (**46**, **47**; R = *t*-Bu) could not be further chlorinated under drastic conditions[91].

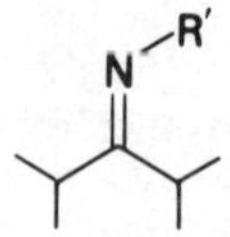

(53)

The reaction of imines with *N*-halosuccinimide requires an initiation period, after which the reaction proceeds smoothly to completion. As expected, and concordant with the proposed reaction mechanism, secondary enamines, i.e. enamines in which stabilizing groups such as a nitrile moiety force tautomerizable substrates to occur exclusively as enamines, react instantaneously with *N*-halosuccinimide (*vide infra*)[115,116]. Imines having no α′-hydrogen atoms, such as **54**, **55** and **58**, are easily chlorinated with NCS in CCl_4[85–90]. Substitution of all available α-hydrogen atoms by chlorine atoms is accomplished without problems. However, methylketimines (**44** and **45**) or imines having an α-CH_2 function, e.g. **54** and **55**, cannot be converted by this method into α-monochloro imines, because the rate of introduction of the first and the second halogen are of the same magnitude. This seems to be a general observation and points to the major limitation of the chlorination procedure of imines with *N*-chlorosuccinimide.

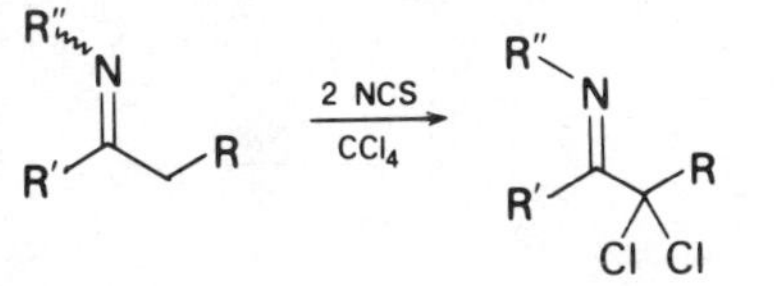

(**54**; R′ = H) (**56**; R′ = H)
(**55**; R′ = Ar) (**57**; R′ = Ar)

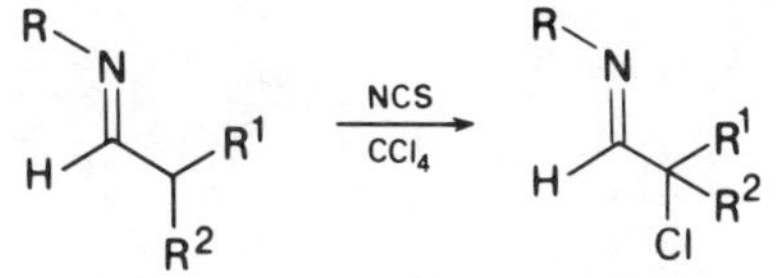

(**58**) (**59**)

The same comments as given for the α-chlorination of imines using NCS are applicable for *N*-bromosuccinimide (NBS). Bromination of imines **44**, **54** and **55** with NBS in CCl_4 yielded α-bromo imines **60**[91], **61**[92] and **62**[88] in good yield. In many cases

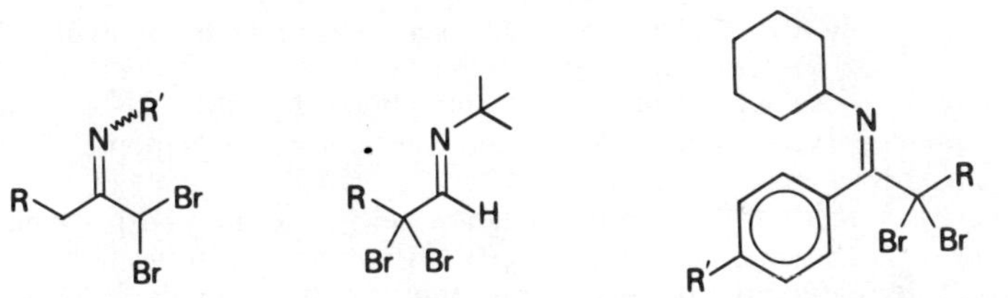

(**60**) (**61**) (**62**)

the use of NBS required the aid of benzoyl peroxide, irradiation or acid catalysis[93], or the combined action of these influences. A variety of classes of imino compounds have been α-brominated with *N*-bromosuccinimide, including oximine benzoates (**63**)[94], nitrones (**65**)[95–97], 2-methoxycarbonyl-1-pyrrolines[98], cyclic imino ethers (imidates) (**67**)[99,100], hydrazone-type compounds (**69**)[101–103], amidines[104] and 1-pyrrolines[448].

Other sources of positive chlorine, which have been used for α-chlorination of amines, are sodium hypochlorite[106,107] and *t*-butyl hypochlorite[107,108]. These reagents converted steroidal methylketimines (**71**) into mixtures of α-halogenated imines, while

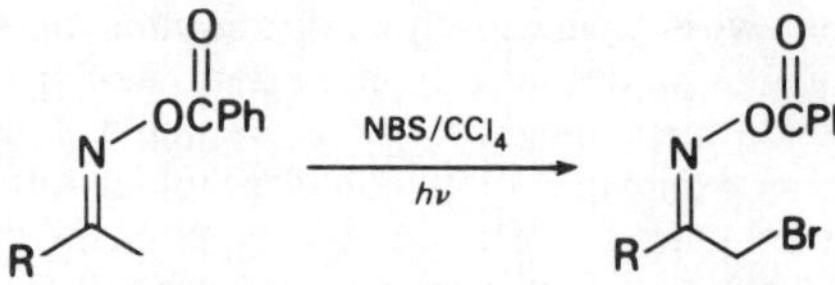

NBS/CCl_4, $h\nu$

(63) (64)

NBS/CCl_4, BPO

(65; R = CN, Ph, *t*-Bu) (66)

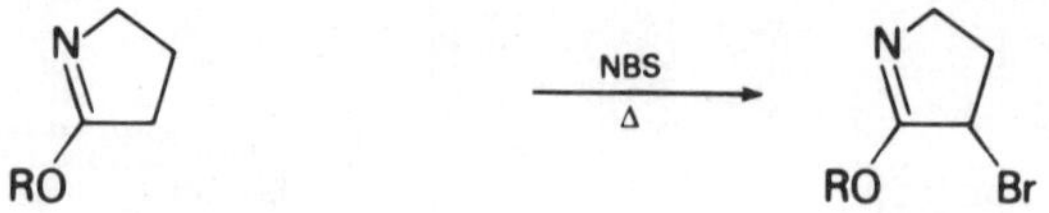

NBS, Δ

(67; R = Me, Et) (68)

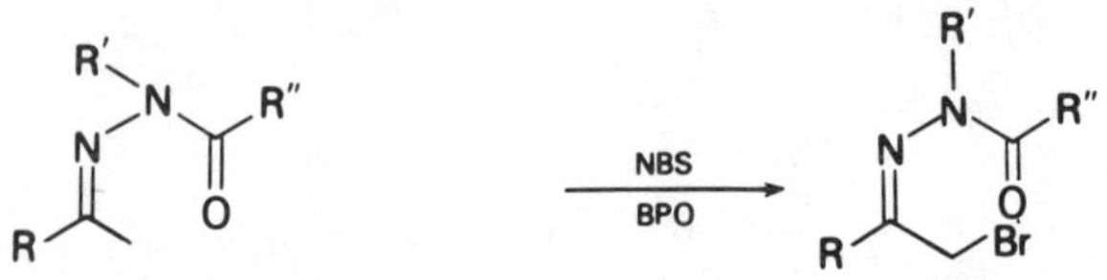

NBS, BPO

(69) (70)

NaOCl or *t*-BuOCl

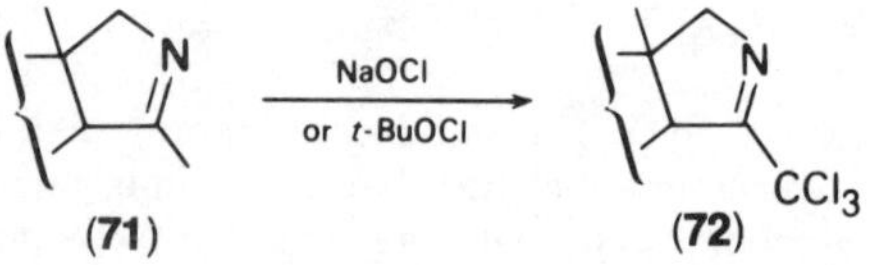

(71) (72)

only the α,α,α-trichloromethyl derivative (**72**) was obtainable in a synthetically useful manner[107]. Monobromination of ketimines can be performed with 2,4,4,6-tetrabromocyclohexadienone, as exemplified for the *N*-*t*-butyl imine of 3,3,5,5-tetramethylcyclohexanone, but the monobromo compound existed in equilibrium with its enamino form[93].

A superior reagent for the α-bromination of hydrazones (**73**) seemed to be phenyltrimethylammonium perbromide (PTAB) in tetrahydrofuran[102,103,111,112,453]. With two equivalents of the brominating reagent α,α′-dibromination occurred[111,112].

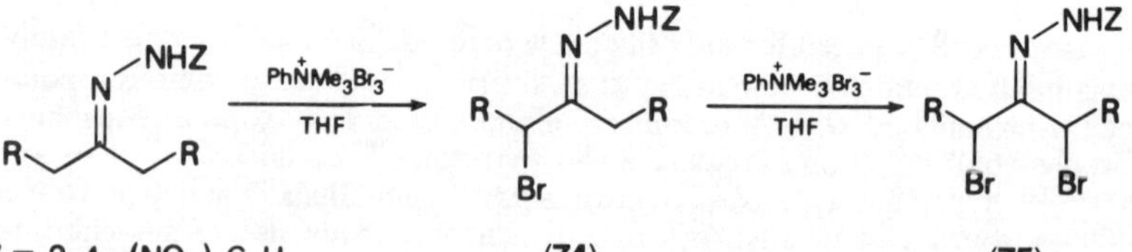

$Ph\overset{+}{N}Me_3Br_3^-$, THF

(73) Z = 2,4-$(NO_2)_2C_6H_3$
Z = Tosyl

(74) (75)

C. Synthesis of α-Halogenated Imino Compounds by Halogenation of Enamines

Only those cases in which secondary enamines (**76**), i.e. enamines having one hydrogen bonded to nitrogen, are converted into α-halo imines will be discussed in this section. Halogenation of other enamines to produce β-halogenated enamines will be discussed elsewhere[113].

(**76**) —Halogenation→ (**1**)

Halogenation of enamines with chlorine or bromine to give α-haloimines has not been amply documented[114,119], while the halogenation with *N*-halosuccinimide has been mainly applied to enamines carrying electron-withdrawing groups[115–118]. α-Cyanoenamines (**77**) were halogenated to produce α-halo imidoyl cyanides (**78**; X = Cl, Br)[115,116], while indoles were converted into 3-bromoindolenines by reaction

(**77**) —NXS, CCl_4→ (**78**) X = Cl, Br

with NBS in carbon tetrachloride[120,121]. Similarly, the conversion of α-cyanoenamines (**77**) into **78** (X = Cl) was accomplished with aqueous sodium hypochlorite[115]. On the other hand, indoles (**79**) were transformed into 3-chloroindolenines (**81**) with sodium hypochlorite[122,123], but it was shown that the reaction proceeded via intermediately formed *N*-chloroindoles (**80**)[124].

(**79**) —NaOCl→ (**80**) → (**81**)

t-Butyl hypochlorite has been proven to be a very efficient reagent for the conversion of indoles into chloroindolenines[119,125–131], and this method found widespread application in the alkaloid field. In the latter field, various indole-type alkaloids have been chlorinated to chloroindolenines such as yohimbine[132,133], ibogaine[134], cleavamine (**82**)[135], 14,15-dehydroquebrachamine[135], voaphylline[135], cacubine[136], tetraphylline[136], hirsutine *N*-oxide[137], pseudoyohimbine *N*-oxide[137], catharanthine[455] and several other examples[138,139,153,215,270,271,373,417,426]. It was recently shown that the chloroindolenines derived from cleavamine, 14,15-dehydroquebrachamine, voaphylline and some related derivatives have their chloro substituent (C-7 position) in a β-orientation, a conclusion which was drawn from detailed investigation of their ^{13}C-NMR spectra[135].

Another source of positive chlorine which was found to be efficient for the

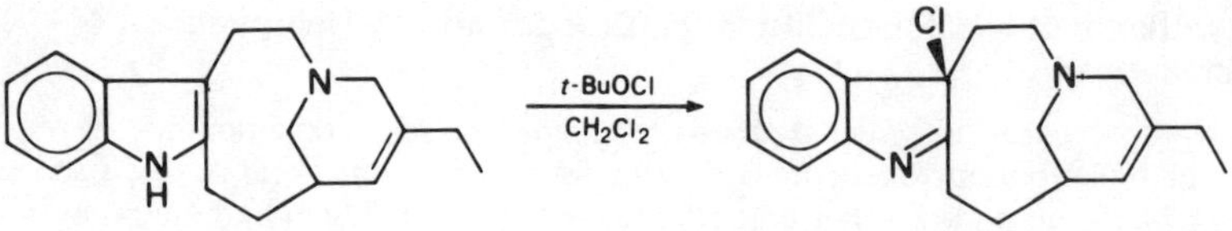

t-BuOCl, CH_2Cl_2

(**82**) Cleavamine (**83**)

conversion of indole alkaloids into chloroindolenines was N-chlorobenzotriazole[125,140], as reported for deserpine, yohimbine, catharanthine and (±)-dihydrocorynantheal[140].

D. Miscellaneous Methods

α-Halo imines carrying electron-withdrawing groups, e.g. alkoxycarbonyl, sulphonyl, acyl, aroyl, etc., at the nitrogen atom have a very electrophilic imino carbon, suitable for various reactions, including cycloadditions.

So-called 'anhydrochloralurethanes' (**86**) were synthesized from carbamate adducts of chloral (**84**) via conversion into the chlorides (**85**) and subsequent dehydrochlorination[141]. A similar methodology was applied to the synthesis of other related

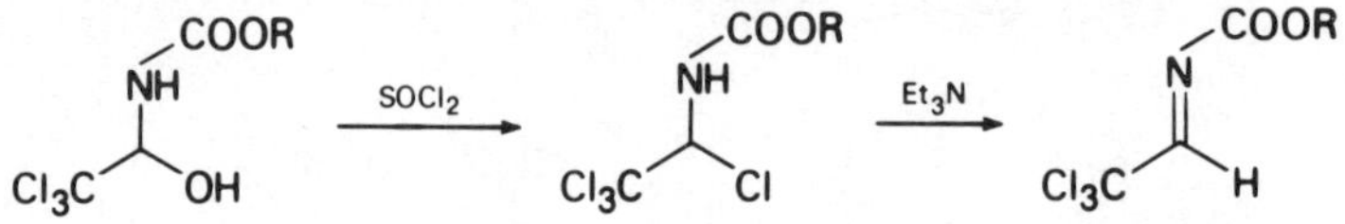

Cl_3C–CH(OH)–NH–COOR (**84**) $\xrightarrow{SOCl_2}$ Cl_3C–CH(Cl)–NH–COOR (**85**) $\xrightarrow{Et_3N}$ Cl_3C–CH=N–COOR (**86**)

classes of N-activated α-halogenated aldimines like α,α,β-trichloroimines (**87** and **88**)[142], N-acetyl-α,α,α-trichloroacetaldimines (**90**)[146], α,α-dichloroaldimines (**89**)[143], and N-sulphonyl-α,α,α-trihaloacetaldimines (**91**, **92** and **93**)[144,145].

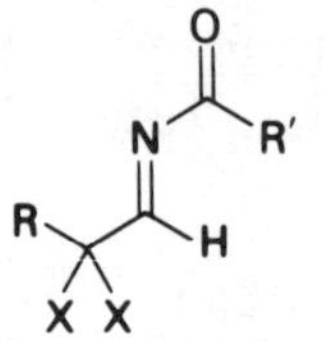

(**87**) R = CH_2Cl; X = Cl, R′ = Me, Et, n-Pr, Ph
(**88**) R = CH_2Cl; X = Cl; R′ = OEt
(**89**) R = alkyl; R′ = Me, Ph; X = Cl
(**90**) R = X = Cl; R′ = Me

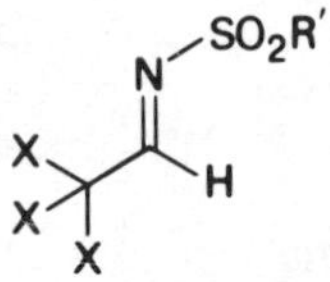

(**91**) X = Cl; R′ = p-MeC_6H_4
(**92**) X = F; R′ = p-MeC_6H_4
(**93**) X = Cl; R′ = NMe_2

Several other papers have reported the synthesis of N-activated α,α,α-trihaloacetaldimines, some of which are shown above[147–152,155].

The reaction of 1,2,2,2-tetrachloroethyl isocyanate (**94**) with alkyl orthoformates, N-silylamines or sulphur trioxide led to compounds **86**, **96** and **97**, respectively[147–151]. In many instances, these N-activated α-halogenated aldimines were postulated as intermediates[146–154]. For instance, sulphinate elimination from **98** under the influence

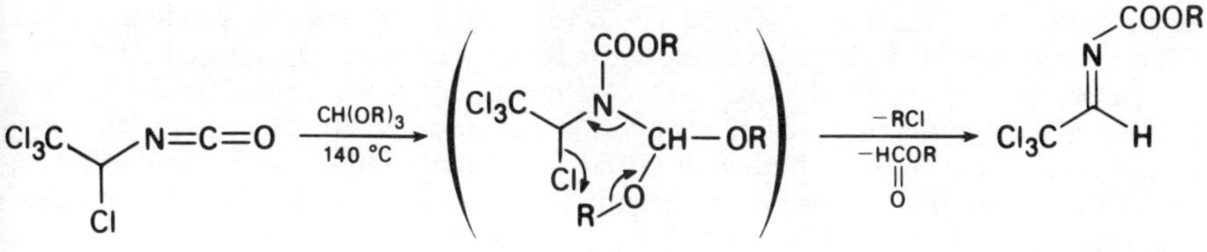

(94) (95) (86)

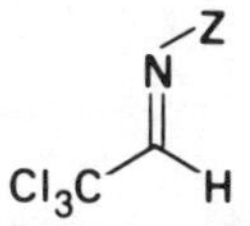

(96) Z = $C(O)NR_2$
(97) Z = SO_2Cl

of vinylmagnesium bromide produced the intermediate imine, to which the Grignard reagent added to give adduct **100**[154].

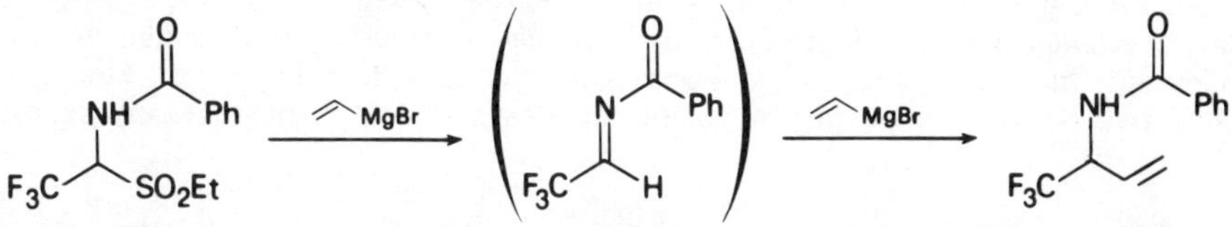

(98) (99) (100)

α-Halogenated oximes are available by the direct oximation of α-halo carbonyl compounds (*vide supra*), but can be obtained by two other general routes, namely the addition of nitrosyl halides to alkenes and the reduction of nitroalkenes.

The Markovnikov addition of nitrosyl chloride to olefins (**101**) yields β-chloronitroso compounds (**102**) which isomerize into α-chloro oximes (provided that at least one olefinic hydrogen is present in the starting alkene)[158–162]. Dimerization of the intermediate β-chloronitroso compound (**102**) is frequently observed, but thermal dissociation or acid-catalysed conversion of the dimer (**104**) into the monomer can generate α-chloro oximes (**103**)[163–164,435]. The addition of nitrosyl chloride to olefins is acid catalysed or can be photo-induced[165]. Simple alkenes[166,167,216,217,449], endocyclic[165,168,170,429,435] and exocyclic alkenes[166] or functionalized alkenes (e.g.

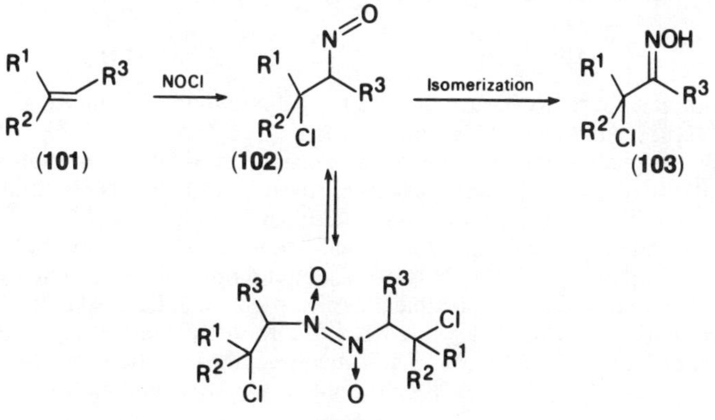

(101) (102) (103)

(104)

acrylonitrile)[169,450] react with nitrosyl chloride in a general mode to produce α-halo oximes. Nitrosyl chloride adds preferentially to the more substituted olefin as illustrated by the reaction of NOCl with a 4:1 mixture of 2-butene (**105**) and 1-butene (**106**) in decalin in the presence of dry hydrogen chloride, to afford the hydrochloride of the oxime of 3-chloro-2-butanone (**107**)[167]. 1-Butene (**106**) remained unaffected under these conditions. Nitrosyl sulphate, in the presence of hydrogen chloride, also converts alkenes into α-chloro oximes[171].

(**105**) 4:1 + (**106**) —NOCl, HCl(gas) / Decalin, 0–5 °C/10 min→ (**107**) NOH.HCl, Cl + (**106**)

Nitrosyl fluoride has been reported to add to alkenes to give unstable α-fluorooximes[418], while steroidal olefins (steroid 5-enes) are known to react with excess NOF at 0°C in dichloromethane or carbon tetrachloride to furnish 5α-fluoro-6-nitrimines (i.e. *N*-nitro-α-fluoro imines)[418]. Another valuable route to α-halo oximes, mainly α-chloro derivatives, entails the reduction of nitroalkenes with stannous chloride in ether in the presence of hydrogen chloride[172,173,221]. Sterically hindered α-chloro oximes are accessible by this method[64,65], but a recent report

$R^1-C(R^2)=C(NO_2)-R^3$ (**108**) —$Sn(II)Cl_2$/ether, dry HCl, −10–0 °C→ $R^1R^2C(Cl)-C(=NOH)-R^3$ (**109**)

claimed an unexpected reduction of nitroölefin (**110**) with stannous chloride in tetrahydrofuran, containing hydrogen chloride, to give the non-halogenated oxime (**111**)[174].

(**110**) NO_2, O —$SnCl_2$. 2 aq, HCl/THF, 75%→ (**111**) NOH, O

An important route to *N*-unsubstituted α-halo imino compounds, e.g. imidoyl cyanides (**116**), amidines (**113**) and imidates (**123**), involves the addition of nucleophilic reagents (amines, cyanide, methoxide) to α-halogenated nitriles. Even sulphur nucleophiles added to the carbon–nitrogen triple bond, as exemplified by the reaction of phosphorus dithioacids with α-chlorinated acetonitriles[436].

All kinds of amines (ammonia, primary and secondary amines) have been shown to add to α-halo nitriles[175–181,190,191,203,204,408]. An equilibrium between isomeric amidines can exist when tautomerism is possible[182]. Alkylations of amidines (**113**) with methyl fluorosulphonate[180] or trimethyloxonium tetrafluoroborate[179] are easily accomplished.

Other approaches to α-haloamidines involved the reaction of β-halogenated α-chloroenamines (**118**)[193,194] or β-halogenated α-cyanoenamines (**119**)[145,177] with primary amines.

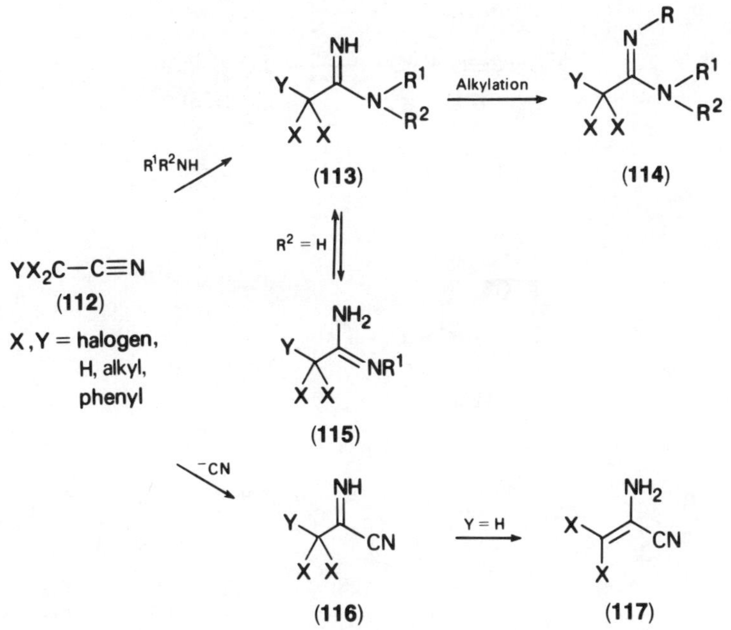

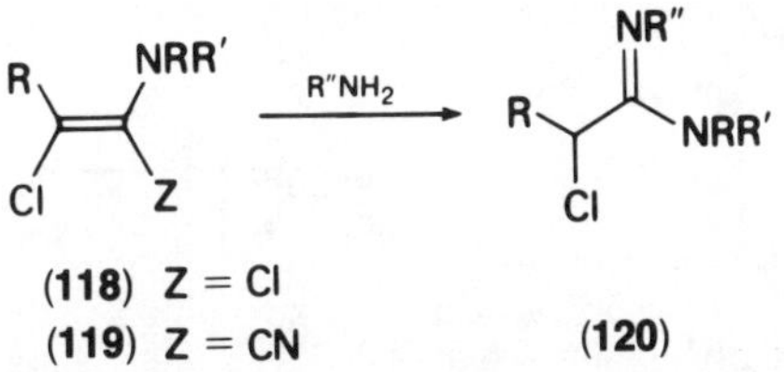

The base-catalysed addition of hydrogen cyanide to α-halo nitriles provides α-halo imidoyl cyanides (**116**)[183,184], which tautomerize to the more stable α-cyanoenamine (**117**) when an hydrogen atom α to the imino function is available[184].

Similarly, base-induced addition of alcohols[183,187,188], including allylic[187,192,425] and propargylic alcohols[189,205,425] to α-halo nitriles (mainly trichloroacetonitrile) to produce α-halogenated imidates (**123**) is a well known reaction. A cyclic functionalized imidate (**122**) was obtained from the reaction of sulphur trioxide with trichloroacetonitrile[192].

The condensation of α-bromo imidoyl chlorides (**125**), prepared from α-bromo carboxylic amides (**124**), with Grignard reagents in ether at low temperature yielded α-bromo ketimines (**126**) in 50–90% yield[195–198].

Some sophisticated α-halo imines in the small ring series have been synthesized by elegant strategies. Dichlorocarbene addition to azidoalkenes (**127**) gave 1-azido-2,2-dichlorocyclopropanes (**129**), which rearranged thermally under nitrogen expulsion to give 3,3-dichloro-1-azetines (**130**)[209]. The azidocyclopropanes (**129**) were also synthesized from aminocyclopropanes (**128**) via magnesium salt formation and treatment with tosyl azide (Anselme reaction[210])[211], and their pyrolysis furnished the four-membered heterocycles **130**[211].

3-Chloroazirines (**133** and **134**) are available from photolysis of β-chlorovinyl azides (**132**), the latter being obtained by iodine azide addition to vinyl chlorides (**131**) and

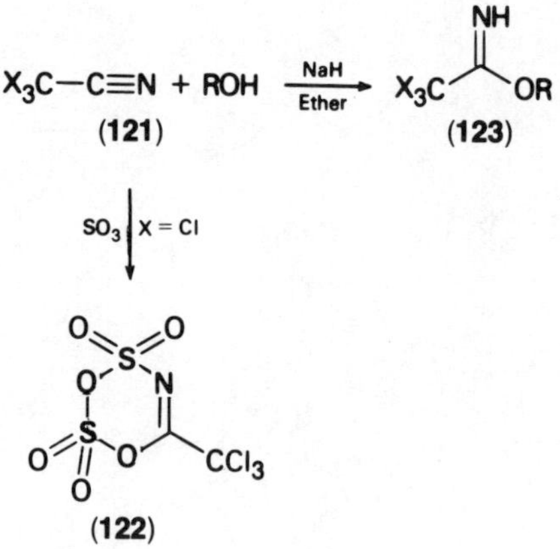

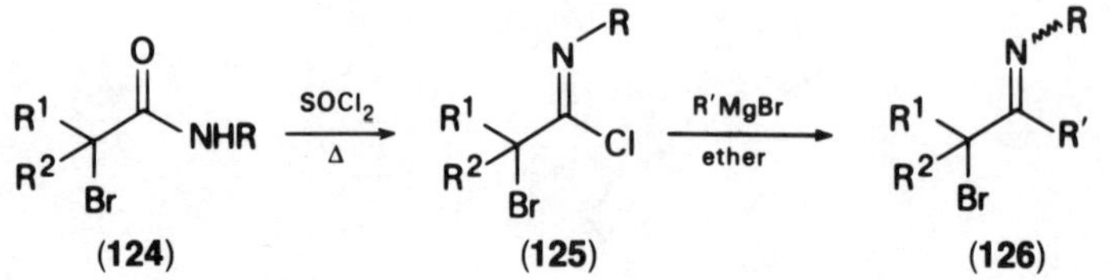

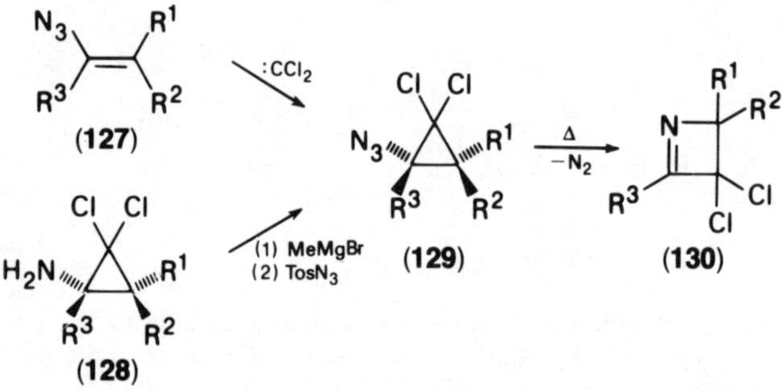

$X_3C{-}C{\equiv}N + ROH \xrightarrow[\text{Ether}]{\text{NaH}} X_3C{-}C(=NH){-}OR$

(**121**) (**123**)

SO_3 | X = Cl

(**122**) [O_2S ring with CCl_3]

(**124**) $R^1R^2C(Br){-}C(=O){-}NHR \xrightarrow[\Delta]{SOCl_2}$ (**125**) $R^1R^2C(Br){-}C(=NR){-}Cl \xrightarrow[\text{ether}]{R'MgBr}$ (**126**) $R^1R^2C(Br){-}C(=NR){-}R'$

(**127**) N_3, R^1, R^3, R^2 $\xrightarrow{:CCl_2}$ (**129**) Cl Cl, N_3, R^1, R^3, R^2 $\xrightarrow[-N_2]{\Delta}$ (**130**) N, R^1, R^2, R^3, Cl, Cl

(**128**) Cl Cl, H_2N, R^1, R^3, R^2 $\xrightarrow{(1)\ MeMgBr\ (2)\ TosN_3}$ (**129**)

subsequent base treatment[213,214]. Compound **132** (R = Ph) is photolysed to a 5:1 ratio of **133** and **134**, respectively, in carbon tetrachloride while a 3.3:1 ratio was observed in acetonitrile[214]. However, low temperature (−40°C) photolysis of **132** (R = Ph)

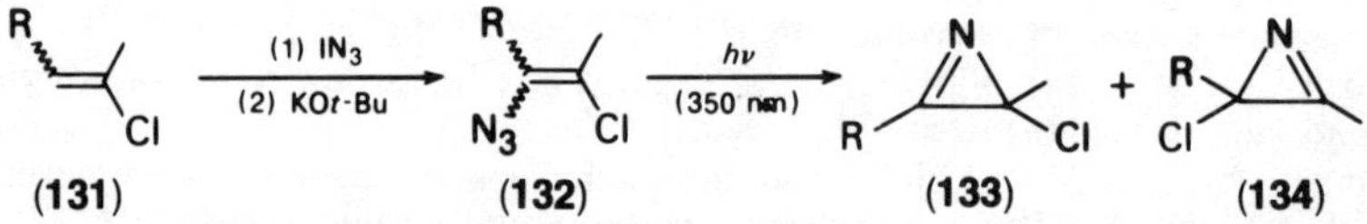

(**131**) R, Cl $\xrightarrow[(2)\ KOt\text{-}Bu]{(1)\ IN_3}$ (**132**) R, N_3, Cl $\xrightarrow[(350\ nm)]{h\nu}$ (**133**) N, R, Cl + (**134**) N, R, Cl

produced 3-chloro-3-methyl-2-phenylazirine (**133**, R = Ph) exclusively. The equilibrium mixture of **133** and **134** can be explained by interconversion via the azacyclopropenyl cation (**135**), but it was reasoned that an alternative mechanism involving a polar bridged transition state (**136**) cannot be excluded.

Many other reports dealing with the synthesis of less general types of α-halogenated imino compounds exist in the literature, some of which are reported in a recent review[1].

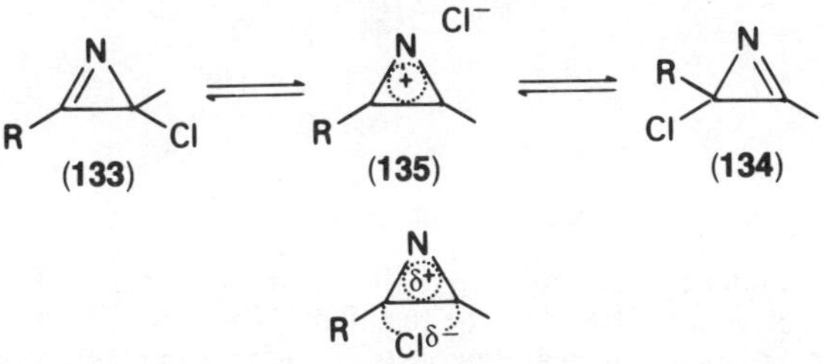

(133) (135) (134)

(136)

III. REACTIVITY OF α-HALOGENATED IMINES

As discussed in the foregoing sections, a great variety of synthetic methods for the synthesis of α-halo imines have become available, especially as a result of efforts in the last decade. Because of these efforts, many useful transformations of α-halo imines have been performed and it will be demonstrated here that their reactivity constitutes, among other things, a broadening of the possibilities for the widely used chemistry of α-halo carbonyl compounds. Indeed, α-halo imines can be regarded as masked α-halo carbonyl compounds and hence very specific transformations of α-halo imines, which cannot be executed with α-halo carbonyl derivatives, may be carried out. Simple hydrolysis of the resulting imines provides the carbonyl compounds. This strategy is outlined in the following scheme by means of an example. Dehydrohalogenation of α-halo aldehydes (**137**) to form α,β-unsaturated aldehydes (**138**) is not applicable in a synthetically useful manner[454], but this transformation is easily accomplished via the corresponding α-bromo *N*,*N*-dimethylhydrazone **139**, which is subsequently dehydrohalogenated in the same reaction; finally, acidic hydrolysis affords the desired unsaturated aldehydes (**138**)[219]. Many other applications will follow in the forthcoming text.

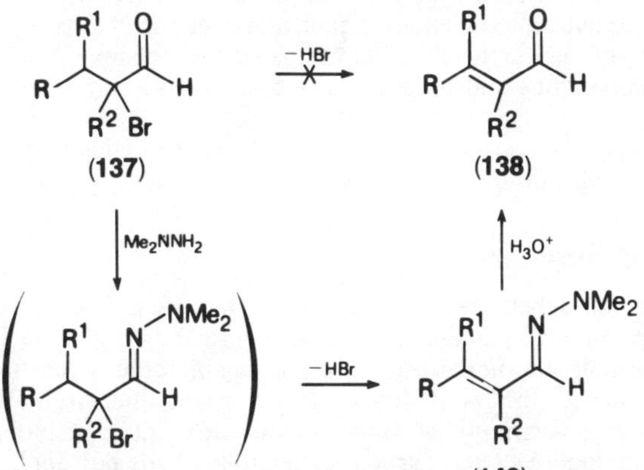

(137) (138)

(139) (140)

α-Halogenated imino compounds and α-halogenated carbonyl compounds are related substances in which the heteroatom determines the difference in reactivity. Also allylic halides can be compared in this context, in that the heteroatom is replaced by carbon. The difference in reactivity between compounds **141**, **142** and **143** is mainly based on the difference in electronegativity between oxygen, nitrogen and carbon.

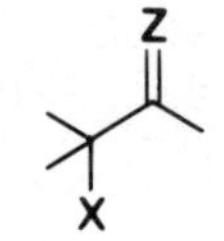

Z

X

(**141**) Z = O
(**142**) Z = NR
(**143**) Z = CR^1R^2

X = halogen

Nitrogen holds an intermediate position in this series and it is therefore expected that its reactivity will be situated between the reactivity of α-halo carbonyl compounds and allylic halides. Many reactions will demonstrate the intermediate character of α-halo imines (*vide infra*).

The reactivity of α-halo carbonyl compounds has already received considerable attention in the literature and, in an accompanying chapter in this book, some general trends and novel developments in this field will be discussed. Allylic halides (**143**; Z = CR^1R^2) can be considered as the carbon analogues of α-halo imines and α-halo carbonyl derivatives and their chemistry is well known, mainly because of its various nucleophilic substitutions, e.g. S_N1, S_N2, S_N2', etc.[222,223].

When combining an imino function and a halide into an α-halo imino system, one can expect a reactivity which depends on one or other of these functional groups or one can expect a greater versatility of the system by the combined interaction of the halide and the imine. In several aspects, the reactivity of α-halo imines parallels the reactivity of α-halo carbonyl compounds. Reactions such as rearrangements via three-membered heterocycles, elimination, nucleophilic substitution, addition to the carbon–heteroatom bond, Favorskii-type rearrangements, elimination–addition, etc., are frequently encountered. These possibilities have recently been treated in detail[2]. The decreased electronegativity of nitrogen as compared to oxygen lowers the electrophilic character of the imino carbon atom and reduces the acidity of the α-protons. These two fundamental characteristics account for a substantial decrease in reactivity of α-halo imines with respect to α-halo carbonyl compounds. The drop in reactivity permits other reactions to become more important. As already discussed above, the infrequently encountered elimination reaction of α-halo carbonyl derivatives will be shown to be an important characteristic of α-halogenated imines. In the α-halo carbonyl series, this reaction can usually not compete with other reactions such as α-deprotonations and following reactions, substitutions, rearrangements via epoxides, etc.

The discussion of the reactivity of α-halo imines will be divided into several sections, each one dealing with different pertinent reaction types.

A. Nucleophilic Substitutions

Many nucleophilic substitutions of α-halo carbonyl compounds have been reported in the literature, but this reaction cannot be regarded as a general feature of these substrates as the substitution pattern in the starting material is determinative in these cases. For example, the well known nucleophilic substitutions of phenacyl halides[224,225] by a variety of nucleophilic reagents, including nitrogen[226], oxygen[227,228], sulphur[229,230] and carbon nucleophiles[231], is not applicable that much to α-substituted and α,α-disubstituted[232,242] phenacyl halides (secondary and tertiary derivatives) as only good nucleophiles (e.g. azide) were found to substitute the latter tertiary α-halo ketones. Other nucleophiles, such as methoxide in methanol, were reported to react with aromatic secondary and tertiary α-halogenated ketones, such as 1-aryl-2-halo-1-alkanones[240,241] and 1-aryl-2,2-dichloro-1-alkanones[233], via an epoxide rearrangement. Several mechanistic propositions concerning the pronounced

S_N2 reactivity of phenacyl halides have been formulated in the literature [228,234–237]. Introduction of one or two α-alkyl substituents in phenacyl halides drastically reduced the S_N2 reactivity[238]. It seemed that steric factors determine this behaviour, although it was reported that nucleophilic substitutions of α-halo carbonyl compounds are almost unaffected for steric reasons[239]. All these arguments can be considered when overlooking the chemical behaviour of α-halo imines towards nucleophilic substitution. The reduced electronegativity of nitrogen as compared to oxygen is responsible for a less positively induced imino carbon, thus resulting in a decreased repulsive effect of the latter with the adjacent positively induced halocarbon atom. Due to the latter feature, α-halo imines show a reasonable tendency to give α-substitution, despite its decreased general reactivity. Since no mechanistic details for substitutions of α-halo imines are available, distinction between a classical S_N2-type displacement[243], or cases in which considerable positive charge develops in the transition state[247], or displacement on an ion pair intermediate, is, at present, difficult[245].

Strong nucleophiles, e.g. thiolates, gave α-substitution of α-halogenated imines (**144**) to afford **145**[87,116,246,247], but with other nucleophiles competition with other reactions frequently occurred.

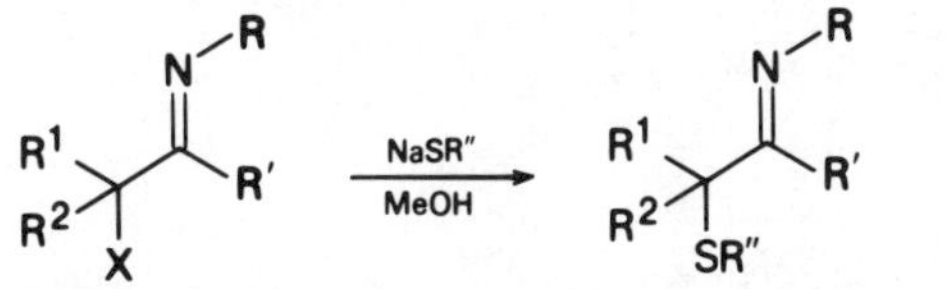

(**144**) X = Cl, Br
R′ = H, CN

NaSR″ / MeOH →

(**145**) R′ = H, CN
R″ = alkyl, Ph

Alkoxides in the corresponding alcohol often yield α-alkoxy imines. *N*-Cyclohexyl α,α-dichloromethylketimines (**46**) gave α,α-dimethoxymethylketimines (**146**) exclusively when refluxed with concentrated methanolic sodium methoxide solution for a prolonged period[80]. Similarly, *N*-aryl α,α-dimethoxyketimines (**147**) were obtained but a Favorskii-type rearrangement to α,β-unsaturated imidates was a competing reaction (*vide infra*)[82,248].

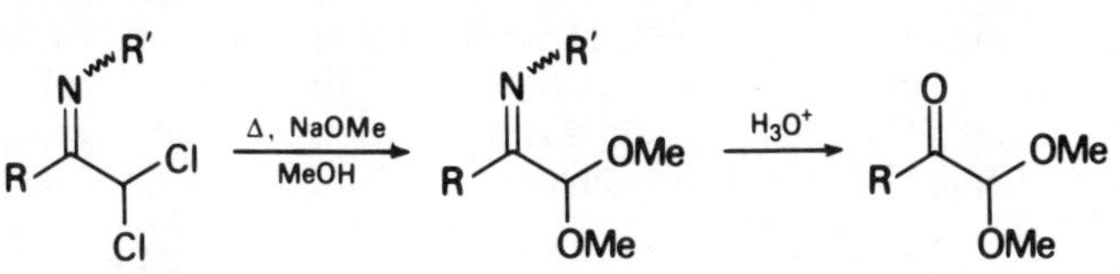

(**46**) R′ = C_6H_{11}
(**47**) R′ = aryl

Δ, NaOMe / MeOH →

(**146**) R′ = C_6H_{11}
(**147**) R′ = aryl

H_3O^+ →

(**148**)

Haloindolenines (**149**) are readily converted into α-alkoxyindolenines (**150**) by treatment with alkoxides, because of the stabilizing effect of the aryl substituent on the halogenated carbon atom; the aryl group participates in the resonance stabilization of

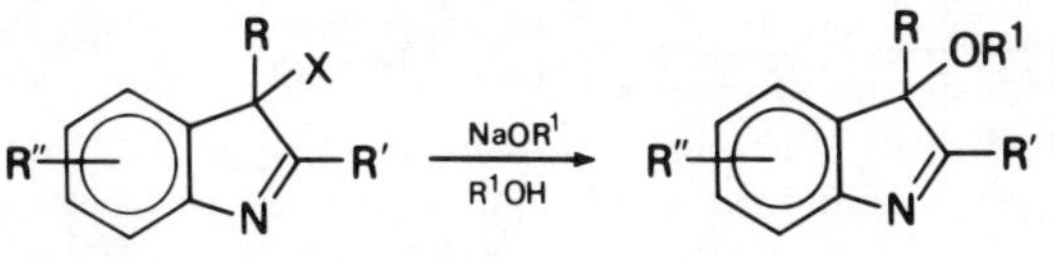

(**149**) X = Cl, Br

$NaOR^1$ / R^1OH →

(**150**)

the developing carbonium ions during nucleophilic substitution. When treated with cold base, the products are alkoxyindolenines (**150**)[119,122,125] while at elevated temperature[121,122,125,132,133,138] or with mild acid[129,249,250] the product is a rearranged spiro compound (see Section III.E).

Alcoholysis of the bromo- or chloroindolenines derived from tetrahydrocarbazole or 2,3-dimethylindole produced the alkoxyindolenine **150**[119,251], but the fact that α-bromoindolenines did not react with methanol in the presence of triethylamine strongly suggested that the methanolysis of the haloindolenines is an acid-catalysed process and thus probably proceeded via a transient *N*-protonated bromoindolenine (**151**)[119].

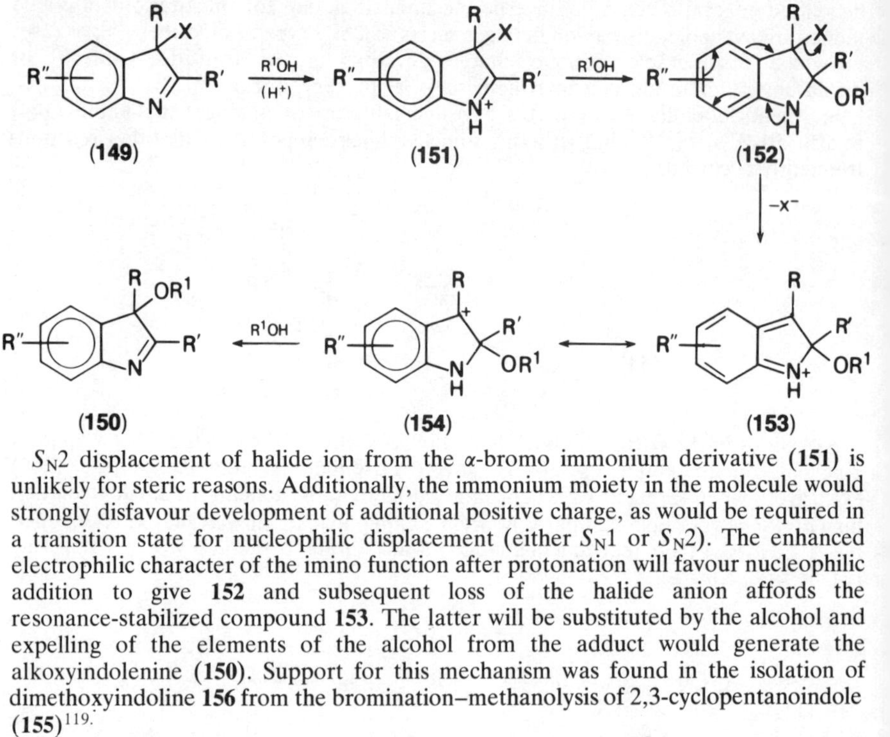

S_N2 displacement of halide ion from the α-bromo immonium derivative (**151**) is unlikely for steric reasons. Additionally, the immonium moiety in the molecule would strongly disfavour development of additional positive charge, as would be required in a transition state for nucleophilic displacement (either S_N1 or S_N2). The enhanced electrophilic character of the imino function after protonation will favour nucleophilic addition to give **152** and subsequent loss of the halide anion affords the resonance-stabilized compound **153**. The latter will be substituted by the alcohol and expelling of the elements of the alcohol from the adduct would generate the alkoxyindolenine (**150**). Support for this mechanism was found in the isolation of dimethoxyindoline **156** from the bromination–methanolysis of 2,3-cyclopentanoindole (**155**)[119].

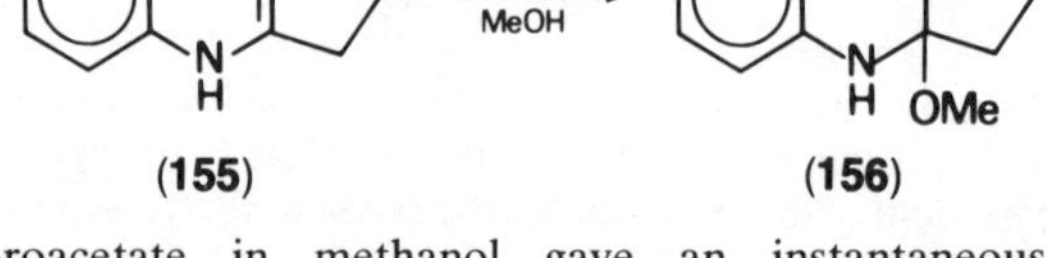

Silver trifluoroacetate in methanol gave an instantaneous reaction with 3-chloro-2,3-dimethylindole[122], but it was recently shown that the reaction also proceeded without the aid of silver salts[119].

When an α'-hydrogen is present in α-halo imino systems, tautomerism to allylic halides (**158**) is possible and these substances produce a delocalized carbonium ion (**159**), which is trapped by the solvent. Depending on the stabilizing effect of the substituents in **159**, the solvolysis leads to one or other (or both) of the two α-methoxy

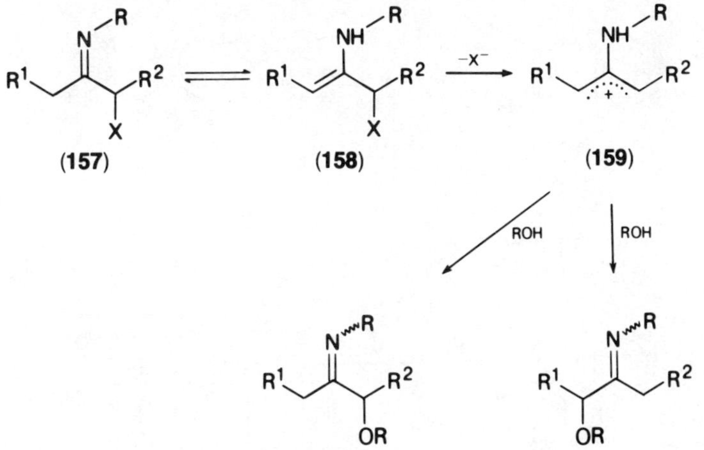

(157) (158) (159)

(160) (161)

ketimines. Many papers about chloroindolenines have dealt with this topic[122,125,128,130,132,249–252,373]. The conversion of secondary *N*-phenyl-1,1-dichloromethylketimines (**162**) with sodium methoxide in methanol under reflux into *N*-phenyl-1,3-dimethoxymethylketimines (**164**) was explained by a solvolysis mechanism (additionally, nucleophilic substitution and Favorskii-type rearrangement occurred) via an enamine allylic halide[248]. The intermediacy of α-chloro-α′-methoxyketimine (**163**) was substantiated by spectral evidence[248].

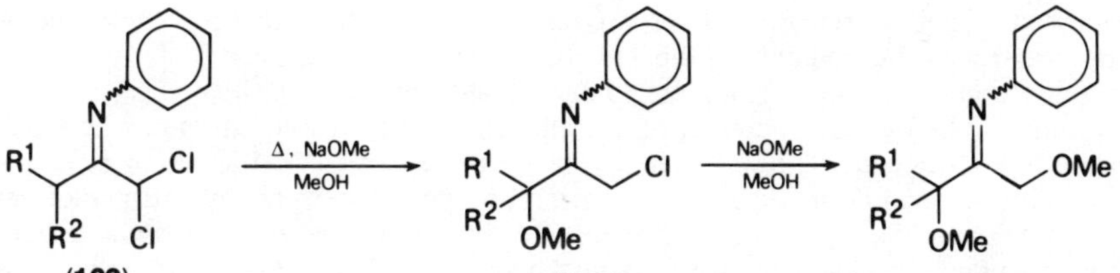

(162) (163) (164)

Similarly to the solvolyses in the chloroindolenine series (*vide supra*), the presence of an α-phenyl substituent in α-chloro aldimines (**165**) is of major importance in determining the course of the reaction. With methoxide in methanol, α-chloro aldimines (**165**) afforded α-methoxy aldimines (**166**) exclusively, while α-methoxyacetals (**167**) were produced in methanol, indicating methanolysis via α-methoxy aldimines (**166**)[87]. Silver ion-assisted alcoholysis of the α-bromo tosylhydrazone of 14-bromodaunomycine (**31**) proceeded smoothly at room temperature, giving rise to α-alkoxy tosylhydrazones[51] (**168**; R = Me, Et, *i*-Pr).

Of course, questions arise here concerning the structure and the stability of a carbonium ion at the α-carbon of imines. No mechanistic studies have been directed hitherto towards the identity of α-imidoyl carbonium ions. The terminology 'α-imino carbonium ion' is incorrect as positional labelling in carbonium ions assigns α to the charge-centre carbon atom. Analogously, in the oxygen series, the well-known species α-keto carbonium ions are better referred to a α-acyl carbonium ions[427,428]. As

(165) R = CH_3, Ph → NaOMe/MeOH → (166)

(165) → MeOH → (166) → MeOH (HCl) → (167)

(31) → ROH, CH_2Cl_2/ether, CF_3SO_3Ag, RT = 1 h → (168) R = Me, Et, *i*-Pr

discussed above, electronic effects reduce markedly the stability of α-imidoyl carbonium ions, but, similar to the case of α-acyl carbonium ions[427], the electronic configuration of an imino group is capable of stabilizing the positive charge on the adjacent carbon by overlapping of the vacant orbital of the carbonium ion with either the occupied lone pair orbital of nitrogen or the π-orbital of the imino function. Accordingly, the intermediacy of an azirinium species, formed by intramolecular nucleophilic halide displacement, seems to be attractive and warrants serious consideration in mechanistic explanations. Quantitative data of α-acyl carbonium ions only very recently became available[427,428], but the corresponding nitrogen analogues have only been postulated as intermediates (*vide supra*).

Other examples of nucleophilic substitutions using oxygen nucleophiles entailed sodium acetate in acetic acid[253], intramolecular substitution of α,α-dichloroimidates by aryloxides[254,255], α-substitutions with silver nitrate[95–97,116], sodium nitrite[96,116], hydroxide[98], bicarbonate[96] or phenoxide[98]. The bromination in acetic acid and

(78) X = Cl, Br → $NaNO_2$ or $AgNO_3$ (DMSO or MeCN) → (169) Z = ONO, ONO_2

subsequent hydrolysis of tetrahydrocarbazole also provided an example of α-hydroxylation[256].

α-Substitutions with amino compounds are not frequently reported. Ordinary aliphatic α-halo imines show a complete lack of reactivity towards amines. Chloromethyl imino compounds seem to be the substrates of choice for α-aminations, as demonstrated by reactions of α-brominated diazines (**170**)[74,258], α-chloro amidines[257] and α-bromo hydrazones[259,260]

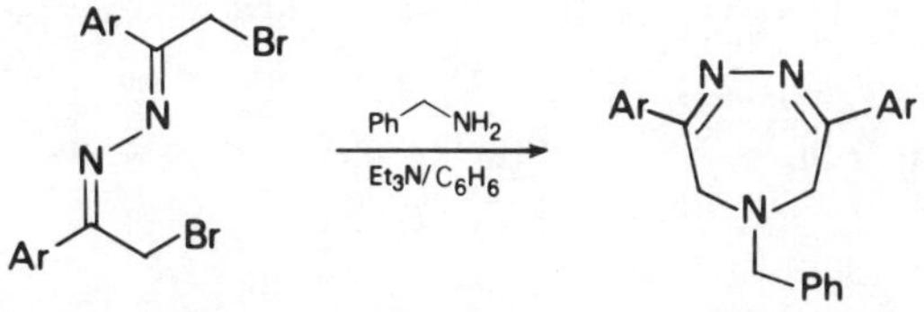

The introduction of an amino substituent α to an oxime can be accomplished by substituting an α-bromo oxime (**172**) with potassium phtalimide in acetonitrile in the presence of crown-18 ether, after which the α-phtalimido oxime (**173**) is subjected to hydrazinolysis in ethanol, the resulting α-amino oxime (**174**) being used as a key intermediate for the construction of 11-oxahomofolic acid, a potential antitumour agent[156].

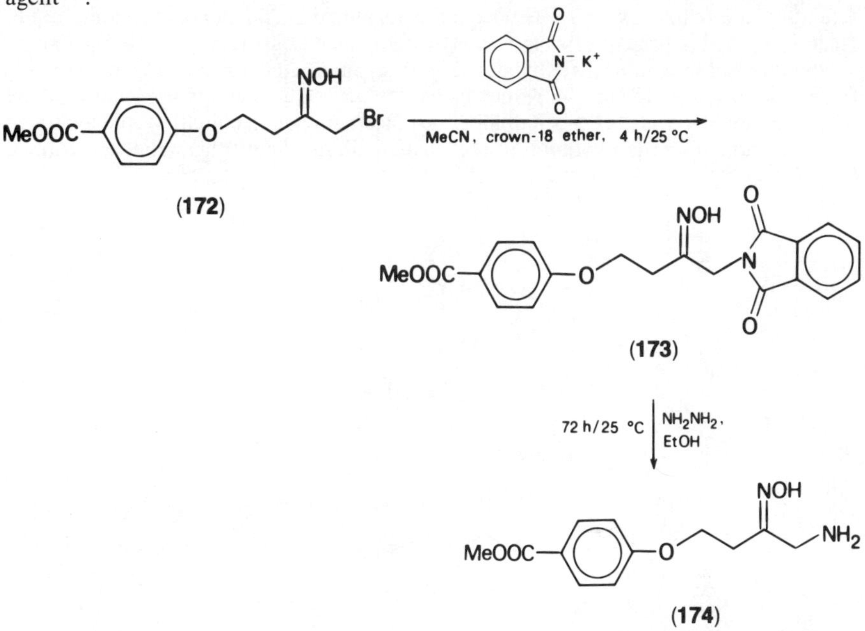

Azide ion, usually in acetone, acetonitrile or acetic acid, converts α-halogenated imino derivatives into unstable α-azido imines[87,253,261]. Phosphorus-containing nucleophiles like trialkyl phosphites do not react with α-chloro aldimines (**59**), but are known to substitute trichloroacetimidates and trichloroacetamidines at the α-position[262] (or at the imino-nitrogen atom[263] in analogy to the Arbuzov or Perkow reaction of α-halo carbonyl compounds[264]).

Triphenylphosphine easily substituted the protected α-bromo oximes (**175**), the α-substituted derivatives (**176**) subsequently yielding oximes (**177**)[157]. The latter underwent ring closure under basic conditions to give five-membered heterocycles (**178**) which were converted into azirines (**179**) by thermolysis[157]. When the group replacing the halogen is sensitive to nucleophilic reagents, intramolecular nucleophilic

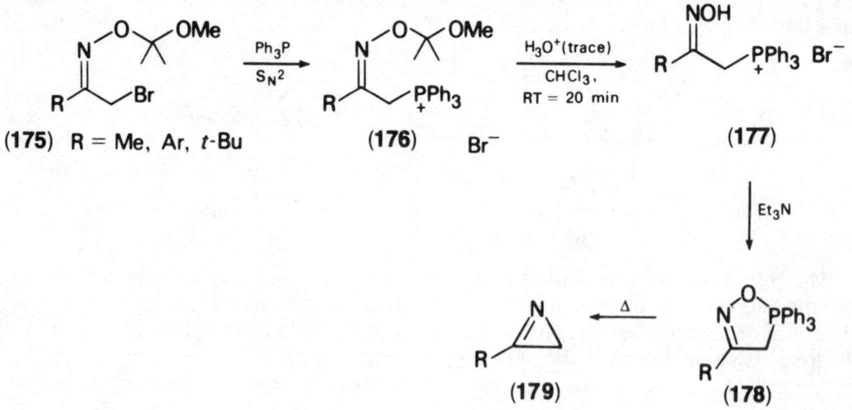

(**175**) R = Me, Ar, *t*-Bu (**176**) (**177**) (**178**) (**179**)

attack by the oxime oxygen can take place to afford *O,N*-heterocyclic compounds. According to this principle, α-chloromethylketoximes (**180**) reacted with phosphines or dimethylsulphoxonium methylide to give α-substituted oximes **181** or **183** and further heterocycles **182** or **184**, respectively[265–268]. It has not been stated whether these reactions involved direct nucleophilic displacement or elimination of hydrogen chloride and subsequent addition of the nucleophile to the nitrosoölefin thus formed (*vide infra*).

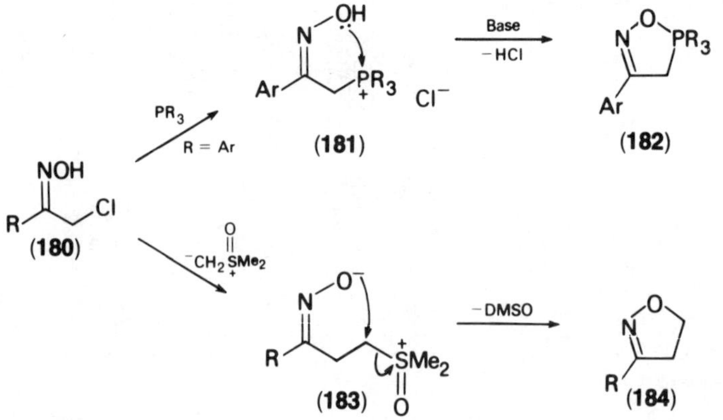

(**180**) (**181**) (**182**) (**183**) (**184**)

Finally, some displacements by direct attack of the nucleophile (iodide, thiophenolate, triphenylphosphine) at the halogen in chloroindolenines were reported to yield the parent indoles[253]. Nucleophilic substitutions involving carbon nucleophiles are included in the next section.

As pointed out above, α-halogenated oximes (**187**) are known to react with nucleophiles to yield the corresponding α-substituted oximes (**191**), but the reaction involves elimination to a nitrosoölefin (**189**) and Michael-type addition of the nucleophile to the latter intermediate (**189**). A similar type of elimination–addition is known for α-halogenated hydrazones (**188**), which are transformed by nucleophiles

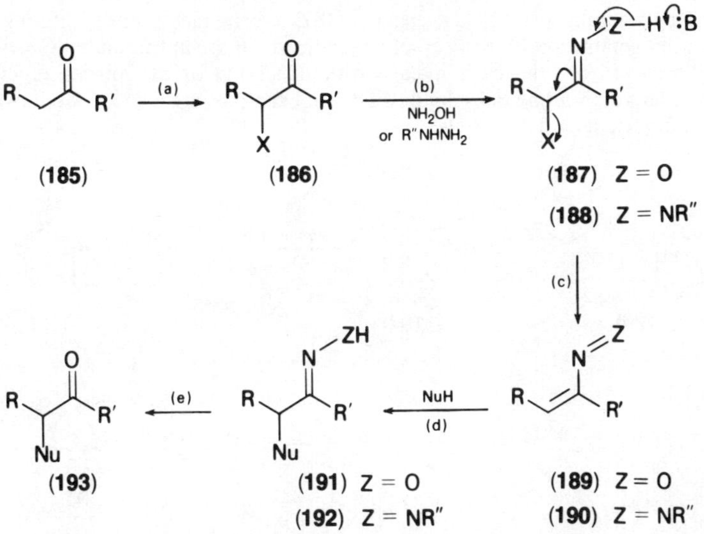

(185) (186) (187) $Z = O$ (188) $Z = NR''$

(193) (191) $Z = O$ (192) $Z = NR''$ (189) $Z = O$ (190) $Z = NR''$

into α-substituted hydrazones (**192**) via the intermediacy of azoalkenes (**190**). As outlined in the accompanying scheme, carbonyl compounds (**185**) are transformed into α-substituted derivatives (**193**) by a sequence involving (a) halogenation, (b) oximation or hydrazone formation, (c) elimination of hydrogen halide to form a nitrosoölefin (**189**) or an azoalkene (**190**), (d) addition of the nucleophile and (e) hydrolysis. Steps (c) and (d) are usually performed in one treatment when the nucleophile displays basic properties.

Secondary amines have been widely used to substitute α-halo oximes[94,165,168,172,287–289,409], but also primary amines[287,290,449] and ammonia[287] gave α-amino oximes. α-Alkoxylations, usually α-methoxylations, proceed smoothly for

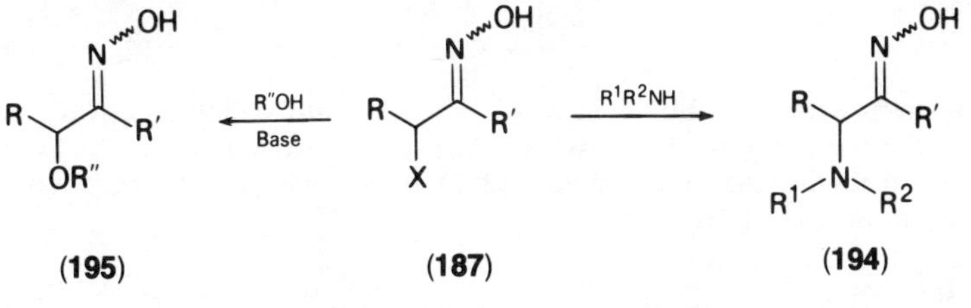

(195) (187) (194)

certain substrates with alcohols[172,291,409], but are facilitated when bases, e.g. triethylamine[165] or alkoxides[166,168,291], are used.

Other nucleophilic reagents such as sulphur nucleophiles[165,447], cyanide ion[293], sodium borohydride[297], sodium nitrite[287], sodium nitrate[287] and sodium azide[287] also provided elimination–addition reactions of α-halo oximes to generate α-substituted oximes (**191**). Carbon–carbon bond formation[447] was accomplished using carbanions derived from ethyl acetoacetate[295], 3-phenyl-2-isoxazolin-5-one[295], diethyl malonate[287] and 2,4-pentanedione[287], while Grignard reagents gave the α-alkylated oximes[287,295]. The generality of such reactions was shown by the reaction of α-bromo oxime (**196**) with the lithium enolate **197**, upon which cyclization resulted[69], and by the substitution of an α-bromo cyclohexanone oxime derivative with 1-lithio-1-butyne, which furnished the α-(1-butynyl)oxime[68]. In similar fashion, α-bromo oximes were alkylated by enamines[296]. α-Halo oximes (**187**) are in fact useful synthons for the base-induced generation of nitrosoölefins (**189**), which are apt to undergo a variety of cycloadditions. Either the carbon–carbon double bond or the nitroso function can participate as a dienophile in cycloaddition, but examples are also known in which the nitrosoölefin acts as a heterodiene.

(**196**) (**197**) 45 min THF, −78 °C (**198**)

(**187**) Base (**189**) (**199**) (**200**) (**201**)

Applications of cycloadditions in which transient nitrosoölefins operate as dienophiles are the reaction of α-chloro oximes (**180**) with cyclopentadiene in the presence of sodium carbonate, the initial adducts (**202**) being transformed spontaneously into *cis*-fused oxazine derivatives (**203**)[292].

(**180**) Na_2CO_3 (**202**) (**203**)

The nitroso function of nitrosoölefin (**205**), generated from chloral oxime (**204**) and sodium bicarbonate, underwent cycloaddition with cyclopentadiene to give oxazine derivative **206**, which rearranged to the tricyclic compound **207**[292,297]. An example in which the intermediate nitrosoalkene acts as a heterodiene in cycloadditions was found recently with α-chloro oximes (**208**), carrying an electron-withdrawing substituent[298,451]. Related reactions are the cyclocondensations of α-chloronitrones[299–305], e.g. **211**, with cycloalkenes (**212**) and alkenes[411] to give bicyclic adducts (**213**). This reaction was recently shown to be applicable also with ketones and α-chloro nitrones to generate 4*H*-1,5,2-dioxazinium salts[456].

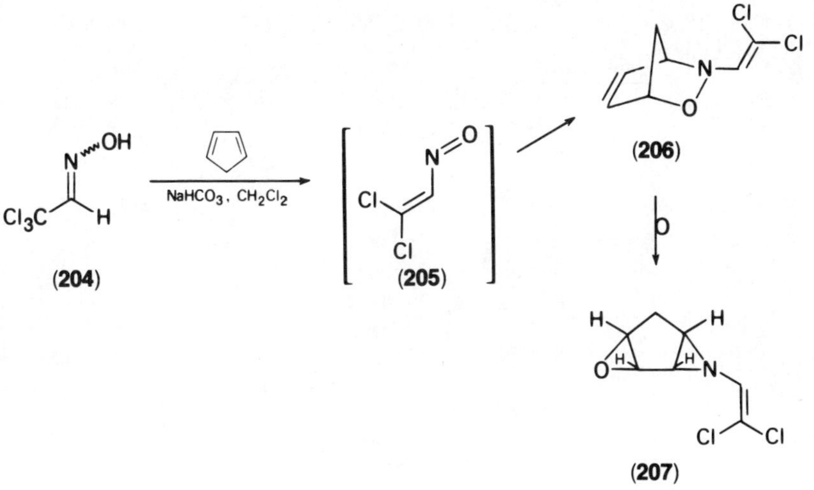

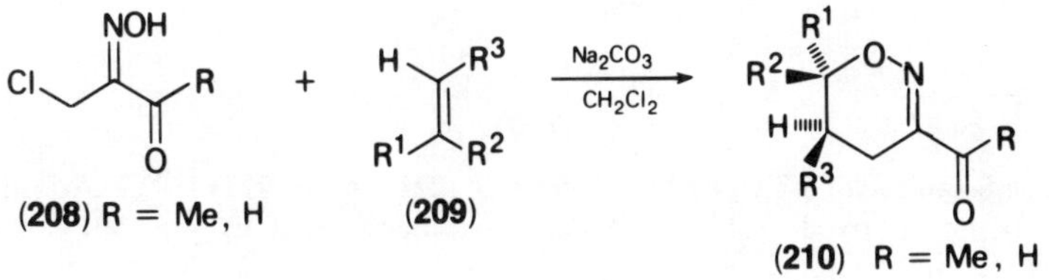

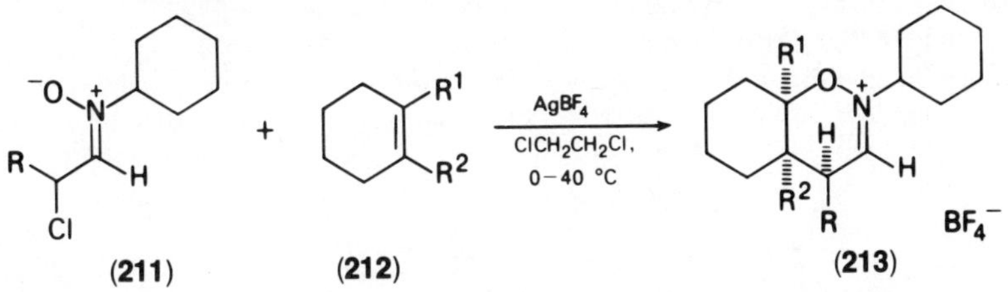

According to the general principle of base-promoted elimination–addition of α-halo hydrazones (**188**), a wide variety of α-substituted hydrazones (**192**) are obtainable. Azoalkenes (**190**) have been isolated in many cases[8–10,47,48,111,306,307–309,317–319,334] and their stereochemistry was investigated to some extent[308,317,434]. However, the *in situ* preparation of azoalkenes is most often applied[18,23,46,310]. Addition of nucleophiles, e.g. acetate[23,46], amines[23,46,307], organocopper reagents[49,50,309], Grignard reagents[325–327] or carbanions[112] to azoalkenes (isolated or prepared *in situ*) provided α-substituted hydrazones (**192**) but also cycloaddition products, in which azoalkenes act as diene or dienophile, have been reported (see, for instance, **218**)[308,311,312]. α-Halogenated hydrazone-type compounds (**219**), contained in a ring system, have been used for the synthesis of α-vinylcarboxylic acids (**220**)[101], α,β-unsaturated carboxylic acids[77,313–316]

and 1,5-diazabicyclo[3.3.0]octadienediones (**221** and **222**)[76,286]. Additionally, tosylazoalkenes derived from aldehydes have been used to generate alkylidene carbenes[453].

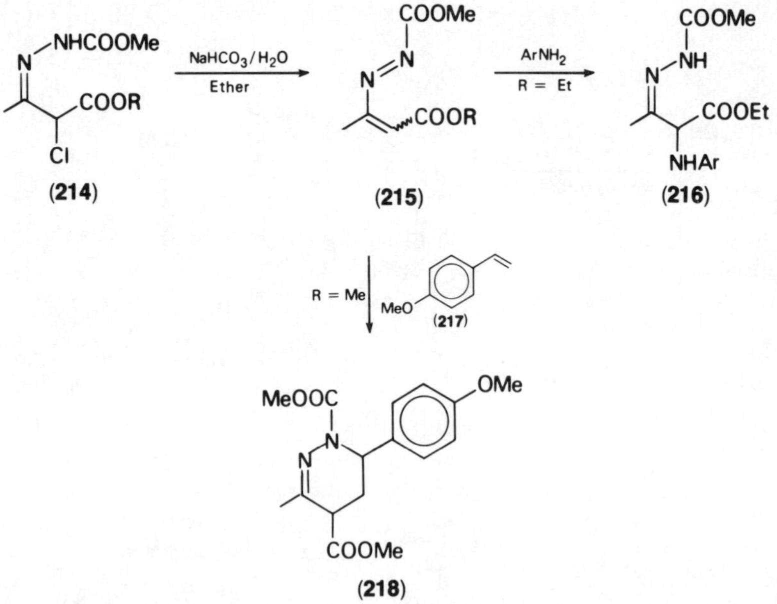

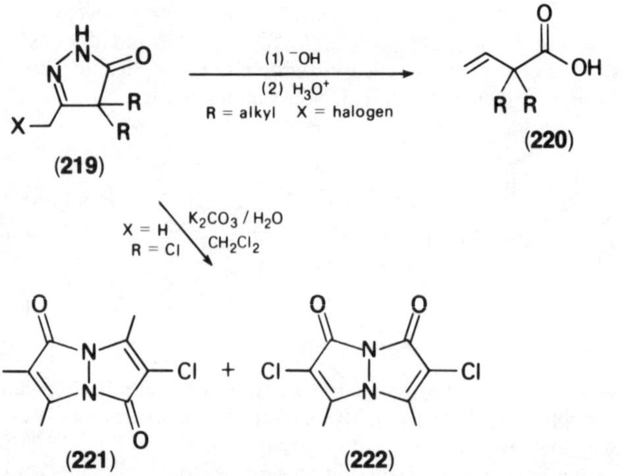

B. Carbon–Carbon Bond Formation

Because of the importance of carbon–carbon bond formation in synthetic organic chemistry, emphasis to this topic is given in a separate section covering reactions of α-halogenated imino compounds with carbanions, cyanide ion and organometallic reagents.

1. Reactions of α-halogenated imino compounds with carbanions

Few reports have been published on the reaction of α-halo imines with carbanions. Tertiary α-chloro aldimines (**59**) are completely resistent to reaction with carbanions derived from active methylene functions[87], but α-halomethyl imino derivatives proved to be more successful in nucleophilic substitutions. Chloroacetone semicarbazone gave α-substitution with the active methylene compound **224** and the resulting product was subjected to ring closure with hydrogen chloride in alcoholic medium to afford functionalized pyrroles (**226**)[109].

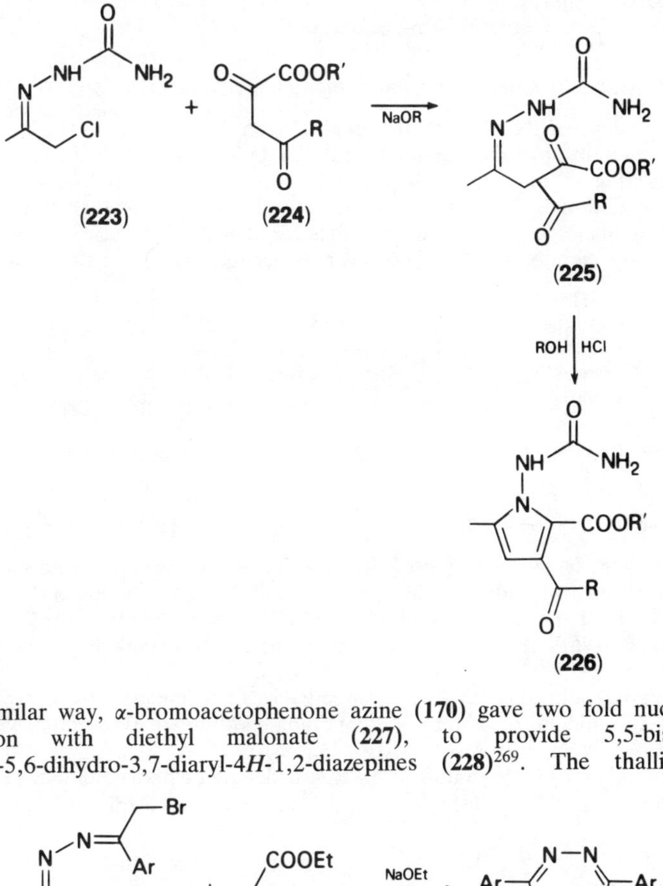

O NH NH2 N Cl + O COOR' R O NaOR O N NH NH2 O COOR' R O

(**223**) (**224**) (**225**)

ROH HCl

O NH NH2 N COOR' R O

(**226**)

In a similar way, α-bromoacetophenone azine (**170**) gave two fold nucleophilic substitution with diethyl malonate (**227**), to provide 5,5-bis(ethoxycarbonyl)-5,6-dihydro-3,7-diaryl-4*H*-1,2-diazepines (**228**)[269]. The thallium salt

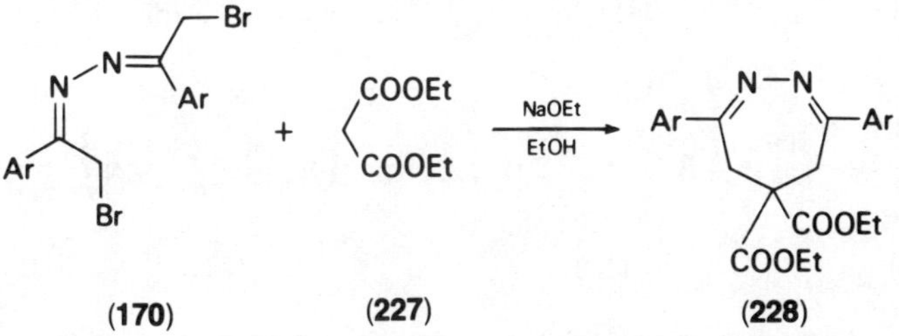

Br N N Ar Ar Br + COOEt COOEt NaOEt EtOH N N Ar Ar COOEt COOEt

(**170**) (**227**) (**228**)

of diethyl malonate in benzene reacted with the chloroindolenine derived from tetrahydrocarbazole (**229**) to produce the nucleophilic addition product (**230**), which rearranged into compound **231**[271]. Such a rearrangement was applied to a

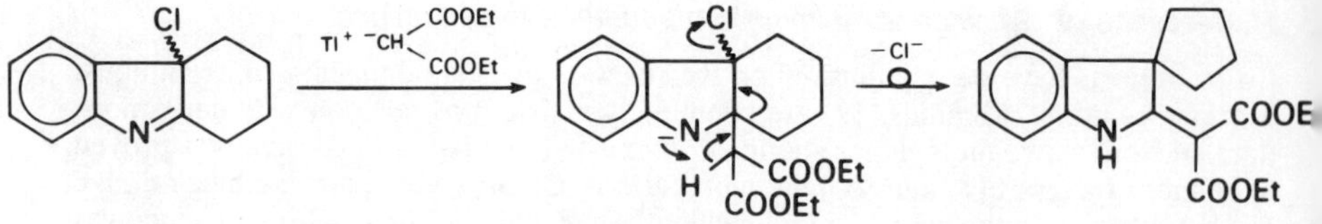

(229) (230) (231)

synthesis leading to the alkaloid vincadifformine[153,271]. The aforementioned base-induced coupling reaction of α-iodomethylketimines (**40**)[78] was similarly observed with 2-bromomethyl-1,3-oxazine derivatives[430].

2. *Reactions of α-halogenated imino compounds with cyanide ion*

Although nucleophilic displacement of α-halogens by cyanide ion are known[272], the preferred reaction of cyanide is addition to the imino function[273,284], and eventually further reaction of intermediately formed adducts[274,275]. α-Chloro aldimines (**59**), α,α-dichloro aldimines (**232**) and α,α,α-trichloro aldimines (**233**) react with potassium cyanide in methanol to give α-cyanoenamines **235**, **236** and **237**, respectively[276–278]. The last-mentioned compound (**237**), however, is accompanied by several by-products,

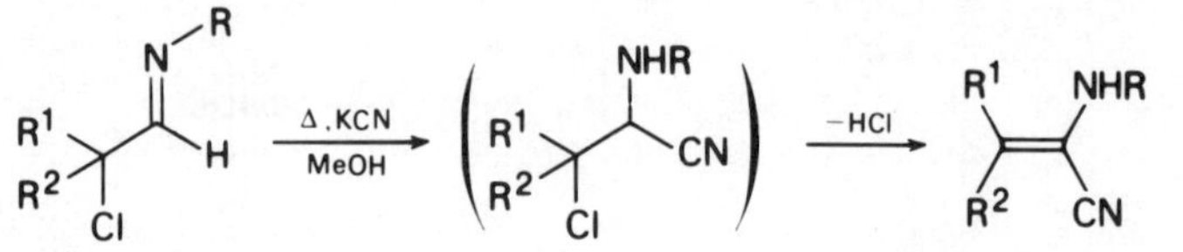

(**59**) R^1, R^2 = alkyl (**234**) (**235**) $R^1 = R^2$ = alkyl

(**232**) R^2 = Cl (**236**) R^2 = Cl

(**233**) $R^1 = R^2$ = Cl; R = *t*-Bu (**237**) $R^1 = R^2$ = Cl; R = *t*-Bu

originating from further reaction of this reactive α-cyanoenamine with methanol[278]. The reaction involves addition of cyanide to the imino function and subsequent dehydrochlorination of the transient adduct (**234**). α-Cyanoenamines (**235**) were found to be valuable synthons as they can be transformed into trialkylketenimines[279] and carboxylic amides[280,457].

The preferred addition of cyanide to an imine was demonstrated by incorporation of ^{14}C-labelled cyanide into α-chloro imidoyl cyanide (**78**; X = Cl) on reflux with $K^{14}CN$ in methanol[116].

With potassium cyanide in dimethyl sulphoxide, α,α-dichloro aldimines (**238**) gave 2-amino-5-cyanopyrroles (**241**) by a sequence involving elimination of hydrogen

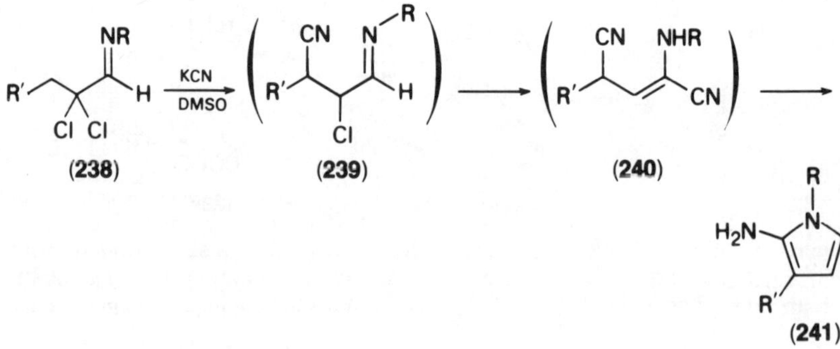

(**238**) (**239**) (**240**)

(**241**)

chloride, Michael addition, α-cyanoenamine formation and ring closure[280]. When non-displaceable halogens were incorporated in the starting α-halo imine, e.g. *N*-activated trifluroacetaldimines (**242**), the adduct **243** was easily isolated and could be hydrolysed into trifluoroalanine (**244**)[285].

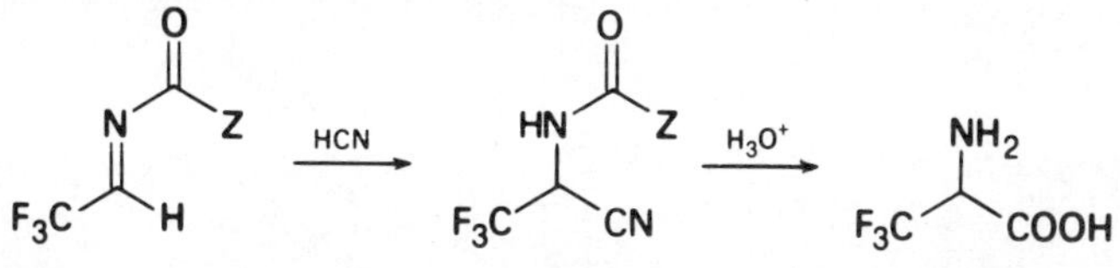

(**242**) Z = C_6H_5CO
Z = $PhCH_2CO$ (**243**) (**244**)

The behaviour of cyanide ion towards α-halo imines parallels the reactivity of this nucleophile towards α-halo carbonyl compounds as the latter are also known to give nucleophilic addition to the carbonyl function, but the adduct is dehydrohalogenated in a different way, namely by intramolecular nucleophilic substitution to produce α-cyano epoxides[231,438–440]. All the above-mentioned examples in the imine series belong to the class of α-halo aldimines and give rise to 1,2-dehydrohalogenation, but aliphatic α-halo ketimines behave like α-halo ketones in that they undergo 1,3-dehydrohalogenation, generating α-cyano aziridines[91].

3. Reaction of α-halogenated imino compounds with organometallic reagents

Grignard reagents usually add to the carbon–nitrogen double bond of *N*-activated α-halogenated imines (**245** and **246**)[146,154,320]. Aliphatic α-bromo aldimines (**247**) with

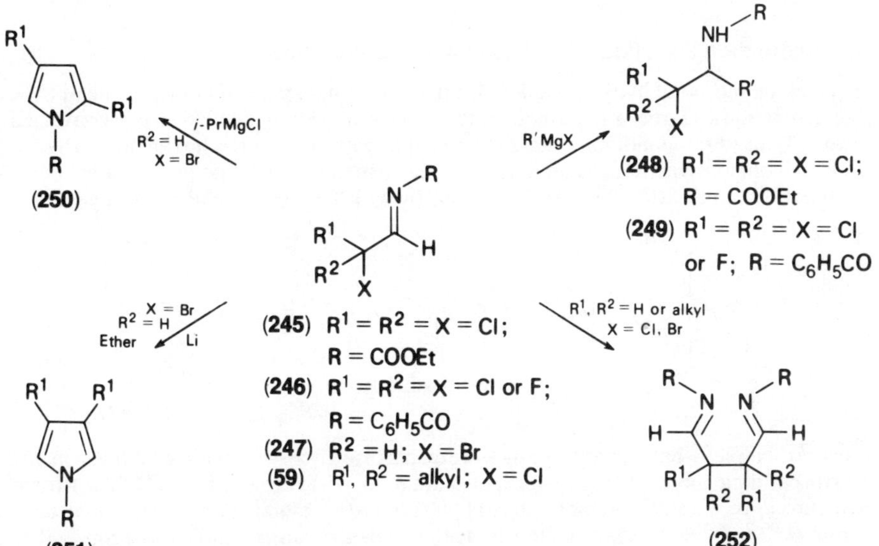

(**250**) (**251**)

(**245**) $R^1 = R^2 = X = Cl$; R = COOEt
(**246**) $R^1 = R^2 = X =$ Cl or F; R = C_6H_5CO
(**247**) $R^2 = H$; X = Br
(**59**) R^1, R^2 = alkyl; X = Cl

(**248**) $R^1 = R^2 = X = Cl$; R = COOEt
(**249**) $R^1 = R^2 = X = Cl$ or F; R = C_6H_5CO

(**252**)

isopropylmagnesium chloride produced 1,2,4-trisubstituted pyrroles (**250**), while 1,3,4-trisubstituted pyrroles (**251**) were obtained with lithium in ether[321,323]. On the other hand, coupling reactions to produce 1,4-diimines (**252**) were observed when secondary and tertiary α-haloimines reacted with sodium in liquid ammonia or with methylmagnesium iodide[87,322]. Metallation of α,α-dibromoaldimines (**253**) with butyllithium in tetrahydrofuran produced lithiated α-bromo aldimines (**254**), which underwent protolysis with methanol at low temperature to give the first non-conjugated secondary β-halo enamines (**255**), while α-alkylation of anions (**254**) was accomplished with allyl bromide[324].

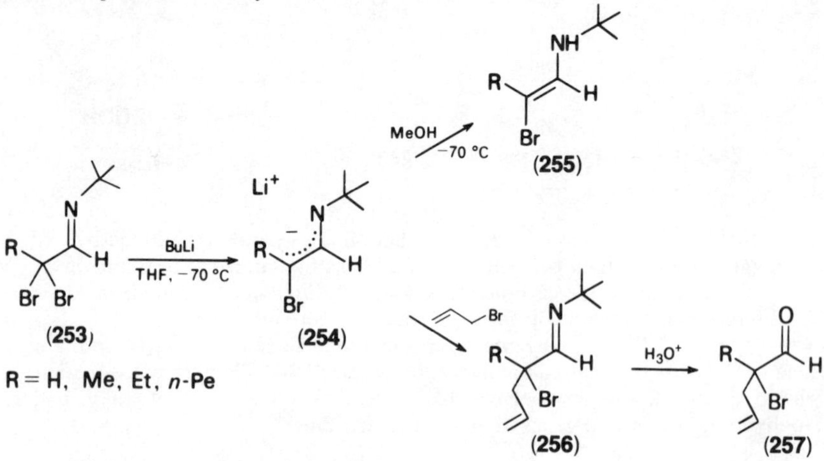

BuLi, THF, −70 °C
MeOH, −70 °C
H_3O^+
(**253**) (**254**) (**255**) (**256**) (**257**)
R = H, Me, Et, *n*-Pe

As mentioned above organometallic reagents, e.g. organocopper compounds[49,50,309], alkyllithium compounds[69] and Grignard reagents[102,103,287,295,325–327], substitute halogens α with respect to an oxime or an hydrazone moiety, but the reaction proceeds through intermediacy of nitrosoölefins or azoalkenes.

C. Elimination Reactions of α-Halo Imino Compounds

The elimination of hydrogen halide from simple α-halogenated imines is one of their basic reactions, as recently demonstrated. α-Chloro aldimines (**258**) were converted into α,β-unsaturated aldimines (**259**) by reaction with sodium methoxide in methanol, but a competitive rearrangement via an α-methoxyaziridine was noticed for isobutyraldimines (**258**; R^1 = H; R^2 = Me) by which α-aminoacetals (**260**) resulted[87].

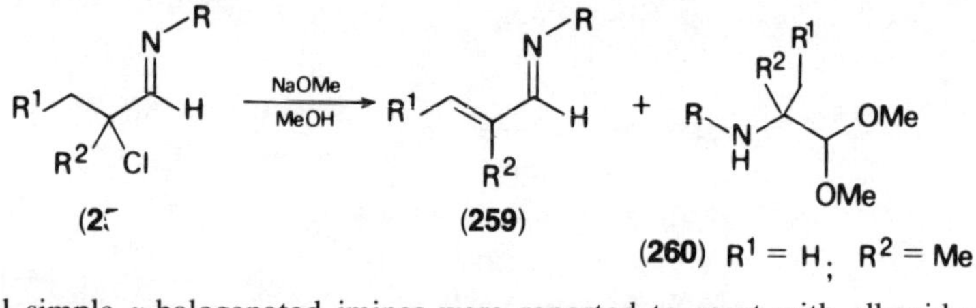

NaOMe, MeOH
(**2** (**259**)
(**260**) R^1 = H; R^2 = Me

Several simple α-halogenated imines were reported to react with alkoxides in the corresponding alcohol to give initial elimination of hydrogen halide[86,90]. α-Bromo hydrazone-type compounds (**261**) suffered elimination of hydrogen bromide[20–23,46,328–330] when heated in acetic acid, and this method was proposed to introduce a double bond at C_4–C_5 in 3-ketosteroids[20–22].

Finally, 3-bromo-2-cyano-1-pyrroline 1-oxides (**65**; R = CN)[95–97] and α-halogenated imidoyl cyanides (**78**)[116] readily underwent base-induced elimination reactions to afford the corresponding α,β-unsaturated imino compounds.

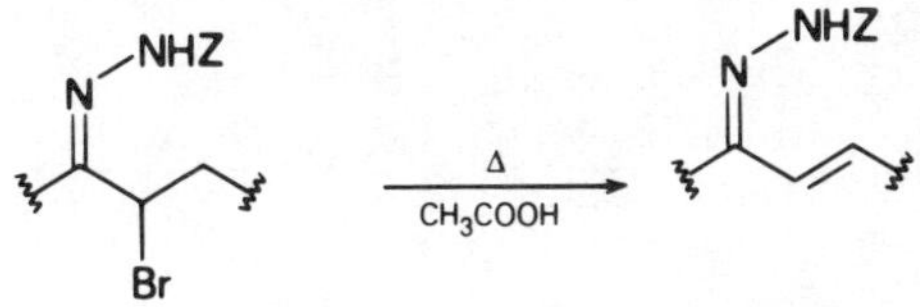

(261) Z = Ar, CONH$_2$ → (Δ, CH$_3$COOH) → (262)

D. Nucleophilic Additions to *N*-Activated α-Halogenated Imino Compounds

Imino compounds (**263**), especially aldimines, having an electron-withdrawing *N*-substituent and α-perhalogenation, add nucleophiles at the imino function under very mild conditions[141,143,145,154,335,336,410]. The high electrophilic character of the imino

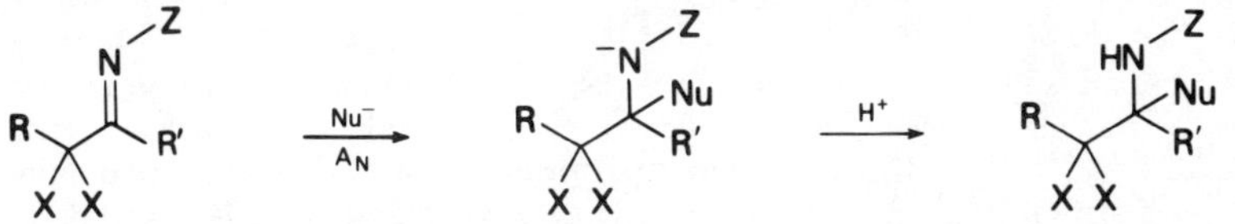

(263) → (Nu⁻, A$_N$) → (264) → (H⁺) → (265)

(**263**) X = halogen
R = halogen or alkyl
R′ = usually H
Z = COOR″, COR″, SO$_2$R″

(**264**)

(**265**)
Nu = OR^1, OAr, NR^{1}R^2, NHR^1, SR, SAr, OH, OOCR^1, NHCOR^1, NHNHR^1

carbon originates from the inductive effect of the α-halogens and the mesomeric effect of the *N*-activating group. The extreme form of the polarization in, for example, *N*-acyl α-halo imines (**266**) can be expressed as dipolar structure (**267**), clearly indicating the tendency to give Michael-type additions to such heteroenones. Various

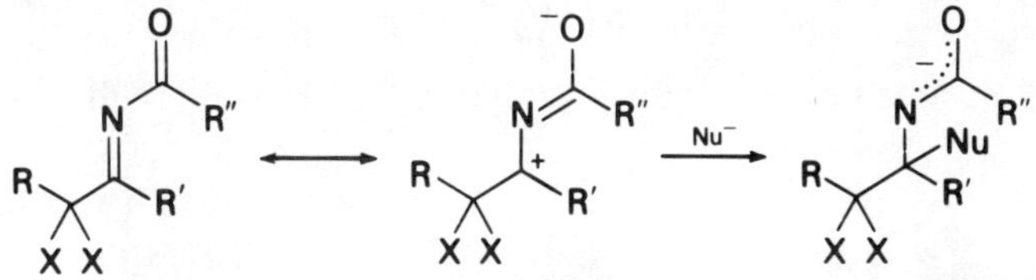

(**266**) ⟷ (**267**) → (Nu⁻) → (**268**)

kinds of nucleophiles, have been reported to give stable adducts (**265**; R′ = H), among others alcohols[141,143,154,337], amines[141,143,145,335,336], thiols[143,335,336], hydrogen sulphide[335], water[335], phenols[335,336], carboxylic acids[143,335,336], amides[143,335,336] and hydrazines[143,336]. This tendency for nucleophilic addition to the imino function is at maximum for α-halogenated aldimines (**266**; R′ = H) and originated from the decreased steric hindrance as compared to the ketimine case (**266**; R′ ≠ H). An analogous aptitude for nucleophilic addition is well known for α-halogenated aldehydes, which practically always react by such an initial addition reaction[443,444]. Similarly, several examples with α-halo ketimines[37,117,118,247,338–341] or imidate derivatives[452] have been found. However, the nucleophilic addition can be followed by an elimination of hydrogen halide (**270**)[278,333,343], an expelling of a leaving group (R′) connected to the imino carbon (**269**)[342] or a haloform-type reaction (**271**)[333].

Of major importance is the reaction of α-halogenated imino compounds with mixed metal hydrides (usually lithium aluminium hydride), which add to the imino function

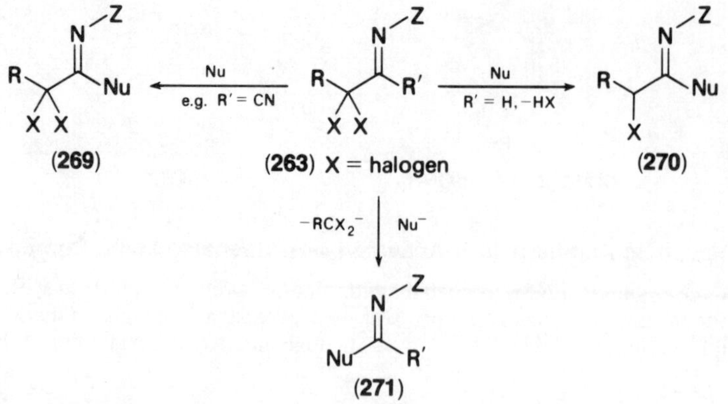

(269) (263) X = halogen (270)

(271)

in a very general way. When the halogen is displaceable (X = Cl, Br), the reaction proceeds further by intramolecular nucleophilic attack giving aziridines[85,143,321,344,345,347]. If the reaction is performed with α,α-dihalogenated imines (**272**; R^2 = X = Cl)[85,143,345,346], the intermediate α-haloaziridine (**274**; R^2 = Cl) is transformed into the final aziridines (**276**) by expelling of a halide anion to generate an azirinium halide (**275**), which is stereospecifically attacked from the less hindered side (most remote from substituent R^3) to give *cis*-aziridines[346]. Ring opening, however, of transient α-haloaziridines by hydrides has also been encountered[106,346,348].

These results are in sharp contrast to the reactivity of α-halo carbonyl compounds towards mixed hydrides, from which only β-halohydrins and/or alcohols result.

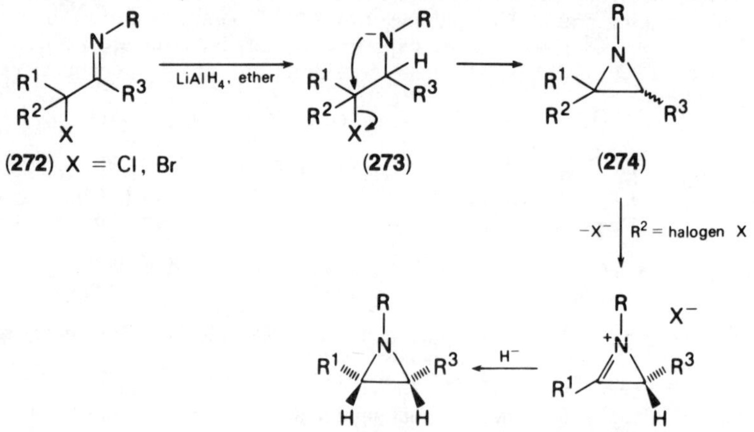

(272) X = Cl, Br (273) (274)

(276) (275)

When α-fluorinated imino compounds (**227**) react with mixed metal hydrides, the exclusive addition reaction leads to β-fluorinated amino compounds (**278**)[37,38,349–351], a

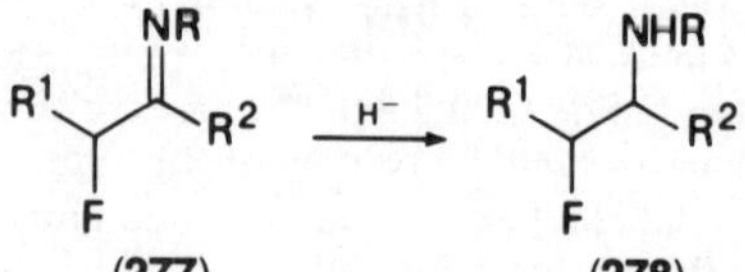

(277) (278)

reaction which was also observed with some α-chloro- or α-bromo imino compounds[211,278,351,352].

E. Rearrangement of α-Halogenated Imino Compounds

Three types of more or less frequently encountered rearrangements of α-halo imines will be discussed here, namely the Favorskii-type rearrangement, the rearrangement via activated aziridines and the Wagner–Meerwein-type rearrangement of chloroindolenines. Finally, a single case of the Beckmann rearrangement of an α-bromo oxime will be discussed.

1. The Favorskii-type rearrangement

The base-induced skeletal rearrangement of α-halo ketones to afford carboxylic acid derivatives, known as the Favorskii rearrangement[353–360], has also been encountered with α-halo imines. Quast and coworkers performed the first transformation of an α-halo ketimine (**279**) into a carboxylic amide (**281**) via a two-step sequence, which could be accounted for in terms of the Favorskii rearrangement[195,361]. 1,3-Dehydrobromination of α-bromo ketimine (**279**) was obtained with potassium *t*-butoxide to generate cyclopropylidene amines (**280**), which underwent hydroxide-induced opening to give amide (**281**). The opening of the nitrogen analogues of cyclopropanones (**280**) is directed by the stability of the intermediate anion. Accordingly, **280** is opened via path *a*, giving rise to the more branched carboxylic amide (**281**). No trace of the alternative route *b* was observed. Another

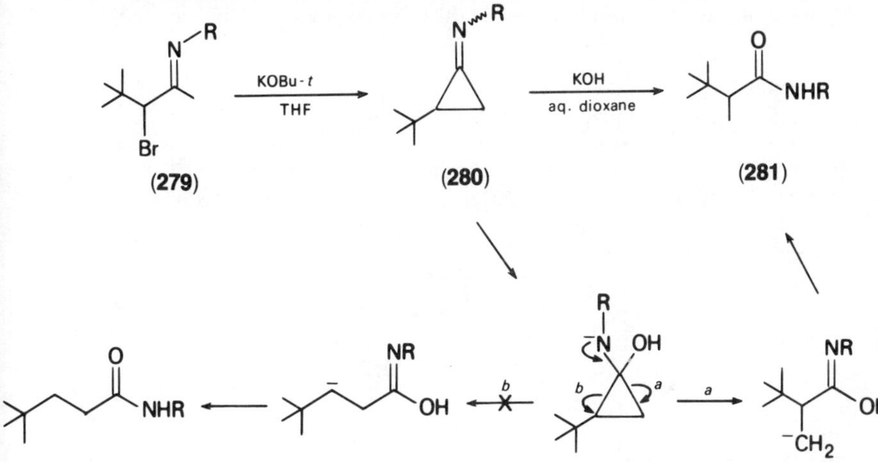

Favorskii-type rearrangement was observed by reaction of *N*-aryl α,α-dichloromethylketimines (**286**) with sodium methoxide in methanol, affording α,β-unsaturated imidates (**289**) via transient cyclopropylideneamines (**287**)[82,248]. The reaction of primary derivatives (**286**; $R^2 = H$) was shown to be stereospecific and gave rise to a regiospecific opening of the chlorocyclopropylideneamine (**287**)[82,248]. Secondary derivatives (**286**; $R^2 \neq H$) also afforded a regiospecific opening of the transient cyclopropylideneamine (**287**) because of the directive aid of the chloride anion expulsion. The latter reaction was not stereospecific in that a mixture of *E*- and

(286) → NaOMe/MeOH → (287) → ⁻OMe → (288) → −Cl⁻ → (289)

Z-imidates (**289**) was produced[248]. Side reactions, e.g. nucleophilic substitution and solvolysis of **286** with sodium methoxide in methanol could be avoided by working in ethereal medium[248].

2. *Rearrangement of α-halo imines via activated aziridine intermediates*

As exemplified for mixed metal hydrides (*vide supra*), the addition of nucleophiles to the imino function of α-halo imines can be followed by intramolecular attack with halide expulsion, by which aziridines result. In the case of nucleophiles other than hydride, but most often with alkoxides (or alcohol), the aziridine thus formed is a very reactive species and undergoes alcoholysis when the reaction is carried out in an alcohol. α-Chloro isobutyraldimines (**290**), when treated with sodium methoxide in methanol, are converted into α-alkylaminoacetals (**292**) and α,β-unsaturated aldimines[87]. The latter competitive elimination reaction was removed by working in methanol only. The intermediacy of α-methoxyaziridines (**291**) was established by trapping these transient species (see dipolarophilic form **293**) with the ambident thiocyanate or cyanate anions[362,363], resulting in 2-imidazolidinethiones (**294**) and 2-Functionalized 3-chloroindolenines (**303**) rearranged in protic solvents into

(290) → MeOH/NaOMe → (291) → MeOH → (292)

(293) → KZCN, Z = O,S → (294) Z = S; (295) Z = O

nitrogen atom were observed with α,α-dichloroaldimines (**56**)[86], α-monochloroaldimines[87] and α-halo immonium halides[352,364–369]. These transpositions are completely comparable to the alkoxide-induced rearrangements of α-halo carbonyl compounds (**186**) to α-hydroxyacetals (**297**) via intermediate alkoxyepoxides (**296**)[419–424].

(186) → ⁻OR″/R″OH → (296) → R″OH → (297)

3. Rearrangement of chloroindolenine derivatives

Chloroindolenines, obtained by reaction of indoles with *t*-butyl hypochlorite, yield rearranged iminoethers, e.g. **299** (and/or oxindoles), upon treatment with base (methoxide, hydroxide) at elevated temperature, while α-substitutions were noticed when treated with cold base. This general rearrangement was applied to simple

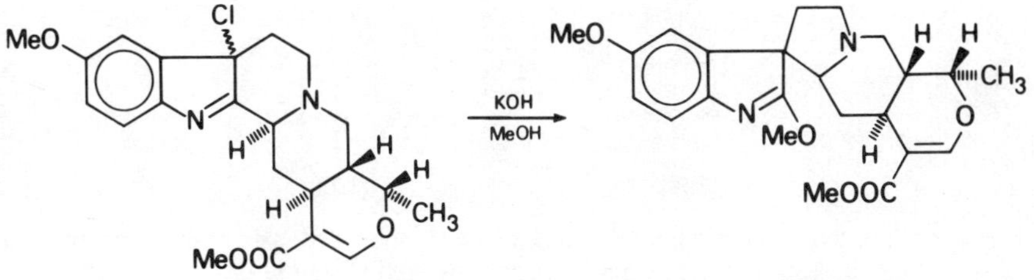

KOH / MeOH

(**298**) (**299**)

indolenine derivatives, like the chloroindolenine derived from tetrahydrocarbazole, and more complex indole-type alkaloids[125–128,132,133,138,249,250,270,271,370]. From the mechanistic point of view, the rearrangement of chloroindolenines (**300**) was explained by initial nucleophilic addition of methoxide at the C=N double bond, followed by Wagner–Meerwein-type rearrangement to give **302**. This transformation requires *cis* disposition of the chlorine atom and the methoxy group in adduct **301**.

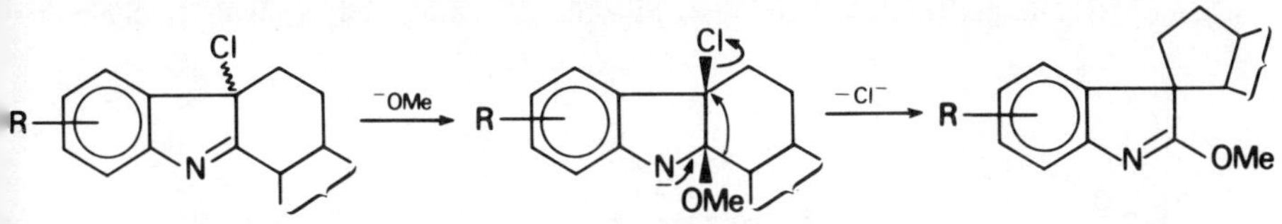

$^{-}$OMe; $-$Cl^{-}

(**300**) (**301**) (**302**)

Not only under basic conditions but also under neutral conditions were rearrangements of chloroindolenine and related alkaloids observed[250]. 2-Functionalized 3-chloroindoenine (**303**) rearranged in protic solvents into

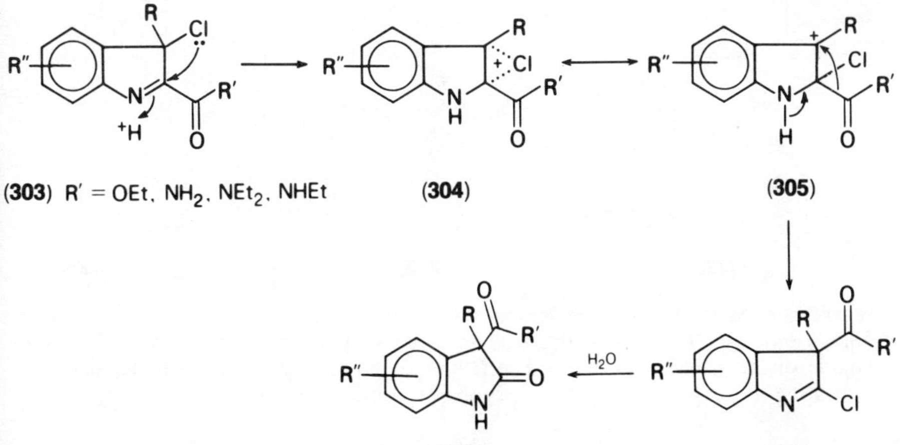

(**303**) R′ = OEt, NH_2, NEt_2, NHEt (**304**) (**305**)

H_2O

(**307**) (**306**)

oxindoles with migration of the functional group from the 2-position to the 3-position[129]. In this case a carbonium ion (**305**), mesomeric with chloronium ion **304**, is involved and migration of the carbonyl group furnishes imidoyl chloride (**306**), which leads to **307** upon hydrolysis.

Many other migrations of substituents from the 2-position to the 3-position, starting from 3-chloroindolenines by treatment in acidic medium, are known[88,121,253,371,372];

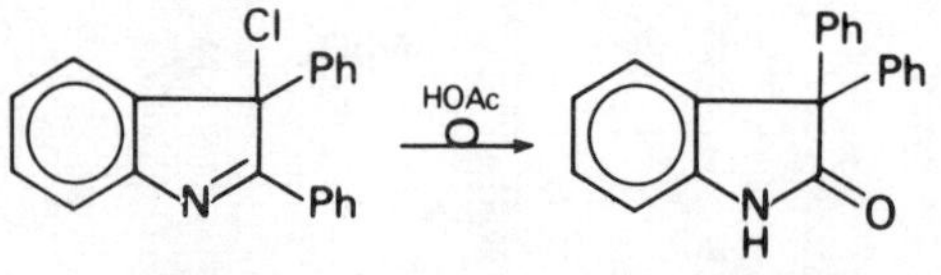

(**308**) (**309**)

one example is given here, namely the spontaneous rearrangement of chloroindolenine (**308**) in acetic acid solution into oxindole (**309**).

4. Beckmann rearrangement of α-bromo oximes

The reaction of *para*-substituted phenacyl bromide oximes (**310**) with triphenylphosphine in acetonitrile at room temperature produced imidoyl bromides (**311**) and triphenylphosphine oxide[226]. The rearrangement is explained by addition of the phosphine to the imino function after which debrominated oxime derivative (**314**) is formed via a series of tranformations, visualized by the arrows in the accompanying scheme. Compound **314** is then susceptible to the well known Beckmann

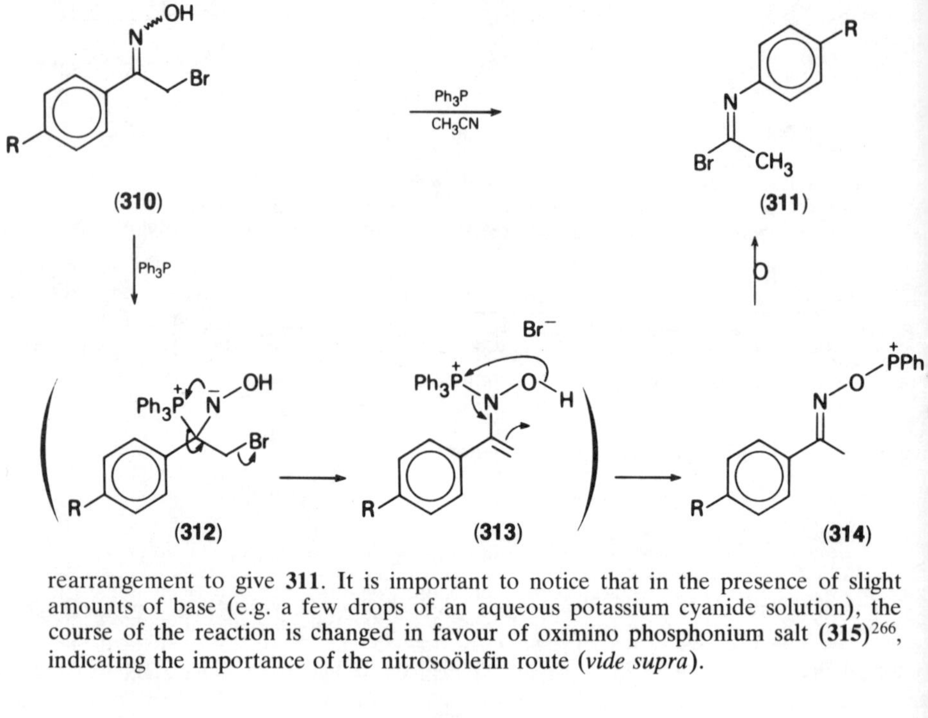

(**310**) (**311**)

(**312**) (**313**) (**314**)

rearrangement to give **311**. It is important to notice that in the presence of slight amounts of base (e.g. a few drops of an aqueous potassium cyanide solution), the course of the reaction is changed in favour of oximino phosphonium salt (**315**)[266], indicating the importance of the nitrosoölefin route (*vide supra*).

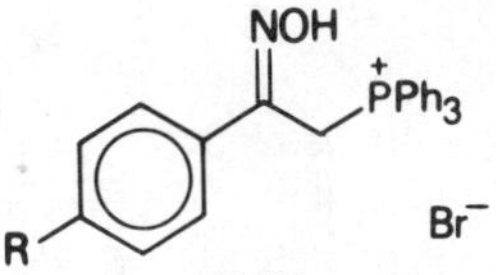

(315)

F. Cycloadditions

Numerous cycloadditions have been reported with α-halo imines, practically all of them α-perhalogenated and having an activating *N*-substituent, e.g. alkoxycarbonyl, tosyl, benzoyl, etc. α-Halo imines having a carbonyl substituent directly bonded to nitrogen can act as dienophile or as heterodiene in Diels–Alder reactions. *N*-Alkoxycarbonyl and *N*-tosyl chloralimines (**86** and **91**) gave cycloadducts with dienes, e.g. acyclic and cyclic 1,3-dienes, functionalized 1,3-dienes, etc.[144,396–399,437]. The *N*-acetyl analogue **316** reacted with 2,3-dimethylbutadiene as dienophile to give adduct **319**, but adduct **320** was also isolated, indicating that **316** acted as a heterodiene[399]. On the other hand, *N*-alkoxycarbonyl chloralimines (**86**) behaved exclusively as heterodienes towards electron-rich alkenes, e.g. ketene acetals, resulting in an oxazine derivative (**321**), which was hydrolysed to carbamate derivative **322**[400].

α-Perfluorinated ketimines such as hexafluoroacetone imines especially have been found to give a variety of cycloadditions, but these reactions are not dealt with here in detail because of lack of space.

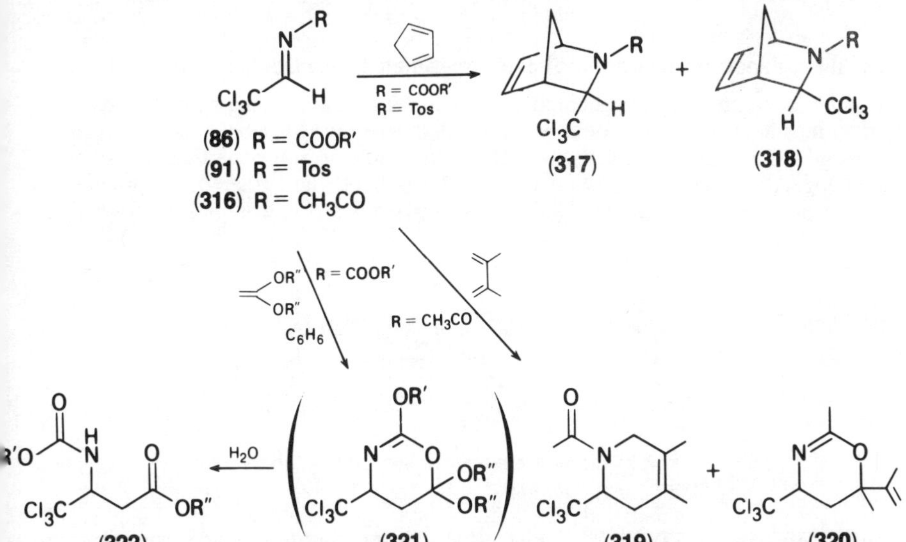

(**86**) R = COOR′
(**91**) R = Tos
(**316**) R = CH_3CO

(317) (318) (319) (320) (321) (322)

Only some examples of the various possible types of cycloadditions of α-perfluorinated imines are reported in the following scheme, together with some leading references in this area[2,41,42,154,374–395,401,458]. Among the reagents found to give cycloadditions to α-perfluorinated imines are included nitriles, nitrones, enol ethers, carbenes, ketenes, alkenes, ketones, ynamines and isonitriles.

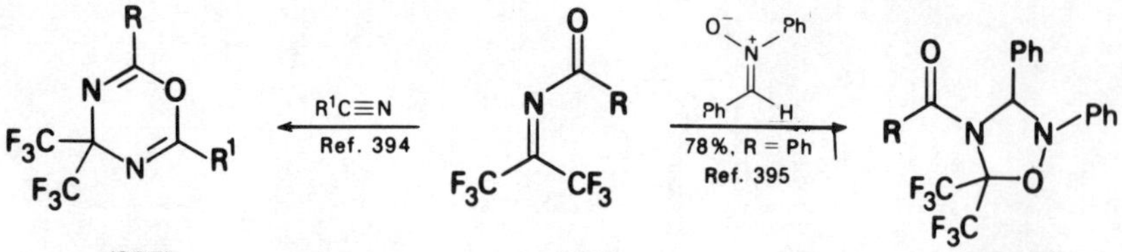

(325) ← $R^1C{\equiv}N$, Ref. 394 ← (323) → [Ph–CH=N(O)–Ph nitrone], 78%, R = Ph, Ref. 395 → (324)

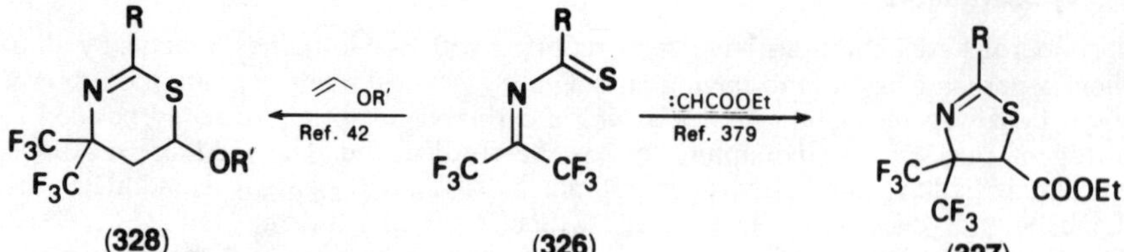

(328) ← $CH_2{=}CHOR'$, Ref. 42 ← (326) → :CHCOOEt, Ref. 379 → (327)

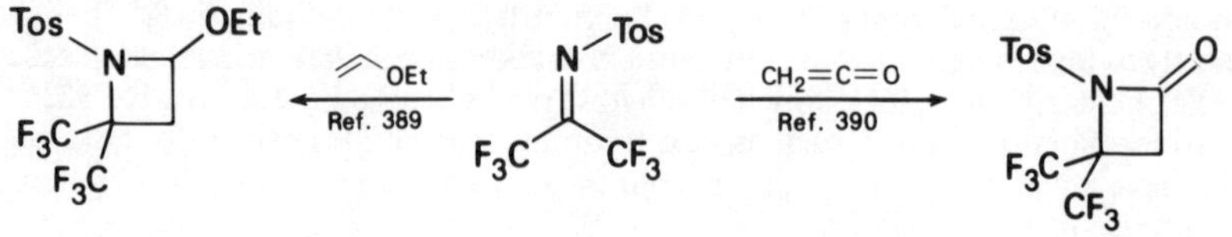

(331) ← $CH_2{=}CHOEt$, Ref. 389 ← (329) → $CH_2{=}C{=}O$, Ref. 390 → (330)

G. Miscellaneous Reactions of α-Halogenated Imino Compounds

A large number of α-halogenated imino compounds have been used for the synthesis of various heterocyclic compounds. A detailed description has been given in a recent review[2]. This survey will be limited to mentioning some particular examples. α-Halogenated thiosemicarbazones, e.g. **331** and **332**, and related compounds are known to give intramolecular cyclizations to pyrazoles[57,402], 2-iminothiazolines[52,53,55]

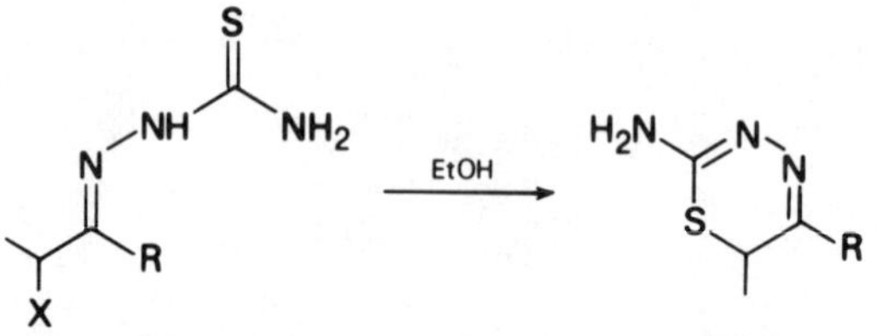

EtOH

(**331**) X = Cl; R = Me
(**332**) X = Br; R = H

(**333**)

and thiadiazines (**333**)[52–55,403], while α-halo oximes produced thiazoles (**335**) and thiadiazines on reaction with thiourea and dithiocarbazic acid, respectively[216,217].

As pointed out already in the section describing cycloadditions, perfluoroacetone imines (**336**, **323**, **326**) are versatile substrates for syntheses of nitrogen heterocycles as they are known to give rise to imidazoles (**337**)[404], oxazoles (**338**)[404], thiazoles (**339**)[404], dithiazolines, thiaselenazolines and diselenazolines[405–407,459].

This section will be closed by discussing briefly the acidic hydrolysis of α-halo imines as a path in the specific α-halogenation of certain ketones. The acid-catalysed

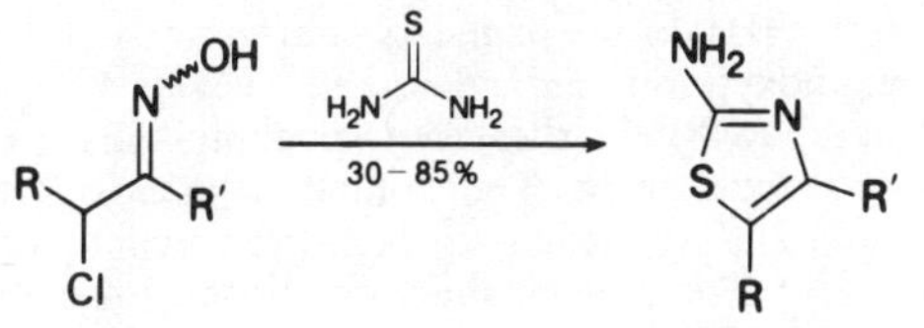

(334) (335)

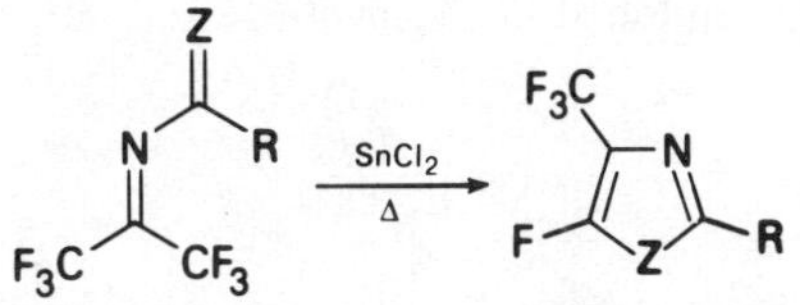

(336) Z = NR′ (337) Z = NR′
(323) Z = O (338) Z = O
(326) Z = S (339) Z = S

hydrolysis of α-halo imines affords α-halo carbonyl compounds. Direct halogenation of carbonyl compounds can be problematic from the viewpoint of regioselectivity, but in some cases the halogenation of carbonyl compounds via imination, subsequent halogenation and hydrolysis offers a complementary method. Via this three-step sequence, the regiospecific mono- and dihalogenation of methyl ketones (**340**) has been accomplished[79,84], but in cases where no competitive α′-halogenation can take place,

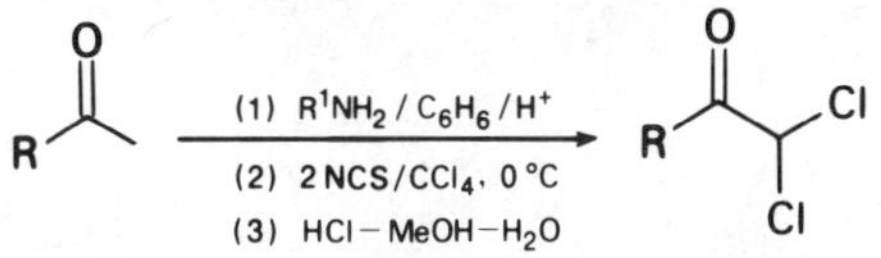

(1) $R^1NH_2 / C_6H_6 / H^+$
(2) $2NCS/CCl_4$, 0 °C
(3) $HCl-MeOH-H_2O$

(340) (341)

this method has been developed as the synthesis of choice for α,α-dichloroaldehydes (**344**)[92], α,α-dibromoaldehydes (**345**)[92] and α,α-dichloroalkyl aryl ketones (**346**)[88]. However, the direct chlorination of aldehydes (**342**) or alkyl aryl ketones (**343**) with

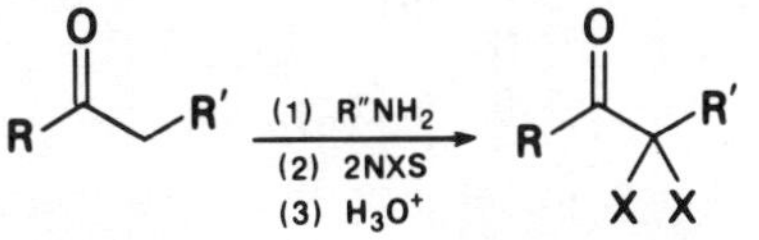

(1) $R''NH_2$
(2) 2NXS
(3) H_3O^+

(342) R = H (344) R = H; X = Cl
(343) R = Ar (345) R = H; X = Br
(346) R = Ar; X = Cl

chlorine in dimethylformamide has been shown very recently to be an improved synthetic method for the preparation of α,α-dichlorocarbonyl derivatives **344** and **346**[445,446].

IV. PROPERTIES AND APPLICATIONS OF α-HALOGENATED IMINO COMPOUNDS

Several α-halo imino compounds have been found to have pharmaceutical and phytopharmaceutical properties, while they have also been used for the synthesis of medical products. α-Halo oxime derivatives e.g. 17β-acetoxy-2α-chloro-3-(p-nitrophenoxy)imino-5α-androstane (**347**)[220,431] and 1,3-dichloroacetone oxime acetate (**348**)[414], displayed postimplantive antifertility activity and slimicide activity, respectively. The contact acaricide Tranid (**350**) belongs to the important class of oxime carbamates, which were recently successfully applied in various pest controls[432,433]. The fungicidal activity of the 2,4-dinitrophenylhydrazone of chloroacetaldehyde (**351**) is well known, the compound being referred to as 'Fungicide 1763' in pesticide science[442]. Very recently, a great variety of α-halogenated hydrazones (**352**) have been found to display fungicidal activity[460].

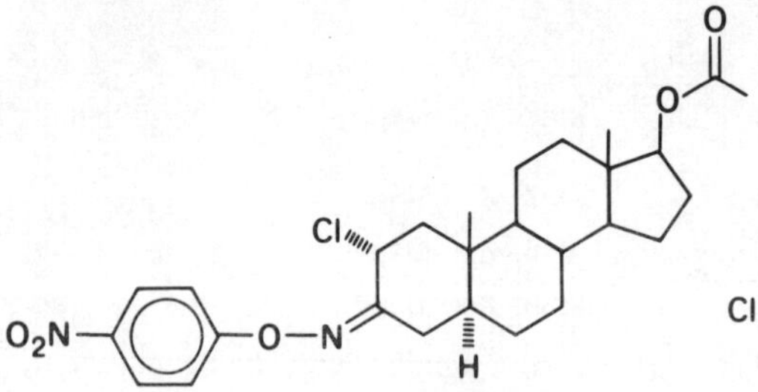

(**347**)

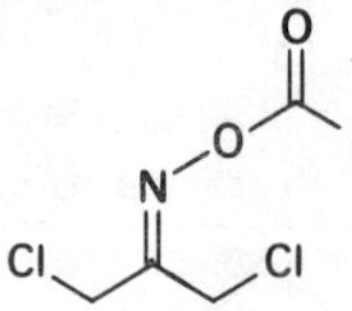

(**348**)

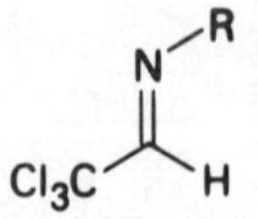

(**349**) R = alkyl
(**204**) R = OH

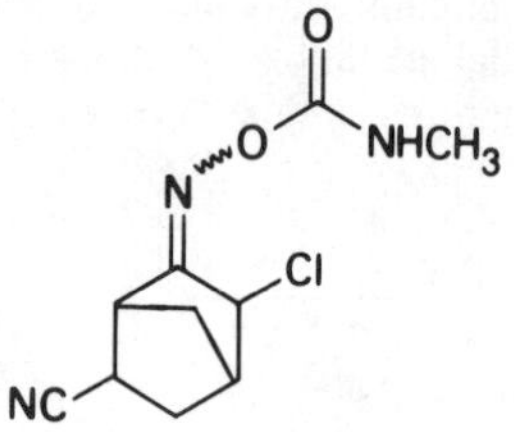

(**350**)

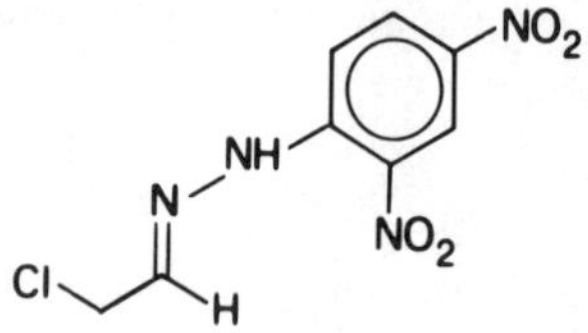

(**351**)

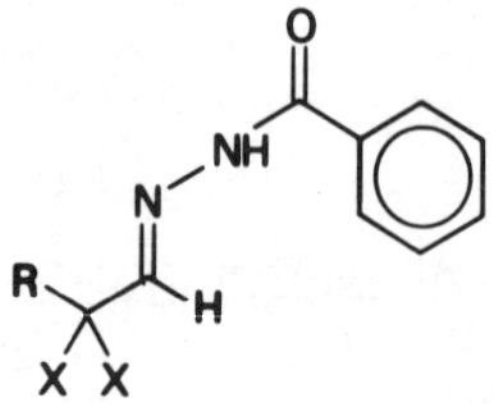

(**352**) R = H, F, Cl, Br, CH_3, $CHClCH_3$
X = Cl, Br

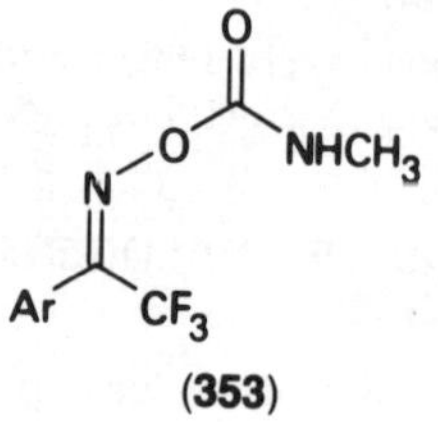

(**353**)

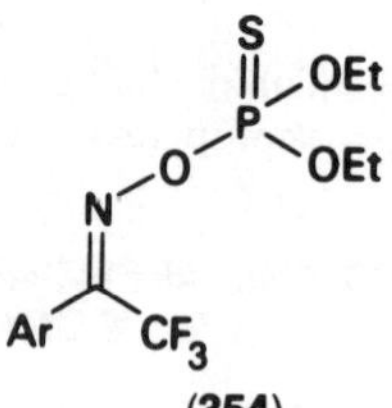

(**354**)

The insecticidal activity of a large number of trifluoroacetophenone oxime carbamates (**353**) and trifluoroacetophenone oxime thiophosphates (**354**) was recently evaluated[441]. Some of these compounds were good aphicides and some were strong cholinesterase inhibitors[441].

Trichloroacetamidines (**113**; X = Y = Cl) showed enzyme inhibitory activity besides bacterial mutagenic and inotropic activity[180,203]. α-Bromoacetimidate acted as cross-linking agent for proteins[415], while choral imines (**349**) were found to be useful as compounding agents for rubber[332]. Imino derivatives of chloral, e.g. **349**, and oxime (**204**), have been proposed as insecticides[331,416]. Besides these industrial uses, α-halo imines have been used for the synthesis of medicinal products, among others antispasmodics[246], vasodilators[246], convulsants[126], glycine antagonists[126], radioprotective agents[204] and products having gastric antisecretory activity[181]. An area of major importance is the transformation of various indole-type alkaloids via their chloroindolenines[132–139,153,215,270–272,417], but also their application as intermediates in the total synthesis of steroids[69] and antitumour agents, e.g. 11-oxahomofolic acid[156] and anthracyclines[51], deserves attention.

V. REFERENCES

1. N. De Kimpe and N. Schamp, *Org. Prep. Proced. Int.*, **11**, 115 (1979).
2. N. De Kimpe, R. Verhé, L. De Buyck and N. Schamp, *Org. Prep. Proced. Int.*, **12**, 49 (1980).
3. R. W. Layer, *Chem. Rev.*, **63**, 489 (1963).
4. *The Chemistry of the Carbon–Nitrogen Double Bond* (Ed. S. Patai), Interscience, London (1970).
5. E. F. Janetzky and P. E. Verkade, *Rec. Trav. Chim. Pays-Bas*, **65**, 691 (1946).
6. Y. A. Shaikh, *Org. Prep. Proced. Int.*, **8**, 293 (1976).
7. J. Wolinsky, R. O. Hutchins and T. W. Gibson, *J. Org. Chem.*, **33**, 407 (1968).
8. I. Migaichuk and I. Khaskin, *Zh. Prikl. Khim.* (Leningrad), **52**, 946 (1976); *Chem. Abstr.*, **91**, 39092 (1979).
9. H. Pauly and I. Rossback, *Ber.*, **32**, 2000 (1899).
10. H. Pauly and C. Boehm, *Ber.*, **33**, 919 (1900).
11. J. G. Aston and R. B. Greenburg, *J. Amer. Chem. Soc.*, **62**, 2590 (1940).
12. J. Wolinsky, J. J. Hamsher and R. O. Hutchins, *J. Org. Chem.*, **35**, 207 (1970).
13. C. L. Stevens, P. Blumbergs and M. Munk, *J. Org. Chem.*, **28**, 331 (1963).
14. M. Miocque, C. Combet-Farnoux, J.-F. Givardeau and H. Galons, *Compt. Rend. C.* **282**, 469 (1976).
15. P. Catsoulacos and A. Hassner, *J. Org. Chem.*, **32**, 3723 (1967).
16. P. Catsoulacos, *Bull. Soc. Chim. Fr.*, 642 (1976).
17. J. Van Alphen, *Rec. Trav. Chim. Pays-Bas*, **64**, 109 (1945).
18. B. T. Gillis an J. D. Hagerty, *J. Amer. Chem. Soc.*, **87**, 4576 (1965).
19. L. Caglioti, G. Rossini and F. Rossi, *J. Amer. Chem. Soc.*, **88**, 3865 (1966).
20. V. R. Mattox and E. C. Kendall, *J. Amer. Chem. Soc.*, **70**, 882 (1948).
21. C. Djerassi, *J. Amer. Chem. Soc.*, **71**, 1003 (1949).
22. V. R. Mattox and E. C. Kendall, *J. Amer. Chem. Soc.*, **72**, 2290 (1950).
23. F. Ramirez and A. F. Kirby, *J. Amer. Chem. Soc.*, **74**, 4331 (1952).
24. O. Hess, *Ann.*, **232**, 234 (1886).
25. J. Culman, *Ann.*, **258**, 235 (1890).
26. S. Bodforss, *Ber.*, **52**, 1762 (1919).
27. S. Bodforss, *Ber.*, **72**, 468 (1939).
28. D. Y. Curtin and E. W. Tristam, *J. Amer. Chem. Soc.*, **72**, 5238 (1950).
29. K. H. Ongania and J. Schantl, *Monats. Chem.*, **107**, 481 (1976).
30. J. Schantl, *Monats. Chem.*, **108**, 599 (1977).
31. J. Schantl, *Monats. Chem.*, **108**, 325 (1977).
32. P. A. Reddy, S. Singh and V. R. Srinivasan, *Indian J. Chem.. B*, **14**, 793 (1976).
33. L. Duhamel and J. Y. Valnot, *Compt. Rend. C.* **286**, 47 (1978).

34. T. Kametani and O. Umezawa, *Chem. Pharm. Bull.* (Tokyo), **14**, 369 (1966).
35. T. Kametani and O. Umezawa, *Yakugaku Zasshi*, **85**, 514, 518 (1965); *Chem. Abstr.*, **63**, 6851d, 6959h (1965).
36. A. Roedig, F. Hagedorn and G. Märkl, *Chem. Ber.*, **97**, 3322 (1964).
37. Y. Zeifman, N. Gambaryan and I. Knunyants, *Izv. Akad. Nauk SSSR, Ser. Khim.*, 450 (1965).
38. W. Pirkle and J. Hanske, *J. Org. Chem.*, **42**, 2436 (1977).
39. E. W. Abel and C. A. Burton, *J. Fluorine Chem.*, **14**, 105 (1979).
40. K. Burger, J. Albanbauer and M. Eggersdorfer, *Angew. Chem.*, **87**, 816 (1975).
41. K. Burger, J. Albanbauer and W. Foag, *Angew. Chem.*, **87**, 816 (1975).
42. K. Burger, R. Ottlinger and J. Albanbauer, *Chem. Ber.*, **110**, 2114 (1977).
43. L. Duhamel, P. Duhamel and J. Y. Valnot, *Compt. Rend. C.* **271**, 1471 (1970).
44. L. Duhamel and J.-C. Plaquevent, *Tetrahedron Lett.*, 2285 (1977).
45. O. L. Brady, *J. Chem. Soc.*, 757 (1931).
46. F. Ramirez and A. F. Kirby, *J. Amer. Chem. Soc.*, **75**, 6026 (1953).
47. A. Dondoni, G. Rossini, G. Mossa and L. Caglioti, *J. Chem. Soc. B*, 1404 (1968).
48. L. Caglioti, P. Grasselli, F. Morlacchi and G. Rossini, *Chem. Ind.*, 25 (1968).
49. C. A. Bunnell and P. L. Fuchs, *J. Org. Chem.*, **42**, 2614 (1977).
50. W. E. Fristad, Y.-K. Han and L. A. Paquette, *J. Organometal. Chem.*, **174**, 27 (1979).
51. P. Masi, A. Suarato, P. Giardino, L. Bernardi and F. Arcamone, *Il Farmaco, Ed. Sci.*, **34**, 907 (1979).
52. H. Beyer, W. Lässig and E. Bulka, *Chem. Ber.*, **87**, 1385 (1954).
53. H. Beyer, W. Lässig, E. Bulka and D. Behrens, *Chem. Ber.*, **87**, 1392 (1954).
54. H. Beyer, W. Lässig and U. Schultz, *Chem. Ber.*, **87**, 1401 (1954).
55. H. Beyer and G. Wolter, *Chem. Ber.*, **89**, 1652 (1956).
56. H. Beyer and T. Pyl, *Chem. Ber.*, **89**, 2556 (1956).
57. H. Beyer and G. Badicke, *Chem. Ber.*, **93**, 826 (1960).
58. S. C. De, *J. Indian Chem. Soc.*, **3**, 30 (1926).
59. S. C. De and P. C. Rakshit, *J. Indian Chem. Soc.*, **13**, 509 (1936).
60. A. Hantzsch and W. Wild, *Ann.*, **289**, 285 (1896).
61. R. Scholl and G. Matthaiopoulos, *Ber.*, **29**, 1550 (1896).
62. N. Tokuro and R. Oda, *Bull. Int. Phys. Chem. Res.* (Tokyo), **22**, 844 (1943).
63. P. Ciattoni and L. Rivolta, *Chim. Ind.* (Rome), **49**, 1186 (1967).
64. K. Wieser and A. Berndt, *Angew. Chem.*, **87**, 73 (1976).
65. W. Ahrens, K. Wieser and A. Berndt, *Tetrahedron*, **31**, 2829 (1975).
66. D. Bertin, J. Peronnet and J. Caumartin, French Patent 2,029,890, Appl. 30 Jan 1969; *Chem. Abstr.*, **75**, 35134 (1971).
67. H. Brintzinger and R. Titzmann, *Chem. Ber.*, **85**, 344 (1952).
68. E. J. Corey, M. Petrzilka and Y. Ueda, *Tetrahedron Lett.*, 4343 (1975).
69. W. Oppolzer, M. Petrzilka and K. Bättig, *Helv. Chim. Acta*, **60**, 2964 (1977).
70. J. Turcan, *Bull. Soc. Chim. Fr.*, 486 (1932).
71. M. A. Berg, *Bull. Soc. Chim. Fr.*, 637 (1925).
72. J. Turcan, *Bull. Soc. Chim. Fr.*, 283 (1936).
73. D. S. Malament and N. Levi, *J. Org. Chem.*, **40**, 3285 (1975).
74. O. Tsuge, M. Tashiro, K. Kamata and K. Hokama, *Org. Prep. Proced. Int.*, **3**, 289 (1971).
75. E. Aufderhaar and W. Seelinger, *Justus Liebigs Ann. Chem.*, **701**, 166 (1967).
76. E. M. Kosower, B. Pazhenchevsky and E. Hershkowitz, *J. Amer. Chem. Soc.*, **100**, 6516 (1978).
77. A. Silveira, Jr, Y. R. Mehra and M. A. Atwell, *J. Org. Chem.*, **42**, 3892 (1977).
78. M. Larchevêque, G. Valette, T. Cuvigny and H. Normant, *Synthesis*, 256 (1975).
79. W. Coppens and N. Schamp, *Bull. Soc. Chim. Belg.*, **81**, 643 (1972).
80. N. De Kimpe and N. Schamp, *Bull. Soc. Chim. Belg.*, **83**, 507 (1974).
81. N. De Kimpe, N. Schamp and W. Coppens, *Bull. Soc. Chim. Belg.*, **84**, 227 (1974).
82. N. De Kimpe and N. Schamp, *Tetrahedron Lett.*, 3779 (1974).
83. N. De Kimpe and N. Schamp, *Bull. Soc. Chim. Belg.*, **84**, 235 (1975).
84. J. F. W. Keana and R. R. Schumaker, *Tetrahedron*, **26**, 5191 (1970).
85. N. De Kimpe, R. Verhé, L. De Buyck and N. Schamp, *Synthetic Commun.*, **5**, 269 (1975).
86. N. De Kimpe, R. Verhé, L. De Buyck and N. Schamp, *Bull. Soc. Chim. Belg.*, **84**, 417 (1975).

87. N. De Kimpe, R. Verhé, L. De Buyck, H. Hasma and N. Schamp, *Tetrahedron*, **32**, 2457 (1976).
88. N. De Kimpe, R. Verhé, L. De Buyck and N. Schamp, *Synthetic Commun.*, **8**, 75 (1978).
89. N. De Kimpe, R. Verhé, L. De Buyck and N. Schamp, *J. Org. Chem.*, **43**, 2933 (1978).
90. N. De Kimpe, R. Verhé, L. De Buyck, Sunari Tukiman and N. Schamp, *Tetrahedron*, **35**, 789 (1979).
91. N. De Kimpe, Unpublished results.
92. R. Verhé, N. De Kimpe, L. De Buyck and N. Schamp, *Synthesis*, 455 (1975).
93. H. Quast and A. Heublein, *Tetrahedron Lett.*, 3317 (1975).
94. H. P. Fischer and C. A. Grob, *Helv. Chim. Acta*, **45**, 2528 (1962).
95. D. S. C. Black, N. A. Blackman and R. F. Brown, *Tetrahedron Lett.*, 3423 (1975).
96. D. S. C. Black, N. A. Blackman and R. F. Brown, *Aust. J. Chem.*, **32**, 1785 (1979).
97. D. S. C. Black and N. A. Blackman, *Aust. J. Chem.*, **32**, 1795 (1979).
98. J. Häusler and U. Schmidt, *Justus Liebigs Ann. Chem.*, 1881 (1979).
99. T. Yamazaki, K. Matoba, S. Imoto and M. Terashima, *Chem. Pharm. Bull.* (Tokyo), **24**, 3011 (1976).
100. Y. Yamada, T. Emori, S. Kinoshita and H. Okada, *Agr. Biol. Chem.*, **37**, 649 (1977).
101. L. A. Carpino and G. S. Rundberg, Jr, *J. Org. Chem.*, **34**, 1717 (1969).
102. M. Flammang, *Compt. Rend. C.* **283**, 593 (1976).
103. M. Flammang, *Compt. Rend. C.* **286**, 671 (1978).
104. M. Z. Kirmani and S. R. Ahmed, *Indian J. Chem. B*, **15**, 892 (1977).
105. B. Erni and H. G. Khorana, *J. Amer. Chem. Soc.*, **102**, 3888 (1980).
106. A. Picot and X. Lusinchi, *Tetrahedron Lett.*, 679 (1974).
107. A. Picot, M. Dendane and X. Lusinchi, *Tetrahedron*, **32**, 2899 (1976).
108. M. Poutsma and P. Ibarbia, *J. Org. Chem.*, **34**, 2849 (1969).
109. O. Migliara, S. Petruso and V. Spiro, *J. Heterocyclic Chem.*, **16**, 833 (1979).
110. N. I. Korotkikh, A. Y. Chervinskii, S. N. Baranov, L. M. Kapkan and O. P. Shvaika, *Zh. Org. Khim.*, **15**, 962 (1979); *Chem. Abstr.*, **91**, 74124 (1979).
111. G. Rosini and G. Baccolino, *J. Org. Chem.*, **39**, 826 (1974).
112. S. Cacchi, D. Misiti and M. Felici, *Synthesis*, 147 (1980).
113. N. De Kimpe and N. Schamp, *Org. Prep. Proced. Int.*, **13** 241 (1981).
114. A. I. Fetell and H. Feuer, *J. Org. Chem.*, **43**, 1238 (1978).
115. N. De Kimpe, R. Verhé, L. De Buyck, J. Chys and N. Schamp, *Synthetic Commun.*, **9**, 901 (1979).
116. N. De Kimpe, R. Verhé, L. De Buyck, J. Chys and N. Schamp, *Bull. Soc. Chim. Belg.*, **88**, 695 (1979).
117. C. Shin, Y. Sato and J. Yoshimura, *Bull. Chem. Soc. Japan*, **49**, 1909 (1976).
118. C. Shin, Y. Sato, H. Sugiyama, K. Nanjo and J. Yashimura, *Bull. Chem. Soc. Japan*, **50**, 1788 (1977).
119. G. I. Dmitrienko, E. A. Gross and S. F. Vice, *Canad. J. Chem.*, **58**, 808 (1980).
120. H. Tohru, E. Mamoru, T. Masakatsu and N. Masako, *Heterocycles*, **2**, 565 (1974).
121. T. Hino, M. Endo, M. Tonozuka, Y. Hashimoto and M. Nakagawa, *Chem. Pharm. Bull.* (Tokyo), **25**, 2350 (1977).
122. P. G. Gassman, G. A. Campbell and G. Metha, *Tetrahedron*, **28**, 2749 (1972).
123. M. De Roza, *Chem. Commun.*, 482 (1975).
124. M. De Roza and J. L. T Alonso, *J. Org. Chem.*, **43**, 2639 (1978) and references cited therein.
125. R. J. Owellen, *J. Org. Chem.*, **39**, 69 (1974).
126. F. M. Hershenson, K. A. Prodan, R. L. Kochman, J. L. Bloss and C. R. Mackerer, *J. Med. Chem.*, **20**, 1448 (1977).
127. F. M. Hershenson, L. Swenton and K. A. Prodan, *Tetrahedron Lett.*, 2617 (1980).
128. R. J. Owellen and C. A. Hartke, *J. Org. Chem.*, **41**, 102 (1976).
129. A. Walser, J.-F. Blount and R. I. Fryer, *J. Org. Chem..*, **38**, 3077 (1973).
130. S. Sakai, E. Yamanaka and L. Dolby, *J. Pharm. Soc. Japan*, **97**, 309 (1977).
131. J. Y. Laronze, J. Laronze, D. Royer, J. Lévy and J. Le Men, *Bull. Soc. Chim. Fr.*, 1215 (1977).
132. N. Finch and W. Taylor, *J. Amer. Chem. Soc.*, **84**, 1319 (1962).
133. N. Finch and W. Taylor, *J. Amer. Chem. Soc.*, **84**, 3871 (1962).
134. G. Büchi and R. E. Manning, *J. Amer. Chem. Soc.*, **88**, 2532 (1966).

135. E. Wenkert, E. W. Hagaman, N. Wang and N. Kunesch, *Heterocycles*, **12**, 1439 (1979).
136. F. Titeux, L. Le Men-Olivier and J. Le Men, *Bull. Soc. Chim. Fr.*, 1473 (1976).
137. N. Aimi, E. Yamanaka, M. Ogawa, T. Kohmoto, K. Mogi and S. Sakai, *Heterocycles*, **10**, 73 (1978).
138. E. Wenkert, J. S. Bindra, C.-J. Chang, D. W. Cochram and D. E. Rearick, *J. Org. Chem.*, **39**, 1662 (1974).
139. D. Herlem and F. Khuong-Huu, *Tetrahedron*, **35**, 633 (1979).
140. K. V. Lichman, *J. Chem. Soc. C*, 2539 (1971).
141. H. Ulrich, B. Tucker and A. A. R. Sayigh, *J. Org. Chem.*, **33**, 2887 (1968).
142. H. Zinner, W. Siems, D. Kuhlman and G. Erfurt, *J. Prakt. Chem.*, **316**, 54 (1974).
143. N. De Kimpe, R. Verhé, L. De Buyck, W. Dejonghe and N. Schamp, *Bull. Soc. Chim. Belg.*, **85**, 763 (1976).
144. G. Kresze and R. Albrecht, *Chem. Ber.*, 97, 490 (1964).
145. B. Drach, A. Martynyuk, G. Miskevich and O. Lobanov, *Zh. Org. Khim.*, **13**, 1404 (1977).
146. H. E. Zaugg, *Synthesis*, 49 (1970).
147. F. Weygand, W. Steglich and F. Fraunberger, *Angew. Chem.*, **79**, 822 (1967).
148. A. D. Sinitsa, N. A. Parkhomenko and E. A. Stukalo, *Zh. Obshch. Khim.*, **47**, 2077 (1977).
149. A. D. Sinitsa, S. V. Bonadyk and L. N. Markovskii, *Zh. Org. Khim.*, **15**, 2003 (1979).
150. A. D. Sinitsa, N. A. Parkhomenko and S. V. Bonadyk, *Zh. Org. Khim.*, **12**, 974 (1976).
151. A. D. Sinitsa, S. V. Bonadyk and L. N. Markovskii, *Zh. Org. Khim.*, **13**, 721 (1977).
152. F. Weygand, W. Steglich, I. Lengyel, F. Fraunberger, A. Maierhofer and W. Oettmeier, *Chem. Ber.*, **99**, 1944 (1966).
153. M. E. Kuehne, US Patent 4,154,943 (Cl. 546–51; C07D487/16), 15 May 1979, Appl. 865,657, 29 Dec. 1977; *Chem. Abstr.*, **91**, 91816 (1979).
154. F. Weygand and W. Steglich, *Chem. Ber.*, **98**, 487 (1965).
155. E. Schmidt and K. Kuehlein, Ger. Offen., 2,645,280 (Cl. C07C145/00) 13 Apr. 1978, Appl. 7 Oct 1976; *Chem. Abstr.*, 89, 5908 (1978).
156. M. G. Nair, C. Saunders, S.-Y. Chen, R. L. Kisliuk and Y. Gaumont, *J. Med. Chem.*, **23**, 59 (1980).
157. A. Hassner and V. Alexanian, *J. Org. Chem.*, **44**, 3861 (1979).
158. *Houben-Weyl's Methoden der Organischen Chemie*, Vol. X/4, p. 92, G. Thieme, Stuttgart (1968).
159. O. V. Schickh and H. Metzger, Ger. Offen. 1,082,253, 25 May, 1960 (Cl.12o); *Chem.. Abstr.*, **55**, 17547c (1961).
160. J. Schmidt, *Ber.*, **35**, 3729 (1902).
161. N. Thorne, *J. Chem. Soc.,* 2587 (1956).
162. N. Thorne, *J. Chem. Soc.*, 4271 (1956).
163. M. Ohno, M. Okamoto and K. Nukada, *Tetrahedron Lett.*, 4047 (1965).
164. S. N. Danilov and K. A. Ogloblin, *J. Gen. Chem. USSR*, **22**, 2167 (1952).
165. M. Ohno, N. Naruse, S. Torimitsu and I. Terasawa, *J. Amer. Chem. Soc.*, **88**, 3168 (1966).
166. O. Wallach, *Ann.*, **306**, 278 (1899); **332**, 305 (1904); **360**, 26 (1908); **374**, 198 (1910); **389**, 185 (1912); **414**, 257 (1918).
167. M. Nishi, T. Komukai and K. Oikawa, Japan Kokai Tokyo Koho, 79, 24,809 (Cl.C07C131/00), 24 Feb. 1979, Appl. 77/89,789, 28 Jul. 1977; *Chem. Abstr.*, **91**, 38928 (1979).
168. M. Ohno, N. Naruse, S. Torimitsu and M. Okamoto, *Bull. Chem. Soc. Japan*, **39**, 1119 (1966).
169. K. A. Ogloblin and V. P. Semenov, *Zh. Org. Khim.*, **1**, 1361 (1965); *Chem. Abstr.*, **64**, 588a (1966).
170. P. Ciattoni, L. Rivolta and C. Divo, *Chim. Ind.* (Rome), **46**, 875 (1964).
171. A. Nenz, Belg. Patent. 627932 (1962); *Chem. Abstr.*, **60**, 14406c (1964).
172. A. Dornow, H. D. Jorden and A. Müller, *Chem. Ber.*, **94**, 67 (1961).
173. A. Dornow and A. Müller, *Chem. Ber.*, **93**, 41 (1960).
174. M. J. Haire, *J. Org. Chem.*, **45**, 1310 (1980).
175. R. L. Shriner and F. W. Neumann, *Chem. Rev.*, **35**, 351 (1944).
176. *Houben-Weyl's Methoden der Organischen Chemie*, 4th edn, Vol. XI(2), p. 39, G. Thieme, Stuttgart (1958).
177. B. S. Drach and G. N. Miskevich, *Zh. Org. Khim.*, **13**, 1398 (1977).

178. I. C. Grivas and A. Taurius, *Canad. J. Chem.*, **36**, 771 (1958).
179. W. Kantlehner, U. Dinkeldein and H. Bredereck, *Justus Liebigs Ann. Chem.*, 1354 (1979).
180. W. S. Saari, M. B. Freedman, J. R. Huff, S. W. King, A. W. Raab, S. J. Bergstrand, E. L. Engelhardt, A. Scriabine, G. Morgan, A. Morris, J. M. Stavorski, R. M. Noll and D. E. Duggan, *J. Med. Chem.*, **21**, 1283 (1978).
181. W. A. Bolhofer, C. N. Habecker, A. M. Pietruszkiewicz, M. L. Torchiana, H. I. Jacoby and C. A. Stone, *J. Med. Chem.*, **22**, 295 (1979).
182. A. G. Moritz, *Spectrochim. Acta*, **20**, 1555 (1964).
183. K. Matsumura, T. Saraie and N. Hashimoto, *Chem. Pharm. Bull.* (Tokyo), **24**, 912 (1976).
184. K. Matsumura, T. Saraie and N. Hashimoto, *Chem. Commun.*, 705 (1972).
185. F. Cramer, K. Pawelzik and H. J. Baldauf, *Chem. Ber.*, **91**, 1049 (1958).
186. F. Cramer and H. J. Baldauf, *Chem. Ber.*, **92**, 370 (1959).
187. L. E. Overman, *J. Amer. Chem. Soc.*, **96**, 597 (1974).
188. L. E. Overman, *J. Amer. Chem. Soc.*, **98**, 2901 (1976).
189. L. E. Overman and L. A. Clizbe, *J. Amer. Chem. Soc.*, **98**, 2352 (1976).
190. W. Steinkopf and L. Bohrmann, *Ber.*, **40**, 1635 (1907).
191. V. Shevchenko, V. Kalchenko and A. Sinitsa, *Zh. Obshch. Khim.*, **47**, 2157 (1977).
192. A. A. Michurin, E. A. Lyandaev and I. V. Bodrikov, *Zh. Org. Khim.*, **13**, 222 (1977).
193. A. J. Speziale and R. C. Freeman, *J. Amer. Chem. Soc.*, **82**, 909 (1960).
194. C. T'Kint and L. Ghosez, Unpublished results, mentioned as Ref. 99 by L. Ghosez and J. Marchand-Brynaert in *Advan. Org. Chem.*, **9**, 421 (1976).
195. H. Quast, E. Schmitt and R. Frank, *Angew. Chem.*, **83**, 728 (1971).
196. H. Quast, R. Frank and E. Schmitt, *Angew. Chem.*, **84**, 316 (1972).
197. H. Quast, R. Frank, B. Freundenzeich, P. Schäfer and E. Schmitt, *Jusbus Liebigs Ann. Chem.*, 74 (1979).
198. H. Quast, R. Frank, A. Heublein and E. Schmitt, *Justus Liebigs Ann. Chem.*, 83 (1979).
199. A. Ross and R. N. Ring, *J. Org. Chem.*, **26**, 579 (1961).
200. B. Yamada, R. W. Campbell and O. Vogl, *J. Polym. Sci.*, **15**, 1123 (1977).
201. D. W. Lipp and O. Vogl, *J. Polym. Sci.*, **16**, 1311 (1978).
202. R. W. Campbell and O. Vogl, *Monats. Chem.*, **110**, 453 (1979).
203. R. M. Noll and D. E. Duggan, *J. Med. Chem.*, **21**, 1283 (1978).
204. B. A. Titov, P. G. Zherebchenko, E. A. Krasheninnikova, V. Y. Kovtun and A. V. Terekhov, *Khim.-Farm. Zh.*, **13**, 26 (1979).
205. L. A. Overman, C. K. Marlowe and L. A. Clizbe, *Tetrahedron Lett.*, 599 (1979).
206. J. Wrobel, V. Nelson, J. Sumiejski and P. Kovacic, *J. Org. Chem.*, **44**, 2345 (1979).
207. M. Numazawa and Y. Osawa, *Steroids*, **32**, 519 (1978).
208. A. Silveira, Jr and S. K. Satra, *J. Org. Chem.*, **44**, 873 (1979).
209. A. B. Levy and A. Hassner, *J. Amer. Chem. Soc.*, **93**, 2051 (1971).
210. W. Fischer and J.-P. Anselme, *J. Amer. Chem. Soc.*, **89**, 5284 (1967).
211. J. Harnisch and G. Szeimies, *Chem. Ber.*, **112**, 3914 (1979).
212. K. Bott, *Chem. Ber.*, **108**, 402 (1975).
213. J. Ciabattoni and M Cabell, Jr, *J. Amer. Chem. Soc.*, **93**, 1482 (1971).
214. A. Padwa, T. J. Blacklock, P. H. J. Carlsen and M. Pulwer, *J. Org. Chem.*, **44**, 3281 (1979).
215. F. Sigaut-Titeux, L. Lemen-Olivier, J. Lévy and J. Le Men, *Heterocycles*, **6**, 1129 (1977).
216. J. Beger and P. D. Thong, East Ger. Patent 127,811 (Cl.C07D285/16), 12 Oct. 1977, Appl. 194,777, 14 Sep. 1976; *Chem. Abstr.*, **88**, 121250 (1978).
217. J. Beger and P. D. Thong, East Ger. Patent 127, 813 (C1.C07D277/38), 12 Oct. 1977, Appl. 194, 779, 14 Sep. 1976; *Chem. Abstr.*, **88**, 121164 (1978).
218. N. Barbulescu, S. Moga-Gheorghe, E. Andrei and A. Sintamarian, *Rev. Chim.* (Bucharest), **30**, 598 (1979); *Chem. Abstr.*, **91**, 157206y (1979).
219. L. Duhamel and J. Y. Valnot, *Compt. Rend. C* **286**, 47 (1978).
220. L. S. Abrams, H. S. Weintraub, J. E. Patrick and J. L. McGuire, *J. Pharm. Sci.*, **67**, 1287 (1978).
221. Y. Komeichi, S. Tomioka, T. Iwasaki and K. Watanabe, *Tetrahedron Lett.*, 4677 (1970).
222. A. Streitwieser, *Solvolytic Displacement Reactions*, McGraw-Hill, New York (1962).
223. R. M. Magid, *Tetrahedron*, **36**, 1901 (1980).
224. J. B. Conant, W. R. Kirner and R. E. Hussey, *J. Amer. Chem. Soc.*, **47**, 488 (1925).
225. A. Streitwieser, *Chem. Rev.*, **56**, 600 (1956).

226. J. Sadet, J. Lipszyc, E. Chenu, C. Gansser, G. Deyson, M. Hayat and C. Viel, *Eur. J. Med. Chem.—Chim. Therap.*, **13**, 277 (1978).
227. L. E. S. Barata, P. M. Baker, O. R. Gottlieb and E. A. Ruveda, *Phytochemistry*, **17**, 783 (1978).
228. M. J. Hunter, A. B. Cramer and H. Hibbert, *J. Amer. Chem. Soc.*, **61**, 516 (1939).
229. Y. Nagao, K. Kaneko and E. Fujita, *Tetrahedron Lett.*, 4115 (1978).
230. Y. Nagao, M. Ochiai, K. Kaneko, A. Maeda, K. Watanabe and E. Fujita, *Tetrahedron Lett.*, 1345 (1977).
231. H. Kobler, K.-H. Schuster and G. Simchen, *Justus Liebigs Ann. Chem.*, 1946 (1978).
232. J. H. Boyer and D. Straw, *J. Amer. Chem. Soc.*, **75**, 1642 (1953).
233. N. De Kimpe, R. Verhé, L. De Buyck and N. Schamp, *J. Org. Chem.*, **45**, 2803 (1980).
234. M. J. S. Dewar, *The Electronic Theory of Organic Chemistry*, Clarendon Press, Oxford (1948), p. 73.
235. R. G. Pearson, S. H. Langer, F. V. Williams and W. J. McGuire, *J. Amer. Chem. Soc.*, **74**, 5130 (1952).
236. P. D. Bartlett and E. N. Trachtenberg, *J. Amer. Chem. Soc.*, **80**, 5808 (1958).
237. A. Halvorsen and J. Songstad, *JCS Chem. Commun.*, 327 (1978).
238. W. Reeve, E. L. Caffary and T. E. Keiser, *J. Amer. Chem. Soc.*, **76**, 2280 (1954).
239. J. W. Thorpe and J. Warkentin, *Canad. J. Chem.*, **51**, 927 (1973).
240. T. I. Temnikova and E. N. Kropacheva, *Zh. Obshch. Khim.*, **19**, 1917 (1949); *Chem. Abstr.*, **44**, 1919 (1950).
241. C. L. Stevens, W. Malik and R. Pratt, *J. Amer. Chem. Soc.*, **72**, 4758 (1950).
242. O. E. Edwards and C. Grieco, *Canad. J. Chem.*, **52**, 3561 (1974).
243. U. Miotti and A. Fava, *J. Amer. Chem. Soc.*, **88**, 4274 (1966).
244. C. A. Grob, K. Seckinger, S. W. Tam and R. Traber, *Tetrahedron Lett.*, 3051 (1973).
245. F. G. Bordwell and T. G. Mecca *J. Amer. Chem. Soc.*, **94**, 2119 (1972).
246. T. Kishimoto, H. Kochi and Y. Yaneda, Japan Kokai, 76 32, 569 (Cl.C07D401/12), 19 Mar. 1976, Appl. 74/104,520 10 Sep. 1974; *Chem. Abstr.*, **85**, 177266 (1976).
247. B. S. Drach, G. N. Miskevich and A. P. Martynyuk, *Zh. Org. Khim.*, **14**, 508 (1978).
248. N. De Kimpe and N. Schamp, *J. Org. Chem.*, **40**, 3749 (1975).
249. J. Shavel Jr. and H. Zinnes, *J. Amer. Chem. Soc.*, **84**, 1320 (1962).
250. H. Zinnes and J. Shavel, Jr., *J. Org. Chem.*, **31**, 1765 (1966).
251. G. I. Dmitrienko, *Heterocycles*, **12**, 1141 (1979).
252. S. Sakai and N. Shinma, *J. Pharm. Soc. Japan*, **97**, 309 (1977).
253. Y. Tamura, M. W. Chun, H. Nishida and M. Ikeda, *Heterocycles*, **8**, 313 (1977).
254. W. Gauss and H. Heitzer, *Justus Liebigs Ann. Chem.*, **733**, 59 (1970).
255. K. Dickore, K. Sasse and K. Bode, *Justus Liebigs Ann. Chem.*, **733**, 70 (1970).
256. S. G. P. Plant and M. L. Tomlinson, *J. Chem. Soc.*, 298 (1933).
257. P. Stoss, *Arch. Pharm.*, **310**, 509 (1977).
258. D. A. Trujillo, K. Nishiyama and J.-P. Anselme, *Chem. Commun.*, 13 (1977).
259. M. V. Povstyanoi, E. V. Logachev and P. M. Kochergin, *Khim. Geterotsikl. Soedin.*, (5), 715 (1976); *Chem. Abstr.*, **85**, 94322 (1976).
260. T. Sarawatki and V. Srinivasan, *Tetrahedron*, **33**, 1043 (1977).
261. M. Ikeda, F. Tabusa and Y. Nishimura, *Tetrahedron Lett.* 2347 (1976).
262. V. P. Kukhar and E. I. Sagina, *Zh. Obshch. Khim.*, **49**, 1025 (1979).
263. V. Shevchenko, V. Kalchenko and A. Sinitsa, *Zh. Obshch. Khim.*, **47**, 2157 (1977).
264. (a) F. W. Lichtenthaler, *Chem. Rev.*, **61**, 607 (1961); (b) A. J. Kirby and S. G. Warren, *The Organic Chemistry of Phosphorus*, Elsevier, Amsterdam (1967).
265. G. Gaudiano, R. Modelli, P. P. Ponti, C. Ticozzi and A. Umani-Ronchi, *J. Org. Chem.*, **33**, 4431 (1968).
266. M. Masaki, K. Fukui and M. Ohta, *J. Org. Chem.*, **32**, 3564 (1967).
267. P. Bravo, G. Gaudiano, C. Ticozzi and A. Umani-Ronchi, *Chem. Commun.*, 1311 (1968).
268. P. Bravo, G. Gaudiano and A. Umani-Ronchi, *Gazz. Chim. Ital.*, **97**, 1664 (1967).
269. O. Tsuge, K. Kamara and S. Yogi, *Bull. Chem. Soc. Japan*, **50**, 2153 (1977).
270. M. E. Kuehne and R. Hafter, *J. Org. Chem.*, **43**, 3702 (1978).
271. M. E. Kuehne, D. M. Roland and R. Hafter, *J. Org. Chem.*, **43**, 3705 (1978).
272. R. Child and F. L. Pyman, *J. Chem. Soc.*, 36 (1971).
273. W. J. Middleton and C. G. Krespan, *J. Org. Chem.*, **30**, 1398 (1965).

274. W. J. Middleton and C. G. Krespan, *J. Org. Chem.*, **35**, 1480 (1970).
275. W. J. Middleton and D. Metzger, *J. Org. Chem.*, **35**, 3985 (1970).
276. N. De Kimpe, R. Verhé, L. De Buyck, H. Hasma and N. Schamp, *Tetrahedron*, **32**, 3063 (1976).
277. R. Verhé, N. De Kimpe, L. De Buyck, M. Tilley and N. Schamp, *Tetrahedron*, **36**, 131 (1980).
278. R. Verhé, N. De Kimpe, L. De Buyck, M. Tilley and N. Schamp, *Bull. Soc. Chim. Belg.*, **86**, 879 (1977).
279. N. De Kimpe, R. Verhé, L. De Buyck, J. Chys and N. Schamp, *J. Org. Chem.*, **43**, 2670 (1978).
280. N. De Kimpe, R. Verhé, L. De Buyck, J. Chys and N. Schamp, *Org. Prep. Proced. Int.*, **10**, 149 (1978).
281. For some leading references in the field of tertiary α-cyanoenamines, see Refs 282 and 283.
282. H. Ahlbrecht and D. Liesching, *Synthesis*, 495 (1977).
283. N. De Kimpe, R. Verhé, L. De Buyck and N. Schamp, *Synthesis* 741 (1979).
284. G. Costa, C. Riche and H. P. Husson, *Tetrahedron*, **33**, 315 (1977).
285. F. Weygand, W. Steglich and F. Fraunberger, *Angew. Chem.*, **79**, 822 (1967).
286. E. M. Kosower and B. Pazhenchevsky, *J. Amer. Chem. Soc.*, **102**, 4983 (1980).
287. M. Ohno, S. Torimitsu, N. Naruse, M. Okamoto and I. Sakai, *Bull. Chem. Soc. Japan*, **39**, 1129 (1966).
288. J. H. Smith, J. H. Heidema, E. T. Kaiser, J. B. Wetherington and J. W. Moncrief, *J. Amer. Chem. Soc.*, **94**, 9274 (1972).
289. J. H. Smith, J. H. Heidema and E. T. Kaiser, *J. Amer. Chem. Soc.*, **94**, 9276 (1972).
290. W. Höbold, U. Prietz and W. Pritzkow, *J. Prakt. Chem.*, **311**, 260 (1969).
291. C. L. Stevens and P. M. Pillai, *J. Amer. Chem. Soc.*, **89**, 3084 (1967).
292. (a) E. Francotte, Ph.D. thesis, University of Louvain-La-Neuve, Belgium (1978); (b) E. Francotte, Personal Communication.
293. M. Ohno and N. Naruse, *Bull. Chem. Soc. Japan*, **39**, 1125 (1966).
294. J. H. Smith and E. T. Kaiser, *J. Org. Chem.*, **39**, 728 (1974).
295. A. Dornow and H. D. Jorden, *Chem. Ber.*, **94**, 76 (1961).
296. P. Bravo, G. Gaudiano, P. P. Ponti and A. Umani-Ronchi, *Tetrahedron*, **26**, 1315 (1970).
297. H. G. Viehe, R. Merenyi, E. Francotte, M. Van Meerssche, G. Germain, J. P. Declercq and M. Bodart-Gilmont, *J. Amer. Chem. Soc.*, **99**, 2940 (1977).
298. T. L. Gilchrist and T. G. Roberts, *Chem. Commun.*, 847 (1978).
299. U. M. Kempe, T. K. Das Gupta, K. Blatt, P. Gygax, D. Felix and A. Eschenmoser, *Helv. Chim. Acta*, **55**, 2187 (1972).
300. T. K. Das Gupta, D. Felix, U. M. Kempe and A. Eschenmoser, *Helv. Chim. Acta*, **55**, 2198 (1972).
301. M. Petrzilka, D. Felix and A. Eschenmoser, *Helv. Chim. Acta*, **56**, 2950 (1973).
302. P. Gygax, T. K. Das Gupta and A. Eschenmoser, *Helv. Chim. Acta*, **55**, 2205 (1972).
303. S. Shatzmiller and A. Eschenmoser, *Helv. Chim. Acta*, **56**, 2975 (1973).
304. A. Ruttimann and D. Ginsburg, *Helv. Chim. Acta*, **58**, 2237 (1975).
305. S. Shatzmiller, P. Gygax, D. Hall and A. Eschenmoser, *Helv. Chim. Acta*, **56**, 2961 (1973).
306. F. D. Chattaway and T. E. W. Browne, *J. Chem. Soc.*, 1088 (1931).
307. A. G. Schultz and W. K. Hagmann, *J. Org. Chem.*, **43**, 3391 (1978).
308. S. Sommer, *Tetrahedron Lett.*, 117 (1977).
309. C. A. Sacks and P. L. Fuchs, *J. Amer. Chem. Soc.*, **97**, 7372 (1975).
310. S. R. Sandler and W. Karo, *Organic Functional Group Preparations*, Vol. II, Academic Press, New York (1971), p. 307.
311. J. Schantl, *Monats. Chem.*, **105**, 322 (1974).
312. V. Sprio and S. Plescia, *Ann. Chim.* (Roma), **62**, 345 (1972).
313. L. A. Carpino, *J. Amer. Chem. Soc.*, **80**, 601 (1958).
314. L. A. Carpino, P. H. Terry and S. D. Thatte, *J. Org. Chem.*, **31**, 2867 (1966).
315. A. Silveira, Jr, T. J. Weslowski, T. A. Weil, V. Kumar and J. P. Gillespie, *J. Amer. Oil Chem. Soc.*, **48**, 661 (1971).
316. P. J. Kocienshi, J. M. Ansell and R. W. Ostrow, *J. Org. Chem.*, **41**, 3625 (1976).
317. J. Schantl, *Monats. Chem.*, **103**, 1705 (1972).
318. J. Schantl, *Monats. Chem.*, **103**, 1718 (1972).

319. J. Schantl and P. Karpellus, *Monats. Chem.*, **109**, 1081 (1978).
320. C. Kashima, Y. Aoko and Y. Omote, *JCS Perkin I*, 2511 (1975).
321. L. Duhamel and J.-Y. Valnot, *Tetrahedron Lett.*, 3167 (1974).
322. P. Duhamel, L. Duhamel and J.-Y. Valnot, *Tetrahedron Lett.*, 1339 (1973).
323. L. Duhamel, P. Duhamel and J.-Y. Valnot, *Compt. Rend. C.* **278**, 141 (1974).
324. L. Duhamel and J.-Y. Valnot, *Tetrahedron Lett.*, 3319 (1979).
325. S. Bozzini, S. Gratton, A. Risaliti, A. Stener, m. Calligaris and G. Nardin, *JCS Perkin I*, 1377 (1977).
325. S. Bozzini, S. Gratton, A. Risaliti, A. Stener, M. Calligaris and G. Nardin, *JCS Perkin I*,
327. S. Bozzini, B. Cova, S. Gratton, A. Lisini and A. Risaliti, *JCS Perkin I*, 240 (1980).
328. F. J. McEvoy and G. R. Allen, *J. Med. Chem.*, **17**, 281 (1974).
329. W. V. Curran and A. Ross, *J. Med. Chem.*, **17**, 273 (1974).
330. J. D. Albright, F. J. McEvoy and D. B. Moran, *J.Heterocyclic Chem.*, **15**, 881 (1978).
331. S. C. Dorman, US Patent 2,468,592, Apr. 26, 1949; *Chem. Abstr.*, **43**, 5153f (1949).
332. S. C. Dorman, US Patent 2,468,593, Apr. 26, 1949; *Chem. Abstr.*, **43**, 7499 (1949).
333. D. Borrmann and R. Wegler, *Chem. Ber..*, **100**, 1814 (1967).
334. S. Brodka and H. Simon, *Chem. Ber.*, **102**, 3647 (1969).
335. H. Zinner, W. Siems and G. Erfurt, *J. Prakt. Chem.*, **316**, 443 (1974).
336. H. Zinner, W. Siems and G. Erfurt, *J. Prakt. Chem.*, **316**, 491 (1974).
337. E. Schmidt and K. Kuehlein, Ger. Offen. 2,645,280 (Cl. C07C145/00), 13 Apr. 1978, Appl. 7 Oct. 1976; *Chem. Abstr.*, **89**, 5908 (1978).
338. R. E. Banks, M. G. Barlow and M. Nickkhoamiry, *J. Fluorine Chem.*, **14**, 383 (1979).
339. R. K. Olsen and A. J. Kolar, *Tetrahedron Lett.*, 3579 (1975).
340. Y. G. Balon and V. A. Smirnov, *Zh. Org. Khim.* **14**, 668 (1978).
341. K. Burger, G. George and J. Fehn, *Justus Liebigs Ann. Chem.*, **757**, 1 (1972).
342. B. S. Drach and G. N. Miskevich, *Zh. Org. Khim.*, **14**, 501 (1978).
343. T. Hino, H. Miura, T. Nakamura, R. Murata and M. Nakagawa, *Heterocycles*, **3**, 805 (1975).
344. N. De Kimpe, R. Verhé, L. De Buyck and N. Schamp, *Rec. Trav. Chim. Pays-Bas*, **96**, 242 (1977).
345. N. De Kimpe, N. Schamp and R. Verhé, *Synthetic Commun.*, **5**, 403 (1975).
346. N. De Kimpe, R. Verhé, L. De Buyck and N. Schamp, *J. Org. Chem.*, **45**, 5319 (1980).
347. A. Hassner and A. B. Levy, *J. Amer. Chem. Soc.*, **93**, 5469 (1971).
348. N. De Kimpe, R. Verhé and N. Schamp, *Bull. Soc. Chim. Belg.*, **84**, 701 (1975).
349. K. Burger, J. Albanbauer and F. Manz, *Chem. Ber.*, **107**, 1823 (1974).
350. G. A. Boswell, Jr, *J. Org. Chem.*, **33**, 3699 (1968).
351. M. J. Haire, *J. Org. Chem.*, **42**, 3446 (1977).
352. M. Takeda, M. Inoue, M. Konda, S. Saito and H. Kugito, *J. Org. Chem.*, **37**, 2677 (1972).
353. R. Jacquier, *Bull. Soc. Chim. Fr.*, [5] **17**, D35 (1950).
354. A. S. Kende, *Org. Reactions*, **11**, 261 (1960).
355. A. A. Akhrem, T. K. Ustynyuk and Y. A. Titov, *Russian Chem. Rev.*, **39**, 732 (1970).
356. C. Rappe in *The Chemistry of the Carbon—Halogen Bond* (Ed. S. Patai), Wiley, Chichester (1973), p. 1071.
357. K. Sato and M. Oohashi, *Yuki Gosei Kagaku Kyokai Shi*, **32**, 435 (1974).
358. J. March in *Advanced Organic Chemistry: Reactions Mechanisms and Structure*, McGraw-Hill, New York (1968), p. 804.
359. W. J. le Noble in *Highlights of Organic Chemistry*, Marcel Dekker, New York (1974), p. 864.
360. P. J. Chenier, *J. Chem. Educ.*, **55**, 286 (1978).
361. R. Frank, Ph.D. thesis, University of Würzburg (1971).
362. N. De Kimpe, R. Verhé, L. De Buyck, N. Schamp, J. P. Declercq, G. Germain and M. Van Meerssche, *J. Org. Chem.*, **42**, 3704 (1977).
363. N. De Kimpe, R. Verhé, L. De Buyck and N. Schamp, *Bull. Soc. Chim. Belg.*, **86**, 663 (1977).
364. P. Duhamel, L. Duhamel, C. Collet and A. Haïder, *Compt. Rend. C* **273**, 1461 (1971).
365. L. Duhamel, P. Duhamel, C. Collet, A. Haïder and J-M. Poirier, *Tetrahedron Lett.*, 4743 (1972).
366. L. Duhamel and J.-M. Poirier, *Bull. Soc. Chim. Fr.*, 329 (1975).
367. A. Buzas, C. Retourné, J. P. Jacquet and G. Lavielle, *Heterocycles*, **6**, 1307 (1977).

368. A. Picot and X. Lusinchi, *Tetrahedron*, **34**, 2747 (1978).
369. G. Costa, C. Riche and H. P. Husson, *Tetrahedron*, **33**, 315 (1977).
370. J.-Y. Laronze, J. Laronze-Fontaine, J. Lévy and J. Le Men, *Tetrahedron Lett.*, 491 (1974).
371. T. Hino, H. Yamaguchi, M. Endo and M. Nakagawa, *JCS Perkin I*, 745 (1976).
372. T. Hino and M. Nakagawa, *Heterocycles*, **6**, 1680 (1977).
373. J. P. Kutney, J. Beck, F. Bylsma, J. Cook, W. J. Cretney, K. Fuji, R. Imhof and A. M. Treasurywala, *Helv. Chim. Acta*, **58**, 1690 (1975).
374. K. Burger and R. Ottlinger, *J. Fluorine Chem.*, **11**, 29 (1978).
375. K. Burger, W.-D. Roth, K. Einhellig and L. Hatzelmann, *Chem. Ber.*, **108**, 2737 (1975).
376. K. Burger, H. Schikaneder and A. Meffert, *Z. Naturforsch., B*, **30**, 622 (1975).
377. K. Burger and S. Penninger, *Synthesis*, 524 (1978).
378. R. W. Hoffmann, K. Steinbach and W. Lilienblum, *Chem. Ber.*, **109**, 1759 (1976).
379. K. Burger and R. Ottlinger, *Chem. Zeit.*, **101**, 402 (1977).
380. R. W. Hoffmann, K. Steinbach and B. Dittrich, *Chem. Ber.*, **106**, 2174 (1973).
381. K. Burger, W. Thenn and H. Schickaneder, *J. Fluorine Chem.*, **6**, 59 (1975).
382. K. Burger, J. Fehn and A. Gieren, *Justus Liebigs Ann. Chem.*, **757**, 9 (1972).
383. K. Burger, S. Tremmel and H. Schickaneder, *J. Fluorine Chem.*,. **6**, 471 (1975).
384. K. Burger, W. Thenn and A. Gieren, *Angew. Chem.*, **86**, 481 (1974).
385. K. Burger, W. Thenn, R. Rauk and H. Schickaneder, *Chem. Ber.*, **108**, 1460 (1975).
386. K. Burger and F. Hein, *Justus Liebigs Ann. Chem.*, 133 (1979).
387. N. P. Gambaryan, I. L. Knunyants *et al., Dokl. Akad Nauk SSSR*, **166**, 864 (1966).
388. F. Weygand, W. Steglich, W. Oettmeier, A. Maierhofer and R. S. Loy, *Angew. Chem.*, **78**, 640 (1966).
389. Y. V. Zeifman, N. P. Gambaryan, L. A. Simonyan, R. B. Minasyan and I. L. Knunyants, *Zh. Obshch. Khim.*, **37**, 2476 (1967); *J. Gen. Chem. USSR*, **37**, 2355 (1967).
390. Y. V. Zeifman and I. L. Knunyants, *Dokl. Akad. Nauk. SSSR*, **173**, 354 (1967).
391. I. L. Knunyants, Y. V. Zeifman and N. D. Gambaryan, *Izv. Akad. Nauk. SSSR, Ser. Khim.*, 1108 (1966).
392. I. L. Knunyants, N. P. Gambaryan and R. B. Minasyan, *Izv. Akad. Nauk. SSSR, Ser. Khim.*, 1910 (1965).
393. F. Hein, K. Burger and J. Firl, *JCS Chem. Commun.*, 792 (1979).
394. L. Kryukov, L. Y. Kryukova, M. A. Kurykin, R. N. Sterlin and I. L. Knunyants, *Zh. Vses. Khim. O-va*, **24**, 393 (1979); *Chem. Abstr.*, **91**, 193269 (1979).
395. D. P. Del'tsova, Z. V. Safronova, N. P. Gambaryan, M. Y. Antipin and Y. T. Struchkov, *Izv. Akad. Nauk. SSSR, Ser. Khim.*, (8), 1881 (1978); *Chem. Abstr.*, **89**, 215280 (1978).
396. R. Albrecht and G. Kresze, *Chem. Ber.*, **98**, 1431 (1965).
397. G. R. Krow, C. Pyun, R. Rodebaugh and J. Marakowski, *Tetrahedron*, **30**, 2977 (1974).
398. T. Imagawa, K. Sisido and M. Kawanisi, *Bull. Chem. Soc. Japan*, **46**, 2922 (1973).
399. Y. A. Arbuzov, E. I. Klimova, N. D. Autonova and Y. T. Tomilov, *Zh. Org. Khim.*, **10**, 1164 (1974).
400. T. Akiyama, N. Urasato, T. Imagawa and M. Kawanisi, *Bull. Chem. Soc. Japan*, **49**, 1105 (1976).
401. K. Burger and S. Penninger, *Synthesis*, 526 (1978).
402. H. Beyer, G. Wolter and J. Lemke, *Chem. Ber.*, **89**, 2550 (1956).
403. H. Beyer, W. Lässig and G. Ruhlig, *Chem. Ber.*, **86**, 764 (1953).
404. R. Ottlinger, K. Burger, H. Goth and J. Firl, *Tetrahedron Lett.*, 5003 (1978).
405. K. Burger, J. Albanbauer and W. Strych, *Synthesis*, 57 (1975).
406. K. Burger and R. Ottlinger, *Synthesis*, 44 (1978).
407. K. Burger and R. Ottlinger, *Tetahedron Lett.*, 973 (1978).
408. D. Maytum, *Chem. Brit.*, **14**, 382 (1978).
409. K. Isogai, T. Sasaki, C. Sato, *J. Syn. Org. Chem. Japan*, **36**, 1104 (1978).
410. B. S. Drach, T. P. Popovich, A. A. Kisilenko and O. M. Polumbrik, *Zh. Org. Khim.*, **15**, 31 (1979).
411. V. J. Lee and R. B. Woodward, *J. Org. Chem.*, **44**, 2487 (1979).
412. E. Elkik and H. Assadifar, *Bull. Soc. Chim. Fr.*, 129 (1978).
413. J. Patrick, H. Weintraub and J. McGuire, *Steroids*, **32**, 147 (1978).
414. P. Swered and M. A. Girard, Canadian Patent 1,035,698 (Cl,A01N9/20), 01 Aug. 1978, US Appl. 545,678, 30 Jan. 1975; *Chem. Abstr.*, **90**, 98550 (1979).
415. J. Diopoh and M. Olomucki, *Hoppe-Seyler's Z. Physiol. Chem.*, **360**, 1257 (1979).

416. Shell Intern. Res. Maatsch. N.V., British Patent 963.055, US Appl. 13 May, 1962; *Chem. Abstr.*, **61**, 11252b (1964).
417. N. Langlois, F. Guéritte, Y. Langlois and P. Potier, *J. Amer. Chem. Soc.*, **98**, 7017 (1976).
418. G. A. Boswell, Jr *J. Org. Chem.*, **33**, 3699 (1968).
419. T. I. Temnikova and E. N. Kropacheva, *Zh. Obshch. Khim.*, **19**, 1917 (1949); *Chem. Abstr.*, **44**, 1919 (1950).
420. C. L. Stevens, W. Malik and R. Pratt, *J. Amer. Chem. Soc.*, **72**, 4758 (1950).
421. C. L. Stevens, M. L. Weiner and R. C. Freeman, *J. Amer. Chem. Soc.*, **75**, 3977 (1953).
422. C. L. Stevens and J. J. DeYoung, *J. Amer. Chem. Soc.*, **76**, 718 (1954).
423. T. I. Temnikova and E. N. Kropacheva, *Zh. Obshch. Khim.*, **22**, 1150 (1952); *Chem. Abstr.*, **47**, 6901 (1953).
424. N. De Kimpe, R. Verhé, L. De Buyck and N. Schamp, *J. Org. Chem.*, **45**, 2803 (1980) and references cited therein.
425. L. E. Overman, *Acc. Chem. Res.*, **13**, 218 (1980).
426. N. Kunesch, P.-L. Vaucamps, A. Cavé, J. Poisson and E. Wenkert, *Tetrahedron Lett.*, 5073 (1979).
427. J.-P. Bégué and M. Charpentier-Morize, *Acc. Chem. Res.*, **13**, 207 (1980).
428. X. Creary, *J. Org. Chem.*, **44**, 3938 (1979).
429. Y. L. Chow, K. S. Pillay and H. Richard, *Canad. J. Chem.*, **57**, 2923 (1979).
430. A. I. Meyers, H. W. Adickes, I. R. Politzer and W. N. Beverung, *J. Amer. Chem. Soc.*, **91**, 765 (1969).
431. A. F. Hirsch, G. O. Allen, B. Wong, S. Reynolds, C. Exarhos, W. Brown and D. W. Hahn, *J. Med. Chem.*, **20**, 1546 (1977).
432. R. Wegler in *Chemie der Pflanzenschutz- und Schädlingsbekämpfungsmittel*, Springer-Verlag, Berlin, Heidelberg and New York (1970).
433. B. Unterhalt, *Pharm. Zeit.* **125**, 361 (1980).
434. J. Schantl, *Org. Magn. Resonance*, **12**, 652 (1979).
435. E. W. Della, M. P. Reimerink and B. G. Wright, *Aust. J. Chem.*, **32**, 2235 (1979).
436. M. G. Zimin, N. G. Zabirov, V. N. Smirnov, R. A. Cherkasov, and A. N. Pudovik, *Zh. Obshch. Khim.*, **50**, 24 (1980).
437. T. N. Maksimova, V. B. Mochalin and B. V. Unkovskii, *Khim. Geterotsiklich. Soedin*, 273 (1980)
438. J. Cantacuzène and D. Ricard, *Bull. Soc. Chim. Fr.* 1587 (1967).
439. J. Cantacuzène, M. Atlanti and J. Anibié, *Tetrahedron Lett.* 2335 (1968).
440. D. Ricard and J. Cantacuzène, *Bull. Soc. Chim. Fr.*, 628 (1969).
441. D. D. Rosenfeld and J. R. Kilsheier, *J. Agr. Food Chem.*, **22**, 926 (1974).
442. K. Packer, *Nanogen Index – A Dictionary of Pesticides and Chemical Pollutants* (Ed. K. Packer), Nanogens Int., Freedom, USA (1975).
443. A. Kirrmann, *Bull. Soc. Chim. Fr.*, 657 (1961).
444. F. I. Luknitskii, *Chem. Rev.*, **75**, 259 (1975).
445. L. De Buyck, R. Verhé, N. De Kimpe, D. Courtheyn and N. Schamp, *Bull. Soc. Chim. Belg.*, **89**, 441 (1980).
446. N. De Kimpe, L. De Buyck, R. Verhé, F. Wychuyse and N. Schamp, *Synthetic Commun.*, **9**, 575 (1979).
447. T. L. Gilchrist, D. A. Lingham and T. G. Roberts, *JCS Chem. Commun.*, 1089 (1979).
448. K. H. Pfoertner and J. Foricher, *Helv. Chim. Acta*, **63**, 658 (1980).
449. C. J. Barnett, U.S. 4,199,525 (Cl.260–453RW; C07C119/00), 22 Apr 1980, Appl. 27,627, 6 Apr. 1979; *Chem. Abstr.*, **93**, 94970 (1980).
450. K. A. Ogloblin and A. A. Potekhin, *J. Org. Chem. U.S.S.R.*, **1**, 1370 (1965).
451. T. L. Gilchrist and T. G. Roberts, *JCS Chem. Commun.*, 1090 (1979).
452. A. Roedig, W. Ritschel and M. Fouré, *Chem. Ber.*, **113**, 811 (1980).
453. P. J. Stang and D. P. Fox, *J. Org. Chem.*, **42**, 1667 (1977).
454. D. R. Williams and K. Nishitani, *Tetrahedron Lett.*, 4417 (1980).
455. R. Z. Andriamialisoa, N. Langlois, Y. Langlois and P. Potier, *Tetrahedron*, **36**, 3053 (1980).
456. R. Neidlein, S. Shatzmiller and E. Walter, *Justus Liebigs Ann. Chem.*, 686 (1980).
457. N. De Kimpe, R. Verhé, L. De Buyck and N. Schamp, *Bull. Soc. Chim. Belg.*, **88**, 59 (1979).
458. K. Burger and H. Goth, *Angew. Chem.*, **92**, 836 (1980).

459. K. Burger, R. Ottlinger, H. Goth and J. Firl, *Chem. Ber.*, **113**, 2699 (1980).
460. K. Aoki, T. Shida, S. Kamawawa, M. Ohtsuru and S. Yamazaki, UK Patent Appl., 2,019,402 (Cl.C07C109/10), 31 Oct. 1979; Japan Patent Appl., 78/48,493, 24 Apr. 1978; *Chem. Abstr.*, **93**, 204296 (1980).
461. N. De Kimpe, R. Verhé, L. De Buyck, L. Moëns and N. Schamp, *Syhthesis*, **43**, (1982).

*Note added in proof
A very recent paper described a straightforward general synthesis of α-haloketimines by condensation of α-halocarbonyl compounds with primary amines in the presence of titanium(IV) chloride.[461]

The Chemistry of Functional Groups, Supplement D
Edited by S. Patai and Z. Rappoport
© 1983 John Wiley & Sons Ltd

CHAPTER **14**

Fluorocarbons

B. E. SMART

E. I. du Pont de Nemours and Company, Central Research and Development Department, Wilmington, Delaware 19898, USA

I. INTRODUCTION

The purpose of this chapter is to review recent developments in fluorocarbon chemistry with particular emphasis on the fundamental effects of fluorine substituents on the structure, bonding and reactivity of organic molecules. Although many of the topics covered in this chapter have been discussed for halocarbons in previous volumes of the series 'The Chemistry of Functional Groups', this is the first treatment of fluorocarbons *per se*. (In this chapter, the general terms halo-, halogen, and halide refer to chlorine, bromine, and iodine.)

Space limitations require that this discussion be limited primarily to the chemistry of polyfluorinated hydrocarbons. Unfortunately, the rapidly expanding area of synthetic organofluorine chemistry cannot be covered. The chemistry of fluorinated functional groups containing oxygen, sulphur, nitrogen as well as organometallic chemistry and synthetic methods can be found in several excellent texts[1–6] and reviews[7–10].

II. STRUCTURE AND BONDING

A. Alkanes

1. Carbon–halogen bonds

Fluorine is the most electronegative of the elements. Its electronegativity of 4.0 compares with 3.0 and 2.8 for chlorine and bromine, respectively, on the Pauling scale. Fluorine forms the strongest single bond with carbon of any element. The C—F bond is 43% ionic from electronegativity considerations, and the shorter C—F bond length relative to other carbon–halogen bond lengths (Table 1) is correctly predicted by the Stevenson–Schomaker equation which relates bond contractions to electronegativity differences[11].

Fluorinated methanes display unique changes in bond lengths and bond strengths in proceeding from methane to tetrafluoromethane[12] (Table 1). As the fluorine content increases, the bonds progressively shorten and increase in strength. A similar trend is observed with the fluorinated ethanes. In the series FCH_2CH_2F[13], HCF_2CF_2H[14] and CF_3CF_3[15], the C—F bond lengths are respectively 1.389, 1.350, and 1.326 Å. These trends are not observed in the chlorocarbon analogues.

Several explanations for this phenomenon, all of which propose systematic differences in bonding, have been offered. Pauling[19] invoked double-bond no-bond resonance structures, e.g. **1a** ↔ **1b**, to account for bond shortening in the fluorinated methanes. Twelve such structures can be written for CF_4, six for CHF_3, two for CHF_2 and none for CH_3F. *Ab initio* calculations (4-31G) support this valence bond description[20,21]. In molecular orbital terms, **1a** ↔ **1b** represents π-type donation of a

TABLE 1. Halomethane bond lengths[16] and bond dissociation energies[17]

	X = F		X = Cl		X = Br	
	r(C—F), Å	$D°$(C—F), kcal mol^{-1}	r(C—Cl), Å	$D°$(C—Cl), kcal mol^{-1}	r(C—Br), Å	$D°$(C—Br), kcal mol^{-1}
CH_3X	1.385	109.0[a]	1.782	83.7[a]	1.939	69.2[a]
CH_2X_2	1.358	122	1.772	81	1.934	64
CHX_3	1.332	128.0	1.767	77.7	1.930	62
CX_4	1.317	129.7	1.766	72.9	1.942	56.2

[a]Ref. 18.

$$\mathrm{F{-}\overset{\displaystyle F}{\underset{\displaystyle F}{C}}{-}F} \longleftrightarrow \mathrm{F{-}\overset{\displaystyle F^-}{\underset{\displaystyle F}{C}}{=}F^+}$$

(1a) **(1b)**

lone pair of electrons from fluorine into an acceptor C—F σ-bond. The degree of overlap will depend upon the occupancy of the appropriate acceptor carbon 2p orbital. The calculations show that the occupancy of the acceptor orbital decreases with the number of electron-withdrawing fluorine atoms, and the π-overlap between C and F increases in going from CH_3F to CH_2F_2.

Distinctly different treatments that do not involve fluorine lone pair electrons have been advanced. Peters[22] considered the nature of the C—X bond as X becomes more electronegative and concluded that the fluorine substituent will induce carbon to preferentially give up its less tightly bound p electrons. The C—F bond therefore will have more s-character than usual and will be shorter. The hybridization of carbon was proposed to be $sp^{2.7}$ in CH_3F, $sp^{2.6}$ in CH_2F_2, $sp^{2.5}$ in CHF_3, and $sp^{2.4}$ in CF_4. An alternative hybridization scheme has been reached using valence bond logic[23,24]. Bent[24] argues that by symmetry the C—F bonds in CF_4 must be sp^3 hybridized. In CH_3F, the electronegative fluorine rehybridizes the carbon to give increased p-character in the C—F bond but increased s-character in the C—H bonds. The p-rich C—F bond in CH_3F is therefore longer than the C—F bond in CF_4. Bent's arguments also explain the unusual molecular geometries of fluorinated methanes. For example, the FCF and HCH bond angles in CH_2F_2 are 108.3° and 111.9°, respectively, instead of the normal tetrahedral angle, 109.3° [12]. Rehybridization of the sp^3 carbon toward sp^n ($n > 3$), as suggested by Bent, would shrink the angle between those hybrids with increased p-character, and thus would reduce the FCF bond angle in CH_2F_2 (and conversely would enlarge the HCH bond angle). Peters' explanation does not appear to be consistent with these results.

Ab initio calculations (STO-3G and 4-31G) by Kollman, however, indicate that the carbon hybrids in the C—F and C—H bonds in CH_2F_2 have about 36% and 32% s-character, respectively, and the difficulty in comparing hybridization derived from molecular orbital theory with that inferred from valence bond restrictions has been emphasized[25]. Kollman further gives a molecular orbital analysis that predicts a decrease in the BAB angle for a general AB_2 fragment as B becomes more electronegative.

Electrostatic models which complement the hybridization schemes also have been used to explain semi-quantitatively the bonding in fluorinated alkanes[22,26]. Successive fluorine substitution on carbon will make the C—F bonds more ionic and thus stronger. Photoelectron spectroscopy studies support this contention. For example, the carbon 1s binding energy in fluorinated methanes increases linearly with the number of fluorines bound to carbon[27].

The unusual strengthening of C—F bonds with increasing α-fluorination is not observed for other C—X bonds. In comparing the C—X bond dissociation energies in C_2H_5X with those in C_2F_5X (X = H, Cl, Br, I), substitution of F for H has little effect (Table 2). This is also the case for the methane series (Table 3). The striking exception is again the C—F bond, which is strengthened by 20 kcal mol^{-1} on fluorination (CF_3CF_2—F versus CH_3CH_2—F; CF_3—F versus CH_3—F). There is currently no satisfactory explanation for these results.

The wide range of descriptions for bonding in molecules as 'simple' as the fluoromethanes indicates the uncertainty that remains in understanding the fundamental

TABLE 2. Bond dissociation energies (in kilocalories per mole) for the C—X bond in ethanes[28,29]

X	$D°$ (CF_3CF_2—X)	$D°$ (CH_3CH_2—X)
H	103	98
F	127	108[a]
Cl	85	82
Br	68	68
I	53	53

[a]Estimated.

properties of fluorocarbons. A multitude of often conflicting theoretical treatments have been advanced to explain the unusual chemical and physical properties of fluorocarbons. Unfortunately, there rarely are definitive experimental data from which the merits of one theory over those of another can be established. For the remainder of this chapter, various theoretical propositions will be referred to but they will not be discussed in detail. The reader should consult the appropriate references for full discussions.

TABLE 3. Bond dissociation energies (in kilocalories per mole) for halomethanes[17]

X	$D°$ (CH_3—X)	$D°$ (CH_2F—X)	$D°$ (CHF_2—X)	$D°$ (CF_3—X)
H	104[a]	102.7	103.2	105.9
F	109.0[a]	122	128	129.7
Cl	83.7	89	87.6	85.5
Br	69.2	73	70	69.2

[a]Ref. 18.

2. *Carbon–carbon bonds*

The strengths of C—C bonds generally increase upon fluorination. The C—C bond strength in poly(tetrafluoroethylene) is estimated to be 8 kcal mol^{-1} greater than that in polyethylene[28]. The CF_3—CH_3 and CF_3—CF_3 bonds are respectively 14 and 10 kcal mol^{-1} stronger than the C—C bond in ethane[30,31]. It is curious that CF_3—CF_3 has a longer (1.545 versus 1.536 Å)[15] yet stronger C—C bond than ethane.

Kinetic studies on the unimolecular decomposition of fluorinated cyclobutanes illustrate the increased stability of fluorinated C—C bonds (Table 4). The activation energy for the decomposition of perfluorocyclobutane is 12 kcal mol^{-1} greater than that for cyclobutane. The data for 1,1,2,2-tetrafluorocyclobutane indicate that the favoured decomposition pathway involves cleavage of the $C_{(3)}$—$C_{(4)}$ bond, which is probably only slightly stronger than the C—C bond in cyclobutane itself, and the strong $C_{(1)}$—$C_{(2)}$ bond. The less favourable pathway involves cleavage of two relatively strong bonds, $C_{(1)}$—$C_{(4)}$ and $C_{(2)}$—$C_{(3)}$.

Fluorinated cyclopropanes are exceptions to this pattern. Perfluorocyclopropane is thermally less stable than cyclopropane[37], and its strain energy has been estimated to be 40 kcal mol^{-1} greater than that for cyclopropane[38], even though its C—C bonds (1.505 Å)[39] are slightly shorter than those in the hydrocarbon. (A shorter bond is not necessarily a stronger bond, in terms of bond dissociation energies. The shorter C—C bond in cyclopropane versus propane is an example. For an interesting discussion of this subject, see Greenberg and Liebman[40].)

TABLE 4. Cyclobutane decompositions – Arrhenius parameters

Reaction		log A, s^{-1}	E_a, kcal mol^{-1}	Reference
 □ ⟶	2 $CH_2{=}CH_2$	15.6	62.5	32, 33
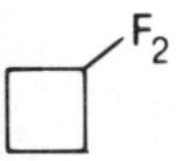 F_2 ⟶	$CH_2{=}CF_2$ $CH_2{=}CH_2$	15.61	67.2	34
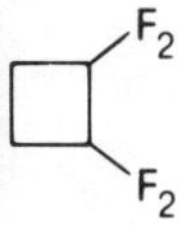 F_2, F_2 ⟶	2 $CH_2{=}CF_2$	15.35	69.8	35
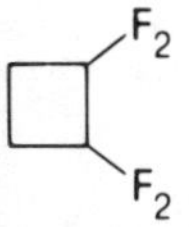 F_2, F_2 ⟶	$CH_2{=}CH_2$ $CF_2{=}CF_2$	15.27	73.6	35
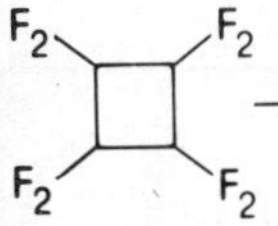 F_2, F_2, F_2, F_2 ⟶	2 $CF_2{=}CF_2$	15.97	74.2	36

The effect of fluorine substituents on the structure and reactivity of cyclopropanes has received considerable theoretical[41–46] and experimental attention[47–51]. Günter predicted a general weakening of the cyclopropane ring bonds upon fluorine substitution; others suggest that fluorocyclopropane is slightly less strained than the hydrocarbon[42] or that bonds adjacent to fluorine are shortened (strengthened) whereas those opposite are lengthened (weakened)[41,46]. Presently, there are no experimental data on the strain energy or C—C bond strengths of a monofluorocyclopropane. Paquette and coworkers[52] found that the semibullvalene isomer **2a** (X = F), was thermodynamically favoured over **2b** (X = F), but the opposite was true for other substituents (X = CH_3; CH_2OCH_3). However, these data reflect only the relative effect of F on a double bond versus a cyclopropane ring, not the absolute effect on a cyclopropane ring (see Section II.B.1.b).

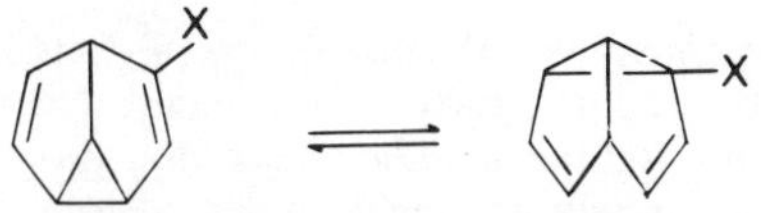

X ⇌ X

(2a) (2b)

Although the effect of a single fluorine substituent is unresolved, geminal fluorine substituents clearly facilitate homolytic cleavage of the cyclopropane ring. The activation energy for the thermal isomerization of **3a** to **3b** (49.7 kcal mol^{-1}) is about 10 kcal mol^{-1} less than that for the hydrocarbon system[50]. The rapid isomerization of the *endo* isomer **4a** to **4b** at 80°C ($\Delta G^{\neq}$ = 28.2 kcal mol^{-1}) also indicates enhanced bond cleavage[53]. The 1,1-difluorocyclopropane isomerizations apparently proceed by preferential $C_{(2)}$—$C_{(3)}$ bond cleavage, which is consistent with the observed $C_{(2)}$—$C_{(3)}$ bond lengthening from microwave spectral studies[51].

TABLE 5. Methylenecyclopropane isomerizations[48,49]

Isomerization	log A	E_a, kcal mol^{-1}	$\Delta G°$
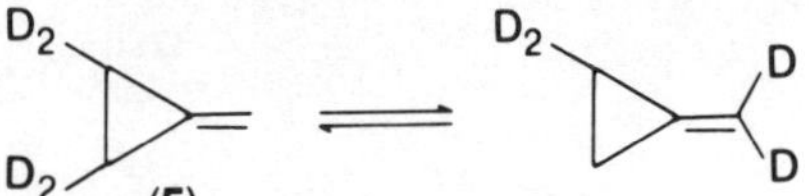 (5)	14.3	41.6	0
 (6)	13.2	38.3 ± 0.4	−1.7
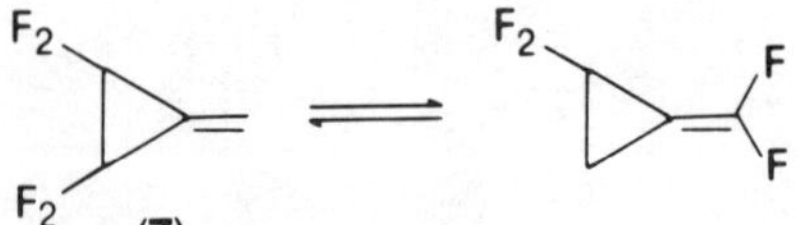 (7)	12.6	29.6 ± 0.5	−7.5[a]

[a]Estimated.

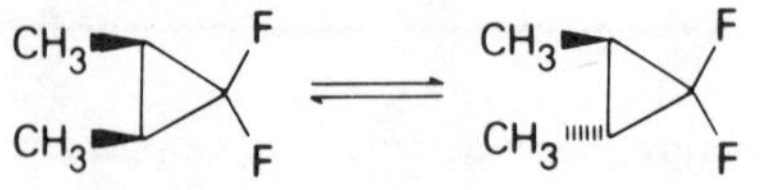

(3a) (3b) 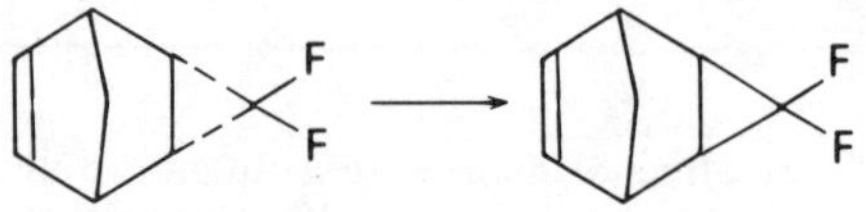 (4a) (4b)

Related studies on methylenecyclopropane isomerizations reveal dramatic *gem*-difluoro substituent effects (Table 5). The relative rates of rearrangement of **5**:**6**:**7** at 150°C are 1:4.4:3.4 × 10^4! A non-additive increase in methylenecyclopropane strain energies upon introducing pairs of fluorine substituents was proposed to explain these rate differences. Furthermore, the equilibrium thermodynamics ($\Delta G°$ values) indicate that cyclopropyl *gem*-difluoro substituents are more destabilizing than vinylic *gem*-difluoro substituents on a methylenecyclopropane.

3. *The* 'gauche *effect*'

1,2-Dihaloethanes exist mainly in the *trans* form (**8**) for Cl, Br and I. 1,2-Difluoroethane, however, exists predominantly in the *gauche* conformation (**9**)[13,54]. This phenomenon has been termed the '*gauche* effect' and it occurs in many structurally similar molecules containing pairs of highly polar bonds[55]. An analogous situation, called the '*cis* effect', occurs in substituted ethylenes (see Section II.B.1.c). Some

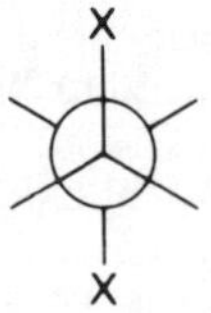

(8) 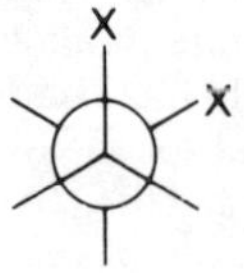 (9)

TABLE 6. Ethane rotational barriers[60]

Ethane	Barrier, kcal mol^{-1}
CH_3CH_3	2.88
CH_3CH_2F	3.30
CH_3CHF_2	3.18
CH_3CF_3	3.48, 3.25[a]
CF_3CH_2F	4.58
CF_3CHF_2	3.51
CF_3CF_3	3.92, 3.7[b]

[a]Ref. 15.
[b]Ref. 61.

special stabilization of the *gauche* conformer that is absent in the *trans* conformer, such as fluorine–fluorine lone pair attraction[56–58], is commonly postulated to explain this effect. Bingham has criticized this analysis and has contended that the *gauche* form is not stabilized but the *trans* conformer is conjugatively destablized, based on simple perturbation theory[59]. This contention has been defended[43] and criticized[56].

Fluorine substitution also has a peculiar effect on the rotational barriers of ethanes (Table 6). The rotational barriers increase in going from CH_3CH_3 to CH_3CH_2F, or from CF_3CH_3 to CF_3CH_2F, but they decrease upon further fluorination. Various theoretical rationales for this effect have been offered[62–65].

B. Unsaturated Compounds

1. Olefins

a. Carbon–fluorine bonds. The C—F bond lengths in fluoroalkenes are of the same order as those in CHF_3 or CF_4 (Table 7). The C—F bond strengths also appear to be comparable (see Section II.B.1.b). The fluoroalkene C=C bond lengths are all shorter than that in $CH_2{=}CH_2$ (1.337 Å). The most striking feature is the geminal FCF bond angle which is close to tetrahedral.

The same arguments used to rationalize the molecular geometries of polyfluorinated alkanes have been made to explain the fluoroolefin geometries. Bernett[38] proposed that the attachment of two electronegative fluorine atoms to normally sp^2 hybridized carbon changes the hybridization to sp^3. Kollman[25] similarly suggested that the decreased FCF angle is a result of electronegativity effects, but calculated about 40% s-character in the carbon hybrid orbital of the C—F bond in $CF_2{=}CH_2$. Epiotis

TABLE 7. Bond angles and bond lengths in fluorinated ethylenes

	$CH_2{=}CHF$[66]	$CH_2{=}CF_2$[67]	$FHC{=}CF_2$[68]	$F_2C{=}CF_2$[69]
r(C—C), Å	1.333	1.315	1.309[a]	1.311
r(C—F), Å	1.348	1.323	1.32[b]; 1.33[c]	1.319
<HCH, deg	120.4	121.8		
<HCF, deg	115.4		116.2	
<FCF, deg		109.3	112.8	112.5

[a]Ref. 69.
[b]C—F = 1.32 Å for $=CF_2$.
[c]C—F = 1.33 Å for =CHF.

and collaborators[56,70] have criticized the theoretical validity of these analyses and have advocated the importance of fluorine lone pair attractions.

Although the precise nature of the bonding in fluoroolefins remains controversial, fluorine substitution clearly can have a pronounced effect on the reactivity of the double bond.

b. Carbon–carbon double bonds. Many addition reactions of *gem*-difluoroolefins to form saturated compounds are more exothermic than the corresponding hydrocarbon reactions[16,71]. The heats of hydrogenation of $CF_2{=}CF_2$, $CF_2{=}CFH$, and $CF_2{=}CH_2$ are respectively about 16, 8 and 4 kcal mol^{-1} greater than that for ethylene[72]. The heat of addition of Cl_2 to $CF_2{=}CF_2$ is nearly 14 kcal mol^{-1} greater; the heat of polymerization is 17 kcal mol^{-1} greater, and the heat of dimerization is 42 kcal mol^{-1} greater than that estimated for ethylene. The activation energy for the dimerization of $CF_2{=}CF_2$ to octafluorocyclobutane is 25 ± 1 kcal mol^{-1}, which compares with 44.3 kcal mol^{-1} calculated for the hypothetical ethylene dimerization. These observations have been attributed to either π-bond destabilization[73] or to an increase in C—F bond strength in the saturated system relative to the olefin[22]. The cumulative experimental evidence strongly favours the former explanation, at least for tri- and tetrafluorinated double bonds. The thermodynamic ΔH values for thermal cyclobutene ring openings are illustrative (Table 8). At 200°C cyclobutene is quantitatively converted to butadiene, whereas the perfluorocarbon reaction (**14** $\rightleftharpoons$ **15**) proceeds quantitatively in the opposite direction under similar conditions. Taken alone, these differences could be explained by either fluorine π-bond destabilization or σ-bond stabilization. However, since the three fluorocyclobutenes (**10**, **14**, **18**) rearrange at similar rates and each involve the same changes in C—F hybridization, but their equilibrium constants markedly differ, fluorine substitution must destabilize the diene. These results also indicate a non-linear relationship between π-bond energy and the number of vinyl fluorines (fluoroolefin heats of hydrogenation also show a non-linear relationship, *vide supra*). Three fluorines on a double bond are uniformly destabilizing, whereas geminal fluorines are only slightly destabilizing (**16** $\rightleftharpoons$ **17**, **18** $\rightleftharpoons$ **19**) or stabilizing (**10** $\rightleftharpoons$ **11**).

Wu and Rodgers[28] have shown that the π-bond dissociation energy, defined as $D_\pi^\circ\,(C{=}C) = DH^\circ\,(CC{-}X) - DH^\circ\,({}^\bullet CC{-}X)$ for $CF_2{=}CF_2$ is 52.3 ± 2 kcal mol^{-1}, 7 kcal mol^{-1} less than that for $CH_2{=}CH_2$ at 59.1 ± 2 kcal mol^{-1}. This value is in good agreement with the thermodynamic data on $CF_2{=}CF_2$. The π-bond dissociation energy for $CF_2{=}CH_2$ at 62.1 ± 1.5 kcal mol^{-1}, however, is 3 kcal mol^{-1} *greater* than that for $CH_2{=}CH_2$[76,77]. This value is consistent with the observed low free radical reactivity of $CF_2{=}CH_2$[78], the failure of $CF_2{=}CH_2$ to cyclodimerize at ordinary pressures[79], and the thermodynamic stability of **11** relative to **10**. The greater heat of hydrogenation of $CF_2{=}CH_2$ relative to $CH_2{=}CH_2$ and the ΔH for the Cope rearrangement equilibrium in equation (1)[75], in contrast, suggest that a *gem*-difluoro group destabilizes the π-bond by 4–5 kcal mol^{-1}.

$$\Delta H \cong -5\ \text{kcal}^{-1}. \qquad (1)$$

Experimental data on the effects of a single vinyl fluorine or vicinal fluorines on C=C bond strengths are scarce. The thermal isomerization of *cis*- to *trans*-$CF_3CF{=}CFCF_3$ has an activation energy of 58.8 kcal mol^{-1}, about 8 kcal mol^{-1} lower than that for 2-butene itself[80]. Similarly, E_a for the isomerization of *cis*-CHF=CHF is about 5 kcal mol^{-1} less than that for CHD=CHD[81]. Based on these limited data, vicinal fluorination weakens double bonds. Epiotis's group noted that 1,1-disubsti-

TABLE 8. Cyclobutene ring openings[74]

Reaction	E_a, kcal mol^{-1}	ΔH, kcal mol^{-1}
cyclobutene → 1,3-butadiene	32.7	−8
(10) → (11)	47.9	[a]
(12) → (13)	—	2.5[b]
(14) → (15)	47.1	11.7
(16) → (17)	—	1.0[b]
(18) → (19)	46.0	0.4

[a]Not reported. At 600°C, $K_{eq}(\mathbf{10} \rightleftharpoons \mathbf{11}) > 200$ compared with $K_{eq}(\mathbf{14} \rightleftharpoons \mathbf{15}) = 7 \times 10^{-3}$.
[b]Ref. 75.

tuted ethylenes are in general more stable than the 1,2-isomers when the substituents are identical[82].

To date, only two equilibrium studies of the effect of monofluorination have been reported, and the results are contradictory. In monofluorobullvalene the isomer with fluorine at the bridgehead **(20)** predominates, as opposed to the bromo- and chlorobullvalenes in which the halogens are at vinylic positions **(21)**[83]. In contrast, a double bond stabilizing parameter of 3.3 kcal mol^{-1} was established for F[84], based on the equilibrium represented in equation (2)[85].

The interpretation of fluorine substituent effects on π-bond dissociation energies is obviously not a simple matter. The generalization that substitution of F for vinylic H in an olefin is destabilizing, which often appears in the fluorocarbon literature, is valid for $CF_2{=}CF_2$ and $CF_2{=}CF{-}$ systems, but is misleading for the lower fluorinated

(20)

(21) X = Br, Cl

$$CH_2{=}CHCH_2F \rightleftharpoons CHF{=}CHCH_3 \quad \Delta H \cong -4 \text{ kcal mol}^{-1}. \tag{2}$$

olefins. The effects of geminal or vinyl fluorine substitution on thermodynamic stability clearly depend on the particular olefin in question.

c. The 'cis *effect' and* ab initio *calculations. cis*-1,2-Difluoroethylene is thermodynamically more stable than the *trans* isomer. This anomaly is not unique to the fluorocarbons; it occurs with several ethylenes substituted vicinally with electronegative atoms (Table 9). This phenomenon, which is unexpected based on steric or dipole–dipole repulsion considerations, has been termed the '*cis* effect'.

Theoretical arguments to explain the '*cis* effect' abound, e.g. resonance structure formulations[86], steric attraction mechanisms[56,57], rehybridization schemes[25], and conjugative destabilization models[59]. The problem is intimately related to the questions of structure and bonding in *gem*-difluoroalkanes and -alkenes, and in 1,2-difluoroalkanes (see Section II.A.3).

The results of *ab initio* SCF calculations on 1,2-difluoroethylenes point out a general caveat concerning the validity of such calculations. Since 1977, four major papers[43,75,87,88] have dealt with the 1,2-difluoroethylene problem and the results are disappointing in that even large basis sets, e.g. 6-311G, have not been able to reproduce the correct sign of the energy difference between the isomers. With the popular 4-31G basis set including geometry optimization, the *trans* isomer is calculated to be more stable by 1.3 kcal mol^{-1}. Only the highest quality calculation with Hartree–Fock wave functions, 6-311G*, though not geometrically optimized, predicts the *cis* isomer to be slightly more stable ($\Delta H = 0.26$ kcal mol^{-1})[88].

Relatively modest basis sets consistently give optimized C—F bond lengths that are too long compared to the experimental values[89,90]. In some cases, increased basis sets correctly decrease the C—F bond lengths, but for the 1,2-difluoroethylenes the C—F bonds become even more overelongated as the basis set increases[87]. It is evident that basis sets for reliable *ab initio* calculations on fluorocarbons must include polarization functions, and it indeed may be necessary to include electron correlation beyond the Hartree–Fock limit. (The reader is reminded that the best value of the dissociation energy for F_2 obtained by the Hartree–Fock method is −1.37 eV[91]; the experimental value being +1.65 eV. When correlation effects are included, positive values close to the experimental ones are obtained[92].) *Ab initio* SCF calculations on fluorocarbons at the STO-3G or 4-31G levels should be treated with circumspection.

TABLE 9. Enthalpy values for the *cis–trans* isomerization of 1,2-dihaloethylenes, XCH=CHX[59]

X	ΔH_0^{298}, kcal mol^{-1}
F	0.93
Cl	0.65
Br	0.32
I	0.00

2. *Fluoroacetylenes*

The shortest reported C—F bond length of 1.279 Å is found in HC≡CF[12]. The C—F bond strength is estimated to be 114 kcal mol^{-1}, considerably stronger than the C—F bond in fluoroethane but comparable to that in a vinyl fluoride.

Although fluorine substitution shortens the triple bond by about 0.01 Å (r(FC≡CH) = 1.198 Å versus r(HC≡CH) = 1.209 Å)[12], the triple bond is appreciably destabilized (Table 10). Monofluoroacetylene[93,94] and chlorofluoroacetylene[95] are dangerously explosive. *t*-Butylfluoroacetylene is extremely reactive and it dimerizes below 0°C[96]. Tetrafluoropropyne, $CF_3C{\equiv}CF$, is considerably more stable than other monofluoroacetylenes, but it is thermodynamically less stable than its allene isomer, $CF_2{=}C{=}CF_2$[97,98]. Difluoroacetylene, claimed to have been prepared by pyrolysis of difluoromaleic anhydride[99] and by other routes[100,101], cannot be isolated.

TABLE 10. Acetylene bond dissociation energies[103]

Acetylene	$D°$ (C≡C), kcal mol^{-1}
HC≡CH	230
FC≡CH	178
ClC≡CH	166
BrC≡CH	202
FC≡CF	171[a]

[a]Calculated, Ref. 104.

Fluorine attached to an sp-hybridized carbon is expected to perturb triple bonds in much the same manner that it is proposed to affect double bonds. If the electronegative fluorine imparts extra p-character in the carbon orbitals of the C—F bond, a normal p-orbital is no longer available for π-bonding[38,73]. Alternatively, if repulsion between the electron pairs on fluorine and those of the π-system (Iπ repulsion) is the source of destabilization, this effect should be maximized with triple bonds[102]. A single fluorine indeed destabilizes the triple bond by over 50 kcal mol^{-1}, whereas its effect on a double bond is much less pronounced. This analysis is obviously an oversimplification since chlorine unexpectedly destabilizes the triple bond more than fluorine, judging from the π-dissociation energies of FC≡CH versus ClC≡CH (Table 10).

3. *Fluorobenzenes*

The C—F bond lengths in fluorobenzenes vary with the degree and position of fluorine substitution (Table 11). Compared with monofluorobenzene, other fluorobenzenes have shorter C—F bonds. The C—F bonds in *o*- and *m*-difluorobenzene are

TABLE 11. C—F bond distances in fluorobenzenes[12]

Arene	C—F bond length, Å[a]
C_6H_5F	1.354
1,2-$C_6H_4F_2$	1.306
1,3-$C_6H_4F_2$	1.304
C_6F_6	1.321

[a]Taken as mean value of r_0.

TABLE 12. C—X bond dissociation energies (in kilocalories per mole) in benzenes and pentafluorobenzenes[105,106]

	X = F	Cl	H
C_6F_5—X	154	130	152
C_6H_5—X	123	94	110

shorter by about 0.05 Å, whereas those in hexafluorobenzene are shorter by only 0.03 Å. This trend suggests some specific interaction between fluorines on different carbons, the exact nature of which has not been delineated.

The C—F bond strengths in aryl fluorides increase with increasing fluorination (Table 12). This pattern is similar to that observed with the alkanes (see Section II.A.1). Unlike the fluoroalkanes, other C—X bond strengths also increase on fluorination. The C—Cl and C—H bonds in C_6F_5—X are stronger than those in C_6H_5—X by an astounding 40 kcal mol^{-1}.

These effects have not been adequately explained and they again emphasize the unpredictable and mysterious nature of the structure and bonding in fluorocarbons.

III. FLUORINE SUBSTITUENT EFFECTS

The electronic effects of fluorine and other halogens on neutral and charged species were reviewed by Modena and Scorrano[107] in 1973. The current perspectives on fluorine and fluoroalkyl substituent effects, including more recent theoretical and experimental data, will be reviewed in this section.

A. Polar and Resonance Effects

The electron-withdrawing effects of fluorine and the halogens in aliphatic compounds in solution are well known. The increases in acidities of carboxylic, phosphoric and sulphonic acids and of alcohols upon halogenation are typical examples[108,109]. Trifluoroacetic acid is two to six times stronger than CCl_3CO_2H and about 10^4 times stronger an acid than CH_3CO_2H in aqueous solution. Perfluoro-*t*-butanol, $(CF_3)_3COH$ ($pK_a = 5.4$), is about as acidic as CH_3CO_2H[110]; hexafluoroacetone hydrate, $(CF_3)_2C(OH)_2$, is quite acidic ($pK_{a_1} = 6.76$)[111], and CF_3SO_3H is the strongest known monobasic organic acid ($H_0 = -14.5$)[112].

A quantitative measure of the polar effect is the substituent constant σ_I which increases in the order Br < Cl < F (Table 13) and, as expected, parallels the increasing electronegativity of F and the halogens. This polar effect is classically transmitted in σ-bonded systems by a relay of the induced bond dipole along the chain of carbon atoms (σ-inductive effect, I_σ). or by a through space electrostatic interaction (field effect)[113,114].

The electron-withdrawing order is reversed in the gas phase and becomes F < Cl < Br[115–118]. For example, in the gas phase the fluoroacetic acids FCH_2CO_2H and F_2CHCO_2H are more acidic than CH_3CO_2H but less acidic than the corresponding bromo- and chloroacetic acids[115]. Polarizability (charge-induced dipole) effects apparently predominate in the gas phase and σ-inductive effects are relatively unimportant[115,119–120]. (The atomic polarizabilities of F, Cl and Br are 0.53, 2.61 and 3.79 Å^3 respectively[121].)

The electronic effects of fluorine substituents attached directly to a π-system are complex. Besides I_σ and the classical field effect, several additional electronic effects are known to be important, especially in aromatic systems. Classical resonance in

TABLE 13. Substituent constants for fluorine, halogens and fluorocarbon groups[122]

Substituent	σ_m	σ_p	σ_I	σ_R	$\mathscr{F}$	$\mathscr{R}$	$E_S{}^a$
Br	0.39	0.23	0.44	−0.19	0.44	−0.17	−1.16
Cl	0.37	0.23	0.47	−0.23	0.41	−0.15	−0.97
F	0.34	0.06	0.52	−0.34	0.43	−0.34	−0.46
CH_2F	0.12	0.11	0.12	−0.02	—	—	−1.48
CHF_2	0.29	0.32	0.32	0.06	—	—	−1.91
CF_3	0.43	0.54	0.42	0.10	0.38	0.19	−2.40
C_2F_5	0.47	0.52	0.41	0.11	0.44	0.11	—
n-C_3F_7	0.47	0.52	0.39	0.11	—	—	—
i-C_3F_7	0.37	0.53	0.48	0.04	0.30	0.25	—
t-C_4F_9	0.35	0.52	0.27	0.26	0.28	0.27	—
C_6F_5	0.34	0.41	0.25	0.02	0.30	0.13	—

[a]Taft E_S values referenced to H by subtracting 1.24; Ref. 123.

which electrons from the substituent are donated back into the ring (**22**) is generally said to explain the *ortho–para* directing effects of halogen substituents. This resonance effect increases in the order Br < Cl < F, as the σ_R° values quantitatively reflect (Table 13). Fluorobenzene, in fact, is more reactive than benzene in some electrophilic substitution reactions. Politzer and Timberlake[124] have questioned the validity of this interpretation. The repulsion of halogen lone pair electrons with the aromatic π-electrons (I_π repulsion) which induces π-charge at the *ortho* and *para* positions (**23**) was claimed to account for the σ_R° values. The magnitude of I_π repulsion increases in the order Br < Cl < F and fluorine I_π repulsion is known to be quite important in α-fluoroanions and radicals (see Sections III.D and E). Although this analysis may be valid, it treats only ground-state effects. Since fluorine resonance has been unequivocally demonstrated in cations (Section III.C), it is also expected to be dominant in electrophilic substitution reactions with relatively late transition states[107].

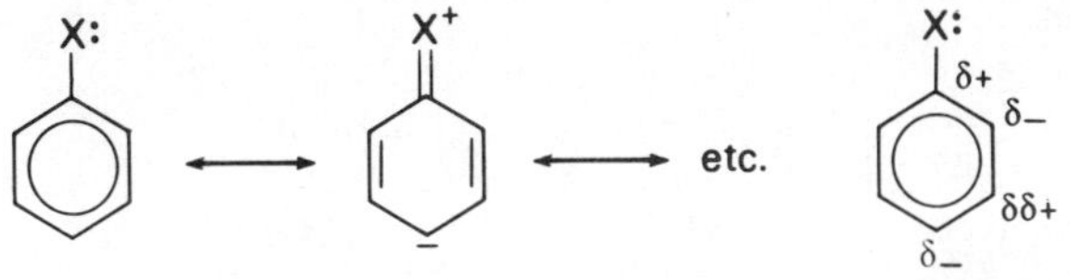

(**22**) (**23**)

Considerable controversy has surrounded the nature of perfluoroalkyl group electronic effects. The positive σ_R° and $\mathscr{R}$ values for CF_3 (Table 13) indicate apparent resonance which was originally attributed to C—F no-bond resonance or negative hyperconjugation (**24**). This concept has since fallen into disrepute[26,125,126]. Holtz[26] has shown that many of the effects attributed to hyperconjugation can be adequately

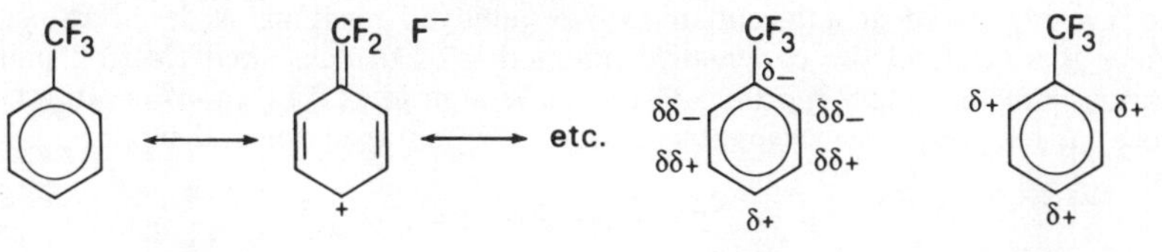

(**24**) (**25**) (**26**)

explained by field-induced π-polarization in which the field effect of the substituent directly polarizes the π system as in **25**. π-Polarization recently has been established as an important mechanism for substituent effects on ^{13}C and ^{19}F NMR chemical shifts in aryl fluorides[127,128]. A related, but conceptually different phenomenon is the π-inductive effect in which the substituent, through its I_σ effect, induces a redistribution of π-electron density in the sense represented by **26**[129]. Since the inductive effects of fluorine and the fluoroalkyl groups are comparable (Table 13), the π-inductive mechanism also may be important for fluoroalkyl groups.

B. Steric Effects

The steric effect of a fluorine substituent is often said to be negligible because the van der Waals radius of F (1.47 Å) is only 0.27 Å greater than that for H[130,131]. This generalization is misleading since situations can arise where steric crowding is so severe that this small difference is enough to require non-bonded overlap of F with an interacting atom, but no overlap between H and the interacting atom. The energetic consequences of this can be substantial since the van der Waals potential function for repulsion rises very steeply in the region of overlapping radii. The largest fluorine steric effect known occurs with **27**, wherein the *meta* ring-flipping rate ratio k_H/k_F is greater than 10^{11} at 25°C[132]. Several other fluorine steric effects in conformational processes are summarized by Förster and Vögtle[133].

Notable fluorine steric effects also are reported for the free radical additions of halogens and polyhalomethanes to norbornenes[134–136]. For example, **28** reacts with CCl_4 to give 73% **29** and 27% *trans* adduct, whereas, norbornene itself gives >95% *trans* adduct **30**. The ring fluorines were proposed to sterically retard *endo* attack by CCl_4 on the free radical intermediate.

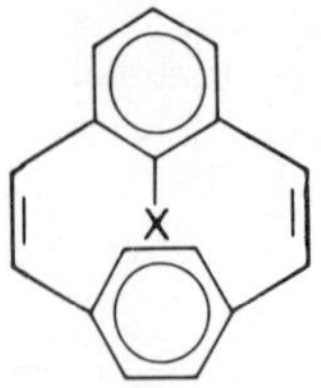

X F2 F2 F2 F2 CCl3 Cl CCl3 Cl

(27) 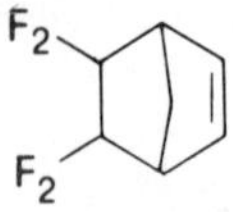 **(28)** 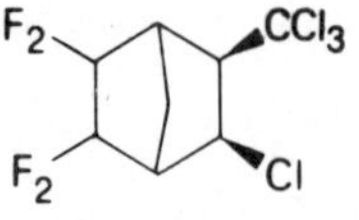 **(29)** 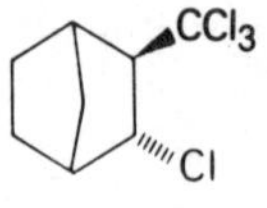 **(30)**

Although the potential steric effect of F is often overlooked, the large steric bulk of CF_3 and other perfluoroalkyl groups is well recognized. From the E_S values (Table 13), CF_3 is even larger than *i*-Pr ($E_S = -1.71$), and *i*-C_3F_7 is suggested to be comparable in size to *t*-Bu[137].

C. Fluorocarbocations

Fluorine directly bonded to the cation centre stabilizes a carbenium ion, whereas fluorine substituted at adjacent or further removed positions is destabilizing. The former is a result of the conjugative interaction of the unshared electron pairs on fluorine with the empty p-orbital of the carbenium ion **(31)** (π(p–p) bonding). This effect has received considerable theoretical[138–142] and experimental support.

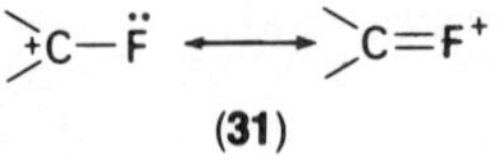

$$>\overset{+}{C}-\ddot{F} \longleftrightarrow >C=F^+$$

(31)

Beauchamp and coworkers[143,144] have shown the following increasing orders of cation stabilities in the gas phase: $^+CH_3 < {}^+CF_3 < {}^+CH_2F < {}^+CHF_2$; $^+CH_2CH_3 \ll {}^+CFHCH_3 \cong {}^+CF_2CH_3$; $^+CH(CH_3)_2 < {}^+CF(CH_3)_2$. The cation CF_3^+ has been generated by the matrix photoionization of trifluoromethyl halides and its IR spectrum is consistent with extensive π(p–p) bonding[145].

Numerous long-lived fluorocarbocations have been observed in solution and much of the experimental data has been reviewed[146,147]. The $CH_3\overset{+}{C}FCH_3$ and $CH_3CF_2^+$ cations have been directly observed as long-lived species; however, no fluorinated methyl or monofluoroalkyl carbenium ions, $R\overset{+}{C}HF$, have been directly observed in solution under non-exchanging conditions. When SbF_5–SO_2ClF was used to ionize CH_3CHF_2, $CH_3\overset{+}{C}Hf$ was formed, but it underwent rapid fluoride exchange with CH_3CHF_2 and SbF_6^-. The decreased stability of $R\overset{+}{C}HF$ compared with $R_2\overset{+}{C}H$ indicates that an alkyl group better stabilizes electron-deficient carbon than does fluorine. Similar attempts to ionize trifluorohalomethanes only gave CF_4. From comparative studies of α-fluoro- and α-haloisopropyl, -cycloalkyl and -2-norbornyl cations in solution, the stability and degree of π(p–p) bonding was found to increase in the order $Br < Cl < F$[146].

No simple β-fluoroalkyl carbocations have been observed in solution. The $FCH_2CH_2^+$ cation is calculated to be less stable than CH_3CHF^+ by 18.3 kcal mol^{-1}, but more stable than the fluoronium ion **32** (X = F) by 11.5 kcal mol^{-1}[140]. Although long-lived halonium ions **32** (X = Cl, Br, I) have been observed in solution, fluoronium ions have never been detected[147]. Heats of formation data from the gas-phase proton affinities of fluoroolefins indicate that FCH_2CHF^+ is at least 20 kcal mol^{-1} less stable than $CH_3CF_2^+$[148]. Unlike β-halocarbenium ions, β-fluorocarbenium ions are clearly not stabilized by bridging.

(32)

The α-CF_3 group in a carbocation is enormously destabilizing relative to CH_3. The $CF_3CH_2^+$ cation is calculated to be less stable than $CH_3CH_2^+$ by 37 kcal mol^{-1}[149]. Trifluoropropene is not protonated by FSO_3H, but is ionized to the 1,1-difluoroallyl cation[150]. Anti-Markovnikov additions of strong acids to $CF_3CH{=}CH_2$, which are often incorrectly quoted as examples of the electron-withdrawing effect of CF_3, actually proceed by the 1,1-difluoroallyl cation and do not involve direct addition to the double bond[151]. The relative rates of acid-catalysed hydration of $PhCR{=}CH_2$ for R = H, CH_3, CF_3 are approximately 1, 2.6×10^2, 2.4×10^{-8} respectively[152].

The pronounced deactivating effect of CF_3 also has been demonstrated in solvolysis reactions. The rate ratios k_H/k_{CF_3} for the solvolysis of $Me_2C(CF_3)OTs$ and PhCMe-$(CF_3)OTs$ in comparison with Me_2CHOTs and PhCHMeOTs in various solvents are 1.1×10^5–2.3×10^6[153].

The only reported long-lived α-CF_3 cations are those which benefit from additional conjugative stabilization, e.g. $Ph\overset{+}{C}R(CF_3)$ (R = Me, c-C_3H_5, Ph)[146]. Magic acid, SbF_5–FSO_3H, in SO_2 at $-60\,^\circ$C will protonate CF_3COCH_3, but not HCF_2COCF_2H or CF_3COCF_3[147]. The H_0 values for the first two protonated ketones are -14.7 and *c*. -17, respectively, which compare with -7.5 for acetone[154].

Polyfluorination also destabilizes delocalized cations. In contrast to the well characterized, long-lived cyclobutenyl, allyl and perchloroallyl cations, the analogous perfluorinated cations are unstable and have not been detected in solution[146]. However, the cations **33**[155], **34**[156], **35**[157] and **36**[158], which benefit from additional resonance or conjugative stabilization, are quite stable.

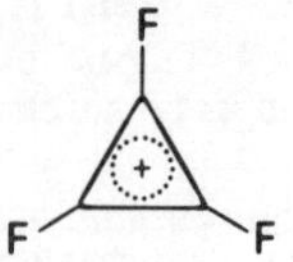

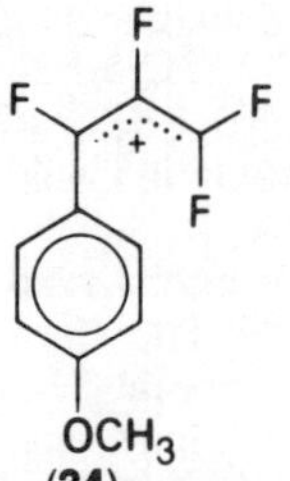

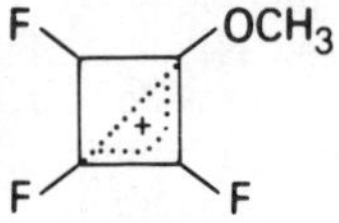

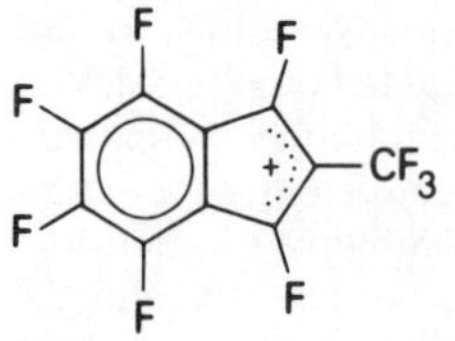

(33) (34) (35) (36)

No fluorovinyl cations have been detected. The cations $CH_2{=}CF^+$ and $FCH{=}CH^+$ are calculated to be less stable than $CH_2{=}CH^+$ by 20.8 and 32 kcal mol^{-1}, respectively[138,139]. The bridged cation **37** is calculated to be 31 kcal mol^{-1} less stable than $FCH{=}CH^+$ [159].

(37)

The combined effects of α- and β-fluorines on carbocation stability imply that fluoroolefins will react regiospecifically with electrophiles as in equation (3). This is the case (see Section IV.B.1). For example, FSO_3H reacts with $CH_2{=}CF_2$ to give exclusively $CH_3CF_2OSO_2F$. Because of the pronounced instability of β-fluorocarbocations, polyfluoroolefins are relatively resistant to electrophilic attack. Tetrafluoroethylene and hexafluoropropene are not protonated by $HF{-}SbF_5{-}SO_2ClF$ at -5°C.

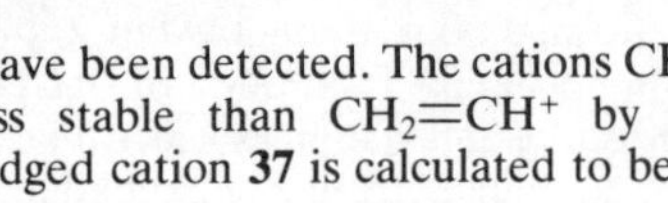

(3)

The effect of F on the reactivity of an olefin towards electrophiles is more difficult to predict. Although a carbocation is stabilized by α-F relative to H because of resonance, the inductive effect of F predominates in the ground state and reduces the nucleophilicity of a double bond. Consequently, F is an activating group only in electrophilic additions to olefins which have very late transition states, and examples of this are rare[160,161]. The activating influence of F relative to H and halogens, however, is quite common in electrophilic aromatic substitutions[107].

D. Fluorocarbanions

β-Fluorine substituents stabilize a carbanion, but α-fluorines can be either stabilizing or destabilizing, depending upon the geometry of the carbanion. The former effect is evident from the remarkable ability of fluoroalkyl groups to enhance the acidity of carbon acids (Table 14). The acidity of $(CF_3)_3CH$ (**41**) is nearly 50 orders of magnitude greater than that of CH_4, and **38** ($pK_a < -2$) is a stronger acid than HNO_3[162].

In addition to polar effects, negative hyperconjugation was claimed to contribute to the unusual stability of β-fluorocarbanions[163]. This concept has been severely criticized and it was concluded that only polar effects are needed to explain the data[26,125,126,164] (see Section III.A). The classic studies by Streitwieser and coworkers[26,126] on the acidities of **39** and **40** are the most cogent evidence against negative hyperconjugation.

TABLE 14. Equilibrium acidities of carbon acids[126], pK_{CsCHA}[a]

CHF_3 30.5	$CHCl_3$ 24.4	$CHBr_3$ 22.7	CHI_3 22.5
CF_3CF_2H 28.2	$(CF_3)_2CFH$ 25.2	$(CF_3)_3CH$ (41) ~21[b]	
CF_3CCl_2H 24.4	CF_3CBr_2H 23.7	CF_3CI_2H 24.1	

[a]Cs ion pair values in cyclohexylamine solution.
[b]Estimated.

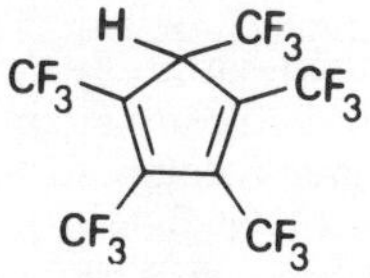

(38)

The pK_a values for **39** and **40**, 20.5 and 18.3 ± 0.3, respectively, are lower than that *calculated* for **41** (*c*. 21). Since the anions of **39** and **40** cannot benefit from hyperconjugation owing to their highly unfavourable anti-Bredt's rule canonical forms, but both **39** and **40** are more acidic than **41**, hyperconjugation was ruled out.

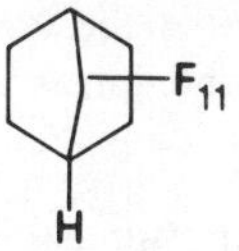

(39) 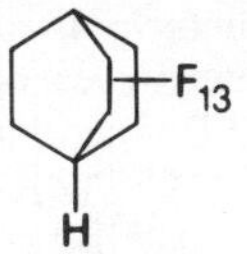 (40) $(CF_3)_3CH$ (41)

Recent experimental and theoretical results have reopened the controversy over negative hyperconjugation. Tatlow and coworkers have measured the relative rates of exchange for **39**, **40** and **41** under the same reaction conditions for the first time; after 450 h at 55°C in neutral D_2O–CD_3COCD_3, the hydrocarbons incorporated 0%, 12% and 70% deuterium, respectively[165]. Compound **41** clearly exchanges more rapidly than **40** and very much more rapidly than **39**. These data indicate some special stabilization in the perfluoro-*t*-butyl anion, and the validity of the estimated pK_a of **41** is in doubt.

Early molecular orbital calculations (CNDO/2) on $FCH_2CH_2^-$ indicated no appreciable energy difference between the conformers **42** and **43**[26]. More recent *ab initio* calculations show that **42** is more stable than **43** by 9–11 kcal mol^{-1} [166–168], indicating appreciable negative hyperconjugation in the anion. The status of fluorine hyperconjugation, at least in carbanions, is now an open question.

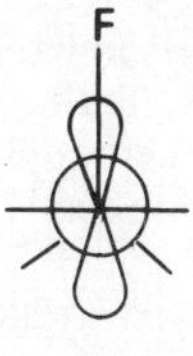

(42) 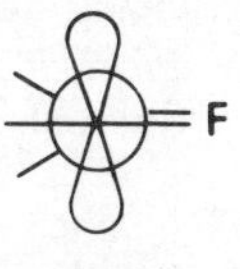 (43)

The stability of α-fluoro- and α-halocarbanions depends upon a balance of polar stabilization (F > Cl > Br > I > H), p–p lone pair repulsion (I_π repulsion), and the carbanion geometry which affects the magnitude of I_π repulsion. From the carbon acidity data (Table 14), CX_3^- and $CF_3CX_2^-$ are stabilized in the order X = I ≃ Br > Cl > F, respectively. The carbanions $(CF_3)_2CX^-$ are known to be stabilized in the order Br > Cl > I > Ph > F > OMe[164]. These data indicate that anion stability is primarily determined by the degree of I_π repulsion (Br < Cl < F). Since I_π repulsion is maximized in planar carbanions, α-F carbanions prefer to be pyramidal[169]. The energy differences between the planar and pyramidal forms of CH_3^- and FCH_2^- are 1.1 and 13.2 kcal mol^{-1}, respectively, from *ab initio* calculations[170]. The calculated inversion barriers for the cyclopropyl and α-fluorocyclopropyl carbanions are 17.4 and 41.8 kcal mol^{-1}, respectively[171], which again illustrates the pronounced effect of F on carbanion geometry.

Although pyramidal α-fluorocarbanions are less stable than α-halocarbanions, they are more stable than their unsubstituted counterparts (CF_3H is about 10^{40} times more acidic than CH_4). In contrast, *planar* carbanions are destabilized by substitution of α-F for H. Examples that reflect this include the decreased C—H acidities of α,α-difluoroacetates, α-fluoronitroalkanes and 9-fluorofluorene compared to their respective unsubstituted systems[107]. A 45 year old case of I_π repulsion is the report that *p*-fluorophenol is less acidic than phenol[172].

Unlike fluorocarbocations, very few stable fluorocarbanions are known. The perfluoroanions **44**–**46**, generated from olefin precursors with CsF in DMF, are rare examples of long-lived anions, directly observable by ^{19}F NMR[173]. A few specially stabilized, isolable perfluorocarbanions, **47**[174,175], **48**[176], **49**[177] and **50**[178] (ylids), are known. The stability of these structures is attributed to the electron-withdrawing β-fluorines; however, if an α-fluorine is present, the carbanions are destabilized by I_π repulsion. For example, $CF_3CF_2\bar{C}F\overset{+}{P}Bu_3$ cannot be detected since it rapidly rearranges to the vinyl phosphorane, $CF_3CF{=}CFP(F)Bu_3$[177]. Similarly, the fluoromethylene ylids, R_3PCFH and R_3PCF_2, are established reagents but cannot be isolated or directly observed[179,180].

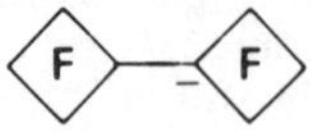

(44) 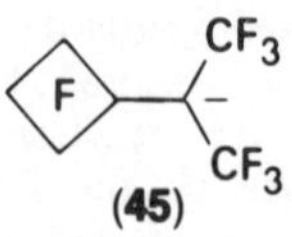 (45) 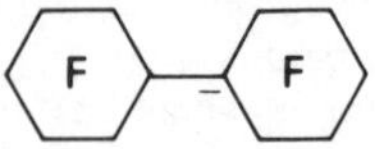 (46)

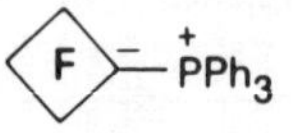

(47) 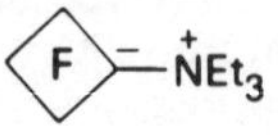 (48) 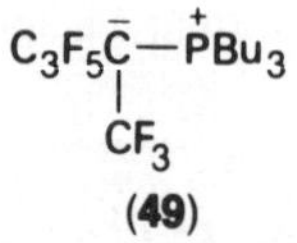 (49) 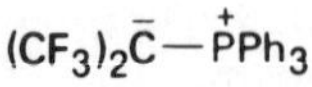 (50)

Several organofluorolithium compounds of varying stability are known, but it is not clear if these derivatives are ionic, covalent or are best described as carbenoids. The bicyclic organolithium **51** is stable in ether above 25°C[181]; **52** decomposes rapidly to perfluoroheptene at −75°C[182], and **53** is stable at low temperatures but eliminates F^- near 0°C to give perfluoro-*t*-butylcarbene[183]. Difluoroallyllithium is unstable at −95°C, but can be trapped *in situ* by R_3SiCl or carbonyl compounds[184]. The synthesis and synthetic utility of polyfluoroalkyllithium and Grignard reagents has been reviewed[185–187].

Halodifluoromethides, XCF_2^-, once considered to have no finite existence[188], are now known to be sufficiently long-lived to be trapped by electrophiles[189,190] and to

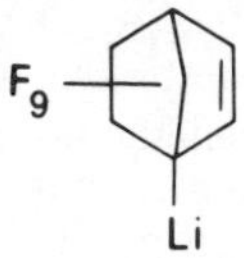

F_9 (bicyclic structure with Li) $CF_3(CF_2)_6Li$ $(CF_3)_3CCF_2Li$

(51) **(52)** **(53)**

undergo H–D exchange[191]. Methyl chlorodifluoroacetate was decarboxylated by 1:1 LiCl/HMPA (hexamethylphosphoric triamide) complex to generate $ClCF_2^-$, or the $ClCF_2Li$/HMPA complex, which was trapped *in situ* with ketones, fluoroolefins, pentafluoropyridine and halogens (e.g. equation 4). The $BrCF_2^-$ anion or its complex was similarly generated and trapped[190].

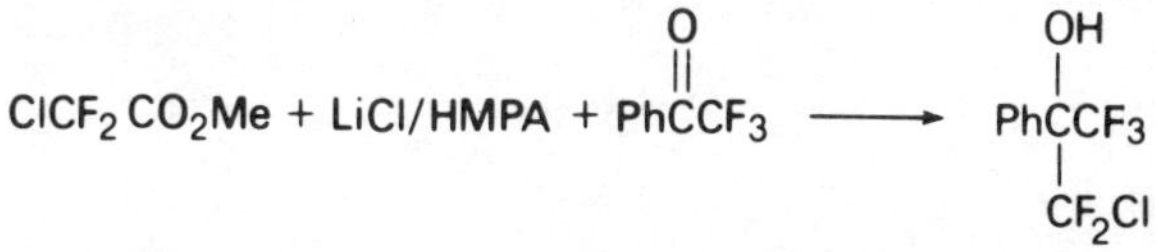

$$ClCF_2CO_2Me + LiCl/HMPA + PhC(=O)CF_3 \longrightarrow PhC(OH)(CF_2Cl)CF_3 \quad (4)$$

Lithium fluoromethylides, generated by conventional proton abstraction or halide exchange with an alkyllithium, are unstable at $-116°C$ and rapidly decompose to give carbenes[192,193]. *Ab initio* calculations on CHF_2Li and CH_2FLi predict rather startling structures for these species, none of which corresponds to the classical tetrahedral structure[194,195]. An inverted carbon structure with $CH_2Li^+F^-$ character is favoured for FCH_2Li.

A wide variety of polyfluorovinyl and polyfluoroaromatic lithium and magnesium derivatives have been prepared by conventional procedures and have been used extensively in organic synthesis[9,196]. These derivatives are generally more stable than the saturated systems. The fluorinated vinyllithiums **54** and **55** do not isomerize during their syntheses and are trapped stereospecifically by electrophiles[197]. The effects of F on the stability of vinyl anions have been discussed[107].

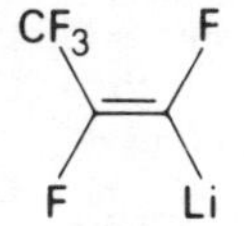

$CF_3(F)C{=}C(F)Li$ **(54)** $Ph(CF_3)C{=}C(F)Li$ **(55)**

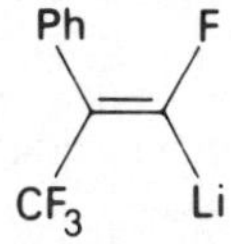

The effects of F and fluoroalkyl groups on the reactivity of an olefin towards nucleophiles are straightforward. Since these substituents both increase the electrophilicity of a double bond and can appreciably stabilize carbanions, fluoroolefins are much more susceptible to nucleophilic attack than their hydrocarbon counterparts (see Sections IV.B.1 and 2). The well known feature of polyfluoroolefins of being attacked regiospecifically at terminal $=CF_2$ groups is consistent with the relative effects of α- and β-fluorines on carbanion stability.

E. Fluororadicals

α-Fluorine substituents have a pronounced effect on the geometry of free radicals. Whereas the $CH_3^\bullet$ radical is planar, or nearly so[198], fluorinated methyl radicals increasingly deviate from planarity in the order $H_2CF^\bullet < HCF_2^\bullet < CF_3^\bullet$ [199]. The $CF_3^\bullet$ radical is nearly pyramidal with a calculated inversion barrier of *c*. 25 kcal mol[200]. A similar increase in pyramidality is observed in the series $CF_3CF_2^\bullet < CH_3CF_2^\bullet <$

$ClCF_2^{\bullet} < CF_3^{\bullet}$ and $(CF_3)_3C^{\bullet} < (CF_3)_2CF^{\bullet} < CF_3CF_2^{\bullet}$, although $(CF_3)_3C^{\bullet}$ appears to be less bent than t-Bu$^{\bullet}$ [201,202].

Fluorine also markedly increases the barrier to pyramidal inversion of cyclopropyl radicals. The radicals **56a** and **56b** (X = F), generated by homolysis of the respective (*Z,Z*) and (*E,E*)-1-bromo-1-fluoro-2,3-dimethylcyclopropanes, are configurationally fixed on the ESR time scale at $-108°C$[203], whereas **56a** and **56b** (X = H) rapidly interconvert under these conditions ($k \geqslant 5 \times 10^9\ s^{-1}$)[204]. The highly stereospecific homolytic reductions of many 1-halo-1-fluorocyclopropanes is a consequence of the high inversion barriers of the intermediate 1-fluorocyclopropyl radicals[205–207].

The first persistent cyclopropyl radicals (**57**; X = H, F) have been reported[208]. These unusual radicals have planar and almost planar configurations, respectively. This was attributed to steric repulsion between the 1-substituent and the t-Bu groups.

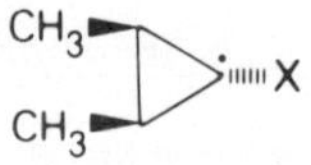

(**56a**) 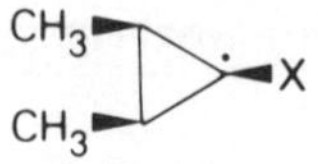 (**56b**) 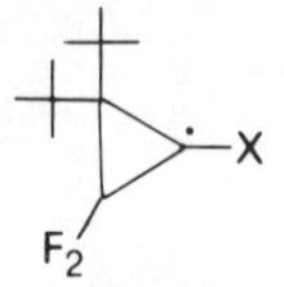 (**57**)

The bending in α-fluororadicals has been attributed to primarily inductive and I_π repulsion effects[200], although other factors have been considered[209,210]. Substitution of H by F will increase the s-character of the singly occupied orbital and, hence, will increase non-planarity. The pyramidal structure also minimizes I_π repulsion between the F lone pair electrons and the electron on C, as is the case for α-fluorocarbanions.

β-Fluorine substituents have little or no effect on the geometry of free radicals. The radicals $CH_3CH_2^{\bullet}$ and $CF_3CH_2^{\bullet}$ are planar and essentially rotate freely on the ESR time scale, unlike $CH_3CF_2^{\bullet}$ and $CF_3CF_2^{\bullet}$ which have substantial barriers to rotation (2.2 and 2.85 kcal mol^{-1}, respectively)[211]. The $FCH_2CH_2^{\bullet}$ radical is calculated to prefer the eclipsed rather than perpendicular conformation[166]; low temperature ESR studies suggest two rapidly interconverting conformations with the β-F approximately 50° from the plane of the p-orbital[212]. In marked contrast, the $ClCH_2CH_2^{\bullet}$ radical strongly favours the perpendicular conformation with the Cl nucleus displaced toward the radical centre. The proposed bridging or σ-bond delocalization for β-haloradicals is negligible for β-fluororadicals[213].

The properties of delocalized radicals are also affected by fluorine substituents. Allyl radicals ordinarily are geometrically very stable; the allyl radical is static on the ESR time scale at 280°C with a $\Delta G^{\neq}$ for internal rotation greater than 17 kcal mol^{-1} [214]. In contrast, the $\Delta G^{\neq}$ values for internal rotation of the terminal fluorines in **58** (X = H, F, Cl) are 7.2, 6.1 and 4.5 kcal mol^{-1}, respectively[215]. The radical **58** (X = Cl) is freely rotating at $-52°C$!

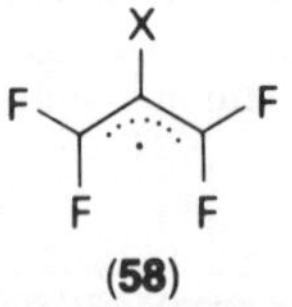

(**58**)

The effect of fluorine substitution on the stability of free radicals is difficult to assess. Directive effects in radical additions to olefins or in atom transfer reactions of substituted alkanes are normally dominated by polar factors and do not reflect relative radical stabilities[216–218]. The C—H and C—X (X = Cl, Br, I) bond dissociation energy data (Tables 2 and 3) are perhaps the best measure of radical stability. These data, within experimental error, indicate that α-fluorine substitution has little or no effect on the stability of methyl or ethyl radicals (e.g. $D°(CF_3—X)$ and $D°(CH_3—X)$ (X = Cl, Br) are nearly identical). The activation parameters for the thermal rearrangement of **59** to **60**, which are practically identical for X = H or F, also indicate that the stability of an allyl radical is not significantly affected by partial fluorination[219].

(59) $\xrightarrow{\Delta}$ → (60)

F. Fluorocarbenes

α-Fluorocarbenes and CH_2 differ in structure, stability and reactivity. The most recent calculated geometries and singlet–triplet (S–T) splittings for simple carbenes are given in Table 15. (To date, only the ground state geometries for CH_2, CHCl, CHF and CF_2 are known accurately.) The fluorocarbenes are all ground-state singlets in contrast to CH_2. The magnitude of the S–T separation in CXY increases as the substituent's electronegativity or π-donating ability increases, although the relative importance of each is debatable[220–224]. The very large S–T energy difference in CF_2 is consistent with the absence of triplet chemistry in the thermal reactions of this species. For example, CF_2 adds stereospecifically to *cis* or *trans* FCH=CFCl at 200°C. In general, simple α-fluorocarbenes stereospecifically cyclopropanate olefins in keeping with their ground-state singlet structures.

α-Fluorocarbenes are more stable and less electrophilic than CH_2 and their corresponding halocarbenes. The electrophilicity of singlet carbenes increases in the order $CF_2 < CHF < CH_2$ and $CF_2 < CCl_2 < CBr_2 < CI_2 < CH_2$. This reflects a balance of inductive and resonance effects, and the predominant π-donation from F lone pair electrons into the singlet carbene's vacant carbon 2p orbital is analogous to π(p–p) bonding in α-fluorocarbocations.

The subjects of carbene spin states[225], carbene reactivity[226] and fluorocarbene chemistry[192,227] have been reviewed. The reviews of Seyferth and Burton also summarize the many procedures for generating fluorocarbenes, including the interesting

TABLE 15. Calculated bond angles and singlet–triplet energy separations for simple carbenes[224]

CXY	CXY angle: Singlet	CXY angle: Triplet	S–T ΔE, kcal mol^{-1}
CH_2	102.4 (102.4)[a]	128.8	12.8
CHBr	102.6	125.6	1.1
CHCl	102.0 (103.4)[a]	123.3	−1.6
CHF	102.2 (101.6)[a]	120.4	−9.2
CCl_2	109.4	125.5	−13.5
CBr_2[b]	110.1	127.3	−7.7
CF_2	104.3 (104.9)[a]	117.8	−44.5

[a]Experimental value.
[b]Ref. 229.

TABLE 16. Carbene selectivity indexes[232]

CX_2	m_{CXY}
$(CF_3)_2C$	-0.01^a
$BrCCO_2Et$	0.29
CBr_2	0.65
PhCBr	0.70
PhCCl	0.83
PhCF	0.89
MeSCl	0.91
CCl_2	1.00
CFCl	1.28
CF_2	1.48
ClCOMe	1.59^b
FCOMe	1.85^b
$C(OMe)_2$	2.22^b

[a]Calculated, this work.
[b]Calculated, Ref. 232.

fluoroethylidenes, $:CFCF_3$ and $:CFCF_2H$[230,231]. The use of diazo compounds and diazirenes as fluorocarbene precursors also has been reviewed[228].

A semiquantitative, empirical correlation of carbenic selectivity towards alkenes has been developed by Moss[232]. The 'carbene selectivity index' m_{CXY}, as defined by Moss, measures the selectivity of a singlet carbene in cyclopropanation reactions at 25°C (Table 16). The larger the m_{CXY} value, the greater is the selectivity of CXY. The fluorocarbenes in each series of structures are uniformly the most selective ($CF_2 > CCl_2 > CBr_2$; PhCF > PhCCl > PhBr) and CF_2 is experimentally the most selective electrophilic carbene known. Carbenes with $1.40 < m_{CXY} < 2.22$, e.g. CH_3OCCl and CH_3OCF, were predicted to be *ambiphilic*, i.e. they act as electrophiles toward electron-rich alkenes and as nucleophiles toward electron-poor alkenes. This prediction has been confirmed for CH_3OCCl. The carbene $(CF_3)_2C$ is predicted to be exceedingly electrophilic and very unselective. Although its relative reactivity with simple alkenes is unknown, this highly reactive species adds to hexafluorobenzene, reacts with CO to give $(CF_3)_2C{=}C{=}O$, inserts into CH bonds, and cyclopropanates olefins[229,233,234]. A frontier molecular orbital formulation of carbenic selectivity which complements the empirical correlation also has been developed and reviewed[232].

Unfortunately, the elegance of these selectivity formulations is somewhat diminished by the recent findings of Giese and coworkers[235]. From competition experiments with $MeCH{=}CMe_2$ and $Me_2C{=}CH_2$ at 20°C, Giese found the expected increase in selectivity $CBr_2 < CClBr < CCl_2 < CFCl < CF_2$. However, at 87°C, the carbenes react with *equal* selectivity and, at 120°C, the selectivity order is exactly reversed. These results suggest that carbenic selectivity is determined predominantly by entropic effects, and the molecular orbital formulations, based only on enthalpic considerations, appear rather dubious.

G. The 'Perfluoroalkyl Effect'

The remarkable influence of perfluoroalkyl groups on the stability of strained molecules and certain functional groups is undoubtedly the most prominent fluorine substituent effect. Perfluoroalkylated compounds are often incredibly stable compared with their corresponding hydrocarbons. Octakis(trifluoromethyl)cyclooctatetraene

(61)[236] only decomposes slightly at 400°C; the valence isomers **62**[237], **63**[238] and **64**[238] have half-lives of 2 h at 360°C, 9 h at 170°C and 29 h at 170°C, respectively. In addition, perfluoroalkyl groups often dramatically change the relative thermodynamic stability of isomers. For example, the valence isomers **63** and **64** are stabilized by over 30 kcal mol^{-1} relative to perfluorohexamethylbenzene in comparison to the corresponding hexamethyl analogues[239]. A particularly pronounced case is the Dewar isomer **65** which is more stable than **66** above 280°C[240]. In fact, **66** is quantitatively converted to **65** by vacuum flow pyrolysis at 400°C!

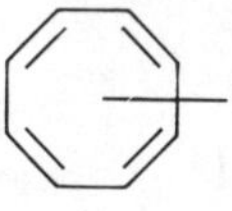

$(CF_3)_8$ **(61)** $(CF_3)_6$ **(62)** $(CF_3)_6$ **(63)** $(CF_3)_6$ **(64)**

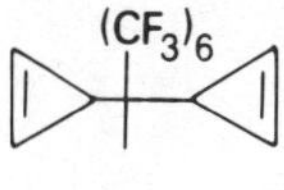

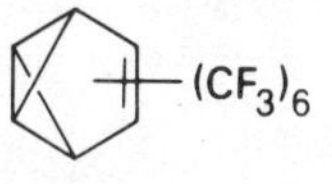

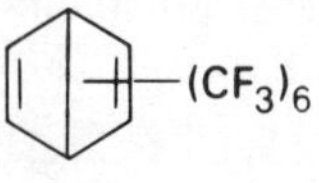

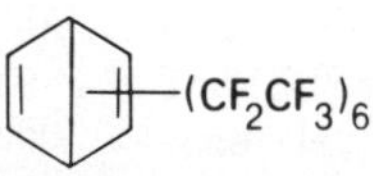

$(CF_2CF_3)_6$ **(65)** $(CF_2CF_3)_6$ **(66)**

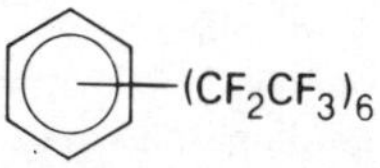

Several examples of stable perfluoroalkylated compounds for which isolable hydrocarbon analogues either are very rare (**67**[241], **68**[242], **69**[243]) or are unknown (**70**[244], **71**[245], **72**[246], **73**[247], **74**[248], **75**[249], **76**[250]) are given below. The successful isolation of perfluoroalkylated valence bond isomers of aromatic and heteroaromatic systems has opened fascinating new areas of chemistry that are only speculative for hydrocarbons (see Section IV.C.2).

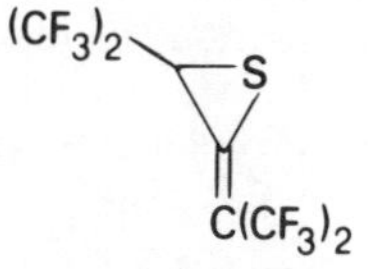

$(CF_3)_2$, S, $C(CF_3)_2$ **(67)** $(CF_3)_2$, O, O **(68)** $(CF_3)_2C{=}C{=}S$ **(69)** CF_3, S, S, CF_3 **(70)**

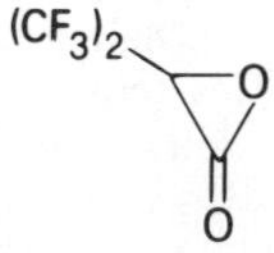

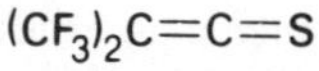

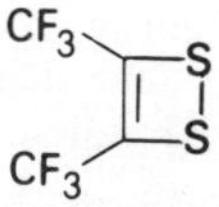

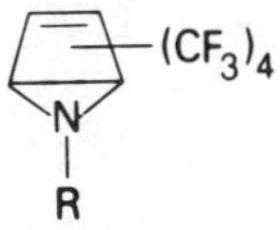

$(CF_3)_4$, N, R **(71)** $(CF_3)_4$, S **(72)** P, P, $(CF_3)_4$ **(73)**

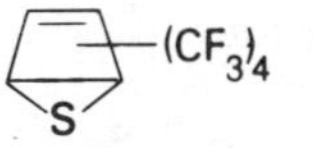

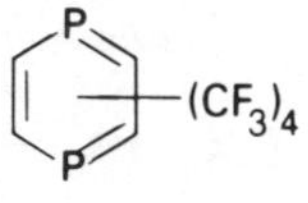

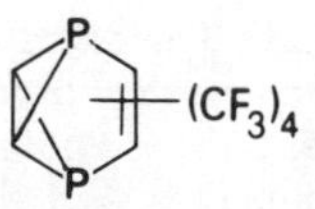

P, P, $(CF_3)_4$ **(74)** N, $(C_2F_5)_5$ **(75)** O, CO, $C_2F_5CCCF_3$ **(76)**

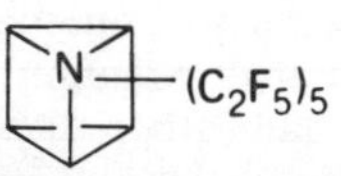

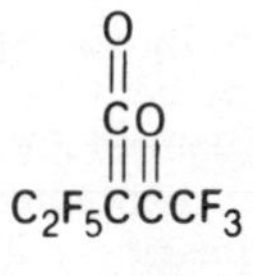

The stabilizing influence of perfluoroalkyl groups on highly strained compounds was dubbed the 'perfluoroalkyl effect' by Lemal and Dunlap[239]. Both kinetic and thermodynamic factors were claimed to be important, although a more recent analysis suggests that the thermal stability of perfluoroalkylated strained rings is kinetic in nature[251].

IV. POLYFLUOROCARBON REACTIONS

The reactions of fluorocarbons have been discussed in several textbooks and in numerous reviews. One text[2] that stresses the physical organic aspects of organofluorine chemistry is especially recommended. Since this work adequately surveys the effects of fluorine substitution on reactivity and mechanism for virtually every important class of organic reactions, only select topics that have not been previously covered in detail will be discussed in the following sections. General trends in fluorocarbon reactivities will be noted but the reader should refer to the cited special topic reviews for comprehensive discussions.

A. Polyfluoroalkanes

1. Nucleophilic displacements

Nucleophilic displacement on saturated carbon is rarely encountered in highly fluorinated systems. This is normally attributed to the shielding of the carbon centre by the surrounding lone pair electrons on fluorine. If reaction does occur, the nucleophile attacks a substituent other than fluorine with apparent displacement of a fluorocarbanion. The classic example is the attack of hydroxide on the electron-deficient iodine in a polyfluoroalkyl iodide to give hydropolyfluoroalkane (equation 5)[252].

$$R_F^{\delta-}I^{\delta+} + OH^- \longrightarrow [R_F^- + IOH] \xrightarrow{\text{Solent}} R_FH, \tag{5}$$

where $R_F{=}CF_3$, C_2F_5, C_3F_7, etc.

Several displacements that formally give products from direct attack on carbon are also proposed to proceed by initial attack on halogen. The reactions of phosphines with fluorohaloalkanes are typical examples (equations 6, 7). Mechanism studies suggest that these reactions proceed via ion pairs which recombine to give the observed products (equation 8)[180,253].

$$CF_3I + PPh_3 \longrightarrow CF_3\overset{+}{P}Ph_3I^-, \tag{6}$$

$$CF_2Br_2 + P(NMe_2)_3 \longrightarrow BrCF_2\overset{+}{P}(NMe_2)_3Br^- \tag{7}$$

$$R_FX + PR_3 \longrightarrow [R_F^-\ R_3\overset{+}{P}X] \longrightarrow R_F\overset{+}{P}R_3X^-, \tag{8}$$

The direct observation of the $CF_3X^{\overline{\cdot}}$ radical anions in ESR studies[254] raises the possibility that many of these reactions proceed by a one-electron transfer process, rather than by direct displacement on halogen. Since $CF_3X^{\overline{\cdot}}$ radical anions are known to decompose to $CF_3^{\cdot}$ and X^-, not to CF_3^- and $X^{\cdot}$, any one-electron transfer mechanism requires the intermediacy of fluoroalkyl radicals. This is certainly reasonable for the phosphine reactions (equation 9). Although this process remains speculative for most polyfluorohalocarbon displacements, there is good evidence for electron transfer in the reactions with Co^I 'supernucleophiles'. Cobalamin, cobaloximes and other Co^I chelates have been reported to react with several polyfluorohalocarbons, including perfluoroalkyl iodides, CF_3Br, CF_2Cl_2, $CFCl_3$ and CF_2HCl, to give the corresponding polyfluoroalkyl–Co^{III} complexes[256–258]. In particular, the $[Co^I(Salphen)]^{-1}$ complex reacted with n-C_3F_7I to give n-$C_3F_7^{\cdot}$ radical reaction products in addition to the expected $C_3F_7Co^{III}$ alkylation product. A general electron transfer scheme was proposed[258].

$$R_FX + PR_3 \longrightarrow [R_FX^{\dot{-}}\ R_3P^{\dot{+}}] \longrightarrow [R_F^{\cdot}\ X\dot{P}R_3] \longrightarrow R_F\overset{+}{P}R_3\ X^-. \quad (9)$$

Weak nucleophiles, notably polyfluoroalkoxides, are reported to displace chloride from α-chloropolyfluoroketones under mild conditions[259]. Since saturated chloropolyfluoroalkanes are inert to these reagents, the unusual reactivity of the ketones was ascribed to special activation by the well known α-effect. Instead of direct attack on $—CF_2Cl$, reversible addition of the nucleophile to the carbonyl carbon followed by displacement through the triangular transition state **78** was suggested.

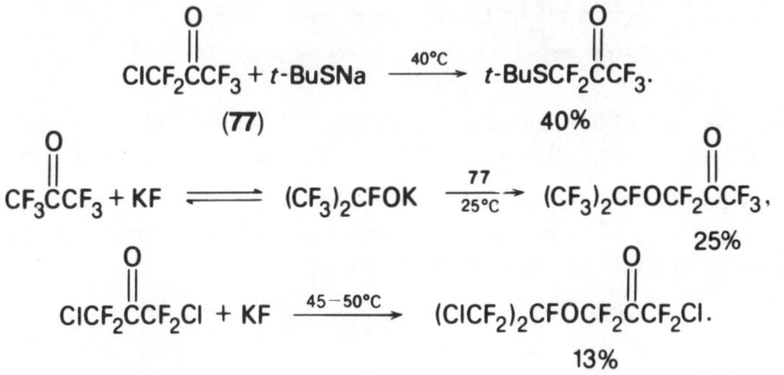

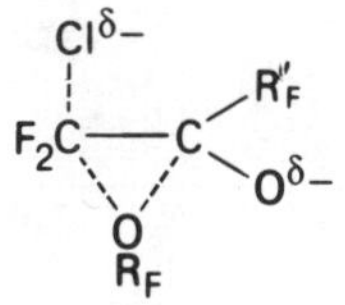

$$ClCF_2C(O)CF_3 + t\text{-}BuSNa \xrightarrow{40°C} t\text{-}BuSCF_2C(O)CF_3.$$

(**77**) 40%

$$CF_3C(O)CF_3 + KF \rightleftharpoons (CF_3)_2CFOK \xrightarrow[25°C]{77} (CF_3)_2CFOCF_2C(O)CF_3,$$

25%

$$ClCF_2C(O)CF_2Cl + KF \xrightarrow{45-50°C} (ClCF_2)_2CFOCF_2C(O)CF_2Cl.$$

13%

$Cl^{\delta-}$ … $F_2C—C(R'_F)(O^{\delta-})$ … $O R_F$

(**78**)

The only examples of direct displacement on fluorinated carbon appear to be nucleophilic attack on fluoroepoxides and intramolecular ring closure of polyfluorohalohydrins to epoxides.

Simple perfluorinated epoxides all readily react with nucleophiles, including H_2O, alcohols, amines, thiols, fluoride and halide ions, to give ring-opened products. The chemistry of the epoxides **79–81** has been reviewed[260]. The high strain energy of these epoxides (**79** and **80** are estimated to be 20–25 kcal mol^{-1} more strained than ethylene oxide[261]), coupled with the excellent fluoroalkoxide leaving groups, undoubtedly contribute to their susceptibility to nucleophilic attack.

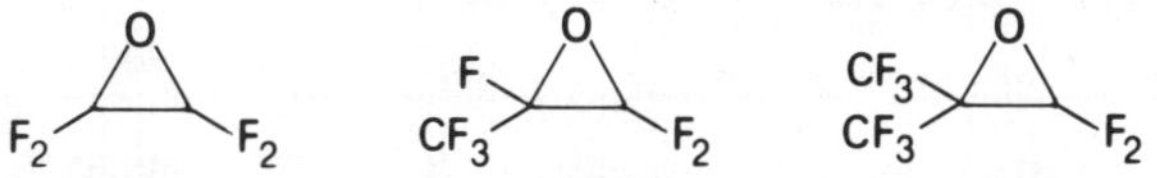

(**79**) (**80**) (**81**)

A curious feature of the ring opening of unsymmetrical fluorinated epoxides is the preferential attack of nucleophiles at the most sterically hindered position in all cases except those involving extremely bulky nucleophiles and very hindered epoxides[260,262]. For example, F^- isomerizes **80** to CF_3CF_2COF, not to CF_3COCF_3. Additional examples are shown in equations (10)[263], (11)[264] and (12)[190].

The specificity of nucleophilic fluoroepoxide ring openings has not been adequately explained. It may be that the stability of the leaving alkoxide governs the positions of

$$C_6F_5OCs + \mathbf{81} \longrightarrow C_6F_5OC(CF_3)_2C(=O)F. \tag{10}$$

$$RMgX + \mathbf{80} \longrightarrow CF_3CFXC(=O)F. \tag{11}$$

$$Ph(ClCF_2)C\overset{O}{—}CF_2 + LiCl \longrightarrow PhCClCOF(CF_2Cl). \tag{12}$$

attack, i.e. late transition states are involved. Based on inductive and hyperconjugative considerations, alkoxide stability is expected to decrease in the order $—CF_2O^- > R_F\overset{|}{C}FO^- > (R_F)_2\overset{|}{C}—O^-$. Nucleophilic attack on **80**, for example, thus gives **82**, not **83**. An even more striking example of this specificity is the attack of nucleophiles exclusively at nitrogen in the oxaziridine **84**[265].

$$\mathbf{80} + Nu^- \longrightarrow CF_3CF(Nu)CF_2O^- \; (\mathbf{82}) \quad CF_3CF(CF_2Nu)O^- \; (\mathbf{83})$$

$$CF_3N\overset{O}{—}CF_2 \; (\mathbf{84}) \xrightarrow[(X^- = F^-,\ R_FO^-)]{X^-} CF_3N(X)—C(=O)F$$

The efficient ring closure of polyfluorohalohydrins to epoxides under relatively mild conditions (Table 17) apparently proceeds by direct displacement of Cl^- from

TABLE 17. Conversion of polyfluorohalohydrins (**85**) to epoxides (**86**)[a]

$$R_FRC(OH)CFXY \; (\mathbf{85}) \longrightarrow R_FRC\overset{O}{—}CFX \; (\mathbf{86})$$

R_F	R	X	Y	Yield, %	Ref.
$ClCF_2$	C_6H_5	F	Cl	80; 65	266, 190
$ClCF_2$	C_6H_5	Cl	Cl	66	266
CF_3	C_6H_5	F	Cl	(79)[b]	262
$ClCF_2$	m-$CF_3C_6H_4$	F	Cl	85	267
$ClCF_2$	C_6F_5	F	Cl	78.5	268
$ClCF_2$	C_6F_5	Cl	Cl	81	268
CF_3	C_6F_5	F	Cl	79	268
$ClCF_2$	CH_3	Cl	Cl	78	266
$ClCF_2$	C_3H_5	Cl	Cl	83	266

[a]In aqueous KOH, 70–100°C.
[b]Epoxide unstable, $C_6H_5C(OH)CF_3CO_2H$ isolated.

$-CF_2Cl$ and $-CFCl_2$ groups. This is most unexpected since these groups generally are quite resistant to nucleophilic attack and the anions of the relatively acidic polyfluorinated alcohols are poor nucleophiles. These disadvantages are seemingly overcome by the favourable entropy of the intramolecular process.

Although many examples of the ring closure of **85** ($R_F = CF_2Cl$) have been reported and confirmed in some cases, the situation with **85** ($R_F = CF_3$) is unsettled. Burton reported that **85** ($R_F = CF_3$; R = Ph, Me, *n*-Bu; X = F, Y = Cl) were recovered unchanged under the same conditions that Knunyants and coworkers[266] claimed reaction with **85** ($R_F = CF_3$; R = Ph; X = F, Y = Cl)[190]. Clearly additional research is needed to resolve this discrepancy and to clarify the effect of the R_F group on this extremely interesting ring closure.

Because of the strong C—F bond, the highly basic F^- leaving group, and the negligible polarizability of the F atom, nucleophilic displacements on alkyl fluorides are notoriously sluggish[269]. Recent gas-phase studies of nucleophilic displacements further indicate a large intrinsic barrier to the displacement of F^-, even when thermodynamic and solution effects have been eliminated[270]. It therefore is not surprising that there are no *bona fide* cases of F^- displacement from $-CF_3$ or $-CF_2H$ groups. Perfluorocarbons, for example, are resistant to hydrolysis below 500°C. The displacement of F^- from $-CF_3$ analogous to the displacement of Cl^- from $-CF_2Cl$ in ketones or halohydrins discussed above has not been reported.

There have been recent claims to the intramolecular displacement of F^- from polyfluoroalkyl groups that warrant comment. Chambers' group reported that the reaction of **87** with F^- gave **89** and a mechanism involving F^- displacement was

(13)

(**87**) (**88**) 79% (**89**)

proposed (equation 13)[271]. A more reasonable alternative which involves F^--catalysed isomerization of the fluorinated double bond followed by an addition–elimination reaction (see Section IV.B.2) is suggested (equation 14). A related intramolecular cyclization has been reported (equation 15)[272].

(14)

The hydrolysis of **90** to give **92** appears to be an unequivocal case of F^- displacement (equation 16)[273]. However, the structure of the product has been questioned and an alternative pathway involving an (unprecedented) 1,3-perfluoroalkyl shift was

(15)

(16)

78%

(**90**) (**91**) (**92**)

proposed (equation 17)[274]. A definitive structure proof of the product is needed to resolve this controversy.

(17)

(**93**)

2. *Exchange reactions*

The most common halofluorocarbon substitution reactions are those involving exchange induced by Lewis acids. Since the commercially important fluorohalomethanes and ethanes are synthesized by exchange reactions, this chemistry has been thoroughly researched and reviewed[1,275–276]. Antimony fluorides, often in combination with HF, are the most widely used catalysts to effect the exchange of Cl by F (the Swarts process). The reactivity of the catalysts increases in the order $SbF_3 < SbF_3 + SbCl_3 < SbF_3 + SbCl_5 < SbF_3Cl_2 < SbF_5$. In general the reactivity of C—Cl bonds decreases in the order $-CCl_3 > CFCl_2 > CF_2Cl$. For example CCl_3CCl_3 is sequentially fluorinated to give $FCCl_2CCl_2F$, CCl_2FCClF_2, and finally $ClCF_2CClF_2$. A carbenium ion mechanism (equation 18) has generally been assumed for these exchange processes, although a four-centre exchange (**94**) cannot be ruled out. Groups that are known to stabilize carbenium ions facilitate exchange. For example, SbF_3 selectively converts $PhCCl_2CCl_3$ and $MeOCCl_2CCl_3$ to $PhCF_2CCl_3$ and $MeOCF_2CCl_3$, respectively.

A carbenium ion mechanism at first would seem to be inconsistent with the observed

$$RCl + SbF_5 \longrightarrow [R^+SbF_5Cl^-] \longrightarrow RF + SbF_4Cl. \quad (18)$$

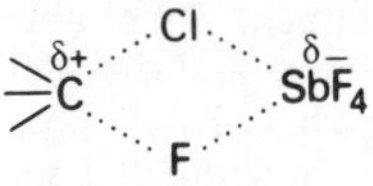

$$\text{>}\overset{\delta+}{C}\cdots Cl \cdots \overset{\delta-}{SbF_4} \cdots F \cdots C$$

(94)

reactivity of $-CF_nCl_{3-n}$ groups since the order of cation stability in solution is $CF_2^+ > CFCl^+ > CCl_2^+$ (see Section III.C). However, the exchange process probably proceeds through tight ion pairs rather than dissociated ions, and Okuhara suggested that the resonance effect (F > Cl) will dominate the inductive effects (F > Cl) only in dissociated ions[277].

Aluminium and boron halides complement the action of antimony catalysts. These catalysts replace F by Cl or Br in halofluorocarbons (equations 19[278], 20[279], 21 and 22[277]), presumably because of the much greater Al—F (159 kcal mol^{-1}) and B—F (180 kcal mol^{-1}) bond strengths compared with the Al—Cl (118 kcal mol^{-1}) and B—Cl (119 kcal mol^{-1}) bonds.

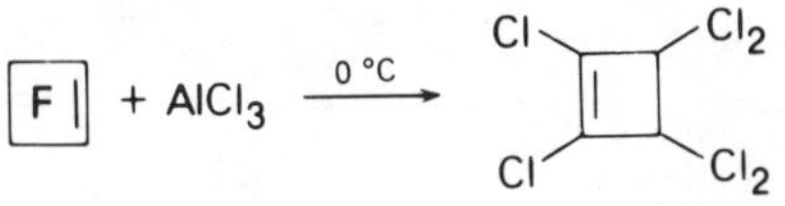

$$\boxed{F\,\|} + AlCl_3 \xrightarrow{0\,°C} \text{(cyclobutene: Cl, Cl, Cl}_2\text{, Cl}_2\text{)} \quad (19)$$

$$\text{(cyclobutene: F}_2\text{, F}_2\text{, F, Me)} + AlCl_3 \longrightarrow \text{(cyclobutene: Cl}_2\text{, Cl}_2\text{, F, Me)} \quad (20)$$

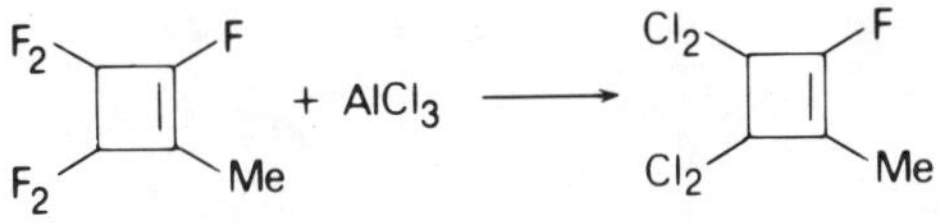

$$CF_2ClCFCl_2 + AlBr_3 \longrightarrow ClCF_2CCl_2Br \quad (21)$$

$$BrCF_2CFClBr + AlCl_3 \longrightarrow BrCF_2CCl_2Br \quad (22)$$

Isomerization (equation 23) and disproportionation (equations 24 and 25) also commonly occur when chlorofluorocarbons are treated with $AlCl_3$[276,277]. The reactions of $ClCF_2CFCl_2$ and other fluorohalocarbons with $AlCl_3$ have been studied in detail to elucidate the factors which control isomerization and substitution[277]. The substitution reactions are proposed to proceed via tight ion pairs, e.g. $[ClCF_2CCl_2^+\ ^-AlFCl_3]$, whereas isomerization occurs via a chain mechanism involving the more stable, dissociated carbenium ion $CCl_3CF_2^+$.

$$ClCF_2CFCl_2 + AlCl_3 \longrightarrow CF_3CCl_3. \quad (23)$$

$$CCl_3F + AlCl_3 \longrightarrow CClF_3 + CCl_4 \quad (24)$$

$$CHClF_2 + AlCl_3 \longrightarrow CHF_3 + CHCl_3. \quad (25)$$

B. Unsaturated Compounds

1. *Nucleophilic, electrophilic and free radical attack*

The electron-deficient double bonds of polyfluoroolefins are much more susceptible to attack by nucleophiles than by electrophiles. Since nucleophilic addition and addition–elimination reactions of fluoroolefins are presumed to proceed via carbanion intermediates[167,280], the effect of substituents on carbanion stability controls the rate and orientation of nucleophilic attack. Thus, nucleophiles attack exclusively at the terminal $CF_2=$ group in fluoroolefins and the olefin reactivity increases in the order $CF_2=CF_2 < CF_2=CFCF_3 \ll CF_2=C(CF_3)_2$ and $CF_2=CF_2 < CF_2=CFCl < CF_2=CFBr$. The latter sequence reflects the importance of I_π repulsion (Section III.D). Steric factors also can be important as the following order of reactivity indicates: $(R_F)_2C=C(R_F)_2 < (R_F)_2C=CFR_F < (R_F)_2C=CF_2$[281,282].

Nucleophilic addition–elimination reactions of fluoroolefins proceed exclusively or predominantly with retention if a good leaving group, e.g. Cl^-, is replaced (equations 26[283], 27 and 28[284]). If the relatively poor F^- leaving group is replaced, the olefin configuration may isomerize, as in the reaction of t-$BuNH_2$ with $CH_3CF=CHCN$[285], although Normant and coworkers reported retention in the replacement of both F and Cl in the reactions of $PhCF=CFCl$ or $BuCF=CFCl$ with EtO^- and BuS^-, and in the conversion of $EtOCF=CFCl$ to $BuC(OEt)=CFCl$ with BuLi[286]. The factors that control the stereochemistry of nucleophilic vinyl substitution have been reviewed[167,280,287,288].

$$(Cl)(CF_3)C=C(Cl)(CF_3) + MeO^- \longrightarrow (MeO)(CF_3)C=C(OMe)(CF_3) \quad (>90\%) \tag{26}$$

$$(Ph)(CF_3)C=C(Cl)(F) + MeO^- \longrightarrow (Ph)(CF_3)C=C(OMe)(F) \quad (96\%) \tag{27}$$

$$(Ph)(CF_3)C=C(F)(Cl) + MeO^- \longrightarrow (Ph)(CF_3)C=C(F)(OMe) \quad (96\%) \tag{28}$$

The principles which govern the reactivity, regiochemistry and product distribution (addition, vinyl substitution or allylic displacement) in nucleophilic attack on acyclic and cyclic fluoroolefins have been covered in detail in Chambers' text[2] and in several reviews[287,289,290] and recent articles[271,291–295]. Ionic additions to fluoroacetylenes have also been reviewed[296].

Highly fluorinated olefins are relatively resistant to electrophilic attack, particularly when one or more perfluoroalkyl groups are present. Hydrofluoroethylenes and halofluoroethylenes will react with a large range of electrophiles and this chemistry has been reviewed[297]. Electrophilic additions normally proceed in accordance with either the polarization of the olefin π-bond or the known effects of substituents on carbenium ion stability (e.g. equations 29[298] and 30[299]), although several exceptions are known, particularly in the ionic additions of halomethanes to fluoroolefins[300].

Numerous electrophilic dimerization, addition and isomerization reactions have been reported, e.g. equations (31)[156], (32)[301] and (33)[302].

$$C_{10}H_{21}CH{=}CF_2 + EtCOCl \xrightarrow{AlCl_3} C_{10}H_{21}CH(CF_2Cl)C(O)Et \qquad (29)$$

$$CF_2{=}CFX + SO_3 \xrightarrow[(X = CF_3,\ H,\ Cl,\ Bu,\ etc.)]{} \underset{O-SO_2}{CF_2CFX} \qquad (30)$$

$$CF_3CF{=}CF_2 + SbF_5 \longrightarrow CF_3CF{=}CFCF(CF_3)_2 \qquad (31)$$

$$\boxed{F\|} + CF_2{=}CF_2 \xrightarrow{SbF_5} \text{(cyclobutene: } F_2, F_2, CF_2CF_3, F) \qquad (32)$$

$$HCF_2CF_2CF{=}CF_2 \xrightarrow{SbF_5} FCH{=}CFCF_2CF_3. \qquad (33)$$

The factors which control the rate and orientation of free radical additions to fluoroolefins are complex and depend upon the nature of the attacking radical as well as the olefin substrate. For example, the ratio of addition rates for $CF_2{=}CF_2/CH_2{=}CH_2$ is 9.5 for the $CH_3^{\bullet}$ radical, but is 0.1 for the $CF_3^{\bullet}$ radical. Similarly, $CH_3^{\bullet}$ preferentially attacks at $CF_2{=}$ in $CHF{=}CF_2$, whereas $CF_3^{\bullet}$ attacks at $CFH{=}$. Tedder and Walton have reviewed in detail the directive effects in radical additions[216], and rules for predicting the rate and preferred orientation have been summarized[303]. The free radical polymerization of fluoroolefins and the properties of many commercially important fluoropolymers also have been reviewed[304].

2. *Perfluorocarbanion chemistry*

The pioneering studies of Miller established that carbanions can be generated by the reaction of F^- with polyfluoroolefins[305,306]. Potassium, caesium or tetralkylammonium fluorides in an aprotic solvent are commonly used and the carbanions generated under these conditions can be trapped with a variety of electrophiles (equations 34[307], 35[308], 36[309], 37[310], 38[311], 39[312], 40[313]). The reaction in equation (40) is particularly interesting since it involves electron transfer to give initially the trityl radical.

$$CF_2{=}CF_2 + KF \longrightarrow [CF_3CF_2^-] \xrightarrow[(2)\ H_2SO_4]{(1)\ CO_2,\ 115\ ^\circ C} \underset{75\%}{CF_3CF_2CO_2H}. \qquad (34)$$

$$(CF_3)_2C{=}CF_2 + KF \xrightarrow{CH_2{=}CHCN} \underset{42\%}{(CF_3)_3CCH_2CH_2CN}. \qquad (35)$$

$$\underset{(1:1:0.06)}{CF_3CF{=}CF_2 + S + KF} \longrightarrow \underset{98\%}{(CF_3)_2C\langle S_2 \rangle C(CF_3)_2}. \qquad (36)$$

$$\boxed{F\|} + KF \xrightarrow{CF_3CO_2Ag} \boxed{F}\text{–}Ag \qquad (37)$$

$$(CF_3)_2C{=}CFC_2F_5 + CsF \xrightarrow{PhCH_2Br} \underset{90\%}{(CF_3)_2C(C_3F_7)CH_2Ph}. \quad (38)$$

$$CF_3CF{=}CF_2 + CF_3\overset{O}{\overset{||}{C}}CF_3 \xrightarrow{CsF} (CF_3)_2CF\overset{O^-Cs^+}{\overset{|}{C}}(CF_3)_2. \quad (39)$$

$$(CF_3)_3C^- Cs^+ + Ph_3CCl \longrightarrow (CF_3)_3CC_6H_4CHPh_2. \quad (40)$$

Fluoride ion also isomerizes and oligomerizes fluoroolefins. Olefins with a terminal $CF_2{=}$ group are particularly susceptible to fluoride attack and are converted to the more stable internal olefins (equations 41[305] and 42[314]). Fluoroallenes are likewise converted into thermodynamically more stable isomers (equations 43[315] and 44[316]).

$$CF_2{=}CFCF_2C_4F_9 \underset{}{\overset{Et_4N^+ F^-}{\rightleftharpoons}} CF_3CF{=}CFC_4F_9. \quad (41)$$

$$PhC(C_2F_5){=}CF_2 \xrightarrow{CsF} PhC(CF_3){=}CFCF_3. \quad (42)$$

$$C_2F_5CF_2{=}C{=}CF_2 \xrightarrow{F^-} CF_3C{\equiv}CC_2F_5 \quad (43)$$

$$(CF_3)_2CFC(CF_3){=}C{=}CF_2 \xrightarrow{F^-} \underset{72\%}{CF_3CF{=}C(CF_3)C(CF_3){=}CF_2} + \underset{18\%}{\text{cyclobutene: } (CF_3)_2,\ F_2,\ CF_3,\ F} \quad (44)$$

Dimers and trimers of $CF_3CF{=}CF_2$ (HFP) are formed relatively easily, but higher oligomers are difficult to produce in fluoride-induced reactions. Higher telomers of $CF_2{=}CF_2$ are more accessible, presumably because of the greater reactivity of the $CF_3CF_2^-$ anion, and highly branched $C_{14}F_{28}$ oligomers have been produced. The chemistry of the fluoride ion and perfluorocarbanions prior to 1968 have been reviewed[317]. Some important advances that postdate Young's review are discussed below.

The oligomerization of $CF_3CF{=}CF_2$ initiated by tertiary amines or F^- normally gives a mixture of dimers **95** and **96**, and trimers **97**, **98** and **99**. Moderately basic trialkylamines, such as $(CF_3CHFCF_2OCH_2CH_2)_3N$, have been recommended for the control of isomerization[318]. With this amine in CH_3CN, the kinetic dimer **95** is formed almost exclusively, but with Et_3N present, trimers **97** (66%), **98** (30%) and **99** are formed instead. The dimer **95** is rearranged to the thermodynamic dimer **96** with triethylenediamine or DABCO in DMSO. Hexamers of HFP have been produced using a mixture of $[(CF_3)_2CHOCH_2CH_2]_3N$ and DABCO in DMSO solvent[319].

$(CF_3)_2CFCF{=}CFCF_3$ **(95)**

$(CF_3)_2C{=}CFC_2F_5$ **(96)**

$(CF_3)_2C{=}C(C_2F_5)CF(CF_3)_2$ **(97)**

$[(CF_3)_2CF]_2C{=}CFCF_3$ **(98)**

$(CF_3)_2CFCF{=}C(CF_3)CF_2CF_2CF_3$ **(99)**

Hexafluoropropene oligomerization catalysed by F^- can also be controlled by judicious choice of reagents and reaction conditions[320]. At 20°C, KF in CH_3CN gave 94% **95**, whereas CsF in CH_3CN gave exclusively **96**. If a crown ether (18-crown-6) was added, KF/CH_3CN also gave only **96**. With CsF in THF plus crown ether, the rate of oligomerization was rapidly increased and only trimers, mostly **97** and **98**, were formed at 20°C in the best yields yet reported; at 130–200°C only trimer **99** was formed. By contrast, the addition of crown ether to similar solvent systems had no effect on the rate, yield or product distribution for the oligomerization of $CF_2{=}CF_2$. The limiting factor in these reactions therefore appears to be the concentrations of olefin in the solvent rather than the concentration of F^-. Recent detailed studies of the dimerization and trimerization of perfluorocyclobutene, -cyclopentene, and -cyclohexene with F^- or pyridine also have appeared[321].

The development of nucleophilic equivalents of Friedel–Crafts alkylation and disproportionation is one of the more interesting advances in fluorocarbanion chemistry. Fluorocarbanions mono- or polyalkylate activated heterocycles depending upon the reaction conditions (equations 45[322], 46[323], 47[322]). Polyalkylation with $(CF_3)_2C{=}CF_2$ is normally thermodynamically controlled, whereas $CF_2{=}CF_2$ alkylations are kinetically controlled. The 2,4,5-isomer **101** cannot be rearranged by F^-, whereas $(CF_3)_2C{=}CF_2$ gives only the 2,4,6-compound **100**. The disproportionation of the 2,4,5-isomer from $CF_3CF{=}CF_2$ has also been reported (equation 48[324]). These rearrangement processes have been shown to be intermolecular and the ease of rearrangement parallels the stability of the migrating anion, i.e. $(CF_3)_3C^- > (CF_3)_2CF^- > CF_3CF_2^-$.

C(CF$_3$)$_3$ C(CF$_3$)$_3$

F + $(CF_3)_2C{=}CF_2$ —CsF, T / Sulpholane→ F (N) + (CF$_3$)$_3$C F C(CF$_3$)$_3$ (N) (45)

(**100**)

(90%, T = 20 °C) (85%, T = 80 °C)

F + $CF_2{=}CF_2$ —CsF / Sulpholane→ C$_2$F$_5$, C$_2$F$_5$, F, C$_2$F$_5$ (N) (46)

(**101**)

F (N, N) + $(CF_3)_2C{=}CF_2$ —CsF / Sulpholane→ (CF$_3$)$_3$C F (N, N) + C(CF$_3$)$_3$ F (N, N) C(CF$_3$)$_3$ (47)

(T = 20 °C) (T = 80 °C)

R$_F$, R$_F$, F, R$_F$ (N) —F$^-$ / 177 °C→ R$_F$ F R$_F$ (N) + R$_F$ F R$_F$ (N) + R$_F$ R$_F$ F R$_F$ (N) (48)

where $R_F = CF(CF_3)_2$.

Hexafluoro-2-butyne (**102**) is readily polymerized by F^- [317], although the initially formed vinyl carbanion can be trapped with very reactive substrates (equation 49)[325]. Chain transfer agents such as $CF_3CF{=}CBrCF_3$, $CHCl_3$ or CCl_2BrCCl_2Br have been used to prepare 1,3-dienes $XC(CF_3){=}C(CF_3)C(CF_3){=}C(CF_3)Y$ (X = F, Y = Br or H; X = Cl, Y = Br) from **102**. This chemistry was used to synthesize **103**, a precursor to tetrakis(trifluoromethyl)cyclobutadiene derivatives[326] (see Section IV.B.3).

$$\text{C}_4\text{F}_4\text{N}_2 \ (\text{F-pyrimidine}) + CF_3C{\equiv}CCF_3 \ (\mathbf{102}) \xrightarrow[\text{Sulpholane}]{\text{CsF, 20 °C}} CF_3C{=}CFCF_3\text{-substituted F-pyrimidine} \ (73\%) \quad (49)$$

$$\mathbf{102} + CF_3CF{=}CBrCF_3 \xrightarrow{\text{CsF}} BrC(CF_3){=}C(CF_3)C(CF_3){=}C(CF_3)F \xrightarrow{100-150\ °C} \mathbf{(103)}$$

3. *Fluorinated cyclobutadienes and derivatives*

Cyclobutadiene and its derivatives are of considerable interest to fluorocarbon as well as hydrocarbon chemists[327]. Both tetrafluoro- (**104**) and tetrakis(trifluoromethyl)-cyclobutadiene (**105**) have been synthetic targets of several laboratories, and **105** has been successfully characterized. The status of **104** is less certain.

(**104**) (**105**)

Kobayashi and coworkers reported that the photolysis of ozonide **106**, prepared from **63**, in an organic matrix at 77 K followed by thawing gave the *syn* dimer of **105** (**107**)[328]. When the photolysis was conducted in the presence of ethyl diazodicarboxylate, adduct **108** was formed in high yield. The postulated intermediacy of **105** was soon confirmed by Masamune[329]. Irradiation of **106** in organic glasses at 77K followed by warming to 135 K gave a yellow solution of **105**. The cyclobutadiene was fully

$$\mathbf{(106)}\ (CF_3)_6 \xrightarrow{h\nu} (CF_3CO)_2O + [\mathbf{105}] \xrightarrow{\Delta} \mathbf{(107)}\ (CF_3)_8$$

$$[\mathbf{105}] \xrightarrow{EtO_2CN{=}NCO_2Et} \mathbf{(108)}\ (CF_3)_4$$

characterized spectroscopically and was shown to exist in a rectangular singlet ground state.

Miller had earlier reported that the treatment of **103** with MeLi in ether at $-125°C$ gave an unstable lithium compound which eliminated LiF to give predominantly **107**[326]. However, the dimer was later shown to be the *anti* isomer **109**[236,328], which makes the intermediacy of **105** highly improbable. Both **107** and **109** isomerized thermally to the very stable but photochemically labile perfluorooctamethylcyclooctatetraene **(61)**[326]. Irradiation of **107** or **109** gave the cubane **110** and the cuneane **111**, and the photolysis of **61** also afforded **110** and **111** as major final products. This is the first example of the conversion of a cyclobutadiene dimer into a cubane.

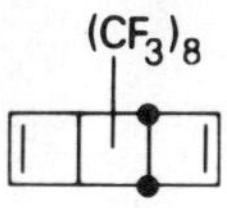

$(CF_3)_8$ $(CF_3)_8$ $(CF_3)_8$

(109) 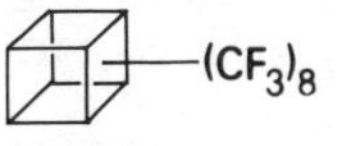 **(110)** 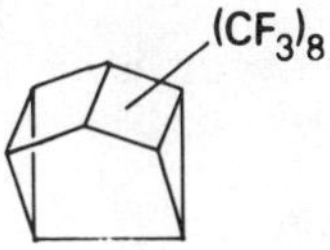 **(111)**

Lemal has presented evidence for the intermediacy of **104**, generated by the photolysis of **112** (equation 50)[330]. Irradiation of **112** in the presence of 500 mm N_2 gave **113**; in the absence of N_2, **114** was produced. Irradiation in the presence of furan gave the adduct **115** in addition to **114**. Lemal originally proposed that the dimerization of the photoproduct **104** formed vibrationally excited **113** which rearranged rapidly to **114**, unless it was collisionally deactivated by an inert gas, but more recent results have undermined this proposal[331,332]. Dimer **113** was found to be predominantly the *anti* isomer, not the required cyclobutadiene *syn* dimer. Furthermore, **114** was shown to be formed first and **113** resulted from the photocyclization of **114**, which was dramatically accelerated by an inert gas. Perfluorocyclopentadienone (**116**), prepared by the vacuum pyrolysis of **112** at 585°C[333], is now implicated in the photolysis of **112**. Photolysis of **116** (2537 Å) also gives perfluorocyclooctatetraene (**114**) directly, but its photolysis is strongly inhibited by an inert gas. A revised mechanism has been proposed to account for these results (equation 51).

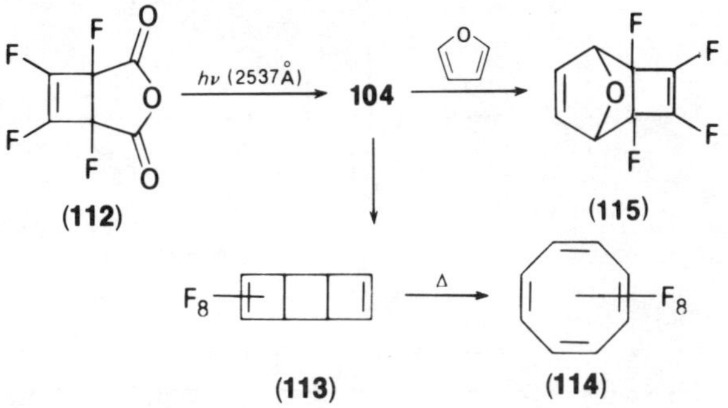

F F O F O F O
$h\nu$ (2537 Å) **104** F F O F F
(112) **(115)** (50) 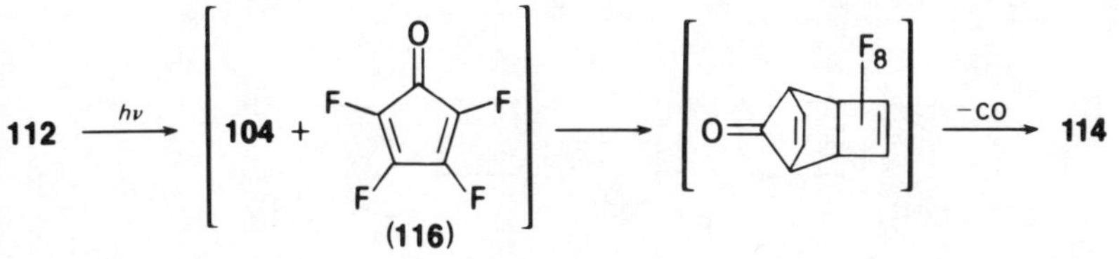
F_8 Δ F_8
(113) **(114)**

O
F F F_8
112 $h\nu$ **104** + O $-CO$ **114** (51)
F F
(116)

Attempts to prepare a π-$(C_4F_4)Fe(CO)_3$ complex in a manner analogous to the synthesis of the hydrocarbon complex were unsuccessful[334]. Cyclobutene **117** (X = Cl) reacted with $Fe_2(CO)_9$ to give only the unstable *cis* complex **118**. Pyrolysis of **117** (X = I) gave **114** in 58% yield.

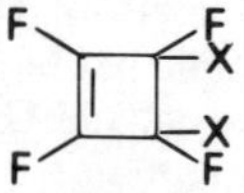

(117) 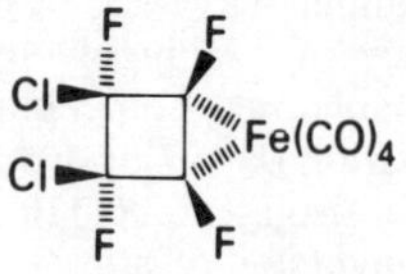 **(118)**

There currently is no direct evidence for the existence of tetrafluorocyclobutadiene.

4. Cycloadditions

The propensity of fluoroolefins to thermally cyclodimerize and cycloadd to olefins and acetylenes to give four-membered rings is one of the most familiar and unusual aspects of fluorocarbon chemistry. Polyfluorochloroolefins are exceptionally reactive in these additions and they characteristically give almost exclusively [2 + 2] cycloadducts with dienes rather than the normal [4 + 2] Diels–Alder adducts. The [2 + 2] cycloadditions are generally not stereospecific and are insensitive to solvent effects. A biradical mechanism for cycloadditions is well established. Head-to-head coupling is normally preferred when unsymmetrical olefins dimerize and the orientation of the product can almost always be predicted on the basis of the most stable biradical intermediate (equations 52, 53, 54). As the difference between the ability of the substituents at each end of the olefin to stabilize a radical diminishes, the cycloaddition becomes less selective. Trifluoroethylene, for example, cyclodimerizes at 260°C to give approximately equal amounts of all four possible hexafluorocyclobutanes. Unlike $CF_2{=}CF_2$ and $CF_2{=}CCl_2$, which give 99% [2 + 2] cycloadducts with butadiene exclusively via biradical mechanisms, $CF_2{=}CFH$ gives only 86% [2 + 2] and 14% [4 + 2] adduct, with most of the [4 + 2] adduct arising from concerted addition.

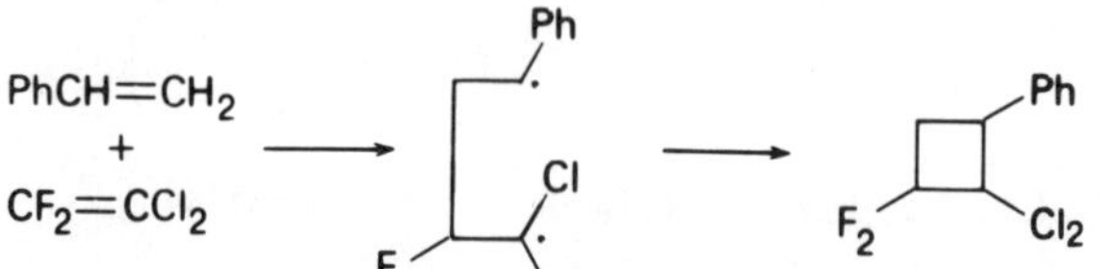

$2\ CF_2{=}CFCl \longrightarrow$ (52)

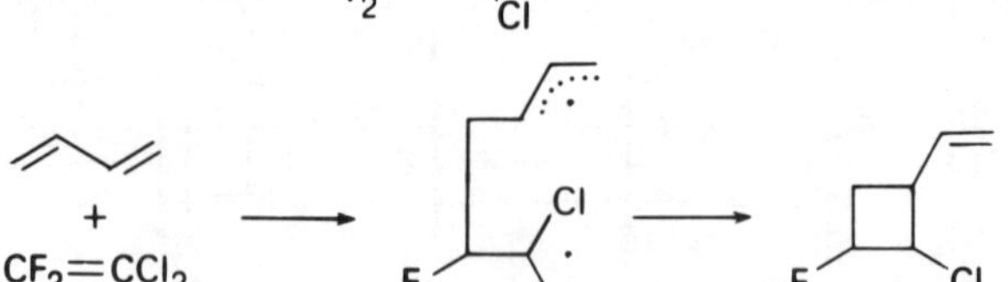

1:1 *cis/trans*

$PhCH{=}CH_2 + CF_2{=}CCl_2 \longrightarrow$ (53)

$+ CF_2{=}CCl_2 \longrightarrow$ (54)

Fluoroolefin cycloadditions, including the elegant mechanism studies of Bartlett and coworkers, have been thoroughly reviewed[335–338]. The Diels–Alder reactions of fluorodienes also have been reviewed[339].

Why do fluoroolefins [2 + 2] cycloadd but their hydrocarbon counterparts generally do not? A particular requirement for biradical reactivity is the presence of a *gem*-difluoro group in the olefin; 1,2-difluoroolefins are much less reactive in this context. The low π-bond energy in $CF_2{=}CFX$ (Section II.B.1.b) and the formation of an unusually strong fluorinated C—C bond (Section II.A.2) are the major driving forces for biradical cycloaddition. (Methylenecyclopropanes, which also have weak π-bonds, are rare examples of hydrocarbons that efficiently undergo thermal [2 + 2] cycloadditions. The analogy between their reactions and those of *gem*-difluoroolefins has been discussed[73,340].) A *gem*-difluoro group, however, is not sufficient for biradical reactivity; $CF_2{=}CH_2$ does not cyclodimerize and no [2 + 2] cycloadditions are known for $(CF_3)_2C{=}CF_2$. The π-bond in a 1,1-difluoroolefin apparently is not weak enough to ensure biradical reactivity (Section II.B.1.b), but if substituents that stabilize radicals are present, the olefin can be quite reactive. For example, at 212°C $CF_2{=}CCl_2$ is approximately 5×10^6 more reactive than $CF_2{=}CH_2$ in 1,2-cycloaddition to butadiene[336].

Fluoroallenes readily undergo [2 + 2] and [2 + 4] cycloadditions[335], and the highly reactive 1,1-difluoro- and monofluoroallenes have been used as mechanistic probes for concertedness in cycloadditions[341–343].

Fluoroacetylenes are powerful Diels–Alder dienophiles and enophiles[296]. Hexafluoro-2-butyne (**102**) is so reactive that it converts *N*-alkylated pyrroles into isolable 7-azanorbornadienes[344]. Gassman and collaborators used **102** to synthesize novel 'inside–outside' carbobicyclic dienes (equation 55)[345]. The reactions of **102** or $CF_3C{\equiv}CH$ with simple olefins generally give *cis* adducts via a concerted [π2s + π2s + σ2s] mechanism. For example **102** reacts suprafacially with isobutylene at 145°C to give the *Z*-diene **119** in 80% yield[346]. In contrast, *trans* adducts are also observed in reactions with allenes (equation 56) and a biradical mechanism has been proposed[347].

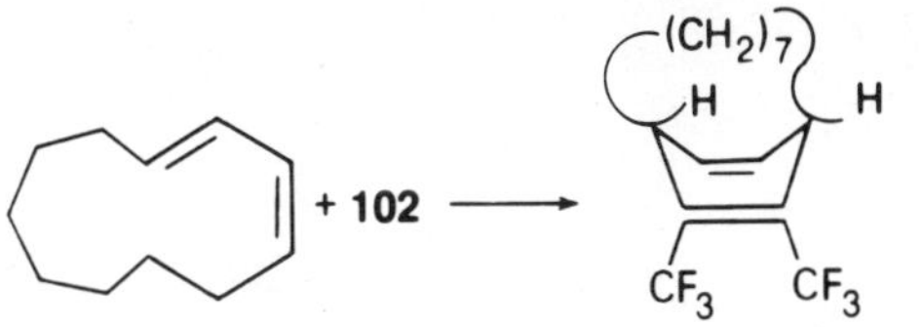

+ **102** → [bicyclic diene with $(CH_2)_7$, H, H, CF_3, CF_3] (55)

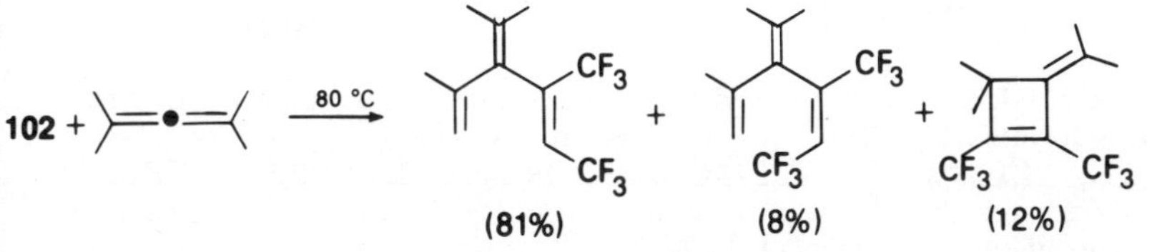

102 + [tetramethylallene] —80 °C→ [diene, CF_3, CF_3] (81%) + [diene, CF_3, CF_3] (8%) + [cyclobutene, CF_3, CF_3] (12%) (56)

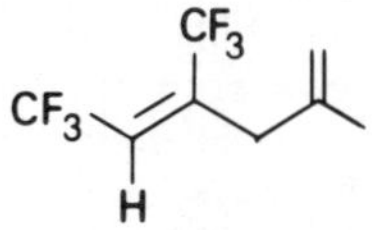

[CF_3, CF_3, H] (**119**)

Dipolar [2 + 2] cycloadditions to fluoroolefins are fairly common. The reaction of SO_3 with fluoroolefins to give β-sultones is one large class of dipolar additions[299] (equation 30, Section IV.B.1). Ynamines, e.g. $MeC{\equiv}CNEt_2$, react with $CF_2{=}CFX$ to give cyclobutenes **120** (X = Cl, Br, CF_3, n-C_5F_{11}) together with α-fluoroenamines **121** (X = CF_3 or n-C_5F_{11}) via an ionic mechanism[348].

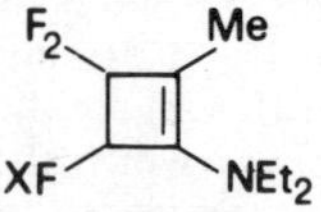

$XCF{=}CFCMe{=}CFNEt_2$

(**120**) (**121**)

Relatively few [2 + 3] dipolar cycloadditions of simple fluoroolefins have been reported. Vinyl fluorides are particularly poor 1,3-dipolarophiles: $CHF{=}CH_2$, $CF_2{=}CH_2$ and $CF_3CF{=}CF_2$ do not react with CF_3CHN_2[349]. Organic azides and $CF_3CF{=}CF_2$ react only under forcing conditions (equation 57)[350], and there is only one claim to isoxazolidine formation from the reaction of perfluoroolefins with nitrones (equation 58)[351]. *C,N*-Diphenylnitrone reacts with perfluoropropene to give surprisingly the β-lactam **122**[352]; pyridine-*N*-oxide reacts at 70–80°C to afford **123**, presumably via the intermediate adduct **124**[353].

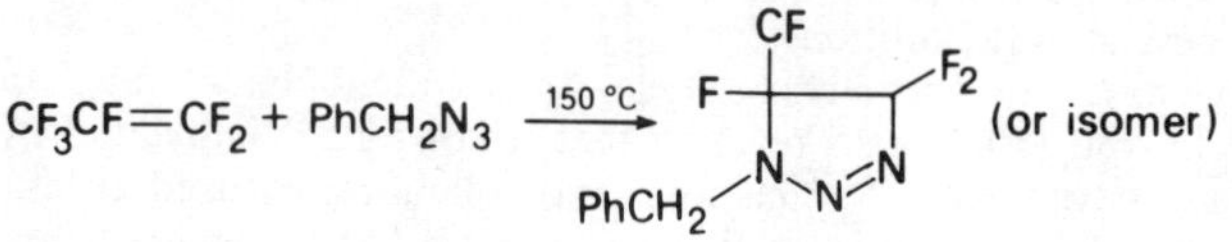

$$CF_3CF{=}CF_2 + PhCH_2N_3 \xrightarrow{150\,°C} \text{(or isomer)} \quad (85\%) \qquad (57)$$

$$CF_3C(R_F){=}CF_2 + PhCH{=}\overset{+}{N}(O^-)CH_2Ph \longrightarrow \qquad (58)$$

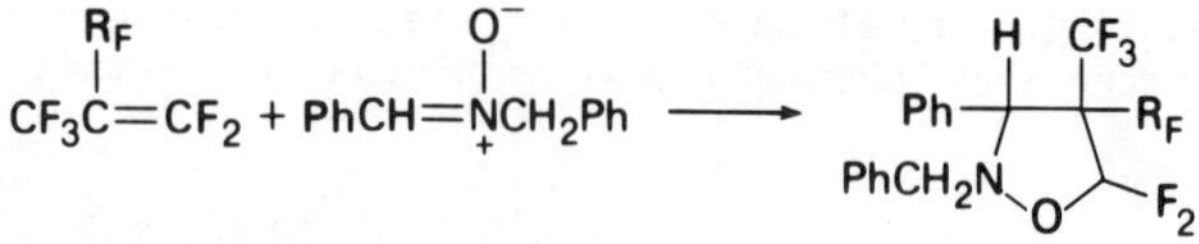

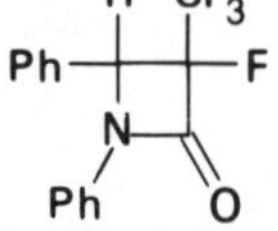

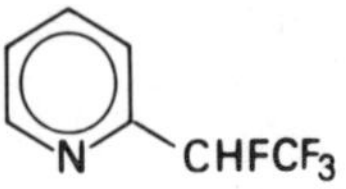

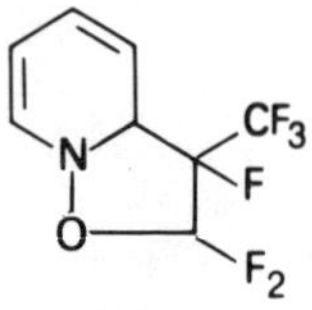

(**122**) (**123**) (**124**)

Perfluoroalkylolefins with no vinyl fluorines are good dipolarophiles. The reactivity of olefins towards CF_3CHN_2 was found to increase in the order $MeCH{=}CH_2 < CH_2{=}CH_2 < CH_2{=}CHCl < CH_2{=}CHBr < CH_2{=}CHCF_3 < CF_3CH{=}CHCF_3$[349]. Nitrile oxides readily add to alkenes, $R_FCH{=}CH_2$, below 80°C to give isolable isoxazolines in excellent yields[354].

Only highly strained vinyl fluorides are especially reactive towards 1,3-dipoles. The Dewar benzene isomer **125** reacted with PhN_3, CH_2N_2, $PhCN \rightarrow O$ and $PhC\overset{+}{N}{-}\overset{-}{N}Ph$ at room temperature or below to give both 1:1 and 1:2 [3 + 2] cycloadducts[355]. From comparative studies of additions to the double bonds in **126**, an olefin reactivity sequence which parallels the double bond electrophilicity was observed: $FC{=}CCF_3 > CF{=}CF > CH{=}CF \gg CF{=}CMe$, $CF{=}COMe$[356].

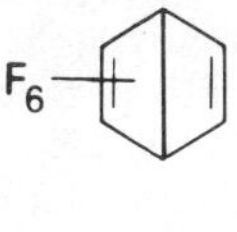

(125) 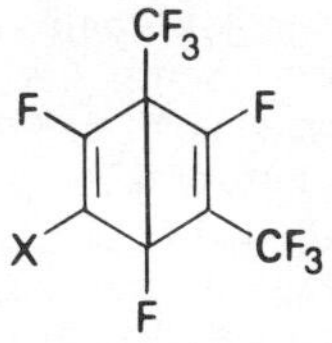 **(126)**

Fluoroacetylenes are also quite reactive dipolarophiles. Diazo compounds[357,358] and nitrile oxides[354,359] readily cycloadd at room temperature or below (equations 59 and 60).

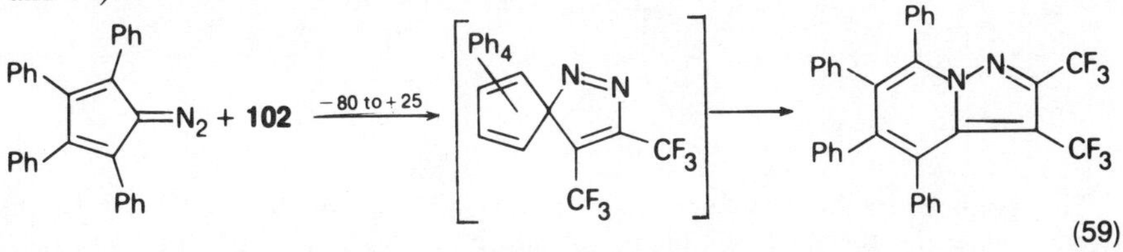

(59)

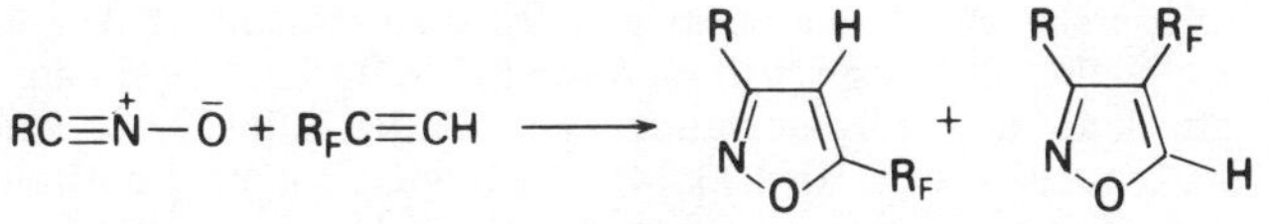

(60)

C. Polyfluoroarenes

The general chemistry of fluoroarenes has been surveyed recently[360], and detailed reviews of nucleophilic substitution in polyfluoroarenes[361] and the chemistry of polyfluoroaromatic[362] heterocycles are available. The effects of the pentafluorophenyl group on the reactivity of organic compounds has also been reviewed[363]. Only more recent mechanism studies of nucleophilic aromatic substitution will be discussed at length in the following section.

1. Nucleophilic and electrophilic attack

Nucleophilic aromatic substitution is the most characteristic reaction of polyfluoroarenes. The rate and orientations of substitution in C_6F_5X derivatives is particularly interesting because nucleophiles attack mainly at the position *para* to X in most cases (X = H, Me, CF_3, NMe_2, SMe, C_6F_5, RC=O, etc.), whereas *meta* attack predominates in only a few cases (X = NH_2, O^-), and comparable *meta* and *para* attack occurs but occasionally (X = OMe, NHMe). The relative rates of substitution, however, correspond to the expected effect of the substituent X in nucleophilic aromatic substitution. For example, the relative reactivities of C_6F_5X towards NaOMe in MeOH at 60°C for X = NMe_2, Me, H, CF_3 are respectively 0.1, 0.63, 1, 4.5×10^2 [361]. Thus, unlike electrophilic aromatic substitution in benzene derivatives, the rates of C_6F_5X reactions vary greatly, but the position of nucleophilic attack is relatively insensitive to the aromatic substituent.

Burdon has rationalized this behaviour based on how substituents affect the stability of the Wheland anion intermediate[364,365]. Resonance hybrid **127** was proposed to best model the transition state for nucleophilic attack; thus the effect of substituents on carbanion stability will determine the position of attack. Since F strongly destabilizes planar carbanions by I_π repulsion (Section II.D), attack *para* to X in C_6F_5X **(128)** is preferred except for those substituents with greater electron repulsion than F ($X = O^-$, NH_2). This I_π-repulsion theory has proved to be highly predictive for several fluorinated benzene systems[361], and the generalization that nucleophilic attack on polyfluoroaromatics is fastest at the position which localizes the least anionic charge on fluorine-bearing carbon atoms has been made[365]. (Polyfluoroaromatic heterocycles are all activated relative to the fluorobenzenes, and pentafluoropyridine, for example, reacts with various nucleophiles exclusively at the 4-position in accordance with the I_π repulsion theory. In general, however, the directing effect of the ring N on substitution appears to be more important than electron-pair repulsion involving F[362].)

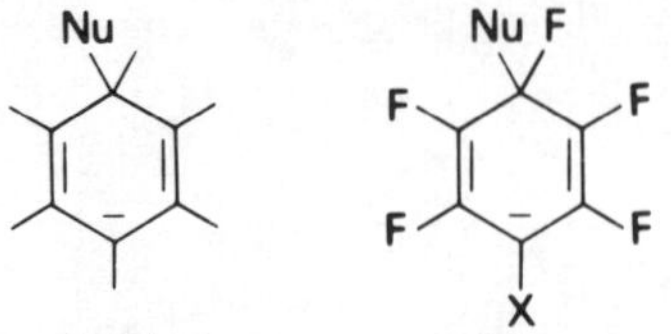

(127) (128)

The recent studies of Chambers and coworkers indicate that the I_π repulsion theory overestimates the deactivating influence of *para*-F in C_6F_5X[366–368]. From relative rate data for nucleophilic attack on fluorinated benzenes (Table 18), *para*-F is only modestly deactivating with respect to H, whereas the fluorines *ortho* and *meta* to the position of attack are markedly activating. (The relative influence of *ortho*:*meta*:*para* F is 57:106:0.43 relative to H, or 133:246:1 relative to the *para* position, calculated from the data in Table 18.) Similar effects were found in the reactions of polyfluoropyridines and chlorofluorobenzenes. Chambers concluded that relatively early transition states are involved in nucleophilic attack on polyfluoroarenes and ground state polarization governs the orientation of attack. Therefore, nucleophilic substitution in C_6F_5X generally occurs *para* to maximize the number of fluorines *ortho* and *meta* to the point of attack, and the effect of the *para* fluorine can be largely ignored.

Compared with aromatic hydrocarbons, polyfluoroarenes are relatively resistant to electrophilic attack. 1,2-Difluoro-1,2,3,4-tetrafluoro-, 1,2,4,5-tetrafluoro- and pentafluorobenzene are not protonated by $FSO_3H–SbF_5$; only the strongest superacid system, $SbF_5–HF–SO_2ClF$, protonates these derivatives. Hexafluorobenzene and $C_6F_5CH_3$ cannot be protonated; both react with $FSO_3H–SbF_5$ to give radical cations. In $DF–SbF_5–SO_2ClF$, $C_6F_5CH_3$ is converted to $C_6F_5CH_2D$ but apparently is not ring protonated[146].

Perfluoroarenium ions have been generated from hexa-1,4-dienes and these ions are converted to radical cations on heating (equations 61, 62)[369,370]. Radical cations are

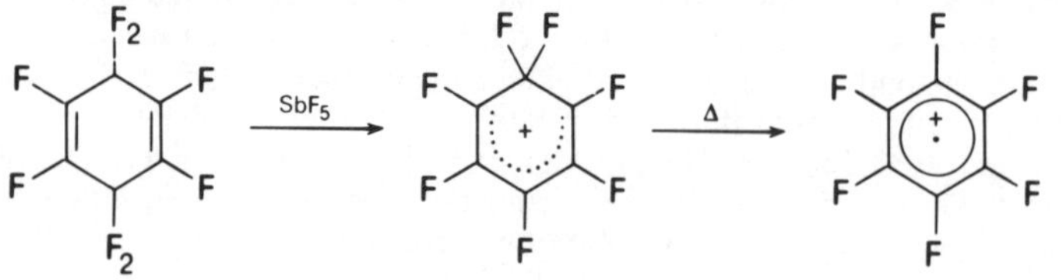

(61)

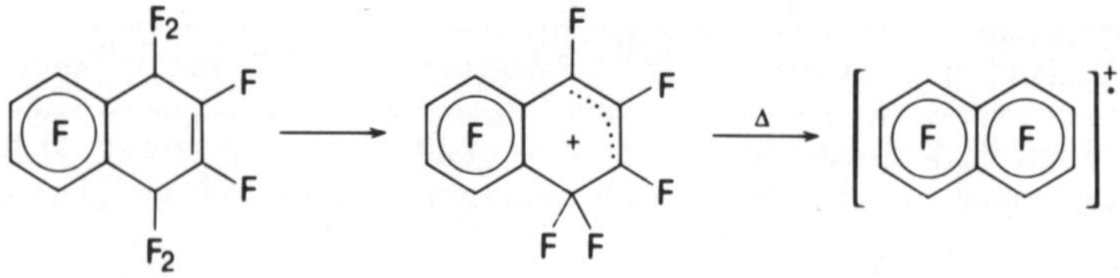

(62)

TABLE 18. Relative second-order rate constants for reactions of polyfluorobenzenes with NaOMe in MeOH at 58.00°C[366]

Polyfluorobenzene[a]	k_{rel}
C_6F_6	1
C_6HF_5 (ring: F, H, F, F, F*, F)	2.34
$C_6H_2F_4$ (ring: F, H, F, H, F*, F)	4.09×10^{-2} [b]
$C_6H_2F_4$ (ring: F, H, H, F, F*, F)	2.21×10^{-2} [b]
$C_6H_2F_4$ (ring: H, F, F, H, F*, F)	1.99×10^{-4} [b]

[a]Position of attack indicated by *.
[b]Corrected for statistical factor.

probably quite common intermediates in electrophilic reactions of polyfluoro-aromatics. The fluorination of arenes with high valency metal fluorides proceeds via radical cations[371], and several radical cations have been produced by the anodic oxidation of polyfluoroarenes[372,373].

Electrophilic substitution with displacement of F^+ obviously is never observed. If perfluoroarenes react at all, they undergo electrophilic addition, not substitution, e.g. equation (63)[374]. Electrophilic substitution of hydrogen in highly fluorinated systems, however, is surprisingly facile (equations 64[375], 65[376]), but little is known about the mechanism of this process.

$$C_{10}F_8 \xrightarrow{NO_2BF_4} \text{(}O_2N, F, F, F, F_2\text{ adduct) (61\%)} \qquad (63)$$

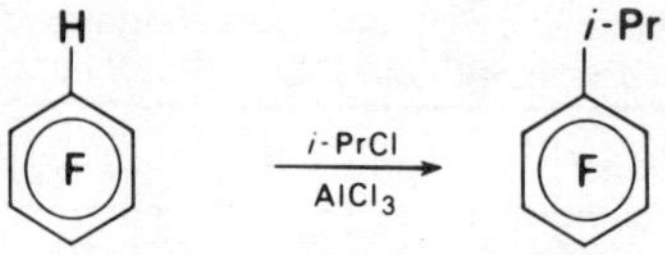

$$\text{C}_6\text{F}_5\text{H} \xrightarrow[\text{AlCl}_3]{i\text{-PrCl}} \text{C}_6\text{F}_5\text{-}i\text{-Pr} \ (40\%) \qquad (64)$$

$$\text{C}_{10}\text{F}_7\text{H} \xrightarrow[20\,^\circ\text{C}]{\text{Br}_2/\text{Fe}} \text{C}_{10}\text{F}_7\text{Br} \ (68\%) \qquad (65)$$

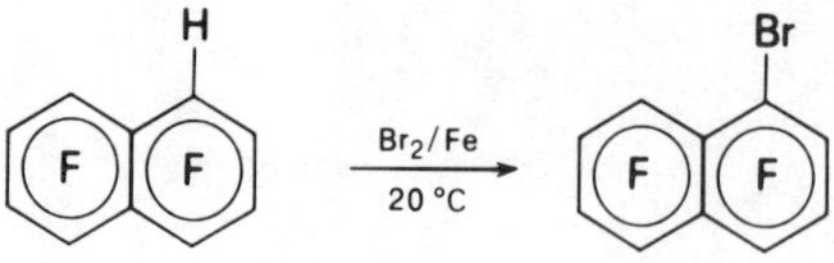

2. Photochemical reactions and valence isomers

The photochemistry of fluorinated aromatic molecules has been a very active research area for the last 15 years. Numerous photoisomerization, photoaddition and photosubstitution reactions have been reported, and the mechanisms involved in all three types of reactions have been studied and compared with those of aromatic hydrocarbons. These subjects have been adequately reviewed[377–381] and the discussion in this section is mainly limited to photocycloadditions of C_6F_6 and to recent research on valence isomers of fluorinated heterocycles.

The course of C_6F_6 cycloadditions to alkenes and alkynes is highly dependent upon reaction conditions and the nature of the unsaturated substrate. *cis*-Cyclooctene reacts photochemically with C_6F_6 to give five 1:1 cycloadducts[382]. The 1,3-adducts **129** and **130** (major) and the 1,2-adduct **131** (minor) are the primary photoproducts. Both free and complexed S_1-C_6F_6 were proposed as the reactive species. In contrast, indene and 1,2-dihydronaphthalene react stereospecifically with C_6F_6 on irradiation (λ = 245 nm) to give only the *endo*-1,2-adducts **132**[383], whereas cyclopentene gives mainly the *exo*-adduct **133**[384]. These apparently are triplet processes since added piperylene inhibited cycloaddition.

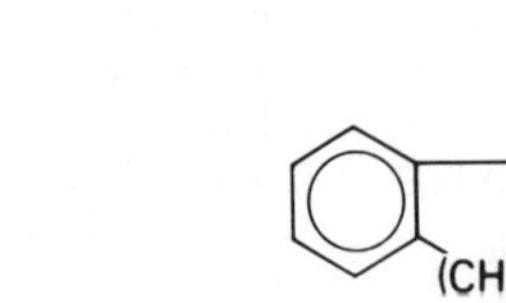

(**129**) 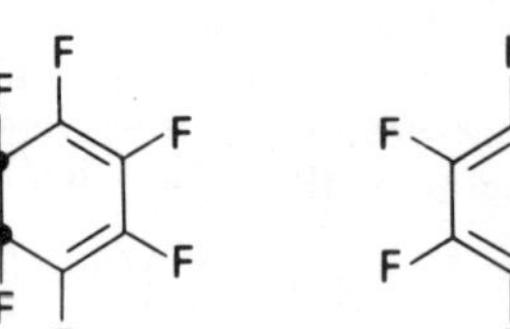 (**130**) 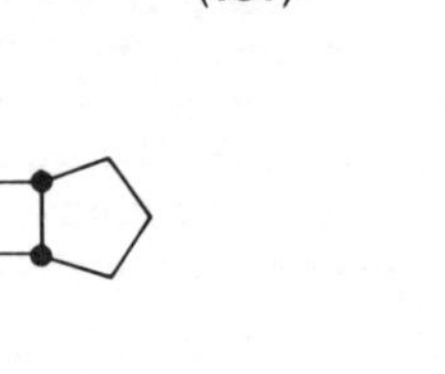 (**131**)

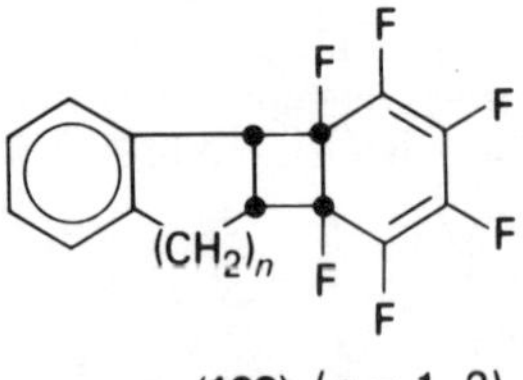

(**132**) (n = 1, 2) 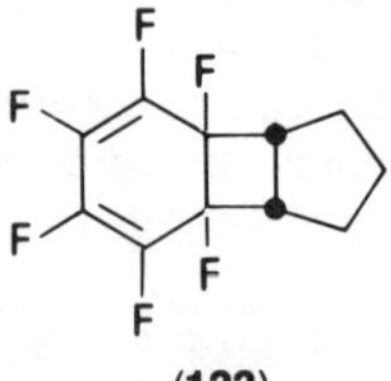 (**133**)

The photocycloaddition of ClCF=CFCl to C_6F_6 has been used in a practical synthesis of perfluorocyclooctatetraene (**114**) (equation 66)[385].

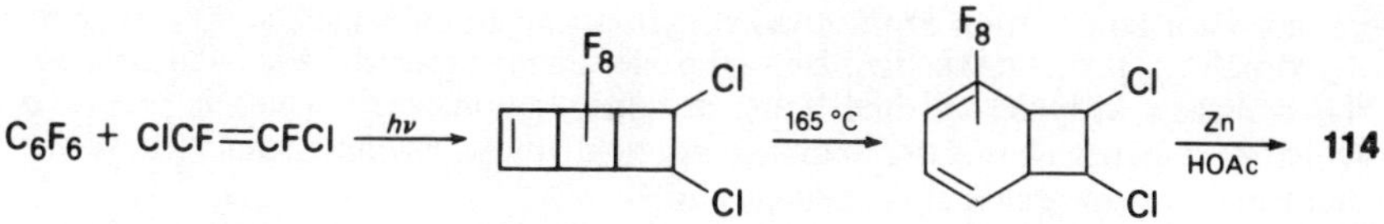

$$C_6F_6 + ClCF{=}CFCl \xrightarrow{h\nu} \text{[F}_8\text{ tricyclic dichloride]} \xrightarrow{165\,°C} \text{[F}_8\text{ bicyclic dichloride]} \xrightarrow[\text{HOAc}]{\text{Zn}} \mathbf{114} \quad (66)$$

Few photocycloaddition reactions of polyfluorinated heterocycles have been reported. Pentafluoropyridine reacts photochemically ($\lambda > 200$ nm) with $CH_2{=}CH_2$ to give 1:1 (**134**) and 2:1 (**135**) adducts, in contrast to pyridine itself which does not cycloadd with olefins[386]. Pentafluoropyridine 2:1 photocycloadducts with cycloalkenes also have been reported[387].

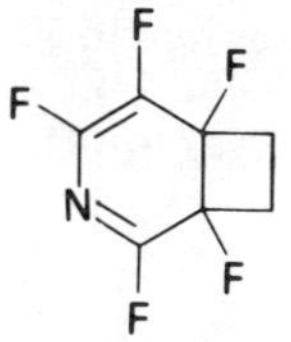

(**134**) 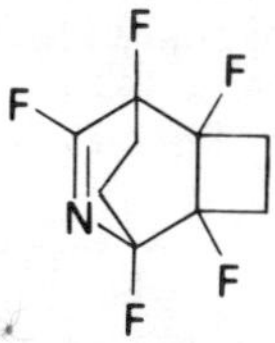 (**135**)

The photoisomerizations of benzene and fluorobenzenes differ markedly[379]. Whereas irradiation (λ = 254 nm) of liquid C_6H_6 gives a mixture of fulvene and benzvalene, but no trace of Dewar benzene, irradiation (λ = 212–265 nm) of C_6F_6 in the liquid or vapour phases gives exclusively hexafluoro-Dewar benzene (**125**). The photochemical transformation of C_6F_6 is a symmetry-allowed process which occurs from the S_2 or S_3 singlet state. The tendency for *para* rather than *meta* bonding is also observed in the photolysis of lower fluorinated benzenes. Perfluoroalkylated benzenes photoisomerize somewhat differently; $C_6(CF_3)_6$ gives the benzvalene **63** and the prismane isomers as well as the Dewar benzene **64** on irradiation. Isolable Dewar and prismane isomers are intermediates in the photochemical isomerization of perfluoro-1,3,5- and perfluoro-1,2,4-trimethylbenzenes.

Photolysis of perfluoropentaalkylpyridines gives stable azaprismanes (**75**) and 1-azabicyclo[2.2.0]hexa-2,5-dienes (Dewar pyridines)[377], whereas various perfluorotetra- and trialkylpyridines give both 1-aza- and 2-aza-Dewar isomers[388], e.g. equation (67). Continuous photolysis of pentafluoropyridine affords isomer **136** in *c*. 1% yield, which re-aromatizes with a half-life of about 5 days at room temperature. Flash photolysis of C_5F_5N gives two transients with half-lives of 22 ms and 3 ms which were identified spectroscopically as the respective novel fulvenes **137** and **138**[389].

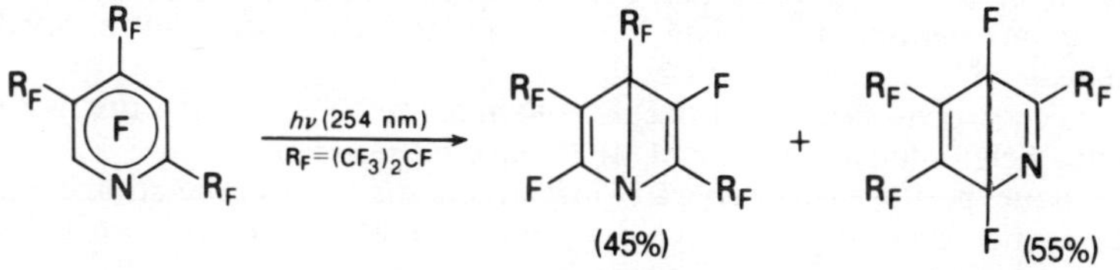

$$\xrightarrow[R_F=(CF_3)_2CF]{h\nu\,(254\text{ nm})} \quad (45\%) \; + \; (55\%) \quad (67)$$

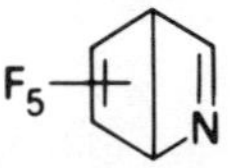

(**136**) 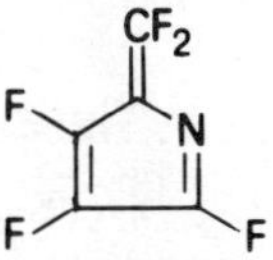 (**137**) 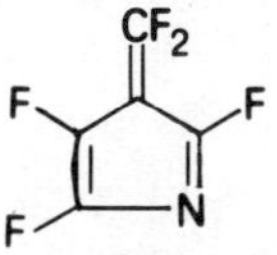 (**138**)

The photoisomerization of fluoroalkyl pyridazines to pyrazines, and in some cases to pyrimidines, proceeds via *para* Dewar species, many of which have been isolated [377]. This contrasts with the thermal isomerization of pyridazines, which is proposed to involve diazabenzvalene intermediates, although these valence isomers have never been isolated or detected spectroscopically[390].

Photolysis of tetra(trifluoromethyl)thiophene at 214 nm or by Hg sensitization at 254 nm afforded **102** and **72**, the only known Dewar thiophene[246]. The half-life for rearomatization is 5.1 h at 160°C; phosphines catalyse its re-aromatization at 25°C[391]. Peroxytrifluoroacetic acid oxidizes **72** to its *exo-S*-oxide **139**[392].

The Dewar thiophene **72** is an exceptionally reactive dienophile and 1,3-dipolarophile[243,393,394]. Azides, including HN_3, react rapidly with **72** at room temperature to give 1,3-dipolar cycloadducts **140**; photolysis of **140** affords a valence bond isomer of 1,4-thiazene (**141**). Compounds **141** are desulphurized by Ph_3P to give the first Dewar pyrroles (**71**). Photolysis of **142** (R = C_6H_{11}, *t*-Bu, H) did not produce the Dewar form, although **142** (R = Ph) was slowly converted to **143**.

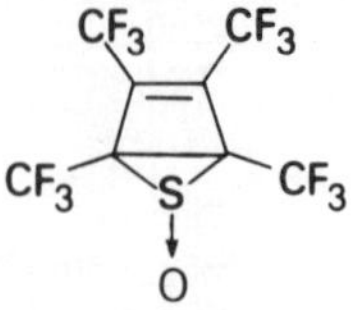

(**139**) 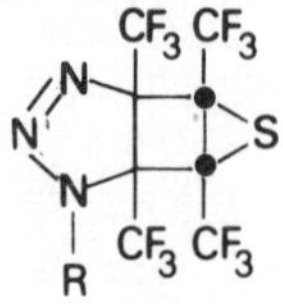 (**140**) 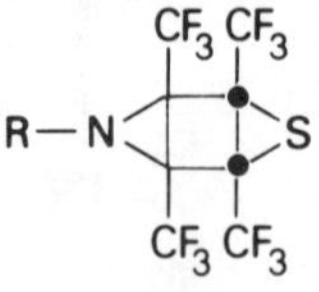 (**141**)  (**142**)

R = Ph, C_6H_{11} *t*-Bu, H

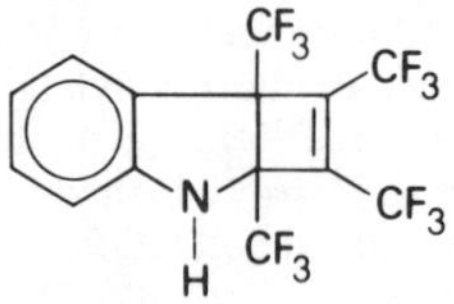

(**143**) 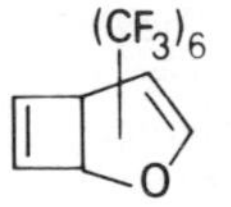 (**144**) 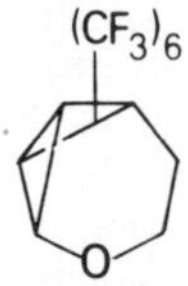 (**145**)

The most intriguing property of the Dewar thiophene system is the remarkably facile degenerate rearrangement of **139**[392,395]. The ^{19}F NMR spectrum of **139** shows a sharp singlet at −95°C that separates into two singlets of equal areas at −150°C. The $\Delta G^{\neq}$ for CF_3 exchange is 6.7 ± 0.1 kcal mol^{-1} at −135.8°C. The parent Dewar thiopene **72** also 'automerizes', but at much higher temperatures with $\Delta G^{\neq} = 22.1 \pm 0.1$ kcal mol^{-1} at 157°C. The relative rate for rearrangement of **139**:**72** is 3×10^{10} at 25°C! The mechanism of exchange which accounts for this remarkable difference is unresolved[392,395,396].

Two valence bond isomers of oxepin, **144** and **145**[397,398], and the first benzvalene containing heteroatoms in the ring (**74**)[248], have been isolated.

The unique 'perfluoroalkyl effect' is responsible for many of the advances made in valence isomer chemistry. The synthesis and chemistry of heteroaromatic valence isomers are now fruitful areas of research.

V. REFERENCES

1. M. Hudlicky, *Chemistry of Organic Fluorine Compounds*, 2nd edn, Ellis Horwood, Chichester (1976).
2. R. D. Chambers, *Fluorine In Organic Chemistry*, Wiley, New York (1973).

3. W. A. Sheppard and C. M. Sharts, *Organic Fluorine Chemistry*, W. A. Benjamin, New York (1969).
4. R. E. Banks, *Fluorocarbons and their Derivatives*, 2nd edn, Macdonald, London (1970).
5. *Organofluorine Chemicals and their Industrial Applications* (Ed. R. E. Banks), Ellis Horwood, Chichester (1979).
6. *Fluorocarbon and Related Chemistry* Vols. I–III (Ed. R. E. Banks and M. G. Barlow), The Chemical Society, London (1971, 1974, 1976).
7. G. A. Boswell, W. C. Ripka, R. M. Scribner and C. W. Tullock, *Org. Reactions*, **21**, 1 (1974).
8. C. M. Sharts and W. A. Sheppard, *Org. Reactions*, **21**, 215 (1974).
9. S. C. Cohen and A. G. Massey, *Adv. Fluorine Chem.*, **6**, 185 (1970).
10. P. M. Treichel and F. G. A. Stone, *Adv. Organometal. Chem.*, **1**, 143 (1964).
11. L. Pauling, *Nature of the Chemical Bond*, 3rd edn, Cornell University Press, Ithaca, N.Y. (1960), pp. 88–102.
12. A. Yokozeki and B. Bauer, *Topics Curr. Chem.*, **53**, 71 (1975).
13. D. Friesen and K. Hedberg, *J. Amer. Chem. Soc.*, **102**, 3987 (1980).
14. D. E. Brown and B. Beagley, *J. Mol. Struct.*, **38**, 167 (1977).
15. K. L. Gallaher, A. Yokozeki and S. H. Bauer, *J. Phys. Chem.*, **78**, 2389 (1974).
16. C. R. Patrick, *Adv. Fluorine Chem.*, **2**, 1 (1961).
17. K. W. Egger and A. T. Cooks, *Helv. Chim. Acta*, **56**, 1516 (1973).
18. S. Furayama, D. M. Golden and S. W. Benson, *J. Amer. Chem. Soc.*, **91**, 7564 (1969).
19. L. Pauling in Ref. 11, pp. 314–316.
20. L. Radom, W. J. Hehre and J. A. Pople, *J. Amer. Chem. Soc.*, **93**, 289 (1971).
21. L. Radom and P. J. Stiles, *Tetrahedron Lett.*, 789 (1975).
22. D. Peters, *J. Chem. Phys.*, **38**, 561 (1963).
23. C. E. Mellish and J. W. Linnett, *Trans. Faraday Soc.*, **50**, 657 (1954).
24. H. A. Bent, *Chem. Revs*, **61**, 275 (1961).
25. P. Kollman, *J. Amer. Chem. Soc.*, **96**, 4363 (1974).
26. D. Holtz, *Prog. Phys. Org. Chem.*, **8**, 1 (1971).
27. T. D. Thomas, *J. Amer. Chem. Soc.*, **92**, 4184 (1970).
28. E.-C. Wu and A. S. Rodgers, *J. Amer. Chem. Soc.*, **98**, 6112 (1976).
29. J. M. Pickard and A. S. Rodgers, *J. Amer. Chem. Soc.*, **99**, 691 (1977).
30. A. S. Rodgers and W. G. Ford, *Int. J. Chem. Kinetics*, **5**, 965 (1973).
31. S. S. Chen, A. S. Rodgers, J. Chow, R. C. Wilhoit and B. J. Zwolinski, *J. Phys. Chem. Ref. Data*, **4**, 441 (1975).
32. H. R. Gerberich and W. D. Walters, *J. Amer. Chem. Soc.*, **83**, 4935 (1961).
33. H. R. Gerberich and W. D. Walters, *J. Amer. Chem. Soc.*, **83**, 4884 (1961).
34. R. T. Conlin and H. M. Frey, *JCS Faraday I.*, **75**, 2556 (1979).
35. R. T. Conlin and H. M. Frey, *JCS Faraday I*, **76**, 322 (1980).
36. J. N. Butler, *J. Amer. Chem. Soc.*, **84**, 1393 (1962).
37. R. A. Mitsch and F. W. Neuvar, *J. Phys. Chem.*, **70**, 546 (1966).
38. W. E. Bernett, *J. Org. Chem.*, **34**, 1772 (1969).
39. J. F. Chiang and W. E. Bernett, *Tetrahedron*, **27**, 975 (1971).
40. A. Greenberg and J. F. Liebman, *Strained Organic Molecules*, Academic Press, New York (1978), pp. 29–55.
41. S. Durmaz and H. Kollmar, *J. Amer. Chem. Soc.*, **102**, 6942 (1980).
42. J. D. Dill, A. Greenberg and J. F. Liebman, *J. Amer. Chem. Soc.*, **101**, 6814 (1979).
43. A. Skancke and J. E. Boggs, *J. Amer. Chem. Soc.*, **101**, 4063 (1979).
44. C. A. Deakyne, L. C. Allen and N. C. Craig, *J. Amer. Chem. Soc.*, **99**, 3895 (1977).
45. H. Günter, *Tetrahedron Lett.*, 5173 (1970).
46. R. Hoffmann and W. P. Stohrer, *J. Amer. Chem. Soc.*, **93**, 6941 (1971).
47. M. E. Jason and J. A. Ibers, *J. Amer. Chem. Soc.*, **99**, 6012 (1977).
48. W. R. Dolbier, Jr, S. F. Sellers, B. H. Al-Sader and B. E. Smart, *J. Amer. Chem. Soc.*, **102**, 5398 (1980).
49. W. R. Dolbier, Jr and T. H. Fielder, Jr, *J. Amer. Chem. Soc.*, **100**, 5577 (1978).
50. W. R. Dolbier, Jr and H. O. Enoch, *J. Amer. Chem. Soc.*, **99**, 4532 (1977).
51. A. T. Perretta and V. W. Laurie, *J. Chem. Phys.*, **62**, 2469 (1975).
52. D. R. James, G. H. Birnberg and L. A. Paquette, *J. Amer. Chem. Soc.*, **96**, 7465 (1974).

53. C. W. Jefford, J. Mareda, J. C. E. Gehret, T. Karbengele, W. D. Graham and U. Burger, *J. Amer. Chem. Soc.*, **98**, 2585 (1976).
54. L. Fernholt and K. Kveseth, *Acta Chem. Scand. A*, **34**, 163 (1980).
55. S. Wolfe, *Acc. Chem. Res.*, **5**, 102 (1972).
56. N. D. Epiotis, W. R. Cherry, S. Shaik, R. Yates and F. Bernardi, *Topics Curr. Chem.*, **70**, 1–242 (1977).
57. A. Liberles, A. Greenberg and J. E. Eilers, *J. Chem. Educ.*, **50**, 676 (1973).
58. R. Hoffmann and R. A. Olofson, *J. Amer. Chem. Soc.*, **88**, 943 (1966).
59. R. C. Bingham, *J. Amer. Chem. Soc.*, **98**, 535 (1976).
60. M. S. Gordon, *J. Amer. Chem. Soc.*, **91**, 3122 (1969).
61. S. Weiss and G. Leroi, *J. Chem. Phys.*, **48**, 962 (1968).
62. T. K. Brunck and F. Weinhold, *J. Amer. Chem. Soc.*, **101**, 1700 (1979).
63. A. Y. Meyer, *J. Mol. Struct.*, **49**, 383 (1978).
64. L. Radom and P. J. Stiles, *Tetrahedron Lett.*, 789 (1975).
65. D. H. Wertz and N. L. Allinger, *Tetrahedron*, **30**, 1579 (1974).
66. D. Bak, D. Christensen, L. Nygard and J. R. Anderson, *Spectrochim. Acta*, **13**, 120 (1958).
67. V. W. Laurie and D. T. Pence, *J. Chem. Phys.*, **38**, 2693 (1963).
68. A. Bhaumik, W. V. F. Brooks and S. C. Dass, *J. Mol. Struct.*, **16**, 29 (1973).
69. J. L. Carlos, Jr, R. R. Karl, Jr and S. H. Bauer, *JCS Faraday II*, **70**, 177 (1974).
70. F. Bernardi, A. Bottini, N. D. Epiotis and M. Guerra, *J. Amer. Chem. Soc.*, **100**, 6018 (1978).
71. E.-C. Wu, J. M. Pickard and A. S. Rodgers, *J. Phys. Chem.*, **79**, 1078 (1975).
72. J. R. Lacher and H. A. Skinner, *J. Chem. Soc. A*, 1034 (1968).
73. P. D. Bartlett and R. C. Wheland, *J. Amer. Chem. Soc.*, **94**, 2145 (1972).
74. H. M. Frey, R. G. Hopkins and I. C. Vinall, *JCS Faraday I*, **68**, 1874 (1972).
75. W. R. Dolbier, Jr, unpublished results (1980).
76. J. M. Pickard and A. S. Rodgers, *J. Amer. Chem. Soc.*, **98**, 6115 (1976).
77. J. M. Pickard and A. S. Rodgers, *J. Amer. Chem. Soc.*, **99**, 695 (1977).
78. J. M. Tedder and J. C. Walton, *Adv. Phys. Org. Chem.*, **16**, 86 (1978).
79. A. Wilson and D. Goldhamer, *J. Chem. Educ.*, **40**, 599 (1963).
80. E. W. Schlag and E. W. Kaiser, Jr, *J. Amer. Chem. Soc.*, **87**, 1171 (1965).
81. P. M. Jeffers, *J. Phys. Chem.*, **78**, 1469 (1974).
82. N. D. Epiotis, J. R. Larson, R. L. Yates, W. R. Cherry, S. Shaik and F. Bernardi, *J. Amer. Chem. Soc.*, **99**, 7460 (1977).
83. J. F. M. Oth, R. Merenyi, H. Rottle and G. Schroder, *Tetrahedron Lett.*, 3941 (1968).
84. J. Hine, *Structural Effects on Equilibria in Organic Chemistry*, Wiley, New York (1975), pp. 270–276.
85. P. I. Abell and P. K. Adolf, *J. Chem. Thermodyn.*, **1**, 333 (1969).
86. K. S. Pitzer and J. L. Hollenberg, *J. Amer. Chem. Soc.*, **76**, 1493 (1954).
87. C. W. Bock, P. George, G. J. Mains and M. Trachtman, *JCS Perkin II*, 814 (1979).
88. J. S. Brinkley and J. A. Pople, *Chem. Phys. Lett.*, **45**, 197 (1977).
89. J. A. Pople in *Modern Theoretical Chemistry*, Vol. 4 (Ed. H. F. Schaefer III), Plenum Press, New York (1977), Chap. 1, pp. 12–18.
90. K. B. Wiberg, *J. Amer. Chem. Soc.*, **102**, 1229 (1980).
91. A. C. Wahl and G. Das, *Adv. Quantum Chem.*, **5**, 26 (1970).
92. T. L. Gilbert and A. C. Wahl, *J. Chem. Phys.*, **55**, 5247 (1971).
93. W. J. Middleton and W. H. Sharkey, *J. Amer. Chem. Soc.*, **81**, 803 (1959).
94. K. M. Smirov and A. P. Tomilov, *Zhur. Vsesoyuz Khim. obshch. D. I. Mendeleeva*, **19**, 350 (1974).
95. S. Y. Delavarenne and A. G. Viehe, *Chem. Ber.*, **103**, 1198 (1970).
96. H. G. Viehe, *Angew Chem. Int. Edn. Engl.*, **4**, 746 (1965).
97. R. E. Banks, M. G. Barlow, W. D. Davies, R. N. Haszeldine and D. R. Taylor, *J. Chem. Soc.*, 1104 (1969).
98. R. E. Banks, M. G. Barlow and K. Mullen, *J. Chem. Soc. C*, 1131 (1969).
99. W. J. Middleton, U.S. Patent 2,831,835 (1958); *Chem. Abstr.*, **52**, 14658 (1958).
100. J. Heicklen and V. Knight, *J. Phys. Chem.*, **69**, 2484 (1965).
101. L. Kevan and P. Hamlet, *J. Chem. Phys.*, **42**, 2255 (1965).

102. R. D. Chambers in Ref. 2, pp. 142–144.
103. E. Kloster-Jensen, C. Pascual and J. Vogt, *Helv. Chim. Acta*, **53**, 2109 (1970).
104. T. C. Ehlert, *J. Phys. Chem.*, **73**, 949, (1969).
105. M. J. Krech, S. J. W. Prize and H. J. Scepians, *Canad. J. Chem.*, **55**, 4222 (1977).
106. M. J. Krech, S. J. W. Prize and W. F. Yared, *Canad. J. Chem.*, **52**, 2673 (1974).
107. G. Modena and G. Scorrano in *The Chemistry of the Carbon–Halogen Bond*, (Ed. S. Patai), Wiley, London (1973), Chap. 6, pp. 301–406.
108. D. F. Loncrini, *Adv. Fluorine Chem.*, **6**, 43 (1970).
109. S. K. De and S. R. Palit, *Adv. Fluorine Chem.*, **6**, 69 (1970).
110. B. L. Dyatkin, E. P. Mochalina and I. L. Knunyants, *Tetrahedron*, **21**, 2291 (1965).
111. J. Hine and N. W. Flachskam, *J. Org. Chem.*, **42**, 1979 (1977).
112. R. D. Howells and J. D. McCown, *Chem. Revs*, **77**, 69 (1977).
113. R. D. Topsom, *Prog. Phys. Org. Chem.*, **12**, 1 (1976).
114. L. S. Levitt and H. F. Widing, *Prog. Phys. Org. Chem.*, **12**, 119 (1976).
115. K. Hiroaka, R. Yamdagni and P. Kebarle, *J. Amer. Chem. Soc.*, **95**, 6833 (1973).
116. R. Yamdagni, T. B. McMahon and P. Kebarle, *J. Amer. Chem. Soc.*, **96**, 4035 (1974).
117. T. B. McMahon and P. Kebarle, *J. Amer. Chem. Soc.*, **99**, 2222 (1977).
118. R. T. McIver, Jr and J. H. Silvers, *J. Amer. Chem. Soc.*, **95**, 8463 (1973).
119. C. S. Yoder and C. H. Yoder, *J. Amer. Chem. Soc.*, **102**, 1245 (1980).
120. R. D. Topsom, *Tetrahedron Lett.*, 403 (1980).
121. S. Fraga, K. M. Saxena and B. W. N. Lo, *Atomic Data*, **3**, 323 (1971).
122. C. Hansch and A. Leo, *Substituent Constants for Correlation Analysis in Chemistry and Biology*, Wiley, New York (1979).
123. S. H. Under and C. Hansch, *Prog. Phys. Org. Chem.*, **12**, 91 (1976).
124. P. Politzer and J. M. Timberlake, *J. Org. Chem.*, **37**, 3557 (1972).
125. R. D. Chambers, J. S. Waterhouse, and D. L. H. Williams, *Tetrahedron Lett.*, 743 (1974).
126. A. Streitwieser, Jr, D. Holtz, G. R. Ziegler, J. O. Stoffer, M. L. Brokaw and F. Guibe, *J. Amer. Chem. Soc.*, **98**, 5229 (1976).
127. W. Adcock and T.-C. Khor, *J. Amer. Chem. Soc.*, **100**, 7799 (1978).
128. W. Kitching, M. Bullpitt, D. Gartshore, W. Adcock, T.-C. Kohr, D. Doddrell and I. D. Rae, *J. Org. Chem.*, **42**, 2411 (1977).
129. A. Pross, L. Radom and R. W. Taft, *J. Org. Chem.*, **45**, 819 (1980).
130. *Biochemistry Involving Carbon–Fluorine Bonds* (Ed. R. Filler), American Chemical Society, Washington, D.C. (1976).
131. M. Schlosser, *Tetrahedron*, **34**, 3 (1978).
132. S. A. Sherrod, R. L. daCosta, R. A. Barnes and V. Boekelheide, *J. Amer. Chem. Soc.*, **96**, 1565 (1974).
133. H. Förster and F. Vögtle, *Angew. Chem. Int. Edn Engl.*, **16**, 429 (1977).
134. B. E. Smart, *J. Org. Chem.*, **38**, 2027 (1973).
135. B. E. Smart, *J. Org. Chem.*, **38**, 2035 (1973).
136. B. E. Smart, *J. Org. Chem.*, **38**, 2039 (1973).
137. W. H. Dawson, D. H. Hunter and C. J. Willis, *JCS Chem. Commun.*, 874 (1980).
138. Y. Apeloig, P. v. R. Schleyer and J. A. Pople, *J. Amer. Chem. Soc.*, **99**, 5901 (1977).
139. Y. Apeloig, P. v. R. Schleyer and J. A. Pople, *J. Amer. Chem. Soc.*, **99**, 1291 (1977).
140. W. J. Hehre and P. C. Hilberty, *J. Amer. Chem. Soc.*, **96**, 2665 (1974).
141. J. Burdon, D. W. Davies and G. delConde, *JCS Perkin II*, 1193 (1976).
142. L. D. Kispert, C. U. Pittmann, Jr, D. L. Allison, T. B. Patterson, Jr, C. W. Gilbert, Jr, C. F. Harris and J. Prather, *J. Amer. Chem. Soc.*, **94**, 5979 (1972).
143. R. J. Blint, T. B. McMahon and J. L. Beauchamp, *J. Amer. Chem. Soc.*, **96**, 1269 (1974).
144. A. D. Williams, P. R. LeBreton and J. L. Beauchamp, *J. Amer. Chem. Soc.*, **98**, 2705 (1976).
145. F. T. Prochaska and L. Andrews, *J. Amer. Chem. Soc.*, **100**, 2102 (1978).
146. G. A. Olah and Y. K. Mo in *Carbonium Ions* (Ed. G. A. Olah and P. von R. Schleyer), John Wiley and Sons, New York (1976), Vol. V, Chap. 36, pp. 2135–2261.
147. G. A. Olah and Y. K. Mo., *Adv. Fluorine Chem.*, **7**, 69 (1973).
148. D. P. Ridge, *J. Amer. Chem. Soc.*, **97**, 5670 (1975).
149. M. N. Paddon-Row, C. Santiago and K. N. Houk, *J. Amer. Chem. Soc.*, **102**, 6561 (1980).
150. P. C. Myhre and G. D. Andrews, *J. Amer. Chem. Soc.*, **92**, 7595 (1970).

151. P. C. Myhre and G. D. Andrews, *J. Amer. Chem. Soc.*, **92**, 7596 (1970).
152. K. M. Koshy, D. Roy and T. T. Tidwell, *J. Amer. Chem. Soc.*, **101**, 357 (1979).
153. K. M. Koshy and T. T. Tidwell, *J. Amer. Chem. Soc.*, **102**, 1216 (1980).
154. G. Levy, *Chem. Commun.*, 1257 (1969).
155. P. B. Sargeant and C. G. Krespan, *J. Amer. Chem. Soc.*, **91**, 415 (1969).
156. R. D. Chambers, A. Parkin, and R. S. Matthews, *JCS Perkin I*, 2107 (1976).
157. B. E. Smart and G. S. Reddy, *J. Amer. Chem. Soc.*, **98**, 5593 (1976).
158. T.-L. Ho and G. A. Olah, *Proc. Natl. Acad. Sci. U.S.A.*, **75**, 4 (1978).
159. I. G. Csizmadia, V. Lucchini and G. Modena, *Theoret. Chim. Acta*, **39**, 51 (1975).
160. P. E. Peterson, R. J. Bopp and M. M. Ajo, *J. Amer. Chem. Soc.*, **92**, 2834 (1970).
161. S. R. Hooley and D. L. H. Williams, *JCS Perkin II*. 1053 (1973).
162. E. D. Laganis and D. M. Lemal, *J. Amer. Chem. Soc.*, **102**, 6633 (1980).
163. S. Andreades, *J. Amer. Chem. Soc.*, **86**, 2003 (1964).
164. K. J. Klabunde and D. J. Burton, *J. Amer. Chem. Soc.*, **94**, 5985 (1972).
165. J. H. Sleigh, R. Stephens and J. C. Tatlow, *J. Fluorine Chem.*, **15**, 411 (1980).
166. A. Pross and L. Radom, *Tetrahedron*, **36**, 1999 (1980).
167. Y. Apeloig and Z. Rappoport, *J. Amer. Chem. Soc.*, **101**, 5095 (1979).
168. R. Hoffmann, L. Radom, J. A. Pople, P. von R. Schleyer, W. J. Hehre and L. Salem, *J. Amer. Chem. Soc.*, **94**, 6221 (1972).
169. J. C. Tatlow, *Chem. Ind.*, 522 (1978).
170. J. W. Burdon, D. W. Davies and G. delConde, *JCS Perkin II*, 1205 (1978).
171. J. Tyrell, V. M. Kolb and C. Y. Meyers, *J. Amer. Chem. Soc.*, **101**, 3497 (1979).
172. G. Baddeley, G. M. Bennett, S. Glasstone and B. Jones, *J. Chem. Soc.*, 1827 (1935).
173. R. D. Chambers, R. S. Matthews and G. Taylor, *JCS Perkin I*, 435 (1980).
174. M. A. Howells, R. D. Howells, N. C. Baenzinger and D. J. Burton, *J. Amer. Chem. Soc.*, **95**, 5366 (1973).
175. N. C. Baenzinger, B. A. Foster, M. A. Howells, R. D. Howells, P. D. Vander Valk and D. J. Burton, *Acta Crystallogr. B*, **33**, 2339 (1977).
176. R. D. Howells, P. D. Vander Valk and D. J. Burton, *J. Amer. Chem. Soc.*, **99**, 4830 (1977).
177. D. J. Burton, S. Shinya and R. D. Howells, *J. Amer. Chem. Soc.*, **101**, 3689 (1979).
178. D. J. Burton and Y. Inouye, *Tetrahedron Lett.*, 3397 (1979).
179. D. J. Burton and P. E. Greenlimb, *J. Org. Chem.*, **40**, 2797 (1975).
180. D. G. Naae, H. S. Kesling and D. J. Burton, *Tetrahedron Lett.*, 3789 (1975).
181. S. F. Campbell, R. Stephens, J. C. Tatlow and W. T. Westwood, *J. Fluorine Chem.*, **1**, 439 (1971/72).
182. P. Johncock, *J. Organometal. Chem.*, **19**, 257 (1969).
183. J. L. Adcock and E. R. Renk, *J. Org. Chem.*, **44**, 3431 (1979).
184. D. Seyferth, R. M. Simon, D. J. Sepelak and H. A. Klein, *J. Org. Chem.*, **45**, 2273 (1980).
185. W. R. Cullen, *Fluorine Chem. Revs*, **3**, 73 (1969).
186. H. Gilman, *J. Organometal. Chem.*, **100**, 83 (1975).
187. D. Denson, C. F. Smith and C. Tamborski, *J. Fluorine Chem.*, **3**, 73 (1973/74).
188. J. Hine, *Divalent Carbon*, Ronald Press, New York (1964), Chap. 3.
189. H. S. Kesling and D. J. Burton, *Tetrahedron Lett.*, 3355 (1975).
190. G. A. Wheaton and D. J. Burton, *J. Org. Chem.*, **43**, 2643 (1978).
191. W. Kimpenhaus and J. Buddrus, *Chem. Ber.*, **109**, 2370 (1976).
192. D. J. Burton and J. L. Hanfeld, *Fluorine Chem. Revs*, **8**, 119 (1977).
193. D. J. Burton and J. L. Hanfeld, *J. Org. Chem.*, **42**, 828 (1977).
194. T. Clark and P. von R. Schleyer, *JCS Chem. Commun.*, 883 (1979).
195. T. Clark and P. von R. Schleyer, *Tetrahedron Lett.*, 4963 (1979).
196. R. D. Chambers and T. Chivers, *Organometal. Chem. Revs*, **1**, 279 (1966).
197. J. L. Hanfeld and D. J. Burton, *Tetrahedron Lett.*, 773 (1975).
198. T. Koenig, T. Bell and W. Snell, *J. Amer. Chem. Soc.*, **97**, 662 (1975).
199. R. W. Fessenden and R. H. Schuler, *J. Phys. Chem.*, **43**, 2704 (1965).
200. F. Bernardi, W. Cherry, S. Sharts and N. D. Epiotis, *J. Amer. Chem. Soc.*, **100**, 1352 (1978).
201. R. V. Lloyd and M. T. Rogers, *J. Amer. Chem. Soc.*, **95**, 1512 (1973).
202. P. J. Krusic and R. C. Bingham, *J. Amer. Chem. Soc.*, **98**, 232 (1976).

203. T. Kawamura, M. Tsumara, Y. Yokomichi, and T. Yonezawa, *J. Amer. Chem. Soc.*, **99**, 8251 (1977).
204. G. Boche and D. R. Schneider, *Tetrahedron Lett.*, 2327 (1978).
205. H. M. Walborsky and P. C. Collins, *J. Org. Chem.*, **41**, 940 (1976).
206. I. Ishihara, *J. Org. Chem.*, **40**, 3264 (1975).
207. T. Ando, K. Wakabayashi, H. Yamanaka and W. Funasaka, *Bull. Chem. Soc. Japan*, **45**, 1576 (1972).
208. V. Malatesta, D. Forrest and K. U. Ingold, *J. Amer. Chem. Soc.*, **100**, 7073 (1978).
209. R. C. Bingham and M. J. S. Dewar, *J. Amer. Chem. Soc.*, **95**, 7180 (1973).
210. R. C. Bingham and M. J. S. Dewar, *J. Amer. Chem. Soc.*, **95**, 7182 (1973).
211. K. S. Chen, P. J. Krusic, P. Meakin and J. K. Kochi, *J. Phys. Chem.*, **78**, 2014 (1974).
212. D. J. Edge and J. K. Kochi, *J. Amer. Chem. Soc.*, **94**, 6485 (1972).
213. L. Kaplan, *Bridged Free Radicals*, Marcel Dekker, New York (1972).
214. P. J. Krusic, P. Meakin and B. E. Smart, *J. Amer. Chem. Soc.*, **96**, 6211 (1974).
215. B. E. Smart, P. J. Krusic, P. Meakin and R. C. Bingham, *J. Amer. Chem. Soc.*, **96**, 7382 (1974).
216. J. M. Tedder and J. C. Walton, *Adv. Phys. Org. Chem.*, **16**, 51 (1978).
217. G. A. Russel in *Free Radicals*, Vol. I (Ed. J. K. Kochi), Wiley, New York (1973), pp. 275–321.
218. M. L. Poutsma in *Free Radicals*, Vol. II (Ed. J. K. Kochi), Wiley, New York (1973), pp. 159–222.
219. W. R. Dolbier, Jr, C. A. Piedrahita and B. H. Al-Sader, *Tetrahedron Lett.*, 2957 (1979).
220. D. Feller, W. T. Borden and E. R. Davidson, *Chem. Phys. Lett.*, **71**, 22 (1980).
221. L. Pauling, *JCS Chem. Commun.*, 688 (1980).
222. W. W. Schoeller, *JCS Chem. Commun.*, 1241 (1980).
223. J. F. Harrison, R. C. Liedtke and J. F. Liebman, *J. Amer. Chem. Soc.*, **101**, 7162 (1979).
224. C. W. Bauschlicher, H. F. Schaefer III and P. J. Bagus, *J. Amer. Chem. Soc.*, **99**, 7106 (1977).
225. P. Gaspar and G. S. Hammond in *Carbenes*, Vol. II (Ed. R. A. Moss and M. Jones, Jr), Wiley, New York (1975), pp. 207–362.
226. R. A. Moss in *Carbenes*, Vol. I (Ed. R. A. Moss and M. Jones, Jr), Wiley, New York (1973), pp. 153–304.
227. D. Seyferth in *Carbenes,* Vol. II (Ed. R. A. Moss and M. Jones, Jr), Wiley, New York (1975), pp. 101–158.
228. C. G. Krespan and W. J. Middleton, *Fluorine Chem. Revs*, **5**, 57 (1971).
229. C. W. Bauschlicher, Jr, *J. Amer. Chem. Soc.*, **102**, 5492 (1980).
230. R. N. Haszeldine, R. Rowland, J. G. Speight and H. E. Tipping, *JCS Perkin I*, 1943 (1979).
231. R. N. Haszeldine, C. Parkinson, P. J. Robinson and W. J. Williams, *JCS Perkin II*, 954 (1979).
232. R. A. Moss, *Acc. Chem. Res.*, **13**, 58 (1980).
233. D. M. Gale, *J. Org. Chem.*, **33**, 2536 (1968).
234. W. Mahler and P. R. Resnick, *J. Fluorine Chem.*, **3**, 451 (1973/74).
235. B. Giese, W. -B. Lee and J. Meister, *Ann. Chem.*, 725 (1980).
236. L. F. Pelosi and W. T. Miller, *J. Amer. Chem. Soc.*, **98**, 4311 (1976).
237. M. W. Grayston and D. M. Lemal, *J. Amer. Chem. Soc.*, **98**, 1278 (1976).
238. D. M. Lemal, J. V. Staros and V. Austel, *J. Amer. Chem. Soc.*, **91**, 3374 (1969).
239. D. M. Lemal and L. H. Dunlap, Jr, *J. Amer. Chem. Soc.*, **94**, 6562 (1972).
240. A-M. M. Dabbagh, W. T. Flowers, R. H. Haszeldine and P. J. Robinson, *JCS Perkin II*, 1407 (1979).
241. W. J. Middleton, *J. Org. Chem.*, **34**, 3201 (1969).
242. W. Adam, J.-C. Liu and O. Rodriguez, *J. Org. Chem.*, **38**, 2270 (1973).
243. M. S. Raasch, *J. Org. Chem.*, **35**, 3475 (1970).
244. J. L. Hencher, Q. Shen and D. G. Tuck, *J. Amer. Chem. Soc.*, **98**, 899 (1976).
245. Y. Kobayashi, A. Ando, K. Kawada and I. Kumadaki, *J. Org. Chem.*, **45**, 2968 (1980).
246. H. A. Wiebe, S. Braslavsky and J. Heicklen, *Canad. J. Chem.*, **50**, 2721 (1972).
247. Y. Kobayashi, S. Fujino, H. Hamana, I. Kumadaki and Y. Hanzawa, *J. Amer. Chem. Soc.*, **99**, 8511 (1977).

248. Y. Kobayashi, S. Fujino, H. Hamana, Y. Hanzawa, S. Morita and I. Kumadaki, *J. Org. Chem.*, **45**, 4683 (1980).
249. M. G. Barlow, R. N. Haszeldine and J. G. Dingwall, *JCS Perkin I*, 1542 (1973).
250. D. C. England, *J. Org. Chem.*, **46**, 147 (1981).
251. A. Greenberg, J. F. Liebman and D. Van Vechten, *Tetrahedron*, **36**, 1161 (1980).
252. J. Banus, H. J. Emeleus and R. N. Haszeldine, *J. Chem. Soc.*, 60 (1951).
253. H. Teichmann, *Z. Chem.*, **14**, 216 (1974).
254. A. Hasegawa, M. Shiotani and F. Williams, *JCS Faraday Discuss.*, 157 (1978).
255. J. T. Wang and F. Williams, *J. Amer. Chem. Soc.*, **102**, 2860 (1980).
256. M. W. Parley, D. G. Brown and J. M. Wood, *Biochemistry*, **9**, 4320 (1970).
257. A. van den Bergen, K. J. Murray and B. O. West, *J. Organometal. Chem.*, **33**, 89 (1971).
258. A. M. van den Bergen and B. O. West, *J. Organometal. Chem.*, **92**, 55 (1975).
259. C. G. Krespan, *J. Org. Chem.*, **43**, 637 (1978).
260. P. Tarrant, G. Allison, K. P. Barthold and E. C. Stump. Jr, *Fluorine Chem. Revs*, **5**, 77 (1971).
261. R. C. Kennedy and J. B. Levy, *J. Fluorine Chem.*, **7**, 101 (1976).
262. B. A. Bekker, G. V. Asratyan, B. L. Dyatkin and I. L. Knunyants, *Tetrahedron*, **30**, 3539 (1974).
263. J. T. Hill, *J. Fluorine Chem.*, **9**, 97 (1977).
264. R. O'B. Watts, C. G. Allison, K. P. Barthold and P. Tarrant, *J. Fluorine Chem.*, **3**, 7 (1973/74).
265. A. Sekiya and D. D. Des Marteau, *J. Org. Chem.*, **44**, 1131 (1979).
266. R. A. Bekker, G. V. Asratyan, B. L. Dyatkin and I. L. Knunyants, *Dokl. Akad. Nauk SSSR Engl. Trans.*, **204**, 439 (1972).
267. R. A. Bekker, G. V. Asratyan and B. L. Dyatkin, *Zhur. Organ. Khim. Engl. Trans.*, **9**, 1663 (1973).
268. R. A. Bekker, G. V. Asratyan and B. L. Dyatkin, *Zhur. Organ. Khim. Engl. Trans.*, **9**, 1658 (1973).
269. R. E. Parker, *Adv. Fluorine Chem.*, **3**, 63 (1963).
270. M. J. Pellerite and J. I. Brauman, *J. Amer. Chem. Soc.*, **102**, 5993 (1980).
271. R. D. Chambers, A. A. Lindley, P. B. Philpot, H. C. Fielding, J. Hutchinson and G. Whittaker, *JCS Perkin I*, 214 (1979).
272. S. Bartlett, R. D. Chambers and N. M. Kelly, *Tetrahedron Lett.*, 1891 (1980).
273. R. D. Chambers, A. A. Lindley, P. D. Philpot, H. C. Fielding, J. Hutchinson and G. Whittaker, *JCS Chem. Commun.*, 431 (1978).
274. K. V. Scherer, Jr and K. Yamanouchi, Abstracts of the Division of Fluorine Chemistry, Second Chemical Congress of the North American Continent, Las Vegas, Nevada, August, 1980, p. 3.
275. A. K. Barbour, L. J. Belf and M. W. Buxton, *Adv. Fluorine Chem.*, **3**, 181 (1963).
276. J. M. Hamilton, *Adv. Fluorine Chem.*, **3**, 117 (1963).
277. K. Okuhara, *J. Org. Chem.*, **43**, 2745 (1978).
278. R. F. Merritt, *J. Amer. Chem. Soc.*, **89**, 609 (1967).
279. J. D. Park and T. S. Croft, *J. Org. Chem.*, **38**, 4026 (1973).
280. S. I. Miller, *Tetrahedron*, **33**, 1211 (1977).
281. R. D. Chambers, A. A. Lindley and H. C. Fielding, *J. Fluorine Chem.*, **12**, 85 (1978).
282. H. H. Evans, R. Fields, R. N. Haszeldine and M. Illingworth, *JCS Perkin I*, 649 (1973).
283. J. D. Park and E. W. Cook, *Tetrahedron Lett.*, 4853 (1965).
284. D. J. Burton and H. C. Krutzsch, *J. Org. Chem.*, **36**, 2351 (1971).
285. J.-C. Chalchat and F. Theron, *Bull. Soc. Chim. Fr.*, 3361 (1973).
286. J. Normant, R. Sauvetre and J. Villieras, *Tetrahedron*, **31**, 891, 897 (1975).
287. Z. Rappoport, *Adv. Phys. Org. Chem.*, **7**, 1 (1969).
288. F. Texier, O. Henri-Rousseau and J. Bourgois, *Bull. Soc. Chim. Fr. Ser. II*, 86 (1979).
289. R. D. Chambers and R. H. Mobbs, *Adv. Fluorine Chem.*, **4**, 50 (1965).
290. J. D. Park, R. J. McMurtry and J. H. Adams, *Fluorine Chem. Revs*, **2**, 55 (1968).
291. C. A. Franz and D. J. Burton, *J. Org. Chem.*, **40**, 2791 (1975).
292. R. D. Chambers, A. A. Lindley and H. C. Fielding, *J. Fluorine Chem.*, **12**, 85 (1978).
293. R. N. Haszeldine, I-ud-D. Mir and A. E. Tipping, *JCS Perkin II*, 565 (1979).
294. S. Bartlett, R. D. Chambers, A. A. Lindley and H. C. Fielding, *JCS Perkin I*, 1551 (1980).

295. W. D. Dmowski, *J. Fluorine Chem.*, **15**, 299 (1980).
296. M. I. Bruce and W. R. Cullen, *Fluorine Chem. Revs*, **4**, 79 (1969).
297. B. L. Dyatkin, E. P. Mochalina and I. L. Knunyants, *Fluorine Chem. Revs*, **3**, 45 (1969).
298. M. Suda, *Tetrahedron Lett.*, **21**, 255 (1980).
299. I. L. Knunyants and G. A. Sokolski, *Angew. Chem. Int. Edn Engl.*, **11**, 583 (1972).
300. O. Paleta, *Fluorine Chem. Revs*, **8**, 39 (1977).
301. G. G. Belen'kii, E. P. Lur'e and L. S. German, *Izv. Akad. Nauk SSSR, Ser. Khim. Engl. Trans.*, 2208 (1976).
302. T. I. Filyakova, G. G. Belen'kii, E. P. Lur'e, A. Ya. Zapevalov, I. P. Kolenko and L. S. German, *Izv. Akad. Nauk SSSR, Ser. Khim. Engl. Trans.*, 635 (1979).
303. J. M. Tedder and J. C. Walton, *Tetrahedron*, **36**, 70 (1980).
304. *Fluoropolymers*, (Ed. L. A. Wall), Wiley, New York (1972).
305. W. T. Miller, J. H. Fried and H. Goldwhite, *J. Amer. Chem. Soc.*, **82**, 3901 (1960).
306. W. T. Miller, W. Frass and P. R. Resnick, *J. Amer. Chem. Soc.*, **83**, 1767 (1961).
307. D. P. Graham and W. B. McCormick, *J. Org. Chem.*, **31**, 958 (1966).
308. N. I. Delyagina, E. Ya. Pervova and I. L. Knunyants, *Izv. Akad. Nauk SSSR, Ser. Khim. Engl. Trans.*, 326 (1972).
309. B. L. Dyatkin, S. R. Sterlin, L. G. Zhuravkova, B. I. Martynov, E. I. Mysov and I. L. Knunyants, *Tetrahedron*, **29**, 2759 (1973).
310. B. L. Dyatkin, B. I. Martynov, L. G. Martynova, N. G. Kisim, S. R. Sterlin, Z. A. Stambrevichute and L. A. Fedorov, *J. Organometal. Chem.*, **57**, 423 (1973).
311. K. N. Makarov, L. L. Gervits and I. L. Knunyants, *J. Fluorine Chem.*, **10**, 157 (1977).
312. J. A. Young and M. H. Bennett, *J. Org. Chem.*, **42**, 4055 (1977).
313. N. I. Delyagina, B. L. Dyatkin, I. L. Knunyants, N. N. Bubnov and B. Ya. Medvedev, *JCS Chem. Commun.*, 456 (1973).
314. D. J. Burton and F. Herkes, *J. Org. Chem.*, **33**, 1854 (1968).
315. R. E. Banks, A. Braithwaite, R. N. Haszeldine and D. R. Taylor, *J. Chem. Soc.*, 454 (1969).
316. P. W. L. Bosbury, R. Fields and R. N. Haszeldine, *JCS Perkin I*, 422 (1978).
317. J. A. Young, *Fluorine Chem. Revs*, **1**, 2 (1967).
318. S. P. von Halasz, F. Kluge and T. Martini, *Chem. Ber.*, **106**, 2950 (1973).
319. T. Martini and S. P. von Halasz, *Tetrahedron Lett.*, 2129 (1979).
320. W. Dmowski, W. T. Flowers and R. N. Haszeldine, *J. Fluorine Chem.*, **9**, 94 (1977).
321. R. D. Chambers, G. Taylor and R. L. Powell, *JCS Perkin I*, 426, 429 (1980).
322. S. L. Bell, R. D. Chambers, M. Y. Gribble and J. R. Maslakiewicz, *JCS Perkin I*, 1716 (1973).
323. R. D. Chambers and M. Y. Gribble, *JCS Perkin I*, 1405 (1973).
324. C. J. Drayton, W. T. Flowers and R. N. Haszeldine, *JCS Perkin I*, 1029 (1975).
325. R. D. Chambers, S. Partington and D. B. Speight, *JCS Perkin I*, 2673 (1974).
326. W. T. Miller, R. J. Hummel and L. F. Pelosi, *J. Amer. Chem. Soc.*, **95**, 6850 (1973).
327. T. S. Balley and S. Masamune, *Tetrahedron*, **36**, 343 (1980).
328. Y. Kobayashi, I. Kumadaki, A. Ohsawa, Y. Hanzawa and M. Honda, *Tetrahedron Lett.*, 300 (1975).
329. S. Masamune, T. Machiguchi and M. Aratani, *J. Amer. Chem. Soc.*, **99**, 3524 (1977).
330. M. J. Gerace, D. M. Lemal and H. Erth, *J. Amer. Chem. Soc.*, **97**, 5584 (1975).
331. M. Grayston, A. C. Barefoot III, W. D. Saunders and D. M. Lemal, Abstracts of the Fourth Winter Fluorine Conference, Daytona Beach, 28 Jan.–2 Feb. 1979, p. 4.
332. A. C. Barefoot III, W. D. Saunders, J. M. Buzby, M. W. Grayston and D. M. Lemal, *J. Org. Chem.*, **45**, 4292 (1980).
333. M. W. Grayston, W. D. Saunders and D. M. Lemal, *J. Amer. Chem. Soc.*, **102**, 413 (1980).
334. M. G. Barlow, M. W. Crawley and R. N. Haszeldine, *JCS Perkin I*, 122 (1980).
335. W. H. Sharkey, *Fluorine Chem. Revs*, **2**, 1 (1969).
336. P. D. Bartlett, *Science*, **159**, 833 (1968).
337. P. D. Bartlett, *Quart. Revs*, **24**, 473 (1970).
338. P. D. Bartlett and J. J.-B. Mallett, *J. Amer. Chem. Soc.*, **98**, 143 (1976) and references therein.
339. D. R. A. Perry, *Fluorine Chem. Revs*, **1**, 253 (1967).

340. B. E. Smart, *J. Amer. Chem. Soc.*, **96**, 929 (1974).
341. W. R. Dolbier, Jr, C. A. Piedrahita, K. N. Houk, R. W. Strozier and R. W. Gandour, *Tetrahedron Lett.*, 2231 (1978).
342. L. N. Domelsmith, K. N. Houk, C. Piedrahita and W. R. Dolbier, Jr, *J. Amer. Chem. Soc.*, **100**, 6908 (1978).
343. W. R. Dolbiėr, Jr and C. R. Burkholder, *Tetrahedron Lett.*, **21**, 785 (1980).
344. J. C. Blazejewski, D. Cantacuzène and C. Wakselman, *Tetrahedron Lett.*, 363 (1975).
345. P. G. Gassman, S. R. Korn, T. F. Bailey and T. H. Johnson, *Tetrahedron Lett.*, 3401 (1979).
346. H.-A. Chia, B. E. Kirk and D. R. Taylor, *JCS Perkin I*, 1209 (1974).
347. B. E. Kirk and D. R. Taylor, *JCS Perkin I*, 1844 (1974).
348. J. C. Blazejewski, D. Cantacuzène and C. Wakselman, *Tetrahedron Lett.*, 2055 (1974).
349. J. H. Atherton and R. Fields, *J. Chem. Soc. C*, 1507 (1968).
350. W. Carpenter, A. Haymaker and D. W. Moore, *J. Org. Chem.*, **31**, 789 (1966).
351. I. L. Knunyants, E. G. Bykhovskaya, V. N. Frosin, I. V. Galakhov and L. I. Regulin, *Zhur. Vsesoyuz Khim. obshch. D. T. Mendeleeva*, **17**, 356 (1972).
352. K. Tada and F. Toda, *Tetrahedron Lett.*, 563 (1978).
353. R. E. Banks, R. N. Haszeldine and J. M. Robinson, *JCS Perkin I*, 1226 (1976).
354. J. Galluci, M. LeBlanc and J. G. Reiss, *J. Chem. Res. (S)*, 192 (1978).
355. M. G. Barlow, R. N. Haszeldine, W. D. Morton and D. R. Woodward, *JCS Perkin I*, 1978 (1973).
356. M. G. Barlow, G. M. Harrison, R. N. Haszeldine, R. Hubbard, M. J. Kershaw and D. R. Woodward, *JCS Perkin I*, 2010 (1975).
357. H. Durr and R. Segio, *Chem. Ber.*, **107**, 2207 (1974).
358. H. Martin and M. Regitz, *Ann. Chem.*, 1702 (1974).
359. Y. Kobayashi, I. Kunadaki and S. Fujino, *Heterocycles*, **7**, 871 (1977).
360. M. M. Boudakian in *Kirk–Othmer's Encyclopedia of Chemical Technology*, 3rd edn, Vol. 10, Wiley, New York (1980), p. 901.
361. L. S. Kobrina, *Fluorine Chem. Revs*, **7**, 1 (1974).
362. G. G. Yakobson, T. D. Petrova and L. S. Kobrina, *Fluorine Chem. Revs*, **7**, 115 (1974).
363. R. Filler, *Fluorine Chem. Revs*, **8**, 1 (1977).
364. J. Burdon, *Tetrahedron*, **21**, 3373 (1965).
365. J. Burdon and I. W. Parsons, *J. Amer. Chem. Soc.*, **99**, 7445 (1977).
366. R. D. Chambers, D. Close and D. L. H. Williams, *JCS Perkin II*, 778 (1980).
367. R. D. Chambers, D. Close, W. K. R. Musgrave, R. S. Waterhouse and D. L. H. Williams, *JCS Perkin II*, 1774 (1977).
368. R. D. Chambers, J. S. Waterhouse and D. L. H. Williams, *JCS Perkin II*, 585 (1977).
369. V. D. Shteingarts, Yu. V. Pozdnyakovich and G. G. Yakobson, *Chem. Commun.*, 1264 (1969).
370. N. M. Bazin, N. E. Akhmetova, L. V. Orlova, V. D. Shteingarts, L. N. Shchegoleva and G. G. Yakobson, *Tetrahedron Lett.*, 4449 (1968).
371. J. Burdon and I. W. Parsons, *Tetrahedron*, **31**, 2401 (1975).
372. R. D. Chambers, C. R. Sargent and M. J. Silvester, *J. Fluorine Chem.*, **15**, 257 (1980).
373. J. Devynck and A. Ben Hadid, *Compt. Rend. C*, **286**, 389 (1979).
374. V. D. Shteingarts, G. G. Yakobson and N. N. Vorozhtsov, *Dokl. Akad. Nauk SSSR Engl. Trans.*, **70**, 1348 (1966).
375. A. G. Budnick, T. V. Senchenko and V. D. Shteingarts, *Zhur. Org. Khim. Engl. Trans.*, **10**, 344 (1974).
376. O. I. Osina and V. D. Shteingarts, *Zhur. Org. Khim. Engl. Trans.*, **10**, 329 (1974).
377. M. Zupan and B. Šket, *Israel J. Chem.*, 17 (1978).
378. G. Kaupp, *Angew. Chem. Int. Edn Engl.*, **19**, 243 (1980).
379. D. Bryce-Smith and A. Gilbert, *Tetrahedron*, **32**, 1309 (1976).
380. W. A. Noyes and K. E. Al-Ani, Jr, *Chem. Revs*, **74**, 29 (1974).
381. L. T. Scott and M. Jones, Jr, *Chem. Revs*, **72**, 181 (1972).
382. D. Bryce-Smith, A. Gilbert, B. H. Orger and P. J. Twitchett, *JCS Perkin I*, 232 (1978).
383. B. Šket and M. Zupan, *JCS Chem. Commun.*, 1053 (1976).
384. B. Šket and M. Zupan, *JCS Chem. Commun.*, 365 (1977).

385. D. M. Lemal, J. M. Buzby, A. C. Barefoot III, M. W. Grayston and E. D. Laganis, *J. Org. Chem.*, **45**, 3118 (1980).
386. M. G. Barlow, D. E. Brown and R. N. Haszeldine, *JCS Perkin I*, 363 (1978).
387. M. G. Barlow, D. E. Brown and R. N. Haszeldine, *JCS Perkin I*, 129 (1980).
388. R. D. Chambers and R. Middleton, *JCS Chem. Commun.*, 154 (1977).
389. E. Ratajczak, B. Sztuba and D. Price, *J. Photochem.*, **13**, 233 (1980).
390. R. D. Chambers, C. R. Sargent and M. C. Clark, *JCS Chem. Commun.*, 445 (1979).
391. Y. Kobayashi, I. Kumadaki, A. Ohsawa and S. Y. Sekine, *Tetrahedron Lett.*, 1639 (1975).
392. J. A. Ross, R. P. Seiders and D. M. Lemal, *J. Amer. Chem. Soc.*, **98**, 4325 (1976).
393. Y. Kobayashi, I. Kumadaki, A. Ohsawa, Y. Sekine and A. Ando, *JCS Perkin I*, 2355 (1977).
394. Y. Kobayashi, A. Ando, K. Kawada, A. Ohsawa and I. Kumadaki, *J. Org. Chem.*, **45**, 2962 (1980).
395. C. H. Bushweller, J. A. Ross and D. M. Lemal, *J. Amer. Chem. Soc.*, **99**, 629 (1977).
396. J. P. Snyder and T. A. Halgren, *J. Amer. Chem. Soc.*, **102**, 2861 (1980).
397. Y. Kobayashi, Y. Hanzawa, Y. Nakanishi and T. Kashiwagi, *Tetrahedron Lett.*, 1019 (1978).
398. Y. Kobayashi, Y. Hanzawa, W. Mityashita, T. Kashiwagi, T. Nakano and I. Kumadaki, *J. Amer. Chem. Soc.*, **101**, 6445 (1979).

The Chemistry of Functional Groups, Supplement D
Edited by S. Patai and Z. Rappoport
© 1983 John Wiley & Sons Ltd

CHAPTER **15**

Xenon halide halogenations

MARKO ZUPAN
Department of Chemistry and 'Jožef Stefan' Institute, 'E. Kardelj' University of Ljubljana, Ljubljana, Yugoslavia

I. INTRODUCTION

Preparation of fluoro-substituted organic molecules represents a different problem from that of other halogen derivatives. The special reaction techniques required have been reviewed in depth several times[1]. In the last few years extensive work has been done in order to find a fluorinating agent which could be a source of 'positive' fluorine[1].

The most convenient reagent for direct fluorination of organic compounds seemed to be elemental fluorine. The low dissociation energy of the F—F bond on the one hand, and the high formation energy of the C—F bond on the other, presented a problem which was partially experimentally resolved by the use of low temperature techniques[1].

Another convenient group of reagents could be the xenon fluorides, which were discovered in 1962[2]. Xenon difluoride is a stable compound which can be easily prepared by a photosynthetic[3] or thermal reaction from fluorine and xenon, and is commercially available as well. Xenon difluoride reacts with water very slowly to give xenon, oxygen and hydrogen fluoride. With anhydrous hydrogen fluoride it forms a stable solution, and with trifluoroacetic acid xenon-fluoro-trifluoroacetates and xenon-di (trifluoroacetates), while with iodo-substituted molecules it leads to iodo(III)difluorides. On the other hand, xenon tetrafluoride is converted into highly shock-sensitive xenon oxides with moisture, while xenon hexafluoride reacts even with quartz or pyrex, also forming xenon oxides. All the necessary precautions given in the literature should be taken into account[4].

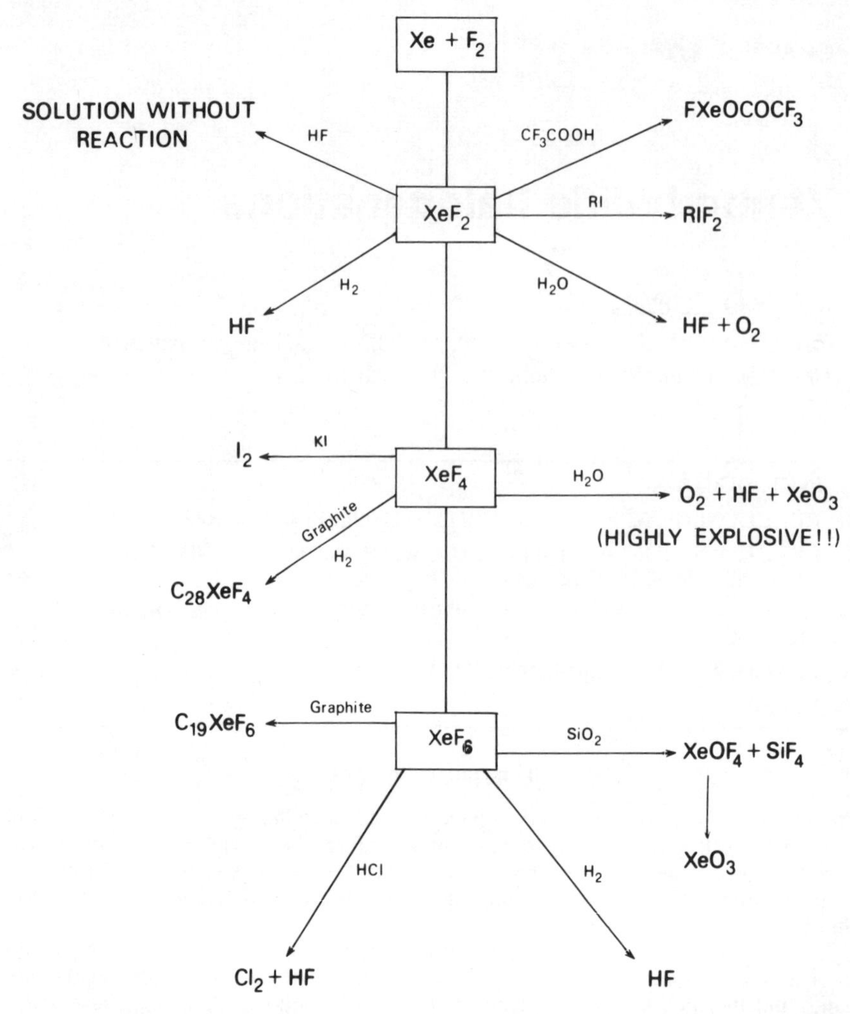

Indeed, xenon difluoride prepared by the thermal reaction may be contaminated with small amounts of xenon tetrafluoride and should also be handled with caution. However, photochemically prepared xenon difluoride can be handled in the same way as other halogenating agents and without the need for moisture exclusion. The usefulness of xenon tetrafluoride and xenon hexafluoride for the fluorination of organic molecules has been increased by the discovery that they form stable intercalates with graphite[5–7].

The existence of $XeCl_2$, $XeBr_2$ and $XeCl_4$ has been spectroscopically established, but they are unstable under ordinary conditions[2,8,9].

II. REACTIONS OF XENON DIFLUORIDE WITH OLEFINS AND ACETYLENES

In 1964 Chernick and coworkers[10] first reported the reactions of xenon difluoride and xenon tetrafluoride with alkenes. They used vacuum techniques and mixed xenon fluorides with alkenes; the reaction mixture was allowed to react at room temperature for 1–12 days. Besides vicinal difluorides, rearranged geminal difluorides were also isolated. The authors suggested a free radical mechanism for fluorination[11].

$$CH_3CH{=}CH_2 \xrightarrow[\text{12 days}]{XeF_2} \underset{(24\%)}{CH_3CHFCH_3} + \underset{(9\%)}{CHF_2CH_3} + \underset{(46\%)}{CHF_2CH_2CH_3} + \underset{(12\%)}{CH_2FCHFCH_3}$$

Nine years later, Zupan and Pollak[12] found that under very simplified conditions, e.g. weighing xenon difluoride in a glass vessel, dissolution in methylene chloride in an open Kel-F tube, etc., xenon difluoride at room temperature readily reacted with phenyl-substituted olefins in the presence of a catalyst, i.e. hydrogen fluoride or trifluoroacetic acid, thus forming vicinal difluorides in high yield. In some cases, the formation of fluoro-trifluoroacetates was also observed in the trifluoroacetic acid-catalysed reaction[12–15].

$$PhR^1C{=}CR^2R^3 \xrightarrow[HF/CH_2Cl_2]{XeF_2} PhCR^1FCR^2R^3F$$

$R^1 = Ph$, $R^2 = R^3 = H$ $\qquad$ $R^1 = R^2 = R^3 = H$
$R^1 = Ph$, $R^2 = CH_3$, $R^3 = H$ $\qquad$ $R^1 = CH_3$, $R^2 = R^3 = H$
$R^1 = Ph$, $R^2 = Ph$, $R^3 = H$

Acid-catalysed liquid-phase fluorine addition with xenon difluoride to a series of *cis* and *trans* phenyl-substituted olefins, e.g. 1-phenylpropene, stilbene and 1-phenyl-2-*t*-butylethylene, resulted in the formation of vicinal difluorides in high yield[13,16] (Scheme 1). The results with olefins as a whole show that the addition of fluorine is clearly a non-stereospecific process. On the basis of the ratios of D,L-*erythro* and D,L-*threo* difluorides being nearly independent of the starting olefin, the absence of free radical inhibition effects (molecular oxygen) on the product distribution, and the fact that the reactions are very strongly catalysed by hydrogen fluoride, the formation of *β*-fluorocarbonium ions was suggested. In the *trans* series of olefins, *anti*-addition of fluorine predominates, while in the *cis* series formation of a mixture of the same composition as in the case of the *trans* series was explained by free rotation about the newly formed C—C single bond in the *β*-fluorocarbonium ion. Indene, which might be considered as a 'cyclic analogue' of *cis*-1-phenylpropene, adds fluorine preferentially *anti*, but not *syn*, as was observed in the fluorination of *cis*-1-phenylpropene[16].

Indene $\xrightarrow[HF]{XeF_2}$ *trans*-1,2-difluoroindane (70%) + *cis*-1,2-difluoroindane (30%)

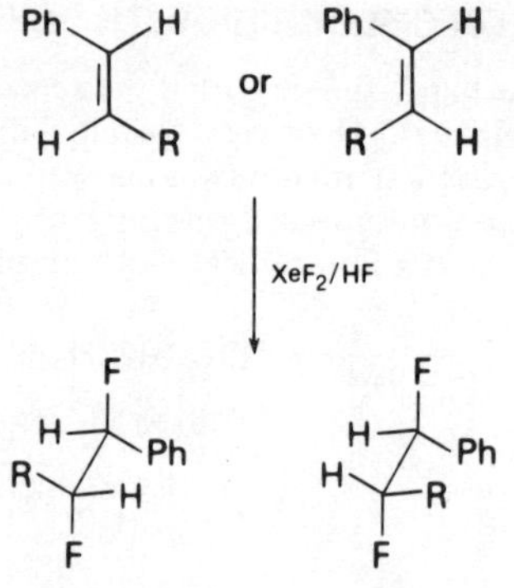

Olefin	R	Relative yields, % D,L-*erythro* **1**	Ratio **1/2**
Trans	CH_3	60	1.50
	Ph	62	1.67
	$C(CH_3)_3$	63	1.70
Cis	CH_3	64	1.78
	Ph	53	1.13
	$C(CH_3)_3$	64	1.78

SCHEME 1. Product distribution in fluorination of substituted 1-phenylethylenes with xenon difluoride in methylene chloride at 25°C[16]

The acid-catalysed liquid-phase fluorine addition to acenaphthylene and 1,2-dihydronaphthalene also resulted in the formation of vicinal difluorides in high yield[17]. The addition again proceeds predominantly via *anti* attack. The reaction with 1,4-dihydronaphthalene produced a complex mixture of naphthalene, 1-fluoronaphthalene, 2-fluoronaphthalene and 2-fluoro-1,2,3,4-tetrahydronaphthalene[17].

The stereochemistry of fluorine addition with xenon difluoride to phenyl-substituted cycloalkenes, e.g. 1-phenylcyclopentene, 1-phenylcyclohexene and 1-phenylcycloheptene, depends on ring magnitude[18]. In the cyclopentene ring there is

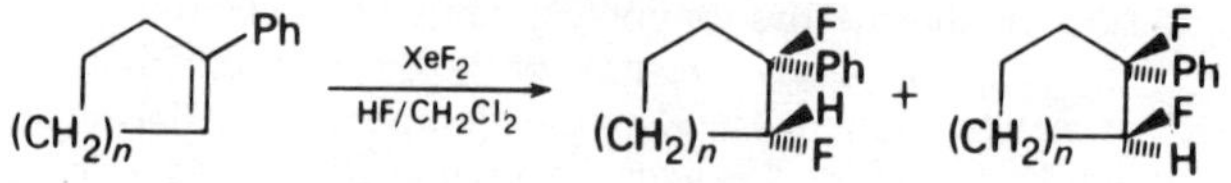

n	Relative yields, %[18] **3**	**4**
1	79	21
2	50	50
3	35	65

preferential *anti*-addition, while in the case of the cycloheptene ring the formation of *syn* difluorides predominates. The stereochemistry of fluorine addition to aryl-substituted cyclohexenes also depends on the substituent on the phenyl ring.

XeF_2/HF, CH_2Cl_2

X = H, *m*-Cl, *p*-OCH_3

The reaction with norbornene has been used as a mechanistic probe to elucidate the reaction mechanism and the stereochemistry of acid-catalysed liquid-phase fluorination with xenon difluoride[19]. The reaction resulted in the formation of seven products: fluoronortricyclane (**5**), 2-*endo*-3-*exo*-difluoronorbornane (**6**), 2-*exo*-7-*anti*-difluoronorbornane (**7**), 2-*endo*-5-*exo*-difluoronorbornane (**8**), 2-*exo*-5-*exo*-difluoronorbornane (**9**), 2-exo-3-*exo*-difluoronorbornane (**10**) and 2-*exo*-7-*syn*-difluoronorbornane (**11**). The product distribution as a function of solvent polarity, concentration of norbornene and the presence of oxygen as an inhibitor of carbon radicals is presented in Table 1. The formation of 2-*endo*-3-*exo*-difluoronorbornane (**6**) and 2-*exo*-3-*exo*-difluoronorbornane (**10**), depending significantly upon the conditions mentioned above, decreases in the presence of oxygen to one-half.

XeF_2

(**5**) (**6**) (**7**)

(**8**) (**9**) (**10**) (**11**)

TABLE 1. Variation in composition of products from addition of xenon difluoride to norbornene at 25 °C[19]

Concentration of norbornene, mg/ml	Solvent	Reaction times, min	Composition, %						
			5	**6**	**7**	**8**	**9**	**10**	**11**
200	CCl_4	120	38	2.5	16	17.5	12	1.5	8
200	$CHCl_3$	60	46	14	13	10.5	5	6.5	4
200	CH_2Cl_2	60	41.5	10.5	16.5	13.5	6	5.5	5
75	CH_2Cl_2	60	37.5	6	18	19	10	4	5.5
75	CH_2Cl_2	3	44	6	17	15.5	9	4	4.5
75	CH_2Cl_2/O_2	60	37.5	3.5	20	20	10	2	6
37.5	CH_2Cl_2	60	38	3.5	19.5	20	11	2	6
25	CH_2Cl_2	60	37	1.5	20	23	12	0.5	6

The fluorination of benzonorbornadiene resulted in the formation of 2-*exo*-7-*syn*-difluorobenzonorbornane, while the fluorination of norbornadiene[19,68] gave 3-*endo*-5-*exo*-difluoronortricyclane, 3-*exo*-5-*exo*-difluoronortricyclane and 2-*exo*-7-*syn*-difluoronorborn-5-ene.

$$\text{norbornadiene} \xrightarrow{XeF_2} \text{3-endo-5-exo-difluoronortricyclane} + \text{3-exo-5-exo-difluoronortricyclane} + \text{2-exo-7-syn-difluoronorborn-5-ene}$$

In the hydrogen fluoride-catalysed liquid-phase fluorine addition with xenon difluoride to 1,1-diphenyl-2-haloethylenes (X = F, Cl, Br) vicinal difluorides were formed[20]. The presence of chlorine or bromine in the olefinic molecule resulted in its decreased reactivity. *trans*-α-Fluorostilbene and α,β-difluorostilbene readily reacted with xenon difluoride, forming trifluoro and tetrafluoro products. *Cis*- and *trans*-α-chloro- or -bromostilbene were converted with XeF_2 to vicinal difluorides and predominance of *syn* addition of fluorine over *anti* was observed. Fluorine addition to *cis*- and *trans*-α-bromostilbene is also accompanied by rearrangements and the production of 1,1-difluoro-2-bromo-1,2-diphenylethane was established[20].

$$Ph_2C{=}CHX \xrightarrow[HF/CH_2Cl_2]{XeF_2} Ph_2CFCHXF$$

$$X = H,\ F,\ Cl,\ Br$$

HF-catalysed fluorination of cyclic enol acetates with xenon difluoride at room temperature resulted in the formation of α-fluorocycloalkanones[29]. The purity of the products and the yields strongly depend on moisture exclusion and on the amount of hydrogen fluoride. In these reactions only catalytic amounts of hydrogen fluoride were used; increasing the amount of hydrogen fluoride results in increasing formation of cycloalkanone. α-Fluorocycloalkanones were formed in yields of 60–85% and were accompanied by 2–17% of cycloalkanone, depending on the reaction conditions and ring magnitude (in the case of larger rings, greater amounts of ketone were observed).

$$\text{1-}OCOCH_3\text{-cycloalkene }(CH_2)_n \xrightarrow[HF]{XeF_2} \text{2-fluorocycloalkanone }(CH_2)_n$$

$$n = 1, 2, 3$$

For acid-catalysed liquid-phase fluorination of various alkenes with xenon difluoride, the mechanism presented in Scheme 2 was suggested[15,16,19]. It involves catalysis with hydrogen fluoride; xenon difluoride behaves as an electrophile in the presence of hydrogen fluoride. In the next step a π-complex is probably formed between this electrophilic species and the olefin, which could be transformed by heterolytic Xe—F bond cleavage into an open β-fluorocarbonium ion intermediate (path A), which can undergo Meerwein–Wagner and hydride rearrangements (in the case of bicyclic olefins), thus forming difluorides after the preferential *anti* attack of a fluorine anion. Another possible explanation put forward for the formation of carbonium ion intermediates was the formation of an ion radical (path B),

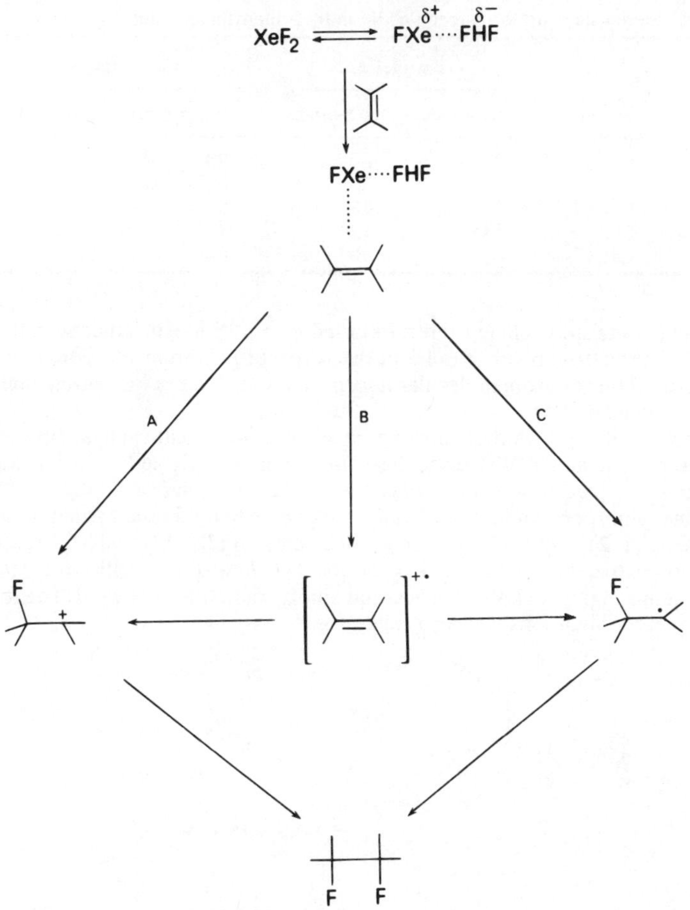

$XeF_2 \rightleftharpoons \overset{\delta+}{FXe}\cdots\overset{\delta-}{FHF}$

FXe····FHF

A B C

F F

SCHEME 2

transforming in the next step by $XeF^{\cdot}$ or XeF_2 to a carbonium ion or to radical species. The formation of a free radical species via homolytic Xe—F bond cleavage, or from an ion radical, was suggested in order to explain the formation of two difluoronorbornanes (**6**, **10**), their amounts depending on the concentration of norbornene and on the presence of oxygen. However, the main intermediates formed in HF-catalysed liquid-phase fluorination of various alkenes with xenon difluoride were of carbonium ion type, but in some cases some free radical intermediates were also present.

Stereochemical results of fluorine addition to phenyl-substituted olefins, and the results obtained in other halogen additions to the same system, are compared in Table 2. The stereochemistry of addition depends on the nature of the reagent. The reasons for these changes are as follows. Since fluorine and chlorine are poor neighbouring atoms, electrophilic additions proceed via open carbonium ions and are non-stereospecific. The tendency toward *syn* addition observed with these reagents is the logical consequence of ion pairing phenomena in non-polar solvents. From the

TABLE 2. Stereochemistry of halogen addition to 1-phenylprop-1-ene

Halogen	Solvent	*Trans* olefin, %		*Cis* olefin, %		Ref.
		D,L-*erythro*	D,L-*threo*	D,L-*erythro*	D,L-*threo*	
F_2	CCl_3F	31	69	78	22	30
Cl_2	CCl_4	38	46	62	29	31
	CH_2Cl_2	55	28	62	22	31
Br_2	CCl_4	88	12	17	83	32
XeF_2	CH_2Cl_2	60	40	64	36	16

comparison of the stereochemical results of halogen addition, it can be seen that xenon difluoride fluorination is very similar in this respect to chlorine addition in methylene chloride, but different from molecular fluorine addition, where ion pairing phenomena are more dominant.

The effect of the catalyst on the mechanism of acid-catalysed liquid-phase fluorination with xenon difluoride has also been investigated[12–15]. Trifluoroacetic acid-catalysed reaction of phenyl-substituted olefins, e.g. *cis* and *trans*-1-phenylpropene and *cis*- and *trans*-stilbene, resulted in the formation of vicinal difluorides **(1, 2)** and fluoro-trifluoroacetates **(12, 13)**. The reaction is non-stereospecific, but D,L-*erythro*- **(12)** and D,L-*threo*-fluoro-trifluoroacetates **(13)** are formed in a highly regiospecific Markownikoff manner in 50% yield. The formation of β-fluorocarbonium ions was also suggested.

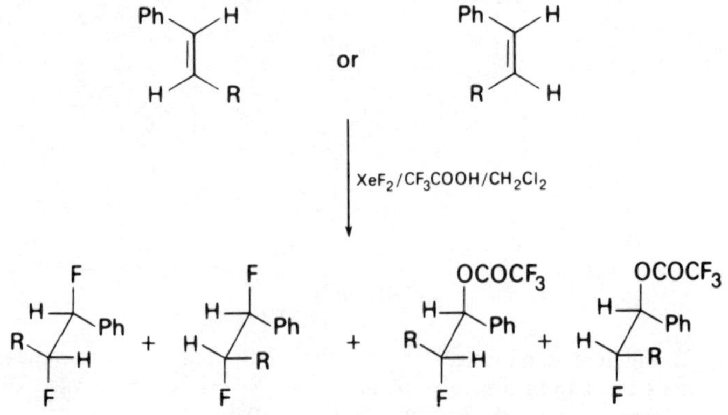

Ph(H)C=C(H)R or Ph(H)C=C(R)H

$XeF_2/CF_3COOH/CH_2Cl_2$

(1) + **(2)** + **(12)** + **(13)**

R = CH_3. Ph

The structure of the organic molecule has an important role in the nature of Xe—F bond cleavage, as was shown in the fluorination of styrene in the presence of trifluoroacetic acid[21]. Room temperature fluorination gave five products:

$$\text{PhCH=CH}_2 \xrightarrow[\text{CF}_3\text{COOH/CH}_2\text{Cl}_2]{\text{XeF}_2} \text{PhCHFCH}_2\text{CF}_3 + \text{PhCHFCH}_2\text{F} + \text{PhCH(OCOCF}_3\text{)CH}_2\text{CF}_3 + \text{PhCH(OCOCF}_3\text{)CH}_2\text{F} + \text{PhCHFCH}_2\text{OCOCF}_3$$

It is known that XeF_2 readily reacts with trifluoroacetic acid giving xenon fluoride trifluoroacetate[33,34], a reagent of electrophilic character, which can react with organic molecules[15]. However, xenon fluoride trifluoroacetate can decompose to $XeF^{\bullet}$ and the trifluoroacetoxy radical and/or to a fluorine radical, a trifluoroacetoxy radical and xenon, while the trifluoroacetoxy radical can further decompose to give a trifluoromethyl radical. Xenon fluoride trifluoroacetate can also further react with trifluoroacetic acid, thus forming xenon bis(trifluoroacetate)[33,34] which after decomposition gives a trifluoroacetoxy radical, decomposing further to a trifluoromethyl radical. The formation of fluoro radicals, trifluoromethyl radicals and trifluoroacetoxy radicals was suggested in order to explain the formation of the five products in the reaction with styrene.

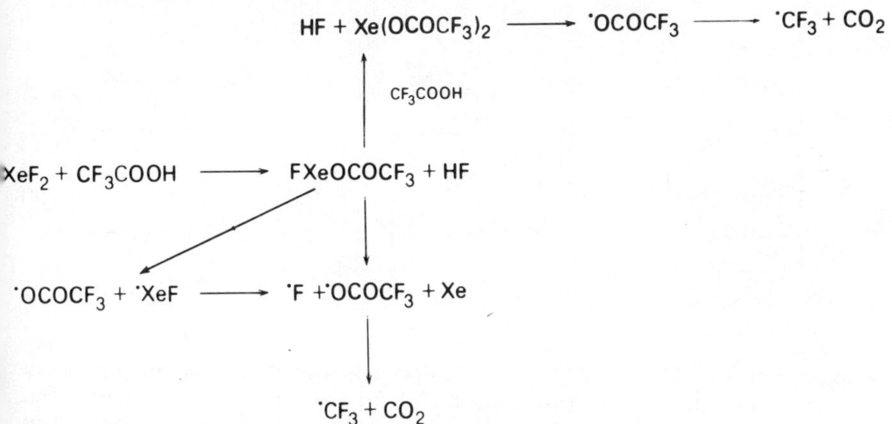

Shackelford and coworkers[22] found that for the fluorination of aliphatic alkenes with xenon difluoride a boron trifluoride etherate proved to be a convenient catalyst. They carried out the reaction in conventional chemical glassware. In the reaction with 1,2-dibromoethene four fluorinated products were formed: 1,1-difluoro-2,2-dibromoethane, 1-fluoro-1-chloro-2,2-dibromoethane, 1-fluoro-2-chloro-1,2-dibromoethane and 1-fluoro-1,2,2-tribromoethane. The fluorination of 2-fluoro-2,2-dinitroethyl vinyl ether gave one major product, identified as 1′,2′-difluoroethyl-2-fluoro-2,2-dinitroethyl ether.

$$\mathrm{CHBr{=}CHBr} \xrightarrow[\mathrm{CH_2Cl_2}]{\mathrm{XeF_2}} \underset{(59\%)}{\mathrm{CHF_2CHBr_2}} + \underset{(18\%)}{\mathrm{CHFBrCHBr_2}} + \mathrm{CHFClCHBr_2} + \mathrm{CHFBrCHClBr}\ (23\%)$$

The utility of XeF_2 as a mild fluorinating agent was confirmed in the case of fluorination reactions of glycals, catalysed by boron trifluoride etherate. The reaction with 3,4,6-tri-*O*-acetyl-1,5-anhydro-2-deoxy-D-*arabino*-hex-1-enitol and the D-*lyxo* epimer gave three products in high yield[23].

The effect of catalyst and reaction temperature on the nature of Xe—F bond cleavage has been examined in detail in the case of norbornene fluorinations[24–26,67]. It was demonstrated that catalyst and temperature play an important role in the product distribution. Shackelford[25] has also made detailed studies of boron trifluoride-catalysed fluorinations of deuterium-labelled norbornene derivatives.

The reaction of xenon difluoride with 1,1-diphenylalkenes in the presence of bromine has also been studied[27]. A mixture of bromine and xenon difluoride remained

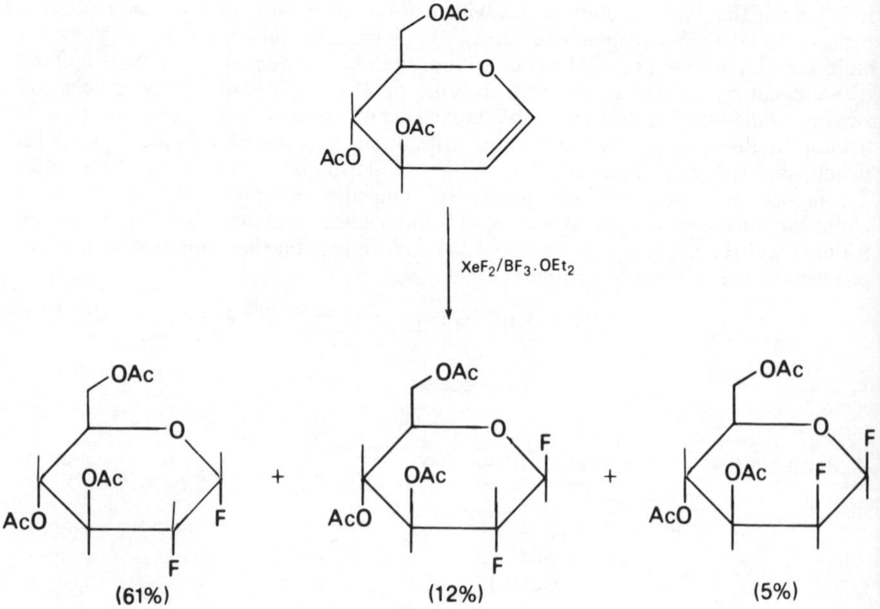

$$\xrightarrow{XeF_2/BF_3.OEt_2}$$

(61%) (12%) (5%)

without reaction, while addition of 1,1-diphenylethylene resulted in immediate gas evolution and 1,1-diphenyl-1-fluoro-2,2-dibromoethane was formed. The reaction occurred in the same way with 1,1-diphenylpropene and 1,1-diphenyl-2-halo-substituted ethylenes (X = F, Cl, Br), whereas it failed in the case of norbornene, *trans*-stilbene, styrene and 2-phenylpropene, resulting in the formation of dibromides only.

Chernick and coworkers[10,11] found that propyne is resistant to fluorine addition with xenon difluoride in a gas-phase reaction. After 100 days at room temperature it gave 2,2-difluoropropane in 33% yield and at least nine other trace products. Zupan and Pollak[28] found that at room temperature xenon difluoride reacted with phenyl-substituted acetylenes in the presence of hydrogen fluoride as catalyst. Less than stoicheiometric amounts of xenon difluoride did not favour difluoroolefin formation and only tetrafluoride and unreacted acetylene were found in the reaction mixture. However, the use of 2.5 equivalents of xenon difluoride led to the formation of tetrafluorides in over 50% yield.

$$PhC{\equiv}CR \xrightarrow[CH_2Cl_2/HF]{XeF_2} PhCF_2CF_2R$$

$$R = Ph, C_3H_7, Me$$

The trifluoroacetic acid-catalysed fluorination of diphenylacetylene is much more complex and produced six products[21]. The formation of fluoro radicals, trifluoromethyl radicals and trifluoroacetoxy radicals formed by decomposition of $FXeOCOCF_3$ was suggested in order to explain the formation of the products.

$$Ph-C\equiv C-Ph \xrightarrow[CF_3COOH/CH_2Cl_2]{XeF_2} Ph(CF_3)C{=}CPh(OCOCF_3) + Ph-CF_2-CF_2-Ph$$

$$+ Ph(CF_3)C{=}CFPh + PhFC{=}CPh(OCOCF_3) + PhC{\equiv}C{-}C_6H_4{-}CF_3$$

(two isomers)

III. REACTIONS OF XENON DIFLUORIDE WITH AROMATIC AND HETEROAROMATIC MOLECULES

Five years after Chernick and coworkers[10] established that xenon difluoride reacted with unsaturated molecules, Filler, Hyman and coworkers[35–38] started fundamental studies on the reactivity of aromatic molecules. On the basis of the effect of the substituent bonded to the benzene ring on the fluorination (Scheme 3), the isolation of biphenyls and the detection of ion radicals by ESR spectroscopy, three possible mechanisms were suggested. The main pathways are presented in Scheme 4. In the first stage polarization of xenon difluoride by hydrogen fluoride was suggested, thus

$$C_6H_5X \xrightarrow{XeF_2/HF/\text{Monel vacuum line}} o\text{-}FC_6H_4X + m\text{-}FC_6H_4X + p\text{-}FC_6H_4X$$

X	ortho	meta	para
OCH_3	30.5	2.5	32.4
CH_3	16.1	2.6	13.7
Cl	16	3.2	46.3
F	11.8	2.8	32.3
CF_3	0	71.7	3.8
NO_2	18.9	50.9	11.4

SCHEME 3.

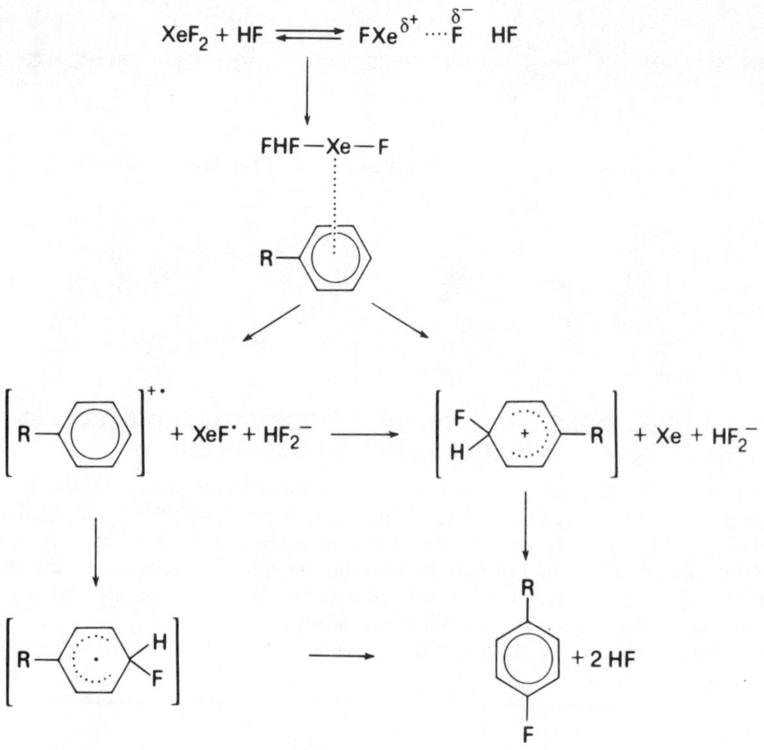

$$XeF_2 + HF \rightleftharpoons FXe^{\delta+} \cdots F^{\delta-} \quad HF \longrightarrow FHF{-}Xe{-}F$$

$$[R{-}C_6H_5]^{+\bullet} + XeF^{\bullet} + HF_2^- \longrightarrow [F,H{-}C_6H_5^{+}{-}R] + Xe + HF_2^-$$

$$\longrightarrow R{-}C_6H_4{-}F + 2\,HF$$

SCHEME 4

forming an electrophilic species, which reacts in the next step to a π-complex with the benzene molecule, transforming further to ion radical species or to a carbonium ion. Turkina and Gragerov[39] studied xenon difluoride reactions with benzene and deuterium-substituted derivatives. For the formation of biphenyls, ion radical[35–38] and radical mechanisms were suggested.

Mackenzie and Fajer[40] have studied vapour-phase fluorination of aromatic molecules and a free radical mechanism was suggested. It was demonstrated that the molar ratio of xenon difluoride to benzene derivative has an important effect on the ratio of substitution process to polymerization; in higher molar ratios, polymerization was preferred.

Filler and coworkers[41] have found that fluorinations of hydroxy, alkoxy and amino substrates do not require external initiation by hydrogen fluoride and have modified the mechanism for those aromatic compounds which are more readily oxidizable than benzene. In these compounds the π-cloud may induce sufficient polarization of the Xe—F bond, while transfer of an electron generates the radical cations.

Filler and coworkers[41] did not find any oxidation products in the fluorination of hydroxy and dihydroxy benzene derivatives. In contrast to these results, Nikolenko and coworkers[42] have shown that pentafluorophenol reacted with xenon difluoride in acetonitrile, forming the peroxide.

It has been demonstrated that benzoic acid and substituted benzoic acids in benzene solution in the presence of xenon difluoride gave various phenyl benzoates[43,44].

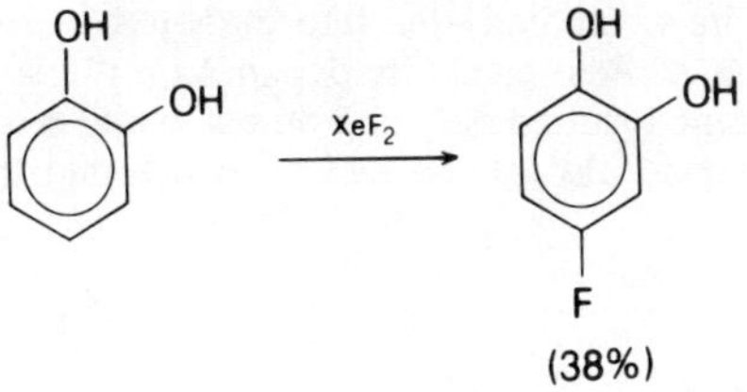

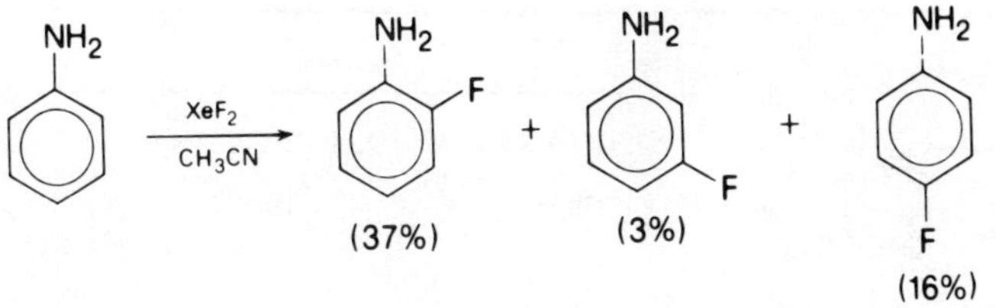

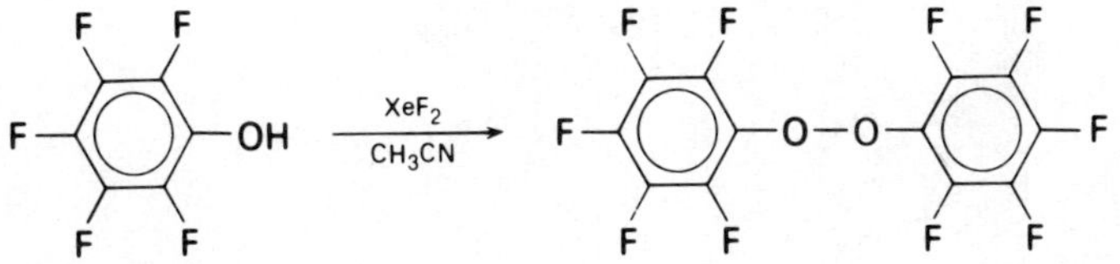

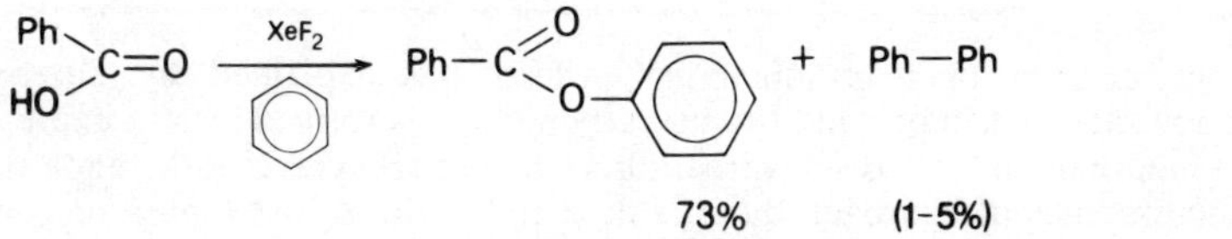

(38%)

(37%) (3%) (16%)

73% (1–5%)

Oxidative properties of xenon difluoride have also been observed in the reaction with 1,2- and 1,4-dihydroxybenzene derivatives, resulting in quinones[45].

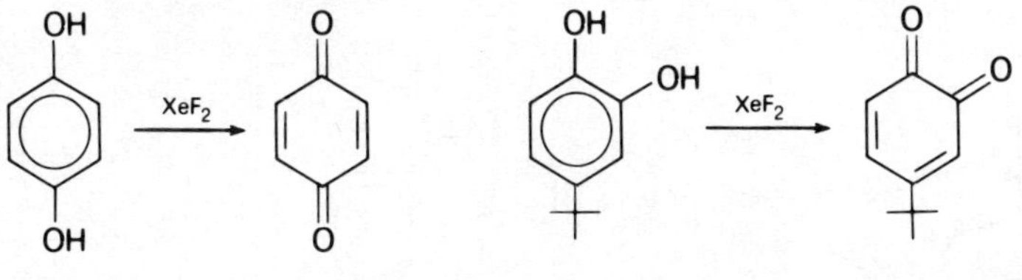

Firnau and coworkers[46] have also found that direct fluorination of L-dopa is not possible, resulting instead in the formation of *o*-quinone, while reaction with L-3-methoxy-4-hydroxyphenylalanine ethyl ester, followed by HBr hydrolysis, gave L-3,4-dihydroxy-6-fluorophenylalanine.

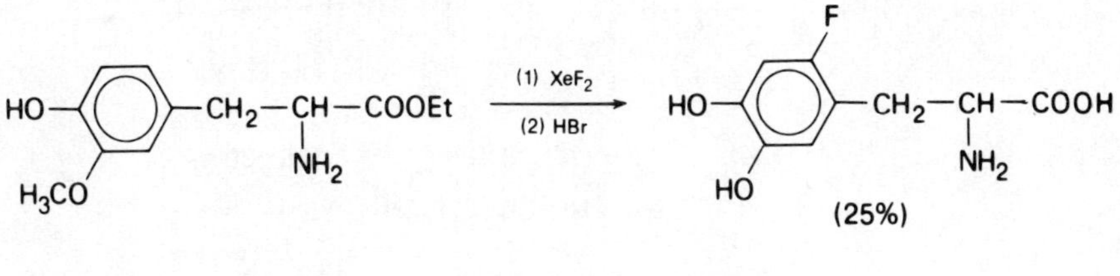

(25%)

Šket and Zupan[47] have studied the fluorination of *ortho*-substituted benzene derivatives and found that regioselectivity depends on the structure of the molecule. The fluorination of indane resulted only in β-attack, while reactions with tetraline and *o*-xylene were accompanied also by α-attack. It was found that fluorination is more

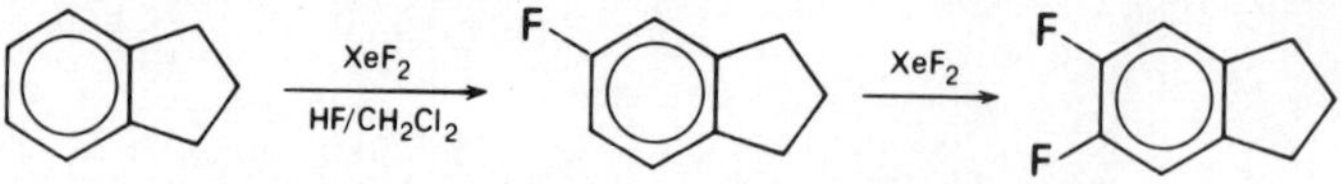

XeF_2, HF/CH_2Cl_2; XeF_2

	β-Attack, %	
	Br_2/AcOH	$XeF_2/HF,CH_2Cl_2$
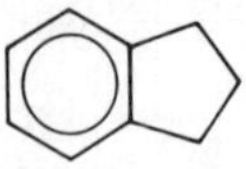 (indane)	78 ± 3	100
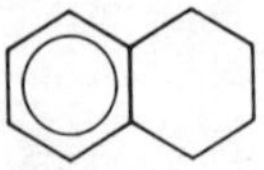 (tetraline)	56 ± 2	70
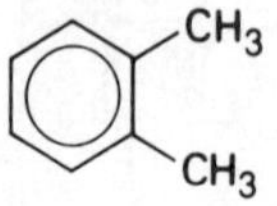 CH_3 CH_3 (o-xylene)	71 ± 3	80

regioselective than bromination[48], which was explained by differences in the stabilization of fluoro and bromo carbonium ions formed after α and β attack.

Fluorination of 9,10-dihydroanthracene and triptycene with xenon difluoride[49] in the presence of hydrogen fluoride occurred at the α- and β-positions, while reaction with acenaphthylene resulted in the formation of 2- and 4-fluoro-substituted products, regioselectivity being very little affected by the nature of the catalyst (HF, BF_3 and pentafluorothiophenol). The reaction must be carried out in much higher dilution than in other cases, otherwise only polymeric material could be isolated.

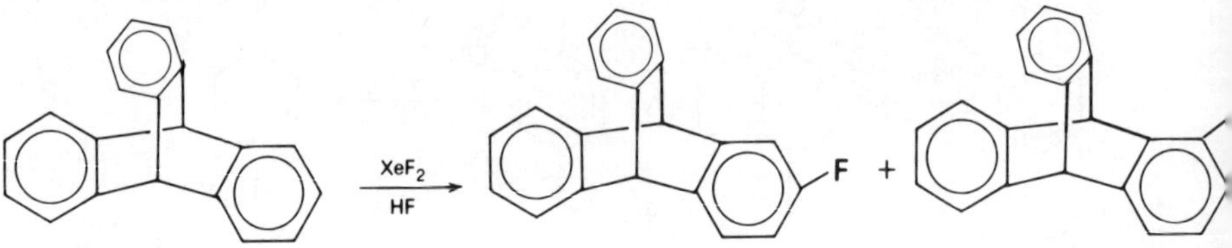

XeF_2, HF → F (71%) + (29%)

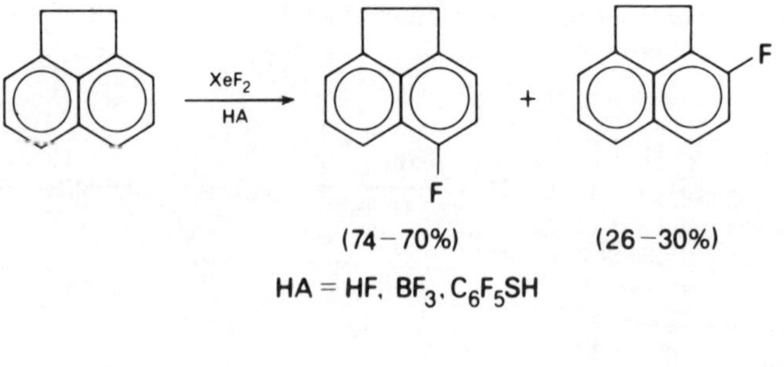

XeF_2, HA → F (74–70%) + F (26–30%)

HA = HF, BF_3, C_6F_5SH

Fluorination of 1,2,3,4-tetrahydro-1,4-methanonaphthalene gave as products 6-fluoro- and 6,7-difluoro-, and as a rearranged product 1-(2,2-difluoroethyl)indane.

The fluorinations of various trimethyl- and tetramethyl-substituted benzene derivatives in the presence of HF catalyst led to ring substitution[50], while the trifluoroacetic acid-catalysed reaction proved to be much more complex, similar to the reactions of styrene and diphenylacetylene[21].

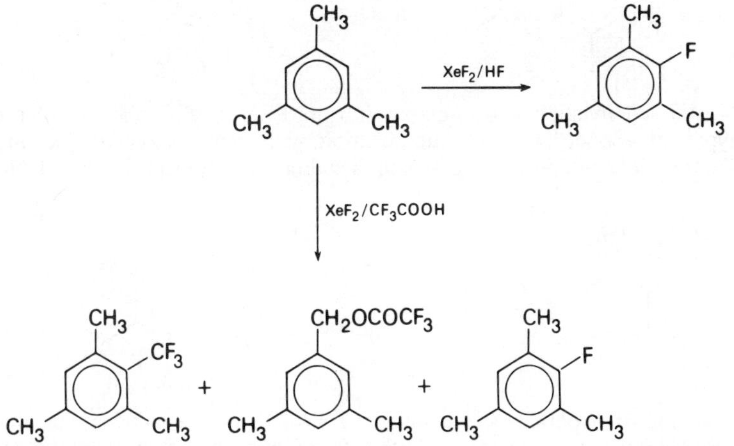
CH3
CH3
CH3
XeF2/HF
CH3
F
CH3
CH3
XeF2/CF3COOH
CH3
CF3
CH3
CH3
+
CH2OCOCF3
CH3
CH3
+
CH3
F
CH3
CH3

HF-catalysed fluorination of hexamethylbenzene[51] resulted in the formation of the fluoromethyl derivative, while the trifluoroacetic acid-catalysed reaction gave the trifluoroacetoxy derivative.

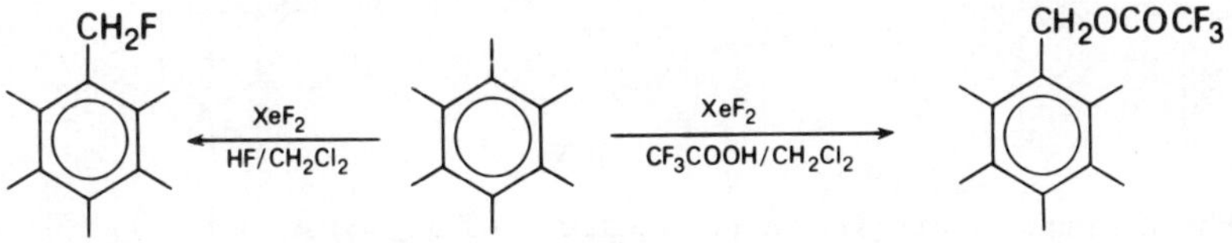
CH2F
XeF2
HF/CH2Cl2
XeF2
CF3COOH/CH2Cl2
CH2OCOCF3

BF_3-catalysed room temperature reactions of xenon difluoride with pentafluoro-substituted benzene derivatives resulted in 1,4- and 1,2-fluorine addition with regiospecificity depending on the substituent. 1-Substituted (H, Cl, Br, C_6F_5) pentafluorobenzenes reacted to give 1-substituted heptafluorocyclohexa-1,4-dienes[52].

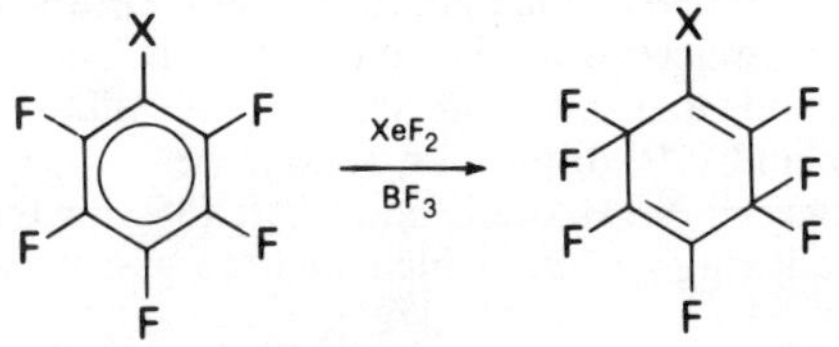
X
F
F
F
F
F
XeF2
BF3
X
F
F
F
F
F
F
F

X = F, Cl, Br, H, C_6F_5

BF_3-catalysed fluorination of octafluoronaphthalene proceeds as 1,4- and 1,2-addition; addition to 2-methoxy and 2-ethoxyheptafluoronaphthalene gave only

1,4-adducts, while reaction with 2-isopropoxyheptafluoronaphthalene resulted in 2-oxooctafluoro-1,2-dihydronaphthalene and 6-isopropoxynonafluoro-1,4-dihydronaphthalene[53].

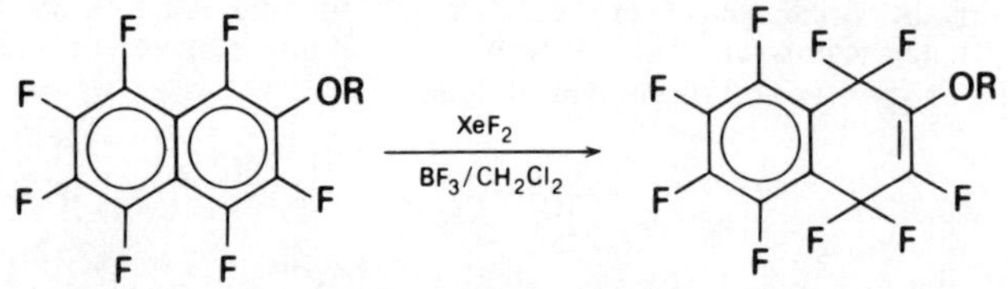

BF_3-catalysed reaction with *n*-alkoxy-substituted pentafluorobenzene[54] produced two types of adduct, while in the reaction with isopropoxypentafluorobenzene, hexafluorocyclohexa-2,5-dien-1-one and hexafluoroacyclohexa-2,4-dien-1-one were formed.

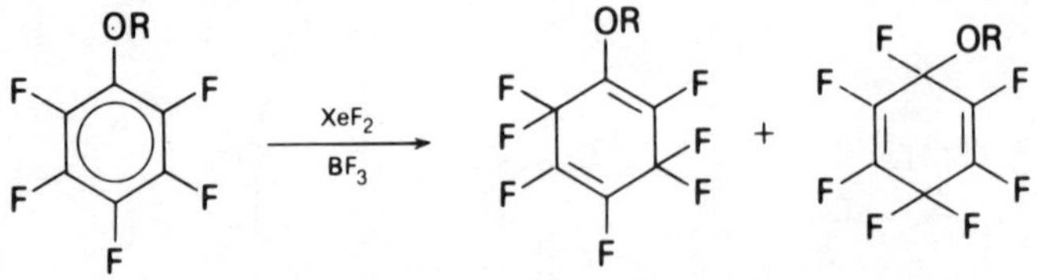

Fluorination of naphthalene[55,56] gave two monofluoro derivatives, the α-product predominating over the β-product. Further fluorination of 1-fluoronaphthalene gave 1,4-difluoronaphthalene[56]. In the fluorination of anthracene three monosubstituted products[55] were formed.

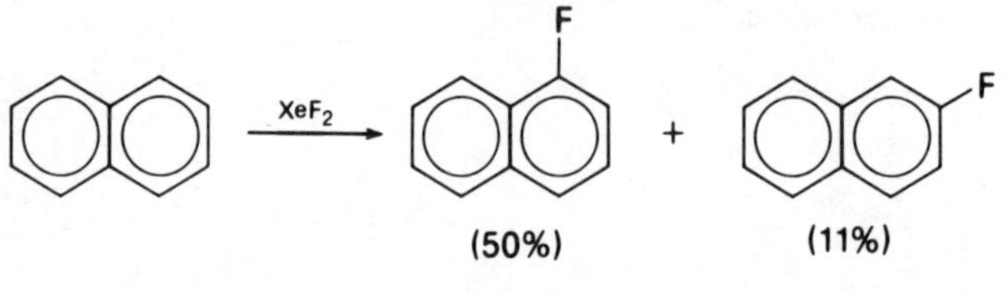

Phenanthrene is well known to undergo addition across the 9,10-positions in chlorination[57] and bromination[58]. The fluorination of phenanthrene has been extensively studied by three groups[55,59,60]. A 3h reaction in the presence of hydrogen fluoride at room temperature gave 9-fluorophenanthrene in 60% yield. Reaction with two molar equivalents of xenon difluoride led to four products: 9-fluorophenanthrene, 9,10-difluorophenanthrene, 9,10,10-trifluoro-9,10-dihydrophenanthrene and 9,9,10,10-tetrafluoro-9,10-dihydrophenanthrene. All efforts to detect the primary addition product were unsuccessful. However, the formation of tri- and tetrafluoro products clearly showed that addition of fluorine with XeF_2 took place with 9-fluorophenanthrene and 9,10-difluorophenanthrene.

Agranat and coworkers[61–63] explored the use of XeF_2 for the preparation of many fluorinated polyaromatic ring systems, which are otherwise very difficult to synthesize, e.g. **14, 15** and **16**.

The fluorination of heteroaromatic molecules with XeF_2 has received much less attention. Yurasova[64] found that reaction with uracil produced 5-fluorouracil. Anand and Filler[65] demonstrated that pyridine reacted with xenon difluoride and 2-fluoro-, 3-fluoro- and 2,6-difluoropyridine were isolated. 8-Hydroxyquinoline was converted to 5-fluoro-8-hydroxyquinoline. Fluorination of imidazo(1,2-b)pyridazine with xenon

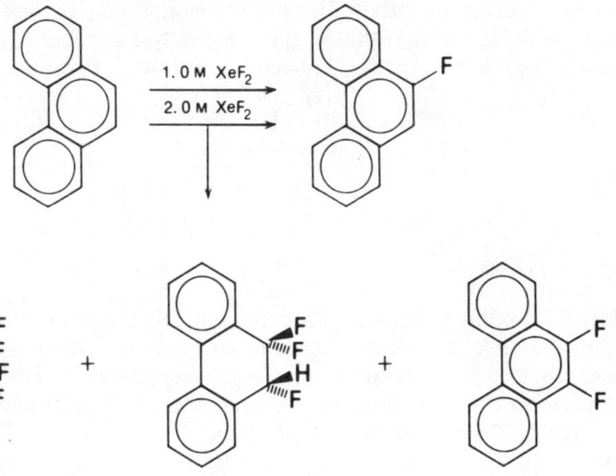

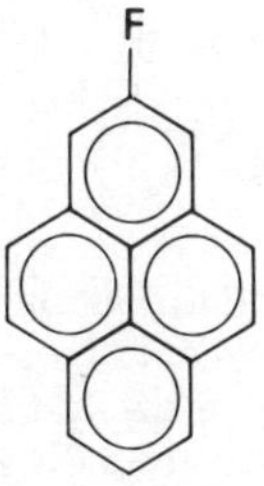

1.0 M XeF_2

2.0 M XeF_2

(14) 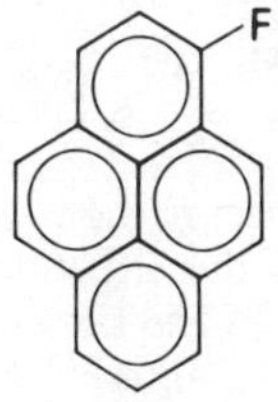 **(15)** 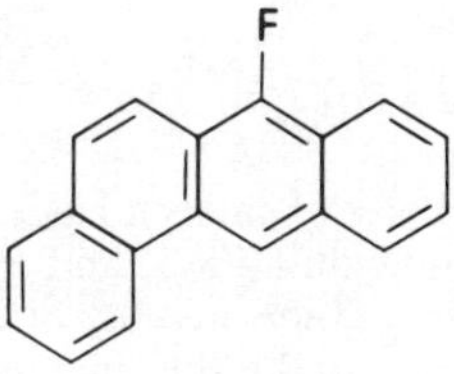 **(16)**

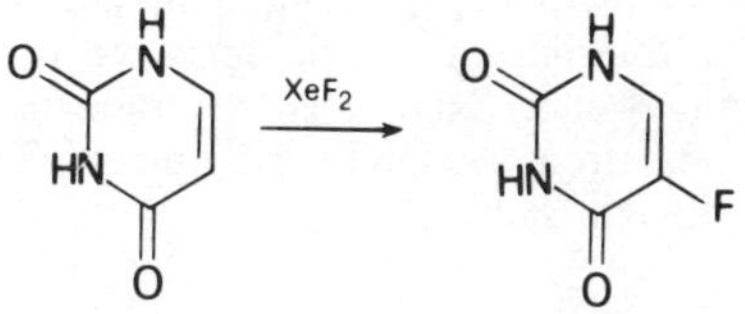

XeF_2

(10%)

XeF_2 / CH_2Cl_2

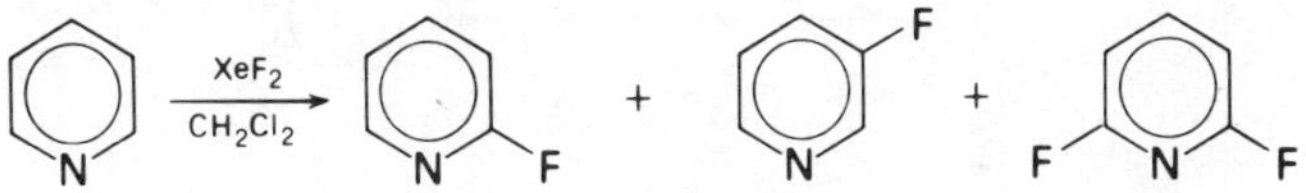

(35%) (20%) (11%)

difluoride gave no fluorinated products[66] and only 3-substituted chloro or bromo derivatives were isolated. Chloro products were formed in CCl_4, $CHCl_3$ and CH_2Cl_2, while bromo products were produced in $CHBr_3$ as solvent and in the presence of HF or trifluoroacetic acid.

IV. REACTIONS OF XENON DIFLUORIDE WITH VARIOUS ORGANIC MOLECULES

Substitution reactions at saturated carbon atoms with xenon difluoride have received little attention. Podkhalyzin and Nazarova[69] found that the reaction of adamantane with xenon difluoride in carbon disulphide resulted in 1-fluoroadamantane. Zajc and Zupan[29] have demonstrated that in the HF-catalysed reaction with acetylacetone the monofluoro product was formed, while further fluorination gave

XeF_2 / CS_2 → F (35%)

3,3-difluoro-2,4-pentanedione. HF-catalysed fluorination of 5,5-dimethylcyclohexa-1,3-dione gave a complex mixture in very low yield. However, it has been discovered that a new catalyst, i.e. the cross-linked polystyrene-4-vinylpyridine complex with boron trifluoride in the presence of cross-linked polystyrene-4-vinylpyridine led to difluoro product in high yield[29].

XeF_2; (P)–C₅H₄N·BF_3/(P)–C₅H₄N; H_3C CH_3 → F F; H_3C CH_3

The choice of catalyst has great influence on the course of the fluorination reaction with xenon difluoride, and depends on the structure of the organic molecule, its reactivity and the stability of the products. The similar influence of catalyst has been pointed out in the fluorination of indan-1,3-dione[70]. HF-catalysed reaction gave one rearranged product in low yield, while Nafion-H-catalysed reaction in the presence of cross-linked polystyrene-4-vinylpyridine resulted in 2,2-difluoroindan-1,3-dione in high yield, while further fluorination in the presence of hydrogen fluoride gave 2,2,3,3-tetrafluorochroman-4-one. (Nafion is the trade name for a perfluorinated resin-sulphonic acid available from the Du Pont Company. The H-form was generated from the commercial potassium salt.)

XeF_2; Nafion-H/(P)–C₅H₄N; XeF_2/HF; XeF_2/HF

Xenon difluoride reactions with sulphides have been studied by three groups[71–74]. Zupan[71] found that phenyl methyl sulphide gave phenyl fluoromethyl sulphide, while further fluorination led to phenyl difluoromethyl sulphide. Reactions with *cis*-2,6-diphenyltetrahydro-1-thio-4-pyrone and thiochroman-4-one produced dehydrogenated products, while reaction with 3,3-dibromothiochroman-4-one gave the 2-fluoro-substituted derivative[72]. In a room-temperature reaction, diphenyl

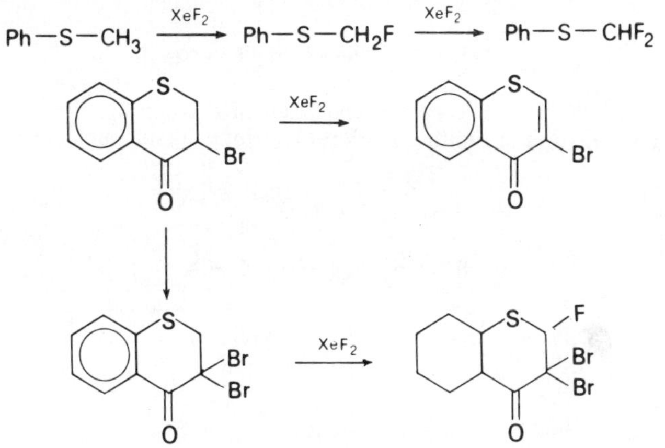

$$Ph{-}S{-}CH_3 \xrightarrow{XeF_2} Ph{-}S{-}CH_2F \xrightarrow{XeF_2} Ph{-}S{-}CHF_2$$

sulphide was converted to sulphone and sulphoxide[72] and no evidence for difluoro- or tetrafluorosulphorane was given. Marat and Janzen[73] have isolated difluorosulphuranes in acetonitrile at low temperature (−20 to −5°C), where decomposition was prevented by $[(CH_3)_3Si]_2NH$. Reactions with thiols led to disulphides. Formation of difluorosulphuranes has been observed from phenyl trifluoromethyl sulphide, while reaction of the selenium analogue also gave the corresponding difluoride[74].

$$Ph{-}S{-}R \xrightarrow[CH_3CN]{XeF_2} Ph{-}SF_2{-}R + X$$

$$Ph{-}SH + XeF_2 \longrightarrow Ph{-}S{-}S{-}Ph$$

$$(CH_3)_3C{-}SH + XeF_2 \longrightarrow (CH_3)_3{-}S{-}S{-}C(CH_3)_3$$

Gibson and Janzen[75] have studied fluorination of various organosilicon compounds and found that the chlorine atom is readily replaced by fluorine.

$$R_3SiCl \xrightarrow[CH_3CN]{XeF_2} R_3SiF + Cl_2$$

$$R = Me,\ Et$$

$$(CH_3)_2SiCl_2 \xrightarrow{XeF_2} (CH_3)_2SiF_2 + Cl_2$$

Alkyl and aryl phosphines give alkyl or aryl hydrodifluorophosphoranes, while reaction with phenyldichlorophosphine results in oxidative fluorination and the substitution of chlorine by fluorine[76]. Gibson and Janzen found that methyliodine(III) difluoride could be prepared by reaction of methyl iodide with xenon difluoride[77]. The reaction has been extended to the preparation of various aryl- and alkyl-substituted

$$R_2PH \xrightarrow[CH_3CN]{XeF_2} R_2P(F)_2\text{—}H + Xe$$

$$Ph_3P \xrightarrow{XeF_2} Ph_3PF_2 + Xe$$

iodo(III) difluorides[76–79]. The easy preparation of this type of fluorinating reagent promoted investigation of the fluorination of various organic molecules[80–88].

$$R\text{—}I \xrightarrow{XeF_2} R\text{—}IF_2$$

$$R = CH_3,\ CH(CH_3)_2,\ C_6H_4X$$

$$\text{(P)—}C_6H_4\text{—}I \xrightarrow[CH_2Cl_2/HF]{XeF_2} \text{(P)—}C_6H_4\text{—}IF_2$$

(P)— = polymer backbone

V. REACTIONS OF OTHER XENON HALIDES

Xenon tetrafluoride and xenon hexafluoride should be handled with great care because of the possible formation of highly explosive xenon oxides, and all their reactions studied so far have been carried out in Monel vacuum lines. The discovery that both compounds form stable intercalates with graphite has increased their utility.

It has been found that the chemical reactivity of the three xenon halides runs in parallel with the increasing amount of fluorine in the molecule[89]. Differences in chemical reactivity with organic substrates were nicely illustrated by their reactions with perfluoropropene[10]; xenon difluoride gave no reaction, xenon tetrafluoride led to perfluoropropane, while in the case of xenon hexafluoride, tetrafluoromethane and

$$CF_4 + CF_3CF_3 \xleftarrow{XeF_6} CF_3CF{=}CF_2 \xrightarrow{XeF_4} CF_3CF_2CF_3$$

hexafluoroethane were formed. An 18 h reaction with ethene gave three products, and a free radical mechanism was suggested[11].

$$CH_2{=}CH_2 + XeF_4 \longrightarrow {}^{\bullet}XeF_3 + {}^{\bullet}CH_2\text{—}CH_2F$$

$${}^{\bullet}CH_2\text{—}CH_2F \longrightarrow FCH_2\text{—}CH_2F \quad (45\%)$$

$${}^{\bullet}CH_2\text{—}CH_2F \longrightarrow CH_2{=}CHF \longrightarrow CH_2F\text{—}CHF_2 \quad (20\%)$$

$${}^{\bullet}CH_2\text{—}CH_2F \longrightarrow CH_3\text{—}\dot{C}HF \longrightarrow CH_3CHF_2 \quad (35\%)$$

For the fluorination of benzene derivatives[11] with XeF_4, which yielded biphenyl, fluorobiphenyl and other products, a free radical mechanism was also suggested. Fluorination of hexafluorobenzene gave perfluorocyclohexane (11%), perfluoro-1,4-cyclohexadiene (20%) and perfluoro-1,3-cyclohexadiene (4%).

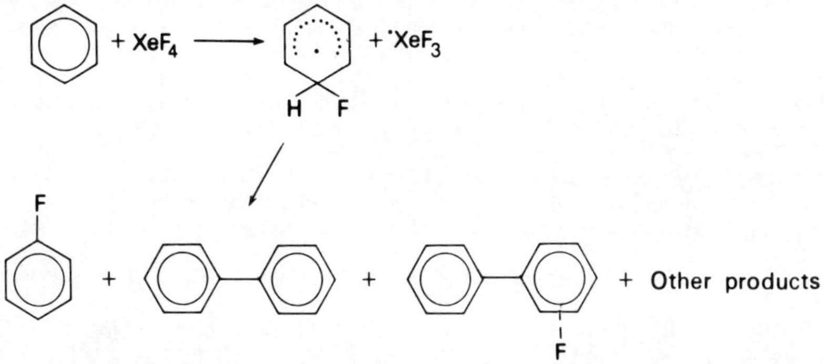

+ XeF_4 ⟶ + $\cdot XeF_3$

H F

F

+ + + Other products

F

Selig, Agranat, Rabinovitz and coworkers have made an important contribution to the use of XeF_4 and XeF_6 for the fluorination of aromatic molecules[5–7,62]. They found that XeF_4, XeF_6 and $XeOF_4$ form stable intercalates with graphite. In the fluorination of benzene, naphthalene, anthracene, phenanthrene, pyrene and other polycyclic aromatic hydrocarbons, fluoro-substituted products were produced. Kagan and

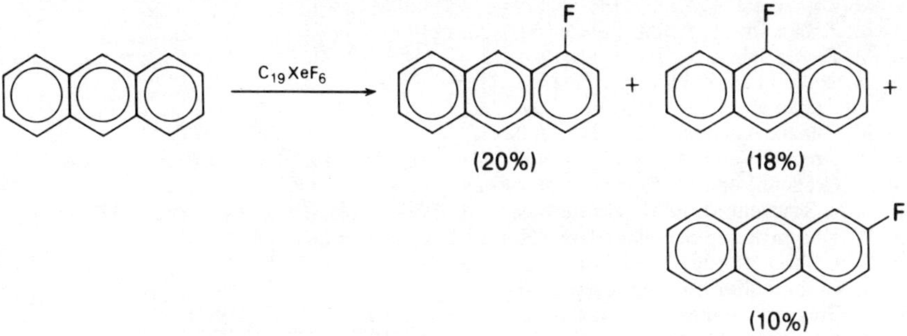

$C_{19}XeF_6$

F F

+ +

(20%) (18%)

F

(10%)

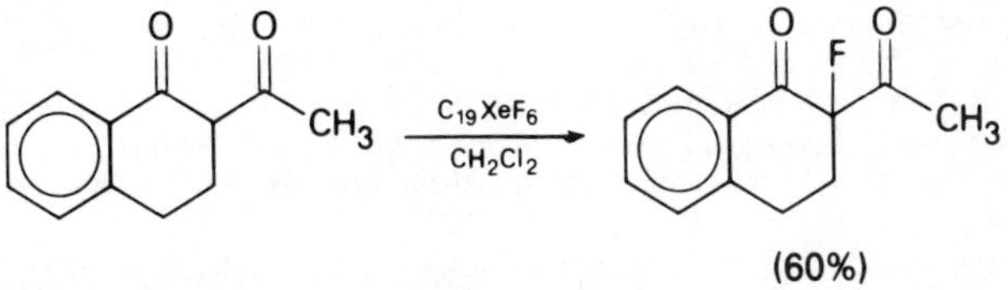

O O O O

F

CH_3 $C_{19}XeF_6$ / CH_2Cl_2 CH_3

(60%)

coworkers[90] found that various β-diketones and β-ketoesters could be fluorinated with $C_{19}XeF_6$, giving monofluoro and difluoro products. Fluorination of uracil and barbituric acid gave monofluoro products in high yield[90].

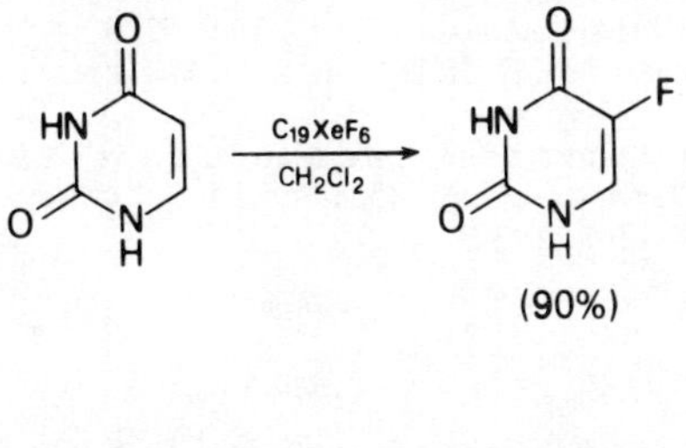

O O

F

HN $C_{19}XeF_6$ / CH_2Cl_2 HN

O N O N

H H

(90%)

VI. REFERENCES

1. W. A. Sheppard and C. M. Sharts, *Organic Fluorine Chemistry*, W. A. Benjamin, New York (1969); R. D. Chambers, *Fluorine in Organic Chemistry*, Wiley, New York (1973); *The Chemistry of the Carbon–Halogen Bond*, Vols. 1 and 2 (Ed. S. Patai), Wiley, Chichester (1973); G. H. Schmid and D. G. Garratt in *The Chemistry of Double-bonded Functional Groups* (Ed. S. Patai), Wiley, Chichester (1977), pp. 725–912; R. Hesse, *Israel J. Chem.*, **17**, 60 (1978); R. Filler, *Israel J. Chem.*, **17**, 71 (1978).
2. *Noble-Gas Compounds* (Ed. H. H. Hyman), The University of Chicago Press, Chicago (1963); J. H. Holloway, *Noble-Gas Chemistry*, Methuen, London, 1968; *Noble Gas Compounds* (Eds. D. T. Hawkins, W. E. Falconer and N. Bartlett), IFI/Plenum, New York (1978).
3. S. M. Williamson, *Inorg. Synthesis*, **11**, 147 (1968).
4. W. E. Falconer and W. A. Sunder, *J. Inorg. Chem.*, **29**, 1380 (1967).
5. H. Selig, M. Rabinovitz, I. Agranat, C.-H. Lin and L. Ebert, *J. Amer. Chem. Soc.*, **98**, 1601 (1976).
6. H. Selig, M. Rabinovitz, I. Agranat, C.-H Lin and L. Ebert, *J. Amer. Chem. Soc*, **99**, 953 (1977).
7. M. Rabinovitz, I. Agranat, H. Selig, C.-H. Lin and L. Ebert, *J. Chem. Res. (S)*, 216 (1977); *J. Chem. Res. (M)*, 2353 (1977).
8. K. Seppelt, *Angew. Chem.*, **91**, 199 (1979).
9. J. H. Holloway, *Educ. Chem.*, 140 (1973).
10. T.-C. Shieh, N. C. Yang and C. L. Chernick, *J. Amer. Chem. Soc.*, **86**, 5021 (1964).
11. T.-C. Shieh, E. D. Feit, C. L. Chernick and N. C. Yang, *J. Org. Chem.*, **35**, 4020 (1970).
12. M. Zupan and A. Pollak, *JCS Chem. Commun.*, 845 (1973).
13. M. Zupan and A. Pollak, *Tetrahedron Lett.*, 1015 (1974).
14. M. Zupan and A. Pollak, *J. Org. Chem.*, **41**, 4002 (1976).
15. M. Zupan and A. Pollak, *Tetrahedron*, **33**, 1017 (1977).
16. M. Zupan and A. Pollak, *J. Org. Chem.*, **42**, 1559 (1977).
17. B. Šket and M. Zupan, *JCS Perkin I*, 2169 (1977).
18. M. Zupan and B. Šket, *J. Org. Chem.*, **43**, 696 (1978).
19. M. Zupan, A. Gregorčič and A. Pollak, *J. Org. Chem.*, **42**, 1562 (1977).
20. A. Gregorčič and M. Zupan, *J. Org. Chem.*, **44**, 1255 (1979).
21. A. Gregorčič and M. Zupan, *J. Org. Chem.*, **44**, 4120 (1979).
22. S. A. Shackelford, R. R. McGuire and J. L. Pflug., *Tetrahedron Lett.*, 363 (1977).
23. W. Korytnyk and S. Valentekovic-Horvat, *Tetrahedron Lett.*, 1493 (1980).
24. S. A. Shackelford, *Tetrahedron Lett.*, 4265 (1977).
25. S. A. Shackelford, *J. Org. Chem.*, **44**, 3485 (1979).
26. A. Gregorčič and M. Zupan, *Bull. Chem. Soc. Japan*, **53**, 1085 (1980).
27. S. Stavber and M. Zupan, *J. Fluorine Chem.*, **10**, 271 (1977).
28. M. Zupan and A. Pollak, *J. Org. Chem.*, **39**, 2646 (1974).
29. B. Zajc and M. Zupan, *JCS Chem. Commun.*, 759 (1980).
30. R. F. Merrit, *J. Amer. Chem. Soc.*, **89**, 609 (1967).
31. R. C. Fahey and C. Schubert, *J. Amer. Chem. Soc.*, **87**, 5172 (1965).
32. R. C. Fahey and H. H. Schneider, *J. Amer. Chem. Soc.*, **90**, 4429 (1968).
33. J. I. Musher, *J. Amer. Chem. Soc.*, **90**, 7371 (1968).
34. M. Eisenberg and D. D. DesMarteau, *Inorg. Nucl. Chem. Lett.*, **6**, 29 (1970).
35. M. J. Shaw, H. H. Hyman and R. Filler, *J. Amer. Chem. Soc.*, **91**, 1563 (1969).
36. M. J. Shaw, J. A. Weil, H. H. Hyman and R. Filler, *J. Amer. Chem. Soc.*, **92**, 5096 (1970).
37. M. J. Shaw, H. H. Hyman and R. Filler, *J. Amer. Chem. Soc.*, **92**, 6498 (1970).
38. M. J. Shaw, H. H. Hyman and R. Filler, *J. Org. Chem.*, **36**, 2917 (1971).
39. M. Y. Turkina and I. P. Gragerov, *Zh. Org. Khim.*, **11**, 340 (1975).
40. D. R. Mackenzie and J. Fajer, *J. Amer. Chem. Soc.*, **92**, 4994 (1970).
41. S. P. Anand, L. A. Quaterman, H. H. Hyman, K. G. Migliorese and R. Filler, *J. Org. Chem.*, **40**, 807 (1975).
42. L. N. Nikolenko, T. I. Yurasova and A. A. Manko, *Zh. Obshch. Khim.*, **40**, 938 (1970).
43. L. N. Nikolenko, L. D. Shustov, T. N. Bocharova, T. I. Yurasova and V. A. Legasov, *Proc. Acad. Sci. USSR*, **204**, 1369 (1972).

44. L. D. Shustov, T. D. Telkovskaya, and L. N. Nikolenko, *Zh. Org. Khim.*, **112**, 2137 (1975).
45. M. Zupan and A. Pollak, *J. Fluorine Chem.*, **7**, 443 (1976).
46. G. Firnau, R. Chirakal, S. Sood and S. Garnett, *Canad. J. Chem.*, **58**, 1449 (1980).
47. B. Šket and M. Zupan, *J. Org. Chem.*, **43**, 835 (1978).
48. J. Vaugham, G. J. Welch and G. J. Wright, *Tetrahedron*, **21**, 1665 (1965).
49. B. Šket and M. Zupan, *Bull. Chem. Soc. Japan*, **54**, 279 (1981).
50. M. Zupan, S. Stavber and B. Šket in *9th International Symposium on Fluorine Chemistry, Avignon, 1979*, p. O-38.
51. M. Zupan, *Chimia*, **30**, 305 (1976).
52. S. Stavber and M. Zupan, *JCS Chem. Commun.*, 969, (1978).
53. B. Zajc and M. Zupan, *Bull. Chem. Soc. Japan*, **54**, in press (1981).
54. S. Stavber and M. Zupan, *J. Org. Chem.*, **46**, 300 (1981).
55. S. P. Anand, L. A. Quaterman, P. A. Christian, H. H. Hyman and R. Filler, *J. Org. Chem.*, **40**, 3796 (1975).
56. M. Rabinovitz, I. Agranat, H. Selig and C.-H. Lin, *J. Fluorine Chem.*, **10**, 159 (1977).
57. P. B. D. de la Mare and R. Koenigsberger, *J. Chem. Soc.*, 5327 (1964).
58. C. C. Price, *J. Amer. Chem. Soc.*, **58**, 1834 (1936); L. Altschuler and E. Berliner, *J. Amer. Chem. Soc.*, **88**, 5837 (1966).
59. M. Zupan and A. Pollak, *J. Org. Chem.*, **40**, 3794 (1975).
60. I. Agranat, M. Rabinovitz, H. Selig, and C.-H. Lin, *Chem. Lett.*, 1271 (1975).
61. E. D. Bergmann, H. Selig, C.-H. Lin, M. Rabinovitz and I. Agranat, *J. Org. Chem.*, **40**, 3793 (1975).
62. I. Agranat, M. Rabinovitz, H. Selig, and C.-H. Lin, *Synthesis*, 267 (1977).
63. I. Agranat, M. Rabinovitz, H. Selig and C.-H. Lin, *Experientia*, **32**, 417 (1976).
64. T. I. Yurasova, *Zh. Obshch. Khim.*, **44**, 956 (1974).
65. P. Anand and R. Filler, *J. Fluorine Chem.*, **7**, 179 (1976).
66. M. Zupan and A. Pollak, *J. Fluorine Chem.*, **8**, 275 (1976).
67. A. Gregorčič and M. Zupan, *Collect. Czech. Chem. Commun.*, **42**, 3192 (1977).
68. A. Gregorčič and M. Zupan, *Tetrahedron*, **33**, 3243 (1977).
69. A. T. Podkhalyzin and M. P. Nazarova, *Zh. Org. Khim.*, **11**, 1568 (1975).
70. B. Zajc and M. Zupan in *VI. Meeting of Chemists of Croatia, Zagreb, 1979*, p. 194.
71. M. Zupan, *J. Fluorine Chem.*, **8**, 305 (1976).
72. B. Zajc and M. Zupan, *JCS Perkin I*, 965 (1978).
73. R. K. Marat and A. F. Janzen, *Canad. J. Chem.*, **55**, 3031 (1977).
74. I. L. Jagupolski and T. I. Savina, *Zh. Org. Khim.* **15**, 438 (1979).
75. J. A. Gibson and A. F. Janzen, *Canad. J. Chem.*, **49**, 2168 (1971).
76. J. A. Gibson, R. K. Marat and A. F. Janzen, *Canad. J. Chem.*, **53**, 3044 (1975).
77. J. A. Gibson and A. F. Janzen, *JCS Chem. Commun.*, 739 (1973).
78. M. Zupan and A. Pollak, *JCS Chem. Commun.*, 715 (1975).
79. M. Zupan and A. Pollak, *J. Fluorine Chem.*, **7**, 445 (1975).
80. M. Zupan and A. Pollak, *Tetrahedron Lett*, 3525 (1975).
81. M. Zupan and A. Pollak, *J. Org. Chem.*, **41**, 2179 (1976).
82. A. Gregorčič and M. Zupan, *Bull. Chem. Soc. Japan*, **50**, 517 (1976).
83. M. Zupan, *Synthesis*, 473 (1976).
84. M. Zupan and A. Pollak, *JCS Perkin I*, 1745 (1976).
85. M. Zupan, *Collect. Czech. Chem. Commun.*, **42**, 266 (1977).
86. A. Gregorčič and M. Zupan, *JCS Perkin I*, 1446 (1977).
87. S. Stavber and M. Zupan, *J. Fluorine Chem.*, **12**, 307 (1978).
88. S. Stavber and M. Zupan, *Bull. Chem. Soc. Japan*, **52**, 925 (1979).
89. N. Bartlett, *Endeavour*, **23**, 3 (1964).
90. S. S. Yemul, H. B. Kagan and R. Setton, *Tetrahedron Lett.*, 277 (1980).

The Chemistry of Functional Groups, Supplement D
Edited by S. Patai and Z. Rappoport
© 1983 John Wiley & Sons Ltd

CHAPTER **16**

The $S_{RN}1$ reaction of organic halides

ROBERT K. NORRIS
Department of Organic Chemistry, The University of Sydney, New South Wales 2006, Australia

I. INTRODUCTION: THE $S_{RN}1$ REACTION

The substitution of a halogen in an organic molecule by a nucleophile may be represented as shown in equation (1), where R represents a wide variety of organic moieties, X is one of the halogens and A^- is an incoming nucleophile. The wide range

$$RX + A^- \longrightarrow RA + X^- \quad (1)$$

of substitution processes which result from variations in the nature of R and changes in solvent, halogen and/or nucleophile, include the S_N1 and S_N2 processes, mainly occurring in saturated systems, and the S_NAr and aryne mechanisms in aromatic systems[1]. The aryne mechanism was proposed in 1953[2] and since then, with the exception of unusual substitution processes with limited general utility such as the multi-step S_N(ANRORC) reaction[3], only one, new, widely applicable substitution reaction has been added to the above well established processes involving organic halides. The basic steps (2)–(5) in this substitution reaction are presented, in general form, in Scheme 1. This sequence of reactions has been variously described as an

$$RX + e \longrightarrow [RX]^{-\bullet} \quad (2)$$

$$[RX]^{-\bullet} \longrightarrow R^{\bullet} + X^- \quad (3)$$

$$R^{\bullet} + A^- \longrightarrow [RA]^{-\bullet} \quad (4)$$

$$[RA]^{-\bullet} + RX \longrightarrow RA + [RX]^{-\bullet} \quad (5)$$

SCHEME 1

electron-transfer chain substitution process[4], or as a substitution reaction involving radical anions and free radicals as intermediates[5]. Bunnett, recognizing the formal similarity between step (3) and the rate-determining unimolecular step in S_N1 reactions, and also appreciating the need for a simple symbolism to represent this reaction[6], suggested the description '$S_{RN}1$', standing for *substitution*, *radical-nucleophilic*, *unimolecular*[7]. This particularly useful designation is used throughout this chapter.

The $S_{RN}1$ reaction has been mentioned briefly in general reviews on the chemistry of radical ions[8,9] and also in other contexts[1,10–12]. Kornblum has extensively reviewed the $S_{RN}1$ reaction in which a nucleofuge is displaced from a saturated carbon atom[4,5], and Bunnett has reviewed aromatic substitution by the $S_{RN}1$ mechanism[6].

II. GENERAL SCOPE OF THE $S_{RN}1$ REACTION

The types of substrates which are known to undergo $S_{RN}1$ reactions comprise two major classes. The first group consists of molecules in which the nucleofuge is directly attached to a saturated carbon, whilst in the second the nucleofuge is attached to the carbon of an unsaturated, usually aromatic, system. Although the first class of substrates invariably contain unsaturated groups of some sort, they will be referred to as belonging to the *saturated system*, whilst the second group will be referred to as the *unsaturated* or *aromatic system*; i.e. the two classes are defined by the nature of the carbon bearing the nucleofuge.

A. Substitution in Saturated Systems

The reactivity of the majority of the substrates in this class usually depends on the presence of a nitro group in either the substrate or the nucleophile[4,5], although exceptions are known[13]. The *o*- and *p*-nitrobenzyl, the sterically hindered *p*-nitrocumyl and other substituted cumyl systems and also various α-nitroalkyl derivatives undergo a variety of $S_{RN}1$ reactions. The discovery of these reactions, their development and a comprehensive set of typical examples have been presented elsewhere in this series[4]. The sterically hindered neopentyl system in the *p*-nitrobenzyl derivative (**1**) has also been shown to undergo $S_{RN}1$ reactions (e.g. reaction 6)[14].

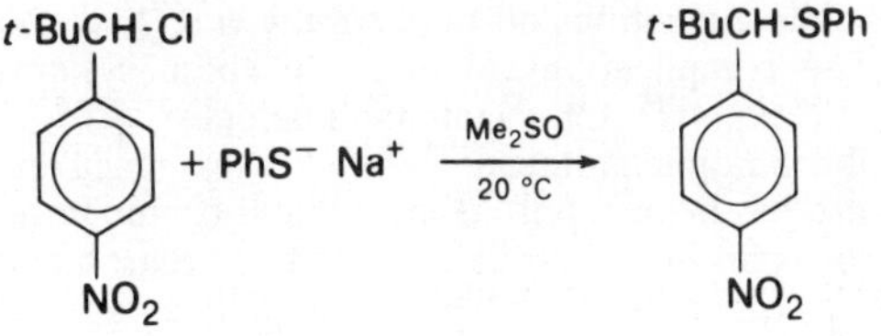

t-BuCH-Cl (1) (NO_2 para) + PhS⁻ Na⁺ →(Me₂SO, 20 °C) t-BuCH-SPh (NO_2 para) (92%) (6)

$$t\text{-BuCH(Cl)}C_6H_4NO_2\ (\mathbf{1}) + PhS^- Na^+ \xrightarrow[20\,^\circ C]{Me_2SO} t\text{-BuCH(SPh)}C_6H_4NO_2\ (92\%) \quad (6)$$

B. Substitution in Aromatic Systems

1. Discovery

Although the $S_{RN}1$ reaction in saturated systems had been demonstrated several years earlier[4,5], it was not until 1970 that an $S_{RN}1$ reaction involving an aromatic halogen compound was discovered. Kim and Bunnett subjected the halotrimethylbenzenes **2–4** and their respective isomers **5–7** to treatment with potassium amide in liquid ammonia[7]. The halides **2**, **3**, **5** and **6** gave the same mixture of trimethylanilines **8** and **9** (**8**/**9** = 0.68), as expected from a reaction proceeding through the common aryne intermediate **10**. However, the iodo compounds **4** and **7** gave appreciably lower

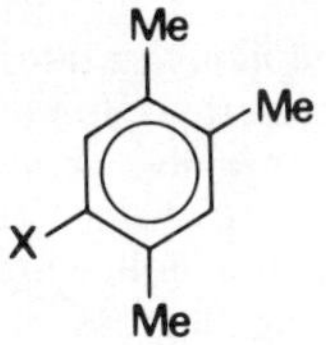

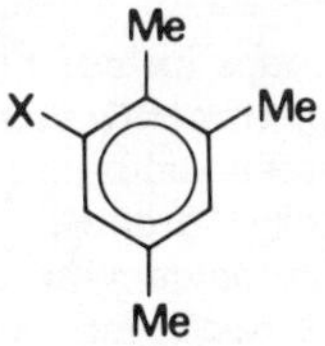

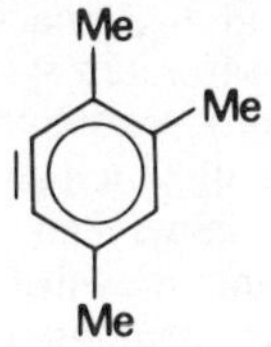

(**2**) X = Cl
(**3**) X = Br
(**4**) X = I
(**8**) X = NH_2

(**5**) X = Cl
(**6**) X = Br
(**7**) X = I
(**9**) X = NH_2

(**10**)

proportions of *cine* substitution products. Product ratios for **8**/**9** of 1.59 and 0.17 respectively were obtained. This observation and its interpretation in terms of the occurrence of an $S_{RN}1$ reaction lead Bunnett and his coworkers into an extensive study of aromatic substitution by the $S_{RN}1$ mechanism[6].

2. Nature of the aromatic moiety

A wide range of aromatic and heteroaromatic systems has been shown to undergo the $S_{RN}1$ reaction. The simple halobenzenes (Scheme 1, R = Ph), readily participate in $S_{RN}1$ reactions[15–39]. The following substituents on the benzene ring do not have a deleterious effect: alkyl[7,15,20–25,34,36] (but see comments on steric effects, Section III.C), alkoxy[7,15,21–24,28,36,40], benzoyl[26,41–44], cyano[42,43], ionized carboxy[23] (but *not* ionized hydroxy)[23,26], dialkylamino[45] and amino groups[45,46]. The nitro group, whose presence is so effective in promoting both $S_{RN}1$ reactions in the saturated system and S_NAr processes, effectively prevents the $S_{RN}1$ reaction on aromatic substrates[23]. Internal $S_{RN}1$ reactions, resulting in cyclic products, also take place on benzene rings bearing both alkyl and alkoxy groups[47,48] or the carboxamido function[49]. The presence of a second halogen on the benzene ring gives rise to products whose nature depends on the halogens, their relative positions on the benzene ring, and also on the nucleophile[21–23,32,36,37,50–53]. The complications involved in these systems are discussed further below (see Section IV.C). Halonaphthalenes[21–23,26,37,39,42,54,55], halobiphenyls[23,26,37,55] and 9-bromophenanthrene[23,26,37,39,55] also react readily under $S_{RN}1$ conditions. Only one case has been reported involving 9-bromoanthracene[23].

The $S_{RN}1$ reaction also occurs in various halogenated heteroaromatic systems. These include pyridines[26,56,57], quinolines[37,43,58–60], isoquinolines[61], thiophenes[62] and pyrimidines[64]. Wolfe and coworkers have been the major contributors in the area of nitrogen-containing heteroaromatic systems[63]. The $S_{RN}1$ mechanism or a related electron transfer process may also be operative in the reaction of 3-bromoquinoline with potassium superoxide[65].

In addition to the $S_{RN}1$ reactions summarized above, in which the halogen is attached to an aromatic system, simple vinyl halides have been found to undergo substitution reactions which appear to belong to the $S_{RN}1$ classification[66].

C. Nature of the Nucleofuge

The nucleofugic group in $S_{RN}1$ reactions is not limited to the halogens. Although reactions involving halide as the nucleofuge are the subject of this chapter, it should be noted that other groups, often poor nucleofuges in S_N1 and S_N2 processes can be involved in $S_{RN}1$ reactions.

In the saturated systems these groups include nitro, arylsulphonyl, azido, aryloxy, pentachlorobenzoyloxy, pyridinium, trimethylammonium, dimethylsulphonium and *p*-toluenesulphonyloxy[4,5,9]. More recent additions to this impressive array are the phenylsulphinyl[67] and arylsulphinyloxy[68] groups.

In aromatic substrates, in addition to the halogens, the following groups have been demonstrated to act as nucleofuges, with varying degrees of success: $(EtO)_2P(=O)O—$[15,17–19,69,70], PhO—[7,15,17,19], ArS—[15,19,24], PhSe—[17,19], $Me_3N^+—$[15,18–20], $Ph_2S^+—$[17,19], and $PhI^+—$[19]. The most useful list of nucleofuges for the aromatic system has been compiled by Rossi and Bunnett[19].

D. Nature of the Nucleophile

The nucleophile in $S_{RN}1$ processes normally performs two functions. First, it may act as a single-electron donor in the initiation step of the reaction (see Section III.A). Secondly, it acts as a radical trap in the key propagating step in the $S_{RN}1$ reaction, namely step (4), Scheme 1. This function of the nucleophile is mandatory or otherwise the reaction fails (see Section III.C).

The range of nucleophiles which are known to participate in $S_{RN}1$ reactions is dependent on the substrate. Different arrays, with some common members, are involved in substitution at saturated and unsaturated carbon.

1. *Saturated systems*

The nucleophiles which engage in substitution reactions at saturated carbon include the anions derived from primary and secondary nitroalkanes, β-keto esters, malonic and monoalkylmalonic esters, monoalkylmalononitriles, α-cyano esters and α-cyano ketones[4,5]. In addition to these ambident nucleophiles, which normally bond through carbon in $S_{RN}1$ reactions, the anions derived from β-diketones[71,72] simple ketones[73] and triethylphosphonoacetate[73] also react in a similar fashion. These latter reactions utilize α-halonitroalkanes as ketone equivalents in condensation reactions, as shown in equations (7)[73] and (8)[72]. Cyanide ion also appears to react, although only one example has been reported[74].

$$\text{(cyclohexenolate } O^-Li^+) + Me_2C(Cl)NO_2 \xrightarrow[10-45\,^\circ C]{THF} \text{2-}(CMe_2NO_2)\text{cyclohexanone} \longrightarrow \text{2-}(=CMe_2)\text{cyclohexanone} \quad (7)$$

$$MeC(O^-Na^+)=CH-C(=O)Ph + \text{1-chloro-1-nitrocyclohexane} \xrightarrow[35\,^\circ C,\ h\nu]{DMF} (MeCO)(COPh)C=\text{cyclohexylidene} \quad (8)$$

Other nucleophiles[4,5], in which the attacking atom is a heteroatom, are nitrite, azide, arenethiolate[75], arenesulphinate and aryloxide, together with certain amines. A recent report adds dialkyl phosphite ion to this list (equation 9)[76].

$$Me_2C(Cl)NO_2 + (MeO)_2PO^- \xrightarrow[-45\text{ to }25\,^\circ C]{THF} Me_2C(NO_2)P(O)(OMe)_2 \quad (9)$$

2. *Aromatic systems*

It is noteworthy that many nucleophiles which react with substrates in the saturated systems are quite unreactive towards aromatic substrates. The anions formed from β-dicarbonyl compounds, such as malonic esters, acetoacetic esters and 2,4-pentanedione, are unreactive towards both halobenzenes[25] and 2-haloquinolines[60,77]. Arenesulphinate and nitromethanide ions also appear to be unreactive[6]. A preliminary report in which phenoxide ion was claimed as an active nucleophile in aromatic $S_{RN}1$ reactions[78] involving halobenzenes has proven to be unreproducible[79]. By means of electrochemical techniques, Savéant[77] has demonstrated that the following nucleophiles are also unreactive towards 2-haloquinolines: CN^-, SCN^-, OH^-, PhO^-, $MeCONMe^-$ and $MeCONH^-$.

Despite the failure of the above anions to participate in aromatic $S_{RN}1$ processes (also see Section III.C), a number of nucleophiles effectively bring about substitution. These nucleophiles can be conveniently classified[6] according to the atom which becomes attached to the aromatic moiety, and this atom may be nitrogen, sulphur, selenium, tellurium, phosphorus or, and this is by far the largest group, carbon.

Amide ion appears to be the only nitrogen-based nucleophile to react in a $S_{RN}1$ process and, unless precluded by substrate structure (e.g. as in equation 10)[7] or by suitable control of reaction conditions, incursion of substitution by the aryne mechanism can take place.

$$\text{2,6-}Me_2C_6H_3I \xrightarrow[\text{liq. }NH_3]{K/KNH_2} \text{2,6-}Me_2C_6H_3NH_2 \quad (10)$$

The divalent, group VIA atoms react readily as arenethiolate[21,24,41–44,61,66,80], alkanethiolate[24,54], phenyl selenide[37,55] and phenyl telluride[37]. Typical examples are given in equations (11)[21] and (12)[37].

The phosphorus-based anions are $(EtO)_2PO^-$ [22,30,32,33,51–53], $PhP(OBu)O^-$ [38], Ph_2PO^- [38], $(EtO)_2PS^-$ [38] and $(Me_2N)_2PO^-$ [38]. These reactions are synthetically useful

$$\text{2-MeO-C}_6\text{H}_4\text{I} + \text{PhS}^- \xrightarrow{h\nu;\ \text{liq. NH}_3} \text{2-MeO-C}_6\text{H}_4\text{SPh}\ (91\%) \tag{11}$$

$$\text{9-bromoanthracene} + \text{PhSe}^- \xrightarrow{h\nu;\ \text{liq. NH}_3} \text{9-(PhSe)anthracene}\ (72\%) \tag{12}$$

since they allow attachment of a phosphorus-based moiety to an aromatic ring. Earlier reports of this type of reaction, for example the reaction of lithium diphenylphosphide with simple phenyl halides[81], have also been shown to be $S_{RN}1$ processes[34]. The examples given in equations (13)[38] and (14)[22] are illustrative of the utility of these reactions.

$$\text{PhI} + (\text{EtO})_2\text{PS}^- \xrightarrow[\text{liq. NH}_3]{h\nu} \text{PhP(S)(OEt)}_2\ (95\%) \tag{13}$$

$$\text{1-iodonaphthalene} + (\text{EtO})_2\text{PO}^- \xrightarrow[\text{liq. NH}_3]{h\nu} \text{1-naphthyl-PO(OEt)}_2\ (93\%) \tag{14}$$

Carbanions involved in reactions with aromatic halides include the enolate ions from acetone[15,17,23,29,40,45,46,54,56,57,59,60,62,66] and/or other ketones[16,25,28,31,35,47,48], esters[48] and amides[39,49]. Cyanomethyl[18,26,36,54] and other α-cyanoalkanide[18] anions, both 2- and 4-picolyl anions[20], and carbanions formed from 1,3-pentadiene, 1-(p-anisyl)propene, indene and fluorene[16] also react. Although, as discussed above, the monoanions from β-diketones are unreactive, the dianions, for example from benzoylacetone[58] or acetylacetone[25], react very satisfactorily. The arylation of all of the above nucleophiles on carbon has been rationalized by a perturbation molecular orbital treatment[82]. This theoretical approach also predicts[82] which of several carbons will be the principal site of arylation in anions such as the dianion of benzoylacetone (equation 15)[58] and pentadienide ion[16].

$$\text{2-chloroquinoline} + \text{CH}_2{=}\text{C(O}^-\text{Li}^+\text{)}{-}\text{CH}{=}\text{C(Ph)O}^-\text{Li}^+ \longrightarrow \text{2-quinolyl-CH}_2\text{COCH}_2\text{COPh}\ (71\%) \tag{15}$$

The yields from many of these $S_{RN}1$ reactions, which are synthetically the arylation of a carbanion, are excellent for enolates which are symmetrical (equations 16 and 17)[23,56], or can form only a single enolate ion (equation 18)[29]. Unfortunately, mixtures, usually reflecting the proportion of the various enolates in the system, often result in unsymmetrical cases (equation 19)[16].

I, MeO, OMe, OMe + $CH_2{=}C(O^-K^+)Me$ $\xrightarrow{h\nu;\ \text{liq. }NH_3}$ CH_2COMe, MeO, OMe, OMe (92%) (16)

N, Br + $Me_2C{=}C(O^-K^+)CHMe_2$ $\xrightarrow{h\nu;\ \text{liq. }NH_3}$ N, $CMe_2COCHMe_2$ (97%) (17)

$$PhBr + CH_2{=}C(O^-K^+)Bu\text{-}t \xrightarrow[Me_2SO]{h\nu} PhCH_2COBu\text{-}t \quad (87\%) \tag{18}$$

$$PhBr + MeCOEt \xrightarrow[\text{liq. }NH_3]{KNH_2,\ h\nu} \underset{(19\%)}{PhCH_2COEt} + \underset{(61\%)}{PhCH(Me)COMe} \tag{19}$$

Further examples, in which the $S_{RN}1$ reaction of a carbanion with an aromatic halide have proved to be of remarkable utility, include the final step in Semmelhack and coworkers' synthesis of cephalotaxine (equation 20)[47], an example of a cyclic variant

O, O, N, I, O, OMe $\xrightarrow[\text{liq. }NH_3]{t\text{-BuOK};\ h\nu}$ O, O, H, N, O, OMe (94%) (20)

of the $S_{RN}1$ reaction[48], and several recent procedures for the synthesis of benzo[*b*]-furans[40], indoles[45,46], oxindoles[49] and 4-azaindoles[57] (e.g. equation 21).

N, Cl, NH_2 + $CH_2{=}C(O^-)Pr\text{-}i$ $\xrightarrow[\text{liq. }NH_3]{h\nu}$ N, N–H, Pr-*i* (61%) (21)

E. Range of Solvents

In the saturated systems, dipolar aprotic solvents such as dimethylformamide (DMF), dimethyl sulphoxide (DMSO) and hexamethylphosphoramide (HMPA) appear to be the solvents of choice, promoting both nucleophilicity and electron-transfer capability, increasingly in the order DMF < DMSO < HMPA, and also being inert to the radical intermediates involved[4,5]. Ethanol has been used successfully in the p-nitrobenzyl[83], α-nitroalkyl[84] and p-nitrocumyl[85] systems. Benzene and dichloromethane allow the $S_{RN}1$ reaction to proceed even more rapidly than in dipolar aprotic solvents when tetrabutylammonium *aci*-nitronates, which are readily soluble in these solvents, are allowed to react with p-nitrobenzyl and p-nitrocumyl systems[86,87].

In the aromatic system Bunnett has made a systematic study of possible solvents[27], and liquid ammonia and dimethyl sulphoxide appear to be the solvents of choice[6], although dimethylformamide and acetonitrile look promising. Acetonitrile also seems to be satisfactory in electrochemically initiated processes[42]. In strong contrast with the saturated system, hexamethylphosphoramide is most unsatisfactory[27].

III. DIAGNOSTIC AND MECHANISTIC FEATURES OF THE $S_{RN}1$ REACTION

A. Initiation

The initiation step of the $S_{RN}1$ reaction, which is formally presented in equation (2), Scheme 1, corresponds to acceptance of a single electron by the substrate to form a radical anion. There is a diversity of sources which can act as one-electron donors in the $S_{RN}1$ process.

Initiation by a solvated electron or by an alkali metal dissolving in liquid ammonia is intimately involved in the discovery of the aromatic $S_{RN}1$ process and in the early development of this reaction[7]. Addition of potassium metal to liquid ammonia containing the aryl halide and nucleophiles such as ketone enolates[15,16,54] and cyanomethyl anion[18,54] proved quite successful. Initiation involving heteroaryl halides was often not as convenient, since competing reduction and other processes (see Section III.D) occurred[56,59,60,62]. The first step in the reaction sequence (22)[16] is another

$$PhBr + CH_2\!\cdots\!CH\!\cdots\!CH\!\cdots\!CH\!\cdots\!CH_2^- \xrightarrow[\text{liq } NH_3]{K} \xrightarrow{H_2/Pd} \underset{(74\%)}{Ph(CH_2)_4CH_3} \tag{22}$$

example of alkali metal-stimulated reaction in liquid ammonia. It is also noteworthy that the reaction of o-haloanisoles, which normally give m-anisidine via an aryne mechanism, with potassium amide in liquid ammonia give only o-anisidine, formed by an $S_{RN}1$ process, when an excess of potassium metal is present[7]. The claim[78] that sodium amalgam in aqueous t-butyl alcohol can stimulate $S_{RN}1$ processes has been recently disputed[79]. The kinetic processes involved in reactions closely allied to the $S_{RN}1$ reaction, brought about by reactions of halobiphenyls with potassium, have been studied by ESR spectroscopy[88]. The only reports of alkali metal-stimulated processes in saturated systems is the initiation of the reaction of p-nitrocumyl chloride with nitrite ion and other related reactions by sodium in hexamethylphosphoramide[89].

Electrochemical initiation of aromatic $S_{RN}1$ processes has proven to be of remarkable utility[77]. A typical example is given in reaction (23)[43], wherein one decided

$$p\text{-}ClC_6H_4S^- + \text{2-chloroquinoline} \xrightarrow[\text{liq. } NH_3]{e^-} \underset{(80\%)}{\text{2-}(p\text{-chlorophenylthio})quinoline} \tag{23}$$

advantage of electrochemical initiation over solvated electron stimulation becomes manifest. The 4′-chloro group in the product is unaffected when judicious choice of electrode reduction potential is made.

Radical anions have also been used as the source of electrons necessary for initiation of $S_{RN}1$ processes, but their use has been restricted mainly to substrates in the saturated systems containing nitro groups. These initiators include the radical anions from naphthalene[71,89], trimesitylborane[89] and nitrobenzene[71], and also sodium superoxide[65,89].

The most common source of single electrons required in step (2) is the nucleophile itself involved in the $S_{RN}1$ reaction, as shown in equation (24) (cf. equations 2–5).

$$RX + A^- \longrightarrow [RX]^{-\cdot} + A^\cdot \tag{24}$$

This mode of initiation seems to be the most common in the saturated systems. These normally have relatively low lying antibonding orbitals (π^*) associated with the nitro group, which can readily accommodate the additional electron. The ease with which nitro compounds form radical ions is well documented[90–94] and their production by single-electron transfer from anions is also well known[95]. The process whereby the electron is transferred is believed to be through some form of charge-transfer complex both in the saturated[96] and aromatic[30] systems. The electron transfer from nucleophile to substrate is often, but not always, light catalysed. This phenomenon has been used as one of the diagnostic tests for occurrence of the $S_{RN}1$ reaction[4–6]. Quantum yields from a quantitative study of the photostimulated reaction of iodobenzene with diethyl phosphite ion were found to be in the range of 20–50[30]; quantum yields in the *p*-nitrocumyl system as high as 6000, and in reaction of α-nitro esters as high as 220, have been observed[96]. In some cases, e.g. *o*-haloiodobenzenes, the $S_{RN}1$ reaction occurs under photostimulation, and quite different processes (see Section IV.C) occur in the dark[53].

Entrainment of $S_{RN}1$ reactions is also possible and may take one of two forms. The addition of a catalytic quantity of a nucleophile which is a potent initiator to a solution of the substrate and an inactive electron-transfer nucleophile, but one which is an adequate radical trap, will promote reaction. Numerous examples of this circumstance have been reported in the saturated system[4,5], but appear to be rare, or perhaps not commonly utilized, in the aromatic series. One example is found in the reaction of 2-chloroquinoline with acetone enolate. The rate of reaction is increased dramatically in the presence of 10 mol% dilithiobenzoylacetone[59]. A second mode of entrainment is observed when small quantities of a more reactive aryl halide is added to a lethargic reaction. A typical example is the immediate increase in the rate of the dark reaction of *p*-bromotoluene with potassium diphenylphosphide upon addition of more reactive *p*-iodotoluene[34].

A single example of radiolytic initiation of an $S_{RN}1$ reaction by β-radiation (from ^{60}Co) has been reported[97]. The homolytic cleavage in reaction (25) has also been suggested[98] as a direct source of the radical component.

$$Me_2C(I)(NO_2) \longrightarrow Me_2\dot{C}(NO_2) + I^\cdot \tag{25}$$

B. Intermediacy and Dissociation of Radical Anions

The dissociation of simple alkyl halides electrochemically[99] or on electron transfer from radical anions[100] appears, in general, to be nearly synchronous with electron transfer and is further complicated by subsequent carbanion formation. Many of the systems which undergo $S_{RN}1$ reactions proceed via radical anions whose presence has been demonstrated independently. The formation and dissociation of the radical anions involved in *p*-nitrobenzyl[101,102] and *p*-nitrocumyl[97] derivatives in the saturated series, and phenyl[77,103–106] and various heteroaryl halides[77,80,104,107,108] have been extensively studied and in general lend support to the thesis that steps (2) and (3) (Scheme 1) should normally be considered as sequential, and not concerted as in equation (26).

$$RX + e \longrightarrow R^{\cdot} + X^- \qquad (26)$$

The actual mechanistic details of dissociation of radical anions appear to have received scant consideration. A rather convenient representation of the dissociation of *p*-nitrobenzyl chloride radical anion is reproduced in equation (27)[4,5] and is described

$$\text{CH}_2\text{Cl (ring radical, } \text{N}^+(\text{O}^-)_2) \longleftrightarrow \text{CH}_2\text{Cl (ring anion, } \text{N}^+(\text{O}^{\cdot})(\text{O}^-)) \xrightarrow{-\text{Cl}^-} \text{CH}_2\text{=ring=N}^+(\text{O}^{\cdot})(\text{O}^-) \longleftrightarrow \text{CH}_2^{\cdot}\text{-C}_6\text{H}_4\text{-NO}_2 \qquad (27)$$

as an internal elimination. Utley and coworkers have suggested a similar process in substrates containing methoxycarbonyl substituents[109]. An alternative proposal by Neta and Behar describes this same dissociation in terms of an intramolecular electron transfer[102] in which the electron is transferred from the nitrophenyl group to the halogen and then dissociation ensues. This latter description may be generalized as shown in equations (28)–(30). The process shown in equation (27) should lead to

$$RX + e \longrightarrow {}^{\cdot -}R{-}X\ (\pi^*) \qquad (28)$$

$${}^{\cdot -}R{-}X \xrightarrow{\text{Slow}} R{-}X^{-\cdot}\ (\sigma^*) \qquad (29)$$

$$R{-}X^{-\cdot} \xrightarrow{\text{Fast}} R^{\cdot} + X^- \qquad (30)$$

planar *p*-nitrobenzylic radicals, and this is generally assumed[4,5]. The process shown in equation (30) could conceivably lead to a pyramidal (sp^3) radical. An experimental demonstration of this hypothesis would be an $S_{RN}1$ reaction proceeding with retention of configuration. Although it has been demonstrated that certain optically active benzylic substrates do racemize when undergoing substitution by the $S_{RN}1$ process[96], other substrates in which collapse of a pyramidal benzylic radical is sterically hindered do undergo $S_{RN}1$ reactions with retention[110].

C. Trapping of Radicals by Anions

The trapping of an anion by a radical to give a radical anion, step (4) of the overall $S_{RN}1$ process (Scheme 1), is in fact formally identical with the reverse of step (3). No examples of trapping with a halide as nucleophile have been demonstrated.

Independent observations of this process have been made in the trapping of phenyl radicals by nitrite ion[111] and *aci*-nitronate ions[112], *p*-nitrophenyl radical by cyanide ion[113], methyl radical by *aci*-nitronate ions[114], and of both *t*-butyl[115] and 3-cyclohexenyl radicals[116] with a variety of nucleophiles.

The bulk of information on the trapping of radicals by anions has, however, been deduced from studies on $S_{RN}1$ processes[4–6,9]. The range of nucleophiles which undergo this process have been discussed above (Section II.D). Whether a given nucleophile will be an efficient trap for a given radical still seems to be in the realm of experimental science. Energy considerations and rationalization based on molecular orbital correlations seem to indicate that, other factors being equal, the basicity of a given site may influence the radical-anion formation step (4), to a greater extent than the stability of the resulting radical ion[117]. Hence the failure of aryl radicals to be trapped by the monoanions of β-diketo compounds, when both dianions of the same compounds and monoanions of simple ketones do act as efficient traps, is rationalized. It is also abundantly clear that the softness of the attacking atom in the nucleophile is important[77], since sulphur- and phosphorus-based nucleophiles are very efficient radical traps.

Somewhat surprisingly, very few data appear in the literature concerning the relative efficiency with which anions trap radicals. Only data from competitive reactions can be relied upon since the relative rates of the initiation step, if nucleophile-induced as in equation (24), can be expected to be nucleophile dependent. Savéant[77] has established a two-tiered qualitative order with PhS^-, $p\text{-}ClC_6H_4S^-$, $(EtO)_2PO^-$, $CH_3COCH_2^-$ and $PhCOCH_2^-$ as *reactive* and CH_2CN^-, Ph_3C^- and lutidine carbanion as *very reactive* towards 2-quinolyl radical. Bunnett and Gloor have shown that the cyanomethyl anion is much more reactive than the enolate of acetone with bromobenzene[18], but with 1-chloronaphthalene Rossi and coworkers found them to react at comparable rates[26]. Komin and Wolfe showed that 2-pyridyl radicals display a preference for combination with tertiary enolates over primary enolates, and furthermore that the reactivity of acetone enolates are counterion dependent[56]. In a recent study, performed in the successful refutation of an $S_{RN}2$ process, whose propagation steps would be (31) and (32), the relative reactivities of diethyl phosphite ion and pinacolone enolate were

$$[PhX]^{-\bullet} + Y^- \longrightarrow [PhY]^{-\bullet} + X^- \tag{31}$$

$$[PhY]^{-\bullet} + PhX \longrightarrow PhY + [PhX]^{-\bullet} \tag{32}$$

determined to be 1.35 ± 0.08 and nucleofuge independent[118]. In this latter study it is suggested that the rates of reaction of both the nucleophiles is approaching the encounter-controlled limit. More recent results by Russell and coworkers[119] in the saturated system again demonstrate that the relative rate of reaction of pairs of nucleophiles is independent of the nucleofuge but reveal that it is critically dependent on the nature of the counterion, free counterion concentration *and* solvent. In the trapping of the 1-methyl-1-nitroethyl radical by a series of nucleophiles (equation 33),

$$Me_2\dot{C}NO_2 + A^- \longrightarrow \left[Me_2C\genfrac{}{}{0pt}{}{\diagup A}{\diagdown NO_2}\right]^{-\bullet} \tag{33}$$

in competitive experiments, it has been found that the anion of diethyl malonate is 10 times as reactive as the anion of 2-nitropropane with K^+[2.2.2]-cryptand in dimethyl sulphoxide but only 0.25 as reactive with lithium as the cation. Similarly, this same ratio of anion reactivities, but with lithium as counterion, varies from greater than 70 in tetrahydrofuran to 0.24 in dimethyl sulphoxide. These latter results[119] clearly show that relative reactivities of nucleophiles with radicals are not open to simple prediction.

The regiochemistry of attack of radicals upon ambident nucleophiles has been rationalized in both the saturated[120] and aromatic systems[82] by application of perturbational molecular orbital theory. The point of attack in the saturated system is often the more sterically hindered atom[4,5], although exceptions are known[14,121]. Hence the kinetically determined products in the reaction of *p*-nitrobenzyl chloride[4,5] and the α-*t*-butyl analogue[14] with the anion of 2-nitropropane are predominantly the

C- and *O*-alkylates respectively. In tertiary *p*-nitrobenzylic substrates reversible *C*- and *O*-alkylation has been invoked to explain the usual predominance of *C*-alkylation[68], although other evidence does not rule out kinetic, sterically controlled, selection of products[14,121].

The problem of regiochemistry aside, the $S_{RN}1$ reaction in both saturated and aromatic systems can be utilized to prepare remarkably congested molecules, and appears to be tolerant of often severe steric hindrance. The preparation of highly branched compounds in the saturated system has been ably exploited by Kornblum and coworkers[4,5]. The aromatic $S_{RN}1$ process is unaffected by *o*-methyl groups (reactions 10[7], and 34[20]) and even when the halogen is flanked by two isopropyl groups some substitution still takes place (reaction 35)[23].

2-Pyridyl-CH_2^- K^+ + 2-Br-1,3,5-trimethylbenzene $\xrightarrow[\text{liq. } NH_3]{h\nu}$ 2-Pyridyl-CH_2-(2,4,6-trimethylphenyl) (87%) (34)

$CH_2{=}C(O^-K^+)CH_3$ + 2-Br-1,3,5-tri(*i*-Pr)benzene $\xrightarrow[\text{liq. } NH_3]{h\nu}$ H-(2,4,6-tri(*i*-Pr)phenyl) (37%) + CH_3COCH_2-(2,4,6-tri(*i*-Pr)phenyl) (16%) (at 55% conversion) (35)

D. Chain Propagation, Termination and Competing Processes

The propagating steps (3) and (4) have been discussed immediately above. The step (5) of Scheme 1, which involves electron transfer between radical anions and neutral molecules has ample precedent[100,122–125]. The chain nature of the $S_{RN}1$ process has been demonstrated[4–6] by measurement of quantum yields in photostimulated reactions (see Section III.A) and by inhibition studies (see Section III.E).

The actual termination processes possible in $S_{RN}1$ reactions have only been marginally considered. Some termination (or pretermination[6]) and other competing reactions which have been suggested are collected in equations (36)–(42).

$$[RX]^{-\bullet} \longrightarrow R^- + X^\bullet \quad (36)$$

$$R^\bullet + A^- \longrightarrow R^- + A^\bullet \quad (37)$$

$$R^\bullet + [RX]^{-\bullet} \longrightarrow R^- + RX \quad (38)$$

$$R^\bullet + e \longrightarrow R^- \quad (39)$$

$$R^\bullet + A^\bullet \longrightarrow RA \quad (40)$$

$$R^\bullet + R^\bullet \longrightarrow R_2 \quad (41)$$

$$R^\bullet + \text{Solv}H \longrightarrow \text{Solv}^\bullet + RH \quad (42)$$

The sense of cleavage of radical ions, which is pertinent to process (36), has been examined in the aromatic system[19]. All the halogen derivatives predominantly cleave according to equation (3), Scheme 1, but the cleavage presented in equation (36) (R = phenyl, X = I) or the electron transfer process (37) (R = phenyl, A = $(EtO)_2PO$) may be a termination step in the reaction of iodobenzene with diethyl phosphite ion[30]. Processes (37)–(39) appear to be common in aromatic systems. Reduced hydrocarbons, resulting from protonation of R^-, appear as products in many $S_{RN}1$ reactions and become significant by-products, if not major products, in alkali metal-stimulated processes[7,15,16,18] (also see Section IV.B), particularly with heteroaromatic substrates[56,59,60,62]. It is not clear which of equations (37) or (38) is operative in the absence of solvated electrons. Termination according to equation (40) is possible but obviously cannot be experimentally verified since the product is that normally produced in $S_{RN}1$ reactions. Varying amounts of symmetrical dimers form in the saturated system[4,5], verifying the occurrence of termination step (41); in benzenoid systems, however, attempts to detect biphenyl have been unsuccessful[30]. The termination step (42), involving hydrogen abstraction, does not appear to interfere with the relatively well stabilized radicals involved in the saturated system. The possibility of phenyl radicals produced in $S_{RN}1$ processes abstracting hydrogens from dimethyl sulphoxide has been considered[27,30] but has been mainly discounted in view of the reported relatively low activity of aryl radicals towards this solvent[126]. The abstraction of hydrogen atoms from acetonitrile and the relative merits of the electron transfer steps (37), (38) and (42) have been discussed and evaluated by Savéant[77].

One significant termination/competing process in the reaction of aryl halides with ketone enolates is β-hydrogen abstraction from the enolate ion by the intermediate aryl radical. The stability of the resulting radical ion, a ketyl of the corresponding α,β-unsaturated ketone (e.g. see reaction 43), and its consequent failure to maintain

$$(H_3C)(Me)C{=}C(O^-)(CHMe_2) + Ar^{\bullet} \longrightarrow \left[(H_2C)(Me)C{\cdots}C(O)(CHMe_2)\right]^{-\bullet} + ArH \qquad (43)$$

the $S_{RN}1$ propagation cycle, would explain the sluggishness of reactions of aryl halides with certain enolates[31]. This process has been noted in cyclic $S_{RN}1$ reactions and to a less significant extent in the reactions of halopyridines[56] and haloquinolines[59,60]. The ultimate fate of the ketyl formed in equation (43) appears to be disproportionation to yield the α,β-unsaturated ketone, which undergoes Michael addition with the original enolate to give an unsymmetrical dimer ($Me_2CHCOCMe_2CH_2CHMeCOCHMe_2$)[31], rather than the symmetrical one originally thought to form[35,59]. A good example of incursion of hydrogen atom abstraction is found in the methoxide ion-induced reactions of some heteroaryl halides in methanol *in the absence* of an efficient radical trap. A radical chain dehalogenation occurs[127].

Another significant competing reaction in $S_{RN}1$ processes is fragmentation of the intermediate radical anion of the product before electron transfer step (5), Scheme 1, can occur. These fragmentations and their consequences are observed in the reactions of phenyl halides with cyanomethyl[18] and alkanethiolate[24] ions and are presented in equations (44) and (45). Similar complications in the reaction of halonaphthalenes with $PhTe^-$ also occur[37]. These reactions do not, in general, interfere in naphthalene[26,54] and heterocyclic systems[54], or in halobiphenyls and halobenzophenones[54].

$$[PhCH_2CN]^{-\bullet} \longrightarrow CN^- + PhCH_2^{\bullet} \longrightarrow (PhCH_2)_2 \qquad (44)$$

$$[PhSEt]^{-\bullet} \longrightarrow Et^{\bullet} + PhS^- \xrightarrow[S_{RN}1]{PhX} Ph_2S \qquad (45)$$

The absence of fragmentation in these latter systems has been rationalized by molecular orbital considerations[26,54,82]. The only example of product radical anion fragmentation leading to alternative products in the saturated system is found in the α,α-dimethyl-5-nitro-2-thenyl system, wherein reaction with benzenethiolate in the presence of benzenethiol leads initially to the expected sulphide but on prolonged reaction leads to the reduced product (Scheme 2)[128].

Suitably substituted allylic substrates have been found to undergo $S_{RN}1$ reactions with allylic rearrangement, and are consequently termed $S_{RN}1'$ reactions[129].

$$\left[O_2N\text{-}C_4H_2S\text{-}CMe_2SPh\right]^{-\bullet} \rightleftharpoons O_2N\text{-}C_4H_2S\text{-}\dot{C}Me_2 + PhS^-$$

$$\downarrow$$

$$O_2N\text{-}C_4H_2S\text{-}CHMe_2 \xleftarrow{PhSH} O_2N\text{-}C_4H_2S\text{-}\bar{C}Me_2 + PhS^{\bullet}$$

SCHEME 2

E. Inhibition of $S_{RN}1$ Reactions

A large number of substances have been found to inhibit the radical chain process involved in $S_{RN}1$ reactions. The use of inhibitors as diagnostics for operation of this mechanism appear in many publications concerning the reaction. The inhibitors normally act in either or both of the following capacities. They may trap radical intermediates or act as electron scavengers[5]. Oxygen is an excellent inhibitor in the saturated system[5] and in some aromatic systems[17,58], but either has no effect[32] or inexplicably accelerates reaction in others[28]. Di-*t*-butyl nitroxide and *m*- and *p*-dinitrobenzene appear to be universally successful inhibitors[5,17,28,30,32,34,56,58–60,64]. Other compounds which have exhibited inhibitory properties are azobenzene[34], tetraphenylhydrazine and 2-methyl-2-nitrosopropane[7,58], benzophenone[28], anthracene and naphthalene[36] and galvinoxyl and *p*-benzoquinone[5].

A typical example of the use of oxygen as a mechanistic probe is to be found in the elucidation of the mechanisms of the reactions in equation (46). The reaction in the

$$O_2N\text{-}C_4H_2S\text{-}CH(Cl)Bu\text{-}t + ArSO_2^- \xrightarrow{Me_2SO} O_2N\text{-}C_4H_2S\text{-}CH(SO_2Ar)Bu\text{-}t \quad (95\text{—}98\%) \qquad (46)$$

5-nitro series is totally inhibited by oxygen and is an $S_{RN}1$ process, whereas the reaction in the 4-nitro series is completely unaffected[130]. The mechanism in the latter case is probably analogous with that found in other 4-nitrothenyl derivatives[131].

IV. SPECIFIC HALOGEN EFFECTS

The only halogen which has been used as a nucleofuge in cumyl systems is chlorine[4,5]. In *p*-nitrobenzyl systems the use of halogens other than chlorine allows successful operation of a competing S_N2 process[4,5]. The reactions of α-chloro-, α-bromo- and

α-iodonitroalkanes have been studied[4,5] but not in a comparative fashion except to the extent that particularly in the bromo[4,71,132] and iodo[133] derivatives nucleophilic attack on the halogen may occur. Accordingly the majority of $S_{RN}1$ reactions of α-halonitroalkanes have been performed with the α-chloro derivatives. The discussion in this section is limited to the effect of variation of halogens in aromatic systems except for a brief discussion of benzylidene dihalides.

A. Relative Reactivities of Monohaloarenes

The relative reactivities in $S_{RN}1$ processes of substrates ArX with a given aromatic nucleus invariably increase as X is changed through the series F, Cl, Br and I. This conclusion has been reached by comparison in a pair-wise competitive fashion of the relative reactivities of aryl halides with nucleophiles[6]. This result may seem expected when one considers the radical anion dissociation step (3) in combination with usual nucleofugalities of the halides, a trend which appears to be supported by electrochemical[104] and other[102,134] studies. Nevertheless, when the kinetic complexity of the $S_{RN}1$ process (Scheme 1) is considered, with its involvement of initiation, dissociation, electron transfer and in addition various termination steps, it is remarkable that the above reactivity order appears to be without exception. Some of the observations on which the relative reactivity order is based include qualitative observations of relative reactivities of individual substrates with a given nucleophile[16,17,21,30,36,39] as well as the following reactivity ratios (symbolized as, for example, k_I/k_{Br})[28] determined by competitive experiments between pairs of halobenzenes in the photostimulated reaction of acetone enolate in liquid ammonia: $k_I/k_{Br} = 8.3$, $k_{Br}/k_{Cl} = 450$ and $k_{Cl}/k_F = 29$[135]. The ratio k_I/k_{Br} has been determined for several nucleophiles. For both diethyl phosphite ion[32] and butyl phenylphosphonite ion[38], k_I/k_{Br} is approximately 1000, for $(Me_2N)_2PO^-$, Ph_2P^- and $(EtO)_2PS^-$, all in liquid ammonia, values of 500, 300 and 45 respectively were obtained[34,38]. For the reaction of pinacolone enolate in dimethyl sulphoxide the ratio k_I/k_{Br} was 6[28]. This very large dependence of the k_I/k_{Br} value on the nucleophile has been interpreted as reflecting a change in the selectivity of the electron-transfer step (5), Scheme 1, from the radical anion $[PhA]^{-\cdot}$ to the halobenzene as the nature of the nucleophile (A^-) is varied[6]. The reaction of six aryl iodides with pinacolone enolate in dimethyl sulphoxide gave a maximum difference in reactivity, *reacting separately*, of almost 400-fold, but a maximum difference from reactions performed *in competition* with bromobenzene of less than twofold. These differences, which again demonstrate the need to consider the many steps in $S_{RN}1$ processes, were attributed to the influence of initiation, propagation and termination steps in the separate reactions as against differences in only the propagating steps (assuming reasonable chain length) in competitive experiments[35].

B. Effect of Halogen Identity on Product Distribution in Monohaloarenes

During initial studies on the utility of the aromatic $S_{RN}1$ process, Bunnett and Rossi[15] observed that in the reaction of acetone enolate with halobenzenes under stimulation by solvated electrons (from potassium in liquid ammonia) not only the expected product phenylacetone (**11**) but also the reduction products 1-phenyl-2-propanol (**12**) and benzene (cf. equation 39) were formed. Furthermore, the co-formation of these by-products was linked to the nature of the halogen, being greatest for fluorine and least for iodine[15]. Recently this result has been quantified and explained in terms of an elaborated $S_{RN}1$ mechanism[136], which includes the termination processes, involving solvated electrons, (47) and (48), and the product-protecting step (49).

$$Ar^{\bullet} \xrightarrow{e} Ar^{-} \xrightarrow{NH_3} ArH \quad (47)$$

$$[ArCH_2COMe]^{-\bullet} \xrightarrow{e} [ArCH_2COMe]^{2-} \xrightarrow{NH_3} ArCH_2CH(O^-)Me \quad (48)$$

$$ArCH_2COMe \xrightarrow{Base} ArCH{=}C(O^-)Me \quad (49)$$

$ArCH_2COMe$

(11) Ar = Ph

(13) Ar = mesityl

$ArCH_2CH(OH)Me$

(12) Ar = Ph

(14) Ar = mesityl

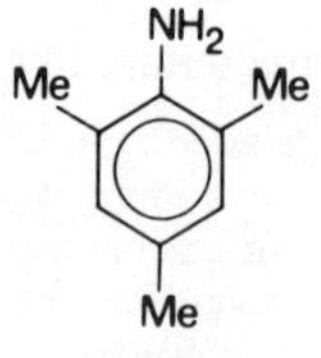

(15)

The proposed explanation describes, in a somewhat military fashion[136], raids of electrons upon aryl halide molecules, which 'annihilate' the electrons with production of halobenzene radical anions. The iodobenzene radical anion fragments rapidly, undergoes the $S_{RN}1$ cycle to give the product **(11)**, which is protected from further reaction by enolate formation (equation 49) before another raiding party of electrons arrives. At the other extreme the fluorobenzene radical ion fragments far more slowly and both the ketyl intermediate, $[\mathbf{11}]^{-\bullet}$, and the phenyl radical are intercepted and reduced by a second wave of electrons as shown in equations (47) and (48).

This concept of product distribution which is dependent on the nature of a leaving group which has already become detached from the substrate at the point of product selection has been termed a 'left group' effect[137]. This concept is further developed in a study of the reaction of the four halomesitylenes in liquid ammonia in the presence of both acetone enolate and amide ions as radical traps. Under stimulation by solvated electrons these reactions gave the ketone **13**, the alcohol **14**, the amine **15** and mesitylene. The ratio **(13 + 14)/15** is constant irrespective of the halogen, whereas the ratio **13/14** increases and the proportion of mesitylene decreases as the halogen is varied from fluorine, through chlorine and bromine to iodine. These results 'demonstrate a very unusual juxtaposition of constancy and sharp variability of product ratios within a single reaction series'[137] and reveal a remarkable effect of halogen variation on the $S_{RN}1$ process.

C. $S_{RN}1$ Reactions of Dihaloarenes

Consideration of the product distributions obtained from the reaction of dihalobenzenes with benzenethiolate[21,50], diethyl phosphite ion[22,32,51–53] and phenyl selenide and phenyl telluride[37], and of 2,6-dihalopyridines with pinacolone enolate[56], has led to both an elaboration of the propagation steps of the $S_{RN}1$ reaction as shown in Scheme 3 (equations 50–55, with Q = C_6H_4 for dihalobenzenes) and to powerful additional evidence in support of the $S_{RN}1$ mechanism[6].

The propagation sequence consisting of steps (50)–(52) will lead to monosubstitution whilst that consisting of steps (50), (51) and (53)–(55) will lead to disubstitution. The overall product distributions in these reactions is decided by the relative rates of steps (52) and (53).

One critical factor in deciding the relative importance of the mono- and disubstitution processes, and also in the selection of which of two different halogens, X and Y, will be lost initially (i.e. act as Y in Scheme 3), is the ease of carbon–halogen bond

$$[X-Q-Y]^{-\cdot} \longrightarrow X-Q^{\cdot} + Y^{-} \quad (50)$$

$$X-Q^{\cdot} + A^{-} \longrightarrow [X-Q-A]^{-\cdot} \quad (51)$$

$$[X-Q-A]^{-\cdot} + X-Q-Y \longrightarrow X-Q-A + [X-Q-Y]^{-\cdot} \quad (52)$$

$$[X-Q-A]^{-\cdot} \longrightarrow A-Q^{\cdot} + X^{-} \quad (53)$$

$$A-Q^{\cdot} + A^{-} \longrightarrow [A-Q-A]^{-\cdot} \quad (54)$$

$$[A-Q-A]^{-\cdot} + [X-Q-Y] \longrightarrow A-Q-A + [X-Q-Y]^{-\cdot} \quad (55)$$

SCHEME 3

cleavage in the intermediate radical ions. This cleavage rate is known to be in the order C—I > C—Br > C—Cl > C—F (see Section IV.A).

Dihalobenzenes in which fluorine is one of the halogens have been found to give only monosubstituted products in which the fluorine is retained[22,23,32,50,53]. The C—F bond does not cleave in either step (50) or (53).

Diiodobenzenes give only disubstituted products[22]. Clearly the cleavage step (53), when X = I, is far more efficient than step (52). In similar fashion, bromoiodobenzenes give mainly disubstituted products accompanied by only small amounts of the monosubstituted products[22,32,51–53]. The absence in these latter two series of significant proportions of monosubstituted products, which normally are less reactive than the starting materials[51], is excellent evidence that the radical anion, and not the neutral form, of the monosubstituted compound is the reactive intermediate.

Both *o*- and *p*-chloroiodobenzene[32,50,53] behave in a similar fashion to the bromoiodobenzenes. *m*-Chloroiodobenzene appears to be at the borderline between mono- and disubstitution; with diethyl phosphite ion, monosubstitution predominates[22], whilst with benzenethiolate ion, disubstitution occurs[50]. An increase in the stability of the radical anion of the monosubstituted compound, caused by the presence of the electron-attracting $PO(OEt)_2$ group in the former case, is invoked to explain this difference[32].

The lower reactivity of *m*-chloroiodobenzene, with diethyl phosphite ion, compared with the *p*-isomer has been correlated with electrochemical results[32].

The *o*-haloiodobenzenes react as described above under light stimulation; in the dark, however, a variety of ionic processes occur[53].

The reactions of both 2,6-dibromo- and 2,6-dichloropyridine with pinacolone enolate give excellent yields of the disubstituted products without accumulation of monosubstituted intermediates[56].

The effect of variation in substrate concentrations on the proportions of mono- and disubstitution have been studied and the results are consistent with a reduction in the rate of bimolecular step (52) whilst the rate of the unimolecular step (53) is unaffected. Consequently the overall rate of reaction decreases but the relative amount of disubstitution increases[52].

D. $S_{RN}1$ and $E_{RC}1$ Reactions of Benzylidene Halides

The anion of 2-nitropropane and *p*-nitrobenzylidene dichloride react readily to produce not only the expected monosubstitution product **16** but also the styrene **17**[138]. This result has been interpreted in terms of a normal $S_{RN}1$ cycle (equations 56–59) operating in competition with a subsequent radical chain elimination process, termed $E_{RC}1$ (equations 60–62 and 64), as summarized in Scheme 4. The dimer of 2-nitropropane (**18**) and its radical anion are intimately involved in the propagation steps of

$p\text{-}O_2NC_6H_4CH(X)CMe_2NO_2$

(**16**) X = Cl

(**19**) X = Br

$RC_6H_4CH{=}CMe_2$

(**17**) R = $p\text{-}O_2N$

(**20**) R = p-NC

(**21**) R = $p\text{-}MeO_2C$

(**22**) R = $m\text{-}MeO_2C$

$$ArCHXY + [CMe_2NO_2]^- \longrightarrow [ArCHXY]^{-\cdot} + \dot{C}Me_2NO_2 \quad (56)$$

$$[ArCHXY]^{-\cdot} \longrightarrow Ar\dot{C}HX + Y^- \quad (57)$$

$$Ar\dot{C}HX + [CMe_2NO_2]^- \longrightarrow [ArCH(X)CMe_2NO_2]^{-\cdot} \quad (58)$$

$$[ArCH(X)CMe_2NO_2]^{-\cdot} + ArCHXY \longrightarrow ArCH(X)CMe_2NO_2 + [ArCHXY]^{-\cdot} \quad (59)$$

or

$$[ArCH(X)CMe_2NO_2]^{-\cdot} \longrightarrow Ar\dot{C}HCMe_2NO_2 + X^- \quad (60)$$

$$Ar\dot{C}HCMe_2NO_2 + [CMe_2NO_2]^- \longrightarrow ArCH{=}CMe_2 + \dot{C}Me_2NO_2 + NO_2^- \quad (61)$$

$$\dot{C}Me_2NO_2 + [CMe_2NO_2]^- \longrightarrow [Me_2C(NO_2)CMe_2NO_2]^{-\cdot} \ (\equiv [\mathbf{18}]^{-\cdot}) \quad (62)$$

$$[\mathbf{18}]^{-\cdot} + ArCHXY \longrightarrow [\mathbf{18}] + [ArCHXY]^{-\cdot} \quad (63)$$

$$[\mathbf{18}]^{-\cdot} + ArCH(X)CMe_2NO_2 \longrightarrow [\mathbf{18}] + [ArCH(X)CMe_2NO_2]^{-\cdot} \quad (64)$$

SCHEME 4

these processes (equations 62–64). In the reactions of both *p*-nitrobenzylidene dichloride (X = Y = Cl) and the corresponding bromide chloride (X = Cl, Y = Br)[139], the $S_{RN}1$ cycle predominates, i.e. step (59) is more efficient than step (60). Furthermore, in the latter case none of the monobromo compound (**19**) could be detected, again demonstrating the preference for C—Br cleavage over C—Cl cleavage in radical anions (e.g. step 56).

In the reaction of *p*-nitrobenzylidene dibromide (X = Y = Br) only the styrene **17** was formed and none of the monobromo compound **19**, which was independently shown to be less reactive than the starting material, could be detected[139]. This result, which finds its parallel in the behaviour of diiodobenzenes (Section IV.C), indicates that the radical anion $[\mathbf{19}]^{-\cdot}$ and not the neutral molecule **19** is the reactive intermediate and that its cleavage (equation 60) is much faster than the electron transfer step (59).

The reactions of *m*-nitrobenzylidene dihalides follow the radical pathways of Scheme 4 but incursion of competitive ionic processes takes place[140]. Moderate yields (35–55%) of the styrenes **20–22** can, however, be obtained on treatment of the appropriately substituted benzylidene dibromides with the lithium salt of 2-nitropropane under somewhat more vigorous conditions (Me_2SO, 60°C) than needed for the nitro derivatives.

V. REFERENCES

1. P. B. D. de la Mare and B. E. Swedlund in *The Chemistry of the Carbon–Halogen Bond* (Ed. S. Patai), Wiley, Chichester (1973), Chap. 7.
2. J. D. Roberts, H. E. Simmons, L. A. Carlsmith and C. W. Vaughan, *J. Amer. Chem. Soc.*, **75**, 3290 (1953).

3. J. de Valk and H. C. van der Plas, *Rec. Trav. Chim. Pays-Bas*, **90**, 1239 (1971); **91**, 1414 (1972); A. P. Kroon and H. C. van der Plas, *Rec. Trav. Chim. Pays-Bas*, **92**, 1020 (1973).
4. N. Kornblum in *The Chemistry of the Functional Groups, Supplement F* (Ed. S. Patai), Wiley, Chichester (1982).
5. N. Kornblum, *Angew. Chem., Int. Edn Engl.*, **14**, 734 (1975).
6. J. F. Bunnett, *Acc. Chem. Res.*, **11**, 413 (1978).
7. J. K. Kim and J. F. Bunnett, *J. Amer. Chem. Soc.*, **92**, 7463, 7464 (1970).
8. N. L. Holy and J. D. Marcum, *Angew. Chem., Int. Edn Engl.*, **10**, 115 (1971); G. A. Russell and R. K. Norris in *Organic Reactive Intermediates* (Ed. S. P. McManus), Academic Press, New York (1973), Chap. 6; G. A. Russell and R. K. Norris, *Rev. React. Species Chem. React.*, **1**, 65 (1973).
9. G. A. Russell, *Proc. 23rd Int. Cong. Pure Appl. Chem.*, **4**, 67 (1971).
10. D. Caine, in *Carbon–Carbon Bond Formation*, Vol. 1 (Ed. R. L. Augustine), Marcel Dekker, New York (1979), Chap. 2.
11. R. O. C. Norman, *Chem. Ind.*, 874 (1973).
12. J. F. Bunnett, *Acc. Chem. Res.*, **5**, 139 (1972); *J. Chem. Educ.*, **51**, 312 (1974).
13. N. Kornblum and M. J. Fifolt, *J. Org. Chem.*, **45**, 360 (1980).
14. R. K. Norris and D. Randles, *Aust. J. Chem.*, **29**, 2621 (1976); **32**, 1487 (1979).
15. R. A. Rossi and J. F. Bunnett, *J. Amer. Chem. Soc.*, **94**, 683 (1972).
16. R. A. Rossi and J. F. Bunnett, *J. Org. Chem.*, **38**, 3020 (1973).
17. R. A. Rossi and J. F. Bunnett, *J. Org. Chem.*, **38**, 1407 (1973).
18. J. F. Bunnett and B. F. Gloor, *J. Org. Chem.*, **38**, 4156 (1973).
19. R. A. Rossi and J. F. Bunnett, *J. Amer. Chem. Soc.*, **96**, 112 (1974).
20. J. F. Bunnett and B. F. Gloor, *J. Org. Chem.*, **39**, 382 (1974).
21. J. F. Bunnett and X. Creary, *J. Org. Chem.*, **39**, 3173 (1974).
22. J. F. Bunnett and X. Creary, *J. Org. Chem.*, **39**, 3612 (1974).
23. J. F. Bunnett and J. E. Sundberg, *Chem. Pharm. Bull.*, **23**, 2620 (1975).
24. J. F. Bunnett and X. Creary, *J. Org. Chem.*, **40**, 3740 (1975).
25. J. F. Bunnett and J. E. Sundberg, *J. Org. Chem.*, **41**, 1702 (1976).
26. R. A. Rossi, R. H. de Rossi and A. F. López, *J. Org. Chem.*, **41**, 3371 (1976).
27. J. F. Bunnett, R. G. Scamehorn and R. P. Traber, *J. Org. Chem.*, **41**, 3677 (1976).
28. R. G. Scamehorn and J. F. Bunnett, *J. Org. Chem.*, **42**, 1449 (1977).
29. R. G. Scamehorn and J. F. Bunnett, *J. Org. Chem.*, **42**, 1457 (1977).
30. S. Hoz and J. F. Bunnett, *J. Amer. Chem. Soc.*, **99**, 4690 (1977).
31. J. F. Wolfe, M. P. Moon, M. C. Sleevi, J. F. Bunnett and R. R. Bard, *J. Org. Chem.*, **43**, 1019 (1978).
32. J. F. Bunnett and R. P. Traber, *J. Org. Chem.*, **43**, 1867 (1978).
33. J. F. Bunnett and R. H. Weiss, *Org. Synthesis*, **58**, 134 (1978).
34. J. E. Swartz and J. F. Bunnett, *J. Org. Chem.*, **44**, 340 (1979).
35. R. G. Scamehorn and J. F. Bunnett, *J. Org. Chem.*, **44**, 2604 (1979).
36. R. A. Rossi, R. H. de Rossi and A. B. Pierini, *J. Org. Chem.*, **44**, 2662 (1979).
37. A. B. Pierini and R. A. Rossi, *J. Org. Chem.*, **44**, 4667 (1979).
38. J. E. Swartz and J. F. Bunnett, *J. Org. Chem.*, **44**, 4673 (1979).
39. R. A. Rossi and R. A. Alonso, *J. Org. Chem.*, **45**, 1239 (1980).
40. R. Beugelmans and H. Ginsburg, *JCS Chem. Commun.*, 508 (1980).
41. J. Pinson and J.-M. Savéant, *JCS Chem. Commun.*, 933 (1974).
42. J. Pinson and J.-M. Savéant, *J. Amer. Chem. Soc.*, **100**, 1506 (1978).
43. C. Amatore, J. Chaussard, J. Pinson, J.-M. Savéant and A. Thiebault, *J. Amer. Chem. Soc.*, **101**, 6012 (1979).
44. W. J. M. van Tilberg, C. J. Smit and J. J. Scheele, *Tetrahedron Lett.*, 2113 (1979); 776 (1978).
45. R. B. Bard and J. F. Bunnett, *J. Org. Chem.*, **45**, 1546 (1980).
46. R. Beugelmans and G. Roussi, *JCS Chem. Commun.*, 950 (1979).
47. M. F. Semmelhack, B. P. Chong, R. D. Stauffer, T. D. Rogerson, A. Chong and L. D. Jones, *J. Amer. Chem. Soc.*, **97**, 2507 (1975).
48. M. F. Semmelhack and T. M. Bargar, *J. Org. Chem.*, **42**, 1481 (1977); *J. Amer. Chem. Soc.*, **102**, 7765 (1980).

49. J. F. Wolfe, M. C. Sleevi and R. R. Goehring, *J. Amer. Chem. Soc.*, **102**, 3646 (1980).
50. J. F. Bunnett and X. Creary, *J. Org. Chem.*, **39**, 3611 (1974).
51. J. F. Bunnett and S. J. Shafer, *J. Org. Chem.*, **43**, 1873 (1978).
52. J. F. Bunnett and S. J. Shafer, *J. Org. Chem.*, **43**, 1877 (1978).
53. R. R. Bard, J. F. Bunnett and R. P. Traber, *J. Org. Chem.*, **44**, 4918 (1979).
54. R. A. Rossi, R. H. de Rossi and A. F. López, *J. Amer. Chem. Soc.*, **98**, 1252 (1976).
55. A. B. Pierini and R. A. Rossi, *J. Organometal. Chem.*, **144**, C12 (1978).
56. A. P. Komin and J. F. Wolfe, *J. Org. Chem.*, **42**, 2481 (1977).
57. R. Beugelmans, B. Boudet and L. Quintero, *Tetrahedron Lett.*, 1943 (1980).
58. J. F. Wolfe, J. C. Greene and T. Hudlicky, *J. Org. Chem.*, **37**, 3199 (1972).
59. J. V. Hay, T. Hundlicky and J. F. Wolfe, *J. Amer. Chem. Soc.*, **97**, 374 (1975).
60. J. V. Hay and J. F. Wolfe, *J. Amer. Chem. Soc.*, **97**, 3702 (1975).
61. J. A. Zoltewicz and T. M. Oestreich, *J. Amer. Chem. Soc.*, **95**, 6863 (1973).
62. J. F. Bunnett and B. F. Gloor, *Heterocycles*, **5**, 377 (1976).
63. J. F. Wolfe and D. R. Carver, *Org. Prep. Proced. Int.*, **10**, 225 (1978).
64. D. A. de Bie, H. C. van der Plas and B. Geurtsen, *JCS Perkin I*, 1363 (1974); E. A. Oostveen and H. C. van der Plas, *Rec. Trav. Chim. Pays-Bas*, **98**, 441 (1979).
65. T. Yamaguchi and H. C. van der Plas, *Rec. Trav. Chim. Pays-Bas*, **96**, 89 (1977).
66. J. F. Bunnett, X. Creary and J. E. Sundberg, *J. Org. Chem.*, **41**, 1707 (1976).
67. G. A. Russell and J. M. Pecoraro, *J. Amer. Chem. Soc.*, **101**, 3331 (1979).
68. N. Kornblum, P. Ackermann and R. T. Swiger, *J. Org. Chem.*, **45**, 5294 (1980).
69. R. A. Rossi and J. F. Bunnett, *J. Org. Chem.*, **37**, 3570 (1972).
70. R. A. Rossi and J. F. Bunnett, *J. Org. Chem.*, **38**, 2314 (1973).
71. G. A. Russell, R. K. Norris and E. J. Panek, *J. Amer. Chem. Soc.*, **93**, 5839 (1971).
72. G. A. Russell, B. Mudryk and M. Jawdosiuk, *Synthesis*, **62** (1981).
73. G. A. Russell, M. Jawdosiuk and F. Ros, *J. Amer. Chem. Soc.*, **101**, 3378 (1979).
74. N. Kornblum and M. M. Kestner, unpublished data cited in Ref. 5.
75. W. R. Bowman and G. D. Richardson, *JCS Perkin I*, 1407 (1980).
76. G. A. Russell and J. Hershberger, *JCS Chem. Commun.*, 216 (1980).
77. J.-M. Savéant, *Acc. Chem. Res.*, **13**, 323 (1980).
78. S. Rajan and P. Sridaran, *Tetrahedron Lett.*, 2177 (1977).
79. R. A. Rossi and A. B. Pierini, *J. Org. Chem.*, **45**, 2914 (1980).
80. F. Ciminale, G. Bruno, L. Testaferri, M. Tiecco and G. Martelli, *J. Org. Chem.*, **43**, 4509 (1978).
81. A. M. Aguiar, H. J. Greenberg and K. E. Rubenstein, *J. Org. Chem.*, **28**, 2091 (1963).
82. R. A. Rossi, R. H. de Rossi and A. F. López, *J. Org. Chem.*, **41**, 3367 (1976).
83. R. C. Kerber, G. W. Urry and N. Kornblum, *J. Amer. Chem. Soc.*, **87**, 4520 (1965).
84. G. A. Russell and W. C. Danen, *J. Amer. Chem. Soc.*, **88**, 5663 (1966).
85. N. Kornblum, T. M. Davies, G. W. Earl, G. S. Greene, N. L. Holy, R. C. Kerber, J. W. Manthey, M. T. Musser and D. H. Snow, *J. Amer. Chem. Soc.*, **89**, 5714 (1967).
86. B. L. Burt, D. J. Freeman, P. G. Gray, R. K. Norris and D. Randles, *Tetrahedron Lett.*, 3063 (1977).
87. R. K. Norris and D. Randles, *Aust. J. Chem.*, **32**, 2413 (1979).
88. O. P. Marquez, J. Giulianelli, T. C. Wallace and D. H. Eargle, Jr., *J. Org. Chem.*, **41**, 739 (1976).
89. M. M. Kestner, Ph.D. thesis, Purdue University, May 1973; cited in Ref. 5.
90. A. H. Maki and D. H. Geske, *J. Amer. Chem. Soc.*, **82**, 2671 (1960); **83**, 1852 (1961).
91. L. H. Piette, P. Ludwig and R. N. Adams, *J. Amer. Chem. Soc.*, **84**, 4212 (1962).
92. P. H. Rieger and G. K. Fraenkel, *J. Chem. Phys.*, **39**, 1793 (1963).
93. I. Bernal and G. K. Fraenkel, *J. Amer. Chem. Soc.*, **86**, 1671 (1964).
94. S. I. Weissman, T. L. Chu, G. E. Pake, D. E. Paul and J. Townsend, *J. Phys. Chem.*, **57**, 504 (1953).
95. G. A. Russell and E. G. Janzen, *J. Amer. Chem. Soc.*, **84**, 4153 (1962); G. A. Russell, E. G. Janzen and E. T. Strom, *J. Amer. Chem. Soc.*, **86**, 1807 (1964).
96. P. A. Wade, Ph.D. thesis, Purdue University, May 1973; cited in Ref. 5.
97. A. L. Scher and N. N. Lichtin, *J. Amer. Chem. Soc.*, **97**, 7170 (1975).
98. N. Kornblum, M. M. Kestner, S. D. Boyd and L. C. Cattran, *J. Amer. Chem. Soc.*, **95**, 3356 (1973).

99. J. Casanova and L. Eberson, in *The Chemistry of the Carbon–Halogen Bond* (Ed. S. Patai), Wiley, Chichester (1973), Chap. 15.
100. N. L. Holy, *Chem. Revs*, **74**, 243 (1974).
101. J. G. Lawless, D. E. Bartak and M. D. Hawley, *J. Amer. Chem. Soc.*, **91**, 7121 (1969).
102. P. Neta and D. Behar, *J. Amer. Chem. Soc.*, **102**, 4798 (1980).
103. G. J. Gores, C. E. Koeppe and D. E. Bartak, *J. Org. Chem.*, **44**, 380 (1979).
104. C. P. Andrieux, C. Blocman, J.-M. Dumas-Bouchiat and J.-M. Savéant, *J. Amer. Chem. Soc.*, **101**, 3431 (1979).
105. C. P. Andrieux, C. Blocman, J.-M. Dumas-Bouchiat, F. M'Halla and J.-M. Savéant, *J. Amer. Chem. Soc.*, **102**, 3806 (1980).
106. F. M'Halla, J. Pinson and J.-M. Savéant, *J. Amer. Chem. Soc.*, **102**, 4120 (1980).
107. K. Alwair and J. Grimshaw, *JCS Perkin II*, 1150 and 1811 (1973).
108. R. D. Chambers, D. T. Clark, C. R. Sargent and F. G. Drakesmith, *Tetrahedron Lett.*, 1917 (1979).
109. J. P. Coleman, Naser-ud-din, H. G. Gilde, J. H. P. Utley, B. C. L. Weedon and L. Eberson, *JCS Perkin II*, 1903 (1973).
110. R. K. Norris and R. J. Smyth-King, *JCS Chem. Commun.*, 79 (1981); R. J. Smyth-King, M.Sc. thesis, University of Sydney, November 1980.
111. A. L. J. Beckwith and R. O. C. Norman, *J. Chem. Soc. B*, 403 (1969).
112. G. A. Russell and A. R. Metcalfe, *J. Amer. Chem. Soc.*, **101**, 2359 (1979).
113. D. E. Bartak, W. C. Danen and M. D. Hawley, *J. Org. Chem.*, **35**, 1206 (1970).
114. M. McMillan and R. O. C. Norman, *J. Chem. Soc. B*, 590 (1968).
115. D. Y. Myers, G. G. Stroebel, B. R. Ortiz de Montellano and P. D. Gardner, *J. Amer. Chem. Soc.*, **96**, 1981 (1974).
116. D. Y. Myers, G. G. Stroebel, B. R. Ortiz de Montellano and P. D. Gardner, *J. Amer. Chem. Soc.*, **95**, 5832 (1973).
117. L. M. Tolbert, *J. Amer. Chem. Soc.*, **102**, 3531 (1980).
118. C. Galli and J. F. Bunnett, *J. Amer. Chem. Soc.*, **101**, 6137 (1979).
119. G. A. Russell, F. Ros and B. Mudryk, *J. Amer. Chem. Soc.*, **102**, 7601 (1980).
120. R. F. Hudson, *Angew. Chem., Int. Edn Engl.*, **12**, 36 (1973).
121. R. K. Norris and D. Randles, *J. Org. Chem.* (1982); D. Randles, Ph.D. thesis, The University of Sydney, October 1979.
122. L. M. Dorfman, *Acc. Chem. Res.*, **3**, 224 (1970).
123. M. Szwarc, *Acc. Chem. Res.*, **5**, 169 (1972).
124. S. I. Weissman, *Z. Electrochem.*, **64**, 47 (1960); M. T. Jones and S. I. Weissman, *J. Amer. Chem. Soc.*, **84**, 4269 (1962).
125. R. L. Ward, *J. Chem. Phys.*, **32**, 410 (1960).
126. R. F. Bridger and G. A. Russell, *J. Amer. Chem. Soc.*, 3754 (1963); however, also see B. Helgee and V. D. Parker, *Acta Chem. Scand. B*, **34**, 129 (1980).
127. J. A. Zoltewicz, T. M. Oestreich and A. A. Sale, *J. Amer. Chem. Soc.*, **97**, 5889 (1975).
128. P. J. Newcombe and R. K. Norris, *Aust. J. Chem.*, **31**, 2463 (1978).
129. S. D. Barker and R. K. Norris, *Tetrahedron Lett.*, 973 (1979).
130. P. J. Newcombe and R. K. Norris, unpublished data; P. J. Newcombe, Ph.D. thesis, University of Sydney, September 1980.
131. P. J. Newcombe and R. K. Norris, *Aust. J. Chem.*, **32**, 2647 (1979).
132. E. E. van Tamelen and G. van Zyl, *J. Amer. Chem. Soc.*, **71**, 835 (1949).
133. J. J. Zeilstra and J. B. F. N. Engberts, *Rec. Trav. Chim. Pays-Bas*, **92**, 954 (1973).
134. M. Anbar and E. J. Hart, *J. Amer. Chem. Soc.*, **86**, 5633 (1964).
135. J. F. Bunnett and X. Creary, unpublished data; cited in Ref. 6.
136. R. R. Bard, J. F. Bunnett, X. Creary and M. J. Tremelling, *J. Amer. Chem. Soc.*, **102**, 2852 (1980).
137. M. J. Tremelling and J. F. Bunnett, *J. Amer. Chem. Soc.*, **102**, 7375 (1980).
138. D. J. Girdler and R. K. Norris, *Tetrahedron Lett.*, 2375 (1975); D. J. Freeman and R. K. Norris, *Aust. J. Chem.*, **29**, 2631 (1976).
139. D. J. Freeman, R. K. Norris and S. K. Woolfenden, *Aust. J. Chem.*, **31**, 2477 (1978).
140. R. K. Norris and R. J. Smyth-King, *Aust. J. Chem.*, **32**, 1949 (1979).
141. R. K. Norris, unpublished work.

The Chemistry of Functional Groups, Supplement D
Edited by S. Patai and Z. Rappoport
© 1983 John Wiley & Sons Ltd

CHAPTER **17**

Reactions involving solid organic halides

E. HADJOUDIS
Solid State Chemistry Laboratory, 'Demokritos' Nuclear Research Centre, Athens, Greece

I. INTRODUCTION

In spite of the scarcity, and the lack of uniformity, of information available about reactions involving solid organic halides, which makes difficult a systematic approach to the subject, the decisive factors for writing this chapter were the new trends and strategies offered by such reactions in the growing field of solid state organic chemistry.

The basic differences between reactions in the solid state and those in fluid and amorphous systems are based on three aspects of crystalline order. First, all the molecules in a given crystal occur in a unique conformation (in some cases in a small number of conformations). Second, the geometries relating different molecules, and the types of possible intermolecular approaches, are limited in the crystal but not in other phases. Third, the relative arrangement of molecules – the molecular packing – absent in fluid or amorphous systems, dominates the chemical behaviour of the solid. These three aspects are closely interrelated and one may not be able to determine which factor is responsible for the particular chemical behaviour of the crystal. In general the conformational effect will be important in monomolecular reactions and the crystal structure and molecular packing effects in bimolecular and polymolecular ones.

In order to accomplish the aims of this chapter in the best possible way, let us divide the reactions involving solid organic halides into five classes: thermal solid state reactions; photochemical solid state reactions; gas–solid reactions; reactions at solid surfaces; miscellaneous reactions. The above reactions have been reviewed in general, but none of these reviews has treated specifically their chemical and mechanistic aspects in terms of the halide participation. In this light the following chapter will treat reactions which involve solid organic halides as starting material, as intermediate, or as product.

II. THERMAL SOLID STATE REACTIONS

A. Aromatic Diazonium Salts

In a study of the structure and solid state chemistry of the ionic salt 3-carboxynaphthalenediazonium bromide (**1a**) Gougoutas and Johnson[1] reported that replacement of the diazonium group by the indigenous bromide ion occurs in quantitative yield in the crystalline matrix. Later, Gougoutas[2] studied the crystal structure and solid state behaviour of the corresponding crystalline diazonium iodide, **1b**, and

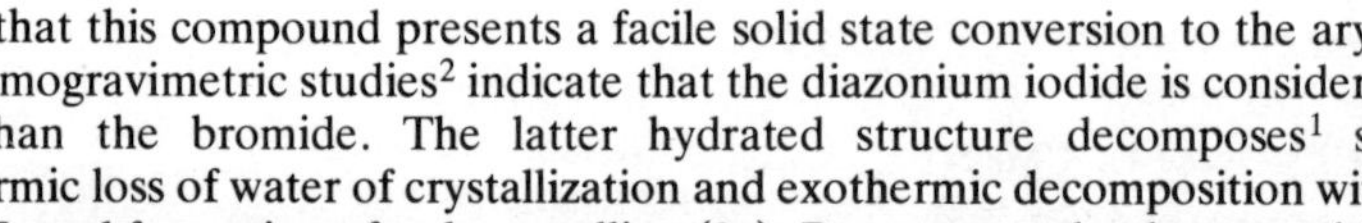

N_2^+ X^- —Solid→ X ... COOH + N_2↑ (1)

COOH

(**1a**) X = Br (**2**)

(**1b**) X = I

showed that this compound presents a facile solid state conversion to the aryl iodide, **2b**. Thermogravimetric studies[2] indicate that the diazonium iodide is considerably less stable than the bromide. The latter hydrated structure decomposes[1] stepwise: endothermic loss of water of crystallization and exothermic decomposition with evolution of N_2 and formation of polycrystalline (**2a**). By contrast, the decomposition of **1b** to **2b** proceeds with evolution of *both* N_2 and water of crystallization. This difference in solid state chemistry of the triclinic structure **1a** and of the orthorhombic structure **1b** is attributed to the fundamentally different packing of these molecules of hydrogen bonding which in both structures link each halide ion to two water molecules, and each of the latter to two halide ions.

The corresponding tetrafluoroborate (X = BF_4) is more stable and such derivatives

$$\text{(3-carboxy-2-naphthalenediazonium)}\ BF_4^- \xrightarrow{\text{Solid}} \text{3-fluoro-2-naphthoic acid} + N_2\uparrow + BF_3\uparrow \quad (2)$$

serve, through solid state decomposition with evolution of N_2 and BF_3, as practical laboratory precursors of aryl fluorides[3].

B. Decarboxylation

p-Aminosalicylic acid hydrochloride can undergo relatively facile dehydrochlorination. On this basis, it was suggested[4] that the decarboxylation of this compound involves initial loss of HCl, followed by decarboxylation, sublimation of *m*-aminophenol, and finally hydrochlorination of the *m*-aminophenol product (equation 3).

$$\text{4-}NH_3Cl\text{-2-}OH\text{-}C_6H_3COOH \longrightarrow \text{4-}NH_2\text{-2-}OH\text{-}C_6H_3COOH + HCl \longrightarrow [\text{4-}NH_2\text{-2-}OH\text{-}C_6H_4(H)(COO^-)]^+ \longrightarrow \text{3-}NH_2C_6H_4OH + CO_2 \xrightarrow{HCl} \text{3-}NH_3Cl\text{-}C_6H_4OH \quad (3)$$

One explanation of the solid state reaction is that it proceeds through the crystal by addition of a carboxylate proton either to ArCOOH or $ArCOO^-$, as does the reaction in aqueous solution.

C. Biologically Important Molecules

Choline chloride (**3**) crystallizes in a stable orthorhombic modification and an unstable, high temperature, cubic modification referred as α-form and β-form[5,6]. The α-form is the most ionizing radiation-sensitive compound known (G factor for radical formation about 2 and G factor for radiolysis as high as 55 000, four orders of magnitude higher than G-values commonly observed for primary processes of radiation damage in organic molecules) while the high temperature β-form is normally radiation sensitive. The difference in radiation sensitivity between the two crystalline forms is probably related to the crystal packing of the two forms. This is supported by the fact that choline chloride in solution is normally radiation sensitive ($G = 3$) and choline iodide shows no elevated reactivity. Probably, the chain reaction leading to the overall reaction of equation (4) is favoured with the particular geometrical arrangement of the α-form, since the sensitivity of choline chloride crystals to ionizing radiation is greatly reduced at elevated temperatures[7] and the stabilization is associated with the phase transition from the orthorhombic to the face-centred cubic form[8]. An examination of

$$\underset{(\mathbf{3})}{[(CH_3)_3NCH_2CH_2OH]^+\ Cl^-} \xrightarrow{\text{Solid}} [(CH_3)_3NH]^+\ Cl^- + CH_3CHO \quad (4)$$

the packing in the α-form indicates that the sensitivity in radiation may be due to the fact that the propagation proceeds rapidly through the stacks of choline molecules, like a topochemically controlled polymerization.

This phenomenon was studied in detail by the group led by Lemmon[9,10], and it was found that choline bromide was the only other similar compound that decomposed at a comparable rate.

A number of nitrogen mustards exhibit antitumour activity and there has been considerable interest in 5-[3,3-bis(2-chloroethyl)-1-triazeno]pyrazole-4-carboxamide (**4**) and 5-[3,3-bis(2-chloroethyl)-1-triazeno]imidazole-4-carboxamide (**5**). These compounds showed significant activity in experimental tumours but have been disappointing in many clinical trials[6]. The reason for these features lies in the fact that these two anticancer agents undergo solid state cyclization reactions which render them nearly useless because of the difficulty involved in stabilizing these compounds during storage. Thus **4** undergoes spontaneous ring closure to form the compound 1-(2-chloroethyl)-3-(4-carbamoylpyrazol-3-yl)-Δ^2-1,2,3-triazolinium chloride (**6**) (equation 5) and compound **5** cyclizes similarly to **7** (equation 6). The structures of the products **6** and **7** were established by crystal structure determinations[11,12].

(**4**) —Solid, Δ→ (**6**) (5)

(**5**) —Solid, Δ→ (**7**) (6)

The solid state reactions raise the question of whether the bis-2-chloroethyl-1-triazenes crystallize with one chloroethyl group in a conformation favourable to ring closure such as the bipyrimidal geometry of the $N_2 \cdots CH_2CH_2Cl$ complex which is favourable to an S_N2 reaction. A favourable conformation (**8**) to this ring closure was proposed[13] but the geometry at the reaction centre is not perfectly trigonal bipyramidal, with the chloride anion leaving as the nitrogen attacks because of the suggested N–H···Cl hydrogen bond. Thus, in this series it seems that the most important crystal structure is that of the starting material. Finally, it should be stressed that, of practical importance here, are experiments directed towards crystallizing these bis-2-chloroethyltriazenes in unreactive solid state modifications in order to be useful as clinical agents.

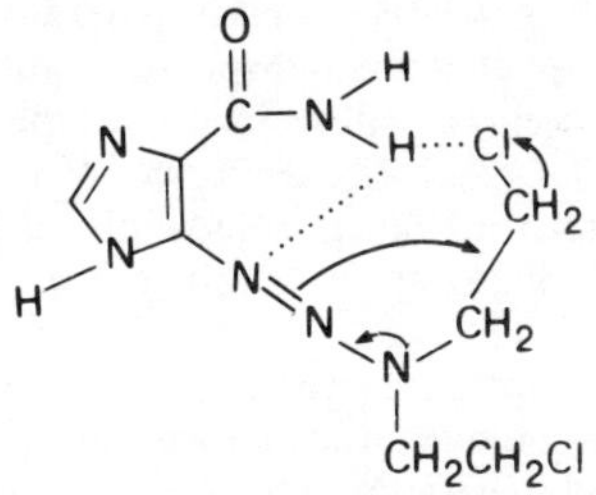

(8)

III. PHOTOCHEMICAL SOLID STATE REACTIONS

A. Anthracene Halides

Cohen, Schmidt and their coworkers in a systematic chemical and crystallographic study[14] have shown that halo-substituted anthracenes, like cinnamic acids[15], fall into at least three packing types (equation 7).

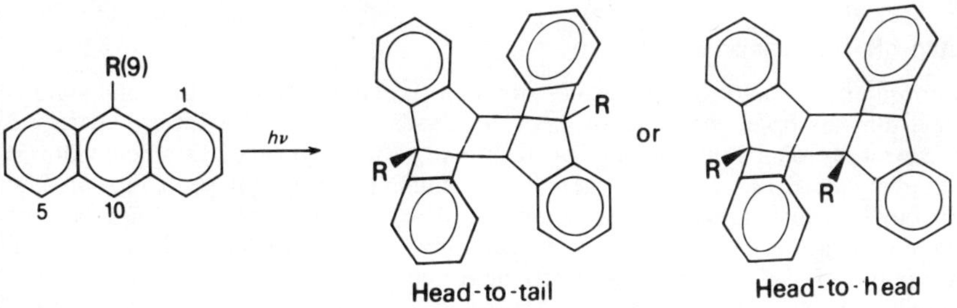

R(9) 1 5 10 $\xrightarrow{h\nu}$ Head-to-tail or Head-to-head (7)

α-type

Isomorphous	9-Cl, 9-Br, 1-Cl	$\xrightarrow{h\nu}$	Head-to-tail dimer

β-type

	1,5-Dichloro, 1-(2,4-Dichlorophenoxycarbonyl)	$\xrightarrow{h\nu}$	Head-to-tail dimer
Space group $P_{2_12_12_1}$, $P_{2_12_12_1}$	9-Cl, 9-CN	$\xrightarrow{h\nu}$	Head-to-tail dimer
$P_{2_12_12_1}$, $P_{2_1/a}$	1,10-Dichloro, 9-Br	$\xrightarrow{h\nu}$	Light-stable

γ-type

	1-Cl, 1-CN	$\xrightarrow{h\nu}$	Light-stable

The α-type units are packed pairwise across centres of symmetry such that the C(9)···C(10′) distances are short (3.6 Å). The monosubstituted anthracenes, whether substituted at position 9 (R = Cl, Br) or elsewhere (1-chloroanthracene, second modification) produce the topochemically expected 1-dimer in low yield.

The anthracene halides which crystallize in the β-type fall into three classes: first, 1,5-dichloroanthracece and 2,4-dichlorophenyl ester of 1-anthroic acid which dimerize to the topochemical head-to-head dimers; secondly, a light-stable group (R = Br, second modification); third, a group yielding the 1-dimer {R = Cl, second modification; CN}.

The γ-types (1-chloroanthracene; 1-cyanoanthracene) have the molecular planes not parallel and the distances between the *meso*-atoms of neighbouring molecules are greater than 5Å. As expected these compounds are light-stable in the crystalline state, though they photodimerize in solution.

The photochemistry of these compounds shows that structural features other than the topochemical ones formulated during the dimerization of cinnamic acids[15], are involved in solid state photochemical dimerizations. Thus, as is seen in equation (7), one form of the dimorphic 9-chloroanthracene and 9-cyanoanthracene yields centric head-to-tail dimers whereas mirror symmetric head-to-head dimers would be expected from the topochemical proformation theory[15]. However, this perverse behaviour of the 9-halo-substituted anthracenes is also found in a mechanism involving crystal imperfection[16–19].

B. Azo-*bis*-isobutyronitrile

A considerable factor of importance in unimolecular reactions in the solid state is the fact that the fragments of dissociative processes cannot readily diffuse apart and also their rotational motions may be restricted. An example of this behaviour is provided by McBride and coworkers[20] who studied the photolysis of azo-*bis*-isobutyronitrile, which crystallizes from methanol in two modifications. Photolysis of either modification generates pairs of cyanopropyl radicals which collapse to give abnormally high yields of the disproportionation products isobutyronitrile and methacrylonitrile, low yields of the symmetrical coupling product tetramethylsuccinodinitrile, and little or none of the unsymmetrical coupling product dimethyl-*N*-(2-cyano-2-propyl) ketenimine, which predominates in solution reaction (equation 8).

$$(CH_3)_2C(CN){-}N{=}N{-}C(CN)(CH_3)_2 \xrightarrow{\text{Solid}} \underset{(5\%)}{(CH_3)_2C(CN){-}C(CN)(CH_3)_2} + \underbrace{(CH_3)_2CH(CN) + CH_2{=}C(CH_3){-}CN}_{(95\%)} \quad (8)$$

$$(CH_3)_2C(CN){-}N{=}N{-}C(CN)(CH_3)_2 \xrightarrow{\text{Solution}} \underset{(40\%)}{(CH_3)_2C(CN){-}C(CN)(CH_3)_2} + \underset{(55\%)}{(CH_3)_2C{=}C{=}NC(CN)(CH_3)_2} + \underbrace{(CH_3)_2CH(CN) + CH_2{=}C(CH_3){-}CN}_{(5\%)}$$

The authors, discussing the relative yields, argue on the basis of their X-ray results[21] that an important factor in determining the molecular packing is the dipole–dipole attraction of the antiparallel nitrile groups. The path leading to the ketenimine, although it possibly might not distort greatly the reaction cavity, is energetically unfavourable because it counteracts the dipole–dipole attraction, while the disproportion reaction can proceed without disrupting the cavity or the nitrile framework. One of the radicals formed can rotate about its C—C—N axis so that a methyl replaces the nitrogen and follows hydrogen transfer (equation 9).

$$(CH_3)_2C(C{\equiv}N)-N{\equiv}N-C(CH_3)_2C{\equiv}N \longrightarrow (CH_3)_2\dot{C}-C{\equiv}N + N{\equiv}N + N{\equiv}C-\dot{C}(CH_3)_2 \longrightarrow CH_2{=}C(CH_3)-C{\equiv}N + N{\equiv}C-CH(CH_3)_2 \quad (9)$$

IV. GAS–SOLID REACTIONS

A. Bromination

Schmitt[22] showed in 1863 that solid cinnamic acid reacts with bromine vapour to give cinnamic acid dibromide (equation 10) while the bromination of anthracene in the solid state was noted in 1870[23]. Many examples of this type of reaction followed

$$C_6H_5CH{=}CHCOOH\ (\text{Solid}) + Br_2\ (\text{Gas}) \longrightarrow C_6H_5CHBrCHBrCOOH\ (\text{Solid}) \quad (10)$$

primarily for synthetic reasons. Thus Buckles and coworkers[24] studied a large group of solid aromatic compounds and found that they react with bromine vapour to yield products in which the *p*-positions of unsubstituted phenyl groups were brominated and in which olefinic double bonds, not highly substituted, added bromine in a way expected in solution bromination by an ionic mechanism. They interpreted the reaction as occurring in either an absorbed face or in a thin film of solution on the surface of the solid. Evidently the work has been directed at understanding the influence of crystal structure upon the course of the reaction. Thus Labes and coworkers[25], following a survey of the reactions of several solid aromatic hydrocarbons with bromine, e.g. naphthalene to 1,4-dibromonaphthalene, anthracene to 9,10-dibromoanthracene and 9,10-dibromo-1,2,3,4-tetrabromotetrahydroanthracene, perylene to 3,9- and 3,10-dibromoperylene, performed a quantitative study of the decomposition of a perylene–bromine charge-transfer complex in the solid state to 3,9-dibromoperylene (equation 11) and viewed it as an intimate, stoicheiometric mixture of reactants

$$\text{perylene}\cdot(Br_2)_{3.5} \longrightarrow \text{3,9-dibromoperylene}\cdot(Br_2)_{1.5} + 2\,HBr \quad (11)$$

TABLE 1. Bromination of solid organic compounds by bromine vapour

Compound	Bromination product
trans-Stilbene	*meso*-1,2-Diphenyl-1,2-dibromoethane
cis-Stilbene	*meso*-1,2-Diphenyl-1,2-dibromoethane
Mesaconic acid	D,L-*erythro*-α,β-Dibromomethylsuccinic acid
Citraconic acid	D,L-*threo*-α,β-Dibromomethylsuccinic acid
trans-Dibenzoylethylene	*meso*-1,4-Diphenyl-2,3-dibromobuta-1,4-dione
cis-Dibenzoylethylene	D,L-1,4-Diphenyl-2,3-dibromobuta-1,4-dione
trans-1,2-Di-*p*-chlorobenzoylethylene	*meso*-1,4-Di-4-chlorophenyl-2,3-dibromobuta-1,4-dione
cis-1,2-Di-*p*-chlorobenzoylethylene	D,L-1,4-Di-4-chlorophenyl-2,3-dibromobuta-1,4-dione
trans-1,2-Di-*p*-methylbenzoylethylene	*meso*-1,4-Di-4-methylphenyl-2,3-dibromobuta-1,4-dione
cis-1,2-Di-*p*-methylbenzoylethylene	*meso*-1,4-Di-4-methylphenyl-2,3-dibromobuta-1,4-dione

ordered at the molecular level in which the course of the reaction reflected the minimum amount of atomic or molecular movement.

Hadjoudis and Schmidt[26–30] demonstrated that solid α,β-unsaturated acids, amides and ketones yield, on exposure to bromine vapour, dibromides in quantitative or near-quantitative yield, even where addition of bromine in solution has been reported to be difficult. Some representative cases are shown in Table 1. In general the addition of bromine proceeds by *trans*-addition to *cis*- and *trans*-substituted ethylenes; the abnormal direction of addition to *cis*-substituted ethylenes and stilbenes was attributed to *cis–trans* isomerization prior to or during addition to *erythro*- or *meso*-dibromides. The solid state *cis–trans* isomerization was demonstrated by treatment of solid olefins with iodine vapour[27]. By performing brominations of several crystalline modifications of the same compound, it was possible to show that the crystal structure plays a role in the course of bromination[26].

After the above studies on polycrystalline materials Panzien and Schmidt[31] performed a topochemically controlled gas–solid reaction in a single crystal which resulted in two enantiomeric dibromides in different yields. These workers used non-chiral disubstituted ethylenes that crystallize in chiral space groups. In a given enantiomorphic single crystal, all molecules adopt a prochiral conformation of a single chirality. Thus, in the heterogeneous bromination of single crystals of 4,4′-dimethylchalcone (space group $P_{2_12_12_1}$) with a gaseous bromine, a 6% excess of one enantiomer was obtained. This excess can rise to 22% depending upon experimental conditions[32,33]. These results have been interpreted by assuming formation of a bromonium ion as an intermediate, formed preferentially from the side of the molecule having the least steric hindrance at the C=C bond.

Lahav and coworkers[34], in order to circumvent drawbacks such as the preparation of large homochiral crystals, employed chiral ethylenes in which *trans* addition of bromine gas could yield two diastereoisomeric dibromides. Thus *trans*-cinnamoylalanine (X = H) and 2-chloro-*trans*-cinnamoylalanine (X = Cl), resolved and racemic, both yield diastereoisomeric products in ratios of 55:45 to 60:40 upon solid state bromination for the first case, while for the second the resolved crystal yields a ratio of almost 50:50 and the racemate a ratio of 60:40, as determined by NMR integration of the methyl groups. The steric influence of the chlorosubstituent probably plays a primary role in determining the configuration of the reaction product.

These experiments show that it is possible to modify the reaction pathways by incorporating a flexible molecule into a rigid crystal lattice.

In the course of gas–solid reactions Naae[35] showed that the ionic addition of

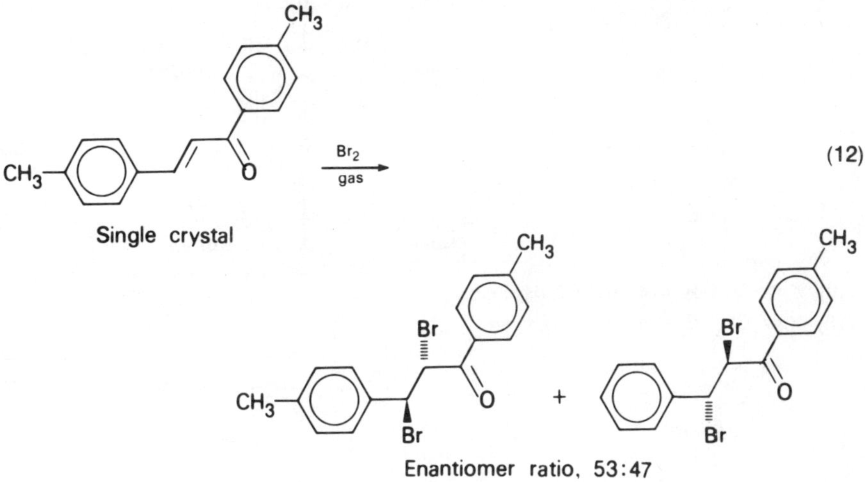

Single crystal

Enantiomer ratio, 53:47

(12)

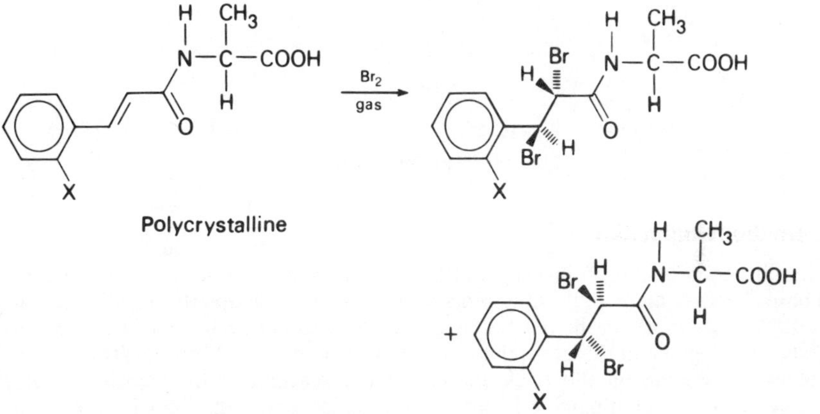

Polycrystalline

(13)

bromine to solid (E)- and (Z)-p-$HOOCC_6H_4CF{=}CFX$ ($X = Cl, CF_3$), leads to the *trans*-dibromo adduct while the radical reactions indicate a slight preference for *cis* addition. For $X = CF_3$, the Z isomer preferentially adds bromine *cis* under either ionic or radical conditions. The E isomer also shows a preference for *cis* addition but the reaction is complicated by competing mechanisms. It should be noted that solution reactions for $X = CF_3$ are mainly non-stereoselective. The reaction of (E)-**10** in the solid state with exposure to more intense light causes a loss of stereospecificity. An explanation of this result may be the two pathways for bromination of (E)-**10**. The chain propagation step, reaction with bromine gas, gives a slight preference to the *cis* adduct. This is in agreement with the 38:12 diastereoisomer ratio for the solid–gas reaction under normal illumination. The competing path is a chain termination step where $Br^{\bullet}$ reacts with the radical to give equal amounts of the diastereoisomers. This path is favoured by the more intense light as this increases the bromine atom concentration. A comparison of the ionic brominations of **9** with **10** demonstrates a difference in mechanism and the author believes that a bromonium ion is involved with **9**, while a non-bridged, open cation is the major intermediate for **10**.

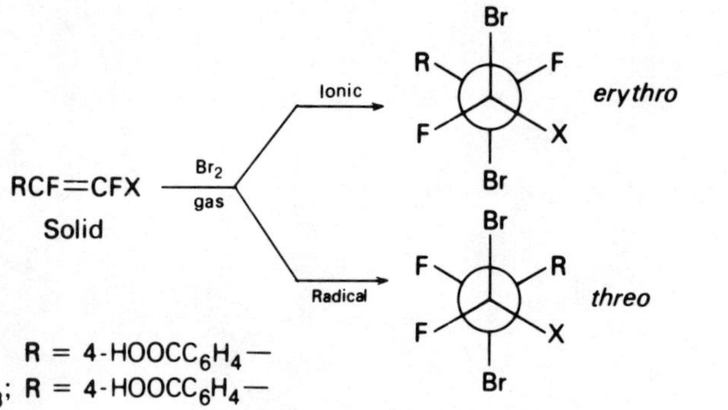

$$\mathrm{RCF{=}CFX}\ (\text{Solid}) \xrightarrow[\text{gas}]{\mathrm{Br_2}} \text{Ionic} \rightarrow \textit{erythro};\quad \text{Radical} \rightarrow \textit{threo} \tag{14}$$

(**9**) X = Cl; R = 4-$HOOCC_6H_4$—
(**10**) X = CF_3; R = 4-$HOOCC_6H_4$—

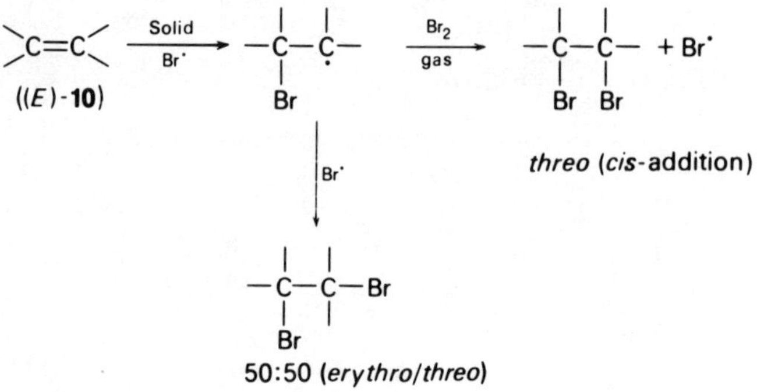

$$\mathrm{>C{=}C<}\ ((E)\text{-}\mathbf{10}) \xrightarrow[\mathrm{Br^{\cdot}}]{\text{Solid}} \mathrm{-C(Br)-\dot{C}-} \xrightarrow[\text{gas}]{\mathrm{Br_2}} \mathrm{-C(Br)-C(Br)-} + \mathrm{Br^{\cdot}} \tag{15}$$

threo (*cis*-addition)

$\mathrm{-C(Br)-\dot{C}-} \xrightarrow{\mathrm{Br^{\cdot}}} \mathrm{-C(Br)-C-Br}$

50:50 (*erythro*/*threo*)

B. Dehydrohalogenation

Schmidt and coworkers[36] investigated the dehydrohalogenation of organic halides with bases because many of these systems were expected to adopt, during elimination, a transition state in which the reacting hydrogen is in an antiperiplanar orientation to the leaving halogen and therefore these reactions were expected to yield *cis* and *trans* ethylenes, depending on the conformation of the reactant in the transition state. Molecules with two eliminatable hydrogens on the β-carbon may populate two conformations, which upon anti-elimination will yield a mixture of ethylenes (equation 16).

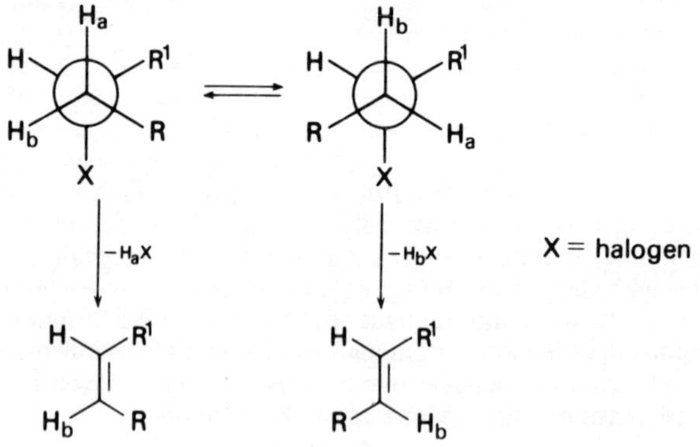

$$\xrightarrow{-\mathrm{H_aX}} \mathrm{H(R^1)C{=}C(H_b)R} \qquad \rightleftharpoons \qquad \xrightarrow{-\mathrm{H_bX}} \mathrm{H(R^1)C{=}C(R)H_b} \qquad \mathrm{X} = \text{halogen} \tag{16}$$

With these guidelines in mind the authors investigated the reactions of *meso*-β,β'-dichloro and dibromo adipic esters with ammonia and gases. The gaseous amine induces double dehydrohalogenation of the solid organic halides and yields practically pure diesters of *trans,trans*-hexa-2,4-dienedioic acids while the β-elimination of the same compounds in solution results in a complex mixture of products such as *trans,trans* and *cis,trans* diene esters and amides. The molecular skeletons of the isomorphous dibromo and dichloro diesters are centrosymmetrical and fully extended[37] while the two enantiotopic hydrogens (H_a) are almost antiperiplanar to the leaving halogens, in agreement with the *anti*-type elimination[38] (equation 17).

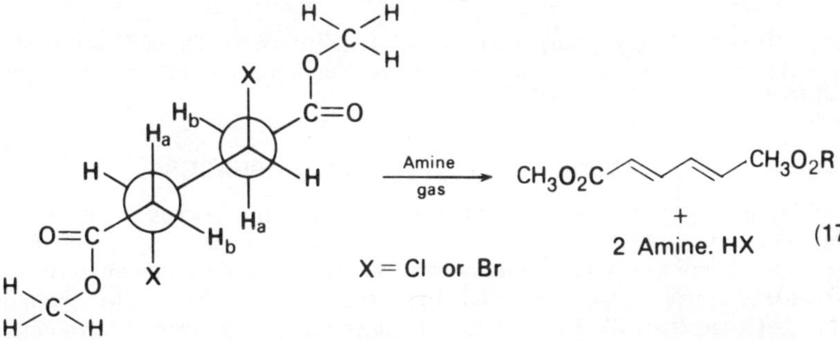

$X = Cl$ or Br + 2 Amine. HX (17)

The β-elimination reactions of the corresponding *meso*-α,α'-dimethyl homologues provided further evidence for this mechanism[39–41]. These molecules have only one eliminatable hydrogen for each halogen and are centrosymmetric in the crystal[40] and both enantiotopic α-hydrogens are antiperiplanar to the leaving hydrogens (equations 18 and 19).

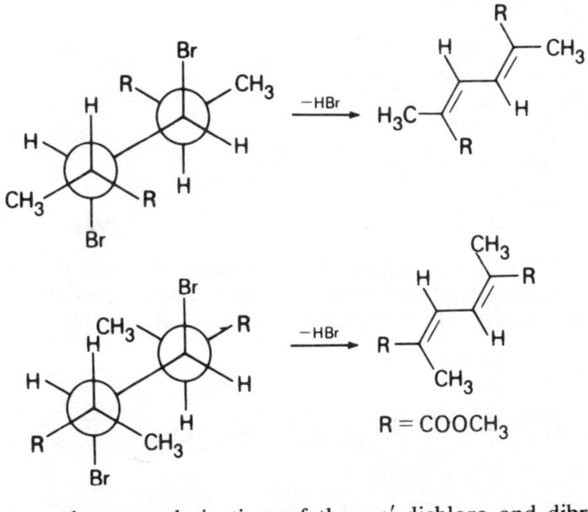

(18)

(19)

$R = COOCH_3$

Similar studies on the *meso* derivatives of the α,α'-dichloro and dibromo adipodinitriles (R = CN) demonstrated an identical mechanism as for the esters.

Crystals of *meso*-2,3-dihalogeno-1,4-dicyanobutanes behave differently, yielding a mixture of the *trans,trans*, *cis,trans* and *cis,cis* muconodinitriles instead of the expected (based on a topochemically controlled reaction from the conformations of

$$\text{Newman projection (X, CN, }H_a, H_b\text{, H)} \xrightarrow{-HX} \text{NC–CH=CH–CH=CH–CN } (cis,cis) + (cis,trans) + (trans,trans) \tag{20}$$

X = Cl or Br

the molecules in the crystal) *cis,cis* isomer (equation 20). This difference was interpreted in terms of a mechanism involving pre-reaction equilibration of rotamers in the solid[40].

V. REACTIONS AT SOLID SURFACES

Kornblum and Lurie[42] investigated the homogeneous and heterogeneous alkylation of phenol and the results illustrate certain constraints imposed on reactions proceeding on the surface of a crystal. This homogeneous alkylation of sodium phenoxide by allyl bromide in ethylene glycol dimethyl ether gave 99% allyl phenyl ether (equation 21) while heterogeneous alkylation gave *o*-allylphenol as the major product (equation 22), and therefore some feature of the surface reaction leads to a strong preference for attack at an *ortho* carbon of the benzene ring.

$$C_6H_5\bar{O}\overset{+}{Na} + CH_2{=}CHCH_2Br \xrightarrow{\text{Homogeneous}} C_6H_5OCH_2CH{=}CH_2 \ (99\%) \tag{21}$$

$$C_6H_5\bar{O}\overset{+}{Na} + CH_2{=}CHCH_2Br \xrightarrow{\text{Heterogeneous}} o\text{-}HOC_6H_4CH_2CH{=}CH_2 \ (54\%) \tag{22}$$

The absence of oxygen alkylation at the crystal surface (equation 22) is explained by the stabilization of the incipient bromide ion by ion pair formation with one of the sodium ions which cluster about the oxygen anion (equation 23). Therefore, in this

$$^{+}Na\ {}^{\delta-}O \cdots {}^{+}Na \cdots Br^{\delta-}\ (\text{cyclohexadienyl}, CH_2CH{=}CH_2) \longrightarrow \text{6-allylcyclohexa-2,4-dienone } (CH_2CH{=}CH_2) \xrightarrow{\text{Fast}} o\text{-}HOC_6H_4CH_2CH{=}CH_2 \tag{23}$$

case, no unfavourable Coulombic repulsion exists between the sodium ions, as it does in the case in solution where the covalent bonding to the oxygen nucleophile dissipates charge on the oxygen, thus depriving the sodium ions in the proximity of the oxygen of one of their counterions (equation 24). Thus, since ion migration in the solid is less ready than in solution, the resulting repulsion between the sodium ions is not relieved.

$$\overset{+}{Na}\ \overset{\delta-}{O}(\cdots CH_2CH{=}CH_2\cdots Br^{\delta-})\ \overset{+}{Na}\text{-}C_6H_5 \longrightarrow C_6H_5OCH_2CH{=}CH_2 \qquad (24)$$

Patchornik and Kraus[43] acylated phenylacetic acid by 'immobilizing' the acid on an insoluble polymeric carrier. Phenylacetic acid was treated with chloromethylated polystyrene (equation 25; Ⓟ = polymer), and the resulting ester was acylated with an acid chloride and trityllithium. The derived polymer was isolated by filtration, washed, dried and treated with HBr in trifluoroacetic acid to give the desired ketone. By running the acylation heterogeneously, self-condensation of the ester was avoided.

$$\text{Ⓟ}{-}CH_2Cl + C_6H_5CH_2COOH \longrightarrow \text{Ⓟ}{-}CH_2OCOCH_2C_6H_5 \xrightarrow[Ph_3CLi]{RCOCl} \text{Ⓟ}{-}CH_2OCOCH(COR)C_6H_5 \xrightarrow[-CO_2]{HBr} RCOCH_2C_6H_5 \qquad (25)$$

As in many solid state reactions, reactions at solid surfaces may have an advantage over their homogeneous counterparts in the case of product isolation.

VI. MISCELLANEOUS REACTIONS

A. Solid State Polymerization

Electron beam irradiation of urea and thiourea clathrate complexes yields very interesting results[5]. Thus the clathrates of urea containing vinyl chloride or acrylonitrile and of thiourea containing vinylidene chloride were found to polymerize[44].

Structural studies show that the canal-type polymer possesses a highly stereoregular structure as compared with the normal poly(vinyl chloride)[45] and therefore the geometric arrangements of monomer molecules in the crystal lattice may be of decisive importance. In connection with this it was found that the cyclic monomer **11** polymerizes only in the crystalline state and that the polymerization of large single crystals leads to the production of fibres whose orientation is determined by the crystal lattice of the monomer[46].

$$BrCH_2-\overset{CH_2Br}{\underset{CH_2-O}{C}}-CH_2 \xrightarrow{\text{Solid}} \left(-CH_2\overset{CH_2Br}{\underset{CH_2Br}{C}}CH_2O-\right)_n \quad (26)$$

(11)

B. Szilard–Chalmers Reactions

When various nuclei capture slow neutrons they emit γ-rays, the momentum of which is balanced by the recoil of the emitting nucleus, which results, almost always, in the rupture of covalent bonds in which the atom of this nucleus participates. If the nucleus is that of a halogen, part of an organic molecule, the new halogen isotope will be radioactive and the radioactivity may be concentrated in an aqueous extract. The percentage of the radioactivity remaining in the organic layer after water extraction is termed 'retention' and the whole phenomenon comprises the well known Szilard–Chalmers reaction.

Libby[47], considering the theory of retention in Szilard–Chalmers reactions, suggested that the radioactive recoil halogen atom X^*, being heavier than the other atoms present in the organic molecules, will have a large probability of transferring most of its energy in a single collision only if it collides with another halogen atom. In this case, the halogen atom X^* and the radical may be retained in the 'cage' of neighbouring molecules to recombine to a molecule similar to the one from which X^* was derived or to escape and decelerate by multiple collisions with light atoms to the point at which another such collision will make X^* be trapped in a solvent cage. Then it will recombine with the radical formed by its last collision to form an organic halide other than the original material. These two processes are shown for the case of methylene bromide in equations (27) and (28). The cage effect operates differently in

High energy

$$\begin{aligned} CH_2Br_2 + Br^* &\longrightarrow \dot{C}H_2Br + Br^* + Br \\ \dot{C}H_2Br + Br^* &\longrightarrow CH_2BrBr^* \end{aligned} \quad (27)$$

Low energy

$$\begin{aligned} CH_2Br_2 + Br^* &\longrightarrow \dot{C}HBr_2 + Br^* + H \\ \dot{C}HBr_2 + Br^* &\longrightarrow CHBr_2Br^* \end{aligned} \quad (28)$$

the liquid and in the solid states, and this is reflected in differences in the products obtained from neutron irradiations of liquid and solid samples, as was demonstrated[48] for the cases of n-propyl bromide and isopropyl bromide (Table 2).

Fox and Libby[48] believe that compounds whose yield is enhanced in the solid by factors of less than 2 are formed from collisions with very high energy X^* atoms, while those with factors of more than 2 are formed from collisions with X^* atoms whose energy had already been largely dissipated by previous collisions. This conclusion is based on the idea that high energy collisions produce local melting and that therefore the cage effect is similar to that in the liquid state.

TABLE 2. Neutron irradiation of *n*-propyl bromide and isopropyl bromide[5]

Radioactive bromine, %	Irradiated liquid	Irradiated solid	Solid/liquid
	n-Propyl bromide at −110°C		
$CH_3CH_2CH_2Br$	14.6	26.8	1.8
$CH_3CHBrCH_3$	2.0	3.4	1.7
CH_2BrCH_2Br	5.6	6.2	1.1
$CH_3CHBrCH_2Br$	4.2	17.4	4.1
'Retention'	39.2	88.1	
	Isopropyl bromide at −80°C		
$CH_3CH_2CH_2Br$	4.1	29.9	7.3
$CH_3CHBrCH_3$	10.1	12.8	1.3
CH_2BrCH_2Br	4.4	4.8	1.1
$CH_3CHBrCH_2Br$	5.8	18.2	3.1
'Retention'	35.6	88.7	

VII. MOLECULAR PACKING MODES

A. Acyl Halides

The solid state and gas–solid reactions, the dehydrohalogenation of solid dihalobutane derivatives and the current activity in the area of asymmetric synthesis in chiral crystals, call for more systematic crystallographic work in order to elucidate more precisely the role of the molecular conformations in the solid and the packing arrangement of the ensemble of the molecules as controlling factors of the mechanisms in such reactions. Systematic work along this line of investigation has been undertaken by Leser and Rabinovich[49] who chose to study the molecular packing modes of a number of dicarbonyl halides in relation to the $R{-}C{\lessgtr}^{O}_{X}$ functional group (X = Cl, Br) and its attached residue (R). The available published crystal structure data on acyl halides is scanty (some other acyl halides have been studied by electron diffraction and microwave spectroscopy[50]) but this group is very reactive even in the solid state and therefore can, in principle, participate in solid state reactions.

From the analyses of terephthaloyl chloride[51], muconyl chloride[52], biphenyl-2,2′-dicarbonyl chloride[53], adamantane-1,3-dicarbonyl chloride[54] and terephthaloyl bromide[55] crystal structures, the above authors observed that no single primary interaction dominates the packing modes of these compounds. Hal···O interaction of the type observed in oxyallyl bromide by Groth and Hassel[56] was found only in terephthaloyl chloride. Instead, Hal···Hal interactions and/or antiparallel $\overset{+}{C}{=}\overset{-}{O}$ dipole–dipole interactions were observed while the C—H···O contacts were present in all the crystal structures analysed. The aromatic acyl halides, except biphenyl-2,2′-dicarbonyl chloride, pack with a short (4 Å) axis[57].

The importance of the packing modes in general, and the relation to the chemical reactivity of the crystals, was explored in length by Schmidt and his coworkers[58].

VIII. ACKNOWLEDGEMENT

Professor M. Lahav is thanked for furnishing a preprint of a paper by himself and his coworkers.

IX. REFERENCES

1. J. Z. Gougoutas and J. Johnson, *J. Amer. Chem. Soc.*, **100**, 5816 (1978).
2. J. Z. Gougoutas, *J. Amer. Chem. Soc.*, **101**, 5672 (1979).
3. G. Balz and G. Schiemann, *Chem. Ber.*, **60**, 1186 (1927).
4. C. T. Lin, P. Y. Siew and S. R. Byrn, *JCS Perkin II*, 957 (1978).
5. H. Morawetz in *Physics and Chemistry of the Organic Solid State* (Ed. D. Fox, M. M. Labes and A. Weissberger), Interscience Publishers, New York and London (1963), pp. 287–328.
6. S. R. Byrn, *J. Pharm. Sci.*, **65**, 1 (1976).
7. I. Serbin, *Science*, **126**, 261 (1957).
8. R. L. Collin, *J. Amer. Chem. Soc.*, **79**, 6086 (1957).
9. Y. Tomkiewicz, R. Agarwal and R. M. Lemmon, *J. Amer. Chem. Soc.*, **95**, 3144 (1973).
10. A. Nater, R. Agarwal, L. Marton, V. Subramanyan and R. M. Lemmon, *J. Amer. Chem. Soc.*, **93**, 2103 (1971).
11. D. J. Abraham, J. S. Rutherford and R. D. Rosenstein, *J. Med. Chem.*, **12**, 189 (1969).
12. J. E. Whinnery and W. H. Watson, *Acta Cryst. B*, **28**, 3635 (1972).
13. S. L. Edwards, J. S. Sherfinski and R. E. Marsh, *J. Amer. Chem. Soc.*, **96**, 2593 (1974).
14. G. M. J. Schmidt, *Pure Appl. Chem.*, **27**, 647 (1971) and references cited herein.
15. M. D. Cohen and G. M. J. Schmidt, *J. Chem. Soc.*, 1996 (1964) and following papers.
16. M. D. Cohen, Z. Ludmer, J. M. Thomas and J. O. Williams, *Chem. Commun.*, 1172 (1969).
17. M. D. Cohen, Z. Ludmer, J. M. Thomas and J. O. Williams, *Proc. Roy. Soc. Lond. A* **324**, 459 (1971).
18. J. P. Desvergne, F. Chepko and H. Bouas-Laurent, *JCS Perkin II*, 84 (1978).
19. E.-Z. M. Eleid, S. E. Morsi and J. O. Williams, *JCS Faraday I*, **76**, 2170 (1980).
20. A. B. Jaffe, K. J. Skinner and J. M. McBride, *J. Amer. Chem. Soc.*, **94**, 8510 (1972).
21. A. B. Jaffe, D. S. Mallament, E. P. Slisz and J. M. McBride, *J. Amer. Chem. Soc.*, **94**, 8515 (1972).
22. A. Schmitt, *Ann. Chem.*, **127**, 319 (1863).
23. C. Graebe and C. Liebermann, *Ann. Chem.*, Suppl. 7, 257 (1870).
24. R. E. Buckles, E. A. Hausman and N. G. Wheeler, *J. Amer. Chem. Soc.*, **72**, 2494 (1950).
25. M. M. Labes, H. W. Blakesbee and J. E. Bloor, *J. Amer. Chem. Soc.*, **87**, 4251 (1965).
26. E. Hadjoudis, E. Kariv and G. M. J. Schmidt, *JCS Perkin II*, 1056 (1972).
27. E. Hadjoudis and G. M. J. Schmidt, *JCS Perkin II*, 1060 (1972).
28. E. Hadjoudis, *Israel J. Chem.*, **11**, 63 (1973).
29. E. Hadjoudis, *Israel J. Chem.*, **12**, 981 (1974).
30. E. Hadjoudis in *Reactivity of Solids* (Ed. J. Wood, O. Lindqvist, C. Helgesson and N. G. Vannerberg), Plenum Press, New York (1977), pp. 493–497.
31. K. Panzien and G. M. J. Schmidt, *Angew. Chem. Int. Edn Engl.*, **8**, 608 (1969).
32. B. S. Green and L. Heller, *Science*, **185**, 525 (1974).
33. D. Rabinovich and H. Hope, *Acta Cryst.*, **31**, 5128 (1975).
34. L. Addadi, S. Ariel, M. Lahav, L. Leiserowitz, R. Popovitz-Biro and C. D. Tang, *JCS Specialist Reports*, in press (1980).
35. D. G. Naae, *J. Org. Chem.*, **44**, 336 (1979).
36. G. Friedman, M. Lahav and G. M. J. Schmidt, *JCS Perkin II*, 428 (1974).
37. H. Kauffman, D. Rabinovich and G. M. J. Schmidt, *JCS Perkin II*, 433 (1974).
38. K. Mislow and M. Raban in *Topics in Stereochemistry* (Ed. N. L. Allinger and E. L. Eliel), Wiley, New York (1976), p. 1.
39. G. Friedman, E. Gati, M. Lahav, D. Rabinovich and Z. Shakked, *Chem. Commun.*, 491 (1975).
40. D. Rabinovich and Z. Shakked, *Acta Cryst. B*, **34**, 1176 (1978).
41. D. Rabinovich and Z. Shakked, *Acta Cryst. B*, **34**, 1183 (1978).
42. N. Kornblum and A. P. Lurie, *J. Amer. Chem. Soc.*, **81**, 2705 (1959).

43. A. Patchornik and M. A. Kraus, *J. Amer. Chem. Soc.*, **92**, 7587 (1970).
44. J. F. Brown, Jr and D. M. White, *J. Amer. Chem. Soc.*, **82**, 5671 (1960).
45. S. Krimm, *Plastics Engineers J.*, **15**, 797 (1959).
46. S. Nakashio, M. Kondo, H. Tsuchiya and M. Yamada, *Makromol. Chem.*, **52**, 79 (1962).
47. W. F. Libby, *J. Amer. Chem. Soc.*, **69**, 2523 (1947).
48. M. S. Fox and W. F. Libby, *J. Chem. Phys.*, **20**, 487 (1952).
49. J. Leser and D. Rabinovich, *Acta Cryst. B*, **34**, 2250 (1978).
50. M. Simonetta and P. Beltrame in *The Chemistry of Acyl Halides* (Ed. S. Patai), Wiley, Chichester (1972), p. 1.
51. J. Leser and D. Rabinovich, *Acta Cryst. B*, **34**, 2253 (1978).
52. J. Leser and D. Rabinovich, *Acta Cryst. B*, **34**, 2257 (1978).
53. J. Leser and D. Rabinovich, *Acta Cryst. B*, **34**, 2260 (1978).
54. J. Leser and D. Rabinovich, *Acta Cryst. B*, **34**, 2264 (1978).
55. J. Leser and D. Rabinovich, *Acta Cryst. B*, **34**, 2269 (1978).
56. P. Groth and O. Hassel, *Acta Chem. Scand.*, **16**, 2311 (1962).
57. J. Leser and D. Rabinovich, *Acta Cryst. B*, **34**, 2272 (1978).
58. G. M. J. Schmidt *et al.*, *Solid State Photochemistry* (Ed. D. Ginsburg). Verlag Chemie, Weinheim and New York (1976).

The Chemistry of Functional Groups, Supplement D
Edited by S. Patai and Z. Rappoport
© 1983 John Wiley & Sons Ltd

CHAPTER **18**

Hypervalent halogen compounds

GERALD F. KOSER
Department of Chemistry, The University of Akron, Akron, Ohio 44325, USA

I. INTRODUCTION

Among the organohalogen compounds, those with halogen atoms bound to a single ligand and residing in the +1 oxidation state are the most familiar to organic chemists. However, hypervalent organohalogen species containing a halogen atom surrounded by 10 or 12 valence electrons and attached to two, three, four or five ligands are not uncommon. The capacity for existence in higher oxidation states increases with the size of the halogen atom. Thus hypervalent organoiodine compounds are far more numerous than hypervalent organobromine and organochlorine compounds, and hypervalent organofluorine compounds are unknown.

The very first hypervalent species with at least one carbon ligand bound to a halogen atom was reported in 1886 by Willgerodt, who prepared iodosobenzene dichloride by the reaction of iodobenzene with molecular chlorine[1]. The syntheses of iodosobenzene, iodoxybenzene and iodosobenzene diacetate were reported by the same investigator in 1892[2] and the first iodonium salt, phenyl(*p*-iodophenyl)iodonium bisulphate, prepared by the autocondensation of iodosobenzene in concentrated sulphuric acid, was reported by Hartmann and Meyer in 1894[3].

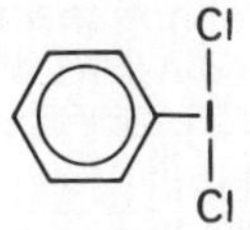

iodosobenzene dichloride

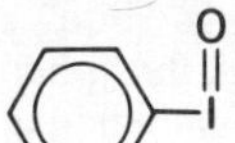

iodosobenzene

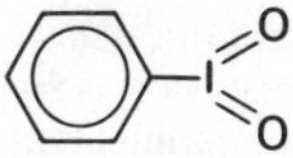

iodoxybenzene

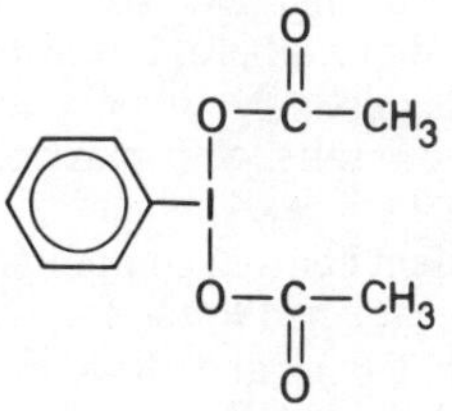

iodosobenzene diacetate

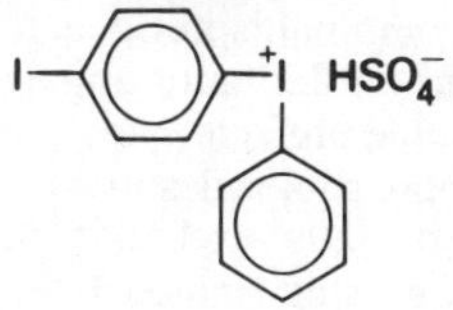

phenyl(*p*-iodophenyl)iodonium bisulphate

The discovery of the 'iodoso' family of compounds initiated a period of intense activity directed toward the synthesis of numerous analogues. In 1914, Willgerodt published a monograph entitled *Die organischen Verbindungen mit mehrwertigem Jod* in which the early work was thoroughly summarized[4]. In this publication, over 480 aryliodoso and aryliodoxy derivatives are tabulated, these having been prepared from approximately 160 iodoarenes. In addition to molecules containing the functional groups present in the compounds already mentioned, Willgerodt's compendium includes aryliodoso difluorides, dinitrates, dibenzoates, hydroxychromates, hydroxyacetates, hydroxynitrates, hydroxyiodates, hydroxyperchlorates, sulphates, chromates and oxydifluorides, some of these being more common than others. The general structures are summarized in Table 1.

Willgerodt also reviewed the iodonium salts and the relatively few aliphatic iodoso compounds that were known up to 1912.

Since that time, the chemistry of hypervalent organoiodine compounds has been reviewed by several authors. Two general articles have appeared, one published by

TABLE 1. Aryliodoso and aryliodoxy derivatives tabulated by Willgerodt[4]

$ArI(OOCCH_3)_2$	$ArI(OOCPh)_2$	$ArI(OH)OCr(=O)_2OI(OH)Ar$
$ArI{=}O$	$ArI(OH)ONO_2$	$ArI(OH)IO_3$
$ArICl_2$	$ArI(OH)OOCCH_3$	$ArI(CrO_4)$
$ArIF_2$	$ArI(OH)OClO_3$	$ArI(SO_4)$
$ArI(ONO_2)_2$	$ArI(OH)OS(=O)_2OI(OH)Ar$	$ArIO_2$
		$ArI(O)F_2$

Sandin in 1943[5] and another by Banks in 1966[6]. As part of a lengthy book chapter on the inorganic chemistry of the halogens, published in 1972, Downs and Adams present a concise overview of polyvalent organohalogen chemistry[7]. In 1956, Beringer and Gindler published an extensive compendium of the physical properties of the polyvalent organoiodine compound known up to that time[8], and, in 1975, Olah published a book entitled *Halonium Ions*[9a]. Finally in 1981, Varvgolis reviewed the chemistry of the (diacyloxyiodo)arenes[9b].

The last decade has witnessed a surge of interest in the chemistry of hypervalent organohalogen compounds. An encyclopaedic coverage of all that has been accomplished in this area is far too ambitious for the space allotted here, but in this chapter we shall attempt to present a review of the chemistry of hypervalent organohalogen compounds not generally regarded as halonium salts, with particular emphasis on some of the more important research of the past 15 years.

In this chapter, the prefix 'organo', when applied to hypervalent halogen compounds, is reserved for those molecules in which at least one ligand is attached to the halogen atom via carbon, thus excluding such species as $I(OOCCF_3)_3$ and $I(OCH_3)_5$. Following the suggested nomenclature of Martin[10], organohalogen(III) compounds are also referred to herein as organohalinanes and organohalogen(V) compounds as perhalinanes. The names (diacetoxyiodo)benzene, (dichloroiodo)benzene, etc., are generally used in preference to iodosobenzene diacetate, iodosobenzene dichloride, etc.

In Table 2, a classification scheme for the diversity of known hypervalent organohalogen compounds is presented, and a number of representative structures are shown. This classification system, similar to that previously suggested by Banks, is based on the oxidation state of and the number of carbon ligands attached to the halogen atom. For discussions of much of the earlier preparative methodology for organoiodine(III) compounds, the reader is directed to review articles already cited.

II. ORGANOIODINANES: STABILITY AND OCCURRENCE

Any discussion of the occurrence of organoiodinanes must first confront the ambiguities associated with the term 'stability'. A stable compound is generally regarded as one that can be kept at room temperature (or above) under atmospheric conditions for reasonable periods of time without noticeable decomposition. The stability of an organoiodinane must be measured against the relative thermodynamic stabilities of the products (especially organoiodine(I) products) into which it can decompose and against the activation energies for the particular modes of decomposition. Furthermore, an organoiodinane might be perfectly stable to thermal degradation at elevated temperatures but not stable to the reactions initiated by atmospheric moisture or oxygen at room temperature.

It will prove expedient to focus first on the dissociative mode of decomposition of organoiodinanes illustrated by the following general reaction (L = ligand). In this

$$\mathrm{R{-}IL_2} \xrightleftharpoons{K_{eq}} \mathrm{R{-}I + L{-}L}; \quad K_{eq} = ([\mathrm{RI}][\mathrm{L_2}])/[\mathrm{RIL_2}] \tag{1}$$

case, those compounds RIL_2 with the smallest equilibrium constants will be the most stable. Some specific examples are given in equations (2)–(5). For most of these reactions, equilibrium measurements have either not been made or the reactions have not yet been directly observed.

In principle, the reaction enthalpies could be approximated by simple bond energy computations. However, the iodine–ligand bond energies of hypervalent organoiodine

TABLE 2. Classes and examples of organohalinanes and organoperhalinanes

(1) Organohalogen(III) compounds

(a) One carbon ligand, one heteroatom ligand

(b) Two carbon ligands

(c) One carbon ligand, two identical heteroatom ligands

(d) One carbon ligand, two different heteroatom ligands

(e) Two carbon ligands, one heteroatom ligand

(f) Three carbon ligands

TABLE 2. (*Continued*)

(2) Organohalogen(V) compounds

(a) One carbon ligand, two heteroatom ligands

(b) One carbon ligand, three heteroatom ligands

(c) One carbon ligand, four heteroatom ligands

(d) Two carbon ligands, two heteroatom ligands

(e) Two carbon ligands, three heteroatom ligands

---

compounds are not known. Furthermore, the bond energies of halogen(III) compounds cannot be confidently estimated from those of the corresponding halogen(I) compounds. For example, the fluorine–bromine bond energy in BrF is 59.42 kcal mol^{-1} while the average bond energy in BrF_3 is 48.2 kcal mol^{-1} [11]. Although the I—L bonds in aryliodinanes are hypercovalent, a useful mnemonic for a qualitative understanding of the stabilities of organoiodinanes relative to their molecular dissociation products is to view them as disalts of general structure $(PhI^{2+})2L^-$. Those ligands L^-, which might be readily oxidized by the hypothetical species PhI^{2+}, should either afford iodinanes of low thermal stability or not give them at all.

For example, the successful isolation of numerous aryldichloroiodinanes ($ArICl_2$) stands in stark contrast to the absence of the corresponding aryldibromoiodinanes ($ArIBr_2$) and aryldiiodoiodinanes ($ArII_2$) in the chemical literature. Only the generation of (dibromoiodo)benzene at low temperature has been claimed and will be

$$\mathrm{PhIX_2} \rightleftharpoons \mathrm{PhI} + \mathrm{X_2} \qquad (2)$$

(X = halogen)

$$\mathrm{PhI(OOCR)_2} \rightleftharpoons \mathrm{PhI} + \mathrm{RC(O)OOC(O)R} \qquad (3)$$

$$\mathrm{PhI(SR)_2} \rightleftharpoons \mathrm{PhI} + \mathrm{RSSR} \qquad (4)$$

$$\mathrm{Ph_3I} \rightleftharpoons \mathrm{PhI} + \mathrm{Ph{-}Ph} \qquad (5)$$

discussed later in this chapter. Apparently, the equilibrium constants for the oxidative addition of molecular bromine and molecular iodine to the iodine atom of iodoarenes are prohibitively low. Since bromide ion and iodide ion are more readily oxidized than chloride ion, it might be expected that the formation of dibromoiodinanes and diiodoiodinanes will be at least less favourable than the formation of dichloroiodinanes, if not altogether prohibited.

The dimercaptoaryliodinanes ($ArI(SR)_2$) are also unknown. This, in conjunction with the known ability of (diacetoxyiodo)benzene to efficiently oxidize thiols to disulphides[12] suggests that, if dimercaptophenyliodinanes are formed as intermediates in such oxidations, the relative stability of iodobenzene and disulphide is sufficiently great to disfavour iodinane isolation (equation 4). Such results are clearly consistent with the disalt mnemonic and the low oxidation potentials of alkyl- and arylthiolate ions.

(Diacetoxyiodo)benzene is commercially available and can be stored for months at room temperature without noticeable decomposition. Triphenyliodine, on the other hand, decomposes readily at $-10°C$ to iodobenzene and biphenyl (equation 5)[13,14]. This, too, can be explained by the disalt model since the oxidation of phenyl anion to phenyl radical by PhI^{2+}, with the ultimate formation of biphenyl, should be more facile than the oxidation of acetate ion to the acetoxy radical eventuating in peroxyacetic anhydride.

The fact that no *acyclic* arylbrominane ($ArBrL_2$) or arylchlorinane ($ArClL_2$) has ever been isolated is likewise consistent with the disalt concept. Such species as $ArCl^{2+}$ and $ArBr^{2+}$ would be expected to suffer electron-transfer reduction by L^- far more easily than ArI^{2+}. This conclusion is borne out indirectly by electrochemical studies on the aryl halides. The anodic oxidation of iodobenzene in aqueous acetic acid has been found to yield (diacetoxyiodo)benzene[15] and, in acetonitrile–benzene–lithium perchlorate, to yield diphenyliodonium perchlorate[16]. Anodic oxidations of chlorobenzene and bromobenzene, on the other hand, afford products derived from oxidation of the aromatic ring[17].

Such arguments must, however, be cautiously applied since organoiodine(III) compounds are not in fact constrained to one mode of decomposition. For example, the thermal decomposition of (diacetoxyiodo)benzene in chlorobenzene proceeds to give the products shown in equation (6)[18]. Two decomposition pathways are evident,

$$\mathrm{PhI(OOCCH_3)_2} \xrightarrow[126.8\,^\circ\mathrm{C}]{\mathrm{PhCl}} \mathrm{CH_3I} + \mathrm{PhI} + \mathrm{PhOCCH_3} + \mathrm{CH_3COOH} + \mathrm{CH_3C_6H_4Cl}\ (o,m,p) + \mathrm{CO_2} \qquad (6)$$

(In $\mathrm{PhOCCH_3}$ the carbon carries =O; the chlorotoluene is drawn as a benzene ring with CH_3 and Cl.)

70.4% 26.0% 74.6% 22.7% 10.8% 72.5%

(mole % per mole of (diacetoxyiodo) benzene)

the major one involving the initial formation of an ion pair $[\mathrm{Ph\overset{+}{I}OAc\ OAc^-}]$ and ultimately leading to iodomethane, acetoxybenzene and carbon dioxide and the minor one involving the initial formation of a radical pair $[\mathrm{Ph\dot{I}OAc\ \dot{O}Ac}]$ and ultimately leading to acetic acid, chlorotoluenes and *some* carbon dioxide.

An investigation of the thermolysis of (ditriphenylacetoxyiodo)benzene (**$\mathrm{PhI(OOCCPh_3)_2}$**) in benzene at 79.8°C revealed the homolytic process to be the dominant mode of decomposition[19].

Equilibrium constants for the dissociation of various (dichloroiodo)arenes to iodoarene and molecular chlorine in solvents of differing polarity (i.e. CCl_4, CH_3COOH, CH_3NO_2) have been measured and are generally quite small[20]. For example, for (dichloroiodo)benzene at 25°C, equilibrium constants of 0.4×10^{-3} M (CH_3NO_2), 2.0×10^{-3} M (CH_3COOH) and 2.0×10^{-2} M (CCl_4) were determined. In the solvent nitromethane, the reaction constant, ρ, was found to be +2.4. However, the complete thermal decomposition of (dichloroiodo)benzene in the solid state leads to the complex mixture of products shown in equation (7) in addition to a 2.1% yield

$$\mathrm{PhICl_2} \xrightarrow[120\,^\circ\mathrm{C}]{\text{solid}} \mathrm{PhCl} + p\text{-}\mathrm{C_6H_4Cl_2} + o\text{-}\mathrm{C_6H_4Cl_2} + \mathrm{PhI} + o\text{-}\mathrm{ClC_6H_4I} + p\text{-}\mathrm{ClC_6H_4I} \qquad (7)$$

0.9% 2.5% 1.7% 27.3% 28.6% 36.8%

of unidentified material[21]. In refluxing benzene, the decomposition products are iodobenzene (67%) and chlorobenzene (30%). Once again, two competing modes of decomposition have been proposed, one involving initial homolysis of an iodine–chlorine bond and another involving initial heterolysis of an iodine–chlorine bond.

The (dichloroiodo)alkanes are rare and those which are known are extremely 'unstable'. (Dichloroiodo)methane, the most stable representative, decomposes at −28°C[22], in contrast to (dichloroiodo)benzene which can be handled conveniently at room temperature. However, this stability comparison does not involve the dissociative equilibria to chlorine and the corresponding iodo compound. (Dichloroiodo)methane decomposes instead to chloromethane and iodine monochloride, probably by nucleophilic heterolysis of a carbon–iodine bond, perhaps by the reaction mechanism shown below. Although nucleophilic heterolysis of the carbon–iodine

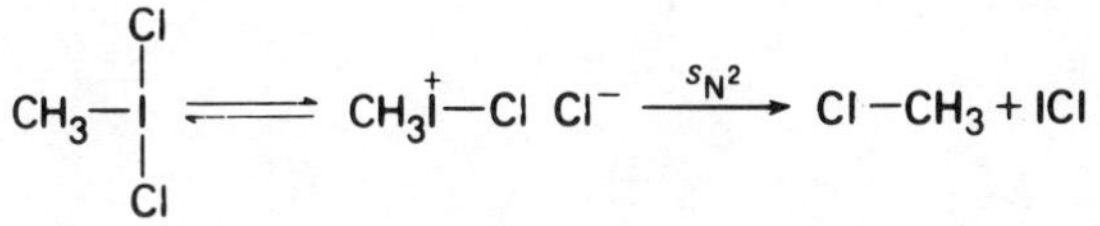

$$CH_3-I(Cl)_2 \rightleftharpoons CH_3\overset{+}{I}-Cl\ Cl^- \xrightarrow{S_N2} Cl-CH_3 + ICl \quad (8)$$

bond in (dichloroiodo)benzene, possibly by the S_NAr mechanism, may well acount for the production of chlorobenzene during its thermolysis, the nucleophilic mode of decomposition would be expected to proceed with much higher activation energy at aryl carbon. The fact that α-(dichloroiodo)sulphones[23,24] and (dichloroiodo)alkenes[8] can be prepared and isolated as stable crystalline compounds at room temperature is consistent with this logic.

Iodosobenzene is another case in point. Its thermal decomposition does not yield iodobenzene and molecular oxygen. Instead, iodosobenzene disproportionates to iodobenzene and iodoxybenzene[6] (equation 9), a common mode of reactivity for compounds in intermediate oxidation states.

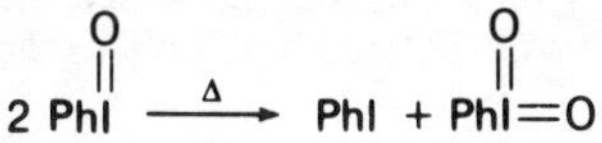

$$2\ PhI{=}O \xrightarrow{\Delta} PhI + PhI({=}O)_2 \quad (9)$$

III. ORGANOIODINE(III) COMPOUNDS: STRUCTURE AND BONDING

A. Molecular Structures of Organoiodinanes

Except for a single crystal X-ray study of a bromonium ylide, no quantitative structural information is available for any organobromine(III) or organochlorine(III) compounds. Fortunately, single crystal X-ray structures of some organoiodine(III) compounds have been published. Among these are solid state structures of (dichloroiodo)benzene (**1**) (published in 1953)[25], (diacetoxyiodo)benzene (**2**) (published in 1977[26] and 1979[27]), bis(dichloroacetoxyiodo)benzene (**3**) (published in 1979)[27], [hydroxy(tosyloxy)iodo]benzene (**4**) (published in 1977)[28], 1-hydroxy-1,2-benziodoxol-3(1*H*)-one (**5**) (published in 1965)[29], 1-acetoxy-1,2-benziodoxol-3(1*H*)-one (**6**) (published in 1972)[30], 1-(*m*-chlorobenzoyloxy)-1,2-benziodoxol-3(1*H*)-one (**7**) (published in 1974)[31], 1-(*o*-iodobenzoyloxy)-1,2-benziodoxol-3(1*H*)-one (**8**) (published in 1972)[32] and 1,2-dichloro-1,2-benziodazol-3(1*H*)-one (**9**) (published in 1975)[33].

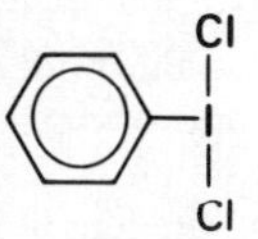

$PhI(Cl)_2$ (**1**) $PhI(OOCCH_3)_2$ (**2**) $PhI(OOCCHCl_2)_2$ (**3**) $PhI(OH)(OTs)$ (**4**)

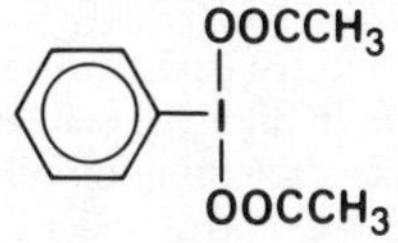

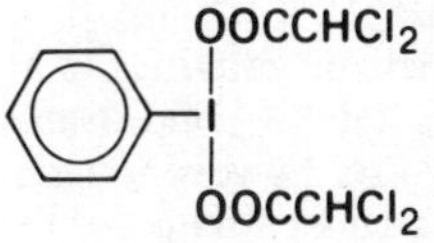

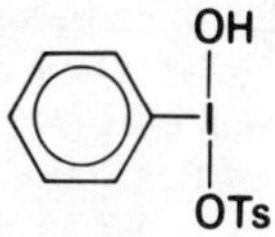

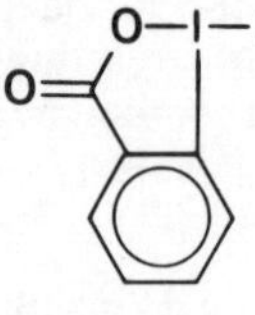

(5) 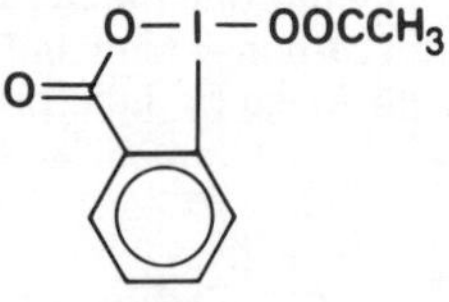 (6) 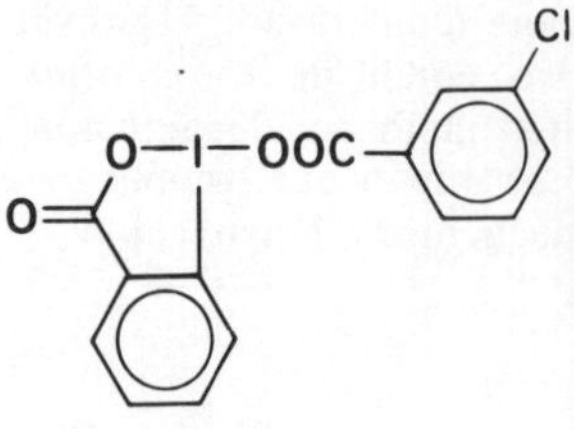 (7)

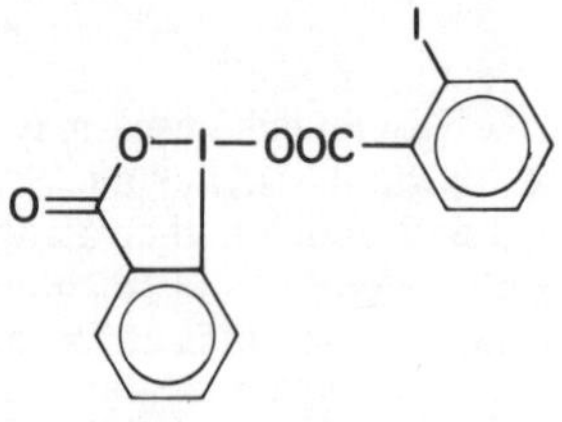

(8) 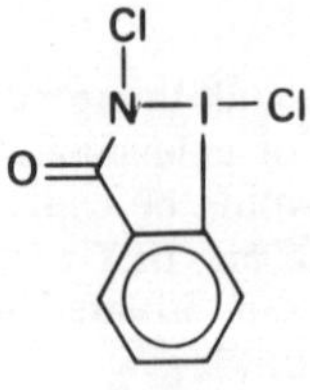 (9)

These compounds share one common structural feature in that all possess one aryl ligand and two heteroatom ligands bound to trivalent iodine. To our knowledge, no X-ray studies of any aliphatic iodinanes or triaryliodinanes have yet appeared in the chemical literature. Single crystal X-ray studies of several diaryliodonium salts have been published, but they will be considered here only as they relate to the structures of compounds **1–9**.

An examination of the bonding parameters of iodinanes **1–9**, summarized in Table 3, reveals several general structural features:

(1) Organoiodine(III) compounds are approximately T-shaped and one must resist the temptation to draw trigonal structures. The same configuration has also been observed with the inorganic halinanes[7] and with diaryliodonium salts.

(2) The most electronegative ligands of organoiodine(III) compounds are invariably colinear; that is, they exhibit structure **A** below and not the isomeric structure **B**, at least in the solid state.

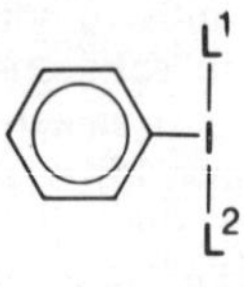

(A) 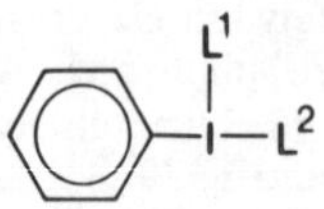 (B)

(3) The carbon–iodine bond distances in acyclic aryliodinanes are approximately equal in length to the sum of the covalent radii of carbon and iodine. Variable carbon–iodine bond distances are observed with cyclic aryliodinanes and probably manifest the electronic requirements associated with the placement of an iodine atom into a five-membered ring.

(4) When the electronegativities of the two heteroligands are nearly equal, both iodine–ligand bonds exhibit bond distances greater than the sum of the appropriate covalent radii. They are, however, significantly shorter than ionic bonds.

(5) When the electronegativities of the heteroligands differ significantly, the length of the bond to the less electronegative (more basic) ligand is very nearly that expected

TABLE 3. Intramolecular bond distances and bond angles for aryliodinanes **1–9**

Structure	Bond	Bond distance, Å	Triad	Bond angle, deg	References
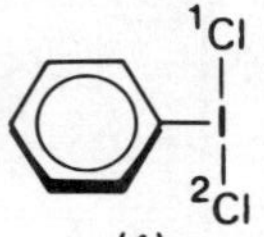 C_6H_5—I(^{1}Cl)(^{2}Cl) **(1)**	I—C	2.00	C—I—$Cl_{(1)}$	86(1)	25
	I—$Cl_{(1)}$	2.54	C—I—$Cl_{(2)}$	86(1)	
	I—$Cl_{(2)}$	2.54			
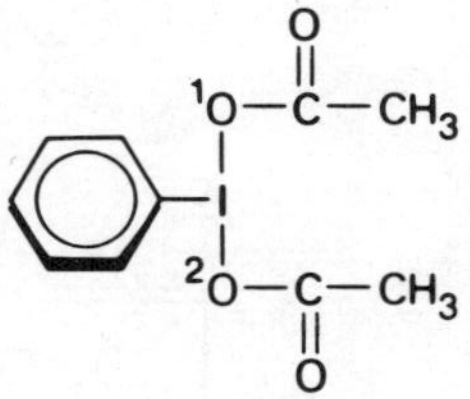 C_6H_5—I(^{1}O—C(=O)—CH_3)(^{2}O—C(=O)—CH_3) **(2)**, 1977 structure	I—C	2.093(9)	C—I—$O_{(1)}$	81.6(4)	26, 27
1979 structure		2.090(6)		81.4(2)	
1977 structure	I—$O_{(1)}$	2.186(9)	C—I—$O_{(2)}$	81.6(4)	
1979 structure		2.159(5)		82.6(3)	
1977 structure	I—$O_{(2)}$	2.143(8)			
1979 structure		2.153(5)			
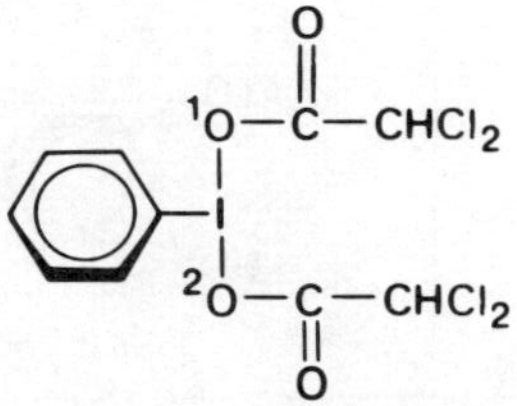 C_6H_5—I(^{1}O—C(=O)—$CHCl_2$)(^{2}O—C(=O)—$CHCl_2$) **(3)**	I—C	2.083(12)	C—I—$O_{(1)}$	86.0(3)	27
	I—$O_{(1)}$	2.136(6)	C—I—$O_{(2)}$	82.5(3)	
	I—$O_{(2)}$	2.163(7)			

TABLE 3 (*Continued*)

Structure	Bond	Bond distance, Å	Triad	Bond angle, deg	References
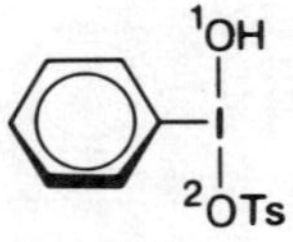 (4)	I—C I—$O_{(1)}$ I—$O_{(2)}$	2.083(4) 1.940(4) 2.473(3)	C—I—$O_{(1)}$ C—I—$O_{(2)}$	92.8(2) 86.0(1)	28
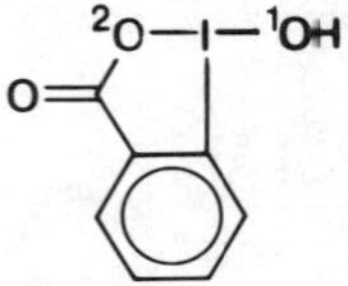 (5)	I—C I—$O_{(1)}$ I—$O_{(2)}$	2.16 2.00 2.30	C—I—$O_{(1)}$ C—I—$O_{(2)}$	89 77	29
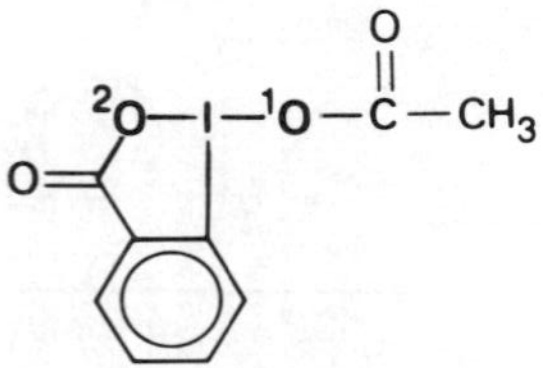 (6)	I—C I—$O_{(1)}$ I—$O_{(2)}$	2.00(3) 2.11(2) 2.13(2)	C—I—$O_{(1)}$ C—I—$O_{(2)}$	83(1) 81(1)	30

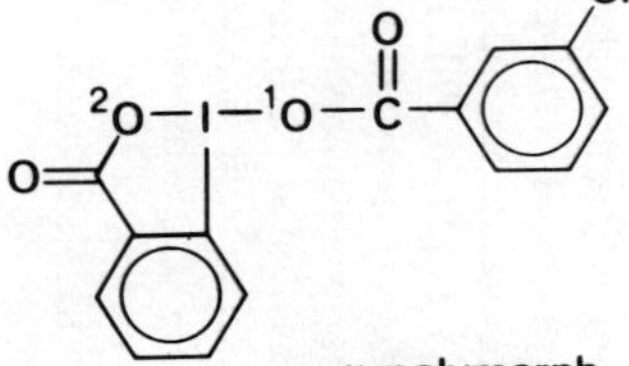

Compound	Bond	Length (Å)	Angle	Degrees	Ref.
(7) α-polymorph	I—C	2.06	C—I—$O_{(1)}$	85	31
	I—$O_{(1)}$	2.13	C—I—$O_{(2)}$	79	
	I—$O_{(2)}$	2.11			
(8) α-polymorph, β-polymorph	I—C	2.14(7)	C—I—$O_{(1)}$	83(8)	32
		2.09(4)		85(2)	
	I—$O_{(1)}$	2.11(7)	C—I—$O_{(2)}$	81(6)	
		2.08(4)		81(2)	
	I—$O_{(2)}$	2.15(6)			
		2.14(4)			
(9)	I—C	2.19	C—I—N	80	33
	I—N	2.06	C—I—Cl	91	
	I—Cl	2.56			

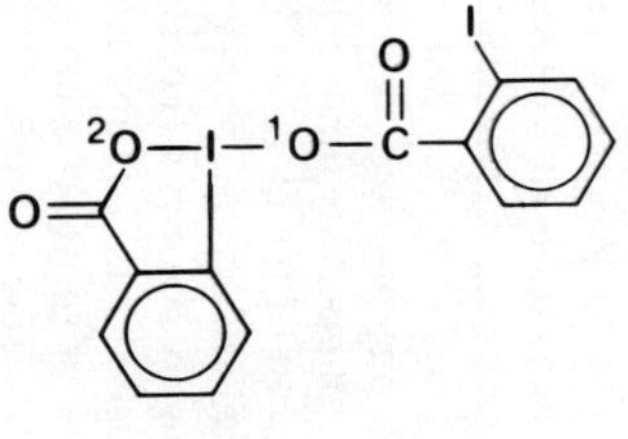

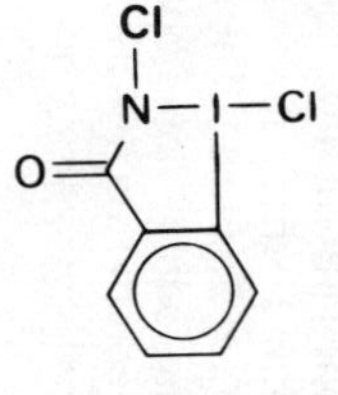

from covalent radii (or perhaps a little shorter). The bond to the more electronegative (less basic) ligand, however, is elongated, even more so than with the symmetrical iodinanes.

(6) With cyclic aryliodinanes, the plane of the benzene ring is constrained to the same plane as that of the L^1—I—L^2 triad. However, in the acyclic aryliodinanes, those planes are almost orthogonal.

(7) In organoiodine(III) compounds with acyloxy ligands, secondary bonding between the carbonyl oxygen atom and the iodine atom appears to be important.

B. Bonding Models for Organohalinanes

The electronic configuration of elemental iodine is $[Kr]4d^{10}5s^25p^55d^0$. Thus, one qualitative model for the bonding in organoiodine(III) compounds is one in which iodine utilizes a set of $5sp^2$–5pd hybrid orbitals in which one of the sp^2 orbitals is singly occupied and used to bind the equatorial ligand and the remaining sp^2 hybrid orbitals are doubly occupied, non-bonded and lie in the equatorial plane. The two axial ligands are covalently bound to a pair of singly occupied pd hybrids, diametrically opposed and perpendicular to the equatorial plane (i.e. $[Kr]4d^{10}\,(5sp^2)^1\,(5sp^2)^2\,(5sp^2)^2\,(5pd)^1\,(5pd)^1$). Overall, the hybrid orbitals are directed toward the corners of a trigonal bipyramid as illustrated below.

An alternative and slightly more quantitative model is one in which the iodine atom in organoiodine(III) compounds is not hybridized at all. In this model, one of the three ligands is bound by normal covalent overlap to the singly occupied 5p orbital at iodine. The remaining ligands are attached, one to each lobe, to one of the doubly occupied 5p orbitals of the iodine atom, resulting in a linear three-centred four-electron bond [3c–4e]. Such covalent bonds are termed 'hypervalent' and are distinct from normal covalent bonds. Although it may be tacitly assumed for this model that the remaining unshared electron pairs and the three ligands are directed toward the corners of a trigonal bipyramid, there is no evidence that this is actually the case and that the unshared pairs are not best described as residing in 5s and 5p orbitals on iodine. Indeed, the periodinanes appear to exhibit square pyramidal structures. The terminology, equatorial and axial, employed herein relates to a presumed trigonal bipyramide geometry about iodine in organoiodinanes.

Such a treatment of the bonding at iodine in organoiodinanes is analogous to the Hückel molecular orbital (HMO) description of allyl anion which predicts a [3c–4e] π-bond with two significant features: (1) the π-bond orders between adjacent carbon atoms are 0.5 and (2) the electronic charge is distributed primarily at the terminal carbon atoms. In the case of the iodinanes, *sigma* overlap of a 5p orbital on the iodine atom with the σ orbitals of two ligands (e.g. the two chlorine ligands in $PhICl_2$) generates three σ molecular orbitals, one bonding, one approximately non-bonding and one antibonding. The four σ electrons occupy the two lower energy levels and lead to σ-bond orders of about 0.5, with most of the electron density residing at the ligand sites (see Figure 1). The entire triad will, of course, be neutral in the organoiodinanes, but the iodine–ligand bonds will be highly polarized (i.e. $X^{\delta-}$—$I^{2\delta+}$—$X^{\delta-}$). This model has been thoroughly discussed by Musher, who has applied it to a variety of hypervalent molecules and who cites the work of those who pioneered its use[34]. It has more

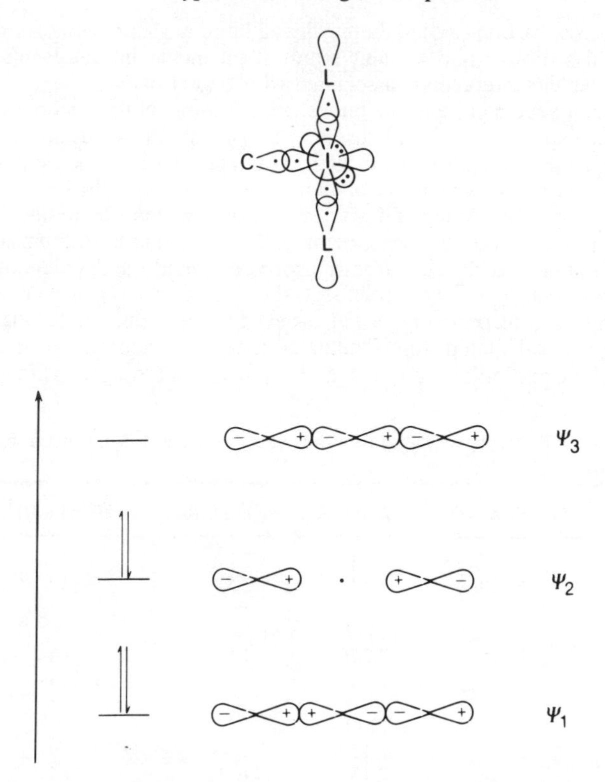

FIGURE 1. Hypervalent bonding model for organoiodinanes in which the iodine atom is bound to one equatorial carbon ligand and two axial, univalent heteroligands. The molecular orbitals for the [3c–4e] bond are shown.

recently been employed by Martin in his work on hypervalent sulphur and halogen compounds (references to his work are given in later sections of this chapter).

It has already been noted that electronegative axial ligands stabilize organohalinanes. This is to be expected from the *inherent* nature of the [3c–4e] bond which places more electron density at the ends of the L—I—L triad than at the centre. Thus, triphenyliodine is less stable than (diacetoxyiodo)benzene because the more electronegative acetoxy ligands can accommodate negative charge better than the phenyl ligands can. It is important to remember that even in symmetrical compounds such as ICl_3 and IPh_3, one covalent iodine–ligand bond and two hypercovalent iodine–ligand bonds (i.e. one [3c–4e] bond) are anticipated, and the chemical behaviour of the axial ligands should be distinct from that of the equatorial ligand.

C. Organoiodinanes: Bond Lengths, Bond Angles and the Hypervalent Bonding Model

The hypervalent bonding scheme, when applied to the organoiodinanes, anticipates the existence of T-shaped molecules with elongated σ-bonds to the axial ligands. The T-configuration is approximately realized for all of the aryliodinanes listed in Table 3, the experimental carbon–iodine–heteroligand bond angles ranging from 77 to 92.8°. Geometric distortions from 90° may manifest secondary intramolecular interactions and intermolecular interactions associated with crystal packing.

A number of years ago, Pauling published an empirical relation between the bond order n, the experimental bond distance $D(n)$, and the computed covalent bond distance $D(l)$ for a chemical bond: $D(n) = D(l) - 0.6 \log n$[35]. Application of this equation to the organoiodinanes is in qualitative accord with the [3c–4e] bond. It has recently been noted by Amey and Martin for example, that the iodine–chlorine bond distances in (dichloroiodo)benzene, being 0.22 Å longer than the sum of the covalent radii of the iodine and chlorine atoms, afford computed bond orders of 0.45[36]. With (diacetoxyiodo)benzene (1977 structure), the computed oxygen–iodine bond orders are 0.47 and 0.56 respectively. Bond orders for the iodine–heteroligand bonds of compounds **1–9**, calculated from Pauling equation, are summarized in Table 4.

In each of the aryliodinanes **1**, **2**, **3**, **6**, **7** and **8**, the heteroligands are either of equal

TABLE 4. Iodine–heteroligand bond orders for aryliodinanes **1–9** calculated from the Pauling equation

Compound	Bond	$D(n)$, Å	$D(l)$, Å	$D(n) - D(l)$, Å	n
1	$I-Cl_{(1)}$	2.54	2.32	0.22	0.43
	$I-Cl_{(2)}$	2.54	2.32	0.22	0.43
2		1977 structure			
	$I-O_{(1)}$	2.186	1.99	0.196	0.47
	$I-O_{(2)}$	2.143	1.99	0.153	0.56
		1979 structure			
	$I-O_{(1)}$	2.159	1.99	0.169	0.52
	$I-O_{(2)}$	2.153	1.99	0.163	0.54
3	$I-O_{(1)}$	2.136	1.99	0.146	0.57
	$I-O_{(2)}$	2.163	1.99	0.173	0.52
4	$I-O_{(1)}$	1.940	1.99	−0.05	1.21
	$I-O_{(2)}$	2.473	1.99	0.483	0.16
5	$I-O_{(1)}$	2.00	1.99	0.01	0.96
	$I-O_{(2)}$	2.30	1.99	0.31	0.30
6	$I-O_{(1)}$	2.11	1.99	0.12	0.63
	$I-O_{(2)}$	2.13	1.99	0.14	0.58
7	$I-O_{(1)}$	2.13	1.99	0.14	0.58
α-polymorph	$I-O_{(2)}$	2.11	1.99	0.12	0.63
8			α-Polymorph		
α-polymorph	$I-O_{(1)}$	2.11	1.99	0.12	0.63
	$I-O_{(2)}$	2.15	1.99	0.16	0.54
β-polymorph			β-Polymorph		
	$I-O_{(1)}$	2.08	1.99	0.09	0.71
	$I-O_{(2)}$	2.14	1.99	0.15	0.56
9	I–N	2.06	2.03	0.03	0.89
	I–Cl	2.56	2.32	0.24	0.40

or comparable basicity and, as a result, the iodine–ligand bonds are of comparable bond order and range from 0.43 to 0.71. However, in each of the compounds **4**, **5** and **9**, the heteroligand basicities differ significantly. The more basic (less electronegative) ligand is bound more tightly to iodine than either ligand of the symmetrical iodinanes while the less basic (more electronegative) ligand is bound less tightly to the iodine atom. for example, in [hydroxy(tosyloxy)iodo]benzene, the heteroligands are HO^- and TsO^- (recall the disalt mnemonic discussed earlier), and the computed I—OH and I—OTs bond orders are 1.21 and 0.16 respectively. This bond order assymmetry is consistent with the [3c–4e] bond and finds precedent in HMO descriptions of [3c–4e] π-bonds. For example, in proceeding from allyl anion to the corresponding enolate system (H_2C⋯CH⋯O)$^-$, both the π-bond order and charge distribution symmetries are lost. The less electronegative (more basic) end of the molecule, $C_{(2)}$, is bound more tightly to the central carbon atom (π-bond order >0.5) while the more electronegative (less basic) oxygen atom is less tightly bound (π-bond order <0.5) to that atom. Furthermore, the electronic charge is shifted largely to the more electronegative (less basic) atom.

At what point an organoiodine(III) compound may be correctly regarded as either hypercovalent or ionic is difficult to assess. Even those molecules generally accepted as iodonium salts may possess hypercovalent character. Alcock and Countryman have recently reported single crystal X-ray structures of diphenyliodonium chloride, bromide and iodide, all of which exist as centrosymmetric dimers in the solid state. They concluded that the iodine–halogen bond distances in all three salts are consistent with a bond order of approximately 0.35[37]. The bond orders, computed from the Pauling relation, are I—Cl (0.053), I—Br (0.053) and I—I (0.052).

X = Cl, Br, I

Furthermore, even if an organohalinane is hypercovalent in the solid state, it *may* exist as an ion pair species in solution. For example, the fact that the proton magnetic resonance (PMR) spectrum of (diacetoxyiodo)benzene, recorded in organic solvents, exhibits a singlet for the methyl groups is consistent with either a hypercovalent structure or an ionic structure $(PhIOAc)^+$ OAc^- in which the acetoxy ligands are in a state of rapid degenerate exchange.

Finally, it is emphasized that geometric parameters of organoiodinanes must certainly be affected by intermolecular interactions in the solid state.

D. Secondary Bonding and Modes of Crystal Packing

The observed molecular bond distances and angles in the aryliodinanes **1–9** are influenced by secondary intramolecular contacts and by intermolecular interactions attending the various modes of crystal packing. The crystal structures of (diacetoxyiodo)benzene (**2**) and bis(dichloroacetoxyiodo)benzene (**3**) have been examined with a particular focus on secondary bonding[27]. In compound **2**, the I—$O_{(1)}$ and I—$O_{(2)}$ bond distances were found to be equal. The I—$O_{(3)}$ and I—$O_{(4)}$ bonds are much longer, but $O_{(3)}$ and $O_{(4)}$ are almost equally displaced from the iodine atom and at a distance significantly less than the sum of the van der Waals' radii of iodine and oxygen. The iodine atom, the four oxygen atoms and $C_{(1)}$ are coplaner, this plane describing a dihedral angle of 75° with respect to the plane of the benzene ring. These weak secondary contacts may account for the small compressions of the C—I—O bond angles from 90°.

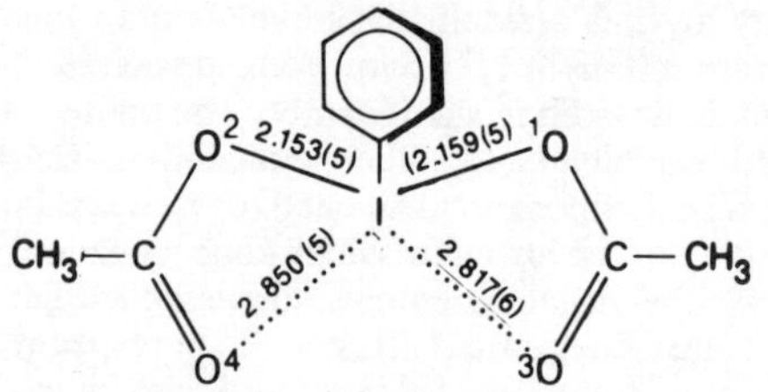

(2)

In compound **3**, the I—$O_{(1)}$ and I—$O_{(2)}$ bond distances are not equal nor are the carbonyl oxygen atoms, $O_{(3)}$ and $O_{(4)}$, symmetrically located about the iodine atom. Molecules of **3** crystallize as centrosymmetric 'dimers' in which the carbonyl oxygen of one of the dichloroacetoxy ligands on each iodine is in secondary intramolecular contact with one of the iodine atoms and in intermolecular contact with the other. The remaining dichloroacetoxy ligand on each iodine atom is unidentate. The I, $C_{(1)}$, $O_{(1)}$, $O_{(2)}$, $O_{(3)}$ and $O_{(4)}$ atoms in **3** exhibit small deviations from a common plane, the dihedral angle of which is 79.2° relative to the plane of the benzene ring.

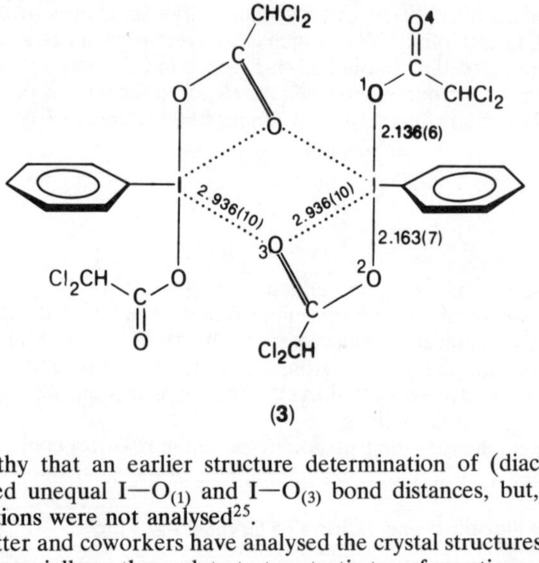

(3)

It is noteworthy that an earlier structure determination of (diacetoxyiodo)benzene (**2**) revealed unequal I—$O_{(1)}$ and I—$O_{(3)}$ bond distances, but, the secondary bonding interactions were not analysed[25].

Gougoutas, Etter and coworkers have analysed the crystal structures of a variety of benziodoxoles, especially as they relate to topotactic transformations. For those benziodoxoles with acyloxy ligands bound to iodine, secondary bonding between the oxygen atom of the acyloxy function and the iodine atom is important. For example, in the α- and β-polymorphs of 1-(*o*-iodo-benzoyloxy)-1,2-benziodoxol-3(1*H*)-one (**8**), the oxygen–iodine(III) intramolecular contacts are at 2.85(6) Å and 2.75(4) Å respectively. Four fundamental intermolecular packing modes have been recognized and related to the 'lactone' carbonyl stretching frequency of the iodole nucleus[38].

E. Solution-phase Studies

The T-configuration about iodine in the organoiodinanes is corroborated by one study of such compounds in solution[39]. The 5-aryl-5*H*-dibenziodoles **10** and **11** exhibit temperature-dependent PMR spectra. In solution (toluene-d_8, benzene, chlorobenzene, tetrahydrofuran, tetrachloroethylene) at low temperatures, the methyl groups of **10** and **11** appear as a pair of singlets which coalesce as the temperature is raised, the coalescence temperatures ranging from 15 to 60°C (decomposition is significant above 40°C). The line shape changes are reversible and are consistent with degenerate thermal isomerization of the aryl ligands about T-shaped iodine. However, the observation of large negative entropies of activation and irreproducible rate constants for the isomerization of the same compound at the same temperature were interpreted as evidence against a unimolecular degenerate isomerization mechanism (equation 10).

CH3 I CH3 ⇌ CH3 I CH3 (10)

R R

(**10**) R = H
(**11**) R = CF_3

F. Hypervalent Bonding in Organoiodoso Compounds

Hypervalent bonding theory may be extended to include divalent ligands as Musher has, for example, in his treatment of the sulphur–oxygen 'coordinate' bond of thionyl

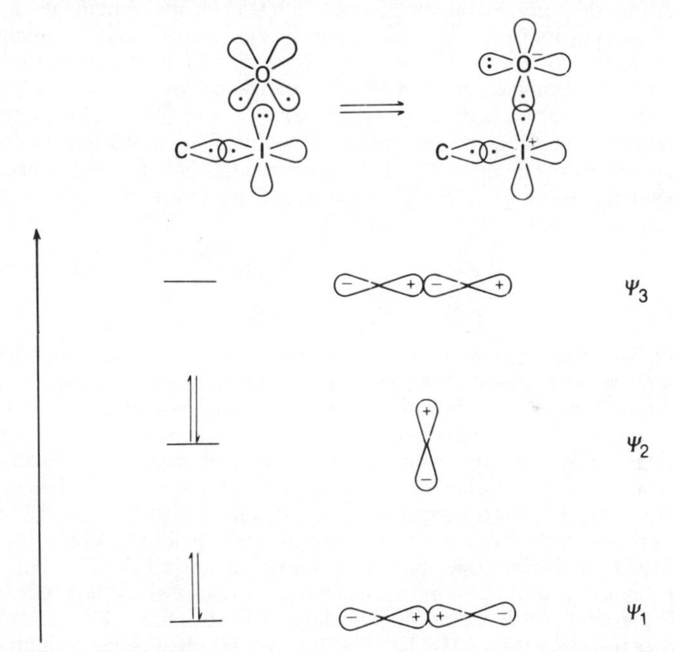

FIGURE 2. Hypervalent bonding model for organoiodinanes in which the iodine atom is bound to one equatorial carbon ligand and one axial divalent ligand (i.e. oxygen). The molecular orbitals for the [2c–4e] bond are shown.

fluoride[34]. Divalent ligands may be pictured as adding to iodo compounds by the interaction of two singly occupied atomic orbitals on the ligand with a doubly occupied 5p orbital on the iodine atom. The *simplest* electronic description of such a bond predicts the existence of three molecular orbitals; one bonding and doubly occupied, one non-bonding, *localized* on the divalent ligand and doubly occupied, and one antibonding and unoccupied.

The result is a non-linear molecule exhibiting a C—I—L bond angle of 90° and a highly polarized I—L bond with considerable positive charge density at iodine and negative charge density at the ligand site. This polarity is an inherent property of the [2c–4e] hypervalent bond, and it is to be expected that ligands of high electronegativity will stabilize such molecules more effectively than ligands of low electronegativity. It is not surprising then, that while many iodosoarenes and iodoxyarenes are known, the corresponding sulphur analogues have not been prepared.

IV. ALKYLIODINANES

A. Occurrence

Tricovalent alkyliodinanes are rare, and we are aware of no reports in the chemical literature describing the *isolation* of iodosoalkanes (R_H—I=O), (diacyloxyiodo)alkanes (R_H—I(OCOR)$_2$) or iodoxyalkanes (R_H—IO_2) where R_H is a hydrocarbon radical. A few (dichloroiodo)alkanes (R_H—ICl_2: R_H = methyl, ethyl, *n*-propyl, *n*-butyl, *i*-propyl, *t*-butyl) have long been known, these having been generated by the action of molecular chlorine on the corresponding alkyl iodides at low temperature, but they are extremely unstable[40]. (Dichloroiodo)methane, a yellow crystalline solid, appears to be the most stable of the known saturated aliphatic dichloroiodinanes, and, as discussed in Section II (equation 8), it decomposes at −28°C to iodine monochloride and chloromethane. (Dichloroiodo)iodomethane (**12**) has also been prepared and exhibits a decomposition point of −11.5°C[40]. The synthesis of (dibromoiodo)methane (**13**) by the oxidative addition of molecular bromine to iodomethane has been described[40]. This compound, obtained as orange-yellow needles which decompose at −45°C, is unique and appears to be the only alkyldibromoiodinane yet observed.

$$ICH_2ICl_2 \qquad\qquad CH_3IBr_2$$
$$(\mathbf{12}) \qquad\qquad\qquad (\mathbf{13})$$

The carbon atom directly bound to the dichloroiodo function in (dichloroiodo)alkanes seems to be significantly more prone to nucleophilic substitution than the carbon atom attached to iodine in the corresponding iodoalkanes (i.e. ICl_2^- is apparently a better leaving group than I^-). This is manifested in the reaction of neopentyl iodide with chlorine, which gives *t*-amyl chloride and 2,3-dichloro-2-methylbutane[41]. The initial formation of (dichloroiodo)neopentane in this reaction and its subsequent ionization-rearrangement to ethyldimethylcarbenium dichloroiodate has been proposed: the products arise either by nucleophilic capture of the cation or via a β-elimination followed by chlorine addition (equation 11).

The stereochemistry of iodine–chlorine exchange in reactions of alkyl iodides with chlorine has been studied[41,42]. Chiral 2-iodooctane, with molecular chlorine in various solvents (e.g. petroleum ether, CH_2Cl_2, CH_2Cl_2/CH_3OH) gives chiral 2-chlorooctane with predominant *inversion* of configuration, the (k_{inv}/k_{ret}) ratio depending on the particular solvent employed. The initial formation of chiral 2-(dichloroiodo)octane followed by nucleophilic displacement of dichloroiodate ion by molecular chlorine,

$$(CH_3)_3CCH_2I \overset{Cl_2}{\rightleftharpoons} (CH_3)_3CCH_2{-}I(Cl)_2 \longrightarrow \left[CH_3{-}\overset{+}{C}(CH_2CH_3)(CH_3) \right] ICl_2^- \quad \begin{cases} \xrightarrow{k_n} CH_3CCl(CH_3)CH_2CH_3 + ICl \\ \xrightarrow{k_e} (CH_3)_2C{=}CHCH_3 \xrightarrow{Cl_2} (CH_3)_2C(Cl)CH(Cl)CH_3 \end{cases} \qquad (11)$$

perhaps via an intermediate ion pair, is one plausible mechanism for this reaction (equation 12). When chiral 2-iodooctane is treated with bromine in petroleum ether

$$CH_3{-}CH(ICl_2){-}(CH_2)_5CH_3 \rightleftharpoons CH_3\overset{+}{C}H(CH_2)_5CH_3\ ICl_2^- \xrightarrow{Cl_2} CH_3{-}CH(Cl)(CH_2)_5CH_3 + ICl + Cl_2 \qquad (12)$$

for 2 h at -78°C and then for 1 h at 0°C, 2-bromooctane is likewise obtained with predominant inversion of configuration ($k_{inv}/k_{ret} = 3.5$). This is consistent with a mechanism involving the intermediate formation and nucleophilic collapse of 2-(dibromoiodo)octane (**14**).

$$CH_3{-}CH(IBr_2)(CH_2)_5CH_3$$

(**14**) (**15**) (**16**)

In 1968, the alicyclic dichloroiodinanes **15** and **16** were synthesized in the hope that they might exhibit some kinetic stability toward nucleophilic species[43]. Both compounds, however, were found to be unstable above 0°C (the decomposition products were not reported), and attempts to hydrolyse them to the corresponding iodoso compounds were unsuccessful.

The ability of electronegative ligands to stabilize iodinanes is well established, and it might be expected that difluoroalkyliodinanes will be more stable than dichloroalkyliodinanes. (Difluoroiodo)methane has been synthesized by treatment of *excess* iodomethane with xenon difluoride in the presence of anhydrous HF at *room temperature*. It has not, however, been isolated free of solvent (equation 13)[44].

$$\underset{\text{(excess)}}{CH_3{-}I} \xrightarrow[HF]{XeF_2} CH_3{-}IF_2 \qquad (13)$$

The addition of alkyl Grignard reagents to diphenyliodonium salts leads to a variety of products[45]. For example, the reaction of ethylmagnesium bromide with diphenyliodonium chloride in ether at -40°C gives benzene, iodobenzene, ethylbenzene, biphenyl and iodoethane. When such additions are conducted below 0°C, yellow-orange precipitates are formed which then decolorize as they are stirred at room temperature. Additions near the ether boiling point are accompanied by a transient flash of colour. The formation of intermediate alkyldiphenyliodinanes and their subsequent homolytic decomposition has been proposed to explain these observations (equation 14).

$$\mathrm{RMgX} + (\mathrm{Ph\overset{+}{I}Ph})\mathrm{X}^- \xrightarrow{\mathrm{Et_2O}} \mathrm{R{-}I(Ph)_2} \xrightarrow{\mathrm{Et_2O}} \mathrm{PhH + PhI + PhPh + PhR + RI} \quad (14)$$

It seems clear that the successful isolation of stable organoiodinanes of general structure $R{-}IL_2$ (where L is a heteroatom ligand) is rendered more probable when the carbon atom bound to iodine is kinetically resistant to nucleophiles. This fact has been corroborated by the isolation at room temperature of crystalline (dichloroiodo)arenes, (dichloroiodo)alkenes[8] and α-(dichloroiodo)sulphones (**17**)[23,24] shown below. How-

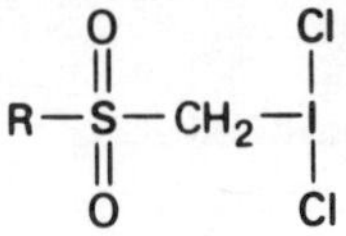

$$\mathrm{R{-}S(=O)_2{-}CH_2{-}ICl_2}$$

(**17**)

ever, the iodososulphones and diacyloxysulphones have not been reported. The disappointing instability of (dichloroiodo)cyclopropane (**15**) may reflect alternate modes of decomposition unique to its structure such as ionic and/or free radical chlorination of the strained carbon–carbon bonds of the cyclopropane ring; the (dichloroiodo)arenes are known to be excellent chlorinating reagents for alkenes and alkanes[6]. The instability of **16** may be partially due to steric interactions between the tertiary bridgehead carbon and the dichloroiodo function which cannot be totally relieved by any angle of rotation about the carbon–iodine bond. It is noteworthy, however, that **16** is significantly more stable than 2-(dichloroiodo)-2-methylpropane, which has been reported to decompose at -100°C.

Recent efforts to prepare aliphatic iodinanes have focused on the capacity of fluorine atoms in the *aliphatic* residue to impart some measure of stability. (Difluoroiodo)trifluoromethane was first reported in 1959, having been prepared by the direct fluorination of iodotrifluoromethane at -80°C[46]. It was described as a 'white, hygroscopic substance', stable to about 0°C, but decomposing at room temperature to the mixture of products shown in equation (15). The preparation of a series of (difluoro)-

$$\mathrm{CF_3{-}I + F_2} \xrightarrow[\mathrm{N_2(g)}]{\mathrm{FCCl_3/-80\ ^\circ C}} \mathrm{CF_3{-}IF_2} \xrightarrow{\text{Room temp.}} \mathrm{CF_3I + CF_4 + IF_5 + I_2} \quad (15)$$

perfluoroalkyliodinanes, including (difluoroiodo)trifluoromethane, by the action of chlorine trifluoride on several perfluoroalkyl iodides was reported in 1969[47]. The stoicheiometry of this ligand transfer reaction is critical. Thus, with an excess of the iodo compound, difluoroiodinanes are obtained, but with an excess of chlorine trifluoride, tetrafluoroperiodinanes are formed instead. The compounds (difluoroiodo)pentafluoroethane (**18**), (difluoroiodo)nonafluoro-*n*-butane (**19**), (difluoroiodo)tridecafluoro-*n*-hexane (**20**) and (difluoroiodo)heneicosafluoro-*n*-decane (**21**) have been synthesized and observed to be moderately stable as long as they are kept from moisture and glass. Compounds **18**, **19** and **20** were usually obtained as low melting solids, but they were sometimes isolated in high melting modifications (m.p. $> 200°C$), while compound **21** was isolated only as a high melting solid. The

$$R_F{-}I\ (\text{excess}) + ClF_3 \longrightarrow R_F{-}IF_2 + Cl_2 \qquad (16)$$

(**18**) $R_F = CF_3CF_2$

(**19**) $R_F = CF_3CF_2CF_2CF_2$

(**20**) $R_F = CF_3(CF_2)_4CF_2$

(**21**) $R_F = CF_3(CF_2)_8CF_2$

high melting materials were suggested to be 'dimeric' and possibly to be iodonium salts of general structure $(R_F)_2\overset{+}{I}\ \overset{-}{I}F_4$. Although the (difluoro)perfluoroalkyliodinanes were reported to undergo hydrolysis, no evidence was presented that the corresponding perfluoroiodosoalkanes ($R_F{-}I{=}O$) were formed.

Perfluoroalkyliodinanes with ligands other than halide attached to the iodine atom have also been prepared. When heptafluoro-*n*-propyl iodide is treated with sulphur tetrafluoride, under rather stringent conditions, (difluoroiodo)heptafluoropropane (**22**) is obtained. The oxidation of the iodide with 80% hydrogen peroxide in trifluoroacetic acid–trifluoroacetic anhydride gives (ditrifluoroacetoxyiodo)heptafluoropropane (**23**) in 65% yield. When **23** is subjected to hydrolysis with cold, aqueous sodium bicarbonate, perfluoroiodosopropane (**24**) is obtained as a light yellow powder which 'decomposes rapidly on storage'[48] (equation 17).

$$CF_3CF_2CF_2{-}I \xrightarrow[\text{autoclave, 2 h, 70 °C}]{SF_4} CF_3CF_2CF_2{-}IF_2\ (\mathbf{22})$$

$$CF_3CF_2CF_2{-}I \xrightarrow[CF_3CO_2H/CF_3COOCOCF_3]{80\ °C,\ H_2O_2} CF_3CF_2CF_2{-}I(OC(O)CF_3)_2\ (\mathbf{23}) \qquad (17)$$

$$(\mathbf{23}) \xrightarrow{H_2O/NaHCO_3} CF_3CF_2CF_2I{=}O\ (\mathbf{24})$$

(Difluoroiodo)heptafluoropropane has been prepared under much milder conditions from heptafluoro-*n*-propyl iodide and xenon difluoride[49] (equation 18).

$$CF_3CF_2CF_2I + XeF_2 \xrightarrow{\text{Neat, 20°C, } N_2 \text{ atmosphere}} CF_3CF_2CF_2IF_2 + Xe \quad (18)$$

The attachment of fluorine atoms to α-carbon in fluoroalkyliodinanes is not essential for their isolation. Thus, the oxidation of fluoroiodoalkanes of general structure $H(CF_2CF_2)_nCH_2I$ (n = 1, 2, 3) with trifluoroperacetic acid gives the corresponding bis(trifluoroacetoxy)iodinanes **25**, **26** and **27** while similar oxidation of the diiodofluoroalkanes, $ICH_2(CF_2)_nCH_2I$ (n = 3, 4), gives the tetra(trifluoroacetoxy)iodinanes **28** and **29**[50]. When compounds **25** and **29** are subjected to the action of sulphur tetrafluoride, the corresponding fluoroiodinanes **30** and **31** are formed[50]. The hydrolysis of compounds **25–27** does not yield the corresponding iodoso compounds but leads instead to the *meso*-oxy derivatives **32**.

$H(CF_2CF_2)_nCH_2I(OOCCF_3)_2$

(25) $n = 1$

(26) $n = 2$

(27) $n = 3$

$(CF_3COO)_2ICH_2(CF_2)_nCH_2I(OOCCF_3)_2$

(28) $n = 3$

(29) $n = 4$

$HCF_2CF_2CH_2IF_2$

(30)

$F_2ICH_2(CF_2)_4CH_2IF_2$

(31)

$$H(CF_2CF_2)_nCH_2\underset{\displaystyle CF_3COO}{\underset{|}{I}}-O-\underset{\displaystyle OOCCF_3}{\underset{|}{I}}CH_2(CF_2CF_2)_nH$$

(32)

Finally, several 'iodoso' derivatives of trifluoromethyl iodide, in addition to the iododifluoride, have been reported and they are tabulated below.

$CF_3I{=}O$ [51] $CF_3I(OOCCF_3)_2$ [52] $CF_3I(ONO_2)_2$ [53] $CF_3I(Cl)ONO_2$ [53]

$CF_3I(F)ONO_2$ [53]

B. Alkyliodinanes as Intermediates in Oxidations of Iodoalkanes

Although the non-fluorinated iodosoalkanes and (diacyloxyiodo)alkanes have thus far eluded isolation, they have been implicated as transient intermediates in the oxidation reactions of various alkyl iodides. Three fundamental reaction manifolds have been recognized for alkyliodoso intermediates[54–57].

(1) *Nucleophilic substitution*. The diacyloxyiodo and iodoso functions render the carbon atom to which they are attached hyperelectrophilic. Primary alkyliodoso compounds, when generated in the presence of nucleophiles (Nu^-), are extremely susceptible to S_N reactions.

$$R{-}I \xrightarrow{\langle O \rangle} R{-}\overset{O}{\overset{\|}{I}} \rightleftharpoons R^+\ IO^- \xrightarrow{Nu^-} [R{-}Nu + IO^-] \quad (19)$$

(2) *Elimination*. Those alkyliodoso intermediates which contain β-carbon–hydrogen bonds and which are sterically or electronically deactivated toward nucleophilic attack, eliminate hypoiodous acid and afford alkenes.

$$\begin{matrix} \text{H} & \text{I} \\ | & | \\ >\text{C} - & \text{C}< \end{matrix} \xrightarrow{\langle O \rangle} \begin{matrix} \text{H} & \text{I}{=}\text{O} \\ | & | \\ >\text{C} - & \text{C}< \end{matrix} \longrightarrow [\text{HOI}] + >\text{C}{=}\text{C}< \qquad (20)$$

(3) *Isomerization with loss of hydroiodic acid*. In rare instances, the oxidation of an alkyl iodide eventuates in ketone formation. Such reactions probably proceed by the isomerization of alkyliodoso intermediates to alkyl hypoiodites and subsequent elimination of hydrogen iodide across the carbon–oxygen bond.

$$>\overset{\text{H}}{\overset{|}{\text{C}}}-\text{I} \xrightarrow{\langle O \rangle} >\overset{\text{H}}{\overset{|}{\text{C}}}-\overset{\text{O}}{\overset{\|}{\text{I}}} \longrightarrow >\overset{\text{H}}{\overset{|}{\text{C}}}-\text{O}-\text{I} \longrightarrow >\text{C}{=}\text{O} + \text{HI} \qquad (21)$$

1. Oxidative substitution

The alkyl iodides are apparently inert to such reagents as hydrogen peroxide, ozone, periodate, *t*-butylhydroperoxide, *N*-chlorosuccinimide, and $NaOCl/Bu_4N^+$ HSO_4^- in $CHCl_3/H_2O$[54,57]. *m*-Chloroperbenzoic acid (*m*-CPBA), on the other hand, has proven to be an excellent oxidizing reagent for these compounds. The oxidation of primary iodoalkanes with *m*-CPBA has been recently studied by several research groups. It proceeds under remarkably mild conditions (CH_2Cl_2, room temperature or below) to give the corresponding primary alcohols in good yield and, in some cases, primary alkyl *m*-chlorobenzoates[54–57]. Some of these reactions are summarized in Table 5.

The oxidative replacement of iodine by the hydroxy group is thought to involve the initial formation of iodosoalkanes which suffer nucleophilic displacement of hypoiodite ion by water, the water being introduced as a contaminant in the *m*-CPBA. The stoicheiometry of the oxidation of *n*-heptyl iodide with *m*-CPBA, chosen as a model reaction, has been thoroughly studied[55]. In dichloromethane, to which water had been deliberately added, *n*-heptyl alcohol was formed quantitatively within 8 min at 27°C. The by-products were identified as *m*-chlorobenzoic acid (*m*-CBA), molecular iodine, iodine pentoxide and molecular oxygen. The iodine and iodine pentoxide were consistently produced in a 2:1 mole ratio, thus providing good evidence for the existence of hypoiodite ion (or hypoiodous acid) in the reaction medium. The disproportionation of hypoiodite ion to iodide and iodate ions is well known, and, in the presence of *m*-CPBA, iodide ion would be oxidized to iodine while the dehydration of any iodic acid formed would eventuate in iodine pentoxide. It was noted that, while a 1.33:1.00 mole ratio of *m*-CPBA to alkyl iodide is theoretically required for complete oxidation, alcohol yields were generally not optimal unless *at least* two equivalents of *m*-CPBA were present. This observation and the formation of molecular oxygen were attributed to a side reaction involving the decomposition of *m*-CPBA. The overall reaction scheme is summarized in equations (22)–(24).

$$n\text{-}C_7H_{15}I + m\text{-CPBA} \longrightarrow n\text{-}C_7H_{15}I{=}O + m\text{-CBA} \qquad (22)$$

$$H_2O + n\text{-}C_7H_{15}I{=}O \longrightarrow n\text{-}C_7H_{15}OH + \text{HOI} \qquad (23)$$

$$3\ \text{HOI} \longrightarrow (2\ \text{HI}) + (\text{HIO}_3) \longrightarrow 0.5\ H_2O + 0.5\ I_2O_5 \qquad (24)$$

$$(2\ \text{HI}) \xrightarrow{m\text{-CPBA}} m\text{-CBA} + I_2 + H_2O$$

When the *m*-CPBA is rigorously dried prior to use, the yield of *n*-heptyl alcohol from 1-iodoheptane decreases with a corresponding increase in the yield of *n*-heptyl *m*-chlorobenzoate.

TABLE 5. Oxidation of iodoalkanes with *m*-chloroperbenzoic acid

Substrate	Reaction conditions	Reaction time, min	Products	Yield, %	Reference
$CH_3(CH_2)_{14}CH_2I$	CH_2Cl_2, 0°C	–	$CH_3(CH_2)_{14}CH_2OH$	62	54
$PhCH_2CH_2CH_2I$	CH_2Cl_2, 0°C	–	$PhCH_2CH_2CH_2OH$	75	54
$PhCH_2CH_2I$	CH_2Cl_2, 0°C	–	$PhCH_2CH_2OH$	73	54
$PhCH_2CD_2I$	CH_2Cl_2, 0°C	–	$PhCH_2CD_2OH$, $PhCD_2CH_2OH$	–	54
$CH_3(CH_2)_5CH_2I$	$CH_2Cl_2(H_2O)$, 27°C	8	$CH_3(CH_2)_5CH_2OH$	100	55
$CH_3(CH_2)_5CH_2I$	CH_2Cl_2, 27°C	6	$CH_3(CH_2)_5CH_2OH$	90	55
			$CH_3(CH_2)_5CH_2OC_6H_4Cl$-*m*	10	55
$CH_3(CH_2)_6CH_2I$	CH_2Cl_2, 20°C	1	$CH_3(CH_2)_6CH_2OH$	95	56, 57
$CH_3(CH_2)_2CH_2I$	CH_2Cl_2, 20°C	5	$CH_3(CH_2)_2CH_2OH$	100	57
$CH_3(CH_2)_3CH_2I$	CH_2Cl_2, 20°C	5	$CH_3(CH_2)_3CH_2OH$	100	57

The nucleophilic displacement of hypoiodite ion from intermediate iodosoalkanes has been observed in oxidations of secondary alkyl iodides and with nucleophiles other than water[56,57]. For example, the reaction of *n*-octyl iodide with *m*-CPBA in methanol gives methyl *n*-octyl ether in 91% yield. When acetic acid is the solvent, both *n*-octanol (24%) and *n*-octyl acetate (72%) are formed. With 2-iodooctane as the starting material, 2-methoxyoctane and 2-acetoxyoctane are obtained in methanol and acetic acid, respectively.

$$CH_3(CH_2)_6CH_2I + m\text{-CPBA} \begin{cases} \xrightarrow{CH_3OH} CH_3(CH_2)_6CH_2OCH_3 \\ \xrightarrow{CH_3COOH} CH_3(CH_2)_6CH_2OH + CH_3(CH_2)_6CH_2OAc \end{cases} \quad (25)$$

$$CH_3CH(I)(CH_2)_5CH_3 + m\text{-CPBA} \begin{cases} \xrightarrow{CH_3OH} CH_3CH(OCH_3)(CH_2)_5CH_3 \\ \xrightarrow{CH_3COOH} CH_3CH(OAc)(CH_2)_5CH_3 \end{cases} \quad (26)$$

The stereochemistry of these oxidative displacement reactions has been studied, and they proceed either with inversion of configuration or retention of configuration depending on the nature of the substrate employed. The oxidation of S(+)-2-iodooctane with *m*-CPBA gives *R*(−)-2-octanol in 56% yield, among other products, and thus proceeds with configurational inversion[57]. On the other hand, the *m*-CPBA oxidations of *trans*-1-iodo-2-methoxycyclohexane and *trans*-1-azido-2-iodocyclohexane either in dichloromethane or in aqueous *t*-butyl alcohol and *trans*-1-hydroxy-2-iodocyclohexane and *trans*-1-bromo-2-iodocyclohexane in aqueous *t*-butyl alcohol all give the corresponding alcohols with retention of configuration[57]. The retention of configuration observed with the cyclohexane substrates has been explained on the assumption that oxidative displacement proceeds with neighbouring group participation.

$$\underset{S(+)}{CH_3CH(I)(CH_2)_5CH_3} + m\text{-CPBA} \longrightarrow \underset{R(-)}{CH_3CH(OH)(CH_2)_5CH_3} + [HOI] \quad (27)$$

$$trans\text{-}1\text{-iodo-}2\text{-X-cyclohexane} + m\text{-CPBA} \xrightarrow{\text{Solvent}} trans\text{-}2\text{-X-cyclohexanol} + [HOI] \quad (28)$$

$X = OCH_3, N_3, OH, Br$

The oxidations of *trans*-1-acyloxy-2-iodocyclohexanes with *m*-CPBA are characterized by alcohol formation with *inversion* of configuration. Once again, the involvement of the acyloxy functions in the displacement process is deemed likely. However, attack of the nucleophile on the resulting cyclic 1,3-dioxolan 2-ylium ions would be directed to the carbenium ion centre, eventuating in products with *cis* stereochemistry[57].

$$\text{trans-2-iodocyclohexyl ester } (\text{O–C(=O)–R}) + m\text{-CPBA} \xrightarrow{-[\text{HOI}]} \text{dioxolanylium ion (R)} \xrightarrow{H_2O} \text{hydroxy-dioxolane (OH, R)} \longrightarrow \text{cis-2-hydroxycyclohexyl ester } (\text{OH}, \text{O–C(=O)–R}) \qquad (29)$$

Epoxide formation is observed when *trans*-1-hydroxy-2-iodocyclohexane and 3α-iodo-5α-androstan-2β-ol are oxidized with 2.2 equivalents of *m*-CPBA in dichloromethane. In both cases, the products are formed in quantitative yield within 5 min[57].

$$\text{trans-2-iodocyclohexanol (I, OH)} + m\text{-CPBA} \xrightarrow{-[\text{HOI}]} \text{cyclohexene oxide (O)} \qquad (30)$$

2. Oxidative elimination

While the oxidations of primary alkyl iodides with *m*-CPBA generally proceed cleanly to the corresponding primary alcohols, mixtures of products are obtained from similar oxidations of the few secondary alkyl halides studied which are not substituted at β-carbon.

For example, the reaction of cyclohexyl iodide with *m*-CPBA has been found to yield cyclohexanol, cyclohexanone and cyclohexene oxide[55]. Under similar conditions,

$$\text{cyclohexyl iodide} + m\text{-CPBA} \xrightarrow[27\ ^\circ C]{CH_2Cl_2} \underset{54\%}{\text{cyclohexanol (OH)}} + \underset{14\%}{\text{cyclohexanone (O)}} + \underset{32\%}{\text{cyclohexene oxide (O)}} \qquad (31)$$

2-iodooctane is converted into 2-octanol, 2-octyl *m*-chlorobenzoate, 2-octanone and 2-octene oxide. The epoxides are thought to arise via the β-elimination of hypoiodous acid from intermediate iodosoalkanes and the subsequent oxidation of the alkenes thus formed with *m*-CPBA.

$$CH_3CH(I)(CH_2)_5CH_3 + m\text{-CPBA} \xrightarrow[40\ ^\circ C]{CH_2Cl_2} \underset{31\%}{CH_3CH(OH)(CH_2)_5CH_3} + \underset{16\%}{CH_3CH(OCOC_6H_4Cl\text{-}m)(CH_2)_5CH_3}$$
$$+ \underset{29\%}{CH_3C(=O)(CH_2)_5CH_3} + \underset{24\%}{CH_3CH\overset{O}{—}CH(CH_2)_4CH_3} \qquad (32)$$

The oxidation of cyclohexyl iodide with peracetic acid has been reported to give cyclohexyl acetate, *vic*-diacetoxycyclohexane and 1-acetoxy-2-iodocyclohexane, the product yields depending on the reaction conditions[58]. The latter compound is thought

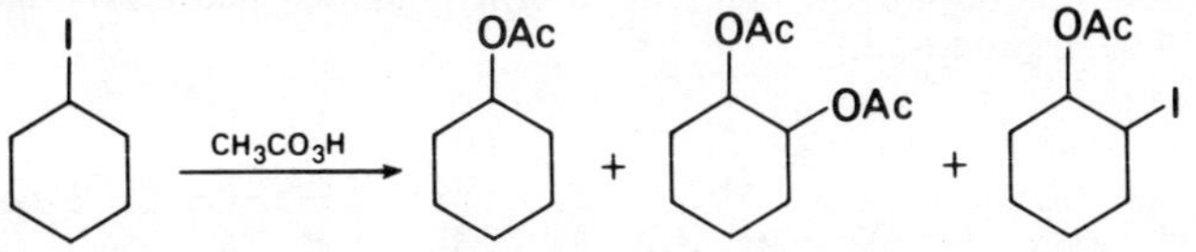

$$\text{C}_6\text{H}_{11}\text{I} \xrightarrow{CH_3CO_3H} \text{C}_6\text{H}_{11}\text{OAc} + 1,2\text{-(AcO)}_2\text{C}_6\text{H}_{10} + 1\text{-AcO-2-I-C}_6\text{H}_{10} \quad (33)$$

to arise by the addition of acetyl hypoiodite to cyclohexene in a secondary reaction following the initial elimination process. The overall reaction scheme is shown in equations (34)–(37).

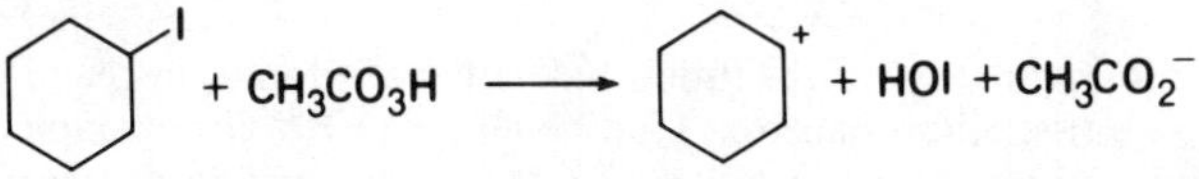

$$\text{C}_6\text{H}_{11}\text{I} + CH_3CO_3H \longrightarrow \text{C}_6\text{H}_{11}^+ + HOI + CH_3CO_2^- \quad (34)$$

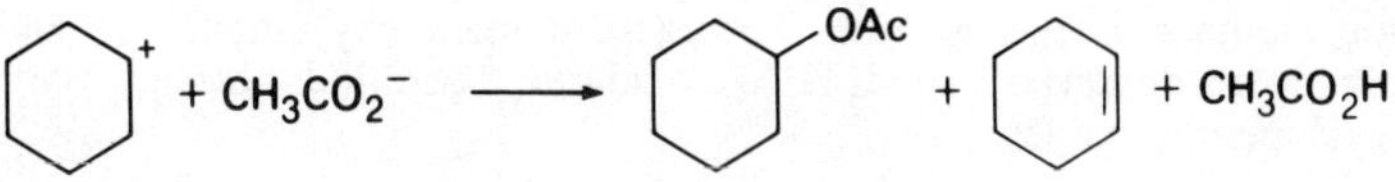

$$\text{C}_6\text{H}_{11}^+ + CH_3CO_2^- \longrightarrow \text{C}_6\text{H}_{11}\text{OAc} + \text{C}_6\text{H}_{10} + CH_3CO_2H \quad (35)$$

$$HOI + CH_3CO_2H \rightleftharpoons CH_3\overset{O}{\overset{\|}{C}}-O-I + H_2O \quad (36)$$

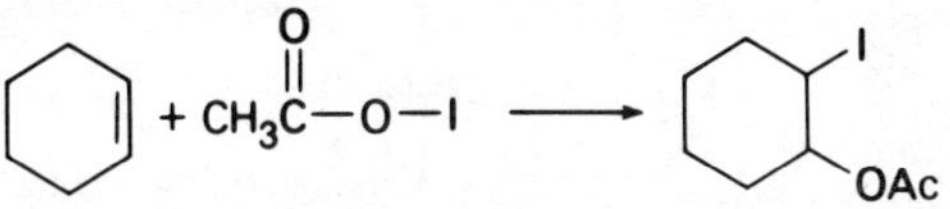

$$\text{C}_6\text{H}_{10} + CH_3\overset{O}{\overset{\|}{C}}-O-I \longrightarrow 1\text{-I-2-AcO-C}_6\text{H}_{10} \quad (37)$$

Clean elimination reactions occur when 2-phenylsulphonyl-2-iodopropane and ethyl 2-iododecanoate are treated with *m*-CPBA (equations 38 and 39). With 2-phenylsulphonyl-2-iodo-3-phenylbutane, the oxidative elimination proceeds in regiospecific fashion to give the vinyl sulphone having the least substituted double bond[54] (equation 40). The mechanism of the α-iodosulphone elimination reaction has

$$Ph-SO_2-C(I)(CH_3)-CH_3 + m\text{-CPBA} \xrightarrow{CH_2Cl_2} Ph-SO_2-C(CH_3)=CH_2 \quad (38)$$

87%

$$CH_3CH_2O\overset{O}{\overset{\|}{C}}CH(I)CH_2(CH_2)_6CH_3 + m\text{-CPBA} \xrightarrow{CH_2Cl_2} CH_3CH_2O\overset{O}{\overset{\|}{C}}CH=CH(CH_2)_6CH_3 \quad (39)$$

81%

$$Ph-SO_2-C(I)(CH_3)-CH(CH_3)Ph \xrightarrow{m\text{-CPBA}} Ph-SO_2-C(=CH_2)-CH(CH_3)Ph + [HOI] \quad (40)$$

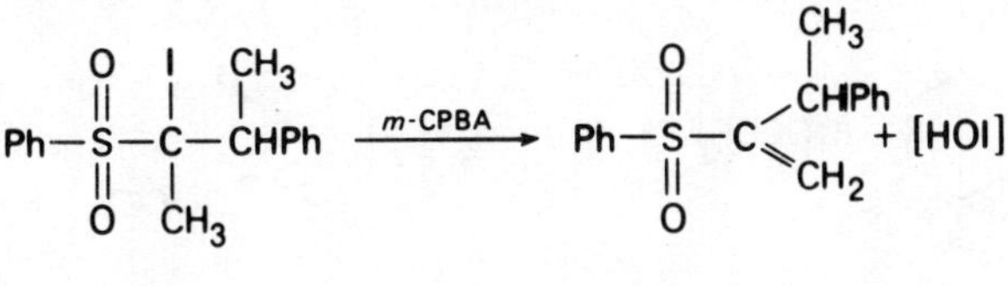

been probed with an appropriately designed tetralin reactant, and the elimination has been found to occur exclusively with *syn* stereochemistry[54]. In view of the fact that the phenylsulphonyl and carboethoxy substituents would inhibit carbenium ion formation at the α carbon, a pericyclic *syn* elimination of hypoiodous acid from α-iodoso intermediates has been proposed.

$$\text{H}(\text{I})\text{C}{-}\text{C}{-}\text{SO}_2\text{Ph} \xrightarrow{\langle O \rangle} \text{H}(\text{I}{=}\text{O})\text{C}{-}\text{C}{-}\text{SO}_2\text{Ph} \not\longrightarrow \text{H}{-}\text{C}{-}\overset{+}{\text{C}}{-}\text{SO}_2\text{Ph} + \text{IO}^- ; \quad \longrightarrow \text{HOI} + {>}\text{C}{=}\text{C}{<}^{\text{SO}_2\text{Ph}} \qquad (41)$$

Oxidative elimination and substitution has also been found in reactions of several iodoalkanes with tris(trifluoroacetoxy) iodine ($I(TFA)_3$)[59]. For example, the reaction of $I(TFA)_3$ with 2-iodobutane in ether affords 2-iodo-3-trifluoroacetoxybutane, but, when dichloromethane is the solvent, 2,3-bis(trifluoroacetoxy)butane is obtained. Similar results were observed with $I(TFA)_3$ oxidations of 2-iodooctane, cyclohexyl iodide and cyclopentyl iodide.

$$\text{CH}_3\text{CH(I)CH}_2\text{CH}_3 + \text{I(OOCCF}_3)_3 \xrightarrow{\text{Et}_2\text{O}} \text{CH}_3\text{CH(I)CH(OOCCF}_3)\text{CH}_3 \quad 45{:}55\ (\textit{erythro}:\textit{threo})$$

$$\xrightarrow{\text{CH}_2\text{Cl}_2} \text{CH}_3\text{CH(OOCCF}_3)\text{CH(OOCCF}_3)\text{CH}_3 \quad (45{:}55\ \textit{meso}:\text{racemic}) \qquad (42)$$

It seems likely that these reactions proceed via initial ligand transfer from $I(TFA)_3$ to the iodoalkanes to give (ditrifluoroacetoxyiodo)alkanes followed by β-elimination and subsequent addition of trifluoroacetyl hypoiodite to the resulting alkenes. The 1-iodo-2-trifluoroacetoxyalkanes thus produced might then suffer oxidative displacement of the iodine atom when the solvent is dichloromethane. The formation of *trans*-1-iodo-2-trifluoroacetoxycyclohexane from cyclohexyl iodide in ether and predominantly *cis*-1,2-ditrifluoroacetoxycyclohexane from the same substrate in dichloromethane is consistent with such a mechanism.

$$\text{C}_6\text{H}_{11}\text{I} + \text{I(OOCCF}_3)_3 \rightleftharpoons \text{IOOCCF}_3 + \text{C}_6\text{H}_{11}\text{I(OOCCF}_3)_2 \qquad (43)$$

$$\text{C}_6\text{H}_{11}\text{I(OOCCF}_3)_2 \longrightarrow \text{cyclohexene} + \text{CF}_3\text{COOH} + \text{IOOCCF}_3 \qquad (44)$$

$$\text{cyclohexene} + \text{IOOCCF}_3 \longrightarrow \textit{trans}\text{-1-I-2-OOCCF}_3\text{-cyclohexane} \qquad (45)$$

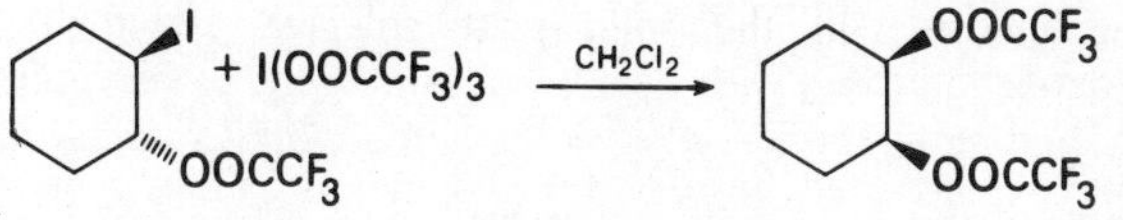

$$\text{(trans-2-iodocyclohexyl trifluoroacetate)} + I(OOCCF_3)_3 \xrightarrow{CH_2Cl_2} \text{(cis-1,2-bis(trifluoroacetoxy)cyclohexane)} \qquad (46)$$

The simple oxidative displacement of iodine from iodoalkanes in the presence of $I(TFA)_3$ has been observed in a couple of instances. For example, treatment of *n*-octyl iodide with $I(TFA)_3$ in ether gives *n*-octyl trifluoroacetate (15%) and *n*-octyl ethyl ether (35%). It has been suggested that such species as $I(TFA)_{3-n}(OEt)_n$ may exist in solution[59].

V. CYCLIC STRUCTURES AND SOME UNUSUAL ORGANOHALINANES

A. Cyclic Structures and Aryliodinane Stabilities

Cyclic aryliodinanes are often observed to be more stable than their acyclic analogues. For example, triphenyliodine (Ph_3I) prepared from diphenyliodonium chloride and phenyllithium in ether at −80°C, undergoes facile homolytic decomposition at −10°C[13,14]. The 5-aryl-5*H*-dibenziodoles **10**, **11**, **33**, **34** and **35**, on the other hand, can be synthesized at 0°C and persist for several hours to several days at room temperature[39,60,61]. The unusual thioiodinanes **36** and **37** exhibit a similar contrast in thermal stability[62].

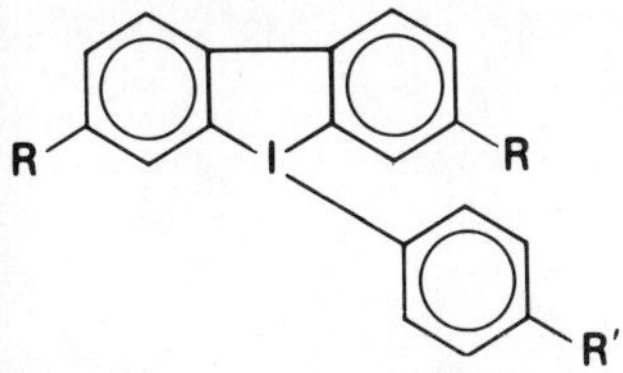

(10) R = Me; R′ = H
(11) R = Me; R′ = CF_3
(33) R = H; R′ = H
(34) R = H; R′ = Cl
(35) R = H; R′ = Me

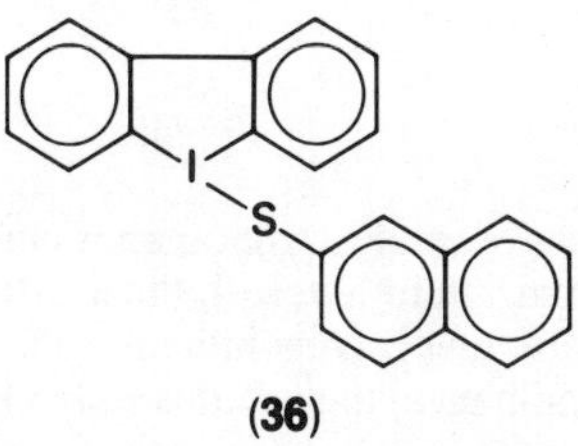

(36)

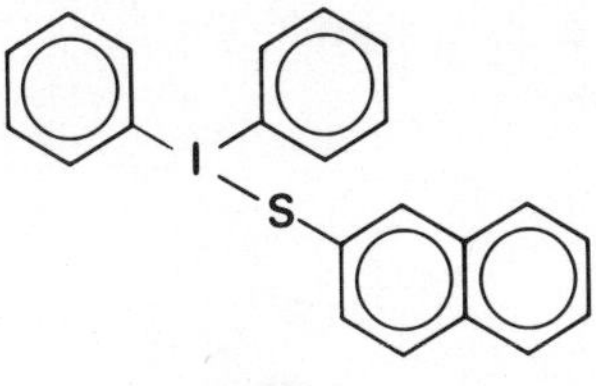

(37)

The oxidations of iodoarenes with functional groups *ortho* to the iodine atom frequently eventuate in iodine(III) heterocycles. For example, '*o*-iodosobenzoic acid'[29,63], '*o*-iodosophenylacetic acid'[64] and '*o*-iodosophenylphosphoric acid'[65] exhibit the cyclic structures **5**, **38** and **39** instead of the isomeric acyclic structures with a free iodoso function.

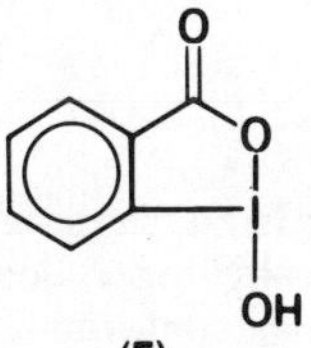

(5) 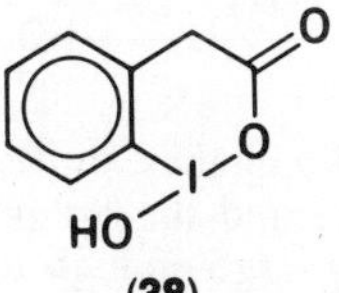 **(38)** 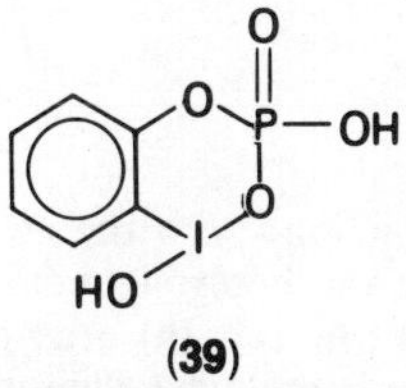 **(39)**

The reaction of *o*-iodobenzoic acid with molecular chlorine gives *o*-(dichloroiodo)benzoic acid. However, this is an unstable compound which eliminates hydrogen chloride in the solid state to give 1-chloro-1,2-benziodoxol-3(1*H*)-one[66].

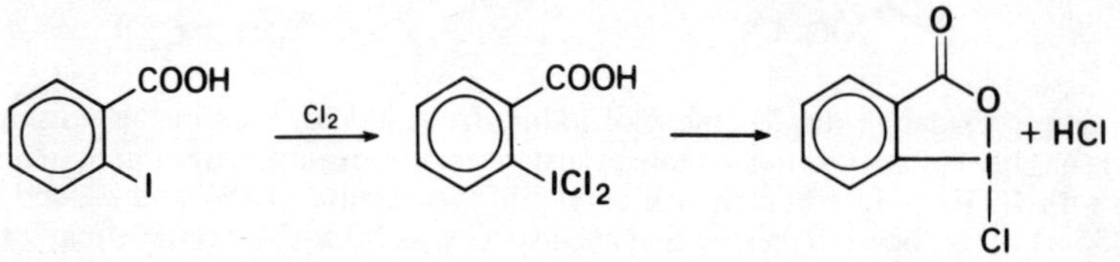

(47)

The direct oxidation of 2-iodoisophthalic acid and 2-iodo-*m*-benzenediacetic acid with peracetic acid likewise affords the cyclic products **40** and **41** instead of the corresponding (diacetoxyiodo)arenes[67]. Even the oxidation of *o*-diiodobenzene with peracetic acid yields the cyclic product **42** with an I—O—I bridge instead of its acyclic isomer **43**[68].

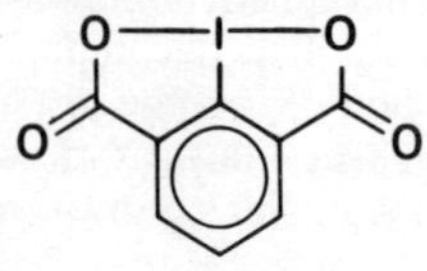

(40) 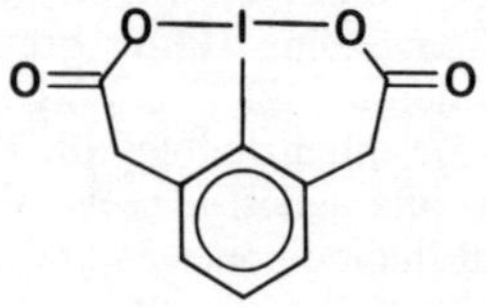 **(41)** 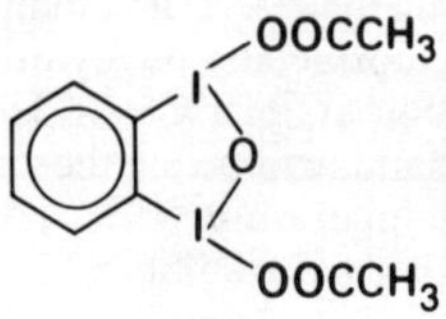 **(42)**

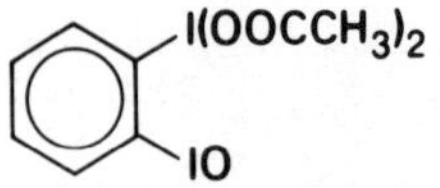

(43)

The increased stability of cyclic aryliodinanes may result from the conjugative overlap of a p-orbital lone pair on iodine with the π-orbitals of the aromatic nucleus. Since the plane of the aromatic ring is constrained to the same plane as that of the L—I—L function in the cyclic iodinanes, the orbitals will at least be favourably aligned for such an interaction. With the acyclic aryliodinanes, on the other hand, steric interactions between the heteroligands on iodine and the *ortho* carbon–hydrogen bonds of the benzene ring (cf. **1a**) are apparently large enough to favour a conformation **(1b)** in which the aromatic nucleus is nearly orthogonal to the L—I—L function, at least in the solid state. Conjugative overlap between the π-orbitals of the ring and the 5p orbital on the iodine atom is eliminated in such conformations.

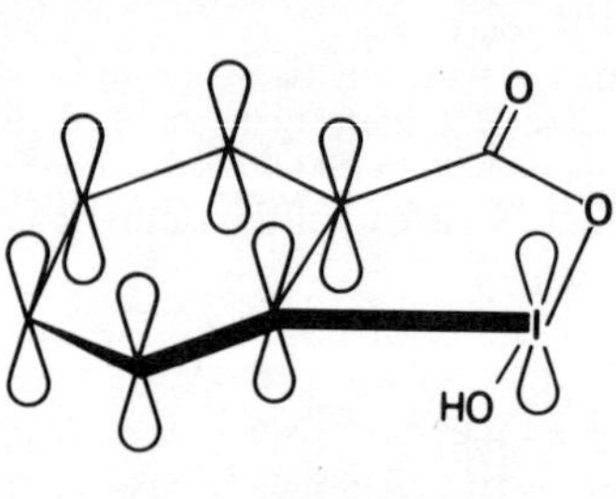

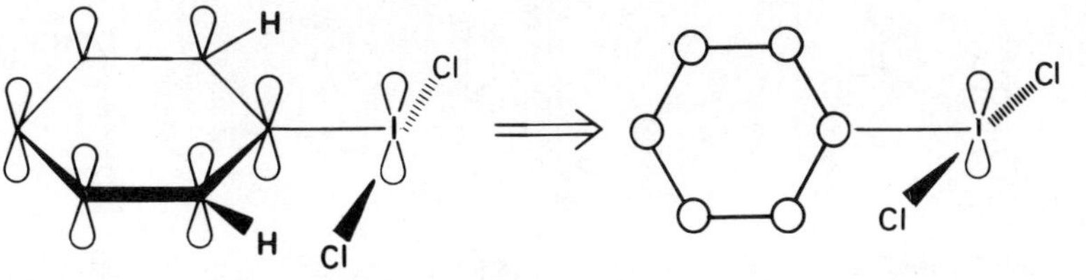

(1a) (1b)

B. Stabilization of Organoalkoxyiodinanes, Organobromoiodinanes and Organoalkoxybrominanes with Cyclic Structures

Prior to 1979, (dibromoiodo)methane (CH_3IBr_2) and (dibromoiodo)benzene ($PhIBr_2$) were the only examples of organobromoiodinanes in the chemical literature. The former compound was reported to decompose at −45°C while the latter was not isolated[69], and an attempt by later investigators to reproduce its synthesis was unsuccessful[36].

Organoiodine(III) compounds with iodine-bound alkoxy ligands are also rare. The production of acetone, benzene, phenyl isopropyl ether, and iodobenzene from diphenyliodonium tetrafluoroborate and sodium isopropoxide in isopropanol (equation 48) very likely proceeds via isopropoxydiphenyliodine, but the iodinane has not been isolated[70]. Cyclic dialkoxyiodinanes have been implicated as intermediates in the

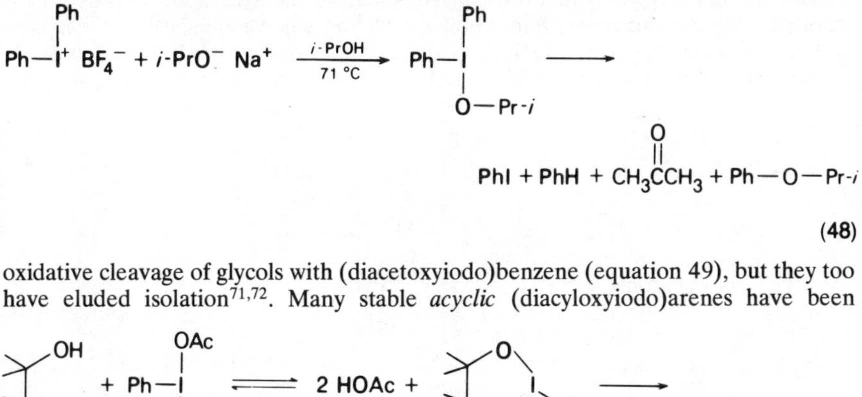

$$Ph_2I^+ BF_4^- + i\text{-}PrO^- Na^+ \xrightarrow[71\,°C]{i\text{-}PrOH} Ph_2I{-}O{-}Pr\text{-}i \longrightarrow PhI + PhH + CH_3C(O)CH_3 + Ph{-}O{-}Pr\text{-}i \qquad (48)$$

oxidative cleavage of glycols with (diacetoxyiodo)benzene (equation 49), but they too have eluded isolation[71,72]. Many stable *acyclic* (diacyloxyiodo)arenes have been

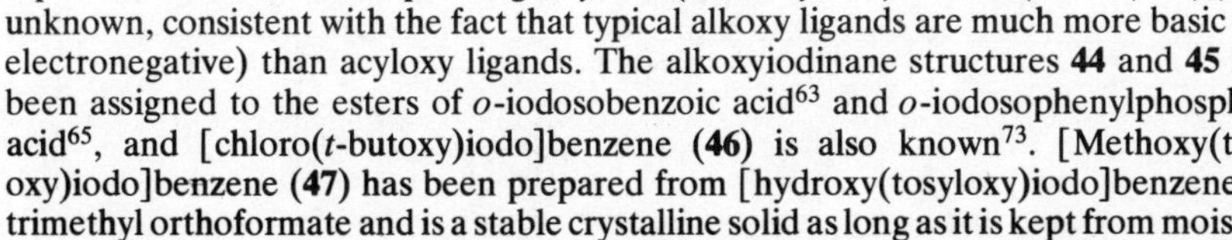

$$Me_2C(OH){-}C(OH)Me_2 + Ph{-}I(OAc)_2 \rightleftharpoons 2\,HOAc + \text{cyclic } Me_4C_2O_2I{-}Ph \longrightarrow 2\,Me_2C{=}O + PhI \qquad (49)$$

reported, but the corresponding *acyclic* (dialkoxyiodo)arenes ($Ar{-}I(OR)_2$) are unknown, consistent with the fact that typical alkoxy ligands are much more basic (less electronegative) than acyloxy ligands. The alkoxyiodinane structures **44** and **45** have been assigned to the esters of *o*-iodosobenzoic acid[63] and *o*-iodosophenylphosphoric acid[65], and [chloro(*t*-butoxy)iodo]benzene (**46**) is also known[73]. [Methoxy(tosyloxy)iodo]benzene (**47**) has been prepared from [hydroxy(tosyloxy)iodo]benzene and trimethyl orthoformate and is a stable crystalline solid as long as it is kept from moisture, in the presence of which it hydrolyses rapidly back to starting material[74]. All of these stable monoalkoxyiodinanes contain two heteroligands of drastically different basicities and probably derive their stability from highly asymmetric [3c–4e] bonds.

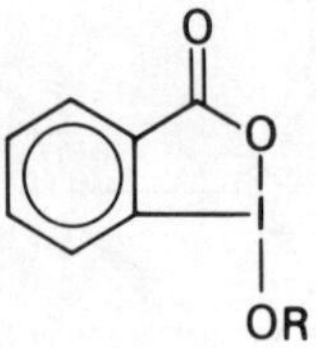

(44) 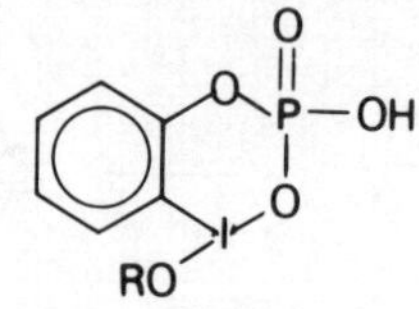 (45) 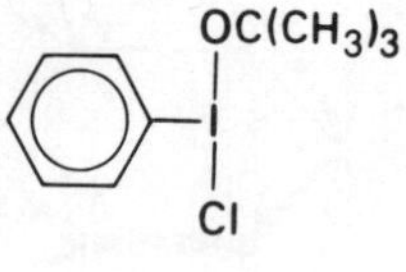 (46)

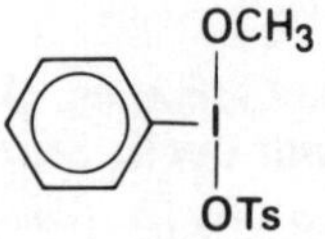

(47)

Martin and his coworkers, exploiting highly electronegative alkoxy ligands in conjunction with cyclic structures to full advantage, have recently *isolated* the first stable arylbromoiodinanes, the first dialkoxyiodinanes with an aryl ligand and the first arylbrominane.

Treatment of 2-(2-iodo-4-methylphenyl)hexafluoropropan-2-ol first with potassium hydride and then with bromine gives 1-bromo-1,3-dihydro-5-methyl-3,3-bis (trifluoromethyl)-1,2-benziodoxol (**48**), as a bright yellow solid[36]. The oxidation of the alcohol alternatively with trifluoromethyl hypofluorite and with *t*-butyl hypochlorite or chlorine gives the analogous fluoroiodinane **49** and chloroiodinane **50**. These compounds were described as 'stable crystalline solids which may be handled in the atomosphere without decomposition' and were even found to be stable above their

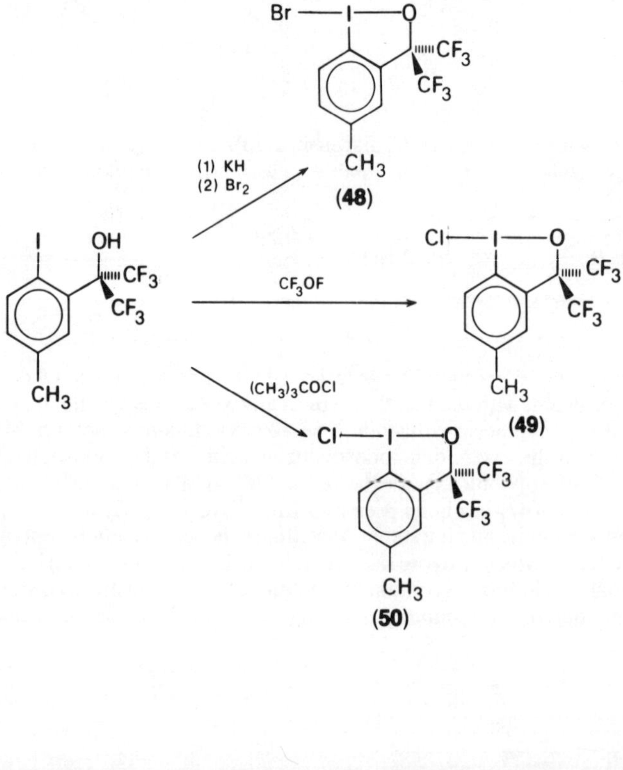

(1) KH (2) Br_2 → (48); CF_3OF → (49); $(CH_3)_3COCl$ → (50) (50)

melting points[36]. The bromo- and chloroiodinanes **51** and **52** have also been prepared and exhibit lower melting points than the perfluoromethyl analogues **48** and **50**. The bromoiodinane **48** is also more stable toward hydrolysis than the bromoiodinane **52**[36].

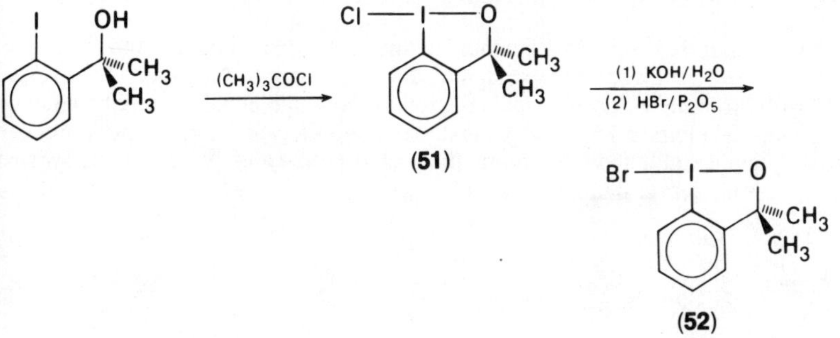

$(CH_3)_3COCl$ (1) KOH/H_2O (2) HBr/P_2O_5

(**51**) (**52**) (51)

The reactions of bromoiodinanes **48** and **52** with the potassium salt of (hexafluoro)cumyl alcohol eventuate in the stable dialkoxyiodinanes **53** and **54**[36].

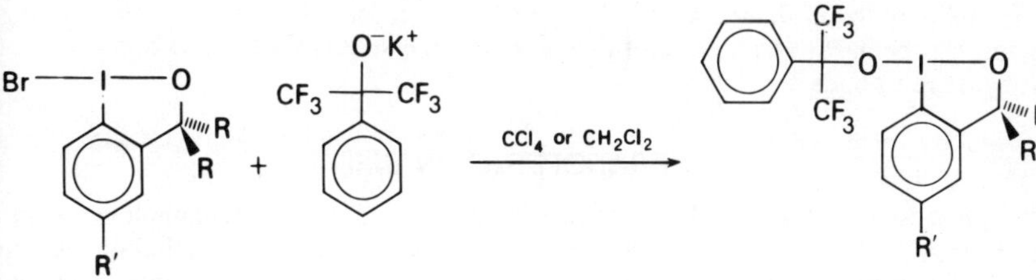

CCl_4 or CH_2Cl_2 (52)

(**48**) $R = CF_3$, $R' = CH_3$

(**52**) $R = CH_3$, $R' = H$

(**53**) $R = CF_3$, $R' = CH_3$

(**54**) $R = CH_3$, $R' = H$

Both of these compounds undergo degenerate ligand exchange with potassium (hexafluoro)cumylate in carbon tetrachloride solution, the former more rapidly than the latter. An associative mechanism for ligand exchange involving the intermediate formation of a trialkoxyaryliodate species was proposed (equation 53). The increased

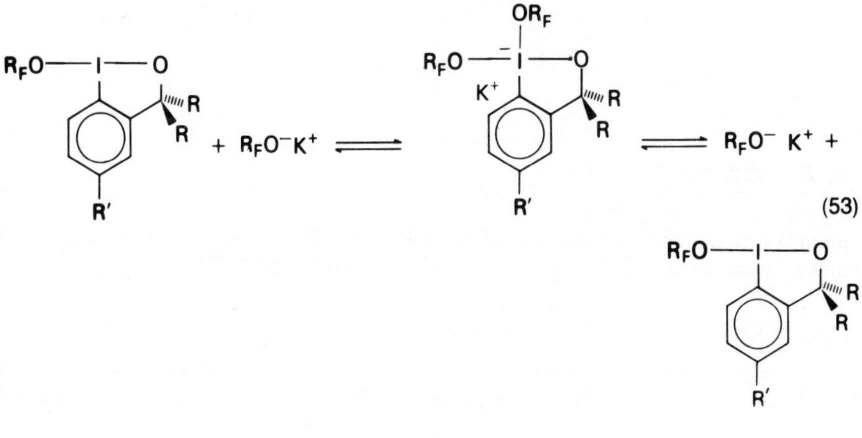

(53)

reactivity of **53** compared to **54** toward ligand exchange may manifest the greater ability of trifluoromethyl groups over methyl groups to stabilize the trialkoxyaryliodate ion. A similar tetraaryliodate species had earlier been proposed to rationalize the observation of ligand exchange in reactions of iodonium salts with aryllithium species[75].

Those molecular features which stabilize organobromoiodinanes and dialkoxyiodinanes have been utilized in the design of the first organobrominane. The reaction of diol **55** with bromine trifluoride yields the remarkably stable dialkoxyarylbrominane **56**[76]. This material melts at 153–154°C, is stable indefinitely at room temperature, can be sublimed, and is stable to water, dilute hydrochloric acid and dilute sodium hydroxide. It is also a moderate oxidizing agent.

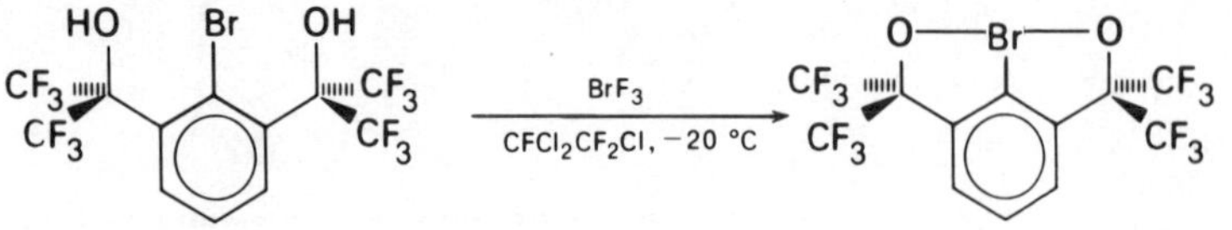

HO Br OH; CF_3 CF_3 CF_3 CF_3 **(55)** $\xrightarrow[CFCl_2CF_2Cl,\ -20\ °C]{BrF_3}$ O—Br—O; CF_3 CF_3 CF_3 CF_3 **(56)** (54)

The ability of the iodinanes **48**, **50**, **51** and **52** to function as free radical halogenating reagents has been investigated, and they exhibit high selectivity toward benzylic carbon–hydrogen bonds[77].

VI. ORGANOPERHALINANES

Although iodoxybenzene has been known for nearly a century, organohalogen compounds with the halogen atom in the +5 oxidation state, the organoperhalinanes, are much less common than the organohalinanes, and only those of the iodine subclass have been isolated. The known organoiodine(V) compounds exhibit one of the four general structures, **I–IV**, given below, where L represents a heteroatom ligand and where the I—L bonds may be either hypercovalent or ionic depending on the nature of the molecule. In some cases, L^1 and L^2 may be introduced via a bidentate ligand. While the general structures are illustrated with aryl groups, some organoperiodinanes are known in which the organic ligand is a perfluoroalkyl group.

$Ar—IO_2$ **(I)** $Ar_2I(=O)L$ **(II)** $ArI(=O)L^1L^2$ **(III)** $ArIL^1L^2L^3L^4$ **(IV)**

A. Bonding and Structure

Hypervalent bonding theory conveniently accommodates the organoiodine(V) compounds according to the principles already discussed in Sections IIIB and IIIF. This is illustrated as follows for the general periodinane structures **I–IV**.

1. *Organoperiodinanes of structural class I*

The expectations of hypervalent bonding theory are roughly corroborated by a single crystal X-ray study of *p*-chloroiodoxybenzene[78]. For this compound the

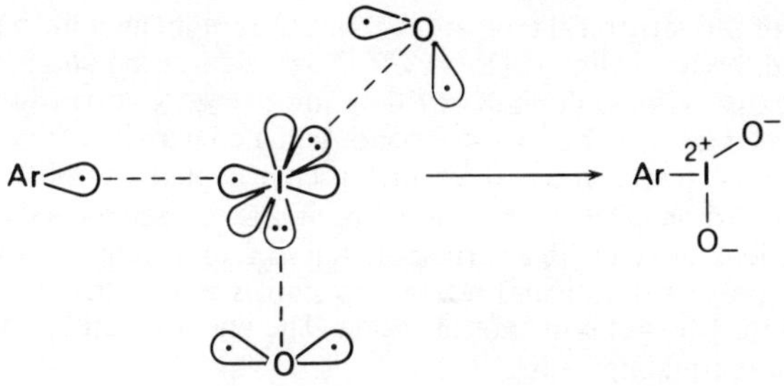

Expected bond angles:

$C{-}I{-}O_{(1)}$	90°
$C{-}I{-}O_{(2)}$	90°
$O_{(1)}{-}I{-}O_{(2)}$	90°

$C{-}I{-}O_{(1)}$, $C{-}I{-}O_{(2)}$ and $O_{(1)}{-}I{-}O_{(2)}$ bond angles have been determined to be 94°, 95° and 103° respectively. The iodine–oxygen bond distances are 1.60 and 1.65 Å, considerably shorter than the sum of the covalent radii for the iodine and oxygen atoms as perhaps owing to strong electrostatic attractions between the charged nuclear centres or to some π-overlap of the p-orbital on oxygen with a 'vacant' 5d orbital on iodine, an interaction not accounted for by the bonding model employed here. The iodoxy function in *p*-chloroiodoxybenzene is bisected by the plane of the aromatic ring. There are also some rather surprising departures from the expected molecular symmetry which probably manifest intermolecular contacts associated with crystal packing. In particular, the C—C—I bond angles are 109° and 129°, and the chlorine atom lies 14° out of the aromatic plane.

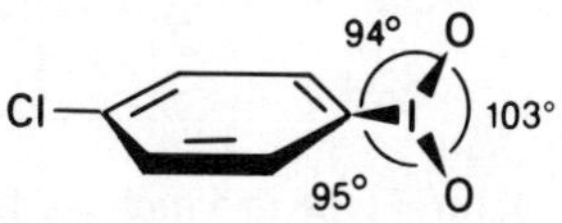

2. Organoperiodinanes of structural class II

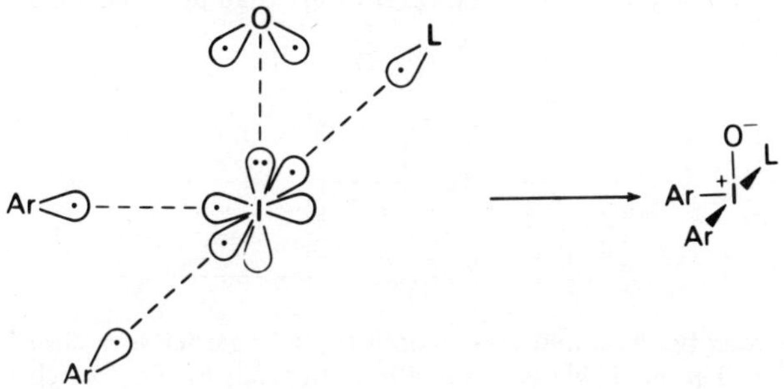

Expected bond angles:

C—I—C	90°	$C_{(1)}{-}I{-}L$	90°
$C_{(1)}{-}I{-}O$	90°	$C_{(2)}{-}I{-}L$	180°
$C_{(2)}{-}I{-}O$	90°	O—I—L	90°

Compounds of this structural type are thought to contain an ionic bond between the iodine atom and the heteroligand ($L^- = CF_3CO_2^-$, IO_3^-, etc.) and have been dubbed 'diaryliodosyl salts'. Although no X-ray data are available to verify this, it is not an unreasonable assumption. The [3c–4e] bond, being constructed from an iodine atom and two ligands of significantly different basicity, should exhibit a high electronic asymmetry. The carbon–iodine bond should be shortened and the iodine–heteroligand bond should be lengthened with a corresponding shift of negative charge density to the more electronegative heteroligand relative to similar molecules of class III in which both ligands of the [3c–4e] bond are the same. The linear nature of the [3c–4e] bond, however, should be maintained.

3. *Organoperiodinanes of structural class III*

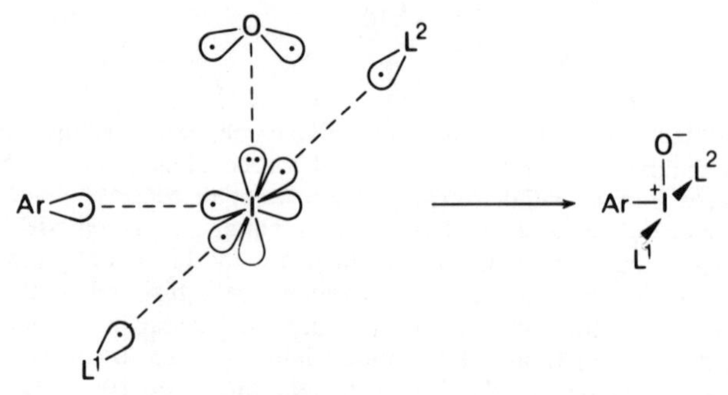

Expected bond angles:

C—I—L^1	90°	L^1—I—O	90°
C—I—L^2	90°	L^2—I—O	90°
C—I—O	90°	L^1—I—L^2	180°

Those molecules of this structural type in which the heteroligands are either identical or of comparable basicity, would be expected to contain two elongated I—L hypercovalent bonds with bond orders of about 0.5.

A recent single crystal X-ray study has revealed unequivocally that '*o*-iodoxybenzoic acid' possesses the heterocyclic structure **57** rather than the isomeric acyclic structure

(57)

with a free iodoxy function and may, therefore, be regarded as a class III compound with $L^1 \neq L^2$.[79] The molecule is essentially flat except for $O_{(3)}$ which lies above the molecular plane. The I—$O_{(1)}$ and I—$O_{(2)}$ bond distances are 2.324(3) Å and 1.895(3) Å respectively, and the C—I—$O_{(1)}$, C—I—$O_{(2)}$ and $O_{(1)}$—I—$O_{(2)}$ bond angles are 75.6(1)°, 87.0(2)° and 162.6(1)° consistent with the existence of an electronically asymmetric [3c–4e] bond comprised of a central iodine atom and two ligands of considerably different basicity. The iodine–oxygen 'double bond' is roughly perpendicular to the molecular plane; the $O_{(1)}$—I—$O_{(3)}$, $O_{(2)}$—I—$O_{(3)}$ and C—I—$O_{(3)}$ angles are 85.5(2)°, 98.2(2)° and 101.6(2)° respectively and the I—$O_{(3)}$ bond distance is 1.784(3) Å. Thus '*o*-iodoxybenzoic acid' is a chiral molecule. The asymmetric units of the unit cell are dimeric, the two molecules being held together by pairwise intermolecular I---O secondary contacts and each molecule being of the same absolute configuration. Thus each crystal of *o*-iodoxybenzoic acid' is enantiomerically pure.

4. *Organoperiodinanes of structural class IV*

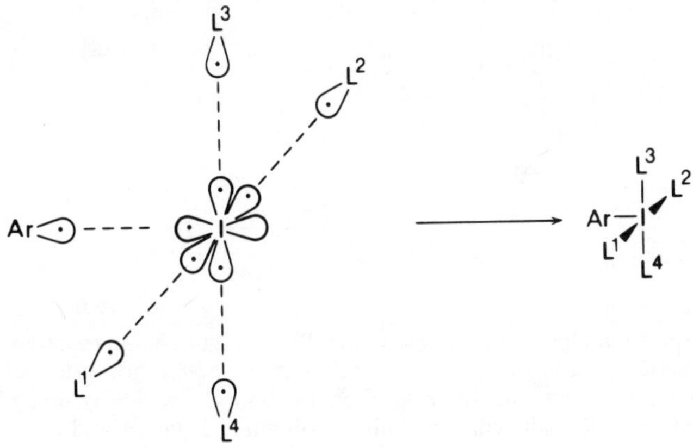

Expected bond angles

Ar—I—$L^{1,2,3,4}$	90°
L^1—I—L^2	180°
L^3—I—L^4	180°
L^1—I—$L^{3,4}$	90°
L^2—I—$L^{3,4}$	90°

These molecules are expected to exhibit square pyramidal structures, a consequence of one normal covalent bond between the carbon ligand and the iodine atom and two orthogonal [3c–4e] bonds constructed from iodine and heteroligand pairs. When the heteroligands are of comparable basicity, the I—L bonds should be hypercovalent, but when they are not or when a heteroligand is replaced by a carbon ligand, the appropriate I—L bonds may develop considerably more ionic character.

B. Diaryliodosyl Salts

In the presence of hydroxide ion, iodoxybenzene condenses with itself to give diphenyliodosyl hydroxide (**58**) as an unstable amorphous solid[80]. When solutions of **58**, generated *in situ*, are subjected to the action of trifluoroacetic and acetic acids, diphenyliodosyl trifluoroacetate (**59**) and diphenyliodosyl acetate (**60**) precipitate, the latter compound as its monohydrate[81]. The trifluoroacetoxy function in **59** exhibits infrared absorption at 1640 cm^{-1}. This is consistent with its formulation as an ionic compound since diphenyliodonium trifluoroacetate ($Ph_2\overset{+}{I}\ {}^{-}O_2CCF_3$) shows a corresponding peak at 1650 cm^{-1} whereas [bis(trifluoroacetoxy)iodo]benzene ($PhI(OOCCF_3)_2$), the hypercovalent analogue, exhibits carbonyl absorption at 1700 and 1740 cm^{-1}.

$$2\ PhIO_2 \xrightarrow{OH^-} Ph_2I^+(O)\bar{O}H\ \textbf{(58)} \xrightarrow{CF_3COOH} Ph_2I^+(O)\ {}^-OOCCF_3\ \textbf{(59)} \xrightarrow{M^+X^-} Ph_2I^+(O)\ X^-\ (X = F, Cl, Br)$$

$$\textbf{(58)} \xrightarrow{CH_3COOH} Ph_2I^+(O)\ {}^-OOCCH_3 \cdot H_2O\ \textbf{(60)} \qquad (55)$$

As expected for an onium salt, compound (**59**) undergoes anion metathesis reactions with alkali metal salts. For example, when a solution of **59** in hot water is treated with saturated aqueous potassium fluoride, diphenyliodosyl fluoride is precipitated. The chloride and bromide salts can be similarly obtained from **59** and aqueous sodium chloride or sodium bromide. Iodide ion, on the other hand, reduces the diphenyliodosyl cation to the diphenyliodonium ion.

Di-*p*-tolyliodosyl salts ($X^- = CF_3COO^-, F^-, IO_3^-$) have been similarly prepared[81].

Diphenyliodosyl trifluoroacetate reacts with sulphur tetrafluoride in dichloromethane at 0°C to give diphenyl(difluoroiodosyl) trifluoroacetate (**61**) in 78% yield[82]. The ^{19}F NMR spectrum of **61** in dichloromethane exhibits two singlets, one for the trifluoromethyl group at +13.12 ppm (relative to benzotrifluoride) and one for iodine-bound fluorines at +22.62 ppm. Under more stringent conditions (CH_2Cl_2, 65°C, 2 h), compound **59** is converted by sulphur tetrafluoride into diphenyl(difluoroiodosyl) fluoride (**62**). The ^{19}F NMR spectrum of **62** in nitromethane exhibits a singlet at +89.25 p.p.m. (relative to benzotrifluoride) which may be assigned to the fluoride ion and a singlet at +24.5 ppm for the two fluorine atoms attached to iodine, consistent with the presence of electronically symmetric and asymmetric [3c–4e] bonds in this molecule. It is noteworthy that the three fluorine ligands were not in a state of rapid dynamic exchange under the conditions of the NMR experiment.

$$Ph_2I^+(O)\ \bar{O}OCCF_3 \xrightarrow{SF_4(CH_2Cl_2,\ 0\ ^\circ C)} Ph_2I^+F_2\ {}^-OOCCF_3\ \textbf{(61)}$$

$$Ph_2I^+(O)\ \bar{O}OCCF_3 \xrightarrow{SF_4(CH_2Cl_2,\ 65\ ^\circ C)} Ph_2I^+F_2\ F^-\ \textbf{(62)} \qquad (56)$$

C. Bis(acyloxy)iodosylarenes

(Difluoro)iodosylbenzene, prepared by the action of aqueous hydrofluoric acid on iodoxybenzene, was first reported in 1901[83]. The synthesis of (difluoro)iodosyltrifluoromethane has also been achieved[51].

A rather recent development is the addition of the bis(acyloxy) analogues to this class of compounds. Various iodoxyarenes, when treated with trifluoroacetic acid in the presence of trifluoroacetic anhydride, are converted quantitatively into the corresponding bis(trifluoroacetoxy)iodosylarenes[84] (equation 57). The iodoxyarenes react similarly with acetic acid/acetic anhydride· or acetic anhydride alone to give bis(acetoxy)iodosylarenes[84,85] (equation 58).

$$\mathrm{ArIO_2} \xrightarrow{\mathrm{CF_3COOH/(CF_3CO)_2O}} \mathrm{Ar{-}I(=O)(OOCCF_3)(CF_3COO)} \qquad (57)$$

$\mathrm{Ar{=}C_6H_5,\ \mathit{p}\text{-}MeC_6H_4,\ \mathit{p}\text{-}NO_2C_6H_4,\ \mathit{m}\text{-}CF_3C_6H_4}$

$$\mathrm{ArIO_2} \xrightarrow{\mathrm{CH_3COOH/(CH_3CO)_2O}} \mathrm{Ar{-}I(=O)(OOCCH_3)(CH_3COO)} \qquad (58)$$

$\mathrm{Ar{=}C_6H_5,\ \mathit{p}\text{-}MeC_6H_4,\ \mathit{p}\text{-}NO_2C_6H_4,\ \mathit{m}\text{-}FC_6H_4,\ \mathit{p}\text{-}FC_6H_4}$

The bis(acyloxy)iodosyl compounds exhibit moderately high melting points and appear to be reasonably stable. The ^{1}H NMR spectrum of bis(acetoxy)iodosylbenzene in acetic anhydride shows the acyloxy ligands to be magnetically equivalent (6H singlet at δ1.89), consistent with the presence of two equivalent iodine–oxygen hypercovalent bonds or with rapid degenerate ligand exchange (see equation 59).

$$\mathrm{Ph{-}I^{+}(=O)(\cdots{}^{-}OOCCH_3)(CH_3COO)} \underset{}{\overset{\text{Solvent}}{\rightleftharpoons}} \mathrm{Ph{-}I^{+}(=O)(OOCCH_3)(\cdots CH_3COO^{-})} \qquad (59)$$

D. Tetrafluoro-, Tetraacyloxy- and Fluoroalkoxyperiodinanes

Iodoxybenzene and *p*-iodoxytoluene react with excess sulphur tetrafluoride at moderately high temperatures to give quantitative yields of (tetrafluoroiodo)benzene and *p*-(tetrafluoroiodo)toluene (equation 60) as white crystalline materials subject to hydrolysis and capable of etching glass[86]. The ^{19}F NMR spectrum of (tetrafluoroiodo)benzene in dichloroethane is characterized by one singlet resonance at -35.97 ppm relative to benzotrifluoride.

$$\mathrm{R{-}C_6H_4{-}IO_2} \xrightarrow{\text{Excess } \mathrm{SF_4}} \mathrm{R{-}C_6H_4{-}IF_4} \qquad (60)$$

R = H, Me

Chlorine trifluoride has emerged as a useful reagent for the synthesis of tetrafluoroperiodinanes from iodo compounds. Reaction stoicheiometries are critical, and chlorine trifluoride must be used in excess or difluoroiodinanes will be obtained instead[47] (equation 61).

$$3\ ArI \xrightarrow{2\ ClF_3} 3\ ArIF_2 + Cl_2 \qquad 3\ ArI \xrightarrow{4\ ClF_3} 3\ ArIF_4 + 2\ Cl_2 \quad (61)$$

Pentafluoroiodobenzene reacts with excess chlorine trifluoride in perfluorohexane at low temperatures to give (tetrafluoroiodo)pentafluorobenzene (equation 62),

$$C_6F_5I + ClF_3 \xrightarrow[-78^\circ \text{ to } -15^\circ]{CF_3(CF_2)_4CF_3} C_6F_5IF_4 \quad (62)$$

observed to be more stable than (tetrafluoroiodo)benzene and (difluoroiodo)benzene[87]. The treatment of pentafluorobromobenzene with chlorine trifluoride does *not* eventuate in perbrominane formation but leads instead to products derived from chlorine and fluorine addition to the aromatic ring.

Various perfluoroalkyl iodides are converted by excess chlorine trifluoride into the corresponding periodinanes. (Tetrafluoroiodo)trifluoromethane (**63**) is a white solid

$$F_3C{-}IF_4$$

(**63**)

subject to volatilization at 20°C and sensitive to moisture[88,89]. The ^{19}F NMR spectrum of **63** is that of an A_3X_4 spin system; $\delta_x - 32.4$ ppm and $\delta_a - 56.1$ ppm ($FCCl_3$ reference), $J_{ax} = 18$ Hz. The observation of F—C—I—F coupling is consistent with a square pyramidal structure for **63** in which the trifluoromethyl ligand is at the axial site and in which all four I—F bonds are hypercovalent, instead of an ionic structure in which the fluoride ligands are in a state of rapid degenerate exchange. A solution of **63** in fluorotrichloromethane showed no evidence of decomposition after 12 h at 20°C, but its decomposition in the solid state does occur within 4 h at that temperature, the products being molecular iodine, trifluoromethyl iodide, iodine pentafluoride and tetrafluoromethane. The following decomposition scheme has been proposed[89].

$$\underset{(\mathbf{63})}{CF_3IF_4} \longrightarrow CF_4 + IF_3 \quad (63)$$

$$IF_3 \longrightarrow IF + IF_5 \quad (64)$$

$$IF \longrightarrow I_2 + IF_5 \quad (65)$$

$$CF_3IF_4 \xrightarrow{I_2/IF_5 \text{ or } IF} CF_3I + IF_5 \quad (66)$$

(Tetrafluoroiodo)trifluoromethane undergoes an interesting metathesis reaction with methoxytrimethylsilane or dimethoxydimethylsilane in fluorotrichloromethane at 20°C giving a series of methoxyperiodinanes (equation 67) isolated as unstable liquids[89]. The number of fluoride ligands replaced depends on the reaction time. For

$$CF_3-IF_4 \xrightarrow{(MeO)_2SiMe_2} CF_3-I(F)_{4-n}(OMe)_n,\ n = 0, 1, 2, 3, 4 \quad (67)$$

example, after 30 min (methoxytrifluoroiodo)trifluoromethane is obtained, but after 9 h (tetramethoxyiodo)trifluoromethane is formed. The ^{19}F NMR spectrum of (dimethoxydifluoroiodo)trifluoromethane (**64**) is that of an A_3X_2 spin system (J_{ax} = 22 Hz). The coupling constant is thought to be more consistent with the square pyramidal structure shown below in which the methoxy ligands are *cis* rather than the corresponding isomeric structure in which those ligands are *trans*.

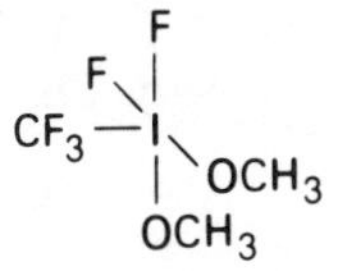

F
F
CF_3—I
OCH_3
OCH_3

(**64**)

Among the (tetrafluoroiodo)perfluoroalkanes, thermal stability appears to increase with the size of the perfluoroalkyl group. Solutions of the perfluorobutyl and perfluoroisopropyl analogues were reported to be stable for weeks, although the solvent and temperature were not specified[47].

Organoperiodinanes with four acyloxy ligands have also been reported[85]. Various iodoxyarenes, when treated with trifluoroacetic anhydride in the absence of trifluoroacetic acid, are converted quantitatively into the corresponding [tetrakis(trifluoroacetoxy)iodo]arenes (equation 68). The action of perfluorobutyric acid in combination with perfluorobutyric anhydride on iodoxybenzene and its 3-fluoro and 4-nitro derivatives likewise affords [tetrakis(perfluorobutoxy)iodo]arenes.

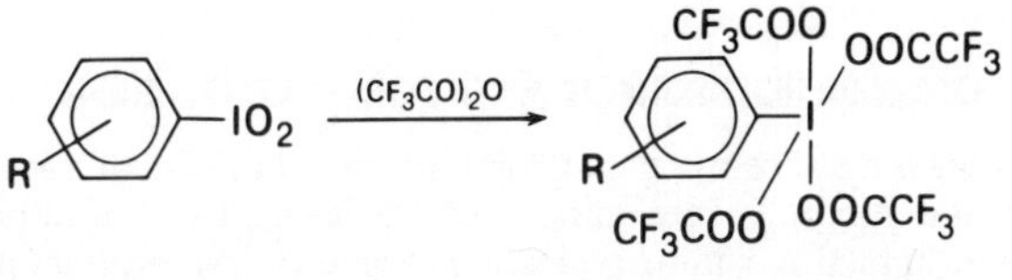

$$R{-}C_6H_4{-}IO_2 \xrightarrow{(CF_3CO)_2O} R{-}C_6H_4{-}I(OOCCF_3)_4 \quad (68)$$

R = H, 4-Me, 4-Cl, 3-F, 4-F, 3-CF_3, 3-NO_2, 4-NO_2

The preparation of two stable organoalkoxytrifluoroperiodinanes has recently been achieved [10,90]. The oxidation of 2-iodo-5-methylhexafluorocumyl alcohol with excess trifluoromethyl hypofluorite proceeds in Freon-113 at −20°C to give 1,1,1-trifluoro-3,3-bis(trifluoromethyl)-5-methyl-3*H*-1,2-benziodoxole (**65**) in 95% yield. The ^{19}F NMR spectrum of **65** exhibits a two fluorine doublet at 13.20 ppm (ϕ scale, $CFCl_3$ reference) and a one fluorine triplet at 41.57 ppm (J = 116, Hz), consistent with the indicated square pyramidal placement of the ligands about the iodine atom. The benziodoxole **65** is stable in contact with water for at least 4 days, but it does hydrolyse in aqueous potassium hydroxide to give a white crystalline product thought to be the iodoxy alcohol **66** or its cyclic tautomer **67** or a mixture of the two. Various amines, when treated first with **65** and then with water are converted into aldehydes. Benzyl alcohol and isopropyl alcohol are also oxidized by **65** to the corresponding aldehyde and ketone.

The trifluoromethyl hypofluorite oxidation of 2-iodocumyl alcohol affords the benziodoxole **68** in 87% yield. This compound is much more reactive toward water than **65**, its hydrolysis in the presence of atmospheric moisture being complete within 4 h at 25°C.

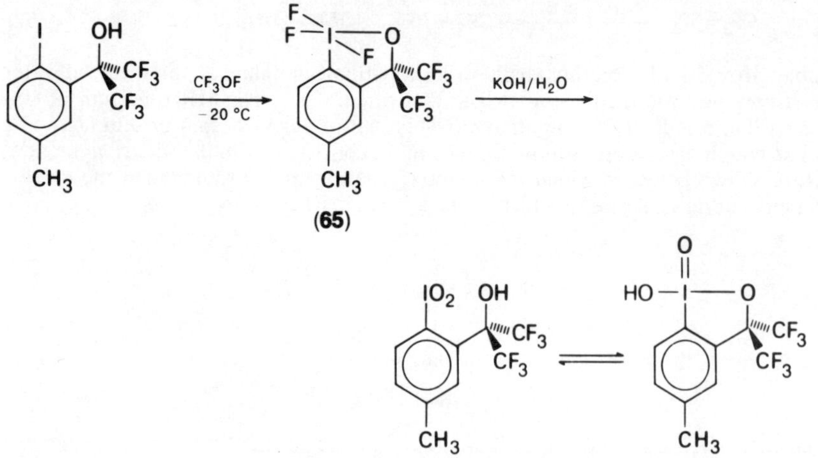

(65)

(66) (67) (69)

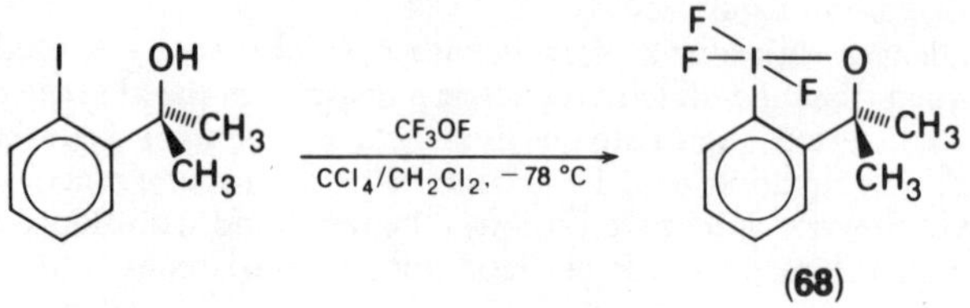

(68) (70)

VII. ORGANOIODINANES AND TOPOTACTIC REACTIONS

The benziodoxoles have recently engendered interest because of their involvement as both reactants and products in a variety of *topotactic* transformations. Solid state reactions of single crystal reactants typically proceed to give product phases which are either amorphous or polycrystalline. Topotactic reactions, on the other hand, are those in which single crystals of reactant afford products in ordered single crystal phases with preferred lattice orientations relative to the lattice directions of the reactant (i.e. there is a reticular correspondence)[91]. The solid state behaviour of bis(aroyl)peroxides of general structure **69** has been studied in detail by Gougoutas and his co-workers[92]. These compounds undergo several primary and secondary chemical reactions in the solid state characterized by specific topotaxis. The primary manifolds include peroxide isomerization to (benzoyloxy)benziodoxoles and peroxide hydrolysis to '*o*-iodosobenzoic acid' and a substituted benzoic acid (equation 71). The initially formed (benzoyloxy)benziodoxoles are subject, in turn, to secondary hydrolysis and photochemical reduction. The topotactic manifestations are diverse and not necessarily predictable. For example, while the thermal isomerization of the 2,2′-diiodoperoxide (**70**) to the corresponding benziodoxole proceeds with a single mode of topotactic alignment, the analogous transformation of the 2-chloro-2′-iodoperoxide (**69**, R = 2-Cl) displays four distinct modes of topotactic alignment. The crystal morphology is usually retained in these reactions. A detailed knowledge of molecular and crystal structures is essential for a clear understanding of topotactic reactions, and accordingly, a number of organoiodinane structures have been elucidated in the course of these studies.

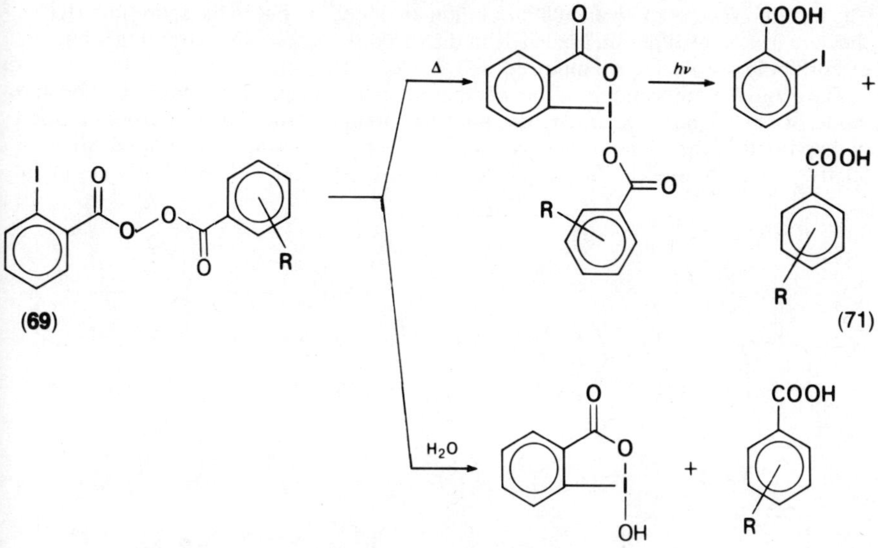

I O O O O R (69)

Δ → O O I O O R — $h\nu$ → COOH I + COOH R (71)

H_2O → O O I OH + COOH R

A. Topochemistry of Bis(*o*-iodobenzoyl)peroxide

1. Crystal structures of alpha and beta polymorphs of 1-(o-iodobenzoyloxy)-1,2-benziodoxol-3(1H)-one

The synthesis of 1-(*o*-iodobenzoyloxy)-1,2-benziodoxol-3(1*H*)-one (**8**) and its subsequent recrystallization from organic solvents affords two monoclinic polymorphs (designated α and β) (equation 72). The **8α** modification exists as acicular crystals with

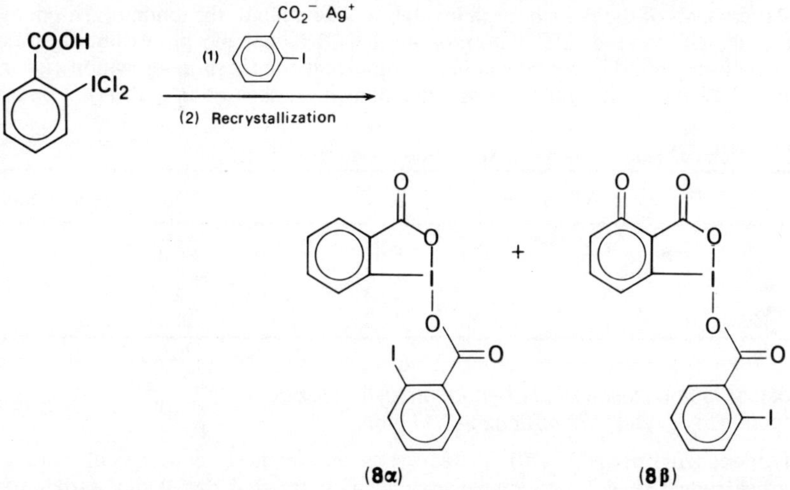

COOH ICl_2 — (1) $CO_2^-\ Ag^+$, I; (2) Recrystallization →

O O I O O I (**8α**) + O O O I O O I (**8β**) (72)

an ill defined melting point owing to its conversion, upon heating, to the β-polymorph. With rapid heating (~25°C min^{-1}), **8α** melts at approximately 180°C. The **8β** modification crystallizes in prisms exhibiting well defined faces and melting at 190°C.

The infrared spectra of **8α** and **8β** in solution are identical, but in the solid state (KBr), they are distinctly different, especially in the carbonyl region. The unit cell parameters of both polymorphs are summarized in Table 6.

The *α* and *β* modifications of **8** are distinguished both in molecular conformation and mode of crystal packing. In **8α**, the univalent iodine atom and the carbon–oxygen double bond of the exocyclic acyloxy ligand exist in a transoid configuration while, in **8β**, they are *cis* to one another. The *α*-polymorph crystallizes with two molecules in the

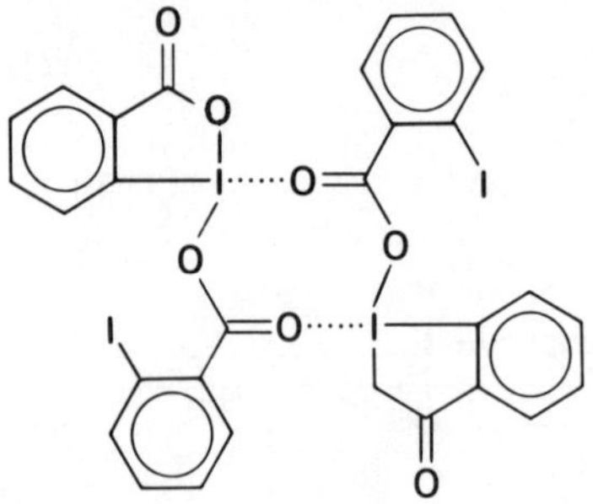

(8α)

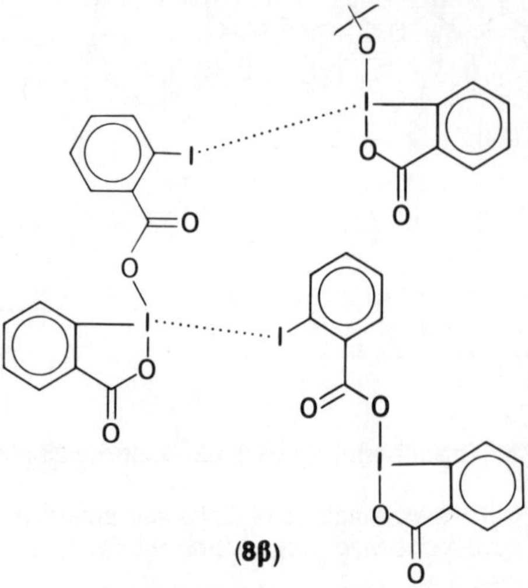

(8β)

asymmetric unit of the unit cell, the two molecules being mutually coordinated through a pair of iodine(III)–oxygen intermolecular contacts about a pseudo inversion centre rendered possible by the *trans* conformation discussed above. The intermolecular contacts of the *β*-polymorph involve coordination of the iondine(I) atom of one molecule with the iodine(III) atom of another. Both polymorphs exhibit layered crystal structures, interlayer contacts being important for the *β*-modification (layer separation 3.42 Å) but not for the *α*-modification (layer separation 4.21 Å).

TABLE 6. Unit cell parameters of **8α**, **8β**, **70** and *o*-iodobenzoic acid

	a, Å	*b*, Å	*c*, Å	*β*, °	*Z*	Space group
8α	4.21	30.86	22.52	93.3	8	*Cc*
8β	8.03	12.58	13.74	91.6	4	$P2_1/c$
70	13.05	4.21	15.47	121.1	2	*Pc*
o-Iodobenzoic acid	4.32	15.08	11.29	91.5	4	$P2_1/c$

2. *Topotactic isomerization of bis(o-iodobenzoyl) peroxide to 1-(o-iodobenzoyloxy)-1,2-benziodoxol-3(1 H)-one*

Bis(*o*-iodobenzoyl)peroxide (**70**) crystallizes as acicular monoclinic crystals with a short lattice repeat of 4.21 Å, approximately twice the van der Waals' radius of univalent iodine[93]. The approximately flat molecules exhibit a molecular conformation in which both iodine(I) atoms are *cis* to their respective carbon–oxygen double bonds and both *o*-iodobenzoyl groups are locked in a transoid configuration about the central oxygen–oxygen bond. When allowed to stand for several weeks at room temperature or overnight at 110°C, single crystals of **70** are converted topotactically (in a single alignment) and with morphological retention to single crystals of **8** exclusively in its α-modification[91]. This is a remarkable rearrangement since it requires 180° *ring flips* of half of the aryl moieties in the starting peroxide; i.e. in half of the *o*-iodobenzoyloxy groups the iodine (I)/C=O conformation disposition changes from cisoid to transoid, but not via a simple bond rotation. Since the short lattice repeat in **70**, a reflection of the layer separation in the crystal, is only 4.21 Å, there does not

(70) —Δ, Solid state→ (8α) (73)

appear to be sufficient space for such ring flips, and this is a surprising result. The topotaxis is such that the short lattice repeat of **8α** (*a*-axis) is aligned parallel to the short lattice repeat of **70** (*b*-axis), but since their unit cell symmetry axes are not aligned, the crystals of **8α** are conservatively twinned[94].

3. *Topotactic conversion of the alpha polymorph of 1-(o-iodobenzoyloxy)-1,2-benziodoxol-3(1*H*)-one to o-iodobenzoic acid*

The **8α** pseudomorph (i.e. a daughter crystal with the same morphology as the parent crystal), formed topotactically from **70**, is subject, in turn, to slower topotactic conversion to a monoclinic single crystal phase of *o*-iodobenzoic acid, the short lattice repeat (4.32 Å) of which is aligned parallel to the 4 Å repeats of **8α** and **70**[91]. It was originally thought that *o*-iodobenzoic acid was formed via atmospheric hydrolysis of **8α**. The expected by-product, 1-hydroxy-1,2-benziodoxol-3(1*H*)-one (**5**), went

(8α) —H_2O, ?→ o-IC_6H_4COOH + (5)
(8α) —$h\nu$→ 2 o-IC_6H_4COOH (74)

undetected and was presumed to be located in amorphous regions of the product crystal. This thinking was later revised with the discovery that various benziodoxoles are subject to topotactic photochemical reduction to benzoic acids on continuous exposure to X-radiation. Even though striking similarities exist between the crystal structure of the starting peroxide and the *o*-iodobenzoic acid thus formed, the direct solid state conversion of peroxide to acid has not been observed, and the acid is formed after the peroxide phase is no longer evident (equation 74).

Under non-photochemical conditions, other benziodoxoles have been observed to undergo topotactic hydrolysis to **5** and the corresponding carboxylic acid[95].

B. Topochemistry of 2-Iodo-2′-chlorobis(benzoyl)peroxide

1. *Thermal and photochemical conversion of 2-iodo-2′-chlorobis (benzoyl) peroxide to polymorphs of 1-(o-chlorobenzoyloxy)-1,2-benziodoxol-3(1H)-one*

Unsymmetrical analogues of **70** in which one of the iodine atoms is replaced either by chlorine or bromine are isostructural with **70** and undergo similar topotactic isomerizations to the corresponding (benzoyloxy)benziodoxoles. However, their solid state behaviour is more complicated. Single monoclinic crystals of 2-iodo-2′-chlorobis-(benzoyl)peroxide (**71**)[93], when allowed to stand for *ca* 4 weeks at room temperature, are converted topotactically to monoclinic 1-(*o*-chlorobenzoyloxy)-1,2-benziodoxol-3(1*H*)-one (**72**), identical to the only polymorph obtainable from solvent recrystallizations[92,96]. However, while the solid state isomerization of **70** to **8α** manifests only one topotactic alignment, the isomerization of **71** to **72** proceeds along four distinct

(**71**) —Δ, Solid state→ (**72**) (monoclinic) (75)

topotactic manifolds in an approximately 20:2:1:1 distribution. In each topotactic alignment, the short repeat (7.403 Å, *a*-axis) of the benziodoxole unit cell is aligned with the short repeat (4.07 Å, *b*-axis) of the peroxide unit cell, but, since the symmetry axes of the reactant and product unit cells are *not* aligned, all four topotactic modes of alignment are attended by conservative twinning. As with the **8α** polymorph obtained from **70**, the halogen(I) atom in **72** and the oxygen atom of the carbon–oxygen double bond of the exocyclic ligand adopt a transoid conformation. Like **8α**, **72** crystallizes in dimeric units characterized by two iodine(III)–oxygen intermolecular contacts. Once again, it is evident that a 180° ring flip is required for this isomerization despite the closely packed (4 Å) peroxide layers. Examination of the unit cell parameters given in Table 7 reveals that benziodoxole **72** is *not*, however, isostructural with benziodoxole **8α**.

The solid state topotactic isomerization of monoclinic peroxide **71** likewise proceeds under the influence of continuous X-radiation or ultraviolet light and eventuates in a

TABLE 7. Unit cell parameters for **71**, **72**, **72′** and the solid solution of 2-iodo- and 2-chlorobenzoic acids

	a, Å	b, Å	c, Å	α, deg	β, deg	γ, deg	Z	Space group
71	13.02	4.07	15.39	90	120.9	90	2	Pc
72	7.403	13.88	13.12	90	102.1	90	4	$P2_1/a$
72′	13.86	4.00	13.89	100	114	84	2	$P\bar{1}$
Solid solution of benzoic acids	14.99	4.06	26.41	90	119	90	8	$C2/c$

novel single crystal phase in a single topotactic alignment and unique to photoactivation[92]. The photochemical product, **72′**, exhibits the same transoid conformation as the thermal product, and like **72** crystallizes in dimeric units characterized by two iodine(III)–oxygen intermolecular contacts. However, **72′** is a triclinic polymorph of **72** with a short lattice repeat very nearly the same as and aligned with that of the starting peroxide.

2. *Topotactic photoreduction of the triclinic polymorph of 1-(o-chlorobenzoyloxy)-1,2-benziodoxol-3(1*H*)-one*

When the triclinic polymorph **72′** is subjected to prolonged X-radiation, it is ultimately transformed topotactically to a new monoclinic single crystalline phase identified as a solid solution of *o*-iodobenzoic acid and *o*-chlorobenzoic acid, the crystal structure of which is the same as that observed for pure *o*-chlorobenzoic acid and unlike that of the conventional crystal structure of *o*-iodobenzoic acid[92]. Typical total irradiation times are 50 h for the complete conversion of peroxide **71** to benziodoxole **72′** and 480 h for the complete formation of benzoic acids. The topotaxis of the photoreduction is such that the short repeat (4.06 Å, *b*-axis) of the benzoic acids single crystal phase is aligned with the short repeats of both **71** and **72′**.

3. *Topotactic hydrolysis of 2-iodo-2-chlorobis(benzoyl) peroxide*

The direct topotactic hydrolysis of peroxide **71** to *o*-chlorobenzoic acid and '*o*-iodosobenzoic acid' is promoted when single crystals are exposed to a humid atmosphere[92]. However, the thermal isomerization of **71** to the benziodoxole **72** competes effectively, and, even after 16 days, the peroxide phase is still present. Analysis of partially hydrolysed crystals of **71** reveals *ten* single crystalline phases present simultaneously: starting peroxide, benziodoxole **72** (four modes of topotactic alignment), 1-hydroxy-1,2-benziodoxol)-3(1*H*)-one (four modes of topotactic alignment) and *o*-chlorobenzoic acid (one mode of topotactic alignment). The four reticular alignments of 1-hydroxy-1,2-benziodoxol-3(1*H*)-one are generated in a 15:7:5:1 distribution and exhibit the short lattice repeat (~4 Å) either parallel or antiparallel to that of the starting peroxide.

C. Generalizations on the Topochemistry of Bis(aroyl)peroxides

The bis(aroyl)peroxides (**69**) may, on the basis of their solid state topochemical behaviour, be divided into two groups: group 1 (R = *o*-iodo, *o*-bromo, *o*-chloro), group 2 (R = *o*-fluoro, hydrogen, *m*-chloro)[97]. The group 1 peroxides, already discussed, undergo moderately facile thermal topotactic isomerizations to the corresponding benziodoxoles and relatively slow topotactic hydrolyses. The group 2 peroxides, on the other hand, undergo facile topotactic hydrolyses and slow topotactic isomerizations.

For example, single monoclinic crystals of 2-iodo-3′-chlorobis(benzoyl)peroxide (**73**), upon exposure to atmosphere moisture at 22°C, afford single crystalline phases of 1-hydroxy-1,2-benziodoxol-3(1*H*)-one and *m*-chlorobenzoic acid each with its own mode of topotactic alignment[98]. The topotactic isomerization of **73** to the corresponding benziodoxole proceeds only at a higher temperature and has been studied at *c.* 55°C[98].

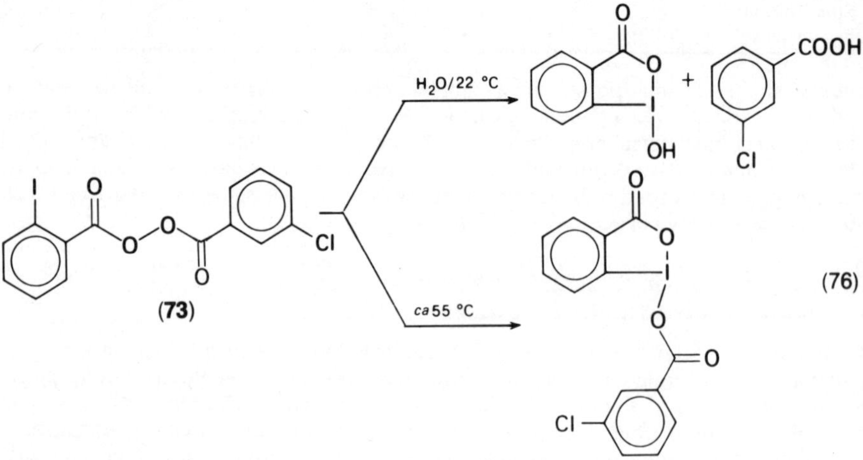

(**73**) $\xrightarrow{H_2O/22\ °C}$ + ; $\xrightarrow{ca\ 55\ °C}$ (76)

D. Topochemistry of Alkoxybenziodoxoles

The solid state hydrolysis and photoreduction of two crystalline polymorphs of 1-methoxy-1,2-benziodoxol-3(1*H*)-one (**74**) have been studied by Etter[99,100]. The photoreduction of the α-polymorph affords four oriented crystalline phases of *o*-iodobenzoic acid, each of which exhibits conservative twinning, while the photoreduction of the β-polymorph eventuates in a powdered phase of *o*-iodobenzoic acid. Hydrolyis of both polymorphs affords multiple product phases of '*o*-iodosobenzoic acid' in various topotactic alignments relative to the parent lattices. The behaviour of **74α** and **74β** on hydrolysis is, however, distinct and, in the case of **74β**, the distribution of oriented product phases depends significantly on the rate at which hydrolysis is allowed to occur.

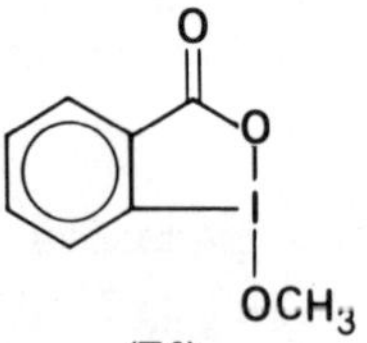

(**74**) 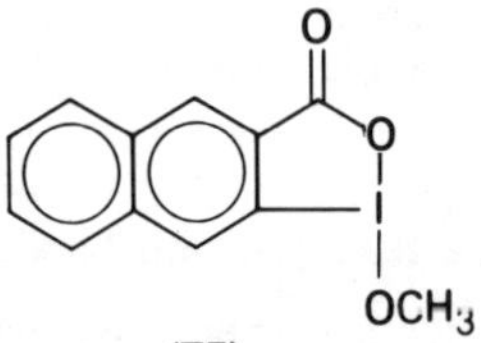 (**75**)

The topotactic photoreduction of the naphthalene analogue **75** has likewise been observed[101].

VIII. HALONIUM YLIDES

A. Occurrence and Synthesis

1. *Halonium ylides in carbene reactions*

The formation of halonium ylides by direct electrophilic attack of *singlet* carbenes on the lone-pair electrons of organic halides seems reasonable. The facility for such a process must certainly depend upon the affinity of the particular carbene for the

$$R-\ddot{X}: + :C(R^1)(R^2) \longrightarrow R-\ddot{X}^{+}-\bar{C}R^1R^2 \longleftrightarrow R-\ddot{X}=CR^1R^2 \quad (77)$$

particular halide and upon the stability of the halonium ylide in question. One might expect, for example, that those carbenes which are hard electrophiles (R^1, R^2 highly electronegative, high energy LUMO) might be more reactive toward chlorine lone pairs than toward iodine lone pairs and those carbenes which are soft electrophiles to exhibit the reverse selectivity. Carbenes, such as bis(carbomethoxy)carbene, with two substituents capable of stabilizing negative charge at α-carbon should give ylides of higher thermodynamic stability than carbenes, such as methylene, lacking such substituents. For a given substituent pair, the order of stability among halonium ylides should be iodonium > bromonium > chloronium for at least two reasons: (1) this trend parallels the ability of halogen atoms to accommodate positive charge and (2) multiple bonding between carbon and halogen, if it occurs, should parallel the ability of the halogen atom to 'expand' its valence octet. Finally, with the well known halonium salts taken as models, those halonium ylides with R = aryl should be more stable than those with R = alkyl.

In 1954, it was reported that the thermal decomposition of ethyl diazoacetate in benzal chloride gives ethyl α-chlorocinnamate in 24% yield[102]. In a footnote, the authors credited a referee with the suggestion that the reaction might proceed through the chloronium ylide **76** formed by electrophilic attack of carboethoxycarbene at a

$$PhCHCl_2 \xrightarrow[\Delta]{N_2CHCO_2Et} PhCH(Cl)-\ddot{C}l^{+}-\bar{C}HCO_2Et \; (\mathbf{76}) \longrightarrow \longrightarrow PhCH{=}C(Cl)CO_2Et + HCl \quad (78)$$

chlorine lone pair. Although the specific term 'ylide' was not applied, this is probably the first proposal for the existence of halonium ylides in the chemical literature.

In 1968, the photolysis of 3,5-di-*tert*-butylbenzene-1,4-diazooxide (**77**) in brominated and iodinated solvents to give significant yields of 3,3′,5,5′-tetra-*tert*-butyldiphenoquinone (**78**) was described[103,104]. For example, while the irradiation of **77** in benzene gives an 89% yield of 2,6-di-*tert*-butyl-4-phenylphenol (**79**), a product of solvent incorporation, and less than 1% of **78**, similar irradiation of **77** in iodobenzene gives a 55% yield (by isolation) of **78**. The formation of **78** was proposed to proceed via intermediate halonium ylides of general structure **80**, analogy for the conversion of **80**

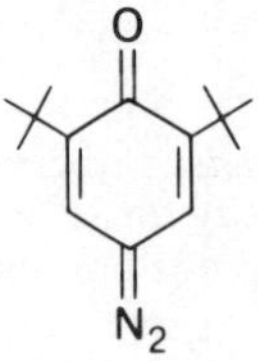

(77) 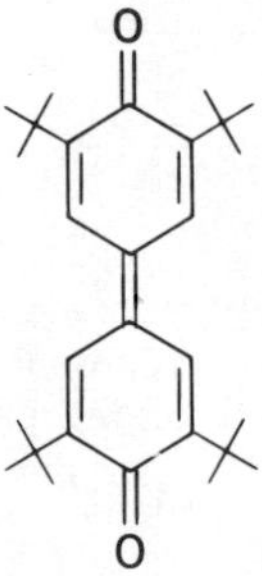 (78) 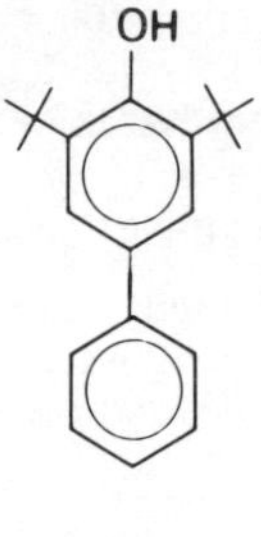 (79) 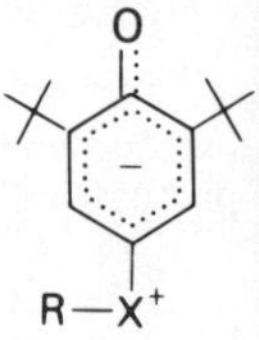 (80)

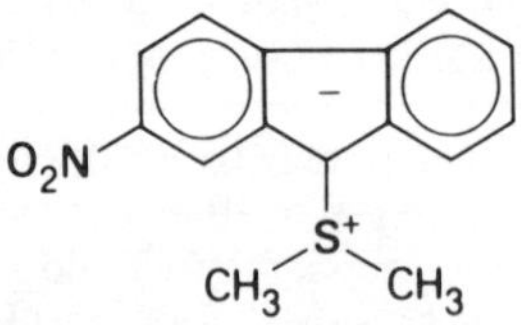

(81) 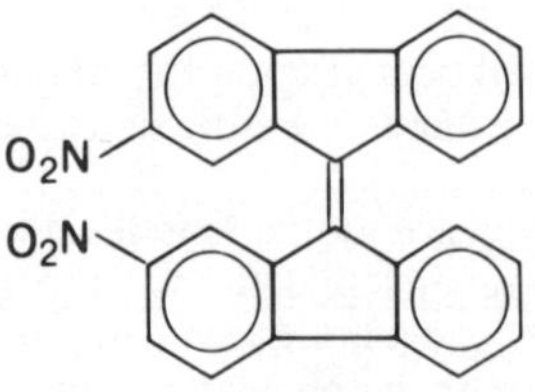 (82)

to **78** being drawn from the literature. 2-Nitrofluorenylidenedimethylsulphurane (**81**) was known, for example, to give a similar 'dimer', **82**, upon its thermolysis in nitromethane[105]. It was subsequently reasoned that the photolysis of **77** in the presence of 2,6-diisopropyl-4-bromophenol should generate the bromonium ylide **83**, which by proton transfer might be in equilibrium with its tautomer, **83'**. Should such a proton transfer proceed more rapidly than 'decomposition', a mixture of three diphenoquinones would be expected. When the experiment was conducted, in hexafluorobenzene as solvent, this was indeed the case. In addition, the ylide hydrobromide, **84**, was isolated in 15% yield, although the source of hydrogen bromide was not established (equation 79). When **84** was treated with triethylamine in CH_2Cl_2, all three diphenoquinones were obtained, consistent with the ylide hypothesis.

The intervention of intermediate halonium ylides in carbon–halogen insertion reactions of singlet carbenes with organic halides has been recognized[104,106–109]. In particular, the action of bis(carbomethoxy)carbene and carboethoxycarbene on allyl bromide and several allylic chlorides has been studied in detail[108,109]. When the carbenes are generated by *direct* photolysis from the corresponding diazo compounds, carbon–halogen insertion is favoured over π-bond addition, but, when the carbenes are generated under conditions of photosensitization, C—X insertion does not compete favourably with π-bond addition. With crotyl chloride as the substrate, the C—X bond insertion of singlet bis(carbomethoxy)carbene proceeds with 100% allylic inversion while the C—X bond insertion of singlet carboethoxycarbene proceeds with retention and inversion and in lower overall yield[108]. Furthermore, singlet $:C(CO_2CH_3)_2$ attacks C—Br bonds twice as fast as it attacks C—Cl bonds. These insertion reactions have been interpreted as Stevens rearrangements and [2,3] sigmatropic transpositions from intermediate halonium ylides. This is illustrated in equation (80) for carboethoxycarbene and crotyl chloride. The observation that such carbenes, when generated in the presence of allylic sulphides, gave *stable* sulphonium ylides which undergo similar rearrangements upon heating provides analogy for this mechanism. The fact that carbon–halogen insertion reactions of methylene and other carbenes have been shown

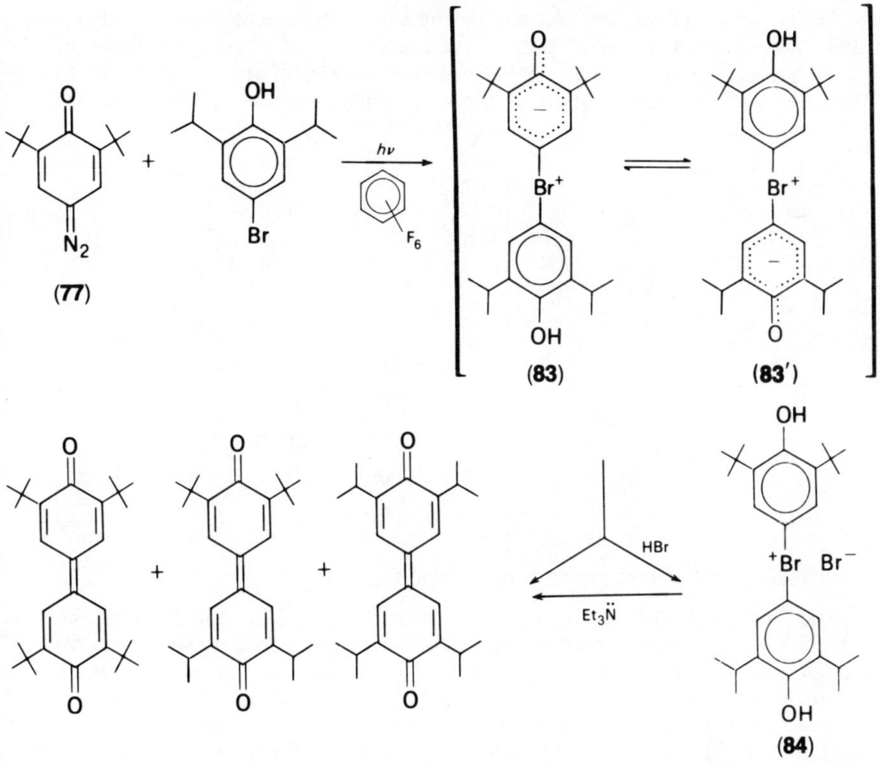

(77) + Br-substituted phenol $\xrightarrow[C_6F_6]{h\nu}$ [(83) $\rightleftharpoons$ (83′)]

HBr / $Et_3\ddot{N}$

(84) Br^-

(79)

Stevens rearrangement: $CH_3CH{=}CHCH_2C(CO_2Et)(H){-}Cl$

$[\ddot{C}HCO_2Et]$

[2.3]-rearrangement: $CH_2{=}CH{-}CH(CH_3){-}CH(Cl)CO_2Et$

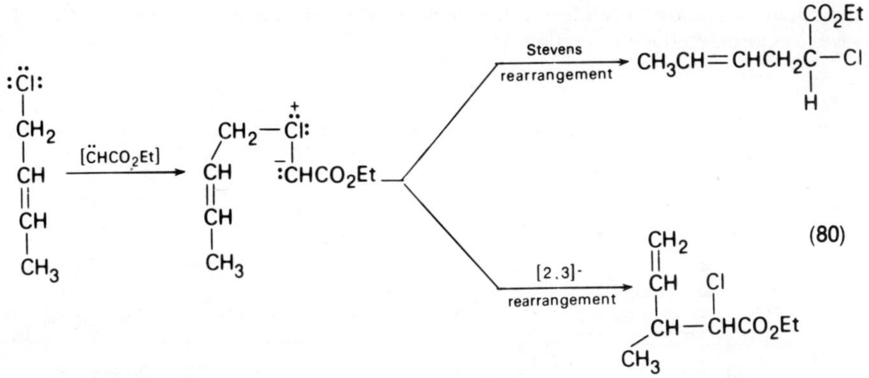

(80)

to involve free radical intermediates does not preclude the possible intervention of halonium ylides along the reaction coordinate. The formation of radical pairs from halonium ylides is a feasible process.

$$R{-}\ddot{X}{:} + \ddot{C}H_2 \rightleftharpoons R{-}\overset{+}{\ddot{X}}{-}\overset{-}{\ddot{C}}H_2 \longrightarrow R^{\cdot}\ ({:}\overset{+}{\dot{X}}{-}\overset{-}{\ddot{C}}H_2 \longleftrightarrow {:}\ddot{X}{-}\dot{C}H_2) \quad (81)$$

We are aware of only one reaction in which stable halonium ylides are obtained via the carbene pathway. The thermal decomposition of dicyanodiazoimidazole in iodobenzene, bromobenzene, chlorobenzene, *p*-chlorotoluene and *m*-fluoroiodobenzene leads to the isolation of the corresponding halonium ylides **85–89**[110,111].

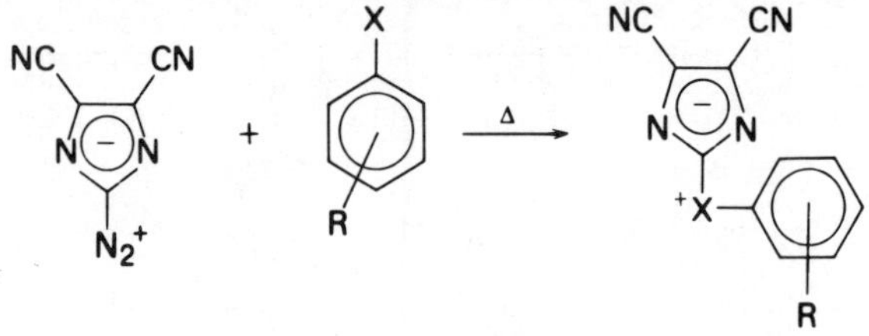

NC CN N(−)N N_2^+ + X R —Δ→ NC CN N(−)N ^+X R (82)

(**85**) X = I, R = H
(**86**) X = Br, R = H
(**87**) X = Cl, R = H
(**88**) X = Cl, R = *p*-Me
(**89**) X = I, R = *m*-F

2. *Iodonium ylides from β-dicarbonyl compounds*

The first stable halonium ylide was reported by Neiland and his coworkers in 1957[112]. In an attempt to prepare 2-fluorodimedone, they treated dimedone **(90)** with (difluoroiodo)benzene in chloroform. However, when the solvent was evaporated and the residual material treated with base, phenyldimedonyliodone **(91)** was obtained instead. The condensation of dimedone with iodosobenzene in chloroform, benzene or acetone likewise afforded phenyldimedonyliodone. The iodonium ylide was described as a 'white felt-like substance' which forms a monohydrate with water and iodonium salts with hydrogen chloride and nitric acid. The thermal decomposition of phenyl(2-dimedonyl)iodonium chloride and the thermolysis of **91** in aqueous hydrochloric acid give 2-chlorodimedone (equation 83).

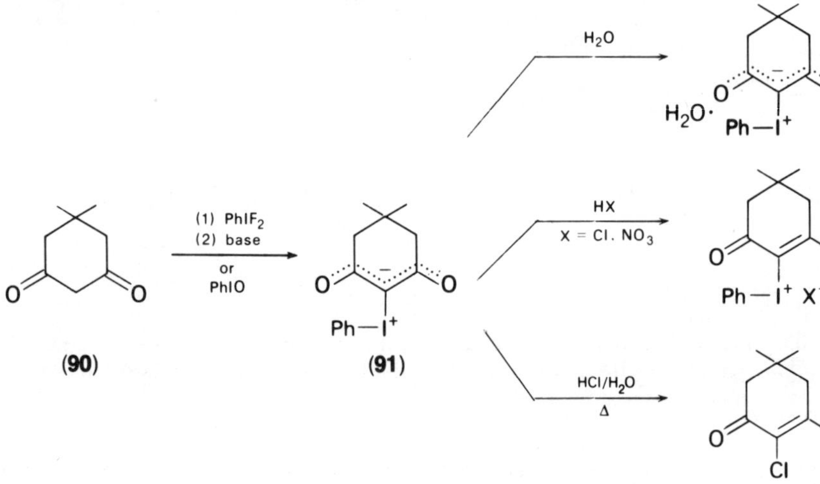

O O (**90**) —(1) $PhIF_2$ (2) base or PhIO→ O O Ph—I^+ (**91**)

—H_2O→ O O $H_2O\cdot$ Ph—I^+

—HX, X = Cl, NO_3→ O OH Ph—I^+ X^-

—HCl/H_2O, Δ→ O OH Cl (83)

A general synthesis of iodonium ylides by the condensation of β-dicarbonyl compounds with (diacetoxyiodo)arenes in alcoholic base (OH^- or OR^-) was subsequently developed[113,114]. In those preparations utilizing iodosoarenes instead of (diacetoxyiodo)arenes, the addition of base to the reaction medium is unnecessary. The condensation is illustrated in equation (84).

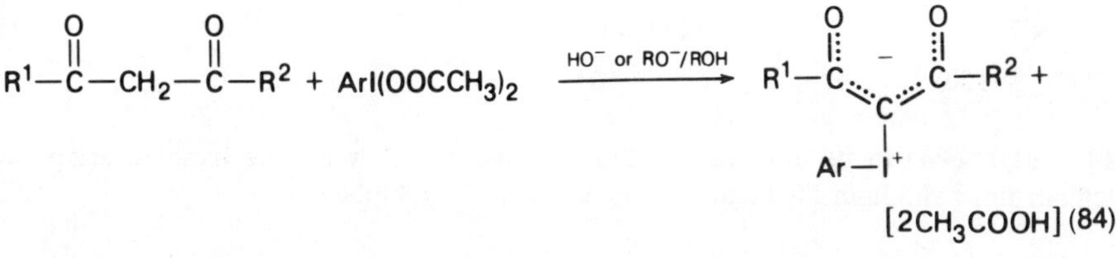

$$R^1{-}\overset{O}{\overset{\|}{C}}{-}CH_2{-}\overset{O}{\overset{\|}{C}}{-}R^2 + ArI(OOCCH_3)_2 \xrightarrow{HO^- \text{ or } RO^-/ROH} R^1{-}C(O){\cdots}\overset{-}{C}(I^+{-}Ar){\cdots}C(O){-}R^2 + [2CH_3COOH] \quad (84)$$

For example, the condensations of 1,3-indanedione[113,115], dibenzoylmethane[113,116], 5-phenyl-1,3-cyclohexanedione[117], benzoylacetone[113,116], ethyl acetoacetate[113,118], dimethyl and diethyl malonate[113,119], isopropylidene malonate[120], barbituric acid and *N,N*-dimethylbarbituric acid[121] with (diacetoxyiodo)benzene give the corresponding iodonium ylides **92–99**.

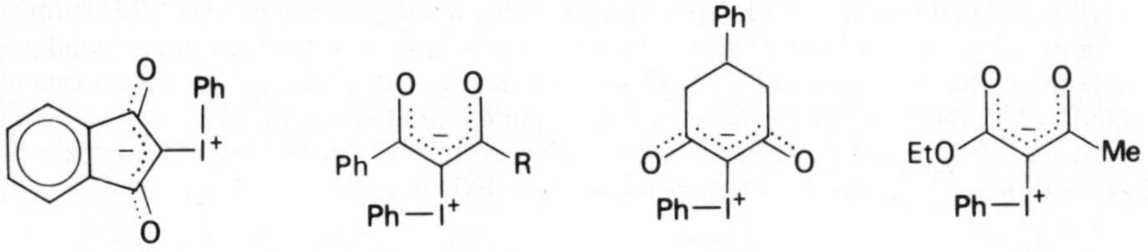

(**92**) (**93**) R = Me, Ph (**94**) (**95**)

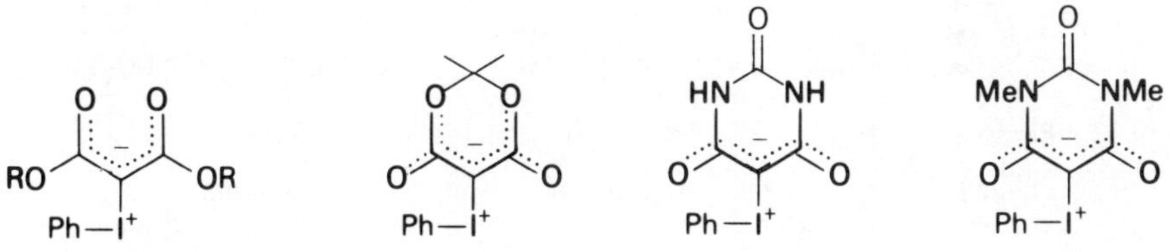

(**96**) R = Me, Et (**97**) (**98**) (**99**)

Analogous iodonium ylides with nitro[122], methyl[123,124] and methoxy groups[125] at various positions in the aromatic ring have likewise been prepared from appropriate (diacetoxyiodo)arenes and iodosoarenes.

3. *Iodonium ylides from phenols*

Phenol and resorcinol undergo oxidative decomposition in the presence of (diacetoxyiodo)benzene[113]. However, electron-withdrawing substituents on the phenol nucleus encourage the formation of iodonium ylides. Thus, the condensations of *p*-nitrophenol and *p*-carboethoxyphenol with (diacetoxyiodo)benzene in acetic acid and of 2-chloro-4-nitrophenol with *o*-(diacetoxyiodo)chlorobenzene in acetic acid afford the iodonium ylides **100**, **101** and **102** respectively,[126,127]. Similar condensations of (diacetoxyiodo)benzene with *p*-hydroxybenzaldehyde and *p*-hydroxyacetophenone did not eventuate in the isolation of the corresponding iodonium ylides, but they were detected in the crude products by ^{1}H NMR analysis[126]. The iodonium ylides

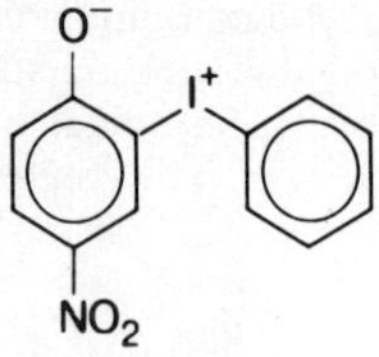

(**100**) 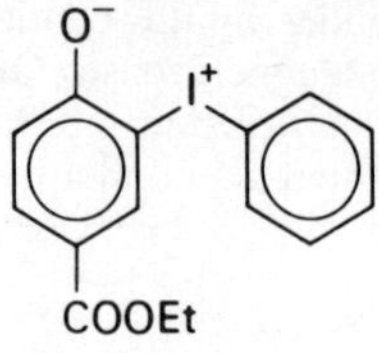 (**101**) 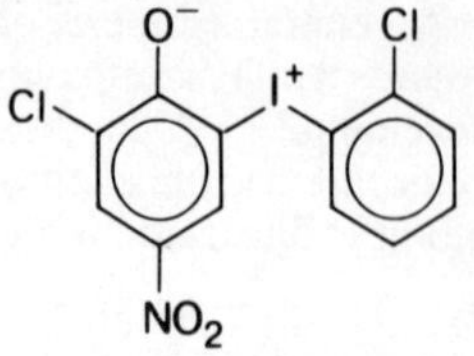 (**102**)

100–102 appear to be moderately stable solids, and they can be handled at room temperature although they can be isomerized to diaryl ethers.

4. Iodonium ylides from cyclopentadienes

In view of the relatively low pK_a of cyclopentadiene, its treatment with (diacetoxyiodo)benzene in the presence of base might reasonably be expected to afford phenyliodonium cyclopentadienylide. However, only 'dark products', presumably derived from oxidative decomposition of cyclopentadiene, are obtained[113]. When the cyclopentadiene nucleus is substituted with electron-withdrawing groups, ylide formation proceeds, and at least 23 such ylides are now known[128–130]. For example, condensations of the potassium salts of 2,3,4-tricyanocyclopentadiene, 2,3,5-tricyanocyclopentadiene and 2,5-dicarboethoxy-3,4-dicyanocyclopentadiene in acetic acid at 20°C and 2,3,5-triformylcyclopentadiene in methanol at −5°C with (diacetoxyiodo)benzene afford the iodonium cyclopentadienylides **103–106**[128,129].

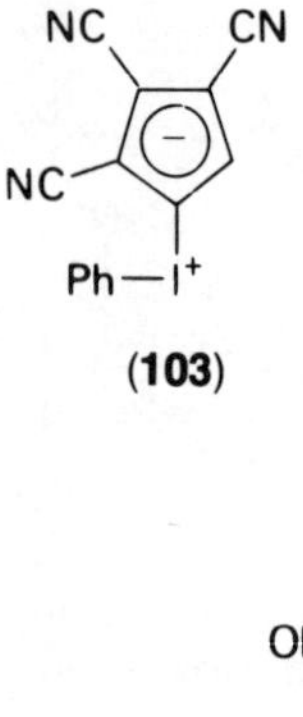

(**103**) 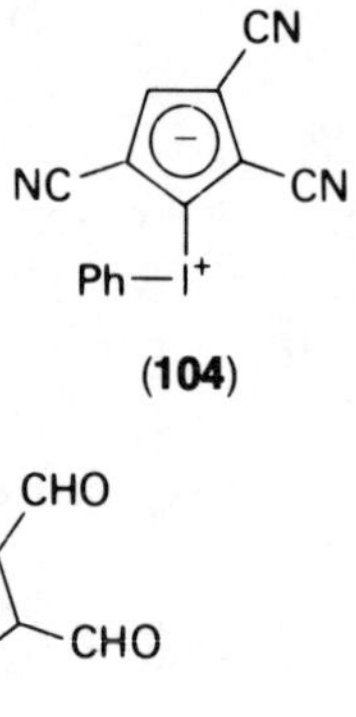 (**104**) 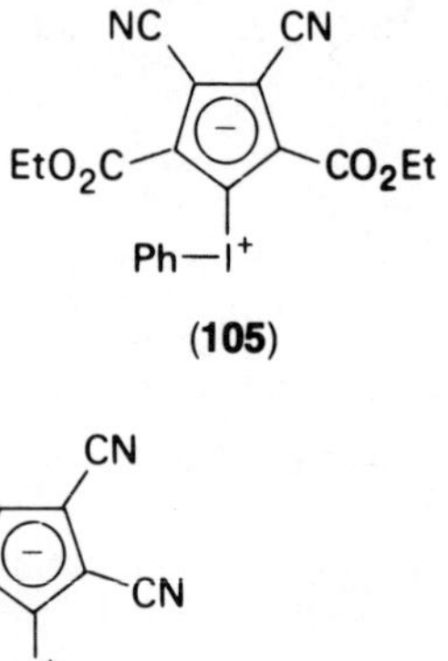 (**105**)

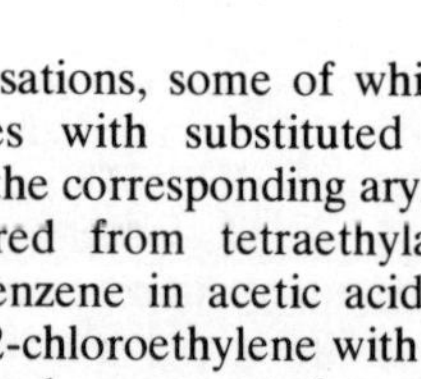

(**106**) 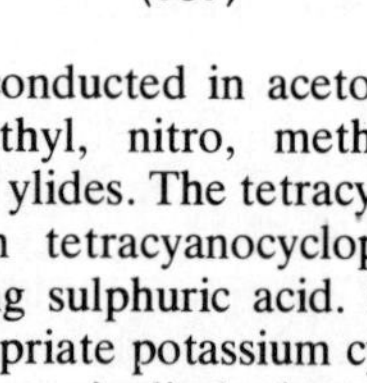 (**107**)

Similar condensations, some of which were conducted in acetonitrile at 0°C, of cyclopentadienides with substituted (e.g. methyl, nitro, methoxy) (diacetoxyiodo)arenes give the corresponding aryliodonium ylides. The tetracyano analogue **107** has been prepared from tetraethylammonium tetracyanocyclopentadienide and (diacetoxyiodo)benzene in acetic acid containing sulphuric acid. Condensations of 1-diacetoxyiodo-2-chloroethylene with the appropriate potassium cyclopentadienides in acetonitrile have been reported to give the chlorovinyliodonium ylides **108–110**[129]. These compounds are unique since they are the only known stable iodonium ylides with an aliphatic ligand attached to iodine.

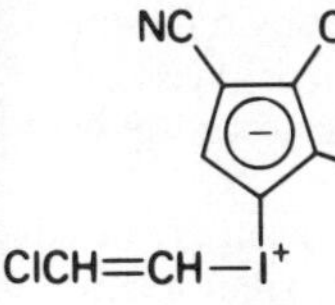

(108) 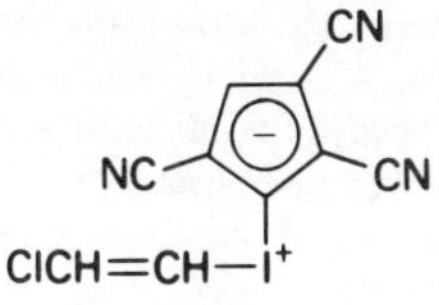 (109) 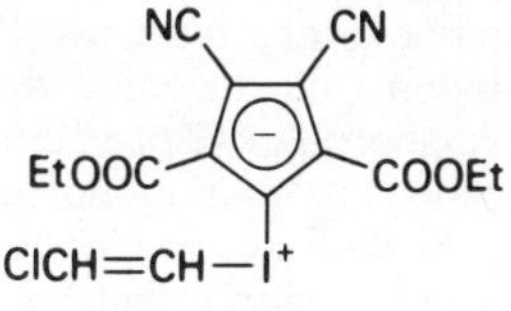 (110)

5. Iodonium ylides from monocarbonyl compounds

It seems clear that the successful preparation of stable iodonium ylides by carbon–iodine condensation reactions is rendered more probable when the 'carbanion' portion of such molecules is highly delocalized. Indeed, no halonium ylides derived from unactivated monocarbonyl compounds have yet been isolated. The reactions of various monoketones (e.g. acetone, acetophenone, cyclohexanone) in acetic acid–sulphuric acid with (diacetoxyiodo)benzene give α-acetoxyketones[131–134] (equation 85). [Hydroxy(tosyloxy)iodo]benzene reacts likewise with monoketones (e.g. acetone, cyclopropyl methyl ketone, acetophenone, 2-thienyl methyl ketone) in acetonitrile to give α-tosyloxyketones in good yield[135] (equation 86). These α-functionalization reactions

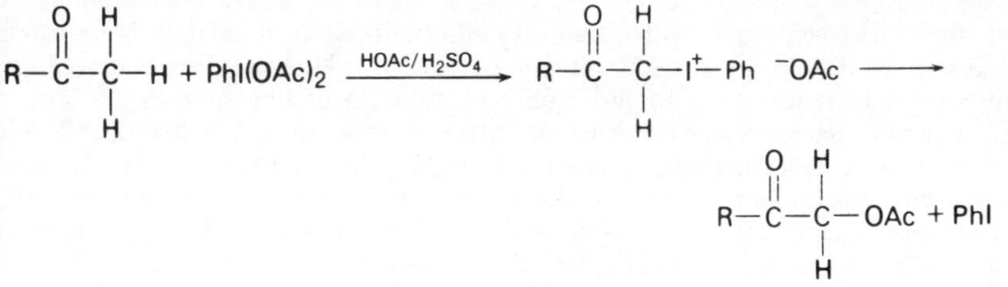

$$\mathrm{RC(O)CH_2H + PhI(OAc)_2 \xrightarrow{HOAc/H_2SO_4} RC(O)CH_2{-}I^{+}{-}Ph\ {}^{-}OAc \longrightarrow RC(O)CH_2{-}OAc + PhI} \quad (85)$$

$$\mathrm{RC(O)CH_2H + PhI(OH)OTs \xrightarrow{CH_3CN} RC(O)CH_2{-}I^{+}{-}Ph\ {}^{-}OTs \longrightarrow RC(O)CH_2{-}OTs + PhI} \quad (86)$$

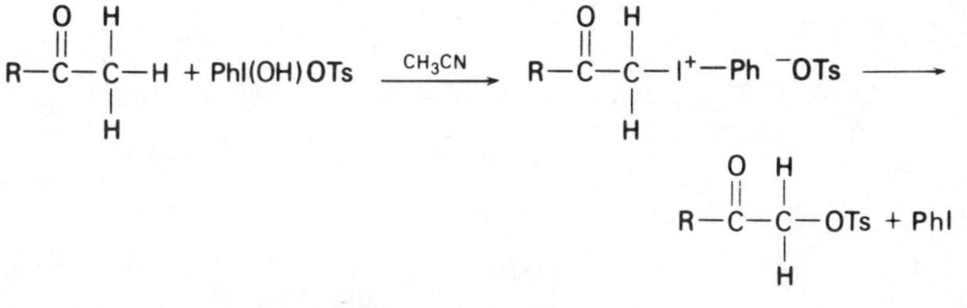

may proceed through intermediate phenyl(α-keto)iodonium acetates and tosylates, and should it prove possible to isolate such species, their successful deprotonation should afford monoketoiodonium ylides. The condensation of dimedone with [hydroxy(tosyloxy)iodo]benzene in acetonitrile at room temperature gives phenyldimedonyliodonium tosylate, which, upon thermolysis, affords 2-tosyloxydimedone (equation 87), thus providing evidence that such intermediates might intervene in the

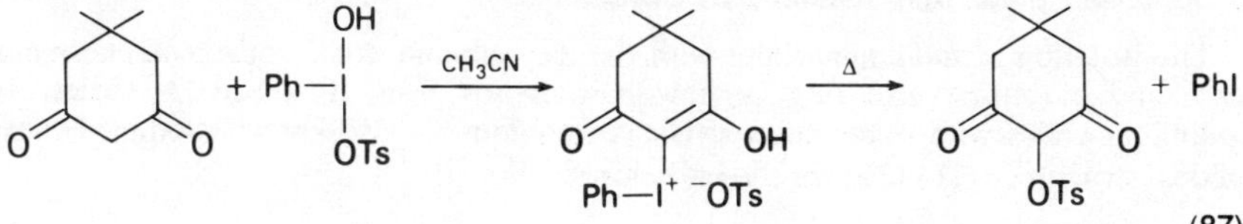

$$\text{dimedone} + \mathrm{Ph{-}I(OH)OTs} \xrightarrow{CH_3CN} \text{(enol, } \mathrm{Ph{-}I^{+}\ {}^{-}OTs}\text{ at C-2)} \xrightarrow{\Delta} \text{2-OTs-dimedone} + \mathrm{PhI} \quad (87)$$

monoketone reactions. Furthermore, the iodonium tosylate yields phenyldimedonyliodone when treated with base[135]. A similar observation of the α-hydroxylation of isophorone, cyclohexanone and acetophenone with iodosobenzene has recently been reported, but no iodonium ylides were obtained[136].

Several ketoiodonium ylides, **111–114**, have been prepared by the action of (diacetoxyiodo)benzene and *p*-(diacetoxyiodo)nitrobenzene on α-nitroacetophenone and methyl α-nitroacetate in ether at 20°C, and they decompose rapidly on storage[137,138]. What stability they do enjoy must certainly be due, in part, to the nitro group bound to the carbanion centre. The aryldinitromethylides **115** and **116** have also

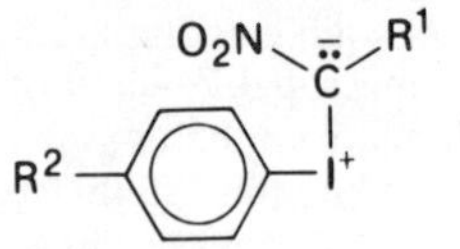

(**111**) R^1 = COPh, R^2 = H
(**112**) R^1 = COOMe, R^2 = H
(**113**) R^1 = COPh, R^2 = NO_2
(**114**) R^1 = COOMe, R^2 = NO_2
(**115**) R^1 = NO_2, R^2 = H
(**116**) R^1 = NO_2, R^2 = NO_2

been synthesized and are comparable in stability to ylides **111–114**[137,138]. The condensation of benzoylacetonitrile with (diacetoxyiodo)benzene in methanolic potassium hydroxide gives the α-cyanoketoiodonium ylide **117**[139,140]. However, it was described as not being sufficiently stable to give a reliable carbon/hydrogen analysis.

The action of (diacetoxyiodo)benzene on various α-cyanoesters in methanol at 20°C and in methanolic potassium hydroxide at 0°C has also been investigated[141]. Various products are obtained, but iodonium ylides are not among them. For example, methyl 2-cyano-2-phenylacetate is converted to the diester **118** and the ether **119** while methyl 2,3-dicyano-3,3-diphenylpropanoate gives the ketenimine **120** in 85% yield.

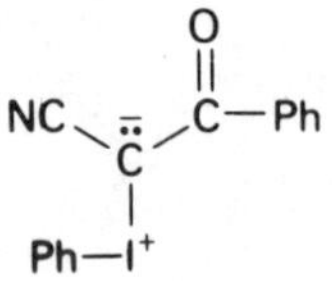

(**117**) 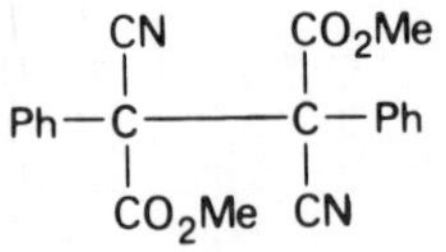 (**118**) 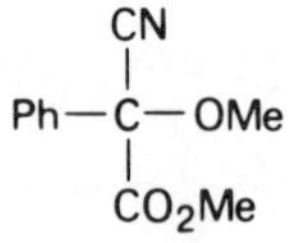 (**119**)

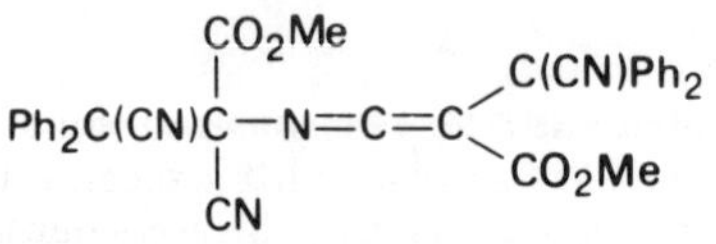

(**120**)

6. Iodonium ylides from heterocyclic compounds

The isolation of iodonium ylides with the dicyanoimidazole, isopropylidene malonate and barbituric acid ring systems has already been discussed. A variety of iodonium ylides with other heterocyclic 'carbanions' is also known. Some of these ylides, structures **121–126**, are shown below[142,143].

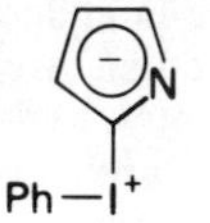

(121)  (122) 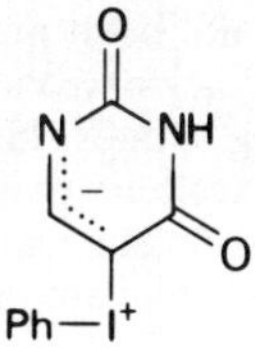 (123) 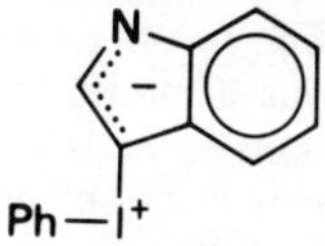 (124)

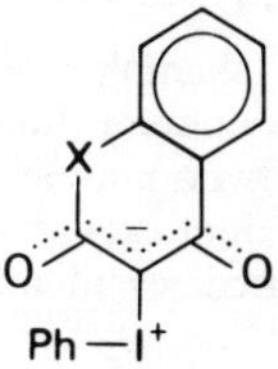

(125) X = O
(126) X = NH

7. *Iodine–nitrogen ylides*

Very few iodonium ylides are known in which the negatively charged centre is something other than a 'carbanion', and they are confined to ylides derived from sulphonamides. The reaction of methanesulphonamide with (diacetoxyiodo)benzene in pyridine–ether gives the mesyliminoiodinane **127**[144]. Similar condensations of *p*-toluenesulphonamide with (diacetoxyiodo)benzene, *p*-(diacetoxyiodo)chlorobenzene and *p*-(diacetoxyiodo)toluene in methanolic potassium hydroxide eventuate in the iminoiodinanes **128–130**[145].

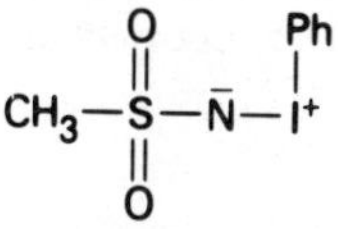

(127)

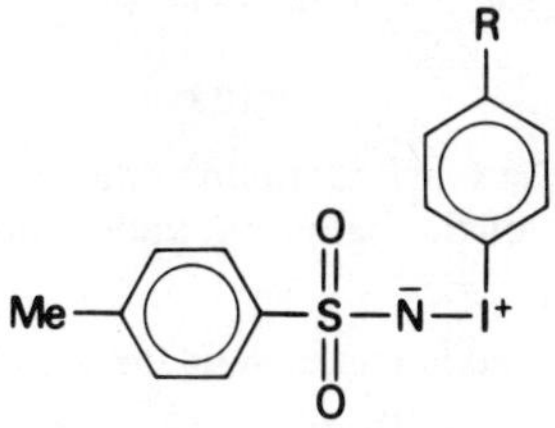

(128) R = H
(129) R = Cl
(130) R = Me

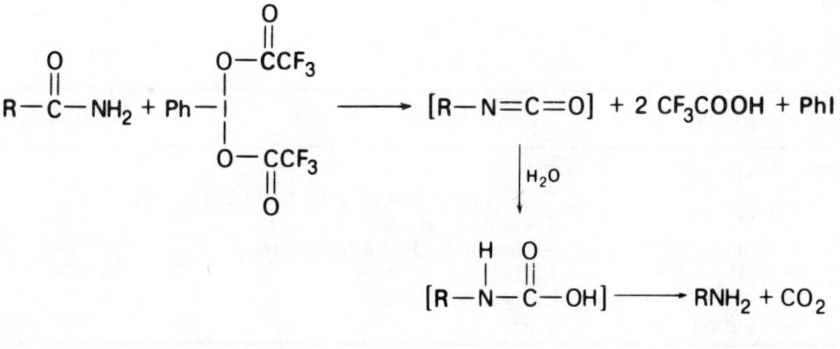

$$R{-}C(=O){-}NH_2 + Ph{-}I(OCOCF_3)_2 \longrightarrow [R{-}N{=}C{=}O] + 2\ CF_3COOH + PhI$$
$$\xrightarrow{H_2O} [R{-}NH{-}C(=O){-}OH] \longrightarrow RNH_2 + CO_2 \qquad (88)$$

Iodonium ylides have not been prepared from carboxamides or amines. Reactions of carboxamides with (diacyloxyiodo)arenes proceed by formal amide decarbonylation to the corresponding amines. Indeed, bis(trifluoroacetoxyiodo)benzene in aqueous acetonitrile is an excellent reagent for such conversions. These reactions are thought to involve the oxidative rearrangement of carboxamides to isocyanates, hydrolysis of the isocyanates to carbamic acids and decarboxylation of the carbamic acids[146] (equation 88).

8. *Halonium ylides from alkanes*

No examples of iodonium ylides with simple alkyl carbanion centres or with alkyliodonium functions have been isolated. However, dimethylbromonium and dimethyliodonium hexafluoroantimonates have been observed to undergo hydrogen–deuterium exchange in D_2SO_4 at 30°C. This exchange is thought to proceed via the intermediacy of methylbromonium methylide **(131)** and methyliodonium methylide **(132)**[147].

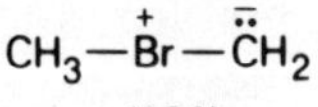

$$CH_3-\overset{+}{Br}-\overset{-}{\ddot{C}}H_2 \quad \textbf{(131)} \qquad\qquad CH_3-\overset{+}{I}-\overset{-}{\ddot{C}}H_2 \quad \textbf{(132)}$$

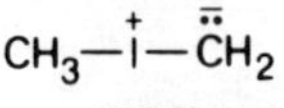

9. *Iodonium ylides from alkenes*

Although, no aryliodoniumethenylides ($Ar-\overset{+}{I}-\overset{-}{C}=C<$) have yet been prepared, a reaction has been reported which may involve the intermediate formation of such an ylide. An attempt to metathesize (β-phenylethynyl)phenyliodonium chloride to the corresponding tetrafluoroborate salt by its treatment with aqueous fluoroboric acid led instead to (2-chloro-2-phenylethenyl)phenyliodonium tetrafluoroborate[148] (equation 89). It seems plausible that this reaction proceeds by Michael addition of chloride

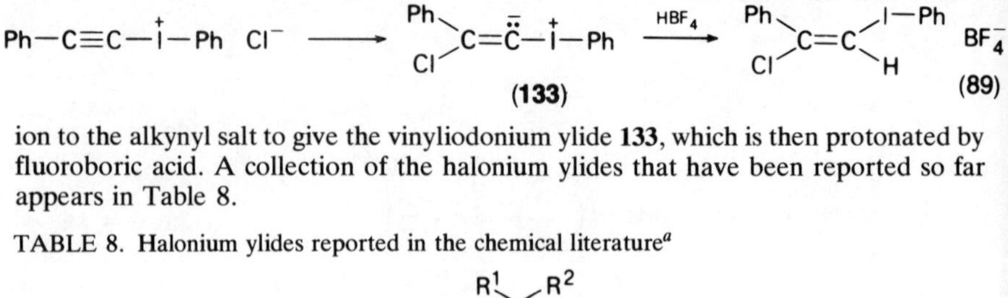

$$Ph-C\equiv C-\overset{+}{I}-Ph\ \ Cl^- \longrightarrow \underset{\textbf{(133)}}{(Ph)(Cl)C=\overset{-}{\ddot{C}}-\overset{+}{I}-Ph} \xrightarrow{HBF_4} (Ph)(Cl)C=C(H)-\overset{+}{I}-Ph\ \ BF_4^- \qquad (89)$$

ion to the alkynyl salt to give the vinyliodonium ylide **133**, which is then protonated by fluoroboric acid. A collection of the halonium ylides that have been reported so far appears in Table 8.

TABLE 8. Halonium ylides reported in the chemical literature[a]

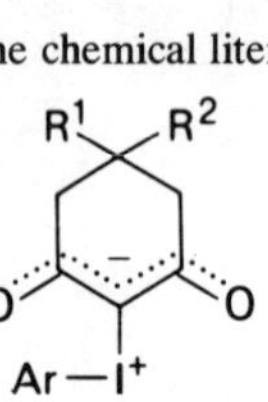

(General structure: 5-R^1,5-R^2-cyclohexane-1,3-dione 2-ylide with $Ar-I^+$ at C-2)

R^1	R^2	Ar	Reference
Me	Me	C_6H_5	112
Me	Me	2-$NO_2C_6H_4$, 3-$NO_2C_6H_4$, 4-$NO_2C_6H_4$	122
Me	Me	2,4,6-$Me_3C_6H_2$, 4-MeC_6H_4	124
Me	Me	4-$MeOC_6H_4$, 3-Cl-4-$MeOC_6H_3$	125
H	H	C_6H_5	124
H	C_6H_5	C_6H_5	117
H	α-furyl	C_6H_5	149

TABLE 8 (*Continued*)

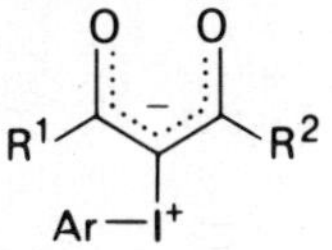

R^1	R^2	Ar	Reference
C_6H_5	C_6H_5	C_6H_5	113, 116
C_6H_5	C_6H_5	4-$MeOC_6H_4$, 3-Cl-4-$MeOC_6H_3$	125
C_6H_5	Me	4-$MeOC_6H_4$	125
C_6H_5	Me	C_6H_5	113, 116
C_6H_5	CF_3	C_6H_5	150
MeO	MeO	C_6H_5	119
MeO	MeO	2-$NO_2C_6H_4$, 3-$NO_2C_6H_4$	122
MeO	MeO	4-$MeOC_6H_4$	125
EtO	EtO	C_6H_5	113, 119
EtO	EtO	4-$MeOC_6H_4$	125
EtO	Me	C_6H_5	113, 118
EtO	Me	4-$MeOC_6H_4$	125
MeO	Me	C_6H_5	118
Me	Me	C_6H_5	150
CF_3	α-thienyl	C_6H_5	150
C_6H_5	CN	C_6H_5	140

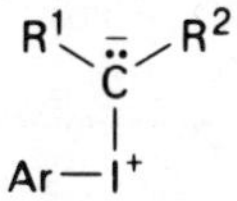

R^1	R^2	Ar	Reference
NO_2	COC_6H_5	C_6H_5	137
NO_2	CO_2Me	C_6H_5	137
NO_2	NO_2	C_6H_5, 4-$NO_2C_6H_4$	138
NO_2	COC_6H_5	4-$NO_2C_6H_4$	138
NO_2	CO_2Me	4-$NO_2C_6H_4$	138

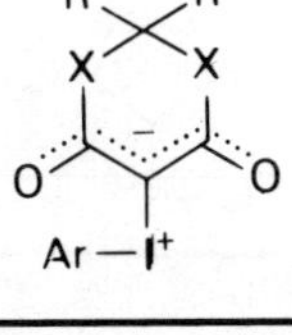

X	R^1	R^2	Ar	Reference
O	Me	Me	C_6H_5, 4-$MeOC_6H_4$, 2-$NO_2C_6H_4$	120
NH	O		C_6H_5, 4-$MeOC_6H_4$	121
NMe	O		C_6H_5	121

TABLE 8. (*Continued*)

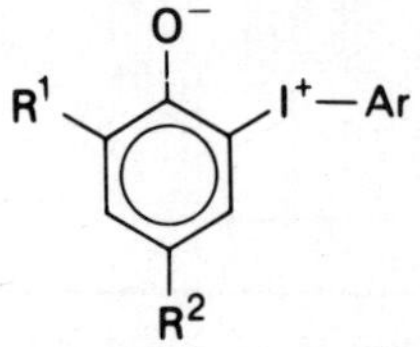

R^1	R^2	Ar	Reference
H	NO_2	C_6H_5	126
H	CO_2Et	C_6H_5	126
Cl	NO_2	$2\text{-}ClC_6H_4$	127

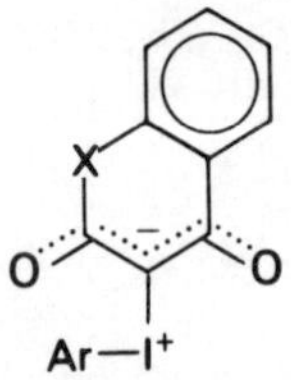

X	Ar	Reference
O	C_6H_5	143
NH	C_6H_5, $4\text{-}MeC_6H_4$, $4\text{-}MeOC_6H_4$	143
NMe	C_6H_5	143
NPH	C_6H_5	143

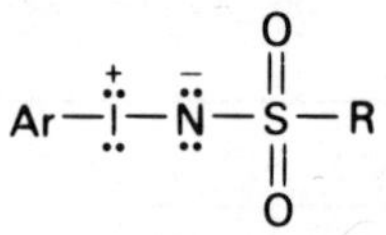

R	Ar	Reference
Me	C_6H_5	144
$4\text{-}MeC_6H_4$	C_6H_5, $4\text{-}ClC_6H_4$, $4\text{-}MeC_6H_4$	145

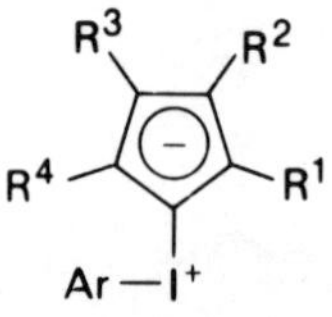

R^1	R^2	R^3	R^4	Ar	Reference
CO_2Et	CN	CN	CO_2Et	C_6H_5, $2\text{-}MeC_6H_4$, $4\text{-}MeC_6H_4$, $4\text{-}MeOC_6H_4$ $4\text{-}NO_2C_6H_4$, $4\text{-}PhN{=}NC_6H_4$, $2,4,6\text{-}Me_3C_6H_2$	129
CHO	H	CHO	CHO	C_6H_5, $2,4,6\text{-}Me_3C_6H_2$, $2\text{-}Me\text{-}3,5\text{-}(NO_2)_2C_6H_2$	129
H	CN	CN	CN	C_6H_5, $4\text{-}MeOC_6H_4$, $4\text{-}NO_2C_6H_4$	129
CN	H	CN	CN	C_6H_5	129
CN	CN	CN	CN	C_6H_5	129

TABLE 8. (*Continued*)

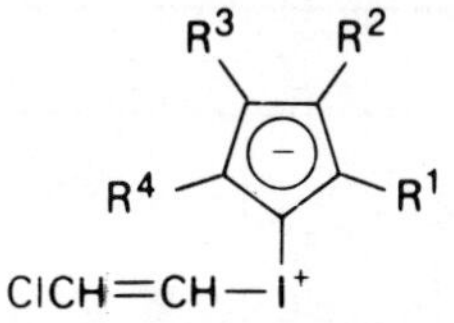

R^1	R^2	R^3	R^4	Reference
H	CN	CN	CN	129
CN	H	CN	CN	129
CO_2Et	CN	CN	CO_2Et	129

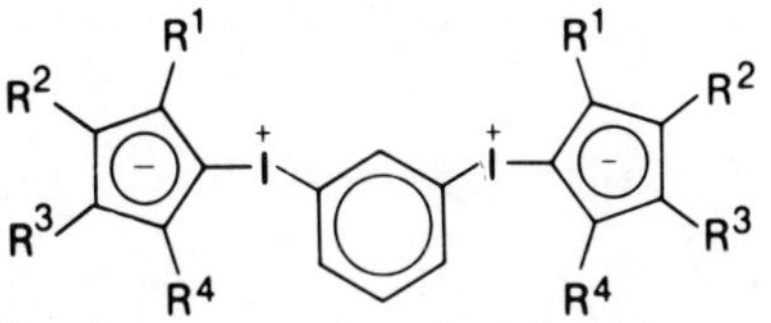

R^1	R^2	R^3	R^4	Reference
H	CN	CN	CN	130
CO_2Et	CN	CN	CO_2Et	130

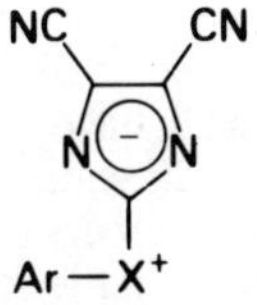

Ar	X	Reference
C_6H_5	I, Br, Cl	110, 111
$4\text{-}MeC_6H_4$	Cl	110, 111
$3\text{-}FC_6H_4$	I	110, 111

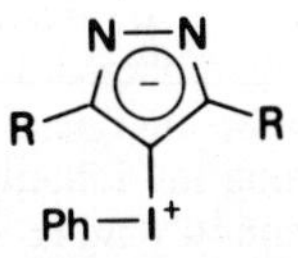

R	Reference
H	142, 151
Me, Ph	151

TABLE 8. (*Continued*)

Structure	Reference
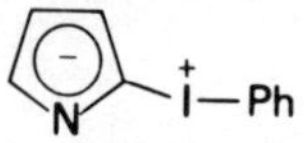 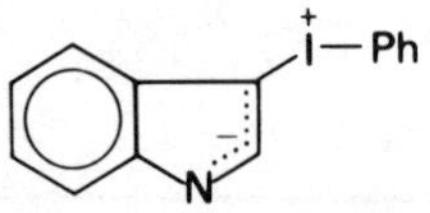 N, $\overset{+}{I}$—Ph (−) $\overset{+}{I}$—Ph, N (−)	113, 142
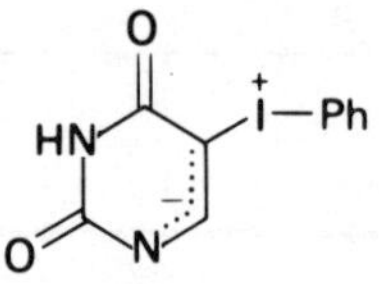 O, HN, $\overset{+}{I}$—Ph, O, N (−)	142
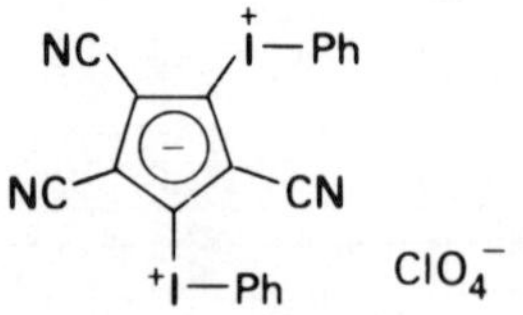 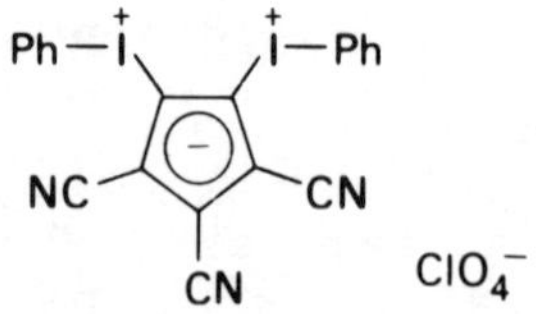 NC, $\overset{+}{I}$—Ph, NC, CN, ^{+}I—Ph (−) ClO_4^- Ph—$\overset{+}{I}$, $\overset{+}{I}$—Ph, NC, CN, CN (−) ClO_4^-	130

[a]*Chem. Abstr.*, **85**, 20763j (1976)[123] lists various aryliodonium ylides wherein Ar = *o*-tolyl, *p*-tolyl, 2,4-xylyl and 3,4-xylyl. However, it is not clear in the abstract which ylides were actually prepared, and the original document was not obtained by this author.

B. Thermal Reactions

1. Stability

Among those halonium ylides which have been isolated, there are considerable variations in thermal stability. Thus, while some can be kept for days at room temperature without significant decomposition, others decompose within minutes or hours at room temperature. Unfortunately, for most iodonium ylides, the products of thermal decomposition other than iodoarenes have not been characterized.

While the pK_a values of the carbon acids corresponding to the carbanion moiety of iodonium ylides may be used in a qualitative way to predict the likelihood of ylide occurrence, caution must be exercised in relating such pK_a values to relative ylide stabilities. For example, the pK_a values of acetylacetone and ethyl acetoacetate are 9.0 and 10.7 respectively[152], and it might be expected that phenyliodonium diacetylmethylide (**134**) should enjoy greater stability than phenyliodonium(carboethoxy)acetylmethylide (**95**). In fact, the reverse is true. Similarly, the pK_a of dinitromethane (3.6) is lower than that of dimedone (5.6), yet phenyliodonium dinitromethylide (**113**) is far less stable than phenyldimedonyliodone (**91**). Obviously, ylide stabilities will depend on the activation energies for particular modes of decomposition and those modes are diverse. Among the iodonium ylides derived from β-dicarbonyl compounds, the cyclic ones are generally more stable than their acyclic analogues.

It is difficult to generalize the effect of substituents on iodonium ylide lifetimes since the same substituent variation may operate in opposite directions in different ylide subclasses. For example, *p*-nitrophenyliodonium dinitromethylide (**116**) is more stable to thermal decomposition than phenyliodonium dinitromethylide (**113**), but the reverse is true for phenyldimedonyliodone and its *p*-nitro analogue.

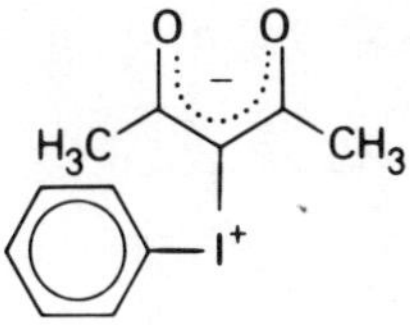

(134) 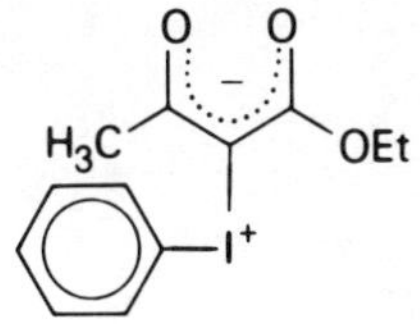 (95) 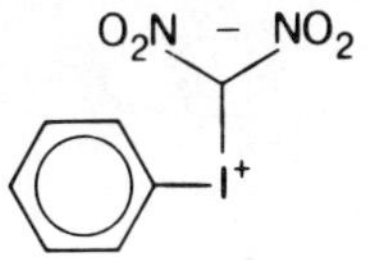 (113)

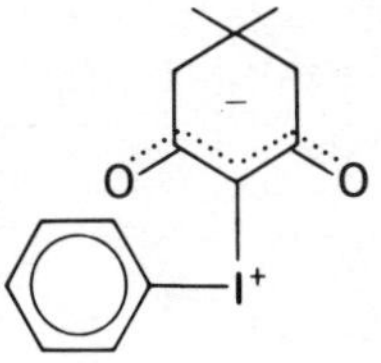

(91) 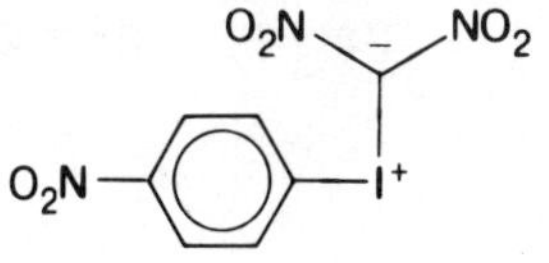 (116) 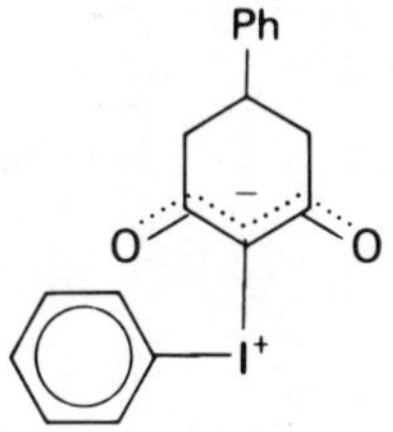 (94)

2. Rearrangement

Iodonium ylides derived from cyclohexanediones are subject to an intramolecular rearrangement in which the aryl group bound to iodine migrates to an oxygen atom in a formal [1,4] sigmatropic transposition. Phenyldimedonyliodone (**91**) isomerizes to phenyl 2-iododimedonyl ether in various solvents[153,154]. In refluxing toluene, the ether is formed in 47% yield and is accompanied with the formation of iodobenzene (40%). Rearrangement to the corresponding ethers likewise occurs with *o*-nitrophenyldimedonylidone (95% yield in hot DMF) and *p*-nitrophenyldimedonyliodone (64% yield in hot 1,2-dichloroethane) and to a lesser extent with the *m*-nitro analogue[122]. Similar isomerizations have been observed with *p*-tolyl- and mesityldimedonyliodones, but no information on yields has been published[124]. On the other hand the *p*-methoxy derivative is *not* converted to the corresponding ether, even in boiling DMF[125]. these observations are qualitatively consistent with an S_NAr mechanism for rearrangement (equation 90).

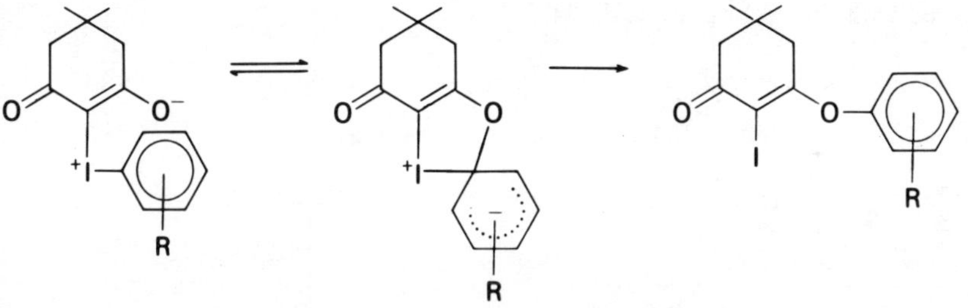

(90)

Other iodonium ylides which undergo this rearrangement include the phenylcyclohexanedionate ylide **94**[117,155], the phenoxyiodonium ylides **100–102** (equation 91) and the heterocyclic malonyl ylides **125**, **126**, **135–138** (equation 92), the last-mentioned group with particular efficiency[126,127,143]. It is interesting that **138** isomerizes to the corresponding ether in 92% yield despite the presence of a *p*-methoxy substituent on the aromatic nucleus in contrast to the inertness of *p*-methoxyphenyldimedonyliodone.

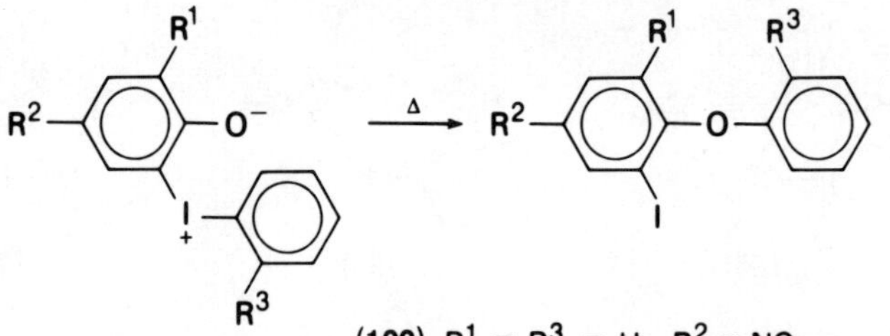

(91)

(100) $R^1 = R^3 = H$, $R^2 = NO_2$
(101) $R^1 = R^3 = H$, $R^2 = CO_2Et$
(102) $R^1 = R^3 = Cl$, $R^2 = NO_2$

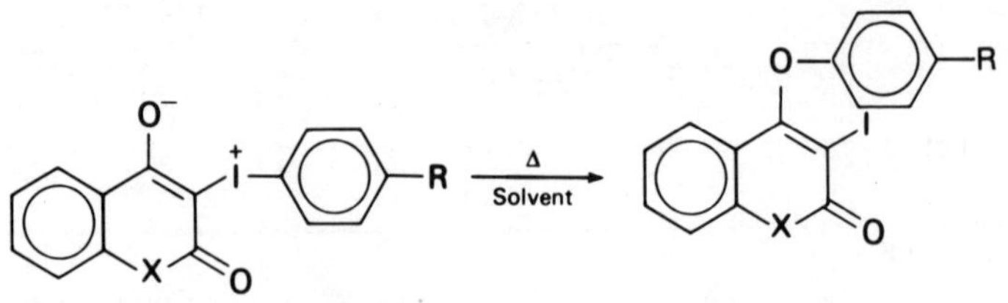

(92)

(125) X = O, R = H
(126) X = NH, R = H
(135) X = NPh, R = H
(136) X = NMe, R = H
(137) X = NH, R = Me
(138) X = NH, R = OMe

Not all cyclic dicarbonyl ylides are subject to this aryl migration reaction. It has been specifically noted, for example, that the *o*-nitrophenyl malonate ylide **139** does not isomerize in boiling ethanol over a 12 h period, although the dimedone analogue is rearranged within a few minutes[120]. Such a rearrangement has likewise not been observed with the barbiturate ylide **98**[121].

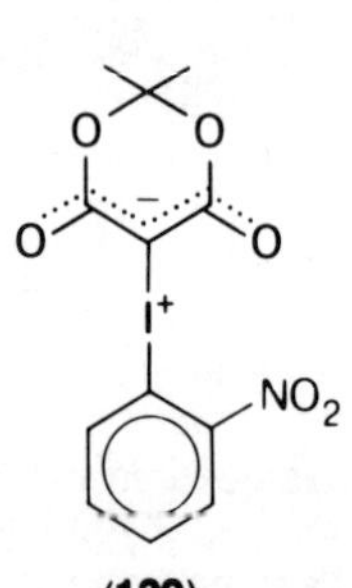

(139) 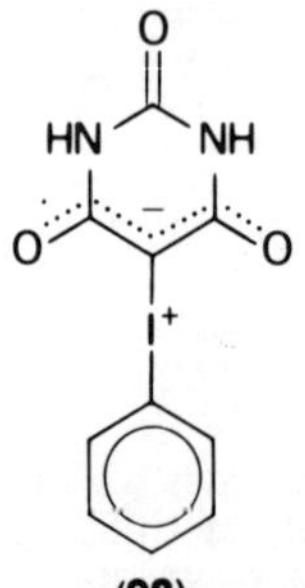 (98)

Formal 1,3 migration of the phenyl group from the halonium centre to nitrogen has been observed with the dicyanoimidazole halonium ylides **85**–**89**[110,111] (equation 93). It is noteworthy that while the thermal decomposition of dicyanodiazoimidazole in

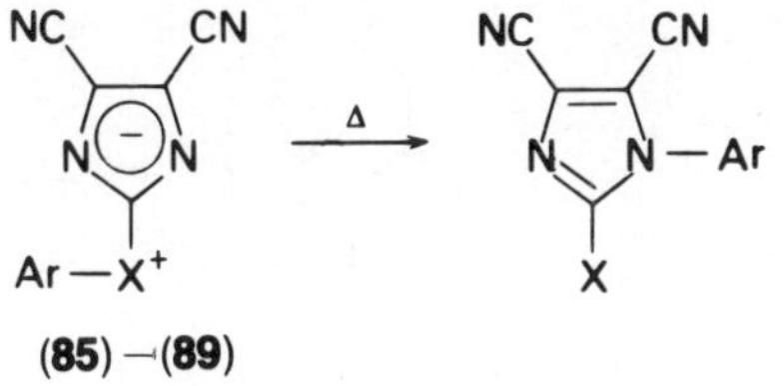

(85)—(89) (93)

1,2-dichloroethane does not afford a stable chloronium ylide, the imidazole **140** is formed in 81% yield, probably by intramolecular nucleophilic collapse of an intermediate chloronium ylide (i.e. by formal migration of a β-chloroethyl group).

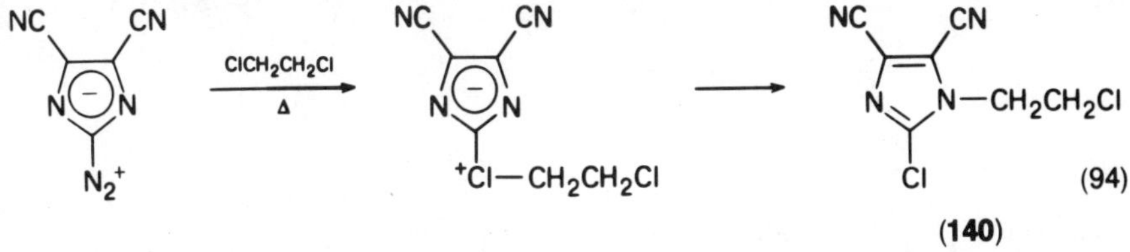

(140) (94)

3. *Cleavage*

Except for the observation that iodoarene formation often attends iodonium ylide decompositions, detailed product studies have thus far been confined to only a few iodonium ylides.

Phenyldimedonyliodone (**91**) has been studied in the most detail, and its thermal rearrangement to phenyl 2-iododimedonyl ether and fragmentation to iodobenzene has already been discussed. Metal catalysts strongly influence the thermal reactions of **91**. For example, when **91** is heated in pyridine, phenyl 2-iododimedonyl ether and iodobenzene are formed just as they are in toluene, although the ratio of ether to iodobenzene is much greater. However, when the ylide is heated in aqueous methanol in the presence of pyridine and silver nitrate, the major product is the silver salt of 2-iododimedone, and not much iodobenzene is formed[153,154]. The presence of pyridine is important in this reaction, although its mode of action in facilitating the rupture of the phenyl–iodine bond has yet to be elucidated. In refluxing ethanol, **91** undergoes intramolecular isomerization, replacement of iodobenzene by the solvent, and reductive cleavage at the dimedonate–iodine bond, as evidenced by the formation of phenyl 2-iododimedonyl ether, ethyl dimedonyl ether and dimedone in addition to iodobenzene[124] (equation 95). The presence of Cu(I) speeds up and alters the decomposition of **91** in ethanol. In the presence of <10 mol % cuprous chloride, **91** reacts at room temperature to give 2-iododimedone, a small quantity of phenyl 2-iododimedonyl ether, phenetole, ethyl 4,4-dimethyl-2-oxocyclopentanecarboxylate and iodobenzene[124] (equation 96).

The formation of dimedone in the uncatalysed reaction and the β-ketoester in the catalysed reaction is thought to be consistent with a carbenic mode of ylide fragmentation. Loss of iodobenzene from the ylide would presumably afford 4,4-dimethylcyclohexane-2,6-dionylidene initially in its singlet state (equation 97). Rapid intersystem crossing (ISC) in the incipient carbene facilitated by the proximal heavy iodine atom of iodobenzene would eventuate in dimedone. Stabilization of the singlet carbene by coordination with copper in the catalysed reaction has been suggested. The β-ketoester then arises via Wolff rearrangement of the carbene and subsequent addition of ethanol to the resulting ketene[124].

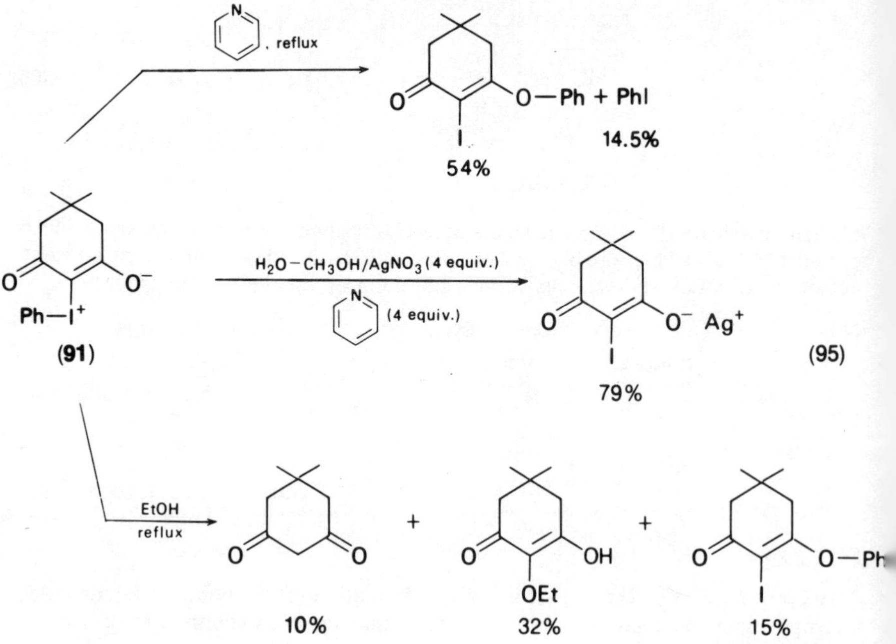

pyridine, reflux → O–Ph + PhI
54% 14.5%

$H_2O-CH_3OH/AgNO_3$ (4 equiv.), pyridine (4 equiv.) → $O^- Ag^+$
Ph—I^+
(91)
79% (95)

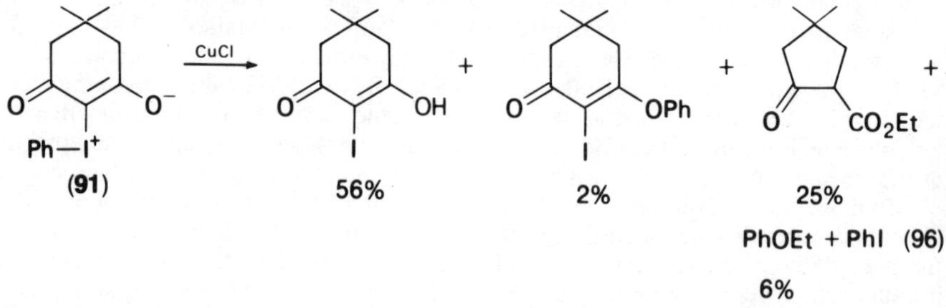

EtOH reflux → OH, OEt, O—Ph
10% 32% 15%

CuCl → OH, OPh, CO_2Et
Ph—I^+
(91)
56% 2% 25%
PhOEt + PhI (96)
6%

The iodonium ylide **94** reacts analogously to **91** in neat pyridine and in methanolic silver nitrate containing pyridine, but its copper(I)-promoted decomposition has not been studied[117,155].

Phenyliodonium dibenzoylmethylide (**93**) exhibits reactivity patterns similar to those of **91** but not identical. In pyridine, for example, ylide **93** is transformed into the corresponding pyridinium ylide with loss of iodobenzene[116,156]. When **93** is heated in refluxing ethanol, the major modes of decomposition are reductive cleavage and carbenic rearrangement, the respective products being dibenzoylmethane and ethyl α-benzoylphenylacetate[124]. In the presence of cuprous chloride in ethanol, the reductive cleavage of **93** does not occur while carbenic rearrangement emerges as the major process, consistent with the selective copper(I) stabilization of singlet bis(benzoyl)methylene[124]. Tetrabenzoylethylene is also formed in the copper(I)-promoted decomposition of **93**.

The thermolyses of the tosyliminoiodinanes **127** and **130** in benzene at 130°C have also been studied. In addition to iodobenzene, the major products are *p*-toluenesul-

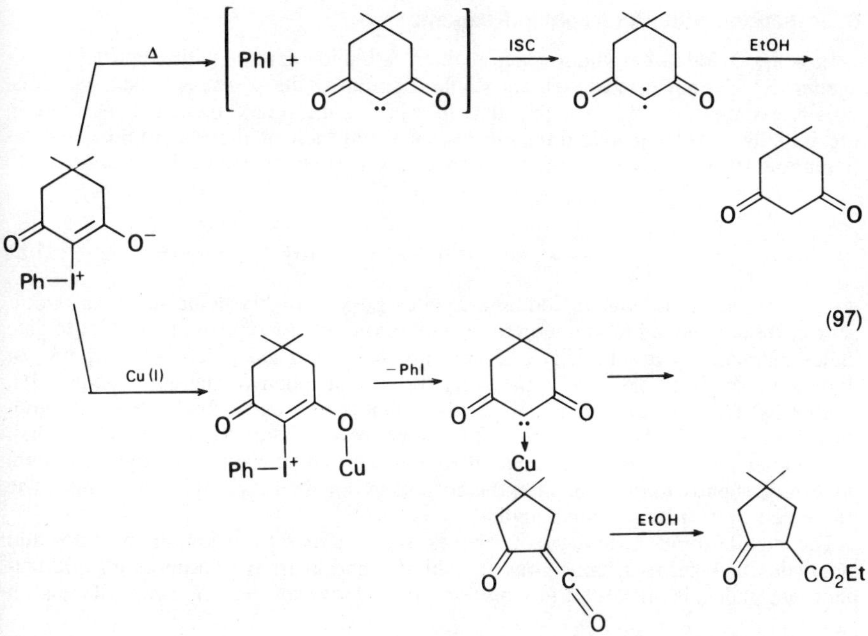

(97)

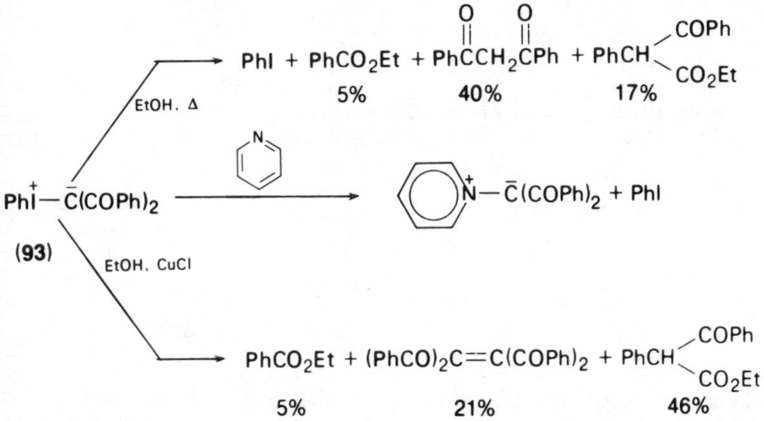

PhI + $PhCO_2Et$ + PhC(O)CH_2C(O)Ph + PhCH(COPh)CO_2Et
5% 40% 17%

$Ph\overset{+}{I}-\overset{-}{C}(COPh)_2$ (93) → (EtOH, Δ; pyridine; EtOH, CuCl)

$\overset{+}{N}-\overset{-}{C}(COPh)_2$ + PhI (98)

$PhCO_2Et$ + $(PhCO)_2C{=}C(COPh)_2$ + PhCH(COPh)CO_2Et
5% 21% 46%

phonamide from **130** and methanesulphonamide from **127**[144]. The sulphonamides are thought to be derived from the corresponding triplet nitrenes generated from the ylides as iodobenzene departs.

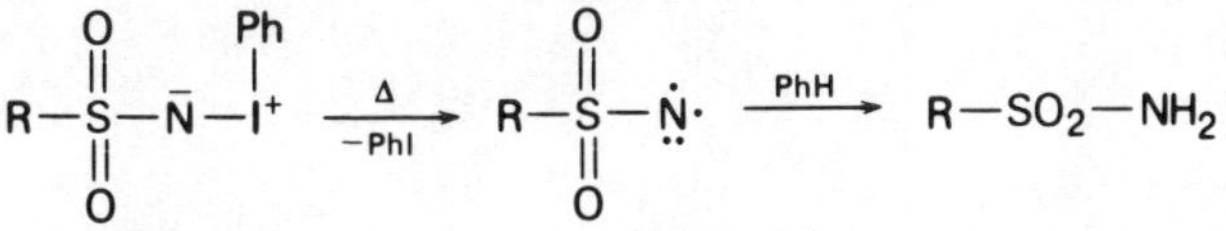

$$R-SO_2-\bar{N}-\overset{+}{I}Ph \xrightarrow[-PhI]{\Delta} R-SO_2-\dot{N}\cdot \xrightarrow{PhH} R-SO_2-NH_2 \qquad (99)$$

R = Me, *p*-tolyl

C. Reactions with Electrophilic Reagents

In general, iodonium ylides react at their carbanion centres with electrophilic reagents, $E^+ Z^-$, sometimes yielding stable iodonium salts. However, such reactions usually eventuate in cleavage products formed via the replacement of iodoarene in the initially generated iodonium ion by the counterion of the electrophilic species (equation 100). In assessing their reactivity towards electrophiles, it is important to

$$\mathrm{Ar{-}\overset{A:^-}{\overset{|}{I^+}}} \xrightarrow{E^+Z^-} \mathrm{Ar{-}\overset{A{-}E}{\overset{|}{I^+}}\ Z^-} \longrightarrow \mathrm{ArI + Z{-}A{-}E} \qquad (100)$$

remember that most known iodonium ylides possess highly delocalized carbanion centres directly bound to an iodonium centre, structural features that conspire to give ylides of relatively low basicity and nucleophilicity. For example, while the pK_a of dimedone in water is 5.25, the pK_a values of phenyldimedonyliodone (**91**), 2-dimethylsulphoniodimedone, 2-pyridiniodimedone and 2-trimethylammoniodimedone are 0.72, 0.46, 1.21 and 2.22, respectively[157]. Thus, **91** is far less basic than dimedonate ion, less basic than the corresponding pyridinium and trimethylammonium ylides and slightly more basic than the corresponding dimethylsulphonium ylide. The pK_a of 2-(*o*-nitrophenyl)dimedonyliodone is 0.11[157].

The possibility of formation of alkenes and oxiranes from iodonium ylides and aldehydes and ketones, analogous to their formation from phosphonium and sulphonium ylides, is illustrated in equation (101). However, we are aware of no such

$$\mathrm{Ar{-}\overset{+}{I}{-}\ddot{C}^{-}\!\!<} + \mathrm{R{-}\overset{O}{\overset{\|}{C}}{-}R'} \longrightarrow \mathrm{Ar{-}\overset{+}{I}{-}\overset{^-O{-}\overset{R}{\overset{|}{C}}{-}R'}{\overset{|}{C}}{-}} \begin{cases} \longrightarrow \mathrm{ArI} + \text{oxirane } (>C{-}O{-}C(R)(R')) \\ \longrightarrow \mathrm{Ar{-}I{-}C{-}C(R)(R'){-}O} \text{ (ring)} \longrightarrow \mathrm{ArIO} + \mathrm{>C{=}C(R')(R)} \end{cases} \qquad (101)$$

reactions in the chemical literature and suspect that they will require localized iodonium ylides. In those few known cases where iodonium ylides have been treated with carbon electrophiles, they react through oxygen instead of carbon. Furthermore, the iodonium ylides are subject to thermal decomposition and rearrangement in organic solvents, and it is not likely that their reactions with mild electrophiles such as iodomethane and benzaldehyde can be pushed by significantly increasing the reaction temperatures.

1. *Brønsted acids*

The reactions of a variety of iodonium ylides with Brønsted acids have been investigated and some selected examples follow. With some ylides, 'stable' iodonium salts are obtained if the reaction conditions are not too severe. It has already been noted that phenyldimedonyliodone (**91**) reacts with hydrogen chloride in ether/chloroform at 0°C to give phenyl(2-dimedonyl)iodonium chloride, but, in refluxing concentrated hydrochloric acid, it cleaves, to iodobenzene and 2-chlorodimedone[112]. It has also been converted to phenyl(dimedonyl)iodonium nitrate by reaction with fuming nitric acid in ether[112] (equation 102). Phenyliodonium dinitromethylide likewise yields

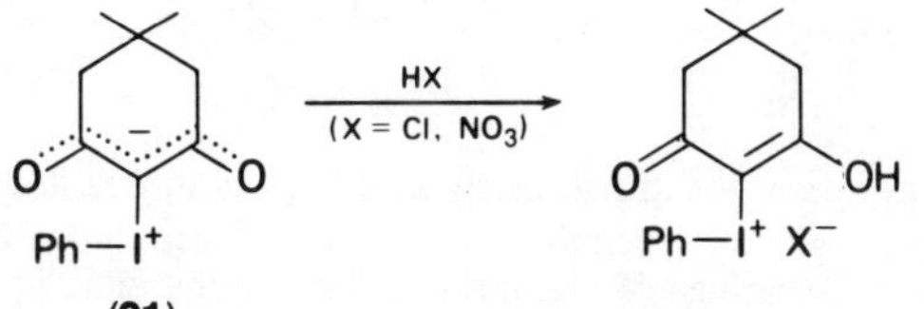

$$\textbf{(91)}\ \xrightarrow[(X = Cl,\ NO_3)]{HX}\ \text{Ph—I}^+\ X^- \text{ salt} \qquad (102)$$

phenyl(dinitromethyl)iodonium chloride and fluorosulphate upon treatment with hydrogen chloride and fluorosulphuric acids in ether, and the *p*-nitro analogue has been converted to the corresponding iodonium chloride (equation 103)[138]. The indole

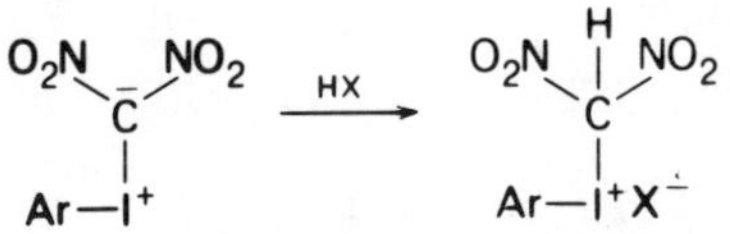

$$Ar\text{—}I^+\text{—}\bar{C}(NO_2)_2 \xrightarrow{HX} Ar\text{—}I^+X^-\text{—}CH(NO_2)_2 \qquad (103)$$

$Ar = C_6H_5,\ X = Cl^-,\ FSO_3^-,\ Ar = 4\text{-}NO_2C_6H_4,\ X = Cl^-$

iodonium ylide (**124**) reacts with trifluoroacetic and *p*-toluenesulphonic acids to give the corresponding iodonium trifluoroacetate and tosylate. However, when those iodonium salts are treated with the more nucleophilic fluoride, chloride, bromide and iodide ions, iodobenzene is displaced and the corresponding 3-haloindoles are obtained[158] (equation 104). With some iodonium ylides, direct cleavage reactions with

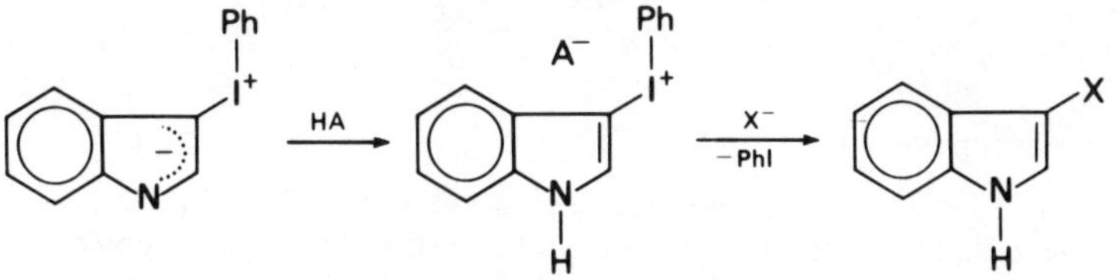

$$\textbf{(124)} \xrightarrow{HA} A^- \text{ salt} \xrightarrow[-PhI]{X^-} \text{3-X-indole} \qquad (104)$$

$A^- = CF_3CO_2^-,\ OTs^-$ $\qquad X = F, Cl, Br, I$

Brønsted acids are observed. For example, phenyliodonium dibenzoylmethylide affords substituted dibenzoylmethanes with hydrochloric, trichloroacetic and *p*-nitrobenzoic acids in methanol[116] (equation 105). The barbiturate ylide **98** is simi-

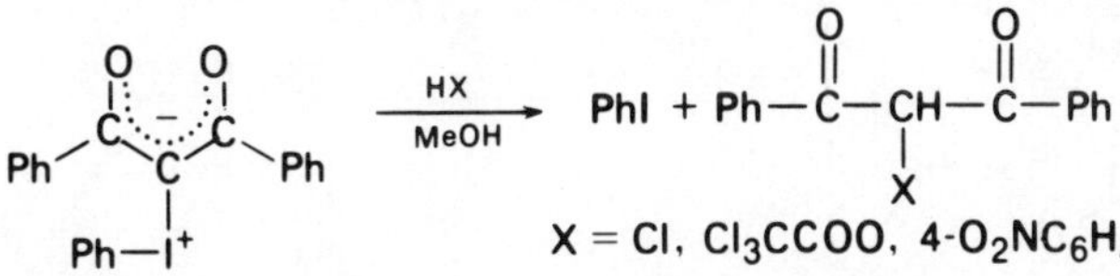

$$PhC(O)\bar{C}(I^+Ph)C(O)Ph \xrightarrow[MeOH]{HX} PhI + Ph\text{—}C(O)\text{—}CH(X)\text{—}C(O)\text{—}Ph \qquad (105)$$

$X = Cl,\ Cl_3CCOO,\ 4\text{-}O_2NC_6H_4COO$

larly converted to halobarbituric acids with hydrochloric and hydrobromic acids in ethanol[121] (equation 106). The *p*-toluenesulphonate salts of amino acids have been observed to react with iodonium ylides in one of two ways, the aliphatic acids promoting ylide cleavage via the carboxyl function and the aromatic acids causing cleavage

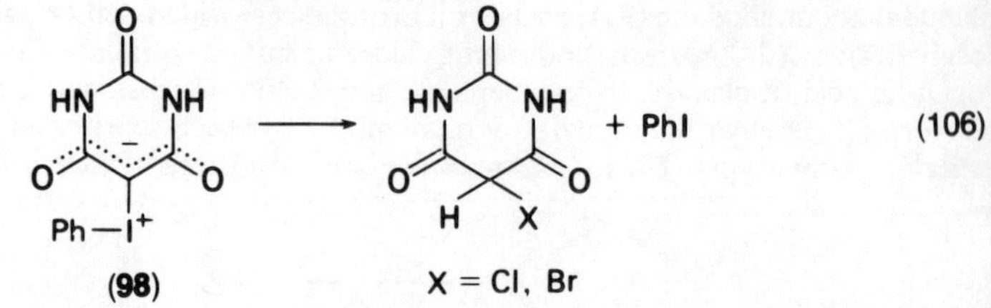

$$\text{(98)} \longrightarrow \text{5-H-5-X-barbituric acid} + PhI \quad X = Cl, Br \tag{106}$$

through the amino group[159,160]. Both of these reaction modes are shown in equation (107) for the cleavage of phenyliodonium dicarbomethoxymethylide with the ammonium tosylate derivatives of alanine and *o*-aminobenzoic acid.

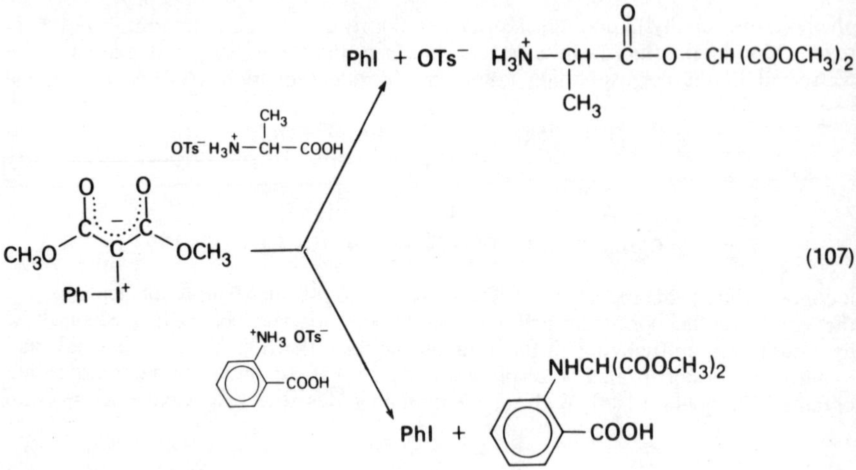

$$(CH_3OCO)_2C^{-}-I^{+}Ph \xrightarrow{OTs^{-}\,H_3\overset{+}{N}-CH(CH_3)-COOH} PhI + OTs^{-}\; H_3\overset{+}{N}-CH(CH_3)-C(=O)-O-CH(COOCH_3)_2$$

$$(CH_3OCO)_2C^{-}-I^{+}Ph \xrightarrow{o\text{-}(^{+}NH_3\,OTs^{-})C_6H_4COOH} PhI + o\text{-}(NHCH(COOCH_3)_2)C_6H_4COOH \tag{107}$$

2. *Carbon electrophiles*

The action of electrophilic carbon species on iodonium ylides has, thus far, been little investigated. The ylides **91** and **94** have been observed to undergo *O*-acylation and *O*-alkylation reactions with benzoyl chloride and triethyloxonium tetrafluoroborate eventuating in stable iodonium salts[117,161] (equation 108). No examples of direct

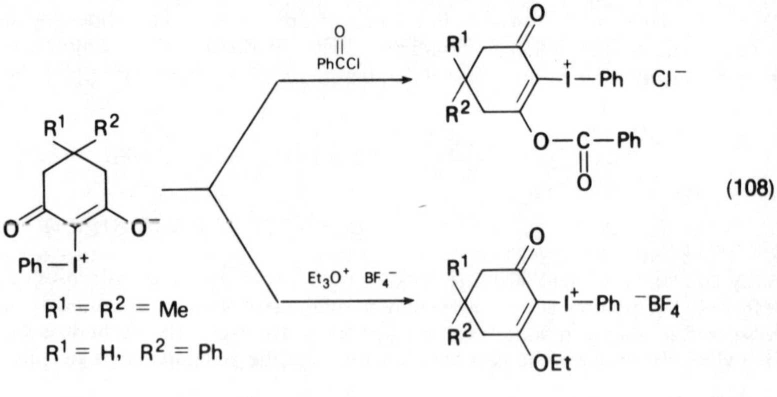

$$\text{ylide} \xrightarrow{PhCCl(=O)} \text{3-(benzoyloxy)-2-(}\overset{+}{I}-Ph\text{)cyclohexenone } Cl^{-}$$

$$\text{ylide} \xrightarrow{Et_3O^{+}\,BF_4^{-}} \text{3-OEt-2-(}\overset{+}{I}-Ph\text{)cyclohexenone } ^{-}BF_4 \tag{108}$$

$R^1 = R^2 = Me$

$R^1 = H,\ R^2 = Ph$

acylation and alkylation of iodonium ylides at the ylidic carbon, however, to our knowledge, have yet been published.

Diphenylketene reacts smoothly with **91** in dichloromethane to give iodobenzene, phenyl 2-iododimedonyl ether, the ketene acetal **141** and the lactone **142**[162]. This reaction very likely proceeds by initial nucleophilic attack of the ylide oxygen atom at the highly electrophilic carbonyl group of diphenylketene to give an intermediate iodonium betaine. Formal displacement of iodobenzene from vinyl carbon in the betaine either through the oxygen atom or the carbon atom of the carbanion moiety would give the observed products. Although several polar mechanisms may be envisioned for the displacement process, a one-electron transfer (ET) from the carbanion centre of the betaine to the iodonium centre and subsequent homolytic collapse of the resulting diradical species is an attractive alternative[162] (equation 109).

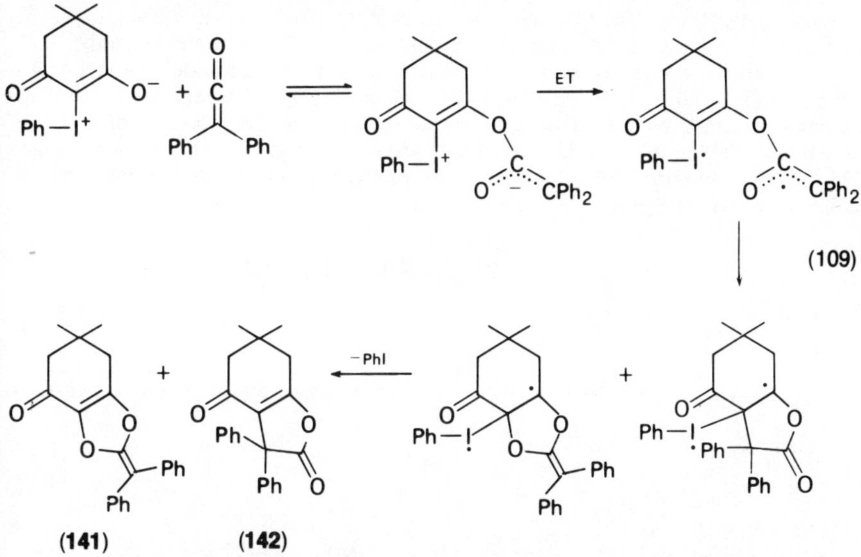

(109)

(141) **(142)**

It is noteworthy that 2-diazodimedone and 2-dimethylsulphoniodimedonate are *inert* to diphenylketene under the same reaction conditions, a fact which may reflect the higher reduction potentials of the diazonium and dimethylsulphonium functions in the intermediate betaines. The phenylation of a variety of carbanions with diaryliodonium salts has been observed and is believed to involve a similar electron-transfer process[163–167].

3. *Electrophilic halogen species*

Phenyliodonium dinitromethylide reacts with various electrophilic halogen species with loss of iodobenzene[138]. With molecular bromine and chlorine in ether, dihalodinitromethanes are obtained in high yields (equation 110). The same products are formed when the ylide is treated with *N*-bromo- and *N*-chlorosuccinimides. These reactions very probably proceed by the initial formation of iodonium salts and their

$$Ph-\overset{+}{I}-\bar{C}(NO_2)_2 \xrightarrow{X_2/Et_2O} PhI + X_2C(NO_2)_2 \qquad (110)$$

$X = Br, Cl$

$$O_2N{-}\overset{NO_2}{\underset{Ph-I^+}{C}}{:}^- + X{-}X \longrightarrow O_2N{-}\overset{NO_2}{\underset{Ph-I^+}{C}}{-}X\ X{:}^- \begin{cases} \xrightarrow[\text{cleavage}]{\text{Nucleophilic}} X_2C(NO_2)_2 + PhI \\ \xrightarrow[\text{Electrophilic cleavage}]{X_2} X_2C(NO_2)_2 + [PhIX_2] \end{cases} \quad (111)$$

subsequent decomposition under the reaction conditions by nucleophilic *and* electrophilic cleavage of the carbon (sp^3)–iodine bond (equation 111). The electrophilic cleavage process would presumably involve the formation of (dihaloiodo)benzenes as the *initially* formed by-products, although none were isolated. Evidence that the electrophilic heterolysis of the carbon–iodine bond of the iodonium salt is a viable process is provided by studies on the stable fluorosulphate salt. For example, when the salt is treated with tetramethylammonium bromide in acetonitrile, bromodinitromethane is obtained in 100% yield along with iodobenzene (100%). Cleavage with molecular bromine in ether on the other hand proceeds with the formation of bromodinitromethane (27%), dibromodinitromethane (67%) and iodobenzene. With hydrogen chloride in acetonitrile, the fluorosulphate salt gives only dinitromethane and (dichloroiodo)benzene (equation 112).

$$\underset{FSO_3^-}{Ph\overset{+}{I}{-}CH(NO_2)_2} \begin{cases} \xrightarrow{Me_4N^+\ Br^-} BrCH(NO_2)_2 + PhI \\ \xrightarrow[Et_2O]{Br_2} BrCH(NO_2)_2 + Br_2C(NO_2)_2 + PhI \\ \xrightarrow[CH_3CN]{HCl} H_2C(NO_2)_2 + PhICl_2 \end{cases} \quad (112)$$

Consistent with this interpretation is the fact that phenylsulphenyl chloride reacts with phenyliodonium dinitromethylide to give both chloro(thiophenyl)dinitromethane (45%) and bis(thiophenyl)dinitromethane (20%). The ylide likewise reacts with hydrogen chloride in acetonitrile to give an 85% yield of dinitromethane (equation 113).

$$Ph\overset{+}{I}{-}\overset{-}{C}(NO_2)_2 \begin{cases} \xrightarrow[Et_2O]{PhSCl} \underset{Cl^-}{Ph\overset{+}{I}{-}\overset{SPh}{C}(NO_2)_2} \xrightarrow[-PhI]{PhSCl} (PhS)_2C(NO_2)_2 + Cl(PhS)C(NO_2)_2 \\ \xrightarrow[CH_3CN]{HCl} \underset{Cl^-}{Ph\overset{+}{I}{-}CH(NO_2)_2} \xrightarrow{HCl} H_2C(NO_2)_2 \end{cases} \quad (113)$$

D. Transylidation Reactions

The ability of iodonium ylides to undergo transylidation reactions with nucleophilic species appears to be fairly general. Most investigations of such reactions have so far focused on the generation of nitrogen and sulphur ylides, particularly those in the pyridinium, sulphonium and thiocarbonyl subclasses, but a few phosphonium and arsonium ylides have also been prepared. The limitation seems to be only that transylidations with other heteroatom species have not been tried. In some cases, transylidation reactions of iodonium ylides require a catalyst (e.g. TsOH, Cu(II)), but most proceed with reasonable efficiency without one. In Table 9, a number of these reactions are summarized along with appropriate literature references.

Among the known aryliodonium cyclopentadienylides, the bis(carboethoxy)-bis(cyano) derivative **105** has been studied most extensively. Direct transylidations with such diverse nucleophiles as cyclic and acyclic sulphides, cyclic and acyclic thioureas,

TABLE 9. Transylidation reactions of iodonium ylides

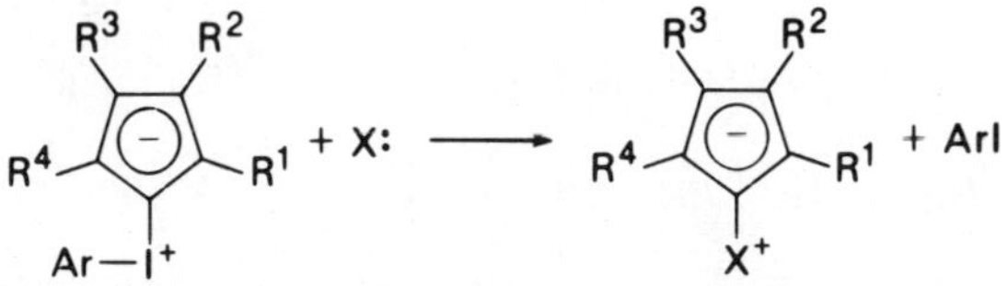

Aryl group	R^1	R^2	R^3	R^4	X:	Yield, %	Ref.
Phenyl	COOEt	CN	CN	COOEt	Ph_2S	61	168
Phenyl	COOEt	CN	CN	COOEt	$(p\text{-}MeOC_6H_4)_2S$	94	168
Phenyl	COOEt	CN	CN	COOEt	PhSMe	77	168
Phenyl	COOEt	CN	CN	COOEt	Et_2S	91	168
Phenyl	COOEt	CN	CN	COOEt	$(n\text{-}Pr)_2S$	91	168
Phenyl	COOEt	CN	CN	COOEt	 S	76	168
Phenyl	COOEt	CN	CN	COOEt	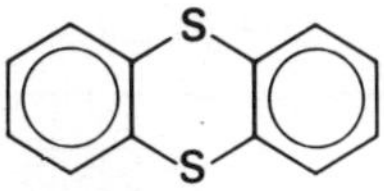 S S	67	168
Phenyl	COOEt	CN	CN	COOEt	$(Ph)_2Se$	24	168
Phenyl	COOEt	CN	CN	COOEt	$(H_2N)_2C{=}S$	47	168
Phenyl	COOEt	CN	CN	COOEt	$(Me_2N)_2C{=}S$	75	168
Phenyl	COOEt	CN	CN	COOEt	$(Ph)_3P$	65	168
Phenyl	COOEt	CN	CN	COOEt	$(Ph)_3As$	90	168
Mesityl	COOEt	CN	CN	COOEt	$(n\text{-}Pr)_2S$	98	168
p-Anisyl	COOEt	CN	CN	COOEt	$(n\text{-}Pr)_2S$	86	168
p-Nitrophenyl	COOEt	CN	CN	COOEt	$(n\text{-}Pr)_2S$	27	168
p-Anisyl	COOEt	CN	CN	COOEt	$(Ph)_3P$	68	168
p-Anisyl	COOEt	CN	CN	COOEt	$(Ph)_3As$	87	168
p-Nitrophenyl	COOEt	CN	CN	COOEt	$(Ph)_3As$	88	168
Phenyl	H	CN	CN	CN	$(H_2N_2)_2C{=}S$	8	169
Phenyl	CN	H	CN	CN	$(MeNH)_2C{=}S$	40	169
Phenyl	CN	H	CN	CN	$(Me_2N)_2C{=}S$	46	169

TABLE 9. (*Continued*)

Aryl group	R^1	R^2	R^3	R^4	X:	Yield, %	Ref.
Phenyl	CN	H	CN	CN	1-H-3-Me-imidazolidine-2-thione (H–N, N–Me, C=S)	49	169
Phenyl	CN	H	CN	CN	1,3-Me₂-imidazolidine-2-thione (Me–N, N–Me, C=S)	1.5	169
Phenyl	CN	H	CN	CN	1,3-Me₂-benzimidazole-2-thione (Me–N, N–Me, C=S)	75	169
Phenyl	H	CN	CN	CN	$(i\text{-Pr})_2N$, $(i\text{-Pr})_2N$ cyclopropenethione (C=S)	71	171
Phenyl	CN	H	CN	CN	$(i\text{-Pr})_2N$, $(i\text{-Pr})_2N$ cyclopropenethione (C=S)	86	171

$$\mathrm{Ar{-}I^{+}{-}\bar{C}(R^1)(R^2)} + \mathrm{X{:}} \longrightarrow \mathrm{ArI} + \mathrm{\overset{+}{X}{-}\bar{C}(R^2)(R^1)}$$

Aryl group	R^1	R^2	X:	Yield, %	Catalyst	Ref.
Phenyl	CN	COPh	Pyridine	58		140
Phenyl	CN	COPh	Quinoline	20		140
Phenyl	CN	COPh	Isoquinoline	83		140
Phenyl	NO_2	NO_2	Me_2S	3	79% with Cu(*acac*)$_2$	172
Phenyl	NO_2	COPh	Me_2S	70		172
Phenyl	NO_2	COOMe	Me_2S	25		172
Phenyl	NO_2	NO_2	Pyridine	20		172
Phenyl	NO_2	NO_2	$(H_2N)_2C{=}S$	85		172
p-Nitrophenyl	NO_2	NO_2	Me_2S	21.5		172
p-Nitrophenyl	NO_2	COPh	Me_2S	50		172
p-Nitrophenyl	NO_2	COOMe	Me_2S	18		172
p-Nitrophenyl	NO_2	NO_2	Pyridine	20		172
p-Anisyl	COOMe	COOMe	Pyridine	61	Cu(*acac*)$_2$	125
p-Anisyl	COOEt	COOEt	Pyridine	91	Cu(*acac*)$_2$	125

TABLE 9. (*Continued*)

Aryl group	R^1	R^2	X:	Yield, %	Catalyst	Ref.
p-Anisyl	COMe	COPh	Pyridine	62	Cu(*acac*)$_2$	125
p-Anisyl	COMe	COOEt	Pyridine	74	Cu(*acac*)$_2$	125
p-Anisyl	COPh	COPh	Pyridine	61	Cu(*acac*)$_2$	125
p-Anisyl	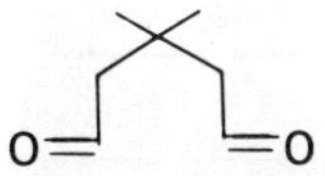 O=	=O	Pyridine	70	Cu(*acac*)$_2$	125
Phenyl	COOMe	COOMe	Pyridine	48	68% with Cu(*acac*)$_2$	156
Phenyl	COOEt	COOEt	Pyridine	48	80% with Cu(*acac*)$_2$	156
Phenyl	COMe	COOMe	Pyridine	68	79% with Cu(*acac*)$_2$	156
Phenyl	COMe	COOEt	Pyridine	56	75% with Cu(*acac*)$_2$	156
Phenyl	COPh	COPh	Pyridine	13	55% with Cu(*acac*)$_2$	156
Phenyl	COPh	COMe	Pyridine	28	59% with Cu(*acac*)$_2$	156
Phenyl	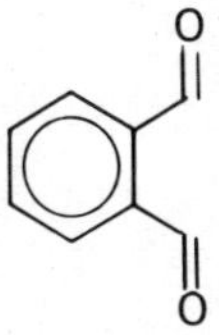 O	O	Pyridine	53	64% with Cu(*acac*)$_2$	156
Phenyl	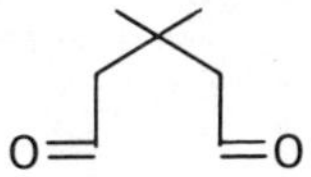 O=	=O	Pyridine	0	40.5% with Cu(*acac*)$_2$	156
Phenyl	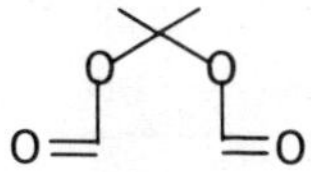 O= O	O =O	Me_2S	44	TsOH	120
Phenyl	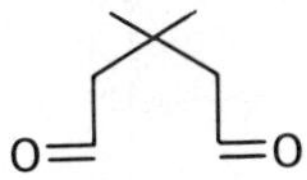 O=	=O	Quinoline	45	CuCl	157
Phenyl	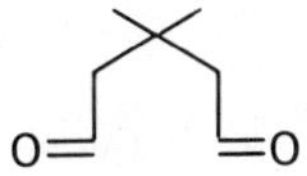 O=	=O	Me_2S	75	TsOH	157
Phenyl	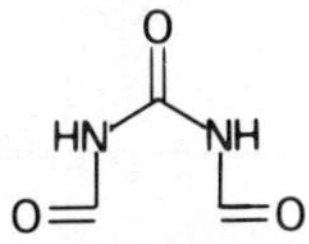 O; HN; O=	NH; =O	Me_2S	80	TsOH	121
Phenyl	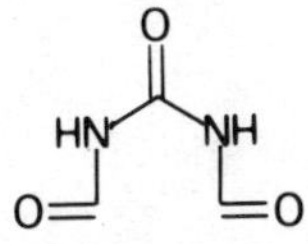 O; HN; O=	NH; =O	Pyridine	70	TsOH	121

TABLE 9. (*Continued*)

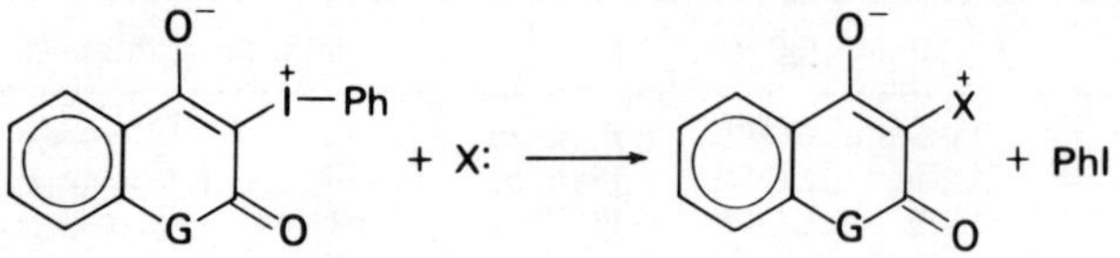

G	X:	Yield, %	Catalyst	Ref.
O	Pyridine	25	TsOH	143
NH	Pyridine	42		143
NPh	Pyridine	38		143
NPh	Isoquinoline	90	TsOH	143

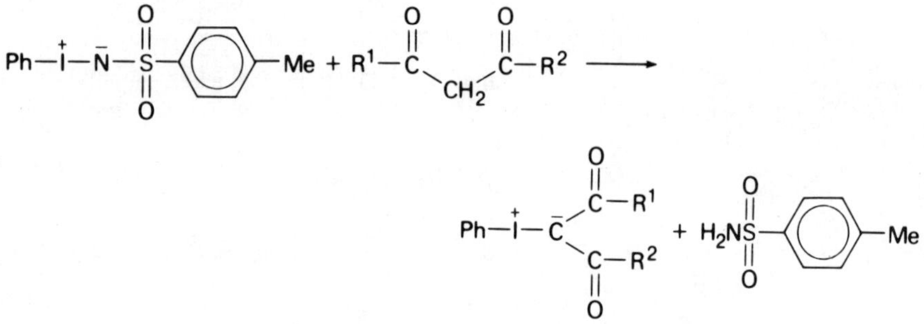

R^1	R^2	Yield, %	Ref.
Me	Phenyl	70	150
Phenyl	Phenyl	65	150
Me	OEt	47	150
CF_3	Phenyl	80	150
CF_3	α-Thienyl	48	150
Me	Me		150
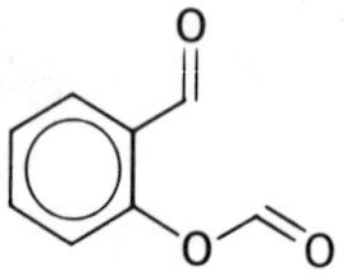		77	150
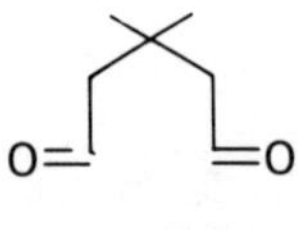		76	150
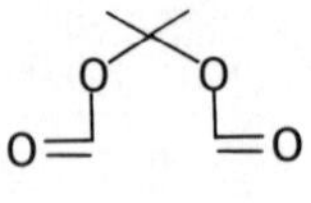		85	150

TABLE 9. (*Continued*)

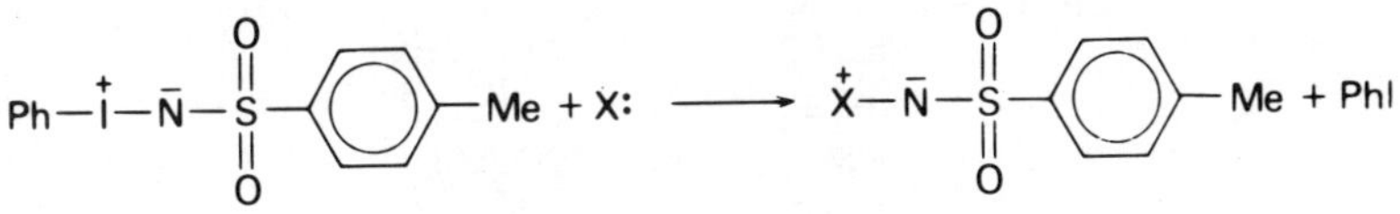

$$\mathrm{Ph{-}\overset{+}{I}{-}\bar{N}{-}SO_2{-}C_6H_4{-}Me + X\!: \longrightarrow \overset{+}{X}{-}\bar{N}{-}SO_2{-}C_6H_4{-}Me + PhI}$$

X:	Yield, %	Ref.
PhSMe	49	145
$(Ph)_3P$	69	145
$Me_2S{=}O$	100	145

diphenylselenide, triphenylphosphine and triphenylarsine having been observed, a few of which are illustrated in equation (114)[168].

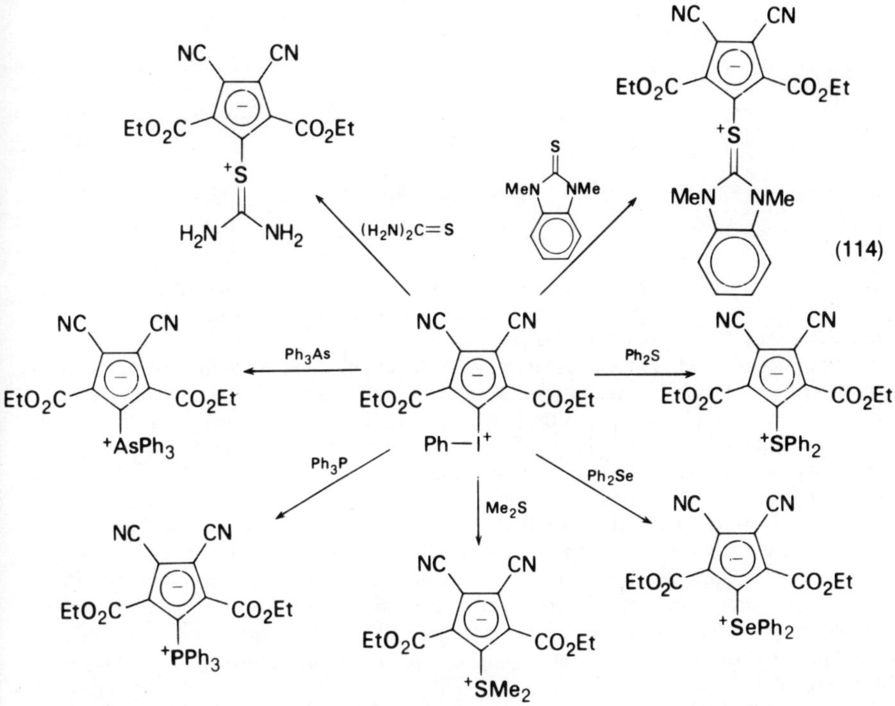

(114)

Aryliodonium ylides derived from α-cyano- and α-acetylacetophenone, methyl and ethyl acetoacetate, dimethyl and diethyl malonate, 1,3-indanedione, dimedone and barbituric acid react with pyridine to give the corresponding pyridinium ylides[121,140,156] (equation 115). Although a catalyst is not essential for most of these reactions, the presence of copper acetylacetonate in catalytic amounts does afford higher pyridinium ylide yields, in a few cases dramatically so. For example, product yields of 13, 28 and 0% respectively were observed in the direct pyridine transylidations of phenyliodonium dibenzoylmethylide, phenyliodonium acetylbenzoylmethylide and phenyldimedonyliodone respectively, in refluxing 1,2-dichloroethane. However, in the pres-

$$\text{Ar}\overset{+}{\text{I}}-\overset{-}{\text{C}}\text{R}^1\text{R}^2 + \text{C}_5\text{H}_5\text{N} \longrightarrow \text{ArI} + \text{C}_5\text{H}_5\text{N}^+-\overset{-}{\text{C}}\text{R}^1\text{R}^2 \qquad (115)$$

R^1	R^2
CN	COPh
COMe	COPh
COOR	COMe
COMe	COOR
CPh	COPh
NO_2	NO_2
(phthaloyl, C_6H_4(C=O)(C=O))	
(dimedone, $O=C-CH_2-CMe_2-CH_2-C=O$)	
(barbituric, $O=C-NH-CMe_2-NH-C=O$)	

ence of 0.1 mol % of the copper catalyst, those yields were increased to 55, 59 and 40.5%. It seems plausible that these reactions proceed by nucleophilic capture of initially generated copper carbenoids by pyridine[156].

The effect of substituents attached to the aromatic nucleus of the iodonium ylides on the efficiency of their transylidation reactions has not been studied in a systematic way. It has been found, however, that the copper(II)-mediated reactions of various *p*-methoxyphenyliodonium ylides afford pyridinium ylides in approximately the same yields as those obtained with their phenyliodonium counterparts[125]. Phenyliodonium dinitromethylide and its *p*-nitro analogue likewise react with pyridine to give the same yield of pyridinium dinitromethylide[172]. It has already been noted that phenyldimedonyliodone isomerizes to phenyl 2-iododimedonyl ether in pyridine and that copper acetonylacetonate diverts this reaction to the transylidation manifold. However, *o*-nitrophenyldimedonyliodone and *p*-nitrophenyldimedonyliodone rearrange to the corresponding ethers even in the presence of a copper catalyst[122]. The *m*-nitro analogue, on the other hand, gives mostly pyridinium ylide under the influence of copper(II).

Although pyridine has been most commonly employed as the nitrogen component in the preparation of nitrogen ylides from iodonium ylides, quinoline, isoquinoline, 4-cyanopyridine and 4,4-dipyridyl have also been utilized[173].

The propensity of the iodonium ylides derived from 2-hydroxycoumarin and carbostyril for intramolecular isomerization has already been discussed. Even so, these ylides are converted to the corresponding pyridinium ylides in pyridine, the former under the catalytic influence of *p*-toluenesulphonic acid[143] (equation 116).

The aryliodonium ylides of dinitromethane, α-nitroacetophenone, methyl nitroacetate, isopropylidene malonate, dimedone and barbituric acid all afford dimethylsulphonium ylides on their metatheses with dimethyl sulphide[120,121,157,172] (equation 117).

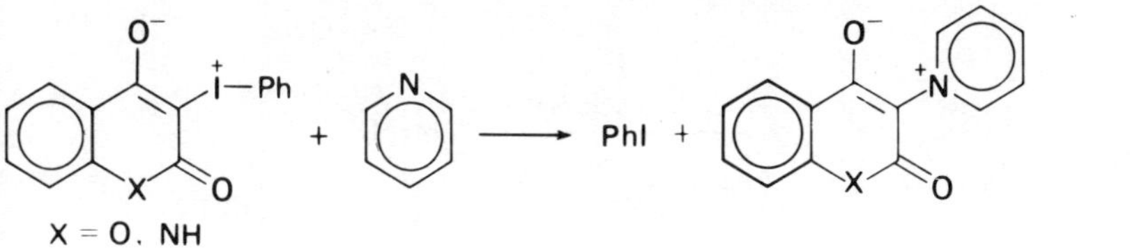

$$\text{(3-phenyliodonio-4-oxido-2-oxo-chromene, X = O, NH)} + \text{pyridine} \longrightarrow PhI + \text{(3-pyridinio-4-oxido-2-oxo-chromene)} \qquad (116)$$

X = O, NH

$$Ar\overset{+}{I}-\overset{-}{C}(R^1)(R^2) + CH_3-S-CH_3 \longrightarrow ArI + (CH_3)_2\overset{+}{S}-\overset{-}{C}(R^1)(R^2) \qquad (117)$$

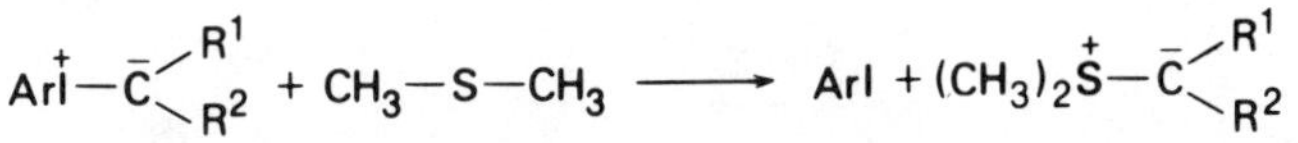

R^1	R^2
NO_2	NO_2
NO_2	$COOCH_3$
NO_2	COPh

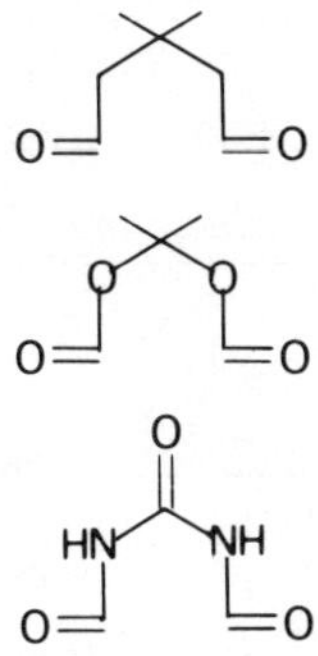

With the cyclic ylides, either *p*-toluenesulphonic acid or copper acetonylacetonate were employed as catalysts.

(*N*-Tosyliminoiodo)benzene has been used to synthesize various iodonium ylides by transylidation reactions with β-dicarbonyl compounds[150]. Two of these, 3-phenyliodonio-1,1,1-trifluoro-4-phenyl-2,4-butanedionate and 3-phenyliodonio-1,1,1-trifluoro-4-thienyl-2,4-butanedionate, shown below, have not been prepared by the standard condensation method from (diacetoxyiodo)benzene. The imino-iodinane also reacts with triphenylphosphine, dimethyl sulphide and dimethyl sulphoxide to give the corresponding iminophosphorane, iminosulphurane and iminodimethylsulphurane oxide[145] (equation 118).

Transylidation reactions of iodonium ylides derived from β-dicarbonyl compounds are not confined to heteronucleophiles, but also proceed with active methylene compounds in methanol solution in the absence of base. Such metathesis reactions allow at least a qualitative assessment of relative iodonium ylide stabilities. For example, when phenyliodonium dibenzoylmethylide is treated with dimedone, phenyldimendonyliodone is obtained[174] (equation 119).

Phenyldimedonyliodone (**91**) reacts with phenyl isothiocyanate in dichloromethane to give phenyl 2-iododimedonyl ether, the spirosulphide **143** and 2-(2-benzothiazolyl)dimedone (**144**), a quite unexpected mixture of products in view of the reactions of **91** with diphenylketene and phenyl isocyanate[162,175]. With the formation of the stable thiocarbonyl ylide **145** from the iodone and thiourea providing an

$$PhI^+-\bar{N}Ts \xrightarrow{CF_3COCH_2COPh} PhI^+-\bar{C}(COCF_3)(COPh)$$

$$PhI^+-\bar{N}Ts \xrightarrow{CF_3COCH_2CO(2\text{-thienyl})} PhI^+-\bar{C}(CO(2\text{-thienyl}))(COCF_3)$$

$$PhI^+-\bar{N}Ts \xrightarrow{Ph_3P} Ph_3P{=}NTs$$

$$PhI^+-\bar{N}Ts \xrightarrow{Me_2S} Me_2S{=}NTs$$

$$PhI^+-\bar{N}Ts \xrightarrow{Me_2S{=}O} Me_2S(O){=}NTs \quad (118)$$

$$PhI^+-\bar{C}(COPh)_2 + \text{5,5-dimethylcyclohexane-1,3-dione} \xrightarrow{CH_3OH} PhCOCH_2COPh + \text{(91)} \quad (119)$$

(120)

(91) $\xrightarrow{PhN=C=S}$ 2-iodo-3-phenoxy-5,5-dimethylcyclohex-2-enone 8% + (143) 14% + (144) 36%

(91) $\xrightarrow{(H_2N)_2C=S}$ (145) 67%

Ph—I⁺

(**91**)

(**143**)

14%

HO

(**144**)

36%

H_2N NH_2

(**145**)

67%

analogy (equation 120), it seems likely that this reaction proceeds by the initial transylidation of **91** with the heterocumulene to give the intermediate ylide **146**, which collapses either by intramolecular cyclization to give the benzothiazole or by cleavage of phenyl isocyanide to give the thio derivative **147**. Transylidation of **147** with another molecule of phenyldimedonyliodone to give the thiocarbonyl ylide **148** and cyclization of that ylide would afford the spiro sulphide **143** (equation 121). The proposed cyclization of **148** should be much more favourable than the corresponding cyclization of **145** since, in **148**, the positive charge is highly localized on the sulphur atom.

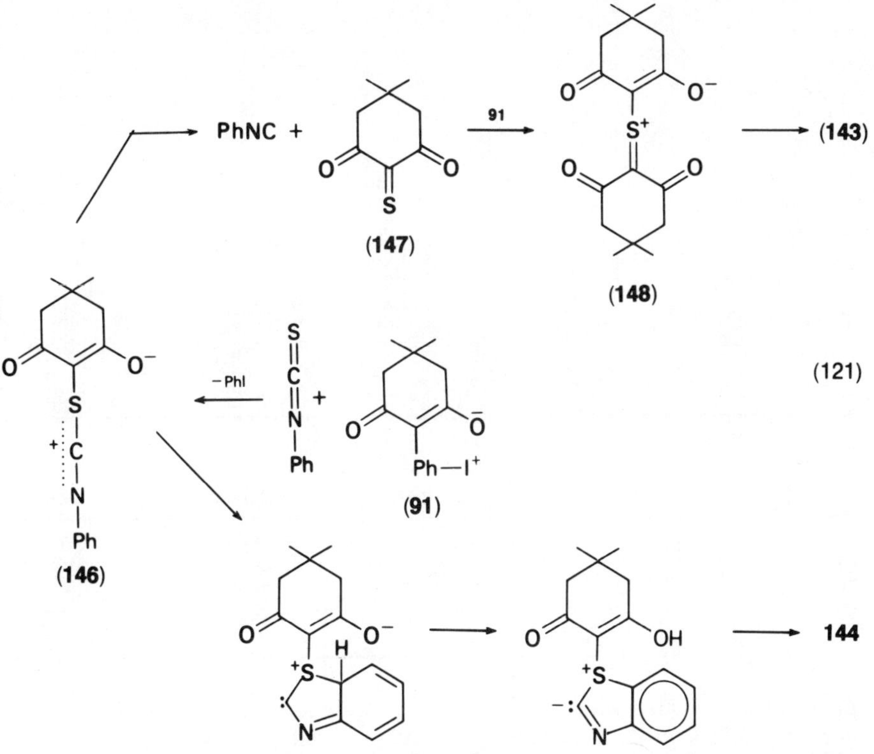

(121)

An attempt to prepare authentic **147** by the action of hydrogen sulphide on **91** was unsuccessful, but the spiro sulphide **143** was isolated in 40% yield. Methyl isothiocyanate likewise reacts with **91** in dichloromethane to give a 26% yield of the spiro sulphide and a 36% yield of phenyl 2-iododimedonyl ether[175].

E. Reductive Cleavage

Phenyldimedonyliodone reacts rapidly with various thiophenols in dichloromethane at ice-bath temperature[176]. With thiophenol, the dominant reaction mode is that of oxidation–reduction, as evidenced by the formation of iodobenzene (99%), diphenyl disulphide (73%), and dimedone (70%). However, phenyl 2-dimedonyl sulphide, a product of substitution, is also formed in 15% yield (equation 122). When electron-withdrawing groups are present in the *para* position of the thiophenols, the yield of substitution product decreases while the yield of oxidation product (ArSSR) increases,

$$\text{(122)}$$

R	Yields, %			
	Iodobenzene	Disulphide	Dimedone	Sulphide
NO_2	97	80.5	43	0
Cl	99	82.5	86	10
H	99	73	70	15
CH_3	100	72	62	32
OCH_3	94	65	46	43

the reverse being true with electron-donating substituents. The yields of dimedone, however, do not conform to any obvious pattern (equation 122).

These results have been rationalized by a mechanism involving initial protonation of **91** by the thiophenol to give a *hydrogen-bonded* ion pair and subsequent one-electron transfer from the thiophenolate ion to the iodonium ion to give a *hydrogen-bonded* radical pair. Collapse of the radical pair would ultimately afford the product of substitution while radical diffusion into the bulk solvent would eventuate in diphenyl disulphide. Electron-withdrawing substituents should weaken the hydrogen bond and permit diffusion to compete with substitutive collapse (equation 123).

That the simultaneous presence of the phenyl 2-dimedonyliodonium ion and the thiophenoxide ion is a prerequisite for electron transfer is indicated by the following control studies: (1) when the iodonium ylide is treated with thioanisole, a model for thiophenol, no reaction occurs; (2) when phenyl 2-(3-ethoxy-5,5-dimethyl-2-cyclohexanonyl)iodonium tetrafluoroborate, a model for the protonated ylide, is treated with thiophenol, again no reaction occurs; (3) when the iodonium tetrafluoroborate is mixed with sodium thiophenoxide, a rapid reaction ensues and affords the mixture of products shown in equation (124).

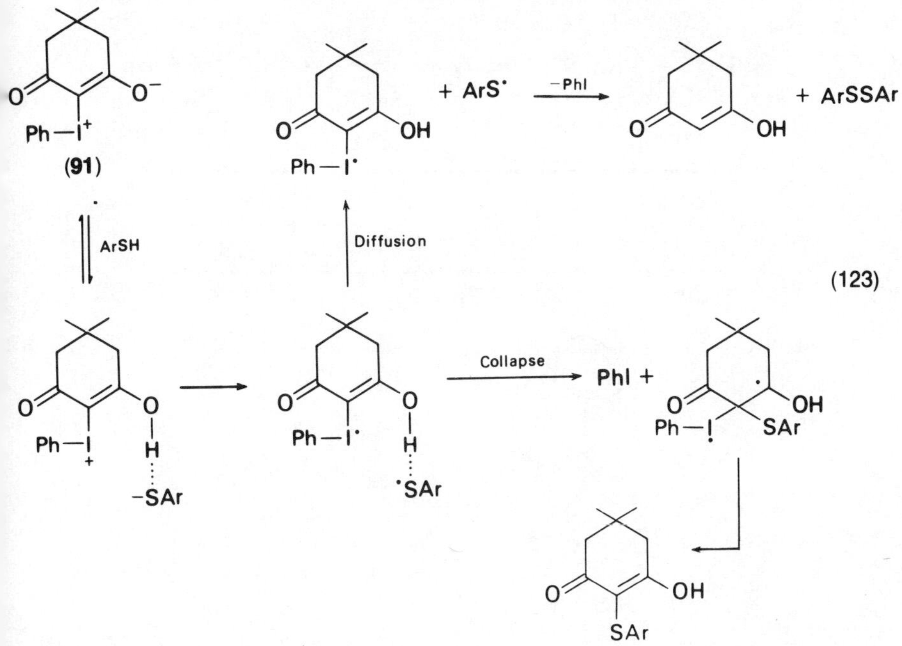

(123)

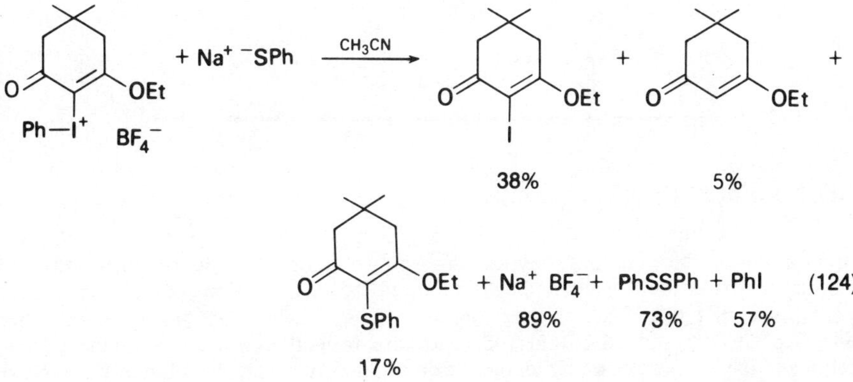

(124)

It is interesting to note that neither 2-diazodimedone nor 2-(dimethylsulphonio)dimedonate reacts under the same conditions with thiophenol. It may be that these compounds are not sufficiently basic to provide threshold concentrations of the conjugate acids via protonation by thiophenol or that the reduction potentials of the conjugate acids are too high to permit electron transfer.

The reactions of **91** with methanethiol and hydrogen sulphide have been studied in some detail and exhibit reactivity patterns similar to those found with the thiophenols[176].

Several iodonium ylides have been observed to undergo efficient reductive cleavage reactions with sulphurous acid or sodium bisulphite in aqueous ethanol. The reactions and product yields are summarized below[177].

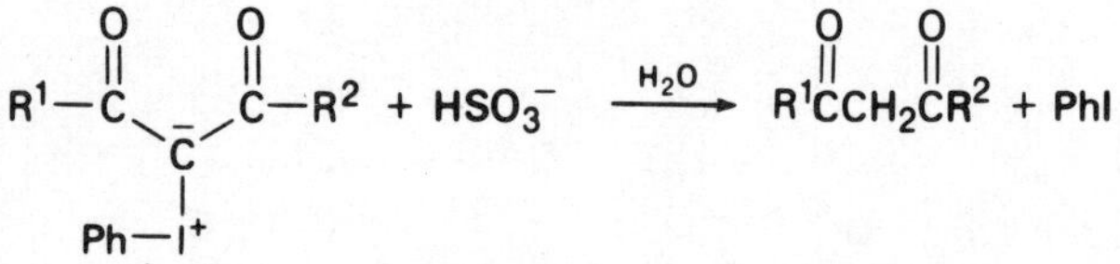

$$R^1-\overset{O}{\overset{\|}{C}}-\underset{Ph-I^+}{\underset{|}{\bar{C}}}-\overset{O}{\overset{\|}{C}}-R^2 + HSO_3^- \xrightarrow{H_2O} R^1\overset{O}{\overset{\|}{C}}CH_2\overset{O}{\overset{\|}{C}}R^2 + PhI$$

R^1	R^2	Yield, %	
		$NaHSO_3$	H_2SO_3
C_6H_5	C_6H_5	73	68
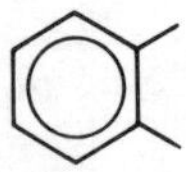		63	
		90	81
O, HN, NH			98
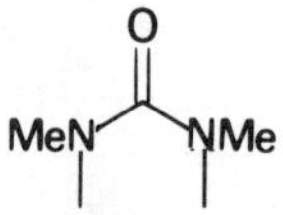 O, MeN, NMe			92

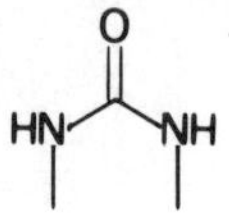

F. Molecular Structure

Among the known halonium ylides, single crystal X-ray structures of phenylbromonium 3,4-dicyanoimidazolylide (**86**)[178], 2-(*o*-chlorophenyliodonium)-4-nitro-6-chlorophenoxide (**102**)[127] and phenyliodonium 2,5-dicarboethoxy-3,4-dicyanocyclopentadienylide (**105**)[179] have been published. The C—$\overset{+}{X}$—C bond angles are 99.3° in **86**, 97.8° in **102** and 98.65° in **105**. They are roughly consistent with the expectations of hypervalent bonding theory which predicts the existence of highly polarized [2c–4e] bonds between the halogen atoms and the divalent 'ylidene' ligands, these

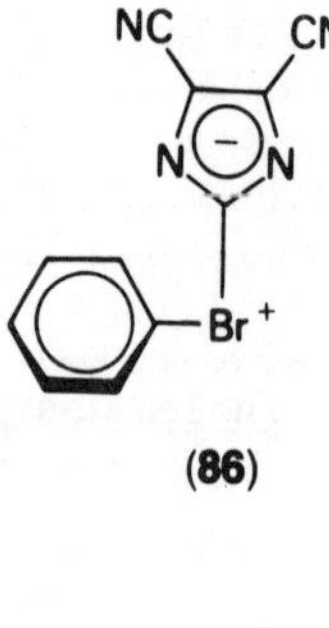

NC CN N N Br$^+$

(**86**)

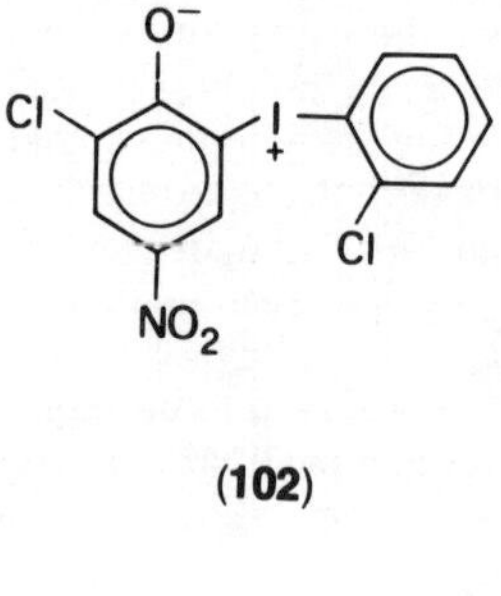

O$^-$ Cl I$^+$ Cl NO$_2$

(**102**)

NC CN EtO$_2$C CO$_2$Et I$^+$

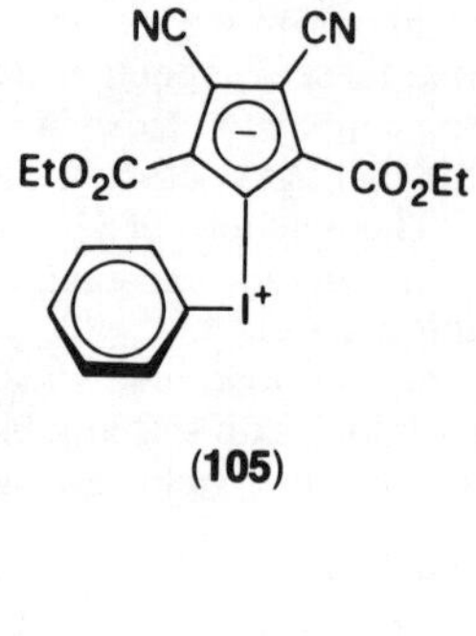

(**105**)

being at right angles to the covalent bonds between the halogen atoms and the aryl ligands. The bond distances between the ylidic carbon and the halogen atom are 1.89 Å in **86**, 2.08(1) Å in **102** and 2.078(5) Å in **105** and may be compared with the carbon–iodine and carbon–bromine distances of 2.10 Å and 1.91 Å computed from single covalent radii. Thus, very little carbon–halogen bond shortening in halonium ylides is evident, and the dipolar structure formulations shown seem to be much more likely than canonical structures exhibiting carbon–halogen double bonds. The five-membered carbocyclic ring in ylide **105** is very nearly a regular pentagon, and judging from the observed carbon–nitrogen and carbon–oxygen bond distances, there seems to be little delocalization of negative charge onto the cyano and carboethoxy functions. Significant charge delocalization onto the cyano groups in **86** is likewise not indicated. The carbon–oxygen bond distance in **102** (1.26 Å) was observed to be shorter than the average carbon–oxygen bond distance (1.36 Å) for a sampling of phenols, indicative of some conjugation of the oxyanion function with the ring. The geometric relationship of the ring planes in **102** was not specified, but in ylides **86** and **105** the five- and six-membered rings are very definitely not coplaner. Finally, the $C_{(1)}—C_{(2)}—I$ angle in **102** is 104.2°, this compression from the ideal angle of 120° probably manifesting some electrostatic attraction between the oxyanion and iodonium centres.

Ultraviolet and infrared measurements on a number of iodonium ylides are described in some of the papers already cited and are generally consistent with dipolar ylide structures in which the 'carbanion' centres are delocalized[180].

IX. REFERENCES

1. C. Willgerodt, *J. Prakt. Chem.*, **33**, 154 (1886).
2. C. Willgerodt, *Chem. Ber.*, **25**, 3494 (1892).
3. C. Hartmann and V. Meyer, *Chem. Ber.*, **27**, 426 (1894).
4. C. Willgerodt, *Die Organischen Verbindungen mit Mehrwertigem Jod*, F. Enke, Stuttgart (1914).
5. R. B. Sandin, *Chem. Rev.*, **32**, 249 (1943).
6. D. F. Banks, *Chem. Rev.*, **66**, 243 (1966).
7. A. J. Downs and C. J. Adams in *Comprehensive Inorganic Chemistry*, Vol. 2 (Executive Ed. A. F. Trotman-Dickenson), Pergamon Press, Oxford (1973), pp. 1564–1573.
8. F. M. Beringer and E. M. Gindler, *Iodine Abstr. Rev.*, **3**, 1956.
9. (a) G. A. Olah, *Halonium Ions*, John Wiley and Sons, New York (1975).
   (b) A. Varvgolis, *Chem. Soc. Rev.*, **10** (3), 377 (1981).
10. R. L. Amey and J. C. Martin, *J. Amer. Chem. Soc.*, **100**, 300 (1978).
11. Reference 7, pp. 1487, 1491.
12. K. K. Verma, J. Ahmed, M. P. Sahasrabuddhey and S. Bose, *J. Indian Chem. Soc.*, **54**, 699 (1977).
13. G. Wittig and M. Rieber, *Ann. Chem.*, **562**, 187 (1949).
14. G. Wittig and K. Clauss, *Ann. Chem.*, **578**, 136 (1952).
15. F. Fichter and P. Lotter, *Helv. Chim. Acta*, **8**, 438 (1925).
16. L. L. Miller and A. K. Hoffmann, *J. Amer. Chem. Soc.*, **89**, 593 (1967).
17. L. Eberson, *J. Amer. Chem. Soc.*, **89**, 4669 (1967).
18. J. E. Leffler and L. J. Story, *J. Amer. Chem. Soc.*, **89**, 2333 (1967).
19. J. E. Leffler, D. C. Ward and A. Burduroglu, *J. Amer. Chem. Soc.*, **94**, 5339 (1972).
20. R. M. Keefer and L. J. Andrews, *J. Amer. Chem. Soc.*, **80** 5350 (1958).
21. E. B. Merkushev and V. S. Raida, *J. Org. Chem. USSR (Engl. Transl.)*, **10**, 404 (1974).
22. J. Thiele and W. Peter, *Ann. Chem.*, **369**, 119 (1909).
23. J. L. Cotter, L. J. Andrews and R. M. Keefer, *J. Amer. Chem. Soc.*, **84**, 4692 (1962).
24. O. Exner, *Coll. Czechoslov. Chem. Commun.*, **24**, 3562 (1959).
25. E. M. Archer and T. G. D. van Schalkwyk, *Acta Cryst.*, **6**, 88 (1953).
26. C.-K. Lee, T. C. W. Mak and W.-K. Li, *Acta Cryst. B*, **33**, 1620 (1977).
27. N. W. Alcock, R. M. Countryman, S. Esperas and J. F. Sawyer, *J.C.S. Dalton*, 854 (1979).

28. G. F. Koser, R. H. Wettach, J. M. Troup and B. A. Frenze, *J. Org. Chem.*, **41**, 3609 (1976).
29. E. Shefter and W. Wolf, *J. Pharm. Sci.*, **54**, 104 (1965).
30. J. Z. Gougoutas and J. C. Clardy, *J. Solid State Chem.*, **4**, 226 (1972).
31. J. Z. Gougoutas and L. Lessinger, *J. Solid State Chem.*, **9**, 155 (1974).
32. J. Z. Gougoutas and J. C. Clardy, *J. Solid State Chem.*, **4**, 230 (1972).
33. D. G. Naae and J. Z. Gougoutas, *J. Org. Chem.*, **40**, 2129 (1975).
34. J. I. Musher, *Angew. Chem. Int. Ed.*, **8**, 54 (1969).
35. L. Pauling, *The Nature of the Chemical Bond*, 3rd edn, Cornell University Press, Ithaca, N.Y. (1960), p. 255.
36. R. L. Amey and J. C. Martin, *J. Org. Chem.*, **44**, 1779 (1979).
37. N. W. Alcock and R. M. Countryman, *J. C. S. Dalton*, 217 (1977).
38. M. C. Etter, *J. Solid State Chem.*, **16**, 399 (1976).
39. H. J. Reich and C. S. Cooperman, *J. Amer. Chem. Soc.*, **95**, 5077 (1973).
40. Reference 4, pp. 247–248.
41. F. M. Beringer and H. S. Schultz, *J. Amer. Chem. Soc.*, **77**, 5533 (1955).
42. E. J. Corey and W. J. Wechter, *J. Amer. Chem. Soc.*, **76**, 6040 (1954).
43. J. B. Dence and J. D. Roberts, *J. Org. Chem.*, **33**, 1251 (1968).
44. J. A. Gibson and A. F. Janzen, *J.C.S. Chem. Commun.* 739 (1973).
45. F. M. Beringer, J. W. Dehn, Jr and M. Winicov, *J. Amer. Chem. Soc.*, **82**, 2948 (1960).
46. M. Schmeisser and E. Scharf, *Angew. Chem.*, **71**, 524 (1959).
47. C. S. Rondestvedt, Jr, *J. Amer. Chem. Soc.*, **91**, 3054 (1969).
48. V. V. Lyalin, V. V. Orda, L. A. Alekseeva and L. M. Yagupol'skii, *J. Org. Chem. USSR (Engl. Transl.)*, **6**, 317 (1970).
49. I. I. Maletina, V. V. Orda, N. N. Aleinikov, B. L. Korsunskii and L. M. Yagupol'skii, *J. Org. Chem. USSR (Engl. Transl.)*, **12**, 1364 (1976).
50. V. V. Lyalin, V. V. Orda, L. A. Alekseeva and L. M. Yagupol'skii, *J. Org. Chem. USSR (Engl. Transl.)*, **8**, 1027 (1972).
51. D. Naumann, L. Deneken and E. Renk, *J. Fluorine Chem.*, **5**, 509 (1975); *Chem. Abstr.*, **83**, 57982Q (1975).
52. D. Naumann and J. Baumanns, *J. Fluorine Chem.*, **8**, 177 (1976); *Chem. Abstr.*, **85**, 159290W (1976).
53. D. Naumann, H. H. Heinsen and E. Lehmann, *J. Fluorine Chem.*, **8**, 243 (1976); *Chem. Abstr.*, **86**, 4867m (1977).
54. H. J. Reich and S. L. Peake, *J. Amer. Chem. Soc.*, **100**, 4888 (1978).
55. T. L. MacDonald, N. Narasimhan and L. T. Burka, *J. Amer. Chem. Soc.*, **102**, 7760 (1980).
56. R. C. Cambie, B. G. Lindsay, P. S. Rutledge and P. D. Woodgate, *J.C.S Chem. Commun.*, 919 (1978).
57. R. C. Cambie, D. Chambers, B. G. Lindsay, P. S. Rutledge and P. D. Woodgate, *J.C.S. Perkin I*, 822 (1980).
58. Y. Ogata and K. Aoki, *J. Org. Chem.*, **34**, 3978 (1969).
59. M. Linskeseder and E. Zbiral, *Ann. Chem.*, 1039 (1977).
60. K. Clauss, *Chem. Ber.*, **88**, 268 (1955).
61. F. M. Beringer and L. L. Chang, *J. Org. Chem.*, **36**, 4055 (1971).
62. J. W. Greidanus, W. J. Rebel and R. B. Sandin, *J. Amer. Chem. Soc.*, **84**, 1504 (1962).
63. G. P. Baker, F. G. Mann, N. Sheppard and A. J. Tetlow, *J., Chem Soc.*, 3721 (1965).
64. J. E. Leffler, L. K. Dyall and P. W. Inward, *J. Amer. Chem. Soc.*, **85**, 3443 (1963).
65. J. E. Leffler and H. Jaffe, *J. Org. Chem.*, **38**, 2719 (1973).
66. L. J. Andrews and R. M. Keefer, *J. Amer. Chem. Soc.*, **81**, 2374 (1959).
67. W. C. Agosta, *Tetrahedron Lett.*, 2681 (1965).
68. W. Wolf, E. Chalekson and D. Kobata, *J. Org. Chem.*, **32**, 3239 (1967).
69. J. Thiele and W. Peter, *Chem. Ber.*, **38**, 2842 (1905).
70. J. J. Lubinkowski, J. W. Knapczyk, J. L. Calderon, L. R. Petit and W. E. McEwen, *J. Org. Chem.*, **40**, 3010 (1975).
71. R. Criegee and H. Beucker, *Ann. Chem.*, **541**, 218 (1939).
72. K. H. Pausacker, *J. Chem. Soc.*, 107 (1953).
73. D. D. Tanner and G. C. Gidley, *Canad. J. Chem.*, **46**, 3537 (1968).
74. G. F. Koser and R. H. Wettach, *J. Org. Chem.*, **45**, 4988 (1980).

75. F. M. Beringer and L. L. Chang, *J. Org. Chem.*, **37**, 1516 (1972).
76. T. T. Nguyen and J. C. Martin, *J. Amer. Chem. Soc.*, **102**, 7382 (1980).
77. R. L. Amey and J. C. Martin, *J. Amer. Chem. Soc.*, **101**, 3060 (1979).
78. E. M. Archer, *Acta Cryst.*, **1**, 64 (1948).
79. The author wishes to thank Dr Jack Z. Gougoutas (University of Minnesota) for a copy of a manuscript detailing the structure of *o*-iodoxybenzoic acid prior to publication.
80. I. Masson, E. Race and F. E. Pounder, Jr, *J. Chem. Soc.*, 1669 (1935).
81. F. M. Beringer and P. Bodlaender, *J. Org. Chem.*, **33**, 2981 (1968).
82. V. V. Lyalin, V. V. Orda, L. A. Alekseeva and L. M. Yagupol'skii, *J. Org. Chem. USSR (Engl. Transl.)*, **8**, 215 (1972).
83. R. F. Weinland and W. Stille, *Chem. Ber.*, **34**, 2631 (1901).
84. I. I. Maletina, N. V. Kondratenko, V. V. Orda and L. M. Yagupol'skii, *J. Org. Chem. USSR (Engl. Transl.)*, **14**, 809 (1978).
85. L. M. Yagupol'skii, I. I. Maletina, N. V. Kondratenko and V. V. Orda, *Synthesis*, 574 (1977).
86. L. M. Yagupol'skii, V. V. Lyalin, V. V. Orda and L. A. Alekseeva, *J. Gen. Chem. USSR (Engl. Transl.)*, **38**, 2714 (1968).
87. J. A. Berry, G. Oates and J. M. Winfield, *J.C.S. Dalton,* 509 (1974).
88. O. R. Chambers, G. Oates and J. M. Winfield, *J.C.S. Chem. Commun.*, 839 (1972).
89. G. Oates and J. M. Winfield, *J.C.S. Dalton*, 119 (1974).
90. R. L. Amey and J. C. Martin, *J. Amer. Chem. Soc.*, **101**, 5294 (1979).
91. J. Z. Gougoutas, *Pure Appl. Chem.*, **27**, 305 (1971).
92. J. Z. Gougoutas and D. G. Naac, *J. Phys. Chem.*, **82**, 393 (1978).
93. J. Z. Gougoutas and J. C. Clardy, *Acta Cryst*. B, **26**, 1999 (1970).
94. J. Z. Gougoutas, *Israel J. Chem.*, **10**, 395 (1972).
95. J. Z. Gougoutas, K. H. Chang and M. C. Etter, *J. Solid State Chem.*, **16**, 283 (1976).
96. J. Z. Gougoutas and D. G. Naae, *J. Solid State Chem.*, **16**, 271 (1976).
97. J. Z. Gougoutas and L. Lessinger, *J. Solid State Chem.*, **7**, 175 (1973).
98. J. Z. Gougoutas and L. Lessinger, *J. Solid State Chem.*, **12**, 51 (1975).
99. M. C. Etter, *J. Amer. Chem. Soc.,* **98**, 5326 (1976).
100. M. C. Etter, *J. Amer. Chem. Soc.*, **98**, 5331 (1976).
101. Personal communication from J. Z. Gougoutas.
102. G. D. Gutsche and M. Hillman, *J. Amer. Chem. Soc.*, **76**, 2236 (1954).
103. W. H. Pirkle and G. F. Koser, *Tetrahedron Lett.*, 3959 (1968).
104. W. H. Pirkle and G. F. Koser, *J. Amer. Chem. Soc.*, **90**, 3598 (1968).
105. A. W. Johnson, *Ylide Chemistry*, Academic Press, New York (1964), pp. 258, 321.
106. W. Kirmse, *Carbene Chemistry*, 2nd ed. Academic Press, New York (1971), pp. 442–447.
107. A. P. Marchand and N. MacBrockway, *Chem. Rev.*, **74**, 431 (1974); see pp. 442–446.
108. W. Ando, S. Kondo and T. Migata, *J. Amer. Chem. Soc.*, **91**, 6516 (1969).
109. W. Ando, S. Kondo, K. Nakayama, K. Ichibori, H. Kohoda, H. Yamato, I. Imai, S. Nakaido and T. Migita, *J. Amer. Chem. Soc.*, **94**, 3870 (1972).
110. W. A. Sheppard and O. W. Webster, *J. Amer. Chem. Soc.*, **95**, 2695 (1973).
111. W. A. Sheppard, U.S. Patent 3,914,247 (Cl. 260–309; CO7D), 21 October 1975; *Chem. Abstr.*, **84**, P59470r (1976).
112. E. Gudrinietse, O. Neilands and G. Vanag, *J. Gen. Chem. USSR (Engl. Transl.)*, **27**, 2777 (1957).
113. O. Neilands and G. Vanag, *Proc. Acad. Sci. USSR (Engl. Transl.)*, *Chem. Sect.*, **141**, 1232 (1961).
114. O. Neilands and G. Vanag, *Proc. Acad. Sci. USSR (Engl. Transl.), Chem. Sect.*, **131**, 425 (1960).
115. O.Neilands, M. Sile and B. Karele, *Latv. PSR Zinat. Akad. Vestis, Kim. Ser.*, 217 (1965); *Chem. Abstr.*, **63**, 13166f (1965).
116. O. Neilands, *J. Org. Chem. USSR (Engl. Transl.)*, **1**, 1888 (1965).
117. O. Neilands and G. Vanag, *J. Gen. Chem. USSR (Engl. Transl.)*, **31**, 137 (1961).
118. O. Neilands and B. Karele,*J. Org. Chem. USSR (Engl. Transl.)*, **2**, 491 (1966).
119. O. Neilands and B. Karele, *J. Org. Chem. USSR (Engl. Transl.)*, **1**, 1884 (1965).
120. O. Neilands and B. Karele, *J. Org. Chem. USSR (Engl. Transl.)*, **7**, 1674 (1971).
121. O. Neilands and D. E. Neiman, *J. Org. Chem. USSR (Engl. Transl.)*, **6**, 2522 (1970).

122. B. Karele and O. Neilands, *J. Org. Chem. USSR (Engl. Transl.)*, **4**, 627 (1968).
123. B. Karele and O. Neilands, *VINITI*, USSR, 4123 (1972); *Chem. Abstr.*, **85**, 20763j (1976).
124. Y. Hayasi, T. Okada and M. Kawanisi, *Bull. Chem. Soc. Japan*, **43**, 2506 (1970).
125. F. Karele and O. Neilands, *J. Org. Chem. USSR (Engl. Transl.)*, **4**, 1755 (1968).
126. P. B. Kokil and P. M. Nair, *Tetrahedron Lett.*, 4113 (1977).
127. S. W. Page, E. P. Mazzola, A. D. Mighell, V. L. Himes and C. R. Hubbard, *J. Amer. Chem. Soc.*, **101**, 5858 (1979).
128. K. Friedrich and W. Amann, *Tetrahedron Lett.*, 3689 (1973).
129. K. Friedrich, W. Amann and H. Fritz, *Chem. Ber.*, **111**, 2099 (1978).
130. K. Friedrich and W. Amann, *Tetrahedron Lett.*, 2885 (1977).
131. F. Mizukami, M. Ando, T. Tanaka and J. Imamura, *Bull. Chem. Soc. Japan*, **51**, 335 (1978).
132. S. C. Pati and B. R. Dev, *Indian J. Chem.*, **17A**, 92 (1979).
133. V. Mahalingam and N. Venkatasubramanian, *Indian J. Chem.*, **18B**, 94 (1979).
134. V. Mahalingam and N. Venkatasubramanian, *Indian J. Chem.*, **18B**, 95 (1979).
135. G. F. Koser, A. G. Relenyi, A. N. Kalos, L. Rebrovic and R. H. Wettach, *J. Org. Chem.*, **47**, 2487 (1982).
136. R. M. Moriarty, S. C. Gupta, H. Hu, D. R. Berenschot and K. B. White, *J. Amer. Chem. Soc.*, **103**, 686 (1981).
137. V. V. Semenov, S. A. Shevelev and A. A. Fainzil'berg, *Bull. Acad. Sci. USSR (Engl. Transl.), Chem. Div.* **25**, 2459 (1976).
138. V. V. Semenov, S. A. Shevelev and A. A. Fainzil'berg, *Bull. Acad. Sci. USSR (Engl. Transl.), Chem. Div.*, **27**, 2080 (1978).
139. R. A. Abramovitch and I. Shinkai, *J.C.S. Chem. Commun.* 569 (1973).
140. R. A. Abramovitch, G. Grins, R. B. Rogers and I. Shinkai, *J. Amer. Chem. Soc.*, **98**, 5671 (1976).
141. A. Seveno, G. Morel, A. Foucaud and E. Marchand, *Tetrahedron Lett.*, 3349 (1977).
142. B. Karele, S. Kalnina, I. Grinberga and O. Neilands, *Nov. Issled. Obl. Khim. Khim. Teknol., Mater. Nauchno-Tekh. Konf. Professorsko-Prepod. Sostava Nauchn. Rab. Khim. Fac. RPI (Russ).* 19–20 (1973); *Chem. Abstr.*, **82**, 4204k (1975).
143. T. Kappe, G. Korbuly and W. Stadlbaur, *Chem. Ber.*, **111**, 3857 (1978).
144. R. A. Abramovitch, T. D. Bailey, T. Takaya and V. Uma, *J. Org. Chem.*, **39**, 340 (1974).
145. Y. Yamada, T. Yamamoto and M. Okawara, *Chem. Lett., Chem. Soc. Japan*, 361 (1975).
146. A. S. Radhakrishna, M. E. Parham, R. M. Riggs and G. M. Loudon, *J. Org. Chem.*, **44**, 1746 (1979).
147. G. A. Olah, Y. Yamada and R. J. Spear, *J. Amer. Chem. Soc.*, **97**, 680 (1975).
148. F. M. Beringer and S. A. Galton, *J. Org. Chem.*, **30**, 1930 (1965).
149. O. Neilands and J. Polis, *Latv. PSR Zinat. Akad. Vestis. Kim. Ser.*, 192 (1963); *Chem. Abstr.*, **60**, 5427e (1964).
150. B. Adamsone, D. E. Prikule and O. Neilands, *J. Org. Chem. USSR (Engl. Transl.)*, **14**, 2416 (1978).
151. B. Karele, S. Kalnina, I. Grinberga and O. Neilands, *Khim. Geterotsikli Soedin*, 245 (1973); *Chem. Abstr.*, **78**, 136161w (1973).
152. J. Hine, *Structural Effects on Equilibria in Organic Chemistry*, John Wiley and Sons, New York (1975), pp. 183, 180.
153. O. Neilands, G. Vanag and E. Gudrinietse, *J. Gen. Chem. USSR (Engl. Transl.)*, **28**, 1256 (1958).
154. O. Neilands and G. Vanag, *Proc. Acad. Sci. USSR (Engl. Transl.)*, **130**, 19 (1960).
155. O. Neilands and G. Vanag, *Proc. Acad. Sci USSR (Engl. Transl.)*, **131**, 325 (1960).
156. B. Karele and O. Neilands, *J. Org. Chem. USSR (Engl. Transl.),* **2**, 1656 (1966).
157. S. V. Kalnin' and O. Neilands, *J. Org. Chem. USSR (Engl. Transl.),* **7**, 1668 (1971).
158. M. S. Ermolenko, V. A. Budylin and A. N. Kost, *Khim. Geterotsikl Soedin,* 933 (1978); *Chem. Abstr.*, **89**, 146707v (1978).
159. D. E. Neiman and O. Neilands, *J. Org. Chem. USSR (Engl. Transl.),* **6**, 1015 (1970).
160. D. E. Neiman and O. Neilands, *J. Org. Chem. USSR (Engl. Transl.)*, **6**, 633 (1970).
161. O. Neilands and G. Vanag, *Zhur. Obshchei Khim*, **30**, 510 (1960); *Chem. Abstr.*, **54**, 24469b (1960).
162. G. F. Koser and S.-M. Yu, *J. Org. Chem.*, **40**, 1166 (1975).

163. F. M. Beringer, P. S. Forgione and M. D. Yudis, *Tetrahedron*, **8**, 49 (1960).
164. F. M. Beringer and P. S. Forgione, *J. Org. Chem.*, **28**, 714 (1963).
165. F. M. Beringer, S. A. Galton and S. J. Huang, *J. Amer. Chem. Soc.*, **84**, 2819 (1962).
166. F. M. Beringer and S. A. Galton, *J. Org. Chem.*, **28**, 3417 (1963).
167. F. M. Beringer, W. J. Daniel, S. A. Galton and G. Rubin, *J. Org. Chem.*, **31**, 4315 (1966).
168. K. Friedrich, W. Amann and H. Fritz, *Chem. Ber.*, **112**, 1267 (1979).
169. P. Gronski and K. Hartke, *Chem. Ber.*, **111**, 272 (1978).
170. P. Gronski and K. Hartke, *Tetrahedron Lett.*, 4139 (1976).
171. K. Nakasuji, K. Nishino, I. Murata, H. Ogoshi and Z. Yoshida, *Angew. Chem. Int. Edn Engl. Transl.*, **16**, 866 (1977).
172. V. V. Semenov and S. A. Shevelev, *Bull. Acad. Sci. USSR (Engl. Transl.), Chem. Div.*, **27**, 2087 (1978).
173. V. Kokars, V. Kampars and O. Neilands, *Latv. PSR Zinat. Akad. Vestis, Kim. Ser.*, 734 (1975); *Chem. Abstr.*, **85**, 62480v (1976).
174. D. E. Prikule and O. Neilands, *J. Org. Chem. USSR (Engl. Transl.)*, **7**, 2537 (1971).
175. G. F. Koser and S.-M. Yu, *J. Org. Chem.*, **41**, 125 (1976).
176. G. F. Koser, S.-M. (Yu) Linden and Y.-J. Shih, *J. Org. Chem.*, **43**, 2676 (1978).
177. D. E. Prikule and O. Neilands, *J. Org. Chem. USSR (Engl. Transl.)*, **13**, 1033 (1977).
178. J. L. Atwood and W. A. Sheppard, *Acta Cryst. B*, **31**, 2638 (1975).
179. U. Drück and W. Littke, *Acta Cryst. B*, **34**, 3092 (1978).
180. S. Valtere, O. Neilands and A. Burkevica, *Latv. PSR Zinat. Akad. Vestis, Kim. Ser.*, 345 (1973); *Chem. Abstr.* **79**, 91174e (1973).

The Chemistry of Functional Groups, Supplement D
Edited by S. Patai and Z. Rappoport
© 1983 John Wiley & Sons Ltd

CHAPTER **19**

Synthesis and reactivity of α-halogenated ketones

ROLAND VERHÉ and NORBERT DE KIMPE
Laboratory of Organic Chemistry, Faculty of Agricultural Sciences, State University of Gent, Coupure 533, B-9000 Gent, Belgium

I. INTRODUCTION

Although much information on the synthesis and the chemistry of α-halogenated carbonyl compounds is scattered throughout the literature, there appear to be few comprehensive sources of information in this important area, with the exception of a short chapter dealing with the preparation of halogenated ketones in Houben-Weyl's *Methoden der organischen Chemie*[1,2]. In addition, the Favorskii rearrangement of α-haloketones has been reviewed by several authors[3–8], while the reactivity of α-haloketones towards nucleophiles was described by Tchoubar in 1955[9]. The past two decades has seen a considerable expansion in synthetic procedures and mechanistic studies on the reactivity of α-halogenated ketones. It is our hope that putting together a survey of the widely scattered information on the synthesis and reactivity of α-haloketones will focus new attention on the broad potential of these compounds in synthetic and mechanistic organic chemistry.

The presentation of this chapter is divided into two major sections. The first part deals with the synthetic methods for the preparation of α-haloketones. In the second section the reactivity will be considered, with emphasis on preparative applications, although some mechanistic interpretations of the results will be treated in some important cases. The section on reactivity has been subdivided according to the nature of the nucleophile, e.g. oxygen, nitrogen or carbon nucleophiles, and not on the basis of the reaction type, e.g. substitution, elimination.

This chapter has been restricted to halogenated ketones which carry one or more halogen atoms at the α-carbon atom to a carbonyl function, excluding compounds derived from diketones, β-keto esters and quinones. Other α-halogenated carbonyl compounds such as aldehydes, esters and acids will not be treated in this chapter.

II. SYNTHESIS OF α-HALOGENATED KETONES

While a number of reviews have been published during the last decade on the preparation of α-fluoroketones[10–13], practically no general review deals with new syntheses of α-chloro-, α-bromo- and α-iodoketones[14]. The syntheses of α-fluoro, α-chloro-, α-bromo- and α-iodoketones are treated separately and the procedures are classified according to the starting substrates. Some procedures, using the same class of reagents, are described separately for each class of haloketones.

A. Synthesis of α-Fluoroketones

1. *α-Fluoroketones from ketones and their derivatives*

Conventional methods for the synthesis of α-fluoroketones by direct fluorination of ketones often give rise to side reactions and are therefore of limited use (equation 1).

$$R^1CH_2COCH_2R^2 \xrightarrow{\text{'F'}} R^1CH(F)COCH_2R^2 + \text{polyfluorinated and degradation products} \quad (1)$$

For example, treatment of acetone with fluorine yields a complex mixture of fluoroacetone, hexafluoroacetone and degradation products such as trifluoroacetyl chloride, tetrafluoromethane and carbonyl difluoride[15]. The direct action of perchloryl fluoride on ketones has also met with little success because of degradation reactions.

The reactions of a variety of fluorinating agents on derivatives of ketones appear to be more advantageous. Potential synthetic interest may be found in the reaction of perchloryl fluoride with enol ethers[16], enol esters[17], enamines[18] and lithium enolates[230]. 1-Ethoxycyclohexene (**1**) gives 2-fluorocyclohexanone (**3**) via 1-ethoxy-1,2-difluorocyclohexane (**2**) on treatment with perchloryl fluoride in pyridine at 0°C (equation 2)[16]. The enamines of 3-oxo steroids are transformed into 2α-fluoro-3-oxo steroids on treatment with perchloryl fluoride followed by hydrolysis of the intermediate fluoroenamines.

OC_2H_5 $\xrightarrow[\text{Pyridine}]{FClO_3}$ C_2H_5O F F $\xrightarrow{H_3O^+}$ O F (2)

(**1**) (**2**) (**3**)

2α-Fluorocholestan-3-one (**5**) is formed on treatment of 3-(*N*-pyrrolidinyl)-2-cholestene (**4**) with this reagent in benzene in 72% yield (equation 3)[19].

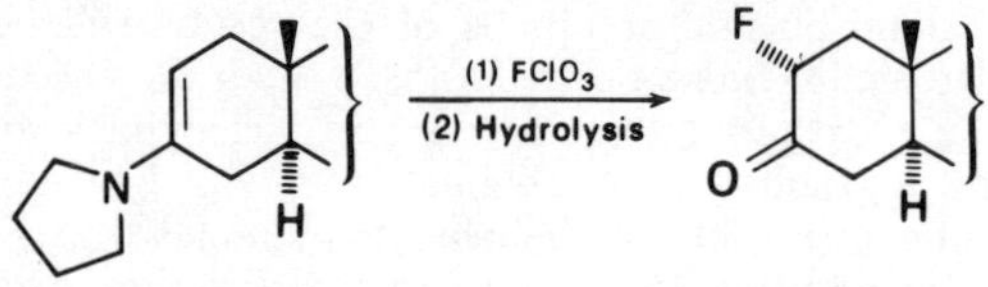

(1) $FClO_3$ (2) Hydrolysis (3)

(**4**) (**5**)

When fluorinated steroidal enamines (**6**) are treated with perchloryl fluoride, 2,2-difluoro compounds (**7**) and 2,2,4-trifluoro compounds (**9**) become accessible (equation 4)[20,21]. A related process for the synthesis of α-fluoroketones employs lithium enolates of ketones and perchloryl fluoride in tetrahydrofuran. In this manner ω-fluoroacetophenone is obtained in 44% yield[22].

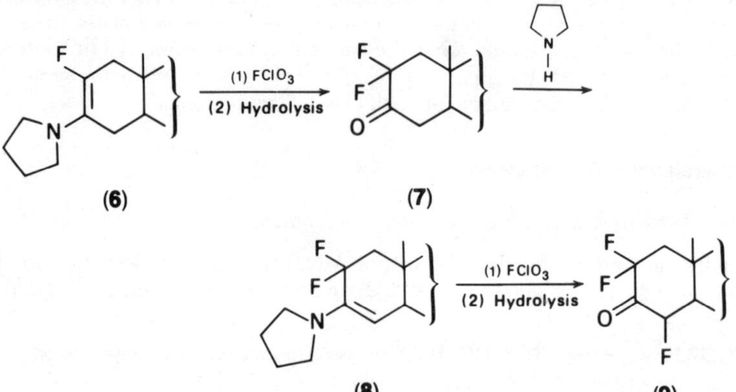

(1) $FClO_3$ (2) Hydrolysis

(**6**) (**7**) (4)

(1) $FClO_3$ (2) Hydrolysis

(**8**) (**9**)

Fluorination of ketones with perchloryl fluoride is also performed via intermediate methoxalyl ketones[23] (i.e. — COCOOMe) and hydroxymethylene ketones[24]. 2α-Fluorohydrocortisone is synthesized from the sodium salt of 20-ethylenedioxy-2-methoxalyl-Δ^4-pregnentriol-(11β, 17α, 21)-3,20-dione[23] and 2α-fluorotestosterone from the sodium salt of 2-hydroxymethylenetestosterone[24].

Recently, a new and powerful method for the α-fluorination of carbonyl compounds was developed which utilizes trifluoromethyl hypofluorite with silyl enol ethers, as exemplified by the preparation of 2-fluorocyclohexanone (**3**) in 70% yield (equation 5)[25]. A similar method with enol acetates is used by Rozen, by passing fluorine

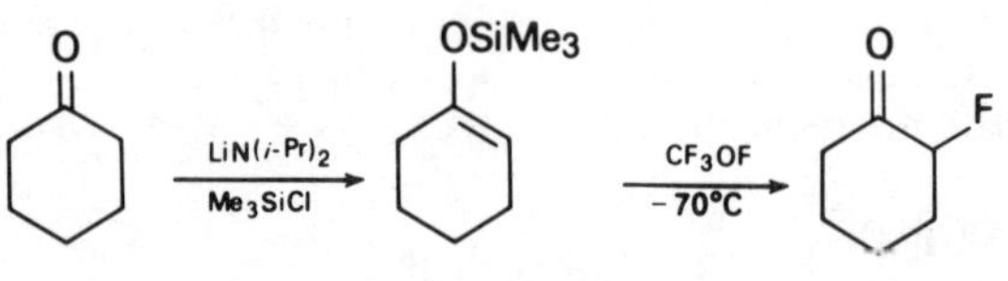

$LiN(i\text{-}Pr)_2$, Me_3SiCl; $OSiMe_3$; CF_3OF, $-70°C$ (5)

(**10**) (**11**) (**3**)

into a suspension of sodium trifluoroacetate in Freon at −75°C. A considerable portion of the oxidizing ability of this solution is due to the presence of pentafluoroethyl hypofluorite (CF_3CF_2OF) and other oxidizing compounds of the perfluoroxyfluoride type. 2-Fluoro-1-tetralone (**13**) can be obtained by this procedure in 85% yield by starting from the enol acetate of 1-tetralone (**12**) (equation 6)[26].

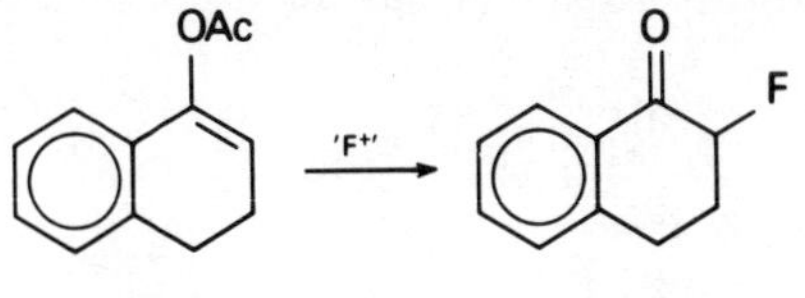

(6)

(**12**) (**13**)

In attempts to react enol acetates with molecular fluorine, no α-fluoroketones could be isolated from the complicated reaction mixtures. Geminal α,α-difluoroketones are formed by decomposition of geminal difluorocyclopropanes (**14**), prepared by difluorocarbene addition to enol acetates. Reaction of these cyclopropanes with sodium hydroxide in methanol provides α,α-difluoroketones (**15**) in addition to other products (equation 7)[27,28]. The corresponding dichloro- and dibromocyclopropanes exhibit completely different pathways, resulting in the formation of halogenated enones.

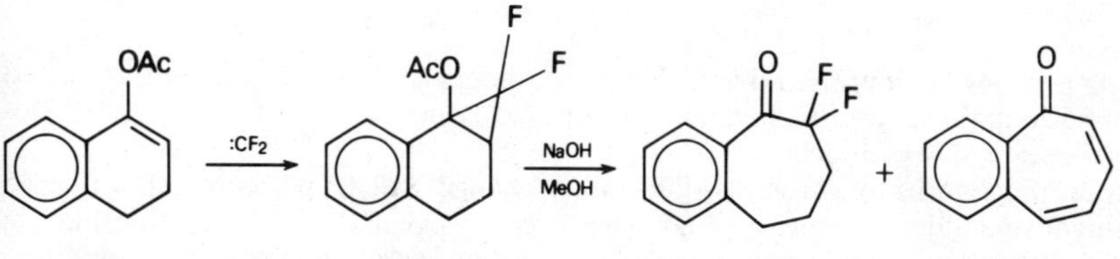

(7)

(**12**) (**14**) (**15**) (**16**)

Finally, the action of a Lewis acid on α-fluorinated amines (**17**), easily obtained by addition of secondary amines to fluorinated alkenes, produces fluorinated immonium salts (**18**), which on arylation with electron-rich aromatic compounds and subsequent hydrolysis furnish α-halo-α-fluoroacetophenones (**20**) (equation 8)[29].

$$\underset{(\mathbf{17})}{XCHF{-}CF_2{-}NR_2} \rightleftharpoons XCHF{-}\overset{F}{\overset{|}{C}}{=}\overset{+}{N}R_2\ F^- \xrightarrow{BF_3} \underset{(\mathbf{18})}{XCHF{-}\overset{F}{\overset{|}{C}}{=}\overset{+}{N}R_2\ BF_4^-} \xrightarrow{ArH}$$

$$\underset{\substack{(\mathbf{19})\\ X=Cl,Br}}{Ar{-}\overset{{}^+NR_2\ BF_4^-}{\overset{\|}{C}}{-}CHFX} \xrightarrow{H_3O^+} \underset{(\mathbf{20})}{ArCOCHXF} \qquad (8)$$

2. *α-Fluoroketones from α-haloketones by halogen exchange*

The exchange of a chlorine atom in α-chlorinated ketones by fluorine on treatment with hydrogen fluoride only takes place when there is no possibility of hydrogen chloride elimination, such as in perchloroketones and chloroacetone. Better results are obtained with potassium fluoride[30,161] and potassium hydrogen fluoride (KHF_2)[31] (equation 9). Excellent results of bromine–fluorine exchange are obtained by the use

$$R^1R^2C(X)COR^3 \xrightarrow[KHF_2]{KF\text{ or}} R^1R^2C(F)COR^3 \qquad (9)$$

of mercuric fluoride[32,33]. If a chlorine atom is also present in the molecule, it is retained. 1-Aryl-2,2-difluoro- and 1-aryl-2-chloro-2-fluoro-1-alkanones are prepared by this procedure in moderate yields. Another method involves the use of silver tetrafluoroborate in ether. However, this method is not applicable to primary bromoketones or to chloroketones (equation 10). The method does not seem to have a broad scope since several side products, mainly α,β-unsaturated ketones, are formed,

$$R^1R^2C(Br)H{-}COR^3 \xrightarrow{AgBF_4} R^1R^2C(F)H{-}COR^3 + AgBr + BF_3 \qquad (10)$$

making isolation on a preparative scale rather laborious. When the reaction is carried out in nucleophilic solvents (methanol, acetic acid), α-methoxy- and α-acetoxyketones are isolated as side products.

Other procedures of bromine–fluorine exchange utilize potassium fluoride in dimethylformamide, glycerine and diethylene glycol[33–35], silver fluoride in acetonitrile–water[36], thallium fluoride[37] and pyridinium poly(hydrogen fluoride) used in conjunction with mercuric oxide[38].

3. α-Fluoroketones from α-diazoketones

Fluoromethyl ketones are easily formed when diazomethyl ketones, prepared by condensation of acid chlorides with diazomethane, are treated with hydrogen fluoride[39,40] or pyridinium poly(hydrogen fluoride)[38] (equation 11). α,α-Difluoro-

$$R{-}COCl + CH_2N_2 \longrightarrow RCOCHN_2 + HCl$$
$$R{-}COCHN_2 + HF \longrightarrow RCOCH_2F + N_2 \qquad (11)$$

ketones (e.g. **22**) are obtained by fluorination of diazoketones such as diazocamphor (**21**) with trifluoromethyl hypofluorite; additionally, minor amounts of α-fluoro-α-trifluoromethoxyketones (e.g. **23**) were isolated[41]. In the case of **21** a rearrangement also occurs, leading to the formation of a fluorotricyclanone (**24**) as another side product (equation 12).

$$\textbf{(21)} \xrightarrow{CF_3OF} \textbf{(22)} + \textbf{(23)} + \textbf{(24)} \qquad (12)$$

(21) (**22**) (**23**) (**24**)

4. α-Fluoroketones from carboxylic acid derivatives

Reaction of fluorinated carboxylic acids and derivatives with organometallic reagents usually gives rise to α-fluoroketones. Treatment of trifluoroacetic acid with phenyllithium in ether at −65°C affords ω,ω,ω-trifluoroacetophenone[42], while the reaction of lithium trifluoroacetate with butyllithium yields 1,1,1-trifluoro-2-hexanone[43].

The condensation of organomagnesium compounds with α-fluorinated esters gives satisfactory yields of α-fluoroketones (equation 13)[44,45]. Condensation of

$$R^1CH(F)COOR^2 + R^3MgX \longrightarrow R^1CH(F)COR^3 \tag{13}$$

α-fluoronitriles (**25**) with Grignard reagents affords α-fluoroketones in high yields (equation 14)[46,47]. ω-Fluoroacetophenone is produced in good yield by the Friedel–Crafts method, provided that the reaction with fluoroacetyl chloride is carried out rapidly[48].

$$\underset{(\mathbf{25})}{R^1CH(F)CN} \xrightarrow[(2)\ H_2O]{(1)\ R^2MgX} R^1CH(F)COR^2 \tag{14}$$

An α-fluoroketone (**29**) is formed during the hydroxide-catalysed hydrolysis of an α-fluoro-β-keto ester (**27**), while under the same circumstances the difluoro-β-keto ester (**31**) is transformed into the 1,3-difluoroketone (**32**) (equation 15)[49].

(**26**) COOEt —$FClO_3$→ (**27**) (65%) EtOOC, F + (**28**) (10%) F, COOEt

(**27**) —^-OH→ (**29**) F

(**30**) F, COOEt —$FClO_3$→ (**31**) F, F COOEt —^-OH→ (**32**) F, F (15)

5. α-Fluoroketones from α-functionalized epoxides

A general synthesis of α-fluorocarbonyl compounds is developed from fluorocyanohydrins (**34**), obtained by the simultaneous action of hydrogen fluoride and boron trifluoride on epoxynitriles (**33**). Decomposition (**34**) with silver nitrate in the presence of an equimolecular amount of ammonia gives rise to the formation of α-fluoroketones in moderate yields (equation 16)[50]. Thermal isomerization of

$$\underset{(\mathbf{33})}{R^1R^2C\overset{O}{—}C(R^3)CN} \xrightarrow[\text{Ether}]{BF_3/HF} \underset{(\mathbf{34})}{R^1R^2C(F)—C(R^3)(OH)CN} \xrightarrow[NH_3]{Ag^+} R^1R^2C(F)COR^3 \quad (16)$$

α-fluoroepoxides (**36**), prepared by epoxidation of fluorinated olefins (**35**), gives rise to the formation of α-fluoroketones by migration of the fluorine atom (equation 17)[51].

$$\underset{(\mathbf{35})}{R^1—CH{=}CFR^2} \xrightarrow{\text{Epoxidation}} \underset{(\mathbf{36})}{R^1CH\overset{O}{—}CFR^2} \xrightarrow{\Delta} R^1CH(F)COR^2 \quad (17)$$

6. α-Polyfluoroketones by condensation reactions

α-Polyfluoroketones are produced by several condensation reactions, e.g. Friedel–Crafts, Hoesch, Claisen, Knoevenagel and aldol condensations. These types of reaction are undoubtedly the most suitable for perfluoroketone synthesis. Much of the literature concerning the various aspects of this topic has been covered elsewhere[2] and will not be repeated here.

B. Synthesis of α-Chloroketones

1. Synthesis of α-chloroketones from ketones and their derivatives

The preparation of α-chloroketones starting from ketones and their derivatives can be achieved by various procedures. The choice of method is dependent upon the nature of the ketone and the degree of chlorination wanted. Therefore no general procedure seems to be available for the synthesis of a given chlorinated ketone. The substitution pattern in the starting ketone determines the method to be employed, as will be demonstrated below.

a. Chlorination with chlorine. In general, reaction of aliphatic ketones with chlorine most commonly affords higher chlorinated products (equation 18).

$$R^1CH_2COCH_2R^2 \xrightarrow{Cl_2} R^1CH(Cl)COCH_2R^2 + R^1CH_2COCH(Cl)R^2 + R^1CH_2COCCl_2R^2 + R^1CH(Cl)COCH(Cl)R^2 + R^1CCl_2COCH_2R^2 + \text{Polychlorinated ketones} \quad (18)$$

During the monochlorination of acetone, minor amounts of dichloroacetone are always isolated. However, good results for the monochlorination of acetone and 3-pentanone are possible when the chlorination is carried out in aqueous solutions of calcium carbonate and calcium chloride; using this procedure 2-butanone furnishes a mixture of 75% 3-chloro-2-butanone and 25% 1-chloro-2-butanone[52].

Further chlorination of monochloroacetone at 100–140°C in the presence of iodine, antimony pentachloride and ferric trichloride gives a mixture of 1,1,1,3-tetrachloroacetone and 1,1,3,3-tetrachloroacetone in a 1:4 ratio[53]. Chlorination of acetone in carbon tetrachloride at 50–70°C gives pentachloroacetone, which is transformed into hexachloroacetone by further chlorination in the presence of antimony trisulphide and iodine[54].

Hexachloroacetone is also formed during chlorination in an acetic acid–sodium acetate medium[55].

In general, α-perchloroketones are produced in very good yields in the presence of light without catalysis[56]. Photochlorination in the gas phase only affords α-substituted ketones[57].

The degree of chlorination in alicyclic ketones is strongly dependent upon the reaction medium. Cycloalkanones are monochlorinated in the α-position in acetic acid[58], water[59,60], methanol[61], or dichloromethane[62], while α,α'-dichloro compounds are produced upon further treatment with chlorine, except for α-tetralone, of course, where 2,2-dichloro-α-tetralone is obtained[62] (equation 19). Tetrachloro- and hexachlorocyclohexanone are formed when the chlorination is carried out in the presence of rhodium(III) chloride and iridium(IV) chloride[63], respectively.

$$\text{cyclo-}(CH_2)_n\text{C=O} \xrightarrow{Cl_2} \text{2-Cl-cyclo-}(CH_2)_n\text{C=O} \xrightarrow{Cl_2} \text{2,}n'\text{-Cl}_2\text{-cyclo-}(CH_2)_n\text{C=O} \quad (19)$$

Chlorine in dimethylformamide seems to be a powerful reagent for the substitution of α-protons in aldehydes and ketones[64,65]. Usually all the α-protons are rapidly replaced at 50–90°C, except in aliphatic ketones, where the last α-proton is substituted only at 120°C, because of the sterically hindered enolization[66]. A clean conversion of cyclopentanone (**37**) into 2,2,5,5-tetrachlorocyclopentanone (**38**) is obtained using this procedure at 20–30°C, when a continuous excess of chlorine is maintained during the course of the reaction.

Several intermediate α-chlorinated cyclopentanones are dehydrochlorinated in dimethylformamide (DMF), yielding chlorinated 2-cyclopentenones which are further chlorinated to afford penta- (**39**) and hexachlorocyclopentanone (**40**). Chlorination of cyclopentanone (**37**) in DMF at 120°C gives a mixture of the isomeric perchlorocyclopentenones (**41**) and (**42**) (equation 20)[67].

Chlorination of cyclopentanone with chlorine in dichloromethane and carbon tetrachloride is not a synthetically useful method as rather complex mixtures of mono-, di- and trichloro derivatives are formed[67]. However, 2,2,3-trichlorocyclopentanone can be prepared via chlorination of 2-chloro-2-cyclopentenone; the latter compound is produced upon treatment of 2-cyclopentenone with chlorine in carbon tetrachloride[68].

Direct chlorination of 2-methylcyclohexanone with chlorine yields 2-chloro-2-methylcyclohexanone as the major product, besides *cis*- and *trans*-6-chloro-2-methylcyclohexanone and substantial amounts of the 2,6-dichloro compound[69]. Treatment of cyclohexanones with chlorine in dimethylformamide results in substitution of all the α-hydrogens (equation 21)[66].

Aryl alkyl ketones are mostly monochlorinated in the aliphatic chain using solutions of chlorine in acetic acid, methanol or carbon tetrachloride at low temperatures[70].

At 60°C ω,ω-dichloroacetophenone is produced[71], which in turn is converted into ω,ω,ω-trichloroacetophenone in the presence of sodium acetate on further treatment

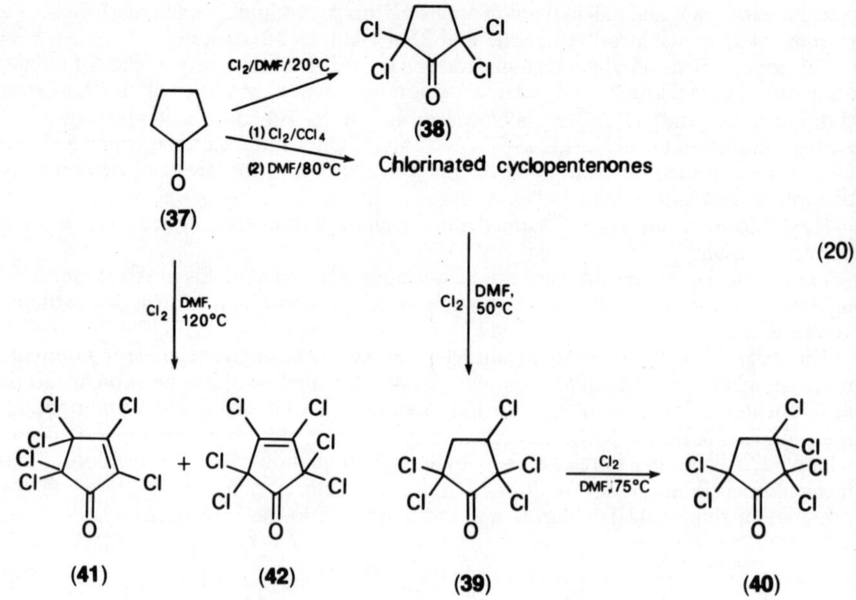

(20)

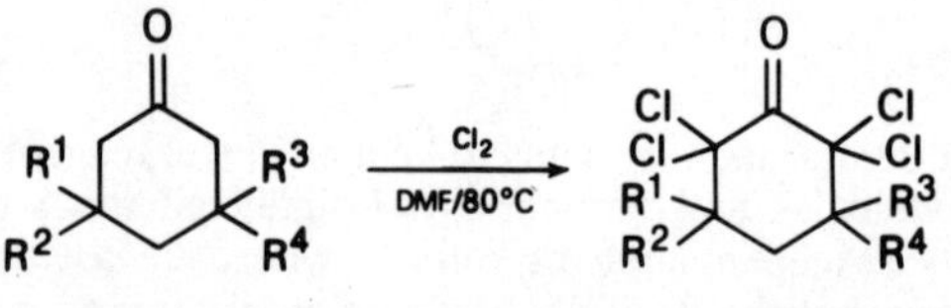

(21)

with chlorine[72]. Surprisingly, the higher homologues, 2,2-dichloro-1-aryl-1-alkanones, could only be prepared by chlorination in dimethylformamide at 100°C[73], with the exception of 2,2-dichloropropiophenone, which is also formed during the chlorination of propiophenone in a solution of sodium acetate in acetic acid[74] (equation 22).

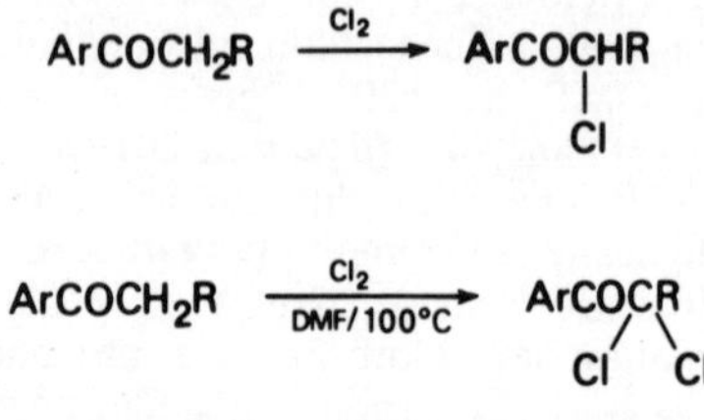

$$ArCOCH_2R \xrightarrow{Cl_2} ArCOCHR(Cl)$$

(22)

$$ArCOCH_2R \xrightarrow[DMF/100°C]{Cl_2} ArCOCR(Cl)(Cl)$$

Chlorination of enamines has been used for the preparation of α-chloroketones. A procedure for the regiospecific synthesis of chloromethyl ketones **(43)** via immonium salts is described by Carlson[75]. By regioselective deprotonation of these salts, mixtures of tautomeric enamines, derived from methyl ketones, are transformed into the less sterically hindered enamines, which upon reaction with chlorine and subsequent hydrolysis yield chloromethyl ketones **(43)** (equation 23). Enamines react with chlorine in ether at −78°C under exclusion of oxygen and moisture to give the isolable α-chloroimmonium halides **(44)**, after which acid hydrolysis leads to α-chloroketones (equation 24)[76].

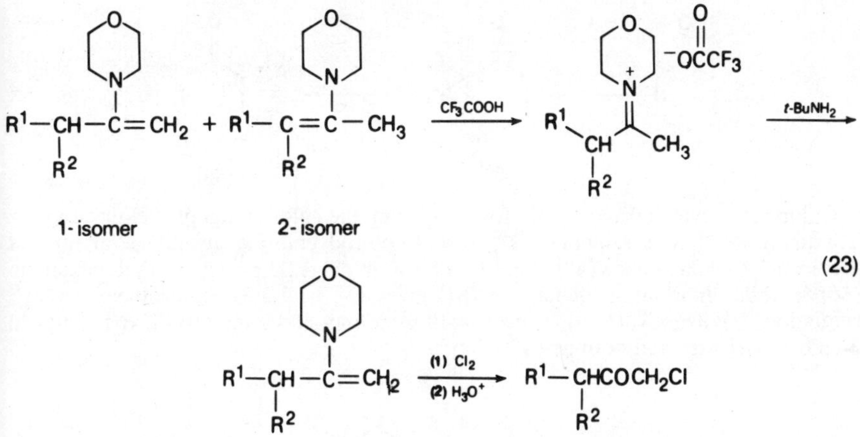

$$R^1-CH(R^2)-C(NC_4H_8O)=CH_2 + R^1-C(R^2)=C(NC_4H_8O)-CH_3 \xrightarrow{CF_3COOH} R^1(R^2)CH-C(=\overset{+}{N}C_4H_8O)-CH_3\ \ {}^{-}OCOCF_3 \xrightarrow{t\text{-}BuNH_2}$$

1-isomer 2-isomer

$$R^1-CH(R^2)-C(NC_4H_8O)=CH_2 \xrightarrow[(2)\ H_3O^+]{(1)\ Cl_2} R^1-CH(R^2)COCH_2Cl \quad (23)$$

(**43**)

$$\text{cyclic } C(NR_2)=CH\ (CH_2)_n \xrightarrow[\text{Ether}/-78\,^\circ C]{Cl_2} C(=\overset{+}{N}R_2\ Cl^-)-C(H)(Cl)\ (CH_2)_n \xrightarrow{H_3O^+} C(=O)-C(H)(Cl)\ (CH_2)_n \quad (24)$$

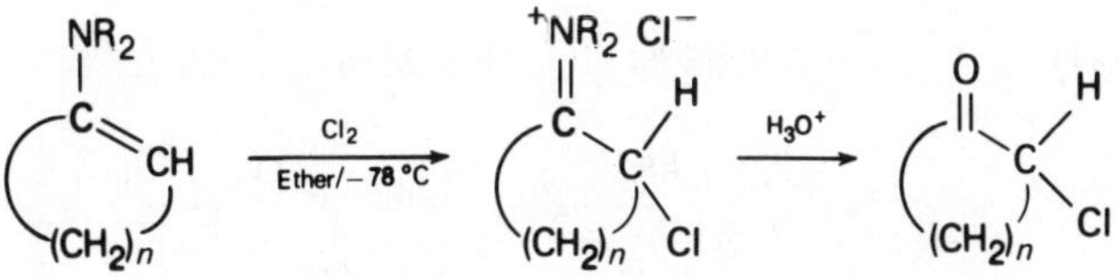

(**44**)

b. Chlorination with sulphuryl chloride and selenium oxychloride. As in the case of the chlorination with chlorine, secondary hydrogens are more easily substituted than primary hydrogens and tertiary hydrogens more easily than secondary hydrogens on treatment with sulphuryl chloride. Hydrogens in the α-position next to a carbonyl function react with sulphuryl chloride at room temperature without any catalysts[77]. Chloroacetone[78], 3-chloro-3-methyl-2-butanone[79], 2-chloro-2-methylcyclohexanone[80], 2-chloropropiophenone[33] and 1-benzoyl-1-chlorocyclohexane[81] are prepared in high yields by treatment of the corresponding ketones with sulphuryl chloride (equation 25).

$$R^1CH_2COR^2 \xrightarrow{SO_2Cl_2} R^1CH(Cl)COR^2 \quad (25)$$

Reaction of ketones with two moles of sulphuryl chloride generally leads to mixtures of products. For example, from acetone at 30°C a mixture of 72% 1,1-dichloro-, 6% 1,3-dichloro- and 20% 1,1,3-trichloroacetone is produced, while from 2-butanone a mixture of 42% 3,3-dichloro-, 7% 1,1-dichloro- and 46% 1,3-dichloro-2-butanone is obtained[77]. Chlorination of cyclopentanone with an excess of sulphuryl chloride affords a mixture of 2,2-dichloro- and 2,5-dichlorocyclopentanone, while in the case of cyclohexanone only 2,2-dichlorocyclohexanone (**45**) is isolated when the reaction is carried out in dichloromethane or in acetic acid at 20°C[82]. Heating of α,α-dichlorocycloalkanones in acetic acid–hydrogen chloride results in rearrangement of a chlorine atom with formation of α,α′-dichloro compounds (**46**)[83], but this rearrangement is not applicable to the acyclic series (equation 26).

$$\text{cyclohexanone} \xrightarrow{SO_2Cl_2} \underset{(45)}{\text{2,2-dichlorocyclohexanone}} \xrightarrow{HCl/HOAc} \underset{(46)}{\text{2,6-dichlorocyclohexanone}} \quad (26)$$

Chlorinated cyclohexanones are formed during the chlorination of cyclohexenones. Treatment of 2-cyclohexenone (**47**) with sulphuryl chloride affords a mixture of 2-chloro-2-cyclohexenone (**48**), 2,2,3-trichloro- (**49**) and 2,3,6-trichlorocyclohexanone (**50**)[84], while chlorination of flavone (**51**) gives rise to 2,3,3-trichloroflavone (**52**)[85] (equation 27). Thionyl chloride reacts with 6-methyl- and 7-methoxyflavone to yield 3-chloro derivatives in both cases[85].

$$\underset{(47)}{\text{2-cyclohexenone}} \xrightarrow{SO_2Cl_2} \underset{(48)\ (10\%)}{\text{2-chloro-2-cyclohexenone}} + \underset{(49)\ (35\%)}{\text{2,2,3-trichlorocyclohexanone}} + \underset{(50)\ (55\%)}{\text{2,3,6-trichlorocyclohexanone}}$$

$$\underset{(51)}{\text{flavone}} \xrightarrow{SO_2Cl_2} \underset{(52)}{\text{2,3,3-trichloroflavone}} \quad (27)$$

During the chlorination of methyl ketones with selenium oxychloride, the intermediate dichloroselenium compounds (**53**) are decomposed thermally to furnish α-chloroketones (equation 28)[86].

$$RCOCH_3 \xrightarrow{SeOCl_2} \underset{(53)}{(RCOCH_2)_2SeCl_2} \xrightarrow{\Delta} RCOCH_2Cl \quad (28)$$

c. Chlorination with hypochlorites. Methyl ketones react with sodium hypochlorite in aqueous alkaline solution to give intermediate trichloromethyl ketones which are further transformed into chloroform and carboxylic acids (i.e. the so-called haloform reaction) (equation 29). Trichloromethyl ketones are isolated when acetophenones are used as substrates[87].

$$RCOCH_3 \xrightarrow{NaOCl} RCOCCl_3 \xrightarrow{H_2O/OH^-} CHCl_3 + RCOOH \quad (29)$$

Alkyl hypochlorites react easily with ketones; chloroacetone and ω-chloroacetophenone are prepared in good yields using ethyl hypochlorite[88]. *t*-Butyl hypochlorite seems to be an excellent reagent for the chlorination of steroidal ketones[89,90]. By the latter method, 2-chloro-3-cholestanone (**55**) is prepared from 3-cholestanone (**54**), while in pregnantrione derivatives (**56**) chlorination takes place at the 4-position (equation 30).

(54) $\xrightarrow{t\text{-BuOCl}}$ (55) (30)

(56) $\xrightarrow{t\text{-BuOCl}}$ (57)

d. Chlorination with N-*chlorosuccinimide.* Direct chlorination of ketones with *N*-chlorosuccinimide (NCS) is not a potential method for the synthesis of α-chloroketones because the reaction rate is often too slow and in most cases mixtures of reaction products are formed. Treatment of 2-heptanone with NCS in the presence of benzoyl peroxide gives a mixture of several mono-, di- and trichloro derivatives which are difficult to separate[91].

However, NCS is an excellent chlorinating agent of the corresponding *N*-analogues of ketones and enol ethers, i.e. ketimines and enamines, yielding α-chlorinated ketimines and β-chlorinated enamines. This subject has been reviewed elsewhere[92]. The last-mentioned compounds are potential sources for α-haloketones by a simple hydrolysis procedure. 1,1-Dichloromethyl ketones **(60)** are prepared by chlorination of *N*-cyclohexyl methyl ketimines **(58)** with two equivalents of NCS in carbon tetrachloride at 0°C, followed by hydrolysis in acidic medium[93–95] (equation 31). By

$$R^1R^2CHCOCH_3 \xrightarrow[H^+]{C_6H_{11}NH_2} R^1R^2CHC(=N\text{-}C_6H_{11})CH_3 \; (58) \xrightarrow[CCl_4/0°C]{2\ NCS}$$

$$R^1R^2CHC(=N\text{-}C_6H_{11})CHCl_2 \; (59) \xrightarrow[H_2O/CH_3OH]{H^+} R^1R^2CHCOCHCl_2 \; (60) \qquad (31)$$

$ArCOCCl_2R$

(**61**)

the same procedure 1-Aryl-2,2-dichloro-1-alkanones (**61**) have been successfully synthesized[96]. A similar method has been developed in which steroidal *N*-(β-hydroxyethyl)methylketimines (**62**) are treated with NCS in ether at 25°C, followed by mild acidic hydrolysis leading to the corresponding α-chloromethyl ketones (**63**) (equation 32). However, application of this halogenation method to 2-pentanone yields a mixture 1-chloro-, 3-chloro-, 1,1-dichloro- and 1,1,1-trichloro-2-pentanone[97].

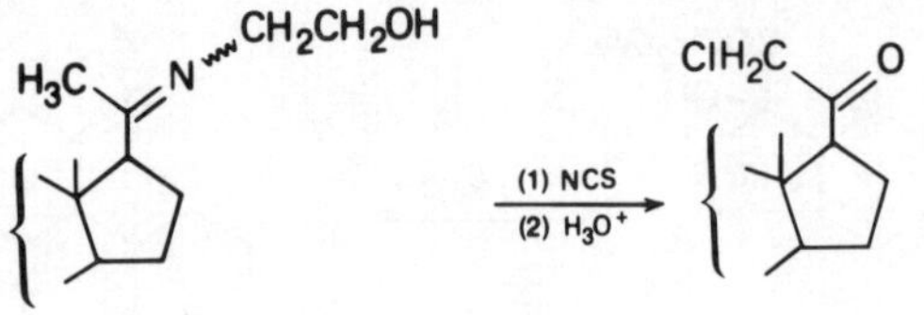

CH_2CH_2OH H_3C N ClH_2C O (1) NCS (2) H_3O^+ (32)

(**62**) (**63**)

Chlorination of the pyrrolidine enamines derived from 2-methylcyclohexanone (**64**) (which exists as a 9:1 mixture of two isomers) with NCS and subsequent hydrolysis gives 2-chloro-2-methylcyclohexanone (**65**), while the isomeric 6-chloro isomer (**66**) is not formed (equation 33)[98].

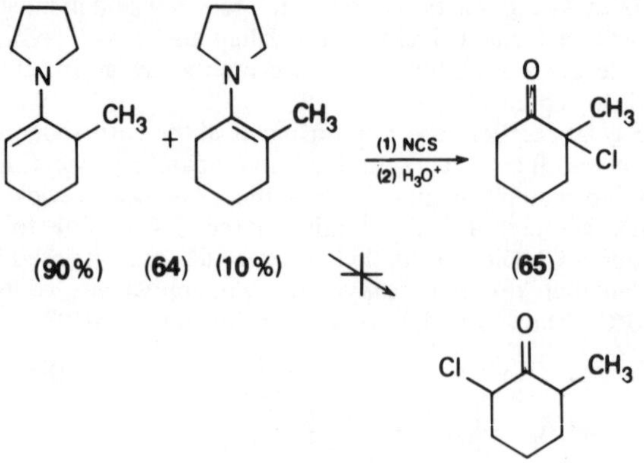

N CH_3 + N CH_3 (1) NCS (2) H_3O^+ O CH_3 Cl

(**90**%) (**64**) (10%) (**65**) (33)

O Cl CH_3

(**66**)

e. Chlorination with cupric and ferric chlorides. Cupric chloride is known as a chlorination catalyst but it has also been used for the preparation of chloroacetone from acetone[99,100]. Cyclohexanone and its methyl derivatives react with a large excess of cupric chloride in 50% aqueous acetic acid or 50% aqueous dioxan to give dichloro and trichloro derivatives of 1,2-cyclohexanediones (**67**, **68**)[101] (equation 34).

A convenient synthetic method consists of the reaction of silyl enol ethers (**69**) with cupric or ferric chlorides (equation 35)[102]. The mechanism involves a vinyloxy radical, generated from the collapse of the copper(II) or iron(III) enolate which is formed initially. The selection of specific solvents is important; for cupric chloride dimethylformamide must be used, while acetonitrile is the solvent of choice for ferric

$$\text{cyclohexanone} \xrightarrow[\text{50\% aq. HOAc or 50\% aq. dioxan}]{CuCl_2} \textbf{(67)} \xrightarrow{CuCl_2} \textbf{(68)} \quad (34)$$

$$\underset{\textbf{(69)}}{R^1C(OSiMe_3){=}CHR^2} \xrightarrow[FeCl_3/CH_3CN]{CuCl_2/DMF\ \text{or}} R^1COCHClR^2 \quad (35)$$

chloride. This method possesses the interesting feature that α-chlorination of unsymmetrical ketones can be performed regiospecifically and that extra double bonds are left intact. (Note that this does not occur when ferric chloride is used, but only with cupric chloride.)

f. Miscellaneous chlorination agents. Several other reagents or procedures of minor importance have been used for α-chlorination of ketones and their applications are strongly dependent upon the substrate. Treatment of acetophenones and aryl benzyl ketones with phenylchloroiodonium chloride gives rise to α-monochlorination, but reaction of 2-butanone with this reagent affords a mixture of 3-chloro- and 3,3-dichloro-2-butanone[103]. Pyridine hydrochloride perchloride[89] and phosphorus pentachloride[104] have also been used for α-chlorination of ketones.

Ketones possessing α-hydrogens are easily chlorinated with a system consisting of carbon tetrachloride, powdered potassium hydroxide and *t*-butanol, but subsequent rapid reactions generally lead to the formation of a variety of products such as Favorskii rearrangement products, α-hydroxy ketones and cleavage products[105,106]. Ketones with only one α-hydrogen, no α′-hydrogens and a sterically blocked carbonyl function such as **70** are especially suitable substrates and are easily converted into α-chloroketones (e.g. **71**), which are resistant to further reaction (equation 36). The

$$\underset{\textbf{(70)}}{2,4,6\text{-}(CH_3)_3C_6H_2COCH(CH_3)_2} \xrightarrow[t\text{-BuOH}]{CCl_4-KOH} \underset{\textbf{(71)}}{2,4,6\text{-}(CH_3)_3C_6H_2COCCl(CH_3)_2} \quad (36)$$

α-chlorination of ketones with this reagent involves the reaction of enolate anions with carbon tetrachloride in a discrete electron transfer/chlorine atom transfer step proceeding through a radical anion–radical pair (RARP) mechanism. As shown in equation (37), the formation of $Cl_3C:^-$ in the chlorination step leads to the generation of $:CCl_2$ as well as of $\dot{C}Cl_3$.

Hexachloroacetone acts as a source of positive chlorine in its reaction with enamines giving α-chloroketones after acid hydrolysis (equation 38)[107]. This reaction results in regioselective α-chlorination because of the availability of either α- or α′-enamines, thus making routes to 6-chloro-2-alkyl- or 6-chloro-3-alkylcyclohexanones quite feasible. For example, 6-methyl-1-pyrrolidinocyclohexene is transformed into a mixture of *cis*-6-chloro-2-methyl-, *trans*-6-chloro-2-methyl- and 2-chloro-2-methylcyclohexanone in a 93:6:1 ratio. The pyrrolidine enamines of 2-methylcyclohexanone

$$-\overset{O}{\overset{\|}{C}}-\underset{H}{\overset{|}{C}}- \xrightarrow{B^-} -\overset{O}{\overset{\|}{C}}\cdots\overset{-}{C} \xrightarrow{CCl_4} \left[-\overset{O}{\overset{\|}{C}}\cdots C\ (CCl_4)^{-\cdot}\right] \longrightarrow -\overset{O}{\overset{\|}{C}}-\overset{Cl}{\overset{|}{C}}- + Cl_3C:^- \quad (37)$$

$$Cl_3C:^- \longrightarrow :CCl_2 + Cl^- \qquad Cl_3C:^- \xrightarrow{CCl_4} 2Cl_3C^{\cdot} + Cl^-$$

$$R^1C(:NR^3_2)=CHR^2 + Cl-CCl_2-C(=O)-CCl_3 \xrightarrow[(2)\ H_3O^+\ (3)\ NaHCO_3]{(1)\ THF/-78-0^{\circ}C} R^1COCH(Cl)R^2 \quad (38)$$

react with sulphuryl chloride or NCS to give primarily the 2-chloro-2-methyl isomer.

Reaction of enamines with dimethyl(succinimido)sulphonium chloride (**72**) yields 2-amino-1-cycloalkenylsulphonium chlorides (**73**), which decompose into chlorinated enamines (**74**) under expulsion of dimethyl sulphide. Hydrolysis of these chlorinated enamines gives rise to α-chlorocycloalkanones (equation 39)[108].

$$\text{1-(dimethylamino)cycloalkene } (CH_2)_n + \text{(72)} \longrightarrow \text{(73)}: \overset{+}{S}(CH_3)_2\ Cl^- \xrightarrow{-(CH_3)_2S} \text{(74)}: \text{2-chloro enamine} \xrightarrow{H_3O^+} \text{2-chlorocycloalkanone} \quad (39)$$

A highly convenient electrolysis procedure for the preparation of α-halogenated ketones from enol acetates, enol ethers and silyl enol ethers has been developed (equation 40)[109]. The method consists of an electrolysis with halide salts in an undivided cell.

Reaction of dichlorocarbene with dioxolane (**75**) derivatives gives rise to dichloromethyl-1,3-dioxolanes (**76**) under phase transfer catalysis. These acetals are

$$\text{cyclo-}(CH_2)_n\text{-}C(OR)=CH \xrightarrow[NH_4Cl/MeCN]{-2e} \text{cyclo-}(CH_2)_n\text{-}C(=O)\text{-}CHCl \qquad (40)$$

excellent starting materials for syntheses of α,α-dichloromethyl ketones (equation 41)[110].

$$\underset{(75)}{\text{CH}_2\text{O}\text{-}\text{CH}_2\text{O}\text{>}C(R)H} \xrightarrow{:CCl_2} \underset{(76)}{\text{CH}_2\text{O}\text{-}\text{CH}_2\text{O}\text{>}C(R)CHCl_2} \xrightarrow{H_3O^+} RCOCHCl_2 \qquad (41)$$

A generally applicable synthesis of unsymmetrical α-chloroketones involves the chlorination of β-oxoalkylidenephosphorane (**77**) with iodobenzene dichloride followed by alkaline hydrolysis of the intermediate triphenylphosphonium chlorides (**78**) (equation 42)[111].

$$\underset{(77)}{Ph_3P=C(R^1)COR^2} \xrightarrow{PhICl_2} \underset{(78)}{Ph_3P^+\text{-}C(R^1)(Cl)COR^2}\ Cl^- \xrightarrow{^-OH} R^1CH(Cl)COR^2 + Ph_3PO \qquad (42)$$

2. *Synthesis of α-chloroketones from alcohols and phenols*

It is obvious that chlorination agents, which are capable of oxidizing alcohols to ketones, will give rise to chlorinated ketones using secondary alcohols as starting materials. The chlorination of isopropanol and 2-octanol, respectively, with chlorine gives 1,1,1,3-tetrachloroacetone and 1,1,1,3,3-pentachloro-2-octanone[112], while 2-chloro- and 2,2,6,6-tetrachlorocyclohexanone could be obtained from cyclohexanol in high yield[113,114]. Sterols are simultaneously oxidized and chlorinated upon treatment with hypochlorites[115].

Other reagents substitute hydroxy functions for chlorine atoms. 2-Hydroxytropolone (**79**) is transformed into 2-chlorotropolone (**80**) upon treatment with thionyl chloride in benzene[116] and 3-chloro-3-phenyl-*trans*-2-decalone (**82**) is formed from the 3-hydroxy derivative (**81**) by reaction with thionyl chloride in carbon tetrachloride[117] (equation 43). Hydroxyl functions in the side chain of steroids are easily substituted for chlorine by the action of arylsulphonyl chlorides[118].

The chlorination of phenols and halophenols with chlorine gives rise to polychlorinated cyclohexanones and cyclohexenones[119–122]. For example 2,4,4,6-tetrachloro-2,5-cyclohexadienone (**84**) and 2,2,4,5,6,6-hexachloro-3-cyclohexenone (**85**) are formed on chlorination of 2,4,6-trichlorophenol (**83**) (equation 44)[123].

1-Aryl-1-chloro-2-propanones (**87**) are formed from 1-aryl-2,2-dichloro-1-propanols (**86**) in generally good yields (69–90%) by an acid-catalysed rearrangement involving a 1,2-chlorine shift (equation 45)[124–126].

Studies on the acid-catalysed trifluoroacetolysis of 1-(*o*-chlorophenyl)-2,2-dichloro-1-propyl trifluoroacetate (**88**) indicate that the rearrangement takes place through the intermediacy of a halonium ion (equation 46).

$$\textbf{(79)} \xrightarrow[\text{Benzene}]{SOCl_2} \textbf{(80)}$$

$$\textbf{(81)} \xrightarrow[CCl_4]{SOCl_2} \textbf{(82)} \qquad (43)$$

$$\textbf{(84)} \xleftarrow[HOAc/H_2O]{Cl_2/80^{\circ}C} \textbf{(83)} \xrightarrow[HCl-HOAc]{Cl_2/20^{\circ}C} \textbf{(85)} \qquad (44)$$

$$\textbf{(86)} \xrightarrow{H_2SO_4} \textbf{(87)} \qquad (45)$$

$$\underset{\textbf{(88)}}{Ar-CH(OCOCF_3)-CCl_2-CH_3} \longrightarrow Ar-CH-CH(Cl)-CH_3 \; (\text{bridging } Cl^{+}) \xrightarrow{TFA}$$

$$Ar-CH(Cl)-C(Cl)(OCOCF_3)-CH_3 \xrightarrow[H_2SO_4]{TFA} Ar-CH(Cl)-C(=O)-CH_3 \qquad (46)$$

Another procedure for the synthesis of chloromethyl ketones involves the reaction of trichlorosilyl ethers with butyllithium, generating dichlorolithium compounds (**90**), which upon heating furnish chlorinated silyl enol ethers (**91**). Hydrolysis yields α-chloroketones (equation 47)[127].

It is evident that α-chloroketones are easily formed by oxidation of the corresponding β-chloro alcohols. However, β-chlorinated alcohols are not accessible in a general way and are mostly prepared by reduction of α-chloroketones (*vide infra*).

$$R^1R^2CHCH(OSiMe_3)CCl_3 \ (\mathbf{89}) \xrightarrow[-70\,°C]{n\text{-BuLi}} R^1R^2CHCH(OSiMe_3)CCl_2Li \ (\mathbf{90}) \xrightarrow{\Delta} R^1R^2CH(OSiMe_3)C{=}CHCl \ (\mathbf{91}) \xrightarrow{H_3O^+} R^1R^2CHCOCH_2Cl \quad (47)$$

The oxidation is exemplified by the conversion of 3,3-dichlorobicyclo[2,2,2]octan-2-ol (**92**) into the corresponding ketone (**93**) on treatment with chromic acid (equation 48)[128].

$$(\mathbf{92}) \xrightarrow[H_2SO_4]{CrO_3} (\mathbf{93}) \quad (48)$$

3. *Synthesis of α-chloroketones from α-diazoketones*

A general procedure for the preparation of α-chloroketones consists of decomposition of diazoketones in the presence of hydrogen chloride or chlorine yielding chloromethyl[129] and dichloromethyl ketones[130–131], respectively (equation 49). Excellent results are obtained for chloroacetone[132], 1,1,1,3-tetrachloroacetone[133]

$$RCH_2COCl \xrightarrow{CH_2N_2} RCH_2COCHN_2 \begin{cases} \xrightarrow{HCl} RCH_2COCH_2Cl \\ \xrightarrow{Cl_2} RCH_2COCHCl_2 \end{cases} \quad (49)$$

and chloromethyl benzyl ketone[134] on treatment of diazoketones, derived respectively from acetyl chloride, trichloroacetyl chloride and phenylacetyl chloride, with hydrogen chloride. 3,3-Dibromo-1,1-dichloro-2-butanone is synthesized by the reaction of chlorine in ether with the diazoketone derived from 2,2-dibromopropionyl bromide[131].

4. *Synthesis of α-chloroketones from alkenes and alkynes*

Addition of nitrosyl chloride to alkynes affords chlorinated nitroso compounds which upon acid hydrolysis yield monochlorinated ketones in excellent yields (equation 50)[135].

$$\underset{H}{>}C{=}C< \xrightarrow{NOCl} H{-}\underset{NO}{C}{-}\underset{Cl}{C}< \rightleftharpoons \left[H{-}\underset{NO}{C}{-}\underset{Cl}{C}<\right]_2 \xrightarrow{H^+} -\underset{O}{\overset{\|}{C}}{-}\underset{Cl}{C}< \qquad (50)$$

Oxidation of di- or trisubstituted olefins with chromyl chloride in acetone provides an efficient preparation of α-chloroketones (equation 51). For example, 2-chlorocyclododecanone is prepared by this method from *trans*-cyclododecene in 90% yield[136].

$$R^1CH{=}CR^2R^3 \xrightarrow{CrO_2Cl_2} R^1CO\underset{Cl}{C}R^2R^3 \qquad (51)$$

Several mono- and dichlorocyclobutanones have been synthesized by cycloadditions of chloro- or dichloroketenes to olefins. The dehydrohalogenation of 2-haloalkanoyl chlorides with triethylamine generates the chloroketene *in situ*, which in turn adds rapidly to dienes. 7-Chlorobicyclo[3,2,0]hept-2-en-6-ones (**94**) have been prepared in such a way by addition of chloroketenes to cyclopentadiene (equation 52)[137].

$$R\underset{Cl}{C}HCCl(=O) \xrightarrow{Et_3N} \underset{Cl}{\overset{R}{>}}C{=}C{=}O \xrightarrow{C_5H_6} \text{exo-alkyl} + \text{endo-alkyl} \qquad (52)$$

exo-alkyl (**94**) *endo*-alkyl

Chloro(2,2,2-trichloroethyl)ketene gives higher yields of [2 + 2] cycloadducts and a large variety of monochlorocyclobutanones (**95**) can readily be prepared (equation 53)[138].

$$CCl_3CH_2CHClCOCl \xrightarrow{Et_3N} \left[\underset{Cl}{\overset{CCl_3CH_2}{>}}C{=}C{=}O\right] \xrightarrow{(CH_3)_2C{=}CH_2} \qquad (53)$$

(**95**)

The cycloaddition of dichloroketene, generated *in situ* from trichloroacetyl chloride with triethylamine or with activated zinc in the presence of phosphorus oxychloride, constitutes a useful method for the synthesis of 2,2-dichlorocyclobutanone derivatives[139]. Styrene is converted into 2,2-dichloro-3-phenylcyclobutanone (**96**) in 87% yield (equation 54)[140]. Also silyl enol ethers seem to be suitable substrates for the preparation of functionalized cyclobutanones[141], while the adducts of indene and cyclopentadienes are valuable precursors in the synthesis of tropolones[142].

2,2-Dichlorocyclobutanones easily undergo regioselective one-carbon ring expansion by reaction with diazomethane, yielding 2,2-dichlorocyclopentanones (e.g. **97**). The presence of α-chloro substituents accelerates this reaction. Epoxide formation is not significant, probably because of the strained nature of the four-membered ring[143].

$$PhCH{=}CH_2 + [Cl_2C{=}C{=}O] \longrightarrow \textbf{(96)} \xrightarrow{CH_2N_2} \textbf{(97)} \tag{54}$$

Terminal alkynes can be converted into dichloromethyl ketones by treatment with hypochlorous acid (equation 55)[144–146]. 1,1-Dichloroacetone, 1,1-dichloro-3,3-

$$R{-}C{\equiv}CH \xrightarrow{HOCl} [R{-}C(OH)_2{-}CHCl_2] \xrightarrow{-H_2O} RCOCHCl_2 \tag{55}$$

dimethyl-2-butanone and ω,ω-dichloroacetophenone are obtained from propyne, 3,3-dimethyl-1-butyne and phenylacetylene, respectively. Treatment of phenylacetylene with chlorine in methanol gives 1-phenyl-1,1-dimethoxy-2,2-dichloroethane[147].

α,α-Dichloroketones are also prepared by reaction of acetylenes with *N*-chlorosuccinimide in methanol, followed by hydrolysis of the resulting dichlorodimethyl acctals (equation 56)[148].

$$R^1C{\equiv}CR^2 \xrightarrow[CH_3OH]{NCS} R^1C(OMe)_2{-}CR^2Cl_2 \xrightarrow{H_3O^+} R^1COCCl_2R^2 \tag{56}$$

5. *Synthesis of α-chloroketones from epoxides*

Ring opening of α-chloroepoxides, prepared by treatment of *gem*-dichloroalcohols with bases, gives rise to several halogenated ketones under various conditions (equation 57)[149,150].

$$R^1R^2C(O)CR^3Cl \begin{cases} \xrightarrow{\Delta} R^1R^2C(Cl){-}COR^3 \\ \xrightarrow{BF_3} R^1COC(R^2)(R^3)Cl \\ \xrightarrow{HCl} R^1R^2C(Cl){-}C(R^3)(OH)Cl \longrightarrow R^1R^2C(Cl){-}COR^3 \end{cases} \tag{57}$$

Neat thermal rearrangement of chlorinated epoxides normally gives rise to the formation of α-chloroketones, while on treatment with boron trifluoride a rearrangement takes place with formation of the isomeric chloroketone. Bifunctional epoxides also afford α-chloroketones on thermal or acid-catalysed isomerization (equation 58)[33,151,209].

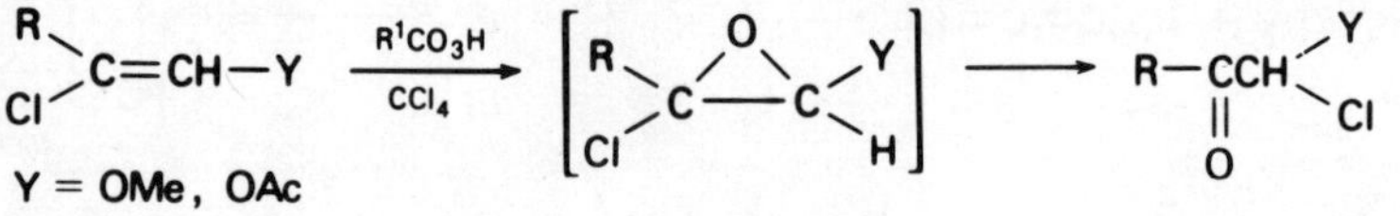

$$\text{R(Cl)C=CH-Y} \xrightarrow[\text{CCl}_4]{\text{R}^1\text{CO}_3\text{H}} \left[\text{R(Cl)C(-O-)C(Y)H}\right] \longrightarrow \text{R-C(=O)-CH(Y)Cl} \quad (58)$$

Y = OMe, OAc

A stereospecific chlorine migration occurs when a *cis–trans* mixture of 1-chloro-4-methylcyclohexene oxides (**98**) gives exclusively *trans*-2-chloro-4-methylcyclohexanone (**99**) on heating, while a zinc chloride-catalysed rearrangement gives rise to a mixture of the *cis* and the *trans* isomers (equation 59)[152]. If a hydride shift occurred, the other isomer (**100**) would be produced.

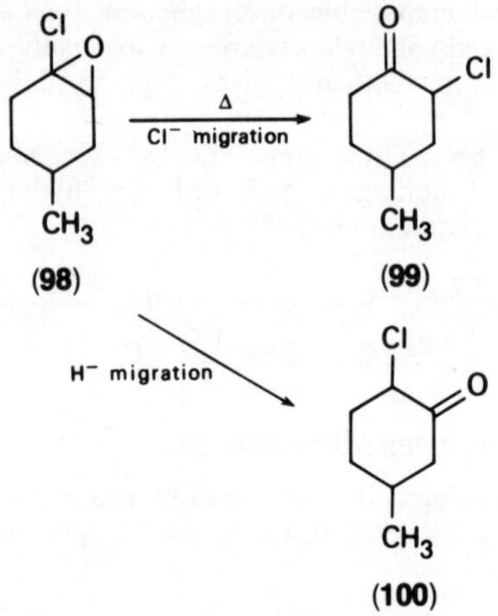

(59)

Δ, Cl^- migration; H^- migration

(**98**) (**99**) (**100**)

However, thermal rearrangement of 2-chlorobicyclo[2,2,1]hept-2-ene *exo*-oxide (**101**) gives rise to two major products, *exo*-3-chlorobicyclo[2,2,1]heptan-2-one (**102**, 38%) and *exo*-2-chlorobicyclo[2,2,1]heptan-7-one (**103**, 35%), while 2-chlorobicyclo-[2,2,2]oct-2-ene oxide (**104**) produces 89% 3-chlorobicyclo[2,2,2]octan-2-one (**105**) (equation 60)[153,154].

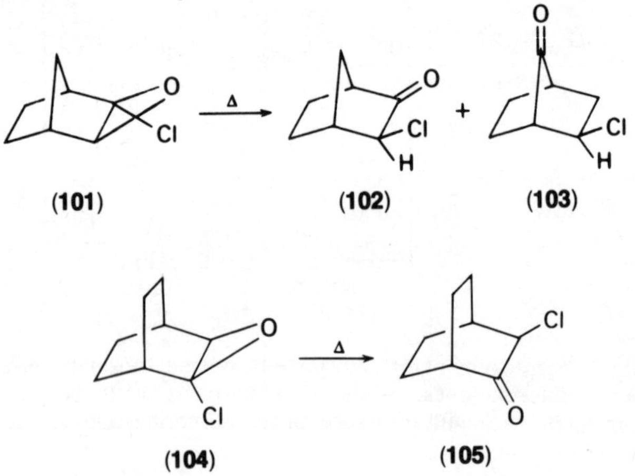

(60)

(**101**) Δ→ (**102**) + (**103**)

(**104**) Δ→ (**105**)

It has been proved in the case of α-chlorostyrene oxides that such thermal rearrangements occur by disrotatory $C_\beta-O$ bond heterolysis to yield an α-acylcarbenium chloride ion pair (equation 61)[155]. Upon heating 2,3-dichloro-

(61)

epoxides rearrange into α,α-dichloroketones, but the reaction course is strongly dependent upon the substitution pattern of the epoxide ring (equation 62)[156].

$R^1CCl_2COR^2$ $R^1COCCl_2R^2$ $R^1R^2C(Cl)COCl$ (62)

Ring opening of glycidonitriles (formed by condensation of a ketone with an α-halonitrile) with anhydrous hydrogen chloride leads to chlorinated cyanohydrins, which in turn upon treatment with sodium hydroxide expel hydrogen cyanide, yielding α-chloroketones (equation 63)[157].

$R^1COR^2 + NCCH(Cl)CHR^3R^4 \longrightarrow$ (63)

Another excellent conversion of epoxides into α-chloroketones involves the reaction of chlorodimethylsulphonium chloride (generated *in situ* by reacting molecular chlorine with dimethyl sulphide at −20°C) with epoxides in the presence of a tertiary amine (equation 64). For example, 2-chlorocyclohexanone is formed from cyclohexene oxide in 83% yield[158].

(1) $Cl\overset{+}{S}(CH_3)_2Cl^-$ (2) Et_3N (64)

6. *Synthesis of α-chloroketones from carboxylic acids and their derivatives*

Chlorination of diketene gives rise to the unstable γ-chloroacetoacetic acid chloride which decomposes in aqueous medium to yield monochloroacetone[159]. Dichloromethylketones have been synthesised by hydrolysis of lactone derivatives[160], by treatment of α,α-dichloroesters with Grignard reagents[149], and by the action of dichloromethyllithium on esters[162] (equation 65).

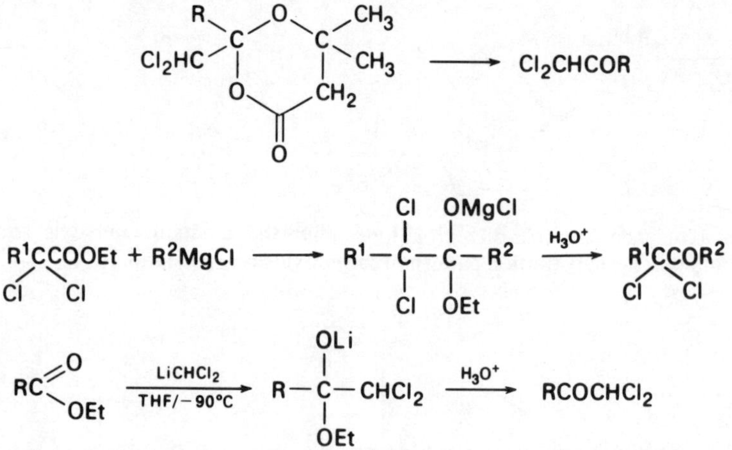

$$\text{2-R-2-}Cl_2HC\text{-6,6-dimethyl-1,3-dioxan-4-one} \longrightarrow Cl_2CHCOR$$

$$R^1CCl_2COOEt + R^2MgCl \longrightarrow R^1\text{—}CCl_2\text{—}C(OMgCl)(OEt)\text{—}R^2 \xrightarrow{H_3O^+} R^1CCl_2COR^2 \qquad (65)$$

$$RC(=O)OEt \xrightarrow[THF/-90°C]{LiCHCl_2} R\text{—}C(OLi)(OEt)\text{—}CHCl_2 \xrightarrow{H_3O^+} RCOCHCl_2$$

Acylation of alkynes with α,β-unsaturated acid chlorides provides 5-chloro-2-cyclopentenones **(106)** (equation 66)[163].

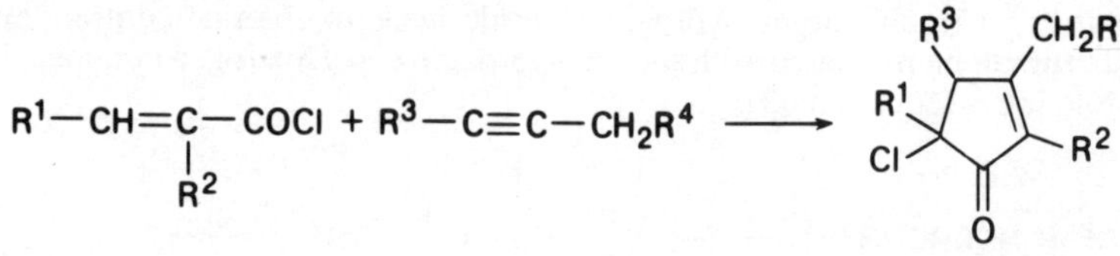

$$R^1\text{—}CH{=}C(R^2)\text{—}COCl + R^3\text{—}C{\equiv}C\text{—}CH_2R^4 \longrightarrow \textbf{(106)} \qquad (66)$$

(106): 5-R^1-5-Cl-4-R^3-3-CH_2R^4-2-R^2-cyclopent-2-enone

7. *Synthesis of α-chloroketones from aromatic amines*

Aromatic amines are converted into polychlorinated cyclohexanone compounds upon treatment with chlorine in acetic acid. For example, *p*-toluidine **(107)** gives 2,2,3,4,5,6,6-heptachloro-4-methylcyclohexanone **(108)** (equation 67)[164,165].

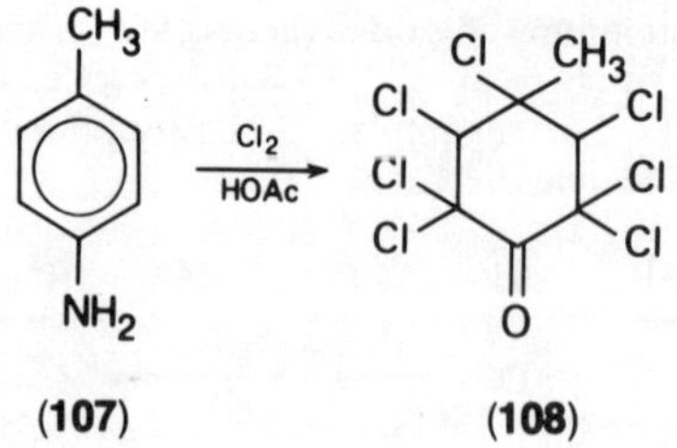

$$\textbf{(107)}\ \xrightarrow[HOAc]{Cl_2}\ \textbf{(108)} \qquad (67)$$

C. Synthesis of α-Bromoketones

The synthesis of α-bromoketones can be achieved by methods similar to those mentioned for the preparation of α-chloroketones, in addition to typical procedures and reagents for the introduction of bromine atoms in ketones.

1. *Synthesis of α-bromoketones from ketones and their derivatives*

a. Bromination with bromine. Treatment of ketones with bromine gives rise to substitution of at least one α-hydrogen. However, some side reactions take place during the bromination of ketones which are not encountered during the chlorination[166–169].

The bromination of ketones with bromine is a reversible process and the debrominated ketones are regenerated by reaction of the bromoketones with the liberated hydrogen bromide (equation 67a). In order to shift the equilibrium to the right,

$$RCOCH_3 + Br_2 \rightleftharpoons RCOCH_2Br + HBr \quad (67a)$$

precautions have to be taken to evaporate the hydrogen bromide or to take it up by an acceptor. In principle all brominated ketones are reducible by hydrogen bromide, but with varying ease depending on their structure. More 'positive' bromine atoms are reduced more rapidly. The tendency to debromination is directly related to the difficulty of introducing more than one bromine atom on a carbon atom in the presence of hydrogen bromide; trapping of the liberated hydrogen bromide is necessary.

Besides reduction, disproportionation reactions also take place, with the consequence that during the reaction of a ketone with an equimolecular amount of bromine some dibromoketone is always produced (equation 68). The monobromo:dibromo

$$2\,RCOCH_2Br \rightleftharpoons RCOCH_3 + RCOCHBr_2 \quad (68)$$

compound ratio is dependent upon the solvent and reaction time. ω,ω,ω-Tribromoacetophenone, with highly 'positive' bromine atoms, is able to brominate acetophenone, yielding phenacyl bromide (equation 69)[4].

$$PhCOCBr_3 + PhCOCH_3 \longrightarrow PhCOCH_2Br \quad (69)$$

Another side reaction occurring during the synthesis of bromoketones consists of a rearrangement of a bromine atom under the influence of hydrogen bromide. α,α-Dibromoketones rearrange to 1,3-dibromo compounds, but geminal dibromoketones are formed when the bromination is carried out in the presence of potassium acetate[170]. Bromomethyl cyclohexyl ketone (**107a**) is transformed to 1-acetyl-1-bromocyclohexane (**108a**) (equation 70)[171]. In the case of 1-bromo-3-phenyl-2-propanone

$$RCH_2COCH_2Br \xrightarrow{Br_2} RCH_2COCHBr_2 \xrightarrow{HBr} RCH(Br)COCH_2Br$$

$$\text{c-}C_6H_{11}COCH_2Br\ (\mathbf{107a}) \xrightarrow{HBr} \text{1-Br-c-}C_6H_{10}COCH_3\ (\mathbf{108a}) \quad (70)$$

(**109**) an equilibrium is established between both isomers (**109** and **110**)[171]. In addition, the solvent seems to have a great influence on the position of substitution, as exemplified by the bromination of 1,1-diphenyl-2-propanone (**111**) (equation 71)[168].

$$PhCH_2COCH_2Br \rightleftharpoons PhCHBrCOCH_3$$

(**109**) (**110**)

$$Ph_2CHCOCH_3 \text{ (111)} \xrightarrow[CCl_4]{Br_2} Ph_2CBrCOCH_3 \text{ (112)} \xrightarrow{HBr} Ph_2CHCOCH_2Br \text{ (113)}$$

$$Ph_2CHCOCH_3 \text{ (111)} \xrightarrow[Br_2]{HOAc} Ph_2CHCOCH_2Br \text{ (113)} \quad (71)$$

Various procedures have been employed for the bromination of aliphatic ketones with bromine. Direct treatment gives very impure products since the liberated hydrogen bromide tends to promote the formation of condensation products and resinous materials next to by-products which are formed during side reactions. These difficulties are minimized by bromination in an inert atmosphere, by the use of acetic acid as solvent or in the presence of calcium carbonate, potassium acetate and potassium chlorate. Bromination of acetone in a mixture of acetic acid and water provides bromoacetone in a 44% yield together with 2,2-dibromo- and 1,3-dibromoacetone[172]. Better results are obtained when the bromination is carried out in an aqueous solution in the presence of potassium chlorate[173]. Further bromination of bromoacetone gives 1,1,3-tribromo- and 1,1,3,3-tetrabromoacetone, while reaction of acetone with an excess of bromine yields pentabromoacetone[174].

Bromination of alkyl methyl ketones always leads to the formation of isomeric compounds. Normally bromo-substitution of methylene groups is faster than of methyl groups, but the rate is nearly identical for methylene and methine moieties (equation 72). For example, acid-catalysed bromination in the presence of potassium chlorate

$$CH_3COCH_2R \xrightarrow{Br_2} CH_3COCHBrR + BrCH_2COCH_2R$$

$$CH_3COCHR^1R^2 \xrightarrow{Br_2} CH_3COCBrR^1R^2 + BrCH_2COCHR^1R^2 \quad (72)$$

affords mixtures of 73% 3-bromo- and 27% 1-bromo-2-butanone from 2-butanone and 63% 3-bromo- and 37% 1-bromo-2-pentanone from 2-pentanone while 3-methyl-2-butanone gives rise to 76% 3-bromo-3-methyl- and 24% 1-bromo-3-methyl-2-butanone[175–177]. Pinacolone can be converted into the mono- and the dibromo compounds when the bromination is carried out in ether[178], while tribromopinacolone is obtained in a refluxing carbon tetrachloride–water mixture in the presence of mercuric chloride[179].

Monobromination of unsymmetrical aliphatic ketones is rarely a regiospecific reaction and seems to be strongly dependent upon the solvent used. While in carbon tetrachloride, ether and acetic acid the substitution mostly occurs at the most substituted α-carbon atom, yielding mixtures of reaction products[180], bromination in methanol preferentially takes place at the less substituted carbon atom, as illustrated by the bromination of 3-methyl-2-butanone (**114**) (equation 73). The latter procedure constitutes an excellent method for the preparation of bromomethyl ketones, not readily accessible previously by direct bromination[181].

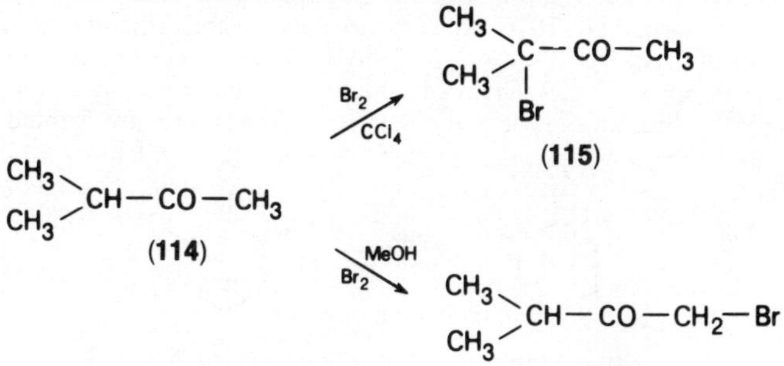

$$(CH_3)_2CH-CO-CH_3 \ (\mathbf{114}) \xrightarrow[CCl_4]{Br_2} (CH_3)_2CBr-CO-CH_3 \ (\mathbf{115})$$

$$(CH_3)_2CH-CO-CH_3 \ (\mathbf{114}) \xrightarrow[Br_2]{MeOH} (CH_3)_2CH-CO-CH_2-Br \ (\mathbf{116}) \qquad (73)$$

Bromination of ketones with bromine in carbon tetrachloride during irradiation with a 100 W tungsten lamp in the presence of 1,2-epoxycyclohexane gives monobromoketones in which bromine has entered exclusively the more highly substituted α-position or the benzylic position (equation 74). The extent of substitution α to the

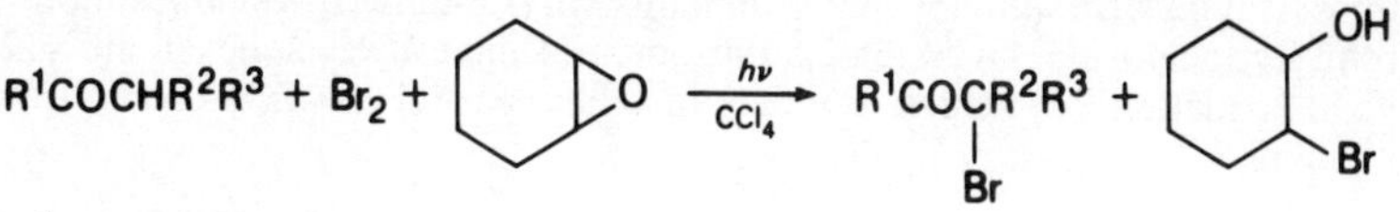

$$R^1COCHR^2R^3 + Br_2 + \text{1,2-epoxycyclohexane} \xrightarrow[CCl_4]{h\nu} R^1COCBrR^2R^3 + \text{2-bromocyclohexanol} \qquad (74)$$

R^1 = Me, Et

R^2 = alkyl, phenyl

R^3 = H, alkyl

carbonyl group plays a decisive role in the reaction. Ketones with a secondary or a benzylic α-carbon atom are brominated at this position exclusively. With less substituted ketones (2-butanone, acetone) the reaction takes a different course, providing a mixture of 2-bromocyclohexane, 2-bromocyclohexanol and the starting ketone[182]. Under the same reaction conditions a regiospecific introduction of bromine occurred at the C_{17} of 5α- and 5β-pregnane-3,20-dione (**117**). The selectivity of these reactions is due to the epoxide which, by scavenging the hydrogen bromide produced during the reaction, inhibits any ionic acid-catalysed bromination of the ketones.

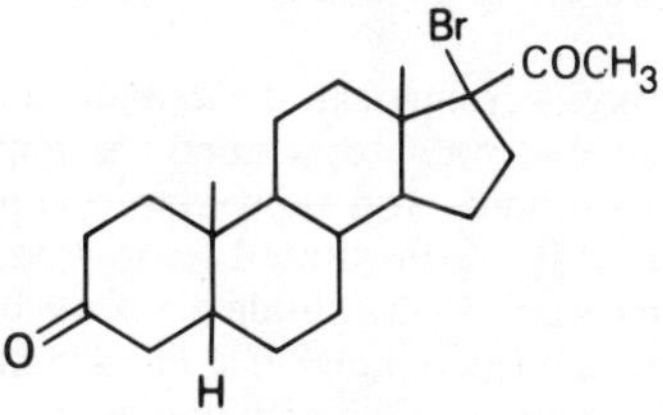

(**117**)

Treatment of aliphatic ketones with two equivalents of bromine in acetic acid or ether results in the formation of stereoisomeric α,α′-dibromoketones and no geminal dibromo compounds are formed (equation 75)[183,184].

$$R^1CH_2COCH_2R^2 \xrightarrow{Br_2} R^1CH(Br)COCH(Br)R^2 \quad (75)$$

Monobromocycloalkanones, prepared from cycloalkanones with bromine in an acetic acid–water mixture at 50–70°C, are very air-sensitive and difficult to purify[185–186].

Bromination of 2-chloro-, 2-cyano- or 2-fluorocyclohexanone (**118**) in carbon tetrachloride in the presence of calcium carbonate takes place mainly at the 6-position (equation 76)[187]. Stereoisomeric α,α'-dibromocycloalkanones are formed by bromi-

(**118**) $\xrightarrow{Br_2}$ (**119**) (76)

X = Cl, CN, F

nation with two equivalents of bromine in acetic acid or anhydrous ether[183]. Reaction of cyclohexanol with bromine in acetic acid containing 15% hydrogen bromide yields 20% crystalline *cis*-2,6-dibromocyclohexanone and 80% of the *trans* compound, which decomposes upon distillation[188]. Dibromination of 4,4-dimethylcyclohexanone (**120**) in carbon tetrachloride gives *cis*-2,6-dibromo-4,4-dimethylcyclohexanone (**121**) in 66% yield (equation 77). Upon standing in ether, partial *cis–trans* isomerisation is

(**120**) $\xrightarrow[CCl_4]{Br_2}$ (**121**) (77)

observed[189]. The *cis* isomers have higher melting points and higher infrared carbonyl stretching frequencies and are more polar as well as less soluble than the *trans* analogues, which are considered to be conformationally more mobile.

Neat bromination of cyclohexanone afford tetrabromocyclohexanone[190], while bromination of cyclohexadecanone with 3.5 mole equivalents of bromine in dichloromethane at 25–30°C gives 2,2,15-tribromocyclohexadecanone in 92% yield[191].

Bromination of aryl alkyl ketones can be carried out selectively and ω-bromo, ω,ω-dibromo- and ω,ω,ω-tribromoacetophenone are synthesised in high yields from acetophenone[192–194]. Monobromo- and dibromopropiophenone are obtained from propiophenone, although for the disubstituted compound to be obtained the bromination must be performed in carbon tetrachloride at reflux temperature under irradiation and in the presence of benzoyl peroxide[33,195]. The bromination of indanone can be carried out selectively. Reaction in ether with one molar equivalent of bromine yields the 2-bromo compound while 2,2-dibromoindanone is obtained upon treatment with two molar equivalents of bromine in chloroform[196]. In general, α,α-dibrominated alkyl aryl ketones are not easily accessible due to exchange processes in the presence of hydrogen bromide[197,198].

During the addition of bromine to α,β-unsaturated ketones which yields α,β-dibromoketones, precautions have to be taken in order to avoid decomposition. The reaction has to be carried out very slowly and at low temperature (0°C) as exemplified by the preparation of 3,4-dibromo-3-methyl-2-butanone[199] and 3,4-dibromo-4-phenyl-2-butanone[200].

Alkoxybromination occurs when α,β-unsaturated ketones are treated with bromine in an alcohol, giving rise to α-bromo-β-alkoxy ketones[1] (equation 78)[201].

$$R^1CH{=}CHCOR^2 \xrightarrow[R^3OH]{Br_2} R^1CH(OR^3){-}CH(Br)COR^2 \qquad (78)$$

2-Bromo-2-cyclohexenone (**123**) is prepared by treatment of 2-cyclohexenone (**122**) with bromine in collidine, the initial adduct being dehydrobrominated (equation 79)[202].

$$\textbf{(122)} \xrightarrow[\text{Collidine}]{Br_2} \textbf{(123)} \qquad (79)$$

Treatment of isophorone (**124**) with an excess of bromine in carbon tetrachloride gives 2,4,6-tribromo-3-bromomethyl- (**125**) and 2,6,6-tribromo-3-dibromomethyl-5,5-dimethyl-2-cyclohexenone (**126**), respectively at 0 and 25°C. Further treatment of **125** affords the pentabromocyclohexenone (**127**) (equation 80)[202].

$$\textbf{(124)} \xrightarrow[CCl_4]{Br_2} \begin{cases} 0^\circ C: \textbf{(125)} \xrightarrow[CCl_4]{Br_2} \textbf{(127)} \\ 25^\circ C: \textbf{(126)} \end{cases} \qquad (80)$$

Just as in the preparation of α-chloroketones, various ketone derivatives serve as substrates for the synthesis of bromoketones. The bromination of trimethylsilyl enol ethers with bromine in carbon tetrachloride at −20°C represents an excellent method for the regiospecific introduction of a bromine atom into aliphatic and cyclic ketones (equation 81)[203].

Treatment of enol acetates with bromine gives rise to α-bromoketones: 2-bromo-1-phenyl-1-propanone and 2-bromocycloalkanones are prepared according to this procedure (equation 82)[204].

$$R^1R^2CHCOR^3 \xrightarrow[NEt_3/DMF]{ClSi(CH_3)_3} R^1R^2C{=}C(OSi(CH_3)_3)R^3 \xrightarrow{Br_2} R^1R^2C(Br){-}COR^3 \quad (81)$$

(128) $\xrightarrow[NEt_3/DMF]{ClSi(CH_3)_3}$ (129) $\xrightarrow{Br_2}$ (130)

$$R^1CH{=}C(OAc){-}CH_3 \xrightarrow{Br_2} R^1{-}CH(Br){-}C(Br)(OAc){-}CH_3 \xrightarrow{R^2OH} R^1CH(Br)COCH_3 + CH_3COOR^2 \quad (82)$$

Bromination of enamines constitutes a suitable method for the preparation of bromoketones and the procedures mentioned for the synthesis of α-chloroketones are also applicable here[75,76].

b. Bromination with N-*bromo compounds.* The use of *N*-bromo compounds in the preparation of α-bromocarbonyl compounds, first reported by Schmid and Karrer[205] using *N*-bromosuccinimide (NBS), shows the advantage that neither hydrogen bromide nor free bromine are present during the reaction, with the consequence that side reactions are largely eliminated. Monobromination occurs smoothly and geminal dibromination rarely takes place, so that brominated ketones which are not available by the bromine method can be synthesized. However, the reaction rate is much slower with *N*-bromo compounds and in some cases no bromination occurs at all[206].

Numerous examples of monobromination of aliphatic and acyclic ketones with *N*-bromosuccinimide are known, mostly in the presence of initiators (benzoyl peroxide (BPO), azo-isobutyronitrile) and/or illumination[207–208]. Geminal dihaloketones (**133**, **135**) are formed when α-chloroketones, such as α-chlorocyclohexanone and α-fluoropropiophenone, are treated with NBS (equation 83)[33,187,209].

Introduction of a bromine atom in α,β-unsaturated ketones takes place at the allylic position and not at the carbon atom next to the carbonyl function[210]. Bromination of isophorone (**124**) with NBS gives rise to 4-bromoisophorone (**136**) (equation 84)[202]. Other *N*-bromo compounds, e.g. *N*-bromophthalimide, *N*-bromoacetamide, *N*-bromotolylsulphonylamide, 3-bromo- and 1,3-dibromo-5,5-dimethylhydantoin have been used less frequently as brominating agents[211]. Besides bromination of ketones with these reagents, derivatives such as enol acetates, enol ethers and enamines have also been treated and excellent yields of bromoketones are obtained, as in the case of the corresponding chloro derivatives[98,212].

However, the bromination of ketimines and subsequent hydrolysis turns out not to be a useful method for the preparation of dibromoketones. Reaction of methylketimines with NBS in CCl_4 gives rise to α,α-dibromoketimines, but hydrolysis of the latter compounds provides a mixture of α,α-dibromo and α,α'-dibromoketones[213]. On the other hand, hydrolysis of *N*-1-(2,2-dibromo-1-phenylalkylidene)cyclohexylamines affords a mixture of mainly 1-aryl-2,2-dibromo-1-alkanones and 1-aryl-1,2-alkanediones[96].

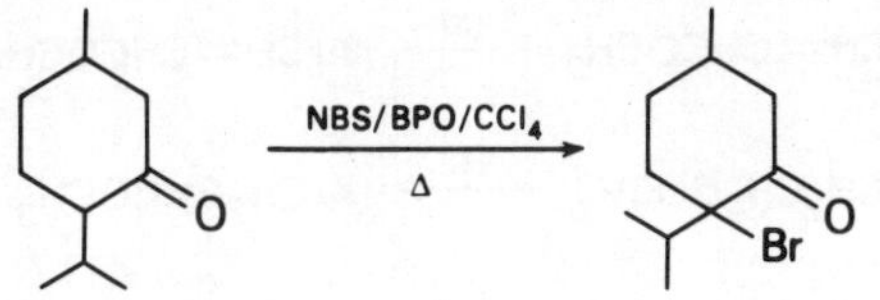

NBS/BPO/CCl_4, Δ

(131) → (132)

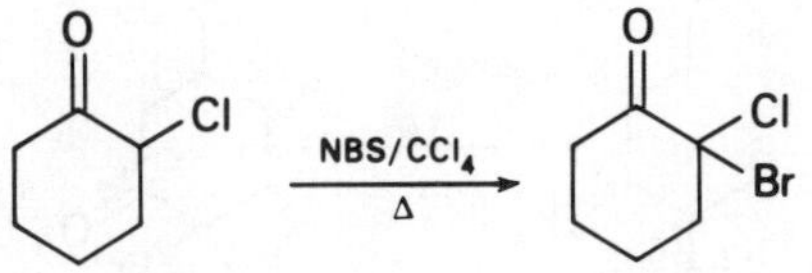

NBS/CCl_4, Δ

(118) → (133) (83)

$$PhCOCH(F)CH_3 \xrightarrow[\Delta]{NBS/CCl_4} PhCOC(F)(Br)-CH_3$$

(134) → (135)

$$R^1CH_2CH{=}CHCOCH_2R^2 \xrightarrow{NBS} R^1CH(Br)CH{=}CHCOCH_2R^2$$

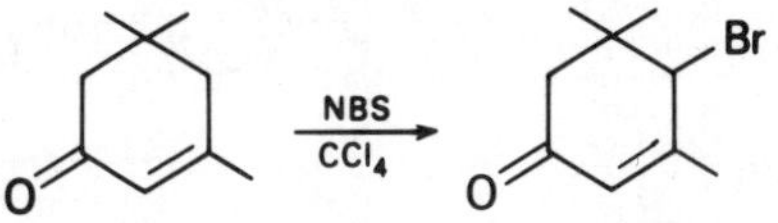

(84)

NBS, CCl_4

(124) → (136)

c. *Miscellaneous brominating agents.* Copper(II) bromide is an excellent reagent for the preparation of α-monobromoketones when the reaction is carried out in refluxing chloroform-ethyl acetate[214,215] (equation 85) (R^1, R^2 = alkyl, phenyl, $-(CH_2)_n-$).

$$R^1COCH_2R^2 \xrightarrow[CHCl_3/EtOAc]{CuBr_2} R^1COCH(Br)R^2 + CuBr + HBr \quad (85)$$

Selective bromination of C—H α to a carbonyl function can also be achieved by pyridinium hydrobromide perbromide[216], tetrazolium perbromide[217], phenyltrimethylammonium perbromide[218], 2,4-diamino-1,3-thiazole hydroperbromide[219] and 2-carboxyethyltriphenylphosphonium perbromide (**137**)[220].

The last-mentioned compound, which is conveniently prepared by heating triphenylphosphine and acrylic acid in 49% hydrobromic acid followed by treatment with bromine in acetic acid, selectively gives monobromination at the α-position of a keto function even in the presence of double bonds. In the case of unsymmetrically substituted ketones, α-bromination occurs predominantly at the most substituted carbon atom due to the preferred enolization in that direction (equation 86).

$$(C_6H_5)_3\overset{+}{P}CH_2CH_2COOH\ Br_3^-$$

(**137**)

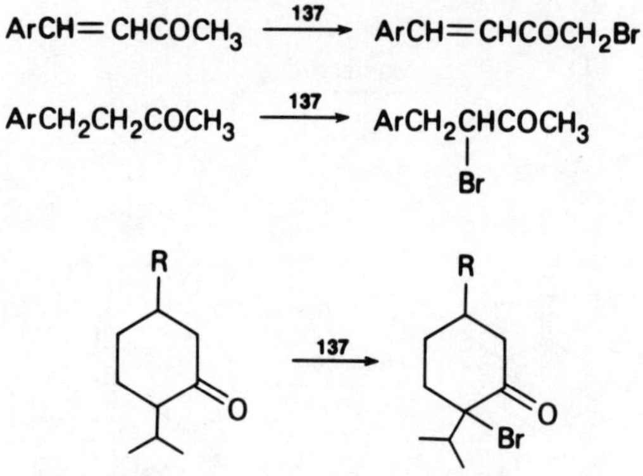

$$ArCH{=}CHCOCH_3 \xrightarrow{137} ArCH{=}CHCOCH_2Br$$

$$ArCH_2CH_2COCH_3 \xrightarrow{137} ArCH_2CH(Br)COCH_3 \qquad (86)$$

Recently, an insoluble regenerable brominating polymer has been developed. This polymer is easily and safely prepared from the macroreticular anion exchange resin Amberlyst-A26 Cl^- (Rohm and Haas Co.) (equation 87).

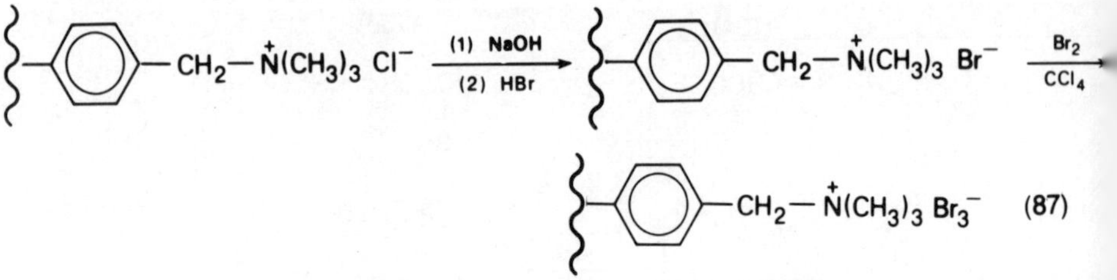

$$\text{P}-C_6H_4-CH_2-\overset{+}{N}(CH_3)_3\ Cl^- \xrightarrow[(2)\ HBr]{(1)\ NaOH} \text{P}-C_6H_4-CH_2-\overset{+}{N}(CH_3)_3\ Br^- \xrightarrow[CCl_4]{Br_2} \text{P}-C_6H_4-CH_2-\overset{+}{N}(CH_3)_3\ Br_3^- \qquad (87)$$

(138)

Amberlyst-A26 bromide form is converted into the perbromide form (**138**) by treating with a carbon tetrachloride solution of bromine. Unsymmetrical ketones are selectively brominated at the more highly substituted position, in the presence of a free radical initiator and methyloxirane as scavenger of the hydrobromic acid. α,β-Unsaturated ketones are completely converted into the corresponding dibromo saturated adducts. In the reaction of steroidal ketones, bromination occurs mainly from the less hindered side of the molecule (equation 88). The advantage of this reagent consists of the ease of operation, the work-up conditions and the capability for regeneration[221–222].

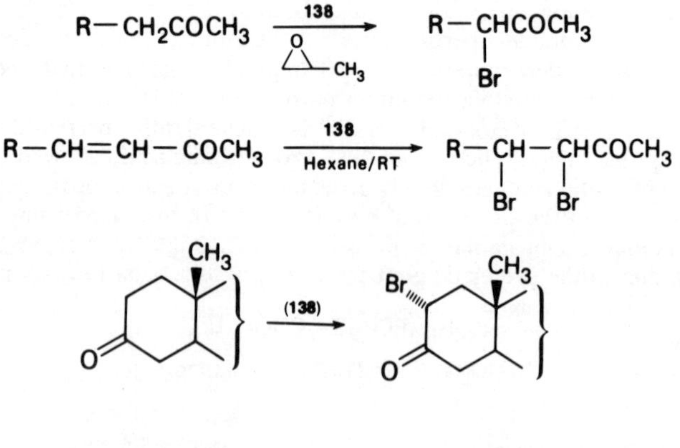

$$R-CH_2COCH_3 \xrightarrow[\text{methyloxirane}]{138} R-CH(Br)COCH_3$$

$$R-CH{=}CH-COCH_3 \xrightarrow[\text{Hexane/RT}]{138} R-CH(Br)-CH(Br)COCH_3 \qquad (88)$$

Brominated active methylene functions are also able to act as bromonium ion sources. Monobromo- and dibromomalononitrile are suitable reagents for the bromination of active methylene functions, but they are not active enough to introduce a bromine atom at the α-carbon of monocarbonyl functions[223,224].

An excellent reagent, however, seems to be 5,5-dibromo-2,2-dimethyl-4,6-dioxo-1,3-dioxan (**138a**), which is able to monobrominate saturated aldehydes and ketones and the α′-carbon atom of α,β-unsaturated ketones with high selectivity (equation 89)[225].

$$2\ R^1R^2CH-CO-R^3 + \textbf{(138a)} \longrightarrow 2\ R^1R^2C(Br)-CO-R^3 + \text{2,2-dimethyl-4,6-dioxo-1,3-dioxan} \quad (89)$$

(**138a**)

In addition, 2-bromo-2-cyano-*N,N*-dimethylacetamide is also effective for the synthesis of α-monobromoketones[223], while selective monobromination of α,β-unsaturated ketones has been performed by the action of 2,4,4,6-tetrabromocyclohexa-2,5-dienone without affecting the double bond or any allylic position (equation 90)[226].

$$R-CH_2-CH{=}CH-CO-CH_3 + \text{2,4,4,6-tetrabromocyclohexa-2,5-dienone} \longrightarrow R-CH_2-CH{=}CH-\overset{O}{\overset{\|}{C}}-CH_2Br + \text{2,4,6-tribromophenol} \quad (90)$$

Dioxan dibromide has been used for bromination in the side chain of electron-rich hydroxy- and methoxy-substituted acetophenones, which often suffer nuclear bromination with other brominating agents[227].

Another procedure involves the attack of bromodimethylsulphonium bromide (**138b**) on enamines followed by hydrolysis of the intermediate brominated immonium salt (equation 91)[158]. Sodium hypobromite is not a suitable reagent for the preparation of bromoketones. Methyl ketones give the haloform reaction; e.g. propiophenone has been oxidized to benzoic and acetic acid[228].

$$\text{(138b)} + Br{-}\overset{+}{S}(CH_3)_2 \longrightarrow \text{[2-bromocyclohexylidene morpholinium]}^{+}\,Br^{-} + S(CH_3)_2 \xrightarrow{H_3O^+} \text{2-bromocyclohexanone} \qquad (91)$$

2. *Synthesis of α-bromoketones from α-diazoketones*

Monobromo- and dibromoketones are produced by decomposition of diazoketones with hydrogen bromide and bromine, respectively (equation 92)[130,229]. Sometimes minor amounts of the corresponding chloroketones are produced when starting from acid chlorides (X = Cl). Therefore acid bromides are preferentially used for the preparation of the diazoketones[231].

$$RCOX \xrightarrow{CH_2N_2} RCOCHN_2 \begin{cases} \xrightarrow{HBr} RCOCH_2Br \\ \xrightarrow{Br_2} RCOCHBr_2 \end{cases} \qquad (92)$$

3. *Synthesis of α-bromoketones from epoxides*

Most of the procedures mentioned in the section dealing with the synthesis of α-chlorinated ketones using epoxides as substrates are also applicable for the preparation of α-bromoketones. Nevertheless, some specific methods using epoxides are available for the synthesis of the corresponding bromoketones.

Photocatalytic bromination of epoxides in carbon tetrachloride yields ketones directly, exclusively monobrominated at the less substituted α-carbon atom (equation 93)[232].

$$\text{RHC}\overset{O}{-\!\!-}\text{CH}_2 + Br_2 \xrightarrow{CCl_4/h\nu} RCOCH_2Br + HBr \qquad (93)$$

R = alkyl, phenyl

The majority of epoxides tested react with a stoichiometric amount of bromine to produce only the bromoketones and no bromohydrins, the latter arising from ring cleavage of the epoxide by the generated hydrogen bromide. However, cyclohexene oxide and styrene oxide give, besides the monobromo-, also the dibromoketones and the bromohydrins. The photocatalysis is indispensable and the choice of the solvent is critical. In ether only the two bromohydrins are formed.

The mechanism must involve a free radical hydrogen abstraction process followed by a fast rearrangement to an α-oxoalkyl radical (equation 94).

$$\text{R(H)}\overset{\text{O}}{\text{C—C}}\text{H}_2 + {}^{\bullet}\text{Br} \longrightarrow \text{R}\dot{\text{C}}\overset{\text{O}}{\text{—C}}\text{H}_2 + \text{HBr}$$

$$\downarrow$$

$$\text{R—CO—}\dot{\text{C}}\text{H}_2 \xrightarrow{\text{Br}_2} \text{R—COCH}_2\text{Br} \qquad (94)$$

Reaction of epoxysulphonyl compounds, now conveniently available from α-chlorosulphones and aldehydes under phase transfer conditions with magnesium dibromide in ether at room temperature, affords α-bromo carbonyl compounds (equation 95)[233].

$$\text{R}^1\text{R}^2\text{C=O} + \text{PhSO}_2\text{CH(R}^3\text{)Cl} \longrightarrow \text{PhSO}_2\text{—C(R}^3\text{)}\overset{\text{O}}{\text{—}}\text{C(R}^1\text{)R}^2 \xrightarrow{\text{MgBr}_2}$$

$$\text{PhSO}_2\text{—C(O}^-\text{)(R}^3\text{)—C(R}^2\text{)(R}^1\text{)—Br} \longrightarrow \text{PhSO}_2^- + \text{R}^3\text{—C(=O)—C(R}^2\text{)(R}^1\text{)—Br} \qquad (95)$$

$$\text{R}^1, \text{R}^2 = \text{H, alkyl, phenyl}$$

The epoxysulphone route is more general than the α-chloroepoxide route[149] and can be carried out easily. For example, 1-bromo-1-phenyl-2-propanone can be obtained in a yield greater than 95%. Ring cleavage of nitroepoxides with hydrogen bromide also gives rise to the formation of α-bromoketones (equation 96)[234].

$$\text{R}^1\text{R}^2\text{C}\overset{\text{O}}{\text{—}}\text{C(R}^3\text{)NO}_2 \xrightarrow{\text{HBr}} \text{R}^1\text{R}^2\text{C(Br)—COR}^3 \qquad (96)$$

4. Synthesis of α-bromoketones from miscellaneous substrates

Some of the procedures already mentioned in the section concerning the synthesis of α-chloroketones can be utilized for the preparation of bromoketones.

Dibromomethyl ketones are formed by the action of dibromomethyllithium on esters[162]. Jones oxidation and oxidation with pyridinium chlorochromate of brominated cycloalkanols also give excellent results, as in the synthesis of *cis*-2,8-dibromocyclooctanone[235].

Another method uses vinyl esters which are transformed into dibromoesters upon addition of bromine. These esters spontaneously decompose into acyl bromides and

$$R^1CO-O\underset{R^2}{C}=CH_2 \xrightarrow{Br_2} R^1CO-O\overset{Br}{\underset{R^2}{C}}-CH_2Br \longrightarrow R^1COBr + R^2COCH_2Br \quad (97)$$

α-bromoketones (equation 97)[236]. 1-Bromo-2-heptanone is obtained from 2-acetoxy-1-heptene in 75% yield. Pyrolytic elimination of α-bromo-β-hydroxysulphoxides, obtained by reaction of aldehydes with lithiobromomethyl phenyl sulphoxide, produces α-bromomethyl ketones in excellent yields (equation 98)[237].

$$R-CHO + Ph\overset{O}{\overset{\|}{S}}-\underset{Li}{CHBr} \xrightarrow{THF/-78^{\circ}C} R-\underset{OH}{CH}-\underset{Br}{CH}-SOPh \xrightarrow[160^{\circ}C]{Diglyme} RCOCH_2Br \quad (98)$$

D. Synthesis of α-Iodoketones

α-Iodoketones are usually prepared by treatment of ketones with iodine in the presence of a strong base[238], by exchange reaction of chloro- or bromoketones with inorganic iodides[212,239,240], by treatment of ketones and their enol acetates with *N*-iodosuccinimide[241], by action of iodine(I) chloride on enol acetates[212] and by decomposition of diazoketones in the presence of iodine[242]. Several of these procedures suffer from disadvantages such as condensation and decomposition reactions and the availability and stability of substrates and reagents.

During the last few years successful methods have been developed for the synthesis of α-iodoketones. The reaction of enol acetates with thallium(I) acetate–iodine[243] and the oxidation of alkenes with silver chromate–iodine[244] gives α-iodoketones in moderate yields. However, thallium(I) acetate is highly toxic, and, using the latter method, only 1-iodo-2-alkanones can be prepared from terminal alkenes.

Cyclic α-iodoketones are obtained directly by oxidation of olefin–iodine complexes with pyridinium dichromate (PDC), but the reaction failed with linear olefins (equation 99)[245].

$$\text{cyclohexene} \xrightarrow[(2)\ PDC]{(1)\ I_2} \text{2-iodocyclohexanone (O, I)} + \text{1-chloro-2-iodocyclohexane (Cl, I)} \quad (99)$$

Terminal alkynes react with iodine in methanol in the presence of silver nitrate to give mainly α,α-diiodoketones together with diiodoalkenes and iodoalkynes (equation 100)[246].

$$RC\equiv CH \xrightarrow[CH_3OH]{I_2,\ AgNO_3} RCOCHI_2 + RC(I)=CHI + RC\equiv C-I \quad (100)$$

At present the most general method consists of the sequential treatment of enol silyl ethers with silver acetate–iodine followed by triethylammonium fluoride. High yields of α-iodo carbonyl compounds are reported (equation 101)[247]. The mechanism can be envisioned as occurring with initial formation of an iodonium ion followed by acetate attack.

$$\text{Me}_2\text{C}{=}\text{C(OSiMe}_3\text{)CH(Me)}... \xrightarrow[\text{I}_2]{\text{AgOAc}} [\text{iodonium}]^+ \xrightarrow{^-\text{OAc}} \text{(OSiMe}_3\text{)(OAc) iodo adduct} \xrightarrow{\text{Et}_3\overset{+}{\text{N}}\text{H F}^-} \alpha\text{-iodoketone} \quad (101)$$

Iodoketones are relatively unstable and are not widely used in synthesis. Therefore the reactivity of α-iodoketones will not be discussed in the following sections because of their limited applicability.

E. Mechanisms of α-Halogenation of Ketones

As already pointed out during the discussion of the various procedures for the preparation of α-halogenated ketones, the halogenation occurs according to three different types of mechanisms[248,249].

(1) In the presence of acids an electrophilic attack of the halogen on the enol takes place and subsequent loss of a proton from the intermediate oxonium ion leads to the α-haloketone (equation 102). For sufficiently high halogen concentrations, the rate-limiting step is the enolization while the rate of halogenation seems to be independent of the nature and concentration of the halogen.

$$R^1{-}\overset{\overset{\displaystyle O}{\|}}{C}{-}\underset{\underset{\displaystyle R^2}{|}}{\overset{\overset{\displaystyle H}{|}}{C}}{-}R^3 \underset{}{\overset{(H^+)}{\rightleftharpoons}} R^1{-}\overset{\overset{\displaystyle OH}{|}}{C}{=}\underset{\underset{\displaystyle R^2}{|}}{C}{-}R^3 \xrightarrow{X_2}$$

$$\left[R^1{-}\overset{\overset{\displaystyle ^+OH}{\|}}{C}{-}\underset{\underset{\displaystyle R^2}{|}}{\overset{\overset{\displaystyle X}{|}}{C}}{-}R^3 \right] X^- \xrightarrow{-HX} R^1{-}\overset{\overset{\displaystyle O}{\|}}{C}{-}\underset{\underset{\displaystyle R^2}{|}}{\overset{\overset{\displaystyle X}{|}}{C}}{-}R^3 \quad (102)$$

In the halogenation of unsymmetrical ketones, the substitution position is determined by the relative ease of formation of the isomeric enols. Consequently, the predominant isomer produced on halogenation of a ketone is that in which the halogen enters the more highly substituted α-position, because enol formation is enhanced by the presence of α-alkyl substituents and by other substituents which stabilize the enol.

However, the presence of an α-halo atom results in a decrease of the rate of enol formation and the substitution of each successive halogen atom becomes more difficult.

(2) In base-catalysed halogenations the halogen reacts with the enolate anion rather than with the enol. The rate of enolate formation is retarded by alkyl substituents and enhanced by α-halogen substituents. Therefore, base-catalysed halogenation is not suitable for the preparation of α-monohaloketones (equation 103).

(3) The halogenation can be carried out via halogen radicals, but further introduction of halogens proceeds via an ionic mechanism under the influence of the generated hydrogen halide.

$$R^1COCH\begin{matrix}R^2\\R^3\end{matrix} \underset{}{\overset{\text{Base}}{\rightleftharpoons}} R^1-\overset{O^-}{\overset{|}{C}}=C\begin{matrix}R^2\\R^3\end{matrix} \xrightarrow{X_2} R^1CO\underset{X}{\underset{|}{C}}\begin{matrix}R^2\\R^3\end{matrix} \qquad (103)$$

In the traditional mechanism for halogenation of ketones, which involves halogenation of a reactive enol or enolate, the observed rate of halogenation is independent of the halogen concentration and the nature of the halogen, when the halogen concentration is sufficiently high. Under these conditions the rate of deuteration should also be equal to the rate of halogenation. At sufficiently low halogen concentrations, the reaction between the enol or enolate and the halogen becomes rate determining and the observed rates become dependent on both the nature and the concentration of the halogen species.

At very low halogen concentration and high acidity, Bell demonstrated that the rate of the addition of the halogen to the enol form becomes slower in comparison with the enolization[250]. Nearly the same observations were made when the kinetics of the iodination, bromination and chlorination of acetone, diethyl ketone and diisopropyl ketone were studied at $[X_2] = 10^{-7}$–10^{-5} M (equation 104)[251]. The apparent rate

$$-\underset{\underset{O}{\|}}{C}-\overset{|}{\underset{|}{C}}-H + (H^+) \underset{k_{-1}}{\overset{k_1}{\rightleftharpoons}} -\underset{\underset{OH}{|}}{C}=\overset{|}{\underset{|}{C}} + (H^+) \underset{k_2}{\overset{X_2}{\longrightarrow}} -\underset{\underset{O}{\|}}{C}-\overset{|}{\underset{|}{C}}-X \qquad (104)$$

$$k_{II} = \frac{k_1}{k_{-1}}\,k_2 = K_E k_2$$

constants k_{II} for iodination, bromination and chlorination are approximately equal and k_2 is rate controlling only at very low concentrations of halogen (diffusion-controlled kinetics). The order of magnitude of such limiting rate constants of 10^9 M^{-1} s^{-1} leads to new values for K_E in solution, much smaller than those reported earlier[252].

After many years of unchallenged acceptance, the enolization mechanism for halogenation of carbonyl compounds was questioned by two groups. Rappe has postulated no less than five different mechanisms of halogenation for 2-butanone and other related ketones. Of these five reactions two are acid catalysed, two base catalysed and one is a free radical mechanism[253–255]. Rappe has claimed, for example, that base-catalysed bromination of 2-butanone can result in a ratio of monohalides (3-Br/1-Br = 7–7.5) quite different from that predicted on the basis of relative exchange rates (CH_2/CH_3 = 0.6–0.7). In addition there is an apparent 20–30-fold (at pH 5.5–7) and a fivefold difference (at pH 12) in the reaction rates for bromine and iodine. In view of these results a mechanism is postulated which involves a reaction of unenolized ketone with hypohalite anions.

Sytilin also claimed that the initial rate of bromination of acetone is dependent upon the concentration of bromine[256]. However, a few years later several groups proved independently that there is no reliable evidence to suggest that the base-catalysed halogenation of unsymmetrical ketones proceeds by alternative non-enolic halogenation routes other than by a traditional enolization mechanism[257–260].

III. REACTIVITY OF α-HALOGENATED KETONES

The interest in the reactivity of halogenated carbonyl compounds has grown since the discovery in 1895 of the Favorskii rearrangement, and numerous reports have dealt with theoretical studies and synthetic applications of α-haloketones. Therefore it is extremely difficult to review all the reactions in which α-haloketones are important intermediates. The main focus will be upon the reactivity with nucleophiles and bases, although other important reactions and transformations will be treated selectively.

A. Reactivity of α-Haloketones towards Nucleophilic Agents and Bases

1. Introduction

On treatment of an α-haloketone with various nucleophiles and/or bases, the attack can take place at six possible electrophilic sites:

```
         (6)  (5)        (4)
          H    H    O    H
          |    |(2) ||   |
  R1 —— C —— C —— C —— C —— R4
       (β)|    |(α) (1)  |(α')
          R2   X(3)      R3
```

The nucleophile is able to attack the carbon of the carbonyl function (position 1), the carbon atom carrying the halogen atom (position 2) and the halogen atom (position 3). In addition, due to the presence of two polar electron-withdrawing groups, namely the carbonyl function and the halogen atom, the hydrogen atoms in the α-, α′- and β-positions also become susceptible to attack by nucleophiles or bases (positions 4, 5, 6).

Theoretically, the following types of reaction can be envisioned during the reaction of an α-haloketone with a nucleophilic reagent. Besides nucleophilic substitution (a), elimination (b) and reduction (c), a nucleophilic addition to the carbonyl (d) can take place, followed by a nucleophilic intramolecular substitution (e) with formation of an epoxide which is able to give further reactions. In addition a Favorskii rearrangement, via an intermediate cyclopropanone, with formation of carboxylic acid derivatives is an alternative route (f) (equation 105).

In most cases it is very difficult to predict which reaction type will occur on treatment of an α-haloketone with a nucleophile. This complexity is mainly due to the following factors:

(1) Several reaction pathways are often occurring simultaneously, resulting in mixtures of reaction products.

(2) The same reagent gives rise to different reaction products with different ketones.

(3) The same ketone may show completely different reaction pathways with very similar nucleophilic reagents.

(4) The reaction is strongly dependent upon the reaction conditions (solvent, temperature, etc.).

(5) Structurally similar ketones which are substituted with different halogens give different reaction products with the same reagents.

(6) The reaction products can undergo further transformations during the reaction, such as rearrangement, oxidation and dimerization, while the starting α-haloketones can also be transformed into different ketones which then react further, giving rise to unexpected compounds.

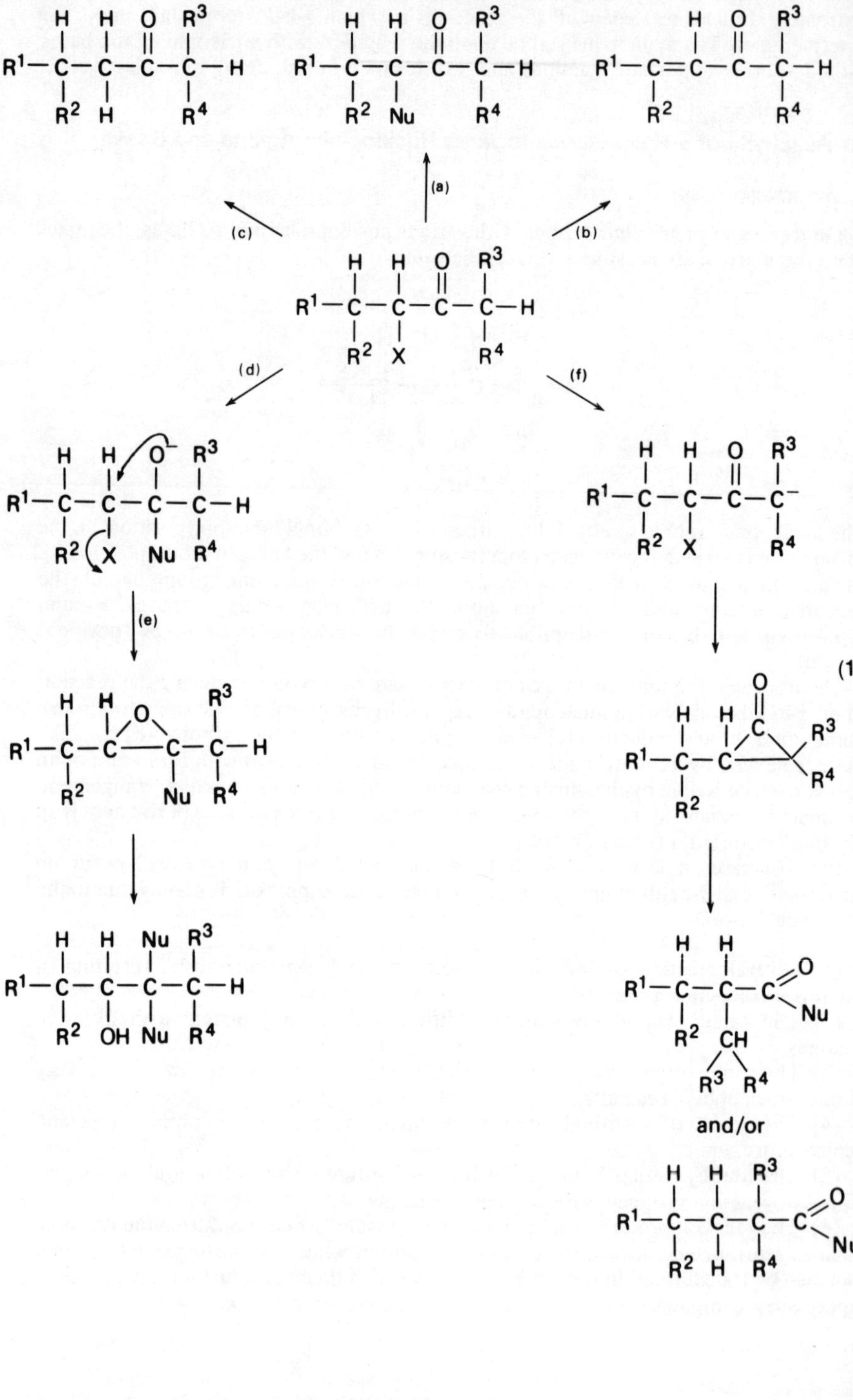

(a) (b) (c) (d) (e) (f)

and/or

(105)

2. The effect of the carbonyl function on the relative reactivity of α-halocarbonyl compounds with alkyl halides

The enhanced reactivity of α-halogenated ketones relative to the corresponding alkyl halides in bimolecular nucleophilic substitution is well known[261–265] and is illustrated in Table 1.

TABLE 1. Relative reactivities[a] of α-halo carbonyl compounds, alkyl halides and benzyl halides in nucleophilic substitution

Reaction	n-C_3H_7X	$PhCH_2X$	XCH_2COOEt	CH_3COCH_2X	$PhCOCH_2X$	Ref,
R—Cl + KI/acetone	1	197	1720	35 700	105 000	261
R—Cl + $S_2O_3^-$/water	1	–	220	1400	1600	263
R—Cl + $^-$OAc/ methanol	1	–	28	198	228	263
R—Br + pyridine/ methanol	1	286	56	208	406	261
R—Br + thiourea/ methanol	1	300	640	–	10 700	262
R—Cl + $^-N_3$/ methanol	1	–	33	210	276	263
R—Cl + $^-$OCN/ methanol	1	–	75	156	176	263
R—Cl + $^-$SCN/ methanol	1	–	83	401	770	263

[a]Relative reactivity to C_3H_7.

Ester, cyano and related groups also show this powerful activating effect, but surprisingly the sulphonyl group is deactivating, although the carbonyl and the sulphonyl groups exert the same inductive and resonance effects as expressed by their σ-constants[264].

It is also noteworthy that the activating effect of the carbonyl function is still operative when the group is situated at the β- or γ-carbon atom; $PhCOCH_2CH_2Cl$ and $PhCOCH_2CH_2CH_2Cl$ are respectively 80 and 230 times as reactive as n-butyl chloride[261]. Various mechanisms have been postulated to explain the enhancement of reactivity due to the presence of the carbonyl function. Hughes[266] ascribes the reactivity to the inductive effect of the carbonyl group which enhances the polarity of the carbon–halogen bond by increasing the electron deficiency at the α-carbon atom. The more polar the C—X bond, the faster is the reaction of nucleophiles in bimolecular substitution. Baker[267] has proposed a mechanism in which the first and rate-determining step is the addition of the basic reagent to the carbonyl function, followed by a rapid intramolecular displacement (equation 106).

The isolation of stable epoxides in the reaction of an α-haloketone with sodium methoxide and the evidence that these epoxides are reactive intermediates leading to other products[268] gives rise to another explanation by Pearson and coworkers[262] (equation 107). If any of the steps of the first reaction is slow, then this mechanism is in agreement with the second-order kinetics. A key point is that the reagent B′ is not necessarily the same as B.

The interaction between the carbonyl group and the nucleophile is mainly

$$R-\overset{O}{\overset{\|}{C}}-CH_2X + B \xrightarrow{\text{Slow}} \left[R-\overset{O^-}{\overset{|}{\underset{B^+}{\underset{|}{C}}}}-CH_2X \xrightarrow{\text{Fast}} R-\overset{O^-}{\overset{|}{\underset{B^+}{\underset{|}{C}}}}-CH_2-X \right] \xrightarrow{\text{Fast}} R-\overset{O}{\overset{\|}{C}}CH_2B^+ + X^- \quad (106)$$

$$R-\overset{O}{\overset{\|}{C}}-CH_2X + B \longrightarrow R-\overset{O^-}{\overset{|}{\underset{B^+}{\underset{|}{C}}}}-CH_2X \longrightarrow R-\underset{B^+}{\underset{|}{C}}\overset{O}{\frown}CH_2 + X^-$$

$$R-\underset{B^+}{\underset{|}{C}}\overset{O}{\frown}CH_2 + B' \xrightarrow{\text{Fast}} R-\overset{O}{\overset{\|}{C}}CH_2B' + B \quad (107)$$

electrostatic and the high S_N2 reactivity is due to polarization interaction caused by the smaller steric requirement of RCO as compared to RCH_2.

Another interpretation by Dewar[269] and Winstein and coworkers[270] is based upon neighbouring group orbital overlap with the adjacent electron-deficient carbon atom. The transition state for the substitution of α-haloketones is envisaged as including partial bonding of the reagent with the *p*-orbital of the carbonyl carbon.

An alternative explanation is that substitution products are formed via an enolization–solvolysis mechanism as expressed in equation (108)[271].

$$RCH_2COCH_2X \rightleftharpoons RCH{=}\overset{OH}{\overset{|}{C}}-CH_2X \longrightarrow$$

$$RCH\cdots\overset{OH}{\overset{|}{\underset{X^-}{\underset{+}{C}}}}\cdots CH_2 \xrightarrow{B^-} RCH_2COCH_2B + R\underset{B}{\underset{|}{C}}HCOCH_3 \quad (108)$$

Many cases have been reported in which this mechanism is operative. An example is the reaction of 1-chloro-3-phenylmercapto-2-propanone (**139**) with acetic acid in the presence of potassium acetate, which yields 1-acetoxy-1-phenylmercapto-2-propanone (**140**) and the thiol ester (**141**) (equation 109)[272].

Another example involving solvolysis of an enol allylic chloride is responsible for the formation of the α-alkoxyketone (**143**) and the α-hydroxyketone (**144**) from the chloroketone (**142**) (equation 109)[273].

A fast rate of substitution could result from fast enolization–solvolysis and comparison with the rates of the corresponding alkyl halides is worthless. Several of

$$C_6H_5SCH_2COCH_2Cl \xrightarrow[HOAc]{CH_3COOK} C_6H_5SCH(OCOCH_3)COCH_3 + CH_3CH(OCOCH_3){-}COSC_6H_5$$

(139) (140) (141)

$$PhCH_2COCHClCH_3 \xrightarrow[\text{Acid or dilute base}]{75\%\ H_2O/MeOH} PhCH{=}C(OH){-}CH(Cl)CH_3 \longrightarrow$$

(142)

$$PhCH_2COCH(OMe)CH_3 + PhCH_2COCH(OH)CH_3 \qquad (109)$$

(143) (76%) (144) (24%)

the proposed rationales are in contradiction to experimental data, for example with the substitution of α-haloketones by weakly basic nucleophiles.

The explanation of Hughes[266] fails in these cases where electron-withdrawing α-substituents other than carbonyl should show the same rate enhancement. However, α-halogen, α-alkoxy and α-sulphonyl substituents cause a substantial decrease in the rate of nucleophilic substitution of alkyl halides; α-halosulphones and α-halonitro compounds are quite unreactive. Baker's mechanism[267] does not fit the observation that the rate of substitution of α-haloketones is dependent upon the nature of the halogen. Bromoketones react faster than the corresponding chloro compounds, phenacyl bromide being 120 times as fast as phenacyl chloride with thiourea in methanol. Although a number of epoxides have been isolated or are shown to be important intermediates in reactions of α-haloketones (especially α-halobenzyl ketones react with alkoxides to yield α-hydroxyacetals), cleavage of epoxide intermediates leading to substitution products is dependent on the system and the experimental conditions. Arguments against an epoxide intermediate are presented in the reaction of haloketones with weakly basic nucleophilic reagents and against a rate-determining addition of the reagent to the carbonyl group[262]. Lutz showed that optically active desyl chloride undergoes exchange and racemization at the same rate with radioactive Cl^-, while for an epoxide mechanism an exchange without racemization is predicted[274]. However, Turro and coworkers proved that α-methoxy ketones are formed from α-bromoketones via an epoxide mechanism[275].

Thorpe and Warkentin interpret the bimolecular substitution of α-haloketones with acetate and azide ion, which are remarkably insensitive to steric hindrance, in terms of a normal S_N2 transition state, not involving either special alignment of entering and leaving groups with the π-orbital of the carbonyl function (conjugation) or covalent interaction between nucleophile and carbonyl carbon (bridging) for reaction of conformationally mobile systems. Conformationally fixed systems, on the other hand, may be affected by such factors. *Trans*-4-*t*-butyl-2-chloro cyclohexanone is 61 times more reactive than the *cis*-isomer in reaction with acetate ion. Activation parameters support the statement that only those α-haloketones which are set up for conjugation and bridging show substitution according to a different pathway from that operating in the corresponding reactions of alkyl halides[276].

TABLE 2. Relative reactivity of phenacyl bromide and methyl iodide in nucleophilic substitution

Nucleophile	$\frac{k_2(PhCOCH_2Br)}{k_2(MeI)}$
Cl^-	110
$(NH_2)_2CS$	63
NCS^-	30.5
$NCSe^-$	26.6
Pyridine	7.07
Ph_3P	4.44
Me_2S	3.22
Ph_3As	2.15
Me_2NPh	0.37
Et_3N	0.14

A recent report of Halvorsen and Songstad concerning comparison of second-order rate constants for reactions of phenacyl bromide and methyl iodide with various nucleophiles in acetonitrile reveals that the rate enhancement due to the carbonyl group is not a general effect but is dependent upon the nucleophile (Table 2).

Apparently, reactions with ionic nucleophiles tend to involve a 'tight' transition state (containing a mainly sp^2 hybridized central carbon atom). On the other hand, reactions with uncharged nucleophiles (amines) react via an 'early' transition state (an sp^3 hybridized central carbon atom) where no conjugation with the α-carbonyl group is possible. In the first case, the α-carbonyl function does exert a significant influence upon reaction rates due to its $+E$ effect, while in the latter case a decrease of the reaction rate of phenacyl bromide is observed in comparison with methyl iodide[277]. Considering all these results it is reasonable to postulate that any favourable effect exerted by a carbonyl function on the nucleophilic reactivity stems in part from the absence of rate-retarding steric effects (the enhancement is much lower for propionyl or butyryl functions in comparison with an acetyl function[33]), coupled with a mildly rate-enhancing inductive effect.

3. *Reaction of α-haloketones with oxygen nucleophiles and bases*

a. Reaction with inorganic oxygen nucleophiles. The products of the reaction of α-halogenated ketones with oxygen nucleophiles and bases are strongly dependent upon the substrate, the nature of the nucleophile and the reaction conditions. Besides substitution reactions, eliminations and rearrangements are also occurring, with the result that in many cases the outcome of the reaction cannot be predicted and that several reaction pathways take place simultaneously, resulting in a mixture of products. Several examples are known in which hydrolysis of α-haloketones with hydroxide and carbonate solutions in various solvents (water, alcohols, ether, dioxan, etc.) gives rise to α-hydroxyketones[277–282]. For example, α-hydroxycyclohexanone is formed in 76% yield from 2-chlorocyclohexanone with an aqueous solution of potassium carbonate[283], while ω-hydroxyacetophenone is obtained by boiling ω-chloroacetophenone in water[284].

However, during the hydrolysis of α-haloketones, side reactions, and especially Favorskii rearrangements and elimination reactions, are able to occur.

While treatment of 2-bromocyclododecanone (**145**) with aqueous potassium hydroxide gives 78% 2-hydroxycyclododecanone (**146**), the reaction in isopropanol results in the formation of the Favorskii rearrangement product, i.e. cycloundecanecarboxylic acid (**147**)[285]. The same phenomenon is observed during the reaction of halogenated aryl cyclohexyl ketones (**148**) with potassium hydroxide (equation 110)[286,287]. The nature of the halogen also plays an important role in the

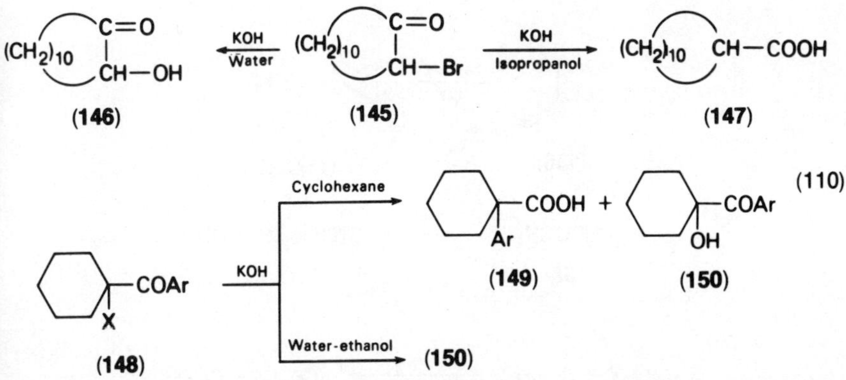

$(CH_2)_{10}$ C=O CH—OH (**146**) ← KOH, Water — $(CH_2)_{10}$ C=O CH—Br (**145**) — KOH, Isopropanol → $(CH_2)_{10}$ CH—COOH (**147**)

(**148**) COAr, X — KOH → Cyclohexane → (**149**) COOH, Ar + (**150**) COAr, OH; Water-ethanol → (**150**) (110)

reaction course. Reaction of α-chlorinated dicyclohexyl ketone (**151**) with potassium hydroxide in dioxan gives mainly a carboxylic acid (**152**), while the corresponding bromo compound shows a completely different reaction resulting in a debromination (equation 111)[288]. The initially formed α-hydroxyketones sometimes undergo further

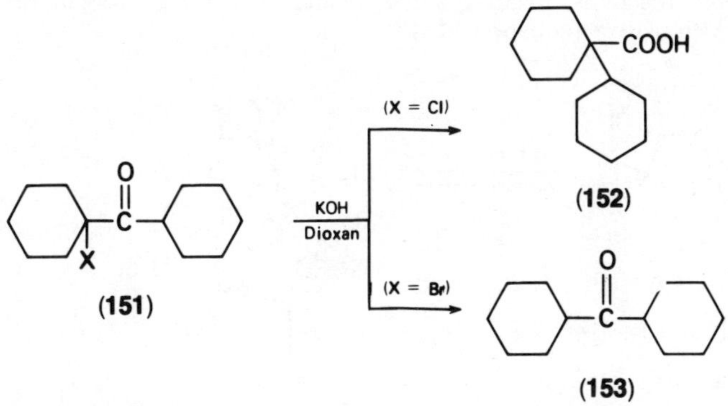

(**151**) — KOH, Dioxan → (X = Cl) (**152**) COOH; (X = Br) (**153**) (111)

reactions under the basic reaction conditions, such as isomerization[289], oxidation with formation of diketones[290], benzylic rearrangement with formation of α-hydroxycarboxylic acids[291] and dimerization[292].

Numerous examples are known of Favorskii rearrangements of α-halogenated ketones with metal hydroxides[293], carbonates[294], bicarbonates[295] and silver nitrate[296] in water, as depicted above (equation 112). In some cases, such as polyhalogenated cycloalkanones, even treatment with water gives rise to Favorskii products[293].

$$\text{(154)} \xrightarrow[H_2O]{NaOH} \text{(155)}$$

$$\text{(156)} \xrightarrow[H_2O]{Na_2CO_3} \text{(157)} \qquad (112)$$

$$\underset{(158)}{Cl_2CHCOCH_2Cl} \xrightarrow{KHCO_3} \underset{(159)}{ClCH{=}CHCOOH}$$

$$\underset{(160)}{(CH_3)_2C(Br){-}C(O){-}Ar} \xrightarrow[C_2H_5OH-H_2O]{AgNO_3} \underset{(161)}{(CH_3)_2C(Ar){-}COOH}$$

Next to the Favorskii rearrangement, the Grob fragmentation is frequently encountered in reactions of halogenated cycloalkanones such as polychlorocyclopentanones with base[162] (equation 113)[64].

$$\text{(162)} \xrightarrow[\text{Acetone-water}]{NaHCO_3} \text{(163)} \longrightarrow \underset{(164)}{HOOC{-}CCl_2{-}CH_2{-}C(X){=}CCl_2} \qquad (113)$$

X = H, Cl

A similar fragmentation reaction leading to dichlorinated acids takes place during the synthesis of geminal dichloroketones with the system carbon tetrachloride–potassium hydroxide–*t*-butyl alcohol[106]. Also, in the reaction of carvone tribromide **(165)** with sodium hydroxide in water or ether, little or no Favorskii rearrangement occurs; instead the carbonyl group is attacked, leading to compounds

166, **167** and **168** by a Grob fragmentation and to an epoxide (**169**). However, in the reaction of the *trans* isomer (**171**) with sodium hydroxide, the Favorskii rearrangement prevails with formation of unsaturated esters (after treatment with diazomethane) and a lactone (equation 114)[297].

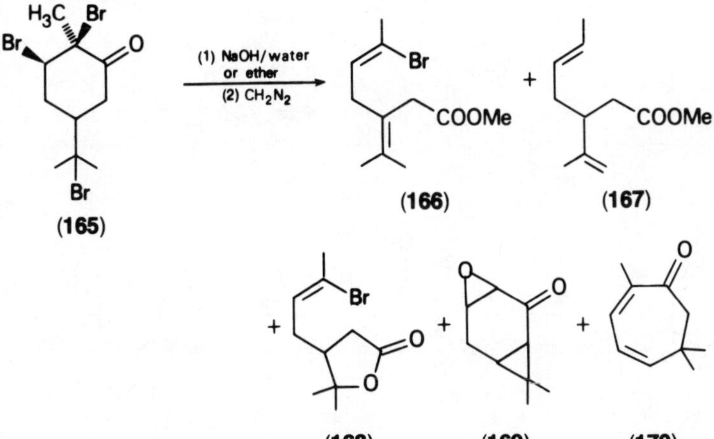

(1) NaOH/water or ether; (2) CH_2N_2

(**165**) (**166**) (**167**) (**168**) (**169**) (**170**)

(114)

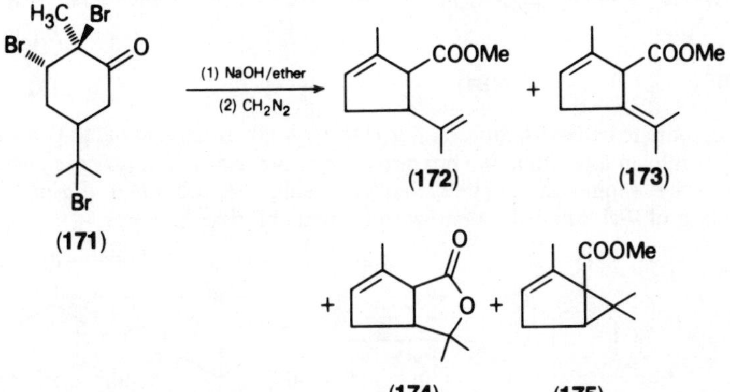

(1) NaOH/ether; (2) CH_2N_2

(**171**) (**172**) (**173**) (**174**) (**175**)

α-Hydroxycycloalkenones (**177**, **179**) are prepared from α-brominated cycloalkanones (**176**, **178**) by hydrolysis with aqueous sodium hydroxide (equation 115)[298,299].

The reaction of 2,2-dihalo-1-arylalkanones (**180**) with hydroxide ion takes a completely different course, yielding α-hydroxycarboxylic acids (**182**) via a benzilic rearrangement of intermediate α-diketones (**181**) (equation 116)[300].

Although a number of dehydrohalogenation reactions by the action of carbonates on α-haloketones have been reported, this reaction has little synthetic value due to rearrangements and aldol condensations already mentioned. In most cases lithium carbonate in dimethylformamide or dimethyl sulphoxide has been used in the elimination reactions.

Cyclohexenones are formed from α-halocyclohexanones. For example, 5-*t*-butyl-2-cyclohexenone (**184**) is produced from the chloro compound (**183**) using

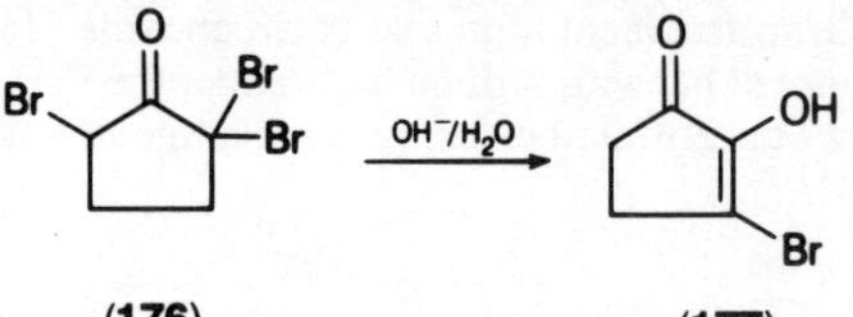

(176) (177)

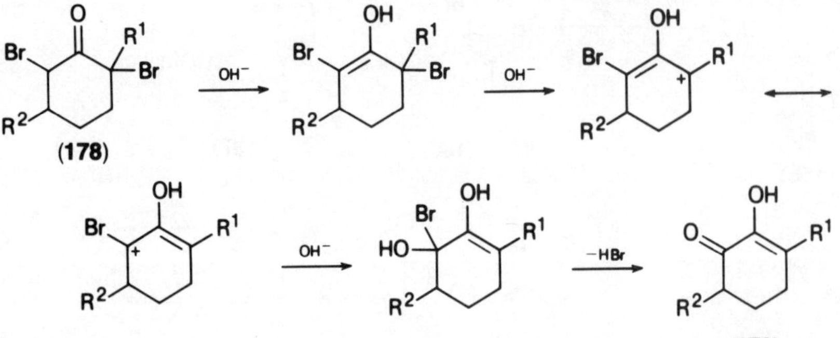

(178) (179) (115)

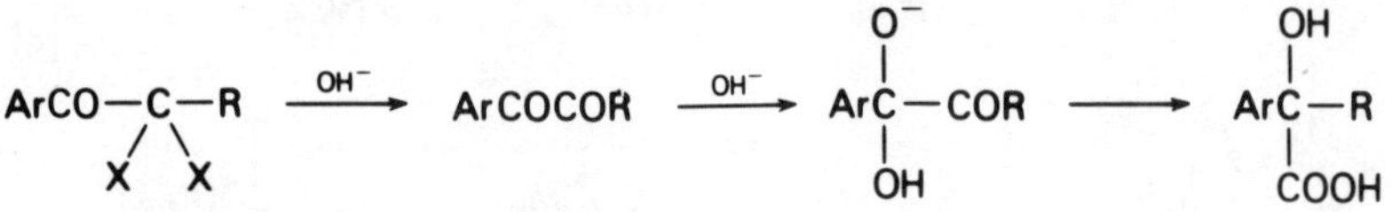

(180) (181) (182) (116)

lithium carbonate in dimethylformamide (DMF), while treatment of **(183)** with lithium chloride results in formation of a mixture of the isomeric cyclohexenones (**184**, **185**) in a ratio 3.5:1 (equation 117)[301]. Similar results are obtained during dehydrochlorination of 9-chloro-1-decalone with lithium chloride[302].

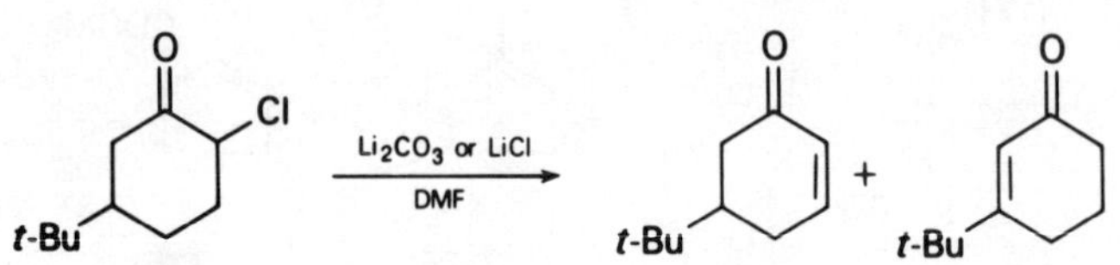

(183) (184) (185) (117)

Dehydrohalogenation readily occurs using alkali carbonates in DMF or dimethyl sulphoxide (DMSO) in reactions with γ,δ-unsaturated α-haloketones, producing dienones (equation 118)[303].

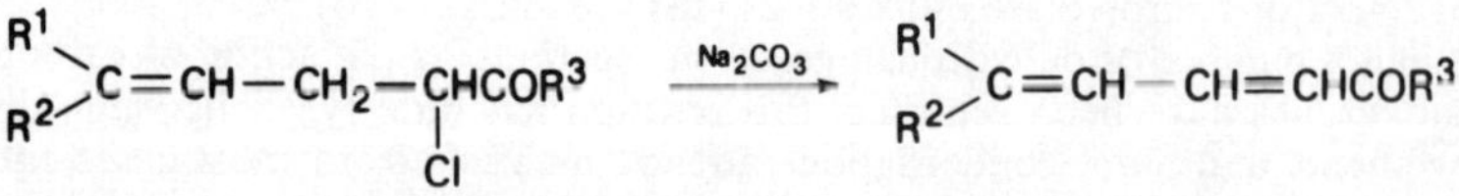

$$R^1R^2C{=}CH{-}CH_2{-}CHClCOR^3 \xrightarrow{Na_2CO_3} R^1R^2C{=}CH{-}CH{=}CHCOR^3 \quad (118)$$

Dehydrobromination of α,α'-dibromocycloalkanones occurs easily and 4,4-tetramethyl-2,5-cyclohexadienone (**187** and 2,4,6-cycloheptatrienones (**189**) are formed from 2,6-dibromo-4,4-dimethylcyclohexanone (**186**) and 2,2,7-tribromo-

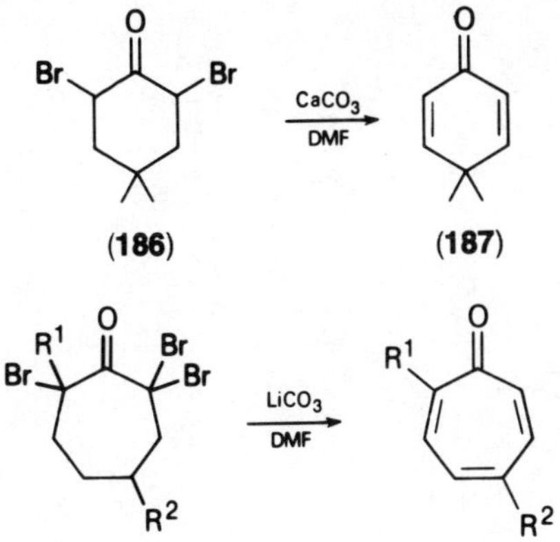

(186) (187)

(119)

(188) (189)

cycloheptanone (**188**), respectively (equation 119)[189,305]. However, the reaction of α,α′-dibromocyclopentanone (**190**) with two equivalents of sodium hydrogen carbonate in DMF affords 2-bromocyclopent-2-enone in high yield (equation 120)[306].

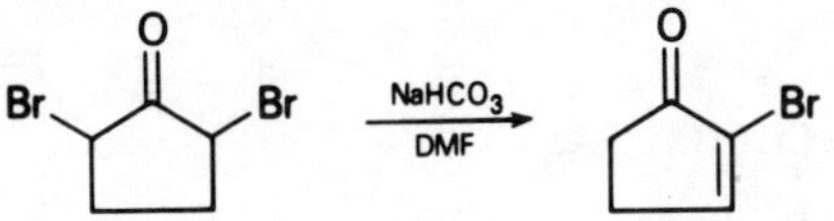

(120)

(190) (191)

Hydroxide-catalysed cyclization takes place when α-bromo-*o*-acyloxyaryl alkyl ketones (**192**) and α-bromo-β-methoxydihydrochalcones (**194**) are treated with aqueous sodium hydroxide, yielding respectively 3-substituted chromone epoxides (**193**)[307] and aurones (**195**)[308]. The bromohydrins (**196**) are cyclized to chalcone epoxides (**197**)[309] by reaction with potassium carbonate in aqueous *t*-butanol (equation 121).

Finally, treatment of bromoketones with sodium hydroxide or potassium carbonate in the presence of peroxides provides olefins as major products via intermediate cyclopropanones (equation 122)[310].

b. Reaction with organic oxygen nucleophiles and bases. There is no doubt that the reaction of α-haloketones with alkoxides is the most profoundly investigated in the field of the reactivity of α-halogenated carbonyl compounds. Nevertheless, prediction of the reaction products of α-haloketones with oxygen nucleophiles turns out to be very puzzling. Nearly all the reaction pathways proposed in the introduction to this section can take place, and the reaction outcome is dependent upon the nature of substrate and reaction conditions.

A typical example of the complexity of this type of reaction was given by Turro during study of the isomeric pair of α-bromo-2-butanone and α-bromo-3-methyl-2-butanone with sodium methoxide. The reaction products consist of mixtures of Favorskii esters, α-hydroxy- and α-methoxyketones (**198, 199, 199a**) (equation 123)[275]. The ester formation is favoured in ether, while methoxy ketones are the dominant products in methanol. Turro proved that the Favorskii esters are formed via a cyclopropanone intermediate and the hydroxy- and methoxyketones are generated through epoxy ethers, which subsequently decompose directly or upon work-up.

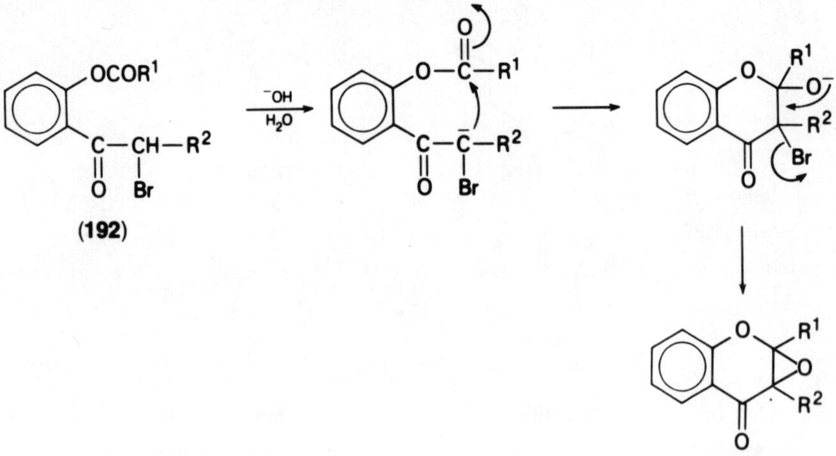

(192)

(193)

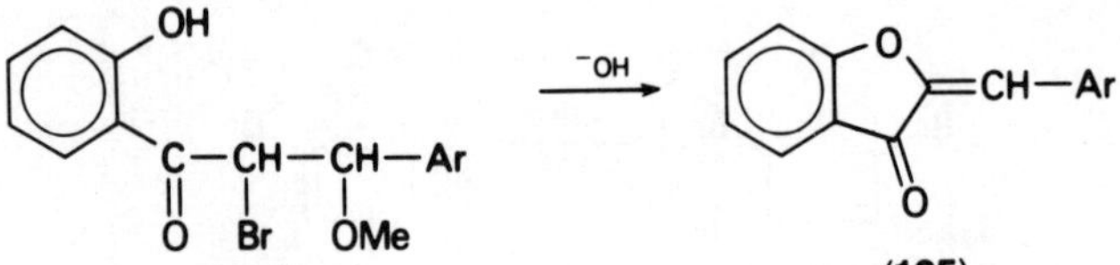

(194) (195) (121)

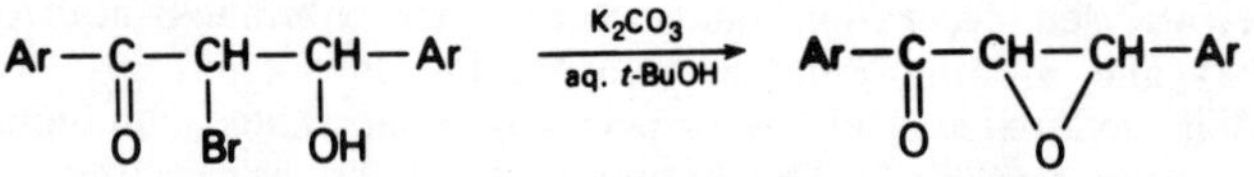

(196) (197)

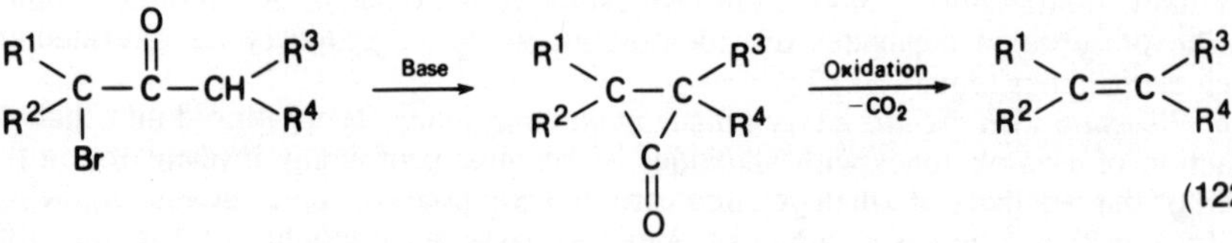

(122)

Increasing the degree of substitution at the carbon atom to which the halogen is attached usually favours the Favorskii rearrangement by lowering the rate of side reactions, whereas substitution on the α'-atom hinders rearrangement[310]. Replacement of chlorine by bromine favours Favorskii rearrangement most of all[229]. The competition between substitution, epoxide formation and Favorskii rearrangement is illustrated by the following examples, which emphasize the important influence of the structure of the substrate.

Treatment of 3-bromo-3-methyl-2-butanone (**200**) with sodium methoxide gives the ester (**201**)[312,313], while in compounds without α'-hydrogen atoms such as brominated alkyl aryl ketones (**202**), epoxide formation (**203**) predominates (equation 124)[314].

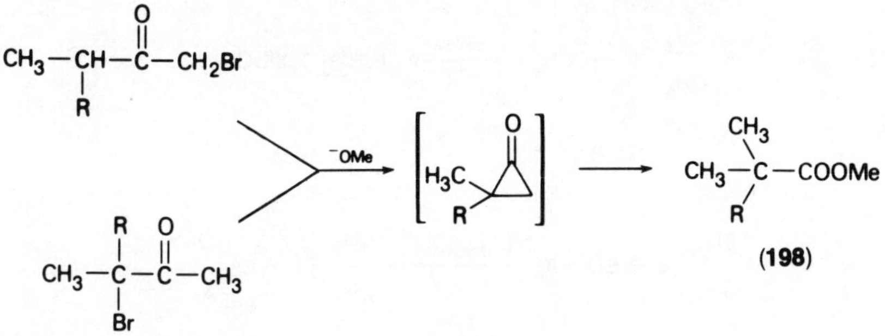

$CH_3-CH(R)-C(=O)-CH_2Br$ and $CH_3-C(R)(Br)-C(=O)-CH_3$ $\xrightarrow{^-OMe}$ [cyclopropanone intermediate: H_3C, R] $\longrightarrow$ $CH_3-C(CH_3)(R)-COOMe$

(**198**)

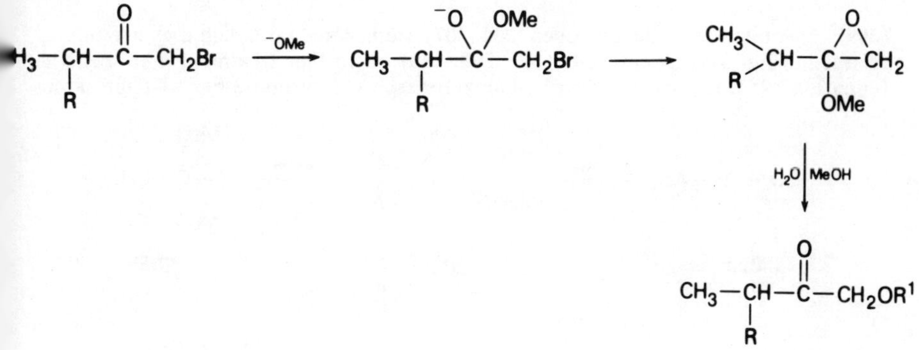

$CH_3-CH(R)-C(=O)-CH_2Br \xrightarrow{^-OMe} CH_3-CH(R)-C(O^-)(OMe)-CH_2Br \longrightarrow$ (CH₃)(R)CH–C(OMe) epoxide $-CH_2$ $\xrightarrow{H_2O \mid MeOH} CH_3-CH(R)-C(=O)-CH_2OR^1$

$R^1 = H, Me$

(**199**)

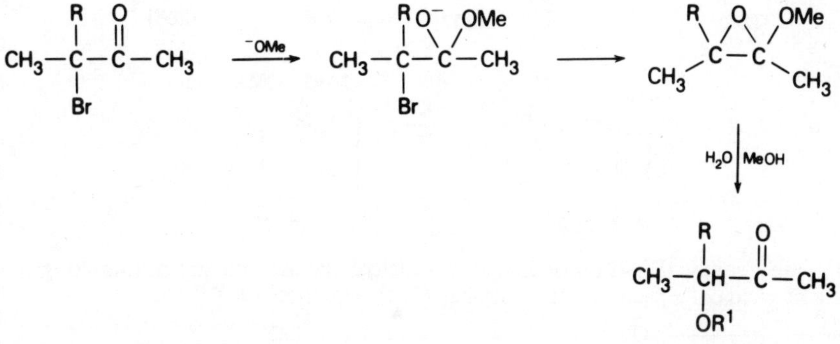

$CH_3-C(R)(Br)-C(=O)-CH_3 \xrightarrow{^-OMe} CH_3-C(R)(Br)-C(O^-)(OMe)-CH_3 \longrightarrow$ epoxide [R, CH_3]C–C[OMe, CH_3] $\xrightarrow{H_2O \mid MeOH} CH_3-CH(OR^1)(R)-C(=O)-CH_3$ (123)

$R^1 = H, Me$

(**199a**)

Nucleophilic addition at the carbonyl function followed by intramolecular nucleophilic substitution yielding methoxy epoxides seems to be the most favourable pathway in the reaction of ketones without α′-hydrogen atoms such as in halogenated aryl benzyl ketones (**203a**)[315,316], tetralones (**206**)[317] and steroidal α-bromoketones

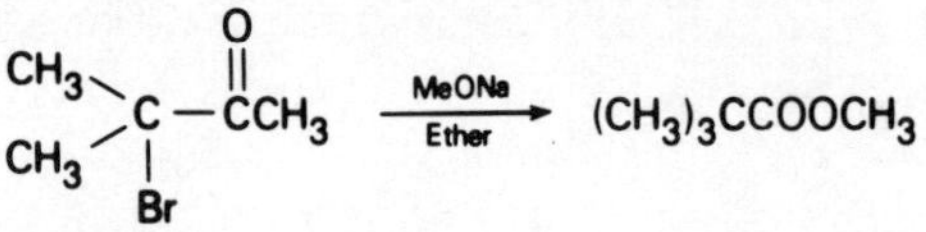

$$(CH_3)_2C(Br)-COCH_3 \xrightarrow[\text{Ether}]{\text{MeONa}} (CH_3)_3CCOOCH_3$$

(200) (201)

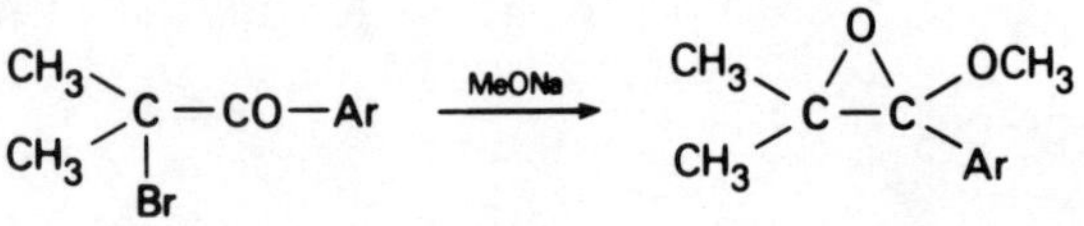

$$(CH_3)_2C(Br)-CO-Ar \xrightarrow{\text{MeONa}} (CH_3)_2C\overset{O}{-}C(OCH_3)Ar \tag{124}$$

(202) (203)

(209)[318]. A number of the epoxides (**204**, **207**) are isolated, although they are readily converted into α-hydroxyacetals (**205**, **208**, **210**) on further treatment with alcohols (equation 125). However, tertiary α-haloketones, e.g. 2-bromo-2-benzyl-1-tetralones

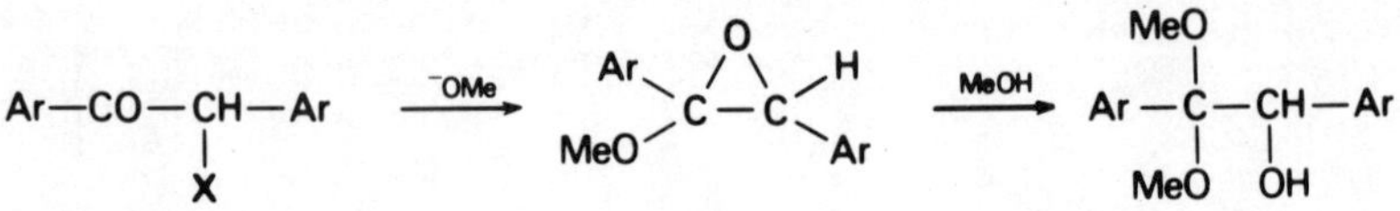

$$Ar-CO-CH(X)-Ar \xrightarrow{^-OMe} Ar(MeO)C\overset{O}{-}CH(Ar) \xrightarrow{\text{MeOH}} Ar-C(MeO)_2-CH(OH)-Ar$$

(203a) (204) (205)

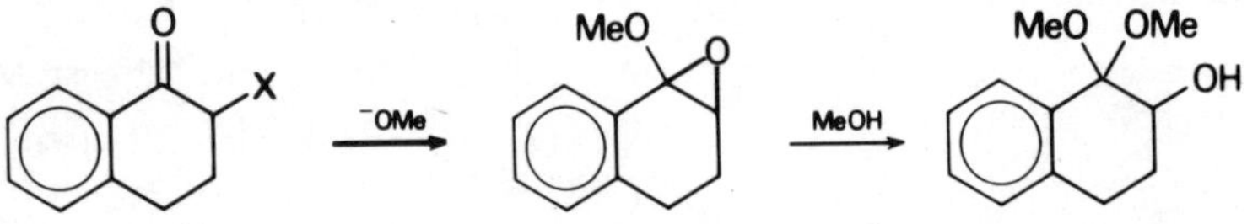

$$(206) \xrightarrow{^-OMe} (207) \xrightarrow{\text{MeOH}} (208) \tag{125}$$

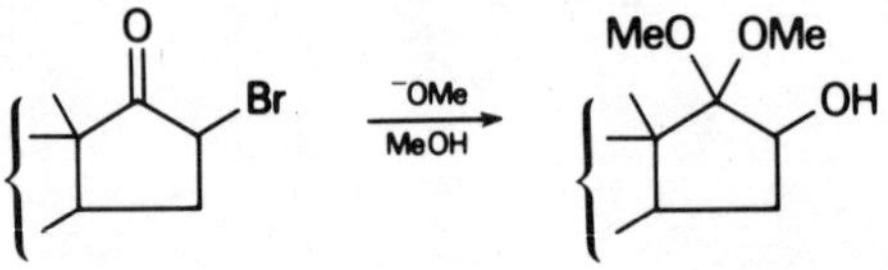

$$(209) \xrightarrow[\text{MeOH}]{^-OMe} (210)$$

(**211**), are readily dehydrobrominated with alcoholic sodium methoxide to give excellent yields of α,β-unsaturated ketones (**212**) (equation 126)[319,320].

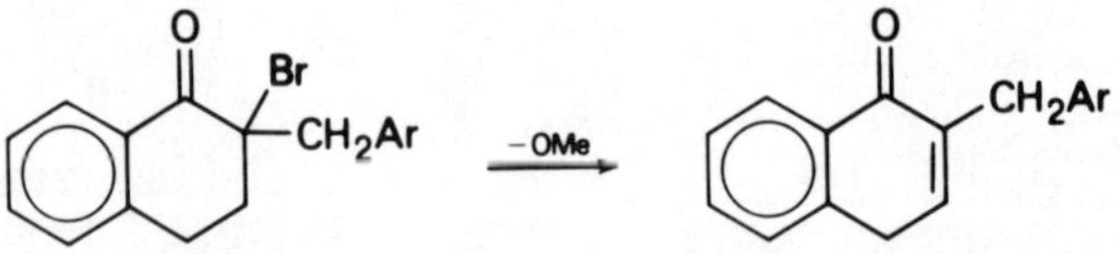

$$(211) \xrightarrow{^-OMe} (212) \tag{126}$$

In the field of α-halocycloalkanones the reaction products with alkoxides are also strongly dependent upon the substrate. While 2-chlorocyclohexanone gives rise to the

Favorskii ester (**213**)[321], treatment of 6-phenyl-2-chlorocyclohexanone produces the substitution product (**214**) in different solvents[322] (equation 127). On the other hand,

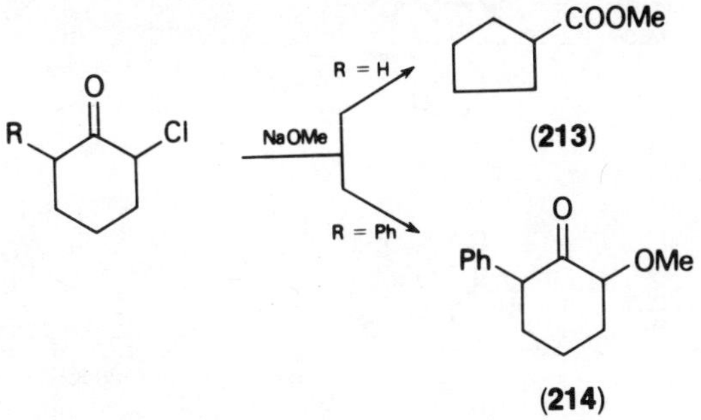

NaOMe; R = H → (**213**) COOMe; R = Ph → (**214**) Ph, OMe (127)

2-chloro-2,6,6-trimethylcyclohexanone (**215**) gives a stable epoxy ether (**216**)[323], while 9-chloro-1-decalone (**217**) produces the rearranged substitution product (**218**)[324] (equation 128).

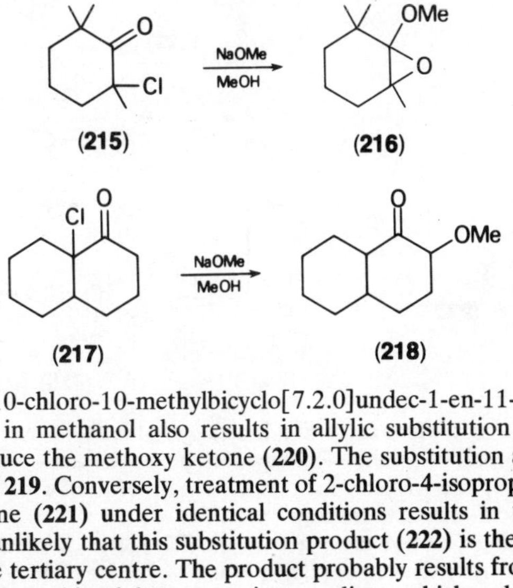

(**215**) $\xrightarrow[\text{MeOH}]{\text{NaOMe}}$ (**216**)

(**217**) $\xrightarrow[\text{MeOH}]{\text{NaOMe}}$ (**218**) (128)

Treatment of 10-chloro-10-methylbicyclo[7.2.0]undec-1-en-11-one (**219**) with sodium methoxide in methanol also results in allylic substitution rather than ring contraction to produce the methoxy ketone (**220**). The substitution apparently occurs through the enol of **219**. Conversely, treatment of 2-chloro-4-isopropylidene-2,3,3-trimethylcyclobutanone (**221**) under identical conditions results in the unrearranged product. It seems unlikely that this substitution product (**222**) is the result of a direct displacement at the tertiary centre. The product probably results from an elimination proceeding through a bicyclobutanone intermediate which adds methoxide to produce **222** (equation 129)[325].

Not only does the nature of the substrate play an important role in the reaction pathway. The halogen atom also influences the reaction, as illustrated in the case of the triaryl ketone (**223**) where the chloro compound gives a methoxy epoxide (**224**) and the bromo compound a methoxyketone (**225**) (equation 130)[316].

Other important factors controlling the reaction course are the reaction conditions and especially the nature of the solvent and the concentration of the nucleophile. In the reaction of α-chlorocyclohexanone or 2-bromo-5-methyl-5-phenylcyclohexanone with sodium methoxide in methanol, the yield of the Favorskii esters has been found to increase markedly at the expense of the α-methoxyepoxide and the α-methoxyketone on increasing the methoxide concentration. The increased yield can be attributed

(219) $\xrightarrow[MeOH]{NaOMe}$ (220)

(221) $\xrightarrow[MeOH]{NaOMe}$ (222) (129)

(223) $\xrightarrow{NaOMe}$ X = Cl: (224); X = Br: (225) (130)

partly to a positive salt effect favouring ionization of halide ion from the enolate ion. 2-Chloro- and 2-bromo-4-methyl-4-phenylcyclohexanone are much less subject to this concentration effect due to steric factors: 40% yield of the Favorskii ester is obtained even at low methoxide concentrations[326].

Substituted α-chlorobenzyl methyl ketones (**226**) also give mixtures of Favorskii esters (**227**) and α-hydroxyacetals (**228**) (equation 131)[327]. The yield of the Favorskii ester increases from 9% at 0°C with 0.05 M sodium methoxide to 61% with 2 M sodium methoxide at 63°C. This is believed to be a consequence of a 2–3 kcal mol^{-1}

$$\mathrm{Ar{-}CH(Cl)COCH_3} \xrightarrow[MeOH]{NaOMe} \mathrm{ArCH_2CH_2COOCH_3} + \mathrm{Ar{-}CH(OH){-}C(CH_3)(OCH_3){-}OCH_3} \quad (131)$$

(226) (227) (228)

Ar = C_6H_5, p-$NO_2C_6H_4$, p-ClC_6H_4, p-$MeOC_6H_4$, m-$MeOC_6H_4$, p-MeC_6H_4, m-MeC_6H_4, p-FC_6H_4

higher activation energy for the Favorskii reaction. The yield of the ester is increased to 68% for Ar = p-$MeOC_6H_4$ and is decreased to 0% for Ar = p-$NO_2C_6H_4$. Similar effects are observed in the reactions of 3-chloro-1-phenyl-2-butanone (**228a**) and 1-chloro-1-phenyl-2-butanone (**231**) with variable concentrations of methoxide ion yielding mixtures of ester (**229**) and α-methoxyketone (**230**) (equation 132)[328,329].

$$PhCH_2C(=O)CH(Cl)CH_3\ (\mathbf{228a}) \xrightarrow[MeOH]{NaOMe} PhCH_2CH(CH_3)COOCH_3\ (\mathbf{229}) + PhCH_2C(=O)CH(OCH_3)CH_3\ (\mathbf{230})$$

[⁻OMe], M	229, %	230, %
0.05 (inverse addition)	0	100
0.02	50	50
2	100	0

$$PhCH(Cl)C(=O)CH_2CH_3\ (\mathbf{231}) \xrightarrow{NaOMe} PhCH_2CH(CH_3)COOCH_3\ (\mathbf{229}) + PhCH_2C(=O)CH(OCH_3)CH_3\ (\mathbf{230}) \qquad (132)$$

[⁻OMe], M	229, %	230, %
0.05	70	30

1-Halo-1,1-diphenyl-2-propanones (**232**) react with 0.05 M sodium methoxide to give essentially quantitative yields of Favorskii ester (**233**), while under inverse addition (addition of the nucleophile to the substrate) and low concentration of methoxide ion a mixture of **233**, **234**, and **235** is formed. Reaction of 3-chloro-1,1-diphenyl-2-propanone under the same conditions gives the same product distribution, while α-bromo-1,1,3-triphenyl-2-propanones (**236**, **237**) yield 1,3-diphenyl-2-indanone (**238**) (equation 133)[330].

The mechanisms leading to the various reaction products have been elucidated by Bordwell[273,322,325–330]. The Favorskii rearrangement of α-chloroarylpropanones and α-chloroarylbutanones falls into two classes. The series $ArCHClCOCH_3$ and $ArCH_2COCH_2Cl$ react with methoxide by way of reversible carbanion (enolate ion) formation followed by rate-limiting halide release, while for systems like $ArCH_2COCHXCH_3$, $PhCHXCOCH_2CH_3$ and $PhCH_2COCHXPh$, halide ion release is greatly accelerated and proton removal becomes rate limiting. The alkoxy ketones are formed through solvolysis of intermediate enol allylic chlorides. The various pathways are depicted in equation (134). The same mechanism is observed during the methanolysis of 3-chloro-1,3-diphenyl-2-propanone with lutidine or lutidine–lutidine $\cdot H^+$ buffer, yielding exclusively the α-methoxyketone[328]. No reaction is observed when chloroacetone is treated under similar circumstances[331].

The distribution of the products obtained by reaction of dichlorinated methyl ketones (**239**, **240**) with sodium methoxide in methanol is strongly dependent upon the

$$Ph_2C(X)C(=O)CH_3 \xrightarrow[MeOH]{NaOMe} Ph_2CHCH_2COOCH_3 + Ph_2C(OMe)COCH_3 + o\text{-}(COPh)C_6H_4CH_2COOCH_3$$

(232) → (233) (48%) + (234) (17%) + (235) (19%)

$$Ph_2C(Br)C(=O)CH_2Ph \ \text{(236)} \quad \text{or} \quad Ph_2CHC(=O)CH(Br)Ph \ \text{(237)} \xrightarrow[MeOH]{NaOMe} \text{1,3-diphenyl-2-indanone (238)} \qquad (133)$$

structure of the ketone. Primary dichloromethyl ketones (R^2 = H) give the normal *cis* acrylic esters (**241**), together with chloromethyl esters (**242**) whose amount increases with the increase in the bulk of the R^1 group, while the secondary dichloromethyl ketones (**240**) afford small amounts of methyl esters (**242**) but variable amounts of methoxyketones (**243**). The stereospecificity is complete for primary ketones and in the secondary derivatives the ratio between the *cis* and *trans* acrylic esters depends on the difference between both alkyl substituents (bulkiness) and on the chlorine substitution (1,1- and 1,3-dichloroketones) (equation 135)[95]. The reaction takes a completely different course when α,α-dihaloalkyl aryl ketones (**244**) react with sodium alkoxides to produce a mixture of isomeric α,α-dialkoxy ketones (**245**, **246**) in variable ratios[33,209]. In the cases where R^1 = H and R^1 = *t*-Bu small amounts of alkyl benzoates (**247**) are detected (equation 136).

2-Chloro-2-fluoro and 2-bromo-2-fluoro compounds give rearrangement to **246** exclusively while α,α-difluoroketones show exclusive reduction of the carbonyl function. It is reasonable that the dichloroketones react by an initial nucleophilic addition and subsequent intramolecular nucleophilic attack with halide displacement, furnishing α-halo-α′-alkoxyepoxide intermediates. The latter compounds rapidly rearrange spontaneously to α-halo-α-alkoxy ketones which further give rise to α,α-dialkoxy ketones by a direct route or via the hemiacetal. Alternatively, the latter product can be deprotonated by the alkoxide, after which intramolecular nucleophilic substitution yields the α,α′-dialkoxyepoxides. The diactivated epoxides are then opened at both sides to produce the final isomeric α,α-dialkoxy ketones **245** and **246** (equation 137). The reaction mechanism was supported by the synthesis of reaction intermediates[33,209]. The intermediacy of α-chloro-α′-methoxyoxiranes seems to be reasonable as compounds of this type have been observed during the reaction of tetrachlorocyclopentanone (**38**) which gives **248** and **249** (equation 138)[64].

In addition, stable α-chloro-α′-methoxy oxiranes (**251**) have been isolated during the reaction of tetrachlorocyclohexanones (**250**) with sodium methoxide in methanol (equation 139)[332].

$$\mathrm{PhCHClCOCH_2CH_3}\ (\mathbf{231}) \xrightarrow{MeO^-} \mathrm{PhCHCl{-}C(O^-){=}CHCH_3} \xrightarrow{-Cl^-} \mathrm{Ph{-}CH{\cdots}C(O^-){\cdots}CH{-}CH_3}\ (\mathbf{229}) \xleftarrow{-Cl^-} \mathrm{PhCH{=}C(O^-){-}CHClCH_3} \xleftarrow{MeO^-} \mathrm{PhCH_2COCHClCH_3}\ (\mathbf{228a})$$

$$(\mathbf{229}) \rightleftharpoons \text{2-phenyl-3-methylcyclopropanone} \xrightarrow{MeO^-} \mathrm{PhCH_2CH(CH_3)COOCH_3}$$

$$\mathrm{PhCHCl{-}C(OH){=}CHCH_3} \rightleftharpoons \mathrm{PhCH{\cdots}C^{+}(OH){\cdots}CHCH_3\ Cl^-} \rightleftharpoons \mathrm{PhCH{=}C(OH){-}CHCl_3Cl}$$

$$\mathrm{PhCH{\cdots}C^{+}(OH){\cdots}CHCH_3\ Cl^-} \xrightarrow{MeOH} \mathrm{PhCH_2COCH(OCH_3)CH_3}\ (\mathbf{230}) \qquad (134)$$

(the enolates are interconverted with the enols by $+H^+$ / $-H^+$)

$$\mathrm{R^1R^2CHCOCHCl_2}\ (\mathbf{239}),\ \mathrm{R^1R^2CClCOCH_2Cl}\ (\mathbf{240}) \xrightarrow[MeOH]{^-OMe} \mathrm{R^1R^2C{=}CHCOOMe}\ (c/t)\ (\mathbf{241}) + \mathrm{R^1R^2C(CH_2Cl)COOMe}\ (\mathbf{242}) + \mathrm{R^1R^2C(OMe)COCH_2Cl}\ (\mathbf{243}) \qquad (135)$$

α,α′-Dibromocycloalkanones react with sodium methoxide via the intermediacy of epoxides to yield α-hydroxy acetals (**253**)[306], although earlier investigations claimed the formation of a Favorskii rearrangement product[333]. The acetals are formed by addition of methanol to the carbonyl group and substitution of one bromine with

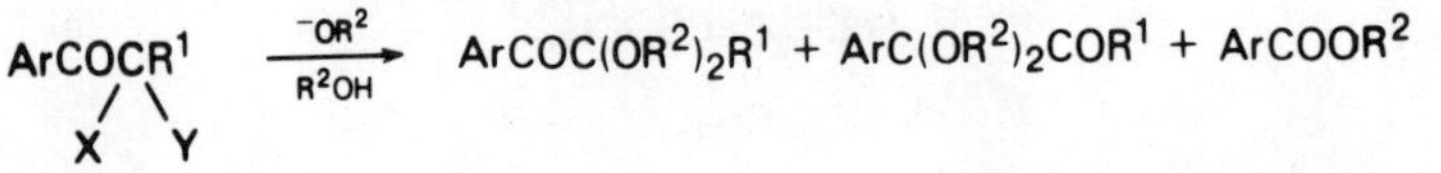

$$\underset{(244)}{\mathrm{ArCOCR^1XY}} \xrightarrow[R^2OH]{^-OR^2} \underset{(245)}{\mathrm{ArCOC(OR^2)_2R^1}} + \underset{(246)}{\mathrm{ArC(OR^2)_2COR^1}} + \underset{(247)}{\mathrm{ArCOOR^2}} \quad (136)$$

X, Y = Cl, Br, F
R^1 = H, alkyl
R^2 = Me, Et, *i*-Pr

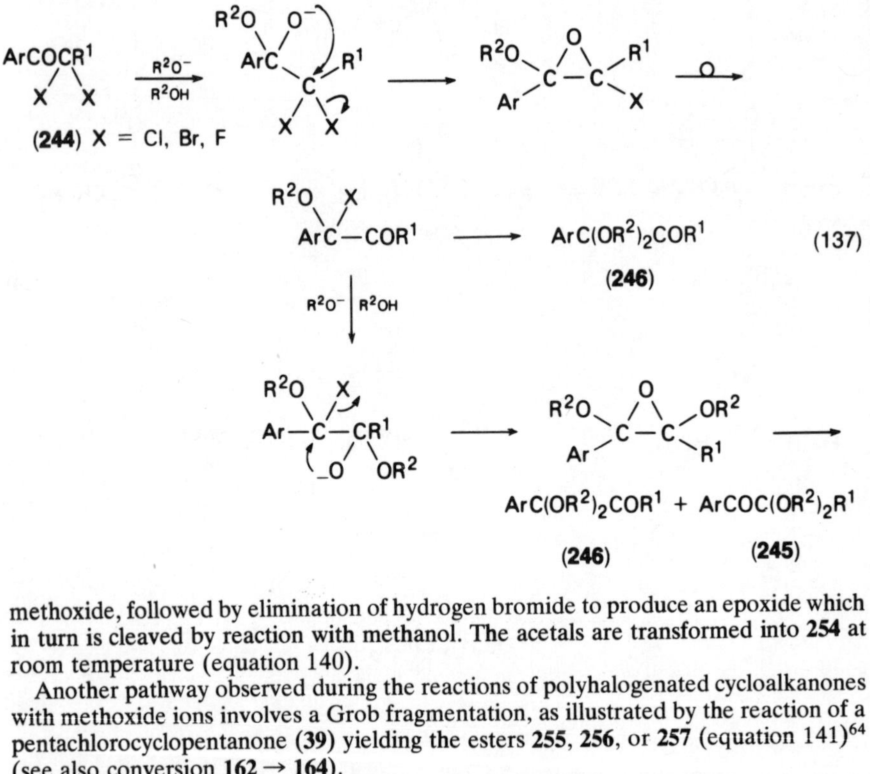

$$\underset{(244)\ X = Cl, Br, F}{\mathrm{ArCOCR^1X_2}} \xrightarrow[R^2OH]{R^2O^-} \mathrm{ArC(OR^2)(O^-)-CR^1X_2} \longrightarrow \mathrm{R^2O(Ar)C\overset{O}{-}C(R^1)X} \xrightarrow{\Delta} \mathrm{ArC(OR^2)X-COR^1} \longrightarrow \underset{(246)}{\mathrm{ArC(OR^2)_2COR^1}} \quad (137)$$

$$\mathrm{ArC(OR^2)X-COR^1} \xrightarrow{R^2O^-\,|\,R^2OH} \mathrm{Ar-C(OR^2)X-CR^1(O^-)(OR^2)} \longrightarrow \mathrm{R^2O(Ar)C\overset{O}{-}C(OR^2)R^1} \longrightarrow \underset{(246)}{\mathrm{ArC(OR^2)_2COR^1}} + \underset{(245)}{\mathrm{ArCOC(OR^2)_2R^1}}$$

methoxide, followed by elimination of hydrogen bromide to produce an epoxide which in turn is cleaved by reaction with methanol. The acetals are transformed into **254** at room temperature (equation 140).

Another pathway observed during the reactions of polyhalogenated cycloalkanones with methoxide ions involves a Grob fragmentation, as illustrated by the reaction of a pentachlorocyclopentanone (**39**) yielding the esters **255**, **256**, or **257** (equation 141)[64] (see also conversion **162** → **164**).

Monofluoroketones show a completely different route in the reaction with sodium alkoxides in ether at −60°C. The exclusive reaction product constitutes of a ketol (**259**) produced by aldol condensation (equation 142)[334].

Finally, reaction of α-bromoketones (**260**, **263**) with methanol in the presence of silver hexafluoroantimonate affords substitution products (**262**) via an α-keto-carbenium ion (**261**) and Favorskii ester (**265**) via an intermediate hemiacetal (**264**) (equation 143)[335].

Reaction of the chlorinated bicyclic ketones (**266**, **267**) with potassium *t*-butoxide (strong base, poor nucleophile) in *t*-butanol shows a regioselective elimination–rearrangement via a zwitterionic intermediate leading to a bicyclic enone (**268**) system (equation 144)[336].

Cyclopropanone derivatives are produced when specific α-bromoketones are

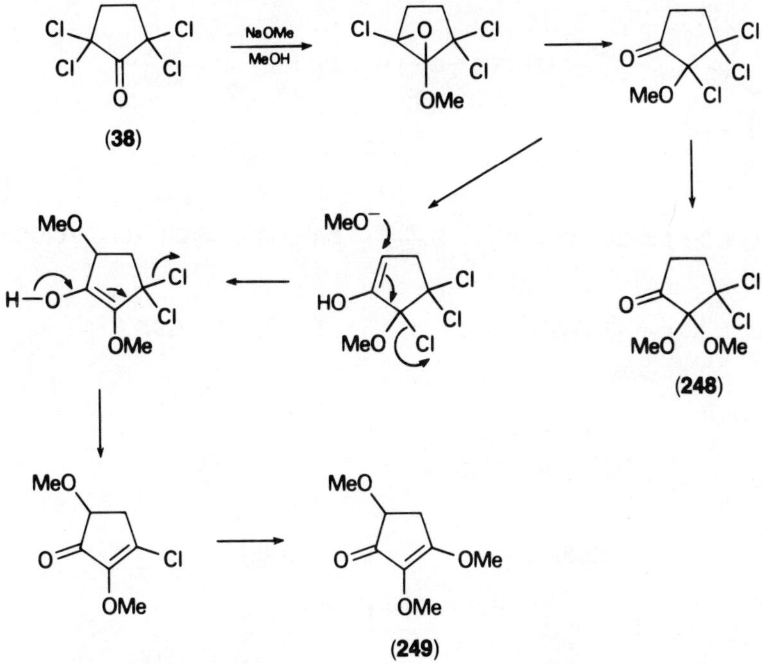

(38) (248) (249) (138)

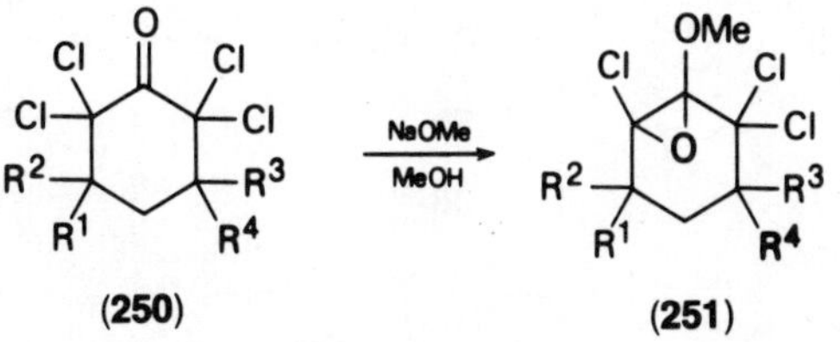

(250) (251) (139)

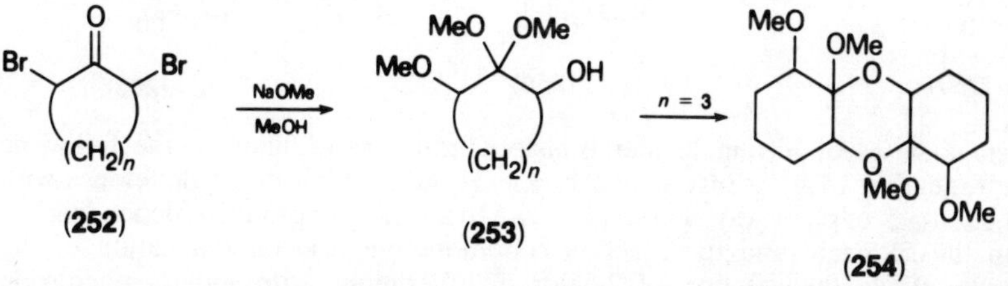

(252) (253) (254) (140)

treated with potassium *t*-butoxide in THF. An example is the formation of 2,3-di-*t*-butylcyclopropanone **(271)** from α-bromodineopentyl ketone **(270)**. When the reaction is carried out in *t*-butanol, not the *t*-butyl ester, but the corresponding acid **(272)** is formed, presumably from the action of adventitious hydroxide in the *t*-butoxide (equation 145)[337,338]. In addition, reaction of α,α′-dibromodineopentyl ketone **(273)** with potassium *t*-butoxide in THF provides di-*t*-butylcyclopropene **(274)** in 80% yield (equation 146)[339,340].

Addition of a solution of tris(chloroacetonyl)methane **(275)** in THF to a solution of potassium *t*-butoxide in *t*-butanol affords a direct entry to the triasterane structure

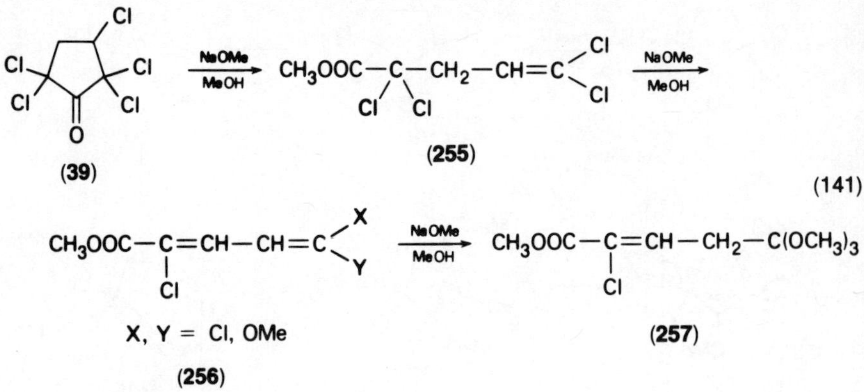

$$\text{(39)} \xrightarrow[\text{MeOH}]{\text{NaOMe}} CH_3OOC{-}CCl_2{-}CH_2{-}CH{=}CCl_2 \;\text{(255)} \xrightarrow[\text{MeOH}]{\text{NaOMe}} \tag{141}$$

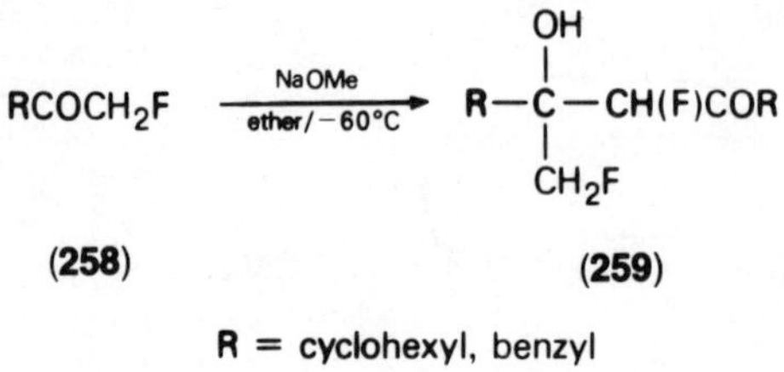

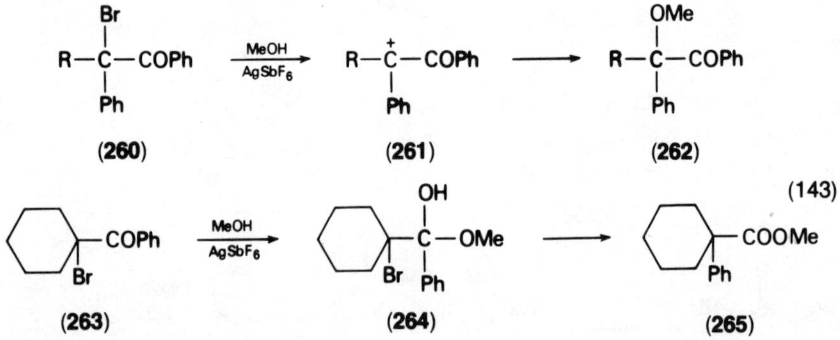

$$CH_3OOC{-}C(Cl){=}CH{-}CH{=}CXY \;\; \text{(256)} \xrightarrow[\text{MeOH}]{\text{NaOMe}} CH_3OOC{-}C(Cl){=}CH{-}CH_2{-}C(OCH_3)_3 \;\; \text{(257)}$$

X, Y = Cl, OMe

$$RCOCH_2F \;\text{(258)} \xrightarrow[\text{ether}/-60°C]{\text{NaOMe}} R{-}C(OH)(CH_2F){-}CH(F)COR \;\text{(259)} \tag{142}$$

R = cyclohexyl, benzyl

$$R{-}C(Br)(Ph){-}COPh \;\text{(260)} \xrightarrow[\text{AgSbF}_6]{\text{MeOH}} R{-}\overset{+}{C}(Ph){-}COPh \;\text{(261)} \longrightarrow R{-}C(OMe)(Ph){-}COPh \;\text{(262)} \tag{143}$$

$$\text{(263)} \xrightarrow[\text{AgSbF}_6]{\text{MeOH}} \text{(264)} \longrightarrow \text{(265)}$$

(263): 1-bromocyclohexyl–COPh; (264): 1-bromocyclohexyl–C(OH)(OMe)Ph; (265): 1-phenylcyclohexyl–COOMe

(276) via a series of intramolecular transformations, as outlined in the following scheme (equation 147)[341]. Just as in the case of the reaction of α-haloketones with alkoxides, attack of phenoxide anions gives rise to a variety of products, depending not only on the substrate and the reaction conditions but also on the nature of the phenoxide. While the reaction of 2-chlorocyclohexanone with sodium phenoxide affords the α-phenoxy ketone **(277)**, treatment of the same ketone with sodium (2-isopropyl-5-methyl)phenoxide provides a mixture of the substitution product **(277a)** and the Favorskii ester **(278)** (equation 148)[342–344]. Hypothetically, the formation of 2-phenoxycyclohexanone may occur by, first, a S_N2 attack at the α-carbon (path 1), second, a S_N2' attack at $C_{(2)}$ in the enol (path 2), third, an attack at either the α- or α'-carbon of a symmetrical cyclopropanone intermediate (path 3). A decision between the various possibilities is offered by the use of [1,2-^{14}C]-2-chlorocyclohexanone and it is proved that only path 3 is consistent with the results (equation 149)[345].

(266) (267) $\xrightarrow[t\text{-BuOH}]{t\text{-BuO}^- \ K^+}$ [intermediate] → (268) (not (269)) (144)

t-BuCH$_2$C(=O)CH(Br)Bu-t (270) $\xrightarrow{t\text{-BuO}^-}$ THF: (271); t-BuOH: t-BuCH$_2$CH(Bu-t)COOH (272) (145)

t-BuCH(Br)C(=O)CH(Br)Bu-t (273) $\xrightarrow[\text{THF}/-70\,^{\circ}\text{C}]{t\text{-BuOK}}$ (274) (146)

The reaction of 1-chlorocyclohexyl methyl ketone (**279**) with sodium phenoxide in phenol gives a mixture of the substitution and elimination products (**280** and **281**) together with the Favorskii ester (**282**) (equation 150)[346]. On the other hand, α,α′-dibromocycloalkanones (**252**) provide a single reaction product, 2-phenoxy 2-cycloalkenones (**283**) with sodium phenoxide in methanol or DMF (equation 151)[306]. The reaction of chloroacetone with activated phenols in the presence of potassium carbonate and potassium iodide in DMF gives substitution products[347], but 1,3-dichloroacetone reacts with phenols under similar conditions to give mixtures

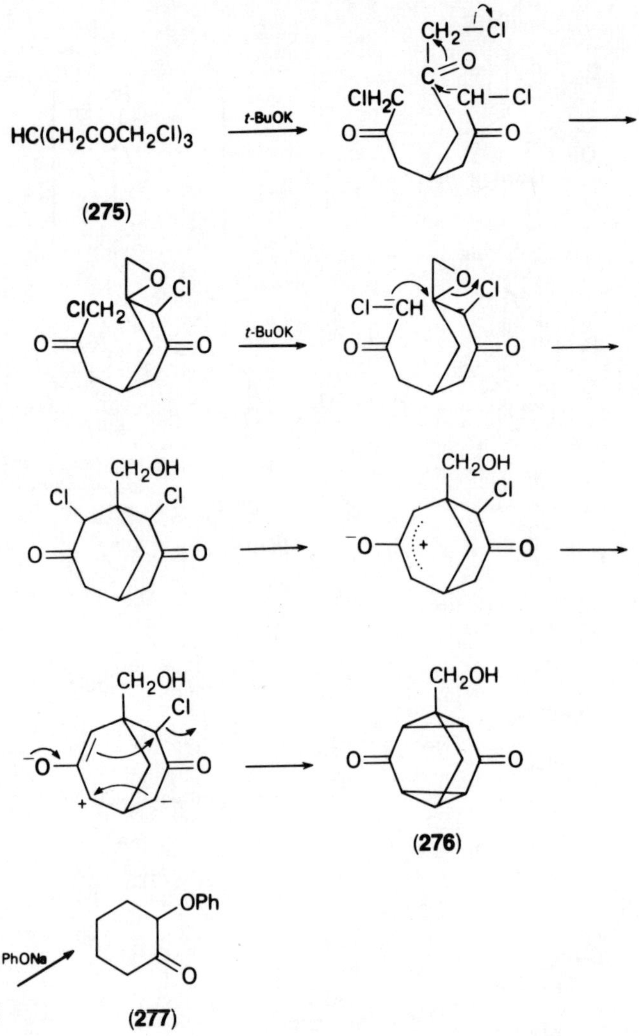

$HC(CH_2COCH_2Cl)_3$ (275) $\xrightarrow{t\text{-BuOK}}$ → → (147)

$\xrightarrow{t\text{-BuOK}}$ →

CH_2OH → → (276)

(148)

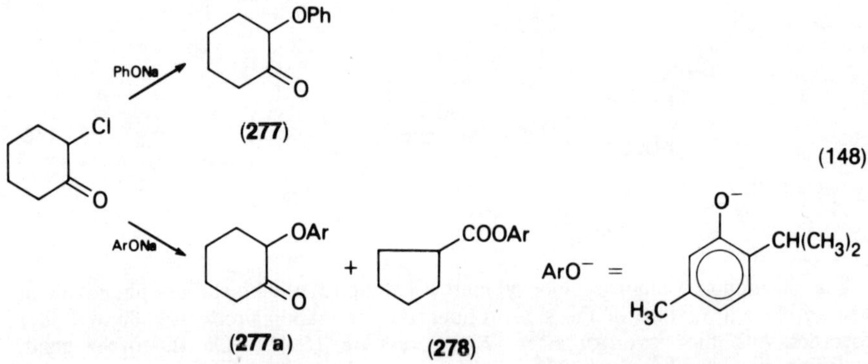

PhONa → (277)

ArONa → (277a) + (278) $ArO^- =$

where 1,1-bis(aryloxy)acetone is the major product, while the more acidic *p*-nitrophenol (which is ionized under the reaction conditions) provides the expected 1,3-disubstituted compound[348]. Reaction of the dibromoketones (**284**) with catechol gives rise to the formation of 1,4-benzodioxan derivatives (**285**) (equation 152)[349].

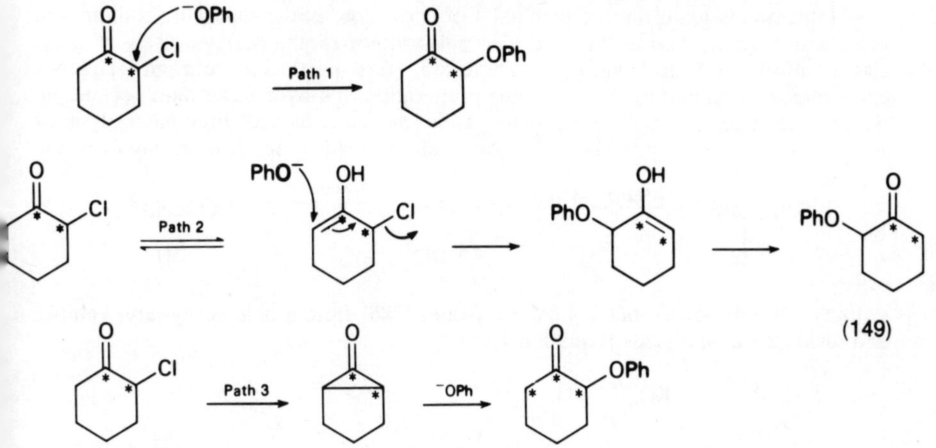

Path 1

Path 2

Path 3

(149)

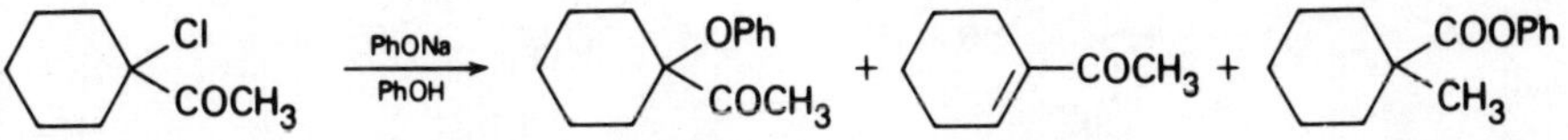

(279) (280) (281) (282) (150)

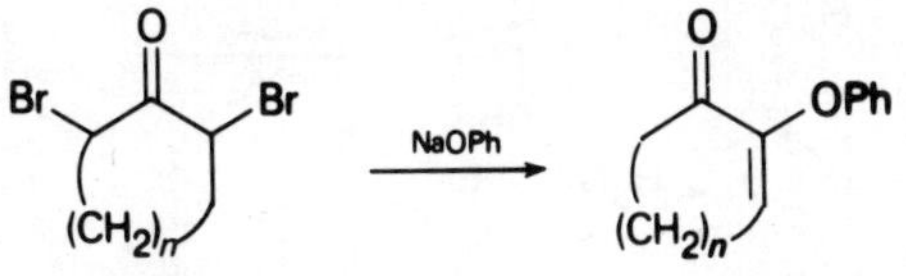

(252) (283) (151)

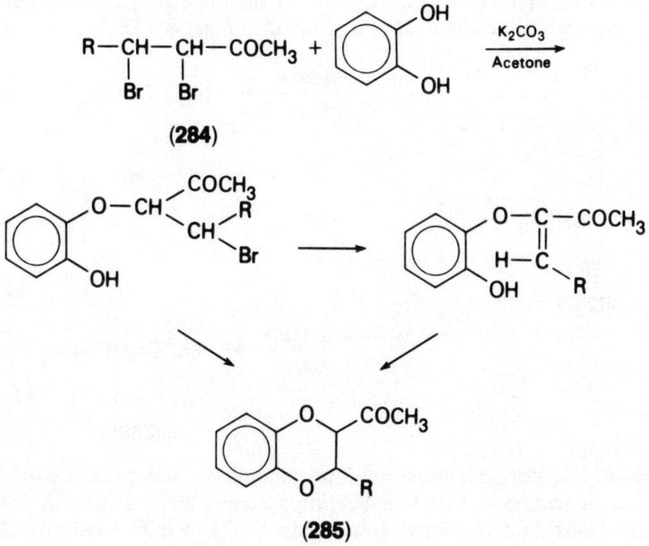

(284)

(285) (152)

α-Haloketones readily react with salts of carboxylic acids, especially sodium and potassium formate and acetate, to give substitution products. Hydrolysis of these esters affords α-hydroxyketones. Therefore, this particular reaction sequence constitutes the method of choice for the preparation of α-hydroxyketones because no major side reactions are taking place as in the case of the direct hydrolysis of α-haloketones (equation 153)[350–352]. Another interesting application involves the

$$R^1COCHR^2(Br) \xrightarrow{R^3COO^- \; M^+} R^1COCHR^2(OCOR^3) \xrightarrow{\text{Hydrolysis}} R^1COCHR^2(OH) \qquad (153)$$

synthesis of 2,3-dihydro-6*H*-1,4-oxazin-2-ones (**286**) from α-halomethyl aryl ketones and protected amino acids (equation 154)[353].

$$ArCOCH_2Br + KOOC{-}CHR{-}NH{-}CbZ \xrightarrow{DMF} ArCOCH_2O{-}CO{-}CHR{-}NH{-}CbZ \xrightarrow[HOAc]{HBr} ArCOCH_2O{-}CO{-}CHR{-}NH_3^+\;Br^- \xrightarrow[HOAc]{NaOAc} \textbf{(286)} \qquad (154)$$

A number of isomerizations are observed during the reaction of α-haloketones with carboxylate anions. Treatment of 2-bromo-1-phenyl-1-propanone (**287**) with acetate ion followed by hydrolysis provides 1-hydroxy-1-phenyl-2-propanone (**288**) while with formate ion the normal product (**289**) is formed (equation 155)[352].

$$PhCOCHCH_3(Br)\ \textbf{(287)} \xrightarrow[(2)\ H_3O^+]{(1)\ KOAc/MeOH} PhCH(OH)COCH_3\ \textbf{(288)}$$

$$PhCOCHCH_3(Br)\ \textbf{(287)} \xrightarrow[(2)\ H_3O^+]{(1)\ HCOOK/MeOH} PhCOCH(OH)CH_3\ \textbf{(289)} \qquad (155)$$

Acetolysis of 1-chloro-3,3-diphenyl-2-propanone in the presence of potassium acetate gives 1-acetoxy-3,3-diphenyl-2-propanone (**291**), 1-acetoxy-1,1-diphenyl-2-propanone (**293**) and 1-phenyl-2-indanone (**292**), while 1-chloro-1,1-diphenyl-

2-propanone (**290**) produces **293** exclusively. The reaction mechanism suggests the intervention of an allylic carbonium ion capable of capturing nucleophilic species at both $C_{(1)}$ and $C_{(3)}$ (equation 156)[354].

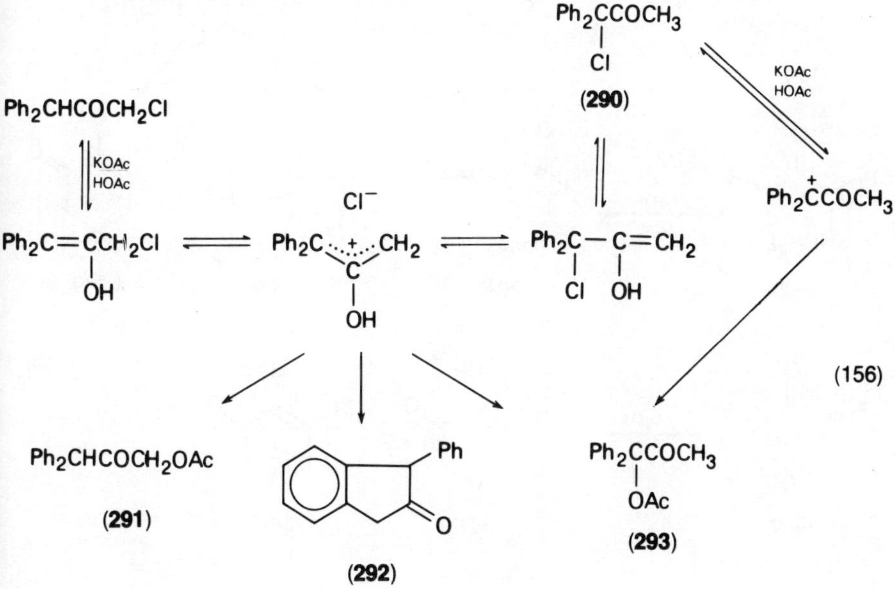

$$Ph_2CHCOCH_2Cl \xrightleftharpoons[HOAc]{KOAc} Ph_2C{=}C(OH)CH_2Cl \rightleftharpoons [Ph_2C{\cdots}C^{+}(OH){\cdots}CH_2]\,Cl^- \rightleftharpoons Ph_2C(Cl){-}C(OH){=}CH_2 \rightleftharpoons Ph_2C(Cl)COCH_3\ (\mathbf{290}) \xrightleftharpoons[HOAc]{KOAc} Ph_2\overset{+}{C}COCH_3 \tag{156}$$

Products: $Ph_2CHCOCH_2OAc$ (**291**); 1-phenyl-2-indanone (**292**); $Ph_2C(OAc)COCH_3$ (**293**)

Other examples of cine substitution are illustrated below (equation 157)[355,356]. Rearrangement also occurs when 2,6-dibromo-4,4-dimethylcyclohexanone (**186**)

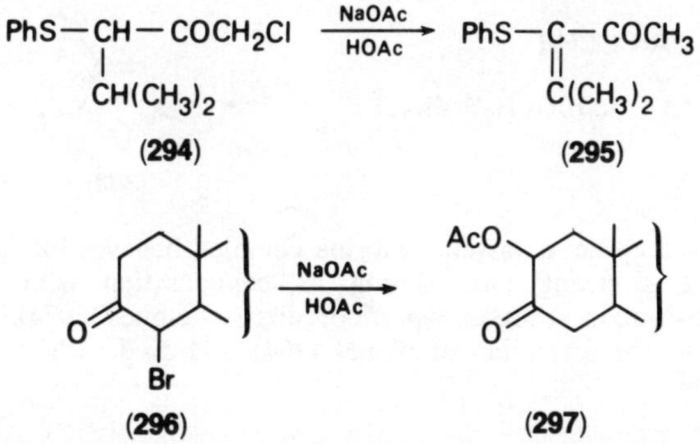

$$PhS{-}CH(CH(CH_3)_2){-}COCH_2Cl\ (\mathbf{294}) \xrightarrow[HOAc]{NaOAc} PhS{-}C({=}C(CH_3)_2){-}COCH_3\ (\mathbf{295}) \tag{157}$$

$$(\mathbf{296}) \xrightarrow[HOAc]{NaOAc} (\mathbf{297})$$

reacts with sodium acetate in acetic acid yielding **298**. The mechanism involves bromine substitution followed by a 1,3-hydrogen bromide elimination together with an acyl migration[189]. Similar results are obtained in the cases of α,α′-dibromocycloalkanones (**252**) and tribromotetrahydro-4*H*-pyran-4-ones (**300**) with acetate anions[357] (equation 158).

Bromoketones react with bromoacetic acid in the presence of triethylamine to give the substitution products, which, via intermediate phosphonium salts, can be cyclized to α,β-unsaturated lactones (**303**) (equation 159)[358].

(186) $\xrightarrow[\text{HOAc}]{\text{NaOAc}}$ $\xrightarrow{-H^+}$ $\xrightarrow{-Br^-}$ (298)

(252) $\xrightarrow[\text{DMF}]{\text{NaOAc}}$ (299)

(300) $\xrightarrow{\text{AgOAc}}$ (301) + (302) (158)

$$\text{ArCOCHR(Br)} \xrightarrow[\text{Et}_3\text{N}]{\text{BrCH}_2\text{COOH}} \text{ArCOCHR(OCOCH}_2\text{Br)} \xrightarrow{\text{PPh}_3}$$

$$\text{ArCOCHR(OCOCH}_2\text{P}^+\text{Ph}_3\cdot\text{Br}^-) \xrightarrow{\text{Et}_3\text{N}} \textbf{(303)} \quad (159)$$

The hexafluoroacetone–potassium fluoride complex behaves like a weak oxygen nucleophile and a strong base during its condensation with α-haloketones. Nucleophilic substitution produces α-perfluoroalkoxy ketones (**303a**). Abstraction of a proton leads to the formation of diones (**304**) and cyclic ethers (**305** and **306**) (equation 160)[359].

4. Reaction of α-haloketones with nitrogen nucleophiles and bases

a. Reaction of α-haloketones with amines. Amines have been widely used to substitute α-haloketones. Numerous examples are known in which ammonia, primary and secondary amines produce α-aminoketones, while treatment with tertiary amines gives rise to ammonium salts (equation 161)[360]. Geminal diaminoketones are formed when α,α-dichloroketones react with an excess of amines[361].

Aminoketones are rather unstable compounds, and it is therefore advisable to isolate them as salts of strong acids.

$$PhCOCH_2Br + (CF_3)_2CO/KF \xrightarrow{\Delta} PhCOCH_2OCF(CF_3)_2$$

(**303a**)

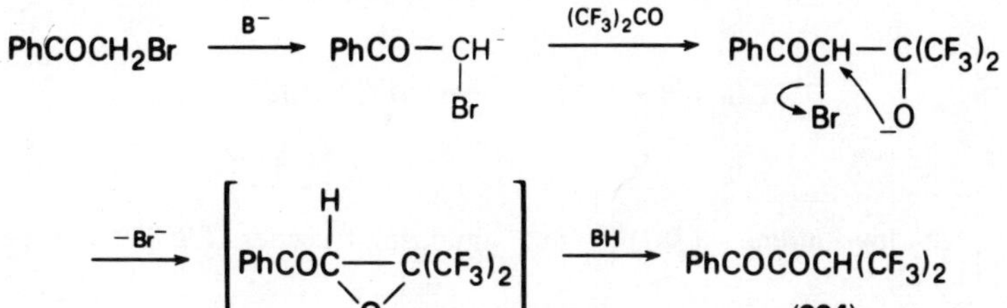

$$PhCOCH_2Br \xrightarrow{B^-} PhCO{-}\underset{Br}{CH^-} \xrightarrow{(CF_3)_2CO} PhCOCH(Br){-}C(CF_3)_2O^- \qquad (160)$$

$$\xrightarrow{-Br^-} \left[PhCOC(H){-}C(CF_3)_2 \text{ (epoxide)} \right] \xrightarrow{BH} PhCOCOCH(CF_3)_2$$

(**304**)

(**305**)

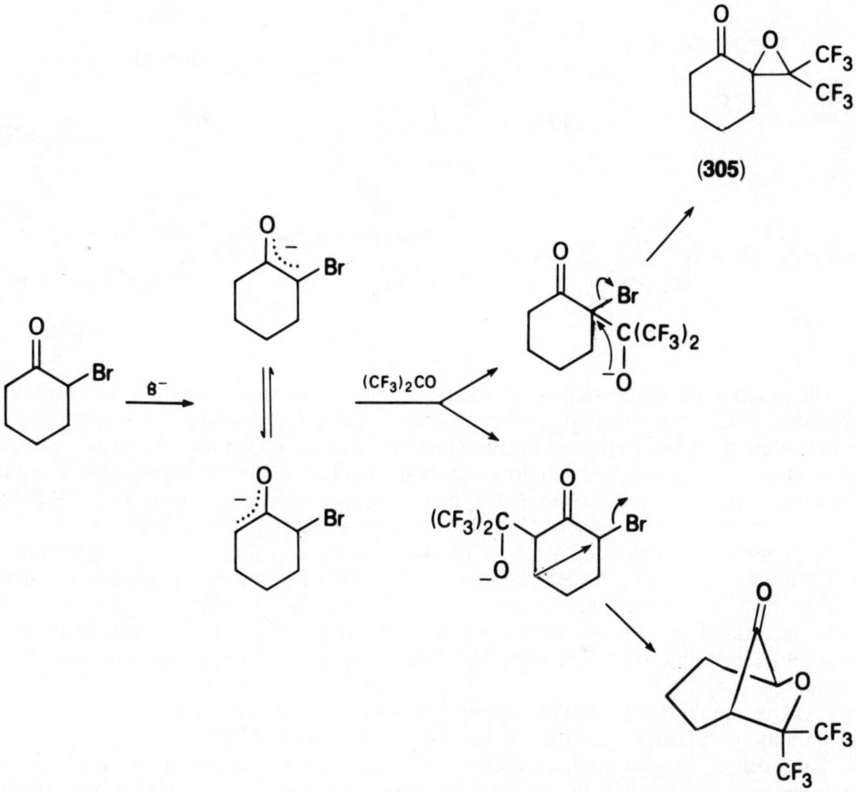

(**306**)

Primary aminoketones are also synthesized by introduction of a protected amino function using the urotropine[362] and the phthalimide method[363] or by hydrolysis of *N*-benzylaminoketones[364]. Substitution reactions of α-haloketones with heterocyclic amines often give rise to cyclized products; e.g. reaction of bromoketones with 4-aminopyrimidines **(306a)** and 3-amino-1,2,4-triazines **(308)** affords

$$R^1COCHR^2(X) + R^3R^4NH \longrightarrow R^1COCHR^2(NR^3R^4)$$

$$R^1COCHR^2(X) + NR^5_3 \longrightarrow R^1COCHR^2(\overset{+}{N}R^5_3)\ X^- \quad (161)$$

imidazo[1,2-c]pyrimidines (**307**)[365] and imidazolotriazines (**309**)[366], respectively (equation 162).

$$R^1COCHR^2(Br) + (\mathbf{306a}) \xrightarrow[\Delta]{EtOH} (\mathbf{307})$$

$$R^1COCHR^2(Br) + (\mathbf{308}) \longrightarrow \text{intermediate (N–CH(R}^2\text{)–COR}^1\text{, =NH)} \xrightarrow{\Delta} (\mathbf{309}) \quad (162)$$

The reaction between α-haloketones and amines is not always a simple substitution reaction. While the reaction of 2-bromo-2-methyl-1-aryl-1-propanones (**160**) with morpholine gives rise to the substitution product (**310**), reaction with the stronger base piperidine affords the elimination–addition product (**311**)[367]. From aniline and 1-bromo-1-phenyl-2-propanone (**312**) and 2-bromo-1-phenyl-1-propanone (**313**) a mixture of the α-aminoketones (**314**, **315**) was obtained (equation 163)[368].

In general, α-halogenated ketones of the primary and secondary type (primary, $ClCH_2CO$; secondary, RCHClCO; tertiary, R^1R^2CClCO) are expected to give substitution, but tertiary ketones are able to give elimination products. High yields of α,β-unsaturated ketones are reported when α-haloketones are treated with pyridine, quinoline, collidine and *N,N*-dimethylaniline, especially with cyclic α-haloketones (equation 164)[369–371].

α,α′-Dibromoketones provide cyclopropenones (**316**) by double 1,3-dehydrobromination on treatment with tertiary amines (equation 165)[372].

The ordinary course of the reaction of α-haloketones with pyridines, resulting in dehydrohalogenation and displacement, is often apparently accompanied by varying amounts of reduction and double bond rearrangement products as illustrated for 2β-bromocholestan-3-one (**317**) (equation 166)[373,374].

Favorskii rearrangement amides are frequently encountered when α-haloketones are treated with ammonia[375], primary amines[376,377] and secondary amines[322] (equation 167). The reaction course is dependent on the kind of base and solvent used. *Cis*-carvone tribromide (**165**) undergoes a Favorskii rearrangement to afford an iminolactone (**320**) when treated with primary amines in methanol but suffers

$ArCOC(CH_3)_2Br$ (**160**) $\xrightarrow{\text{Morpholine}}$ $ArCOC(CH_3)_2$–N(morpholino) (**310**)

$ArCOC(CH_3)_2Br$ (**160**) $\xrightarrow{\text{Piperidine}}$ $ArCOCH(CH_3)-CH_2-N$(piperidino) (**311**)

$PhCH(Br)COCH_3$ (**312**) + $PhCOCH(Br)CH_3$ (**313**) $\xrightarrow{PhNH_2}$ $PhCOCH(NHPh)CH_3$ (**314**) + $PhCH(NHPh)COCH_3$ (**315**) (163)

2-R-2-Br-cyclohexanone $\xrightarrow[\Delta]{R^1R^2R^3N}$ 2-R-cyclohex-2-enone + $R^1R^2R^3\overset{+}{N}H\ Br^-$ (164)

$R-CH(Br)-CO-CH(Br)-R \xrightarrow[\Delta]{Et_3N}$ 2,3-R₂-cyclopropenone (**316**) (165)

(**317**) $\xrightarrow{\text{pyridine}}$ pyridinium Br^- salt + enone + enone + ketone (166)

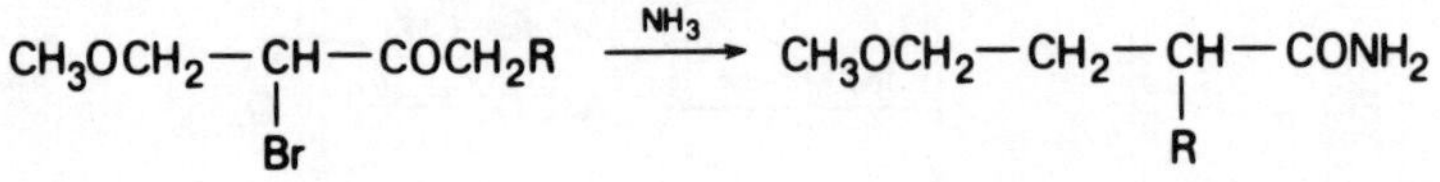

$$CH_3OCH_2-\underset{\displaystyle Br}{\underset{|}{CH}}-COCH_2R \xrightarrow{NH_3} CH_3OCH_2-CH_2-\underset{\displaystyle R}{\underset{|}{CH}}-CONH_2$$

(**318**) (**319**)

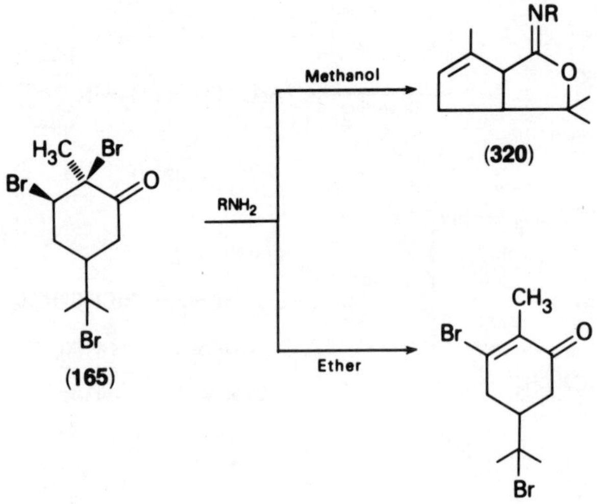

(**165**) $\xrightarrow{RNH_2}$ Methanol → (**320**); Ether → (**321**) (167)

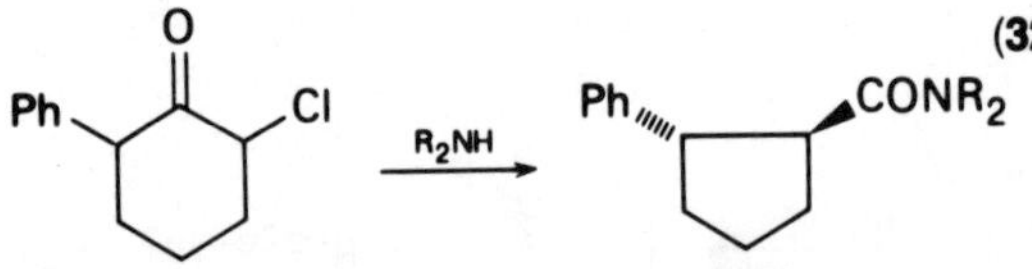

(**322**) (via R_2NH)

dehydrobromination when the reaction is conducted in ether[377]. The configuration also plays an important role, as the *trans* isomer yields the lactone **320** in both solvents. Similar solvent effects are reported during the action of amines on α,α'-dibromocycloalkanones, e.g. **252** and **300a**[306,307] (equation 168). In both cases the enamino ketones are the predominant products in polar aprotic solvents such as HMPA, whereas the Favorskii rearrangement products predominate in ether.

Another type of frequently occurring reaction consists of an addition of the amine to the carbonyl function (e.g. of **202**) followed by an intramolecular substitution yielding an aminooxirane (**329**) which can rearrange into an α-hydroxyketimine (**330**) (equation 169)[378].

Formation of ketimines[405,406] normally does not take place when α-haloketones are treated with primary amines except for α-fluorinated ketones, and especially high yields of trifluoroketimines (**331**) are easily obtained (equation 170)[379,380]. However, we recently developed a general method for the preparation of α-halogenated ketimines (**332**) by condensing α-haloketones with primary amines in ether using titanium tetrachloride as condensing agent (equation 171)[381]. This method is also applicable to dihalo- and trihaloketones[381].

The formation of compounds containing C=N bonds will be discussed in the section dealing with reactions of carbonyl reagents. In general[586], the condensation of α-halogenated ketones with secondary amines does not afford β-halogenated enamines

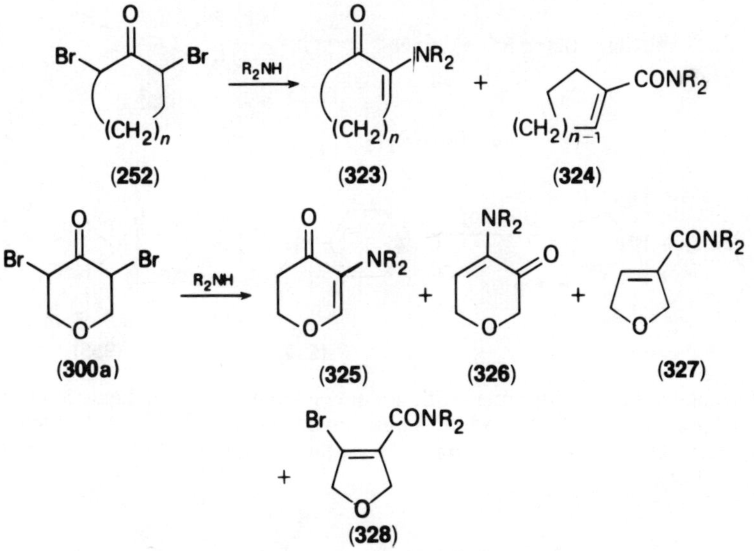

(252) —R_2NH→ (323) + (324)

(300a) —R_2NH→ (325) + (326) + (327) + (328) (168)

R_2NH = morpholine, piperidine, pyrrolidine

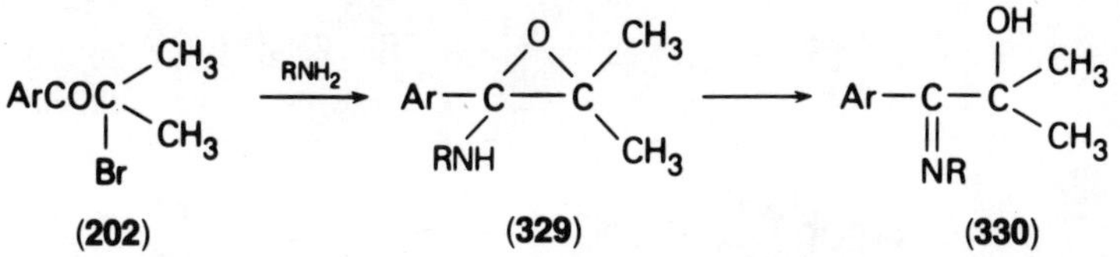

(202) —RNH_2→ (329) → (330) (169)

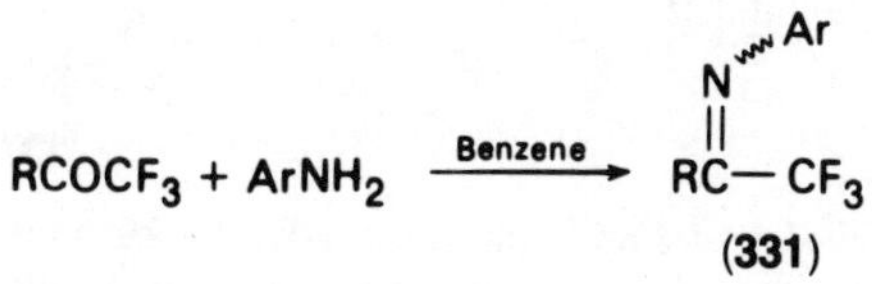

$RCOCF_3 + ArNH_2$ —Benzene→ $RC(=N\text{-}Ar)-CF_3$ (331) (170)

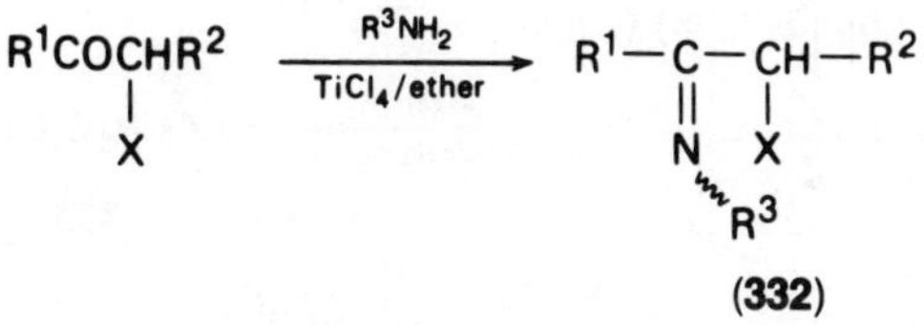

$R^1COCHXR^2$ —R^3NH_2, $TiCl_4$/ether→ $R^1-C(=NR^3)-CHX-R^2$ (332) (171)

R^1, R^2 = H, alkyl, aryl

R^3 = alkyl, aryl

(333) except when the reaction is carried out in the presence of metal chlorides such as $AsCl_3$, $SbCl_3$, $FeCl_3$ and $TiCl_4$[382–383], or by the use of tris(*N*,*N*-dialkylamino)arsines[384] (equation 172).

In addition, reaction of 2-chlorocyclohexanone with pyrrolidine at −100°C in the presence of magnesium sulphate produces an enamine **(334)** with the chlorine atom in allylic position and minor amounts of a bicyclic compound **(335)** (equation 173)[385].

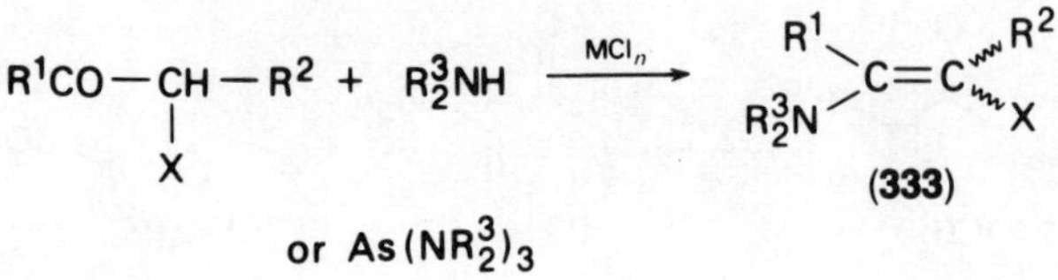

$$R^1CO-CH(X)-R^2 + R^3_2NH \xrightarrow[\text{or } As(NR^3_2)_3]{MCl_n} R^1(R^3_2N)C{=}C(R^2)X \quad \textbf{(333)} \qquad (172)$$

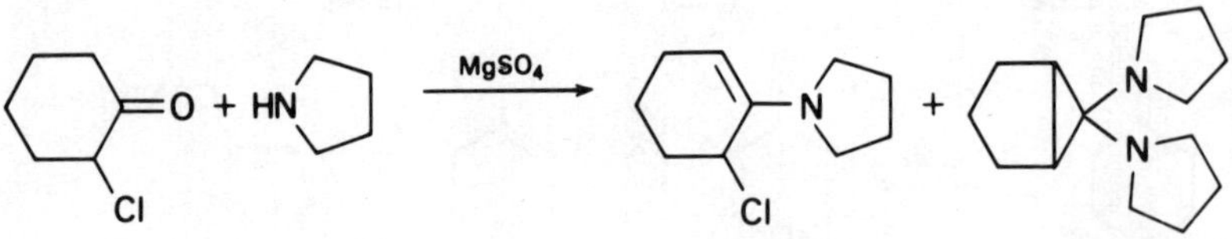

$$\text{2-chlorocyclohexanone} + \text{pyrrolidine} \xrightarrow{MgSO_4} \textbf{(334)} + \textbf{(335)} \qquad (173)$$

Chloromethyl ketones also react with lithium or sodium amide in liquid ammonia to produce oxazolines **(336)**[386], while stable epoxyamines **(337)** are formed when brominated alkyl aryl ketones are treated with the lithium salt of ethyleneimine[387,388] (equation 174).

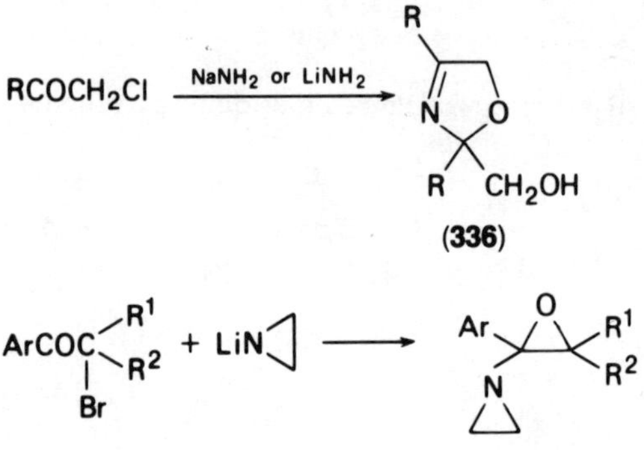

$$RCOCH_2Cl \xrightarrow{NaNH_2 \text{ or } LiNH_2} \textbf{(336)}$$

$$ArCOC(Br)R^1R^2 + LiN\!\!\vartriangleleft \longrightarrow \textbf{(337)} \qquad (174)$$

b. Reaction of α-haloketones with enamines. α-Bromoketones react with enamines of methyl ketones to provide immonium salts which upon hydrolysis afford 1,4-dicarbonyl compounds **(338)** (equation 175)[389].

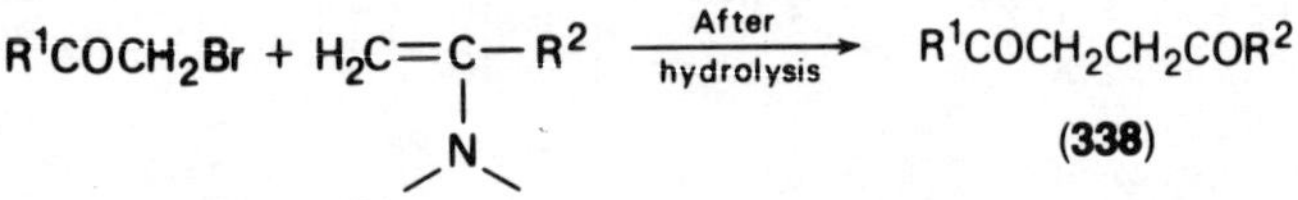

$$R^1COCH_2Br + H_2C{=}C(NMe_2)-R^2 \xrightarrow[\text{hydrolysis}]{\text{After}} R^1COCH_2CH_2COR^2 \quad \textbf{(338)} \qquad (175)$$

Cycloaddition of β-amino-α,β-unsaturated carboxylic acids and derivatives with α-haloketones gives rise to pyrrole compounds **(339)** (Hantzsch synthesis)[390], while reaction with β-aminovinyl thioketones **(340)** give 2-acylthiophenes **(341)**[391] (equation 176).

c. Reaction of α-haloketones with amides, thioamides and derivatives. Various heterocyclic compounds have been synthesized by the reaction of α-haloketones with amides, thioamides, urea, thiourea, amidines, guanidines and sulphonylamides. Reaction of α-bromoketones with amides produces oxazoles **(342)**[392–393], while reaction with alkynyl thioamides **(343)** gives 1,3-oxathiazoles **(344)**[394] (equation 177).

A cyclodehydrohalogenation leading to 2-azetidinones **(346)** is observed when

$$R^1COCH(Br)R^2 + R^3C(NH_2){=}CHCOY \longrightarrow \textbf{(339)} \quad (Y = H,\ OR,\ NHR) \qquad (176)$$

$$R^1COCH_2Br + ArC(=S){-}CH{=}C(R^2){-}N< \ \textbf{(340)} \xrightarrow{NEt_3} \textbf{(341)}$$

$$R^1COCH(Br)R^2 + R^3CONH_2 \xrightarrow{110^\circ C} \textbf{(342)} \qquad (177)$$

$$R^1COCH_2Br + R^2{-}C{\equiv}C{-}C(=S){-}NHR^3 \ \textbf{(343)} \xrightarrow{NEt_3} \textbf{(344)}$$

anilides of α-bromoketones (**345**) are subjected to the action of various bases (equation 178)[395].

$$Ar{-}NHCOCH_2CH(Br)COPh \ \textbf{(345)} \xrightarrow{Base} \textbf{(346)} \qquad (178)$$

Reaction of α-bromoketones with amidines or formimidates constitutes an excellent method for the synthesis of imidazole derivatives (**347**) (equation 179)[396,375]. However, a pyrimidine ring (**350**) is formed during the reaction of

$$R^1COCH(Br)R^2 + HN{=}C(R^3){-}NH_2 \longrightarrow \textbf{(347)} \qquad (179)$$

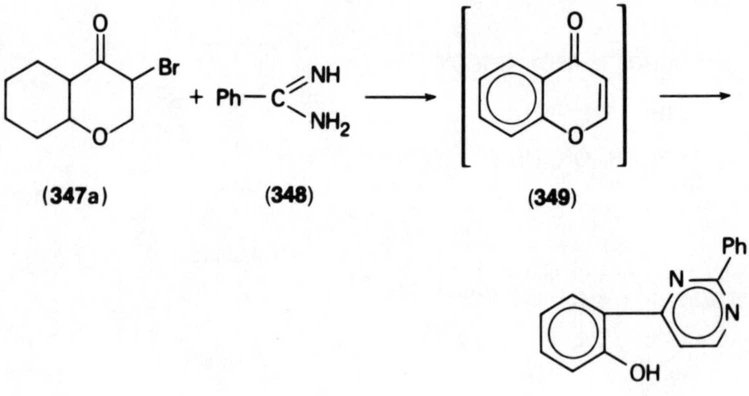

(347a) + (348) → [(349)] → (350) (180)

α-bromochromanone (**347a**) with benzamidine (**348**) probably via an intermediate chromone (**349**) (equation 180)[397].

Another synthesis of imidazole compounds (**351**) involves the condensation of phenacyl bromides with guanidine using bromine in methanol as the condensing agent (equation 181)[398]. On the other hand, reaction of an α-haloacetone with

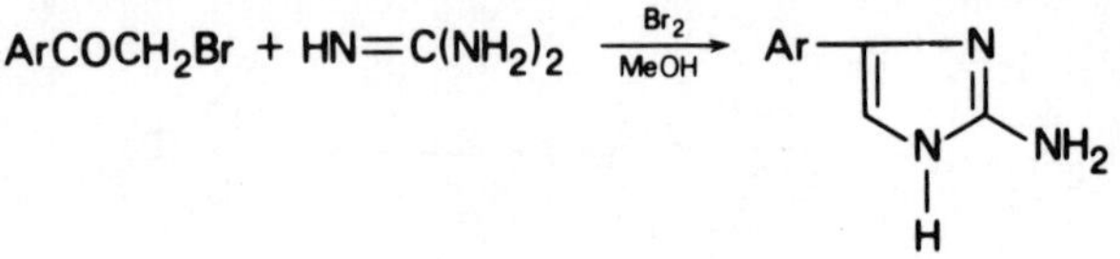

$$ArCOCH_2Br + HN{=}C(NH_2)_2 \xrightarrow[MeOH]{Br_2} (351) \quad (181)$$

isothiosemicarbazones (**352**) gives rise to a competitive formation of imidazoles (**353**) and triazole compounds (**354**). The ratio is dependent upon the nature of the halogen and the reaction temperature (equation 182)[399].

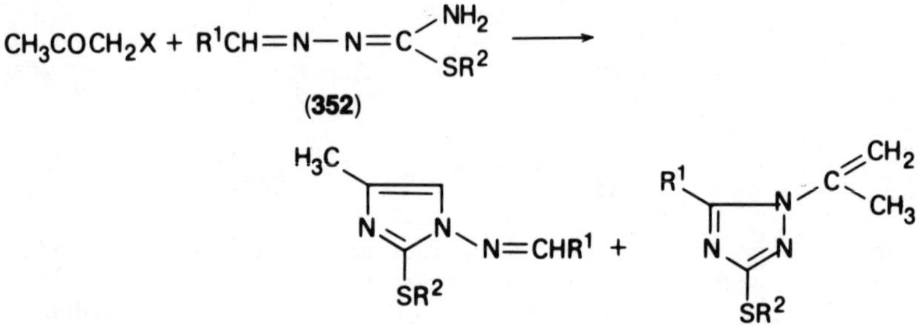

$$CH_3COCH_2X + R^1CH{=}N{-}N{=}C(NH_2)SR^2 \longrightarrow (353) + (354) \quad (182)$$

(352)

The Hantzsch reaction of α-haloketones with thioamides, thiourea and dithiocarbamates affords thiazolium derivatives (**357**). By isolation of intermediate thiazolines it is proved that the first step in this reaction is a direct substitution of the halogen atom and not an addition of the nitrogen atom at the carbonyl function (equation 183)[400–402].

$$R^1COC(X)(R^2)R^3 + R^4NHC(=S)R^5 \longrightarrow R^2R^3C(SC(R^5)=\overset{+}{N}HR^4)C(=O)R^1\ X^- \ \textbf{(355)} \longrightarrow \qquad (183)$$

$$\textbf{(356)} \xrightarrow[-H_2O]{R^3 = H} \textbf{(357)}$$

Substitution of α-bromoketones with *N*-phenyltrifluoromethanesulphonamides (**358**) under mild conditions gives rise to the formation of α-iminoketones (**359**) which in turn can further be converted into pyrazines (**360**) (equation 184)[403–404].

$$R^1COCHR^2(Br) + PhNHSO_2CF_3\ \textbf{(358)} \xrightarrow[DMF]{K_2CO_3} R^1COCHR^2N(H)(Ph)SO_2CF_3 \xrightarrow{-CF_3SO_2H}$$

$$R^1COC(=NPh)R^2\ \textbf{(359)} \xrightarrow[(2)\ O_2/CH_3OH/CH_3ONa]{(1)\ NH_2CH_2CH_2NH_2} \textbf{(360)} \qquad (184)$$

d. Reaction of α-haloketones with carbonyl reagents. The reactivity of α-halocarbonyl compounds towards the usual carbonyl identification reagents has been reviewed by De Kimpe and coworkers in reports dealing with the synthesis and reactivity of α-halogenated imino compounds[405–406]. Therefore, only the most typical reactions will be covered in this section. Except for 2,4-dinitrophenylhydrazones of α-haloketones, which are easily formed when prepared in aqueous methanol in the presence of sulphuric acid[407], α-haloimino compounds are not generally available by the condensation of α-haloketones with carbonyl reagents, due to the reactivity of the imino compounds which lead to further reactions under the normal reaction conditions. Consequently, only a limited number of α-halohydrazones[408], semicarbazones[409] and oximes[410–412] have been obtained by the direct condensation route (equation 185).

$$R^1COCHR^2(X) + H_2N{-}R^3 \longrightarrow R^1C(=N{\sim}R^3){-}CHR^2(X) \qquad (185)$$

R^3 = OH, NH_2, NHR, NHCOR, NHCSR, NHAr, NHTos

The most frequently encountered side reactions are the formation of azoalkenes (**361**) by 1,4-dehydrohalogenation[413,414], nitrosoolefins (**362**)[410] and the formation of diimino compounds[415], as illustrated in the following examples (equation 186). In

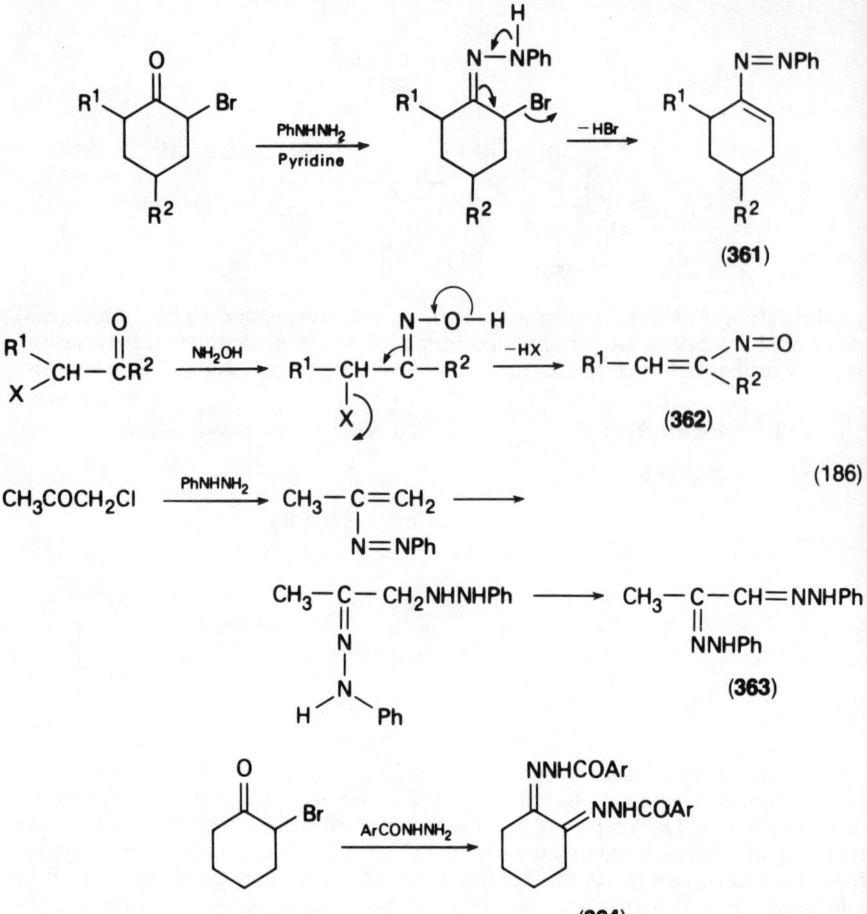

(186)

(**361**)

(**362**)

(**363**)

(**364**)

addition, the initially formed α-haloimino compounds are able to undergo ring closure reactions to yield a variety of heterocyclic products. Reaction of phenacyl bromides with *N,N*-dimethylhydrazine and phenylhydrazine gives rise to the formation of pyrazoles (**365**)[416] and tetrahydropyridazines (**366**)[417], respectively. The reaction of dibromoketones with hydroxylamine and hydrazine furnishes isoxazoles (**367**)[418] and pyrazolidinones (**368**)[419] (equation 187).

1,2,4-Triazines (**369**)[420,421], 1,3,4-thiadiazines (**370**)[422] and thiazolines (**371**)[423] are formed when α-haloketones are treated respectively with acylhydrazines, thioacylhydrazines and thiosemicarbazide (equation 188).

e. Reaction of α-haloketones with sodium azide. Reaction of α-haloketones with sodium azide produces α-azidoketones, which on pyrolysis afford α-iminoketones via nitrene intermediates (equation 189)[304,424,425]. Action of sodium azide on chalcone dibromides furnishes α-azidochalcones (**372**) and isoxazoles (equation 189a)[418].

$$ArCOCH_2Br + H_2NN(CH_3)_2 \longrightarrow [ArCOCH{=}NH] \longrightarrow \textbf{(365)}$$

$$ArCOCH_2Br + H_2NNHPh \longrightarrow \textbf{(366)}$$

$$ArCOCH(Br){-}CH(Br)Ph + NH_2OH \longrightarrow \textbf{(367)} \qquad (187)$$

$$R^1R^2C(Br){-}CO{-}C(Br)R^3R^4 + NH_2NH_2 \longrightarrow \textbf{(368)}$$

$$R^1COCH(X)R^2 + H_2NNHCOR^3 \longrightarrow \textbf{(369)}$$

$$ArCOCH(Br)R^1 + H_2NNHCSR^2 \longrightarrow \textbf{(370)} \qquad (188)$$

$$R^1COCH(Cl)R^2 + H_2NCSNHNH_2 \longrightarrow \textbf{(371)}$$

$$R^1COC(X)R^2R^3 \xrightarrow{N_3^-} R^1CO{-}C(N_3)R^2R^3 \xrightarrow{\Delta} R^1COC(R^2){=}NR^3 \qquad (189)$$

$$\text{ArCOCH(Br)CH(Br)—Ph} \xrightarrow{N_3^-} \text{ArCO—C(Br)=CH—Ph} \xrightarrow{HN_3} \text{ArCO—CH(Br)—CH(N}_3\text{)—Ph} \tag{189a}$$

$$\text{ArCO—CH(Br)—CH(N}_3\text{)—Ph} \rightarrow \text{ArCO—CH(N}_3\text{)—CH(N}_3\text{)—Ph} \rightarrow \text{ArCO—C(N}_3\text{)=CH—Ph}\ \textbf{(372)}$$

$$\text{ArCO—CH(Br)—CH(N}_3\text{)—Ph} \rightarrow \text{ArCO—CH=C(N}_3\text{)—Ph} \xrightarrow{-N_2} \text{5-Ar-3-Ph-isoxazole}\ \textbf{(373)}$$

5. *Reaction of α-haloketones with sulphur nucleophiles*

a. Reaction of α-haloketones with inorganic sulphur compounds. Reaction of α-haloketones with sodium hydrogen sulphide gives rise to α-mercaptoketones (**374**) in 50–80% yield. However, when the reaction temperature is higher than 0°C, sulphides (**375**) can be generated (equation 190)[426,427]. α,α′-Dimercaptoketones (**376**) (isolated

$$R^1COC(X)R^2R^3 + NaSH \longrightarrow R^1COC(SH)R^2R^3\ \textbf{(374)} \xrightarrow{\Delta} R^1COC(R^2)(R^3)—S—C(R^2)(R^3)COR^1\ \textbf{(375)} \tag{190}$$

as the cyclic dimers (**377**)) are also produced on treatment of α,α′-dihaloketones with sodium hydrogen sulphide (equation 191)[428].

$$ClCH_2COCH_2Cl \xrightarrow{NaSH} HSCH_2COCH_2SH\ \textbf{(376)} \rightleftharpoons \text{2,5-bis(mercaptomethyl)-2,5-dihydroxy-1,4-dithiane}\ \textbf{(377)} \tag{191}$$

3-Thietanones (**378**) have been synthesized by reaction of α,α′-dibromoketones and sodium hydrogen sulphide, together with minor amounts of dithiols (**379**), 1,2-dithiolan-4-ones (**380**) and polycondensates (**381**) (equation 192)[429]. Mercaptomethyl aryl ketones are also formed when aryl α-halomethyl ketones are treated with hydrogen sulphide in pyridine[430].

$$R^1R^2C(Br)-CO-C(Br)R^3R^4 \xrightarrow{^-SH} R^1R^2C(SH)-CO-C(Br)R^3R^4 \xrightarrow{\text{Base}} \textbf{(378)}$$

$$\longrightarrow R^1R^2C(SH)-CO-C(SH)R^3R^4 \ \textbf{(379)} + \textbf{(380)}$$

$$\longrightarrow R^1R^2C(SH)-CO-CR^3R^4-[S-CR^1R^2-CO-CR^3R^4]_n-SH \ \textbf{(381)} \qquad (192)$$

Reaction of α-haloketones with sodium sulphide affords sulphides (**375**)[431]. However, 2-chlorocyclohexanone produces a tricyclic compound (**382**) via aldol-type condensation of the intermediate sulphide in an inert atmosphere, while in the presence of oxygen a disulphide (**383**) is formed[432] (equation 193).

$$R^1COCR^2R^3X \xrightarrow{Na_2S} R^1COCR^2R^3-S-CR^2R^3COR^1 \ \textbf{(375)}$$

$$\text{2-chlorocyclohexanone} \xrightarrow{Na_2S} \begin{cases} \xrightarrow{N_2} \text{bis(2-oxocyclohexyl) sulphide} \longrightarrow \textbf{(382)} \\ \xrightarrow{O_2} \textbf{(383)} \end{cases} \qquad (193)$$

Finally, α-ketothiocyanates (**384**) are formed by the reaction of α-haloketones with potassium thiocyanate (equation 194)[433].

$$RCOCH_2Br \xrightarrow{KSCN} \underset{\mathbf{(384)}}{RCOCH_2SCN} \tag{194}$$

b. Reaction of α-haloketones with organic sulphur nucleophiles. Treatment of α-haloketones with mercaptans in the presence of bases mainly gives α-substitution to afford α-alkylthio and α-arylthio ketones (**385**) (equation 195)[434–436].

$$R^1COC(X)(R^2)R^3 + R^4SH \xrightarrow{Base} \underset{\mathbf{(385)}}{R^1CO-C(R^2)(SR^4)-R^3} \tag{195}$$

Other reactions involve dehalogenation[437] and transformation of the initially formed mercaptans into the corresponding disulphides[438]. 1,4-Dithienes (**386**) are prepared by cyclocondensation of α-haloketones with 1,2-ethanedithiol in the presence of acids (equation 196)[439,440].

$$R^1COCHR^2(X) + HSCH_2CH_2SH \xrightarrow{H^+} \mathbf{(386)} \tag{196}$$

The reaction of chloroacetone with mercaptoacetamide proceeds smoothly to give 3-hydroxy-5-methyl-1,4-thiazine (**387**) (equation 197), while with phenacyl bromide the substitution product is formed[441].

$$CH_3COCH_2Cl + HSCH_2CONH_2 \longrightarrow \mathbf{(387)} \tag{197}$$

Thioacids and their derivatives react readily with α-haloketones. Treatment with thioacids in the presence of ammonium acetate in refluxing acetic acid gives 1,3-thiazoles (**388**) (equation 198)[442]. Reaction of thioacid salts normally gives rise to

$$R^1COCHR^2(Br) + HSCOR^3 \xrightarrow[HOAc]{NH_4OAc} \mathbf{(388)} \tag{198}$$

substitution products (**389**)[443]. A useful synthetic application of this reaction is the formation of selenocarboxylates (**389**) on treatment with selenoacids. Selenium elimination from **389** with strong bases yields 1,3-diketones (**390**) (equation 199)[444]. *S*-potassium hydrazino monothio- and dithioformate (**391**) react with α-haloketones to

$$R^1COCH_2Br + R^2COYK \longrightarrow R^1COCH_2YCOR^2 \ \mathbf{(389)} \xrightarrow[t\text{-BuOK}]{Y = Se} R^1COCH_2COR^2 \ \mathbf{(390)} \qquad Y = S, Se \quad (199)$$

form acylmethyl (hydrazino)thioformates (**392**) which can be cyclized to 1,3-thiazolin-2-ones (**393**) or 1,3,4-thiadiazin-2-ones (**394**) dependent upon the substitution pattern of the ketone (equation 200)[445]. Dilithium salts of thioacids (**395**) also

$$R^1CH(X)COR^2 + KS{-}C({=}Y){-}NHNHR^3 \ \mathbf{(391)} \longrightarrow R^2C({=}O){-}CH(R^1){-}S{-}C({=}Y){-}NHNHR^3 \ \mathbf{(392)} \longrightarrow \mathbf{(393)} \text{ or } \mathbf{(394)} \qquad Y = O, S \quad (200)$$

react with α-chloroketones to yield β-hydroxy thioacids (**396**) which are cyclized to thiolactones (**397**) upon action with triethylamine, while unsaturated thiolactones (**398**) are formed with sodium hydride in DMF (equation 201)[446].

$$R^1COCH(Cl)R^2 + R^3CH{=}C(OLi)(SLi) \ \mathbf{(395)} \xrightarrow{THF} R^1C(OH)(R^2CHCl){-}CH(R^3)C(O)SH \ \mathbf{(396)}$$

$$\mathbf{(396)} \xrightarrow[R^2, R^3 = H]{NaH, DMF} \mathbf{(398)} \qquad \mathbf{(396)} \xrightarrow{THF, NEt_3} \mathbf{(397)} \quad (201)$$

6. *Reaction of α-haloketones with carbon nucleophiles*

a. Reaction of α-haloketones with cyanides. α-Haloketones have been shown to give a variety of reactions when treated with sodium or potassium cyanide[447]. Earlier reports claimed the formation of α-cyanoketones[448]; however, it is proved later that in most cases the reaction products are α-cyanoepoxides (**400**) formed via an addition–substitution mechanism[311,449,450] (equation 202).

$$R^1COCHXR^2 \xrightarrow{^-CN} R^1C(OH)(CN)CHXR^2\ (\mathbf{399}) \longrightarrow R^1(NC)C\overset{O}{—}CHR^2\ (\mathbf{400}) \xrightarrow{\Delta} R^1COCH(CN)R^2 \qquad (202)$$

$$R^1COCHXR^2 \xrightarrow{^-CN} R^1COCH(CN)R^2$$

The α-cyanoketones could either be formed via direct substitution or via thermal rearrangement of the cyanoepoxides. Only when R^1 is a *t*-butyl or an aryl substituent are α-cyanoketones produced[451–452]. On the other hand, α-fluoroketones undergo cyanation at the carbonyl function with formation of cyanohydrins (**399**) without substitution of the halogen[453]. Cyanohydrins (**399**) are also produced by condensation with hydrocyanic acid at 0°C in the presence of potassium cyanide as condensing agent[454].

Reaction of α-haloketones with tetraethylammonium cyanide in dichloromethane also gives rise to cyanoepoxides (**400**) which upon heating at 80–135°C in the presence of ammonium cyanide rearrange into α-cyanoketones[455].

Reinvestigation of the reaction of chloroacetone with alkali cyanides in aqueous solution at room temperature shows another route leading to an enaminoketone (**401**)[456] and not to the tetrahydrofuran (**402**) as proposed earlier[457] (equation 203).

$$CH_3COCH_2Cl \xrightarrow{^-CN} CH_3CO\bar{C}HCN \xrightarrow{CH_3COCH_2Cl} CH_3COCH(CN)CH_2COCH_3 \xrightarrow{^-CN} CH_3COCH(C{\equiv}N)CH_2C(CN)(O^-)CH_3 \longrightarrow \mathbf{(401)} \qquad (203)$$

(**401**): 2-amino-3-acetyl-5-methyl-5-cyano-4,5-dihydrofuran (H_2N, CH_3CO, CH_3, CN)

(**402**): 2-hydroxy-2,5-dimethyl-5-cyanotetrahydrofuran (H_3C, HO, CH_3, CN)

Another reaction involves the formation of a cyclopropane (**404**), induced by a Favorskii-type rearrangement on treatment of 1-chloro-3-phenyl-2-propanone (**403**) with alkali cyanides in the cold (equation 204)[458].

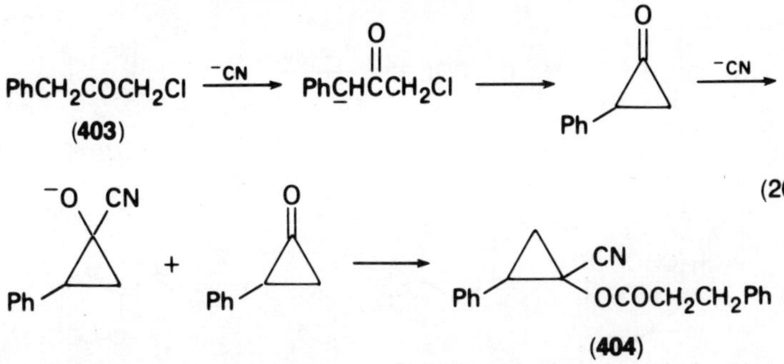

$PhCH_2COCH_2Cl$ (**403**) $\xrightarrow{^-CN}$ $Ph\underline{C}HCOCH_2Cl$ ⟶ [2-phenylcyclopropanone] $\xrightarrow{^-CN}$

(204)

[^-O, CN-substituted phenylcyclopropane] + [2-phenylcyclopropanone] ⟶ [Ph-cyclopropane bearing CN and $OCOCH_2CH_2Ph$] (**404**)

b. Reaction of α-haloketones with carbanions, ylides and enolates. The reaction of α-haloketones with diethyl sodium malonate and ethyl sodium acetoacetate produces exclusively substitution products via an S_N2 reaction[459–461].

The reaction course, however, is influenced strongly by the temperature. While diethyl (2-oxocyclohexyl)malonate (**405**) is formed during the reaction of 2-chlorocyclohexanone with diethyl sodium malonate in refluxing benzene[460], 6-[bis (ethoxycarbonyl) methyl]bicyclo[3.1.0]hexan-6-ol (**406**) is isolated at 0–25°C via a malonate anion-induced Favorskii-type rearrangement[462] (equation 205).

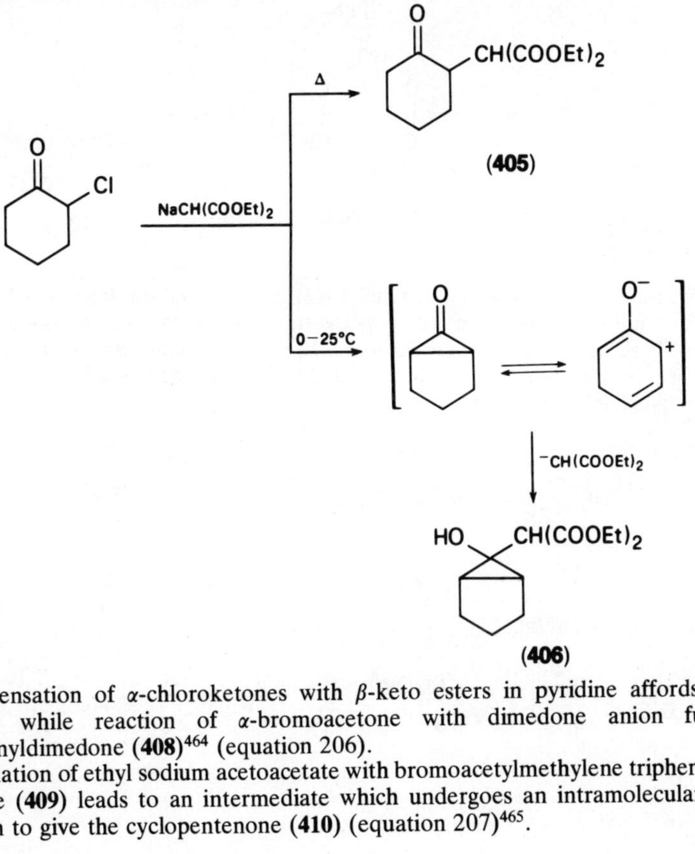

2-chlorocyclohexanone $\xrightarrow{NaCH(COOEt)_2}$ (Δ) ⟶ 2-[$CH(COOEt)_2$]cyclohexanone (**405**)

(0–25°C) ⟶ [bicyclic cyclopropanone ⇌ oxyallyl cation] $\xrightarrow{^-CH(COOEt)_2}$ HO, $CH(COOEt)_2$ bicyclic product (**406**)

(205)

Condensation of α-chloroketones with β-keto esters in pyridine affords furans (**407**)[463] while reaction of α-bromoacetone with dimedone anion furnishes 2-acetonyldimedone (**408**)[464] (equation 206).

Alkylation of ethyl sodium acetoacetate with bromoacetylmethylene triphenylphosphorane (**409**) leads to an intermediate which undergoes an intramolecular Wittig reaction to give the cyclopentenone (**410**) (equation 207)[465].

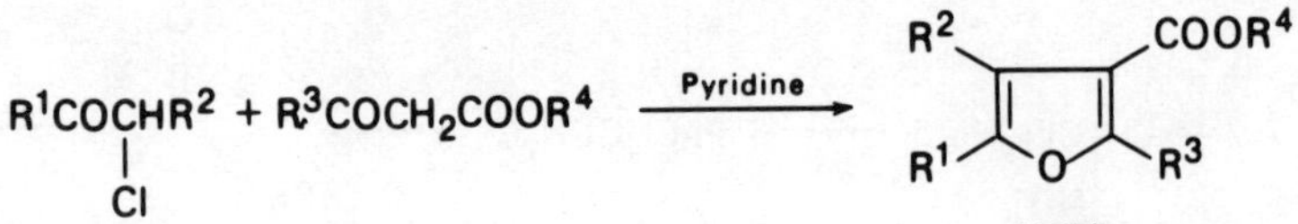

$$R^1COCHR^2(Cl) + R^3COCH_2COOR^4 \xrightarrow{\text{Pyridine}} \textbf{(407)}$$

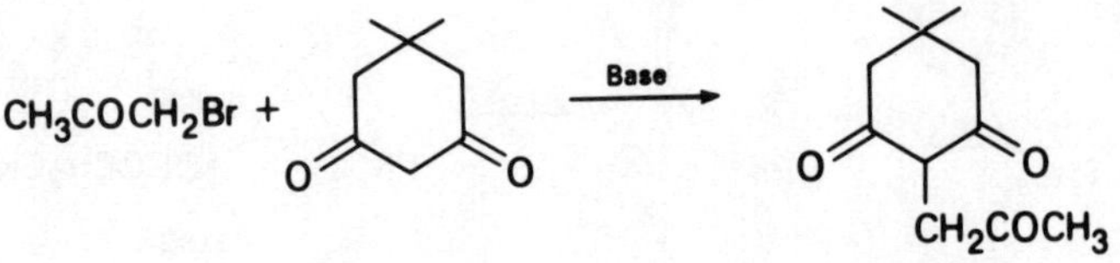

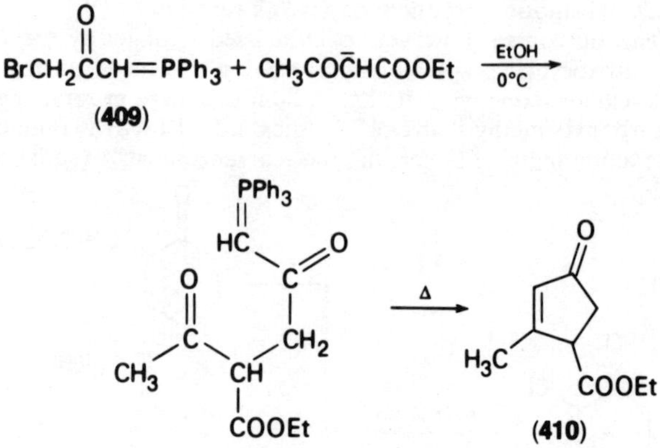

$$CH_3COCH_2Br + \text{5,5-dimethylcyclohexane-1,3-dione} \xrightarrow{\text{Base}} \textbf{(408)} \quad (206)$$

$$BrCH_2C(=O)CH{=}PPh_3 \ \textbf{(409)} + CH_3CO\bar{C}HCOOEt \xrightarrow[0°C]{EtOH} \xrightarrow{\Delta} \textbf{(410)} \quad (207)$$

α,β-Unsaturated esters and cyanides **(412)** are formed in high yields by the Emmons–Wadsworth reaction of α-haloketones with the corresponding phosphoranes[466]. Knoevenagel condensation of α-haloketones with active methylene functions gives electrophilic allylic halides **(413)** by the action of titanium tetrachloride–pyridine[467,468] (equation 208).

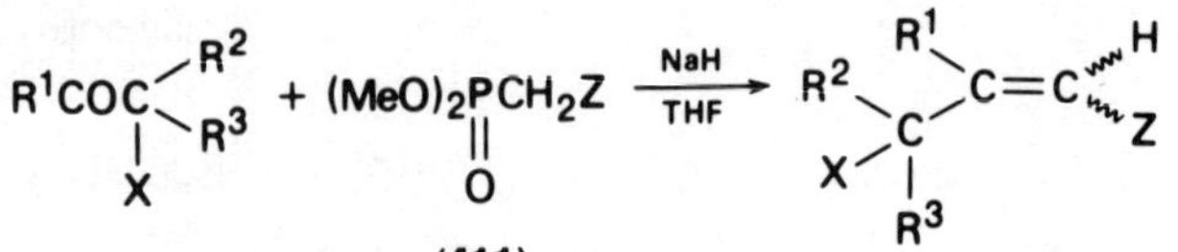

$$R^1COCXR^2R^3 + (MeO)_2P(O)CH_2Z \ \textbf{(411)} \xrightarrow[THF]{NaH} R^2R^3XC(R^1)C{=}CHZ \ \textbf{(412)}$$

$Z = COOR^4$, CN

$$R^1COCXR^2R^3 + CH_2(COOR^4)Z \xrightarrow[\text{Pyridine}]{TiCl_4} R^2R^3XC(R^1)C{=}C(COOR^4)Z \ \textbf{(413)} \quad (208)$$

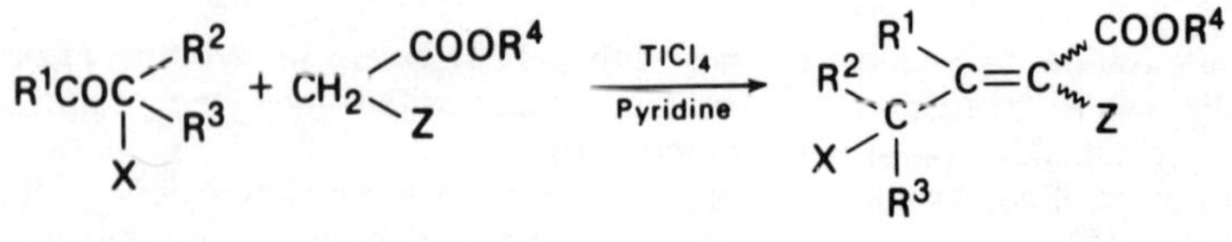

$Z = COCH_3$, $COOR^4$, CN

Another reaction type, namely aldol condensation yielding α,β-enones (e.g. **414**, **415**), is reported when α-chloroketones are treated with pyridine in the presence of titanium tetrachloride (equation 209)[469].

$$ClCH_2COCH_3 \xrightarrow[\text{Pyridine}]{TiCl_4} ClCH_2-C(CH_3)=C(COCH_3)Cl \quad \textbf{(414)}$$

$$R^1R^2C(Cl)-COCH_3 \xrightarrow[\text{Pyridine}]{TiCl_4} R^1R^2C(Cl)-C(CH_3)=CH-C(=O)-C(Cl)R^1R^2 \quad \textbf{(415)} \qquad (209)$$

Reaction of α-haloketones with dimethylsulphoxonium methylide (**416**) results in cyclopropanation. First, the halogen is displaced by the methylide to give the intermediate salt. According to path *a* the salt can be converted to an olefin which reacts successively with the methylide (**416**) to give the cyclopropane (**417**). An alternative

$$R^1COCHXR^2 + {}^-CH_2\overset{+}{S}(O)(CH_3)_2 \; \textbf{(416)} \longrightarrow R^1COCHR^2CH_2\overset{+}{S}(O)(CH_3)_2\;X^-$$

Path *a* (−DMSO, −HX): $\longrightarrow R^1-C(=O)-C(R^2)=CH_2 \xrightarrow[-\text{DMSO}]{+\textbf{416}} R^1CO(R^2)\text{-cyclopropane} \; \textbf{(417)}$

Path *b* (−DMSO, +**416**): $\longrightarrow R^1COCHR^2CH_2CH_2\overset{+}{S}(O)(CH_3)_2\;X^- \xrightarrow{-HX} R^1CO\bar{C}R^2CH_2CH_2\overset{+}{S}(O)(CH_3)_2 \xrightarrow{-\text{DMSO}} \textbf{(417)} \qquad (210)$

route (path *b*) involves a nucleophilic displacement of the salt to afford an homologous salt which in turn affords a cyclopropane (**417**) by an intramolecular displacement (equation 210)[470,471]

Intramolecular cyclopropanation takes place when α-haloketones, carrying electron-withdrawing groups in the γ-position, are treated with strong bases, providing, for example, nitrocyclopropanes (**419**)[472]. 1-Halocyclopropyl methyl ketones (**421**) may be obtained by simply heating the appropriate 3,5-dihalo-2-pentanones (**420**) with potassium fluoride as base in diethylene glycol[473] (equation 211).

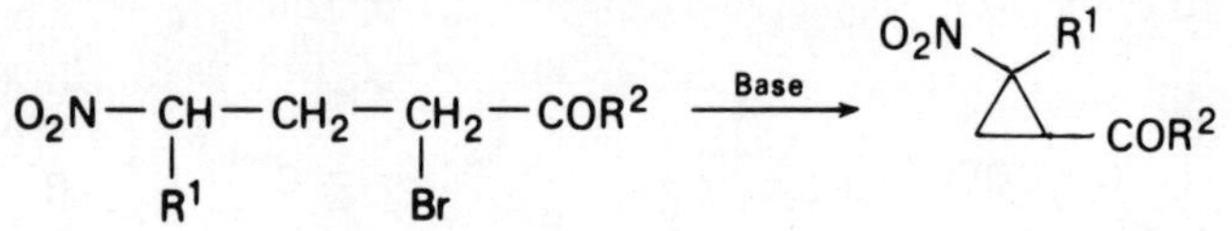

$$O_2N-CH(R^1)-CH_2-CH(Br)-COR^2 \xrightarrow{\text{Base}} \text{1-}O_2N\text{-1-}R^1\text{-2-}COR^2\text{-cyclopropane}$$

(**418**) (**419**) (211)

$$X^1-CH_2-CH_2-CH(X^2)-COCH_3 \xrightarrow{\text{KF}} \text{1-}COCH_3\text{-1-}X^2\text{-cyclopropane}$$

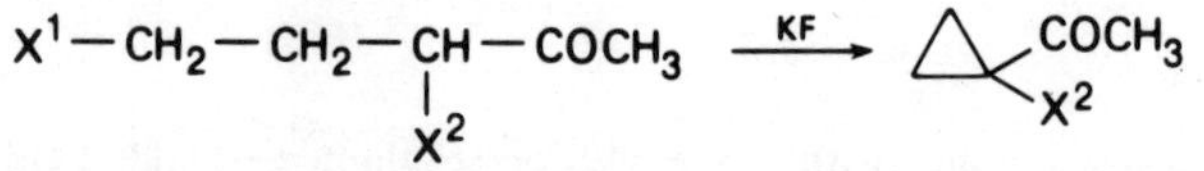

(**420**) (**421**)

X^1, X^2 = halogen

Lithium enolates (**422**) react smoothly with certain α-halocarbonyl derivatives, e.g. bromoacetylmethylene triphenylphosphorane (**409**), which constitutes a useful annelation reagent, to give cyclopentenones (**423**) (equation 212)[465]

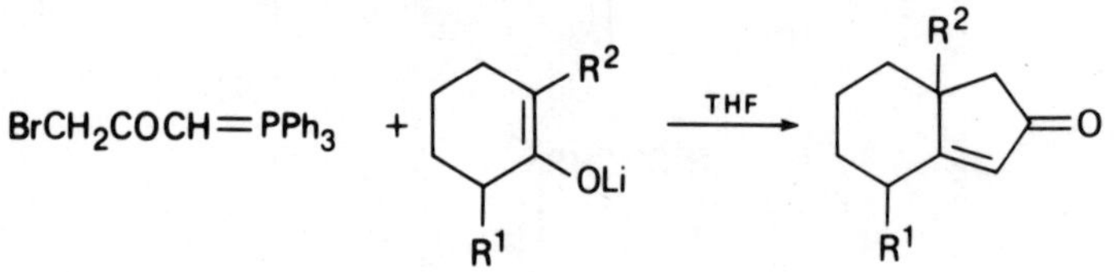

$$BrCH_2COCH{=}PPh_3 + (\mathbf{422}) \xrightarrow{\text{THF}} (\mathbf{423}) \quad (212)$$

(**409**) (**422**) (**423**)

7. *Reaction of α-haloketones with organometallic reagents*

a. Reaction of α-haloketones with Grignard reagents. α-Haloketones react readily with Grignard reagents to afford mainly magnesium salts of halohydrins (which can be hydrolysed to the parent halohydrins) and rearranged ketones in variable proportions (equation 213)[474–478].

The majority of the rearrangements of halomagnesium derivatives of halohydrins can be accounted for by considering them to be pinacol-like rearrangements induced by an electrophilic attack of the MgX group on the neighbouring halogen atom (route A). A second way consists of an internal nucleophilic substitution (route B) (equation 214). Whether mechanism A or B is followed will be determined by structural factors. Route A should be favoured when the halogen atom is secondary or tertiary, when the migrating group R can participate in the process and contribute to the resonance stabilization of the transition state and when the —X and —OMgX moieties are in a *cis* relationship to one another. When the halogen atom is secondary or tertiary, route A seems always to be followed, but when it is primary the nature of the R^3 and R^4 groups directs the course of the rearrangement, as illustrated below (equation 215)[475].

$$R^1R^2C(X)-C(=O)R^3 + R^4MgX \longrightarrow R^1R^2C(X)-C(R^3)(R^4)OMgX \xrightarrow{H_3O^+} R^1R^2C(X)-C(R^3)(R^4)OH$$

$$\longrightarrow R^1R^2R^3C-C(=O)-R^4 + R^1R^2R^4C-C(=O)-R^3 \qquad (213)$$

$$R^1R^2C(X)-C(R^4)(R^3)-O-Mg-X \xrightarrow{\text{Route A}} \begin{cases} \xrightarrow{R^4 \text{ migration}} R^1R^2R^4C-C(=O)-R^3 \\ + \\ \xrightarrow{R^3 \text{ migration}} R^1R^2R^3C-C(=O)-R^4 \end{cases} \qquad (214)$$

$$R^1R^2C(X)-C(R^3)(R^4)-O-Mg-X \xrightarrow{\text{Route B}} R^1R^2C\overset{+}{\underset{|}{O}}(^-MgX_2)C(R^3)R^4 \text{ (epoxide ring)} \xrightarrow{-MgX_2} R^1R^2C \underset{O}{—} CR^3R^4 \text{ (epoxide)}$$

$$CH_3COCH_2Cl + RMgBr \longrightarrow CH_3-C(R)(OMgBr)-CH_2Cl \longrightarrow \left[CH_3-C(R)\underset{O}{—}CH_2\right] \longrightarrow$$

$$CH_3-CH(R)-CHO \xrightarrow{R^4MgBr} CH_3-CH(R)-CH(R^4)-OH \qquad (215)$$

R^4 = alkyl, benzyl

$$CH_3COCH_2Cl + ArMgBr \longrightarrow CH_3C(Ar)(OMgBr)-CH_2Cl \longrightarrow CH_3COCH_2Ar$$

The influence of stereochemical factors is illustrated by the rearrangement, via the halomagnesium derivative, of *cis*-1-methyl-2-chlorocyclohexanol, yielding mostly 2-methylcyclohexanone and a small amount of acetylcyclopentane, while the *trans* isomer exclusively affords acetylcyclopentane[474].

Reaction of cyclic α-chloroketones with arylmagnesium bromide gives rise to α-arylketones (**424**)[479], while with vinylmagnesium chloride 1,2-divinylcycloalkanols (**425**) are formed, except for α-chlorocyclobutanone which furnishes 1-cyclopropyl-4-penten-1-one (**426**) (equation 216)[480].

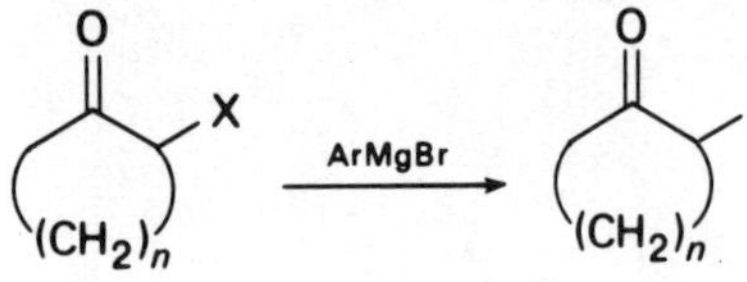

ArMgBr
$(CH_2)_n$
(**424**)

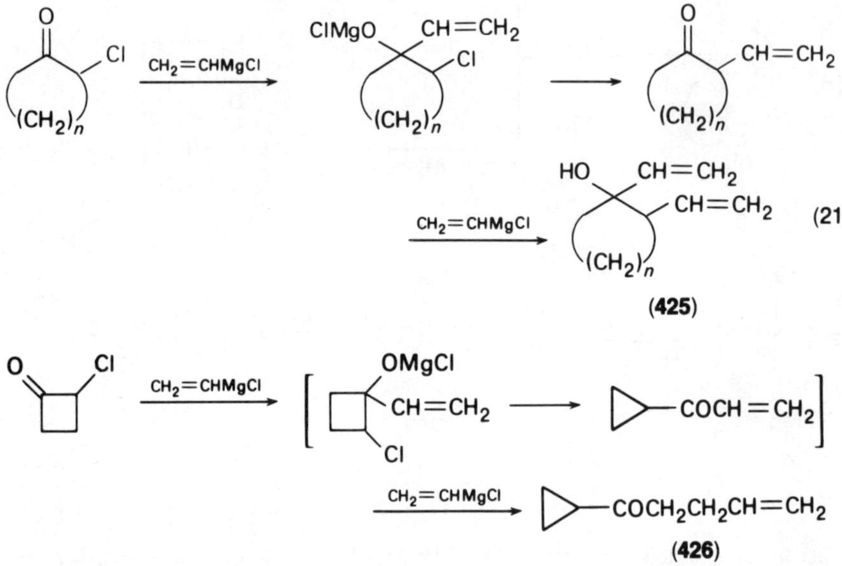

$CH_2{=}CHMgCl$
ClMgO $CH{=}CH_2$ Cl
$(CH_2)_n$
$CH{=}CH_2$
HO $CH{=}CH_2$ $CH{=}CH_2$ (216)
(**425**)

$CH_2{=}CHMgCl$
OMgCl $CH{=}CH_2$ Cl
$COCH{=}CH_2$
$COCH_2CH_2CH{=}CH_2$
(**426**)

A 'one flask' synthesis of olefins has been described by the reaction of α-chloroketones with Grignard reagents and further treatment with lithium metal at −60°C (equation 217)[481].

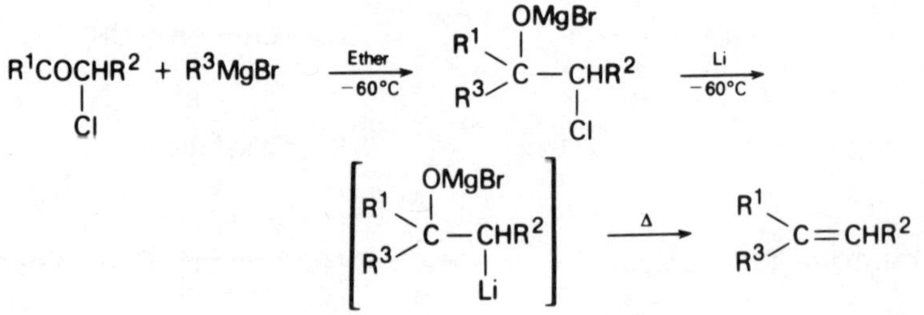

$$R^1COCHR^2(Cl) + R^3MgBr \xrightarrow[-60°C]{Ether} R^1R^3C(OMgBr)-CHR^2(Cl) \xrightarrow[-60°C]{Li} [R^1R^3C(OMgBr)-CHR^2(Li)] \xrightarrow{\Delta} R^1R^3C{=}CHR^2 \quad (217)$$

1-Arylcyclopropanols (**427**) are produced when 1,3-dichloro-2-propanone first reacts with arylmagnesium halides and the product is subsequently treated with ethylmagnesium bromide in the presence of ferric chloride (equation 218)[482,483].

$$ClCH_2COCH_2Cl \xrightarrow{ArMgBr} ClCH_2C(Ar)(OMgBr)-CH_2Cl \xrightarrow[FeCl_3]{EtMgBr} \text{1-Ar-1-OH-cyclopropane } (\mathbf{427}) \qquad (218)$$

Reaction of phenacyl halides with Grignard reagents gives dibenzyl ketones or deoxybenzoins depending upon the aromatic substitution pattern and the reaction conditions which determine the relative migratory aptitudes of the aryl and phenyl groups[484].

1-Aryl-2,2-dichloro-1-alkanones rearrange with methylmagnesium iodide to highly sterically hindered alcohols (**428**). The mechanism involves two pseudo-pinacol-type rearrangements of the carbonyl adducts (equation 219)[485,486].

$$ArCOCCl_2R \xrightarrow[Ether]{CH_3MgI} Ar-C(OMgI)(CH_3)-CCl_2R \longrightarrow CH_3COC(R)(Cl)-Ar \xrightarrow{CH_3MgI}$$

$$CH_3-C(CH_3)(OMgI)-C(R)(Ar)Cl \longrightarrow CH_3COC(CH_3)(R)-Ar \xrightarrow{CH_3MgI} (CH_3)_2C(OH)-C(CH_3)(Ar)R \; (\mathbf{428}) \qquad (219)$$

Finally, non-addition reactions of α-haloketones and Grignard reagents (R′MgX) also occur which result in the formation of halomagnesium enolates with elimination of R′H or R′X[487,488].

Reaction of α,α,α-trichloroketones with isopropylmagnesium chloride gives, after hydrolysis, a mixture of alcohols and α,α-dichloroketones, the latter compounds resulting from intermediate magnesium enolates (equation 220)[488]. Magnesium

$$RCOCCl_3 \xrightarrow{i\text{-PrMgCl}} \begin{cases} \xrightarrow{-C_3H_6} RCH(OMgCl)CCl_3 \xrightarrow{H_3O^+} RCH(OH)CCl_3 \\ \xrightarrow{-i\text{-PrCl}} R-C(OMgCl)=CCl_2 \xrightarrow{H_3O^+} RCOCHCl_2 \end{cases} \qquad (220)$$

enolates are stable compounds possessing high nucleophilic reactivity. They can be prepared by reaction of α-haloketones with magnesium (equation 220a)[487].

$$R^1COCH(X)R^2 + Mg \longrightarrow R^1C(OMgBr)=CHR^2 \qquad (220a)$$

b. Reaction of α-haloketones with organolithium compounds. Reaction of α-haloketones with alkyllithium derivatives normally gives rise to halohydrins, which can be converted into epoxides by the action of bases (equation 221)[477].

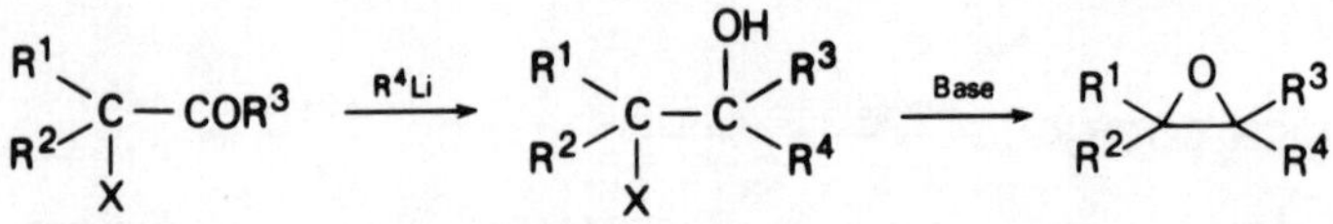

$$R^1R^2C(X)-COR^3 \xrightarrow{R^4Li} R^1R^2C(X)-C(OH)R^3R^4 \xrightarrow{Base} \text{epoxide } R^1R^2C(O)CR^3R^4 \quad (221)$$

A very useful synthetic procedure for alkylation of α-chloroketones utilizes halohydrin formation by the action of alkyllithium compounds, followed by addition of Grignard reagents and thermal decomposition of the resulting magnesium salts into the α-alkylated ketones as illustrated below (equation 222)[489]. However, application of the same sequence to 2-chlorocyclohexanone gives rise to a mixture of 32% 2-methylcyclohexanone and 22% 2-acetylcyclopentane[489].

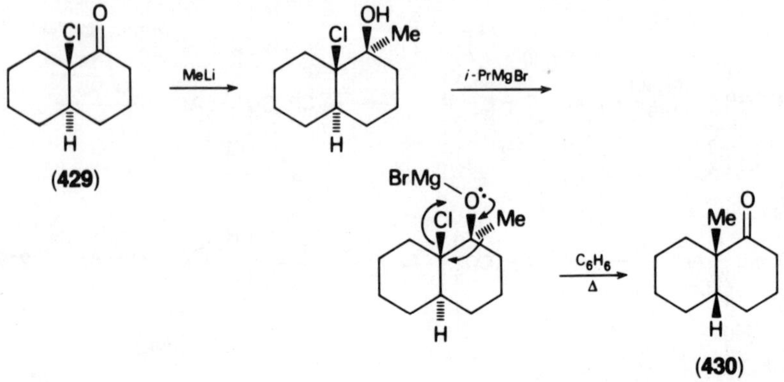

$$(429) \xrightarrow{MeLi} \text{chlorohydrin} \xrightarrow{i\text{-}PrMgBr} \text{BrMg alkoxide} \xrightarrow[\Delta]{C_6H_6} (430) \quad (222)$$

The reaction of α-chlorocycloalkanones with aryllithium reagents also proceeds to the formation of α-arylketones in high yields[490]. The alkylation of α-bromoketones with alkyllithium cuprates allows the regiospecific introduction of a primary, secondary or tertiary alkyl group on the ketone at the site initially brominated. Two concomitant mechanisms, halogen–metal exchange and nucleophilic substitution occur. While these two mechanisms co-exist in substitution by primary and secondary alkyl groups, only nucleophilic substitution seems possible in the case of tertiary alkyl groups (equation 223)[491,492].

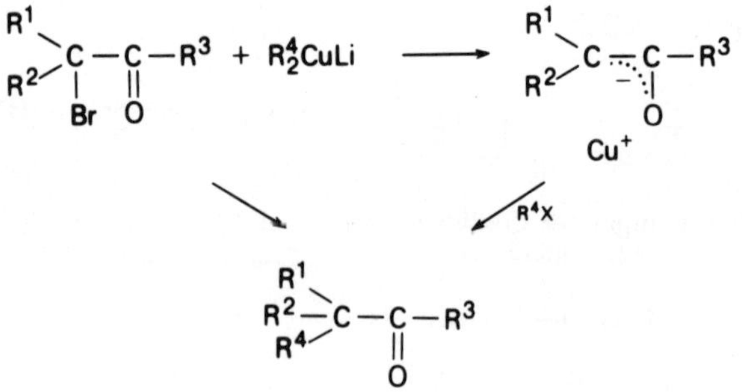

$$R^1R^2C(Br)-C(=O)-R^3 + R^4_2CuLi \longrightarrow R^1R^2C{=}C(O^-)-R^3\ Cu^+ \xrightarrow{R^4X} R^1R^2R^4C-C(=O)-R^3 \quad (223)$$

Complications arise during the action of di-*t*-butyllithium cuprate on α-bromoketones possessing hydrogen atoms at the β-position: products from the halogen–metal exchange are obtained together with alkylation not at the α-position but at the β-position. The latter reaction is explained by a dehydrobromination yielding an α,β-unsaturated ketone, followed by a 1,4-addition (equation 224)[493,494].

$$R^1C(=O)-C(R^2)(Br)-CH(R^3)(R^4) \xrightarrow{t\text{-}Bu_2CuLi} [R^1C(O^-)=C(R^2)-CH(R^3)(R^4)]\ Cu^+ \begin{cases} \xrightarrow{H_2O} R^1COCH(R^2)CHR^3R^4 \\ \xrightarrow{R^5X} R^1COC(R^2)(R^5)-CHR^3R^4 \end{cases}$$

$$\downarrow -HBr$$

$$\left[R^1C(=O)-C(R^2)=C(R^3)(R^4)\right] \xrightarrow{t\text{-}Bu_2CuLi} [R^1-C(O^-)=C(R^2)-C(R^3)(R^4)(Bu\text{-}t)]\ Cu^+ \begin{cases} \xrightarrow{H_2O} R^1COCH(R^2)C(R^3)(R^4)Bu\text{-}t \\ \xrightarrow{R^5X} R^1COC(R^2)(R^5)-C(R^3)(R^4)Bu\text{-}t \end{cases} \quad (224)$$

α,α′-Dibromoketones also react with dialkyllithium cuprates leading initially to enolates in which internal displacement of bromide produces cyclopropanones as in the Favorskii rearrangement. Further reaction with the organocopper reagent affords a new enolate and exposure of the latter to various electrophiles yields α-substituted ketones (equation 225)[495].

$$R^1CH(Br)C(=O)CH(Br)R^2 \xrightarrow{R^3_2CuLi} \left[\text{2-}R^1\text{-3-}R^2\text{-cyclopropanone}\right] \xrightarrow{R^3_2CuLi}$$

$$(R^3)(R^1)CH-C(O^-)=CH(R^2)\ M^+ \xrightarrow{E^+} (R^3)(R^1)CH-C(=O)-CH(R^2)-E \quad (225)$$

$$E = H, D, Me$$

Enolate formation is reported when α,α-dichloro- or α,α,α-trichloroketones are treated with isopropyllithium. These enolates are stable at −75°C and are hydrolysed to α-chloroketones. Reaction with aldehydes provides *C*-alkylation, while enol acetates are obtained with acetic anhydride (equation 226)[496].

$$R^1CO{-}C(Cl)_2{-}R^2 \xrightarrow[-i\text{-}PrCl]{i\text{-}PrLi} [R^1{-}C(O^-){=}C(Cl)R^2]\,Li^+ \longrightarrow \begin{cases} \xrightarrow{H_2O} R^1COCHR^2(Cl) \\ \xrightarrow{R^3CHO} R^1C(=O){-}C(Cl)(R^2){-}CH(OH)R^3 \\ \xrightarrow{Ac_2O} R^1{-}C(OAc){=}C(Cl)R^2 \end{cases} \quad (226)$$

R^1 = alkyl, phenyl
R^2 = alkyl, phenyl, Cl

c. Reaction of α-haloketones with organoboron compounds. α-Bromoketones are alkylated by 9-alkyl-9-borabicyclo[3.3.1]nonanes via their enolates but this reaction is more sensitive to steric hindrance than the analogous reaction using cuprates (equation 227)[497,498].

$$R^1COCHR^2(Br) \xrightarrow{t\text{-BuOK}} [R^1C(O^-){=}C(Br)R^2]\,K^+ \xrightarrow{R^3_3B} \left[R^3{-}\bar{B}(R^3)_2{-}C(R^2)(Br){-}C(=O)R^1\ K^+\right] \longrightarrow$$

$$(R^3)(R^2)C{=}CR^1(OBR^3_2) \xrightarrow{t\text{-BuOH}} R^1C(=O){-}CH(R^2)(R^3) + t\text{-BuOBR}^3_2 \quad (227)$$

Reaction of α-bromoketones with alkynyltrialkylborates (**431**) gives intermediate vinylboranes (**432**) which upon hydrolysis or oxidation afford olefins (**433**) and 1,4-diketones (**434**), respectively (equation 228)[499].

$$R^2_3B + LiC{\equiv}CR^3 \longrightarrow \underset{(\mathbf{431})}{[R^2_3\bar{B}C{\equiv}CR^3]\,Li^+} \xrightarrow{R^1COCH_2Br} \underset{(\mathbf{432})}{(R^1COCH_2)(R^3)C{=}C(R^2)(BR^2_2)} \quad (228)$$

$$(\mathbf{432}) \xrightarrow{H^+} \underset{(\mathbf{433})}{(R^1COCH_2)(R^3)C{=}C(R^2)(H)} \qquad (\mathbf{432}) \xrightarrow{[O]} \underset{(\mathbf{434})}{R^1COCH_2CH(R^3)COR^2}$$

d. Reaction of α-haloketones in the presence of metal complexes. Oxyallyl cations (**435**) are generated when α,α′-dibromoketones are treated with sodium iodide in acetonitrile[500], mercury[500], zinc–copper couple[500,501], iron carbonyl compounds[502], zinc and triethylborate[503] and copper powder with sodium iodide[504]. The formation of these cations, for example with diiron nonacarbonyl, can be envisioned by initial reduction of the dibromide producing an iron enolate (L = Br^-, CO, solvent, etc), which eliminates a bromide ion to form the oxyallyl cations (equation 229)[505].

$$R^1R^2C(Br){-}C(=O){-}C(Br)R^1R^2 \xrightarrow{Fe_2(CO)_9} \left[R^1R^2C{=}C(O^-){-}C(Br)R^1R^2\right]Fe(II)L_n \xrightarrow{-Br^-} \left[R^1R^2C\overset{+}{\cdots}C(O^-)\cdots CR^1R^2\right]Fe(II)L_n \qquad (229)$$

(**435**)

Oxyallyl species serve as highly versatile synthons for the construction of carbocyclic frameworks by cycloaddition with alkenes[505], dienes[505], furans[506,507], pyrrols[508] and enamines[509] (equation 230).

Attempted formation of substituted cyclopentanones from oxyallyl species and 2π-systems failed; however, allenic compounds (**443**) and tetrahydrofurans (**445**) are isolated from the reactions of α,α′-dibromoketones with acetylenes (**442**) and 1,1-dimethoxyethylene (**444**), respectively (equation 231)[510].

2-(*N*-Alkylimino)cyclobutanones (**446**) are produced on reaction of α,α′-dibromoketones with a copper/isonitrile complex (equation 232)[511].

One of the observed side reactions during the formation of oxyallyl cations is the formation of reduction products. Debromination of α,α′-dibromoketones with a zinc–copper couple in methanol yields the parent ketone and an α-methoxy ketone. The latter is suggested to arise from a selective 2-oxyallyl cation which is produced in an S_N1 reaction[512]. Debromination in DMF gives rise to **447** and reductive coupling products (**448**) (equation 233)[513]. Similar debromination and coupling reactions are obtained with dicobalt octacarbonyl under phase transfer conditions[514]. α-Monohaloketones also give a variety of reactions with various organometallic compounds.

Iron pentacarbonyl reacts with α-haloketones in refluxing 1,2-dimethoxyethane, followed by treatment with water, to give 1,4-diketones (**449**), reduced monoketones (**450**) and β-epoxy ketones (**451**) (equation 234)[515] α-Bromoketones react with zinc in ether to provide organozinc compounds which on further reaction with alkyl chloroformates and aldehydes afford respectively β-keto esters (**452**) and β-hydroxy ketones (**453**) (equation 235)[516,517].

An efficient and regiospecific aldol condensation is reported which consists of a coupled attack on the α-haloketone by dialkylaluminium chloride and zinc generating an aluminium enolate regiospecifically. The enolate is sufficiently reactive to cause a facile addition to carbonyl compounds to give β-hydroxy ketones (**453**) (equation 236)[518].

During the debromination of α,α-dibromocamphor (**454**) with diethyl zinc in refluxing benzene, an α-elimination occurs to produce an α-keto carbene (**455**) which leads to the formation of (**456**) (equation 237)[519].

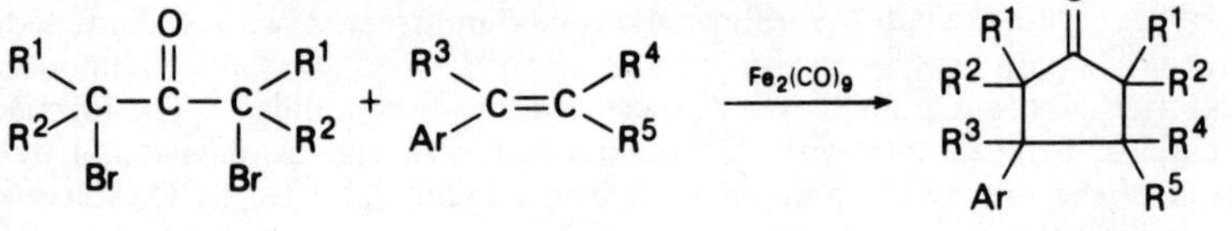

$Fe_2(CO)_9$

(436)

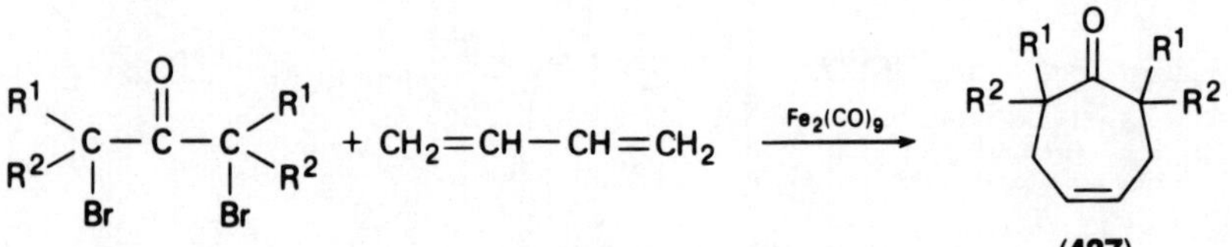

$Fe_2(CO)_9$

(437)

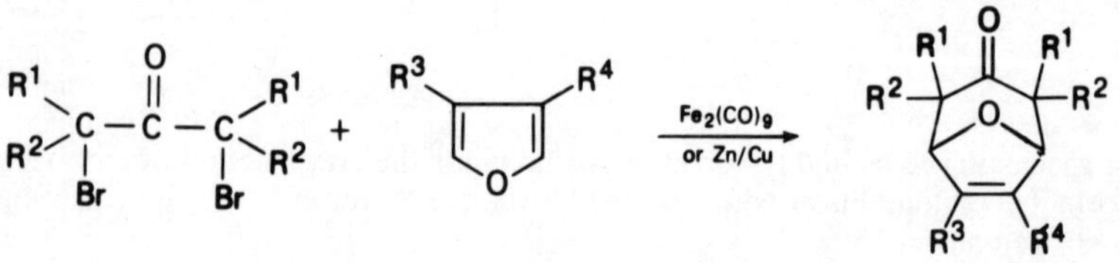

$Fe_2(CO)_9$ or Zn/Cu

(438)

(230)

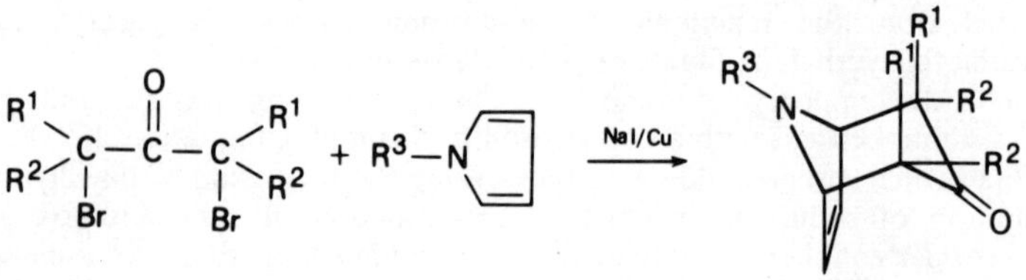

NaI/Cu

(439)

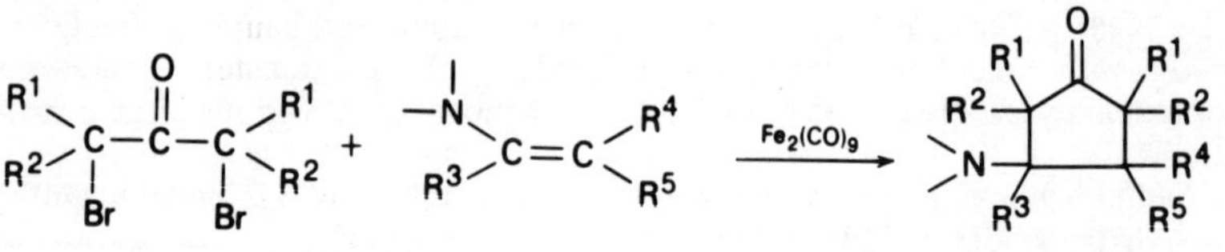

$Fe_2(CO)_9$

(440)

$R^2 = H$

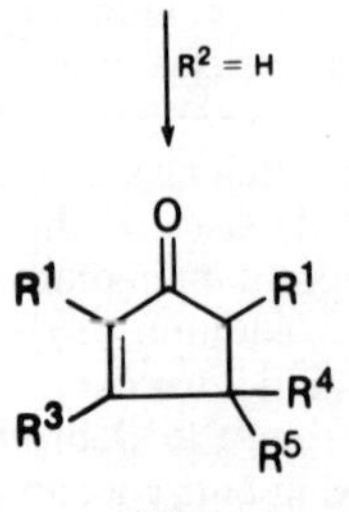

(441)

$$\underset{(442)}{\text{RCH(Br)COCH(Br)R}} + \text{HC}{\equiv}\text{CCH}_2\text{OH} \xrightarrow{\text{Zn/Cu}} \underset{(443)}{\text{R}{-}\text{C(COCH}_2\text{R)}{=}\text{C}{=}\text{CHCH}_2\text{OH}} \qquad (231)$$

$$\underset{(444)}{\text{RCH(Br)COCH(Br)R}} + \text{H}_2\text{C}{=}\text{C(OCH}_3)_2 \xrightarrow{\text{Zn/Cu}} (445)$$

(445): 2,2-(H₃CO)(OCH₃)-tetrahydrofuran with =CHR and R substituents

$$\text{R}^1\text{R}^2\text{C(Br)}{-}\text{CO}{-}\text{C(Br)R}^1\text{R}^2 + \text{R}^3\text{NC} \xrightarrow{\text{Cu}} (446) \qquad (232)$$

$$\text{R}_2\text{C(Br)}{-}\text{CO}{-}\text{C(Br)R}_2 \xrightarrow[\text{DMF}]{\text{Zn/Cu}} (447) + \underset{(448)}{\text{R}_2\text{CHCOC(R}_2)\text{C(R}_2)\text{COCHR}_2} \qquad (233)$$

(447): 2-N(CH₃)₂-1,3-dioxolane with =CR₂ and R, R substituents

$$\text{R}^1\text{COC(Br)R}^2\text{R}^3 \xrightarrow[(2)\ \text{H}_2\text{O}]{(1)\ \text{Fe(CO)}_5} \underset{(449)}{[\text{R}^1\text{COCR}^2\text{R}^3]_2} + \underset{(450)}{\text{R}^1\text{COCHR}^2\text{R}^3} + \underset{(451)}{\text{R}^1\text{COCR}^2\text{R}^3{-}\text{C(R}^1)(\text{O})\text{CR}^2\text{R}^3} \qquad (234)$$

$$\text{R}^1\text{COC(Br)R}^2\text{R}^3 + \text{Zn} \xrightarrow{\text{Ether}} \text{R}^1\text{COC(ZnBr)R}^2\text{R}^3 \qquad (235)$$

$$\text{R}^1\text{COC(ZnBr)R}^2\text{R}^3 \xrightarrow{\text{ClCOOR}^4} \underset{(452)}{\text{R}^1\text{COCR}^2\text{R}^3\text{COOR}^4}$$

$$\text{R}^1\text{COC(ZnBr)R}^2\text{R}^3 \xrightarrow{\text{R}^5\text{CHO}} \underset{(453)}{\text{R}^1\text{COCR}^2\text{R}^3{-}\text{CH(OH)R}^5}$$

$$\mathrm{R^1COC(X)R^2R^3} \xrightarrow{\text{Zn, Al}} \mathrm{R^1\!-\!C(OAl)\!=\!CR^2R^3} \xrightarrow{\mathrm{R^4CHO}} \mathrm{R^1\!-\!CO\!-\!CR^2R^3\!-\!CH(OAl)\!-\!R^4} \longrightarrow \underset{(453)}{\mathrm{R^1COCR^2R^3\!-\!CH(OH)R^4}} \quad (236)$$

$$(454) \xrightarrow[\text{Benzene}]{\mathrm{Et_2Zn}} (455) \longrightarrow (456) \quad (237)$$

8. *Reaction of α-haloketones with complex metal hydrides*

Reaction of α-haloketones with sodium borohydride results in reduction of the carbonyl function with formation of halohydrins (equation 238)[520–522]. Reduction with

$$\mathrm{R^1COC(X)R^2R^3} \xrightarrow[\text{or } \mathrm{LiAlH_4}]{\mathrm{NaBH_4}} \mathrm{R^1\!-\!CH(OH)C(X)R^2R^3} \xrightarrow{\mathrm{LiAlH_4}} \mathrm{R^1\!-\!CH(OH)\!-\!CHR^2R^3} \quad (238)$$

lithium aluminium hydride provides the same halohydrins[523,524], while the reduction of phenacyl halides gives mixtures of 1-aryl-1-ethanols and 1-aryl-2-halo-1-ethanols[525]. However, if the aromatic ring is strongly electron-releasing, e.g. compound **457**, the aromatic ring migrates to give primary alcohols (**458**)[526,527] and not secondary alcohols (**459**) as proposed previously[528] (equation 239).

9. *Reaction of α-haloketones with phosphorus compounds*

The reactions between α-haloketones and trivalent phosphorous compounds are of rather a complex nature[529]. Dialkyl phosphites normally react at the carbonyl group to give α-hydroxy and/or epoxy phosphonate esters (**460**, **461**) (equation 240)[530].

Trialkyl phosphites react with α-haloketones yielding enol phosphates (**462**) (Perkow reaction) and/or β-ketophosphonates (**463**) (Arbusov reaction)[531–534]. In general the attack of phosphites can take place at four positions (equation 241): (1) attack on the carbon atom carrying the halogen, which gives rise either to an enol phosphate (**462**) or to a β-ketophosphonate (**463**); (2) attack on the carbonyl oxygen; (3) attack on the carbonyl carbon, furnishing an epoxy phosphonate (**461a**) or a vinyl

(457) $\xrightarrow{LiAlH_4}$ → → (458) (239)

$ArCOCR^1R^2(Br) \xrightarrow[\times]{LiAlH_4} ArCH_2CR^1R^2(OH)$ (459)

$R^1COCR^2R^3X \xrightarrow{(R^4O)_2PO^-}$ (461)

$\downarrow (R^4O)_2POH$ (460) $\xrightarrow{^-OH}$ (461) (240)

phosphate (**462**); (4) attack on the halogen, leading to the enol phosphate (**462**) via an halophosphonium enolate.

No general agreement has been reached concerning the mechanisms of the Perkow and the Arbusov reactions, and further work is still in progress to substantiate the nature of the intermediates[532,535]. The ratio between the Perkow and Arbusov reaction products is dependent upon the nature of the substrate and the phosphite and the reaction conditions employed. Substitution at the α-carbon atom e.g. by alkyl groups and by strongly electron-withdrawing groups, promotes the Perkow reaction, while a

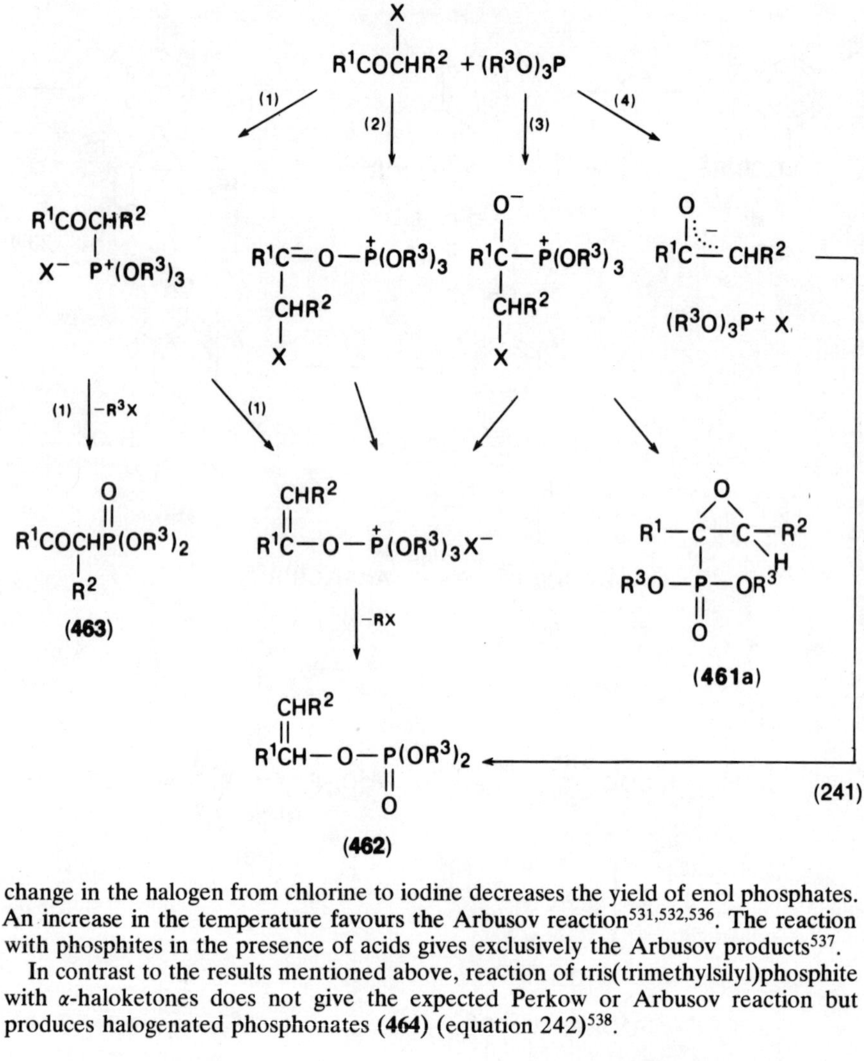

$$R^1COCHXR^2 + (R^3O)_3P \xrightarrow{(1),(2),(3),(4)}$$

(1) $R^1COCHR^2(P^+(OR^3)_3)\ X^-$; (2) $R^1C(=O-\overset{+}{P}(OR^3)_3)CHR^2X$; (3) $R^1C(O^-)(\overset{+}{P}(OR^3)_3)CHR^2X$; (4) $R^1C(O)\cdots CHR^2\ (R^3O)_3P^+\ X$

$$\xrightarrow[-R^3X]{(1)} R^1COCH(R^2)P(O)(OR^3)_2 \quad \textbf{(463)}$$

$$R^1C(=CHR^2)-O-\overset{+}{P}(OR^3)_3X^- \xrightarrow{-RX} R^1CH(=CHR^2)-O-P(O)(OR^3)_2 \quad \textbf{(462)}$$

$$R^1-C(O)C-R^2(H)\ (\text{epoxide}),\ R^3O-P(O)-OR^3 \quad \textbf{(461a)}$$

(241)

change in the halogen from chlorine to iodine decreases the yield of enol phosphates. An increase in the temperature favours the Arbusov reaction[531,532,536]. The reaction with phosphites in the presence of acids gives exclusively the Arbusov products[537].

In contrast to the results mentioned above, reaction of tris(trimethylsilyl)phosphite with α-haloketones does not give the expected Perkow or Arbusov reaction but produces halogenated phosphonates (**464**) (equation 242)[538].

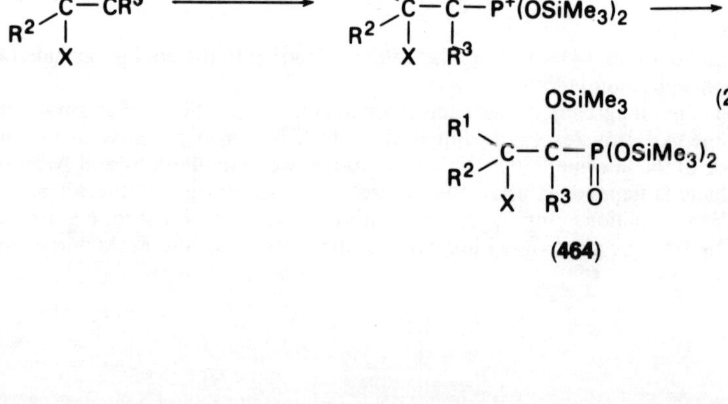

$$R^1R^2CX-C(O)R^3 \xrightarrow{P(OSiMe_3)_3} R^1R^2CX-C(^-O)(R^3)-P^+(OSiMe_3)_2(O-SiMe_3) \longrightarrow R^1R^2CX-C(OSiMe_3)(R^3)-P(O)(OSiMe_3)_2 \quad (242)$$

(464)

Treatment of chlorinated acetophenones with monoalkyl phosphinites under the standard Perkow reaction conditions gives enol phosphonates (**465**) (equation

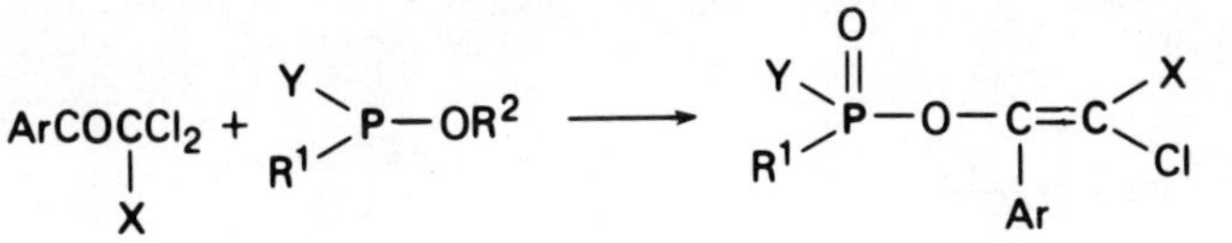

$$ArC(X)Cl_2CO + (Y)(R^1)P{-}OR^2 \longrightarrow (Y)(R^1)P(=O){-}O{-}C(Ar){=}C(X)Cl \quad (243)$$

(**465**)

$X = H, Cl$

$Y = OR^2, NR^2_2, SR^2$

243)[539,540]. On the other hand, reaction of dialkyl phosphinites produces the Perkow (**466**) and/or the Arbusov (**467**) reaction product. With alkyl di-*t*-butylphosphinites only the Arbusov reaction takes place (equation 244)[541,542].

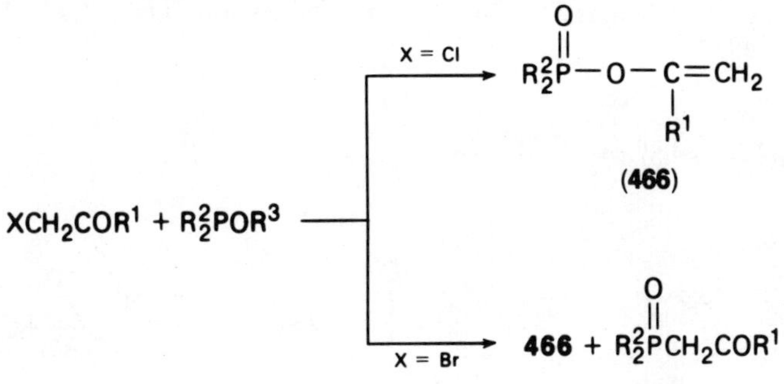

$$XCH_2COR^1 + R^2_2POR^3 \xrightarrow{X = Cl} R^2_2P(=O){-}O{-}C(R^1){=}CH_2 \; (\mathbf{466})$$

$$XCH_2COR^1 + R^2_2POR^3 \xrightarrow{X = Br} \mathbf{466} + R^2_2P(=O)CH_2COR^1 \; (\mathbf{467}) \quad (244)$$

α-Chloroketones react with tertiary phosphines to give phosphonium salts (**468**), while enol phosphonium salts (**469**) are generated in the case of α-bromoketones (equation 245)[543].

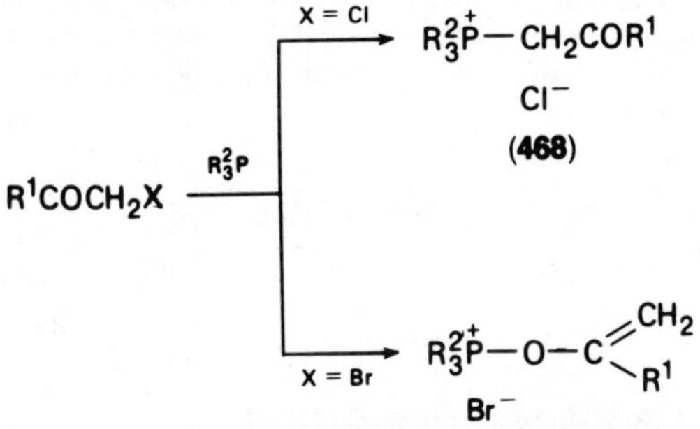

$$R^1COCH_2X \xrightarrow[X = Cl]{R^2_3P} R^2_3\overset{+}{P}{-}CH_2COR^1 \; Cl^- \; (\mathbf{468})$$

$$R^1COCH_2X \xrightarrow[X = Br]{R^2_3P} R^2_3\overset{+}{P}{-}O{-}C(R^1){=}CH_2 \; Br^- \; (\mathbf{469}) \quad (245)$$

Phosphorous acid vinyl esters (**470**, **471**) are produced on treatment of α-chloroketones with phosphorous oxytrichloride[544] and alkyl dichlorophosphinites[545], respectively, in the presence of triethylamine (equation 246).

Epoxy phosphonates (**473**) have been prepared by the action of sodium alkoxide on a dialkyl phosphonate and an α-haloketone. When the reaction is performed in the

$$R^1COCH(R^2)Cl \xrightarrow{Et_3N} \begin{cases} \xrightarrow{POCl_3} Cl_2P(=O)-O-C(R^1)=CR^2 \quad \textbf{(470)} \\ \xrightarrow{(R^3O)PCl_2} (R^3O)(Cl)P(=O)-O-C(R^1)=CR^2Cl \quad \textbf{(471)} \end{cases} \tag{246}$$

presence of triethylamine, the intermediate halohydrins (**472**) could be isolated (equation 247)[546,547].

$$R^1COCHR^2(X) + (R^3O)_2P(=O)H \xrightarrow{NaOR^3} R^1-C(OH)(P(=O)(OR^3)_2)-CHR^2X \; \textbf{(472)} \longrightarrow \text{epoxide } R^1(P(=O)(OR^3)_2)C(-O-)CR^2H \; \textbf{(473)} \tag{247}$$

Monofluorophosphoranes (**474**), especially methyltri(*n*-butyl)fluorophosphorane, are used in exchange reactions to prepare α-fluoroketones. In addition, cyclopropanes (**475**) are produced, due to the high basicity of the reagent which causes the formation of an α-keto carbene by dehydrohalogenation (equation 248)[548].

$$RCOCH_2X + R^1_3(Me)PF \; \textbf{(474)} \xrightarrow[-70^\circ C]{THF} RCOCH_2F + \text{cyclopropane-1,2,3-(COR)}_3 \; \textbf{(475)} \tag{248}$$

B. Miscellaneous Reactions of α-Haloketones

1. *Electrophilic reactions of α-haloketones and their derivatives*

Friedel–Crafts reactions of certain aryl activated α chloroketones with aromatics in the presence of aluminium chloride gives rise to α-aryl ketones[549,550]. Another method for the synthesis of α-aryl ketones consists of an insertion reaction of an α-keto-carbenoid generated from α,α-dibromoketones in the presence of zinc[551] (equation 249).

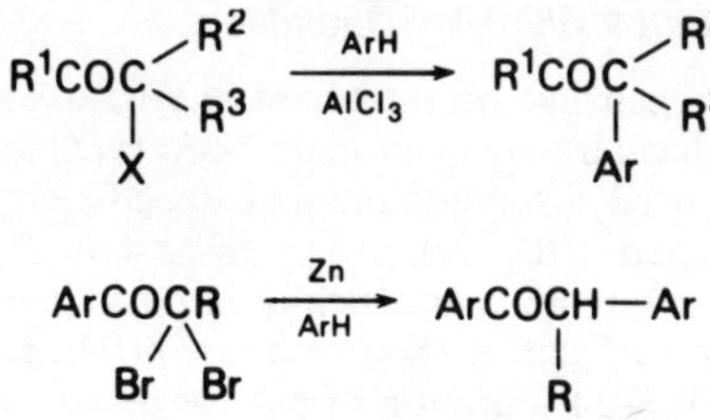

$$R^1COC(X)R^2R^3 \xrightarrow[AlCl_3]{ArH} R^1COC(Ar)R^2R^3$$

$$ArCOCRBr_2 \xrightarrow[ArH]{Zn} ArCOCH(R)-Ar \quad (249)$$

Brominated phenylhydrazones (**476**) are obtained by treatment of α-bromoketones with diazonium salts (equation 250)[552].

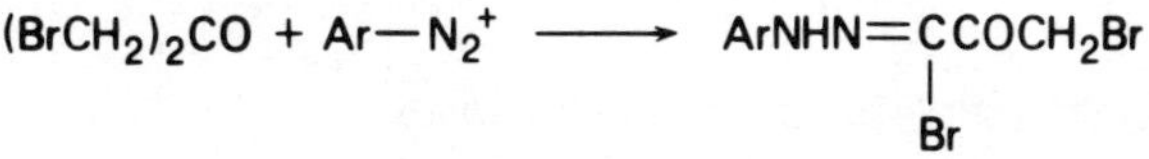

$$(BrCH_2)_2CO + Ar-N_2^+ \longrightarrow ArNHN{=}C(Br)COCH_2Br \quad (250)$$

(**476**)

Finally, α-chloroketones have been converted to the corresponding enol acetates (**477**) by acylation of the intermediate chloroenolates, generated by reaction with a suspension of sodium methoxide in ether at $-50°C$. This procedure takes advantage of the fact that in unsymmetrical ketones the C—H bond adjacent to both the carbonyl group and the chlorine atom is significantly more acidic (by 2 pK_a units) than the C—H bond adjacent only to a carbonyl group[553]. Certain α-haloketones also undergo rapid reaction with lithium diisopropyl amide (LDA) to produce enol acetates (**478**) in the presence of acetic anhydride in competition with reduction products (**479**) via hydride transfer[554]. Enolate formation also takes place when α-chloroketones are treated with trimethylchlorosilane in the presence of tertiary amines to yield trimethylsilyl enol ethers (**480**)[555] (equation 250a).

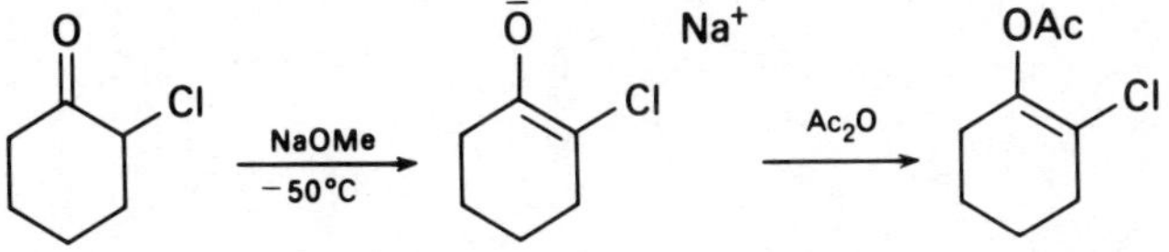

$$\text{2-chlorocyclohexanone} \xrightarrow[-50°C]{NaOMe} \text{2-chlorocyclohexenolate } (O^-\ Na^+) \xrightarrow{Ac_2O} \text{1-acetoxy-2-chlorocyclohexene}$$

(**477**)

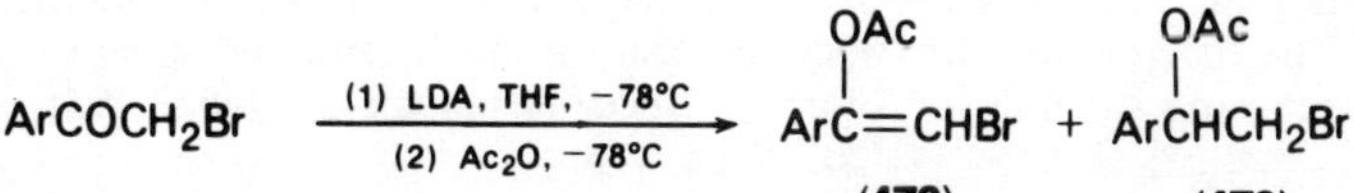

$$ArCOCH_2Br \xrightarrow[(2)\ Ac_2O,\ -78°C]{(1)\ LDA,\ THF,\ -78°C} ArC(OAc){=}CHBr + ArCH(OAc)CH_2Br \quad (250a)$$

(**478**) (**479**)

$$R^1COCH(Cl)R^2 + Me_3SiCl \xrightarrow[DMF]{Et_3N} R^1-C(OSiMe_3){=}C(Cl)R^2$$

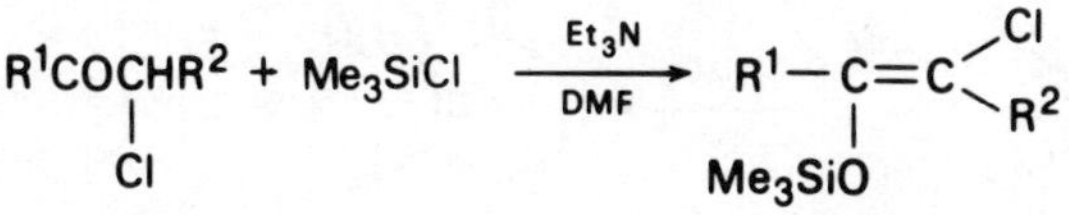

(**480**)

2. *Reaction of α-haloketones with alkali fluorides*

The halogen–fluorine exchange on treatment of α-haloketones with fluoride anion to give α-fluoroketones has already been mentioned (*vide supra*). Another synthetic application involves a desilylbromination and specifically places a double bond between the carbon attached to the carbonyl group and the β-atom to which the silicon atom has originally been bound. Synthetically important α-methylene ketones and lactones have been prepared by using this procedure in which the base- and acid-stable silyl function masks the α,β-unsaturation of enones (equation 251)[556].

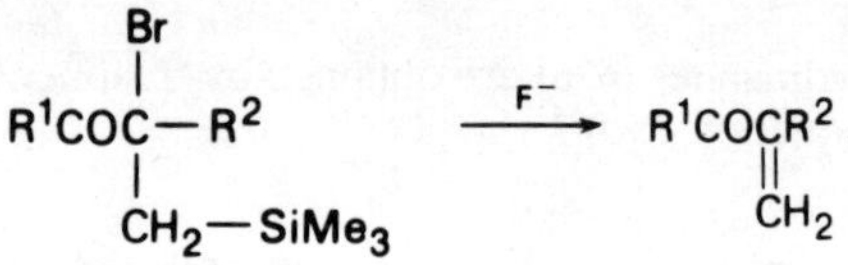

$$R^1COC(Br)(R^2)CH_2{-}SiMe_3 \xrightarrow{F^-} R^1COC(=CH_2)R^2 \qquad (251)$$

3. *Acid-catalysed rearrangement of α-haloketones*

Numerous examples of α-haloketone rearrangements into the α′-isomers in the presence of acids are known. They occur via two mechanisms: (a) a cationic halogen path and (b) an anionic halogen path (equation 252)[557].

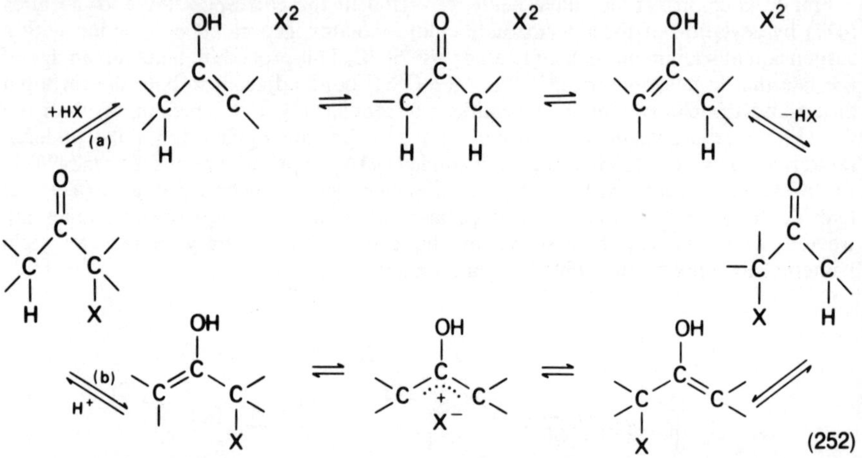

(252)

Equilibration of *trans*-carvone tribromide (**171**) with hydrogen bromide in acetic acid at 0°C gives a mixture of 45% *trans*- and 55% *cis*-carvone tribromide; this isomerization involves the exclusive exchange of a halo substituent β to the carbonyl group[558]. The reaction of 2-bromocyclohexanone in concentrated sulphuric acid also provides rearranged enols of bromo-1,2-cyclohexanediones (**481**, **482**) (equation 253)[559].

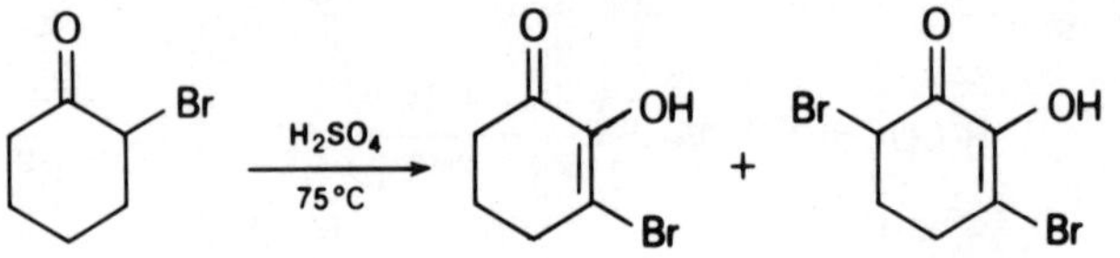

(253)

(**481**) (**482**)

4. Formation of α-acylcarbenium ions from α-haloketones

Silver salts of superacids are able to ionize α-halocarbonyl compounds to α-acylcarbenium ions, and silver hexafluoroantimonate in dichloromethane strikingly promotes this ionization[560,561] (for a recent review dealing with α-acylcarbenium ions, see Ref. 562). In a few cases the reaction leads to the formation of oxonium salts (equation 254).

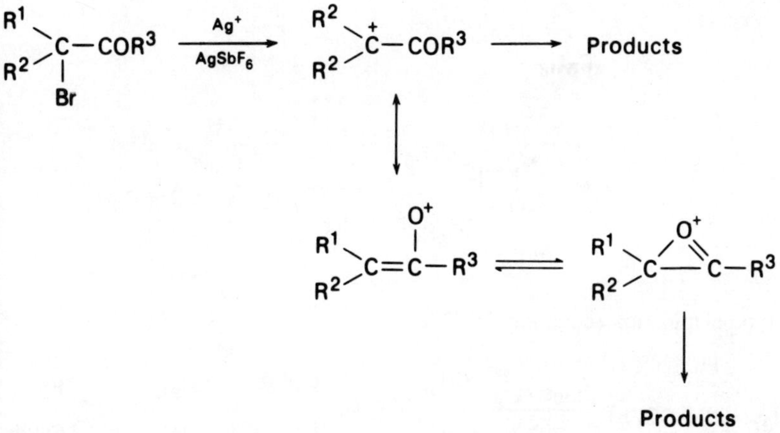

$$R^1R^2C(Br)-COR^3 \xrightarrow[AgSbF_6]{Ag^+} R^1R^2\overset{+}{C}-COR^3 \longrightarrow \text{Products}$$

$$\updownarrow$$

$$R^1R^2C{=}C(O^+)-R^3 \rightleftharpoons R^1R^2C{-}C(-R^3)\ \text{(three-membered ring with } O^+\text{)} \longrightarrow \text{Products} \qquad (254)$$

Acylcarbenium ions are very reactive species which are able to undergo the following reactions: (a) nucleophilic substitution (equation 255)[335,563]

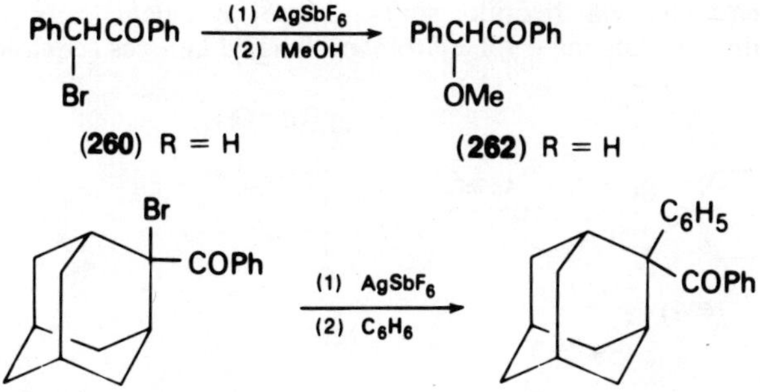

$$\underset{\textbf{(260)}\ R = H}{PhCH(Br)COPh} \xrightarrow[(2)\ MeOH]{(1)\ AgSbF_6} \underset{\textbf{(262)}\ R = H}{PhCH(OMe)COPh}$$

$$\underset{\textbf{(483)}}{\text{2-bromo-2-adamantyl-COPh}} \xrightarrow[(2)\ C_6H_6]{(1)\ AgSbF_6} \underset{\textbf{(484)}}{\text{2-}C_6H_5\text{-2-adamantyl-COPh}} \qquad (255)$$

(b) hydride shift (equation 256)[561]

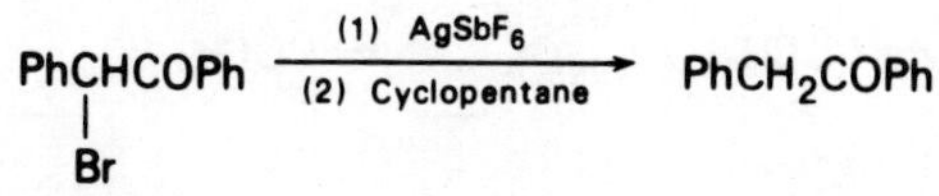

$$\underset{\textbf{(260)}\ R = H}{PhCH(Br)COPh} \xrightarrow[(2)\ \text{Cyclopentane}]{(1)\ AgSbF_6} \underset{\textbf{(485)}}{PhCH_2COPh} \qquad (256)$$

(c) Wagner–Meerwein-type rearrangement (equation 257)[561].

$$\underset{\textbf{(486)}}{H_5C_6(H)C(COC_6H_5)(Br)-C(H)(CH_3)} \xrightarrow[C_6D_6]{AgSbF_6} \left[\text{spiro-cyclopropane cation: } H, H_3C, H, C(=\ddot{O})-C_6H_5\right] \longrightarrow$$

$$\left[C_6D_6 \; ; \; H_3C, H, H, C_6H_5 \text{ oxetenium}, C_6H_5, SbF_6^-\right] \longrightarrow \underset{\textbf{(487)}}{D_5C_6(H)C(H)(CH_3)-C(H)(C_6H_5)CO-C_6H_5} \qquad (257)$$

(d) $E1$-type elimination (equation 258)[561]

$$\underset{\textbf{(488)}}{CH_3CH_2C(Ph)(Br)-COPh} \xrightarrow[20°C]{AgSbF_6} \underset{\textbf{(489)}}{CH_3(H)C{=}C(COPh)(Ph)} + \underset{\textbf{(490)}}{CH_3(H)C{=}C(Ph)(COPh)} \qquad (258)$$

When the structure of the precursor permits, α-acylcarbenium ions are transformed into oxonium salts via hydride shifts. These ambident salts enable further functionalization as illustrated for α-bromocyclohexyl ketones (equation 259)[562].

$$\underset{\textbf{(491)}}{R-C_6H_{10}(Br)COCH_3} \xrightarrow{AgSbF_6} \underset{\textbf{(492)}}{{}^{+}O{=}C-CH_3 \text{ bicyclic}, R}$$

$$\textbf{(492)} \xrightarrow{CH_3OH} \underset{\textbf{(493)}}{R-C_6H_8-COCH_3}; \quad \textbf{(492)} \xrightarrow{H_2O} \underset{\textbf{(494)}}{R(HO)C_6H_{10}-COCH_3}; \quad \textbf{(492)} \xrightarrow{CH_3MgBr} \underset{\textbf{(495)}}{O-C(CH_3)(CH_3) \text{ bicyclic}, R} \qquad (259)$$

5. Photochemistry of α-haloketones

There are a few reports on the study of the photochemistry of α-haloketones involving different types of photoprocesses. α-Halocyclohexanones on irradiation in cyclohexane have been found to give competing radical and ionic photo-behaviour. The principal photoprocess is homolytic β-cleavage of the carbon–halogen bond to afford radical products (**496**, **497**), accompanied by ionic products such as cyclohexenones (**498**) but in much lower proportions (equation 260)[564].

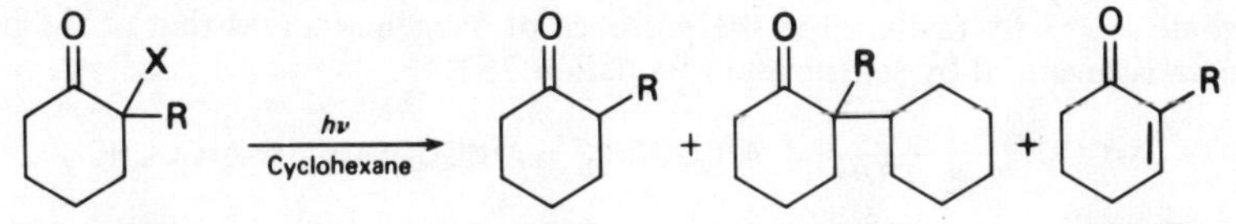

R = H, Me — $h\nu$, Cyclohexane → (**496**) + (**497**) + (**498**) (260)

Favorskii-type ring contraction takes place on irradiation of bicyclic α-chloroketones in methanol. A mechanism involving photoionization of chloride, followed by ring contraction to acylium ions which are trapped by solvent, is suggested (equation 261)[565,566].

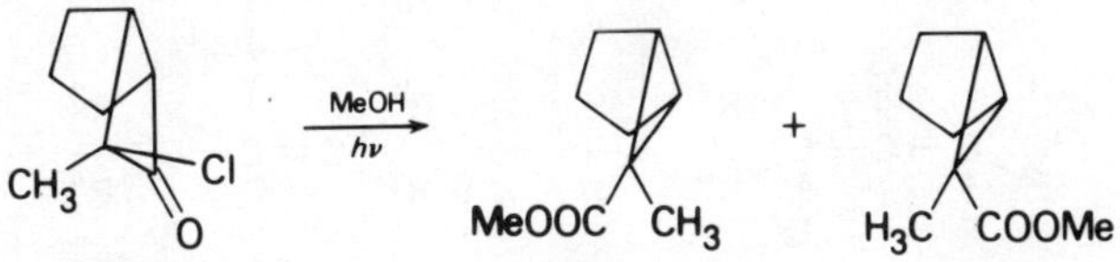

(**499**) — MeOH, $h\nu$ → (**500**) + (**501**) (261)

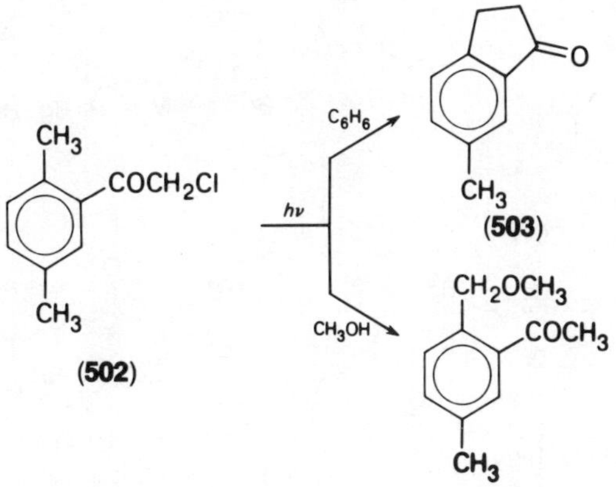

(**502**) — $h\nu$ — C_6H_6 → (**503**); CH_3OH → (**504**) (262)

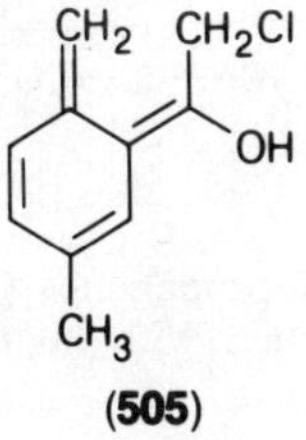

(**505**) 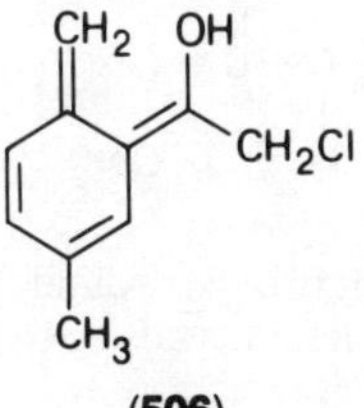 (**506**)

Irradiation of 2,5-dimethyl-α-chloroacetophenone (**502**) in benzene yields 6-methylindan-1-one (**503**) while in methanol the photosolvolysis product (**504**) is obtained. These transformations arise from photoenol intermediates (**505**, **506**) (equation 262)[567].

The behaviour of α-chloro aryl ketones without *ortho* methyl groups is quite different, involving reduction and rearrangement[568]. Irradiation of α,α,α-trichloroacetophenone in methanol affords the alcoholysis products methyl benzoate and methyl benzoylformate along with α,α-dichloroacetophenone. Formation of the benzoate is greatly favoured in the presence of oxygen whereas that of the benzoyl formate is favoured by sensitization (equation 263)[569].

$$ArCOCCl_3 \xrightarrow[CH_3OH]{h\nu} ArCOOCH_3 + ArCOCOOCH_3 + ArCOCHCl_2 \qquad (263)$$

6. *Electrochemistry of α-haloketones*

Electrochemical reduction of α,α′-dibromoketones in acetic acid affords a mixture of the parent ketones and α-acetoxy ketones via an enol allylic bromide intermediate. The same results are obtained when the reduction is carried out by ultrasonically dispersed mercury (equation 264)[570,571].

$$R^1R^2C(Br){-}CO{-}C(Br)R^3R^4 \xrightarrow{2e^-} R^1R^2C{=}C(O^-){-}C(Br)R^3R^4 \xrightarrow{M^+} R^1R^2C{=}C(OM){-}C(Br)R^3R^4 \quad M = H, Hg, Br$$

$$\longrightarrow [R^1R^2C{\cdots}C(OH){\cdots}CR^3R^4]^+ Br^- \xrightarrow{HOAc} R^1R^2CH{-}CO{-}C(OAc)R^3R^4$$

$$\longrightarrow R^1R^2CH{-}CO{-}C(Br)R^3R^4 \xrightarrow{(1)\ 2e^-;\ (2)\ H^+} R^1R^2CH{-}CO{-}CHR^3R^4 \qquad (264)$$

Electroreduction of α,α′-dihaloketones to cyclopropanones (**507**) (isolated as the hemiacetals or hemiacylals (**508**)) is accomplished with highly alkylated ketones (equation 265)[572].

(265)

(507) (508)

7. *Dehalogenation of α-haloketones*

Various reagents are able to effect dehalogenation of α-haloketones such as zinc in acetic acid[573], metal carbonyls $Mo(CO)_6$[574], $Fe(CO)_5$[575] and $HFe(CO)_5^-$[576], transition metals in low valency state[577], triphenylphosphine[578], pyridine followed by sodium dithionite[579], tri-*n*-butyltin hydride[580], lithium iodide–boron trifluoride[581], sodium iodide–amine–sulphur dioxide[582], cerium triodide[583], sodium iodide–chlorotrimethylsilane[584] and sodium iodide in aqueous acid–THF[585] (equation 266).

(266)

IV. REFERENCES

1. W. Hahn, in *Methoden der Organischen Chemie*, Vol. 5/3 (Ed. Houben-Weyl), Georg Thieme Verlag, Stuttgart (1962), p. 1018.
2. O. Bayer, in *Methoden der Organischen Chemie*, Vol. 7/2c (Ed. Houben-Weyl), Georg Thieme Verlag, Stuttgart (1977), p. 2147.
3. R. Jacquier, *Bull. Soc. Chim. France*, **17**, D35 (1950).
4. A. S. Kende, *Org. Reactions*, **11**, 261 (1960).
5. A. A. Akhrem, T. K. Ustynyuk and Y. A. Titov, *Russ. Chem. Rev*., **39**, 732 (1970).
6. C. Rappe, in *The Chemistry of the Carbon–Halogen Bond* (Ed. S. Patai), Wiley, Chichester (1973), p. 1071.
7. K. Sato and M. Oohashi, *Yuki Gosei Kagaku Kyokai Shi*, **32**, 435 (1974).
8. P. J. Chenier, *J. Chem. Educ*., **55**, 286 (1978).
9. B. Tchoubar, *Bull. Soc. Chim. France*, 1363 (1955).
10. N. P. Gambaryan, E. M. Rokhlin, Y. V. Zeifman, C. Ching-Yun and I. L. Knunyants, *Angew. Chem*., **78**, 1008 (1966).
11. Y. A. Cheburkov and I. L. Knunyants, in *Fluorine Chemistry Reviews*, Vol. 1 (Ed P. Tarrant), Edward Arnold, London and Marcel Dekker, New York (1961), p. 107.
12. R. D. Chambers, *Fluorine in Organic Chemistry*, Wiley, New York (1973).
13. M. Hudlicky, *Chemistry of Organic Fluorine Compounds*, Ellis Horwood, Chichester (1976).
14. *Rodd's Chemistry of Carbon Compounds*, 1 C/D Suppl., Elsevier, Amsterdam (1973).
15. N. Fukuhara and L. A. Bigelow, *J. Amer. Chem. Soc*., **63**, 788 (1941).
16. S. Nakanishi, K. Morita and E. V. Jensen, *J. Amer. Chem. Soc*., **81**, 5259 (1959).
17. B. M. Bloom, V. V. Bogert and R. Pinson, *Chem. Ind*., 1317 (1959).
18. R. Joly and J. Warnant, *Bull. Soc. Chim. France*, 569 (1961).
19. R. B. Gabbard and E. V. Jensen, *J. Org. Chem*., **23**, 1406 (1958).
20. S. Nakanishi, *J. Med. Chem*., **7**, 108 (1964).
21. S. Nakanishi and E. V. Jensen, *Chem. Pharm. Bull*. (Tokyo), **25**, 3395 (1977).

22. M. Schlosser and G. Heinz, *Chem. Ber.*, **102**, 1944 (1969).
23. H. M. Kissman, A. M. Small and M. J. Weis, *J. Amer. Chem. Soc.*, **81**, 1262 (1959).
24. J. Edwards and H. J. Ringold, *J. Amer. Chem. Soc.*, **81**, 5262 (1959).
25. W. J. Middleton and E. M. Bingham, *J. Amer. Chem. Soc.*, **102**, 4846 (1980).
26. S. Rozen and Y. Menahem, *Tetrahedron Lett.*, 725 (1979).
27. P. Crabbé, A. Cervantes, A. Cruz, E. Galazzi, J. Iriarte and E. Velarde, *J. Amer. Chem. Soc.*, **95**, 6655 (1973).
28. P. Crabbé, J.-L. Luche, J. C. Damiano, M.-J. Luche and A. Cruz, *J. Org. Chem.*, **44**, 2929 (1979).
29. C. Wakselman and M. Tordeux, *JCS Chem. Commun.*, 956 (1975).
30. E. D. Bergman, *J. Chem. Soc.*, 3457 (1961).
31. E. Cherbuliez, A. de Picciotto and J. Rabinowitz, *Helv. Chim. Acta*, **43**, 1144 (1960).
32. B. Modarai and E. Khoshdel, *J. Org. Chem.*, **42**, 3527 (1977).
33. N. De Kimpe, R. Verhé, L. De Buyck and N. Schamp, *J. Org. Chem.*, **45**, 2803 (1980).
34. A. J. Fry and Y. Migron, *Tetrahedron Lett.*, 3357 (1979).
35. T. E. Gough, W. S. Lin and R. E. Woolford, *Canad. J. Chem.*, **45**, 2529 (1967).
36. R. A. Shepard and A. A. Loiselle, *J. Org. Chem.*, **23**, 2012 (1958).
37. P. C. Ray, A. C. Goswami and A. C. Ray, *J. Indian Chem. Soc.*, **12**, 93 (1935).
38. G. A. Olah, J. T. Welch, Y. D. Vankar, M. Nojima, I. Kerekes and J. A. Olah, *J. Org. Chem.*, **44**, 3872 (1979).
39. R. R. Fraser, J. E. Millington and F. L. M. Pattison, *J. Amer. Chem. Soc.*, **79**, 1959 (1957).
40. E. D. Bergman and R. Ikan, *Chem. Ind.*, 394 (1957).
41. J. Leroy and C. Wakselman, *JCS Perkin I*, 1224 (1978).
42. T. F. McGrath and R. Levine, *J. Amer. Chem. Soc.*, **77**, 3656 (1955).
43. A. F. Blum, H. V. Donnand and H. D. Zook, *J. Amer. Chem. Soc.*, **77**, 4406 (1955).
44. E. Elkik and H. Assadi-Far, *Bull. Soc. Chim. France*, 991 (1970).
45. E. Elkik, R. Dahan and A. Parlier, *Compt. Rend. Acad. Sci. Paris C*, **286**, 353 (1978).
46. E. T. McBee, O. R. Pierce and D. D. Meyer, *J. Amer. Chem. Soc.*, **77**, 917 (1955).
47. E. D. Bergmann, S. Cohen, E. Hoffmann and J. Randmeier, *J. Chem. Soc.*, 3452 (1961).
48. F. Bergmann and A. Kalmus, *J. Amer. Chem. Soc.*, **76**, 4137 (1954).
49. H. Machleidt, *Liebigs Ann. Chem.*, **667**, 24 (1963).
50. J. Cantacuzène and D. Ricard, *Bull. Soc. Chim. France*, 1587 (1967).
51. E. Elkik and M. Le Blanc, *Bull. Soc. Chim. France*, 870 (1971).
52. R. Justoni, *Chimica e Ind.*, **24**, 93 (1942).
53. Z. Bankowska, *Rocz. Chem.*, **32**, 739 (1958).
54. Fr. Pat., 837,741 (1938); *Chem. Abstr.* **33**, 58655 (1939).
55. E. G. Edwards, D. P. Evans and H. B. Watson, *J. Chem. Soc.*, 1944 (1937).
56. M. Geiger, E. Usteri and C. Gränacher, *Helv. Chim. Acta*, **34**, 1340 (1951).
57. A. Bruylants and J. Houssiau, *Bull. Soc. Chim. Belg.*, **61**, 492 (1952).
58. P. D. Bartlett and R. H. Rosenwald, *J. Amer. Chem. Soc.*, **56**, 1992 (1934).
59. H. W. Wanzlick, G. Gollmer and H. Milz, *Chem. Ber.*, **88**, 72 (1955).
60. M. S. Newman, M. D. Farbman and H. Hipsher, *Org. Synth.*, Coll. Vol. III, 188 (1955).
61. G. Hesse and F. Urbanek, *Liebigs Ann. Chem.*, **604**, 54 (1957).
62. C. L. Stevens, J. J. Beereboom and K. G. Rutherford, *J. Amer. Chem. Soc.*, **77**, 4590 (1955).
63. H. Brintzinger and H. Orth, *Monatsh. Chem.*, **85**, 1015 (1954).
64. L. De Buyck, R. Verhé, N. De Kimpe and N. Schamp, *Bull. Soc. Chim. Belg.*, **89**, 307 (1980).
65. L. De Buyck, R. Verhé, N. De Kimpe, D. Courtheyn and N. Schamp, *Bull. Soc. Chim. Belg.*, **89**, 441 (1980).
66. L. De Buyck, R. Verhé, N. De Kimpe, D. Courtheyn and N. Schamp, *Bull. Soc. Chim. Belg.*, **90**, 837 (1981).
67. L. De Buyck, N. De Kimpe, R. Verhé, D. Courtheyn and N. Schamp, *Bull. Soc. Chim. Belg.*, **89**, 1043 (1980).
68. M. F. Guimon, G. Pfister-Guillouro, F. Metras and J. Petrissans, *J. Mol. Struct.*, **33**, 239 (1976).
69. P. A. Peters, R. Ottinger, J. Reisse and G. Chiurdoglu, *Bull. Soc. Chim. Belg.*, **77**, 407 (1968).
70. H. Korten and R. Scholl, *Chem. Ber.*, **34**, 1902 (1901).

71. J. G. Aston, J. D. Newkirk, D. M. Jenkins and J. Dorsky, *Org. Synth*., Coll. Vol. III, 538 (1955).
72. J. G. Aston, J. D. Newkirk, J. Dorsky and D. M. Jenkins, *J. Amer. Chem. Soc*., **64**, 1415 (1942).
73. N. De Kimpe, L. De Buyck, R. Verhé, F. Wychuyse and N. Schamp, *Synthetic Comm*., **9**, 575 (1979).
74. M. Ballester and J. Riera, *Anales Real Soc. Espan. Fis. Quim.* (Madrid), **56**, 897 (1960).
75. R. Carlson, *Acta Chem. Scand*., **B32**, 646 (1978).
76. W. Seufert and F. Effenberger, *Chem. Ber*., **112**, 1670 (1979).
77. D. P. Wyman and P. R. Kaufman, *J. Org. Chem*., **29**, 1956 (1964).
78. E. R. Buchman and H. Sargent, *J. Amer. Chem. Soc*., **67**, 401 (1945).
79. P. Delbaere, *Bull. Soc. Chim. Belg*., **51**, 1 (1942).
80. E. W. Warnhoff, D. G. Martin and W. S. Johnson, *Org. Synth*., **37**, 8 (1957).
81. C. L. Stevens and E. Farkas, *J. Amer. Chem. Soc*., **74**, 619 (1952).
82. Dang Quoc Quan, *Compt. Rend. Acad. Sci. Paris C*, **264**, 320 (1967).
83. F. Caujolle and Dang Quoc Quan, *Compt. Rend. Acad. Sci. Paris C*, **265**, 269 (1967).
84. M. F. Grenier-Loustalot, P. Iratcabal and F. Métras, *Synthesis*, 33 (1976).
85. J. R. Merchant and D. V. Rege, *Tetrahedron*, **27**, 4837 (1971).
86. J. P. Schaefer and F. Sonnenberg. *J. Org. Chem*., **28**, 1128 (1963).
87. G. C. Finger, F. H. Reed, E. W. Maynert and A. M. Weiner, *J. Amer. Chem. Soc*., **73**, 151 (1951).
88. S. Goldsmidt, R. Enders and R. Dirsch, *Chem. Ber*., **58**, 572 (1925).
89. J. J. Beereboom, C. D. Djerassi, D. Ginsburg and L. F. Fieser, *J. Amer. Chem. Soc*., **75**, 3503 (1953).
90. U.S. Pat. 2,714,601 (1952); *Chem. Abstr.* **50**, 8753 (1956).
91. N. P. Buu-Hoï and P. Demerseman, *J. Org. Chem*., **18**, 649 (1953).
92. N. De Kimpe and N. Schamp, *Org. Prep. Proced. Int*., **11**, 115 (1979).
93. W. Coppens and N. Schamp, *Bull. Soc. Chim. Belg*., **81**, 643 (1972).
94. N. De Kimpe, N. Schamp and W. Coppens, *Bull. Soc. Chim. Belg*., **84**, 227 (1975).
95. N. Schamp, N. De Kimpe and W. Coppens, *Tetrahedron*, **31**, 2081 (1975).
96. N. De Kimpe, R. Verhé, L. De Buyck and N. Schamp, *Synthetic Commun*., **8**, 75 (1978).
97. J. F. W. Keana and R. R. Schumaker, *Tetrahedron*, **26**, 5191 (1970).
98. I. J. Borowitz, E. W. Casper, R. K. Crouch and K. Y. Yee, *J. Org. Chem*., **37**, 3873 (1972).
99. J. K. Kochi, *J. Amer. Chem. Soc*., **77**, 5274 (1955).
100. E. M. Kosower, W. J. Cole, G. S. Wu, D. E. Cardy and E. Meisters, *J. Org. Chem*., **28**, 630 (1963).
101. J. T. Satoh and K. Nishizawa, *JCS Chem. Commun*., 83 (1973).
102. Y. Ito, M. Nakatsuka and T. Saegusa, *J. Org. Chem*., **45**, 2022 (1980).
103. A. S. Dneprovskii, I. V. Krainyuchenko and T. I. Temnikova, *Zh. Org. Khim*., **14**, 1514 (1978); *Chem. Abstr.* **88**, 146557 (1978).
104. A. Favorskii, *Russ. Phys.-Chem. Ges*., **44**, 1339 (1912).
105. C. Y. Meyers, A. M. Malte and W. S. Matthews, *J. Amer. Chem. Soc*., **91**, 7512 (1969).
106. C. Y. Meyers and V. M. Kolb, *J. Org. Chem*., **43**, 1985 (1978).
107. F. M. Laskovics and E. M. Schulman, *J. Amer. Chem. Soc*., **99**, 6672 (1977).
108. E. Vilsmaier, W. Tröger, W. Sprügel and K. Gagel, *Chem. Ber*., **112**, 2997 (1979).
109. S. Torrii, T. Inokuchi, S. Misima and T. Kobayashi, *J. Org. Chem*., **45**, 2731 (1980).
110. K. Steinbeck, *Tetrahedron Lett*., 1103 (1978).
111. E. Zbiral and M. Rasberger, *Tetrahedron*, **25**, 1871 (1969).
112. A. Brochet, *Ann. Chim*., **10**, 135 (1897).
113. R. E. Meyer, *Helv. Chim. Acta*, **16**, 1291 (1933).
114. R. Riemschneider, *Z. Naturf*., **8b**, 161 (1953).
115. R. H. Levin, *J. Amer. Chem. Soc*., **75**, 502 (1953).
116. W. E. Doering and L. H. Knose, *J. Amer. Chem. Soc*., **74**, 5683 (1952).
117. S. Vickers and E. E. Smissman, *J. Org. Chem*., **40**, 749 (1975).
118. U.S. Pat. 2,684,968 (1953); *Chem. Abstr.* **49**, 11031k (1955).
119. C. H. R. Elston, *J. Chem. Soc*., 367 (1948).
120. L. Denivelle and R. Fort, *Bull. Soc. Chim. France*, 1837 (1956).
121. K. Fries and K. Schimmelschmidt, *Liebigs Ann. Chem*., **484**, 297 (1930).

122. L. E. Forman and W. C. Sears, *J. Amer. Chem. Soc.*, **76**, 4977 (1954).
123. P. Svec, A. M. Sörensen and M. Zbirovsky, *Org. Prep. Proced. Int.*, **5**, 209 (1973).
124. B. L. Jensen, S. E. Burke, S. E. Thomas and W. H. Klausmeier, *Tetrahedron Lett.*, 2639 (1977).
125. B. L. Jensen and P. E. Peterson, *J. Org. Chem.*, **42**, 4052 (1977).
126. B. L. Jensen, S. E. Burke and S. E. Thomas, *Tetrahedron*, **34**, 1627 (1978).
127. J. Villieras, C. Bacquet and J. F. Normant, *J. Organometal. Chem.*, **97**, 355 (1975).
128. R. M. McDonald and R. N. Steppel. *J. Org. Chem.*, **35**, 1250 (1970).
129. W. E. Bachman and W. S. Struve, *Org. Reactions*, **II**, 48 (1948).
130. A. Roedig and R. Maier, *Chem. Ber.*, **86**, 1467 (1953).
131. C. Rappe and B. Albrecht, *Acta Chem. Scand.*, **20**, 253 (1966).
132. R. E. van Atta, H. D. Look and P. J. Elving, *J. Amer. Chem. Soc.*, **76**, 1185 (1954).
133. C. Gränacher, E. Usteri and M. Geiger, *Helv. Chem. Acta*, **32**, 713 (1949).
134. W. D. McPhee and E. Klingsberg, *Org. Synth.*, Coll. Vol. III, 119 (1955).
135. B. W. Ponder and D. R. Walker, *J. Org. Chem.*, **32**, 4136 (1967).
136. K. B. Sharpless and A. Y. Teranishi, *J. Org. Chem.*, **38**, 185 (1973).
137. W. T. Brady and R. Roe, *J. Amer. Chem. Soc.*, **92**, 4618 (1970).
138. P. Martin, H. Greuter and D. Bellus, *J. Amer. Chem. Soc.*, **101**, 5853 (1979).
139. D. A. Bak and W. T. Brady, *J. Org. Chem.*, **44**, 107 (1979).
140. L. R. Krepski and A. Hassner, *J. Org. Chem.*, **43**, 2879 (1978).
141. L. R. Krepski and A. Hassner, *J. Org. Chem.*, **43**, 3173 (1978).
142. B. M. Trost, *Acc. Chem. Res.*, **7**, 85 (1974).
143. A. E. Greene and J.-P. Deprés, *J. Amer. Chem. Soc.*, **101**, 4003 (1979).
144. N. Wittorf, *J. Russ. Phys.-Chem. Ges.*, **32**, 88 (1900).
145. A. Favorskii, *J. Prakt. Chem.*, **51**, 533 (1895).
146. F. Strauss, *Chem. Ber.*, **63**, 1868 (1930).
147. E. L. Jackson, *J. Amer. Chem. Soc.*, **56**, 977 (1934).
148. S. F. Reed, *J. Org. Chem.*, **30**, 2195 (1965).
149. A. Kirrmann and R. Nouri-Bimorghi, *Bull. Soc. Chim. France*, 3213 (1968).
150. A. Kirrmann, P. Duhamel and R. Nouri-Bimorghi, *Liebigs Ann. Chem.*, **691**, 33 (1966).
151. J. J. Riehl, P. Casara and A. Fougerousse, *Compt. Rend. Acad. Sci. Paris C*, **279**, 113 (1974).
152. R. N. McDonald and T. E. Tabor, *J. Amer. Chem. Soc.*, **89**, 6573 (1967).
153. R. N. McDonald and T. E. Tabor, *J. Org. Chem.*, **33**, 2934 (1968).
154. R. N. McDonald and R. N. Steppel, *J. Org. Chem.*, **35**, 1250 (1970).
155. R. N. McDonald and R. C. Cousins, *J. Org. Chem.*, **45**, 2976 (1980).
156. K. Griesbaum, R. Kibar and P. Pfeffer, *Ann. Chem.*, 214 (1975).
157. G. Stork, W. S. Worrall and J. J. Pappas, *J. Amer. Chem. Soc.*, **82**, 4315 (1960).
158. G. A. Olah, Y. D. Vankar and M. Arvanaghi, *Tetrahedron Lett.*, 3653 (1979).
159. A. B. Boese, *Ind. Eng. Chem.*, **32**, 20 (1940).
160. E. E. Blaise, *Bull. Soc. Chim. France*, 728 (1915).
161. J. Leroy, *J. Org. Chem.*, **46**, 206 (1981).
162. C. Bacquet, J. Villieras and J. F. Normant, *Compt. Rend. Acad. Sci. Paris C*, **278**, 929 (1974).
163. C. Rabiller and G. J. Martin, *Tetrahedron*, **34**, 3281 (1978).
164. T. Zincke, W. Schneider and W. Emmerich, *Liebigs Ann. Chem.*, **328**, 286 (1903).
165. T. Zincke and O. Preiss, *Liebigs Ann. Chem.*, **417**, 203 (1918).
166. F. Kröhnke and F. Timmler, *Chem. Ber.*, **69**, 615 (1936).
167. F. Kröhnke, *Chem. Ber.*, **69**, 921 (1936).
168. C. L. Stevens and C. T. Lenk, *J. Org. Chem.*, **19**, 538 (1954).
169. M. S. Newman, *J. Amer. Chem. Soc.*, **73**, 4993 (1951).
170. F. Kröhnke and O. Lüderitz, *Chem. Ber.*, **83**, 60 (1950).
171. M. Gaudry and A. Marquet, *Bull. Soc. Chim. France*, 4173 (1969).
172. P. A. Levene, *Org. Synth.*, Coll. Vol. II, 88 (1943).
173. J. F. Norris, *Ind. Eng. Chem.*, **11**, 828 (1919).
174. F. Weygand and U. Schmied-Kowarzik, *Chem. Ber.*, **82**, 335 (1949).
175. J. R. Catch, D. F. Elliott, D. H. Hey and E. R. H. Jones, *J. Chem. Soc.*, 272 (1948).
176. H. M. E. Cardwell and A. E. H. Kilner, *J. Chem. Soc.*, 2430 (1951).

177. A. A. Sacks and J. G. Aston, *J. Amer. Chem. Soc.*, **73**, 3902 (1951).
178. J. H. Boyer and D. Straw, *J. Amer. Chem. Soc.*, **74**, 4506 (1952).
179. G. H. Hill and E. L. Kropa, *J. Amer. Chem. Soc.*, **55**, 2511 (1933).
180. Y. Yasor, M. Gaudry and A. Marquet, *Bull. Soc. Chim. France*, 2735 (1973).
181. M. Gaudry and A. Marquet, *Tetrahedron*, **26**, 5611 and 5617 (1970).
182. V. Calo, L. Lopez and G. Pèsce, *JCS Perkin I*, 501 (1977).
183. H. M. R. Hoffman and J. G. Vinter, *J. Org. Chem.*, **39**, 3921 (1974).
184. B. Föhlisch and W. Gottstein *Liebigs Ann Chem.*, 1768 (1979).
185. R. M. Acheson, *J. Chem. Soc.*, 4236 (1956).
186. R. Belcher, W. Hoyle and T. S. West, *J. Chem. Soc.*, 2744 (1958).
187. D. Ricard and J. Cantacuzène, *Bull. Soc. Chim. France*, 628 (1969).
188. E. J. Corey, *J. Amer. Chem. Soc.*, **75**, 3297 (1953).
189. F. G. Bordwell and K. M. Wellman, *J. Org. Chem.*, **28**, 2544 (1963).
190. O. Wallach, *Liebigs Ann. Chem.*, **343**, 41 (1905).
191. T. Kato, H. Kondo and H. Miyake, *Bull. Chem. Soc. Japan*, **53**, 823 (1980).
192. R. M. Cowper and L. H. Davidson, *Org. Synth.*, Coll. Vol. II, 480 (1943).
193. F. Kröhnke, *Chem. Ber.*, **83**, 56 (1950).
194. S. Winstein, T. L. Jacobs, G. B. Linden, D. Seymour, E. F. Levy, B. F. Day, J. H. Robson, R. B. Henderson and W. H. Florsheim, *J. Amer. Chem. Soc.*, **68**, 1831 (1946).
195. S. Wolfe, W. R. Pilgrim, T. F. Garrard and P. Chamberlain, *Canad. J. Chem.*, **49**, 1099 (1971).
196. H. O. House and W. C. McDaniel, *J. Org. Chem.*, **42**, 2155 (1977).
197. K. S. Warren, O. K. Neville and E. C. Hendley, *J. Org. Chem.*, **28**, 2152 (1963).
198. A. Iovchev and S. Spasov, *Monatsh. Chem.*, **98**, 2326 (1967).
199. R. B. Wagner, *J. Amer. Chem. Soc.*, **71**, 3216 (1949).
200. N. H. Cromwell and R. Benson, *Org. Synth.*, Coll. Vol. III, 105 (1955).
201. R. E. Buckles, R. Filler and L. Hilfman, *J. Org. Chem.*, **17**, 233 (1952).
202. R. Verhé, N. Schamp, L. De Buyck and R. Van Loocke, *Bull. Soc. Chim. Belg.*, **84**, 371 (1975).
203. L. Blanco, P. Arnice and J. M. Conia, *Synthesis*, 194 (1976).
204. P. Z. Bedoukian, *J. Amer. Chem. Soc.*, **67**, 1430 (1945).
205. H. Schmid and P. Karrer, *Helv. Chim. Acta*, **29**, 573 (1946).
206. J. D. Billimoria and N. F. Maclagan, *J. Chem. Soc.*, 3067 (1951).
207. E. J. Corey, *J. Amer. Chem. Soc.*, **75**, 2303 (1953).
208. L. A. Paquette and R. F. Doehner Jr, *J. Org. Chem.*, **45**, 5105 (1980).
209. N. De Kimpe, R. Verhé, L. De Buyck and N. Schamp, *Tetrahedron Lett.*, 2257 (1980).
210. C. Djerassi, *Chem. Rev.*, **43**, 271 (1948).
211. O. O. Orazi and J. Meseri, *Ann. Soc. Quim. Arg.*, **37**, 192 (1949).
212. C. Djerassi and C. T. Lenk, *J. Amer. Chem. Soc.*, **75**, 3494 (1953).
213. N. De Kimpe, unpublished results.
214. L. C. King and G. K. Ostrum, *J. Org. Chem.*, **29**, 3459 (1964).
215. D. P. Bauer and R. S. Macomber, *J. Org. Chem.*, **40**, 1990 (1975).
216. C. Djerassi and C. R. Scholz, *J. Amer. Chem. Soc.*, **70**, 417 (1948).
217. D. Jerchel and H. Fischer, *Liebigs Ann. Chem.*, **590**, 224 (1954).
218. W. S. Johnson, J. D. Bass and K. L. Williamson, *Tetrahedron*, **19**, 861 (1963).
219. L. Forlani, *Synthesis*, 487 (1980).
220. V. W. Armstrong, N. H. Chishti and R. Ramage, *Tetrahedron Lett.*, 373 (1975).
221. S. Cacchi and L. Caglioti, *Synthesis*, 64 (1979).
222. A. Bongini, G. Cainelli, M. Contento and F. Manescalchi, *Synthesis*, 143 (1980).
223. T. Hata, *Bull. Chem. Soc. Japan*, **37**, 547 (1964).
224. M. Sekiya, K. Ito and K. Suzuki, *Tetrahedron*, **31**, 231 (1975).
225. R. Bloch, *Synthesis*, 140 (1978).
226. V. Calo, L. Lopez, G. Pèsce and P. E. Todesco, *Tetrahedron*, **29**, 1625 (1973).
227. S. J. Pasaribu and L. R. Williams, *Aust. J. Chem.*, **26**, 1327 (1973).
228. R. Levine and J. R. Stephens, *J. Amer. Chem. Soc.*, **72**, 1642 (1950).
229. R. B. Loftfield and L. Schaad, *J. Amer. Chem. Soc.*, **76**, 36 (1954).
230. M. Schlosser and G. Heinz, *Chem. Ber.*, **102**, 1944 (1969).
231. J. R. Catch, D. H. Hey, E. R. H. Jones and W. Wilson, *J. Chem. Soc.*, 278 (1948).

232. V. Calo, L. Lopez and D. S. Valentino, *Synthesis*, 139 (1978).
233. F. de Reinach-Hirtzbach and T. Durst, *Tetrahedron Lett.*, 3677 (1976).
234. N. A. Sokolov, I. G. Tishchenko and N. V. Kovganko, *Zh. Org. Khim.*, **16**, 281 (1980); *Chem. Abstr.* **93**, 25853 (1980).
235. J. Wolinsky, J. Thorstenson and T. A. Killinger, *J. Org. Chem.*, **43**, 875 (1978).
236. S. J. Slanina, G. F. Hennion and I. A. Nieuwland, *J. Amer. Chem. Soc.*, **58**, 891 (1936).
237. V. Reutrakul, A. Tiensripojamarn, K. Kusamran and S. Nimgirawath, *Chem. Lett.*, 209 (1979).
238. H. J. Ringold and G. Stork, *J. Amer. Chem. Soc.*, **80**, 250 (1958).
239. P. L. Julian and W. J. Karpel, *J. Amer. Chem. Soc.*, **72**, 362 (1950).
240. P. D. Gardner and W. J. Horton, *J. Org. Chem.*, **19**, 213 (1954).
241. C. Djerassi, J. Grossman and G. H. Thomas, *J. Amer. Chem. Soc.*, **77**, 3826 (1955).
242. L. Wolff, *Liebigs Ann. Chem.*, **394**, 40 (1912).
243. R. C. Cambie, R. C. Hayward, J. L. Jurlina, P. S. Rutledge and P. D. Woodgate, *J. Chem. Soc.*, 126 (1978).
244. G. Cardillo and M. Shimizu, *J. Org. Chem.*, **42**, 4268 (1977).
245. R. D'Ascoli, M. D'Auria, L. Nucciarelli, G. Piancatelli and A. Scettri, *Tetrahedron Lett.*, 4521 (1980).
246. V. L. Heasby, D. F. Shellhamer, L. E. Heasley, D. B. Yaeger and G. E. Heasley, *J. Org. Chem.*, **45**, 4649 (1980).
247. G. M. Rubottom and R. C. Mott, *J. Org. Chem.*, **44**, 1731 (1979).
248. E. S. Gould, *Mechanism and Structure in Organic Chemistry*, Holt, Rinehart and Winston, New York (1959), p. 372.
249. H. O. House, *Modern Synthetic Reactions*, W. A. Benjamin, Menlo Park, Calif. (1972), p. 459.
250. R. P. Bell and G. G. Davis, *J. Chem. Soc.*, 902 (1964).
251. J. Toullec and J. E. Dubois, *Tetrahedron*, **29**, 2851, 2859 (1973).
252. G. Schwarzenbach and C. Wittwer, *Helv. Chim. Acta*, **30**, 669 (1947).
253. C. Rappe and W. H. Sachs, *J. Org. Chem.*, **32**, 4127 (1967).
254. C. Rappe, *Acta Chem. Scand.*, **22**, 219 (1968).
255. C. Rappe, *Acta Chem. Scand.*, **23**, 2305 (1969).
256. M. S. Sytilin, *Zh. Fiz. Khim.*, **41**, 1200 (1967); *Chem. Abstr.* **69**, 46429z (1968).
257. C. G. Swain and R. P. Dunlap, *J. Amer. Chem. Soc.*, **94**, 7204 (1972).
258. J. W. Thorpe and J. Warkentin, *Canad. J. Chem.*, **50**, 3229 (1972).
259. R. A. Cox and J. Warkentin, *Canad. J. Chem.*, **50**, 3233 (1972).
260. A. C. Knipe and B. G. Cox, *J. Chem. Soc.*, 1391 (1973).
261. J. B. Conant, W. R. Kirner and R. E. Hussey, *J. Amer. Chem. Soc.*, **47**, 488 (1925).
262. R. G. Pearson, S. H. Langer, F. V. Williams and W. J. McGuire, *J. Amer. Chem. Soc.*, **74**, 5130 (1952).
263. A. J. Sisti and S. Lowell, *Canad. J. Chem.*, **42**, 1897 (1964).
264. F. G. Bordwell and W. T. Brannen, *J. Amer. Chem. Soc.*, **86**, 4645 (1964).
265. A. Streitwieser Jr, *Solvolytic Displacement Reactions*, McGraw-Hill, New York (1962), p. 28.
266. E. D. Hughes, *Quart. Rev.* (London), **5**, 245 (1951).
267. J. W. Baker, *J. Chem. Soc.*, 848 (1938).
268. C. L. Stevens, W. Malik and R. Pratt, *J. Amer. Chem. Soc.*, **72**, 4758 (1950).
269. M. J. S. Dewar, *The Electronic Theory of Organic Chemistry*, Clarendon Press, Oxford (1949), p. 73.
270. S. Winstein, E. Grunwald and H. W. Jones, *J. Amer. Chem. Soc.*, **73**, 2700 (1951).
271. G. Richard, *Bull. Soc. Chim. France*, **5**, 286 (1938).
272. V. Rosnati, F. Sannicolo and G. Zecchi, *Tetrahedron Lett.*, 599 (1970).
273. F. G. Bordwell and M. W. Carlson, *J. Amer. Chem. Soc.*, **91**, 3951 (1969).
274. R. P. Lutz, *J. Amer. Chem. Soc.*, **90**, 3788 (1968).
275. N. J. Turro, R. B. Gagosian, C. Rappe and L. Knutsson, *Chem. Commun.*, 270 (1969).
276. J. W. Thorpe and J. Warkentin, *Canad. J. Chem.*, **51**, 927 (1973).
277. A. Halvorsen and J. Songstad, *JCS Chem. Commun.*, 327 (1978).
278. S. H. McAllister, W. A. Bailey and C. M. Bouton, *J. Amer. Chem. Soc.*, **62**, 3210 (1940).
279. J. Cologne and J. C. Dubin, *Bull. Soc. Chim. France*, 1180 (1960).

280. C. L. Stevens and E. Farkas, *J. Amer. Chem. Soc.*, **74**, 5352 (1952).
281. R. N. McDonald and P. A. Schwab, *J. Amer. Chem. Soc.*, **85**, 4004 (1963).
282. C. A. Buehler, H. A. Smith, K. V. Nayak and T. A. Magee, *J. Org. Chem.*, **26**, 1573 (1961).
283. P. D. Bartlett and G. F. Woods, *J. Amer. Chem. Soc.*, **62**, 2933 (1940).
284. R. Weidenhagen and R. Herrmann, *Chem. Ber.*, **68**, 1955 (1935).
285. W. Ziegenbein, *Chem. Ber.*, **94**, 2989 (1961).
286. B. Tchoubar, *Compt. Rend. Acad. Sci. Paris*, **234**, 2544 (1952).
287. E. E. Smissman and J. L. Diebold, *J. Org. Chem.*, **30**, 4005 (1965).
288. M. Kopp and B. Tchoubar, *Bull. Soc. Chim. France*, 84 (1952).
289. I. Elphimoff-Felkin and B. Tchoubar, *Compt. Rend. Acad. Sci. Paris*, **238**, 1425 (1954).
290. G. Richard, *Bull. Soc. Chim. France*, 286 (1938).
291. R. B. Loftfield and L. Schaad, *J. Amer. Chem. Soc.*, **76**, 35 (1954).
292. R. Jacquier, *Bull. Soc. Chim. France*, 83, (1950).
293. G. Hesse and F. Urbanek, *Chem. Ber.*, **91**, 2733 (1958).
294. J. M. Conia and J. R. Salaun, *Acc. Chem. Res.*, **5**, 33 (1972).
295. C. Rappe and L. Knutsson, *Acta Chem. Scand.*, **73**, 4702 (1964).
296. A. C. Cope and E. S. Graham, *J. Amer. Chem. Soc.*, **73**, 4702 (1964).
297. J. Wolinsky and R. O. Hutchins, *J. Org. Chem.*, **37**, 3294 (1972).
298. C. Rappe, *Acta Chem. Scand.*, **19**, 270 (1965).
299. M. Utaka, S. Matsushita and A. Takeda, *Chem. Lett.*, 779 (1980).
300. K. S. Warren, O. K. Neville and E. C. Hendley, *J. Org. Chem.*, **28**, 2152 (1963).
301. P. Moreau and E. Casadevall, *Compt. Rend. Acad. Sci. Paris C*, **272**, 801 (1971).
302. M. Hanack, C. E. Harding and J.-L. Deroque, *Chem. Ber.*, **105**, 428 (1972).
303. Ger. Pat. 1,262,955 (1966); *Chem. Abstr.* **69**, 57710p (1968).
304. Ger. Pat. 2,819,264 (1978); *Chem. Abstr.* **90**, 103450 (1979).
305. G. Jones, *J. Chem. Soc. C*, 1230 (1970).
306. K. Sato, S. Inoue, S.-I. Kuranami and M. Ōhashi, *JCS Perkin I*, 1666 (1977).
307. J. A. Donnelly and D. E. Maloney, *Tetrahedron*, 2875 (1979).
308. J. A. Donnelly, M. J. Fox and T. C. Sharma, *Tetrahedron*, **35**, 875 (1979).
309. J. A. Donnelly, M. F. Cox and T. C. Sharma, *Tetrahedron*, **35**, 1987 (1979).
310. A. A. Sachs and J. G. Aston, *J. Amer. Chem. Soc.*, **73**, 3902 (1951).
311. O. E. Edwards and C. Grieco, *Canad. J. Chem.*, **52**, 3561 (1974).
312. J. A. Aston, J. T. Clarke, K. A. Burgess and R. B. Greenburg, *J. Amer. Chem. Soc.*, **64**, 300 (1942).
313. T. Oda, *Chem. Lett.*, 957 (1977).
314. C. L. Stevens and J. Tazuma, *J. Amer. Chem. Soc.*, **76**, 215 (1954).
315. C. L. Stevens, M. L. Weiner and R. C. Freeman, *J. Amer. Chem. Soc.*, **75**, 3977 (1953).
316. C. L. Stevens and J. J. De Young, *J. Amer. Chem. Soc.*, **76**, 718 (1954).
317. C. L. Stevens, J. J. Beereboom and K. G. Rutherford, *J. Amer. Chem. Soc.*, **77**, 4590 (1955).
318. A. Hassner and P. Catsoulacos, *J. Org. Chem.*, **31**, 3149 (1966).
319. A. Hassner and N. H. Cromwell, *J. Amer. Chem. Soc.*, **80**, 901 (1958).
320. N. H. Cromwell and R. P. Ayer, *J. Amer. Chem. Soc.*, **81**, 133 (1959).
321. R. B. Loftfield, *J. Amer. Chem. Soc.*, **73**, 4707 (1951).
322. F. G. Bordwell and J. Almy, *J. Org. Chem.*, **38**, 571 (1973).
323. B. Goyau and F. Rouessoc, *Bull. Soc. Chim. France*, 590 (1978).
324. S. Vichers and E. E. Smissman, *J. Org. Chem.*, **40**, 749 (1975).
325. W. T. Brady and A. D. Patel, *J. Org. Chem.*, **39**, 1949 (1974).
326. F. G. Bordwell and J. G. Strong, *J. Org. Chem.*, **38**, 579 (1973).
327. F. G. Bordwell and R. G. Scamehorn, *J. Amer. Chem. Soc.*, **90**, 6751 (1968).
328. F. G. Bordwell and M. W. Carlson, *J. Amer. Chem. Soc.*, **92**, 3370 (1970).
329. F. G. Bordwell and M. W. Carlson, *J. Amer. Chem. Soc.*, **92**, 3377 (1970).
330. F. G. Bordwell and R. G. Scamehorn, *J. Amer. Chem. Soc.*, **93**, 3410 (1971).
331. A. W. Fort, *J. Amer. Chem. Soc.*, **84**, 2620, 4979 (1962).
332. L. De Buyck, unpublished results.
333. M. Mousseron, R. Jacquier and A. Fontaine, *Compt. Rend. Acad. Sci. Paris*, **232**, 1562 (1951).
334. E. Elkik and H. Assadifar, *Bull. Soc. Chim. France II*, 129 (1978).

335. D. Baudry and M. Charpentier-Morize, *Tetrahedron Lett.*, 3013 (1973).
336. H. E. Zimmerman and R. J. Pasteris, *J. Org. Chem.*, **45**, 484, 4876 (1980).
337. J. F. Pazos, J. G. Pacifici, G. O. Pierson, D. B. Slove and F. D. Greene, *J. Org. Chem.*, **39**, 1990 (1974).
338. P. S. Wharton and A. R. Fritzberg, *J. Org. Chem.*, **37**, 1899 (1972).
339. J. Ciabattoni, E. C. Nathan, A. E. Feiring and P. J. Kocienski, *Org. Synth.*, **54**, 97 (1974).
340. R. Breslow, T. Eicher, A. Krebs, R. A. Peterson and J. Posner, *J. Amer. Chem. Soc.*, **87**, 1320 (1965).
341. E. Herranz and F. Serratosa, *Tetrahedron Lett.*, 3335 (1975).
342. M. Mousseron and R. Jacquier, *Compt. Rend. Acad. Sci. Paris*, **229**, 374 (1949).
343. M. Kopp, *Bull. Soc. Chim. France*, 628 (1954).
344. M. Mousseron and R. Jacquier, *Bull. Soc. Chim. France*, 689 (1949).
345. W. B. Smith and C. Gonzalez, *Tetrahedron Lett.*, 5751 (1966).
346. M. Charpentier-Morize, M. Mayer and B. Tchoubar, *Bull. Soc. Chim. France*, 529 (1965).
347. F. Bohlmann and G. Fritz, *Tetrahedron Lett.*, 95 (1981).
348. J. Hill, *J. Chem. Soc. C*, 462 (1970).
349. V. Rosnati and A. Salimbeni, *Gazz. Chim. Ital.*, **107**, 271 (1977).
350. P. A. Levine and A. Walti, *Org. Synth.*, Coll. Vol. II, 5 (1943).
351. E. B. Reid, R. B. Fortenbauch and H. R. Patterson, *J. Org. Chem.*, **15**, 579 (1950).
352. K. V. Auwers, H. Ludewig and A. Müller, *Liebigs Ann. Chem.*, **526**, 143, 158 (1936).
353. V. Caplar, A. Lisini, F. Kajfez, D. Kolbah and V. Sunjic, *J. Org. Chem.*, **43**, 1355 (1978).
354. P. Beltrame, V. Rosnati and F. Sannicolo, *Tetrahedron Lett.*, 4219 (1970).
355. S. Gladiali, M. P. Porcu, V. Rosnati, A. Saba, F. Soccolini and A. Selva, *Gazz. Chim. Ital.*, **107**, 293 (1977).
356. R. B. Warneboldt and L. Weiler, *Tetrahedron Lett.*, 3413 (1971).
357. K. Sato, M. Ohashi, E. Aoki and Y. Murai, *J. Org. Chem.*, **42**, 3713 (1977).
358. S. F. Krauser and A. C. Watterson, *J. Org. Chem.*, **43**, 3400 (1978).
359. C. Wakselman and J. Leroy, *J. Fluorine Chem.*, **12**, 101 (1978).
360. D. Mayer, in *Methoden der Organischen Chemie*, Vol. 7/2C (Ed. Houben-Weyl), Georg Thieme Verlag, Stuttgart (1977), p. 2253.
361. M. Kerfanto, A. Brault, F. Venien, J.-M. Morvan and A. Le Rouzic, *Bull. Soc. Chim. France*, 196 (1975).
362. D. G. Holland and E. D. Amstutz, *Recl. Trav. Chim. Pays-Bas*, **83**, 1047 (1964).
363. L. P. Ellinger and A. A. Goldberg, *J. Chem. Soc.*, 266 (1949).
364. Belgian Pat. 621,456 (1962); *Chem. Abstr.* **59**, 9835d (1963).
365. G. K. Rogulchenko, I. A. Mazur and P. M. Kochergin, *Farm. Zh.* (Kiev), **4**, 29 (1976); *Chem. Abstr.* **85**, 192657 (1976).
366. M. V. Povstyanoi, V. P. Kruglenko and P. M. Kochergin, *Ukr. Khim. Zh.*, **42**, 1166 (1976); *Chem. Abstr.* **86**, 89770 (1977).
367. N. H. Cromwell and P. H. Hess, *J. Amer. Chem. Soc.*, **83**, 1237 (1961).
368. P. L. Julian, E. W. Meyer, A. Magnani and W. Cole, *J. Amer. Chem. Soc.*, **67**, 1203 (1945).
369. E. W. Warnhoff and W. S. Johnson, *J. Amer. Chem. Soc.*, **75**, 494 (1953).
370. Ger. Pat. 1,262,995 (1966); *Chem. Abstr.* **69**, 57710p (1968).
371. O. I. Sorokin, *Izv. Akad. SSSR*, 460 (1961); *Chem. Abstr.* **55**, 22310c (1961).
372. K. T. Potts and I. S. Baum, *Chem. Rev.*, **74**, 189 (1974).
373. E. W. Warnhoff, *J. Org. Chem.*, **27**, 4587 (1962).
374. E. W. Warnhoff and D. R. Marshall, **32**, 2000 (1967).
375. H. J. Sattler, H. G. Lennartz and W. Schunack, *Arch. Pharm.* (Weinheim), **312**, 107 (1979).
376. J. Wolinsky, R. O. Hutchins and T. W. Gibson, *J. Org. Chem.*, **33**, 407 (1968).
377. J. Wolinsky, J. J. Hamsher and R. O. Hutchins, *J. Org. Chem.*, **35**, 207 (1970).
378. C. L. Stevens, P. Blumbergs and M. Munk, *J. Org. Chem.*, **28**, 331 (1963).
379. Y. Zeifman, N. Gambaryan and I. Knunyants, *Izv. Akad. Nauk SSSR, Ser. Khim.*, 450 (1965); *Chem. Abstr.* **64**, 6554f (1965).
380. W. Pirkle and J. Hauske, *J. Org. Chem.*, **42**, 2436 (1977).
381. N. De Kimpe, R. Verhé, L. De Buych, L. Moëns and N. Schamp, *Synthesis*, 43 (1982).
382. L. Duhamel, P. Duhamel and J.-M. Poirier, *Tetrahedron Lett.*, 4237 (1973).

383. L. Duhamel, P. Duhamel and J.-M. Poirier, *Bull. Soc. Chim. France*, 221 (1972).
384. L. Duhamel and J.-M. Poirier, *J. Org. Chem.*, **44**, 3585 (1979).
385. D. Cantacuzène and M. Tordeux, *Tetrahedron Lett.*, 4807 (1971).
386. C. Combet-Farnoux, J. F. Girardeau and H. Galens, *Compt. Rend. Acad. Sci. Paris*, **282**, 469 (1976).
387. C. L. Stevens and P. M. Pillai, *J. Org. Chem.*, **37**, 173 (1972).
388. C. L. Stevens, J. M. Cahoon, T. R. Potts and P. M. Pillai, *J. Org. Chem.*, **37**, 3130 (1972).
389. L. Nilsson and C. Rappe, *Acta Chem. Scand.*, **B30**, 1000 (1976).
390. T. Kato, T. Chiba, M. Noda and M. Sasaki, *Heterocycles*, **10**, 261 (1978).
391. J. C. Meslin, Y. T. N'Guessan, H. Quiniore and F. Tonnard, *Tetrahedron*, **31**, 2679 (1975).
392. H. Bredereck and R. Gompper, *Chem. Ber.*, **87**, 700 (1954).
393. P. B. Terentiev, A. N. Kost, N. P. Lomakina and V. G. Kartev, *Org. Prep. Proced. Int.*, **6**, 145 (1974).
394. W. Ried and L. Kaiser, *Liebigs Ann. Chem.*, 958 (1975).
395. R. F. Abdulla and J. C. Williams, *Tetrahedron Lett.*, 997 (1980).
396. G. Kempter, J. Spindler, H. Fiebig and G. Sarodnick, *J. Prakt. Chem.*, **313**, 977 (1971).
397. M.-C. Dubroeucq, F. Rocquet and F. Weiss, *Tetrahedron Lett.*, 4401 (1977).
398. J. P. Nath and G. N. Mahapatra, *Indian J. Chem. B*, **19**, 526 (1980).
399. C. Yamazaki, *Tetrahedron Lett.*, 1295 (1978).
400. A. N. Mirskova, G. G. Levkovskaya, I. D. Kalikhman and M. G. Voronkov, *Zh. Org. Khim.*, **15**, 2301 (1979); *Chem. Abstr.* **92**, 128792 (1980).
401. A. Babadjamian, J. Metzger and M. Chanon, *J. Heterocycl. Chem.*, **12**, 643 (1975).
402. A. Babadjamian, R. Gallo, J. Metzger and M. Chanon, *J. Heterocycl. Chem.*, **13**, 1205 (1976).
403. R. J. Bergeron and P. G. Hoffman, *J. Org. Chem.*, **44**, 1835 (1979).
404. R. J. Bergeron and P. G. Hoffman, *J. Org. Chem.*, **45**, 161 (1980).
405. N. De Kimpe, R. Verhé, L. De Buyck and N. Schamp, *Org. Prep. Proced. Int.*, **12**, 49 (1980).
406. N. De Kimpe and N. Schamp, *Org. Prep. Proced. Int.*, **11**, 115 (1979).
407. F. Ramirez and A. F. Kirby, *J. Amer. Chem. Soc.*, **75**, 6026 (1953).
408. A. G. Schultz and W. K. Hagman, *J. Org. Chem.*, **43**, 3391 (1978).
409. H. Beyer and G. Bodicke, *Chem. Ber.*, **93**, 826 (1960).
410. E. J. Corey, M. Petrzilka and Y. Ueda, *Tetrahedron Lett.*, 4343 (1975).
411. W. Oppolzer, M. Petrzilka and K. Bättig, *Helv. Chim. Acta*, **60**, 2964 (1977).
412. A. Hassner and V. Alexanian, *J. Org. Chem.*, **44**, 3861 (1979).
413. S. Bozzini, B. Cova, S. Gratton, A. Lisini and A. Risaliti, *JCS Perkin I*, 240 (1980).
414. J. Schantl, *Monatsh. Chem.*, **108**, 325 (1977).
415. N. I. Korotkikh, A. Y. Chervinskii, S. N. Baranov, L. M. Kapkan and O. P. Shvaika, *Zh. Org. Khim.*, **15**, 962 (1979); *Chem. Abstr.* **91**, 74124 (1979).
416. M. Koga and J.-P. Anselme, *Chem. Commun.*, 53 (1973).
417. D. Y. Curtin and E. W. Tristam, *J. Amer. Chem. Soc.*, **72**, 5238 (1950).
418. T. Patonay, M. Rakosi, G. Litkei, T. Mester and R. Bognar, in *Proceedings of the 5th Hungarian Bioflavonoid Symposium, Matrafured, Hungary*, 227 (1977).
419. R. J. Crawford and H. Tokunaga, *Canad. J. Chem.*, **52**, 4033 (1974).
420. T. V. Saraswath and V. R. Srinivasan, *Tetrahedron*, **33**, 1043 (1977).
421. W. R. Mallory and R. W. Morrison, *J. Org. Chem.*, **45**, 3919 (1980).
422. E. Bulka and W. D. Pfeiffer, *J. Prakt. Chem.*, **318**, 971 (1976).
423. H. Beyer and G. Wolter, *Chem. Ber.*, **89**, 1652 (1956).
424. J. H. Boyer and D. Straw, *J. Amer. Chem. Soc.*, **75**, 1642 (1953).
425. G. Pasquet, D. Boucherot, W. R. Pilgrim and B. Wright, *Tetrahedron Lett.*, 931 (1980).
426. F. Asinger, W. Schäfer, M. Baumann and H. Römgens, *Liebigs Ann. Chem.*, **672**, 103 (1964).
427. G. Geiseler and F. Stache, *Chem. Ber.*, **94**, 337 (1961).
428. L. Schotte, *Ark. Kemi*, **5**, 533 (1953).
429. B. Föhlisch and W. Gottstein, *Liebigs Ann. Chem.*, 1768 (1979).
430. K. Nagata, *Chem. Pharm. Bull.* (Tokyo), **17**, 661 (1969).
431. J. Gierer and B. Alfredson, *Chem. Ber.*, **90**, 1240 (1957).

432. D. Martinez and A. Hiller, *Zeit. Chem.*, **16**, 320 (1976).
433. Z. M. Ivanova, T. V. Kim, I. E. Boldeskul and Y. G. Gololobov, *Zh. Obshch. Khim.*, **49**, 1464 (1979); *Chem. Abstr.* **91**, 157197 (1979).
434. H. Horstmann, in *Methoden der Organischen Chemie*, Vol. 7/2c (Ed. Houben-Weyl), Georg Thieme Verlag, Stuttgart (1977), p. 2352.
435. B. M. Trost, W. C. Vladuchick and A. J. Bridges, *J. Amer. Chem. Soc.*, **102**, 3548 (1980).
436. Y. Nagao, M. Ochiai, K. Kaneko, A. Maeda, K. Watanabe and E. Fujita, *Tetrahedron*, 1345 (1977).
437. G. Buchmann and R. Schmuck, *J. Prakt. Chem.*, **28**, 141 (1965).
438. M. Oki, W. Fanakoshi and A. Nakamura, *Bull. Chem. Soc. Japan*, **44**, 828 (1971).
439. I. G. Mursakulov, F. F. Kerimov, N. K. Kasumov, E. A. Ramazanov and N. S. Zefirov, *Azerb. Khim. Zh.*, 93 (1979); *Chem. Abstr.* **91**, 56925b (1979).
440. G. Giusti and G. Schembri, *Compt. Rend. Acad. Sci. Paris C*, **287**, 213 (1978).
441. H. Sokol and J. J. Ritter, *J. Amer. Chem. Soc.*, **70**, 3517 (1948).
442. P. Dubs and R. Stuessi, *Synthesis*, 696 (1976).
443. T. Terasawa and T. Okada, *J. Org. Chem.*, **42**, 1163 (1977).
444. H. Ishihara and Y. Hirabayashi, *Chem. Lett.*, 1007 (1978).
445. G. Ege, P. Arnold and R. Noronha, *Liebigs Ann. Chem.*, 656 (1979).
446. A. M. Sarpeshkar, G. J. Gossick and J. Wemple, *Tetrahedron Lett.*, 703 (1979).
447. D. T. Mowry, *Chem. Rev.*, **42**, 189, 204 (1948).
448. A. E. Matthews and W. R. Hodginson, *Chem. Ber.*, **15**, 2679 (1882).
449. E. P. Köhler and F. W. Brown, *J. Amer. Chem. Soc.*, **55**, 4299 (1933).
450. R. Justoni, *Gazz. Chim. Ital.*, **69**, 378 (1939).
451. H. Behringer, M. Ruff and R. Wiedenmann, *Chem. Ber.*, **97**, 1737 (1964).
452. O. Widman and E. Wahlberg, *Chem. Ber.*, **44**, 2067 (1911).
453. A. Y. Yakubovich, N. A. Bogolovskii, E. P. Pravova and S. M. Rozenshtein, *Zh. Obshch. Khim.*, **28**, 2288 (1958); *Chem. Abstr.* **55**, 13302h (1961).
454. C. D. Hurd and C. H. Rector, *J. Org. Chem.*, **10**, 441 (1945).
455. H. Kobler, K.-H. Schuster and G. Simchen, *Justus Liebigs Ann. Chem.*, 1946 (1978).
456. H. Galons, C. Combet-Farnoux, J.-F. Girardeau and M. Miocque, *Compt. Rend. Acad. Sci. Paris C*, **286**, 663 (1978).
457. R. Justoni and M. Terruzzi, *Gazz. Chim. Ital.*, **78**, 155, 166 (1948).
458. H. Galons, J. F. Girardeau and C. Combet-Farnoux, *Bull. Soc. Chim. France*, 936 (1977).
459. M. M. Schemiakan, M. V. Kolozov, Y. A. Arbusov, V. V. Onoprienco and Y.-Y. Hsieh, *Zh. Obshch. Khim.*, **30**, 545 (1960); *Chem. Abstr.* **54**, 24576b (1960).
460. F. Ebel, F. Huber and A. Brunner, *Helv. Chim. Acta*, **12**, 16 (1929).
461. R. T. Lalonde, N. Muhammad, C. F. Wong and E. R. Sturiale, *J. Org. Chem.*, **45**, 3664 (1980).
462. T. Sakai, E. Amano, A. Kawabata and A. Takeda, *J. Org. Chem.*, **45**, 43 (1980).
463. O. Campos and J. M. Cook, *J. Heterocycl. Chem.*, **14**, 711 (1977).
464. S. R. Ramadas and S. Padmanabhan, *Curr. Sci.*, **48**, 52 (1979).
465. H.-J. Altenbach, *Angew. Chem.*, **91**, 1005 (1979).
466. R. Verhé and R. Thierie, unpublished results.
467. R. Verhé, D. Courtheyn, N. De Kimpe, L. De Buyck, R. Thierie, L. Van Caenegem and N. Schamp, *Org. Prep. Proced. Int.*, **13**, 13 (1981).
468. W. Lehnert, *Tetrahedron*, **29**, 635 (1973).
469. R. Verhé, N. De Kimpe, L. De Buyck, R. Thierie and N. Schamp, *Bull. Soc. Chim. Belg.*, **89**, 563 (1980).
470. P. Bravo, G. Gaudiano, C. Ticozzi and A. Umani-Ronchi, *Tetrahedron Lett.*, 4481 (1968).
471. J. A. Donnelly, M. J. Fox and J. G. Hoey, *JCS Perkin I*, 2629 (1979).
472. A. S. Sopova, N. W. Perekalin, O. I. Jurczenko and G. M. Arnautova, *Zh. Org. Khim.*, **5**, 858 (1969); *Chem. Abstr.* **71**, 38394k (1969).
473. L. Fitjer, *Synthesis*, 189 (1977).
474. M. Tiffeneau, *Bull. Soc. Chim. France*, 612 (1945) and references cited therein.
475. T. A. Geissman and R. Akawie, *J. Amer. Chem. Soc.*, **73**, 1993 (1951).
476. J. J. Riehl, A. Smolikiewicz and L. Thil, *Tetrahedron Lett.*, 1451 (1974).
477. J. W. Cornforth, R. H. Cornforth and K. K. Mathew, *J. Chem. Soc.*, 112 (1959).
478. E. Elkik, M. Le Blanc and A. Vailatti, *Compt. Rend. Acad. Sci. Paris C*, **270**, 246 (1970).

479. A. S. Hussey and R. R. Herr, *J. Org. Chem*., **24**, 843 (1959).
480. T. Kato, H. Kondo, M. Nishino, M. Tanaka G. Hata and A. Miyake, *Bull. Chem. Soc. Japan*, **53**, 2958 (1980).
481. J. Barbuenga, M. Yus and P. Bernad, *Chem. Commun*., 847 (1978).
482. C. H. De Puy, G. M. Dappen, K. L. Eilers and R. A. Klein, *J. Org. Chem*., **29**, 2813 (1964).
483. H. C. Brown and C. G. Rao, *J. Org. Chem*., **43**, 3602 (1978).
484. R. L. Huang, *J. Chem. Soc*., 4089 (1957).
485. N. De Kimpe, R. Verhé, L. De Buyck and N. Schamp, *Tetrahedron Lett*., 955 (1978).
486. N. De Kimpe, R. Verhé, L. De Buyck and N. Schamp, *Bull. Soc. Chim. Belg*., **88**, 719 (1979).
487. J. Cologne and J. Grenet, *Bull. Soc. Chim. France*, 1304 (1954).
488. J. Villieras and B. Castro, *Bull. Soc. Chim. France*, 1189 (1970).
489. A. J. Sisti and A. C. Vitale, *J. Org. Chem*., **37**, 4090 (1972).
490. H. H. Ong, V. B. Anderson, J. C. Wilker, T. C. Spaulding and L. R. Meyerson, *J. Med. Chem*., **23**, 726 (1980).
491. J.-E. Dubois, C. Lion and C. Moulineau, *Tetrahedron Lett*., 177 (1971).
492. J.-E. Dubois and C. Lion, *Tetrahedron*, **31**, 1227 (1975).
493. J.-E. Dubois, P. Fournier and C. Lion, *Tetrahedron Lett*., 4263 (1975).
494. J.-E. Dubois, P. Fournier and C. Lion, *Bull. Soc. Chim. France*, 1871 (1976).
495. G. H. Posner and J. J. Sterling, *J. Amer. Chem. Soc*., **95**, 3076 (1973).
496. R. Nouri-Bimorghi, *Bull. Soc. Chim. France*, 1876 (1975).
497. H. C. Brown, M. M. Rogic, M. W. Rathke and G. W. Kabalka, *J. Amer. Chem. Soc*., **91**, 2150 (1969).
498. J.-J. Katz, J.-E. Dubois and C. Lion, *Bull. Soc. Chim. France*, 683 (1977).
499. A. Pelter, K. J. Gould and C. R. Harrison, *JCS Perkin I*, 2428 (1976).
500. R. C. Cookson, M. J. Nye and G. Subrahmanyam, *J. Chem. Soc. C*, 473 (1967).
501. H. M. R. Hoffmann, D. R. Joy and A. K. Suter, *J. Chem. Soc. B*, 57 (1968).
502. R. Noyori, S. Makino and H. Takaya, *J. Amer. Chem. Soc*., **93**, 1272 (1971).
503. H. M. R. Hoffmann and M. N. Iqbal, *Tetrahedron Lett*., 4487 (1975).
504. H. M. R. Hoffmann, R. Chidgey and G. Fierz, *Angew. Chem*., **13**, 444 (1974).
505. R. Noyori, Y. Hayakawa, M. Funakura, H. Takaya, S. Murai, R. Kobayashi and S. Tsutsumi, *J. Amer. Chem. Soc*., **94**, 7202 (1972).
506. R. Noyori, Y. Baba, S. Makino and H. Takaya, *Tetrahedron Lett*., 1741 (1973).
507. D. I. Rawson, B. K. Carpenter and H. M. R. Hoffman, *J. Amer. Chem. Soc*., **101**, 1786 (1979).
508. A. P. Cowling and J. Mann, *JCS Perkin I*, 1564 (1978).
509. Y. Hayakawa, K. Yokoyama and R. Noyori, *J. Amer. Chem. Soc*., **100**, 1799 (1978).
510. A. P. Cowling and J. Mann, *Chem. Commun*., 1006 (1978).
511. Y. Ito, M. Asada, K. Yonezawa and T. Saegusa, *Synth. Commun*., **4**, 87 (1974).
512. H. M. R. Hoffmann, T. A. Nour and R. H. Smithers, *Chem. Commun*., 963 (1972).
513. H. M. R. Hoffmann, K. E. Clemens, E. A. Schmidt and R. H. Smithers, *J. Amer. Chem. Soc*., **94**, 3201 (1972).
514. H. Alper, K. D. Logbo and H. des Abbayes, *Tetrahedron Lett*., 2861 (1977).
515. H. Alper and E. C. H. Keung, *J. Org. Chem*., **37**, 2566 (1972).
516. F. G. Saitkulova, T. P. Kadyrmatova, G. G. Abashev and I. I. Lapkin, *Izv. Vyssh. Uchebn. Zaved., Khim. Khim. Teckhnol*., **20**, 1078 (1977); *Chem. Abstr.* **87**, 201013 (1977).
517. I. I. Lapkin, F. G. Saitkulova, G. G. Abashev, V. V. Fotin, *Izv. Vyssh. Uchebn. Zaved., Khim. Khim. Tekhnol.*, **23**, 793 (1980; *Chem. Abstr.*, **94**, 15339g (1981).
518. K. Maruoka, S. Hashimoto, Y. Kitagawa, H. Yamamoto and H. Nozaki, *J. Amer. Chem. Soc*., **99**, 7705 (1977).
519. L. T. Scott and W. D. Cotton, *J. Amer. Chem. Soc*., **95**, 2708 (1973).
520. J. Wolinsky, J. H. Thorstenson and T. A. Killinger, *J. Org. Chem*., **43**, 875 (1978).
521. H. Sugihara, K. Ukawa, A. Miyake, K. Itoh and Y. Sanno, *Chem. Pharm. Bull*., 394 (1978).
522. W. Sucrow, A. Fehlauer and U. Sandmann, *Z. Naturforsch*., **32b**, 1072 (1977).
523. H. Bodot, J.-A. Braun and J. Fedière, *Bull. Soc. Chim. France*, 3253 (1968).
524. J. Sauleau, *Bull. Soc. Chim. France*, 474 (1978).
525. L. W. Trevoy and W. G. Brown, *J. Amer. Chem. Soc*., **71**, 1675 (1949).

526. L. H. Schwartz and R. V. Flor, *J. Org. Chem.*, **34**, 1499 (1969).
527. J. Bergman and J. E. Bäckvall, *Tetrahedron*, **31**, 2063 (1975).
528. A. A. Volod'kin, N. V. Portnykh and V. V. Ershov, *Bull. Acad. Sci. U.S.S.R., Div. Chem. Sci.*, 1352 (1962).
529. A. J. Kirby and S. G. Warren, *The Organic Chemistry of Phosphorus*, Elsevier, Amsterdam (1967), p. 117.
530. K. V. Nikonorov, I. D. Neklesova, E. A. Gurylev, M. A. Kudrina, V. A. Nikonenko, I. S. Iraidova and N. N. Anisimova, *Zh. Obshch. Khim.*, **46**, 560 (1976); *Chem. Abstr.* **85**, 5795 (1976).
531. P. A. Chopard, V. M. Clark, R. F. Hudson and A. J. Kirby, *Tetrahedron*, **21**, 1961 (1965).
532. L. Toke, I. Petnehazy and G. Szakal, *J. Chem. Res. (M)*, 1975 (1978).
533. E. M. Gaydou, G. Buono and R. Frèze, *Bull. Soc. Chim. France*, 2284 (1973).
534. I. J. Borowitz, S. Firstenberg, E. W. R. Casper and R. K. Crouch, *J. Org. Chem.*, **36**, 3282 (1971).
535. E. M. Gaydou and J. B. Bianchini, *Canad. J. Chem.*, **54**, 3626 (1976).
536. F. Kienzle and P. Rosen, *Helv. Chim. Acta*, **62**, 442 (1979).
537. T. K. Gazizov, Y. I. Sudarev and A. N. Pudovik, *Zh. Obshch. Khim.*, **46**, 2383 (1976); *Chem. Abstr.* **85**, 4817 (1976).
538. M. Sekine, K. Okimoto and T. Hata, *J. Amer. Chem. Soc.*, **100**, 1001 (1978).
539. R. Malinowski and M. Mikolajczyk, *Pr. IPO*, **6**, 95 (1974).
540. R. Malinowski and J. Legocki, *Polish J. Chem.*, **53**, 2149 (1979).
541. O. Dahl and F. K. Jensen, *Acta Chem. Scand. B*, **30**, 863 (1975).
542. O. Dahl, *JCS Perkin I*, 947 (1978).
543. I. J. Borowitz and R. Virkhaus, *J. Amer. Chem. Soc.*, **85**, 2183 (1963).
544. T. V. Kim, Z. M. Ivanova and Y. G. Gololobov, *Zh. Obshch. Khim.*, **48**, 1967 (1978); *Chem. Abstr.* **88**, 190990 (1978).
545. Y. G. Gololobov, L. F. Kasukhin, G. V. Pesotskaya, V. S. Petrenko, T. V. Kim and Z. M. Ivanova, *Zh. Obshch. Khim.*, **48**, 1974 (1978); *Chem. Abstr.* **90**, 18194 (1979).
546. B. Springs and P. Haake, *J. Org. Chem.*, **41**, 1165 (1976).
547. B. Springs and P. Haake, *J. Org. Chem.*, **42**, 472 (1977).
548. J. Leroy, J. Bensoam, M. Humiliere, C. Wakselman and F. Mathey, *Tetrahedron*, **36**, 1931 (1980).
549. G. Richard, *Bull. Soc. Chim. France*, 286 (1938).
550. E. M. Schultz and S. Mickey, *Org. Synth.*, Coll. Vol. III, 343 (1960).
551. L. T. Scott and W. D. Cotton, *J. Amer. Chem. Soc.*, **95**, 5416 (1973).
552. S. N. Kukota, V. P. Borisenko, V. N. Bodnar, N. I. Zhuravskaya, M. O. Lozinskii, *Fiziol. Akt. Veskchestva*, **10**, 32 (1978); *Chem. Abstr.*, **90**, 6212a (1979).
553. D. J. Cooper and L. N. Owen, *J. Chem. Soc. C*, 533 (1966).
554. C. Kowalski, X. Creary, A. J. Rollin and M. C. Burke, *J. Org. Chem.*, **43**, 2601 (1978).
555. H. O. House, W. F. Fischer, M. Gall, T. E. McLaughlin and N. P. Peet, *J. Org. Chem.*, **36**, 3429 (1971).
556. I. Fleming and J. Goldhill, *JCS Perkin I*, 1493 (1980).
557. E. Warnhoff, M. Rampersad, P. Sundara Raman and F. W. Yerhoff, *Tetrahedron Lett.*, 1659 (1978).
558. J. Wolinsky, J. J. Hamsher and R. O. Hutchins, *J. Org. Chem.*, **35**, 207 (1970).
559. I. V. Zavarzin, T. A. Klimova, M. M. Krayushkin, V. V. Sevost'yanova and S. S. Novikov, *Izv. Akad. Nauk SSSR, Ser Khim.*, 868 (1978); *Chem. Abstr.* **89**, 42537 (1978).
560. J. P. Bégué, D. Bonnet, M. Charpentier-Morize and C. Pardo, *Tetrahedron*, **31**, 2505 (1975).
561. D. Baudry and M. Charpentier-Morize, *Nouv. J. Chim.*, **2**, 255 (1978).
562. J. P. Bégué and M. Charpentier-Morize, *Acc. Chem. Res.*, **13**, 207 (1980).
563. J. P. Bégué and M. Malissard, *Tetrahedron*, **34**, 2094 (1978).
564. P. C. Purohit and H. R. Sonawane, *Tetrahedron*, **37**, 873 (1981).
565. G. Jones II and L. P. McDonnell, *J. Amer. Chem. Soc.*, **98**, 6203 (1976).
566. B. E. Kaplan and A. L. Hartwig, *Tetrahedron Lett.*, 4855 (1970).
567. W. R. Bergmark, *Chem. Commun.*, 61 (1978).
568. J. C. Anderson and C. B. Reese, *Tetrahedron Lett.*, 1 (1961).

569. Y. Izaua, H. Tomioka, M. Natsume, S. Beppu and H. Tsujii, *J. Org. Chem.*, **45**, 4835 (1980).
570. A. J. Fry and A. T. Lefor, *J. Org. Chem.*, **44**, 1270 (1979).
571. A. J. Fry and G. S. Ginsburg, *J. Amer. Chem. Soc.*, **101**, 3928 (1979).
572. W. J. M. van Tilborg, R. Plomp, R. de Ruiter and C. J. Smit, *Recl. Trav. Chim. Pays-Bas*, **99**, 206 (1980).
573. H. E. Zimmerman and A. Mais, *J. Amer. Chem. Soc.*, **81**, 3644 (1959).
574. H. Alper and L. Patter, *J. Org. Chem.*, **44**, 2568 (1979).
575. T. H. Luh, C. H. Lai, K. L. Lei and S. W. Tam, *J. Org. Chem.*, **44**, 641 (1979).
576. H. Alper, *Tetrahedron Lett.*, 2257 (1975).
577. J. E. McMurry, *Acc. Chem. Res.*, **7**, 281 (1974).
578. I. J. Borowitz and L. I. Grossman, *Tetrahedron Lett.*, 471 (1962).
579. Tse-Lok Ho and C. M. Wong, *J. Org. Chem.*, **39**, 562 (1974).
580. D. A. Bock and W. T. Brady, *J. Org. Chem.*, **44**, 101 (1979).
581. J. M. Townsend and T. A. Spencer, *Tetrahedron Lett.*, 137 (1971).
582. G. A. Olah, G. D. Vankar and A. P. Fung, *Synthesis*, 59 (1979).
583. Tse-Lok Ho, *Synth. Commun.*, **9**, 241 (1979).
584. G. A. Olah, M. Arvanaghi and Y. D. Vankar, *J. Org. Chem.*, **45**, 3531 (1980).
585. A. L. Gemal and J. L. Luche, *Tetrahedron Lett.*, 3195 (1980).
586. N. De Kimpe and N. Schamp, *Org. Prep. Proced. Int.*, **13**, 241 (1981).